ILLUSTRATED FLORA

Illustrated Flora

OF THE

PACIFIC STATES

WASHINGTON, OREGON, AND CALIFORNIA

BY

LEROY ABRAMS

PROFESSOR OF BOTANY, STANFORD UNIVERSITY

IN FOUR VOLUMES

Vol. I

OPHIOGLOSSACEAE TO ARISTOLOCHIACEAE

FERNS TO BIRTHWORTS

STANFORD UNIVERSITY PRESS
STANFORD, CALIFORNIA

Stanford University Press
Stanford, California

Printed in the United States of America
ISBN 0-8047-0003-6
Original edition 1923
Last figure below indicates year of this printing:
84 83 82 81 80 79 78 77 76

CONTENTS OF VOLUME I

Seed-Bearing Plants

APPROXIMATE EQUIVALENTS OF THE METRIC SYSTEM

1 millimetre (mm.) = 1/25 of an inch.
3 millimeters = 1/8 of an inch.
1 centimetre (cm.) = 2/5 of an inch.
5 centimeters = 2 inches.
1 decimetre (dm.) = 4 inches.
1 metre (m.) = 40 inches or 3 1/3 feet.
300 meters = 1000 feet.

INTRODUCTION

The aim of the present work is to furnish an authentic reference book that will be of greatest service not only to the trained botanist but to everyone interested in the native plant life of the Pacific States. It is patterned after the classical work of Britton and Brown,—"Illustrated Flora of the Northern United States and Canada,"—and is the second comprehensive illustrated flora published in this country. Every species of fern, flower, tree, and shrub known to grow wild in the Pacific States is illustrated and described.

The Physical Features of the Pacific States.

The three Pacific States—Washington, Oregon, and California—form the western boundary of the United States extending along the shore of the Pacific from 49° to 32° 35′ north latitude, a distance of 1,300 miles. They have an average width of about 250 miles, and an area of 324,123 square miles.

The outstanding physical features are the two great mountain systems, the Cascade-Sierra Nevada and the Coast Ranges, which in general parallel the coast. Of these the former is much the more prominent, and exerts the greater influence on climatic conditions, forming a remarkably distinct divide between the Pacific slope on the west and the interior or Great Basin region on the east. The Pacific slope has an oceanic climate with mild, rainy winters and cool (especially near the coast), dry summers, while east of the divide the climate is continental, with cold winters and hot summers.

Rainfall is greatest in the Northwest, diminishing southward and toward the interior. From northwestern California to the Puget Sound region the average annual rainfall is 75 to 100 inches; in the San Francisco Bay region 20 to 25 inches, and at San Diego 10 inches. East of the Cascade-Sierra Nevada Divide it is much reduced, producing semi-arid conditions in eastern Washington and Oregon, and typical desert conditions in southeastern California, where the annual rainfall is only 2 to 5 inches.

Floral Features of the Pacific States.

The great coniferous forests of the Northwest and the mountains of California are the most prominent floral feature. Nowhere are there more extensive forests, more majestic trees, or greater variety of cone-bearing species. In the foothills and coastal valleys of California chaparral and oaks replace the conifers, the open park-like groves of the latter contributing much to the charm of the region. Finally, forming a striking contrast to the forests, are the bunch-grass plains of eastern Washington and Oregon, the sage-brush tracts of southeastern Oregon and adjacent California, and the weird cacti and yuccas of the southern California deserts.

Life Zones.

The usual manual method of noting geographical distribution by states is inadequate in the Pacific region, where, owing to the diversity of topography and climate, plant distribution is unusually complex. As an aid therefore in designating distribution, Merriam's system of Life Zones has been used. These zones as represented in the Pacific States are as follows:

BOREAL REGION
- *Arctic-Alpine Zone*
- *Hudsonian Zone*
- *Canadian Zone*

AUSTRAL REGION
- *Transition Zone*
- *Upper Sonoran Zone*
- *Lower Sonoran Zone*

The Arctic-Alpine Zone occupies the high altitudes of the mountains above timber-line. On Mt. Rainier, Washington, timber-line is about 7,500 feet altitude; on Mt. Shasta, northern California, about 9,500 feet, while on Mt. San Gorgonio, southern California, which is 11,425 feet in elevation, trees extend to the summit, although a few Arctic-Alpine plants are present. The zone is characterized by the absence of trees, and the dwarf (often matted) habit of the shrubs and perennial herbs. Among the characteristic species are: *Festuca alpina, Carex breweri, Salix nivalis, Oxyria didyna, Ranunculus eschscholtzii, Draba breweri, Smelowskia calycina, Saxifraga tolmiei, Lutkea pectinata,* and *Dryas octopetala.*

The Hudsonian and Canadian Zones include the subalpine forests, and are not always readily separable, especially in the Sierra Nevada where the former is poorly developed. The Hudsonian is the uppermost timber zone extending from timber-line to 2,000 or 3,000 feet below. Characteristic trees are: *Pinus albicaulis, Pinus balfouriana, Abies lasiocarpa, Chamaecyparis nootkatensis,* and *Tsuga mertensiana.*

The Canadian Zone comprises the lower portion of the subalpine forests. In the Cascades of Washington it is usually between 3,000 and 5,000 feet altitude, and in the central Sierra Nevada between 6,500 and 9,000 feet. Among the characteristic trees are: *Pinus monticola, Pinus contorta murrayana, Pinus jeffreyi, Abies amabilis, Abies magnifica,* and *Populus tremuloides.* A narrow coastal strip extending from southern Alaska to northern California, in which the dominating tree is *Picea sitchensis,* is to be considered as belonging to the Canadian Zone. The presence of this boreal flora is undoubtedly due to the cold, foggy summer months.

The Transition Zone is represented in the Pacific States by two clearly-defined areas, the Humid Transition Area and the Arid Transition Area. The Humid Transition includes the redwood belt of northern California and the great coniferous forest belt of western Oregon and Washington, where the dominating tree is *Pseudotsuga taxifolia.* The Arid Transition is the yellow pine belt (*Pinus ponderosa*) of western North America. In Washington and northern Oregon it is restricted to the eastern side of the Cascade Divide, but in southern Oregon and California it extends to the inner Coast Ranges. The most characteristic tree is *Pinus ponderosa,* but other typically Arid Transition trees are: *Libocedrus decurrens, Pinus lambertiana, Abies concolor,* and *Quercus kelloggii.* The bunch-grass plains of eastern Washington and eastern Oregon also belong to this zone.

The Upper Sonoran Zone comprises the oak and chaparral belts of the California foothills and valleys, and the sage-brush and piñon belts east of the Cascade-Sierra Nevada Divide. In the California Upper Sonoran the most important characteristic trees and shrubs are: *Pinus sabiniana, Quercus lobata, Quercus douglasii, Quercus dumosa, Aesculus californica, Adenostoma fasciculatum, Ceanothus cuneatus,* and *Ceanothus divaricatus.* In eastern Oregon and Washington the most conspicuous plant is the sage-brush, *Artemisia tridentata.* Other characteristic lignescent plants are: *Eriogonum microtheca, Grayia spinosa, Purshia tridentata,* and *Chrysothamnus nauseosus.* On the desert slopes of the California mountains the Upper Sonoran represents the piñon-juniper belt, of which the two most conspicuous plants are *Pinus monophylla,* and *Sabina californica.*

The Lower Sonoran Zone is the typical desert flora of the Mojave and Colorado Deserts, but it also includes the San Joaquin Valley and the inner coastal valleys of southern California. Among the characteristic trees and shrubs are: *Yucca brevifolia, Yucca mohavensis, Populus fremonti, Salix gooddingii, Atriplex lentiformis, Olneya tesota, Fouquieria splendens,* and various species of cylindrical opuntias.

The Floral Elements.

Boreal Element. The Boreal Zones,—Arctic, Hudsonian, and Canadian,—have a flora essentially boreal in origin. With scarcely an exception the genera are widely distributed throughout the subarctic and cool temperate regions of the northern hemisphere. Many of the species are spread over most of northern North

America, or are represented in the eastern region by closely related species. The presence of these boreal plants on the high mountain ranges of the Pacific States, a thousand miles or more south of their normal range, is partly due, no doubt, to the Glacial Period.

Great Basin Element. The flora east of the Cascade-Sierra Nevada Divide is chiefly of the Great Basin element, which has developed since the humid conditions of Miocene and Pleiocene times and the devastating Glacial Period. The principal source, at least for the Upper Sonoran Zone, has been from the south.

Mexican Element. "While the common genera familiar to us all were evolving in the moist temperate climates to the north, and spreading over northern North America, Europe and Asia by means of land connections that have disappeared, drought-resisting plants were taking form on the great arid plateau of Mexico. Here originated the cacti, yuccas, agaves, and most of the other genera peculiar to the American deserts. At the end of the Glacial Period this Mexican flora pushed northward into the western United States, following increased aridity. It is this ancient Mexican element that gives the unique character to our desert vegetation."*

California Element. For a continental area the Pacific slope, especially that lying within the Upper Sonoran Zone, has an unusual number of endemic genera and species. Reasons for the unique character of the flora are to be found in the climatic conditions, both of the present and the past, and in the isolation brought about by the climatic and physical barriers prohibiting direct communication with the eastern part of the continent. Throughout Tertiary Time, while the present-day floras were developing, the Pacific Ocean, responding much less readily to the climatic changes taking place on the main continental areas, has acted as a great thermostat along the Pacific Coast. The climate of the Pacific States has therefore undergone less change than in other parts of the continent. The climate of today, at least in the immediate coastal region, cannot be very different in temperature from that of Cretaceous Time. *Sequoia, Torreya,* and other genera still flourish, mere remnants of the great cretaceous forests that spread over the northern hemisphere, even to the Arctic Circle. This theory of the earlier climatic and floral conditions on the Pacific Coast has been strengthened recently by the discovery of fossil remains of *Sequoia sempervirens, Quercus chrysolepis, Arbutus menziesii,* and other living species peculiar to the Pacific region, in Pleiocene rocks of central California, showing that the flora and therefore the climate just prior to the Glacial Period was much like that of the present time.

Illustrations.

In the preparation of the illustrations, drawings have been made natural size, either from fresh material or herbarium specimens, and under the direct supervision of the author or the special contributor. All of the figures are from these original drawings except those of such species as are illustrated in Britton and Brown's "Illustrated Flora of the Northern United States and Canada." Privilege to use these has been kindly granted by Dr. Britton, and is hereby gratefully acknowledged.

The numerals placed on the figures indicate the amount of reduction: for example, the numeral ⅗ indicates that the figure is three-fifths the natural size; but where enlarged details of essential parts have been added no indication of magnification has been attempted.

Mr. W. S. Atkinson has made most of the drawings for the families contributed by the author, also those of the *Equisetaceae, Selaginellaceae, Juncaceae,* and *Salix;* Mrs. Rose E. Gamble has made those of the *Ophioglossaceae* and *Polypodiaceae;* Dr. Norma Pfeiffer those of the *Isoetaceae;* Mrs. Mary W. Gill those of the *Poaceae;* Miss Mary E. Eaton those of the *Cyperaceae;* and Mrs. Helen

*Abrams, LeRoy, "The Deserts and Desert Floras of the West," in *Nature and Science on the Pacific Coast.*

Edwards Bacon most of the *Melanthaceae,* as well as a number in other families contributed by the author.

Scientific Names.

The scientific names of genera and species adopted are in accordance with the code of nomenclature recommended by the Nomenclature Committee of the Botanical Society of America. The fundamental principle of the code is that a specific name is based upon a type specimen and a generic name upon a type species. The code also recognizes the necessity of separating nomenclatorial questions from those that have to do with the interpretation of scientific data, for so long as the eligibility of a name depends upon the status of some other genus or species, stability of nomenclature is impossible. These two fundamentals as well as priority are recognized in all modern codes dealing with the nomenclature of animals and plants.

As an aid to the pronunciation of the scientific names the accented syllable is marked. When the vowel of the accented syllable is short the acute (′) accent is used, when it is long or broad the grave (\`).

English Names.

In this initial attempt to apply an English name to every plant species in the Pacific States it is realized that some well established local names may have been omitted. Selecting the most suitable common name has not been an easy task, but when a name has appeared in the local manuals or popular floras it has been retained whenever possible. Special attention has also been given Indian and Spanish-Californian names.

Assistance.

To the following contributors to this volume the author gratefully acknowledges his indebtedness. Much of the merit of the work is due to their hearty coöperation: Mr. William R. Maxon for the text of the *Pteridophyta,* except the family *Isoetaceae;* Dr. Norma Pfeiffer for the text of the *Isoetaceae;* Professor A. S. Hitchcock for the text of the *Poaceae;* Dr. N. L. Britton for the text of the *Cyperaceae,* except *Carex;* Mr. K. K. Mackenzie for the text of *Carex;* Dr. C. R. Ball for the text of *Salix,* and Mr. F. V. Coville for assistance in preparing the text of the *Juncaceae.*

Mr. and Mrs. G. F. Ferris have read the proof-sheets and assisted in the preparation of the index. Dr. J. H. Barnhart has given valuable assistance in bibliographical questions. Miss Alice Eastwood, Dr. W. L. Jepson, Mr. S. B. Parish, Mr. Carl Purdy, Dr. P. A. Rydberg, Professor J. C. Nelson, and Mr. M. W. Gorman have given valuable suggestions and have cordially assisted the author in many ways.

For the loan of herbarium material or privilege of study the author is deeply indebted to the custodians of the various herbaria not only of the Pacific States but also of the leading botanical institutions of the eastern United States and Europe.

Finally, to Dr. N. L. Britton, Director of the New York Botanical Garden; to the Board of Trustees of Stanford University, and to President Ray Lyman Wilbur, who have jointly made an illustrated flora of the Pacific States financially possible the author and all interested in native plants must ever be deeply grateful.

LeRoy Abrams

Stanford University
May 15, 1923

PREFACE TO THE 1940 REPRINT

This reprint of Volume I is without change in text other than the following:

1. A few typographical errors have been corrected.
2. As the final volume will have comprehensive keys to families and also comprehensive indices to the English and Latin names, the keys to the families in the individual volumes have been omitted and the indices have been confined to the Latin names of the families and genera.
3. Since the publication of this volume in 1923, two International Congresses have been held, one in 1930, the other in 1935. These Congresses greatly modified and modernized the original International Rules of Nomenclature adopted at Vienna in 1905 and brought the International Rules and the so-called American Code so nearly in accord that all the leading American institutions have substituted them for the American Code. We therefore have made the following changes in generic names to conform with the *nomina conservanda* of the International Rules: *Cystopteris* for *Filix* (p. 6), *Torreya* for *Tumion* (p. 51), *Elodea* for *Philotria* (p. 103), *Sorghum* for *Holcus* (p. 108), *Hierochloë* for *Torresia* (p. 124), *Holcus* for *Notholcus* (p. 163), *Cynodon* for *Capriola* (p. 174), *Lamarckia* for *Achyrodes* (p. 196), *Glyceria* for *Panicularia* (p. 211), *Luzula* for *Juncoides* (p. 369), *Narthecium* for *Abama* (p. 372), *Brodiaea* for *Hookera* (p. 405), *Chlorogalum* for *Laothoe* (p. 413), *Camassia* for *Quamasia* (p. 415), *Smilacina* for *Vagnera* (p. 453), *Maianthemum* for *Unifolium* (p. 454), *Epipactis* for *Serapias* (p. 478), *Spiranthes* for *Ibidium* (p. 479), *Calypso* for *Cytherea* (p. 482), *Arceuthobium* for *Razoumofskya* (p. 529).
4. The following changes have been made in the specific names to correspond with the change in rules: *Dryopteris Linnaeana* for *D. Dryopteris* (p. 13), *Pseudotsuga taxifolia* for *P. mucronata* (p. 64), and *Glyceria leptostachya* Buckl. for *Panicularia Davyi* Merr.
5. The International Rules state that specific names should be capitalized if they are derived from the name of a person, as *Calochortus Greenei,* or from the name of a genus, as *Sitanion Hystrix,* or from an aboriginal name, as *Camassia Quamash.* As this rule does not change any of the specific names it has not been considered advisable to make the numerous alterations in the plates that would be necessary to capitalize all the specific names involved. However, in the remaining volumes all such specific names will be capitalized according to the rules.

L. R. A.

STANFORD UNIVERSITY
June 15, 1940

ILLUSTRATED FLORA

Phylum PTERIDÓPHYTA.*

FERNS AND FERN-ALLIES.

Terrestrial, epiphytic, or (rarely) aquatic plants, exhibiting a life cycle of two well-marked phases, Sporophyte and Gametophyte, the so-called alternation of generations. The former is the conspicuous one; the plant, known as a fern or fern ally, is usually differentiated into root, stem, and leaf, is provided with vascular tissues, and bears spores asexually, these either alike (plants homosporous) or of two very unlike kinds called microspores and megaspores (plants heterosporous). The germinating spore produces the Gametophyte, or minute, inconspicuous sexual stage (prothallium). In the homosporous series the prothallia are similar, but may be monoecious or dioecious; in the heterosporous series they are dioecious, the ones developing from the microspores producing only male reproductive organs (antheridia), and those from the megaspores only female reproductive organs (archegonia). Fertilization consists in the impregnation of an egg cell or archegone by the coiled motile male element (spermatozoid). The resulting growth is the Sporophyte, or conspicuous asexual stage.

The pteridophytes first appeared in the early part of the Palaeozoic Era, and reached the height of their development in Carboniferous Time, when they formed the dominating type of vegetation. There are about 7,000 living species, of which more than three-fourths are restricted to tropical regions.

Family 1. OPHIOGLOSSÀCEAE.

ADDER'S-TONGUE FAMILY.

Sporophytes herbaceous, consisting of a short fleshy rhizome, bearing numerous fibrous, usually fleshy roots and one or several leaves. Leaves (fronds) erect (in our genera), not jointed, consisting of a simple to variously compound, sessile or stalked sterile blade and, in the fertile leaves, a stalked sporebearing spike or panicle (sporophyll), borne on a short to elongate common stalk. Sporangia developed partly from subepidermal tissue (plants eusporangiate), naked, opening by a transverse slit. Spores uniform (plants homosporous), yellowish. Gametophytes (prothallia) hypogean, tuberlike, usually lacking chlorophyll and associated with an endophytic mycorhiza.

Five genera, the two following widely distributed and with numerous species, the others monotypic.

Sterile blade simple, with reticulate veins; sporangia united in two rows in a simple, slender, fleshy spike. 1. *Ophioglossum.*
Sterile blade 1–4 times pinnately divided, with free veins; sporangia globose, all distinct, borne in a spike or panicle. 2. *Botrychium.*

1. OPHIOGLÓSSUM [Tourn.] L. Sp. Pl. 1062. 1753.

Small herbaceous terrestrial plants. Rhizome short, hypogean, usually erect, terminating in the erect exposed bud for the following season. Leaves erect in vernation, 1–4, erect, glabrous, fleshy, arising at the side of the apical bud; common stalk short to elongate, slender; sterile blade simple, sessile or short-stalked, linear-lanceolate to ovate or reniform, with reticulate venation, the areoles simple or with free and anastomosing veinlets; sporophyll a simple slender long-stalked spike, the large globose sporangia marginal, coalescent in 2 ranks, transversely dehiscent. [Name Greek, meaning the tongue of a snake, alluding to the form of the narrow sporophyll.]

About 45 species, of wide distribution in both hemispheres. Besides the following, 3 others occur in the United States. Type species, *Ophioglossum vulgatum* L.

Fronds usually solitary and much more than 10 cm. long; sterile blade obtuse or acutish, 1–5 cm. broad, translucent. 1. *O. vulgatum.*
Fronds usually two and less than 10 cm. long; sterile blade acute, often apiculate, 5–10 mm. broad, opaque. 2. *O. californicum.*

---

*Text (except *Isoetaceae*) contributed by WILLIAM R. MAXON.

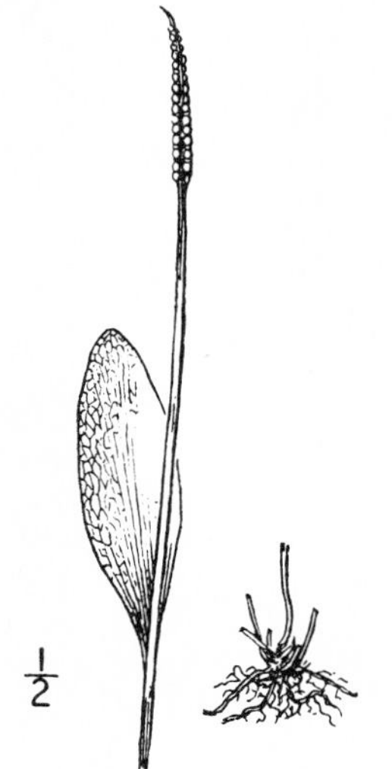

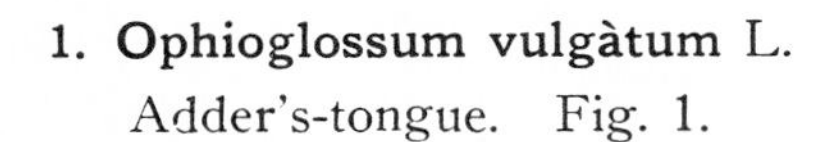

1. Ophioglossum vulgàtum L.

Adder's-tongue. Fig. 1.

Ophioglossum vulgatum L. Sp. Pl. 1062. 1753.
Ophioglossum arenarium E. G. Britton, Bull. Torrey Club **24**: 555. *pl. 318, pl. 319, f. 3.* 1897.
Ophioglossum alaskanum E. G. Britton, Bull. Torrey Club **24**: 556. *pl. 319, f. 5.* 1897.

Rhizome cylindrical, rather slender; fronds usually solitary, 10–40 cm. long; common stalk mostly epigean, one-third to two-thirds the total length of the plant; sterile blade usually sessile, flat, oblique, lanceolate, oblanceolate or spatulate, elliptical, oblong, or ovate, 2.5–12 cm. long, 1–5 cm. broad, rounded or obtuse, sometimes acutish, translucent when dry, the middle areoles long and narrow, the outer ones successively shorter, hexagonal, with or without included veinlets; sporophyll 12–30 cm. long, the spike 2–4 cm. long, 1.5–3.5 mm. broad, apiculate; sporangia 10–30 (50) pairs.

Moist meadows, pastures, and thickets, mainly in the Transition Zone; Maine and Quebec to Alaska, south to the Gulf States and Washington; also in Eurasia. Type locality: in Europe.

2. Ophioglossum califórnicum Prantl.

Western Adder's-tongue. Fig. 2.

Ophioglossum californicum Prantl, Bericht. Deutsch. Bot. Ges. 1: 351. 1883.

Rhizome cylindric, stout, 1–2 cm. long; fronds usually two, 3–11 cm. long; common stalk nearly all hypogean, 0.5–3 cm. long; sterile blade sessile or subsessile, usually conduplicate (at least in the long-tapering basal part), oblique, elliptic, lance-elliptic, or oblanceolate, 1.5–5 cm. long, 5–10 mm. broad, acute, frequently apiculate, opaque; areoles narrow, with or without included veinlets; sporophyll 1.5–9 cm. long, the spike 0.5–2 cm. long, 1–3 mm. broad, conspicuously apiculate; sporangia 8–15 pairs.

Among grasses in moist, stony situations, Upper Sonoran Zone; Amador County and high mesas near San Diego, California; also in Lower California (Ensenada). Type locality: San Diego, California.

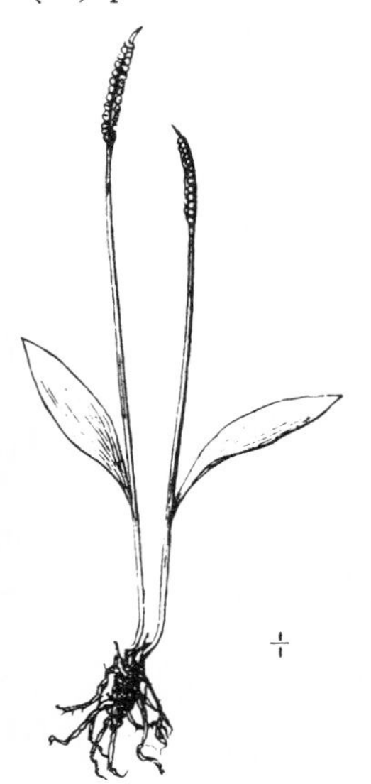

2. BOTRÝCHIUM Swartz, Journ. Bot. Schrad. 1800[2]: 110. 1801.

Fleshy terrestrial plants. Rhizomes hypogean, erect, relatively small, bearing few to many fleshy often corrugated coarse roots, and frequently sheathed at the apex by the persistent bases of old leaves; fronds 1–3; common stalk wholly or partially hypogean, erect, short to elongate; sterile blade erect or bent down in vernation, sessile to long-stalked, 1–4 times pinnately or ternately divided or compound, free-veined; sporophyll a long-stalked simple spike or 1–5-pinnate panicle, erect, the large globose sporangia distinct, sessile or nearly so, borne in 2 rows on the ultimate divisions; spores copious, sulphur yellow. Bud for the following year borne at the apex of the rhizome, enclosed within the base of the common stalk, either wholly concealed or visible along one side. [Name alluding to the grape-like arrangement of the sporangia.]

About 25 species, chiefly natives of the temperate regions of both hemispheres, mostly found in the United States. Type species, *Osmunda lunaria* L.

Buds devoid of hairs.
 Common stalk about one-half to wholly hypogean.
  Rhizome slender, short; sterile blade distinctly stalked; sporophyll long-stalked, erect in vernation. — 1. *B. simplex.*
  Rhizome stout, elongate; sterile blade sessile; sporophyll sessile or short-stalked, the tip recurved in vernation. — 2. *B. pumicola.*
 Common stalk almost wholly epigean.
  Sterile blade bent down only at the tip in vernation, oblong to triangular-oblong; sporophyll straight or with the apex decurved in vernation. — 3. *B. lunaria.*
  Sterile blade wholly bent down in vernation, triangular; sporophyll entirely bent down in vernation. — 4. *B. lanceolatum.*
Buds hairy.
 Common stalk mostly epigean; bud of the following season exposed along one side; leaf tissue very thin, with flexuous epidermal cells. — 5. *B. virginianum.*
 Common stalk all or nearly all hypogean; bud wholly enclosed; leaf tissue fleshy, with straight epidermal cells.
  Blades coriaceous in drying, 10–30 cm. broad, broadly triangular or pentagonal, the segments crenulate. — 6. *B. silaifolium.*
  Blades membrano-herbaceous in drying, 15–40 cm. broad, rounded-triangular or irregular in outline, lax, the segments minutely denticulate. — 7. *B. californicum.*

1. **Botrychium símplex** Hitchc.

Little Grape-fern. Fig. 3.

Botrychium simplex Hitchc. Am. Journ. Sci. **6**: 103. *pl. 8.* 1823.

Plants slender, 3–20 cm. high, the rhizome usually short and slender, with numerous roots; fronds erect, 2.5–18 cm. long, the common stalk short, usually at least half hypogean; sterile blade (straight or slightly bent in vernation) distinctly stalked (0.5–2.5 cm.), orbicular, rounded-deltoid, or deltoid-ovate, 1–5 cm. long, 0.5–2.5 cm. broad, entire or incised, or pinnately divided with entire or radially incised, cuneate or flabelliform segments, or in luxuriant specimens the blades often ternately divided, the divisions pinnately divided, with incised segments; sporophyll (erect in vernation) long-stalked, usually two-thirds the height of the plant, 2–15 cm. long, simple to 2-pinnate, lax.

Grassy meadows and open slopes, chiefly in the Transition Zone; British Columbia to southern California and Nevada, and in the Rocky Mountains to Colorado, east to Quebec and New England; also in Europe. Type locality: Conway, Massachusetts.

2. **Botrychium pumícola** Coville.

Oregon Moonwort. Fig. 4.

Botrychium pumicola Coville in Underw. Nat. Ferns ed. 6, 69. 1900.

Plants stout, fleshy, 8–22 cm. high, the rhizome erect, stout, elongate (2–8 cm. long), copiously radicose; fronds one or sometimes two, erect, 6–14 cm. long, the common stalk hypogean, 4–9 cm. long, thickly sheathed with the stems of old fronds; sterile blade (the apex bent down in vernation) strongly glaucous, sessile, triangular, 2–4 cm. long, 1.5–4 cm. broad, ternately divided, the middle division the largest, broadly oblong to rounded-deltoid, the lateral ones similar or rhombic-oblong, all pinnately parted, the segments closely imbricate, sublunate to flabelliform, broadly crenate to incised, or the larger ones radially cleft into cuneiform lobes; sporophyll with the tip recurved in vernation, sessile or short-stalked, equalling or surpassing the sterile blade.

Fine pumice gravel of open slopes, Hudsonian Zone; Crater Lake National Park, Oregon, altitude about 2500 meters, the type locality.

3. **Botrychium lunària** (L.) Swartz.

Moonwort. Fig. 5.

Osmunda lunaria L. Sp. Pl. 1064. 1753.

Botrychium lunaria Swartz, Journ. Bot. Schrad. **1800**[2]: 110. 1801.

Plants usually stout, 4–30 cm. high, the rhizome erect, rather slender; fronds erect, the common stalk mostly epigean, 3–14 cm. long; sterile blade (the tip bent down and clasping the sporophyll in vernation), sessile or nearly so, oblong to triangular-oblong, rounded at the apex, 1–12 cm. long, 1–5 cm. broad, once pinnately divided, the segments usually close or even imbricate, flabelliform to lunate or reniform, excised at the base below, the outer margins subentire to radially incised, or sometimes cleft into cuneiform lobes; sporophyll straight or with the apex decurved in vernation, at maturity usually surpassing the sterile blade, 2–16 cm. long, the panicle stoutish, 1–3-pinnate.

Moist meadows and open fields, Boreal region; Alaska to Labrador and Newfoundland, southward to Vermont, Michigan, Minnesota and in the mountains to Colorado and southern California. Type locality, European.

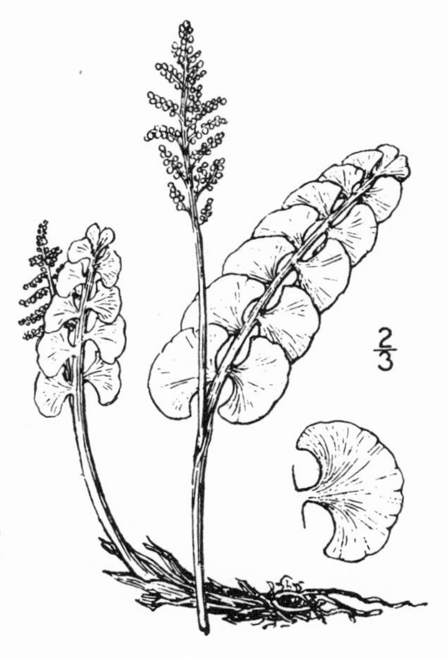

4. Botrychium lanceolàtum (S. G. Gmel.) Ångstr. Lance-leaved Grape-fern. Fig. 6.

Osmunda lanceolata S. G. Gmel. Nov. Comm. Acad. Sci. Petrop. **12:** 516. 1768.
Botrychium lanceolatum Ångstr. Bot. Not. **1854**: 68. 1854.

Plants rather stout, 5–30 cm. high, the rhizome erect; fronds erect, the common stalk mostly epigean, 3–25 cm. long; sterile blade (wholly bent down in vernation) sessile, triangular (usually broadly so), acute, 1–6 cm. long, 1–8 cm. broad, once or twice pinnately divided, the larger primary divisions linear-lanceolate to oblong, or the basal ones oblong-ovate to triangular or triangular-lanceolate, acute, often inequilateral (extended below); lesser divisions oblique, variable, the segments oblong or ovate to linear-lanceolate, blunt or acutish, subentire to obliquely incised; sporophyll entirely bent down in vernation, stout, at maturity equalling or slightly exceeding the sterile blade, sessile or short-stalked, 1–4-pinnate, diffuse, the larger divisions ascending and often subequal.

Grassy, often rocky slopes, Arctic-Alpine Zone; Alaska to Washington, Colorado, and Quebec. Type locality, European.

5. Botrychium virginiànum (L.) Sw. Virginia Grape-fern. Fig. 7.

Osmunda virginiana L. Sp. Pl. 1064. 1753.
Botrychium virginianum Swartz, Journ. Bot. Schrad. **1800**[2]: 111. 1801.

Plants 10–75 cm. high, erect, the common stalk slender, almost wholly epigean, comprising one-half to two-thirds the length of the plant; bud pilose, both the sporophyll and blade wholly bent down; sterile blade deltoid, spreading, sessile, 5–40 cm. broad, nearly as long, ternate, the short-stalked primary divisions 1–2-pinnate; segments numerous, membranous, variable in color and dissection, the ultimate ones mostly oblong, blunt, and toothed at apex; sporophyll long-stalked, panicles 2–3-pinnate, usually with slender divisions.

Moist rich woods, chiefly in the Transition Zone; British Columbia to Labrador, southward to Oregon and nearly throughout the United States; also in Mexico and Eurasia. Type locality: Virginia. The Pacific Coast plants are var. *occidentale* Butters and var. *europaeum* Ångstr., distinguished largely upon characters of the sporangia.

6. Botrychium silaifòlium Presl. Leathery Grape-fern. Fig. 8.

Botrychium silaifolium Presl, Rel. Haenk. **1**: 76. 1825.
Botrychium occidentale Underw. Bull. Torrey Club **25**: 538. 1898.

Rhizome short, with numerous long, thick, strongly corrugated roots; fronds 1 or 2, 20–50 cm. long, fleshy, coriaceous in drying; bud silky-pubescent; common stalk short and stout, entirely hypogean; sterile blade long-stalked, broadly triangular or pentagonal, 10–30 cm. broad, nearly as long, subternately compound, the larger pinnae 2 or 3 times pinnately divided, the basal pair large and nearly 3-pinnate, subpentagonal, the ultimate lateral segments oval or ovate to obovate, cuneate, adnate or decurrent, with irregularly crenulate margins, the terminal segments larger, rhombic; sporophyll long-stalked, stout, 15–45 cm. long, the panicle 2–5-pinnate, very diffuse.

Moist meadows and borders of woods, Boreal Zones; Alaska to northern California, east to Wisconsin and northern New England. Type locality: Nootka Sound, British Columbia.

7. Botrychium califórnicum Underw.
California Grape-fern. Fig. 9.

Botrychium californicum Underw. Torreya 5: 107. 1905.

Rhizome short, with numerous very thick, corrugated roots; fronds lax, 30–70 cm. long; bud hairy; common stalk of medium length, very stout; sterile blade very long-stalked, drooping, rounded-triangular or irregular in outline, usually obtuse, 15–40 cm. broad, nearly as long, ternately compound, the primary divisions 3 or 4 times pinnately or subternately divided; segments large, flat, ovate to rhombic, obtuse or acutish, membrano-herbaceous in drying, the margins minutely denticulate; sporophyll stout, long-stalked, 25–60 cm. long, the panicle 3–5-pinnate, diffuse.

Infrequent, in moist mountain meadows and thickets, in the Transition Zone; northern California, southward in the Sierra Nevada to Tulare County, at middle elevations. Perhaps an extreme, luxuriant form of the last. Type locality: Quincy, California.

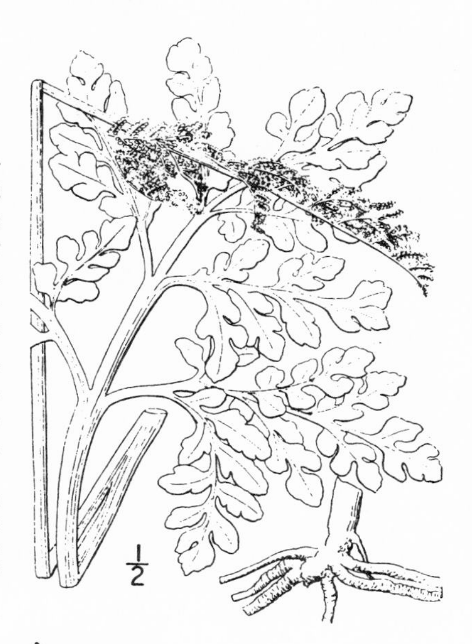

Family 2. POLYPODIÀCEAE.

FERN FAMILY.

Leafy plants of various habit, the rhizomes horizontal and creeping or shorter and oblique or erect, often stout. Fronds pendent to spreading or erect, usually stalked; blades simple to several times pinnatifid or pinnate, or decompound, coiled in vernation. Sporangia borne promiscuously on the under side of wholly fertile blades, or upon slender or contracted, partially foliose or non-foliose blades or parts of blades, or, as in most of our species, in clusters (sori) upon the backs of ordinary foliaceous blades; long-stalked, provided with an incomplete vertical ring of thickened hygroscopic cells (the annulus) and opening transversely. Sori either with or without a membranous protective covering (indusium). Prothallia green.

About 150 genera and 6,000 species, widely distributed, including by far the greater number of living ferns.

Sori dorsal upon the veins, separate, not marginal.
  Indusium wholly or partially inferior.
    Indusium wholly inferior, the divisions stellate or spreading. 1. *Woodsia.*
    Indusium attached by its base at one side, hood-shaped, thrust back by the ripening sporangia. 2. *Cystopteris.*
  Indusium, if present, superior.
    Sori round to oval.
      Stipes jointed to the rhizome; blades pinnatifid or pinnatisect; indusia wanting. 3. *Polypodium.*
      Stipes continuous with the rhizome; blades 1–3-pinnate; indusia present in most species.
        Indusium orbicular, centrally peltate. 4. *Polystichum.*
        Indusium (if present) roundish-reniform, attached at its sinus. 5. *Dryopteris.*
    Sori oblong or linear to lunate or hippocrepiform (roundish in *Athyrium americanum*).
      Venation partially areolate, the large tumid sori borne in a chain-like row close to the midribs. 6. *Woodwardia.*
      Venation wholly free; sori small, oblique.
        Blades small, evergreen; rhizome scales with dark-walled cells; sori oblong to linear, straight or nearly so. 7. *Asplenium.*
        Blades large, delicate; rhizome scales with thin-walled cells; sori mostly lunate to hippocrepiform or roundish. 8. *Athyrium.*
Sori marginal or submarginal (borne at or near the apex of the veins) or in a few cases the sporangia decurrent on the veins or completely covering them.
  Sporangia following the veins throughout. 9. *Pityrogramma.*
  Sporangia borne at or near the apex of the veins.
    Fronds dimorphous, the fertile blades with contracted linear segments.
      Blades linear, pinnatisect or simply pinnate, the segments of the fertile ones with a true intramarginal indusium. 10. *Struthiopteris.*
      Blades broader, 1–3-pinnate, the divisions of the fertile ones with a broadly revolute indusiform margin. 11. *Cryptogramma.*
    Fronds (fertile and sterile) alike or nearly so.
      Plants large, coarse; sporangia borne on a veinlike receptacle connecting the ends of the veins; indusium double, the inner one minute, concealed. 12. *Pteridium.*
      Plants mainly small; sporangia not borne on a special receptacle; indusia (if any) single.
        Sporangia borne on the under side of sharply reflexed membranous lobes. 13. *Adiantum.*
        Sporangia not borne on the back of reflexed lobes.
          Vein-ends distinctly thickened; proper indusium often present. 14. *Cheilanthes.*
          Vein-ends scarcely or not at all enlarged; proper indusium invariably wanting.
            Margin of segments widely reflexed or revolute, usually modified; blades glabrous or nearly so. 15. *Pellaea.*
            Margin of segments narrowly or not at all revolute; blades variously hairy, scaly, or ceraceous beneath. 16. *Notholaena.*

1. WOÒDSIA R. Br. Prodr. Fl. Nov. Holl. 1: 158. 1810.

Small ferns of rocky situations, the short-creeping or ascending rhizomes densely tufted. Fronds numerous, fasciculate, erect or spreading, in several species the stipes jointed above the base and separable; blades linear to lance-ovate, 1–2-pinnate, the segments lobed or pinnatifid, glabrous, variously hairy, or paleaceous; veins free. Sori roundish, separate, often confluent with age; indusia inferior in attachment, either roundish and early cleft into lacerate divisions, or deeply stellate, the filiform divisions mostly concealed beneath the sporangia or inflexed and partially covering them. [Name in honor of Joseph Woods (1776–1864), an English architect and botanist.]

About 35 species, mainly of temperate and boreal regions. Six species besides the two following occur in the United States. Type species, *Polypodium ilvense* L.

Blades glandular-puberulent and bearing flattish, septate, whitish hairs. 1. *W. scopulina.*
Blades glabrous; plants smaller. 2. *W. oregana.*

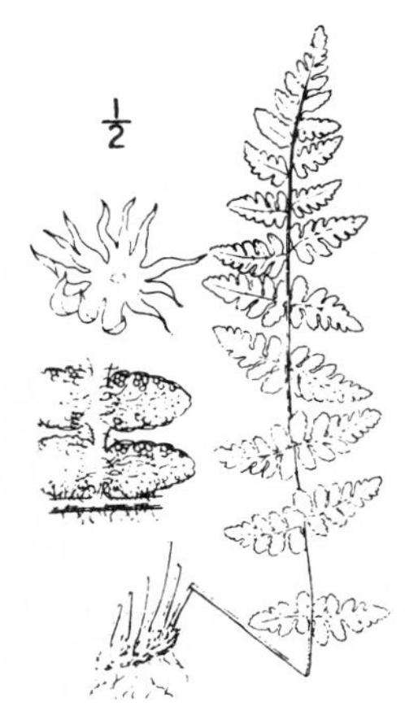

1. Woodsia scopulìna D. C. Eaton.

Rocky Mountain Woodsia. Fig. 10.

Woodsia scopulina D. C. Eaton, Canad. Nat. II. 2: 91. 1865.

Rhizomes short-creeping, forming large tufts, clothed with thin, pale, brown, ovate-oblong scales. Fronds very numerous, tufted, 10–40 cm. long; stipes 3–15 cm. long, inflated, stramineous or golden brown from a castaneous chaffy base; blades linear to oblong-lanceolate, short-acuminate, 6–25 cm. long, 1.5–6 cm. broad, 1–2-pinnate; pinnae numerous, oblong to deltoid-ovate, acute or acutish, the 5–9 pairs of serrate or deeply toothed oblong segments lightly joined, or the larger ones free; surfaces (especially the lower) glandular-puberulent and bearing few to many, flat, septate, whitish hairs; sori supramedial; indusia deeply cleft, the divisions spreading, lacerate, with filamentous tips; leaf tissues membro-herbaceous, dull green.

Crevices and tálus of cliffs, in the Transition Zone; Alaska to Quebec (Gaspé County), Ontario, South Dakota, Colorado, and Utah, and in the Sierra Nevada sparingly to Tulare County, California; also in West Virginia and North Carolina. Type locality: Rocky Mountains (40° N. Lat.).

2. Woodsia oregána D. C. Eaton.

Oregon Woodsia. Fig. 11.

Woodsia oregana D. C. Eaton, Canad. Nat. II. 2: 90. 1865.

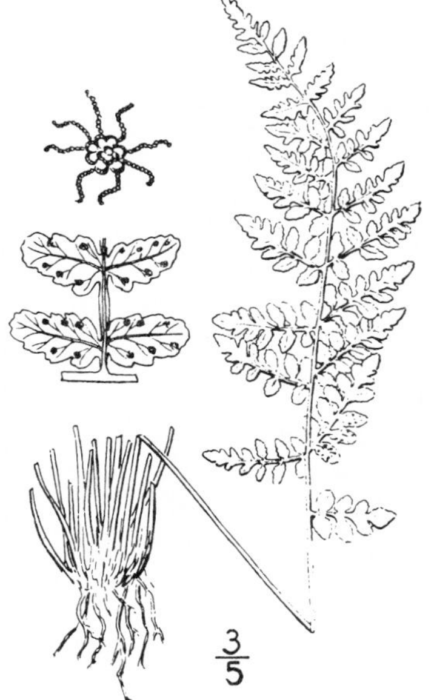

Rhizomes short-creeping, rather slender, tufted, densely paleaceous, the scales thin, pale brown, often dark-striped. Fronds many, erect, fasciculate, 6–25 cm. long; stipes 1.5–12 cm. long, slender, stramineous from a castaneous base; blades linear to lance-oblong, acutish, 5–13 cm. long, 1–3.5 cm. broad, pinnate, glabrous; pinnae 6–12 pairs, mostly deltoid-oblong (the lower ones deltoid, distant), acute or acutish, subpinnate at the base; segments few, apart, mostly oblong, obtuse, crenate or crenately lobed, the larger ones constricted at the base and nearly free, the others decurrent and rather broadly joined; leaf tissue delicately herbaceous, the teeth commonly reflexed and partially concealing the supramedial sori; indusia minute, divided nearly to the base into a few short, turgid, moniliform segments.

Crevices of dryish cliffs and rock slopes, chiefly in the Arid Transition Zone; British Columbia to South Dakota, Nebraska, New Mexico, and Arizona, and in the Sierra Nevada to southern California (San Bernardino Mountains); also in eastern Quebec. Type locality: Dalles, Oregon.

2. CYSTÓPTERIS Bernh. Schrad. Neues Journ. Bot. 1^2: 26. 1806.

[Filix Adans. Fam. Pl. 2: 20, 558. 1763.]

Small, delicate ferns of shaded rocky or alluvial situations, with slender creeping rhizomes. Fronds erect to recurved-spreading, more or less succulent, the stipes slender, not jointed to the rhizome; blades 1–4-pinnate, delicate, the fertile ones commonly less foliose and longer-stalked than the sterile. Sori roundish, dorsal, separate; indusium membranous, hoodlike, attached at the inner side of the broad base and partly under the sorus, early thrust back by the sporangia and partly concealed, withering, the sori thus appearing naked with age.

About ten species, chiefly of temperate and boreal regions. Two besides *F. fragilis* occur in North America. Type species, *Polypodium bulbiferum* L.

1. **Cystopteris frágilis** (L.) Bernh. Brittle-fern. Fig. 12.

Polypodium fragile L. Sp. Pl. 1091. 1753.
Filix fragilis Gilib. Exerc. Phyt. 558. 1792.
Cystopteris fragilis Bernh. Schrad. Neues Journ. Bot. 1[2]: 27. *pl. 2, f. 9.* 1806.

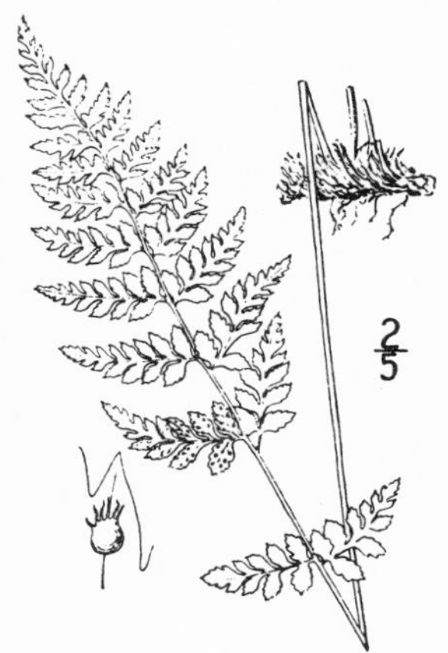

Rhizome creeping, paleaceous toward the apex, the scales thin, ovate, acuminate. Fronds few or several, erect-spreading, usually clustered; stipes about as long as the blades, stramineous to light castaneous, delicate, fragile; blades extremely variable, mostly oblong-lanceolate to lance-ovate, acuminate, about 10–25 cm. long, nearly or quite 2-pinnate; pinnae deltoid to deltoid-oblong or oblong-ovate, acutish to acuminate, glabrous; pinnules few or many, spreading, mostly ovate or oblong-lanceolate, pinnatifid or incised, acutish, mostly decurrent and joined, membranous; veinlets mostly excurrent to marginal teeth; sori small; indusia roundish or ovate-acuminate, deeply convex, the apex often toothed.

Rocky woods and low moist situations, Humid Transition to Arctic Zones; Alaska to Labrador and Newfoundland, south to southern California, the Mexican Border region, Oklahoma, Alabama and Georgia; also in Eurasia and tropical America. Polymorphic. Type locality, European.

3. **POLYPÒDIUM** [Tourn.] L. Sp. Pl. 1082. 1753.

Mostly shade-loving species, of various habitat, commonly epiphytic, the rhizomes usually slender and creeping, often widely so. Fronds uniform or subdimorphous, usually articulate to knob-like prominences of the rhizome; blades simple, 1–3-pinnate, or variously pinnatifid, glabrous, pubescent, or paleaceous; venation free or variously areolate. Sori round to elliptical, relatively large, dorsal or sometimes terminal, invariably non-indusiate. [Name Greek, meaning many feet, alluding to the numerous knoblike prominences of the rhizome.]

Several hundred species, mainly of tropical and subtropical regions. A few additional species occur in other parts of the United States. Type species, *Polypodium vulgare* L.

Segments rigidly coriaceous, the midribs scaly at first; sori very large, crowded against the midrib. 1. *P. scouleri.*
Segments membranous to herbaceous, not scaly; sori smaller, apart from the midrib.
  Blades mostly less than 15 cm. long; segments oblong to elliptical; sori medial, few. 2. *P. hesperium.*
  Blades mostly 15 to 50 cm. long; segments linear; sori inframedial, numerous.
    Blades usually lanceolate; segments linear-attenuate, mostly dilatate; veins free, oblique-spreading, mostly translucent. 3. *P. glycyrrhiza.*
    Blades oblong to deltoid or deltoid-ovate; segments oblong-linear, mostly decurrent; veins casually or regularly areolate, oblique, mostly opaque. 4. *P. californicum.*

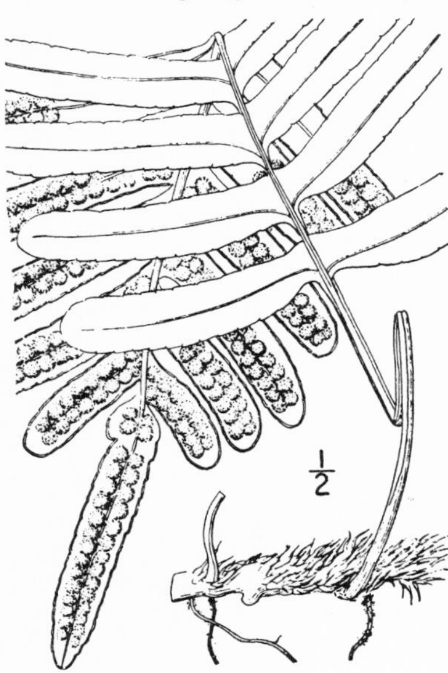

1. **Polypodium scoùleri** Hook. & Grev.

Coast Polypody. Fig. 13.

Polypodium scouleri Hook. & Grev. Icon. Fil. 1: *pl. 56.* 1829.
Polypodium pachyphyllum D. C. Eaton, Am. Journ. Sci. II. 22: 138. 1856.
Polypodium carnosum Kell. Proc. Calif. Acad. 2: 88. *f. 24.* 1861.

Rhizome creeping, woody, 6–10 mm. thick, laxly paleaceous, naked with age, whitish-pruinose; scales dark castaneous, attenuate from a deltoid-ovate base, about 1 cm. long, denticulate. Fronds few, mostly 15–70 cm. long; stipes stout, rigid, shorter than the blade, naked; blades deltoid-ovate, mostly 10–40 cm. long, pinnatisect; segments 2–14 pairs, linear to linear-oblong, obtuse, 7–20 mm. broad, spreading, adnate, slightly decurrent, very rigidly coriaceous, the cartilaginous border crenate to obscurely crenulate-serrate; midribs elevated, deciduously scaly beneath; veins joined in a single series of areoles; sori very large, crowded against the midrib, mostly confined to the upper segments.

On mossy tree-trunks, cliffs, and rock-slopes, Humid Transition Zone; along the coast from Santa Cruz County, California, to Vancouver Island; also on Guadalupe Island, Lower California. Type locality: Columbia River region.

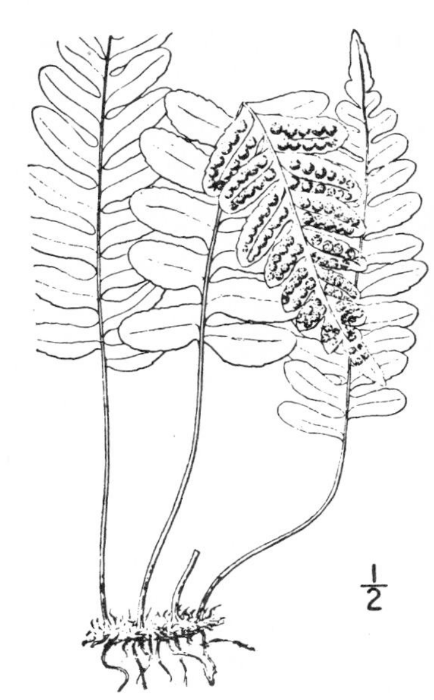

2. **Polypodium hespèrium** Maxon.

Western Polypody. Fig. 14.

Polypodium hesperium Maxon, Proc. Biol. Soc. Washington **13**: 200. 1900.

Rhizome creeping, firm, about 5 mm. thick, densely paleaceous; scales ovate, long-acuminate, 3–5 mm. long, pale to dark castaneous, denticulate. Fronds rather close, ascending, mostly 10–24 cm. long; stipes nearly or quite as long as the blades, stramineous, naked; blades deltoid-oblong to linear-oblong (rarely linear), usually long-acuminate to attenuate, 7–18 cm. long, pinnatifid nearly to the naked rachis; segments spreading, alternate, narrowly oblong to oval, elliptical, or subspatulate, usually rounded-obtuse, crenate to obscurely crenate-serrulate; veins mostly twice-forked and translucent; sori broadly oval, medial; leaf tissue membrano-herbaceous, light to yellowish green, or subglaucous.

Cliffs and rock-slopes, chiefly in the Canadian Zone; Yukon to South Dakota, northern New Mexico, Arizona, and southern California (San Bernardino Mountains); also in Lower California (San Pedro Martir). Type locality: Coyote Cañon, Lake Chelan, Washington.

3. **Polypodium glycyrrhìza** D. C. Eaton.

Licorice Fern. Fig. 15.

Polypodium vulgare occidentale Hook. Fl. Bor. Am. **2**: 258. 1840.
Polypodium falcatum Kell. Proc. Calif. Acad. **1**: 20. 1854, not L. f. 1781.
Polypodium glycyrrhiza D. C. Eaton, Am. Journ. Sci. II. **22**: 138. 1856.
Polypodium occidentale Maxon, Fern Bull. **12**: 102. 1904.

Rhizome creeping, 3–5 mm. thick, often compressed, deciduously paleaceous; scales oblong-ovate, long-acuminate, 5–9 mm. long, rusty brown, thin. Fronds mostly distant, 20–70 cm. long; stipes usually much shorter than the blades, firm, stramineous, naked; blades usually lanceolate, abruptly attenuate or caudate, 15–50 cm. long, 5–17 cm. broad, pinnatisect; segments thin-herbaceous, alternate, linear-attenuate, tapering from the middle or from the dilatate base, often falcate, unevenly curvescent-serrate; basal segments distant, shorter, with rounded sinuses; veins free, oblique-spreading, mostly 3-forked and translucent; sori roundish-oval, inframedial, not extending to the end of the segments.

On rocks, logs, and mossy tree trunks, Humid Transition Zone; Alaska to San Mateo County, California, mainly near the coast. Type locality: southwestern Oregon.

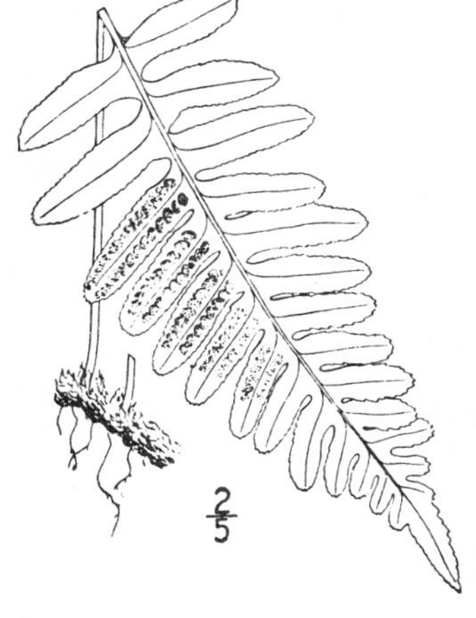

4. **Polypodium califórnicum** Kaulf.

California Polypody. Fig. 16.

Polypodium californicum Kaulf. Enum. Fil. 102. 1824.
Polypodium intermedium Hook. & Arn. Bot. Beechey 405. 1841.

Rhizome creeping, 4–10 mm. thick, deciduously paleaceous; scales deltoid-ovate, long-acuminate, 3–7 mm. long, rusty brown, thin, appressed, imbricate. Fronds ascending, mostly 20–40 cm. long; stipes usually shorter than the blades, stout, stramineous, naked; blades oblong to deltoid or deltoid-ovate, acutish to acuminate, mostly 15–30 cm. long, 7–15 cm. broad, pinnatifid nearly to the rachis; segments membranous to thin-herbaceous, spreading, oblong-linear, usually obtuse or acutish, mostly decurrent (the lower ones slightly apart, excavate below), unevenly serrate to crenate-serrate; veins mostly dark, opaque, oblique, 3 to 5 times forked, producing an irregular series of areoles or only casually joined; sori oval, slightly inframedial.

Rock-crevices or rocky soil, Upper Sonoran and Transition Zones; southern California (common), northward to Butte County, mainly in the coastal region; also in Lower California and on several islands off the California coast. Type locality: California.

4. POLÝSTICHUM Roth, Röm. Arch. Bot. 2[1]: 106. 1799.

Coarse, large to rather small ferns of ledges and rocky woods, the stout woody rhizomes mostly erect to decumbent, copiously paleaceous. Fronds several, rigidly ascending or recurved, borne usually in a crown, the blades similar or subdimorphous, simple to 3-pinnate-pinnatifid, more or less paleaceous, the segments of harsh texture, mainly auriculate, usually with serrate margins, the teeth pungent to spinulose or aristate; veins free. Sori round, dorsal to subterminal; indusia superior, orbicular, centrally peltate, persistent to caducous, or sometimes wanting. [Name Greek, meaning many rows, alluding to the numerous regular rows of sori in *P. lonchitis*.]

A genus of about 100 species, mainly of boreal and temperate regions, agreeing closely in habit. Type species, *Polypodium lonchitis* L.

Blades simply pinnate, the pinnae variously serrate to evenly incised, never pinnately lobed to pinnatifid.
- Stipes very short; pinnae mostly oblong-lanceolate (the lower ones deltoid), serrate-dentate, the teeth conspicuously spreading-spinulose. 1. *P. lonchitis.*
- Stipes much longer; pinnae linear-attenuate, biserrate to incised, the teeth incurved, rigidly long-aculeate. 2. *P. munitum.*

Blades more compound, the pinnae at least pinnately lobed or divided at the base.
- Pinnae pinnately lobed to pinnatifid at the base.
  - Segments very numerous, small, the lobes merely crenulate-dentate. 3. *P. lemmoni.*
  - Segments few, large, pungent to spinulose-aristate.
    - Pinnae inequilateral, deltoid-ovate to deltoid-oblong, the lobes and teeth pungent but not aristate. 4. *P. scopulinum.*
    - Pinnae larger, linear from a broader base, the lobes more numerous, distinctly aristate. 5. *P. californicum.*
- Pinnae nearly or quite pinnate.
  - Blades proliferous below the tip, slightly scaly beneath; teeth long-awned; indusia erose-dentate. 6. *P. andersoni.*
  - Blades not proliferous, copiously scaly beneath; teeth short-awned; indusia long-ciliate. 7. *P. dudleyi.*

1. Polystichum lonchìtis (L.) Roth.

Holly-fern. Fig. 17.

Polypodium lonchitis L. Sp. Pl. 1088. 1753.
Polystichum lonchitis Roth, Röm. Arch. Bot. 2[1]: 106. 1799.
Aspidium lonchitis Swartz, Journ. Bot. Schrad. **1800**[2]: 30. 1801.

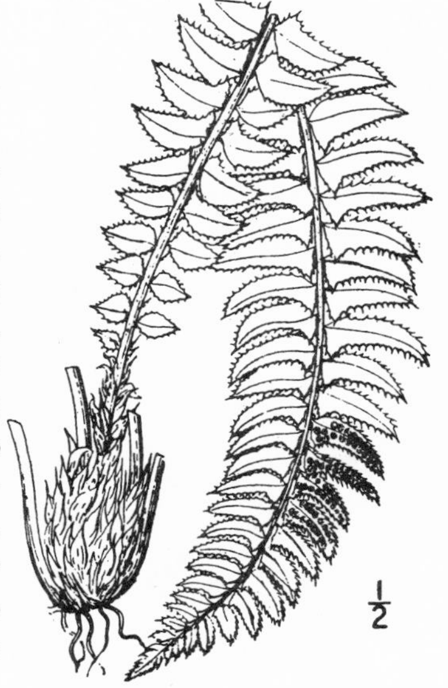

Rhizome large, woody, erect or decumbent. Fronds several, rigidly ascending in a close crown, 15–60 cm. long; stipes 1–6 cm. long, thick, stramineous, densely paleaceous, the scales rusty brown, mostly large, ovate, denticulate; blades linear to narrowly linear-oblanceolate, acuminate, 12–55 cm. long, 2–7 cm. broad, tapering at the base, simply pinnate, the rachis very stout, deciduously paleaceous; pinnae numerous, close, the basal ones deltoid (sometimes minute), equilateral, the middle and upper ones oblong-lanceolate, falcate, acute, auriculate above, cuneate below, unevenly serrate-dentate, the teeth conspicuously spreading-spinulose; sori large, contiguous, usually in 2 rows, nearly medial; indusia large, entire; leaf tissue evergreen, subcoriaceous, fibrillose beneath.

Rocky shaded slopes, in the Boreal region; Alaska to Nova Scotia, southern Ontario, Michigan, and Montana, and in the mountains to Utah, Colorado, and northern California (Castle Lake, Siskiyou County); also in Greenland; Eurasia. Type locality, European.

2. **Polystichum munìtum** (Kaulf.) Presl. Western Sword-fern. Fig. 18.

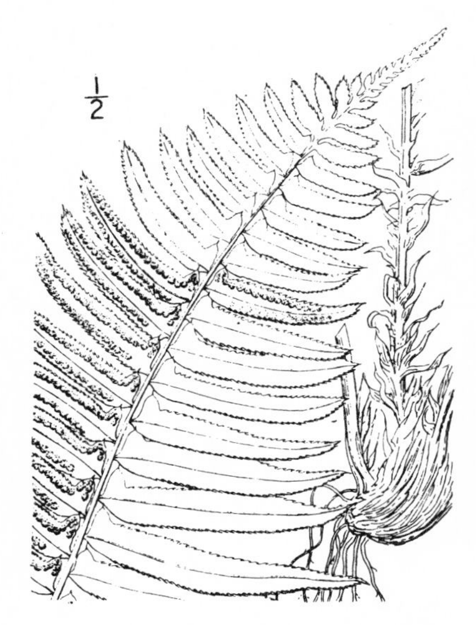

Aspidium munitum Kaulf. Enum. Fil. 236. 1824.
Polystichum munitum Presl, Tent. Pter. 83. 1836.

Rhizome stout, woody, ascending. Fronds several, rigidly ascending in a crown, 30–140 cm. long; stipes stout, 5–60 cm. long, together with the rachis densely paleaceous, the scales large, ascending, bright glossy brown, often dark-centered, ovate to oblong-acuminate, with many small laciniate-ciliate ones interspersed; blades linear-lanceolate, short-acuminate, 25–100 cm. long, 5–25 cm. broad, simply pinnate; pinnae very many, spreading, subfalcate, linear-attenuate, strongly auriculate at the base above, cuneate below, often very chaffy beneath, sharply biserrate to incised, the teeth incurved, rigidly long-aculeate; sori large, borne in a close medial to submarginal row, or sometimes in several crowded rows; indusia papillose-dentate to long-ciliate; leaf tissue coriaceous, evergreen.

Damp wooded slopes, chiefly in the Humid Transition Zone; Alaska to northwestern Montana, northern Idaho, and extreme southern California (Cuyamaca Mountains). Type locality: California. Extremely variable. Attains its best development in the Coast Ranges from the Santa Cruz Peninsula to Washington in a luxuriant form with thick, scurfy, deeply serrate to incised, copiously fertile pinnae, known as the variety *inciso-serratum* (D. C. Eaton) Underw.

2a. **Polystichum munitum ímbricans** (D. C. Eaton) Maxon. Imbricated Sword-fern. Fig. 19.

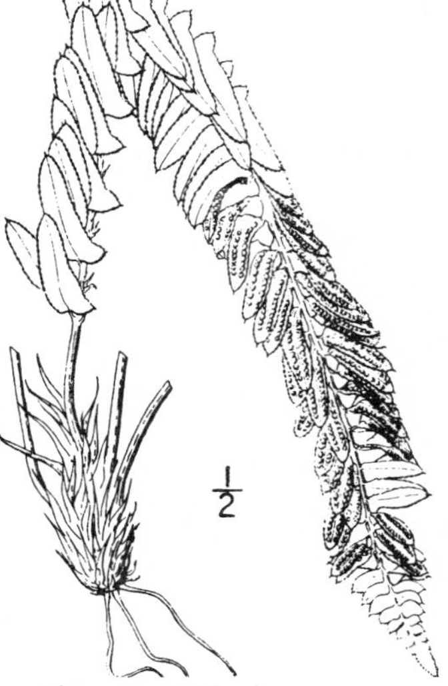

Aspidium munitum imbricans D. C. Eaton, Ferns N. Am. 1: 188. *pl. 25, f. 3.* 1878.
Polystichum munitum imbricans Maxon, Fern Bull. 8: 30. 1900.

Fronds small, 30–50 cm. long; stipes with a conspicuous basal tuft of lance-attenuate, glossy, castaneous scales, nearly naked above; blades usually linear, the stout rachis naked with age; pinnae crowded, obliquely imbricate, acutish, abruptly long-cuspidate; indusia rather thick, never ciliate.

Open rocky slopes, mainly in the Canadian Zone; Vancouver Island and eastern Washington to southern California; less common than the typical form near the coast. A well-marked subspecies, connected with the typical form by numerous intermediate specimens. Type locality: Plumas County, California.

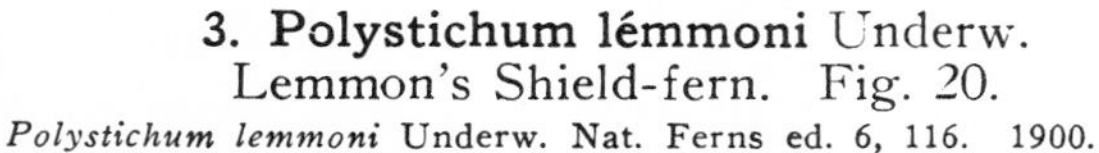

3. **Polystichum lémmoni** Underw. Lemmon's Shield-fern. Fig. 20.

Polystichum lemmoni Underw. Nat. Ferns ed. 6, 116. 1900.

Rhizomes tufted, ascending, stout, thickly covered with old stipe-bases. Fronds ascending, 10–40 cm. long; stipes 3–15 cm. long, stramineous from a very paleaceous, brownish base, the scales mostly large, lanceolate to ovate-acuminate, fulvous; blades linear to narrowly lance-oblong, acutish, 8–28 cm. long, 1.5–5 cm. broad, pinnate-pinnatifid; pinnae numerous, usually imbricate (only the slightly reduced lower ones apart), deltoid-oblong to deltoid-ovate, obtuse or rounded-acutish, pinnately divided at the base, deeply lobed or (often) divided nearly to the tip, the divisions close, spreading, trapeziform-ovate to obliquely oval or obovate, obtuse, crenate or crenately lobed; margins crenulate-dentate; sori borne on the apical pinnae, large, close, 1 or several to the segment; indusia large, thin, bullate, erose-dentate; leaf tissue carnose, herbaceous in drying, venose.

1/2

Moist granitic soil, among loose rocks, Canadian and Hudsonian Zones; Siskiyou and Trinity Counties, northern California, and the Mount Stuart region, Washington; ascribed also to Alaska. Type locality: vicinity of Mount Shasta, California.

4. Polystichum scopulìnum (D. C. Eaton) Maxon.

Eaton's Shield-fern. Fig. 21.

Aspidium aculeatum scopulinum D. C. Eaton, Ferns N. Am. **2**: 125. *pl. 62, f. 8.* 1880.
Polystichum scopulinum Maxon, Fern Bull. **8**: 29. 1900.

Rhizome stout, erect or decumbent, copiously paleaceous; scales light castaneous, linear to ovate-oblong, attenuate, denticulate. Fronds 6–10, erect-spreading, 15–40 cm. long; stipes 3–14 cm. long, stout, sulcate, densely paleaceous at the base, deciduously so (together with the rachis) above; blades linear to narrowly oblong-lanceolate, acuminate, 12–30 cm. long, 2.5–6 cm. broad, pinnate; pinnae numerous, inequilateral, deltoid-ovate to deltoid-oblong, pinnately lobed or divided at the base, less deeply so toward the obtuse to acutish-mucronate apex, the lobes and teeth oblique, pungent, not aristate; lower pinnae usually reduced, distant; sori numerous, large, close, infra-medial, in 2 confluent rows; indusia ample, thin, erose-dentate; leaf tissue coriaceous, deciduously fibrillose-paleaceous beneath.

Dry cliffs and rock crevices, Canadian and Hudsonian Zones; central Washington to eastern Idaho, south to Utah and southern California; also in Gaspé County, Quebec. Type locality: Upper Teton Cañon, Idaho.

5. Polystichum califórnicum (D. C. Eaton) Underw.

California Shield-fern. Fig. 22.

Aspidium californicum D. C. Eaton, Proc. Am. Acad. **6**: 555 1865.
Polystichum californicum Underw. Nat. Ferns ed. 6, 116. 1900.

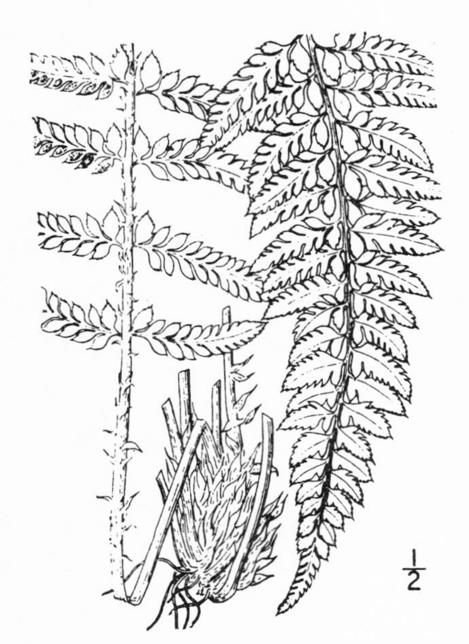

Rhizome stout, suberect. Fronds several, ascending, 30–110 cm. long; stipes 6–35 cm. long, stout, sulcate, nearly naked above the copiously paleaceous base, the scales large, dark castaneous, lance-attenuate to ovate-acuminate; blades linear-oblong to narrowly linear-lanceolate, attenuate, 20–75 cm. long, 5–20 cm. broad, slightly or not reduced at the base, the rachis minutely paleaceous; pinnae numerous, ascending or spreading (the lower ones deflexed), linear from a broader base, obliquely pinnatifid to pinnately incised, the lobes elliptical, decurrent, close, broadly joined, distinctly aristate, as also the teeth of the large basal segments; sori large, several pairs to each segment or lobe; indusia large, erose-ciliate; leaf tissue subcoriaceous, or coriaceous with age, fibrillose beneath.

Creek banks and cañons, in the Humid Transition Zone; coast ranges of Sonoma, Napa, Marin, Santa Clara, and Santa Cruz Counties, California; also in western Washington (Eatonville). Type locality: Santa Cruz Mountains, California.

6. **Polystichum andersòni** Hopkins.

Anderson's Shield-fern. Fig. 23.

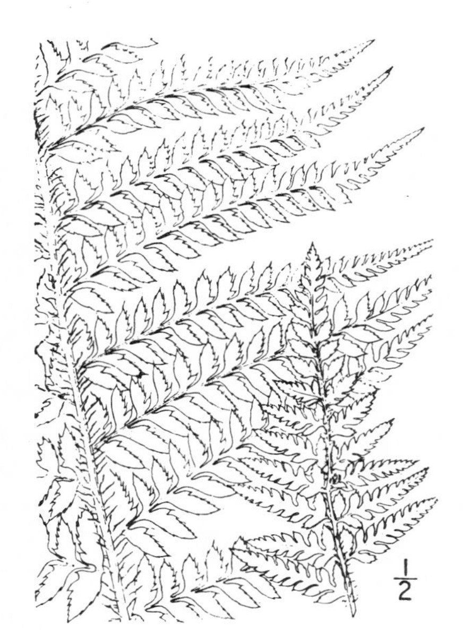

Polystichum andersoni Hopkins, Am. Fern Journ. **3**: 116. *pl. 9.* 1913.

Polystichum jenningsii Hopkins, Ann. Carnegie Mus. **11**: 362. *pl. 37.* 1917.

Rhizome stout, decumbent; scales thin, pale castaneous, large, linear to ovate, denticulate. Fronds in a close crown, ascending, 40–110 cm. long; stipes 5–25 cm. long, sulcate, paleaceous; blades nearly 2-pinnate, narrowly lance-oblong to lance-elliptic, long-acuminate, 35–85 cm. long, 8–20 cm. broad at the middle; rachis sulcate, paleaceous, proliferous below the tip; pinnae numerous, slightly ascending, narrowly triangular, nearly pinnate at the base (the basal segments enlarged), less deeply pinnatifid toward the attenuate tip, minutely paleaceous beneath, glabrate above; segments oblique, elliptical, mostly adnate or semiadnate, decurrent, strongly serrate, the teeth long-awned; sori large, 2–6 pairs, nearly medial; indusia erose-dentate, the teeth gland-tipped.

Moist rocky slopes, usually in alder thickets, Transition Zone; southern British Columbia to Glacier National Park, Montana, and the region of Mount Rainier, Washington. Type locality: Elk River, Strathcona Park, Vancouver Island, British Columbia.

7. **Polystichum dúdleyi** Maxon.

Dudley's Shield-fern. Fig. 24.

Polystichum dudleyi Maxon, Journ. Washington Acad. Sci. **8**: 620. 1918.

Rhizome decumbent, densely paleaceous; scales brown to fulvous, thin, large, linear to ovate, denticulate-ciliolate. Fronds ascending, 40–120 cm. long; stipes 15–45 cm. long, sulcate, chaffy; blades 2-pinnate, oblong-lanceolate to narrowly ovate, long-acuminate to attenuate, 25–75 cm. long, 8–25 cm. broad; pinnae mostly oblique, contiguous, linear to narrowly oblong-lanceolate from a broader unequal base, attenuate, copiously filiform-paleaceous beneath, scantily so above; pinnules obliquely ovate or ovate-oblong, auriculate, the upper basal ones large, obliquely cleft or divided, the others serrate to incised, the curved teeth short-awned; sori numerous, terminal upon the veinlets or nearly so; indusia delicate, long-ciliate.

Rocky slopes of cañons, mainly in the Humid Transition Zone; Santa Cruz Peninsula region, California. Type locality: Santa Cruz Mountains, San Mateo County, California.

5. DRYÓPTERIS Adans. Fam. Pl. **2**: 20, 550. 1763.

[ASPIDIUM Swartz, Journ. Bot. Schrad. **1800²**: 29. 1801, in part.]

Mainly woodland ferns of upright habit, the rhizomes slender, wide-creeping, and nearly naked, to thick, erect, and copiously chaffy. Fronds borne singly or in a crown, usually stipitate, not jointed to the rhizome; fertile and sterile blades usually alike, 1–3-pinnate or decompound (rarely simple), glabrous or variously pubescent, only a few species conspicuously paleaceous; veins free in northern species, connivent to copiously areolate in many tropical and subtropical species. Sori mostly roundish, dorsal, indusiate or not, the indusium (if present) in northern species roundish-reniform, fixed at its sinus. [Name Greek, meaning oak-fern.]

A genus of several hundred species, mainly tropical. Besides the following, numerous species occur in other parts of the United States. Type species, *Polypodium filix-mas* L.

Indusia wanting; blades deltoid.
- Blades 2–3-pinnate below, glabrous and naked; basal pinnae long-stalked, deltoid. 1. *D. dryopteris.*
- Blades pinnate-pinnatifid, hairy beneath, the midribs scaly; basal pinnae subsessile, narrowly ovate. 2. *D. phegopteris.*

Indusia present; blades lanceolate to ovate or oblong (sometimes deltoid in no. 8).
- Blades 1–2-pinnate.
  - Segments entire or nearly so; veins simple or once forked.
    - Blades oblong to ovate-oblong; segments densely puberulous beneath. 3. *D. feei.*
    - Blades lanceolate or oblanceolate, distinctly narrowed at the base; segments glabrous or slightly hairy beneath.
      - Segments 4–5 mm. broad, the margins distinctly hyaline-papillose; veins often forked. 4. *D. oreopteris.*
      - Segments about 2 mm. broad, the margins slightly hyaline; veins mostly simple. 5. *D. oregana.*
  - Segments distinctly toothed; veins twice or several times forked.
    - Pinnae sessile, oblong-lanceolate, the lower basal pinnule usually semicordate and overlying the main rachis; veinlets spreading, all ending in salient spinelike teeth. 6. *D. arguta.*
    - Pinnae mostly short-stalked, deltoid-lanceolate, the basal pinnules symmetrical, apart from the main rachis; veinlets oblique, fewer, ending in oblique usually curved acute teeth. 7. *D. filix-mas.*
- Blades essentially 3-pinnate. 8. *D. dilatata.*

1. Dryopteris linnaeana C. Chr.
Oak-fern. Fig. 25.

Polypodium dryopteris L. Sp. Pl. 1093. 1753.
Phegopteris dryopteris Fée, Gen. Fil. 243. 1852.
Dryopteris linnaeana C. Chr. Ind. Fil. 275. 1905.
Dryopteris dryopteris Christ, Bull. Acad. Internat. Geogr. Bot. **20¹**: 151. 1909.

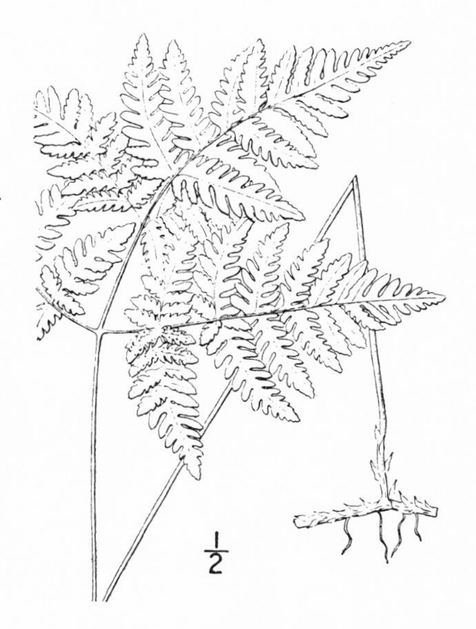

Rhizome cordlike, wide-creeping, the branches bearing a few lax ovate brownish scales. Fronds erect, distant, 20–60 cm. long; stipes slender, very much longer than the blades, stramineous from a chaffy blackish base; blades deltoid, nearly equilateral, 8–25 cm. long and broad, subternate by the enlargement of the basal pinnae, these long-stalked, deltoid, inequilateral, acute, 1–2-pinnate, the larger pinnules pinnately lobed or divided; upper pinnae gradually simpler, sessile or adnate; lobes oblong, rounded-obtuse, entire to serrate-crenate; sori rather small, submarginal, non-indusiate; leaf tissue thin-herbaceous, glabrous.

Moist woods, thickets, and swamps, Canadian and Transition Zones; Alaska to Newfoundland, south to Oregon, Arizona, New Mexico, Kansas, South Dakota, Wisconsin, and the mountains of Virginia; Greenland; Eurasia. Type locality, European.

2. **Dryopteris phegópteris** (L.) C. Chr.

Beech-fern. Fig. 26.

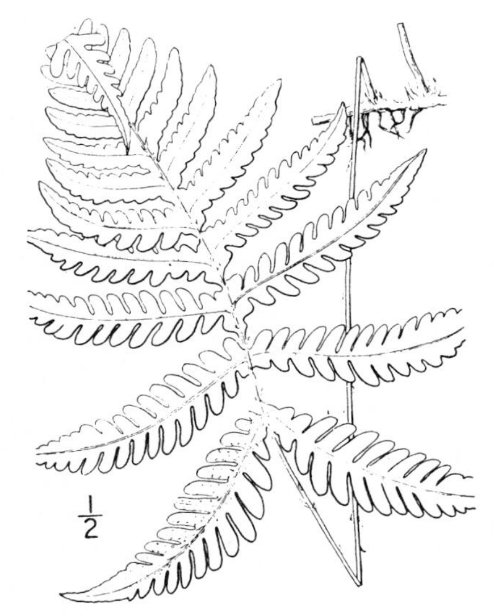

Polypodium phegopteris L. Sp. Pl. 1089. 1753.
Phegopteris polypodioides Fée, Gen. Fil. 243. 1852.
Phegopteris phegopteris Underw.; Small, Bull. Torrey Club **20**: 462. 1893.
Dryopteris phegopteris C. Chr. Ind. Fil. 284. 1905.

Rhizome very slender, wide-creeping; scales few, thin, appressed, pale, ovate. Fronds few, distant, 20–55 cm. long; stipes mostly much longer than the blades, stramineous from a dark base, scantily paleaceous; blades deltoid, 12–25 cm. long, 9–18 cm. broad, long-acuminate, pinnate at the base, the rachis and midribs scantily clothed with small lax rusty lanceolate scales; pinnae mostly close, adnate, horizontal, linear or oblong-lanceolate, attenuate, obliquely pinnatifid, the lower ones distant, subsessile, deflexed and advanced, inequilateral, narrowly ovate; segments thin-herbaceous, close, oblong, rounded-obtuse, entire or crenate, freely whitish-ciliate, white-hairy on both surfaces, chiefly along the veins; sori rather large, submarginal, naked.

Moist, often rocky forests, in the Canadian and Transition Zones; Alaska to Newfoundland, south to Washington (Skamania County), Minnesota, Wisconsin, Ohio, and the mountains of Virginia; Greenland; Eurasia. Type locality, European.

3. **Dryopteris feei** C. Chr.

Downy Wood-fern. Fig. 27.

Aspidium puberulum Fée, Mém. Foug. **10**: 40. 1865, not Gaud. 1827.
Nephrodium puberulum Baker in Hook. & Baker, Syn. Fil. ed. 2, 495. 1874.
Dryopteris feei C. Chr. Ind. Fil. 89, 264. 1905.

Rhizome woody, slender, extensively creeping; scales apical, few, appressed, rusty, linear-attenuate, short-hirsute. Fronds few, distichous, ascending, 50–120 cm. long; stipes stout, stramineous, naked, equaling the blade or not; blades broadly oblong to ovate-oblong, abruptly acuminate-caudate, 30–70 cm. long, 14–40 cm. broad, pinnate-pinnatifid; pinnae spreading, linear-attenuate, cut to within 2 or 3 mm. of the stout, yellowish, elevated, puberulous midrib; segments oblique, narrowly oblong, close, subfalcate, acuminate, the subentire margins closely revolute, short-ciliate; veins 8–11 pairs, simple, oblique, the 2 lower pairs running to the hyaline sinus, together with the leaf-tissue freely puberulous beneath, sparingly so above; sori small, close, supramedial; indusia small, firm, pilose.

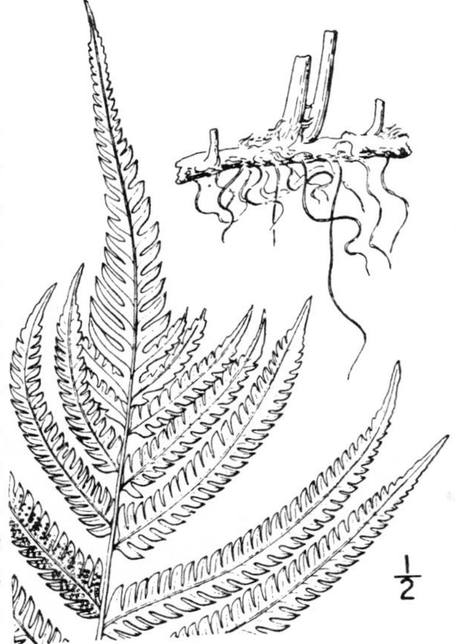

Mountains of southern Arizona and the coastal cañons of southern California (Santa Barbara and Los Angeles Counties); Lower California and Mexico generally, chiefly in the Lower Austral Zone. Type locality: Huatusco, Mexico.

4. **Dryopteris oreópteris** (Ehrh.) Maxon.

Mountain Wood-fern. Fig. 28.

Polypodium montanum Vogler, Diss. Pol. Mont. 1781, not Lam. 1778.
Polypodium oreopteris Ehrh.; Willd. Fl. Berol. Prodr. 292. 1787.
Dryopteris montana Kuntze, Rev. Gen. Pl. **2**: 813. 1891.
Dryopteris oreopteris Maxon, Proc. U. S. Nat. Mus. **23**: 638. 1901.

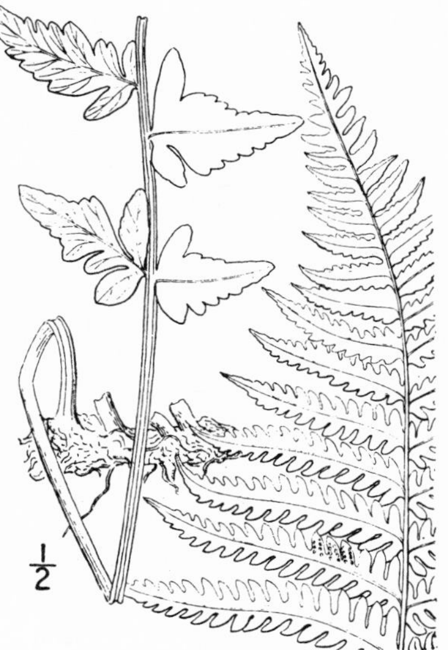

Rhizome rather stout, short-creeping or ascending, scales yellowish brown, thin, ovate-lanceolate. Fronds few, close, terminal, ascending, 40–110 cm. long; stipes very short, sulcate, stramineous from a dark chaffy base; blades narrowly to broadly lanceolate, or oblanceolate, abruptly acuminate, 30–85 cm. long, 10–22 cm. broad at the middle, tapering to a narrow base, pinnate-pinnatifid; pinnae oblong-lanceolate or narrowly deltoid-lanceolate (the basal ones often greatly reduced, ovate or deltoid-ovate), deeply pinnatifid, the midrib minutely hairy or subglabrous beneath; segments spreading, oblong, obtuse, membranous, more or less glandular, subentire, the margins finely hyaline-papillose; veins simple or 1-forked; sori rather small, submarginal; indusia roundish-reniform, toothed, glandular, deciduous.

Rocky mountain-slopes, often along streams, Hudsonian Zone; Alaska to northern Washington (Stehekin River region); Europe; Japan. Type locality, European.

5. **Dryopteris oregàna** C. Chr.

Sierra Wood-fern. Fig. 29.

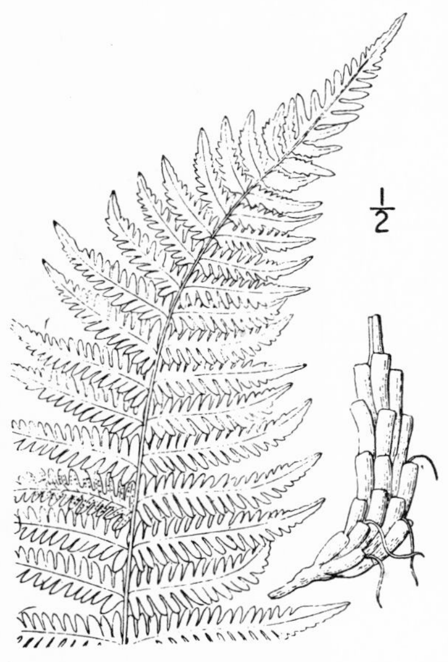

Aspidium nevadense D. C. Eaton, Ferns N. Am. 1: 73. *pl. 10.* 1878, not Boiss. 1838.
Dryopteris nevadensis Underw. Nat. Ferns ed. 4, 113. 1893.
Dryopteris oregana C. Chr. Ind. Fil. 84. 1905.

Rhizome slender, creeping, the apical portion covered with the imbricate stipe-bases of old fronds; scales ovate, yellowish brown, thin, concave, slightly toothed. Fronds few, in a close terminal crown, 50–90 cm. long; stipes slender, stramineous, much shorter than the blade, nearly naked; blades elliptic-lanceolate, acuminate, 40–60 cm. long, 10–16 cm. broad, long-attenuate to a very narrow base, pinnate-pinnatifid; pinnae sessile, horizontal, the larger ones close, linear to oblong-linear, acuminate to subcaudate, the basal ones usually distant, minute; segments close, oblong, obtuse or acutish, slightly oblique, membranous, entire or obscurely crenate, with a few deciduous cilia, minutely resinous-glandular; veins oblique, mostly simple, slightly hairy beneath; sori small, submarginal; indusia minute, long-ciliate, glandular.

Moist meadows and forested slopes, Canadian and Transition Zones; Sierra Nevada of northern California (Mariposa, Butte, Plumas, and Trinity Counties) northward to western Oregon (Josephine and Lane Counties). Type locality: Quincy, Plumas County, California.

6. **Dryopteris argùta** (Kaulf.) Watt.

Coastal Wood-fern. Fig. 30.

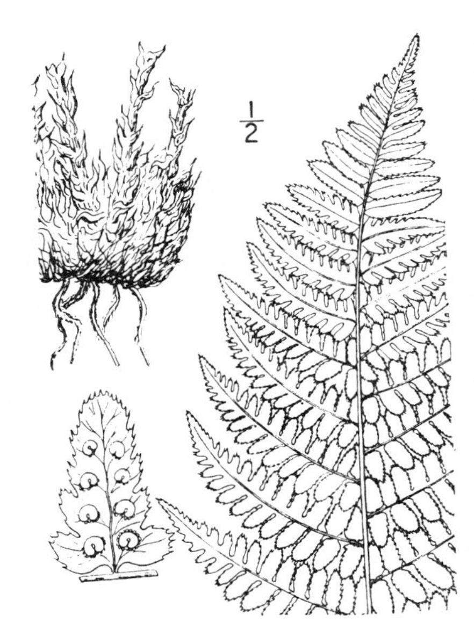

Aspidium argutum Kaulf. Enum. Fil. 242. 1824.
Dryopteris rigida arguta Underw. Nat. Ferns ed. 4, 116. 1893.
Dryopteris arguta Watt, Canad. Nat. II. **3**: 159. 1866.

Rhizome stout, woody, short-creeping; scales bright castaneous, ample, thin, lance-oblong or narrower, attenuate. Fronds several, close, erect, 35–90 cm. long; stipes stout, shorter than the blades, scaly; blades lance-ovate to oblong or deltoid-lanceolate, acuminate, 25–65 cm. long, 10–32 cm. broad, 2-pinnate or nearly so; pinnae spreading, oblong-lanceolate, long-acuminate, the lower ones broadest; pinnules spreading, distant, subcoriaceous, oblong, mostly rounded-obtuse, biserrate to pinnately incised, the lower basal one semicordate and usually overlying the main rachis; veinlets spreading, all ending in salient, often cartilaginous, spinelike teeth; sori large, close; indusia orbicular, firm, strongly convex, with a deep narrow sinus, the margins glandulose.

Rocky ravines and partially shaded slopes, Upper Sonoran and Transition Zones; southern California (common) northward to western Oregon and the bluffs of the Columbia River above Cathlamet, Washington; Santa Catalina Island; ascribed also to British Columbia. Type locality: California.

7. **Dryopteris fìlix-mas** (L.) Schott.

Male-fern. Fig. 31.

Polypodium filix-mas L. Sp. Pl. 1090. 1753.
Aspidium filix-mas Swartz, Journ. Bot. Schrad. **1800**[2]: 38. 1801.
Dryopteris filix-mas Schott, Gen. Fil. 1834.

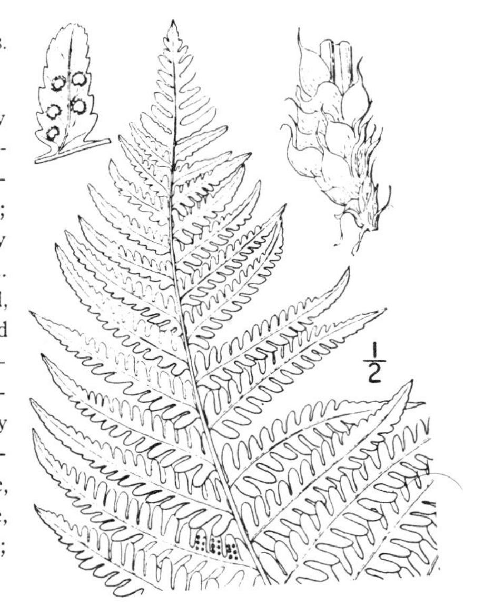

Rhizome stout, woody, erect or decumbent, very chaffy; scales fulvous, large, thin, linear to lanceolate, hair-pointed. Fronds in a close crown, ascending, 30–120 cm. long; stipes short, stout, scaly; blades oblong-lanceolate, short-acuminate, slightly narrowed toward the base, 25–100 cm. long, 10–30 cm. broad, nearly 2-pinnate; pinnae mostly short-stalked, narrowly deltoid-lanceolate, tapering from the broad base, attenuate, the basal ones short, broad, subdeltoid; segments slightly oblique, close, thin-herbaceous, paler and slightly scaly beneath, mostly oblong, rounded-obtuse, adnate, and slightly decurrent (the large basal ones subsessile), usually serrate, the teeth oblique, curved, not spinelike; sori large, usually confined to the basal half of the segment; indusia thin, orbicular-reniform, glabrous.

Rocky woods, Hudsonian and Canadian Zones; Newfoundland to British Columbia, south to Vermont, South Dakota, New Mexico, Arizona, Nevada, and Oregon; known also from a single locality in southern California (San Bernardino Mountains); Eurasia. Type locality, European.

8. Dryopteris dilatàta (Hoffm.) A. Gray.
Spreading Wood-fern. Fig. 32.

Polypodium dilatatum Hoffm. Deutschl. Fl. 2: 7. 1795.
Dryopteris dilatata A. Gray, Man. 631. 1848.
Dryopteris spinulosa dilatata Underw. Nat. Ferns ed. 4, 116. 1893.

Rhizome stout, woody, creeping or ascending, chaffy. Fronds 30–100 cm. long or more, spreading, borne in a complete crown; stipes stout, 15–45 cm. long, clothed with brownish, often darker-centered scales; blades triangular to ovate or broadly oblong, acuminate, 15–90 cm. long, 10–40 cm. broad, nearly or quite 3-pinnate; basal pinnae broadly ovate or triangular, inequilateral, the upper ones lanceolate to oblong; pinnules lanceolate to oblong, acute, the larger ones not decurrent, pinnate or at least pinnately divided; segments herbaceous, obliquely pinnatifid or toothed, the teeth mucronate, straight or falcate, usually not appressed; sori mostly subterminal; indusia glabrous or sparsely glandular.

Rocky woods, Canadian and Humid Transition Zones; Alaska to Newfoundland, south to California (Mendocino, Sonoma, and San Mateo Counties), Montana and New England, and in the mountains to North Carolina and Tennessee; Greenland; Eurasia. Type locality: Germany. Variable in scale and indusium characters and including several very critical forms.

6. WOODWÁRDIA J. E. Smith, Mem. Acad. Sci. Turin 5: 411. 1793.

Coarse, mostly large ferns of low shady situations, the stout woody rhizomes short-creeping to erect. Fronds several or many, in a crown, rigidly ascending or laxly recurved; blades uniform, leafy, once pinnate, the pinnae lobed or coarsely pinnatifid. Sori oblong to linear, straight or slightly curved, borne singly on the outer horizontal veins of a continuous series of elongate areoles lying next to the costa of the segments, sunken, facing inward and occupying the areole, tumid, the ample, elongate, arched indusia at first completely enveloping the sporangia, firm, persistent, at length reflexed; veins arising from the costal areoles branched, either wholly free and excurrent to the margin of the segment or partly joined basally to form 1 or 2 more or less incomplete rows of oblique sterile areoles. [Name in honor of Thomas J. Woodward, an English botanist.]

Four or five species, of North America and Eurasia, only one occurring in the United States. Type species, *Blechnum radicans* L.

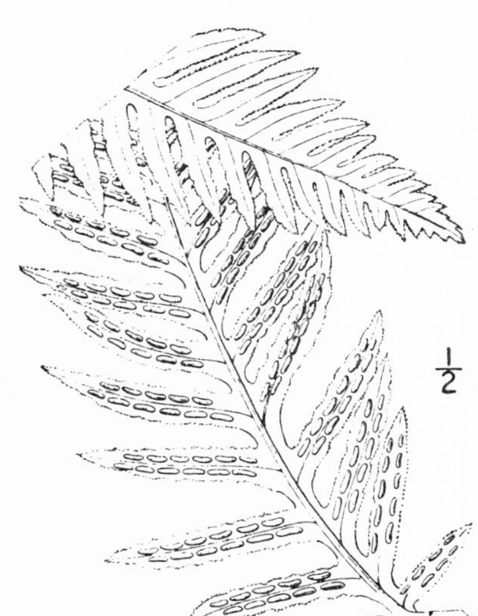

1. Woodwardia chamissòi Brack.
Giant Chain-fern. Fig. 33.

Woodwardia chamissoi Brack. in Wilkes, U. S. Explor. Exped. 16: 138. 1854.
Woodwardia radicans americana Hook. Sp. Fil. 3: 67. 1860.
Woodwardia paradoxa Wright, Gard. Chron. III. 41: 98. 1907.

Rhizome stout, woody, oblique; scales lance-attenuate, 1–3 cm. long, bright castaneous, thin, glossy, entire. Fronds suberect, 1–3 meters long; stipes short, stout, stramineous from a brown base; blades linear-oblong to oblong-ovate or oblanceolate, short-acuminate, 20–50 cm. broad, pinnate, narrow at the base; pinnae very deeply and obliquely pinnatifid, the middle ones up to 30 cm. long, close or imbricate, linear-oblong to ovate, long-acuminate; segments narrowly triangular to linear-attenuate, subfalcate, decurrent, undulate-crenate or shallowly lobed, serrate-spinulose; excurrent veins oblique, 1–2-forked, free, or partly joined basally; leaf tissue herbaceous, paler beneath, the veins resinous-glandular.

Moist shady banks from near sea level to above 1500 meters altitude, Upper Sonoran and Transition Zones; western British Columbia to southern California and Arizona; also in northeastern Nevada. Type locality: California.

7. ASPLÈNIUM L. Sp. Pl. 1078. 1753.

Small to large ferns of forests or rocky ledges, the fronds rosulate to erect-spreading, with creeping to erect rhizomes, the scales rigid, with thick, dark-colored partition cell-walls. Fronds usually uniform, the blades simple to 1–4-pinnate or several times pinnatifid, glabrous, variously pubescent, or slightly palaceous, the rachises dark and lustrous to green and succulent; pinnae or segments articulate or not, entire to variously incised or cleft. Sori oblong to narrowly linear, borne upon the free, usually oblique ultimate veins, usually below their tip; indusia invariably present, lateral, oblong to narrowly linear, usually membranous, often concealed by the sporangia at maturity. [Ancient Greek name, being a supposed remedy for the spleen.]

Nearly 600 species, mainly of tropical and subtropical regions of both hemispheres. Besides the following about 20 species occur in the United States, mainly in Florida and the Appalachian region. Type species, *Asplenium trichomanes* L.

Stipe and rachis dark castaneous or purplish brown throughout, glossy.
  Rachis narrowly yellowish-alate, naked; pinnae crenulate or shallowly crenate, not auriculate; veins mostly forked. 1. *A. trichomanes.*
  Rachis not alate, deciduously fibrillose with tortuous, brown, gland-tipped hairs; pinnae deeply and obliquely crenate, subauriculate; veins mostly simple. 2. *A. vespertinum.*
Stipe reddish brown below, the upper part and the rachis green. 3. *A. viride.*

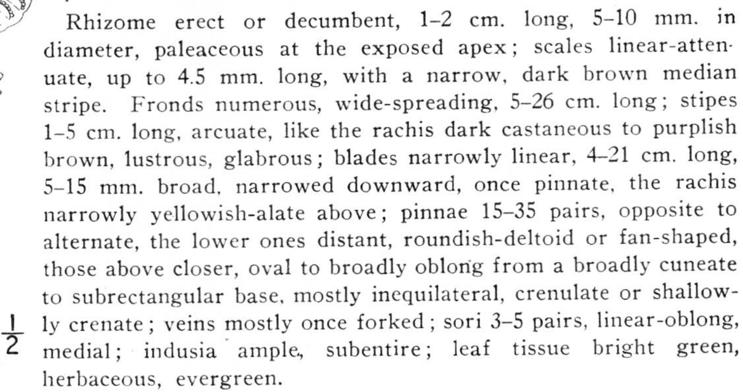

1. Asplenium trichómanes L.

Maidenhair Spleenwort. Fig. 34.

Asplenium trichomanes L. Sp. Pl. 1080. 1753.

Rhizome erect or decumbent, 1–2 cm. long, 5–10 mm. in diameter, paleaceous at the exposed apex; scales linear-attenuate, up to 4.5 mm. long, with a narrow, dark brown median stripe. Fronds numerous, wide-spreading, 5–26 cm. long; stipes 1–5 cm. long, arcuate, like the rachis dark castaneous to purplish brown, lustrous, glabrous; blades narrowly linear, 4–21 cm. long, 5–15 mm. broad, narrowed downward, once pinnate, the rachis narrowly yellowish-alate above; pinnae 15–35 pairs, opposite to alternate, the lower ones distant, roundish-deltoid or fan-shaped, those above closer, oval to broadly oblong from a broadly cuneate to subrectangular base, mostly inequilateral, crenulate or shallowly crenate; veins mostly once forked; sori 3–5 pairs, linear-oblong, medial; indusia ample, subentire; leaf tissue bright green, herbaceous, evergreen.

Crevices of cliffs, usually limestone, chiefly in the Transition and Canadian Zones; Alaska and the Hudson's Bay region to Nova Scotia, south in the mountains to Oregon, the Mexican Border states, Alabama, and Georgia. Type locality, European.

2. Asplenium vespertìnum Maxon.

Western Spleenwort. Fig. 35.

Asplenium vespertinum Maxon, Bull. Torrey Club **27**: 197. 1900.

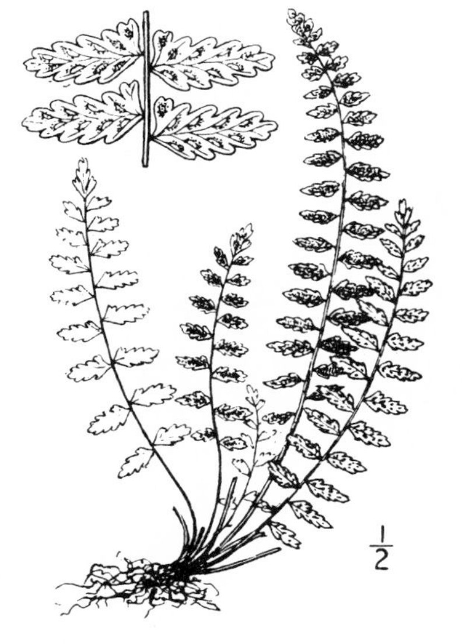

Rhizome erect or ascending, 1–2 cm. long, rather stout; scales linear-acicular, 2–2.5 mm. long, dark brown, rigid. Fronds numerous, closely fasciculate, arcuately ascending, 5–24 cm. long; stipes 1–4 cm. long, stout, purplish brown, lustrous, with a few blackish gland-tipped fibrils; blades linear or oblanceolate-linear, 4–20 cm. long, 1–2.5 cm. broad, once pinnate, the rachis brown, sulcate above, deciduously fibrillose; pinnae 20–30 pairs, subopposite to alternate, the lower ones distant, reduced, short-stalked, reflexed, those above closer, sessile, horizontal, oblong or linear-oblong from an inequilateral, broadly cuneate or rectangular base, subauriculate, deeply and obliquely crenate, the veins (except in the basal lobe) simple; sori 4–6 pairs, short, infra-medial; indusia narrow, firm, crenulate; leaf tissue light or yellowish green, herbaceous.

Moist crevices beneath overhanging rocks, Upper Sonoran Zone; San Gabriel Mountains (Los Angeles County) to the vicinity of San Bernardino, California, southward on the seaward side of the mountains into Lower California. Type locality: National City, San Diego County, California.

3. **Asplenium víride** Huds.

Green Spleenwort. Fig. 36.

Asplenium viride Huds. Fl. Angl. 385. 1762.

Plants tufted, the rhizomes creeping, 2–10 cm. long; scales linear-attenuate, up to 3.5 mm. long, brown, concolorous or with a dark brown basal stripe. Fronds numerous, tufted, 3–20 cm. long, laxly ascending; stipes 1–7 cm. long, weak, stramineous to green above the reddish brown base, scantily fibrillose; blades narrowly linear-lanceolate to oblong-linear, 2–14 cm. long, 5–15 mm. broad, attenuate, once pinnate, the delicate green rachis slightly fibrillose; pinnae 5–25 pairs, the lower ones usually distant, broadly rounded-deltoid, opposite, those above closer, becoming alternate, rhombic to rhombic-oblong or obliquely ovate, obtuse, stalked, broadly cuneate at the inequilateral base, crenate to crenately lobed; basal veins forked, the others forked or simple; sori 2–4 pairs, remote from the margin, soon confluent and concealing the delicate, subentire indusia; leaf tissue bright or yellowish green, soft-herbaceous, glabrous.

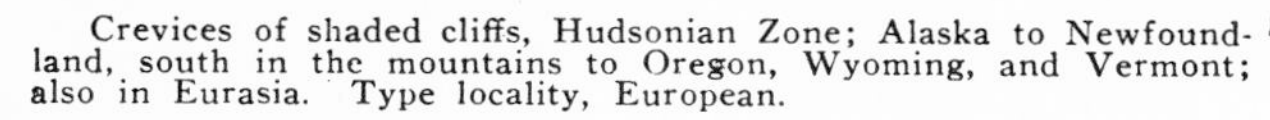
Crevices of shaded cliffs, Hudsonian Zone; Alaska to Newfoundland, south in the mountains to Oregon, Wyoming, and Vermont; also in Eurasia. Type locality, European.

8. **ATHÝRIUM** Roth, Röm. Arch. Bot. 2[1]: 105. 1799.

Medium-sized to large ferns of upright habit, usually growing in moist shaded situations, the rhizomes creeping (often slender) to oblique and densely tufted; rhizome scales membranous, with thin-walled cells. Fronds usually large, erect-spreading, long-stipitate, or the stipe nearly wanting; blades ample, usually elongate, 1–3-pinnate, the segments subentire to incised or pinnatifid, membranous to herbaceous; veins free. Sori dorsal, oblique to the midrib, narrowly oblong, or commonly crossing the vein and recurved, becoming lunate, hippocrepiform, or (rarely) roundish; indusia shaped like the sori, attached along the inner side, facing outward, subentire to fimbriate, delicate, sometimes minute and hidden, rarely wanting. [Name Greek, meaning shieldless, of doubtful application.]

About 150 species, mainly of tropical regions. Four additional species occur in eastern North America. Type species, *Polypodium filix-femina* L.

Blades ample, foliose, the segments mostly close; indusia oblong to lunate or hippocrepiform, fringed with septate cilia. 1. *A. filix-femina.*

Blades skeleton-like, the segments very narrow, oblique, distant; indusia wanting, the sorus roundish. 2. *A. americanum.*

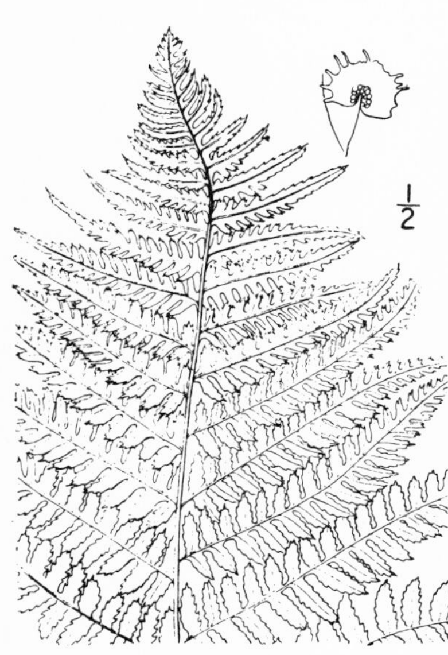

1. **Athyrium fìlix-fémina** (L.) Roth.

Lady-fern. Fig. 37.

Polypodium filix-femina L. Sp. Pl. 1090. 1753.
Athyrium filix-femina Roth, Röm. Arch. Bot. 2[1]: 106. 1799.
Asplenium filix-femina Bernh. Schrad. Neues Journ. Bot. 1[2]: 26. 1806.
γ *Athyrium cyclosorum* Rupr. Beitr. Pflanzenk. Russ. Reich. 3: 41. 1845.
δ *Athyrium sitchense* Rupr. Beitr. Pflanzenk. Russ. Reich. 3: 41. 1845.
Athyrium filix-femina californicum Butters, Rhodora 19: 201. 1918.

Rhizome erect or ascending, stout, paleaceous, the scales lanceolate, up to 1 cm. long, light to dark brown, thin. Fronds up to 2 meters long, erect-arching; stipes short, stramineous, paleaceous at the dark base; blades lanceolate, attenuate in both directions (the lower pinnae distant), 2–3-pinnate; pinnae linear to lance-oblong, acuminate to attenuate, sessile, obliquely spreading; segments crenate, variously incised, or pinnatifid, membranous, glabrous or minutely glandular; sori usually less than 1 mm. long, oblong to lunate or hippocrepiform; indusia sometimes toothed, septate-ciliate.

Forests, moist thickets, and open or brushy slopes, Transition and Canadian Zones; Alaska to southern California and in the Rocky Mountains to Nevada and New Mexico; Eurasia. Type locality, European.

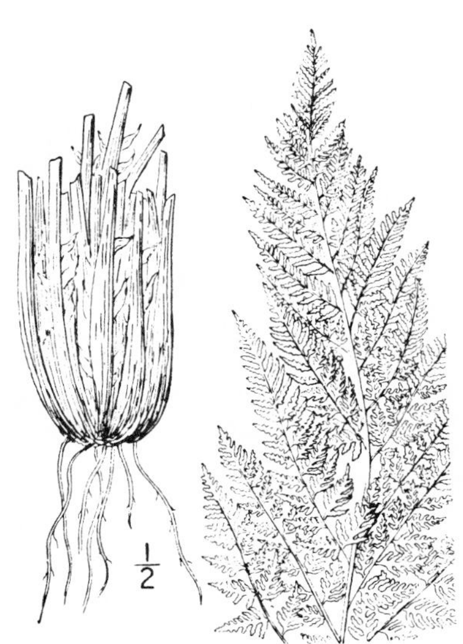

2. Athyrium americànum (Butters) Maxon.
Alpine Lady-fern. Fig. 38.

Athyrium alpestre americanum Butters, Rhodora **19**: 204. 1917.
Athyrium americanum Maxon, Am. Fern Journ. **8**: 120. 1918.

Rhizomes erect or decumbent, branched, forming massive rounded tufts; scales copious, light to dark brown. Fronds close, 20–90 cm. long; stipes short, sparsely paleaceous, stramineous from a dark base; blades linear to oblong-lanceolate, acuminate, 18–65 cm. long, 4–25 cm. broad, usually 2-pinnate-pinnatifid; pinnae oblique, elongate-deltoid, acuminate, the lower ones distant, usually shorter; pinnules delicately herbaceous, stalked, distant, oblique, narrowly deltoid to oblong-ovate or lance-oblong, incised, or the larger ones pinnate, the segments incised or biserrate; sori very numerous, roundish, non-indusiate.

Moist rocky ravines, meadows, or alluvial thickets, Arctic-Alpine Zone; Alaska to Colorado, Nevada, and the southern Sierra Nevada, California (Mineral King); also in Gaspé County, Quebec. Type locality: Rogers Pass, British Columbia.

9. PITYROGRÁMMA Link, Handb. Gewächs. 3: 19. 1833.

[Ceropteris Link, Fil. Sp. 141. 1841.]

Small to medium-sized ferns of dryish banks and rock-ledges, the short-creeping to oblique rhizomes covered with dark rigid linear-lanceolate scales. Fronds erect to drooping, clustered, uniform, not articulate to the rhizome, the stipes dark, glossy, firm; blades 1–3-pinnate, linear to deltoid-pentagonal, covered with a white to deep yellow powder beneath, sometimes glandular above, usually devoid of scales. Sori following the course of the veins, usually confluent, non-indusiate. [Greek, in allusion to the furfuraceous sori.]

About 15 species, mainly of tropical regions, only the following occurring in the United States. Type species, *Acrostichum chrysophyllum* Swartz.

1. Pityrogramma trianguláris (Kaulf.) Maxon.
Gold-fern. Fig. 39.

Gymnogramma triangulare Kaulf. Enum. Fil. 73. 1824.
Gymnopteris triangularis Underw. Nat. Ferns ed. 6, 84. 1900.
Ceropteris triangularis Underw. Bull. Torrey Club **29**: 630. 1902.
Pityrogramma triangularis Maxon, Contr. U. S. Nat. Herb. **17**: 173. 1913.

Rhizome stoutish, short-creeping or ascending; scales linear to lance-attenuate, rigid, brownish, often blackish-carinate. Fronds numerous, clustered, erect, 15–40 cm. long; stipes stout, dark brown, polished, about twice as long as the blades, naked above the base; blades deltoid-pentagonal, 6–18 cm. long, 5–16 cm. broad, pinnate, or 2-pinnate as to the large spreading basal pinnae, these deltoid, strongly inequilateral, with the lower basal segments extended and usually pinnatifid or pinnately divided; other pinnae mostly oblong, pinnately lobed or incised; segments rounded-obtuse, mostly decurrent, coriaceous, essentially glabrous above, yellowish-ceraceous beneath, or the powder rarely white or lacking; sporangia following the short spreading branched veinlets throughout, confluent with age.

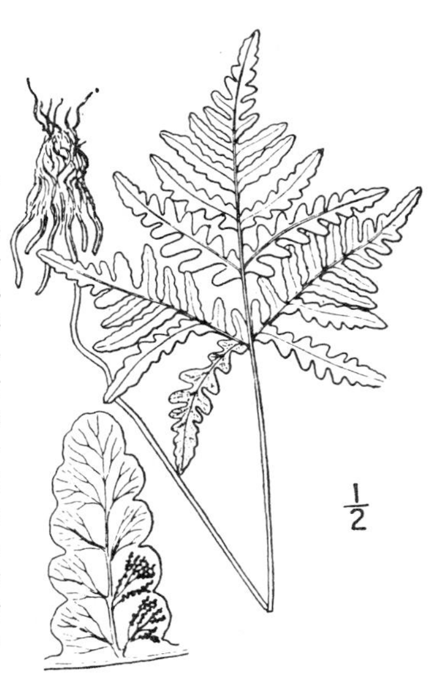

Rocky shaded slopes, Upper Sonoran and Transition Zones; British Columbia (Vancouver Island) to Nevada (Clark County) and southern California, mainly at low altitudes; also in northern Lower California. Type locality: San Francisco Bay region, California.

Pityrogramma triangularis viscòsa (D. C. Eaton) Weatherby, Rhodora **22**: 117. 1920.

Gymnogramme triangularis viscosa D. C. Eaton, Ferns N. Am. **2**: 16. 1879.
Ceropteris viscosa Underw. Bull. Torrey Club **29**: 631. 1902.
Pityrogramma viscosa Maxon. Contr. U. S. Nat. Herb. **17**: 173. 1913.

Stipes reddish brown; blades often small; pinnae few, the upper ones commonly entire, the elongate lower segments of the basal ones usually undulate-crenate only; segments usually few and distant, viscid and often resinous-glandular above, whitish-ceraceous beneath.

Moist ravines, Upper and Lower Sonoran Zones; southwestern California (San Diego, Riverside, and Los Angeles Counties). Type locality: San Diego, California.

Pityrogramma triangularis pállida Weatherby, Rhodora **22**: 119. 1920.

Stipes blackish, glandular and white-farinose above the base; blades thin, nearly 3-pinnate below; lower pinnules of the basal pinnae elongate, deeply pinnatifid or pinnatisect; segments grayish-farinose above with numerous whitish glands, not viscid, white-ceraceous beneath.

Rocky situations, Upper Sonoran Zone; foothills bordering the interior valley region, from Butte County to Tulare County, California. Type locality: Pollasky, Madera County, California.

Pityrogramma triangularis máxoni Weatherby, Rhodora **22**: 119. 1920.

Stipes reddish brown, glossy; blades nearly 3-pinnate below, the lower pinnules of the basal pinnae elongate, very deeply pinnatifid; segments subcoriaceous, sparsely yellowish-glandular above, light yellow or griseous-ceraceous beneath; spores trilobate.

Rocky cañons, Upper Sonoran Zone, desert region of southeastern California and adjacent portions of Arizona, Sonora, and Lower California. Type from the head of Rincon Valley, Arizona, alt. 3500 ft.

10. STRUTHIÓPTERIS Scop. Fl. Carn. 168. 1760.

[LOMARIA Willd. Ges. Naturf. Freund. Berlin Mag. **3**: 160. 1809.]

Terrestrial or epiphytic, mainly forest ferns of various habit, the woody rhizomes scandent to short-creeping (erect in a few tropical species). Fronds cespitose or obliquely imbricate in arrangement, dimorphous, the often narrow blades pinnatisect or simply pinnate, the sterile ones spreading or ascending, numerous, the fertile ones few, long-stipitate, with linear, entire, wholly soriferous segments. Sorus elongate-linear, parallel to the midrib, borne just short of the margin upon a continuous thickish receptacle joining the few short, otherwise free veins running obliquely from the midrib; indusium intramarginal, facing the midrib and at first meeting it, entire to lacerate, often reflexed at maturity, the numerous sporangia covering the lower surface of the segment. [Greek for ostrich and fern.]

Nearly 150 species, mainly of temperate regions, 25 occurring in tropical North America. Only the following, the type species, is found in the United States.

1. Struthiopteris spìcant (L.) Weis.

Deer-fern. Fig. 40.

Osmunda spicant L. Sp. Pl. 1066. 1753.

Blechnum spicant J. E. Smith, Mem. Acad. Sci. Turin **5**: 411. 1793.

Lomaria spicant Desv. Ges. Naturf. Freund. Berlin Mag. **5**: 325. 1811.

Struthiopteris spicant Weis, Pl. Crypt. Gott. 287. 1770.

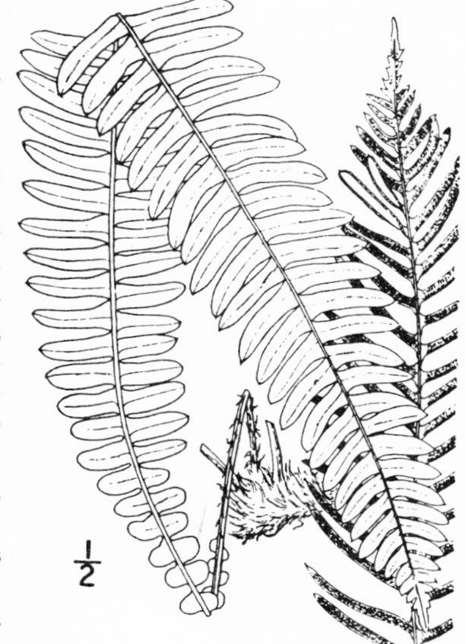

Rhizome short-creeping, 5–10 cm. long, 1–2 cm. thick, woody; scales linear to lance-attenuate, 5–10 mm. long, castaneous. Sterile fronds numerous, in a circular crown, spreading or ascending, 15–100 cm. long, evergreen; stipes 2–30 cm. long, yellowish brown to castaneous; blades linear to linear-oblanceolate, 13–80 cm. long, 2–9 cm. broad, pinnatisect, or pinnate at the long-attenuate base, the basal segments broad, often distant; segments 30–80 pairs, horizontal, subfalcate, mostly linear to linear-oblong, roundish at the apex (the tip acute), dilatate, entire or crenulate, glabrous, paler beneath. Fertile fronds few, central, erect, 40–150 cm. long; stipes long, castaneous; blades similar to the sterile ones in outline, pinnate, with many small basal segments; pinnae distant, mostly very narrowly linear from a dilatate base.

Damp, mostly coniferous forests, in the Humid Transition Zone; Alaska, south in the mountains to the Santa Cruz Peninsula, California, mainly near the coast; Eurasia. Type locality, European.

11. CRYPTOGRÁMMA R. Br. in Richards. Bot. App. Frankl. Journ. 767. 1823.

Small, mainly alpine or boreal ferns of rocky situations. Fronds dimorphous, in most species numerous and densely clustered upon stout ascending rhizomes, the stipes pale, not articulate; blades glabrous, 2–3-pinnate, herbaceous, the sterile ones foliaceous, with numerous, crowded, rather small, flat, obtuse segments, the fertile ones on longer stalks, with fewer, narrower, copiously soriferous segments. Sori marginal or submarginal, in a continuous line at the free ends of the forked veins, confluent; indusia continuous, formed of the broadly revolute or reflexed, modified margin of the blade. [Name Greek, alluding to the sori hidden before maturity.]

Five species, the following, 1 in Chile, and 2 of Eurasia. Type species, *C. acrostichoides* R. Br.

Rhizomes stout, ascending, in massive tufts; fronds very numerous, clustered; blades herbaceous. 1. *C. acrostichoides.*

Rhizomes very slender, creeping; fronds few, scattered; blades membranous, very translucent. 2. *C. stelleri.*

1. Cryptogramma acrostichoìdes R. Br. American Parsley-fern. Fig. 41.

Cryptogramma acrostichoides R. Br. in Richards. Bot. App. Frankl. Journ. 767. 1823.

Rhizomes short-creeping or ascending, in massive tufts, chaffy; scales lance-ovate, attenuate, rusty to dark brown, or striped. Fronds very numerous, closely clustered, the fertile ones 10–30 cm. long, erect, long-stalked, usually much surpassing the short-stalked spreading sterile ones; stipes deciduously scaly below, those of the sterile fronds slender and greenish, and of the fertile ones stouter and stramineous; sterile blades ovate or ovate-lanceolate, 3–12 cm. long, 2–3-pinnate, the rachises compressed, greenish-marginate; pinnae few, mostly close, the ultimate segments crowded, ovate, oblong, or obovate, obtuse, crenate or incised, decurrent, herbaceous, glabrous; fertile blades simpler, the segments fewer, linear-oblong, 6–12 mm. long, about 2 mm. broad, the thin margins widely revolute, nearly meeting, at length expanded; sporangia broadly confluent.

Cliffs and rock-slopes, Canadian and Hudsonian Zones; Alaska to Labrador, southward in the high mountains to southern California (San Bernardino County), Nevada, Utah, northern New Mexico, and the northern shores of Lake Huron. Type locality: Nootka Sound.

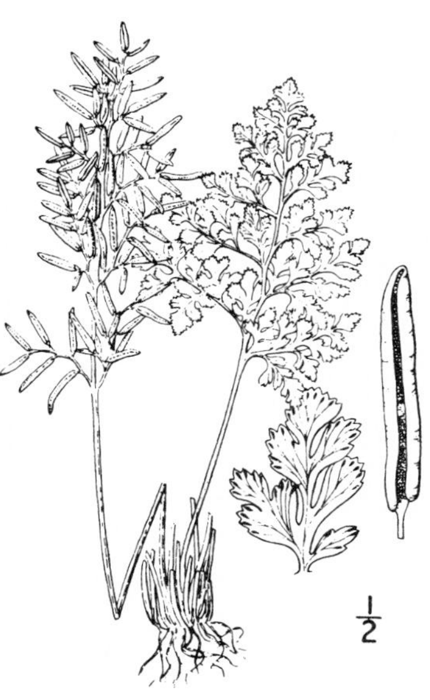

2. Cryptogramma stélleri (S. G. Gmel.) Prantl. Slender Cliff-brake. Fig. 42.

Pteris stelleri S. G. Gmel. Nov. Comm. Acad. Sci. Petrop. **12**: 519. *pl. 12, f. 1.* 1768.
Pteris gracilis Michx. Fl. Bor. Am. **2**: 262. 1803.
Pellaea gracilis Hook. Sp. Fil. **2**: 138. *pl. 133. B.* 1858.
Pellaea stelleri Baker in Hook. & Baker, Syn. Fil. 453. 1868.

Rhizomes very slender, cordlike, creeping; scales few, pale, ovate, appressed. Fronds few, scattered, drooping, 8–20 cm. long, the fertile ones exceeding the sterile; stipes slender, usually longer than the blades, stramineous or yellowish brown from a glossy castaneous base; blades ovate or oblong-ovate, 4–10 cm. long, 2–5 cm. broad, 2-pinnate; pinnae few, the larger ones with pinnately divided basal pinnules; segments of the sterile blades ovate to obovate, cuneate, crenately lobed, adnate-decurrent, those of the fertile blades longer (up to 2 cm.), linear-oblong to lanceolate, mostly acute, distant; indusium membranous, concealing the sporangia; leaf tissue delicately membranous, translucent, the veins conspicuous.

Moist shaded cliffs (usually limestone), Canadian and Transition Zones; Alaska to Labrador and eastern Quebec, south to Washington (Wenatchee Mountains), Montana, Colorado, Iowa, Illinois, and Pennsylvania; also in Asia. Type locality: Siberia.

12. PTERÍDIUM Scop. Fl. Carn. 169. 1760.

Coarse ferns of open or partially shaded, acid-soil situations, the slender, woody, freely branched rhizome wide-creeping underground. Fronds stout, erect to reclining, up to 5 meters long, borne singly, the stout stipes with a feltlike covering at the base, not jointed to the rhizome; blades large, triangular or deltoid-ovate to elongate, pinnately decompound, the ultimate segments entire, toothed, or lobed. Sori linear, marginal, continuous, arising from a transverse veinlike receptacle connecting the ends of the forked free veins; indusium double, the outer one prominent, formed by the reflexed membranous margin of the blade, the inner obscure, delicate, usually minute, borne upon the receptacle, nearly concealed by the sporangia, and facing outward. [Diminutive of *Pteris,* the ancient Greek name of ferns.]

Six or eight widely distributed species, three or four occurring in other parts of North America. Type species, *Pteris aquilina* L.

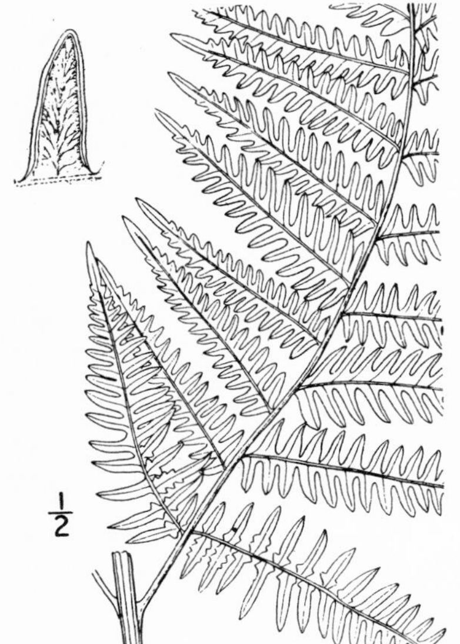

1. Pteridium aquilìnum pubéscens Underw.

Western Bracken. Fig. 43.

Pteris feei Schaffn.; Fée, Mém. Foug. 8: 73. 1857.

Pteridium aquilinum pubescens Underw. Nat. Ferns ed. 6, 91. 1900.

Stipes erect, stout, stramineous from a darker base, 15–100 cm. long; blades oblique, deltoid to deltoid-ovate, 25–100 cm. long, subternately decompound, usually 3-pinnate in the basal part; segments variable, linear-oblong and entire, to hastate or deltoid-oblong and pinnatifid, often dilatate, rigidly herbaceous, pubescent or strongly sericeous-tomentose beneath, slightly hairy or glabrous above; indusia usually narrow, villous, but not often truly ciliate.

Moist woods and thickets or dry open slopes, Alaska to Montana, southward to Mexico. Type locality: not definitely stated. Extremely variable in stature and in nearly all technical characters, these mostly correlated with habitat. A few specimens approach the typical European form.

13. ADIÁNTUM [Tourn.] L. Sp. Pl. 1094. 1753.

Delicate graceful ferns of moist rocky woods and ravines, with thickish suberect to slender wide-creeping, paleaceous rhizomes. Fronds distichous or in several ranks, rigidly ascending to drooping, the stipes firm, dark, usually lustrous; blades simple, 1–3-pinnate, or decompound, usually with dark, polished rachises, extremely variable according to fertility, the pinnules glabrous or variously pubescent, sessile or stalked, often articulate, readily deciduous in many species. Sori appearing marginal, the sporangia borne along and sometimes between the ends of the free forking veins, on the under side of sharply reflexed indusiform marginal lobes of the pinnules or segments. [The ancient name, in allusion to the pinnae repelling rain-drops.]

About 200 species, largely tropical American. In addition to the following, four species occur in other parts of the United States. Type species, *Adiantum capillus-veneris* L.

Blades reniform-orbicular, the two equal divisions spreading, with unilaterally arranged pinnate branches. 1. *A. pedatum aleuticum.*

Blades elongate, with a continuous main rachis.

Fronds erect; pinnules roundish or semicircular, slightly lobed; sori mostly linear, straight or arcuate. 2. *A. jordani.*

Fronds laxly ascending or drooping; pinnules cuneate-obovate to rhombic, deeply incised; sori mostly oblong-lunate. 3. *A. capillus-veneris.*

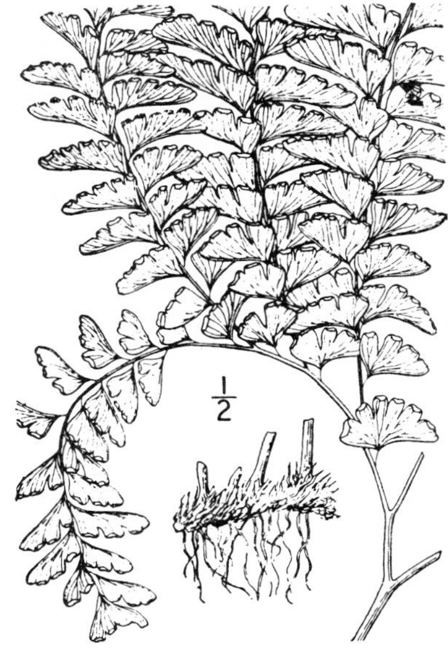

1. Adiantum pedàtum aleùticum Rupr.
Western Maiden-hair. Fig. 44.

Adiantum pedatum aleuticum Rupr. Beitr. Pflanzenk. Russ. Reich. **3**: 49. 1845.

Adiantum pedatum rangiferinum Burgess, Proc. Roy. Soc. Canada **4**[4]: 11. 1887.

Rhizome thickish, short-creeping or oblique, densely paleaceous; scales rich brown, lance-oblong to deltoid-ovate, acuminate, entire. Fronds close, erect, 25–100 cm. long; stipes stout, castaneous to dark brown, much longer than the blades, chaffy at the base; blades usually reniform-orbicular, 12–50 cm. broad; pinnae 5–14, somewhat pedately arranged, arising from the outer side of 2 equal branches surmounting the stipe, pinnate, the larger (middle) ones 12–40 cm. long, 2–5 cm. broad; pinnules close, spreading to very oblique, short-stalked, mostly oblong to deltoid-oblong, the lower margin entire, the upper deeply and often unequally cleft; sori linear to oblong-lunate, solitary on the truncate lobes.

Cliffs and rich rocky or swampy woods, Canadian and Humid Transition Zones; Alaska to the mountains of southern California (San Bernardino County), and along the Rocky Mountains to Utah; also in the Gaspé region, Quebec. Type locality: Aleutian Islands, Alaska. Apparently passing gradually into the common typical form of eastern North America.

2. Adiantum jórdani C. Müll.
California Maiden-hair. Fig. 45.

Adiantum emarginatum D. C. Eaton, Ferns N. Am. **1**: 285. *pl. 38, f. 1–3.* 1879, not Bory, 1810.

Adiantum jordani C. Müll. Bot. Zeit. **1864**: 26. *pl. 1, f. 1.* 1864.

Rhizome creeping, slender, rather densely paleaceous; scales large, rigid, oblique, dark brown, lance-attenuate, entire. Fronds several, rather close, mostly erect, 20–50 cm. long, the stipes nearly as long as the blades, dark brown; blades ovate to deltoid-ovate, 12–30 cm. long, 8–15 cm. broad, 2–3-pinnate at the base, simply pinnate at the apex; pinnae oblique-spreading, stalked; pinnules long-stalked, roundish to transversely oval or semicircular, 8–30 mm. broad, entire below, the outer edge shallowly 2–5-lobed, sharply denticulate if sterile, the numerous veinlets terminating the whitish teeth; sori linear, straight or usually arcuate.

Rocky cañons, Transition Zone; principally at low elevations in the coast ranges of California, from San Diego County northward to southwestern Oregon; rare in the Sierra Nevada of central California. Type locality: Ukiah, California.

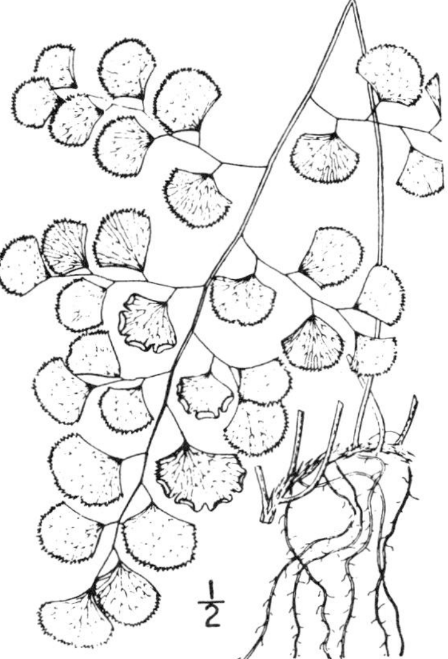

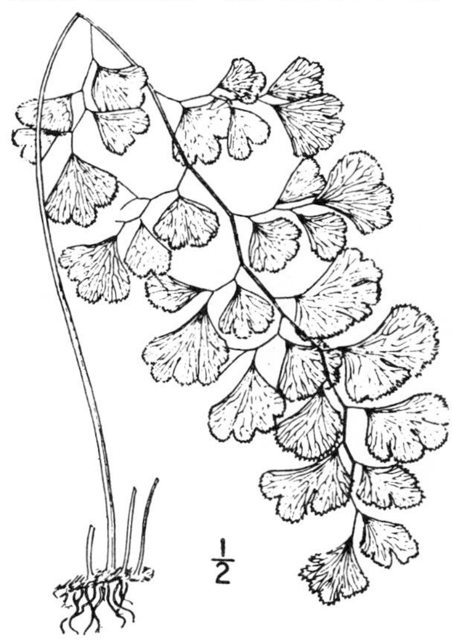

3. Adiantum capíllus-véneris L.
Venus-hair Fern. Fig. 46.

Adiantum capillus-veneris L. Sp. Pl. 1096. 1753.

Adiantum modestum Underw. Bull. Torrey Club **28**: 46. 1901.

Rhizome creeping, cordlike, laxly paleaceous; scales small, thin, light brown, lance-linear, attenuate, entire. Fronds often spaced, laxly ascending or pendent, 20–70 cm. long, the slender purplish-black stipes as long as the blades, or shorter; blades usually ovate or lance-ovate, 12–40 cm. long, 8–22 cm. broad, 2–3-pinnate at the base, gradually simpler above; pinnae laxly spreading, stalked; pinnules stalked, obovate to rhombic or obliquely oblong, 5–30 mm. long, narrowly to broadly cuneate, the outer edge deeply lobed or incised, the margins (if sterile) denticulate, the teeth acutish to long-acuminate; sori solitary on the lobes, mostly oblong-lunate.

Shaded banks, sinkholes, and rocky ravines, Upper and Lower Sonoran Zones; Virginia to Florida, west to Missouri, Utah, southern California, and the Mexican Border region; warm-temperate or subtropical regions of both hemispheres. Type locality, European. Extremely variable according to range, moisture, and degree of fertility.

14. CHEILÁNTHES Swartz, Syn. Fil. 126. 1806.

Mostly small, rock-loving, xerophilous ferns, with glandular-pubescent, tomentose, paleaceous, or (rarely) ceraceous or naked foliage. Fronds uniform, the blades 1–4-pinnate, or 1-pinnate and variously pinnatifid, the segments commonly minute, often beadlike. Sori borne at the enlarged tips of the veins, solitary on minute lobules, or numerous and narrowly confluent. Indusia formed of the revolute or recurved, often thinnish, more or less modified margin of the lobes or segments, or in several species the margin closely reflexed and pouch-like or coherent, giving rise abruptly to a membranaceous, introrse proper indusium. [Name Greek, alluding to the marginal sori.]

About 125 species of temperate and tropical regions. Fourteen species besides the following occur in the United States. Type species, *Cheilanthes micropteris* Swartz.

Blades glabrous and naked; indusia intramarginal, joined continuously to the closely reflexed margin throughout (the sori contiguous) or to minute saccate teeth at either side of separate sori.
- Sori solitary, with separate short, roundish-lunate indusia. — 1. *C. californica.*
- Sori contiguous, with a common narrowly linear indusium. — 2. *C. siliquosa.*

Blades variously paleaceous, tomentose, or hairy; indusia marginal, formed of the nearly unchanged, recurved or reflexed tip or border of the lobes or segments.
- Sori 1 or 2 to each lobe, the indusia separate; segments flattish.
  - Fronds minutely viscid-glandular. — 3. *C. viscida.*
  - Fronds beset with spreading whitish septate hairs. — 4. *C. cooperae.*
- Sori several or many to each lobe or segment, with a common indusium; segments beadlike.
  - Blades wholly devoid of scales. — 5. *C. feei.*
  - Blades paleaceous, as well as villous or tomentose.
    - Segments hairy above.
      - Blades thinly tomentulose above, delicately fibrillose-paleaceous and laxly tomentose beneath. — 6. *C. fibrillosa.*
      - Blades coarsely villous on both surfaces, the rachises bearing also a few lax elongate scales beneath. — 7. *C. parishii.*
    - Segments devoid of hairs above.
      - Blades usually 2-pinnate; segments mostly oblong, densely tomentose beneath. — 8. *C. gracillima.*
      - Blades 3–4-pinnate; segments mostly roundish to oval, imbricate-paleaceous beneath.
        - Fronds many, close; segments irregularly roundish or oval, fertile distally, the lightly crenate margins deeply recurved.
          - Segments devoid of scales above; scales beneath large, much exceeding the segments, whitish to light castaneous. — 9. *C. covillei.*
          - Segments with a few minute pale stellate scales above; scales beneath smaller, more numerous, darker, many reduced to entangled cilia. — 10. *C. intertexta.*
        - Fronds few, 5–10 mm. apart, larger; segments mostly subcordate-orbicular, fertile nearly to the base, the subentire margin closely revolute. — 11. *C. clevelandii.*

1. Cheilanthes califórnica (Hook.) Mett. California Lace-fern. Fig. 47.

Hypolepis californica Hook. Sp. Fil. **2**: 71. *pl. 88. A.* 1852.
Cheilanthes californica Mett. Abh. Senckenb. Ges. Frankfurt **3**: 88. 1859.
Cheilanthes amoena A. A. Eaton, Fern Bull. **5**: 44. 1897.

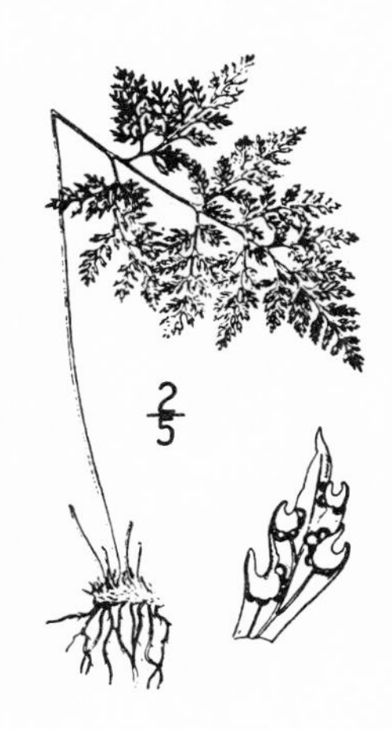

Rhizome short-creeping, thickish; scales acicular, 2–2.5 mm. long, rigid, dark castaneous, with pale narrow borders. Fronds numerous, 6–37 cm. long; stipes 4–28 cm. long, nearly naked, dull castaneous, subflexuous; blades broadly deltoid-ovate or pentagonal (the basal pinnae large, inequilateral, deltoid), 2.5–15 cm. long, 2–12 cm. broad, sub-quadripinnate, the minor rachises greenish-marginate above; segments oblique, linear to elliptic or lance-ovate, decurrent, acute to long-mucronate, callous; sori solitary at the transversely enlarged vein-tips, the ample pale indusium roundish-lunate, adherent to a slender saccate marginal tooth at either side; leaf tissue herbaceous, bright green, glabrous.

Base of cliffs, Transition and Upper Sonoran Zones; chiefly in the coast ranges, Humboldt County to San Diego County, California, and sparingly in the Sierra Nevada from Butte County southward. Type locality: Santa Barbara, California.

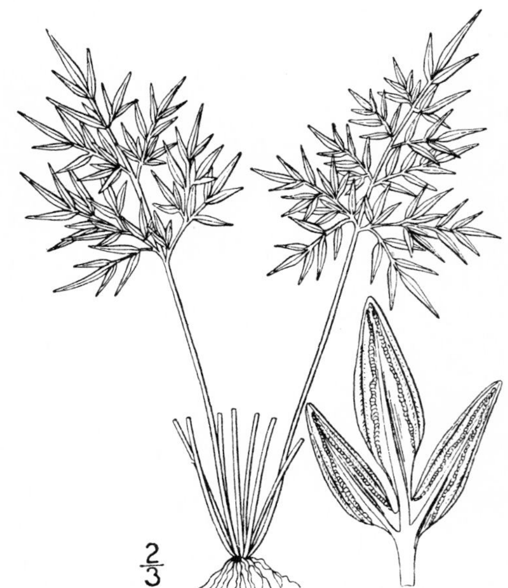

2. Cheilanthes siliquòsa Maxon.
Indian's Dream. Fig. 48.

Onychium densum Brack. in Wilkes, U. S. Explor. Exped. **16**: 120. 1854, not *Cheilanthes densa* Fée, 1852.
Pellaea densa Hook. Sp. Fil. **2**: 150. *pl. 125. B.* 1858.
Cryptogramma densa Diels, in Engl. & Prantl, Pflanzenfam. 1^4: 280. 1899.
Cheilanthes siliquosa Maxon, Am. Fern Journ. **8**: 116. 1918.

Rhizome multicipital, the divisions oblique, cespitose; scales acicular, brownish-castaneous, thick, lustrous, subentire. Fronds many, crowded, 7–30 cm. long; stipes 5–22 cm. long, nearly naked, brownish-castaneous, lustrous, flexuous; blades usually fertile, broadly ovate, deltoid-oblong, or subpentagonal, 1.5–8 cm. long, 3-pinnate below, naked, glabrous; pinnae few, close, oblique, the basal ones broadly triangular, inequilateral; segments narrowly linear, mostly 5–15 mm. long, 1–1.5 mm. broad, mucronate, close; margins sharply revolute, giving rise to a continuous delicate erose-denticulate indusium, the sporangia of adjacent sori crowded transversely in a continuous line; sterile fronds few, the segments broader, callous, serrate.

Crevices of cliffs and rock outcrops, Canadian and Transition Zones; San Luis Obispo and Tulare Counties, California, north in the mountains to Vancouver Island, east to northern Montana, Wyoming, and Utah; also in Gaspé County, Quebec, and Grey County, Ontario. Type locality: Rogue River region, Oregon.

3. Cheilanthes víscida Davenp.
Viscid Lip-fern. Fig. 49.

Cheilanthes viscida Davenp. Bull. Torrey Club **6**: 191. 1877.

Rhizome massive, nodose or multicipital, the divisions subglobose, conspicuously paleaceous; scales tufted, linear, long-attenuate, 6–8 mm. long, less than 1 mm. broad, flaccid, crispate, flexuous, bright castaneous. Fronds numerous, erect, fasciculate, 10–23 cm. long; stipes 4–10 cm. long, slender, fragile, dark brown, beset (like the blades) with sessile or short-stalked viscid glands, the rachises bearing also a few curved moniliform hairs; blades linear-oblong, acute, 5–14 cm. long, 1.5–3.2 cm. broad, 2-pinnate; pinnae spreading, deltoid to ovate-deltoid, the lower 2 or 3 pairs distant, often stalked; pinnules obliquely pinnatifid, with concealed midribs, the segments distant, short, decurrent, shallowly incised, the lobes or teeth minute; sori few-sporangiate, solitary on the ultimate lobes, the roundish or acutish tips reflexed or not; leaf tissue delicately herbaceous, dull grayish green.

Shaded crevices of rocks, Lower and Upper Sonoran Zones; southern California, from the Panamint Mountains to the western borders of the Colorado Desert. Type locality: White Water Cañon, Colorado Desert.

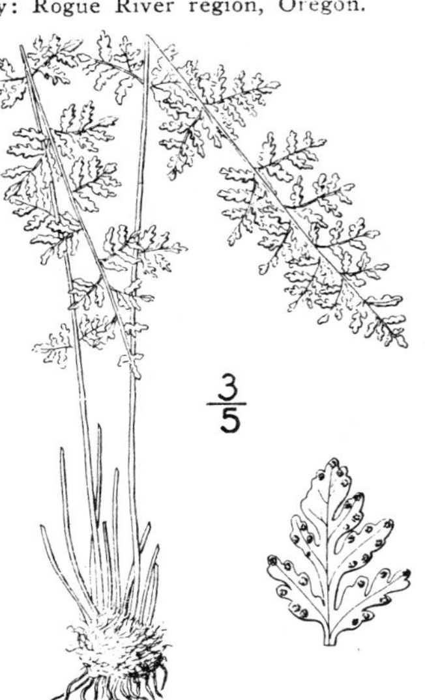

4. Cheilanthes coòperae D. C. Eaton.
Cooper Lip-fern. Fig. 50.

Cheilanthes cooperae D. C. Eaton, Bull. Torrey Club **6**: 33. 1877.

Rhizome short, thick, ascending, noticeably paleaceous at the apex; scales tufted, linear-acicular, 4–7 mm. long, castaneous or darker, translucent and concolorous, or with a dark narrow stripe toward the filiform tip. Fronds numerous, clustered, erect-spreading, 5–22 cm. long; stipes 1–7 cm. long, often thickish, dark castaneous, like the blade thickly beset with white or pale tawny, flattish, septate, mostly gland-tipped hairs; blades narrowly oblong to deltoid-oblong or lance-ovate, acutish, 4–15 cm. long, 2–4.5 cm. broad, 2-pinnate; pinnae mostly ovate-oblong, obtuse or acutish; pinnules pinnately cleft, the larger ones with 2 or 3 pairs of oblique, cuneate lobes, these shallowly bifid or crenate; sori 1 or 2 to a lobe or crenation, the tip recurved or not; leaf tissue delicately herbaceous, grayish green.

Deep crevices of cliffs, Upper Sonoran and Transition Zones; El Dorado County southward to near San Bernardino, California; reported from Mount Shasta. Type locality: near Santa Barbara, California.

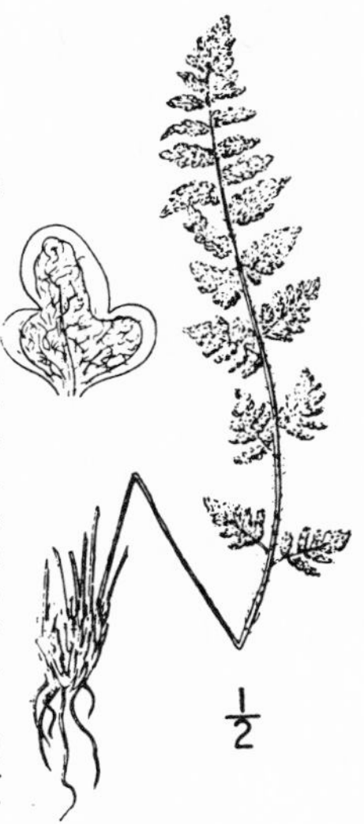

5. **Cheilanthes feèi** Moore.

Slender Lip-fern. Fig. 51.

Myriopteris gracilis Fée, Gen. Fil. 150. *pl. 29, f. 6.* 1852.
Cheilanthes gracilis Riehl; Mett. Abh. Senckenb. Ges. Frankfurt **3**: 80. 1859, not Kaulf. 1824.
Cheilanthes feei Moore, Ind. Fil. xxxviii. 1857.

Rhizome multicipital, the divisions short-creeping or ascending; scales tufted, linear, 5–7 mm. long, cinnamomeous or darker, with a sharp blackish stripe, entire, subflexuous at the tip. Fronds numerous, tufted, erect-spreading, 8–25 cm. long; stipes 3–12 cm. long, very slender, brown, deciduously pilose; blades linear-oblong to ovate or deltoid-ovate, acuminate, 5–13 cm. long, 1.5–4 cm. broad, usually 3-pinnate; pinnae spreading, deltoid to ovate-oblong, thinly villous above with lax, flexuous, whitish hairs, densely but coarsely tomentose with pale fulvous hairs beneath, the rachises similarly clothed; tertiary segments roundish or oval, simple, or the larger ones crenately lobed or divided; margins lightly crenate, narrowly recurved; leaf tissue delicately herbaceous, grayish green.

Ledges and rock crevices, Upper Sonoran and Transition Zones; Illinois and southern Minnesota, westward to British Columbia, Washington (Almota), southern California (Providence Mountains), and the Mexican border region from central Texas westward. Type locality: Hillsboro, Missouri.

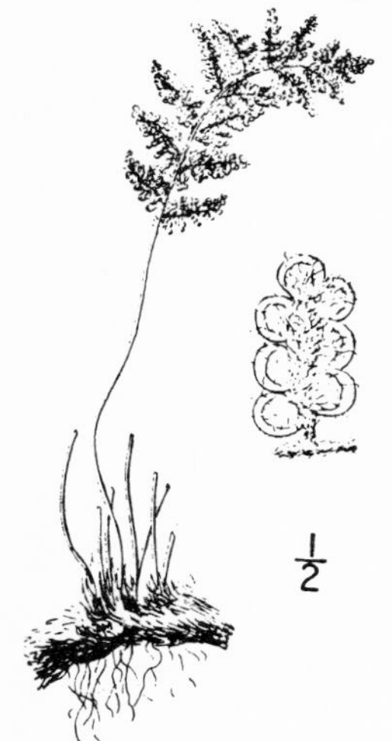

6. **Cheilanthes fibrillòsa** Davenp.

Fibrillose Lip-fern. Fig. 52.

Cheilanthes lanuginosa fibrillosa Davenp. Bull. Torrey Club **12**: 21. 1885.
Cheilanthes fibrillosa Davenp.; Underw. Nat. Ferns ed. 3, 95. 1888.

Rhizome short-creeping, branched, the divisions intricate, densely appressed-paleaceous; scales imbricate, linear-attenuate, hair-pointed, 2.5–3 mm. long, less than 0.5 mm. broad, dark brown, concolorous, rigid, lustrous, subentire. Fronds several, erect; stipes 5–9 cm. long, slender, subflexuous, castaneous, thinly fibrillose-paleaceous; blades narrowly oblong, acuminate, 5–8 cm. long, 2–3 cm. broad, 3-pinnate, thinly tomentulose above with delicate, septate, dirty white hairs, copiously fibrillose-paleaceous on the rachises beneath, the scales pale rusty, linear-attenuate, lax, spreading, tortuous, mingled with loose tomentum and completely covering the pinnae; tertiary segments minute, rounded-cuneate to rhombic, sessile, entire or the largest ones crenately lobed; outer margin crenulate, widely revolute, partially enclosing the few sporangia; leaf tissue very delicately herbaceous, light grayish green.

Among boulders, Upper Sonoran Zone; northern base of San Jacinto Mountain, Riverside County, California. Known only from the type specimens.

7. **Cheilanthes paríshii** Davenp.

Parish Lip-fern. Fig. 53.

Cheilanthes parishii Davenp. Bull. Torrey Club **8**: 61. *pl. 8.* 1881.

Rhizome short-creeping, branched; scales tufted, oblique, linear-lanceolate, attenuate, 3–4 mm. long, castaneous with a sharp blackish stripe, subentire. Fronds erect, 12–15 cm. long; stipes 4–7 cm. long, dark castaneous, deciduously fibrillose-paleaceous and with a few larger whitish scales; blades oblong-lanceolate, 8–10 cm. long, 2–3.5 cm. broad, 3–4-pinnate, coarsely villous on both sides with long, flexuous, hyaline, septate hairs, paleaceous on the rachises beneath, the scales mostly light castaneous, elongate, flexuous, toothed or laciniate, spreading, together with the coarse hairs not wholly covering the pinnae; pinnae inequilateral, deltoid to deltoid-oblong; segments pinnately lobed, or the larger ones with 1 or 2 pairs of free divisions, these minute, rhombic to rounded-cuneate, the terminal ones the larger, mostly 3-lobed; distal margins narrowly recurved, not covering the few sporangia; leaf tissue spongiose-herbaceous, grayish green.

Crevices of rocks, Lower Sonoran Zone; Andreas' Cañon, Palm Springs, east side of San Jacinto Mountain, Riverside County, California. Known only from the type locality.

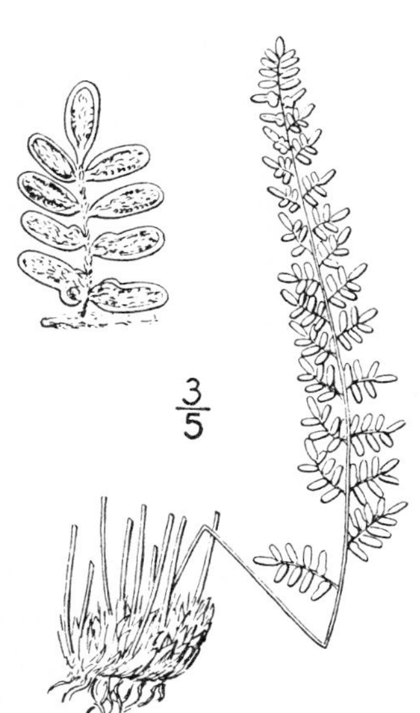

8. Cheilanthes gracíllima D. C. Eaton.

Lace-fern. Fig. 54.

Cheilanthes gracillima D. C. Eaton, in Torr. Bot. Mex. Bound. 234. 1859.

Rhizomes tufted, the numerous short branches close, densely paleaceous; scales light brown in mass, acicular to linear-lanceolate and long-attenuate, 2–3 mm. long. Fronds very numerous, 10–25 cm. long; stipes 5–15 cm. long, dark brown, soon nearly naked; blades linear to oblong-lanceolate, acuminate, 4–13 cm. long, 2-pinnate (rarely 3-pinnate); segments mostly oblong, distant to close, usually oblique, the terminal ones largest; rachises appressed-paleaceous, the scales linear-attenuate from a broader ciliate base, or broader and more numerous in 3-pinnate fronds, gradually more copiously ciliate, the segments densely cinnamomeous-tomentose beneath, a few minute, stellate scales above; margins subentire, deeply recurved, nearly concealing the numerous sporangia; leaf tissue herbaceous, dull or yellowish green.

Rock crevices and ledges, chiefly in the Canadian and arid Transition Zones; Vancouver Island to western Montana, south in the mountains to Nevada and to Mariposa and Marin Counties, California. Type locality: Cascade Mountains, Oregon.

9. Cheilanthes covíllei Maxon.

Coville's Lip-fern. Fig. 55.

Cheilanthes covillei Maxon, Proc. Biol. Soc. Washington **31**: 147. 1918.

Rhizome short-creeping, branched, the divisions short, appressed-paleaceous; scales linear to lanceolate, long-attenuate, 1.5–2.5 mm. long, dark brown to blackish, rigid, subentire. Fronds numerous, 10–30 cm. long; stipes 5–17 cm. long, brown to dark purplish, with small, linear, pale, ascending scales; blades oblong to ovate-deltoid, acuminate, 5–14 cm. long, 3-pinnate, the larger segments ternately to pinnately divided; rachises and under side of pinnae densely paleaceous, the scales imbricate, large, exceeding the roundish or irregularly oval segments, whitish or pale castaneous, cordate-ovate to ovate-lanceolate, attached above the closed sinus of the deeply cordate, long-ciliate base; ultimate rachises fibrillose-paleaceous above, the scales partly recurved upon the segments, these glabrous and naked above; sporangia borne within the deeply recurved, lightly crenate outer border; leaf tissue spongiose-herbaceous.

Rock crevices and rocky slopes, chiefly in the Upper Sonoran and Transition Zones; southern California and adjacent parts of Nevada and Arizona. Type locality: Surprise Cañon, Panamint Mountains, Inyo County, California.

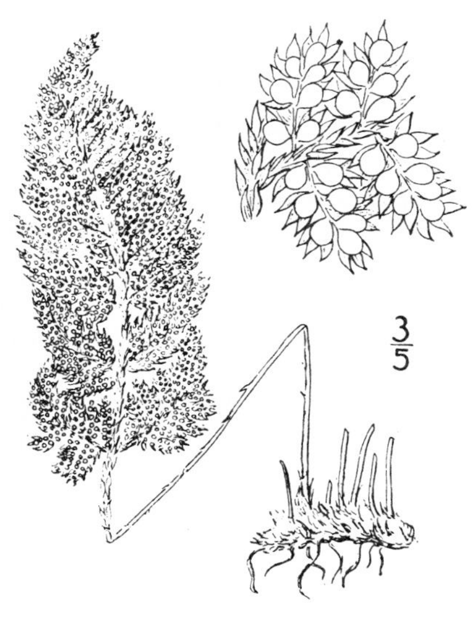

10. Cheilanthes intertéxta Maxon.

Coastal Lip-fern. Fig. 56.

Cheilanthes covillei intertexta Maxon, Proc. Biol. Soc. Washington **31**: 149. 1918.

Similar in general to *C. covillei*, but differing in its smaller shorter-creeping, subnodose rhizomes, its darker green and relatively larger segments, and in vestiture of the blades; segments bearing a few minute, whitish, stellate, subpersistent scales above, thickly covered beneath with numerous bright castaneous to cinnamomeous, imbricate scales, the larger ones (like those of the rachises beneath) mostly deltoid-lanceolate, long-attenuate, sinuate-denticulate nearly throughout, freely long-ciliate at the cordate base, underlaid by successively smaller and more copiously long-ciliate scales, the ultimate ones often reduced to entangled cilia.

Rock crevices, Upper Sonoran Zone; Santa Clara, Sonoma, and Mendocino Counties, California; also in Nevada (Virginia City). Type locality: Black Mountain, Santa Clara County, California.

11. Cheilanthes clevelándii D. C. Eaton.
Cleveland's Lip-fern. Fig. 57.

Cheilanthes clevelandii D. C. Eaton, Bull. Torrey Club **6**: 33. 1875.

Rhizome creeping, woody, slender, with a few short branches, closely imbricate-paleaceous; scales subulate to lance-attenuate, hair-pointed, 2.5–3 mm. long, bright to dark brown, glossy. Fronds erect, 1 cm. or less apart, 15–50 cm. long; stipes stout, 7–24 cm. long, light brown, with a few minute, linear, pale scales; blades linear-lanceolate to deltoid-oblong, acuminate, 7–26 cm. long, 3-pinnate or nearly 4-pinnate; pinnae ascending, falcate, narrowly deltoid to deltoid-oblong, or the lower ones deltoid; rachises and segments densely paleaceous beneath, the scales small, imbricate, castaneous, deltoid-ovate, attached above the closed sinus of the deeply cordate base, erose-dentate, long-ciliate; segments minute, mostly subcordate-orbicular, glabrous and naked above, but with the slender divisions of the rachis scales partly overlying them; sporangia numerous, borne partly within the narrowly revolute border of the segments nearly to the base; leaf tissue spongiose-herbaceous, dull green.

Bases of sheltering rocks, Upper Sonoran Zone; San Diego, Riverside, and Santa Barbara Counties, California; also on Santa Cruz Island, California. Type locality: San Diego County, California.

15. PELLAÈA Link, Fil. Hort. Berol. 59. 1841.

Rather small, rigid, rock-inhabiting plants, with stout and nodose or slender and wide-creeping rhizomes and erect, nearly naked, and glabrous foliage. Fronds uniform or nearly so, the blades 1–4-pinnate, the rachises usually dark colored and lustrous, the segments roundish-oval to linear, minute to large, more or less distinctly articulate; veins free, branched, not thickened at the immersed tips. Sori terminal and subterminal, or sometimes decurrent upon the veins, at length laterally confluent in a broad intramarginal line, nearly or (rarely) not at all concealed by the widely reflexed or revolute, continuous, indusiform margin, the border often modified, thinnish or even membranous. [Name Greek, alluding to the dark-colored stipes.]

About 70 species, mainly of temperate regions. Nine species besides the following occur in the United States. Type species, *Pteris atropurpurea* L.

Blades once pinnate, or the lower pinnae ternately divided.
  Pinnae mostly 2-parted, "mitten-shaped," membranous, the veins evident; stipes corrugate, easily breaking. 1. *P. breweri.*
  Pinnae simple or ternately divided, coriaceous, the veins immersed; stipes not wrinkled.
    Sporangia conspicuously intramarginal, long-decurrent on the veins, not concealed; pinnae oval to cordate-oblong, invariably simple, the fertile ones mostly conduplicate. 2. *P. bridgesii.*
    Sporangia nearer the margin, not long-decurrent, mostly concealed; pinnae mostly simple and linear, plane, the lower ones commonly 3-cleft or 3-divided. 3. *P. suksdorfiana.*
Blades 2–4-pinnate.
  Rhizomes very slender, wide-creeping, with imbricate-secund scales; stipe and rachises flesh-colored; segments obtuse, usually retuse. 4. *P. andromedaefolia.*
  Rhizomes stout, nodose, massive, with densely tufted scales; vascular parts dark or purplish brown; segments mucronulate to rigidly apiculate.
    Blades 2-pinnate, the secondary pinnae (segments) usually simple.
      Rachis of primary pinnae several times as long as the segments. 5. *P. compacta.*
      Rachis of primary pinnae rarely as long as the segments. 6. *P. brachyptera.*
    Blades 2–3-pinnate, the secondary pinnae usually 3-parted, 3-divided, or pinnate. 7. *P. mucronata.*

1. Pellaea bréweri D. C. Eaton.
Brewer's Cliff-brake. Fig. 58.

Pellaea breweri D. C. Eaton, Proc. Am. Acad. **6**: 555. 1865.

Rhizomes ascending or decumbent, massive, covered with the erect crowded stipe-bases of old fronds; scales densely tufted, light castaneous or dark cinnamomeous, almost capillary, 7–10 mm. long, concolorous, thin, lax, twisted. Fronds erect from an arcuate base, crowded, 7–21 cm. long; stipes 3–10 cm. long, stout, bright brown, glossy, naked above the base, transversely corrugate, readily fracturing; blades linear to linear-oblong, acute, 4–16 cm. long, 1.5–3.5 cm. broad, pinnate, glabrous; pinnae 6–12 pairs, mostly 2-parted (the upper lobe the larger), the lobes and simple upper pinnae ovate or deltoid-ovate to oblong-lanceolate from a cuneate to subcordate base, acute or acutish, flat; sporangia terminal and subterminal, confluent, nearly concealed by the thin, whitish, broadly reflexed margin; leaf tissue light green, subglaucous, membranous, the veins usually apparent.

Exposed rocky slopes and clefts of rocks, usually granite, Transition and Canadian Zones; the Sierra Nevada, California, to southern Washington, eastward to Utah, western Wyoming, and Idaho. Type locality: Sierra Nevada, California.

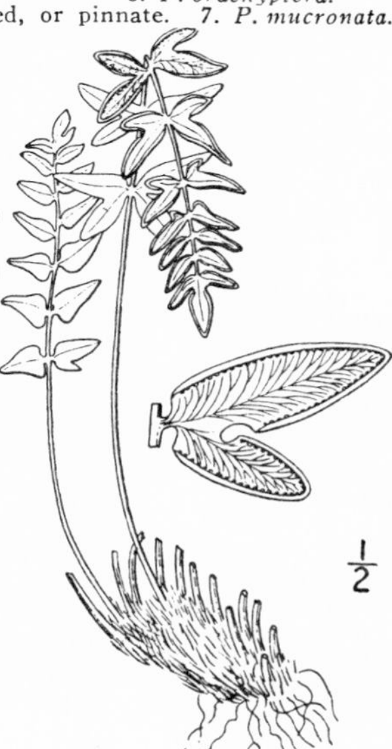

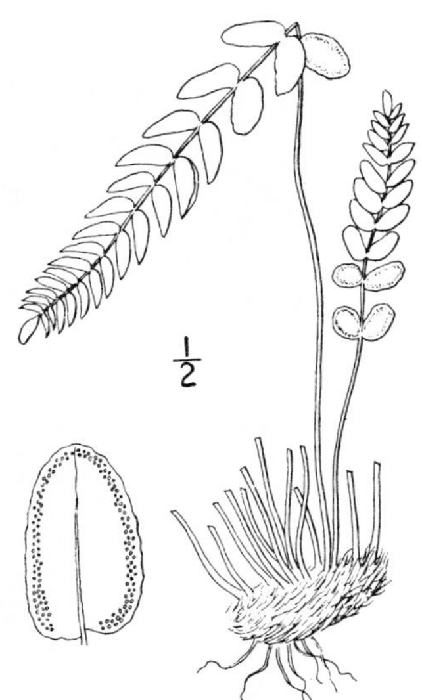

2. Pellaea bridgèsii Hook.

Bridges' Cliff-brake. Fig. 59.

Pellaea bridgesii Hook. Sp. Fil. 2: 238. *pl. 142. B.* 1858.

Rhizome short-creeping, often massive, the divisions short, close; scales tufted, brown in mass, narrowly linear-attenuate, 5–7 mm. long, mostly with a narrow, thick, blackish stripe, the thin pale margins denticulate. Fronds numerous, closely tufted, 8–35 cm. long; stipes 4–21 cm. long, castaneous, lustrous, long-persistent; blades linear to oblong-linear, 4–14 cm. long, 1–2.5 cm. broad, once pinnate, glabrous; pinnae 5–16 pairs and a terminal one, in fully fertile fronds subequal, mostly opposite, sessile, broadly oval to cordate-oblong, usually conduplicate and falcate; sterile pinnae suborbicular, usually flat; sporangia decurrent in a broad intramarginal band, not at all concealed by the narrow, whitish, cartilaginous, reflexed, wrinkled or plane border; leaf tissue coriaceous, grayish green, glaucous.

Rock-slopes and crevices of cliffs, mainly in the Canadian Zone; higher Sierra Nevada, California, from Nevada County southward to Mineral King, Tulare County; also in Boise National Forest, Idaho. Type locality: Sierra Nevada.

3. Pellaea suksdorfiàna Butters.

Suksdorf's Cliff-brake. Fig. 60.

Pellaea glabella simplex Butters, Am. Fern Journ. 7: 84. 1917.
Pellaea suksdorfiana Butters, Am. Fern Journ. 11: 40. 1921.

Rhizomes tufted, the divisions stout, subglobose, 1–1.5 cm. thick; scales closely tufted, dark cinnamomeous in mass, narrowly linear, 5–8 mm. long, thin, flaccid, flexuous, distantly denticulate. Fronds several or numerous, fasciculate, erect, 8–20 cm. long; stipes 2–6 cm. long, castaneous, with a few reduced capillary scales; blades oblong or lance-oblong, 6–14 cm. long, 2–5 cm. broad, pinnate, with a large terminal segment; rachis similar to the stipe; pinnae 3–8 pairs, mostly simple, linear, distant, all but the upper ones stalked; basal pinnae long-stalked, commonly 3-cleft or 3-divided, often deciduous; margins strongly but not widely recurved, the border thin, irregularly repand-crenulate; sporangia borne in a thick submarginal band, partially concealed; leaf tissue light green, glabrous, subchartaceous.

Crevices of limestone cliffs, Transition Zone; British Columbia and Washington, south to Utah and New Mexico. Type locality: Klickitat County, Washington.

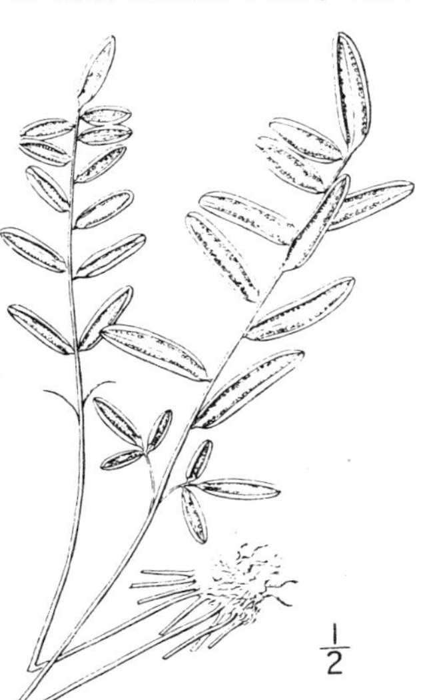

4. Pellaea andromedaefòlia (Kaulf.) Fée.

Coffee-fern. Fig. 61.

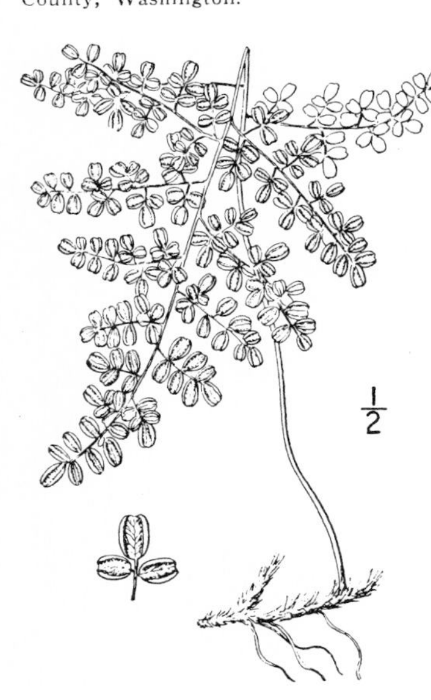

Pteris andromedaefolia Kaulf. Enum. Fil. 188. 1824.
Pellaea andromedaefolia Fée, Gen. Fil. 129. 1852.
Pellaea rafaelensis Moxley, Am. Fern Journ. 5: 107. 1915.

Rhizome very slender, wide-creeping; scales imbricate-secund, 1.5–3.5 mm. long, narrow, tapering to a hair point. Fronds distichous, 5–10 mm. apart, 15–75 cm. long; stipes 6–40 cm. long, flesh-colored, glaucous; blades deltoid to ovate, long-acuminate, 10–40 cm. long, 5–20 cm. broad, 2–4-pinnate (usually 3-pinnate), the rachises glabrous or glandular-pubescent; pinnae subopposite to alternate, the large lower ones often inequilateral, elongate-deltoid to deltoid-ovate; pinnules distant, stalked, the larger ones mostly pinnate; segments distant, oval or elliptical, 3–17 mm. long, short-stalked, mostly retuse, the sterile ones flat, the fertile widely revolute, corrugate at the tips of the spreading veins; sporangia partially concealed; leaf tissue rigidly herbaceous, subglaucous, dull green to reddish purple above, pale or yellowish green beneath.

Dry, stony situations, chiefly in the Upper Sonoran Zone; middle northern California (Mendocino and Tehama Counties) south to Lower California; also in southwestern Oregon (Roseburg). Type locality: California.

5. Pellaea compácta (Davenp.) Maxon.

Desert Cliff-brake. Fig. 62.

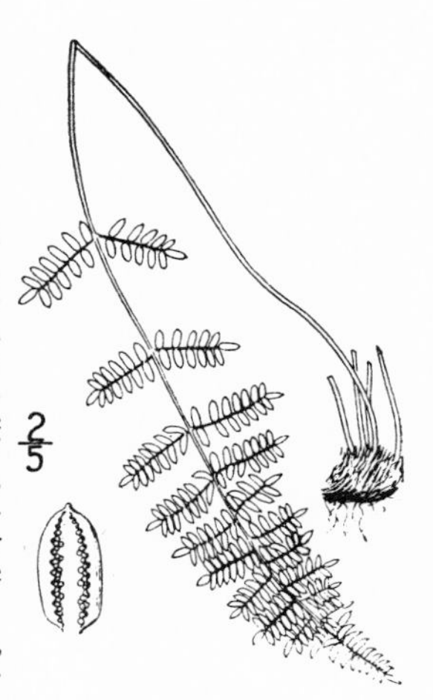

Pellaea wrightiana compacta Davenp. Cat. Davenp. Herb. Suppl. **46.** 1883.
Pellaea compacta Maxon, Proc. Biol. Soc. Washington **30**: 183. 1917.

Rhizome woody, stout, nodose; scales tufted, dark or grayish brown in mass, linear, long-attenuate, 5–7 mm. long, with a broad, black, opaque stripe, the borders pale, thin, denticulate. Fronds mostly fertile, numerous, fasciculate, 20–35 cm. long, the glossy, dark brown, flexuous stipes much longer than the blades, naked; blades narrowly oblong, acuminate, 9–13 cm. long, 2.5–4.5 cm. broad, 2-pinnate; pinnae ascending, the lower ones distant, the others contiguous or crowded; pinnules of the larger pinnae 5–7 pairs, simple, mostly close, sessile or nearly so, the sterile ones oval or oblong, the fertile narrower, broadly revolute, often conduplicate, transversely wrinkled, mucronulate, the margins minutely sinuate-dentate; sporangia nearly concealed within the revolute border; leaf tissue rigidly spongiose-herbaceous, grayish green.

Dry stony slopes, Canadian and Upper Transition Zones; San Bernardino, San Jacinto, San Antonio, and Providence Mountains, southern California. Type locality: San Bernardino Mountains.

6. Pellaea brachýptera (Moore) Baker.

Sierra Cliff-brake. Fig. 63.

Platyloma brachypterum Moore, Gardn. Chron. **1873**: 141. 1873.
Pellaea brachyptera Baker in Hook. & Baker, Syn. Fil. ed. 2, 477. 1874.

Rhizome woody, horizontal, thick, nodose; scales tufted, bright brown in mass, acicular, 5–8 mm. long, rather lax, with a narrow dark brown stripe, the borders thin, pale. Fronds fasciculate, 15–40 cm. long; stipes stout, naked, purplish brown, as long as the blades or longer; blades linear to narrowly linear-oblong, 6–19 cm. long, 1.5–4 cm. broad, 2-pinnate; pinnae numerous, very oblique, the lower 2 or 3 pairs sometimes distant, the others close or imbricate, mostly broader than long, semicircular or broadly rounded-deltoid, the rachis usually much shorter than the subequal segments; segments 5–11, usually close, spreading, simple (rarely 2 or 3-parted), narrowly linear, 6–17 mm. long, sessile, straight or falcate, rigidly short-mucronate, revolute to the middle; sporangia concealed; leaf tissue rigidly spongiose-herbaceous, dull or grayish green, sub glaucous.

Dry rocky slopes, chiefly in the Transition Zone; Sierra Nevada of northern California (Placer County) northward to southwestern Oregon. Type, a cultivated plant from California.

7. Pellaea mucronàta (D. C. Eaton) D. C. Eaton.

Bird's-foot Cliff-brake. Fig. 64.

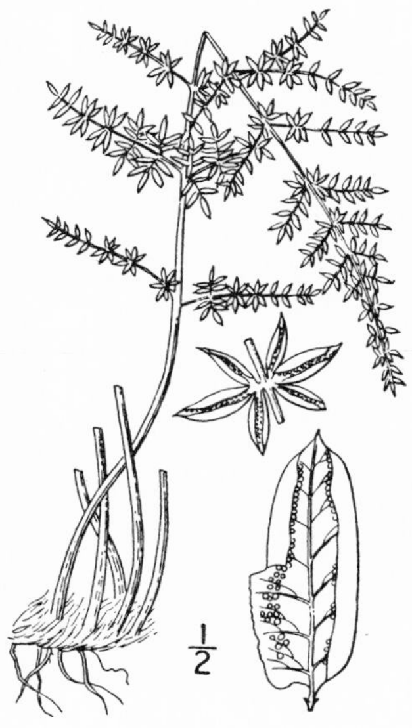

Allosorus mucronatus D. C. Eaton, Am. Journ. Sci. II. **22**: 138. 1856.
Pellaea ornithopus Hook Sp. Fil. **2**: 143. *pl. 116. A.* 1858.
Pellaea mucronata D. C. Eaton in Torr. Bot. Mex. Bound. 233. 1859.

Rhizome woody, thick, nodose; scales castaneous in mass, closely tufted, acicular, 5–10 mm. long, with a narrow, dark brown, opaque stripe, the borders pale, thin, denticulate. Fronds mostly fertile, fasciculate, 15–50 cm. long; stipes 4–20 cm. long, stout, dark purplish-castaneous, rather dull; blades oblong to deltoid-oblong or ovate-lanceolate, abruptly long-acuminate, 8–35 cm. long, 3–15 cm. broad, 2-3-pinnate; pinnae close or distant, linear to lance-linear; pinnules 6–15 pairs, the larger ones normally 3-foliolate or ternately divided (rarely simple), or sometimes fully pinnate, with 5–11 sessile segments; ultimate pinnules and segments 2–6 mm. long, elliptical to linear-oblong, rigidly apiculate, wrinkled, revolute to the middle; sporangia mostly concealed; leaf tissue spongiose-coriaceous, grayish green, subglaucous.

Sunny or sparsely shaded rocky slopes, Upper Sonoran Zone; nearly throughout California, from near sea level to 1500 meters elevation. Type locality: hills near San Francisco Bay, California.

16. NOTHOLAÈNA R. Br. Prodr. Fl. Nov. Holl. 1: 145. 1810.

Small rock-loving, mainly xerophilous ferns, with glandular, densely ceraceous, paleaceous, or variously hairy foliage. Fronds uniform, the blades 1–4-pinnate and variously lobed or pinnatifid, linear to broadly deltoid or pentagonal. Sori usually submarginal, roundish or oblong, borne at or near the tips of the unmodified veins (sometimes decurrent), more or less confluent laterally. Proper indusia wanting, the margin revolute and partially covering the sporangia in some species, flattish in others, the sporangia then somewhat concealed by the hairy, scaly, or ceraceous covering. [Name Greek, meaning spurious cloak, in allusion to the absence of a proper indusium.]

About 60 species, some of them widely distributed. Besides the following, numerous other species occur in the southwestern United States and in tropical America. Type species, *Acrostichum marantae* L.

Blades densely tomentose beneath.
- Stipe and rachis sparsely hirsute; pinnae stalked, coarsely and densely hirsute-tomentose above. 1. *N. parryi.*
- Stipe and rachis at first densely tomentose; pinnae sessile, finely tomentulose above. 2. *N. newberryi.*

Blades glabrous or ceraceous beneath.
- Blades broadly deltoid-pentagonal, ceraceous beneath; segments close, mostly sessile or adnate; margins revolute. 3. *N. californica.*
- Blades oblong-ovate to narrowly triangular, glabrous; segments distant, mostly short-stalked; margins plane. 4. *N. jonesii.*

1. Notholaena párryi D. C. Eaton.

Parry's Cloak-fern. Fig. 65.

Notholaena parryi D. C. Eaton, Am. Nat. 9: 351. 1875.

Rhizomes oblique or short-creeping, tufted, thick, densely paleaceous; scales tufted, subacicular, 4–6 mm. long, cinnamomeous, mostly with a thick narrow brown stripe, entire. Fronds numerous, clustered, erect, 6–18 cm. long; stipes 3–8 cm. long, slender, brown or purplish brown, hirsute; blades linear-oblong to ovate-oblong, acutish, 3–8 cm. long, 1–3 cm. broad, 3-pinnate; pinnae 5–9 pairs, mostly oblong-ovate, delicately stalked, involute with age, the lower ones subdeltoid, distant; segments small, close, roundish to ovate-oblong, crenate or crenately incised, confluent, fragile, closely enveloped by the woolly covering, buff-tomentose beneath, above coarsely hirsute-tomentose, light grayish; sporangia few, black, partly visible with age; leaf tissue very delicately herbaceous, wholly covered.

Crevices of rocks, Sonoran Zones; desert region of southern California to south-central Arizona and southwestern Utah, ascending to 1740 meters in the Panamint Mountains. Type locality: near St. George, Utah.

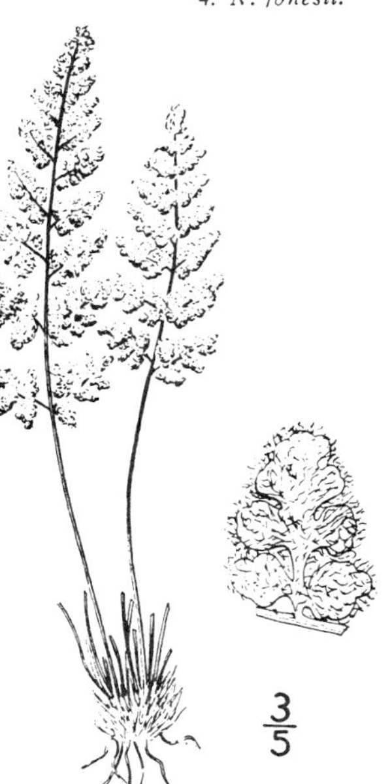

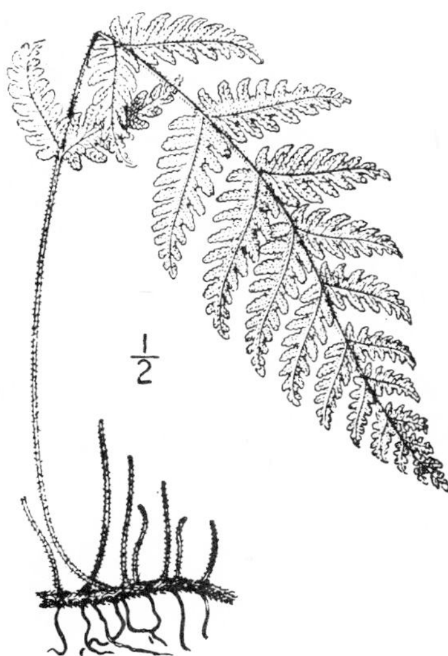

2. Notholaena newbérryi D. C. Eaton.

Cotton-fern. Fig. 66.

Notholaena newberryi D. C. Eaton, Bull. Torrey Club 4: 12. 1873.

Rhizomes slender, creeping, branched, entangled, densely appressed-paleaceous; scales acicular, 2–3 mm. long, rigid, blackish brown, glossy, entire. Fronds clustered, erect, 8–24 cm. long; stipes 4–14 cm. long, wiry, purplish brown, glossy, at first closely tawny-tomentose; blades narrowly oblong to oblong-lanceolate, acuminate, 4–13 cm. long, 1.5–5 cm. broad, 3-pinnate, finely and densely tawny-tomentose beneath, delicately whitish-tomentulose above, the woolly covering persistent throughout; pinnae about 10 pairs, slightly oblique, narrowly triangular to oblong-ovate, mostly inequilateral, sessile, the lower ones distant, subequal, broader; segments minute, close, roundish-obovate, entire, or the larger ones crenately lobed; sporangia protruding and mostly visible with age; leaf tissue membrano-herbaceous.

Rock crevices, mainly in the Upper Sonoran Zone; lower hills of the coast ranges, southern California, from Los Angeles to San Bernardino, south to San Diego; also on San Clemente Island, and on Guadalupe Island, Lower California. Type locality: San Diego.

3. **Notholaena califórnica** D. C. Eaton.
California Cloak-fern. Fig. 67.

Notholaena californica D. C. Eaton, Bull. Torrey Club **10**: *27*, 1883.

Rhizomes nodose-multicipital, densely paleaceous; scales rigidly acicular, 3.5–4.5 mm. long, dark reddish brown to blackish, opaque, denticulate. Fronds several or numerous, fasciculate, erect-spreading, long-stipitate, 4–15 cm. long; stipes 3–12 cm. long, brown to light castaneous; blades broadly deltoid-pentagonal, acutish to acuminate, 2–5 cm. long, 2–4 cm. broad, 3-pinnate as to the large inequilateral deltoid basal pinnae, gradually simpler above, the pinnae and segments all close, the ultimate divisions oblong or deltoid-oblong, rounded-obtuse, inequilateral, sparsely glandular-pulverulent above, beneath rather densely yellowish-ceraceous at maturity, sometimes paler and less ceraceous; rachises brownish, always evident; margins recurved but not concealing the sporangia, these borne in a continuous line; leaf tissue rigidly thick-herbaceous or coriaceous.

Crevices of rocks, Sonoran Zones; Catalina Island, and desert slopes of San Jacinto Mountains, Slover Mountain, and southern San Diego County, California; also in western Arizona (Congress Junction) and Lower California. Type locality: near San Diego.

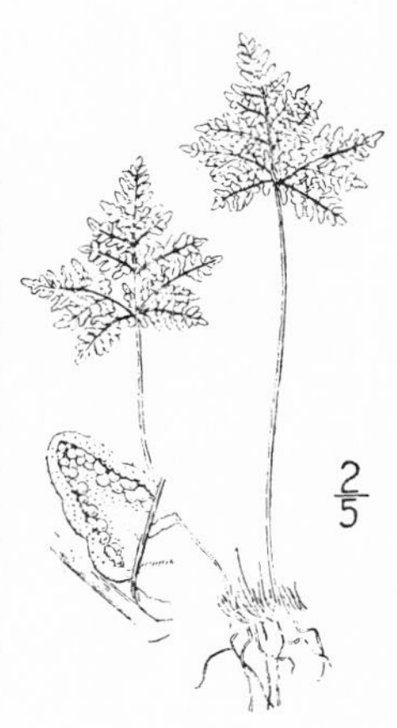

4. **Notholaena jònesii** Maxon.
Jones' Cloak-fern. Fig. 68.

Notholaena jonesii Maxon, Am. Fern. Journ. 7: 108. 1917.

Rhizome short, oblique, conspicuously paleaceous, the scales thin, bright brown, linear, very long-attenuate, with flexuous capillary tips. Fronds several or numerous, depressed-spreading, 3–16 cm. long, the stipes curved, reddish brown; blades 2–9 cm. long, oblong-ovate to narrowly triangular, 2-pinnate or rarely 3-pinnate; pinnae 3–7 pairs, opposite to alternate, spreading, triangular-oblong, with 2–5 pairs of distant, entire to crenately lobed, rounded or subcordate pinnules below the larger terminal segment or, rarely, the larger ones again pinnate; segments fleshy, herbaceous, glabrous, subglaucous, not at all pulverulent, mostly short-stalked; sporangia strongly confluent in a broad submarginal band.

Crevices of rocks, Upper Sonoran Zone; desert slopes of the San Bernardino and San Jacinto Mountains, and in the mountains of Santa Barbara County, California; also in southwestern Utah; rare and local. Type locality: Panamint Cañon, Inyo County, California.

Family 3. **MARSILEÀCEAE.**
Pepperwort Family.

Perennial, herbaceous plants rooting in mud, with slender creeping rhizomes. Leaves long-petioled and with 2- or 4-foliolate blades, or filiform and lacking a blade. Sori borne within bony, ovoid to globose, several-celled, pedunculate sporocarps (conceptacles), these arising from the rhizome near the base of the leaf-stalks or more or less consolidated with them. Spores of two kinds, megaspores and microspores, prothallia from the former producing mainly archegonia and those from the latter antheridia.

Three genera, the two following and *Regnellidium* (monotypic), of South America.

Leaves strongly differentiated into elongate petiole and 4-foliolate blade; sporocarps ovoid. 1. *Marsilea.*
Leaves filiform, lacking a definite blade; sporocarps globose. 2. *Pilularia.*

1. **MARSÍLEA** L. Sp. Pl. 1099. 1753.

Plants of shallow ponds, ditches, and marshy shores. Leaves solitary or subfasciculate, long-petioled; blades 4-foliolate. Sporocarps ovoid, with 2 teeth near the base, 2-celled vertically, with many transverse partitions, at length dehiscent into 2 valves and emitting a mucilaginous band of tissue, this bearing the sori at intervals within membranous envelopes. Sori including both megasporangia and microsporangia, the former few, with solitary megaspores, the latter many, with numerous microspores. [Named in honor of Giovanni Marsigli, an Italian botanist, who died about 1804.]

A genus of very wide distribution, comprising about 55 species, mainly of the Old World. In addition to the following, several occur in the southern United States.

Upper tooth of the sporocarp sharp, conspicuous; megasporangia 12–20 in each sorus. 1. *M. vestita.*
Upper tooth a rounded-tubercle, or obsolete; megasporangia 6–9 in each sorus. 2. *M. oligospora.*

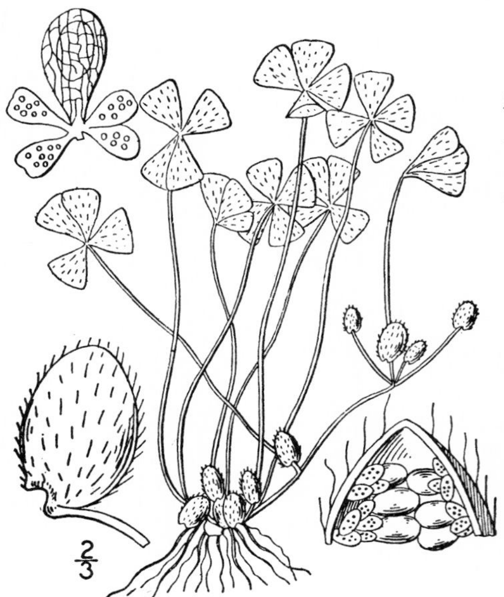

1. Marsilea vestìta Hook. & Grev.

Hairy Pepperwort. Fig. 69.

Marsilea vestita Hook. & Grev. Icon. Fil. **2**: *pl. 159.* 1831.
Marsilea mucronata A. Br. Am. Journ. Sci. II. **3**: 55. 1847.

Rhizomes wide-creeping, clothed with dense tufts of ferruginous or fulvous silky hairs at the nodes. Leaves few or many, arising from the nodes, mostly 5–20 cm. long, ascending; petioles slender, deciduously hairy; blades 1–3 cm. broad; leaflets broadly cuneate, spreading or folding together, 5–15 mm. long, nearly as broad, entire or nearly so, deciduously hairy. Peduncles nearly or quite free from the petiole, short; sporocarps solitary, 4–8 mm. long, 3–6 mm. broad, at first densely hairy; upper tooth rather long, acute, usually curved; lower tooth short, blunt; sori 9–11 in each valve.

Edge of ponds, ditches, and rivers, Upper and Lower Sonoran Zones; British Columbia to southern California, east to South Dakota, Kansas, Oklahoma, and Texas. Type locality: Columbia River region.

2. Marsilea oligóspora Goodding.

Nelson's Pepperwort. Fig. 70.

Marsilea oligospora Goodding, Bot. Gaz. **33**: 66. 1902.

Rhizomes long-creeping or not, the internodes often very short, the plants appearing tufted. Leaves numerous, 3–8 cm. long, erect or ascending; petioles slender; blades 1–2 cm. broad; leaflets broadly cuneate, spreading or folding together, 5–10 mm. long, 3–11 mm. broad, entire, soft-hairy or glabrescent. Peduncles 5–15 mm. long, united basally with the base of the petiole; sporocarps solitary, 4–6 mm. long, 3–5 mm. broad, densely covered with appressed hairs, glabrescent; upper tooth a short rounded tubercle, or obsolete; lower tooth short, blunt; sori 5–8 in each valve.

River banks and bottoms of drying ponds and marshes, chiefly in the Transition Zone; Washington to Montana, south to California (Tulare County), and Wyoming. Type locality: Jackson's Hole, Wyoming.

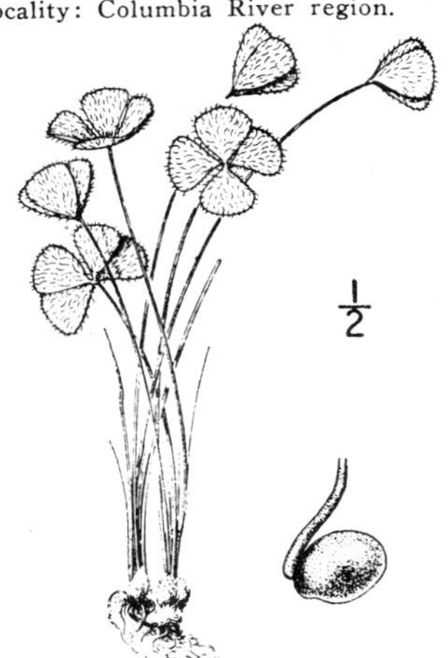

2. PILULÀRIA L. Sp. Pl. 1100. 1753.

Very small, inconspicuous plants of muddy situations, with slender long-creeping rhizomes, the nodes bearing 1 or several minute filiform leaves. Sporocarps globose, short-pedunculate, axillary, longitudinally 2–4-celled, dehiscing by as many valves at the apex; cells (sori) with parietal cushions bearing megasporangia below and microsporangia above, the former with solitary megaspores, the latter with numerous microspores. [Name Latin, meaning a little ball, alluding to the sporocarps.]

A widely distributed genus of 6 species, only the following occurring in North America. Type species, *Pilularia globulifera* L.

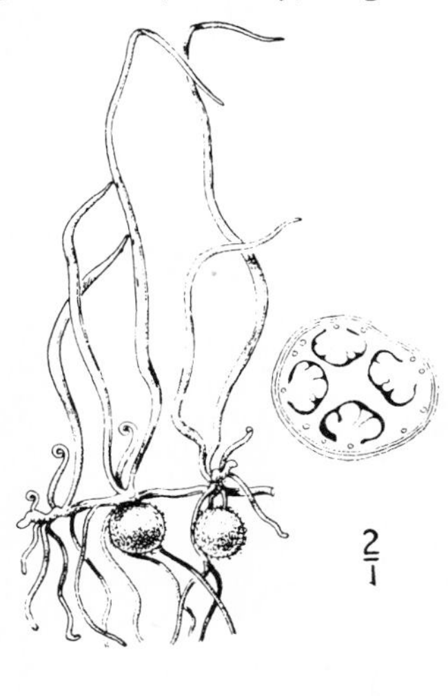

1. Pilularia americàna A. Br.

American Pilularia. Fig. 71.

Pilularia americana A. Br. Monatsb. Kön. Akad. Wiss. Berlin **1863**: 435. 1863.

Rhizome filiform, wide-creeping; leaves setiform, 2–4 cm. long, solitary or not, the nodes rooting beneath. Sporocarps about 2 mm. in diameter, attached laterally to a short descending arcuate peduncle, 2–4-celled (usually 3-celled); megaspores 10–17 in each cell, not constricted in the middle.

Clayey depressions and desiccating pools, mainly in the Upper Sonoran Zone; Oregon (Crook County) to southern California; also in Arkansas. Type locality: Fort Smith, Arkansas.

Family 4. SALVINIÀCEAE.

SALVINIA FAMILY.

Aquatic, floating plants of small size, with a more or less elongate and sometimes branching axis, bearing apparently distichous leaves. Sporocarps (conceptacles) soft and thin-walled, borne 2 or more on a common stalk, 1-celled, having a central often branched receptacle, this bearing either megasporangia containing a solitary megaspore or microsporangia containing numerous microspores.

Two genera of wide distribution, *Salvinia* and the following, each consisting of a few species.

1. AZÓLLA Lam. Encycl. 1: 343. 1783.

Fugacious, reddish or green, mosslike plants, with pinnately branched stems covered with minute imbricate 2-lobed leaves, emitting rootlets beneath. Sporocarps of 2 kinds, borne in pairs beneath the stem, the smaller ones ovoid or acorn-shaped, containing at the base a solitary megaspore and a few minute bodies of doubtful function above it; larger sporocarps globose, with a basal placenta, this bearing many pedicellate microsporangia containing masses of microspores, called massulae. [Greek, signifying killed by drought.]

About 5 species of wide distribution, the following the generic type.

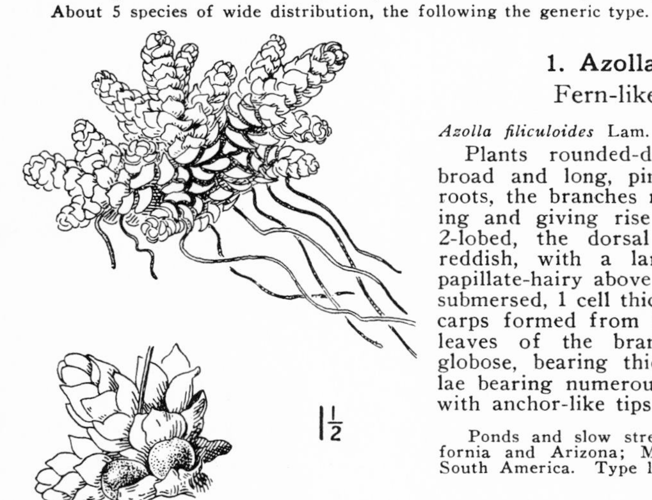

1. Azolla filiculoìdes Lam.

Fern-like Azolla. Fig. 72.

Azolla filiculoides Lam. Encycl. 1: 343. 1783.

Plants rounded-deltoid or elongate, 1–2.5 cm. broad and long, pinnately branched, with solitary roots, the branches mostly 2-pinnate, often separating and giving rise to new plants; leaves deeply 2-lobed, the dorsal lobe rather thick, green or reddish, with a large cavity in the basal part, papillate-hairy above; ventral lobe somewhat larger, submersed, 1 cell thick except in the middle. Sporocarps formed from the ventral lobes of the lowest leaves of the branches, subglobose; megaspores globose, bearing thick cylindrical papillae; massulae bearing numerous rigid linear flattish processes with anchor-like tips.

Ponds and slow streams, Washington, to southern California and Arizona; Mexico southward, nearly throughout South America. Type locality: Straits of Magellan.

Family 5. ISOETÀCEAE.*

QUILLWORT FAMILY.

Perennial plants, aquatic to terrestrial in habitat, with 2–3-lobed corm, bearing branched roots and a rosette of erect or recurved sedge-like leaves. Solitary sporangia sessile in leaf axils, of two sorts, microsporangia and megasporangia, producing respectively microspores and megaspores, which, on germination, develop gametophytes, the former with a single antheridium, the latter with archegonia.

A family consisting of a single genus, world-wide in distribution.

1. ISÒETES L. Sp. Pl. 1100. 1753.

Submerged, amphibious or terrestrial perennials, with a 2–3-lobed short fleshy stem giving rise to a group of elongated, somewhat triangular or quadrangular leaves, with 4 septate longitudinal air-channels, and with or without peripheral strands, with or without stomata. Leaves producing small triangular extension of tissue (ligule) above the single large round or elongated axillary sporangium in basal cavity, more or less encased by membranous extension (velum) on inner face of leaf. Megaspores hemispherical at base, with equatorial ridge, and three other crests joined at apex, with sculptured walls; microspores minute, powdery, usually ovoid. [Name Greek, meaning equal at all seasons.]

About 60 species, of world-wide distribution, of which 20 or more occur in the United States. Type species, *Isoetes lacustris* L.

Submersed or rarely emersed; corms 2-lobed; peripheral strands none.
  Stomata lacking.
    Leaves long; megaspores with irregular network of ridges on basal face, maximum size 640 μ. — 1. *I. occidentalis.*
    Leaves short; megaspores with points or short ridges, maximum size 800 μ. — 2. *I. piperi.*

*Text contributed by NORMA E. PFEIFFER.

Stomata present.
  Leaves coarse; stomata numerous. — 3. *I. flettii.*
  Leaves not coarse; stomata few.
    Megaspores tuberculate. — 4. *I. bolanderi.*
    Megaspores spinose. — 5. *I. braunii.*
Amphibious or terrestrial; peripheral strands usually present.
  Corms 2-lobed; velum incomplete. — 6. *I. howellii.*
  Corms 3-lobed; velum complete.
    Leaves more than 8 cm. long. — 7. *I. nuttallii.*
    Leaves less than 8 cm. long. — 8. *I. orcuttii.*

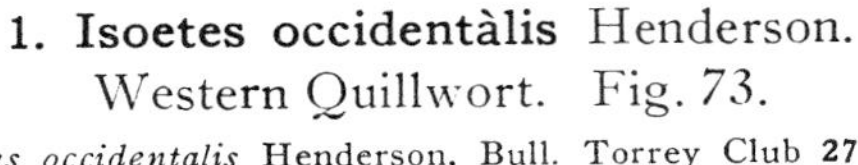

1. Isoetes occidentàlis Henderson.
Western Quillwort. Fig. 73.

Isoetes occidentalis Henderson, Bull. Torrey Club **27**: 358. 1900.

Isoetes lacustris var. *paupercula* Engelm. Trans. St. Louis Acad. Sci. **4**: 377. 1882.

Isoetes paupercula (Engelm.) A. A. Eaton, Proc. U. S. Nat. Mus. **23**: 649. 1901.

Submersed; corm 2-lobed; leaves 9–30, rarely 60, dark green, rigid, gradually tapering, 5–20 cm. long; peripheral strands and stomata lacking; ligule short-triangular; sporangium almost orbicular 5–6 mm. long, about one-third covered by velum, megaspores cream-colored, 525–640 μ in diameter, marked with low conspicuous irregular crests, chiefly simple on apical faces, branching to form irregular network on basal face; microspores 24–42 μ long, spinulose.

Mountain lakes, Canadian Zone; Wyoming, Colorado, Idaho, and California. Type locality: "in ½ foot of water, Lake Coeur d'Alene, Idaho."

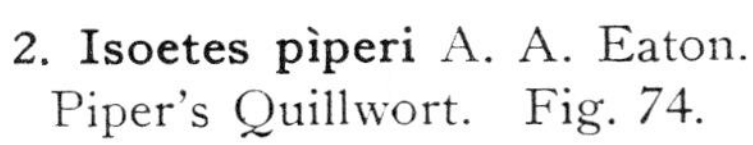

2. Isoetes pìperi A. A. Eaton.
Piper's Quillwort. Fig. 74.

Isoetes piperi A. A. Eaton, Fern Bull. **13**: 51. 1905.

Isoetes howellii var. *piperi* Clute, Fern Allies 246. 1905.

Submersed; corm 2-lobed; leaves 8–26, erect or somewhat spreading, taper-pointed, 3–10 cm., rarely 15 cm., long; peripheral strands and stomata lacking; ligule short triangular; sporangium 4–5 mm. long, one-fifth to two-thirds covered by velum; megaspores white, 540–800 μ in diameter, marked with points or tubercles and short ridges, sometimes serpentine; microspores 29–35 μ, rarely 42 μ long, papillose.

Lakes in Washington, Transition Zone. Type locality: Green Lake, King County, Washington.

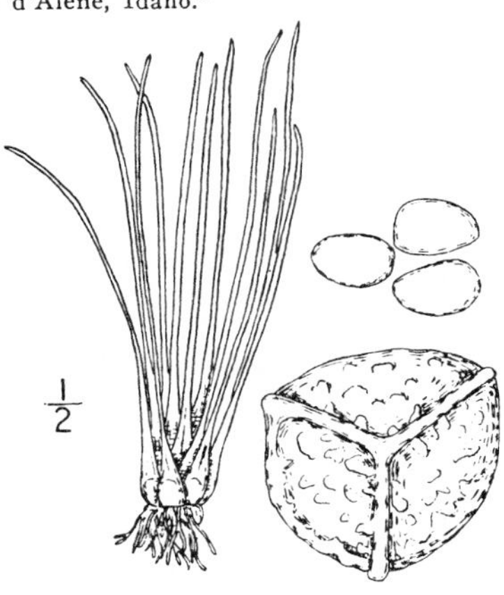

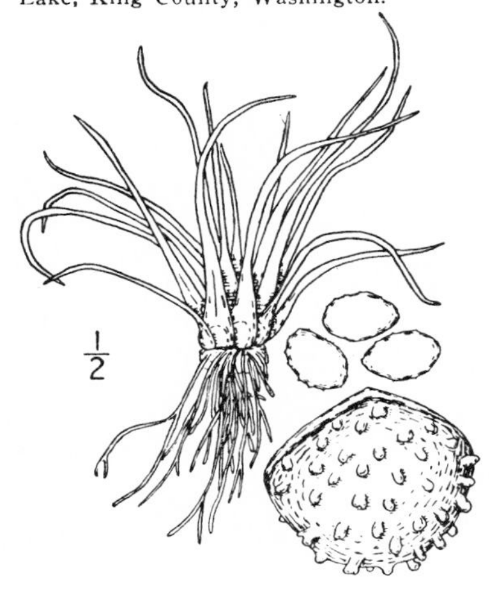

3. Isoetes fléttii (A. A. Eaton) N. E. Pfeiffer.
Flett's Quillwort. Fig. 75.

Isoetes echinospora var. *flettii* A. A. Eaton, Fern Bull. **13**: 51. 1905.

Submersed; corm 2-lobed; leaves 15–30, coarse, tapering, spreading or recurved, 5–8 cm. long; peripheral strands lacking; stomata numerous: ligule blunt-triangular; sporangium oblong, 4–6 mm. long, covered less than one-half by the velum; megaspores 480–560 μ in diameter, marked with coarse low tubercles and short crests, these usually somewhat distant; microspores 29–33 μ long, finely spinulose.

In lakes, Transition Zone; Washington. Type locality: "lake shore in gravel, Spanaway Lake, Pierce County, Washington."

4. Isoetes bolánderi Engelm.

Bolander's Quillwort. Fig. 76.

Isoetes bolanderi Engelm. Am. Nat. **8**: 214. 1874.
Isoetes californica Engelm. in A. Gray, Man. ed. 5, 677. 1867.
Isoetes bolanderi var. *parryi* Engelm. Am. Nat. **8**: 214. 1874.
Isoetes bolanderi var. *sonnei* Henderson, Bull. Torrey Club **27**: 349. 1900.

Submersed; corm deeply 2-lobed; leaves 6–25, bright green, soft, slender, tapering to a fine point, 6–13 or rarely 25 cm. long; stomata not numerous; peripheral strands usually lacking; ligule small, triangular, cordate; sporangia 3–4 mm. long, one-fourth to one-third covered by velum; megaspores white, sometimes bluish, 300–440 μ, rarely 480 μ, in diameter, obscurely or distinctly marked with low tubercles or wrinkles; microspores 23–30 μ long, more or less spinulose.

Mountain lakes and ponds, Transition and Canadian Zones; Wyoming, Colorado, western Idaho, Utah, Arizona, Oregon, Washington, California. Type locality: "in ponds and shallow lakes, Sierra Nevada of California, at altitudes of 5,000–10,000 feet."

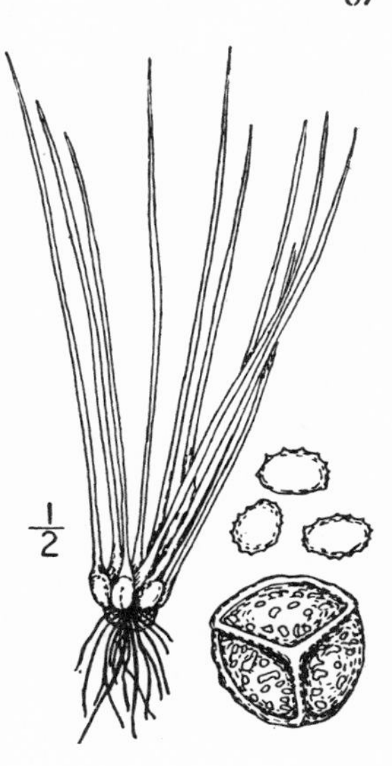

Isoetes bolanderi pygmaea (Englm.) Clute, Fern Allies 228. 1905. (*Isoetes pygmaea* Engelm. Am. Nat. **8**: 214. 1874.) Deeply submersed; corm deeply 2-lobed; leaves 4–15, stout, rather abruptly narrowed at apex, 2–3 cm. long; stomata and peripheral strands lacking; sporangium small, orbicular, with very narrow velum; megaspores white, 360–480 μ in diameter, smoothish or indistinctly marked with short wrinkles and tubercles; microspores 26–30 μ long, smooth to slightly spinulose. Type locality: Mono Pass, Sierra Nevada, California, 7000 ft. altitude.

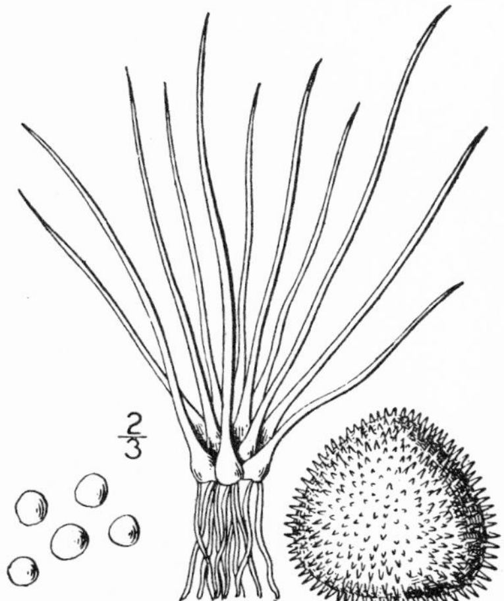

5. Isoetes braùnii Durieu.

Braun's Quillwort. Fig. 77.

Isoetes braunii Durieu, Bull. Soc. Bot. France **11**: 101. 1864.
Isoetes echinospora braunii Engelm. in A. Gray, Man. ed. 5, 676. 1867.
Isoetes echinospora boottii Engelm. in A. Gray, Man. ed. 5, 676. 1867.
Isoetes echinospora muricata Engelm. in A. Gray, Man. ed. 5, 676. 1867.
Isoetes echinospora brittoni Cockerell, Muhlenbergia **3**: 9. 1907.
Isoetes muricata Durieu, Bull. Soc. Bot. France **11**: 101. 1864.

Submersed or in drier seasons emersed; corm 2-lobed; leaves 10–35, straight or recurved, firm, tapering to the apex, 8–25 cm. long; stomata present, usually few, near tip; peripheral strands none; ligule deltoid; sporangia oblong, spotted, 3–5 mm. long, one-half or rarely completely covered by velum; megaspores white, 420–560 μ in diameter, marked with numerous spines, ranging from slender single spines to those toothed or even confluent in short ridges; microspores fawn-colored, 23–33 μ long, smooth to slightly roughened on surface.

In lakes and ponds, Boreal Zones; Labrador, Greenland, eastern Canada, New England, south to New Jersey, Pennsylvania, Ohio, Indiana, Michigan to Minnesota; Washington, Idaho, Utah, Colorado. Type locality: "wiers on Lake Winnipiseogee, New Hampshire."

6. Isoetes howéllii Engelm.

Howell's Quillwort. Fig. 78.

Isoetes howellii Engelm. Trans. St. Louis Acad. Sci. **4**: 385. 1882.
Isoetes nuda Engelm. Trans. St. Louis Acad. Sci. **4**: 385. 1882.
Isoetes underwoodii Henderson, Bot. Gaz. **23**: 124. 1897.
Isoetes melanopoda californica A. A. Eaton, in Gilbert, List N. Am. Pterid. 27. 1901.

Amphibious; corm 2-lobed; leaves 10–30, rarely 50, bright green, erect or slightly outspread, usually slender, but coarser in emersed forms, 5–25 cm. long; stomata numerous; peripheral strands usually 4, sometimes one or more lacking; ligule narrow, elongated-triangular; sporangia frequently brown-spotted, orbicular to oblong, 3–6 mm., rarely 8 mm. long, covered one-third or less by velum; megaspores white, 400–520 μ in diameter, somewhat obscurely marked with simple tubercles and short crests, these isolated or anastomosing, crowded or distant; microspores 25–33 μ long, smoothish to short-spinulose.

Border of lakes and ponds, Transition Zone; western Montana, Idaho, Washington, Oregon, California. Type locality: "on the borders of ponds, The Dalles, Oregon."

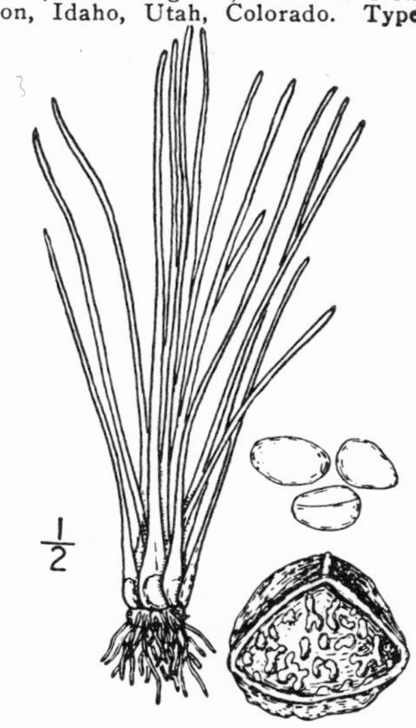

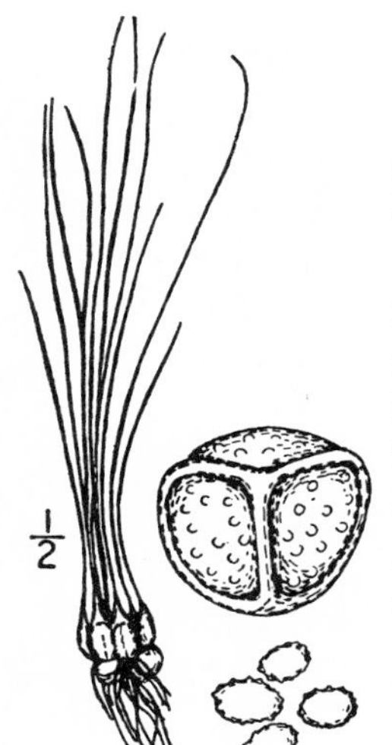

7. **Isoetes nuttállii** A. Br.

Nuttall's Quillwort. Fig. 79.

Isoetes nuttallii A. Br.; Engelm. in Am. Nat. 8: 215. 1874.
Isoetes suksdorfii Baker, Handb. Fern Allies 132. 1887.

Terrestrial in damp places; corm slightly 3-lobed; leaves 12–60, light green, 3-angled, very slender, firm, erect, 8–17 cm. long; stomata numerous; peripheral strands usually 3, sometimes 2, lacking dorsal one; ligule small, triangular; sporangia oblong, 4–7 mm. long, completely covered by velum; megaspores white, chiefly 400–528 μ, rarely smooth, generally densely covered by small, often glistening, distinct papillae on faces between prominent commissural ridges; microspores brown, 25–30 μ long, papillose.

Wet, springy places, Transition Zone; Oregon, Washington, California. Type locality: "on damp flats and springy declivities, Columbia River."

8. **Isoetes orcúttii** A. A. Eaton. Orcutt's Quillwort. Fig. 80.

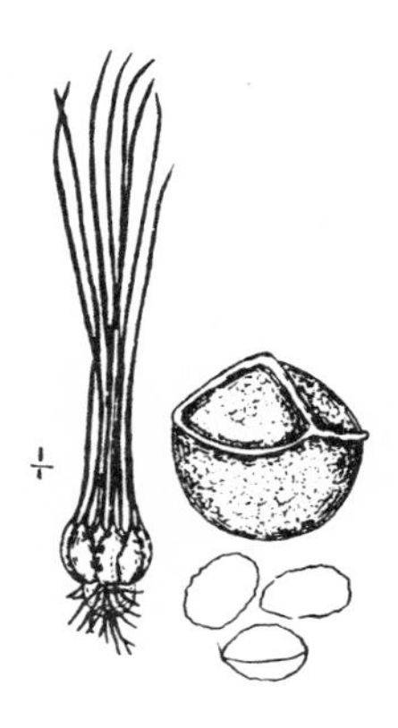

Isoetes orcuttii A. A. Eaton, Fern Bull. 8: 13. 1900.
Isoetes nuttallii orcuttii Clute, Fern Allies 253. 1905.

Terrestrial; corm slightly 3-lobed; leaves 6–14, fine, erect, triangular, 2–8 cm. long; stomata present; peripheral strands none, or two weakly developed; ligule triangular; sporangia orbicular to slightly elongated, 2–5 mm. long, completely covered by velum; megaspores gray at maturity, brownish when wet, 216–360 μ, rarely 480 μ, in diameter, closely marked with numerous small indistinct papillae or almost smooth, glistening; microspores 21–27 μ in length, chiefly spinulose.

California, central and southern, Transition and Upper Sonoran Zones. Type locality: "mesas in low depressions, San Diego."

Family 6. **EQUISETÀCEAE.**

Horsetail Family.

Rush-like plants, mainly of low situations, the rhizomes perennial, blackish, wide-creeping, freely branched, rooting at the nodes. Aerial stems perennial or annual, usually erect, cylindrical, fluted, simple or with whorled branches at the solid sheathed nodes, the internodes usually hollow; surfaces overlaid with silica in the form of tubercles, cross-bands, rosettes, or a smooth covering; stomata arranged in broad bands or regular rows in the grooves. Leaves nodal, minute, united lengthwise to form cylindrical or loose dilated sheaths, their tips (teeth) connivent or free, persistent or deciduous. Strobiles (spikes) terminal, formed of stalked peltate bracts (sporophylls) arranged in close-set circles about a common axis, these bearing a few sporangia beneath; spores uniform, green, provided with 4 hygroscopic bands. Prothallia minute, monoecious or dioecious, green, lobed.

The family consists of the following genus:

1. **EQUISÈTUM** L. Sp. Pl. 1061. 1753.

Character of the family. [Name ancient, signifying horse-tail, in allusion to the copious bushy branching of many species.]

About 25 species, mostly of very wide distribution. Type species, *Equisetum fluviatile* L.

Spikes blunt or barely acute; aerial stems annual, not surviving the winter.
- Aerial stems dimorphous, the fertile ones white, flesh-colored, or brownish, nearly devoid of chlorophyll, the sterile ones green, much branched.
  - Fertile stems 5–25 cm. long, the sheaths with 8–12 teeth; spike 2–3 cm. long; sterile stems 10–60 cm. long, 6–14-furrowed, the branches 3- or 4-angled. 1. *E. arvense.*
  - Fertile stems 25–60 cm. long, the sheaths with 20–30 teeth; spikes 4–8 cm. long; sterile stems 0.5–3 meters long, 20–40-furrowed, the branches 4–6-angled. 2. *E. telmateia.*
- Aerial stems uniform, green.
  - Fertile stems mostly with numerous elongate verticillate branches; plants of swamps and shallow water.

Stems deeply 5–10-angled, slender, with a very small central cavity; sheaths dilated, loose. 3. *E. palustre.*

Stems shallowly 10–30-grooved, stout, with a very broad central cavity; sheaths cylindrical, appressed. 4. *E. fluviatile.*

Fertile stems mostly unbranched above ground, sometimes (toward maturity) developing small verticillate branches; plants of dry or ordinarily moist situations.

Stems very rough, the ridges with sharply projecting cross-bands of silica; limb of sheaths strongly incurved with age; a diffuse tuft of many-branched sterile stems usually borne at the base of the erect fertile ones. 5. *E. funstoni.*

Stems smooth or nearly so, the ridges with or without cross-bands of silica; limb of sheath incurved or not; diffuse basal tuft of branched sterile shoots wanting. 6. *E. kansanum.*

Spikes rigidly apiculate; aerial stems evergreen, persisting 2 or more years.

Stems 5–12-angled; central cavity half the diameter of the stem, or less, or wanting; leaves and teeth not sharply differentiated.

Central cavity one-third to half the diameter of the stem; sheaths 5–12-toothed. 7. *E. variegatum.*

Central cavity wanting; sheaths 3-toothed. 8. *E. scirpoides.*

Stems 16–48-angled; central cavity more than half the diameter of the stem; leaves and teeth sharply differentiated.

Sheaths much longer than broad, dilated above, somewhat funnel-shaped, mostly green, the lower ones with a dark band. 9. *E. laevigatum.*

Sheaths nearly or quite as broad as long, nearly cylindrical, tight, mostly ashy at maturity, with 2 dark bands.

Ridges of the stem with a row of elevated cross-bands of silica; leaves sharply 3-carinate, the central keel rarely grooved. 10. *E. praealtum.*

Ridges of the stem usually with 2 distinct rows of silica tubercles; leaves 4-carinate, the central groove narrow but usually well defined. 11. *E. hiemale californicum.*

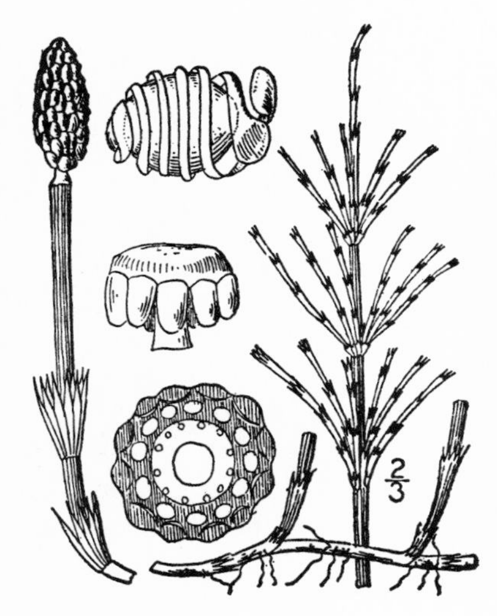

1. Equisetum arvénse L.

Common Horsetail. Fig. 81.

Equisetum arvense L. Sp. Pl. 1061. 1753.

Rhizome slightly angled, felted, tuber-bearing Fertile stems erect, 5–25 cm. long, flesh-colored, smooth, succulent, soon withering; internodes 2–8; sheaths pale, loose, with 8–12 large, brownish, lanceolate teeth; spike lance-ovoid, 2–3 cm. long, pedunculate. Sterile stems prostrate to erect, 10–60 cm. long, green, 6–14-furrowed, roughish, the silex in roundish dots; branches numerous, in dense verticils, usually simple, solid, prominently 3- or 4-angled, their sheaths with triangular-lanceolate, érect, sharp-pointed teeth.

Thickets, open sandy banks, and alluvial situations, rarely in deep shade; Alaska to Labrador and Newfoundland, southward nearly throughout the United States; Greenland; Eurasia. Type locality, European. Extremely variable in habit and size, but offering no permanent varieties.

2. Equisetum telmateìa Ehrh.

Giant Horsetail. Fig. 82.

Equisetum telmateia Ehrh. Hannöv. Mag. 1783: 287. 1783.

Rhizome slightly grooved, felted, tuber-bearing. Fertile stems erect, often clustered, 25–60 cm. long, 1–2.5 cm. thick, white or brownish, smooth, succulent, soon withering; sheaths loose, brownish, membranous, 2–5 cm. long, the lower ones longer than the internodes; teeth 20–30, attenuate, cohering in groups of 2 or 3; spikes stout-pedunculate, 4–8 cm. long, 1–2.5 cm. thick. Sterile stems erect, 0.5–3 meters long, 5–20 mm. thick, white or pale green; internodes 2–9 cm. long, 20–40-furrowed, the ridges smooth; sheaths cylindrical, appressed, with 20–40 teeth, these elongate, with broad hyaline margins, usually coherent in groups; branches in dense spreading verticils, solid, simple, 4–6-angled, the keel very rough with cross-bands of silica.

Swamps and borders of streams, Transition Zone; British Columbia to southern California; Eurasia; north Africa. Type locality, European. Intermediate, partially fertile forms are found occasionally.

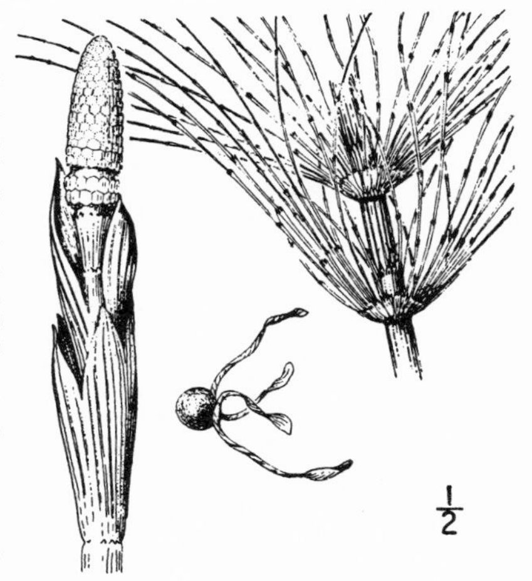

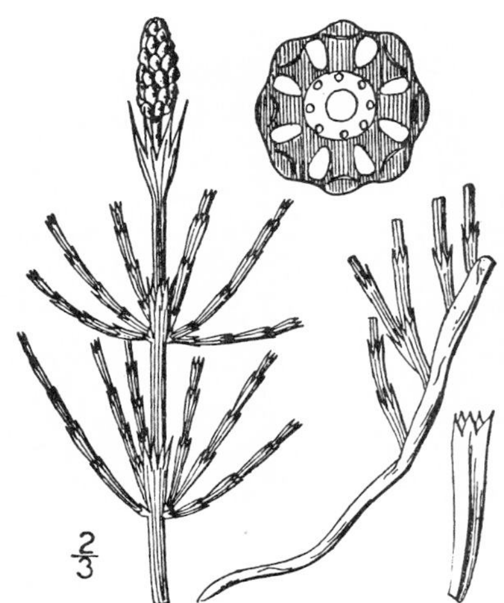

3. Equisetum palústre L.
Marsh Horsetail. Fig. 83.

Equisetum palustre L. Sp. Pl. 1061. 1753.

Rhizome angled, shining, without felt or tubers. Aerial stems conform, annual, slender, erect, 25–90 cm. long, deeply 5–10-angled (the central canal very small), usually naked and nearly smooth below, freely branched and rough above, the ridges sharply elevated but rounded; sheaths rather loose, widened upward, mostly green, the constituent leaflets keeled toward the base and with a distinct carinal groove extending into the teeth, these usually black, lance-subulate, with broad white hyaline margins; branches coarse, in scant verticils, simple, hollow, 4–7-angled, rough with cross-bands of silica, the sheaths similar to those of the stem; spikes short-pedunculate, 1–2.5 cm. long, terminating some of the stems, smaller ones sometimes terminating the branches.

Wet, shady places, Boreal Region; Alaska to Newfoundland, south to Oregon, Illinois, and New Jersey; Europe. Type locality, European.

4. Equisetum fluviátile L.
Swamp Horsetail. Fig. 84.

Equisetum fluviatile L. Sp. Pl. 1062. 1753.
Equisetum limosum L. Sp. Pl. 1062. 1753.

Rhizome shining, rarely tuber-bearing, with a broad central cavity. Aerial stems conform, erect, 60–150 cm. long, 4–6 mm. thick below, smooth, with 10–30 shallow grooves (the central cavity very broad), verticillate branches developed in the upper half or two-thirds, sometimes wanting; sheaths nearly as broad as long, appressed, concolorous; teeth narrowly lanceolate, acute, rigid, erect, distinct, dark, with narrow hyaline margins; lower internodes up to 5 cm. long, yellowish green or flesh-colored, shining, often developing a few 6–8-angled secondary stems, these equaling the main one; verticillate branches 2–15 cm. long, 4- or 5-angled, arcuate-ascending, with rough low-winged angles; sheaths rigid, with sharp, erect or spreading, green or black-tipped teeth; spikes short-pedunculate, cylindric, 1–2 cm. long, terminating the main and secondary stems.

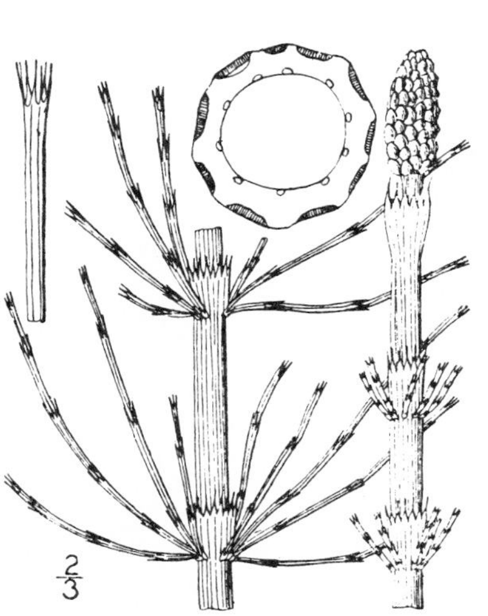

Swamps and in shallow water at the border of ponds and streams, Hudsonian to Transition Zones: Alaska to Labrador, south to Oregon, Wyoming, Nebraska, and Virginia; Eurasia. Type locality, European.

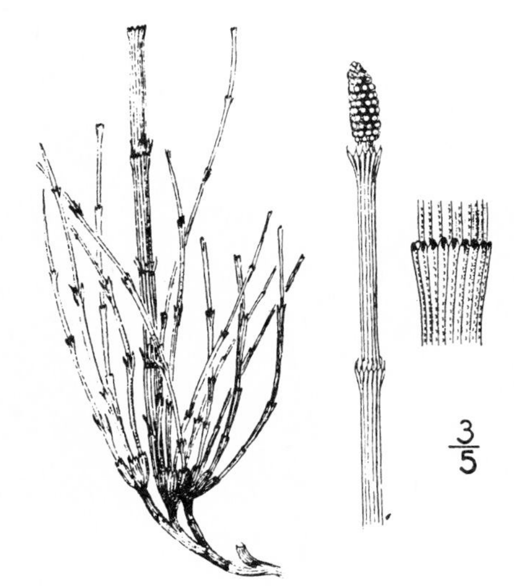

5. Equisetum fúnstoni A. A. Eaton.
California Horsetail. Fig. 85.

Equisetum funstoni A. A. Eaton, Fern Bull. 11: 10. 1903.

Aerial stems up to 75 cm. long, naked, or with elongate, scattering or regular verticillate branches in the upper part, a spreading tuft of numerous branched sterile shoots borne at the base (the winter form); main stems annual, erect, 20–30-angled, very rough, the ridges with many sharply projecting transverse bands of silica; central cavity very broad; sheaths elongate, funnel-form, green (rarely with a dark basal ring), the bases of the elongate coherent teeth small, persistent, horny, strongly incurved with age, forming a very narrow blackish band; uppermost sheaths campanulate, the spike (1–2 cm. long) exserted or not, obtuse or acute.

Damp places, in partial shade, Sonoran Zones; southern California, from Santa Barbara County southeastward; also in the Panamint Mountains, Inyo County, the type locality.

6. **Equisetum kansànum** Schaffn. Summer Scouring-rush. Fig. 86.

Equisetum kansanum Schaffn. Ohio Nat. 13: 21. 1912.

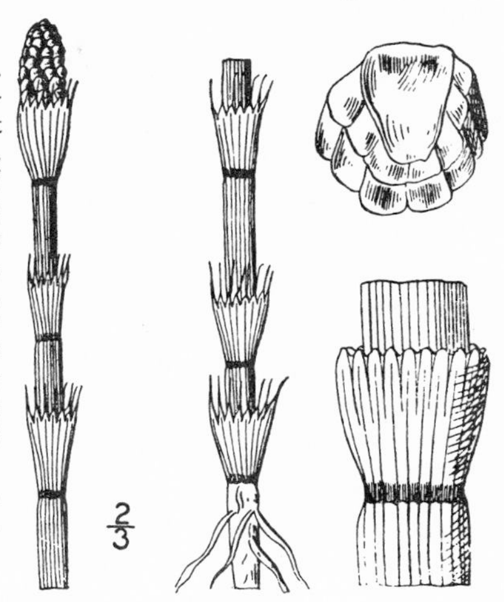

Rhizome dull blackish, naked except at the nodes. Aerial stems annual, single or clustered, erect, usually unbranched, 30–90 cm. long, 2–8 mm. thick, light green, 15–30-grooved, smooth or slightly roughened, the ridges rounded, with or without cross-bands of silica, a single row of stomata borne at each side; central cavity about three-fourths the diameter of the stem; sheaths elongate, green, contracted at the base, dilate upward, with a narrow black terminal band, the constituent leaves keeled in their lower half or two-thirds; teeth partly deciduous, cohering in groups by their whitish membranous borders, the persistent triangular bases rigid, blackish, shining, incurved at the tips or not; spikes sessile or short-exserted, 1–3 cm. long, oblong-ovoid, obtuse.

Ravine banks and moist slopes, commonly in clay, Upper Sonoran and Transition Zones; British Columbia to Ontario, south to southern California, Arizona, New Mexico, Missouri, and Ohio. Type locality: Bloom Township, Clay County, Kansas.

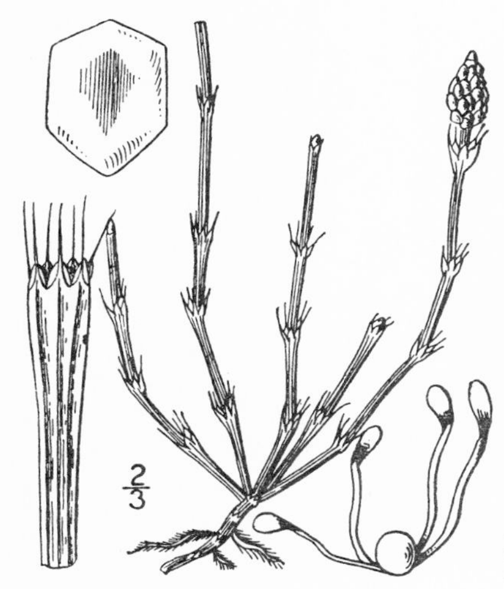

7. **Equisetum variegàtum** Schleich. Northern Scouring-rush. Fig. 87.

Equisetum variegatum Schleich. Cat. Pl. Helvet. 27. 1807.

Aerial stems usually perennial, erect or assurgent, clustered, naked or rarely branched, 15–60 cm. long, 1.5–4 mm. thick, 5–12-angled, rough or smoothish, the ridges grooved, bearing 2 lines of silica tubercles; central cavity about one-third to half the diameter of the stem; sheaths campanulate, green, black or variegated with black above, the constituent leaves distinctly 4-angled, with a deep median groove, this decurrent into the stem ridges and excurrent to the teeth; teeth mostly persistent, ample, connivent by their broad, whitish, membranous borders, the middle portion blackish or dark brown, prolonged into a rough, long-attenuate, usually deciduous tip; terminal sheath large, the spike sessile, 5–10 mm. long, strongly apiculate.

Wet meadows, bogs, and alluvial thickets, Boreal Region; Alaska to Labrador, south to Oregon, Wyoming, and Connecticut; Greenland; Eurasia. Type locality: Switzerland. Several varieties are recognized.

8. **Equisetum scirpoìdes** Michx. Sedgelike Horsetail. Fig. 88.

Equisetum scirpoides Michx. Fl. Bor. Am. 2: 281. 1803.

Aerial stems numerous (10–100), tufted, simple, or slightly branched at the base, green, prostrate or laxly ascending, filiform, about 1 mm. thick, flexuous, 5–15 (25) cm. long, solid at the center, 6-ribbed by the very deep grooving of the 3 angles, the ribs with a regular row of silica tubercles; sheaths loose, becoming black or dark brown, the 3 constituent leaves with a very broad central groove and 2 lateral ones, thus 4-carinate; teeth 3, distinct, persistent, blackish with broad white borders, the tips subulate, fragile; spikes 3–5 mm. long, sessile or short-pedunculate.

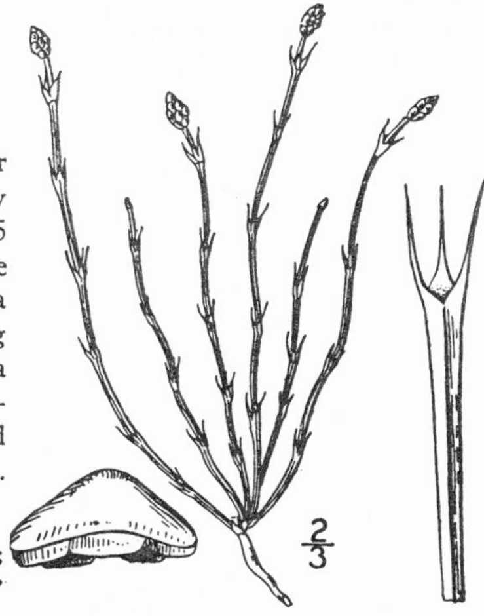

Low fields, swamps, and moist evergreen woods, Boreal Region; Alaska to Labrador, south to Washington, Montana, Michigan, Illinois, and Pennsylvania; Eurasia. Type locality: Canada.

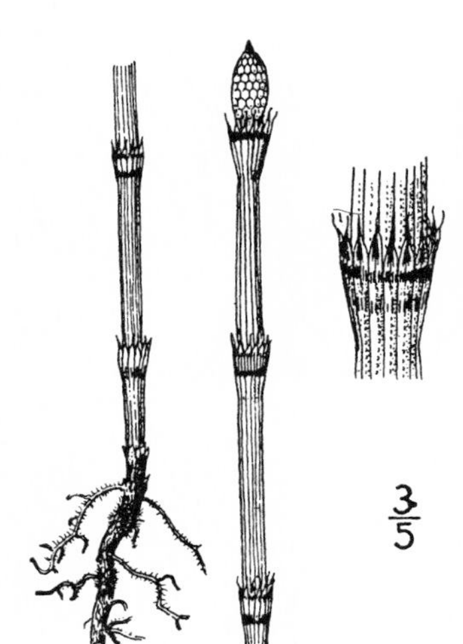

9. **Equisetum laevigàtum** A. Br.

Braun's Scouring-rush. Fig. 89.

Equisetum laevigatum A. Br. Am. Journ. Sci. **46**: 87. 1844.
Equisetum hiemale intermedium A. A. Eaton, Fern Bull. **10**: 120. 1902.
Equisetum intermedium Rydb. Fl. Rocky Mts. 1053. 1917.

Aerial stems usually persisting 2 or more seasons, tufted, erect, simple or at length sparingly branched, 30–120 cm. long, 2–8 mm. thick, 20–30-angled, smoothish or rough, the ridges bearing cross-bands of silica; sheaths much longer than broad, dilated upward, the lower 3 or 4 usually with inferior and apical black rings, the others green, or commonly all the sheaths banded after the first year; leaves centrally keeled below, narrowed upward, the brown-spotted tips contracted and incurved with age, joined by rather broad pale commissures; teeth thin, brownish, flexuous, caducous to persistent, at first cohering in groups by their pale transparent borders; small stems and branches usually very rough, the leaves centrally grooved, with subpersistent teeth; spikes 1–2 cm. long, sharply apiculate.

Damp alluvial thickets and sandy banks, chiefly in the Upper Sonoran Zone; British Columbia to southern California, east to New York, Illinois, Missouri, and Texas. Type locality: near St. Louis, Missouri.

10. **Equisetum praeáltum** Raf.

Prairie Scouring-rush. Fig. 90.

Equisetum praealtum Raf. Fl. Ludovic. 13. 1817.
Equisetum robustum A. Br.; Engelm. Am. Journ. Sci. **46**: 88. 1844.

Aerial stems persisting several seasons, rigidly erect, simple or at length branched, mostly 0.4–2 meters long, 4–12 mm. thick, 16–48-angled, very rough, the ridges with irregular, strongly elevated cross-bands of silica, the tips of these sometimes more prominent; sheaths a little longer than broad, or not, cylindrical, tight, at maturity ashy in the middle with a black basal band and a narrower apical one, at length commonly deciduous in recurved fragments; leaves narrowly linear, sharply 3-carinate, the central keel rarely grooved; teeth articulate, brown centrally, subpersistent, cohering in groups; branches variable, the sheaths with persistent teeth; spikes oval, 1–2.5 cm. long, sharply apiculate.

Moist, usually alluvial situations; British Columbia to Quebec, southward nearly throughout the United States. Type locality: Louisiana.

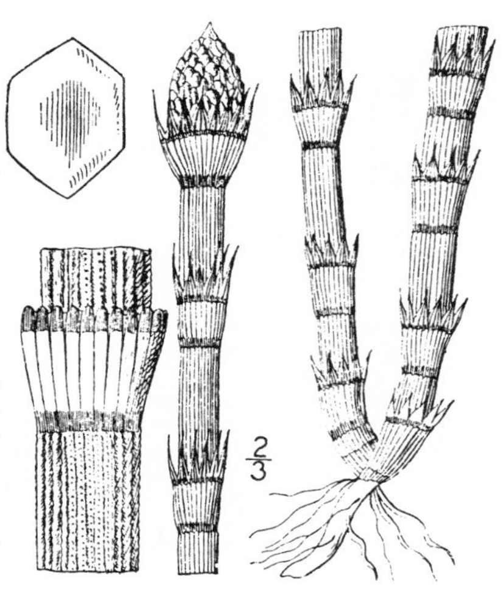

11. **Equisetum hiemàle califórnicum** Milde.

Western Scouring-rush. Fig. 91.

Equisetum hiemale californicum Milde, Nov. Act. Acad. Caes. Leop. Carol. **32²**: 517. 1867.

Aerial stems persisting several seasons, rigidly erect, clustered, naked or at length rarely branched toward the apex, 0.5–2.5 mm. high, 5–15 mm. thick, 25–40-angled, rough, the ridges usually with 2 distinct rows of silica tubercles; sheaths as broad as long, cylindrical or slightly dilated above, variable in coloration, at maturity usually ashy, with a broad or narrow dark brown or blackish band below the middle and a narrow dark band at the top, the constituent leaves with a narrow central groove, 4-carinate; teeth dark brown, mostly cohering in groups by their pale membranous borders, tardily deciduous; spikes exserted or not, 1–3 cm. long, sharply apiculate.

Moist alluvial situations, often in shade; Alaska to California (Santa Clara County), Nevada, Arizona, and New Mexico. Type locality: California.

Family 7. LYCOPODIÀCEAE.

CLUB-MOSS FAMILY

Somewhat moss-like, terrestrial or chiefly epiphytic, rigid to soft-herbaceous plants of upright, trailing, or pendent growth, the stems usually with numerous, apparently alternate branches or repeatedly dichotomous, leafy nearly throughout; leaves small, simple, 1-nerved, continuous with the axis, mostly uniform and multifarious, usually imbricate. Sporangia uniform, 1-celled, 2-valvate, reniform or orbicular, compressed, solitary in the axils of ordinary foliar leaves or of aggregated bractlike sporophylls. Spores uniform, very numerous, minute, globose, smoothish or variously sculptured. Prothallia fleshy, mostly hypogean, monoecious.

Two genera, the following and *Phylloglossum* (monotypic), of Australia.

1. LYCOPÒDIUM L. Sp. Pl. 1100. 1753.

Perennial; leaves arranged in 4–16 ranks, reflexed to ascending, closely appressed, or nearly adnate, often crowded. Sporangia solitary in the axils of some of the nearly uniform leaves along the stem, or in the axils of more or less modified scalelike bracts, these compactly imbricate in sessile or peduncled, terminal spikes (strobiles). Spores copious, sulphur-yellow. [Name from the Greek, meaning wolf's-foot; of doubtful significance.]

About 200 species, mainly of temperate regions. About 10 species occur in other parts of the United States. Type species, *Lycopodium clavatum* L.

Sporophylls not associated in terminal spikes. — 1. *L. selago.*
Sporophylls closely associated in terminal spikes.
  Stems without leafy aerial branches, the peduncles arising directly from the prostrate stem; sporophylls similar to the foliar leaves; sporangia subglobose. — 2. *L. inundatum.*
  Stems with numerous erect or assurgent leafy aerial branches, the spikes terminal upon some of these; sporophylls bractlike, very unlike the foliar leaves; sporangia reniform.
    Ultimate aerial branches flattened, the leaves in 4 rows, conspicuously adnate, those of the under row greatly reduced. — 3. *L. complanatum.*
    Ultimate aerial branches not flattened, the leaves in 5 or more ranks, scarcely adnate, nearly alike.
      Main stem deeply hypogean, horizontal; aerial branches few, treelike. — 4. *L. obscurum.*
      Main stem prostrate or creeping; aerial branches numerous, not treelike.
        Leaves of the ultimate aerial branches in 5 rows. — 5. *L. sitchense.*
        Leaves of the ultimate aerial branches in many ranks.
          Spikes sessile. — 6. *L. annotinum.*
          Spikes pedunculate. — 7. *L. clavatum.*

1. Lycopodium selàgo L. Fir Club-moss. Fig. 92.

Lycopodium selago L. Sp. Pl. 1102. 1753.

Stems rigidly erect from a slender curved base, several times dichotomous, the densely foliaceous branches nearly vertical, usually forming compact level-topped tufts 4–20 cm. high; leaves uniform or nearly so, very numerous, crowded, usually ascending or appressed-imbricate, narrowly deltoid-lanceolate or subulate, acute, lustrous, yellowish to light green (sometimes darker and spreading), usually entire, those bearing the sporangia (below the top) shorter than the others; plants frequently gemmiparous.

Moist, rocky situations, Arctic-Alpine Zone; Alaska to Labrador and Newfoundland, south to Oregon (Mount Hood), Montana, Michigan, northern New York, and northern New England, and in the mountains to North Carolina; Greenland; Eurasia. Type locality, European. Several habital and geographic forms are recognized. The plant, of moist coniferous forests of Oregon and Washington, 20–30 cm. high, with dark, spreading leaves, is apparently a luxuriant state of var. *patens* (Beauv.) Desv.

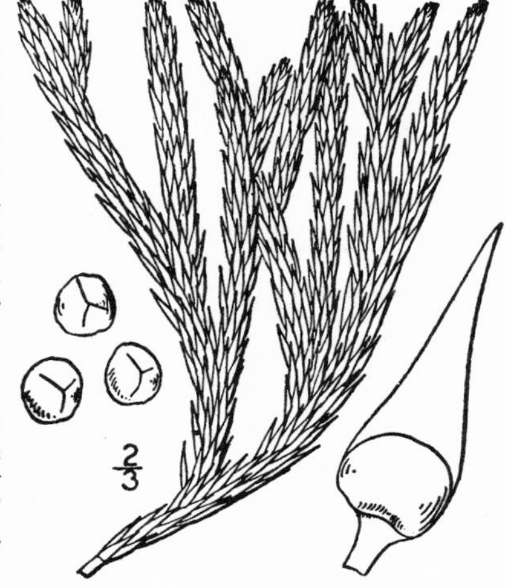

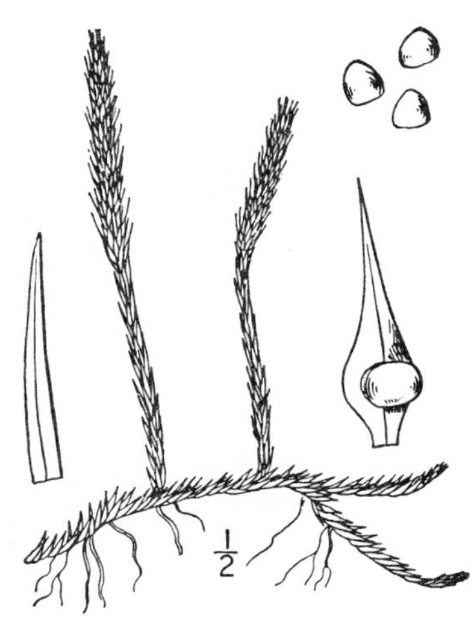

2. Lycopodium inundàtum L.
Bog Club-moss. Fig. 93.

Lycopodium inundatum L. Sp. Pl. 1102. 1753.

Plants small, with simple or 1–2-forked, short, creeping or slightly arched, slender, often lax, leafy stems; stem-leaves linear-lanceolate, attenuate, mostly entire, secund; peduncles long, erect, arising directly from the creeping stem, 0.5–8 cm. long, the leaves spreading or laxly ascending, narrower than those of the stem; spikes 1.5–5 cm. long, the sporophylls similar to the foliar leaves but wider at the base (linear-deltoid), soft, spreading, entire or sometimes toothed just above the base.

Boggy, clayey or sandy situations, Canadian and Transition Zones; Alaska to Idaho and Oregon (Coos County), and from Newfoundland to New Jersey, western Maryland, Illinois, and Ontario; also in Europe. Type locality, European.

3. Lycopodium complanàtum L.
Ground-cedar. Fig. 94.

Lycopodium complanatum L. Sp. Pl. 1104. 1753.

Horizontal stems prostrate and wide-creeping, flattended above, sparingly branched, with numerous erect irregularly branched aerial stems, the branches of these broadly flattened, 2–3-forked, the divisions few, somewhat glaucous, leafy throughout, mostly 2–3 mm. broad; leaves 4-ranked, minute and (excepting those of the under row) imbricate and strongly decurrent, those of the upper row narrow and incurved, of the lateral rows broad, with spreading tips, and of the under row minute, deltoid-cuspidate; peduncles slender, 3–13 cm. long, distantly bracteate, rarely simple, usually once or twice dichotomous near the summit, each branch terminating in a slender cylindric spike 1.5–3 cm. long; sporophylls broadly ovate, acuminate.

In woods and thickets; Alaska to Labrador and Newfoundland, south to northeastern Washington, Idaho, Montana, Ontario, northern New York, Vermont, and Maine; Eurasia. Type locality, European. Replaced in the eastern United States by var. *flabelliforme* Fernald.

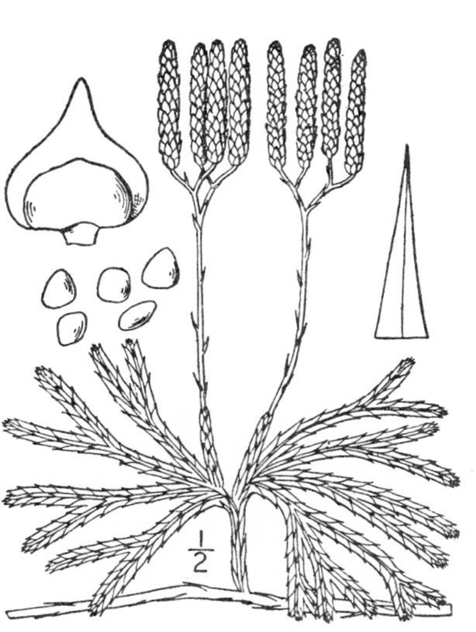

4. Lycopodium obscùrum L. Ground-pine. Fig. 95.

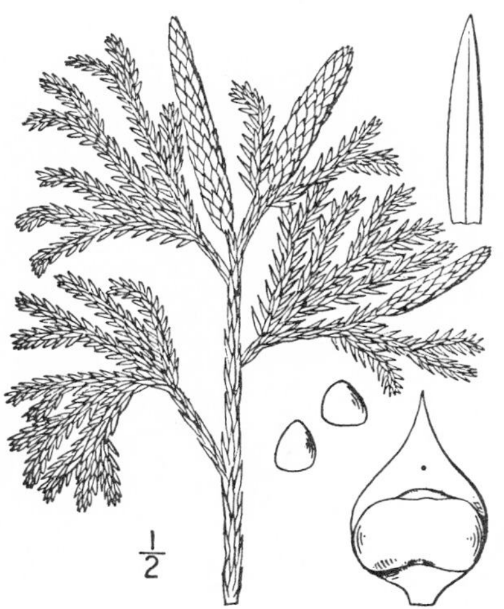

Lycopodium obscurum L. Sp. Pl. 1102. 1753.
Lycopodium dendroideum Michx. Fl. Bor. Am. 2: 282. 1803.
Lycopodium obscurum dendroideum D. C. Eaton, in A. Gray, Man. ed. 6, 696. 1890.

Main stem deeply hypogean, horizontal, elongate, giving rise to a few distant upright aerial branches, these 10–35 cm. high, treelike, with numerous bushy branches; leaves 8-ranked on the lower branches, 6-ranked on the terminal ones, linear-lanceolate, spreading but curved upwards and twisted (especially above), the upper branches thus appearing more or less dorso-ventral, sometimes conspicuously so; spikes 2–5 cm. long, nearly sessile; sporophylls very broadly ovate, abruptly acuminate, the margins scarious, erose.

Moist woods, Canadian and Transition Zones; Alaska to Baffin Land and Newfoundland, south to central Washington, Montana, South Dakota, Indiana, and the mountains of northeastern Alabama, and North Carolina; also in eastern Asia. Type locality: Pennsylvania.

5. **Lycopodium sitchénse** Rupr. Alaskan Club-moss. Fig. 96.

Lycopodium sitchense Rupr. Beitr. Pflanzenk. Russ. Reich. 3: 30. 1845.

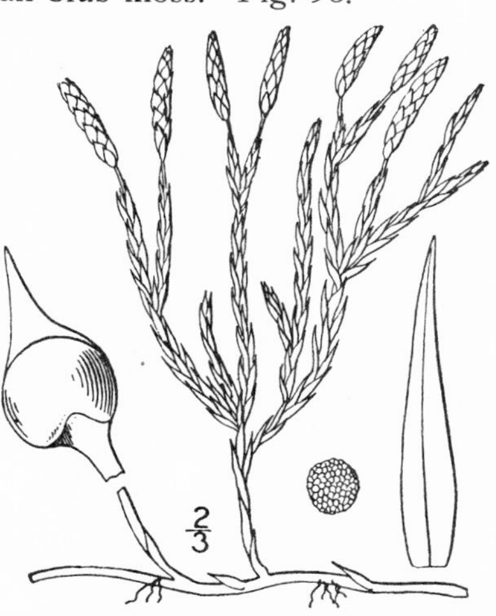

Stems slender, horizontal, nearly superficial, 10–40 cm. long, sending up numerous aerial stems, these several times dichotomous, the branches (with their leaves) glaucous, roundish, vertical, forming compact tufts 4–8 cm. high, with few or numerous stronger projecting fertile branches; leaves of the branchlets 5-ranked, appressed or incurved-ascending, linear, acute, 2–3 mm. long, thick, entire; spikes solitary, cylindric, 0.5–2 cm. long, sessile, or borne upon short minutely bracteate peduncles (up to 1 cm. long); sporophylls acuminate or subulate from an ovate base, greenish, erose.

Moist meadows and coniferous woods, Hudsonian and Arctic Zones; Alaska to Labrador and Newfoundland, south to Oregon, Idaho, western Ontario, New York, and northern New England. Type locality: Alaska.

6. **Lycopodium annótinum** L. Stiff Club-moss. Fig. 97.

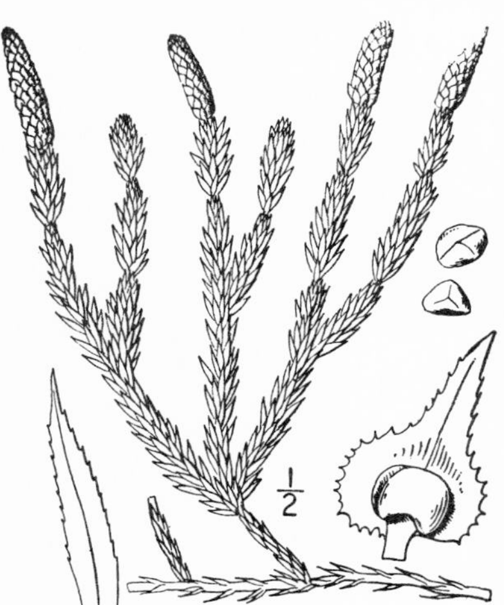

Lycopodium annotinum L. Sp. Pl. 1103. 1753.

Stems prostrate, creeping, often 1 meter long, sparsely leafy, giving rise to numerous erect or ascending branches, these 10–30 cm. high, simple or sparingly forked, leafy throughout; leaves uniform, in 5 ranks, rigid, spreading or reflexed, linear-lanceolate, pungent, 6–10 mm. long, sharply to obsoletely serrate, or (in exposed alpine or boreal situations) the leaves much shorter, thicker, strongly ascending or appressed, subentire; spikes solitary, sessile, oblong-cylindric, 1–2.5 cm. long, the sporophylls broadly ovate, abruptly acuminate–attenuate.

Moist woods or thickets, Boreal Region; Alaska to Labrador and Newfoundland, south to Oregon (Mount Hood), Colorado, Wisconsin, and Maryland; also in Greenland and Eurasia. Type locality, European. The form with thick ascending leaves is usually known as var. *pungens* Desv.

7. **Lycopodium clavàtum** L. Running-pine. Fig. 98.

Lycopodium clavatum L. Sp. Pl. 1101. 1753.

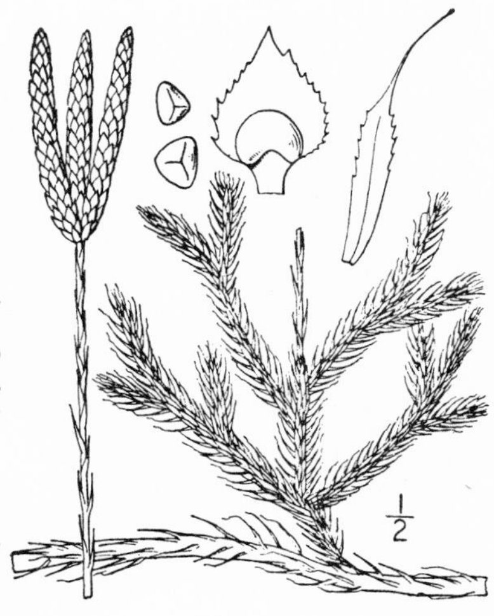

Main stems prostrate, wide-creeping (1–3 meters), branching horizontally, and with numerous very leafy, assurgent, pinnately branched aerial divisions; leaves crowded, many-ranked, incurved-spreading, linear-subulate, bristle-tipped (the bristle commonly deciduous), entire or denticulate; peduncles mostly 3–15 cm. long, with slender whorled or scattered, denticulate, bristle-tipped bracts; spikes 1–4, narrowly cylindric, up to 15 cm. long if solitary, the sporophylls deltoid-ovate, abruptly acuminate, bristle-tipped, the margins scarious, erose-fimbriate.

Moist coniferous woods, brushy slopes, and boggy situations, Boreal Region; Alaska to Newfoundland, south to Oregon, Idaho, Montana, Minnesota, Michigan, and the mountains of North Carolina; nearly cosmopolitan. Type locality, European. The Pacific Coast plant with poorly developed or deciduous terminal bristles upon the leaves of the ascending branches is var. *integerrimum* Spring.

Family 8. SELAGINELLÀCEAE.

SELAGINELLA FAMILY.

Low, depressed or creeping, freely branched, leafy terrestrial plants, our species of moss-like habit. Leaves very numerous, usually imbricate, either all nearly alike, subulate to lanceolate, and arranged radially in many ranks, or of 2 kinds, in 4 ranks, broader, and strongly dorso-ventral. Sporangia in terminal quadrangular sessile spikes of modified leaves (sporophylls), axillary, the larger ones bearing 1–4 large megaspores, the smaller ones very numerous, minute, reddish or orange, powdery microspores.

The family consists of the following genus:

1. SELAGINÉLLA Beauv. Prodr. Aethog. 101. 1805.

Character of the family. [Name a diminutive of Selago, a classical name for some species of Lycopodium.]

A widely distributed genus of about 600 species, of which about 37 occur in the United States. Type species: *Lycopodium selaginoides* L.

Leaves in 4 ranks, those of the 2 upper rows rhombic-ovate, of the 2 lateral rows oval-oblong, rounded-obtuse. 1. *S. douglasii.*
Leaves arranged radially in many ranks, narrowly deltoid-subulate to lanceolate.
  Stems ascending or erect, rooting only at the base. 2. *S. bigelovii*
  Stems creeping or pendent.
    Leaves lacking a terminal seta.
      Plants strongly dorso-ventral, all the divisions flattish; leaves of the lower side thin, lanceolate, imbricate-spreading, those of the upper side smaller, rigid, deltoid-subulate, nearly vertical. 3. *S. eremophila.*
      Plants slightly dorso-ventral, the divisions short, clavate; leaves all appressed-imbricate, herbaceous, linear to broadly subulate. 4. *S. cinerascens.*
    Leaves setigerous.
      Plants lax, usually long-pendent from tree; leaves bright green, wrinkled, lance-attenuate, long-decurrent (up to 1 mm.). 5. *S. oregana.*
      Plants more rigid, terrestrial; leaves somewhat glaucous, not wrinkled, adnate at the base or short-decurrent.
        Branches somewhat dorso-ventral, the leaves of the under ranks largest, obliquely imbricate.
          Leaves broadly subulate, acute, often purplish with age; spikes 5–9 mm. long; sporophylls with 25–30 cilia. 6. *S. hanseni.*
          Leaves linear to lance-subulate, obtuse at apex, not purplish with age; spikes 1–2.5 cm. long; sporophylls with 8–15 cilia. 7. *S. scopulorum.*
        Branches not at all dorso-ventral, the leaves uniform, equally ascending or appressed-imbricate on all sides.
          Setae lutescent, smooth; cilia 2–7 on each side of the leaves, or wanting. 8. *S. watsoni.*
          Setae white or whitish-hyaline, scabrous or serrulate; cilia 8–23 on each side of the leaves.
            Plants cushion-like; branches of the very short creeping stems erect, strongly cespitose, simple; setae milk-white. 9. *S. leucobryoides.*
            Plants forming loose mats; branches numerous, intricate or loosely cespitose, 1–2-pinnate; setae whitish-hyaline.
              Leaves oblong-linear, obtusish, subglaucous, closely appressed-imbricate; setae 0.15–0.6 mm. long; cilia 8–14 on each side, oblique. 10. *S. wallacei.*
              Leaves narrowly deltoid-subulate, strongly glaucous, rigidly ascending, subdistant; setae 0.7–0.9 mm. long; cilia 16–23 on each side, spreading. 11. *S. asprella.*

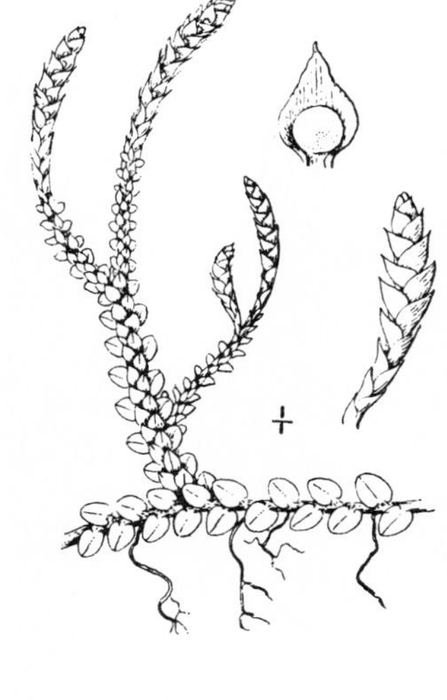

1. Selaginella douglásii (Hook. & Grev.) Spring.
Douglas' Selaginella. Fig. 99.

Lycopodium ovalifolium Hook. & Grev. Icon. Fil. **2**: *pl. 177.* 1831, not Desv. 1813.
Lycopodium douglasii Hook. & Grev. Bot. Misc. **2**: 396. 1832.
Selaginella douglasii Spring, Mém. Acad. Brux. **24**[1]: 92. 1850.

Stems prostrate, wide-creeping, 15–35 cm. long, rooting throughout; main branches alternate, often elongate, rooting or not, distant, with about 3–6 flat assurgent divisions, these 1–3-divided, leafy throughout. Lateral leaves distant, spreading, concave above, yellowish green, oval-oblong, 2–3 mm. long, 1–1.8 mm. broad rounded-obtuse, long-ciliate at the unequally auriculate base; leaves of the upper plane smaller, intricate, narrowly rhombic-ovate, cuspidate. Spikes numerous, sharply quadrangular, 5–15 mm. long; sporophylls closely imbricate, cordate-ovate, acuminate, membranous, sharply carinate.

Moist, shady rocks and banks, Humid Transition Zone; British Columbia and northern Idaho to California. Type locality: Oregon or Washington.

2. Selaginella bigelòvii Underw.
Bushy, or Bigelow's, Selaginella. Fig. 100.

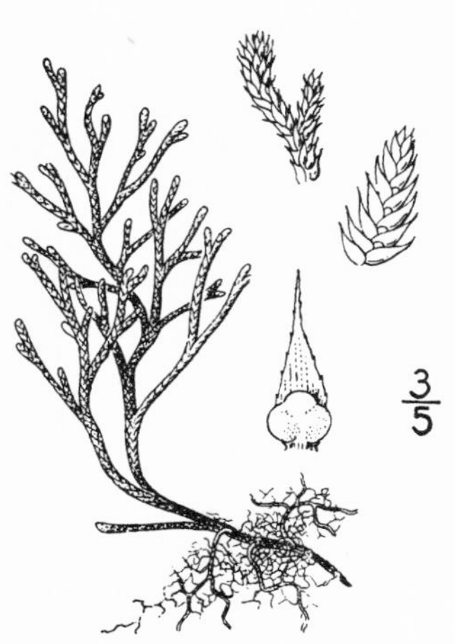

Selaginella bigelovii Underw. Bull. Torrey Club **25**: 130. 1898.

Plants cespitose; stems ascending or erect, 6–20 cm. long, rooting only at the defoliate base, repeatedly branched; branches variable in length, ascending or erect, often crowded or intricate, the ultimate divisions mostly short, terete. Leaves very oblique, rigidly appressed-imbricate, uniform, 1.6–2.4 mm. long (seta included), the blade subulate-attenuate, 1.3–1.9 mm. long, 0.3–0.5 mm. broad, herbaceous, subglaucous, sulcate dorsally nearly to the convex acutish tip, narrowly whitish-marginate; seta 0.25–0.5 mm. long, white, rigid, strongly scrabrous; cilia 8–17 on each side, 0.03–0.06 mm. long, oblique, pungent. Spikes 5–10 mm. long, terminating many of the apical branches, slender; sporophylls mostly deltoid-ovate, long-acuminate, 1.7–2.35 mm. long (seta included), 0.6–1 mm. broad, rigid, carinate; seta 0.3–0.6 mm. long, whitish, rigid, scabrous; cilia 30 or more on each side, very close, oblique, mostly 0.02–0.06 mm. long; megaspores depressed-globose, yellow, up to 0.42 mm. thick, lightly reticulate or rugose-reticulate on all sides; microspores bright orange, about 0.032 mm. thick.

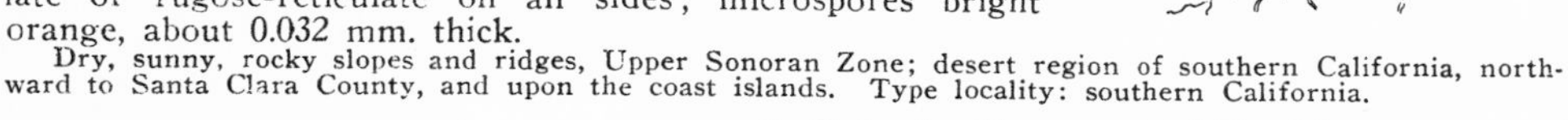
Dry, sunny, rocky slopes and ridges, Upper Sonoran Zone; desert region of southern California, northward to Santa Clara County, and upon the coast islands. Type locality: southern California.

3. Selaginella eremóphila Maxon.
Desert Selaginella. Fig. 101.

Selaginella eremophila Maxon, Smiths. Misc. Coll. **72**[5]: 3. *pl. 2.* 1920.

Selaginella parishii Underw. Bull. Torrey Club **33**: 202. 1906, in part.

Plant wholly prostrate, strongly dorso-ventral, the main stems 5–12 cm. long, rooting sparsely throughout; lower branches subequal, spreading, 2–3-pinnate, the ultimate divisions very short, involute. Leaves crowded, those of the lower side imbricate, thin, lanceolate, acutish, 2 mm. long, 0.5 mm. broad, not setigerous, with about 25 spreading cilia on each side; leaves of upper side rigid, nearly vertical, deltoid-subulate, 1–1.4 mm. long, 0.4–0.47 mm. broad, acutish, not setigerous, with 6–12 weak spreading cilia on each side, these about 0.1 mm. long. Spikes numerous, arcuately ascending, 6-10 mm. long, 1 mm. thick; sporophylls deltoid, acute or acutish, not setigerous, 1.2–1.4 mm. long, 0.9–1 mm. broad, with 12–18 spreading or weakly ascending cilia on each side, these 0.09–0.125 mm. long; megaspores light yellow, 0.36–0.4 mm. thick, deeply reticulate; microspores yellow, about 0.039 mm. thick.

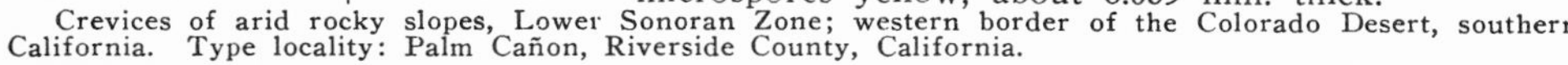
Crevices of arid rocky slopes, Lower Sonoran Zone; western border of the Colorado Desert, southern California. Type locality: Palm Cañon, Riverside County, California.

4. Selaginella cinceráscens A. A. Eaton. Pygmy, or Gray, Selaginella. Fig. 102.

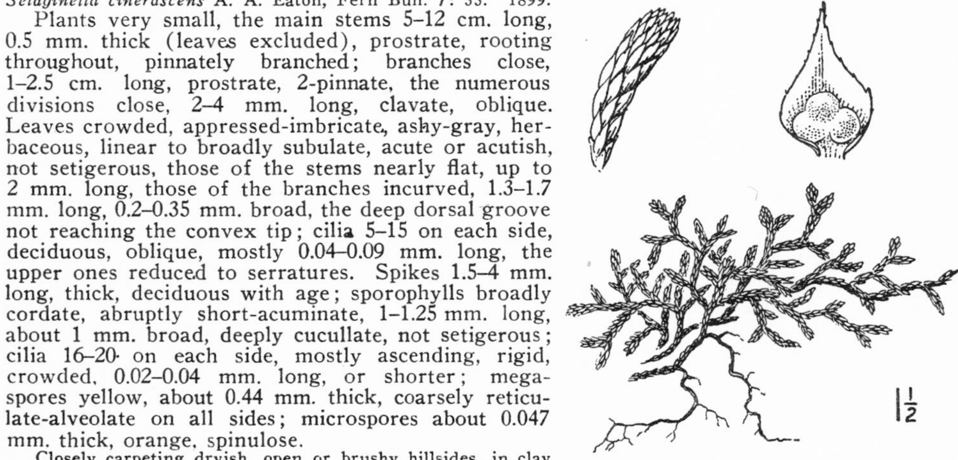

Selaginella cinerascens A. A. Eaton, Fern Bull. **7**: 33. 1899.

Plants very small, the main stems 5–12 cm. long, 0.5 mm. thick (leaves excluded), prostrate, rooting throughout, pinnately branched; branches close, 1–2.5 cm. long, prostrate, 2-pinnate, the numerous divisions close, 2–4 mm. long, clavate, oblique. Leaves crowded, appressed-imbricate, ashy-gray, herbaceous, linear to broadly subulate, acute or acutish, not setigerous, those of the stems nearly flat, up to 2 mm. long, those of the branches incurved, 1.3–1.7 mm. long, 0.2–0.35 mm. broad, the deep dorsal groove not reaching the convex tip; cilia 5–15 on each side, deciduous, oblique, mostly 0.04–0.09 mm. long, the upper ones reduced to serratures. Spikes 1.5–4 mm. long, thick, deciduous with age; sporophylls broadly cordate, abruptly short-acuminate, 1–1.25 mm. long, about 1 mm. broad, deeply cucullate, not setigerous; cilia 16–20 on each side, mostly ascending, rigid, crowded, 0.02–0.04 mm. long, or shorter; megaspores yellow, about 0.44 mm. thick, coarsely reticulate-alveolate on all sides; microspores about 0.047 mm. thick, orange, spinulose.

Closely carpeting dryish, open or brushy hillsides, in clay soil, Upper Austral Zone; coastal belt near San Diego, California, the type locality.

5. Selaginella oregàna D. C. Eaton. Festoon or Oregon Selaginella. Fig. 103.

Selaginella oregana D. C. Eaton in S. Wats. Bot. Calif. **2**: 350. 1880.
Selaginella struthioloides Underw. Bull. Torrey Club **25**: 132. 1898, not *Lycopodium struthioloides* Presl, 1825.

Plants lax, usually pendent, 15–60 (90) cm. long; branches numerous, slender, recurved, loosely intricate. Leaves ascending, loosely imbricate, bright green, lustrous, long-decurrent, narrowly lance-attenuate, 2.35–3.35 mm. long (seta included), 0.47–0.62 mm. broad at the spongiose upper base, herbaceous, with a broad, shallow, dorsal groove; seta 0.15–0.35 mm. long, yellowish green, attenuate, nearly smooth; cilia 5–7 (9) on each side, 0.2–0.6 mm. long, oblique, mostly incurved. Spikes numerous, curved, 1.5–3 cm. long, not sharply differentiated; sporophylls bright green, herbaceous, ovate, long-acuminate, 1.8–2.4 mm. long (seta included), 0.8–1 mm. broad, broadly sulcate; seta 0.08–0.25 mm. long, continuous, yellowish-hyaline or discolored, pungent; cilia 3–7 on each side, 0.015–0.045 mm. long, stout, ascending, often curved; megaspores lenticular, light yellow, 0.345–0.4 mm. broad, coarsely alveolate-reticulate; microspores orange, about 0.034 mm. thick.

Trunks and branches of mossy forest trees, Humid Transition Zone; western Washington (Chehalis and Thurston Counties), southward in the coastal region to Humboldt County, California. Type locality: Port Orford, Oregon.

6. Selaginella hánseni Hieron. Hansen's Selaginella. Fig. 104.

Selaginella hanseni Hieron. Hedwigia **39**: 301. 1900.

Stems prostrate, 5–25 cm. long, rooting throughout, pinnately branched; branches usually close, unequal, the basal ones up to 7 cm. long, 2–3-pinnate, subflabellate, strongly dorso-ventral. Leaves obliquely imbricate (the lower ones of the larger branches often widely so), glaucous, dark ashy or purplish with age, broadly subulate, acute, 2.7–3.5 mm. long (seta included), 0.4–0.5 mm. broad; seta 0.7–1 mm. long, whitish-hyaline, serrulate; cilia 8–16 on each side, mostly 0.06–0.09 mm. long, mostly very oblique, the upper ones appressed. Spikes 5–9 mm. long, numerous; sporophylls deltoid-ovate, acuminate, 2–2.8 mm. long (seta included), the blade 0.9–1.5 mm. broad; seta 0.4–0.7 mm. long, stiff, whitish-hyaline, nearly smooth; cilia 25–30 on each side, rigid, very oblique, the lower ones 0.04–0.07 mm. long, the upper ones dentiform; megaspores about 0.39 mm. thick, orange yellow, slightly rugulose; microspores about 0.039 mm. thick, orange.

Rocky slopes, Transition Zone; mountains of California, from the region of Mount Shasta to Fresno County, mainly in the lower Sierra Nevada. Type locality: Amador or Calaveras County, California.

7. Selaginella scopulòrum Maxon. Rocky Mountain Selaginella. Fig. 105.

Selaginella scopulorum Maxon, Am. Fern Journ. **11**: 36. 1921.

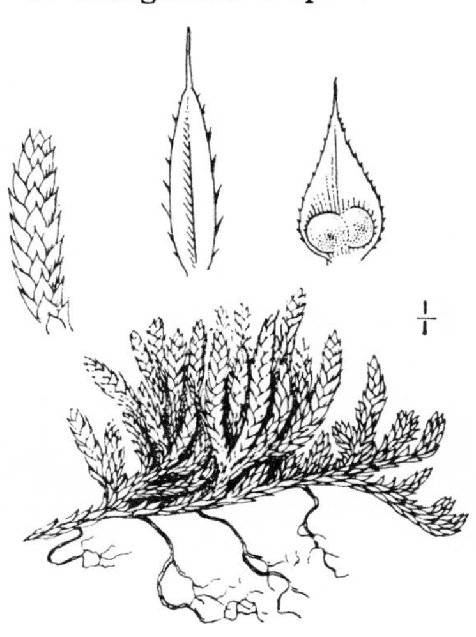

Stems prostrate, short-creeping, 3–6 cm. long, pinnately branched, subcespitose, forming large mats; branches numerous, close, the sterile ones mostly 0.5–1.5 cm. long, ascending, simple or with several very short, oblique divisions. Leaves appressed-imbricate (those on the under side largest), chartaceous, subglaucous, linear to lance-subulate, narrowed to an obtuse apex, 2.35–3.25 mm. long (seta included), 0.3–0.6 mm. broad; seta 0.3–0.6 mm. long, stiff, scabrous, whitish-hyaline from a long lutescent base; cilia 4–8 (11) on each side, 0.06–0.12 mm. long, mostly oblique and incurved. Spikes numerous, 1–2.5 cm. long, slender, suberect; sporophylls coriaceous, deeply concave, broadly ovate, long-acuminate, mostly 2.2–2.8 mm. long (seta included), 1–1.3 mm. broad; seta 0.3–0.55 mm. long, subentire, pungent, whitish-hyaline from a long, lutescent base; cilia usually 8–15 on each side, stout, 0.04–0.06 mm. long, the upper ones dentiform; megaspores subglobose, yellow, about 0.4 mm. thick, foveolate-reticulate; microspores orange, about 0.047 mm. thick.

Open rock-slopes, Canadian to Arctic-Alpine Zone; British Columbia to eastern Oregon (Blue Mountains) and Colorado. Type locality: Cracker Lake, Glacier National Park, Montana.

8. Selaginella wátsoni Underw. Alpine Selaginella. Fig. 106.

Selaginella watsoni Underw. Bull. Torrey Club **25**: 127. 1898.

Plants cespitose, usually densely so; stems prostrate, short-creeping, 2–6 (10) cm. long, rooting throughout; branches numerous, 1–2.5 cm. long, mostly 2-pinnate, rigidly ascending, the divisions short, thick, closely cespitose, equally leafy on all sides. Leaves of the branches uniform, densely appressed-imbricate, thick, linear-oblong, rounded-obtuse, abruptly short-setigerous, 1.8–2.4 mm. long (seta included), 0.5–0.63 mm. broad, strongly cymbiform, with a broad dorsal groove; seta 0.2–0.25 mm. long, stout, lutescent, smooth; cilia 2–7 or wanting, 0.03–0.065 mm. long, slightly oblique, deciduous. Spikes 1–2.5 cm. long, sharply quadrangular; sporophylls appressed-imbricate, ovate, acuminate, 1.8–2.5 mm. long (seta included), 0.9–1.1 mm. broad; seta 0.25–0.35 mm. long, lutescent, stout, rigid, pointed, smooth; cilia 10–15, confined to the basal half, stout, 0.015–0.05 mm. long, often dentiform; megaspores 0.42–0.47 mm. thick, yellow, rugose-reticulate; microspores about 0.039 mm. thick, orange.

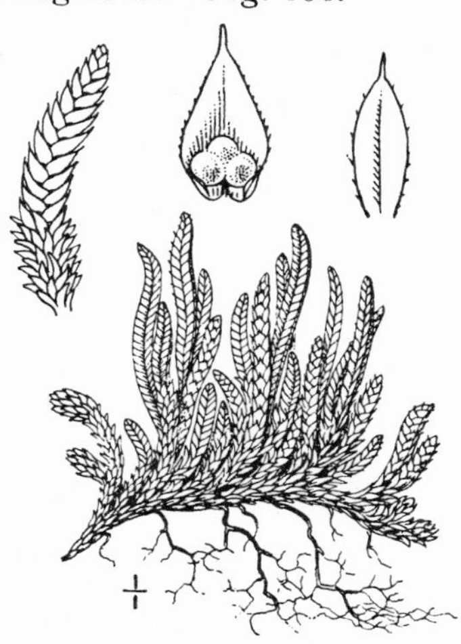

Rocky slopes and cliffs, Hudsonian and Arctic-Alpine Zones; high mountains of California (Nevada County to Riverside County), eastward to Utah, ascending to 3,450 meters. Type locality: Cottonwood Canyon, Utah.

9. Selaginella leucobryoìdes Maxon. Mojave Selaginella. Fig. 107.

Selaginella leucobryoides Maxon, Smiths. Misc. Coll. **72^{5}**: 8. *pl. 5.* 1920.

Plants cushion-like; stems short-creeping, 1–2 cm. long, prostrate, once pinnate; branches thick, erect, strongly cespitose, the sterile ones 2–7 mm. high, densely leafy. Leaves crowded, appressed-imbricate, mostly incurved, glaucous, uniform, linear-subulate, the basal ones up to 3.25 mm. long (seta included), the upper ones shorter, all the leaves thick, rigidly herbaceous, strongly convex dorsally toward the acutish apex, short-setigerous; seta stout, milk-white, 0.12–0.28 mm. long, strongly scabrous, often reflexed; cilia 8–16 on each side, the lower ones spreading, 0.6–0.13 mm. long, the upper ones distant, shorter, ascending. Spikes numerous, close, 5–10 mm. long, about 1.5 mm. thick, erect; sporophylls rigidly appressed-imbricate, deltoid-ovate, long-acuminate, about 2 mm. long, 0.8–1 mm. broad; seta 0.15 mm. long or less, white, rigid, pointed, subentire; cilia and teeth 20–25 on each side, oblique, very short, pungent; megaspores yellow, about 0.47 mm. thick, reticulate; microspores orange, about 0.039 mm. thick.

Crevices of rocky slopes, Upper and Lower Sonoran Zones; Panamint and Providence Mountains, southern California. Type locality: Bonanza Mine, Providence Mountains, California.

10. Selaginella wallácei Hieron. Wallace's Selaginella. Fig. 108.

Selaginella wallacei Hieron. Hedwigia **39**: 297. 1900.
? *Selaginella montanensis* Hieron. Hedwigia **39**: 293. 1900.

Plants loosely cespitose; main stems prostrate, 5–15 (20) cm. long, rooting sparsely throughout; branches numerous, ascending, 1–4 cm. long, freely branched, the divisions short, divaricate, rigid, funiform, often intricate. Leaves chartaceous, rigidly appressed-imbricate on all sides, subglaucous, mostly oblong-linear, narrowed to an obtusish apex; leaves of larger branches 2.5–3.35 mm. long (seta included), 0.45–0.63 mm. broad, with seta 0.3–0.6 mm. long; leaves of smaller branches 1 3–2.5 mm. long (seta included), 0.3–0.45 mm. broad, the seta 0.15–0.3 mm. long, whitish-hyaline, scabrous; cilia 8–14 on each side, oblique, 0.06–0.1 mm. long. Spikes numerous, 1–3 cm. long, slender, curved; sporophylls ovate-deltoid (often narrowly so), 1.8–2.8 mm. long (seta included), 0.75–1.1 mm. broad, abruptly setigerous; seta 0.15–0.45 mm. long, stout, whitish-hyaline, nearly smooth; cilia 10–18 on each side, oblique, incurved, 0.03–0.08 mm. long, or all short and dentiform; megaspores yellow, up to 0.45 mm. thick, tuberculate-reticulate, annulate; microspores about 0.04 mm. thick.

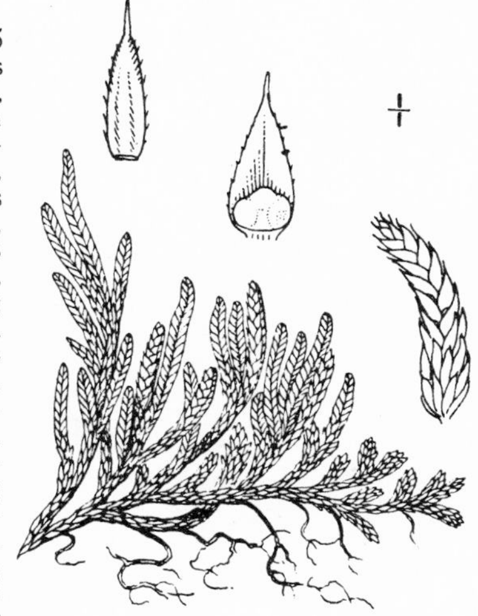

Rock-slopes and bluffs, Humid Transition to Hudsonian Zone; British Columbia to Montana and Oregon, south along the coast to Marin County, California. Type locality: Oregon. Variable.

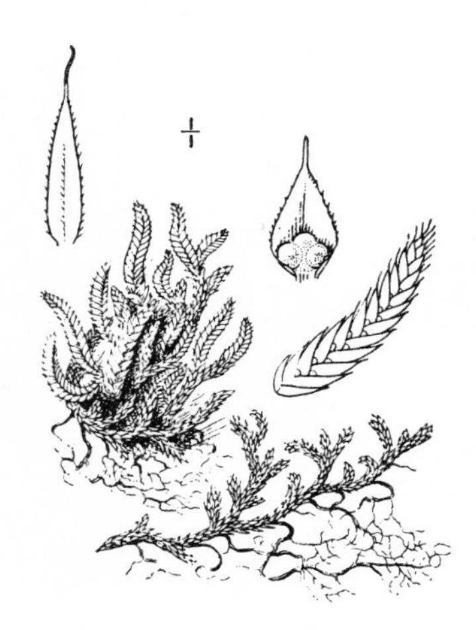

11. Selaginella asprélla Maxon.

Bluish Selaginella. Fig. 109.

Selaginella asprella Maxon, Smiths. Misc. Coll. 72[5]: 6. *pl. 4.* 1920.

Stems loosely matted, 3–6 cm. long, creeping but not prostrate, rooting at intervals; branches laxly ascending, usually intricate, 1–2.5 cm. long, 2-pinnate, the divisions 3–7 mm. long, oblique, slender. Leaves rigidly ascending, subdistant, subimbricate, 2.75–3.2 mm. long (seta included), the blades narrowly deltoid-subulate, strongly glaucous, whitish-marginate; seta 0.7–0.9 mm. long, white-hyaline, subflexuous, serrulate; cilia 16–23 on each side, mostly spreading, 0.05–0.09 mm. long, the upper ones shorter and oblique. Spikes loosely aggregate, apical, 1–2 cm. long, 1.5–2 mm. thick, arcuate; sporophylls laxly imbricate, 2.5–3 mm. long (seta included), the blades narrowly ovate-deltoid, long-acuminate; seta 0.6–0.8 mm. long, whitish, strongly scabrous; cilia 25–35 on each side, mostly 0.03–0.06 mm. long; megaspores yellow, 0.375 mm. thick, lightly reticulate; microspores orange, about 0.033 mm. thick.

Cliffs and rocky cañon floors, Transition Zone; San Antonio and San Bernardino Mountains, southern California. Type locality: Ontario Peak, California.

Phylum SPERMATÓPHYTA.

SEED-BEARING PLANTS.

Plants producing seeds which contain the young plants in a dormant condition until germination. Sporophylls arranged in groups (flowers) of definite or indefinite numbers, heterosporous, those bearing microsporangia (anther-sacs) termed stamens, those producing macrosporangia (ovules) carpels. The gametophytes very much reduced, the female being confined within the macrosporangia, where its egg-cell is fertilized by the spermatozoid of the male gametophyte (pollen-tube).

The seed-bearing plants form the most numerous group in existence, not less than 130,000 species being known.

There are two classes in the subkingdom, which differ from each other as follows:

Ovules and seeds borne on the face of a scale, not enclosed. Class 1. *Gymnospermae.*
Ovules and seeds contained in a closed cavity (ovary). Class 2. *Angiospermae.*

Class 1. *GYMNOSPÉRMAE.*

Ovules (macrosporangia) naked, borne on the flat surface of a sporophyll, this not infolding or uniting to form an ovary, sometimes apparently wanting. Pollen-grains (microspores) dividing at maturity into two or more cells, one of which gives rise to the pollen-tube.

The Gymnosperms are an ancient group, first known in Silurian time, and most numerous in Triassic time. They are now represented by not more than 500 species of trees and shrubs. There are six orders represented by living species: Ginkgoales (only one species extant, *Ginkgo biloba* maiden-hair tree), Cycadales (fern-like tropical or subtropical plants), Araucariales, Taxales, Pinales, and Gnetalés. Only the three last are represented in the flora of the Pacific States. The presence of ciliated spermatozoids in the Ginkgoales and Cycadales, and the well developed archegonia suggest that the Gymnosperms might be more closely related to the Pteridophytes than to the Angiosperms, or, what is more probable, that they represent a distinct offshoot from the former.

Family 1. TAXÀCEAE.

YEW FAMILY.

Slightly resinous and aromatic evergreen trees or shrubs, with fissured or scaly bark, close-grained and durable wood. Leaves stiff, entire, linear or lanceolate, spirally arranged but usually appearing 2-ranked by a twist of the flattened and decurrent petiole. Flowers dioecious, axillary, surrounded by the bud-scales, the staminate composed of many anthers, each bearing several pendent pollen-sacs; the ovulate of a solitary ovule, becoming a bony seed and surrounded by a fleshy aril; cotyledons 2.

Ten genera, the majority of which are in the southern hemisphere, where some species are important forest trees. In North America only the two following genera are found:

Fruit drupe-like, completely enclosed by the fleshy coat; endosperm ruminate; pollen-sacs in a semicircle; leaves very sharp-pointed. 1. *Torreya.*
Fruit surrounded by a scarlet fleshy cup; endosperm uniform; pollen-sacs in a circle around the filament; leaves not sharp-pointed. 2. *Taxus.*

1. TÓRREYA Arn. Ann. Nat. Hist. I. 1: 130. 1838.

Trees with spreading or drooping branches and fissured bark. Leaves appearing 2-ranked, flat, long-pointed. Staminate flowers solitary, forming dense clusters on the lower side of last year's twigs, oval or oblong, pollen-sacs 4; ovulate flowers in pairs on this year's twigs, the ovule surrounded by 4 scales and becoming completely enclosed by the fleshy aril. Fruit maturing the second year, drupe-like with a thick resinous leathery outer coat; seed sharp-pointed; endosperm channeled by the infolding of the thick woody integument. [Name in honor of Dr. John Torrey, American botanist.]

Four species, which are comparatively local but widely separated geographically, there being one each in Florida, California, Japan, and China. Fossils have been found in the arctic region and northern Europe. Type species, *Torreya taxifolia* Arn.

1. Torreya califórnica Torr.

California Nutmeg. Fig. 110.

Torreya californica Torr. N. Y. Journ. Pharm. **3**: 49. 1853.
Torreya myristica Hook. Bot. Mag. III. **10**: *pl. 4780.* 1854.
Tumion californicum Greene, Pittonia **2**: 195. 1891.

Tree 15–35 m. high with a trunk 3–12 dm. in diameter; branches slender, spreading or drooping; bark brown or yellowish brown, 10–15 mm. thick, deeply and broadly furrowed, covered with loose scaly plates. Leaves 25–75 mm. long, 2–3 mm. wide, dark green and glossy above, slightly paler and with a deep, narrow groove beneath, very sharp-pointed, emitting a pungent odor when bruised; staminate flowers about 8 mm. long; fruit ovoid-oblong, 25–35 mm. long, green, with purplish streaks or blotches.

Moist, shaded slopes and along water courses, usually scattered or in small groups, Transition Zone; western slope of the Sierra Nevada, from Tulare to Tehama Counties, and in the Coast Ranges, from Mendocino to Santa Cruz Counties. Type locality: Headwaters of Feather and Yuba Rivers.

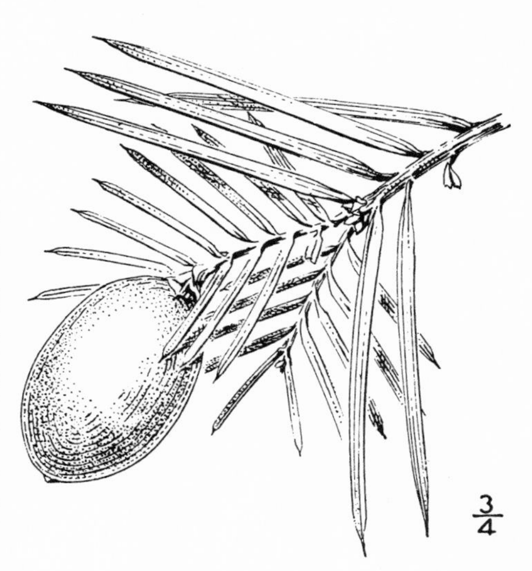

2. TÁXUS [Tourn.] L. Sp. Pl. 1040. 1753.

Trees or shrubs with scaly bark, usually horizontal or drooping branches, and strong elastic durable wood. Leaves persistent for 4–5 years, linear to lanceolate, flat, keeled on the upper surface, stomatiferous beneath, revolute on the margins; staminate flowers globose, of 4–8 stamens; pollen-sacs several, pendent in a circle around the filament; ovulate flowers solitary on rudimentary axillary branches, minute and green, maturing in the autumn. Seed with a bony integument, ovate-oblong, about 8 mm. long, surrounded but free from the thickened gelatinous sweet scarlet cup-shaped aril. [The classical name.]

About 6 closely related species widely distributed over the northern hemisphere, especially in the cooler parts of the north temperate zone. Three species are in North America. Type species, *Taxus baccata* L., the English yew of the gardens.

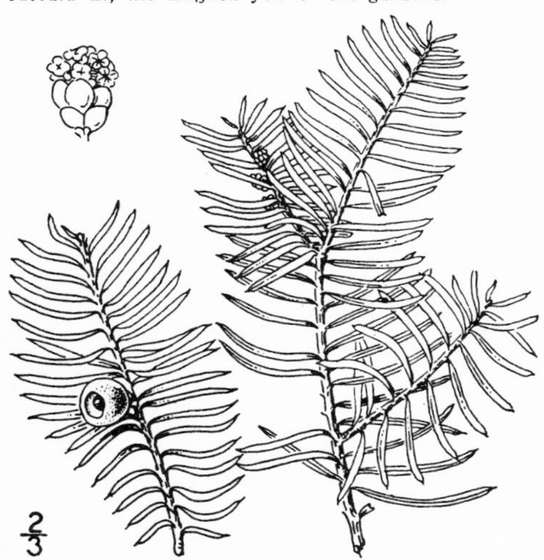

1. Taxus brevifòlia Nutt.

Western Yew. Fig. 111.

Taxus brevifolia Nutt. N. Am. Sylva **3**: 86. *pl. 108.* 1849.
Taxus boursieri Carr. Rev. Hort. **3**: 228. 1854.
Taxus lindleyana Murr. Edinb. New Phil. Journ. II. **1**: 294. 1855.

Tree 10–25 m. high, with straight, often fluted, trunk, 2–12 dm. in diameter, and with slender, spreading or drooping branches; bark about 6 mm. thick, covered with small, dark reddish brown scales. Leaves linear, 12–16 mm. long, about 1–2 mm. wide, acute, deep yellow-green and shining above, much paler beneath, 2-ranked, forming flat sprays, persisting 4–5 years; flowers and fruit as described in the generic description.

In deep, coniferous woods in moist situations, widely distributed, but seldom forming groves, Canadian and Transition Zones; southern Alaska and the Selkirk Mountains, to western Montana, the Santa Cruz Mountains and the western slope of the Sierra Nevada to Tulare County, California. Type locality: in the dense maritime forests of Oregon.

Family 2. PINACEAE.

PINE FAMILY.

Resinous evergreen or rarely deciduous trees or shrubs, with linear, needle-like or scale-like leaves, arranged in spirals. Flowers surrounded at base by the persistent bud scales, monoecious, the staminate consisting of many spirally arranged stamens, with 2 pollen-sacs, the ovulate of many scales, bearing 2 pendent ovules on their inner surface. Fruit a woody cone, maturing the first or second season. Seeds with or without wings; embryo axile in the copious endosperm; cotyledons several.

A family of 8 genera and approximately 150 species widely distributed over the northern hemisphere.

Cones maturing the second or third year; leaves surrounded at base by a persistent or deciduous sheath, in bundles of 2–5 (solitary in one species). 1. *Pinus.*
Cones maturing the first year; leaves not in sheaths, solitary, often appearing 2-ranked.
  Leaves deciduous, forming tuft-like clusters on short spur-like branchlets. 2. *Larix.*
  Leaves persistent, scattered along ordinary branchlets.
    Cones pendulous, their scales persistent.
      Bracts concealed by the scales; branchlets roughened by the woody persistent leaf bases.
        Leaves sessile on the woody stalk-like bases, 4-sided or flattened and stomatiferous above. 3. *Picea.*
        Leaves narrowed to a short petiole, flattened and stomatiferous below or angular. 4. *Tsuga.*
      Bracts exserted beyond the cone-scales, with 3 prominent awns; leaves without woody bases, leaving the branchlets smooth, narrowed to a short petiole. 5. *Pseudotsuga.*
    Cones erect, their scales deciduous; leaves sessile, leaving the branches smooth. 6. *Abies.*

1. PINUS [Tourn.] L. Sp. Pl. 1000. 1753.

Evergreen trees or rarely shrubs, with furrowed or thin, scaly bark, rather hard or often soft wood. Leaves of two kinds, the secondary forming the ordinary foliage leaves, needle-like, borne in clusters of 2–5, or solitary in one species, terminating short rudimentary branchlets in the axils of the scale-like primary leaves, surrounded at base by a deciduous or persistent membranous sheath, persistent from 2–8 years. Staminate flowers clustered at the base of the season's growth, forming a distinct zone on the twig which becomes naked after the flowers have fallen, each flower subtended by an involucre of several scales, composed of many 2-celled anthers; ovulate flowers subterminal, or lateral, solitary, in pairs or clustered, sessile or stalked, consisting of many ovule-bearing scales, subtended by non-accrescent bracts, and bearing 2 pendent ovules. Fruit a cone, maturing the second year, or sometimes the third year, opening and shedding the seeds at maturity, or, in some species, remaining closed and persistent on the branches for years; cone-scales elongated, variously thickened and appendaged at the exposed apex (apophysis). Seeds often with conspicuous membranous wings, their coats crustaceous; cotyledons 3–15, or sometimes more. [The classical Latin name.]

A genus of about 100 species widely scattered over the northern hemisphere. Type species, *Pinus sylvestris* L.

Sheaths of the leaves deciduous; leaves with one fibro-vascular bundle; wood soft, light-colored, close-grained, with thin nearly white sapwood.
  Leaves in 5-leaved clusters.
    Cones with terminal unarmed umbos.
      Seeds much shorter than the wings; cones long-stalked, their scales thin.
        Leaves obtuse at apex; cones 12–25 cm. long; seed-wings acute. 1. *P. monticola.*
        Leaves sharp-pointed; cones 30–45 cm. long; seed-wings rounded at apex. 2. *P. lambertiana.*
      Seeds much longer than the short wings; cones short-stalked, their scales thick, sometimes with pointed umbos.
        Cones opening at maturity, 8–20 cm. long. 3. *P. flexilis.*
        Cones remaining closed, 2–7 cm. long. 4. *P. albicaulis.*
    Cones with dorsal umbos armed with slender prickles; seeds shorter than the wings.
      Cones armed with minute incurved prickles. 5. *P. balfouriana.*
      Cones armed with long slender prickles. 6. *P. aristata.*
  Leaves in 1–4-leaved clusters; cones subglobose, their scales few, much thickened, only the middle seed-bearing; seeds large, wings a mere ring.
    Leaves usually in 4-leaved clusters. 7. *P. quadrifolia.*
    Leaves usually in 1-leaved clusters. 8. *P. monophylla.*
Sheaths of the leaves persistent; leaves with two fibro-vascular bundles; wood usually heavy, rather hard, coarse-grained, generally dark-colored, with often thick pale sapwood.
  Leaves in 5-leaved clusters. 9. *P. torreyana.*
  Leaves in 2-leaved or 3-leaved clusters.
    Leaves in 3-leaved clusters, or sometimes only 2 in a cluster in *P. radiata* and *P. ponderosa.*
      Cone-scales with dorsal slender prickles.
        Cones symmetrical, opening at maturity, deciduous, the basal scales persistent on the branches.
          Leaves yellow-green; branchlets not glaucous; cones 7–15 cm. long. 10. *P. ponderosa.*
          Leaves dull bluish green; branchlets glaucous; cones 15–35 cm. long. 11. *P. jeffreyi.*

Cones unsymmetrical, their outer scales much enlarged and mammillate, remaining closed and persistent on the branches often for many years.
Outer cone-scales with rounded apophysis. 12. *P. radiata.*
Outer cone-scales with prominent, knob-like apophysis. 13. *P. attenuata.*
Cone-scales prolonged into stout, straight or incurved spur-like spines; cones 15–35 cm. long, heavy.
Leaves gray-green and drooping; cones broad-ovate, chocolate brown; seeds longer than the wings. 14. *P. sabiniana.*
Leaves blue-green, erect; cones oblong-conical, light brown; seeds shorter than the wings. 15. *P. coulteri.*
Leaves in 2-leaved clusters; cones small.
Cones nearly symmetrical, opening at maturity, and deciduous, armed with minute prickles. 16. *P. contorta.*
Cones decidedly unsymmetrical, persistent and remaining closed for a number of years; outer scales much enlarged; prickles stout. 17. *P. muricata.*

1. Pinus montícola Dougl.

Western or Mountain White Pine. Fig. 112.

Pinus monticola Dougl.; Lamb. Pinus ed. 2, **3**: 27. *pl. 67.* 1837.
Pinus strobus monticola Nutt. N. Am. Sylva **3**: 118. 1849.
Pinus monticola minima Lemmon, Rep. Calif. State Board Forestry **2**: 70. 1888.
Pinus monticola porphyrocarpa Masters, Journ. Hort. Soc. **14**: 235. 1892.
Strobus monticola Rydb. Fl. Rocky Mts. 13, 1060. 1917.

Tree forming a slender trunk, 8–24 dm. in diameter and 30–50 m. in height, with a narrow short-branched symmetrical crown when growing in dense forests, or scarcely half as tall and forming an irregularly branched crown when growing in open forests; branchlets stout, brownish and puberulent the first year, becoming glabrous the second year; bark 2–4 cm. thick, divided into small nearly square plates. Leaves 5–10 cm. long, in 5-leaved clusters, usually obtuse at apex, bluish green and glaucous, marked by several white bands of stomata on the central side, persistent for 2 or 3 years; staminate flowers 8 mm. long; cones deciduous soon after shedding the seeds in the autumn of the second year, cylindric, 10–20 cm. long, light brown; seeds brown, mottled with black, 8 mm. long; wings 25 mm. long, acute.

A characteristic tree of the Canadian Zone; southern British Columbia eastward to the Continental Divide in northern Montana, southward through the mountains of Washington and Oregon to the southern Sierra Nevada, California. Best developed in northern Idaho and Montana, where it forms extensive forests. Attains an age of 500–600 years. Wood light, soft, fine, and straight-grained, of high commercial value. Type locality: mountains near the Grand Rapids of Columbia River.

2. Pinus lambertiàna Dougl.

Sugar Pine. Fig. 113.

Pinus lambertiana Dougl. Trans. Linn. Soc. **15**: 500. 1827.

Largest of all pines, attaining a height of about 75 m. and a maximum diameter of 4 m., the principal branches usually horizontal, branchlets pubescent. Bark of older trees 5–8 cm. thick, divided into elongated plate-like ridges covered with loose purple-brown or cinnamon-colored scales. Leaves in fives, stout, rigid, sharp-pointed, 7–10 cm. long, marked on the two sides by several rows of stomata, persistent until the second or third year, their resin ducts surrounded by strengthening cells; staminate flowers 8–10 mm. long, light yellow; cones on stalks 5–8 cm. long, shedding the seeds in the autumn and falling the following summer, cylindric, 25–40 cm. long or rarely 55 cm. long, light brown; their scales often 4 cm. wide; seeds 12 mm. long, dark brown or nearly black; wings 25 mm. long, 10–12 mm. wide, rounded at apex.

Mountain slopes, Transition Zone; North Fork of Santiam River, Oregon, southward through the Coast Ranges, and the Sierra Nevada to San Pedro Martir Mountain, Lower California. The Sugar Pine is one of the characteristic trees of the great forest belt of the Sierra Nevada. It reaches its greatest size on the western slope of the central Sierra Nevada, where especially handsome specimens are found associated with the Giant Sequoia. It is a long-lived tree, reaching an age of 500 to nearly 600 years. The wood is light, soft, straight-grained, with a light red-brown heartwood of high commercial value. Wounds in the heartwood often exude a sweet sugar-like substance. Type locality: on the headwaters of the Umpqua River, Oregon.

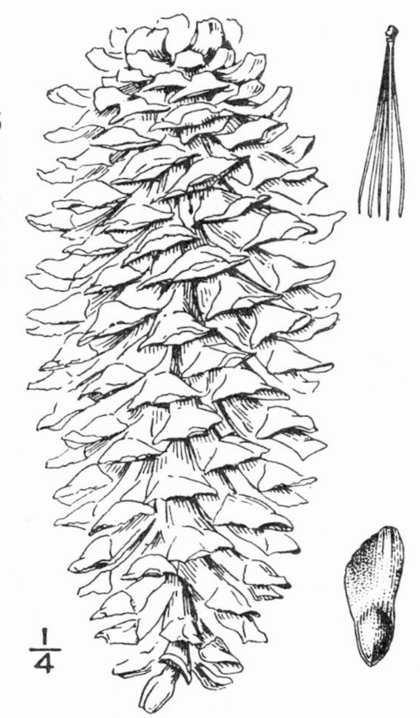

3. **Pinus fléxilis** James. Limber Pine. Fig. 114.

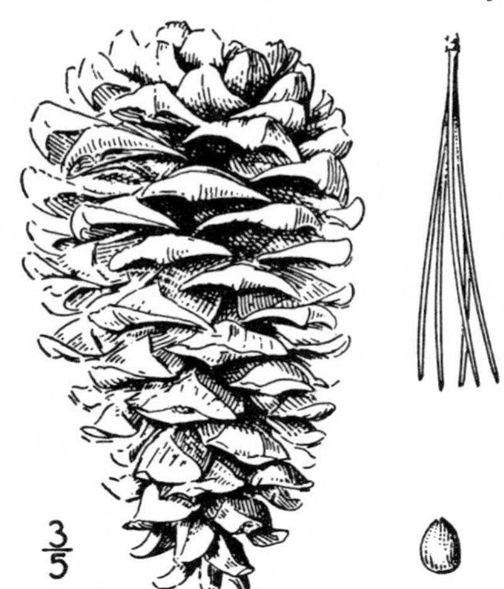

Pinus flexilis James, in Long Exped. 2: 34. 1823.

Tree, 15–25 m. high, with a short stout trunk 1.5–2.5 m. in diameter; young branchlets pubescent, soon glabrous; bark on branches thin, smooth, light gray or silvery white, on older trunks breaking into dark brown plates covered by thin scales, and finally 25–50 mm. thick. Leaves stout, stiff, slightly curved, sharp-pointed, dark green, marked on all sides by 1–4 rows of stomata, persistent until the fifth or sixth year; staminate flowers reddish, oval, 10 mm. long; cones on stout, short stalks, oval or subcylindric, 8–25 cm. long, light brown and more or less tinged with purple, their scales thickened and often incurved at apex, opening at maturity; seeds dark reddish brown, mottled with black, 10–12 mm. long, compressed; wings usually persistent on the scale after the seeds have fallen.

Widely scattered over the Rocky Mountain and Great Basin regions in the Boreal Zones; Alberta and Montana to New Mexico, west to Oregon and California. In Oregon it occurs in the Wallowa and Warner Mountains. In California on the Warner, Panamint, and Inyo Mountains, the eastern slope of the Sierra Nevada south of Mono Pass, and in southern California on Pinos, North Baldy, San Bernardino, San Gorgonio, and San Jacinto Peaks; also on San Pedro Martir Mountain, Lower California. Type locality: Rocky Mountains of Colorado.

4. **Pinus albicaùlis** Engelm. White-bark Pine. Fig. 115.

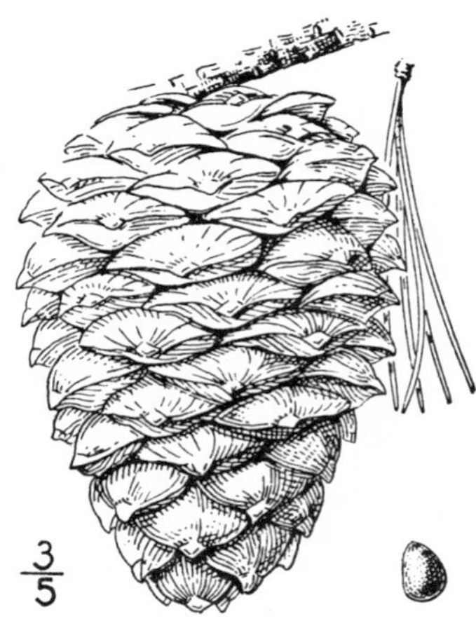

Pinus albicaulis Engelm. Trans. St. Louis Acad. 2: 209. 1863.
Pinus shasta Carr. Trait. Conif. ed. 2, 390. 1867.
Pinus flexilis albicaulis Engelm. in S. Wats. Bot. Calif. 2: 124. 1880.

A low alpine tree with crooked or twisted trunk, 5–15 m. high, or on exposed slopes prostrate, and 0.5–1.5 m. in diameter; branches short and stout, usually erect; bark thin, about 1 cm. thick on old trunks, divided by narrow fissures into thin, whitish or brownish scales. Leaves 35–75 mm. long, dark green, stout, stiff, slightly curved, marked on the back by 1–3 rows of stomata, persistent for 7–8 years; staminate flowers about 10 mm. long; cones oval to subglobose, 35–75 mm. long, purple, remaining closed after maturity and eventually breaking up, their scales often armed with a stout pointed umbo; seeds 10–12 mm. long, dark brown and hard; wings thin, very narrow, remaining on the scale.

A characteristic tree of the Hudsonian Zone, often forming the timber line; ranging from British Columbia and Alberta to northwestern Wyoming and the high mountains of the Pacific Coast as far south as the southern Sierra Nevada. Wood light, soft, close-grained, brittle, light brown. Type locality: Cascade Mountains, Oregon, about latitude 44°

5. **Pinus balfouriàna** Murr. Foxtail Pine. Fig. 116.

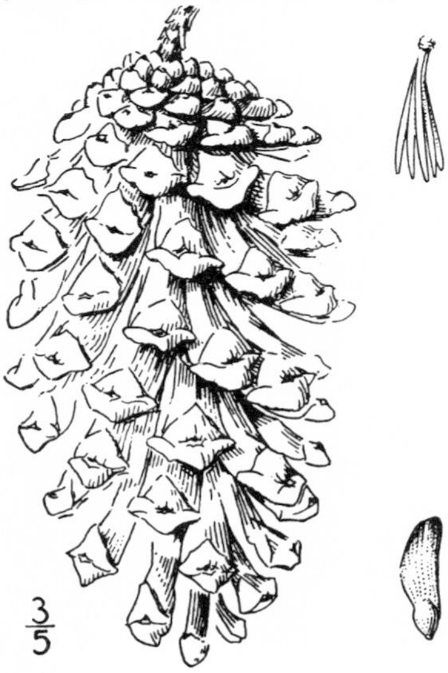

Pinus balfouriana Murr. Rep. Bot. Exped. Oreg. 1. *pl. 3, f. 1.* 1853.

A low tree 7–15 m. high, with a trunk 6–15 dm. in diameter; branches spreading, stout and short; branchlets appearing brush-like, with long, dense clusters of appressed leaves; bark thin, smooth and white, becoming red-brown, 2 cm. thick, divided into broad, flat ridges. Leaves curved, closely appressed to the branches, 25–35 mm. long, stout, stiff, dark blue-green on the back, whitish on the inner surfaces, with many rows of stomata, entire; staminate flowers, 10–12 mm. long; cones 35–75 mm. long, ovoid, sessile, dark reddish brown, their scales thickened and ridged, somewhat 4-sided, armed with minute incurved prickles; seeds pale, mottled with purple, 8 mm. long, their wings 25 mm. long, 6 mm. wide, obliquely narrowed at apex.

A subalpine tree in the Hudsonian Zone, growing on granite at high elevations in the California mountains in two widely separated districts. In the northern district it occurs at altitudes of 6000–8000 ft. on Scott Mountains in Siskiyou County, on Mount Eddy and southward to Yollo Bolly, Trinity County. In the southern district it occurs at altitudes of 10,000–11,000 feet in the southern Sierra Nevada, from Kings River to Olanche Mountain. Type locality: Scott Mountains, California.

3/5

6. **Pinus aristàta** Engelm. Bristle-cone Pine. Fig. 117.

Pinus aristata Engelm. Am. Journ. Sci. II. **34**: 331. 1862.
Pinus balfouriana aristata Engelm. in Wheeler Rep. **6**: 375. 1878.

A small, usually bushy, tree, or occasionally with a trunk 15–20 m. high and 6–9 dm. in diameter; branches short, stout; bark on old trunks 1–2 cm. thick, reddish brown, divided into flat, irregular ridges, with closely appressed scales. Leaves 25–35 mm. long, stout or slender, dark green and shiny on the back, whitish on the inner surfaces; staminate flowers oval, 10–12 mm. long; cones horizontal or somewhat pendent, nearly sessile, ovoid, 7–9 cm. long, dark purplish brown, their scales thickened and ridged, armed with a slender incurved prickle, 6 mm. long; seeds light brown, mottled with black, 8 mm. long and 6 mm. wide.

A subalpine tree, Canadian Zone; extending from Colorado through southern Utah, central and southern Nevada and northern Arizona to the Panamint Mountains, California. Type locality: mountains near Clear Creek (Colorado), 9,000–10,000 feet altitude.

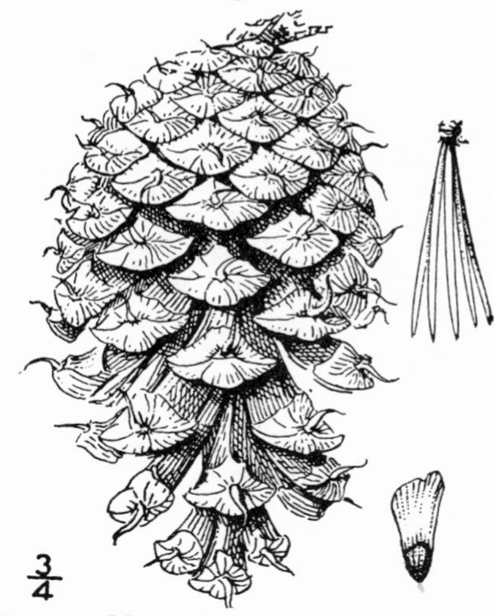

7. **Pinus quadrifòlia** Parry. Parry's Piñon or Nut Pine. Fig. 118.

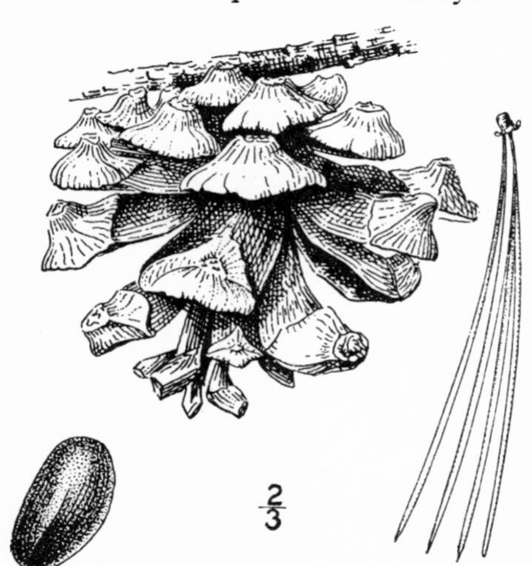

Pinus parryana Engelm. Am. Journ. Sci. II. **34**: 332. 1862, not Gord. 1858.
Pinus quadrifolia Parry; Parl. in DC. Prod. **16²**: 402. 1868, as a synonym.
Pinus quadrifolia Parry; Sudw. U. S. Div. Forest. Bull. **14**: 17. 1897.

A small tree 10–15 m. high with a trunk 5 dm. in diameter; branches spreading, forming a pyramidal or, in age, a flat-topped crown; branchlets soft, pubescent when young; bark dark brown, shallowly divided into broad connected ridges, covered by appressed scales. Leaves 1–5, generally 4, in a cluster, incurved, 25–35 mm. long, pale blue-green on the back, marked by many conspicuous rows of stomata on the inner surfaces; staminate flowers oval, 6 mm. long, cones subglobose, 35–50 mm. long, brown; apex of the scale much thickened, keeled and with a ridged knob; seeds 15 mm. long, their wings 3 mm. high.

Dry mountain slopes of the Upper Sonoran Zone; desert slopes of the Santa Rosa Mountains, Riverside County, California, southward to San Pedro Martir Mountain, Lower California. The large, nut-like seeds are an important food article of the Indians of Lower California. Type locality: near Mountain Springs, San Diego County, California.

8. **Pinus monophýlla** Torr. & Frem. One-leaved Piñon or Nut Pine. Fig. 119.

Pinus monophylla Torr. & Frem. in Frem. Second Rep. 319, *pl. 4.* 1845.
Pinus fremontiana Endl. Syn. Conif. 183. 1847.
Pinus edulis monophylla Torr. in Ives Rep., 28. *pl. 4.* 1860.
Caryopitys monophylla Rydb. Bull. Torrey Club **32**: 397. 1905.

A low tree usually with a branched trunk, 5–15 m. high, in old age, forming a flat-topped, often picturesque, crown; bark 2 cm. thick, divided into narrow, flat ridges. Leaves only 1 or sometimes 2 in a cluster, stout, stiff, incurved, 25–35 mm. long, gray-green; staminate flowers 6 mm. long; cones broadly ovoid or subglobose, 35–55 mm. long, brown, their scales thick, concave, 4-sided, knobbed at apex; seeds 15 mm. long, their wings 8–12 mm. high.

Dry, gravelly slopes in the Upper Sonoran Zone; western slopes of the Wasatch Mountains, Utah, to the eastern slopes of the southern Sierra Nevada, California, southward on the desert slopes to northern Lower California and northern Arizona, usually forming a belt just above the Juniper belt. The seeds are extensively gathered by the Indians for food. Type locality: not indicated.

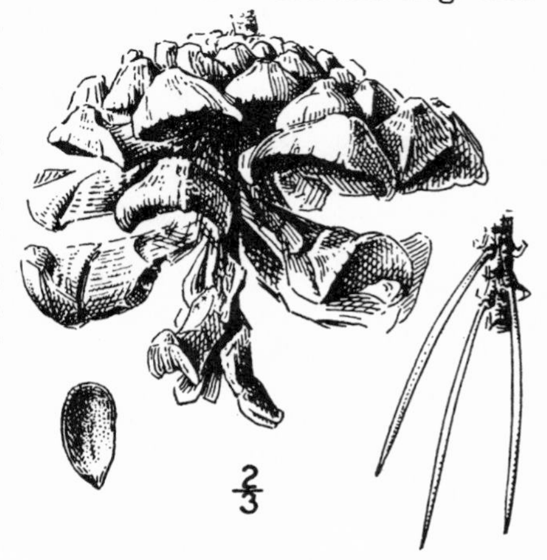

9. **Pinus torreyàna** Parry. Torrey Pine. Fig. 120.

Pinus torreyana Parry, Bot. Mex. Bound. 210. *pl. 58, 59.* 1859.
Pinus lophosperma Lindl. Gard. Chron. **1860**: 46. 1860.

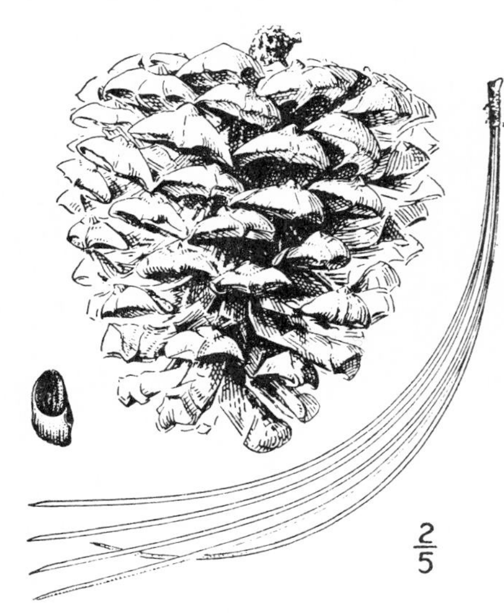

A small tree, 10–20 m. high, with a branched trunk, 4–7 dm. in diameter, and broad, open crown; bark 20–35 mm. thick, red-brown, irregularly divided into broad flat ridges. Leaves in fives, 20–30 cm. long, dark dull green on the back, marked on all surfaces with many rows of stomata, persistent for 3 or 4 years; staminate flowers, cylindric, 25–35 mm. long; cones on elongated stout stalks, broadly ovoid, 10–15 cm. long, chestnut brown, shining, their scales thickened at the apex into a low pyramidal knob, armed with a minute prickle; seeds oval, 20–25 mm. long, with a hard and thick shell; wings nearly surrounding the seeds, about 1 cm. high.

The rarest and most restricted in its distribution of all the American pines. Found in two small colonies in the Upper Sonoran Zone, one on the eastern end of Santa Rosa Island, off the coast of southern California, the other on the mainland near Del Mar, San Diego County. Type locality: Soledad Cañon, San Diego County, California.

10. **Pinus ponderòsa** Dougl. Western Yellow Pine. Fig. 121.

Pinus ponderosa Dougl.; Lawson, Agric. Man. 354. 1836.
Pinus benthamiana Hartweg, Journ. Hort. Soc. Lond. 2: 189. 1847.
Pinus beardsleyi Murr. Edinb. New Phil. Journ. II. 1: 286. *pl. 6.* 1855.
Pinus craigana Murr. Edinb. New Phil. Journ. II. 1: 288. *pl. 7.* 1855.

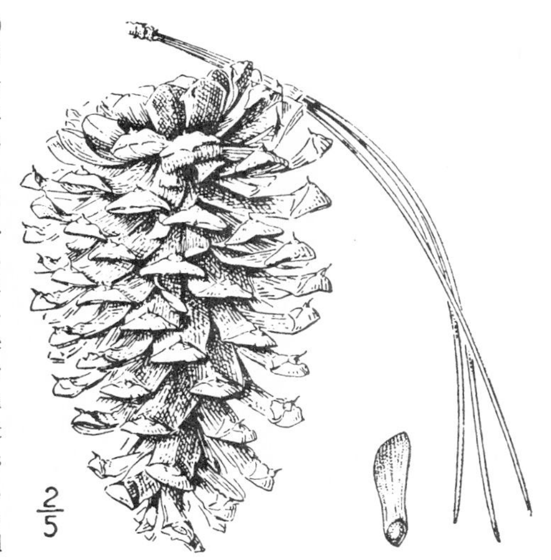

A forest tree, attaining an age of from 300 to 500 years, often 50–75 m. high and 15–25 dm. in diameter; branches short, generally turned upward at the ends and forming a regular spire-like crown, or in open forests in dry situations, often flat-topped; branchlets orange-colored, becoming dark in age, not glaucous; bark on old trees often 7–10 cm. thick, separated into large russet-red plates, usually several inches wide, covered with small concave scales. Leaves in threes, bright yellow-green, forming dense brush-like tufts at the ends of the naked branches, persistent for about 3 years, 12–25 cm. long, stomatiferous on all three sides; staminate flowers in short crowded clusters, cylindric, 2–3 cm. long; cones subterminal, horizontal or somewhat deflexed, short-stalked, oval, 7–15 cm. long, reddish brown, shedding the seeds the second year, and soon falling, their scales thin, light, narrow, thickened at the apex and armed with slender prickles; seeds ovate, acute, 6–7 mm. long, with a thin dark purple more or less mottled shell; wings broadest below the middle, oblique at apex, 25–35 mm. long.

The most characteristic tree of the Arid Transition Zone, forming open park-like forests; British Columbia to Idaho, south through the Pacific States and western Nevada to northern Lower California. Replaced in the Rocky Mountain region by the variety *scopulorum* Engelm. Wood hard, strong, of great commercial value. Type locality: Spokane River, Washington.

11. **Pinus jéffreyi** Murr. Jeffrey or Black Pine. Fig. 122.

Pinus jeffreyi Murr. Rep. Bot. Exped. Oreg. 2. *pl. 1.* 1853.
Pinus deflexa Torr. Bot. Mex. Bound. 209. 1859.
Pinus ponderosa jeffreyi Vasey, Rep. U. S. Dept. Agric. 179. 1875.

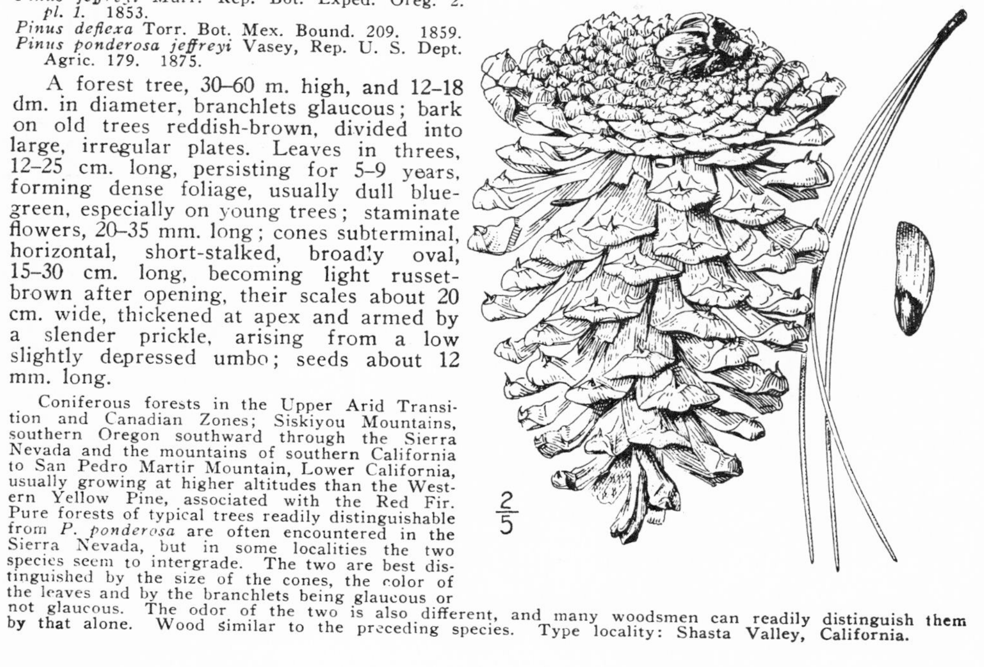

A forest tree, 30–60 m. high, and 12–18 dm. in diameter, branchlets glaucous; bark on old trees reddish-brown, divided into large, irregular plates. Leaves in threes, 12–25 cm. long, persisting for 5–9 years, forming dense foliage, usually dull blue-green, especially on young trees; staminate flowers, 20–35 mm. long; cones subterminal, horizontal, short-stalked, broadly oval, 15–30 cm. long, becoming light russet-brown after opening, their scales about 20 cm. wide, thickened at apex and armed by a slender prickle, arising from a low slightly depressed umbo; seeds about 12 mm. long.

Coniferous forests in the Upper Arid Transition and Canadian Zones; Siskiyou Mountains, southern Oregon southward through the Sierra Nevada and the mountains of southern California to San Pedro Martir Mountain, Lower California, usually growing at higher altitudes than the Western Yellow Pine, associated with the Red Fir. Pure forests of typical trees readily distinguishable from *P. ponderosa* are often encountered in the Sierra Nevada, but in some localities the two species seem to intergrade. The two are best distinguished by the size of the cones, the color of the leaves and by the branchlets being glaucous or not glaucous. The odor of the two is also different, and many woodsmen can readily distinguish them by that alone. Wood similar to the preceding species. Type locality: Shasta Valley, California.

12. **Pinus radiàta** Don.
Monterey Pine. Fig. 123.

Pinus radiata D. Don, Trans. Linn. Soc. **17**: 441. 1836.
Pinus tuberculata D. Don, Trans. Linn. Soc. **17**: 411. 1836.
Pinus insignis Loud. Arb. Brit. **4**: 2265. *f. 2170–2172.* 1838.
Pinus californica Hook. & Arn. Bot. Beechey 393. 1841.

A tree 20–30 m. high, with a trunk 3–10 dm. in diameter, and a symmetrical pyramidal crown or, in age, becoming flat-topped; bark blackish brown, divided into narrow ridges. Leaves in threes or rarely twos, bright, glossy green, 8–15 cm. long, slender, persisting for 3 years; staminate flowers, 12 mm. long, yellow; cones remaining closed and persistent on the trees for many years, short-stalked, deflexed, 7–14 cm. long, light brown and shining, ovoid, bluntly pointed, strikingly unsymmetrical, the lower scales on the outside much thickened into smooth mammillate knobs at apex and armed by slender prickles that usually weather off; seeds black, roughened, 6–7 mm. long; wings brown with longitudinal stripes, 12–18 mm. long.

A maritime species of central California, occurring between Pescadero and Santa Cruz, on the Monterey Peninsula and at San Simeon Bay. The largest area is at Monterey, where it forms an extensive forest, extending inland 6 or 7 miles and southward along the coast about 20 miles. Wood light, soft, weak, and of little commercial value. Type locality: Monterey, California.

Pinus radiata binàta (S. Wats.) Lemmon. West-Am. Cone-Bearer. 42. 1895. This is an insular variety occurring on Santa Rosa Island off the coast of southern California, and on Guadaloupe Island off the coast of Lower California. It is distinguished by the stouter leaves, in clusters of two, and by the deflexed scale tips in the ovulate flowers.

13. **Pinus attenuàta** Lemmon. Knob-cone Pine. Fig. 124.

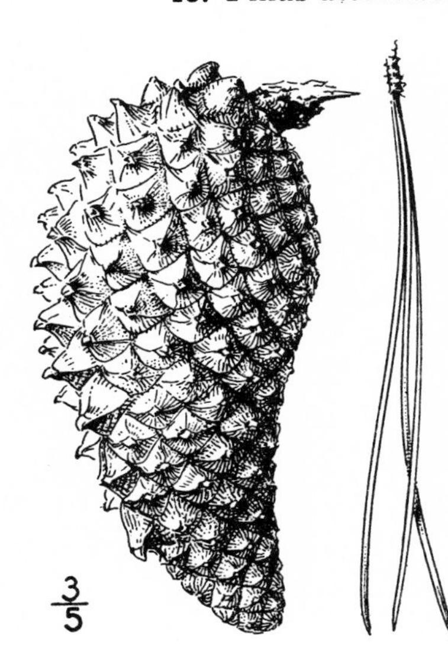

Pinus tuberculata Gord. Journ. Hort. Soc. Lond. **4**: 218. 1849, not Don. 1836.
Pinus californica Hartw. Journ. Hort. Soc. Lond. **2**: 189. 1847, not Hook. & Arn. 1841.
Pinus attenuata Lemmon, Gard. & Forest **5**: 65. 1892.

A small tree, usually 8–15 m. high or occasionally up to 30 m. high, with a trunk 30–60 cm. in diameter, in age forming a straggling crown with sparse foliage, bark on old trunks 6–12 mm. thick, shallowly divided into low ridges and broken into loose scales. Leaves in threes, slender, stiff, pale yellowish green, marked on all sides by stomata, 8–17 cm. long, persisting for 4 or 5 years; staminate flowers, 12 mm. long; cones short-stalked, often clustered in whorls, remaining closed and persistent on the branches for many years, light brown, narrowly ovoid, unsymmetrical, the scales on the outer side enlarged into prominent tranversely flattened knobs, armed with thick flattened incurved prickles; seeds nearly oval, 6 mm. long; wings narrow, 35 mm. long.

Dry, gravelly mountain slopes at low altitudes, in the Upper Sonoran and Transition Zones; extending from the Mackenzie River, Oregon, southward through the Coast Ranges and the Sierra Nevada to the San Bernardino Mountains, southern California. The persistence of the cones has long attracted attention to this tree. They remain unopened for 30 years or more and often may be seen scattered along the trunks half buried in the bark. They seem never to open until the tree or branch dies. The species is therefore well adapted to fire conditions. Cones that have been accumulating for years open and discharge their seeds after a fire, so that a new forest is assured. Type locality: Santa Cruz Mountains, California.

14. **Pinus sabiniàna** Dougl. Digger Pine. Fig. 125.

Pinus sabiniana Dougl. Trans. Linn. Soc. **16**: 747. 1833.

A tree 12–18 m., or occasionally twice as high, with a trunk 10–12 dm. in diameter, divided 5–6 m. above the ground into several large secondary trunks and forming an open crown with scattering branches and sparse foliage; bark 20–35 mm. thick, divided into low, broad ridges, covered by large brown scales. Leaves drooping, 20–30 cm. long, stout, gray-green, marked on all sides by numerous bands of stomata, persisting for 3 or 4 years; staminate flowers cylindric, 3–4 cm. long, cones long-stalked, deflexed, broadly ovoid or sub-globose when open, 15–25 cm. long, chocolate brown, maturing the seeds the second year and slowly shedding them for several months, remaining on the trees for several years; cone-scales ending in stout, flattened, spur-like projections 25 mm. long, those above the middle incurved, the lower recurved; seeds rounded below, slightly compressed toward the apex, 2 cm. long, with a thick, hard, blackish brown shell; wing encircling the seed, about 8 mm. high at apex.

A characteristic tree of the Upper Sonoran Zone in the foothills surrounding the great valley of central California, extending from the head of Sacramento Valley to the Sierra Liebre, Los Angeles County. Wood hard, coarse-grained, of little commercial value. The seeds are an important article of food of the California Indians, hence its name. Type locality: probably near San Juan, California.

15. **Pinus coùlteri** D. Don. Coulter or Big Cone Pine. Fig. 126.

Pinus coulteri D. Don, Trans. Linn. Soc. **17**: 440. 1837.
Pinus macrocarpa Lindl. Bot. Reg. **26**: misc. 61. 1840.
Pinus coulteri diabloensis Lemmon, Bull. Sierra Club **4**: 130. 1902.

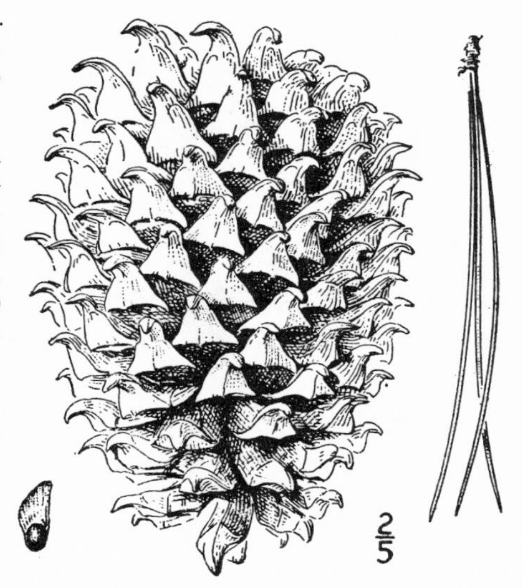

Tree 15–25 m. high with a trunk 4–12 dm. in diameter and a broadly pyramidal or unsymmetrical crown, branchlets stout, densely clothed with large bunches of leaves, very rough; bark blackish brown, deeply divided into wide ridges roughly clothed with scales, 35–50 mm. thick. Leaves not drooping, 15–30 cm. long, stout, deep blue-green, marked by many bands of stomata on all sides, persisting for 3 or 4 years; staminate flowers, cylindric, 25 mm. long; cones long-ovate, 25–30 cm. long, about half as broad when open, buff color, shedding the seeds during the fall and winter of the second year, persisting on the trees for 5 or 6 years; cone-scales ending in stout, incurved, flattened, spur-like projections, often over 25 mm. long; seeds blackish brown, 12–18 mm. long; wings red-brown, 25–30 mm. long.

Inner coast ranges of California from Mt. Diablo southward through the mountains of southern California to San Pedro Martir Mountain, Lower California. Wood soft, coarse-grained, suitable for second-class lumber. Type locality: Santa Lucia Mountains, California.

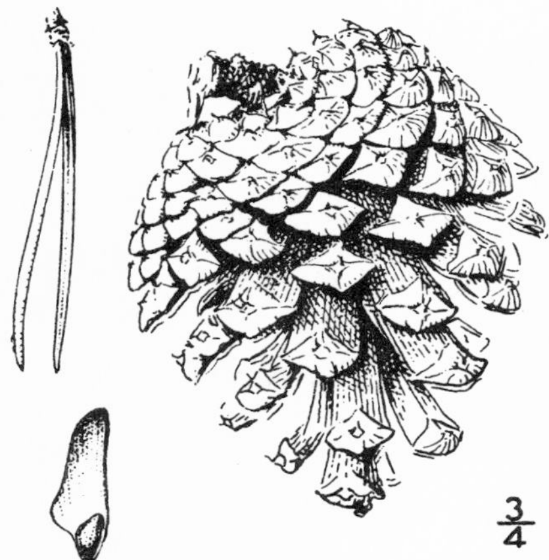

16. **Pinus contórta** Loud. Lodge-pole or Tamarack Pine. Fig. 127.

Pinus contorta Loud. Arb. Brit. **4**: *2292. f. 2210, 2211.* 1838.
Pinus boursieri Carr. Rev. Hort. **1854**: 225. 1854.
Pinus bolanderi Parl. DC. Prod. **16²**: 379. 1869.
Pinus contorta bolanderi Vasey, Rep. U. S. Dept. Agric. **18**: 177. 1875.
Pinus contorta hendersoni Lemmon, West-Am. Cone-Bearers 30. 1895.
Pinus tenuis Lemmon, Erythea **6**: 77. 1898.

A low, scrubby, maritime tree, seldom over 10 m. high, with a short trunk rarely over 4–5 dm. in diameter, and thick branches forming a round-topped, often picturesque crown; bark dark, about 25 mm. thick, becoming rough and ridged, clothed with closely appressed scales; wood light, hard, light red-brown, coarse-grained. Leaves in twos, usually slender, dark green, 3–5 cm. long, densely clothing the branches and persisting for 2 or 3 or rarely more years; staminate flowers, 8 mm. long, orange red; cones ovoid and somewhat oblique, 3–5 cm. long, chestnut brown and shining, opening at maturity and soon falling or persisting for many years; cone-scales thin, concave, armed with long slender prickles; seeds 4 mm. long; wings 8–12 mm. long.

A maritime species extending from Alaska southward along the coast to Albion River, Mendocino County, California. Throughout the greater part of the range this maritime form remains typical, but in western Oregon, where their ranges converge, intergradation between it and the variety seems complete. Type locality: in northwest America, in swampy ground near the sea coast, and abundantly near Cape Disappointment and Cape Lookout.

Pinus contorta murrayàna (Balf.) Engelm. in S. Wats. Bot. Calif. **2**: 126. 1880. (*Pinus murrayana* Balf. Rep. Bot. Oreg. Exped. 2. *pl. 3, 62.* 1853.) Differs from the coastal form mainly in the exceedingly thin bark, which is rarely more than 5–6 mm. thick, neither ridge nor furrowed, but covered by loosely appressed scales. A straight-trunked tree, where not wind-swept, usually 20–25 m. or occasionally 50 m. high, with a trunk 6–10 dm. or rarely 15–18 dm. in diameter; branches slender, forming a long narrow crown.

Usually on the borders of mountain lakes and meadows, in the Canadian Zone; Valley of the Yukon River and the interior plateau region of British Columbia eastward to the eastern slopes of the Rocky Mountains, southward to southern Colorado and along the high mountains of the Pacific Coast to San Pedro Martir Mountain, Lower California. Forms great forests in the Columbia Basin, Montana, and northern Wyoming. In the Sierra Nevada a subalpine species, and especially well developed on the borders of mountain meadows. Type locality: on the Siskiyou Mountains.

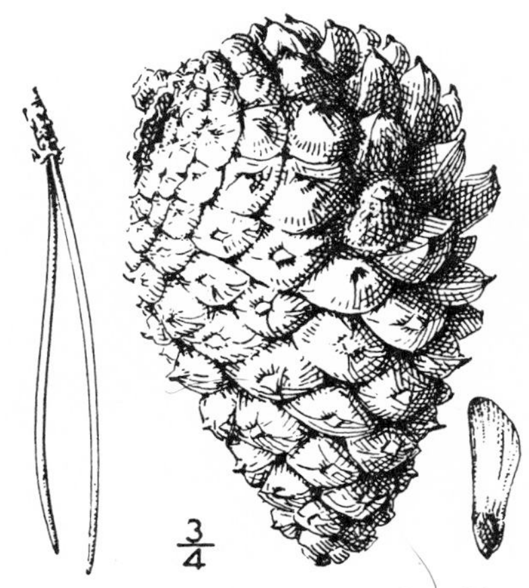

17. Pinus muricàta D. Don.

Prickle-cone or Bishop's Pine. Fig. 128.

Pinus muricata D. Don, Trans. Linn. Soc. 17: 441. 1837.
Pinus edgariana Hartw. Journ. Hort. Soc. Lond. 3: 217. 1848.

A small tree 12–25 m. high with a trunk 3–10 dm. in diameter and a round-topped, compact crown; bark of old trunks 10–15 cm. thick, deeply divided into narrow rounded ridges. Leaves in twos, dark yellow-green, stout, stiff, in dense tufts at the ends of the branches, persisting 2 or 3 years; staminate flowers, 8 mm. long; cones lateral, usually clustered in whorls, sessile, deflexed, 5–7 cm. long, light chestnut brown, unsymmetrical by the much enlarged outer scales which are prolonged into prominent knobs armed with a stout elongated spine, remaining closed for many years, and rarely if ever falling from live trees; seeds 6 mm. long, subtriangular with a thin roughened black shell; wings narrow, nearly 25 mm. long.

A maritime species of the Transition and Upper Sonoran Zones; most abundant and best developed along the coast from Inglenook, Mendocino County, to Point Reyes, California, forming an almost continuous belt. Southward, occurring in small, widely separated groves, Point Pinos Ridge near Monterey, Pecho Mountains, San Luis Obispo County, and Santa Cruz Island; reappearing again in a modified form (var. *anthonyi* Lemmon) between Ensenada and San Quentin, Lower California, and on Cedros Island. Wood light and soft, occasionally cut for lumber. Type locality: near San Luis Obispo, California, "at an elevation of 3000 feet."

2. LÀRIX [Tourn.] Adans. Fam. Pl. 2: 480. 1763.

Tall, pyramidal trees, with slender, remote, horizontal or pendulous branches, thick furrowed or scaly bark and heavy wood. Branchlets slender and often pendulous, covered with short, stout, spur-like lateral twigs. Buds small, nearly globose, covered with broad scales. Leaves deciduous, awl-shaped, triangular or sometimes 4-sided, spirally arranged and remote on leading shoots, densely crowded on the lateral spur-like twigs. Flowers solitary and terminal, the staminate appearing before the leaves, pale yellow, globose to oblong, sessile or stalked, the pistillate appearing with the leaves, their scales orbicular, subtended by longer mucronate bracts. Fruit a woody cone with slightly thickened concave scales, shorter or longer than the bracts. Seeds nearly triangular, rounded on the sides, crustaceous and light brown, shorter than the wings; cotyledons 6. [Name ancient, probably Celtic.]

Ten species confined to the subarctic and cooler parts of the north temperate zone. Besides the following there are two other species in North America, *Larix alaskensis* Wight in Alaska, and *Larix laricina* (Du Roi) Koch, extending from the northern Atlantic coast to the Northwest Territory. Type species, *Larix larix* (L.) Karst.

Leaves triangular; young twigs pubescent, becoming smooth. 1. *L. occidentalis.*
Leaves 4-sided; twigs tomentose. 2. *L. lyallii.*

1. Larix occidentàlis Nutt. Western Larch or Tamarack. Fig. 129.

Larix occidentalis Nutt. N. Am. Sylva 3: 143. *pl. 120.* 1849.
Pinus nuttallii Parl. DC. Prod. 16^2: 412. 1868.

Attaining a maximum height of 80 m. and a diameter of 2–2.5 m., with a short, narrow, open crown and short branches with a sparse foliage, or sometimes with a large crown and long, drooping branches; branchlets pubescent, usually soon glabrous; bark dark colored, divided into large irregularly scaly plates at the base of old trunks, 10–15 cm. thick. Leaves 25–50 mm. long, flatly triangular, distinctly keeled on the inner surface, 14–30 in a cluster; cones oblong, 25–35 mm. long, short-stalked; scales many, nearly entire, sometimes with reflexed margins when open, nearly orbicular, little broader than long; bracts much longer than the scales, abruptly contracted into the mucronate tip; seeds 6 mm. long, pale brown.

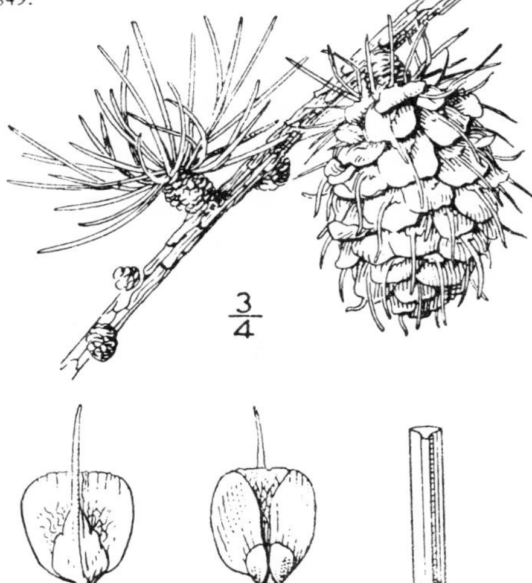

Mountain slopes, Canadian Zone; British Columbia and Montana, southward to the Blue Mountains and the Cascade Mountains, mainly on the eastern slope as far south as Squaw Creek, Oregon. Wood heavy, hard and strong, reddish with nearly white sapwood, very durable in soil and largely used for ties, poles and posts. Type locality: "in the coves of the Rocky Mountains on the western slope toward the Oregon."

2. **Larix lyàllii** Parl. Alpine Larch. Fig. 130.

Larix lyallii Parl. Enum. Sem. Hort. Reg. Mur. Flor. 259. 1863.

Pinus lyallii Parl. in DC. Prod. 16²: 412. 1868.

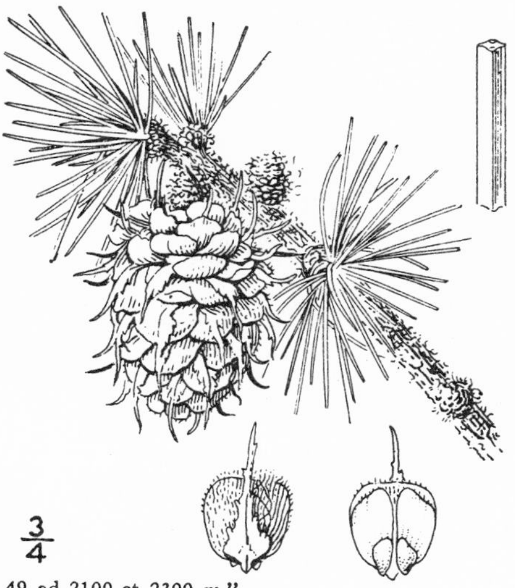

A small tree 15–25 m. high and 5–6 dm. or rarely 12 dm. in diameter, with long, irregular branches, forming an unsymmetrical crown; young branchlets densely covered with a fine white tomentum until the second year; bark rarely over 2 cm. thick, indistinctly furrowed, covered with loose, reddish brown or purplish scales. Leaves 4-angled, short-pointed, stiff, 25–35 mm. long, pale blue-green; cones ovate, acutish, nearly sessile or on slender stalks; scales fringed and covered with matted hairs on the lower surface; bracts much longer than the scales, long-pointed; seeds 3 mm. long.

An alpine tree often growing at timber line, Hudsonian Zone; ranging from the Continental Divide, in western Alberta, to the Cascade Mountains and southward to northern Montana; Nez Perces Pass, Idaho, and Mt. Hood, Oregon. Wood heavy, hard and strong, reddish-brown. Trees 5 dm. in diameter have been found to be 450–500 years old. Type locality: "Cascade Mountains et Galton Ranges Rocky Mountains latitudinis 49 ad 2100 et 2300 m."

3/4

3. **PÌCEA** Link, Abh. Akad. Berlin, 1827: 179. 1827–30.

Tall forest trees with tapering trunks, pyramidal crowns, slender, whorled branches, thin scaly bark and light-colored wood, containing numerous resin ducts. Leaves spirally arranged, extending from all sides of the branchlets or sometimes appearing 2-ranked by the twisting of those on the lower side, persisting for 7–10 years, linear and entire, jointed at the base to a prominent persistent woody base, 4-sided and stomatiferous on all sides, or flattenned and stomatiferous only on the upper side or only on the lower side. Flowers terminal or in the axils of the leaves, the staminate composed of numerous spirally arranged anthers opening lengthwise, surrounded at base by the bud-scales, the ovulate ovate to cylindrical, with numerous round or pointed scales in the axils of accrescent bracts. Cones maturing the first autumn, pendent, ovoid to oblong-cylindric, usually scattered over the upper half of the tree; scales thin, much longer than the bracts, persistent on the axis of the cone. Seeds without resin ducts, much shorter than the wings; cotyledons 4–15. [Name ancient.]

About 18 species of the north temperate zone. Besides the following, four other species occur in North America. Type species, *Picea abies* (L.) Karst.

Leaves 4-sided, with stomata on all 4 sides. — 1. *P. engelmanni.*
Leaves flattened, usually with stomata only on the upper side.
  Branchlets pubescent; cone-scales entire. — 2. *P. breweriana.*
  Branchlets glabrous; cone-scales finely dentate above the middle. — 3. *P. sitchensis.*

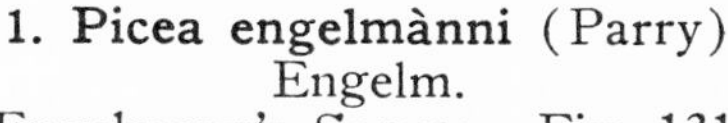

1. **Picea engelmànni** (Parry) Engelm.

Engelmann's Spruce. Fig. 131.

Picea engelmanni Parry; Engelm. Trans. St. Louis Acad. 2: 212. 1863.

Pinus commutata Parl. in DC. Prod. 16²: 417. 1868.

Picea columbiana Lemmon, Gard. & Forest 10: 183. 1897.

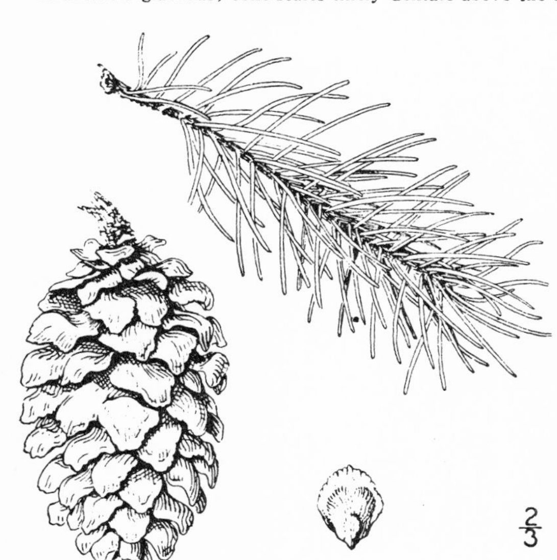

2/3

A forest tree attaining a maximum height of 50 m. with a trunk 1–1.5 m. in diameter, branches spreading in regular whorls, forming a narrow, compact, pyramidal crown, branchlets slender, pubescent for 3–4 years; bark broken into large, loose scales, brownish gray. Leaves ill-scented, 25–30 mm. long, soft and flexible, with acute, callous tips, slender and straight or incurved, with 3–5 rows of stomata on each side, glaucous when young, becoming dark purple; cones sessile or short-stalked, about 5 cm. long, oblong-cylindric, light chestnut-brown and shining, deciduous in the winter after the seeds have fallen; scales thin and flexible, broadest near the middle, gradually narrowed above into the truncate or acute apex, wedge-shaped below, erose-dentate to nearly entire; seeds nearly black, 3 mm. long. Wood weak, light, soft, and close-grained, yellowish white; used for general carpentry.

Mountain slopes of the Canadian Zone; British Columbia southward through the Rocky Mountains to Arizona, and on the eastern slopes of the Cascade Mountains to Oregon. Type locality: Rocky Mountains.

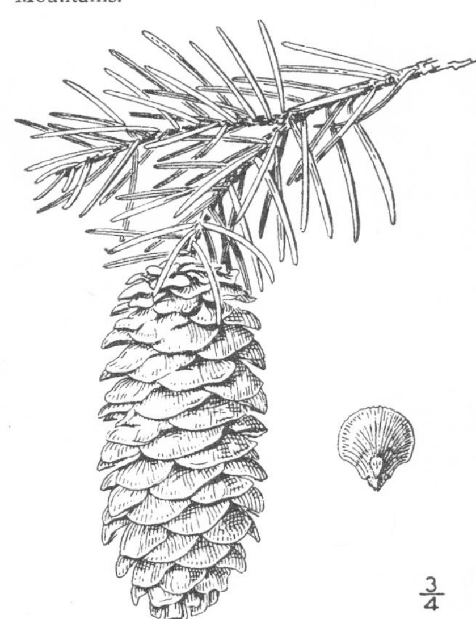

2. Picea breweriàna S. Wats. Brewer's or Weeping Spruce. Fig. 132.

Picea breweriana S. Wats. Proc. Am. Acad. **20**: 378. 1885.

A graceful tree, attaining a maximum height of 35 m. with a trunk 1 m. in diameter, the lower branches horizontal and bearing long, slender, pendulous branchlets, often 2–2.5 m. long and rarely 1 cm. thick; branchlets finely pubescent until the third year; bark 15–20 mm. thick, reddish brown, broken into long, thin, appressed scales. Leaves spreading from all sides of the branchlets, 20–25 mm. long, obtuse, rounded, dark green and shining on the lower surface, flattened and marked by prominent bands of stomata on each side of the midrib on the upper surface; staminate flowers dark purple; cones on slender stalks, oblong, 7–10 cm. long; scales broadly ovate, thin and flat, rounded at the apex with smooth entire margin; bracts oblong, acute, about one-fourth as long as the scales; seeds 3 mm. long, dark brown.

The rarest American spruce, occurring only in the forests of southwestern Oregon and adjacent California, Canadian Zone. Most abundant on the Siskiyou Mountains, but also found on the Oregon Coast Ranges at the head of Illinois River, and in the Klamath and Trinity Mountains, California. Type locality: summit of the Siskiyou Mountains on Happy Camp Trail.

3. Picea sitchénsis (Bong.) Carr. Sitka Spruce. Fig. 133.

Pinus sitchensis Bong. Veg. Sitcha 46. 1832.
Abies trigana Raf. Atlant. Journ. 119. 1832.
Abies falcata Raf. Atlant. Journ. 120. 1832.
Abies menziesii Lindl. in Penny Cyclop. 1: 32. 1833.
Picea sitchensis Carr. Traité Conif. 260. 1855.

A forest tree, often 35 m. in height, with a trunk 10–14 dm. in diameter, or attaining a maximum of 70 m. in height with a trunk 5 m. in diameter; branches horizontal, forming an open pyramidal crown; branchlets glabrous; bark about 15 mm. thick, divided into thin, loose, dark red-brown scales. Leaves 15–25 mm. long, spreading from all sides of the branchlets, acute or acuminate at apex, with long callous tips, rounded and green on the lower surface, flattened and whitened above by two broad bands of stomata; staminate flowers dark red; cones narrowly oblong-oval 6–10 cm. long, yellow to red-brown, their scales oblong, rounded at the apex, thin and stiff, denticulate above the middle; bracts half as long as the scales, lanceolate and finely toothed; seeds pale reddish brown, full and rounded, about 3 mm. long, with narrow oblong wings, 8–12 mm. long.

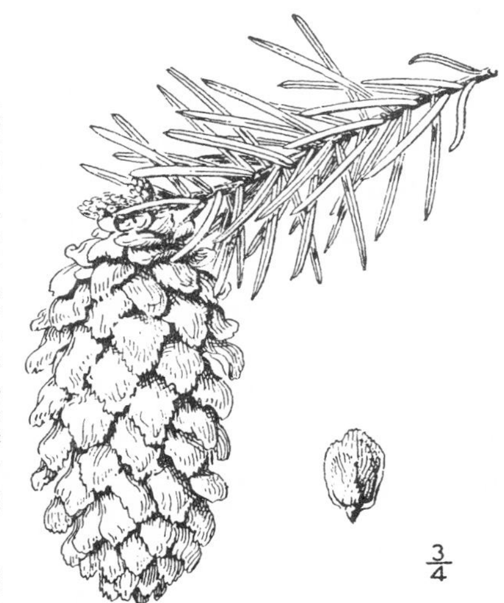

Moist, often swampy, situations, especially in sandy soil, Humid Transition and Canadian Zones; extending along the coast from Kodiak Island, Alaska, to Mendocino County, California. Wood light brown with thick whitish sapwood, light, soft, and weak; used for aeroplanes and shipbuilding. Type locality: Sitka, Alaska.

4. TSÙGA Carr. Trait. Conif. 185. 1855.

Forest trees with a pyramidal crown, nodding leading shoot and slender horizontal branches. Leaves flat or angled, usually appearing 2-ranked by a twist of the petiole, abruptly narrowed to short petioles, on persistent and eventually woody bases, emarginate to acute at the apex, stomatiferous on the lower surface or, in one species, leaves spreading from all sides and stomatiferous on both surfaces. Flowers solitary, the staminate axillary on twigs of the previous year, globose, bearing many small subglobose anthers; the ovulate terminal, erect, their scales rounded, shorter or longer than the bracts. Cones usually pendulous, oval to nearly cylindric, their scales rounded to ovate, thin, entire, much longer than the minute bracts, persistent. Seeds nearly surrounded by their wings, compressed, with resin ducts; cotyledons 3–6. [Name Japanese.]

Eight species confined to the cooler parts of temperate North America and Asia; four are North American. Type species, *Tsuga sieboldi* Carr.

Leaves flat, grooved on the upper surface, stomatiferous on the lower surface, rounded at apex. 1. *T. heterophylla.*
Leaves convex or keeled on the upper surface, stomatiferous on both surfaces, abruptly pointed at apex. 2. *T. mertensiana.*

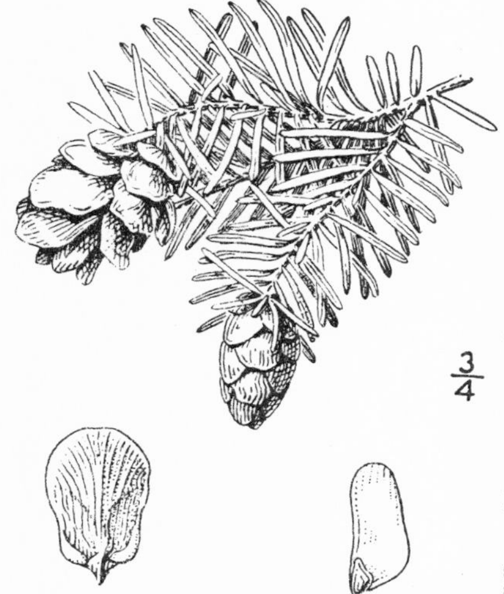

1. Tsuga heterophýlla (Raf.) Sarg.
Western Hemlock. Fig. 134.

Abies heterophylla Raf. Atlant. Journ **1**: 119. 1832.
Abies bridgesii Kell. Proc. Calif. Acad. **2**: 8. 1863.
Abies albertiana Murr. Proc. Hort. Soc. Edinb. **3**: 149. 1863.
Tsuga heterophylla Sarg. Sylva N. Am. **12**: 73. 1898.

A forest tree often 60–70 m. high, with a narrow pyramidal crown and a trunk 2–3 m. in diameter; branchlets pale yellow-brown, pubescent; bark on old trees 25–40 mm. thick, deeply divided into broad, flat ridges. Leaves 2-ranked, flat, 6–20 mm. long, rounded at the apex, distinctly grooved, dark green and shining on the upper surface, marked on the lower surface by two broad bands of 7–9 rows of stomata; staminate flowers yellow; cones sessile, oblong-oval, 20–25 mm. long; scales puberulent, longer than broad; bracts one-fourth the length of the scales, abruptly rounded at the apex and abruptly contracted at the base; seeds 3 mm. long, their wings two or three times as long.

Humid Transition and Canadian Zones; southern Alaska southward along the coast to Mendocino County, California, and eastward through southern British Columbia and Washington to the western slopes of the Bitterroot Mountains, Idaho, and to the western slope of the Cascade Mountains in Oregon. Wood pale yellow-brown, with thin, nearly white sapwood, light, hard, and tough; the strongest and most durable hemlock, and largely used for building purposes. The bark is the principal source for tanning in the Northwest. Type locality: mouth of the Columbia River, Oregon.

2. Tsuga mertensiána (Bong.) Sarg.
Mountain Hemlock. Fig. 135.

Pinus mertensiana Bong. Veg. Sitcha 54. 1832.
Abies pattoniana Murr. Rep. Bot. Exped. Oreg. 1. *pl. 4, f. 2.* 1853.
Abies hookeriana Murr. Edinb. New Phil. Journ II. **1**: 289. *pl. 9, f. 11, 17.* 1855.
Abies williamsonii Newb. Pacif. R. Rep. **6**: 53. *pl. 7, f. 19.* 1857.
Tsuga roezlii Carr. Rev. Hort. 1870: 217. *f. 40.* 1870.
Tsuga mertensiana Sarg. Sylva N. Am. **12**: 77. 1898.
Hesperopeuce mertensiana Rydb. Bull. Torrey Club **39**: 100. 1912.

An alpine forest tree 25–50 m. high, with an open pyramidal crown and a trunk 9–12 dm. in diameter; branches slender, curved and pendent, the branchlets drooping with their tips often curved upward; bark on old trees 25–35 mm. thick, deeply divided into rounded ridges, covered with closely appressed red-brown scales. Leaves spreading from all sides of the branchlets, more or less curved, 15–20 mm. long, blunt-pointed, narrowed toward the base, light blue-green and stomatiferous on both sides, convex above, rounded below; staminate flowers bluish; cones sessile, oblong-cylindric, narrowed toward both ends, 25–75 mm. long; scales very thin, usually as broad as long, gradually narrowed from near the middle toward the wedge-shaped base; bracts one-fourth the length of the scales, sharp-pointed at the apex and wedge-shaped at the base; seeds 3 mm. long, with wings about four times as long.

An alpine species growing on the upper edge of the forests, Hudsonian Zone; southern Alaska southward through the Olympic and Cascade Mountains and the Sierra Nevada to Kings River, California, and eastward to the Selkirk Mountains, northern Idaho, and Montana. Type locality: Sitka.

5. PSEUDOTSÙGA Carr. Trait. Conif. ed. 2, 256. 1867.

Pyramidal trees with long whorled branches, usually drooping branchlets and rough furrowed bark. Leaves flat, appearing 2-ranked by a twist at the base, narrowed into a short petiole, grooved on the upper surface and dark green, paler beneath, with several rows of stomata on either side of the midrib, with 2 lateral resin ducts seen in a cross section of a leaf, falling away after from 6–8 years, leaving a rounded sessile leaf scar. Flowers monoecious, appearing in early spring from buds formed the previous year, surrounded by large membranous scales, the staminate axillary, appearing on the lower side of the branchlets, oblong-cylindrical, nearly or quite sessile, consisting of many short-stalked, rounded anthers opening obliquely and with a short reflexed terminal appendage; the pistillate terminal or subterminal, short-stalked, composed of many small rounded scales, each subtended by a much larger, usually deeply and sharply 3-toothed, bract. Cones maturing the first autumn, oblong-ovoid, with the terminal scales reduced and sterile, the fertile rounded and concave, spreading at maturity; bracts longer than the scales, linear with the midrib extending into a long stiff arm and with a sharp-pointed lobe on each side. Seeds oblong, triangular, shorter than the dark-colored membranous wings; cotyledons 6–12. [Greek and Japanese meaning false hemlock.]

Three species, the third species in Japan. Type species, *Pseudotsuga mucronata* (Raf.) Sudw.

Cones mostly 5–7 cm. long, their scales flexible. — 1. *P. taxifolia.*
Cones 10–15 cm. long, their scales spreading at right angles and very rigid. — 2. *P. macrocarpa.*

1. Pseudotsuga taxifòlia (Lamb.) Britt.

Douglas Fir. Fig. 136.

Pinus taxifolia Lamb. Pinus 1: 51. *pl. 33.* 1803, not Salisb. 1796.
Abies mucronata Raf. Atlant. Journ. 120. 1832.
Abies douglasii Lindl. in Penny Cyclop. 1: 32. 1833.
Pseudotsuga douglasii Carr. Trait. Conif. ed. 2, 256. 1867.
Pseudotsuga lindleyana Carr. Rev. Hort. 1868: 152. 1868.
Pseudotsuga taxifolia Britt. Trans. N. Y. Acad. 8: 74. 1889.
Pseudotsuga mucronata Sudw. Contr. U. S. Nat. Herb. 3: 266. 1895.

A forest tree 60–70 m. high, or sometimes higher, with a trunk 1–3.5 m. in diameter and a broad or narrow pyramidal crown, lower branches drooping, the middle and upper ascending, clothed with numerous drooping branchlets; bark of old trunks 10–50 cm. thick toward the base, deeply fissured, forming broad ridges. Leaves 20–30 mm. long, 2 mm. or less wide, obtuse at the apex, dark yellow-green or sometimes blue-green, usually persistent 8 years; staminate flowers yellow, tinged with red; cones pendent, 5–7 cm. or rarely 10 cm. long; scales rounded, slightly concave, puberulent, flexible; bracts about 6 mm. wide, well exceeding the scales; seeds 6 mm. long; wings a little longer.

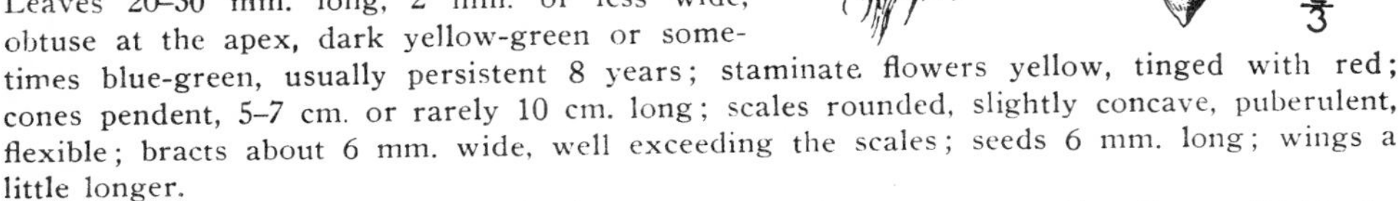

Widely distributed over western North America, Humid and Arid Transition Zones; British Columbia to Central California, New Mexico, Arizona and northern Mexico. Most abundant and growing to the largest size in western Washington and Oregon. The most important lumber tree of the Pacific States. Type locality: mouth of the Columbia River, Oregon.

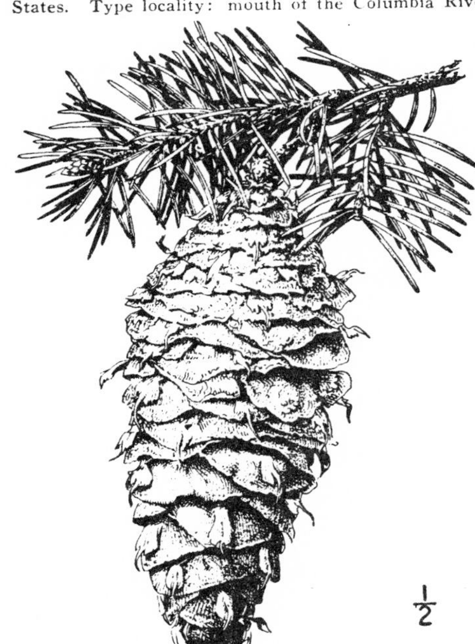

2. Pseudotsuga macrocàrpa (Torr.) Mayr.

Big Cone Douglas Fir. Fig. 137.

Abies douglasii macrocarpa Torr. in Ives Rep. 28. 1861.
Abies macrocarpa Vasey, Gard. Month. 18: 21. 1876.
Pseudotsuga douglasii macrocarpa Englm. in S. Wats. Bot. Calif. 2: 120. 1880.
Pseudotsuga macrocarpa Mayr, Wald. Nordam. 278. 1890.

Tree 15–25 m. high, with a trunk 1.5–2.5 m. in diameter, lower branches elongated and drooping, forming a broad pyramidal crown; bark 8–15 cm. thick, deeply divided into broad rounded ridges. Leaves blue-green, 20–30 mm. long, about 2 mm. wide, acute or tapering at the apex, persistent 4 or 5 years; staminate flowers pale yellow; cones short-stalked, 10–11 cm. long; scales rounded, 40–50 mm. broad, concave, thick and rigid, puberulent; bracts well exceeding the scales, about 8 mm. wide; seeds 12 mm. long, equaling the wings.

Mountains of southern California at 3000 to 6000 feet altitude, Upper Sonoran and Arid Transition Zones; Santa Inez Mountains and Fort Tejon, southward to the Cuyamaca Mountains, San Diego County. Type locality: "mountains near San Felipe," San Diego County, California.

6. ÀBIES [Tourn.] Hill, Brit. Herb. 509. 1756.

Pyramidal trees with light-colored brittle wood, very resinous, smooth or finely roughened bark and slender horizontal branches in regular whorls of 4–6, clothed with once- or twice-branched lateral branchlets forming flattened masses of foliage. Leaves linear, sessile, appearing 2-ranked by a twist at the base, or spreading from all sides of the branchlets, persistent usually 8–10 years, usually flat on sterile branches and young trees, on fruiting branches usually 4-sided and curved upward, stomatiferous on the lower surface or in some species on both sides. Flowers axillary, from buds formed the previous year, surrounded at base by the enlarged bud-scales, the staminate numerous on the lower side of the branchlets, oval or oblong-cylindric, their anthers yellow or scarlet, with knob-like terminal appendages; pistillate on the upper side of the branchlets, usually only near the top of the tree, erect, their scales numerous, rounded, much shorter than the bracts. Cones erect, ovoid to oblong-cylindric, maturing the first autumn; scales thin, incurved at the broad rounded apex, wedge-shaped below and usually narrowed into a short stalk, falling at maturity with the bracts and seeds, leaving the naked axis persistent on the branches. Seeds with large resin ducts, covered on the upper surface by the wing; cotyledons 4–10. [Ancient name of the fir.]

About 25 species confined to the subarctic and temperate regions of the northern hemisphere. Ten species occur in North America. Type species, *Abies picea* (L.) Lindl.

Bracts concealed by the scales or when exserted not furnished with long awl-like tips.
- Leaves dark green above with stomata only on the silvery white surface; bracts concealed by the scales.
  - Bracts gradually narrowed into a slender tip; cones oblong, purple; leaves erect on the branches. — 1. *A. amabilis.*
  - Bracts abruptly narrowed into a short point; cones cylindric, green; leaves 2-ranked forming flat sprays. — 2. *A. grandis.*
- Leaves blue-green with stomata on all sides.
  - Bracts scarcely half the length of the scales; leaves flat.
    - Resin ducts sunken in the leaf tissue (seen in cross section); bracts long-pointed. — 3. *A. lasiocarpa.*
    - Resin ducts next the epidermis; bracts abruptly short-pointed. — 4. *A. concolor.*
  - Bracts three-fourths as long or longer than the scales; leaves 4-sided, usually short and rigid.
    - Bracts shorter than the scales or when exserted much narrower than the scales. — 5. *A. magnifica.*
    - Bracts well exserted and covering the scales. — 6. *A. nobilis.*

Bracts with elongated awl-like tips, 2 or 3 times as long as the scales. — 7. *A. venusta.*

1. Abies amàbilis (Dougl.) Forbes. Amabilis or Lovely Fir. Fig. 138.

Picea amabilis Dougl.; Loud. Arb. Brit. **4**: 2342. 1838.
Abies amabilis Forbes, Pinetum Wob. 125. *pl. 144.* 1839.
Abies grandis densifolia Engelm. Trans. St. Louis Acad. 3: 599. 1878.

A large tree attaining a maximum height of 85 m. with a trunk 1.5–2 m. in diameter; branchlets stout, finely pubescent for several years; bark thin, smooth, and pale, or, on old trees, becoming 5–10 cm. thick near the ground and divided into small plates. Leaves erect on the branches by the curving of those on the lower side, flat, dark green, deeply grooved and shining on the upper surface, silvery white beneath on either side of the prominent midrib, 2–3 cm. long, 1.5–2 mm. wide, recurved on the margins, obtuse or notched at the apex on sterile branches, acute on the fertile branches; staminate flowers bright red, the pistillate narrowly cylindric, dark purple; cones oblong, purple, 9–15 cm. long; scale puberulent, 25–30 mm. wide and nearly as long, gradually narrowed from the rounded apex; bracts scarcely half as long as the scales, oblong-obovate, slightly toothed and tapering to a long tip; seeds 12 mm. long, their wings twice as long.

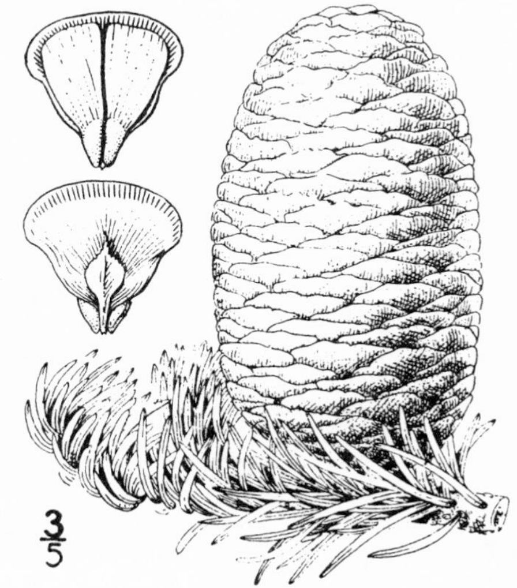

Coniferous forests, Canadian Zone; southern Alaska, southward along the Cascade Mountains and Coast Ranges of British Columbia, Washington and Oregon, reaching its southern limit at Crater Lake, and at Saddle Mountain, about 25 miles south of the Columbia River in the Coast Ranges. Wood close-grained, pale brown, light and weak but hard, and used for interior finishing. Type locality: high mountains, Oregon, near the Cascades of the Columbia.

2. **Abies grándis** Lindl. Grand Fir. Fig. 139.

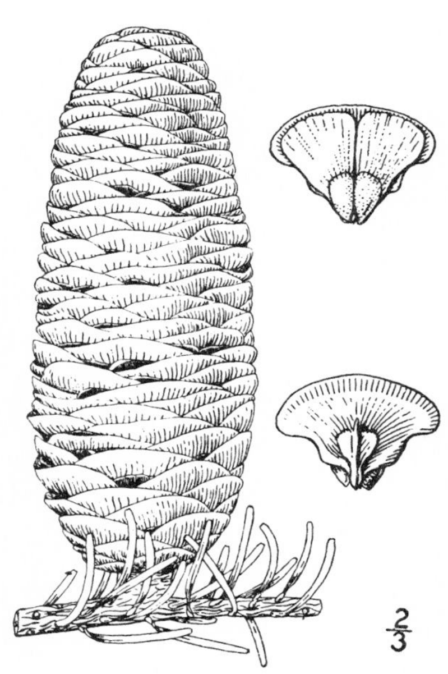

Abies grandis Lindl. in Penny Cyclop. **1**: 30. 1833.
Abies gordoniana Carr. Trait. Conif. ed. 2, 298. 1867.
Abies grandis oregona Beissner, Handb. Conif. 71. 1887.

A tree attaining a maximum height of 90 m. with a trunk 1–1.5 m. in diameter, and long and drooping branches, the branchlets puberulent, becoming glabrous the second year. Leaves on slender branches, 35–50 mm. long, 3 mm. wide, 2-ranked, forming flat sprays, flat, thin, and flexible, deeply grooved, dark green and shining on the upper surface, silvery white on the lower; staminate flowers light yellow, the pistillate greenish; cones long-oblong, 12–18 cm. long, 3–4 cm. thick; scales broadly fan-shaped, 30 mm. wide, about 25 mm. long, puberulent; bracts scarcely half as long as the scales, obcordate, irregularly toothed and short-pointed; seeds pale brown, 8 mm. long, their wings twice as long.

Coniferous forests, mainly in the Canadian Zone; Vancouver Island southward along the coast to Mendocino County, California, and eastward to western Montana, and the Blue Mountains, Oregon. Wood coarse-grained, light brown, light, soft, and not durable; used for interior finishing and for packing boxes. Often called Stinking Fir by lumbermen. Type locality: "low moist valleys in northern California."

3. **Abies lasiocàrpa** (Hook.) Nutt. Alpine Fir. Fig. 140.

Pinus lasiocarpa Hook. Fl. Bor. Am. **2**: 163. 1842.
Abies lasiocarpa Nutt. N. Am. Sylva **3**: 38. 1849.
Abies bifolia Murr. Proc. Hort. Soc. Edinb. **3**: 320. 1863.
Abies subalpina Engelm. Am. Nat. **10**: 555. 1876.
Abies subalpina fallax Engelm. Trans. St. Louis Acad. **3**: 597. 1878.

A tree usually 25–40 m. high or rarely twice as high, with a trunk 0.5–1.5 m. in diameter and short, crowded branches, the lower usually drooping; branchlets stout, rusty pubescent for 3 or 4 years, or rarely glabrous the second year. Leaves crowded, becoming erect by the curving of the lower, 5–10 cm. long, bluish green and shining on both surfaces, rounded or notched at the apex, grooved on the upper side, a cross section shows two resin ducts well within the soft leaf tissue; staminate flowers cylindric, bluish, the pistillate 25 mm. long, scarcely half as thick, dark purple; cones oblong-cylindric, 6–10 cm. long, dark purple; scales usually fan-shaped, about 25 mm. long and scarcely as broad; bracts one-third as long as the scales, oblong-obovate, red-brown, laciniately cut on the margins, abruptly contracted to a slender tip; seeds 6 mm. long; wings dark, nearly as long as the scales.

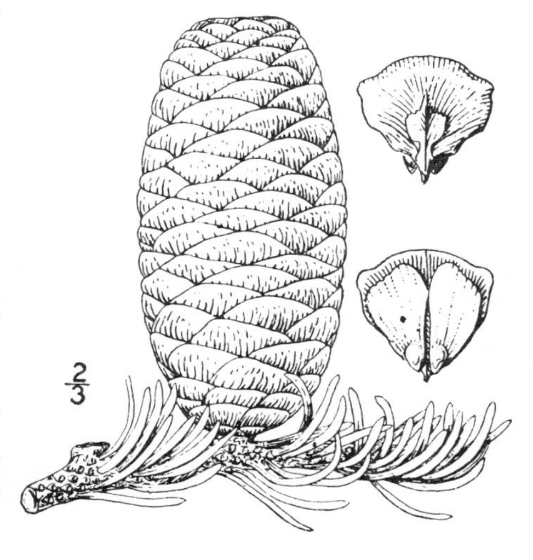

Alpine valleys and mountain slopes, Hudsonian Zone; southeastern Alaska, British Columbia, and western Alberta southward through Washington, Oregon, Idaho, western Montana, and Wyoming to southern Arizona and New Mexico. Wood pale brown, light, soft, and weak; of little commercial value. Type locality: "interior of N. W. America."

4. **Abies cóncolor** Lindl. White Fir. Fig. 141.

Abies concolor Lindl. Journ. Hort. Soc. Lond. **5**: 210. 1850.
Picea lowiana Gord. Pinetum Suppl. 53. 1862.
Abies lowiana Murr. Proc. Hort. Soc. Edinb. **3**: 317. 1863.
Abies grandis concolor Murr. Gard. Chron. II. **3**: 105. 1875.
Abies concolor lowiana Lemmon, West. Am. Cone-Bearers 64. 1895.

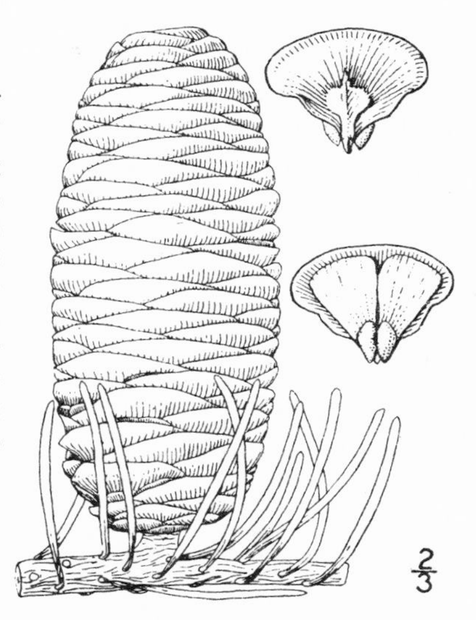

A tree attaining a height of 70–80 m. with a trunk 1.5–2 m. in diameter, and short stout branches forming a narrow spire-like crown, the branchlets glabrous and shining. Leaves crowded, erect on the branches, 5–7 cm. long, usually 3 mm. wide, pale blue-green on both surfaces, on the lower branches flat, straight, rounded to acuminate at apex, on the upper fertile branches curved and usually keeled on the upper surface; staminate flowers rose color or dark red; cones oblong, 7–12 cm. long, scales 30–35 mm. wide, three-fourths as long, fan-shaped, puberulent; bracts thin, scarcely half as long as the scales, emarginate or truncate, ending in a short, slender tip; seeds 8–12 mm. long, dark dull brown; wings twice as long as the seeds.

Common component of the coniferous forests, Arid Transition Zone; northern Oregon to the Rocky Mountains of southern Colorado southward to northern New Mexico and San Pedro Martir Mountain, Lower California. A common tree in the great forest belt of the Sierra Nevada, where it attains its greatest size. Wood pale brown to nearly white, coarse-grained, light, soft, and not durable; used for lumber, packing boxes, and paper pulp. Type locality: mountains of New Mexico.

5. **Abies magnífica** Murr. Red Fir. Fig. 142.

Abies magnifica Murr. Proc. Hort. Soc. Edinb. **3**: 318. 1863.
Abies nobilis magnifica Kell. Trees Calif. 35. 1882.

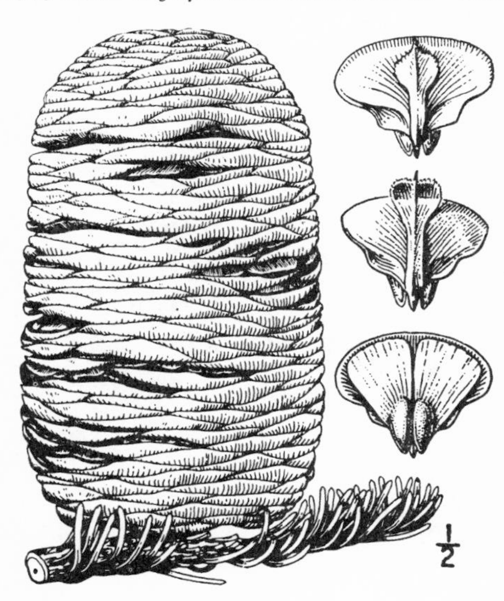

A tree often 60–70 m. high with a trunk 2–3 m. in diameter, clothed with comparatively short branches, the upper ascending, the lower drooping, the branchlets puberulent the first year, finally glabrous and silvery gray. Leaves ribbed on both sides and almost equally 4-sided with stomata on all sides, very glaucous the first year, becoming blue-green, those on the lower branches 2-ranked, round or blunt-pointed, 20–35 mm. long, those on the upper fertile branches curved upward and completely covering the upper side of the branches, acute, 2 cm. long, staminate flowers deep purplish color, the pistillate oblong, reddish brown; cones oblong-cylindric, 15–20 cm. long, scales stalked, 35 mm. wide and a little longer, gradually narrowing toward the heart-shaped base; bracts two-thirds as long as the scales, abruptly contracted to a slender tip, oblong or spatulate; seed deep brown, 2 cm. long with a shining, broadly wedge-shaped wing.

A characteristic tree of the Canadian Zone; Cascade Mountains of southern Oregon southward to the North Coast Ranges of California and through the Sierra Nevada. Wood light red-brown, light, soft and quite durable. Type locality: Sierra Nevada. First discovered by Fremont.

Abies magnifica shasténsis Lemmon. Rep. Calif. State Board Forestry **3**: 145. 1890. (*Abies shastensis* Lemmon, Gard. & Forest, **10**: 184. 1897.) The Shasta fir closely resembles the Red fir, but the bracts are well exserted and yellowish. The general range is that of the species but it is less common. The center bract in Fig. 142 is from the Shasta fir.

6. Abies nóbilis Lindl. Noble Fir. Fig. 143.

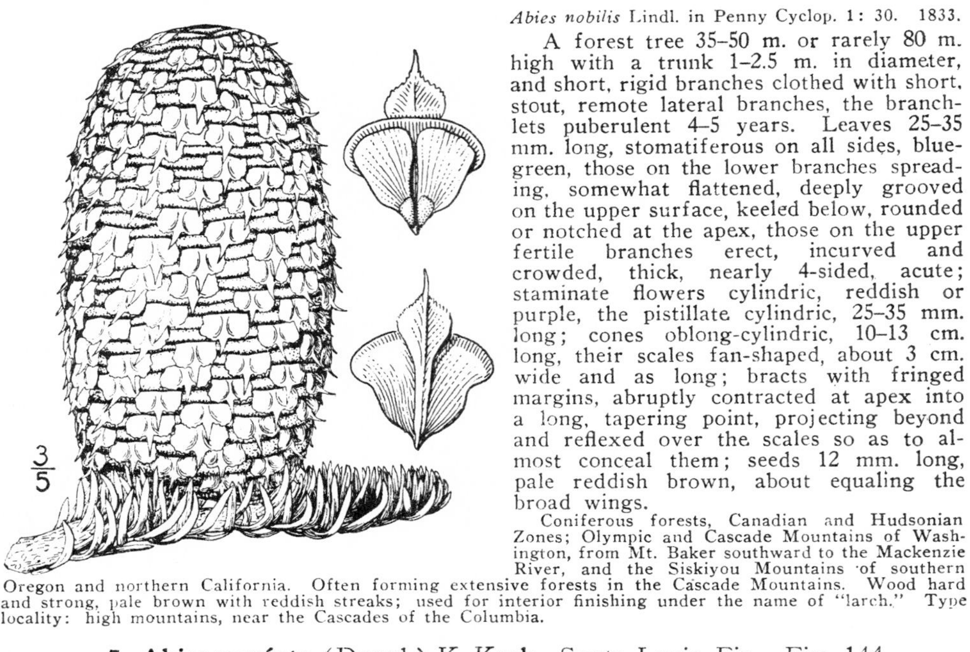

Abies nobilis Lindl. in Penny Cyclop. 1: 30. 1833.

A forest tree 35–50 m. or rarely 80 m. high with a trunk 1–2.5 m. in diameter, and short, rigid branches clothed with short, stout, remote lateral branches, the branchlets puberulent 4–5 years. Leaves 25–35 mm. long, stomatiferous on all sides, blue-green, those on the lower branches spreading, somewhat flattened, deeply grooved on the upper surface, keeled below, rounded or notched at the apex, those on the upper fertile branches erect, incurved and crowded, thick, nearly 4-sided, acute; staminate flowers cylindric, reddish or purple, the pistillate cylindric, 25–35 mm. long; cones oblong-cylindric, 10–13 cm. long, their scales fan-shaped, about 3 cm. wide and as long; bracts with fringed margins, abruptly contracted at apex into a long, tapering point, projecting beyond and reflexed over the scales so as to almost conceal them; seeds 12 mm. long, pale reddish brown, about equaling the broad wings.

Coniferous forests, Canadian and Hudsonian Zones; Olympic and Cascade Mountains of Washington, from Mt. Baker southward to the Mackenzie River, and the Siskiyou Mountains of southern Oregon and northern California. Often forming extensive forests in the Cascade Mountains. Wood hard and strong, pale brown with reddish streaks; used for interior finishing under the name of "larch." Type locality: high mountains, near the Cascades of the Columbia.

7. Abies venústa (Dougl.) K. Koch. Santa Lucia Fir. Fig. 144.

Pinus venusta Dougl. Comp. Bot. Mag. 2: 152. 1836.

Pinus bracteata D. Don. Trans. Linn. Soc. 17: 442. 1837.

Abies bracteata Nutt. N. Am. Sylva 3: 137. 1849.

Abies venusta K. Koch, Dendr. 2²: 210. 1873.

A tree, 30–50 m. high, with a thin, spire-like crown, the trunk sometimes 1 m. in diameter, the branchlets glabrous, glaucous when young; bark near the base 2–3 cm. thick, slightly fissured and broken into appressed scales. Leaves flat, rigid, acuminate and pungent, with a callous tip, dark green above, silvery marked below with broad bands of stomata; staminate flowers pale yellow; pistillate with obcordate yellow-green bracts; cones on stout peduncles, oval 7–10 cm. long; bracts much exserted, with slender, elongated, awn-like tips, often 2–3 cm. long; seeds 6 mm. long, red-brown, nearly as long as the wing.

A unique species, found only in the Santa Lucia Mountains, Monterey County, California. Upper Sonoran and Transition Zones.

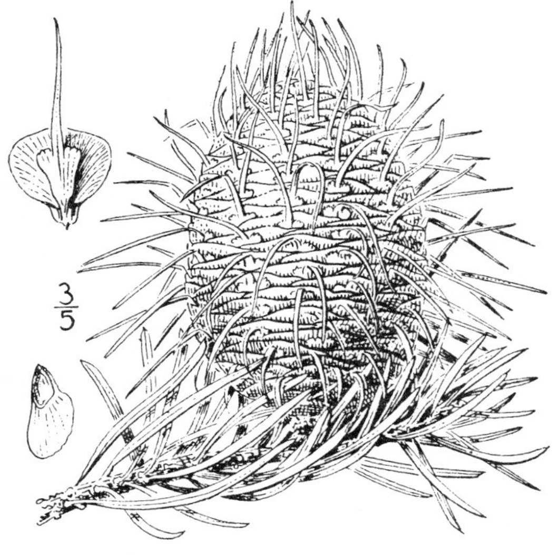

Family 3. TAXODIÀCEAE.

Taxodium Family.

Monoecious trees with linear, spirally arranged leaves decurrent and persistent on the regularly or ultimately deciduous branchlets. Staminate aments terminal on the branchlets or on short lateral branchlets, the stamens spirally arranged, each bearing several pollen-sacs attached to the lower half of the flattened, scale-like connective. Ovulate aments terminal with many spirally arranged ovuliferous scales, each bearing several ovules. Cone woody, the scales thickened toward the summit, usually widely spreading from the axis; seeds erect, small, and irregularly angled; cotyledons 2 or more.

A family of about 8 genera, widely distributed over both hemispheres. Besides the following, *Taxodium*, with 2 species, is found in the southeastern United States and Mexico.

1. SEQUOÌA Endl. Syn. Conif. 197. 1847.

Tall, often massive, forest trees with thick, fibrous, reddish-brown bark; soft, durable, red heartwood, thin whitish sapwood, and short, stout, horizontal branches clothed with slender lateral branchlets. Leaves decurrent at the base, spirally arranged, narrowly linear and spreading in 2 ranks by a twist at the base, or scale-like and somewhat appressed on all sides. Flowers monoecious, the staminate pale yellowish, terminal, ovoid or oblong, surrounded by the ovate bud-scales; anthers numerous, spirally arranged, with slender filaments and a thin ovate connective bearing on the lower half of the inner surface 2–5 pendulous nearly globose pollen-sacs; the ovulate terminal, green, ovoid to oblong, composed of numerous spirally arranged imbricated ovate sharp-pointed scales closely adnate to the shorter rounded ovule-bearing scales, these bearing several erect ovules in 2 rows. Fruit an ovoid or broadly oblong woody cone, maturing the first or second year; cone-scales narrowly wedge-shaped, flattened at the apex and transversely depressed through the middle. Seeds 5–7 to each scale, oblong-ovate, flattened, with a thin, brownish seedcoat produced into lateral wing-like margins; cotyledons 4–6. [Named in honor of Sequoyah, inventor of the Cherokee alphabet.]

Once composed of several species widely distributed over the northern hemisphere, but now reduced to two species and confined to California and southern Oregon. Type species, *Sequoia sempervirens* (Lamb.) Endl.

Leaves of the ordinary foliage linear, spreading in 2 ranks; buds scaly; cones maturing the first year. 1. *S. sempervirens.*
Leaves scale-like, appressed or slightly spreading; buds naked; cones maturing the second year. 2. *S. gigantea.*

1. Sequoia sempervirens (Lamb.) Endl.

Redwood. Fig. 145.

Taxodium sempervirens Lamb. Pinus **2**: 24. 1824.
Schubertia sempervirens Spach, Hist. Veg. **11**: 353. 1842.
Sequoia sempervirens Endl. Syn. Conif. 198. 1847.
Sequoia gigantea Endl. Syn. Conif. 198. 1847.
Gigantabies taxifolia [Nelson] Senelis, Pinaceae. 78. 1866.

A massive tree, usually from 50–80 m. high with a slightly tapering trunk 2.5–5 m. in diameter, strongly buttressed at base, attaining a maximum height of 104 m. and a maximum diameter of 6.7 m. Branches slender on young trees and clothed with slender branchlets spreading and forming flat sprays, long covered by the persistent decurrent leaf bases, on old trees stout and horizontal, forming a narrow, irregular crown. Bark 15–30 cm. thick, divided into broad rounded ridges. Leaves on the leading shoots short, scale-like and more or less appressed, spirally distributed on all sides; those on the ordinary branches 2-ranked, forming flat sprays, linear, sharp-pointed, dark green and shining above, glaucous beneath; staminate flowers ovoid, green, composed of about 12 anthers; the ovulate oblong, composed of 15–20 ovate-orbicular curved and pointed scales, each bearing about 6 ovules; cones maturing the first year, broadly oblong, 25–35 mm. long, red-brown; seeds oblong-lanceolate, light brown, 12 mm. long; the wings about as broad as the body of the seed.

Characteristic tree of the coastal fog belt, Humid Transition Zone; southwestern border of Oregon southward to the Santa Lucia Mountains, Monterey County, California, extending inland rarely more than 20–30 miles. The Redwood is the tallest known tree, 103.6 m. (340 ft.).

2. Sequoia gigántea (Lindl.) Decn.

Giant Sequoia or Big Tree. Fig. 146.

Wellingtonia gigantea Lindl. Gard. Chron. **1853**: 823. 1853, not *Sequoia gigantea* Endl. 1847.
Taxodium washingtonium Winsl. Calif. Farmer, **2**: 58. 1854.
Sequoia wellingtonia Seeman, Bonplandia **3**: 27. 1855.
Sequoia gigantea Decn. Rev. Hort. **1855**: 9. 1855.
Taxodium giganteum Kell. & Behr, Proc. Calif. Acad. **1**: 51. 1855.
Sequoia washingtoniana Sudw. U. S. Div. Forest Bull. **14**: 61. 1897.

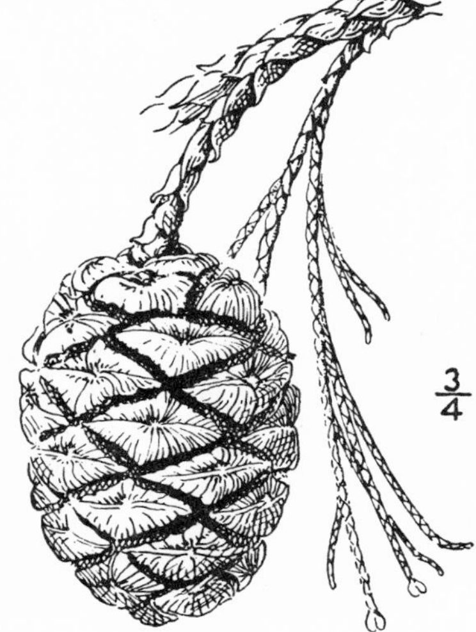

Massive trees, attaining the greatest size and age of any known species. The trunks are usually strongly buttressed, reaching a maximum height of 99 m. and a diameter of 10–12 m., often naked for 50 m. Branches forming a compact, cone-like head, branchlets horizontal with naked buds, becoming reddish brown after the leaves have fallen; bark divided into very broad rounded ridges covered with cinnamon-brown, fibrous scales, 3–6 dm. thick. Leaves decurrent, thickly clothing the branchlets and appressed except at the spreading, awl-like tips, these 3–12 mm. long, rounded on the back and concave on the inner surface, blue-green; staminate flowers borne in great profusion over the whole tree, terminal on the branchlets, oblong, 6 mm. long; the ovulate oblong, consisting, of 25–40 yellow-keeled scales, each bearing 3–7 ovules. Cones broadly oblong, 5–8 cm. long, reddish brown, their scales opening but little when the seeds are discharged, deeply grooved at the apex and often furnished with a reflexed bristle, seeds light brown, about 8 mm. long, linear-lanceolate, flattened and surrounded by wings a little broader than the body and notched at the apex.

Western slopes of the Sierra Nevada in the great forest belt, Arid Transition Zone; ranging from the Middle Fork of the American River to the head of Deer Creek, Tulare County. In the southern part of the range, in the basins of the Kings, Kern, and Tule Rivers, it forms extensive groves, but north of Kings River the groves are small and widely separated. There is much confusion as to the proper specific name, but it seems best to retain the well established name *gigantea*. Type locality: Calaveras Grove. Discovered by Mr. A. T. Dowd, a hunter of Murphy's Camp, in 1853, and made known to science by Wm. Lobb, who sent specimens to Lindley.

The following are the principal groves north of Kings River:

(1) *North Grove:* Near the southern boundary of Placer County on a tributary of the Middle Fork of the American River, about 10 miles east of Michigan Bluff; 6 trees; altitude 5100 ft.

(2) *Calaveras Grove:* In Calaveras County, on the divide at the head of Moran and San Antonio Creeks, at Big Trees Post Office; about 50 trees; altitude 4600 ft. This was the first discovered (1853).

(3) *Stanislaus Grove:* In Tuolumne County, 6 miles southeast of the Calaveras Grove; about 1380 trees; altitude 5000 ft.

(4) *Tuolumne Grove:* Near the south boundary of Tuolumne County in the Yosemite National Park on the Coulterville and Yosemite road, 1½ miles from Crane Flat; about 40 trees; also a few single trees between these and the Merced River.

(5) *Merced Grove:* A few miles southwest of the Tuolumne Grove on the headwaters of the Merced River on the Coulterville road; contains 33 trees.

(6) *Mariposa Grove:* In the Yosemite National Park 16 miles south of the Yosemite Valley. It comprises two groves, the upper with 365 trees and the lower with 182 trees. The largest tree is the Grizzly Giant, 64 feet 3 inches in circumference, 11 feet from the ground.

(7) *Fresno Grove:* About 10 miles southwest of the Mariposa Grove, between Big Creek and Fresno River; contains about 500 trees.

(8) *Dinky Grove:* Fourteen miles east of Pine Ridge, Fresno County, on Dinky Creek, a branch of Kings River; contains about 75 trees. This grove is also called McKinley Grove, and is the most southerly of the isolated groves.

Family 4. **CUPRESSÀCEAE.**

Cypress Family.

Monoecious or dioecious trees or shrubs, with opposite or whorled scale-like (or rarely linear) leaves, decurrent and thickly clothing the branchlets, or jointed at base (in *Juniperus*). Flowers small, terminal on the branchlets or (in *Juniperus*) axillary; stamens with several pollen-sacs attached to the lower half of the thin, shield-like connective; ovuliferous scales, several opposite or whorled, bearing at base 1 to many erect ovules. Cones woody or (in *Juniperus* and *Sabina*) fleshy, the scales shield-shape or imbricated; seeds often angled or winged; cotyledons 2 or more.

A family of about 10 genera widely distributed over both hemispheres.

Fruit a woody cone; leaves scale-like, decussate.
 Cones and their scales oblong, maturing the first year; seeds 2 to each scale.
  Cone-scales 6, the middle pair only fertile; leaves appearing in whorls of four. 1. *Libocedrus.*
  Cone-scales 8–12; leaves obviously in pairs. 2. *Thuya.*
 Cones subglobose, their scales shield- or wedge-shaped.
  Cones maturing the second year; seeds many to each scale. 3. *Cupressus.*
  Cones maturing the first year; seeds few to each scale. 4. *Chamaecyparis.*
Fruit berry-like, formed by the coalition of the scales; ovules 1 or 2.
 Leaves decurrent on the branchlets, usually scale-like; flowers terminal. 5. *Sabina.*
 Leaves jointed at base, not decurrent on the branchlets, awl-shaped; flowers axillary. 6. *Juniperus.*

1. **LIBOCÈDRUS** Endl. Syn. Conif. 42. 1847.

Aromatic evergreen forest trees, with naked buds, flattened branchlets, thin, fibrous bark, and fragrant, straight-grained, durable wood. Leaves scale-like, decurrent, 4-ranked, strongly compressed and keeled, dying and becoming woody before falling. Flowers monoecious, solitary, and terminal; the staminate oblong, of 12–16 subpeltate, broadly ovate-pointed scales, bearing usually 4 pollen-sacs; the ovulate oblong, of 6 acuminate scales, the 2 upper pairs much larger than the lower, only the middle pair ovuliferous. Fruit an oblong cone maturing the first year, the lowest pair of scales ovate, reflexed, the midde pair much larger, oblong, the upper pair united. Seeds 2 to each scale, erect, with 2 unequal wings, the larger nearly as long as the scale; cotyledons 2. [Name Greek, libas and Cedrus, referring to the resinous character.]

Eight species distributed along the Pacific Coast of both North and South America, also in New Zealand, China, and Formosa. Type species, *Libocedrus doniana* Endl.

1. Libocedrus decúrrens Torr.
Incense Cedar. Fig. 147.

Libocedrus decurrens Torr. Smiths. Contr. **6**: 7. 1854.
Thuya craigana Murr. Rep. Bot. Exped. Oreg 2. 1854.
Heyderia decurrens K. Koch, Dendr. **2**: 179. 1873.

Forest tree, often 50 m. high, with an irregularly lobed trunk tapering from a broad base, sometimes 3–3.5 m. in diameter, the upper branches erect, the lower curved downward, branchlets flattened and often vertically placed, 10–15 cm. long, usually deciduous the second or third year; bark 10–25 mm. thick, bright cinnamon-brown, fibrous. Leaves light green, closely adnate except at the tip, 3 mm. long or on leading shoots 10 mm. long, the lateral pair glandless, nearly covering the flattened obscurely pitted inner pair; staminate flowers yellow, ovate, 6 mm. long; the ovulate yellow-green, with ovate acute scales; cones pendulous, maturing the first autumn, oblong, 20–25 mm. long; scales with short, recurved mucro; seeds oblong-lanceolate, 8–10 mm. long, the narrow outer wing scarcely longer, the inner broader and nearly equaling the scale.

A characteristic tree of the Arid Transition Zone, extending from the Santiam River, Oregon, southward through the Cascade Mountains, Sierra Nevada, and Inner Coast Ranges to San Pedro Martir Mountain, Lower California. Wood light reddish brown, close-grained, light and durable in contact with the soil, but often infected with dry rot. It is of considerable commercial value for fencing, shingles, and lead pencils. Type locality: upper waters of the Sacramento River.

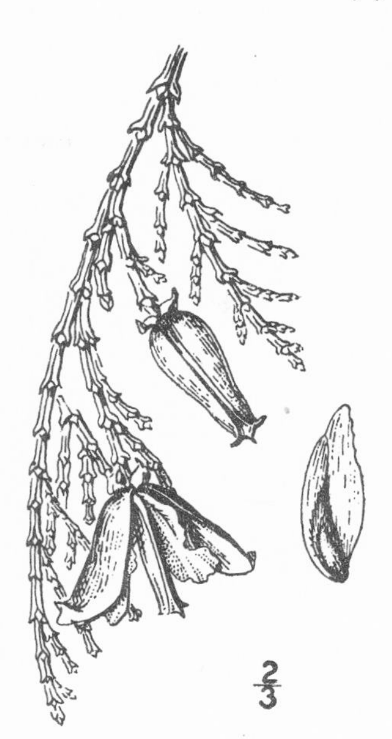

2. THÙYA L. Sp. Pl. 102. 1753.

Aromatic evergreen forest trees with slender, erect or spreading branches forming a pyramidal crown; branchlets pendulous, flattened, forming a flat horizontal frond-like spray; buds naked; bark thin, scaly; wood close-grained, soft, with thin, white sapwood. Leaves scale-like, decussate, the lateral pair compressed and prominently keeled, nearly concealing the other pair. Flowers monoecious, terminal and solitary; the staminate ovate, with 4–6 peltate scales, bearing 2–4 nearly globose pollen-sacs; the ovulate oblong, with 8–12 oblong acute scales and 2 erect ovules to each scale. Cones maturing the first autumn, ovoid-oblong, their scales thin and flexible, oblong, acute, and mucronate at the apex, the 2 or 3 middle pairs fertile. Seeds usually 2, erect, ovate, acute, compressed, light brown, with broad lateral wings distinct at the apex; cotyledons 2. [Name ancient.]

Four species, one in northwestern and one in northeastern America, the other two in China and Japan. Type species, *Thuya occidentalis* L.

1. Thuya plicàta D. Don.
Giant Cedar. Fig. 148.

Thuya plicata D. Don. Hort. Cantab. ed. 6, 249. 1811.
Thuya gigantea Nutt. Journ. Phila. Acad. 7: 52. 1854.
Thuya menziesii Carr. Trait. Conif. 106. 1855.

A handsome tree, attaining a maximum height of 60–70 m. and a diameter of 5 m., the trunk tapering and strongly buttressed at the base, the branchlets bright green and shining on the upper surface, paler on the lower, usually falling the third year; bark 1–2 cm. thick, bright cinnamon-brown, divided into broad, rounded ridges, separating into fibrous shreds. Leaves ovate, short-pointed, bright green and shining, 3 mm. long, obscurely glandular-pitted, those on the leading shoots twice as long, long-pointed, and conspicuously glandular; staminate flowers brownish, about 2 mm. long; cones clustered near the ends of the branches, strongly reflexed, 10–12 mm. long; scales leathery, thickened at the apex and often with a short, stout bristle; seeds 2, or sometimes 3, to each of the central scales, about 6 mm. long; wings divergent at the apex, usually a little longer than the seed.

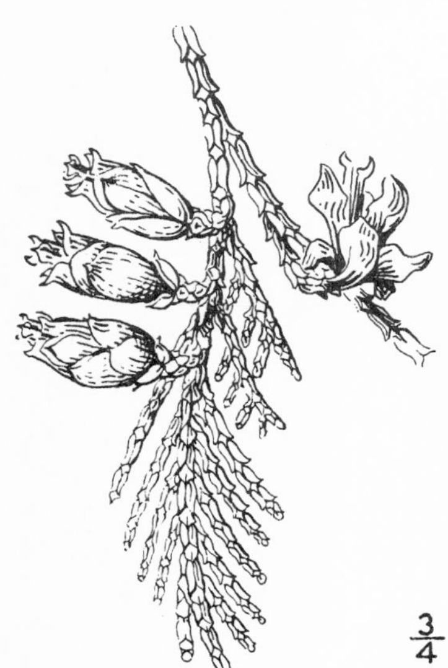

3/4

Moist bottom lands or rarely on dry ridges from sea level to 6000 feet altitude, Humid Transition and Canadian Zones; Portage Bay, Alaska, southward along the coast to Mendocino County, California, and through the Cascade Mountains of Washington and Oregon, and eastward to western Montana. Type locality: Nootka Sound.

3. CUPRÉSSUS [Tourn.] L. Sp. Pl. 1002. 1753.

Resinous aromatic trees with fibrous or exfoliating bark, light brown, durable, and fragrant wood; stout, erect or horizontal branches and naked buds. Leaves scale-like, decussate, ovate, acute, acuminate or rounded at the apex. Flowers monoecious, small, terminal on the branchlets; the staminate rounded to oblong, yellowish, with subpeltate scales bearing 3–5 subtended pollen-sacs; the ovulate green and inconspicuous, the pointed scales spreading and exposing the numerous erect basal ovules; cones maturing the second year, globose to oblong, the scales much thickened into a shield-shaped apex with a prominent central boss or prickle, closely fitting and not overlapping, separating at maturity or remaining closed for years; seeds small, compressed, acutely angled or margined; cotyledons 2–5. [The ancient classical name.]

About 12 species, natives of warm, temperate Asia, southern Europe, southwestern United States and Mexico. Type species, *Cupressus sempervirens* L.

Leaves without glands or with dorsal pits, not resinous glandular; umbos low, crescent-shaped.
  Bark more or less ridged and furrowed, with persistent fibrous scales.
    Seeds reddish brown, often glaucous.
      Leaves without dorsal pits; staminate flowers subglobose. — 1. *C. macrocarpa.*
      Leaves with dorsal pits; staminate flowers oblong. — 2. *C. sargentii.*
    Seeds black or a dark dull brown; usually dwarfed trees. — 3. *C. goveniana.*
  Bark very thin, exfoliating in broad, irregular, and thin plates, leaving a smooth cherry-red surface. — 4. *C. guadalupensis.*
Leaves with conspicuous resinous pits, very fragrant; umbos usually prominent and conical. — 5. *C. macnabiana.*

1. Cupressus macrocàrpa Hartw.

Monterey Cypress. Fig. 149.

Cupressus macrocarpa Hartw.; Gord. Journ. Hort. Soc. Lond. **4**: 296. 1849.
Cupressus hartwegii Carr. Rev. Hort. **1855**: 233. 1855.

A maritime tree 15–25 m. high, with a trunk 0.5–2 m. in diameter; bark 20–25 mm. thick, fibrous; wood close-grained, hard, heavy, strong, and durable. Leaves on short, stout, terete branchlets, about 3 mm. long, obtuse, glandless, obscurely or not at all pitted, persisting for 3 or 4 years; staminate flowers subglobose, usually, with only 2 lateral anthers in each row; the ovulate oblong, with spreading pointed scales; cones solitary or clustered on short, stout stalks, broadly oblong, 25–35 mm. long, puberulent, dull brown; scales 8–12, flat-topped, with a central curved ridge-like umbo, or those of the uppermost scales subconical; seeds about 4 mm. long, light chestnut-brown, irregularly angled and narrowly winged, with a minute white lanceolate hilum.

Occupies a narrow belt on the two promontories (Cypress Point and Point Lobos) bounding Carmel Bay, a few miles south of Monterey, California. Transition Zone.

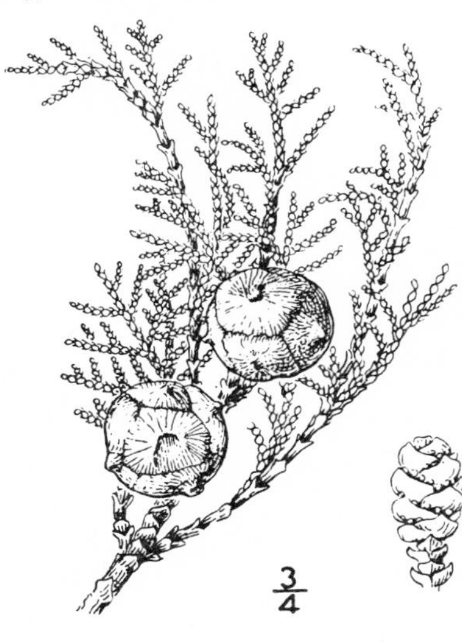

2. Cupressus sargéntii Jepson.

Sargent Cypress. Fig. 150.

Cupressus sargentii Jepson, Fl. Calif. **1**: 61. 1909.

Small tree, usually 3–5 m. or rarely 15 m. high with a short trunk, 20–40 cm. in diameter; branches slender, erect or spreading, forming an open crown; branchlets slender; bark 6–12 mm. thick, dark reddish brown, fibrous. Wood weak, light, and soft, light brown. Leaves on slender terete branchlets, dark green and obscurely glandular or without glands, 2–3 mm. long, acutish, persisting 3–4 years; staminate flowers oblong, with 2–3 làteral anthers in each row; cones subglobose, often clustered, 15–25 mm. long, reddish or purplish brown and shining, puberulent; scales 6–8, with a small upwardly appressed crescent-shaped umbo; seeds reddish brown and glaucous, acutely margined, 3 mm. long.

Dry mountain, usually serpentine, slopes in more or less isolated groves, Upper Sonoran and Transition Zones; Red Mountain, Mendocino County, southward to the Santa Lucia Mountains, Monterey County, California. Type locality: Mount Tamalpais, Marin County.

3. Cupressus goveniàna Gord.
Gowen Cypress. Fig. 151.

Cupressus goveniana Gord. Jour. Hort. Soc. Lond. **4**: 295. 1849.
Cupressus californica Carr. Trait. Conif. 127. 1855.
Cupressus pygmaea Sarg. Bot. Gaz. **31**: 239. 1901.

Small tree, 0.3–5 m. high, or rarely a large tree nearly 30 m. high and 1.5 m. in diameter; bark fibrous, persistent. Leaves on slender 4-sided branchlets, slightly keeled and acutish, glandless; staminate flowers oblong, with 3 lateral anthers in each row; cones subglobose or broadly oblong, 15–20 mm. long, reddish brown and shining, becoming gray-brown with age; scales 6–8 with short, thin-edged, upwardly appressed umbos; seeds black or dark brown, angular or margined, minutely warty, 1–3 mm. long.

First discovered at Monterey on the pine barrens of Huckleberry Hill. It also occurs on the pine barrens or the "White Plains" of Mendocino County, California. Transition Zone.

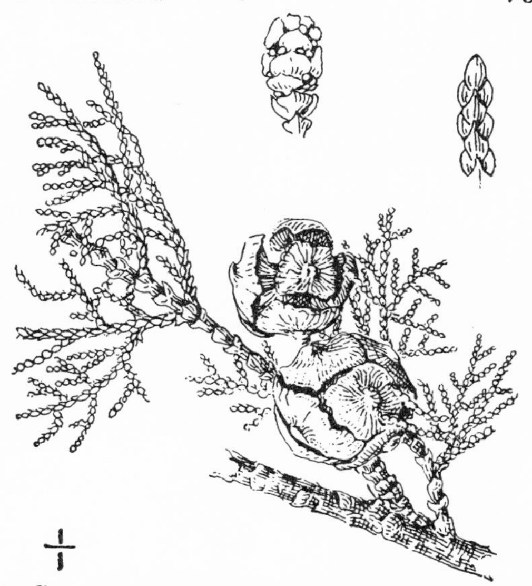

4. Cupressus guadalupénsis S. Wats.
Guadaloupe Cypress. Fig. 152.

Cupressus guadalupensis S. Wats. Proc. Am. Acad. **14**: 300. 1879.
Cupressus macrocarpa guadalupensis Masters, Gard. Chron. III. **18**: 62. *f. 8, 9, 12.* 1895.

Small tree, 5–10 m. high, clothed from the ground with slender, ascending branches, forming a thin, open, conical crown; bark of the trunk separating into extremely thin, reddish brown scales, exfoliating, leaving a smooth, polished, red-brown inner bark. Leaves on slender terete branchlets, light green, acutish, slightly keeled at the tip, furnished with a dorsal pit, but not resinous-glandular; staminate flowers oblong with 3 lateral anthers in each row; cones globose, 20–25 mm. broad, light brown, becoming gray-brown with age; scales 6–8 with central subconical or more or less appressed and crescent-shaped umbos; seeds reddish brown, sharply angled, obscurely warty, 3 mm. long.

Dry mountain slopes, Upper Sonoran Zone; Orange County, to Lower California and Guadaloupe Island. Type locality: Guadaloupe Island.

5. Cupressus macnabiàna Murr.
MacNab Cypress. Fig. 153.

Cupressus macnabiana Murr. Edinb. New. Phil. Jour. II. **1**: 293. *pl. 11.* 1855.
Cupressus glandulosa Henk. & Hochst. Syn. Nadelh. 241. 1865.
Cupressus bakeri Jepson, Fl. Calif. **1**: 61. 1909.

Small tree, 5–10 m., rarely 20 m. high with a trunk up to 6 dm. in diameter; branches usually crooked and forming an irregular open crown; bark thin, gray-brown, fibrous and persistent. Leaves on slender branchlets slightly constricted between the pairs, obtuse, flat and conspicuously glandular on the dorsal surface, very fragrant; staminate flowers subglobose with only 1 or 2 lateral anthers in each row; cones globose, 12–18 mm. broad, brown and shining; scales usually 6; umbos above the middle of the scale, conical, the uppermost pair prominent and often incurved; seeds brown, 3 mm. long, obscurely and sparsely warty.

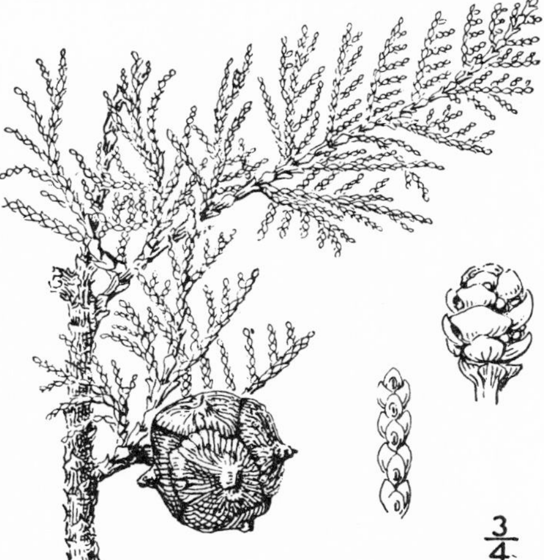

Dry hillsides and cañons, Transition and Upper Sonoran Zones; Shasta and Modoc Counties to Mendocino and Napa Counties, California. Type locality: Near Whiskeytown, Shasta County.

Cupressus macnabiana nevadénsis Abrams (*Cupressus nevadensis* Abrams, Torreya **19**: 92. 1919). Foliage without the characteristic fragrance of the typical form, and branchlets more distinctly 4-sided; cones 20–25 mm. long, light gray-brown.

Dry slopes of igneous rocks, Upper Sonoran Zone; Red Hill, near Bodfish, Piute Mountains, Kern County, California.

4. CHAMAECÝPARIS Spach, Hist. Vég. 11: 329. 1842.

Tall pyramidal trees or rarely shrubs, with thin scaly or deeply furrowed bark, nodding leading shoots, spreading branches, flattened or ultimately terete branchlets, 2-ranked, in a horizontal plane, light-colored, fragrant wood. Leaves scale-like, ovate, acuminate, opposite in pairs, becoming brown and woody in age; flowers minute, terminal, monoecious; staminate flowers oblong, bearing numerous stamens, their filaments short and stout, with connectives broader than long, more or less completely covering the 2 globose pollen-sacs; ovulate flowers subglobose, composed of usually 6 fertile peltate scales, each bearing 2–5 erect ovules; cones small, globose, maturing at the end of the first season, their scales abruptly enlarged and flattened or depressed at the apex, bearing the remnants of the flower as a short prominent projection; seeds 1–5, erect, ovate-oblong, with 2 prominent wings; cotyledons 2. [Name Greek, meaning a low cypress.]

Six species, one on the Atlantic Coast, two on the Pacific Coast of North America, and three in Japan and Formosa. Type species, *Chamaecyparis sphaeroidea* Spach.

Bark thin, about 10–20 mm. thick, divided into flat ridges; branchlets scarcely flattened; leaves glandless or obscurely glandular. 1. *C. nootkatensis.*
Bark thick, often 15–20 cm. thick, divided into broad, rounded ridges; branchlets slender, flattened; leaves conspicuously glandular. 2. *C. lawsoniana.*

1. Chamaecyparis nootkaténsis (Lamb.) Spach. Alaska Cypress. Fig. 154.

Cupressus nootkatensis Lamb. Pinus 2: 18. 1824
Chamaecyparis nootkatensis Spach, Hist. Veg. 11: 333. 1842.

Tree frequently 30–40 m. high, with a trunk 1–2 m. in diameter; branches horizontal, with distichous slightly flattened or terete yellowish branchlets; bark 15–20 mm. thick, irregularly furrowed into flat, brown-gray ridges, these separating on the surface into thin, papery scales. Leaves faintly glandular-pitted or glandless on the back, dark blue-green, long-pointed, closely appressed except on vigorous shoots; staminate flowers oblong, 4 mm. long, composed of 3–10 stamens, with light yellow ovate connectives; ovulate flowers light reddish color with the fertile scales bearing 2–4 ovules. Cone subglobose, 12 mm. broad, dark red-brown, with 4–6 scales, each tipped with a prominent conical projection; seeds 1–4, ovate.

Humid Transition and Canadian Zones; southern Alaska southward through western British Columbia and along the Cascade Mountains of Washington and Oregon as far south as the Santiam River. The wood is clear yellow, with a nearly white sapwood, close-grained, very durable and fragrant. It is used for interior finishing, furniture, and shipbuilding. Type locality: Nootka Sound.

2. Chamaecyparis lawsoniàna (Murr.) Parl. Lawson Cypress. Fig. 155.

Cupressus lawsoniana Murr. Edinb. New Phil. Jour. II. 1: 292. *pl. 10.* 1855.
Cupressus fragrans Kell. Proc. Calif. Acad. 1: 103. 1856.
Chamaecyparis lawsoniana Parl. in DC. Prod. 16^2: 464. 1868.

Tree 25–50 m. high, with a trunk abruptly enlarged at the base and above this 1–3 m. in diameter, and horizontal or pendulous branches bearing slender flattened branchlets; bark dark red-brown, smooth on young trees, on the older often 15–20 cm. thick, separating into large, thin shreds, becoming divided into broad rounded ridges. Leaves bright green above, glaucous below, conspicuously glandular on the back; staminate flowers oblong; stamens about 12, their connectives red; ovulate flowers with pointed scales, each bearing about 4 ovules; cones abundant on the upper branches, globose, 8 mm. broad, reddish brown and somewhat glaucous; scales 8–10, with a flattened or sunken apex, bearing a short conical projection; seeds ovate, slightly flattened, 3 mm. long, chestnut brown, broadly winged.

3/4

Most abundant and reaches its best development on the western slope of the Coast Ranges in southwestern Oregon, Humid Transition Zone; Coos Bay, Oregon, to Mad River, Humboldt County, California. As a lumber tree it is known as Port Orford Cedar, and as an ornamental, Lawson Cypress. Type locality: Sacramento River Cañon, California.

5. SABÌNA Mill. Gard. Dict. Abr. ed. 4. 1754.

Evergreen trees or shrubs with thin, shredded bark, and soft, close-grained, often reddish, wood. Leaves persistent for several years, scale-like or sometimes awl-shaped, opposite or in whorls of 3, decurrent at the base; flowers dioecious or monoecious, terminal; the staminate solitary or clustered, their scales ovate or peltate, bearing 3–6 pollen-sacs; ovulate flowers subglobose, of 2–3 series of scales; ovules 1 or rarely 2 to each scale; cones berry-like by the coalition of the fleshy scales; seeds 1 to several, wingless; cotyledons 2–6. [The ancient classical name of a kind of juniper.]

About 20 species, widely distributed over the northern hemisphere. Type species, *Juniperus sabina* L.

Berries large, red-brown with dry sweet pulp; cotyledons 4–6.
  Leaves glandular-pitted on the back, usually in 3's, rounded at the apex. — 1. *S. californica.*
  Leaves not glandular-pitted, acute or acuminate, usually in 2's. — 2. *S. utahensis.*
Berries rather small, blue or blue-black, with a resinous pulp; cotyledons 2.
  Leaves conspicuously glandular on the back, denticulate. — 3. *S. occidentalis.*
  Leaves not glandular, entire. — 4. *S. scopulorum.*

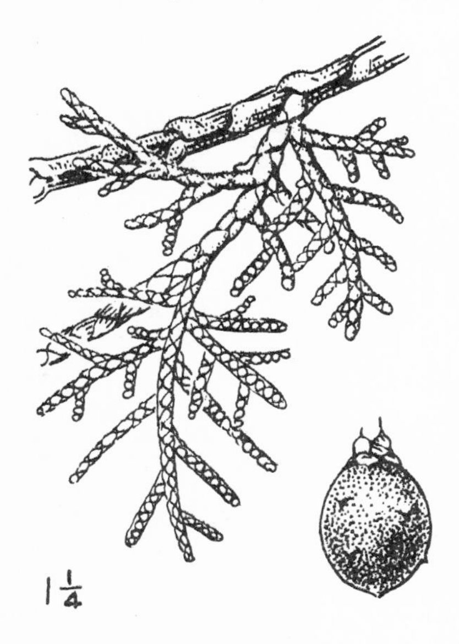

1. Sabina califórnica (Carr.) Antoine.

California Juniper. Fig. 156.

Juniperus californica Carr. Rev. Hort. **1854**: 352. *f. 21.* 1854.
Sabina californica Antoine, Cupress. Gat. 52. *pl. 71, 72.* 1857.

An arborescent shrub with numerous branches forming a broad, rounded head, 2–4 m. high, or rarely a conical tree 12 m. high. Leaves scale-like, closely appressed, usually in 3's, ovate to oblong, 4 mm. long, distinctly glandular-pitted on the back, bluntly pointed; berries at first bluish with a dense bloom, at maturity reddish brown beneath the bloom, globose-oblong, 12–18 mm. long, nearly smooth, the pulp firm, dry, and sweetish; seeds 1 or 2, ovoid, sharp-pointed and angled, 6–9 mm. long, light brown and shining above, dull and yellowish toward the base; cotyledons 4–6.

Dry hillsides, Upper Sonoran Zone; inner Coast Ranges of California, from Lake County to the Tehachapi Mountains, and the western base of the Sierra Nevada in Tulare and Kern Counties. South of the Tehachapi Mountains it is chiefly restricted to the desert slopes of the mountains extending to northern Lower California. Type locality: California; locality not given.

2. Sabina utahénsis (Englm.) Rydb.

Utah Juniper. Fig. 157.

Juniperus californica utahensis Engelm. Trans. St. Louis Acad. **3**: 588. 1877.
Juniperus utahensis Lemmon, Rep. Calif. State Board Forestry 3: 183. *pl. 28, f. 2.* 1890.
Sabina utahensis Rydb. Bull. Torrey Club **32**: 598. 1905.

An arborescent shrub or small tree, rarely over 2.5–3 m. high, with a very crooked trunk, sometimes 6–8 dm. in diameter, but commonly branched at the base, forming rounded clumps, 1.5–2 m. high. Leaves in 2's or rarely in 3's, acute or acuminate, inconspicuously glandular on the back or usually without glands, light yellowish green; staminate flowers with 18–24 stamens; fruit maturing the second year, rounded, or oblong, 6–9 mm. long, reddish brown beneath the bloom when mature, with a thick epidermis and thin dry sweet flesh; seeds 1 or rarely 2, ovate, acute, strongly and abruptly angled, with a conspicuous hilum extending nearly to the apex; cotyledons 4–6.

The arid mountains and table lands of the Great Basin region, Upper Sonoran Zone; southern Wyoming and western Colorado to Nevada, northern Arizona, and the Providence and Panamint Mountains in southeastern California. Type locality: Only general range given.

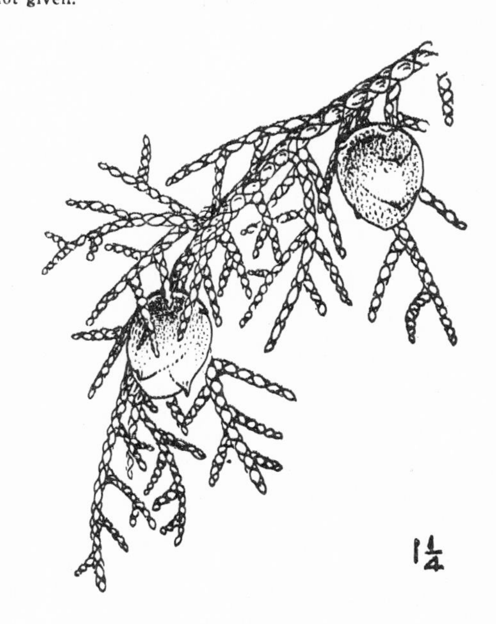

3. **Sabina occidentàlis** (Hook.) Heller. Western Juniper. Fig. 158.

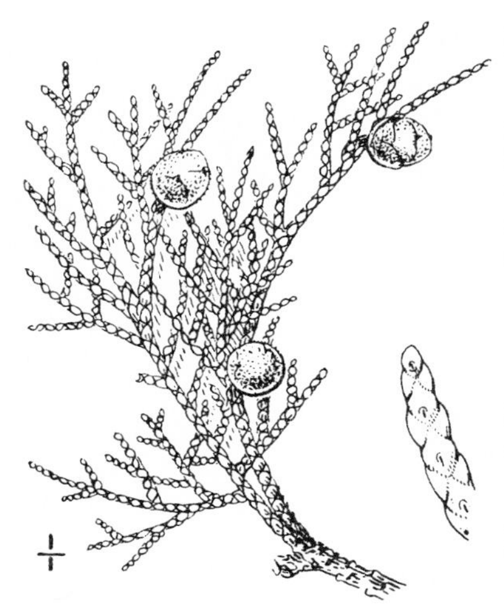

Juniperus occidentalis Hook. Fl. Bor. Am. **2**: 166. 1838.
Sabina occidentalis Heller, Muhlenbergia **1**: 47. 1904.

A tree, usually about 8 m. high, but occasionally 20 m. high, with a trunk 6–12 dm. or rarely 24 dm. in diameter; branches often very large, spreading at right angles and forming a flat top; bark about 20 mm. thick, cinnamon-brown. Leaves in 3's, closely appressed, acute or acuminate, conspicuously glandular and rounded on the back, 3 mm. long, gray-green, the margins slightly denticulate; staminate flowers with 12–18 stamens; berries rounded to oblong, 6–8 mm. long, blue-black at maturity beneath the glaucous bloom; seeds 2 or 3, ovate, acute, rounded and grooved or pitted on the back, 3 mm. long; cotyledons 2.

Mountain slopes, mainly Canadian Zone; western Idaho and Washington, southward through the Cascade Mountains and the Sierra Nevada to the San Bernardino Mountains, southern California. Type locality: "Common on the higher parts of the Columbia at the base of the Rocky Mountains."

4. **Sabina scopulòrum** (Sargent) Rydb. Rocky Mountain Juniper. Fig. 159.

Juniperus scopulorm Sargent Gard. & Forest **10**: 420. 1897.
Sabina scopulorum Rydb. Bull. Torrey Club **32**: 598. 1905.

A tree often 7–15 m. high, with a trunk 1 m. in diameter, or branched at the base into several spreading branches; bark reddish brown, divided into low narrow ridges, becoming shredded on the surface. Leaves in 2's, closely appressed, acute or acuminate, obscurely glandular-pitted, dark green or glaucous; staminate flowers with usually only 6 stamens; berries maturing the second autumn, globose or nearly so, 6–8 mm. long, bright blue beneath the bloom, with a thin epidermis and sweetish flesh; seeds 1 or 2, acute, prominently grooved and angled, 4–5 mm. long; cotyledons 2.

Foothills of the Rocky Mountains, mainly Transition Zone; Alberta, coast of British Columbia and Washington, eastern Oregon, Nevada, northern Arizona, and western Texas. Type locality: not indicated.

6. JUNÍPERUS [Tourn.] L. Sp. Pl. 1038. 1753.

Aromatic trees or shrubs with usually thin, shreddy bark, soft, close-grained, often reddish, wood, and slender branches. Leaves persistent for many years, in whorls of 3, linear-subulate and sharp-pointed, jointed at the base; flowers dioecious or monoecious, axillary, the staminate oblong-ovate, with numerous stamens in 2 or 3 ranks, their connectives ovate or peltate, scale-like, yellow, bearing 2–6 pollen-sacs; the ovulate ovoid, with a few opposite somewhat fleshy scales, each bearing 1 or rarely 2 ovules; cone berry-like, blue, blue-black, or reddish, with a glaucous bloom, formed by the coalition of the flower scales, smooth or roughened by the tips of the original scales; seeds 1–12, ovate, terete or variously angled, wingless; cotyledons mostly 2. [The classical name.]

About 10 species distributed over the northern hemisphere from the Arctic Circle to the mountains of northern Africa, West Indies, Mexico, southern Japan and China. Type species, *Juniperus communis* L.

1. **Juniperus sibirica** Burgsd.

Dwarf Juniper. Fig. 160.

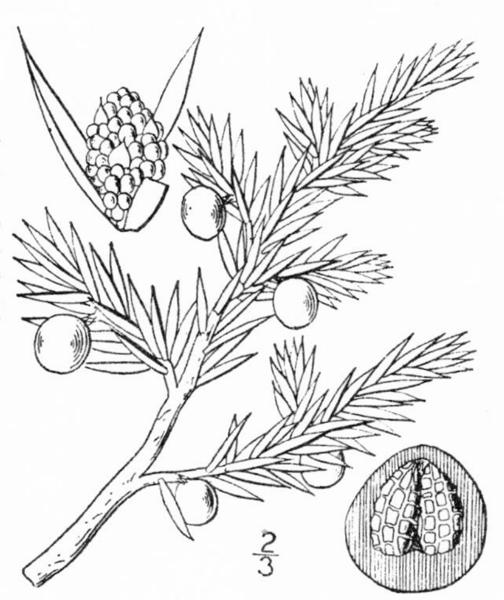

Juniperus sibirica Burgsd. Anleit. Holz. no. 272. 1787.
Juniperus communis montana Ait. Hort. Kew. 3: 414. 1788.
Juniperus nana Willd. Sp. Pl. 4: 854. 1806.
Juniperus communis sibirica Rydb. Contr. U. S. Nat. Herb. 3: 533. 1896.

Low or usually prostrate alpine or arctic shrub, forming patches several feet in diameter. Leaves in whorls of 3, ascending, linear-lanceolate, 6–12 mm. long, pungently acute, dark green, shining and keeled or strongly convex on the lower surface, deeply grooved above and with a broad, white band of stomata; staminate flowers 3–6 mm. long; anther-scales with a short subulate point; berries globose, 7–9 mm. broad, bright blue, covered with a white bloom; seeds, 1–3, ovate, acute, angled, about 3 mm. long.

Widely distributed over the Arctic and alpine regions of the northern hemisphere in the Arctic and Hudsonian Zones. On the Pacific Coast it ranges from Alaska southward to the high altitudes of the central Sierra Nevada. Type locality: in Siberia.

Family 5. **EPHEDRÀCEAE.**

EPHEDRA FAMILY.

Shrubs or small trees with jointed opposite or fascicled branches and scale-like opposite or whorled leaves; flowers unisexual, and usually dioecious, with decussate persistent bracts; stamens monodelphous, subtended by a membranous 2-lobed calyx-like perianth; anthers dehiscent by terminal pores. Ovulate flowers, composed of a solitary erect ovule enclosed in an urn-shaped perianth which becomes hardened in fruit; seed coats 1, separating into 2 and ending in an elongated style-like micropyle; embryo axillary surrounded by copious endosperm; cotyledons 2.

A single genus, inhabiting the arid regions of the northern hemisphere.

1. **EPHÉDRA** [Tourn.] L. Sp. Pl. 1040. 1753.

Desert shrubs, with jointed, equisetum-like branches, and persistent or deciduous scale-like leaves. [Ancient Greek name, used by Pliny for the horse-tail.]

About 15 species distributed over the desert regions of Asia, northern Africa and North America. Type species, *Ephedra distachya* L.

Leaf-scales and bracts in twos, the latter only scarious on the broad margins, sessile.
  Branches bright green, erect and broom-like. — 1. *E. viridis.*
  Branches pale glaucous green, divergent. — 2. *E. nevadensis.*
Leaf-scales and bracts in threes, the latter scarious throughout, clawed.
  Branches not spinose; leaf-scales oblong, obtuse to acutish, soon splitting to the base and recurved. — 3. *E. californica.*
  Branches spinose-tipped; leaf-scales persistent, sheathing, sharp-pointed. — 4. *E. trifurca.*

1. **Ephedra víridis** Coville.

Green Ephedra. Fig. 161.

Ephedra viridis Coville, Contr. U. S. Nat. Herb. 4: 220. 1893.

An erect shrub, 5–10 dm. high, with numerous slender, erect, bright green and broom-like scabrous branches. Leaf-scales opposite, 3–6 mm. long, sheathing for about two-thirds their length, obtuse or toward the ends of the branches tapering at apex, deciduous in age; staminate aments sessile, with 4 or 5 pairs of bracts; fruiting bracts of 4 pairs, round-ovate, sessile; fruit well exserted beyond the bracts, often in pairs and triangular, acutish.

A characteristic species of the Piñon and the upper part of the Juniper belt of the Mojave Desert region, Upper Sonoran Zone; southern Nevada to the vicinity of Fort Tejon and the desert slopes of the San Bernardino Mountains, California. Type locality: near Crystal Springs, Coso Mountains, Inyo County, California. The species of *Ephedra* are often called Mexican Tea.

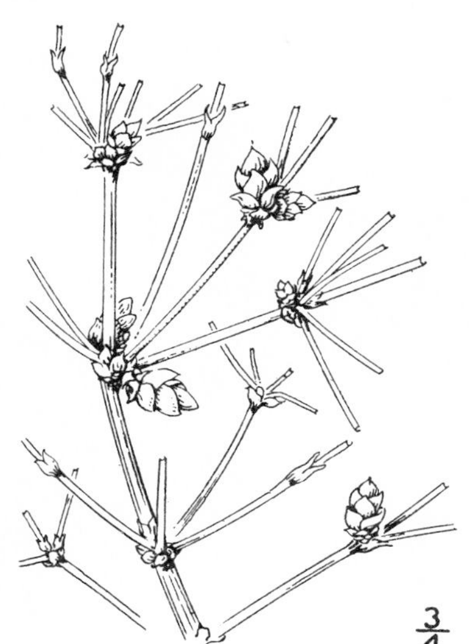

2. **Ephedra nevadénsis** S. Wats.

Nevada Ephedra. Fig. 162.

Ephedra nevadensis S. Wats. Proc. Am. Acad. **14**: 298. 1879.

An erect shrub, 5–10 dm. high, with divergently spreading, glaucous green, scabrous branches. Leaf-scales in 2's, 3–6 mm. long, sheathing to about the middle, obtuse or tapering at the apex, mostly deciduous in age; staminate aments sessile, with 4–6 pairs of bracts; fruiting bracts of 4 or 5 pairs, round-ovate, sessile; fruit exserted, acutish at apex.

Characteristic of the lower part of the Juniper belt and the upper part of the Covillea belt of the desert regions, Upper and Lower Sonoran Zones; Nevada and southern Utah southward through Arizona to northern Mexico, and through the Mojave and Colorado Deserts of California to northern Lower California. Type locality: "Pah Ute Mountains, altitude 5000 ft., and Carson City, Nevada."

3. **Ephedra califórnica** S. Wats.

California Ephedra. Fig. 163.

Ephedra californica S. Wats. Proc. Am. Acad. **14**: 300. 1879.

Low, spreading or suberect shrub, with ternate, smooth or slightly scabrous, pale, glaucous branches, not spinose-tipped. Leaf-scales in 3's, oblong, acutish, sheathing to above the middle but soon splitting to the base and becoming recurved, persistent and turning dark-colored in age; staminate aments sessile, globose, with 4 whorls of ternate bracts; fruiting bracts of 4 or 5 whorls, rather rigid, round-reniform, with a short broad claw; fruit exserted, ovoid, merely acutish at apex, about 15 mm. long.

Juniper belt, Upper Sonoran Zone; Mojave and Colorado Deserts, southward to Lower California, and westward to the coast in the vicinity of San Diego. Type locality: Point Loma, San Diego, Calif.

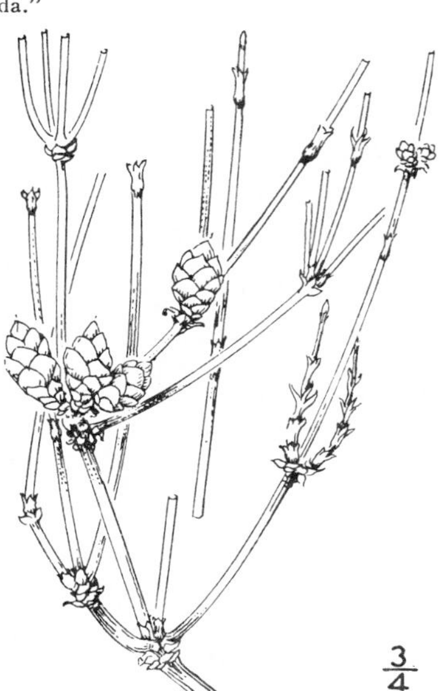

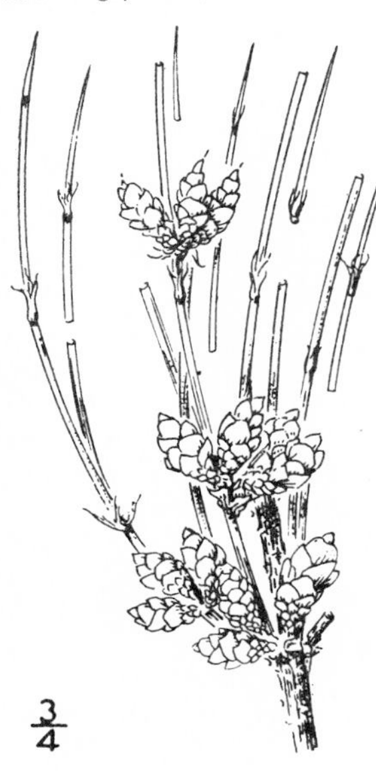

4. **Ephedra trifúrca** Torr.

Three-forked Ephedra. Fig. 164.

Ephedra trifurca Torr. in Emory Rep. 153. 1848.

An erect shrub, 5–10 dm. high, with erect, nearly smooth, pale green, spinose-tipped branches. Leaf-scales in 3's, 5–10 mm. long, sheathing to about the middle, persistent, acuminate, becoming white and shreddy; staminate aments sessile, with 4 or 5 whorls of bracts; fruiting bracts of many whorls, thin and membranous, round-reniform with an elongated, very slender claw; fruit included within the bract, tapering to a prominent beak.

Desert regions, mainly Lower Sonoran Zone; western Texas and southern Utah south to Mexico, and west to the deserts of southern California where it has been found along the Mojave River at Daggett, and on Superstition Mountain, Colorado Desert. Type locality: "From the region between the Del Norte and the Gila, and the hills bordering the latter river to the desert west of the Colorado."

Class 2. *ANGIOSPÉRMAE.*

Ovules borne in an enclosed ovary formed by a single sporophyll (carpel) or by several sporophylls united together. Pollination is effected often through the agency of wind or insects (cross-pollination), or less commonly within the flower itself (self-pollination). The pollen-grains after alighting upon the summit of the ovary (stigma) germinate, sending a tube into the tissue, which, upon entering the ovule, effects fertilization by the fusion of the egg cell of the ovule and the spermatozoid of the pollen-tube.

The angiosperms have been the dominating plants of the world since the Mesozoic Era, and are represented by over 130,000 living species, comprising two distinct sub-classes, as follows:

Cotyledons 1: stem endogenous; leaves commonly parallel-veined. Sub-class 1. MONOCOTYLEDONES.

Cotyledons 2: stem (with rare exceptions) exogenous; leaves commonly net-veined. Sub-class 2. DICOTYLEDONES.

Sub-class 1. *MONOCOTYLEDONES.*

Stem composed of a ground-mass of soft tissue (parenchyma) in which the bundles of wood-cells are irregularly imbedded; no distinction into wood, pith and bark. Leaves usually parallel-veined, mostly alternate and entire, commonly sheathing the stem at the base and often with no distinction of blade and petiole. Flowers mostly 3-merous or 6-merous. Embryo with but a single cotyledon, and the first leaves of the germinating plantlet alternate.

Monocotyledonous plants are first definitely known in Triassic time. They constitute between one-fourth and one-third of the living angiospermous flora.

Family 1. TYPHÀCEAE.

CAT-TAIL FAMILY.

Marsh or aquatic herbaceous plants with creeping rootstocks, fibrous roots and glabrous erect, terete stems. Leaves linear, flat, ensiform, striate, sheathing at the base. Flowers monoecious, densely crowded in terminal spikes, which are subtended by spathaceous, usually fugaceous bracts, and divided at intervals by smaller bracts, which are caducous, the staminate spikes uppermost. Perianth of bristles. Stamens 2–7, the filaments connate. Ovary 1, stipitate, 1–2-celled. Ovules anatropous. Styles as many as the cells of the ovary. Mingled among the stamens and pistils are many bristly hairs, and among the pistillate flowers many sterile flowers with clavate tips. Fruit a minute achene. Endosperm copious.

The family comprises only one genus.

1. TỲPHA [Tourn.] L. Sp. Pl. 971. 1753.

Characters of the family. (Name ancient.)

About 10 species, of temperate and tropical regions. Type species, *Typha latifolia* L.

Spikes with the pistillate and staminate portions usually contiguous, the former without bractlets; stigmas spatulate or rhomboid; pollen-grains in 4's. 1. *T. latifolia.*

Spikes with the pistillate and staminate portions usually distant, the former with bractlets; stigmas linear or oblong-linear; pollen-grains simple. 2. *T. angustifolia.*

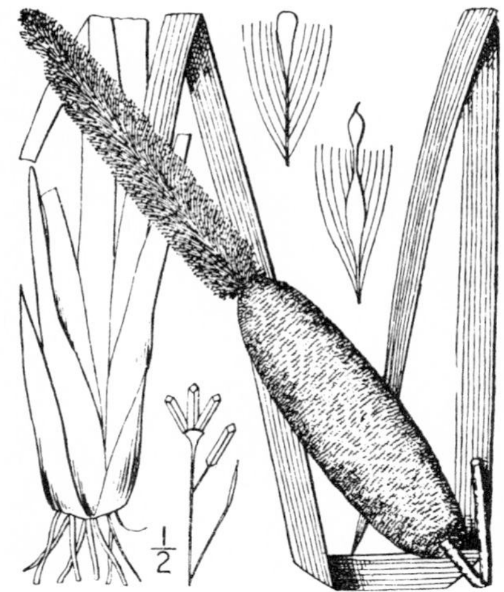

1. **Typha latifòlia** L.

Broad-leaved Cat-tail. Fig. 165.

Typha latifolia L. Sp. Pl. 971. 1753.

Stems stout, 1.5–3 m. tall. Leaves 6–25 mm. broad; spikes dark brown, the staminate and pistillate portions usually contiguous, each 1–3 dm. long and often 25 mm. or more in diameter, the pistillate without bractlets; stigmas rhomboid or spatulate; pollen-grains in 4's; pedicels of the mature pistillate flowers 2–3 mm. long.

In marshes, throughout North America, except the extreme north. Also in Europe and Asia. This is the most common species nearly throughout the Pacific states. Type locality: in Europe.

2. **Typha angustifòlia** L.

Narrow-leaved Cat-tail. Fig. 166.

Typha angustifolia L. Sp. Pl. 971. 1753.
Typha bracteata Greene, Bull. Calif. Acad. 2: 413. 1887.

Stems slender, 1–3 m. tall. Leaves mostly narrower than those of the preceding species, 4–10 mm., or rarely more, wide; spikes light brown, the staminate and pistillate portions usually distant, each 1–2 dm. long, the pistillate at maturity 5–20 mm. in diameter, and provided with bractlets; stigmas linear or linear-oblong; pollen-grains simple; pedicels of the mature pistillate flowers 1 mm. long or less, the perianth-bristles capillary, with one or rarely two chains of cells at the tip.

Widely distributed over the northern hemisphere, also in tropical America and South America. Central and southern California, abundant along irrigating canals in Imperial Valley. *Typha domingensis* Pers., a doubtfully distinct species, has been credited to California, but all the California material seems co-specific and referable to *angustifolia*. Type locality: Europe.

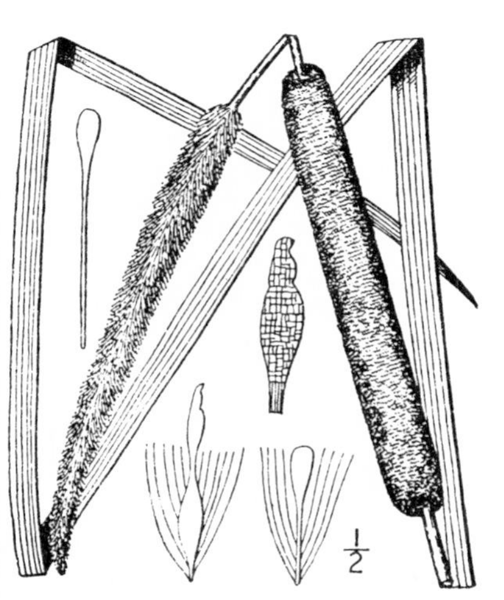

Family 2. **SPARGANIÀCEAE.**

BUR-REED FAMILY.

Marsh or pond herbaceous plants with creeping rootstocks and fibrous roots, erect or floating simple or branched stems, and linear alternate leaves, sheathing at the base. Flowers monoecious, densely crowded in globose heads on the upper part of the stem and branches, the staminate heads uppermost, sessile or peduncled. Spathes linear, immediately beneath or at a distance below the head. Perianth of a few irregular chaffy scales. Stamens commonly 5, their filaments distinct; anthers oblong or cuneate. Ovary sessile, mostly 1-celled, rarely 2-celled. Ovules anatropous. Fruit 1–2-celled and 1–2-seeded, nut-like. Embryo nearly straight, in copious endosperm.

The family comprises only one genus.

1. SPARGÀNIUM [Tourn.] L. Sp. Pl. 971. 1753.

Characters of the family. (Greek, referring to the ribbon-like leaves.)

About 20 species, of temperate and cool regions of both the northern and southern hemispheres. Type species, *Sparganium erectum* L.

Inflorescence branched; achenes sessile, truncate or rounded at apex.
  Achenes cuneate-obpyramidal, nearly as broad as long, truncate or depressed at apex. — 1. *S. eurycarpum.*
  Achenes obovoid, decidedly longer than broad, rounded at the apex.
    Perianth scales dilated toward he base; heads 25–30 mm. broad; achenes 9–10 mm. long. — 2. *S. californicum.*
    Perianth scales narrowed below; achenes 6–8 mm. long; heads 20–25 mm. broad. — 3. *S. greenei.*
Inflorescence simple; achenes stipitate.
  Achenes fusiform, more or less tapering into the beak.
    Leaves, at least the middle ones, triangular-keeled. — 4. *S. simplex.*
    Leaves not keeled, mainly floating. — 5. *S. angustifolium.*
  Achenes obovate, rounded at apex, the slender beak about 1 mm. long. — 6. *S. minimum.*

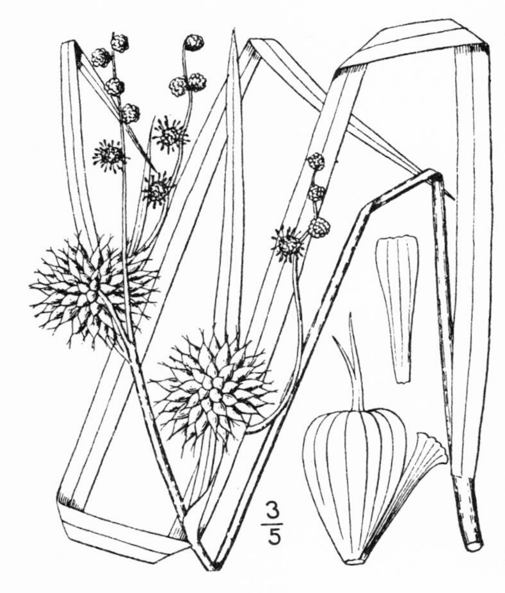

1. Sparganium eurycàrpum Engelm.
Broad-fruited Bur-reed. Fig. 167.

Sparganium eurycarpum Engelm.; A. Gray, Man. ed. 2, 430. 1856.

Stems erect, 5–15 dm. high, stout, branching. Leaves 5–10 dm. long, 7–10 mm. wide, flat, slightly keeled beneath; pistillate heads 2–4 on the stem or branches, sessile or usually peduncled, in fruit 20–25 mm. in diameter; perianth scales long-clawed, nearly as long as the achenes; anthers about 1 mm. long, elliptic-clavate; stigmas often 2, filiform, about 2 mm. long; achenes sessile, cuneate-obpyramidal, 3–5-angled, 6–8 mm. long and nearly as broad, truncate or depressed at the apex, the beak about 3 mm. long.

In marshes and along streams, Austral and Boreal Zones; British Columbia to Newfoundland, south to California, Utah, and Florida. In the Pacific States, mainly east of the Cascade Mountains in Washington and Oregon, and in the interior valleys of California. Type locality: eastern United States.

2. Sparganium califórnicum Greene.
California Bur-reed. Fig. 168.

Sparganium californicum Greene, Bull. Calif. Acad. 1: 11. 1884.

Stem stout, branching, 1–2 m. high. Leaves linear, flat, slightly keeled beneath, the lowest 1–1.5 m. long, staminate heads numerous; pistillate heads 1–2 on a branch, sessile, 25–30 mm. in diameter, compact and hard; achenes sessile, obovoid, 3–5-angled, rounded at the apex, 9–10 mm. long, about 6 mm. broad; beak nearly 2 mm. long; scales long-clawed, rounded and erose at the summit, scarcely wider than the dilated lower portion of the claw; stigmas linear, usually 2.

Marshes, mainly in the Upper Sonoran Zone; central California, from Napa to Monterey Counties. Type locality: Calistoga, Napa County, California.

3. **Sparganium greènei** Morong.

Greene's Bur-reed. Fig. 169.

Sparganium greenei Morong, Bull. Torrey Club **15**: 77. 1888.

Sparganium eurycarpum greenei Graebner, in Engler Pflanzenreich **4**[10]: 13. 1900.

Stem stout, erect, 6–18 dm. high. Leaves 3–15 dm. long, 8–12 mm. wide, keeled; bracts similar but smaller, margins not scarious at base; inflorescence branched; the lower branches erect, with usually 2 pistillate and 6–10 staminate heads; fruiting heads 20–25 mm. in diameter; scales with a cuneate or rhombic blade decidedly broader than the narrow claw; achenes sessile, brown, shining, obovoid with a rounded summit, 7–9 mm. long, about 5 mm. thick, 3–5-angled; beak about 2 mm. long; stigmas 1 or 2, filiform, about 1 mm. long.

In marshes of the Pacific Coast region; Transition to Lower Sonoran Zones; Vancouver Island to northern Lower California. Type locality: Olema, Marin County, California.

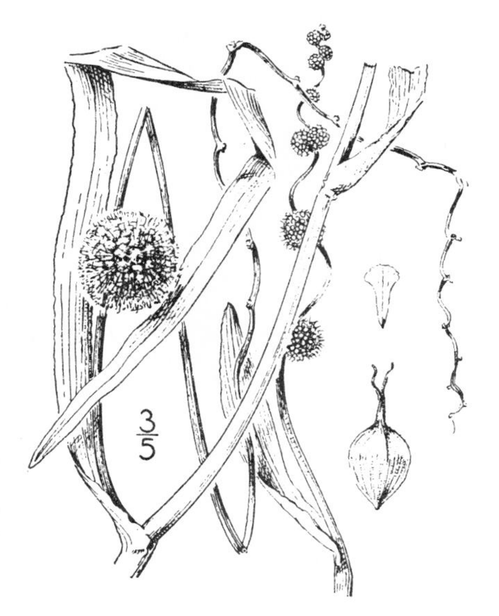

4. **Sparganium símplex** Huds.

Simple-stemmed Bur-reed. Fig. 170.

Sparganium simplex Huds. Fl. Aug. ed. 2, 401. 1788.

Stem rather stout, 4–6 dm. high. Leaves 4–10 dm. long, 8–16 mm. wide, triangular-keeled; bracts flat or slightly keeled; inflorescence usually simple, the pistillate heads 2–5, at least some of them supra-axillary, the lower 1 or 2 peduncled, the staminate heads 4–8; fruiting heads about 15 mm. in diameter; achenes brown or greenish brown, stipitate, 5–6 mm. long, fusiform, often constricted at the middle; stigma linear, about 1 mm. long.

Widely distributed over the cooler regions of the northern hemisphere; British Columbia to Humboldt County, California, mainly in the Canadian Zone. Type locality: Europe.

5. **Sparganium angustifòlium** Michx.

Narrow-leaved Bur-reed. Fig. 171.

Sparganium angustifolium Michx. Fl. Bor. Am. **2**: 189. 1803.

Sparganium simplex angustifolium Torr. Fl. N. Y. **2**: 249. 1843.

Sparganium simplex multipedunculatum Morong, Bull. Torrey Club **15**: 79. 1888.

Sparganium multipedunculatum Rydb. Bull. Torrey Club **32**: 598. 1905.

Stem floating and elongated, or erect and 30–50 cm. high. Leaves narrow 4–10 mm. wide, flat not keeled, dilated and scarious-margined at the base; inflorescence usually simple, the pistillate heads 2–4, the lower 1 or 2 usually on supra-axillary peduncles, the staminate heads usually approximate; fruit heads 7–10 mm. in diameter; achenes stipitate, about 4 mm. long, fusiform, brown; stigma linear, about 1 mm. long.

Margins of slow-running streams and ponds; throughout the cooler regions of North America; ranging from British Columbia to the San Bernardino Mountains, California, mainly in the Canadian Zone. Type locality: Canada.

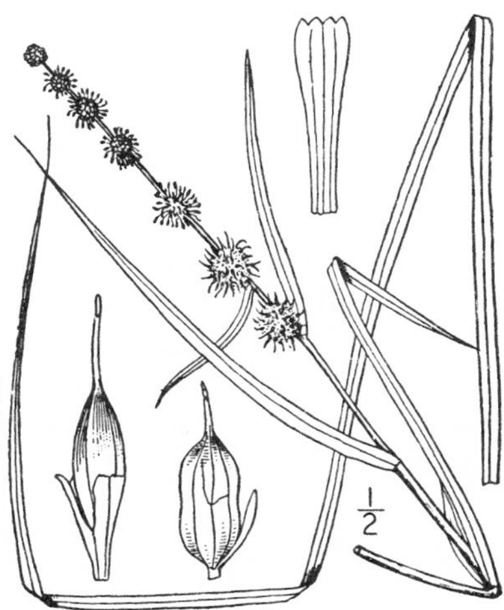

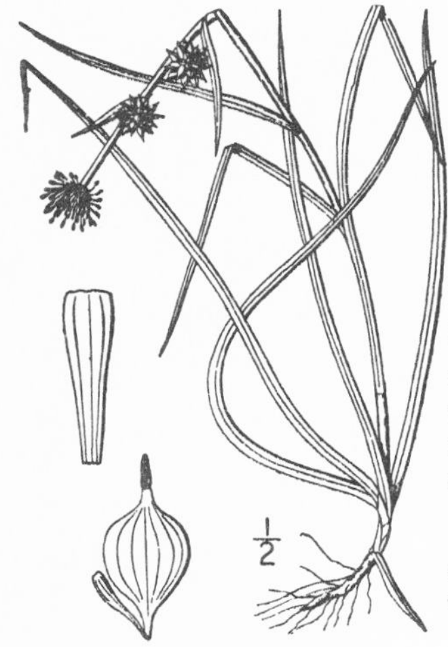

6. Sparganium mínimum Fries.

Small Bur-reed. Fig. 172.

Sparganium minimum Fries, Summa Veg. Scand. 2: 560. 1849.

Stem floating and elongated, or decumbent, ascending or erect, and relatively short. Leaves dark green, flat, narrow, mostly 3–7 mm. wide; the sheaths of the upper leaves dilated but not scarious-margined; inflorescence simple, the pistillate heads solitary or 2 or 3 and placed about equally distant, axillary; fruiting heads about 5 mm. in diameter; achenes short stipitate, about 3 mm. long, broadly ellipsoid, usually constricted below the middle, dull greenish brown; stigma obliquely oblong or oval.

In ponds and streams of the cooler regions of the northern hemisphere; Swan Lake, Vancouver Island, and Mount Rainier. Type locality: Sweden.

Family 3. POTAMOGETONÀCEAE.

Pondweed Family.

Perennial fresh-water plants with floating or submerged leaves or both. Leaves petioled or sessile, capillary or expanded into a proper blade, or rarely reduced to terete phyllodes. Flowers perfect or monoecious, in sessile or peduncled spikes or in small axillary clusters. Perianth none, but flowers sometimes enclosed in a hyaline sheath. Stamens 1–4; anthers extrorse, 1–2-celled, the connective sometimes becoming perianth-like. Pistils of 1–4 distinct carpels. Fruits of small drupelets, sessile or stipitate. Endosperm wanting.

About 4 genera and 70 species of wide geographic distribution, most abundant in north temperate regions.

Flowers perfect; stamens 2 or 4.
  Stamens 4; drupelets sessile. — 1. *Potamogeton.*
  Stamens 2; drupelets stipitate. — 2. *Ruppia.*
Flowers monoecious; stamen 1. — 3. *Zannichellia.*

1. POTAMOGÈTON L. Sp. Pl. 126. 1753.

Leaves alternate or the uppermost opposite, often of two kinds, submerged and floating, the submerged mostly linear, the floating petioled, coriaceous, lanceolate or broader. Stipules membranaceous, more or less united and sheathing, free or connate to the base of the leaf or petiole. Spikes sheathed by the stipules in the bud, mostly raised on a peduncle to the surface of the water. Stamens 4. their connectives sepal-like, opposite the 2-celled anthers, valvate in the bud, orbicular, green or reddish. Ovaries 4, sessile, distinct, 1-celled, 1-ovuled, attenuated into a short recurved style, or with a sessile stigma. Fruit of 4 ovoid or subglobose drupelets, the pericarp usually thin and hard or spongy. Seeds crustaceous, campylotropous, with an uncinate embryo thickened at the radicular end. [Greek, in allusion to the aquatic habit.]

A genus of about 65 species, inhabiting temperate regions. Type species, *Potamogeton natans* L.

Leaves of 2 kinds, floating and submerged, the former coriaceous with a dilated blade, petioled, different in form from the submerged thinner ones.
  Submerged leaves filiform or linear, not over 4 mm. wide.
    Spikes all alike, cylindric.
      Submerged leaves filiform; floating leaves, subcordate at base. — 1. *P. natans.*
      Submerged leaves narrowly linear-lanceolate, 2–4 mm. wide; floating leaves attenuate into the petiole. — 4. *P. epihydrus.*
    Spikes of 2 kinds, the emersed cylindric, many-flowered, the submerged capitate, few-flowered.
      Peduncles of the submerged spikes equaling or exceeding the spikes. — 2. *P. diversifolius.*
      Peduncles of the submerged spikes shorter than the spikes. — 3. *P. dimorphus.*
  Submerged leaves lanceolate to ovate, when approaching linear at least over 4 mm. wide.
    Floating leaves 30–50-nerved. — 5. *P. amplifolius.*
    Floating leaves with fewer nerves.
      Submerged leaves, at least all but the lowest, sessile.
        Submerged leaves serrulate at apex. — 6. *P. angustifolius.*
        Submerged leaves entire.
          Upper leaves and spikes reddish; submerged leaves lanceolate. — 7. *P. alpinus.*
          Upper leaves and spikes green; submerged leaves linear. — 8. *P. heterophyllus.*
      Submerged leaves all petioled. — 9. *P. americanus.*

Leaves all submerged and similar.
  Leaves lanceolate, oblong or ovate.
    Leaves not clasping at base.
      Leaves serrulate only at tip. — 10. *P. lucens.*
      Leaves serrulate throughout. — 11. *P. crispus.*
    Leaves clasping or half clasping at base.
      Stipules conspicuous and persistent; leaves half clasping. — 12. *P. praelongus.*
      Stipules short and inconspicuous. — 13. *P. richardsonii.*
  Leaves narrowly linear or setaceous.
    Stipules free from the leaves.
      Without propagating buds or glands. — 14. *P. foliosus.*
      With propagating buds or glands or both.
        Stems filiform; leaves capillary. — 15. *P. pusillus.*
        Stems much flattened; leaves narrowly linear. — 16. *P. compressus.*
    Stipules adnate to the leaf or petiole.
      Leaves narrowly linear, 1 mm. broad or more.
        Nutlets 1-keeled or without keel; leaves in terminal clusters. — 17. *P. latifolius.*
        Nutlets 3-keeled; leaves 2-ranked. — 18. *P. robbinsii.*
      Leaves capillary, less than 1 mm. wide. — 19. *P. pectinatus.*

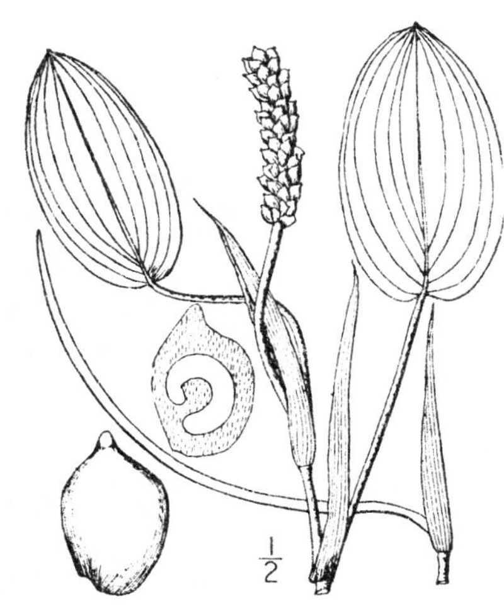

1. Potamogeton nàtans L.

Common Floating Pondweed. Fig. 173.

Potamogeton natans L. Sp. Pl. 126. 1753.

Stems simple or sparingly branched, 6–15 dm. long. Floating leaves 5–10 cm. long, 2.5–5 cm. wide, ovate, oval or elliptic, rounded or subcordate at base, many-nerved, thick, on petioles a little longer to three times as long; submerged leaves reduced to bladeless petioles, commonly perishing and seldom seen at the fruiting period; stipules usually over 5 cm. long, acute, 2-keeled; peduncles 5–10 cm. long, as thick as the stem; spikes cylindric, about 5 cm. long, dense; fruit 4–5 mm. long, narrowly obovoid, turgid, scarcely keeled; style broad and facial; nutlet hard, more or less pitted or impressed on the sides, 2-grooved on the back; embryo forming an incomplete circle, the apex pointing toward the base.

British Columbia and Nova Scotia to New Jersey, Nebraska, and southern California. Also in Europe and Asia. Transition and Boreal Zones. Type locality: in Europe.

2. Potamogeton diversifòlius Raf.

Rafinesque's Pondweed. Fig. 174.

Potamogeton hybridus Michx. Fl. Bor. Am. 1: 101. 1803, not Thuill. 1790.
Potamogeton diversifolius Raf. Med. Rep. II. 5: 354. 1808.

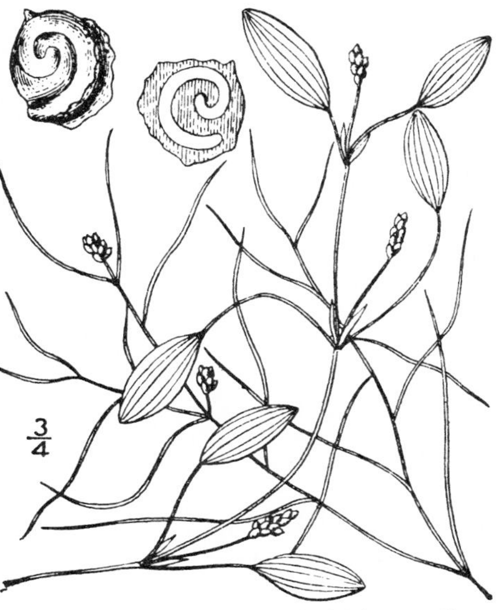

Stems flattened or sometimes terete, much-branched. Floating leaves coriaceous, the largest 2–5 mm. long, 12 mm. wide, oval or elliptic and obtuse, or lanceolate-oblong and acute; petioles shorter or sometimes longer than the blade, filiform or dilated; submerged leaves setaceous, seldom over 0.5 mm. wide, 2.5–8 cm. long; stipules 6–10 cm. long, obtuse or truncate; emersed spikes cylindric, 6–10 mm. long, many-flowered, on peduncles 6–14 mm. long, the submerged globular, few-flowered, on peduncles 4–6 mm. long; fruit cochleate, rarely over 1 mm. long, 3-keeled, the middle keel narrowly winged and usually with 7–12 knob-like teeth on the margin, the lateral keel sharp or toothed; embryo coiled 1½ times.

In still water, Maine to Florida, and Cuba, west to Texas, northern Mexico, and California. Type locality: Carolina.

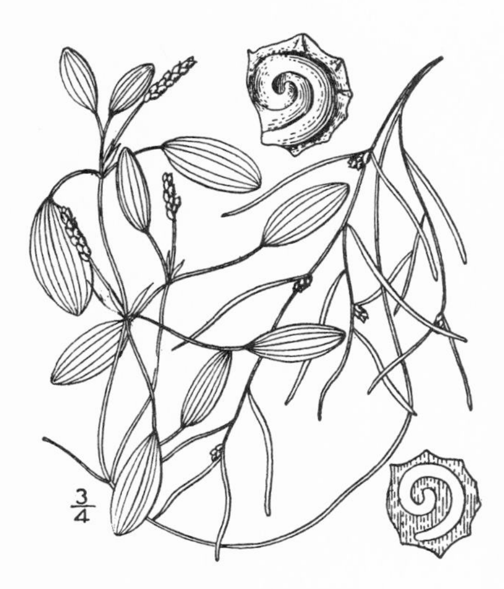

3. Potamogeton dimórphus Raf.

Spiral Pondweed. Fig. 175.

Potamogeton dimorphus Raf. Am. Month. Mag. **1**: 358. 1817.
Potamogeton spirillus Tuckerm. Am. Journ. Sci. II. **6**: 228. 1848.

Stems 1–3 dm. long, compressed, branched, the branches often short and recurved. Floating leaves oval or elliptic, obtuse, the largest about 25 mm. long and 1 cm. wide, with 5–13 nerves deeply impressed beneath; petioles often 25 mm. long; submerged leaves linear, 3–5 cm. long, about 1 mm. wide, mostly 5-nerved; stipules of the upper floating leaves free, those of the submerged leaves adnate to the blade or petiole; emersed spikes 6–10 mm. long, many-fruited, distinctly peduncled; submerged spikes globular and nearly sessile, few-flowered; fruit cochleate, roundish, less than 2 mm. long, flat and deeply impressed on the sides, 3-keeled on the back, the middle keel winged and rarely 4–5-toothed; styles deciduous; embryo spiral, about one and one-half turns.

In still, shallow water, Nova Scotia to Virginia, west to Minnesota and California, where it occurs in the Sierra Nevada and San Jacinto Mountains; Transition Zone. Type locality: Pennsylvania.

4. Potamogeton epihýdrus Raf.

Nuttall's Pondweed. Fig. 176.

Potamogeton epihydrus Raf. Med. Rep. II. **5**: 354. 1808.
Potamogeton nuttallii Cham. & Sch. Linnaea **2**: 226. 1827.
Potamogeton claytoni Tuckerm. Am. Journ. Sci. **45**: 38. 1843.

Stems simple or branched, slender, compressed, 1–2 m. long. Floating leaves elliptic, or obovate, obtuse at the apex, narrowed at base, 3–8 cm. long, 8–24 mm. wide, about 175-nerved; petioles 2–5 cm. long; stipules free from the petioles, obtuse, 2–3 cm. long; submerged leaves sessile, linear or linear-lanceolate, 5-nerved, with a cellular reticulation along the midrib, 6–14 cm. long, 3–4 mm. wide; spikes cylindric, many-flowered, 1.5–6 cm. long; nutlets deeply pitted, 3-keeled, the middle keel sharp; embryo a complete spiral.

In quiet streams and lakes, Boreal and Transition Zones; Newfoundland to British Columbia, south to North Carolina and central California. On the Pacific Coast it is frequent in British Columbia and Washington, but in California it is known only from Yosemite Valley. Type locality: Canada.

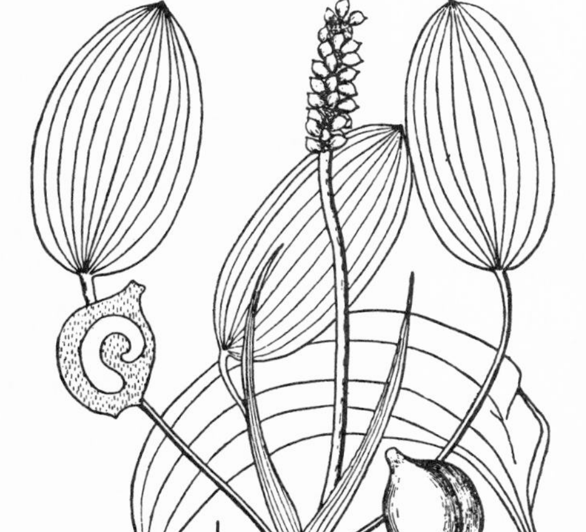

5. Potamogeton amplifòlius Tuckerm.

Large-leaved Pondweed. Fig. 177.

Potamogeton amplifolius Tuckerm. Am. Journ. Sci. II. **6**: 225. 1848.

Stems long, simple or occasionally branched. Floating leaves oval or ovate, abruptly pointed at the apex, rounded at the base, 5–10 cm. long, 3–5 cm. wide, 32–40-nerved; petioles 7–12 cm. long; stipules 2-keeled, 5–8 cm. long; submerged leaves usually petioled, the upper often elliptic or oval, 8–15 cm. long, 3–6 cm. wide, the lower lanceolate, sometimes falcate, often 20 cm. long, 3–5 cm. wide; petioles 1–3.5 cm. long, usually winged; spikes cylindric, 2.5–5 cm. long, densely flowered; peduncles 5–18 cm. long, twice as thick as the petioles; nutlets smooth, 3-keeled, the middle keel larger than the lateral; embryo a complete spiral.

Widely distributed over North America from Maine to Puget Sound, south to Guatemala. On the Pacific Coast ranging from British Columbia to the southern Sierra Nevada, mainly in the Transition Zone. Type locality: Cambridge, Mass.

6. Potamogeton angustifòlius Berch. & Presl. Ziz's Pondweed. Fig. 178.

Potamogeton angustifolius Berch. & Presl, Rost. 19. 1821.
Potamogeton zizii Roth, Enum. 1: 531. 1827.

Stems slender, branching. Floating leaves elliptic, 3–5 cm. long, 12–25 mm. wide, acute or acuminate at apex, sometimes mucronate, narrowed to the acute base, about 15-nerved; petioles shorter than the blades or sometimes longer; submerged leaves sessile or the upper petioled, lanceolate or somewhat spatulate, pellucid, minutely serrulate towards the acute, sometimes cuspidate apex, 5–15 cm. long, 0.5–2.5 cm. wide, 7-nerved, sometimes with a few intermediate secondary nerves; stipules 2-keeled, 15–30 mm. long, obtuse, axillary and free from the petioles; spikes cylindric, 2.5–4.5 cm. long, densely flowered; peduncles as thick as the stem or thicker, 6–12 cm. long; nutlets smooth, 3-keeled at maturity; 3–4 mm. long; apex of the embryo pointing directly toward the base.

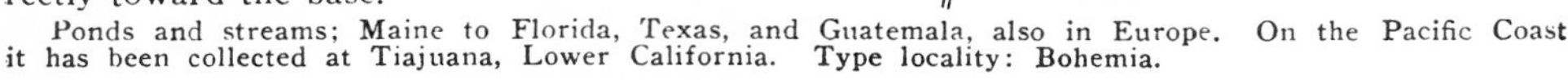

Ponds and streams; Maine to Florida, Texas, and Guatemala, also in Europe. On the Pacific Coast it has been collected at Tiajuana, Lower California. Type locality: Bohemia.

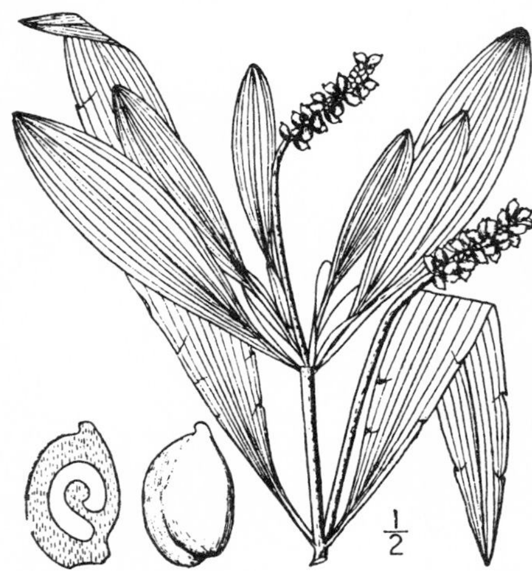

7. Potamogeton alpìnus Balbis. Northern Pondweed. Fig. 179.

Potamogeton alpinus Balbis, Misc. Bot. 13. 1804.
Potamogeton rufescens Schrad.; Cham. Adn. Fl. Ber. 5. 1815.

Whole plant of a reddish tinge; stems simple or branched, somewhat compressed. Floating leaves spatulate or oblanceolate, obtuse at the apex, narrowed at the acute base, 25–35 mm. long, 8–13 mm. wide, coriaceous; petioles as long as the blades or longer; submerged leaves sessile, oblong-lanceolate to linear-lanceolate, 7–25 cm. long, as wide or wider than the floating leaves; stipules axillary and free from the petiole, obscurely 2-keeled, 20–35 mm. long; spikes cylindric, 15–35 mm. long, many-flowered; peduncles 5–18 mm. long, as thick as the stem; nutlets smooth, or pitted when young, 3-keeled, the middle keel obscurely winged and more prominent than the smaller lateral keels; embryo an incomplete spiral, the apex pointing towards the base.

Mainly in the Boreal Zones; Yukon and Labrador to Florida and the southern Sierra Nevada, California. Also in Europe and Asia. Type locality: Lake Chamollet, France.

8. Potamogeton heterophýllus Schreb. Various-leaved Pondweed. Fig. 180.

Potamogeton heterophyllus Schreb. Spicil. Fl. Lips. 21. 1771.
Potamogeton gramineus of various American botanists, not L.

Stem simple, or simply or dichotomously branching, sometimes as long as 4 m. Floating leaves present or sometimes wanting, on petioles usually longer than the blades, ovate or elliptic-ovate, 2–6 cm. long, 7–25 mm. wide, acute or mucronate at apex, narrowed or subcordate at base, coriaceous; submerged leaves sessile or the upper petioled, 15–30 cm. long, 2–18 mm. wide, varying from narrowly linear to lanceolate or oblanceolate, translucent, 3–7-nerved; stipules axillary and free from the petiole, 15–25 mm. long; spikes cylindric, 2–4 cm. long, many-flowered; peduncles mostly twice as thick as the stem, 4–10 cm. long or sometimes becoming much elongated; nutlets slightly compressed or pitted, obscurely 3-keeled; embryo an incomplete spiral, the apex pointing toward the base.

Widely distributed over the northern hemisphere. On the Pacific Coast ranging from British Columbia to the southern Sierra Nevada, California, in the Transition and Boreal Zones. Type locality: Lindenthal, Germany.

9. Potamogeton americànus Cham. & Sch.

Long-leaved Pondweed. Fig. 181.

Potamogeton americanus Cham. & Sch. Linnaea **2**: *226*. 1827.
Potamogeton lonchites Tuckerm. Am. Journ. Sci. II. **6**: *226*. 1848.

Stem terete, much-branched, 1–2 m. long. Floating leaves elliptic, 2–12 cm. long, 5–35 mm. wide, acute at apex, rounded or rarely acutish at base, coriaceous, 17–27-nerved; petioles 4–15 cm. long, about as thick as the peduncles; submerged leaves petioled, linear-lanceolate, 10–30 cm. long, as wide or nearly as wide as the floating leaves, 7-nerved, thin and pellucid, sometimes tinged with red; stipules 2–10 cm. long; spikes cylindric, 2–6 cm. long, many-flowered; peduncles as long as or much longer than the spike, thickening upward; nutlets smooth, 3-keeled, the middle keel prominent, sometimes slightly winged; embryo a complete spiral, the curved apex pointing just inside the base.

Mainly in the Austral Zones; Vermont and Washington, to the West Indies, Mexico, and southern California. Also in Europe and Asia. Type locality: Carolina.

10. Potamogeton lùcens L.

Shining Pondweed. Fig. 182.

Potamogeton lucens L. Sp. Pl. 126. 1753.
Potamogeton proteus lucens Cham. & Schl. Linnaea **2**: 197. 1827.

Stems usually much-branched, thick. Leaves all submerged, petioled or sometimes sessile, lanceolate, elliptic or the uppermost oval, 6–16 cm. long, 2–4.5 cm. wide, acute or acuminate and cuspidate at apex, rarely rounded, rounded or attenuate at base; petioles 0.5–2 cm. long, often winged above the middle; stipules, axillary and free from the petioles, 2–7 cm. long, obtuse, 2-keeled; spikes 4.5–6 cm. long, many-flowered; peduncles as thick or thicker than the stems, 6–14 cm. long; nutlets with a groove running inward from the base, thus apparently pitted; inconspicuously keeled; style scarcely 1 mm. long; embryo a complete spiral.

Widely distributed over the northern hemisphere; on the Pacific Coast extending from British Columbia to southern California. Most common in the valleys and foothills of California. Type locality: Europe.

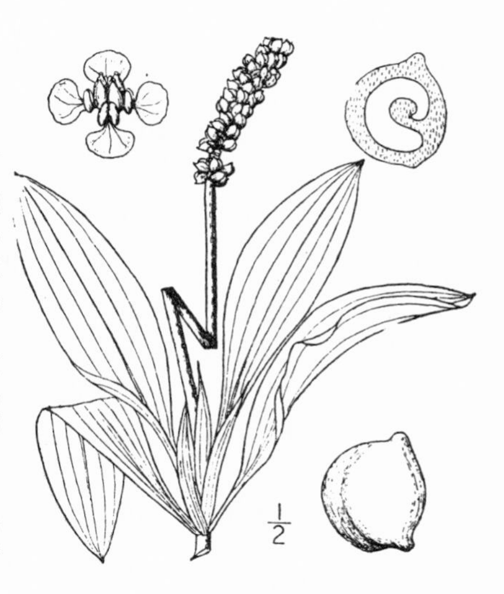

11. Potamogeton críspus L.

Curled-leaved Pondweed. Fig. 183.

Potamogeton crispus L. Sp. Pl. 126. 1753.

Stems branching, compressed. Leaves all submerged, linear-oblong to oblong-lanceolate, 2–5 cm. long, 5–15 mm. wide, sessile or semi-amplexicaul, obtuse at apex, crisped and serrulate, the midrib prominent with 2 fine lateral nerves; stipules scarious, fugacious; propagating buds prominent in the axils of decayed leaves and at the ends of branches; spikes cylindric, 12–18 mm. long; peduncles as thick or thicker than the stems, often recurved in fruit; fruit ovoid, 3-keeled, the middle keel with a spur-like projection near the base; style about 2 mm. long, nearly as long as the body of the fruit; embryo an incomplete spiral, its straight apex pointing directly toward the base.

Santa Ana River, near Corona, southern California; also in the Atlantic States, and Europe. Type locality: Europe.

12. Potamogeton praelóngus Wulf.

White-stemmed Pondweed. Fig. 184.

Potamogeton praelongus Wulf. in Roem. Arch. **3**: 331. 1805.

Stems white, flexuous, flattened, much-branched, sometimes 3 m. long. Leaves all submerged, oblong or oblong-lanceolate, 5–25 cm. long, 15–30 mm. wide, amplexicaul, bright green, with 3–5 main nerves; stipules white, scarious, obtuse, commonly closely embracing the stem; peduncles 1–3 dm. long, erect, straight, about as thick as the stem; spikes cylindric, 1–4 cm. long, thick; nutlets smooth, 1-keeled or sometimes with 2 lateral and less conspicuous keels; embryo an incomplete spiral, the straight apex pointing toward the apex.

Growing in deep water and widely distributed over the cool temperate regions of the northern hemisphere. On the Pacific Coast ranging from British Columbia southward through the Boreal Zones to the southern Sierra Nevada, California. Type locality: Europe.

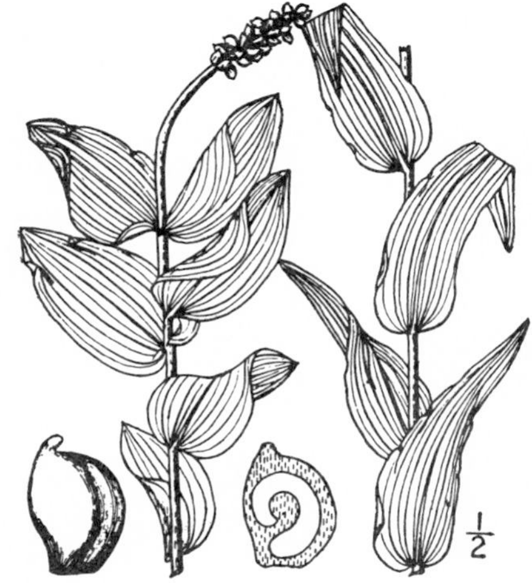

13. Potamogeton richardsònii (A. Benn.) Rydb.

Richardson's Pondweed. Fig. 185.

Potamogeton perfoliatus richardsonii A. Benn. Journ. Bot. **27**: 25. 1889.
Potamogeton richardsonii Rydb. Bull. Torrey Club **32**: 599. 1905.

Stems branched. Leaves all submerged, long-lanceolate from a cordate-clasping base, acuminate, 3–11 cm. long, wavy, pale bright green; stipules conspicuous, at least as shreds; peduncles 5–15 cm. long, thickened upward and somewhat spongy; spikes cylindric, 15–35 mm. long; nutlet irregularly obovoid, 4 mm. long, obscurely 3-carinate on the back, the sides with a shallow indentation that runs into the face; style nearly facial.

Widely distributed over the cool temperate regions of North America. On the Pacific Coast ranging from British Columbia to the Sierra Nevada, California. Mainly confined to the Boreal Zones. Type locality: "along the Great Lakes."

14. Potamogeton foliòsus Raf.

Leafy Pondweed. Fig. 186.

Potamogeton foliosus Raf. Med. Rep. II. **5**: 354. 1808.
Potamogeton pauciflorus Pursh, Fl. Am. Sept. 121. 1814.
Potamogeton foliosus californicus Morong, Bot. Gaz. **10**: 254. 1885.
Potamogeton californicus Piper, Contr. Nat. Herb. **11**: 98. 1906.

Stems flattened, leafy, very rarely with propagating buds. Leaves all submerged, linear, 1.5–8 cm. long, 0.5–2 mm. wide, obtuse or more or less pungent at apex, 3–5-nerved, the midrib prominent; stipules axillary and free from the leaf bases, 15–25 mm. long, obtuse or acute; spikes few-flowered, usually not more than 12 nutlets maturing; peduncles usually 1–2 cm. long, as thick or thicker than the stem; nutlets pitted, 3-keeled, the middle keel winged and more or less dentate; embryo an incomplete spiral, the apex pointing toward the base or outside it.

In rather shallow, still or slowly running water; New Brunswick to British Columbia, south to the West Indies and Mexico. On the Pacific Coast it is found mainly in the Sonoran Zones from eastern Washington to southern California. Type locality: Carolina.

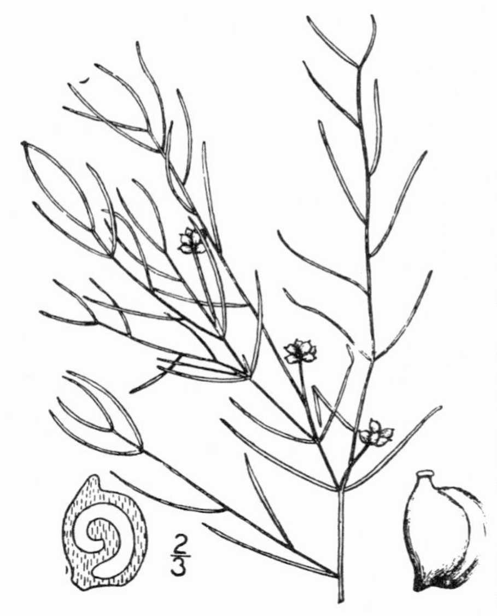

15. Potamogeton pusíllus L.

Small Pondweed. Fig. 187.

Potamogeton pusillus L. Sp. Pl. 127. 1753.

Stems filiform, branching, 3–6 dm. long. Leaves all submerged, narrowly linear, 2–6 cm. long, 0.5–1.5 mm. wide, 1–3-nerved, the lateral nerves often obscure, acute or subacute, furnished with translucent glands on each side at the base; winter propagating buds occasional; stipules short, hyaline, free from the leaves, soon deciduous; spikes interrupted or capitate, 3–10-flowered; peduncles 0.5–3 cm. long; as thick as the stem or thinner, erect or recurved; nutlets smooth or ultimately pitted, distinctly 3-keeled, 1.5–2 mm. long; apex of the embryo incurved and pointing inside of the base.

In ponds and slow running streams, throughout the northern hemisphere. Common on the Pacific Coast from British Columbia to southern California, in both the Boreal and Austral Zones. Type locality: Europe.

16. Potamogeton comprèssus L.

Eel-grass Pondweed. Fig. 188.

Potamogeton compressus L. Sp. Pl. 127. 1753.
Potamogeton zosteraefolius Schum. Enum. Pl. Saell. 50. 1801.

Stems much flattened, sometimes winged, widely branching, often developing propagating buds. Leaves all submerged, linear, 5–20 cm. long, 4–8 mm. wide, grass-like, abruptly pointed, with 3 principal and many fine nerves; stipules scarious, oblong, very obtuse, soon deciduous, free from the leaves; peduncles 4–10 cm. long, as thick as the stems; spikes cylindric, many-flowered, 15–20 mm. long, often interrupted; nutlets smooth, 3-keeled, the lateral keels smaller than the toothed or undulate middle one; embryo a complete spiral, the curved apex pointing inside the base.

Widely distributed over the cool temperate regions of the northern hemisphere, mainly in the Boreal Zones; on the Pacific Coast it ranges from British Columbia south to Honey Lake, Lassen County, California, but it is not common and has been collected in only a few localities. Type locality: Europe.

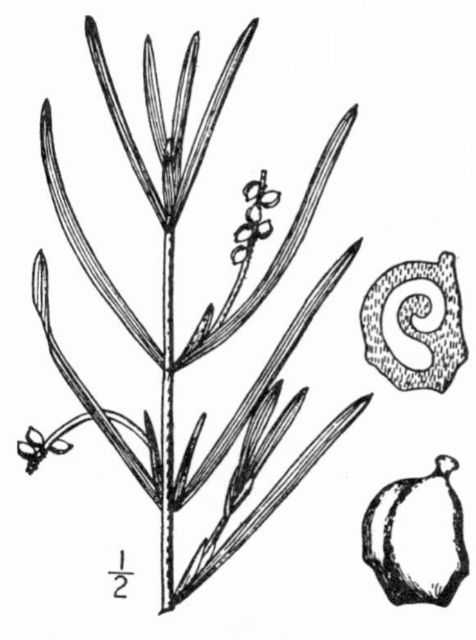

17. Potamogeton latifòlius (Robb.) Morong.

Nevada Pondweed. Fig. 189.

Potamogeton latifolius Morong, Mem. Torrey Club. **3**: 52. 1893.
Potamogeton pectinatus latifolius Robb. in S. Wats. Bot. King Expl. 338. 1871.

Stem whitish, branched. Leaves all submerged, linear, 2–8.5 cm. long, 1–1.5 mm. wide, acute or obtuse at the apex; primary nerves 3–5 with numerous reticulated cross nerves; stipules adnate to the leaf-bases, scarious, the sheath 1–2.5 cm. long, the free part shorter; spikes cylindric, 2–3 cm. long, interrupted below; peduncles whitish, thinner than the stems, 2.5–8 cm. long; nutlets smooth, mostly without keels, rarely with one inconspicuous keel; embryo a complete spiral, the curved apex pointing inside the base.

A local species, restricted to northwestern Nevada and northeastern California, in Modoc and Lassen Counties. Transition and Upper Sonoran Zones. Type locality: Humboldt River, below Humboldt Lake, Nevada.

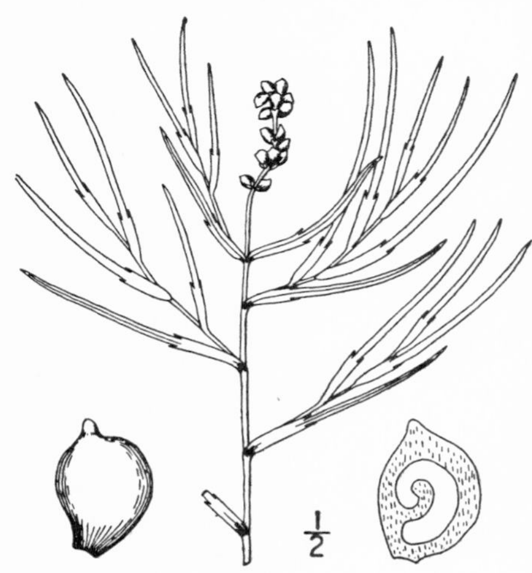

18. Potamogeton robbínsii Oakes.

Robbins' Pondweed. Fig. 190.

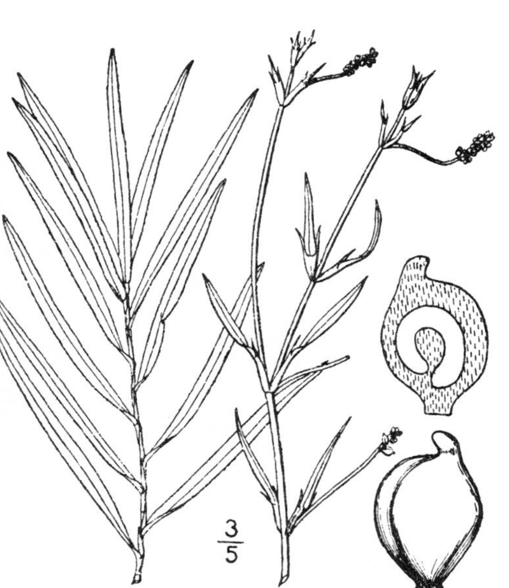

Potamogeton robbinsii Oakes, Mag. Hort. Hovey 7: 180. 1841.

Stems slender, branched and sometimes rooting at the nodes. Leaves all submerged, crowded, linear, 8–13 cm. long, 2.5–4 mm. wide, apiculate or acute at the apex, somewhat clasping at base, minutely serrulate throughout, many-nerved; stipules adnate to the leaf-bases, the sheath 1–1.5 cm. long, the free portion longer; peduncles 3–10 cm. long, a little more slender than the stem; spike slender, 1–2 cm. long, many-flowered but sparingly fruiting; nutlets slightly pitted, 3-keeled, the middle keel more prominent than the lateral; embryo a complete spiral, the apex pointing toward the base or inside it.

In still water in the cool temperate region of North America; New Brunswick, British Columbia, and Pennsylvania to the Pacific Coast, where it is sparingly distributed from British Columbia to Honey Lake, Lassen County, California, in the Boreal and Transition Zones. Type locality: Pondicherry Pond, Jefferson, New Hampshire.

19. Potamogeton pectinàtus L.

Fènnel-leaved Pondweed. Fig. 191.

Potamogeton pectinatus L. Sp. Pl. 126. 1753.
Potamogeton columbianus Suksdorf, Deutsch. Bot. Monatss. 19: 92. 1901.

Stems slender, 0.5–2 m. long, much branched, the branches repeatedly dichotomous. Leaves very narrowly linear or commonly setaceous, attenuate to the apex, 1-nerved with a few transverse veins; stipules adnate half their length, 15–25 mm. long, their sheaths scarious on the margins; peduncles filiform, 5–25 mm. long; the flowers in 2–6 verticels; nutlets obliquely obovoid, rounded on the back, 3–4 mm. long, with two obscure keels; embryo a complete or incomplete spiral, the embryo pointing toward the base or inside it.

In still, fresh or brackish water, and cosmopolitan in its distribution. On the Pacific Coast it is the most widely distributed and abundant species. Type locality: Europe.

2. RÚPPIA L. Sp. Pl. 127. 1753.

Stems capillary, widely branched. Leaves all submerged, very slender, attenuate, 1-nerved with membranous sheaths at the base. Flowers on a capillary spadix-like peduncle, naked, perfect, consisting of 2 sessile anthers, 2-celled, attached by the back to the peduncle, having between them several pistillate flowers with sessile peltate stigmas in 2 sets on opposite sides of the rachis, the whole at first enclosed in the sheathing base of the leaf; in development the peduncle elongates, bearing the pistillate flowers at the end; fertilization takes place at the surface, after which the peduncle coils up. Fruit a small obliquely pointed drupe, pediceled; embryo oval, curved. (Name in honor of Heinrich Bernhard Rupp, a German botanist.)

A genus of three or four species, growing in salt and brackish water all over the world. Type species, *Ruppia maritima* L.

Sheaths of the stipules 10 mm. long or less, the free part shorter. — 1. *R. maritima.*
Sheaths of the stipules 15 mm. long, the free part as long. — 2. *R. occidentalis.*

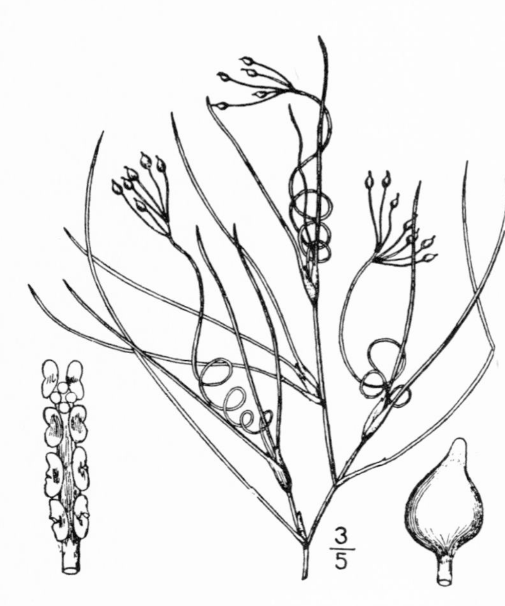

1. Ruppia marítima L.

Ditch-grass. Fig. 192.

Ruppia maritima L. Sp. Pl. 127. 1753.
Ruppia curvicarpa A. Nels. Bull. Torrey Club **26**: 122. 1899.

Stems branching, 6–10 dm. long, often whitish, the internodes irregular, naked, 2.5–8 cm. long. Leaves all submerged, thread-like, 0.3 mm. wide or less, 2–10 cm. long, acute or pungent at apex; stipular sheath, 6–10 mm. long, the free part much shorter or wanting; flowers on a short peduncle which elongates in anthesis and ultimately coils into a loose spiral; anthers 2, sessile, early deciduous, 2-celled; fruit ovoid, equilateral or gibbous and oblique, about 2 mm. long, stipitate; style short and stout, or finely attenuate, straight or hooked; fruiting pedicels 1.5–3 cm. long.

Brackish or salt water along the seashore and in the interior regions; widely distributed over the northern hemisphere. On the Pacific Coast it is found from British Columbia to southern California, occurring in both the Boreal and Austral Zones.

2. Ruppia occidentàlis S. Wats.

Western Ditch-grass. Fig. 193.

Ruppia occidentalis S. Wats. Proc. Am. Acad. **25**: 138. 1890.
Ruppia lacustris Macoun, Cat. Can. Pl. **5**: 372. 1890.

Stems slender, 1 m. long or less, with divaricately ascending branches. Leaves all submerged, filiform, less than 0.3 mm. wide, 7–20 cm. long; stipular sheaths 1.5–3 cm. long, the free part as long; inflorescence and flowers similar to the preceding; fruit pear-shaped, tipped by a straight style; fruiting pedicels less than 2.5 mm. long.

A comparatively little known species, which has been collected only at Kamloops (type locality), British Columbia, in Nebraska, and at Lake Chelan, Washington.

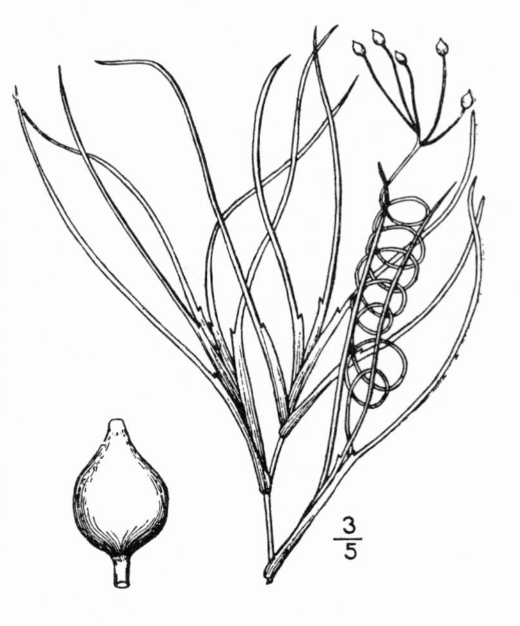

3. ZANNICHÉLLIA L. Sp. Pl. 969. 1753.

Stems capillary, sparsely branched from a creeping rhizome. Leaves all submerged, filiform but flat, 1-nerved. Staminate and pistillate flowers in the same axil, enclosed in the bud by a hyaline spathe-like envelope; staminate solitary, with 2-celled anther on a short pedicel-like filament; pistillate 2–5. Ovary flask-shaped, stipitate at base, tapering into a short style with a broad cup-shaped stigma, its margins angled or dentate. Fruit a flattish falcate nutlet, ribbed or sometimes toothed on the back. (In honor of J. H. Zannichelli 1662–1729, Italian physician and botanist.)

A genus of two or three species of wide distribution. Type species, *Zannichellia palustris* L.

1. **Zannichellia palústris** L.

Horned Pondweed. Fig. 194.

Zannichellia palustris L. Sp. Pl. 969. 1753.

Stems capillary, sparsely branched, from creeping rhizomes. Leaves all submerged, capillary, 2–10 cm. long, acute or almost pungent at the apex, 1-nerved; stipules scarious, free from the leaf-bases, scarcely 2 cm. long; staminate and pistillate flowers in the same leaf-axil, subtended or inclosed by a hyaline bract; staminate flower of a solitary 2-celled anther on a slender filament; pistillate flowers of 2–10 carpels, sessile at first, often pediceled after anthesis; carpels flask-shaped, ribbed or toothed on the margins, or sometimes smooth; style recurved, persistent; stigma cup-shaped, with dentate or angled edges; mature fruit 2–4 mm. long, rarely pitted.

In fresh or brackish water, nearly cosmopolitan in its distribution. Type locality: Europe.

Family 4. **NAIADÀCEAE.**

Naias Family.

Slender, branching, submerged aquatics from fibrous roots. Leaves all submerged, opposite or whorled, spiny-toothed, sheathing at the base. Flowers monoecious or dioecious, axillary, solitary, sessile or pediceled. Staminate with a double perianth; the outer entire or 4-toothed at the apex, the inner one hyaline, adhering to the anthers. Stamen 1, sessile or stalked; anthers 1–4-celled, apiculate or 2-lobed at the summit. Pistillate flowers of a single ovary, tapering into a short style; stigmas 2–4, subulate. Fruit a solitary carpel, sessile, ellipsoidal, with a crustaceous pericarp.

A family with only the following genus.

1. **NÀIAS.** L. Sp. Pl. 1015. 1753.

Slender branching submerged aquatics, with the characters of the family. [Greek, water-nymph.]

A genus of about 10 species living in fresh water, and widely distributed over the world. Type species, *Naias marina* L.

Flowers dioecious; internodes and back of leaf spiny. — 1. *N. marina.*
Flowers monoecious; internodes and back of leaf unarmed.
  Anthers 1-celled; seeds shining, appearing smooth. — 2. *N. flexilis.*
  Anthers 4-celled; seeds dull, reticulate. — 3. *N. guadalupensis.*

1. **Naias marìna** L. Large Naias. Fig. 195.

Naias marina L. Sp. Pl. 1015. 1753.
Naias major All. Fl. Ped. **2**: 221. 1785.
Naias marina californica Rendle, Trans. Linn. Soc. II. **5**: 398. 1889.

Stems branched, sometimes dichotomously, armed with brownish spinulose teeth on the internodes. Leaves opposite or alternate, stiff or recurved, narrowly linear, 1–4.5 cm. long, 0.5–3 mm. wide, with toothed margins, and often dorsally toothed on the midrib, the teeth triangular, apiculate, 1 mm. or more long; leaf-bases sheathing, the sheaths rounded, without teeth or rarely with a few short teeth; flowers dioecious, the staminate 3–4 mm. long; anther 4-celled; pistillate flower usually 3–4 mm. long; stigmas 3, sometimes one shorter than the others; mature fruit apparently tessellated in dried specimens, smooth when fresh.

Growing in quiet fresh water and widely distributed over the world. On the Pacific Coast ranging from Clear Lake, California, to Lower California, but not common. Type locality: Europe.

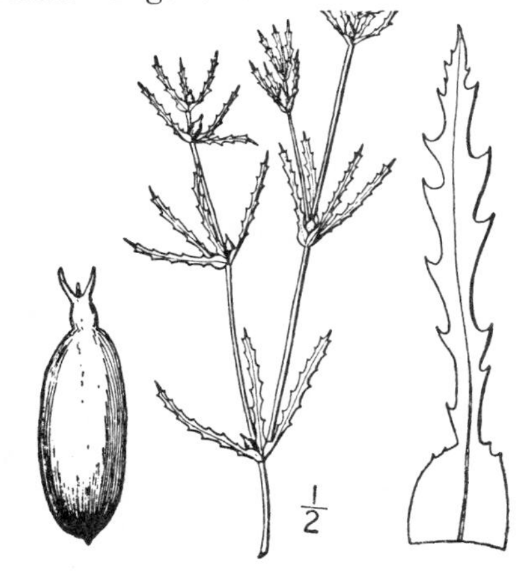

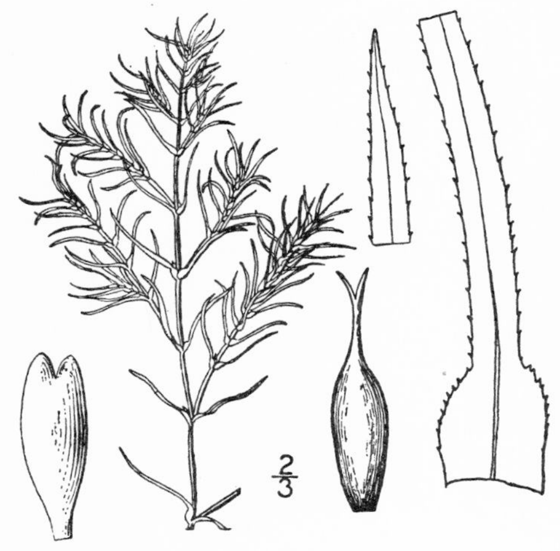

2. Naias fléxilis (Willd.) Rost. & Schmidt.
Slender Naias. Fig. 196.

Caulina flexilis Willd. Abh. Akad. Berlin, 95. 1803.
Naias flexilis Rost. & Schmidt, Fl. Sed. 384. 1824.

Stems branching, slender, 1–2 m. long, or sometimes shorter. Leaves linear, 15–25 mm. long, 1–2 mm. wide, acuminate or abruptly acute, pellucid, with 20–30 minute teeth on each margin, numerous and crowded on the upper parts of the branches; sheaths obliquely rounded, with 5–10 teeth on each margin; flowers monoecious; staminate flower 2.5–3 mm. long, the anther 1-celled; pistillate flower 1–2.5 mm. long, the stigmas 2–4, usually 3, subulate; seed ellipsoid, apparently smooth and shining, but under a strong lens reticulate with minute 6-sided areolae.

In ponds and streams nearly throughout North America and Europe. On the Pacific Coast it ranges from British Columbia to southern California, occurring in the Boreal and Austral Zones, but not common. Type locality: Pennsylvania.

3. Naias guadalupénsis (Spreng.) Morong.
Guadaloupe Naias. Fig. 197.

Caulinia guadalupensis Spreng. Syst. 1: 20. 1825.
Naias microdon A. Br. Bot. Zeit. 26: 510. 1868.
Naias guadalupensis Morong, Mem. Torrey Club 3: 60. 1893.

Stems slender, branching. Leaves all submerged, narrowly linear, 12–25 mm. long, about 0.5 mm. wide, acute at the apex, and usually tipped with 1 or 2 spines, marginal teeth 20–40, inconspicuous, often wanting; leaf-sheaths oblique or rounded, not auriculate, toothed; flowers monoecious; staminate flowers 2–3 mm. long, the anthers 4-celled; pistillate flowers 2–3 mm. long; fruit crowned with 2 or 3 stigmas and usually with 1 or 2 spiny sterile stigmatic processes; seeds ellipsoid, dull, reticulate with 4-sided areolae.

In quiet water, ranging from Pennsylvania to Oregon and southward to Central America. On the Pacific Coast it occurs in Oregon and in the San Francisco Bay region. Type locality: Guadaloupe Island.

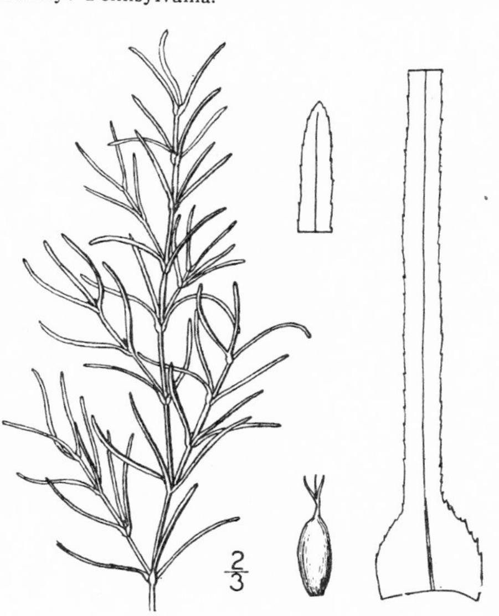

Family 5. ZOSTERÀCEAE.

Eel-grass Family

Submerged marine plants with creeping rootstocks, flattened branching stems and sheathing 2-ranked ribbon-like leaves. Flowers monoecious or dioecious, arranged on a one-sided spadix enclosed in a spathe. Perianth none, but flowers enclosed by a hyaline scale. Staminate flowers a single sessile anther arranged in two rows on the spadix; pollen of slender filaments. Pistillate flower of a single 1-celled ovary, with 2 carpels and 2 slender stigmas.

A family of marine plants comprising two genera.

Flowers monoecious; ovary and fruit ovoid. — 1. *Zostera.*
Flowers dioecious; ovary and fruit heart-shaped. — 2. *Phyllospadix.*

1. ZOSTÈRA L. Sp. Pl. 968. 1753.

Marine plants, wholly submerged, with slender rootstocks and branching compressed stems. Leaves 2-ranked, sheathing at the base, the sheaths with inflexed margins. Spadix linear, contained in a spathe. Flowers monoecious, arranged alternately in 2 rows on the spadix. Staminate flower merely an anther attached to the spadix near its apex, 1-celled; pollen thread-like. Pistillate flower fixed on its back near the middle. Ovary 1; style elongated; stigmas capillary. Mature carpels flask-like, beaked by the persistent style. Seeds ribbed; embryo ellipsoidal.

A genus of about 5 species. The following is the type of the genus.

1. Zostera marìna L. Eel-grass. Fig. 198.

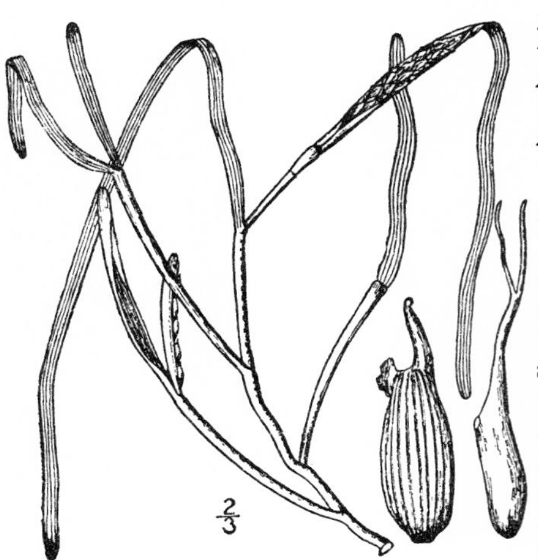

Zostera marina L. Sp. Pl. 968. 1753.
Zostera marina latifolia Morong, Bull. Torrey Club **13**: 160. 1886.
Zostera pacifica S. Wats. Proc. Am. Acad. **26**: 131. 1891.
Zostera oregona S. Wats. Proc. Am. Acad. **26**: 131. 1891.
Zostera latifolia Morong, Mem. Torrey Club **3**²: 63. 1893.

Stems branched, 1–3 m. long, from a thick rootstock. Leaves ribbon-like, obtuse at the apex, 3–15 dm. long, 2–8 mm. wide, with 3–7 principal nerves; spadix 25–60 mm. long; flowers crowded, usually 10–20 of each kind on the spadix, 6 mm. long; the anthers at anthesis escape from the spathe and discharge the glutinous filamentous pollen into the water; ovary somewhat vermiform; the stigmas at anthesis protrude through the spathe and drop off before the anthers of the same spadix open, the ribs showing clearly through the pericarp, seed strongly 20-ribbed, about 3 mm. long and 1 mm. in diameter, truncate at both ends.

In bays and estuaries, usually growing on muddy bottoms; widely distributed in the Atlantic and Pacific Oceans. The common form of the Pacific Coast has broad leaves, 6–12 mm. wide, the broadest with 10–13 principal nerves. Type locality: Europe.

2. PHYLLOSPÀDIX Hook. Fl. Bor. Am. 2: 171. 1838.

Submerged marine plants with thickened rootstocks and slender stems, which bear the inflorescence at the summit or in clusters along the upper part. Leaves linear, sheathing. Flowers dioecious in spathes like those of *Zostera*. Spathes with membranous edges, the back thickened and terminating in long leaf-like appendages. Spadix with a series of short dilated foliaceous flaps, which close over the flower, spreading open at maturity. Staminate flowers of numerous sessile stamens in 2 rows; anthers 1-celled. Pistillate of simple sessile ovaries, attenuate into a short style; stigmas 2, capillary. Fruit beaked by the short persistent style, cordate-sagittate. [Greek, referring to the leaf-like appendages of the spathe.]

A genus of two species, peculiar to the Pacific Coast of North America. Type species, *Phyllospadix sconleri* Hook.

Spadices several, cauline. 1. *P. torreyi.*
Spadix usually solitary, basal. 2. *P. scouleri.*

1. Phyllospadix tórreyi S. Wats.
Torrey's Surf-grass. Fig. 199.

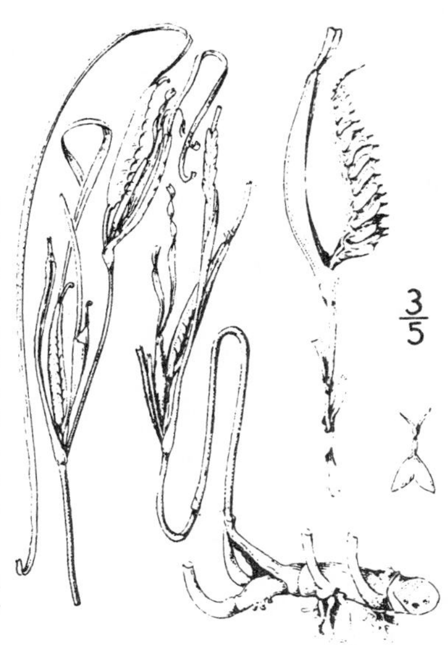

Phyllospadix torreyi S. Wats. Proc. Am. Acad. **14**: 303. 1879.

Stems simple or branched, flat, 3–6 dm. long. Leaves all submerged, linear, flat when young, becoming complicate or terete, 2 m. long or less, 2 mm. wide or less, primary nerves 1–3, or the leaves sometimes nerveless, the base sheathing; spadices cauline, in 2 or 3 pairs, usually about 5 cm. long, each pair subtended by several fugacious bracts; staminate flowers numerous, in 2 rows, consisting of a single 1-celled anther; pistillate flowers numerous, arranged in 2 rows; fruit flask-shaped, 2–3 mm. long, beaked by the persistent style, deeply cordate-sagittate at the base, with 2 projecting wings, and sometimes winged on the back.

Growing among rocks in the surf at low tide mark to about 3 fathoms. Bolinas Bay, California, south to Ensenada, Lower California. Type locality: Santa Barbara.

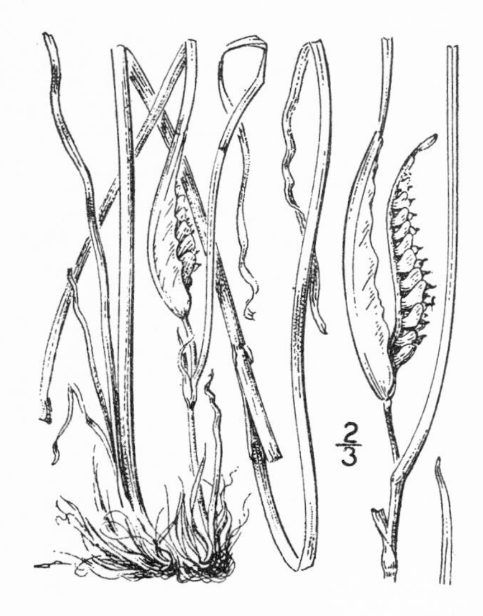

2. Phyllospadix scoùleri Hook.

Scouler's Surf-grass. Fig. 200.

Phyllospadix scouleri Hook. Fl. Bor. Am. 2: 171. 1838.
Phyllospadix serrulatus Rupr.; Asch. Linnaea 35: 169. 1867.

Stem simple or branched, winged, 1–4 dm. long. Leaves all submerged, ribbon-like, flat, 0.5–2 m. long, 2–4 mm. wide, entire or rarely serrulate, obtuse at the apex, sheathing at the base; primary nerves 3; peduncle simple solitary or sometimes in pairs, basal or nearly basal; spathe prominently scarious-winged; staminate spadix about 6 cm. long; anthers about 4.5 mm. long; fruit flask-shaped, about 3.5 mm. long, beaked by the persistent style, deeply cordate-sagittate at the base, projections obliquely striate or costate on the winged margin.

Growing on rocks in the surf; Vancouver Island to Santa Barbara, California; also in Japan on the coast of Hokaido. Type locality: Dundas Island, Columbia River.

Family 6. LILAEÀCEAE.

Flowering Quillwort Family.

Acaulescent, annual marsh plants, with radical, terete, sheathing leaves, the sheaths often containing minute, hyaline scales. Flowers perfect or unisexual in two types of inflorescence, basal and scapose. Perianth none. Basal flowers enclosed by the sheathing leaf-bases, pistillate; carpels solitary, 1-seeded; style long and filamentous, tipped by a capitate stigma; fruit a ribbed caryopsis. Scapose inflorescence of staminate, pistillate and perfect flowers irregularly disposed in a spike; staminate flowers of a single, sessile, 2-celled anther, the connective dilated and bract-like; pistillate flowers of single, 1-seeded carpels, subtended by a bract or bractless, the lower with a long style, the upper with shorter styles, the stigmas papillose; fruit a winged caryopsis; perfect flowers with a stamen and pistil similar to those in the unisexual flowers but united at the base.

A family containing a single monotypic genus of the Pacific regions of North and South America.

1. LILAÈA Humb. & Bonpl. Pl. Aequin. 1: 222. 1808.

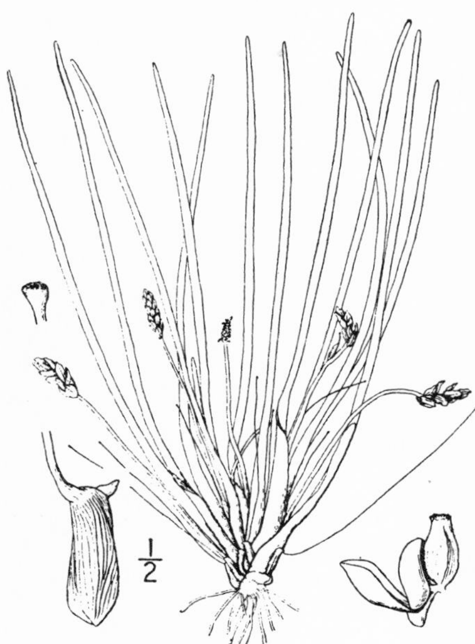

A monotypic genus with the characters of the family. [Name in honor of the French botanist, A. R. Delile. 1778–1850.]

1. Lilaea subulàta Humb. & Bonpl.

Flowering Quillwort. Fig. 201.

Lilaea subulata Humb. & Bonpl. Pl. Aequin. 1: 222. *pl. 63.* 1808.
Heterostylus gramineus Hook. Fl. Bor. Am. 2: 171. 1838.

Acaulescent annual. Leaves terete, fleshy, 5–35 cm. long, 2–4.5 mm. thick, tapering to the acute apex, stipular sheaths hyaline, 4–10 cm. long; basal flowers with styles 10 cm. long or longer, scapose flowers spicate, on terete scapes, 4–20 cm. long; spikes 0.5–2 cm. long, usually many-flowered; anthers nearly sessile, their connective dilated into a thin membranous oblong-cuspidate bract-like appendage; the pistillate flowers of the spike bracted or bractless, the styles of the lower flowers 6 mm. long or longer, much shorter in the upper flowers; mature fruit of the basal flowers, 4–6.5 mm. long, ribbed, those of the scapose flowers smaller, winged.

Growing in mud about lakes, pools, and slow running streams; British Columbia southward through the mountains and lowlands of the Pacific region to Mexico; also in South America. Type locality: Santa Fe de Bogota.

Family 7. SCHEUCHZERIÀCEAE.

Arrow-grass Family.

Marsh herbs with rush-like leaves and small spicate or racemose perfect flowers. Perianth 4–6-parted, its segments in two series, persistent or deciduous. Stamens 3–6, with elongated or short filaments and mostly 2-celled extrorse anthers. Carpels 3–6, 1–2-ovuled, more or less united until maturity, dehiscent or indehiscent. Seeds anatropous. Embryo straight.

A family of 3 genera, and about 14 species, of wide geographic distribution.

Flowers numerous in spike-like racemes terminating scapes; leaves all basal. 1. *Triglochin.*
Flowers few in a loose raceme; leaves cauline. 2. *Scheuchzeria.*

1. TRIGLÒCHIN L. Sp. Pl. 338. 1753.

Marsh plants with radical semiterete fleshy leaves, which have membranous sheaths at the base. Flowers small, perfect in spikes or racemes, on long smooth naked scapes. Perianth-segments 3–6, concave, the 3 inner inserted higher up than the others when present. Stamens 3–6; anthers 2-celled, extrorse, sessile or nearly so, inserted at the base of the segments and deciduous with them. Ovaries 3–6, united or rarely free, 1-celled; style short; stigmas as many as ovaries, plumose. Fruit of 3–6 oblong or ovoid carpels, when ripe separating from the base upward from a persistent central axis, dehiscing by a ventral suture. (Greek, referring to the fruit of the 3-carpeled species.)

A genus of about 12 species, distributed over the temperate and subarctic zones of both hemispheres. Type species, *Triglochin palustris* L.

Carpels 3.
  Fruit linear or clavate, tapering to a subulate base. 1. *T. palustris.*
  Fruit nearly globose. 2. *T. striata.*
Carpels 6; fruit oblong or ovoid, obtuse at base. 3. *T. maritima.*

1. Triglochin palústris L.

Marsh Arrow-grass. Fig. 202.

Triglochin palustris L. Sp. Pl. 338. 1753.

Rootstock short, oblique, with slender fugacious stolons. Leaves linear, shorter than the scapes, 12–30 cm. long, tapering to a sharp tip; ligule very short; scapes 1 or 2, slender, striate, 2–6 dm. high; racemes 12–30 cm. long; pedicels capillary, in fruit 5–7 mm. long, appressed; perianth-segments 6, greenish yellow; anthers 6, sessile; pistil of 3 united carpels, 3-celled, 3-ovuled; stigmas sessile; fruit 6–7 mm. long, about 1.5 mm. thick, linear or clavate, the ripe carpels separating from the axis and hanging suspended from the apex, the axis 3-winged.

Growing in bogs and widely distributed over the subarctic and cool-temperate regions of both hemispheres. On the Pacific Coast found only at Colville, eastern Washington, in the Arid Transition Zone. Type locality: Europe.

2. Triglochin striàta Ruiz & Pav.

Three-ribbed Arrow-grass. Fig. 203.

Triglochin striata Ruiz & Pav. Fl. Per. 3: 72. 1802.
Triglochin triandra Michx. Fl. Bor. Am. 1: 208. 1803.

Rootstocks upright or oblique. Scapes 1 or 2, more or less angular, usually not over 20 cm. high; leaves slender, slightly fleshy, nearly as long or longer than the scapes, 2 mm. or less wide; racemes 3–12 cm. long, spicate; pedicels 1–2 mm. long, not elongating in fruit, spreading; perianth-segments 3, greenish yellow; stamens 3; anthers oval; pistil of 3 united carpels; fruit subglobose or somewhat obovoid, about 2 mm. in diameter, appearing 3-winged when dry by the contracting of the carpels; carpels coriaceous, rounded and 3-ribbed on the back; axis broadly 3-winged.

In saline marshes, along the coast from Maryland to Louisiana, and in California; also in Mexico and South America. In California this species has been found at Santa Cruz, Monterey, and Santa Barbara. Austral Zones. Type locality: Surco, Peru.

3. Triglochin marìtima L.

Seaside Arrow-grass. Fig. 204.

Triglochin maritima L. Sp. Pl. 339. 1753.
Triglochin elata Nutt. Gen. 1: 237. 1818.
Triglochin maritima debilis M. E. Jones, Proc. Calif. Acad. II. 5: 723. 1895.
Triglochin concinna Davy, Erythea 3: 117. 1895.

Rootstock without stolons, often subligneous, the caudex thick, mostly covered with the sheaths of old leaves. Scape stout, nearly terete, 1–7 dm. high; leaves half-cylindric, usually about 2 mm. wide; raceme elongate, often 4 dm. long or more; pedicels decurrent, 2–3 mm. long, slightly longer in fruit; perianth-segments 6, each subtending a large sessile anther; pistil of 6 united carpels; fruit oblong or ovoid, 5–6 mm. long, 2–4 mm. thick, obtuse at the base, with 6 recurved tips at the summit; carpels 3-angled, flat or slightly grooved on the back, or the dorsal edges curved upward and winged, separating at maturity from the hexagonal axis.

In saline marshes along the coast and in interior regions, Boreal and Austral Zones; widely distributed over the northern hemisphere. On the Pacific Coast it ranges from Alaska to Lower California. Type locality: Europe. A slender form with more open racemes has been recognized as distinct; the last two synonyms refer to it.

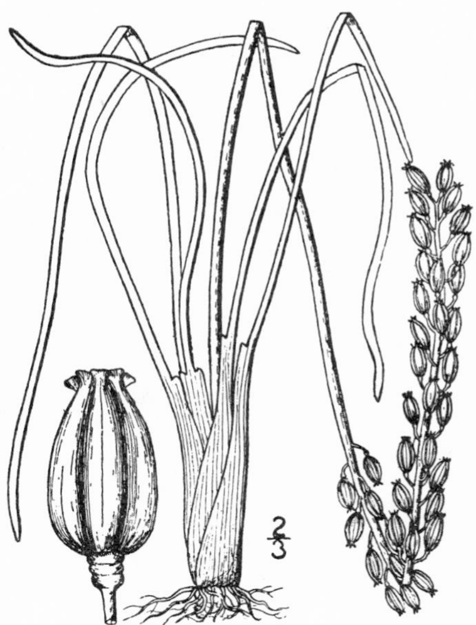

2. SCHEUCHZÈRIA L. Sp. Pl. 338. 1753.

Rush-like bog perennials with creeping rootstocks, and erect leafy stems, the leaves elongate, half-rounded below and flat above, striate, furnished with a pore at the apex and a membranous ligulate sheath at the base. Flowers small, racemose, bracted. Perianth 6-parted, regularly 2-serial, persistent. Stamens 6, inserted at the base of the perianth-segments; filaments elongate; anthers linear, basifixed, extrorse. Ovaries 3 or rarely 4–6, distinct or connate at the base, 1-celled, each cell with 1–3 collateral ovules. Stigmas sessile, papillose or slightly fimbriate. Carpels divergent, inflated, coriaceous, 1- or 2-seeded, follicle-like, laterally dehiscent. Seeds straight or slightly curved, without endosperm. (Name in honor of Johann Jacob Scheuchzer, 1672–1733, Swiss scientist.)

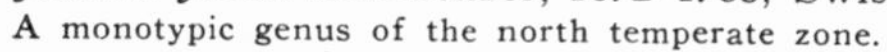

A monotypic genus of the north temperate zone.

1. Scheuchzeria palústris L.

Scheuchzeria. Fig. 205.

Scheuchzeria palustris L. Sp. Pl. 338. 1753.

Leaves 10–40 cm. long, the upper ones reduced to bracts; stems solitary or several, usually clothed at the base with the remains of the old leaves, 10–40 cm. high; sheaths of the basal leaves often 10 cm. long with a ligule 12 mm. long; pedicels spreading in fruit; flowers white, few in a lax raceme; perianth-segments oblong, membranaceous, 1-nerved, 3 mm. long, the inner ones narrower; lower fruiting pedicels 5–25 mm. long; follicles 4–8 mm. long, slightly if at all united at base; seeds oval, brown, with a very hard coat.

Growing in sphagnum bogs and on wet shores, Boreal Zones; widely distributed over the cooler regions of the northern hemisphere. On the Pacific Coast it ranges from Alaska to Sierra County, California, but is infrequent. Type locality: Lapland.

Family 8. ALISMÀCEAE.

Water-plantain Family.

Aquatic or marsh plants, with scapose stems and basal long-petioled sheathing leaves. Inflorescence racemose or paniculate. Flowers regular, perfect, monoecious or dioecious, pediceled, the pedicels in whorls and subtended by bracts. Perianth-segments 6, the outer three small, herbaceous, persistent; the inner three larger and petaloid, deciduous. Stamens 6 or more; anthers 2-celled, extrorse or dehiscing by lateral slits. Ovaries numerous, distinct, on a flat or convex receptacle, 1-celled, 1-ovuled. Carpels becoming achenes in fruit.

A family of about 13 genera and 50 species, of wide distribution in fresh-water swamps and streams.

Carpels borne in one series; achenes verticillate.
  Style lateral; achenes minutely beaked; anthers short; petals entire. 1. *Alisma.*
  Style apical; achenes long-beaked; anthers elongate; petals incised. 2. *Machaerocarpus.*
Carpels borne in several series; achenes capitate.
  Flowers perfect. 3. *Echinodorus.*
  Flowers polygamous or unisexual.
    Lower flowers perfect, upper staminate. 4. *Lophotocarpus.*
    Lower flowers pistillate, the upper usually staminate. 5. *Sagittaria.*

1. ALÍSMA L. Sp. Pl. 342. 1753.

Perennial or rarely annual herbs, with erect or floating basal leaves, several-ribbed, these connected by transverse veinlets. Flowers numerous, in pyramidal panicles, on unequal 3-bracteolate pedicels. Petals white or rose tinged. Stamens 6–9; ovaries few to many, arranged in one whorl on a flat receptacle. Achenes 2–3-ribbed, curved on the back and 1–2-ribbed on the sides. (Greek, the old classical name.)

Type species, *Alisma plantago-aquatica* L. A genus of about 6 or 8 species, widely distributed in temperate and tropical regions.

Achenes longer than wide, grooved on the back, the inner edges not meeting in the whorl; pedicels very slender, ascending. 1. *A. plantago-aquatica.*
Achenes as wide as long, ridged on the back, the inner edges meeting in the whorl; pedicels stout, widely divergent in fruit. 2. *A. geyeri.*

1. Alisma plantàgo-aquàtica L.

Common Water-plantain. Fig. 206.

Alisma plantago-aquatica L. Sp. Pl. 342. 1753.
Alisma subcordatum Raf. Med. Rep. 6: 362. 1808.
Alisma brevipes Greene, Pittonia 4: 158. 1900.

Plants erect, glabrous, the rootstocks becoming bulbous by the sheathing bases of the petioles. Leaves all basal, oblong to ovate or sometimes narrower, 3–15 cm. long, usually abruptly pointed at the apex, cuneate to cordate at the base; the petioles often longer than the blades; scapes 1–10 dm. tall, solitary or several together, the branches and pedicels in whorls of 3–10, variable in length, usually very slender to almost filiform; bracts lanceolate or linear, often acuminate; sepals broadly ovate to suborbicular, obtuse; petals white or pinkish 1–2 mm. long; fruiting heads 3.5–4.5 mm. broad; achenes obliquely obovate, 1.5–2 mm. long, the beak ascending.

Growing in shallow water or in wet ground and widely distributed over the northern hemisphere. Type locality: Europe.

2. Alisma geỳeri Torr.

Geyer's Water-plantain. Fig. 207.

Alisma geyeri Torr. in Nicollet, Rep. Hydrograph. Miss. Riv. 162. 1843.

Plants diffuse, glabrous. Leaves oblong, elliptic, oblong-lanceolate or ovate-lanceolate, or rarely linear, 5–9 cm. long, acute or slightly acuminate at the apex, narrowed at the base; petioles usually longer than the blades; scapes mostly 1–5 dm. long, more or less diffusely spreading, the branches and pedicels relatively stout, the latter widely divergent in fruit; bracts lanceolate; sepals orbicular-ovate, about 2.5 mm. long; petals pink, 2–4 mm. long; fruiting heads 4.5–5.5 mm. broad; achenes suborbicular, about 2 mm. in diameter, ridged on the back, the beak erect or nearly so.

In shallow water or wet ground, mainly Canadian Zone; western New York to North Dakota, Nevada, Oregon, and Washington. In the Pacific States it has been collected near Dalles City, Oregon, and in Klickitat County, Washington. Type locality: near Devil's Lake, North Dakota.

2. MACHAEROCÀRPUS Small. N. Am. Fl. 17: 44. 1909.

Perennial scapose herbs. Leaves erect, ascending or floating, long-petioled, 3–5-veined, not lobed at the base. Flowers perfect, borne in simple panicles. Sepals 3, broad-ribbed, persistent. Petals 3, white or pink, spreading, incised, deciduous. Stamens 6, 2 opposite each sepal; filaments flattened; anthers elongate. Carpels few in one whorl, attached to the conic receptacle by their broad bases; ovules solitary in each carpel. Achenes in one radiant whorl, ribbed on back, faces depressed, beak erect, as long as the achene body or longer.

A monotypic genus peculiar to California, differing from the Old World genus *Damasonium* by the uni-ovulate instead of bi- or multi-ovulate carpels.

1. Machaerocarpus califórnicus (Torr.) Small.

Fringed Water-plantain. Fig. 208.

Damasonium californicum Torr.; Benth. Pl. Hartw. 341. 1857.

Alisma californicum Micheli, in DC. Monogr. Phan. 3: 34. 1881.

Leaves oblong to ovate or linear-oblong, 2.5–6 cm. long, obtuse at the apex, obtuse to subcordate at the base; the petioles, at least some of them, much longer than the blades; scapes usually several together, erect or decumbent, 1–4 dm. long; pedicels loosely spreading or recurving at least in fruit; sepals broadly oblong or ovate-oblong, 4–5 mm. long; petals suborbicular, 8–10 mm. long, more or less irregularly incised; achenes dagger-like, 6–10, horizontally radiating, 7–12 mm. long, the beak subulate.

Shallow water or in mud. Upper Sonoran Zone; California, Modoc County (Honey Lake), through the Sacramento Valley and Sierra foothills to Sierra County; also in the coastal region near Petaluma, Sonoma County. Type locality: Sacramento Valley, near Chico.

3. ECHINÓDORUS Rich.; Engelm. in A. Gray, Man. 460. 1848.

Perennial or annual herbs with long-petioled, elliptic-ovate or lanceolate, often cordate or sagittate leaves, 3–9-ribbed and mostly punctate with dots or lines. Scapes often longer than the leaves. Inflorescence racemose or paniculate. Flowers perfect, in whorls, each whorl with 3 outer bracts and numerous inner bracteoles. Petals white. Receptacle large, convex or globose. Stamens 12–30. Ovaries numerous; style obliquely apical, persistent; stigmas simple. Achenes more or less compressed, ribbed and beaked, forming spinose heads. [Greek, in reference to the prickly fruit.]

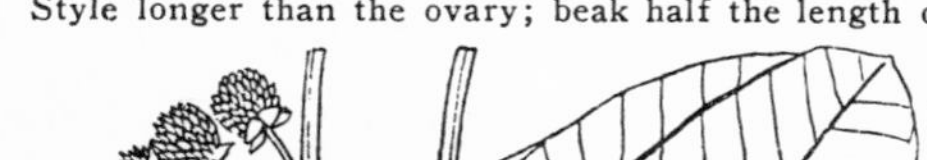

A genus of about 14 species distributed over the temperate and tropical regions of America, Europe, and Africa. Type species, *Alisma rostratum* Nutt.

Style shorter than the ovary; beaks about one-fourth the length of the achene. 1. *E. radicans.*
Style longer than the ovary; beak half the length of the achene. 2. *E. cordifolius.*

1. Echinodorus radìcans (Nutt.) Engelm.

Creeping Bur-head. Fig. 209.

Sagittaria radicans Nutt. Trans. Am. Phil. Soc. II. 5: 159. 1837.

Echinodorus radicans Engelm. in A. Gray, Man. ed. 2, 438. 1856.

Leaves coarse, ovate, obtuse at the apex, cordate at the base, 5–20 cm. long; petioles 1–7 dm. long; scapes elongate, prostrate, creeping, 3–12 dm. long, often solitary; whorls of the inflorescence remote; pedicels 3–12 in a whorl, unequal, 1.5–6 cm. long; bracts linear-lanceolate from dilated bases; sepals ovate or orbicular-ovate, 5–6.5 mm. long, obtuse; petals white, somewhat longer than the sepals; fruiting heads depressed, globose or ovoid, 7–8 mm. in diameter; achenes 2 mm. long, the body falcate, 6–10-ribbed, the beak about one-fourth as long as the body, incurved.

Borders of pools or ditches, Upper and Lower Sonoran Zones; Illinois and Missouri to Florida, Texas and southern California. Type locality: Fort Smith, Arkansas.

2. Echinodorus cordifòlius (L.) Griseb.
Upright Bur-head. Fig. 210.

Alisma cordifolia L. Sp. Pl. 343. 1753.
Echinodorus rostratus Engelm. in A. Gray, Man. ed. 2, 538. 1856.
Echinodorus cordifolius Griseb. Abh. Kon. Gesell. Wiss. Gott. 7: 257. 1857.

Leaves various, broadly ovate to lanceolate, 4–20 cm. long, obtuse or acute at the apex, commonly truncate or cordate at the base, 3–13-veined; petioles usually longer than the blades, angled; scapes solitary or clustered, 1–5 dm. tall or more, simple or branched from the lower whorls of the inflorescence; pedicels 7–15 mm. long, not very variable in length; bracts lanceolate or linear-lanceolate; sepals ovate, 4–5 mm. long, rather acute; petals white, twice as long as the sepals or less, usually broader; fruiting heads bur-like, globose to ovoid, 4–10 mm. in diameter; achenes, 2.5–3 mm. long, the body cuneate, flattish, very prominently ribbed, the beak slender, fully one-half as long as the body.

Borders of ponds and ditches, Austral Zones; Illinois and Missouri to tropical America. On the Pacific Coast it is found from the lower San Joaquin Valley, and Santa Clara County, California, to Lower California. Type locality: Virginia.

4. LOPHOTOCÀRPUS T. Durand, Ind. Gen. Phan. X. 1888.

Annual aquatic or bog plants with basal long-petioled sagittate or cordate leaves, simple erect scapes bearing flowers in several verticils of 2–3 at the summit, the lower perfect, the upper staminate. Petals white. Sepals distinct, enclosing or enveloping the fruit. Receptacle strongly convex. Stamens 9–15, hypogynous, inserted at the base of the receptacle. Pistils numerous with solitary ovules and an elongated persistent style. Achenes winged or crested; embryo horseshoe-shaped. [Greek, referring to the crested achenes.]

An American genus of about 7 species, distributed from the cool temperate to the tropical regions. Type species, *Sagittaria calycina* Engelm.

1. Lophotocarpus califòrnicus J. G. Smith.
California Lophotocarpus. Fig. 211.

Lophotocarpus californicus J. G. Smith, Rep. Mo. Bot. Gard. 11: 146. 1899.

Plants emersed or submerged. Leaves usually erect, mostly 1.5–5 dm. tall, the petioles slender, hastate or sagittate, 2–6 cm., including the divergent basal lobes, these as long as the terminal lobe or shorter; scapes about as long as the leaves, with 3–5 rather approximate whorls; sepals broadly ovate or suborbicular, becoming 6–9 mm. long; fruiting pedicels short and thick, 1–2 cm. long, as long as the internodes or nearly so; fruiting heads 10–12 mm. in diameter; achenes cuneate or obovate-cuneate, nearly 2–5 mm. long, the beak minute, the dorsal wing broad and thick.

Quiet, shallow water, Upper and Lower Sonoran Zones; Lake County, Oregon, to Sacramento Valley and southern California. Type locality: Coyote Creek, Los Angeles County, California.

5. SAGITTÀRIA L. Sp. Pl. 993. 1753.

Perennial aquatic or marsh herbs with tuber-bearing or nodose rootstocks. Leaves with nerves connected by numerous veinlets. Scapes erect, decumbent or floating. Flowers monoecious or dioecious, borne near the summit of the scapes in whorls of threes, pediceled, the staminate usually uppermost, whorls 3-bracted. Perianth-segments 6, the outer three herbaceous, persistent and reflexed or spreading in the pistillate flowers. Stamens numerous, inserted on the convex receptacle; anthers 2-celled, dehiscent by lateral slits. Pistillate flowers with numerous distinct 1-ovuled ovaries and small persistent stigmas. Achenes densely aggregated in globose heads, compressed; seeds curved; embryo horseshoe-shaped. [Latin, in reference to the shape of the leaves.]

A genus of approximately 40 species, native of temperate and tropical regions, but principally American. Type species, *Sagittaria sagittifolia* L.

Sepals of the pistillate flowers ultimately lax or reflexed, not accrescent; filaments glabrous.
Fruiting pedicels slender, ascending; leaves with basal lobes.
Achenes with a conspicuous dorsal wing prominently long-beaked, the beak horizontal. 1. *S. latifolia.*
Achenes minutely or inconspicuously beaked, about equally winged on both faces.
Beak erect; the sides of the achene not winged. 2. *S. cuneata.*
Beak horizontal; the sides of the achene with a prominent wing-margined depression. 3. *S. greggii.*
Fruiting pedicels stout and recurved in fruit; leaves without basal lobes. 4. *S. sanfordii.*
Sepals of the pistillate flowers accrescent and ultimately appressed to the heads; fruiting pedicels stout and recurved; filaments glandular-pubescent. 5. *S. montevidensis.*

1. **Sagittaria latifòlia** Willd.

Broad-leaved Arrow-head. Fig. 212.

Sagittaria latifolia Willd. Sp. Pl. **4**: 409. 1806.
Sagittaria variabilis Engelm. in A. Gray, Man. 461. 1848.
Sagittaria sagittifolia sinensis Brand. Zoe **4**: 217. 1893, not *S. sinensis* Sims. 1833.
Sagittaria esculenta Howell, Fl. NW. Am. 679. 1903.

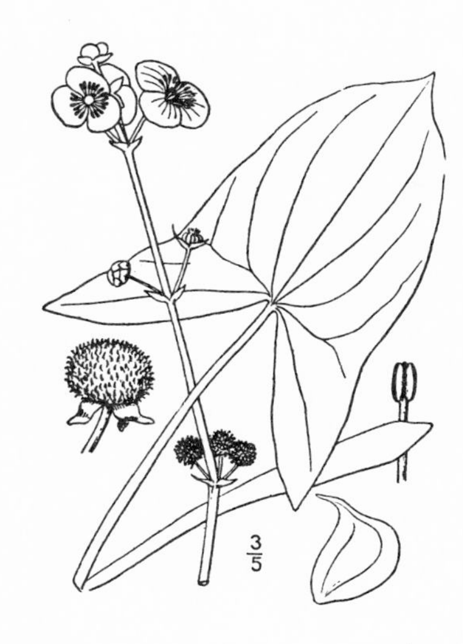

Plants partially or wholly emersed, 2–14 dm. tall. Leaves variable, 10–40 cm. long, the terminal lobe deltoid or ovate to linear; scape erect, angled, stout or slender; bracts ovate, mostly less than 1 cm. long, glabrous, acute or obtusish; whorls of the inflorescence mostly distant; fruiting heads 1.5–3 cm. in diameter; achenes obovate, usually broadly so, about 3 mm. long, broadly winged, especially at the top, the beak horizontal or nearly so.

Shallow water in streams and on margins of lakes, Boreal and Austral Zones; British Columbia to southern California, Atlantic States and Mexico. The tubers are eaten by the Indians and known to them as Wappato. In California, where it grows abundantly on the islands of the lower Sacramento and San Joaquin Rivers, it is known as Tule Potato, and is eaten by the Chinese, who also frequently cultivate it.

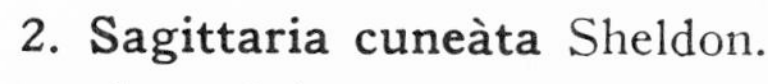

2. **Sagittaria cuneàta** Sheldon.

Arum-leaved Arrow-head. Fig. 213.

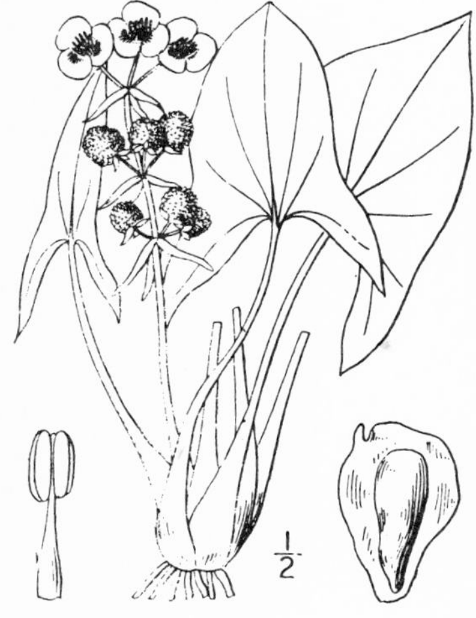

Sagittaria cuneata Sheldon, Bull. Torrey Club **20**: 283. 1893.
Sagittaria arifolia Nutt.; J. G. Smith, Rep. Mo. Bot. **6**: 32. 1894.
Sagittaria paniculata Blankinship, Mont. Agr. Coll. Sci. Stud. **1**: 40. 1905.

Plants emersed and commonly 2–4 dm. tall, or submerged and varying in height with the depth of the water. Leaves hastate to sagittate, 6–18 cm. long, acuminate; phyllodia when present short and linear-lanceolate or elongate and linear-attenuate; scapes as long as the leaves or commonly shorter, weak; whorls of the inflorescence few or numerous; sepals ovate to oblong-ovate, becoming 6–8 mm. long; corolla 2–2.5 cm. broad; fruiting heads 1–1 5 cm. in diameter; achenes obovate, 2–2.5 mm. long, with callous thickened wings, the beak minute, erect over the ventral wing.

Streams and ponds, Canadian Zone; Nova Scotia and Connecticut to the Rocky Mountains and Pacific Coast, where it extends from British Columbia to New Mexico and the San Bernardino Mountains, California. Type locality: East Battle Lake, Otter Trail County, Minnesota.

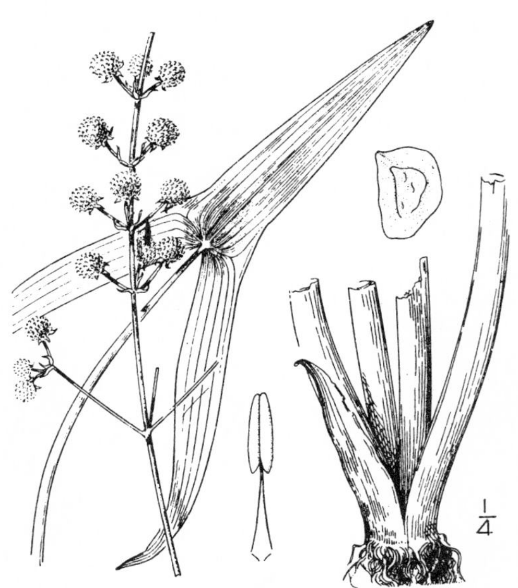

3. **Sagittaria gréggii** J. G. Smith. Gregg's Arrow-head. Fig. 214.

Sagittaria greggii J. G. Smith, Rep. Mo. Bot. Gard. **6**: 43. 1894.

Plants partially emersed, 20–40 cm. tall. Leaves about 1–2 dm. long, the terminal lobe ovate to lanceolate, acute or slightly acuminate, the basal lobes lanceolate or linear-lanceolate, acuminate, about as long as the terminal or longer, submerged leaves bladeless or with an entire linear-lanceolate blade 3–5 cm. long; whorls of the inflorescence several to many; bracts lanceolate or linear-lanceolate, acuminate, 15–30 mm. long; sepals oval to obovate, becoming 6–8 mm. long; corolla mostly 1–2 cm. broad, or more; filaments dilated at base; fruiting heads 8–15 mm. in diameter; achenes obovate to orbicular-obovate, 2–3 mm. long; thick-margined, the sides with a thin-margined depression, the beak minute, horizontal.

Sloughs and slow running waterways, Austral Zones; lower San Joaquin Valley, California, also in Mexico. Type locality: Zamora, Michoacan, Mexico.

4. **Sagittaria sánfordii** Greene. Sanford's Arrow-head. Fig. 215.

Sagittaria sanfordii Greene, Pittonia **2**: 158. 1890.

Plants emersed or partially submerged, erect or ascending. Leaves sometimes represented by 3-sided phyllodia, the petioles spongy, very stout, 2–4 cm. thick at the base, the blades entire, linear-lanceolate to oblong-lanceolate, mostly 10–15 cm. long, acute, narrowed at the base; scapes shorter than the leaves, simple; pedicels 1.5–2 cm. long, becoming stout and recurved in fruit; bracts broad, 5–7 mm. long, united at the base; sepals ovate, 4–6 mm. long; corolla about 2 cm. broad; anthers oblong; fruiting head 12–14 mm. in diameter; achenes broadly cuneate, about 2 mm. long, winged on both margins, the beak short, lateral, ascending, broadly triangular.

In sloughs and slow running waterways, Upper Sonoran Zone; Lower San Joaquin Valley, often growing in large colonies. Type locality: sloughs of the Lower San Joaquin River about Stockton, California.

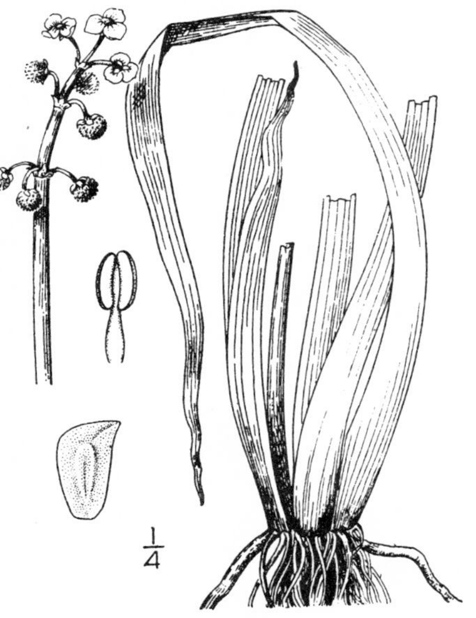

5. **Sagittaria montevidénsis** Cham. & Sch. Montevideo Arrow-head. Fig. 216.

Sagittaria montevidensis Cham. & Sch. Linnaea **2**: 156. 1827.

Plants emersed or partially submerged, 5–17 dm. tall. Leaf-blades hastate or sagittate, 1–6 dm. long, the basal lobes acute or acuminate, more or less divergent, about as long as the terminal; scapes stout, sometimes 6–8 cm. thick at the base; sepals becoming 10–15 mm. long, obtuse, erect in fruit; corolla 2–4 cm. broad, the petals white with a brownish purple spot at the base; filaments glandular-pubescent; fruiting heads 1.5–3 cm. in diameter; achenes cuneate to rhombic-obovate, 2–3 mm. long, winged, the beak slender, horizontal or oblique, the faces unappendaged.

In sloughs and slow running water courses in the Lower San Joaquin Valley, also in the southern Atlantic States. Naturalized from South America. Type locality: Montevideo, Uruguay.

Family 9. **VALLISNERIÀCEAE.**

Tape-grass Family

Submerged or floating, fresh water or marine, perennial herbaceous plants, the leaves various. Flowers regular, mostly dioecious, appearing from an involucre or spathe of 1–3 bracts or leaves. Perianth 3–6-parted, the segments either all petioled or the outer small and herbaceous, the tube adherent to the ovary at its base in the pistillate flowers. Stamens 3–12. Anthers 2-celled. Ovary 1-celled with 3 parietal placentae. Styles 3, with entire or 2-cleft stigmas. Ovules anatropous or orthotropous. Fruit maturing under water, indehiscent. Seeds numerous, without endosperm.

A family of about 6 genera and 25 species.

1. **ELODÈA** Michx. Fl. Bor. Am. **1**: 20. 1803.

Stems submerged, elongated, branching, leafy. Leaves opposite or whorled, usually crowded, 1-nerved, pellucid, minutely serrulate or entire. Flowers dioecious or polygamous, arising from an ovoid or tubular 2-cleft spathe. Staminate flowers solitary, rarely in threes, sessile or nearly so, early separating from the plant and floating on the surface of the water; sepals and petals 3, or the latter sometimes wanting; stamens 9, in two series, the outer of 6 and the inner of 3 stamens. Pistillate flowers solitary, sessile, the perianth-tube prolonged into a long slender pedicel-like tube; sepals and petals 3; stamens none or represented by 3 rudimentary filaments. Hermaphrodite flowers like the pistillate, but with 3–9 stamens. Fruit linear or lanceolate-linear [Name Greek, meaning marshy, in reference to the habitat.]

An American genus of about 10 species. Type species, *Elodea canadensis* Michx.

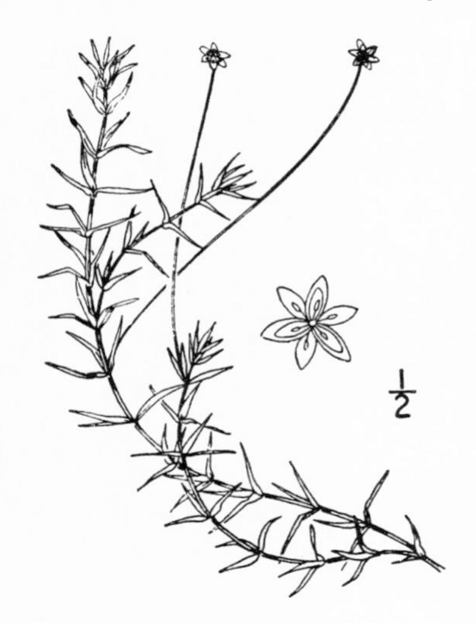

1. **Elodea planchònii** Casp.

Rocky Mountain Waterweed. Fig. 217.

Anacharis canadensis Planch. Ann. Mag. Nat. Hist. II. **1**: 86. 1848, mainly.

Elodea planchonii Casp. Jahrb. Wiss. Bot. **1**: 408. 1858.

Philotria planchonii Rydb Bull. Torrey Club **35**: 462. 1908.

Dioecious aquatic herb, with slender stems, 1–20 cm. long. Leaves in threes or the lower opposite, sessile, oblong to linear, 7–15 mm. long, 1–2 mm. wide, acute or acutish; spathe of the staminate flowers obovoid-clavàte, nearly 1 cm. long, on a peduncle 5–10 mm. long; staminate flowers short-pediceled; sepals elliptic, 5 mm. long; petals wanting; stamens 9; anthers oblong, 3–4 mm. long, subsessile; spathe of the pistillate flowers linear or lanceolate-linear, sessile, 2-cleft at the apex; perianth-tube slender, 3–5 cm. long; sepals and petals linear, about 3 mm. long; stigmas 3, linear; stamens none.

In lakes and ponds, Canadian Zone; Saskatchewan to Washington, south to Colorado, Nevada, and Truckee, California. Type locality: Saskatchewan.

Family 10. **POÀCEAE.***

Grass Family.

Herbs or rarely woody plants, with usually hollow stems (culms) closed at the nodes, and 2-ranked parallel-veined leaves, these consisting of 2 parts, a lower, the sheath, enveloping the culm, its margins overlapping or sometimes grown together, an upper, the blade, usually flat, and, between the two on the inside, a membranaceous hyaline or hairy appendage (the ligule). Flowers perfect (rarely unisexual), small, with no distinct perianth, arranged in spikelets consisting of a shortened axis (rachilla) and 2 to many 2-ranked bracts, the lowest 2 being empty (the glumes, rarely 1 or both of these obsolete), the 1 or more succeeding ones (lemmas) bearing in their axils a single flower, and, between the flower and the rachilla, a second 2-nerved bract (the palea); stamens 1 to 6, usually 3, with very delicate filaments and 2-celled anthers; pistil 1, with a 1-celled 1-ovuled ovary, 2

* Text contributed by A. S. Hitchcock, and published with the permission of the Secretary of Agriculture.

styles, and usually plumose stigmas; fruit a caryopsis with starchy endosperm and a small embryo at the base on the side opposite the hilum. The perianth is usually represented by 2 small hyaline scales (the lodicules) at the base of the flower inside the lemma and palea. The lemma, palea, and inclosed flower constitute the floret. The spikelets are almost always aggregate in an inflorescence at the ends of the main culms or branches. The grain or caryopsis (the single seed and the adherent pericarp) may be free, as in wheat, or permanently inclosed in the lemma and palea, as in the oat. Rarely the seed is free from the pericarp, as in species of *Sporobolus* and *Eleusine*. The culms are solid in our species of *Andropogoneae*. The margins of the sheaths are grown together in species of *Bromus, Festuca, Melica, Panicularia,* and others. The parts of the spikelet may be modified in various ways. The first, and more rarely also the second, glume may be wanting. The lemmas may contain no flower, or even no palea, or may be reduced or rudimentary. The palea is rarely wanting in perfect florets (species of *Agrostis*).

A large family of about 500 genera distributed throughout the world, in all latitudes and altitudes where are found conditions suitable for plant growth, but least abundant in dense tropical forests.

A. Spikelets with 1 perfect terminal floret (disregarding the staminate and neuter spikelets) and (except in *Hilaria*) a sterile or staminate floret below, usually represented by a sterile lemma only, one glume sometimes wanting; articulation below the spikelets, either in the pedicel, in the rachis, or at the base of a cluster of spikelets, the spikelets falling entire, either singly, in groups, or together with joints of the rachis; spikelets, or at least the fruits, more or less dorsally compressed (except in *Hilaria*). Subfamily PANICATAE.

Glumes indurate; fertile lemma and palea hyaline or membranaceous, the sterile lemma (when present) like the fertile one in texture.

Spikelets in pairs, one sessile, the other pedicellate (the pedicellate one sometimes obsolete, rarely both pedicellate); lemmas hyaline. I. ANDROPOGONEAE.

Spikelets in groups of 2–5, the groups falling entire from the continuous axis. II. NAZIEAE.

Glumes membranaceous; fertile lemma and palea indurate or at least firmer than the glumes; sterile lemma like the glumes in texture. III. PANICEAE.

B. Spikelets 1 to many-flowered, the reduced florets, if any, above the perfect florets (except in *Phalarideae*); articulation usually above the glumes, spikelets usually more or less laterally compressed. Subfamily POATAE.

Spikelets with 2 staminate, sterile or rudimentary lemmas unlike and below the fertile lemma; no sterile or rudimentary florets above. V. PHALARIDEAE.

Spikelets without sterile lemmas below the perfect floret.

Spikelets articulate below the glumes, 1-flowered, very flat, the lemma and palea about equal, both keeled, the glumes small or wanting. IV. ORYZEAE.

Spikelets usually articulate above the glumes.

Spikelets sessile on a usually continuous rachis (short-pedicellate in *Leptochloa;* the rachis disarticulating in *Sitanion* and in a few species of allied genera).

Spikelets on opposite sides of the rachis; spike terminal, single. X. HORDEAE.

Spikelets on one side of the rachis; spikes usually more than 1, digitate or racemose. VIII. CHLORIDEAE.

Spikelets pedicellate in open or contracted (sometimes spike-like) panicles.

Spikelets 1-flowered. VI. AGROSTIDEAE

Spikelets 2 to many-flowered.

Glumes as long as the lowest floret, usually as long as the spikelet; lemmas awned from the back (spikelets awnless in *Koeleria, Sphenopholis* and *Trisetum wolfii*). VII. AVENEAE.

Glumes shorter than the first floret (except in *Dissanthelium*); lemmas awnless or awned from the tip (from a bifid apex in *Bromus* and *Triodia*). IX. FESTUCEAE.

Tribe I. ANDROPOGÒNEAE.

Spikelets all fertile, surrounded by copious soft hairs; inflorescence a narrow panicle 1. *Imperata.*

Spikelets unlike, the sessile perfect, the pedicellate staminate or neuter.

Racemes of several joints, silky-villous. 2. *Andropogon.*

Racemes reduced to 1 or few joints, these in a compound panicle, not silky-villous. 3. *Holcus.*

Tribe II. NAZÌEAE.

A single genus. 4. *Hilaria.*

Tribe III. PANÍCEAE.

Spikelets subtended or surrounded by 1 to many bristles (sterile branchlets), these distinct or connate.

Bristles slender, separate, persistent, the spikelets deciduous. 10. *Chaetochloa.*

Bristles united or connate, forming a bur with retrorsely barbed spines, the bur falling with the spikelets. 11. *Cenchrus.*

Spikelets not subtended by bristles.

Glumes awned or mucronate; apex of palea not inclosed by the lemma. 9. *Echinochloa.*

Glumes awnless; apex of palea inclosed by the lemma.

Spikelets in panicles. 8. *Panicum.*

Spikelets in 1-sided spikelike racemes.

First glume and the rachilla joint forming a swollen ringlike callus below the spikelet; racemes several along the main axis. 6. *Eriochloa.*

First glume present or wanting, not forming a ringlike callus; racemes aggregate at the summit of the culm.

Racemes slender, 3–12. 5. *Syntherisma.*

Racemes stout, in pairs. 7. *Paspalum.*

Tribe IV. ORŸZEAE.

A single genus. 12. *Homalocenchrus.*

Tribe V. PHALARÍDEAE.

Lateral florets staminate; spikelets brown and shining. 15. *Torresia.*
Lateral florets neuter; spikelets green or yellowish.
  Lateral florets reduced to small awnless bracts; spikelets much compressed laterally. 13. *Phalaris.*
  Lateral florets consisting of awned hairy sterile lemmas about as long as the fertile floret; spikelets terete. 14. *Anthoxanthum.*

Tribe VI. AGROSTÍDEAE.

Glumes wanting; a low annual. 23. *Coleanthus.*
Glumes present.
  Rachilla articulate below the glumes, these falling with the spikelet.
    Glumes long-awned. 26. *Polypogon.*
    Glumes awnless.
      Rachilla not prolonged behind the palea; panicle dense and spike-like. 22. *Alopecurus.*
      Rachilla prolonged behind the palea; panicle open. 27. *Cinna.*
  Rachilla articulate above the glumes.
    Fruit indurate, terete, awned, the nerves obscure; callus usually well-developed, oblique, bearded.
      Awn trifid, the lateral divisions sometimes short, rarely obsolete (when obsolete no line of demarcation between awn and lemma). 16. *Aristida.*
      Awn simple, a line of demarcation between the awn and lemma.
        Awn persistent, twisted, bent, several to many times longer than the slender fruit; callus sharp-pointed. 17. *Stipa.*
        Awn deciduous, not twisted, sometimes bent, rarely more than 3 or 4 times as long as the plump fruit; callus short, usually obsolete. 18. *Oryzopsis.*
    Fruit thin or firm but scarcely indurate, if firm the nerves prominent or evident; callus not well-developed.
      Glumes longer than the lemma (equaling it only in *Agrostis aequivalvis* and *A. thurberiana*).
        Panicle feathery, capitate, nearly as broad as long; spikelets woolly. 32. *Lagurus.*
        Panicle not feathery; spikelets not woolly.
          Glumes compressed-carinate, abruptly mucronate, stiffly ciliate on the keels; panicle dense, cylindric or ellipsoid. 21. *Phleum.*
          Glumes not compressed-carinate, not ciliate.
            Glumes saccate at base; lemma long-awned; panicle contracted, shining. 29. *Gastridium.*
            Glumes not saccate at base; lemma awned or awnless; panicles open or contracted.
              Florets with hairs at base half as long as the lemma; palea present. 30. *Calamagrostis.*
              Florets naked at base or with short hairs (hairs longer in *A. hallii* with obsolete palea). 28. *Agrostis.*
      Glumes not longer than the lemma, usually shorter (the awn tips longer in *Muhlenbergia racemosa*).
        Lemma awned from the tip or mucronate; 3–5-nerved (lateral nerves obsolete in *M. repens*). 19. *Muhlenbergia.*
        Lemma awnless or awned from the back.
          Lemma bearing a tuft of hairs at the base from the short callus. 31. *Ammophila.*
          Lemmas without hairs at base.
            Fruit at maturity falling from the lemma and palea; seed loose in the pericarp, this usually opening when ripe.
              Inflorescence capitate in the axils of broad bracts. 20. *Crypsis.*
              Inflorescence an open or contracted panicle. 24. *Sporobolus.*
            Fruit not falling from the lemma and palea, remaining permanently inclosed in them; seed adnate to the pericarp; panicles spike-like. 25. *Epicampes.*

Tribe VII. AVÈNEAE.

Spikelets awnless or the upper lemma mucronate. (See *Trisetum wolfii.*)
  Articulation below the glumes; glumes distinctly different in shape, the second widened above. 37. *Sphenopholis.*
  Articulation above the glumes; glumes not much dissimilar in shape. 38. *Koeleria.*
Spikelets awned.
  Florets 2, one perfect, the other staminate.
    Lower floret staminate, the awn twisted, geniculate, exserted. 40. *Arrhenatherum.*
    Lower floret fertile, awnless; awn of upper floret hooked. 33. *Notholcus.*
  Florets 2 or more, all alike except the reduced upper ones.
    Awn arising from between the teeth of a bifid apex, flattened, twisted; inflorescence a simple panicle or reduced to a raceme or even to a single spikelet. 41. *Danthonia.*
    Awn dorsal, not flattened; lemma often bifid at apex.
      Spikelets large, the glumes over 1 cm. long. 39. *Avena.*
      Spikelets less than 1 cm. long.
        Lemmas keeled, bidentate, awned from above the middle. 36. *Trisetum.*
        Lemmas convex, awned from below the middle.
          Apex of lemma acuminate, bidentate; rachilla not prolonged behind the upper floret. 35. *Aspris.*
          Apex of lemma broad, erose; rachilla prolonged behind the upper floret. 34. *Aira.*

Tribe VIII. CHLORÍDEAE.

Spikelets with more than 1 perfect floret.
  Spikes numerous, slender, along an elongate axis. 48. *Leptochloa.*
  Spikes few, thick, digitate or nearly so. 47. *Eleusine.*
Spikelets with only 1 perfect floret, often with additional imperfect florets above.
  Spikelets without additional modified florets, the rachilla sometimes prolonged.
    Rachilla articulate below the glumes.
      Glumes unequal, narrow. 43. *Spartina.*
      Glumes equal, broad and boat-shaped. 46. *Beckmannia.*
    Rachilla articulate above the glumes; spikes digitate; rachilla prolonged. 42. *Capriola.*
  Spikelets with one or more sterile florets above the perfect one.
    Spikes digitate or nearly so. 44. *Chloris.*
    Spikes racemose along the main axis. 45. *Bouteloua.*

Tribe IX. Festùceae.

Lemmas with 5 awn-like lobes; inflorescence a simple erect raceme. 50. *Orcuttia.*
Lemmas awnless or with a single awn.
Tall stout reeds with large plume-like panicles.
Lemma clothed with long silky hairs; rachilla naked. 51. *Arundo.*
Lemma naked; rachilla hairy. 52. *Phragmites.*
Low or moderately tall grasses, rarely over 1.5 meters tall.
Plants dioecious, perennial; lemmas glabrous; grasses of salt or alkaline soil.
Plants low and creeping; spikelets obscure, scarcely differentiated from the short crowded rigid leaves. 49. *Monanthochloë.*
Plants erect from creeping rhizomes; spikelets in a narrow simple exserted panicle. 59. *Distichlis.*
Plants not dioecious (except in a few species of *Poa* with villous lemmas and an annual species of *Eragrostis*).
Spikelets of 2 forms, sterile and fertile intermixed.
Fertile spikelets 2–3-flowered; sterile spikelets with numerous rigid awn-tipped glumes; panicle dense and spikelike. 62. *Cynosurus.*
Fertile spikelets with one perfect floret, long-awned; sterile spikelets with many obtuse glumes; panicles purple or yellow, contracted. 63. *Achyrodes.*
Spikelets all alike in the same inflorescence.
Lemmas 3-nerved.
Lemmas pilose on the nerves, longer than the glumes; callus densely villous. 53. *Triodia.*
Lemmas sometimes pubescent but not pilose on the nerves.
Glumes equaling or exceeding the florets. 54. *Dissanthelium.*
Glumes shorter than the first floret. 55. *Eragrostis.*
Lemmas 5–many-nerved, the nerves sometimes obscure.
Lemmas flabellate; glumes wanting; inflorescence dense, cylindric; low annuals. 56. *Anthochloa.*
Lemmas not flabellate; glumes present; inflorescence not dense and cylindric.
Palea winged; spikelets linear, in a loose raceme. 58. *Pleuropogon.*
Palea not winged; inflorescence mostly paniculate.
Lemmas as broad as long, the margins outspread; florets closely imbricate, horizontally spreading. 60. *Briza.*
Lemmas longer than broad, the margins clasping the palea; florets not horizontally spreading.
Callus of florets bearded; lemmas erose at the summit. 64. *Fluminea.*
Callus not bearded (lemmas cobwebby at base in *Poa*); lemmas not erose (slightly so in *Puccinellia*).
Lemmas keeled on the back.
Spikelets strongly compressed, crowded in 1-sided clusters at the ends of the stiff naked panicle branches. 61. *Dactylis.*
Spikelets not strongly compressed, not in 1-sided clusters.
Lemmas awned from a bifid apex (awnless or nearly so in *Bromus unioloides* and *B. brizaeformis*); spikelets large. 69. *Bromus.*
Lemmas awnless; spikelets small. 65. *Poa.*
Lemmas rounded on the back (slightly keeled toward the summit in *Festuca*).
Glumes papery; lemmas firm, scarious-margined; upper florets sterile, often reduced to a club-shaped rudiment inclosed by the broad upper lemmas; spikelets tawny or purplish, not green. 57. *Melica.*
Glumes not papery; upper florets not unlike the others.
Nerves of lemma parallel, not converging at the summit or but slightly so; lemma awnless, mostly obtuse.
Nerves prominent; glumes nearly as long as the first floret. 66. *Panicularia.*
Nerves faint; glumes much shorter than the first floret. 67. *Puccinellia.*
Nerves of the lemma converging at the summit; lemmas awned or pointed.
Lemmas entire, awned from the tip or pointed (minutely toothed in *F. elmeri*). 68. *Festuca.*
Lemmas awned or awn-tipped from a bifid apex (awnless in *B. brizaeformis*). 69. *Bromus.*

Tribe X. Hórdeae.

Spikelets solitary at each joint of the rachis (occasionally 2 in *Agropyron*).
Spikelets placed edgewise to the rachis, the lateral ones with a single glume.
Spikelets several-flowered. 70. *Lolium.*
Spikelets 1-flowered. 71. *Lepturus.*
Spikelets placed flatwise to the rachis, all with 2 glumes.
Spikelets several-flowered. 74. *Agropyron.*
Spikelets 1-flowered; spikes very slender.
Lemma awnless. 72. *Pholiurus.*
Lemma awned. 73. *Scribneria.*
Spikelets, at least some of them, in twos or threes at each joint of the rachis.
Spikelets 1-flowered, not all alike, in threes, the lateral pair pediceled. 75. *Hordeum.*
Spikelets 2–6-flowered, all alike, usually in twos.
Glumes minute or wanting; spikes loose. 78. *Hystrix.*
Glumes prominent, mostly narrow; spikes dense.
Axis of spike continuous, not disarticulating at maturity; glumes broad or narrow but not greatly elongate. 76. *Elymus.*
Axis of spike disarticulating at maturity; glumes usually setaceous and elongate. 77. *Sitanion.*

1. IMPERÀTA Cyrillo, Pl. Rar. Neap. 2: 26. *pl. 11.* 1792.

Spikelets all alike, awnless, in pairs, unequally pedicellate on a slender continuous rachis, surrounded by long silky hairs; glumes about equal, membranaceous; sterile lemma, fertile lemma and palea thin and hyaline. Perennial slender erect grasses with terminal narrow silky panicles. [Name in honor of the Italian naturalist Ferrante Imperate.]

About 5 species in the warm regions of the world. Type species, *Lagurus cylindricus* L.

1. Imperata hoòkeri Rupr.

Hooker's Imperata. Fig. 218.

Imperata hookeri Rupr.; Anderss. Öfv. Svensk. Vet. Akad. Förh. 12: 160. 1855.

Imperata brevifolia Vasey, Bull. Torrey Club 13: 26. 1886.

Culms erect from creeping rhizomes, 1–1.5 meters tall, glabrous; sheaths glabrous; ligule long-villous; blades elongate, the lower narrowed at the long conduplicate base, 8–15 mm. wide, acuminate, glabrous, the upper shorter, the uppermost much reduced; panicle dense, 15–30 cm. long, pale or tawny or somewhat rose-tinted, soft, silky; spikelets about 3 mm. long, clothed with hairs twice as long.

Desert regions, in the Lower Sonoran Zone; from southern California to western Texas, and south to Jalisco. Sept.–May. Type locality: Texas.

2. ANDROPÒGON L. Sp. Pl. 1045. 1753.

Spikelets in pairs at each node of an articulate rachis, one sessile and perfect, the other pedicellate and staminate, neuter or reduced to the pedicel; glumes of fertile spikelet coriaceous, narrow, awnless, the first rounded, flat or concave on the back, several-nerved, the median nerve weak or wanting; sterile lemma shorter than the glumes, empty, hyaline; fertile lemma hyaline, narrow, entire or bifid, usually bearing a bent and twisted awn from the apex or from between the lobes; palea hyaline, small or wanting; pedicellate spikelet awnless, sometimes staminate and about as large as the sessile spikelet, sometimes consisting of one or more reduced glumes, sometimes wanting, only the pedicel present. Rather coarse perennial grasses with solid culms, the spikelets in racemes, these (in ours) numerous, aggregate on an exserted peduncle, or single or in pairs, the common peduncle enclosed by a spathe-like sheath, these sheaths clustered, forming a compound inflorescence; rachis and pedicels of sterile spikelet often villous. [Greek, man-beard.]

Species about 150 in all warmer parts of the world. Type species, *Andropogon virginicus* L.

Racemes numerous in a pedunculate bractless panicle terminating the culms. 1. *A. saccharoides.*
Racemes in pairs, subtended by a sheathing bract, forming a compound inflorescence. 2. *A. glomeratus.*

1. Andropogon saccharoìdes Swartz.

Plumed Beard-grass. Fig. 219.

Andropogon saccharoides Swartz, Prodr. Veg. Ind. Occ. 26. 1788.

Andropogon argenteus DC. Cat. Hort. Monsp. 77. 1813.

Andropogon barbinodis Lag. Gen. & Sp. Nov. 3. 1816.

Culms tufted, erect or somewhat spreading at base, 60–120 cm. high, glabrous except the densely ascending-hispid nodes; sheaths glabrous; blades 3–6 mm. wide, flat, scabrous above, the upper much reduced; panicle 5–8 cm. long, consisting of several appressed or ascending silky-white racemes, somewhat flabellately aggregate at the summit of the culm; glumes of sessile spikelet 5 mm. long, the awn about 2 cm. long, geniculate at the middle, tightly twisted below the bend, loosely twisted above.

Dry hills in the Transition and Sonoran Zones; from Santa Barbara to San Diego, east to New Mexico and south to the American tropics. Feb.–June. Type locality: Jamaica.

2. Andropogon glomeràtus (Walt.) B. S. P.
Bushy Beard-grass. Fig. 220.

Cinna glomerata Walt. Fl. Carol. 59. 1788.
Andropogon macrourum Michx. Fl. Bor. Am. 1: 56. 1803.
Andropogon glomeratus B. S. P. Prel. Cat. N. Y. 67. 1888.

Culms erect, 50–100 cm. tall, simple below, branched above, the branches erect and again successively branched, forming a club-shaped or corymbiform rather dense, or somewhat interrupted top; sheaths compressed, scabrous; blades 2–5 mm. wide, scabrous, long-acuminate, the basal often as long as the culm; sessile spikelet about 5 mm. long; awn 1–2 cm. long; pedicellate spikelet reduced to a single glume or wanting.

Foothills of San Bernardino Mountains (*Parish, Reed*) and Funeral Mountains (*Coville* and *Funston*), eastward to Florida and New York, and southward to Central America. Type locality: South Carolina.

Andropogon cirràtus Hack. Flora **68**: 119. 1885. A fragmentary specimen of what appears to be this species was collected at Jamacha, Calif. (*Canby*). A perennial with solitary racemes, the sessile spikelet 6 mm. long. West Texas to Arizona and Mexico.

3. SÓRGHUM Moench, Meth. Pl. 207 1794.

Spikelets as in *Andropogon*, but the first glume of sessile spikelet convex; pedicellate spikelet well-developed, usually staminate; racemes reduced to 1–5 joints, these disarticulating tardily, the racemes in a large terminal open or contracted panicle. Annual or perennial, mostly tall grasses with flat blades and terminal panicles. [Name of uncertain origin.]

Species about 20, one Mexican, the others in the Old World, mostly Africa; two cultivated or introduced in America. Type species, *Holcus sorghum* L.

1. Sorghum halepénse (L.) Pers.
Johnson Grass. Fig. 221.

Holcus halepensis L. Sp. Pl. 1047. 1753.
Andropogon halepensis Brot. Fl. Lusit. 1: 89. 1804.
Sorghum halepense Pers. Syn. Pl. 1: 101. 1805.

Culms erect, glabrous, robust, 50–120 cm. tall. producing stout creeping rhizomes; blades flat, 6–15 mm. wide, the midrib prominent, white; panicle 15–25 cm. long, more or less spreading; fertile spikelets about 5 mm. long, the glumes pubescent, becoming glabrate and shining, the awn readily deciduous; staminate spikelets narrow, 4 mm. long on pedicels 3 mm. long, the glumes membranaceous, nerved, glabrous.

Occasional in fields and waste places, California. Common in the southern States; introduced from the Old World. Apr.–Oct. Type locality: Syria.

4. HILÀRIA H. B. K. Nov. Gen. & Sp. 1: 116. *pl. 37.* 1816.

Spikelets sessile in groups of three, the groups falling entire, the central spikelet (next the axis) fertile, 1-flowered, the two lateral spikelets staminate, 2-flowered; glumes coriaceous, those of the 3 spikelets forming a false involucre, in some species connate at the base, more or less unsymmetrical, usually bearing an awn on one side from about the middle; lemma and palea hyaline, about equal in length. Perennial low grasses, the groups of spikelets appressed to the axis in terminal spikes. [Name in honor of the French naturalist Auguste de Saint-Hilaire.]

Species 5, southwestern United States to Central America, all but one found within the limits of the United States. Type species, *Hilaria cenchroides* H. B. K.

Culms felty, pubescent. 1. *H. rigida.*
Culms glabrous or slightly puberulent. 2. *H. jamesii.*

1. Hilaria rígida (Thurb.) Benth.

Galleta. Fig. 222.

Pleuraphis rigida Thurb. in S. Wats. Bot. Calif. 2: 293. 1880.
Hilaria rigida Benth.; Scribn. Bull. Torrey Club 9: 33, 86. 1882.

Culms numerous, rigid, felty-pubescent, glabrate and scabrous above, 50–100 cm. tall; leaves felty or glabrous, usually woolly at the top of the sheath; blades spreading, 2–5 cm. long, or longer on sterile shoots, 2–4 mm. wide, more or less involute, acuminate into a rigid coriaceous point; spikelets about 8 mm. long; glumes of central spikelet broadened upward from a narrow base, woolly-ciliate, several-awned from the tip, a stronger dorsal awn from about the middle; lemma 3-nerved, enclosing the palea and a rudimentary second floret, the nerves villous on the back, extending into delicate awns between the ciliate lobes of the apex; lateral spikelets similar, narrower, the glumes less awned at the tip, the second floret similar to the first.

Mojave and Colorado Deserts, Lower Sonoran Zone; east to Arizona and south into Sonora. Feb.–June. Type locality: southern California.

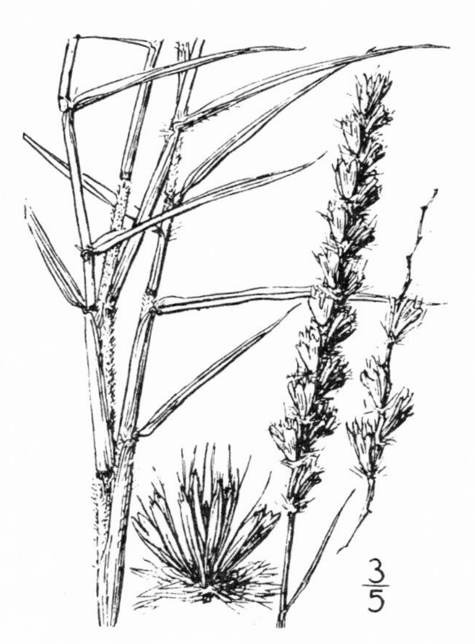

2. Hilaria jàmesii (Torr.) Benth.

James' Galleta. Fig. 223.

Pleuraphis jamesii Torr. Ann. Lyc. N. Y. 1: 148. *pl. 10.* 1824.
Hilaria jamesii Benth. Journ. Linn. Soc. Bot. 19: 62. 1881.

Culms glabrous, the nodes villous; sheaths glabrous or slightly scabrous, sparingly villous around the short membranaceous ligule; blades mostly 2–5 cm. long, 2–4 mm. wide, rigid, soon involute, the upper reduced; spikelets 6–8 mm. long, long-villous at base; glumes of central spikelet pubescent, cuneate, 2-lobed, the lobes 2- or 3-awned, the central nerve between, extending from below the middle into an awn somewhat longer than the others, the awns all minutely plumose; lemma erose at apex, glabrous, 3-nerved, the nerves parallel, the central extending into a short awn; glumes of lateral spikelets narrow, pubescent, the first unsymmetrical, 5-nerved, the second nerve on one side extending into a dorsal awn from below the middle, the apex unequally 2-lobed, the sinus extending down about half-way to the point of departure of the awn, the lobes minutely ciliate; second glume 5-nerved, awnless, entire, ciliate, conduplicate around the floret; lemma as in fertile spikelet; stamens 2.

Deserts of Inyo County, Lower Sonoran Zone; east to Wyoming and Texas. May. Type locality: the sources of the Canadian River.

5. SYNTHERÍSMA Walt. Fl. Carol. 76. 1788.

[Digitaria Hall, Hist. Stirp. Helv. 2: 244. 1768, not Adans. 1763, nor Heist. 1759.]

Spikelets solitary or in twos or threes, subsessile or short-pediceled, alternate in 2 rows on one side of a 3-angled winged or wingless rachis; spikelets lanceolate or elliptic, plano-convex; first glume minute or wanting; second glume equaling the sterile lemma or shorter; fertile lemma cartilaginous, the hyaline margins pale. Annual or sometimes perennial erect or prostrate grasses, the slender racemes digitate or aggregate at the summit of the culms. [Greek, crop-making.]

Species about 75 in the warmer parts of the world. Type species, *Syntherisma praecox* Walt.

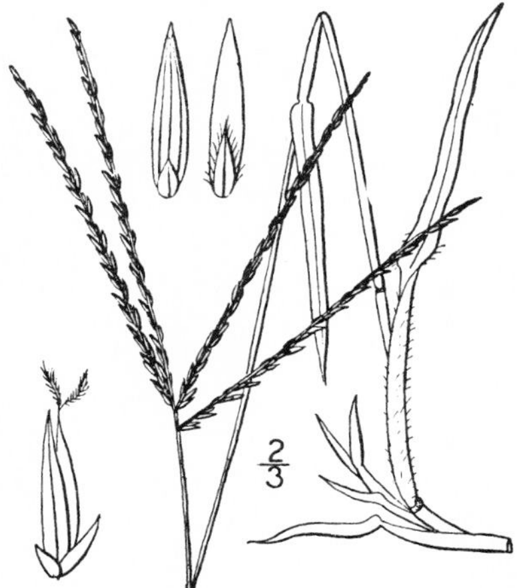

1. Syntherisma sanguinàlis (L.) Dulac.

Crab-grass. Fig. 224.

Panicum sanguinale L. Sp. Pl. 57. 1753.
Digitaria sanguinalis Scop. Fl. Carn. ed. 2. 1: 52. 1772.
Syntherisma praecox Walt. Fl. Carol. 76. 1788.
Syntherisma sanguinalis Dulac, Fl. Haut. Pyr. 77. 1867.

Annual, usually much-branched at base; culms 30–60 cm. long, geniculate-spreading, or creeping and rooting at the nodes, the flower-stalks ascending; sheaths more or less papillose-hirsute; blades lax, 6–10 cm. long, 4–10 mm. wide, often pilose; racemes 3–12, subdigitate, 5–10 cm. long; spikelets in pairs, 3–3.5 mm. long, usually appressed-pubescent between the smooth or scabrous nerves; pedicels angled; first glume minute; second glume about half as long as the spikelet; fruit lead color.

A weed in cultivated soil; occasional on the Pacific slope; common in the warmer parts of America; a native of the Old World. June–Sept. Type locality, European.

Syntherisma ischaèmum (Schreb.) Nash (*S. humifusa* Rydb.) has been found at Walla Walla. Plant glabrous; first glume wanting, second and sterile lemma glandular-puberulent; fruit dark brown.

6. ERIÓCHLOA H. B. K. Nov. Gen. & Sp. 1: 94. *pls. 30, 31.* 1816.

Spikelets dorsally compressed, more or less pubescent, solitary or sometimes in pairs, short-pediceled or subsessile in 2 rows on one side of a narrow, usually hairy rachis, the pedicels often clothed with long stiff hairs, the back of the fertile lemma turned from the rachis; first glume represented by a prominent ring below the spikelet; second glume and sterile lemma about equal, acute or acuminate, the lemma usually inclosing a hyaline palea or sometimes a staminate flower; fertile lemma indurate, minutely papillose-rugose, mucronate or awned, the margins slightly inrolled. Annuals or perennials, with terminal panicles consisting of several or many spreading or appressed racemes, usually rather closely arranged along the main axis. [Greek, wool-grass.]

Species about 15 in the warmer parts of the world, mostly in America. Type species, *Eriochloa distachya* H. B. K.

1. Eriochloa aristàta Vasey.

Awned Eriochloa. Fig. 225.

Eriochloa aristata Vasey, Bull. Torrey Club 13: 229. 1886.

Plants annual; culms 30–100 cm. tall, smooth; blădes flat, 10–20 cm. long, 3–10 mm. wide, smooth; panicle 8–12 cm. long; spikes several, appressed, 2–3 cm. long, the rachis more or less villous; spikelets lanceolate, appressed-silky, 5 mm. long excluding the awn-point of the second glume, the pedicels with a tuft of stiff hairs near the apex; fruit oval, about half as long as the sterile lemma, apiculate.

Alluvial land along the Colorado River at Fort Yuma. Sept.–Dec. Type locality: Chihuahua.

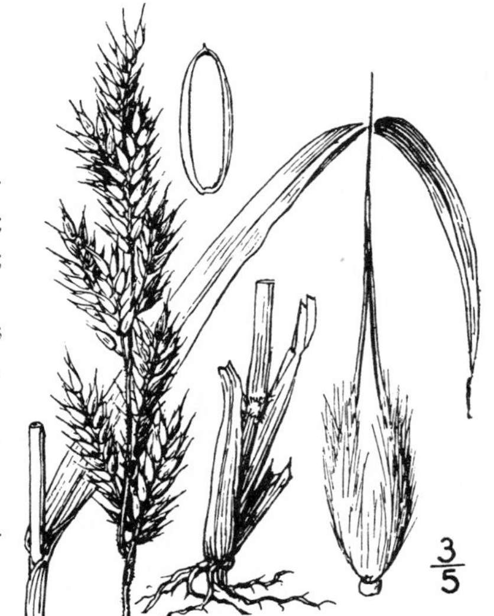

7. PÁSPALUM L. Syst. Nat. ed. 10. 2: 855. 1759.

Spikeless plano-convex, subsessile, solitary or in pairs, in two rows on one side of a narrow or dilated rachis, the back of the fertile lemma toward it; first glume wanting or occasionally present in a few spikelets; second glume and sterile lemma about equal; fertile lemma obtuse, chartaceous-indurate, the margins inrolled. Mostly perennials, with 1 to many spikelike racemes, these single or paired at the summit of the culms or racemosely arranged along the main axis. [Greek name for some grass.]

Species numerous, probably as many as 200, widely distributed in the warmer parts of both hemispheres. Type species, *Paspalum dimidiatum* L.

Racemes 2, digitate. — 1. *P. distichum.*
Racemes several to many, racemose, forming a panicle.
 Racemes several to many, laxly spreading; spikelets about 3.5 mm. long. — 2. *P. dilatatum.*
 Racemes numerous, erect or suberect; spikelets scarcely over 2 mm. long. — 3. *P. larranagae.*

1. Paspalum dístichum L.
Knot-grass. Fig. 226.

Paspalum distichum L. Syst. Nat. ed. 10. **2**: 855. 1759.

Culms erect from a decumbent rooting base, with numerous creeping rhizomes, glabrous, or the nodes pubescent, 30–60 cm. tall; sheaths glabrous or sometimes pubescent; blades flat, glabrous, rarely pubescent, 5–10 cm. long, the upper shorter; racemes 2, paired or closely approximate, sometimes a third below the second, more or less pilose at base, slender, ascending or appressed, usually 2.5–5 cm. long; spikelets elliptic, 3 mm. long; first glume occasionally present and nearly as long as spikelet, glabrous; second glume pubescent; sterile lemma glabrous.

Along the seacoast and especially along irrigation ditches and wet places in the interior, California, and occasionally along the Columbia River. June–Sept. Type locality: Jamaica.

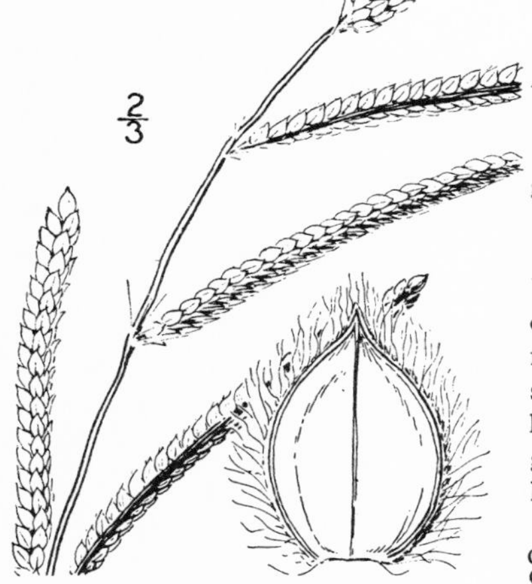

2. Paspalum dilatàtum Poir.
Dallis Grass. Fig. 227.

Paspalum dilatatum Poir. Lam. Encycl. **5**: 35. 1804.

Culms few to many, tufted, erect, or geniculate at base, 1 meter or more tall, glabrous; sheaths loose, glabrous or the lowermost softly pubescent; ligule about 3 mm. long; blades flat, 5–12 mm. wide, commonly elongate, pilose on the upper surface at the base, otherwise glabrous; panicle 10–20 cm. long, the common axis slender; racemes mostly 5–10, densely flowered, 5–10 cm. long, the rachis narrowly winged; spikelets in pairs, ovate, abruptly pointed, 3–3.5 mm. long; 2–2.3 mm. wide, flat, the margins of the second glume and sterile lemma, especially the glume, with long silky white hairs.

Ballast, Portland, Oregon, and roadsides, Los Angeles, California. Introduced from South America; cultivated in the Gulf States and abundantly naturalized; rare on the Pacific Coast. July–Oct. Type locality: Argentina.

3. Paspalum larranàgai Arechav.
Vasey Grass. Fig. 228.

Paspalum virgatum pubiflorum Vasey, Bull. Torrey Club **13**: 167. 1886.

Paspalum larranagai Arechav. Anal. Mus. Nac. Montevid. **1**: 60. *pl. 2.* 1894.

Paspalum vaseyanum Scribn. U. S. Dept. Agr. Div. Agrost. Bull. **17**: 32. *f. 328.* 1899.

Resembles *P. dilatatum,* culms usually more robust; lower sheaths hirsute, the upper hirsute on the margins; panicles commonly more than 15 cm. long, racemes 10–25 (rarely fewer); spikelets 2–2.2 mm. long, the silky hairs longer and more copious than in *P. dilatatum.*

Along irrigation ditches, California. Introduced from South America; open low ground from Texas to South Carolina, spreading rapidly. Aug.–Oct. Type locality: Uruguay.

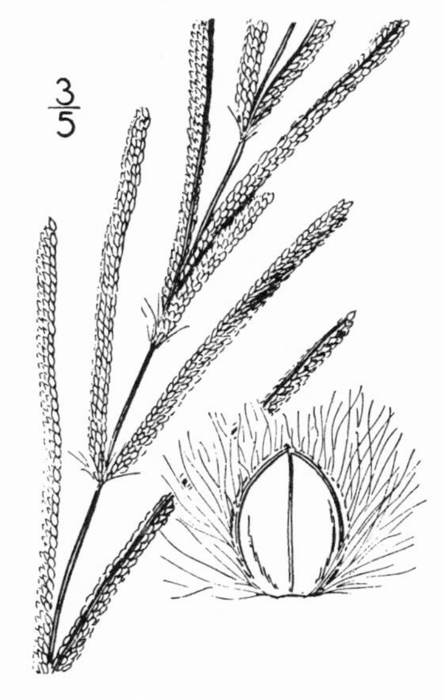

8. PÁNICUM L. Sp. Pl. 55. 1753.

Spikelets more or less compressed dorsally, arranged in open or compact panicles, rarely in racemes; glumes 2, herbaceous, nerved, usually very unequal, the first often minute, the second typically equaling the sterile lemma, the latter of the same texture and simulating a third glume, bearing in its axil a membranaceous or hyaline palea and sometimes a staminate flower, the palea rarely wanting; fertile lemma chartaceous-indurated, typically obtuse, the nerves obsolete, the margins inrolled over an inclosed palea of the same texture. Annual or perennial grasses of various habit. [An ancient Latin name for the Italian millet, *Chaetochloa italica.*]

Species probably about 500, confined to the warmer regions of both hemispheres. Type species, *Panicum miliaceum* L.

Plants annual.
  Fruit (mature fertile lemma) transversely rugose. 1. *P. arizonicum.*
  Fruit smooth.
    First glume not over one-fourth the length of the spikelet, truncate or broadly triangular; sheaths smooth. 2. *P. dichotomiflorum.*
    First glume as much as half the length of the spikelet, acute or acuminate; sheaths hispid.
      Panicle erect.
        Panicle more than half the length of the entire plant.
          Spikelets 2–2.5 mm. long. 3. *P. capillare.*
          Spikelets about 3 mm. long. 4. *P. barbipulvinatum.*
        Panicle not more than one-third the entire height of plant; spikelets 2.7–3.3 mm. long. 5. *P. hirticaule.*
      Panicle drooping. 6. *P. miliaceum.*
Plants perennial.
  Spikelets 6–7 mm. long. 7. *P. urvilleanum.*
  Spikelets less than 4 mm. long.
    Spikelets not turgid nor strongly nerved, pubescent, not over 2.6 mm. long.
      Spikelets not over 2 mm. long.
        Blades glabrous on both surfaces. 8. *P. lindheimeri.*
        Blades pubescent, at least below.
          Plants more or less pubescent but not velvety.
            Vernal blades pubescent above.
              Upper surface of blades appressed-pubescent; autumnal form erect or ascending 9. *P. huachucae.*
              Upper surface of blades pilose; autumnal form decumbent-spreading. 10. *P. pacificum.*
            Vernal blades glabrous above. 11. *P. occidentale.*
          Plants velvety-pubescent. 12. *P. thermale.*
      Spikelets about 3 mm. long. 13. *P. shastense.*
    Spikelets turgid, strongly nerved, sparsely hispid, spikelets 3.2–3.3 mm. long. 14. *P. scribnerianum.*

1. Panicum arizónicum Scribn. & Merr. Arizona Panicum. Fig. 229.

Panicum arizonicum Scribn. & Merr. U. S. Dept. Agr. Div. Agrost. Circ. 32: 2. 1901.

Annual; culms usually branching from the base, glabrous except below the panicle, 20–60 cm. tall; nodes sometimes slightly pubescent; sheaths glabrous to strongly papillose-pubescent; blades 5–15 cm. long, 6–12 mm. wide, glabrous or papillose-hispid beneath; panicles long-exserted, finely pubescent and copiously papillose-hirsute, 7–20 cm. long, the branches solitary, ascending, few-flowered; spikelets nearly 4 mm. long, obovate-elliptical, abruptly pointed, densely hirsute to glabrous, borne on very short appressed branchlets.

Open ground, Sonoran Zones; Jamacha, California, to western Texas and south to Oaxaca. July. Type locality: Arizona.

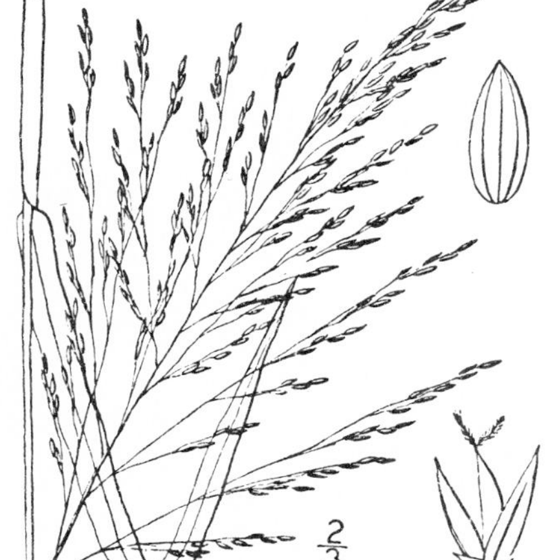

2. Panicum dichotomiflòrum Michx. Smooth Witch-grass. Fig. 230.

Panicum dichotomiflorum Michx. Fl. Bor. Am. 1: 48. 1803.

Annual, usually much-branched from a geniculate base, smooth throughout; culms rather succulent, 60–90 cm. tall; blades 10–50 cm. long, about 1 cm. wide; panicles 10–30 cm. long, finally spreading; spikelets 2.5 mm. long, narrowly oblong-ovate, acute, faintly 7-nerved, the first glume short, truncate, about one-fourth the length of the spikelet.

Low ground and cultivated soil. Fresno (*Bioletti*). Common in eastern United States. Type locality: Allegheny Mountains.

3. Panicum capilláre L.

Old-Witch Grass. Fig. 231.

Panicum capillare L. Sp. Pl. 58. 1753.

Annual, erect, 30–60 cm. tall; foliage papillose-hispid; blades 10–25 cm. long, 5–10 mm. wide; panicle large and diffuse, often half the length of the entire plant, included at the base until maturity, the whole panicle finally breaking away and rolling before the wind; spikelets 2 mm. long, elliptic; first glume acute, half as long as spikelet, 5-7-nerved.

Open ground, cultivated soil, and river banks, California (Pine Grove, *Hansen; Yreka, Butler*). A common weed in eastern United States. August. Type locality: Virginia.

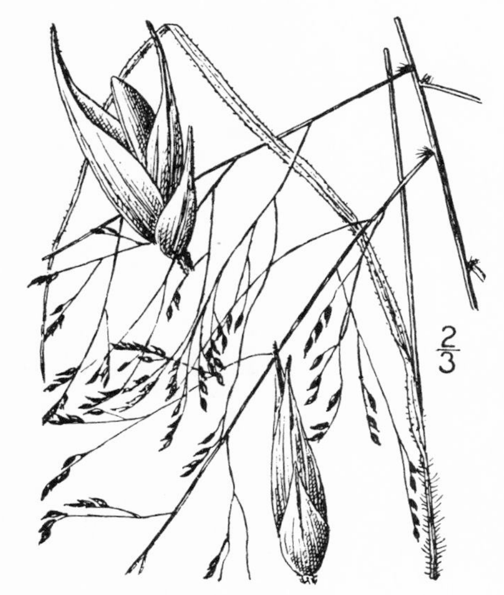

4. Panicum barbipulvinàtum Nash.

Western Witch-grass. Fig. 232.

Panicum capillare occidentale Rydb. Contr. U. S. Nat. Herb. 3: 186. 1895.

Panicum capillare brevifolium Vasey; Rydb. & Shear, U. S. Dept. Agr. Div. Agrost. Bull. 5: 21. 1897, not *P. brevifolium* L.

Panicum barbipulvinatum Nash, in Rydb. Mem. N. Y. Bot. Gard. 1: 21. 1900.

Similar in habit to *P. capillare*, usually stouter, the blades shorter, less pubescent and more crowded toward the base of the plant; spikelets 2.5–3.5 mm. long, acuminate, larger on the average than in *P. capillare*.

Open ground and cultivated fields, in the Transition and Upper Sonoran Zones; common at moderate altitudes throughout our area and eastward to the Great Plains. July–Oct. Type locality: Yellowstone Park.

5. Panicum hirticàule Presl.

Mexican Witch-grass. Fig. 233.

Panicum hirticaule Presl, Rel. Haenk. 1: 308. 1830.

Annual, erect or nearly so, 15–60 cm. tall, more or less papillose-hispid throughout, especially on the sheaths; blades 4–12 mm. wide, often sparsely hispid toward the often cordate base; panicle 7–15 cm. long, scarcely one-third the height of the entire plant, open, the branches ascending; spikelets 3 mm. long, acuminate, usually reddish brown.

Open ground and sandy soil, Sonoran Zones; California (Jamacha, *Canby; Fort Yuma*, Arizona, *Parish*), eastward to Texas and south to northern South America. July–Oct. Type locality: Acapulco, Mexico.

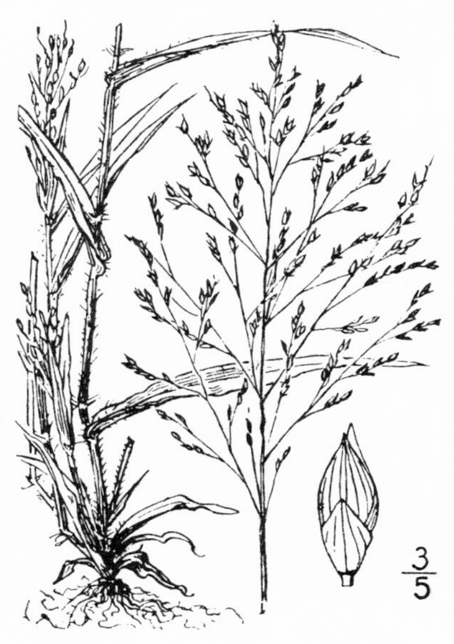

6. Panicum miliàceum.
Broom-corn Millet. Fig. 234.

Panicum miliaceum L. Sp. Pl. 58. 1753.

Annual, as much as 1 meter high; culms and leaves more or less papillose-hispid; panicle 10–30 cm. long, usually nodding, rather compact, the numerous branches ascending, scabrous, spikelet-bearing toward the summit; spikelets 5 mm. long, ovate, acuminate, strongly many-nerved, first glume half the length of the spikelet, acuminate.

A native of the Old World, cultivated in some parts of the United States as a forage plant under the names broom-corn millet, hog millet, and proso millet. Scattered specimens have been found in California. Aug.–Oct. Type locality: India.

7. Panicum urvilleànum Kunth.
Desert Panicum. Fig. 235.

Panicum urvilleanum Kunth, Rév. Gram. 2: 403. 236. *pl. 115.* 1831.
Panicum urvilleanum longiglume Scribn. U. S. Dept. Agr. Div. Agrost. Bull. 17 (ed. 2): 49. 1901.

Plants robust, 60–100 cm. tall, perennial from creeping rhizomes; culms solitary or few in a tuft, the nodes densely bearded but usually hidden by the harshly villous sheaths; blades 4–6 mm. wide, flat, tapering to a long involute-setaceous point; panicle about 30 cm. long, open; spikelets 6–7 mm. long, densely silvery- or tawny-villous; first glume acuminate, two-thirds to nearly as long as spikelet.

Sandy deserts, southern California and Arizona, also in southern South America. Mar.–June. Type locality: Chile; of the variety *longiglume*, San Jacinto, California.

8. Panicum lindheìmeri Nash.
Lindheimer's Panicum. Fig. 236.

Panicum lindheimeri Nash, Bull. Torrey Club 24: 196. 1897.
Panicum funstoni Scribn. & Merr. U. S. Dept. Agr. Div. Agrost. Circ. 35: 4. 1901.

Vernal culms stiffly ascending or spreading, 30–60 cm. high, glabrous, or the lower portion somewhat pubescent; leaves glabrous except the ciliate margin of the lower part of the blades (sheaths rarely with scattered hairs), ligule a ring of hairs 4–5 mm. long; panicle 5–8 cm. long, open; spikelets 1.5 mm. long, obovate, obtuse, turgid, pubescent; autumnal form stiffly spreading or radiate-prostrate, with tufts of short appressed branches at the nodes; blades reduced, involute-pointed, often conspicuously ciliate at the base.

Open ground and grass land, California (Three Rivers, *Coville & Funston*; Sacramento, *Michener*), and eastward; common in the eastern States. August. Type locality: Texas; of *P. funstoni*, Three Rivers, California.

9. Panicum huachùcae Ashe.

Hairy Panicum. Fig. 237.

Panicum huachucae Ashe, Journ. Elisha Mitchell Soc. **15**: 51. 1898.

Vernal form usually stiffly upright, more or less harsh-pubescent throughout; culms 30–60 cm. long, the nodes bearded; ligule of stiff hairs about 4 mm. long; panicle 5–8 cm. long, the axis and usually the branches pilose; spikelets about 1.6–1.8 mm. long, obovate, turgid, pubescent; autumnal form stiffly erect, the reduced branches fascicled, the crowded blades ascending.

Open ground, California (San Bernardino Mountains, *Abrams*), and eastward; common in the Eastern States. June–Aug. Type locality: Huachuca Mountains, Arizona.

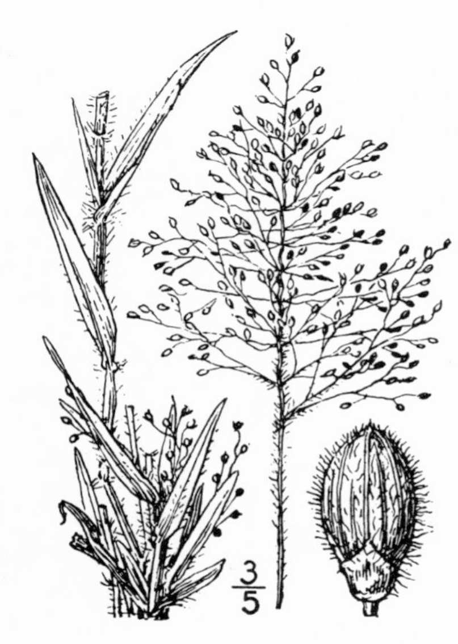

10. Panicum pacíficum Hitchc. & Chase.

Pacific Panicum. Fig. 238.

Panicum pacificum Hitchc. & Chase, Contr. U. S. Nat. Herb. **15**: 229. *f. 241.* 1910.

Vernal form light green, more or less papillose-pilose throughout, 30–60 cm. tall; ligule ciliate, about 4 mm. long; spikelets 1.8–2.0 mm. long, obtuse, pubescent; autumnal form prostrate-spreading, repeatedly branching from the upper and middle nodes. Distinguished from *P. occidentale* by the more copious pubescence throughout, more leafy culms, and, in the autumnal phase, by the branching habit.

Sandy shores and slopes, and moist crevices in rocks, British Columbia to San Bernardino Mountains. June–Aug. Type locality: Castle Crag, California. This and allied species have been included under *P. dichotomum* in Pacific Coast botanies.

11. Panicum occidentàle Scribn.

Western Panicum. Fig. 239.

Panicum occidentale Scribn. Rep. Mo. Bot. Gard. **10**: 48. 1899.

Vernal form yellowish green; culms slender, 10–20 cm. long, spreading, sparsely pubescent; leaves tending to be clustered toward the base; sheaths sparsely pubescent; ligule ciliate, about 4 mm. long; blades glabrous or nearly so above, appressed-pubescent beneath; panicle 5–8 cm. long, open; spikelets 1.8 mm. long, pubescent; autumnal form branching from the lower nodes, forming a spreading tussock; leaves and panicles reduced.

Peat bogs and moist soil, common from British Columbia to southern California. June–Aug. Type locality: Nootka Sound.

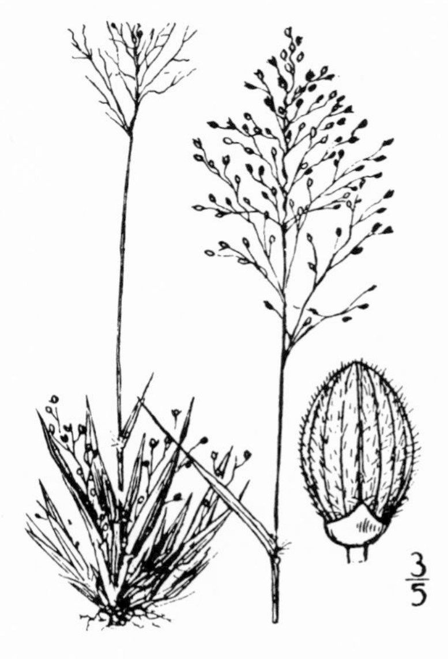

12. Panicum thermàle Boland.
Hot Spring Panicum. Fig. 240.

Panicum thermale Boland. Proc. Calif. Acad. **2**: 181. 1862.

Vernal form grayish green, densely tufted, velvety-villous, 10–30 cm. tall; culms ascending or spreading; nodes bearded; ligule about 3 mm. long; blades densely villous on both surfaces; spikelets about 2 mm. long, pubescent; autumnal form widely spreading, repeatedly branching, forming a dense cushion.

Wet saline soil mostly in vicinity of hot springs, Washington (Granite Falls, *Smith*); Oregon (Rogue River, *Nelson*); California (Sonoma County; vicinity of Lassen Peak); Yellowstone Park, and northward. June–Aug. Type locality: Sonoma County, California.

13. Panicum shasténse Scribn. & Merr.
Shasta Panicum. Fig. 241.

Panicum shastense Scribn. & Merr. U. S. Dept. Agr. Div. Agrost. Circ. **35**: 3. 1901.

Vernal form 30–40 cm. tall, papillose-pilose throughout; ligule 2–3 mm. long, sparse; spikelets 2.4–2.6 mm. long, papillose-pubescent; autumnal form spreading, with geniculate nodes and elongate arched internodes, rather sparingly branching from the middle nodes.

Meadows, Castle Crag, California, the type locality and only known station (*Greata, Hitchcock*).

14. Panicum scribneriànum Nash.
Scribner's Panicum. Fig. 242.

Panicum scribnerianum Nash, Bull. Torrey Club **22**: 421. 1895.

Vernal form erect, 30–60 cm. tall, sheaths papillose-hispid; ligule about 1 mm. long; blades 5–8 cm. long, 6–12 mm. wide, firm, rounded and ciliate at base, glabrous above, often pubescent beneath; panicles 5–8 cm. long; spikelets 3.2 mm. long, turgid, blunt, sparsely hispid or nearly glabrous, strongly nerved; autumnal form branching from the middle and upper nodes, the branches longer than the internodes, late in the season producing crowded branchlets with ascending not greatly reduced blades and small, partially included panicles from their upper nodes.

Dry prairies and open ground, British Columbia to California (Castle Crag, *Hitchcock*), and east to the Atlantic Coast. June–Aug. Type locality: Pennsylvania. This species was included under *P. scoparium* in the Botany of California.

9. ECHINÓCHLOA Beauv. Ess. Agrost. 53. *pl. 11. f. 2.* 1812.

Spikelets, plano-convex, often stiffly hispid, subsessile, solitary or in irregular clusters on one side of the panicle branches; first glume about half the length of the spikelet, pointed; second glume and sterile lemma equal, pointed, mucronate, or the glume short-awned and the lemma long-awned, sometimes conspicuously so, inclosing a membranaceous palea, and sometimes a staminate flower; fertile lemma plano-convex, smooth and shining, abruptly pointed, the margins inrolled below, flat above, the apex of the palea not inclosed. Coarse often succulent annuals or sometimes perennials with compressed sheaths, linear flat blades, and rather compact panicles composed of short densely flowered racemes along the main axis. [Greek, hedgehog-grass.]

Species about 15 in the warm regions of both hemispheres. Type species, *Panicum crusgalli* L.

Spikelets long-awned. 1. *E. crusgalli.*
Spikelets awnless or short-awned (the awn usually not longer than the spikelet).
  Branches more or less compound; glumes firm in texture, strongly hispid; spikelets about 4 mm. long. 1a. *E. crusgalli zelayensis.*
  Branches simple; glumes rather soft, sparingly hispid; spikelets about 3 mm. long. 2. *E. colonum.*

1. Echinochloa crusgálli (L.) Beauv. Barnyard Grass. Fig. 243.

Panicum crusgalli L. Sp. Pl. 56. 1753.
Echinochloa crusgalli Beauv. Ess. Agrost. 53, 161. 1812.

Culms stout, rather succulent, branching from the base or erect, usually 30–100 cm. tall, sometimes larger; leaves glabrous; panicle dense, 10–25 cm. long, consisting of several erect or spreading, or even drooping racemes; spikelets green or purple, about 3 mm. long, exclusive of awns, densely and irregularly crowded in 3 or 4 rows; awns usually longer than the spikelet.

Fields and cultivated soil, especially along irrigating ditches. Common in eastern and southern United States and in the warmer parts of both hemispheres. Introduced and not common on the Pacific Slope. Aug.–Sept. Type locality, European.

1a. Echinochloa crusgalli zelayénsis (H. B. K.) Hitchc. Mexican Barnyard Grass. Fig. 244.

Oplismenus zelayensis H. B. K. Nov. Gen. & Sp. **1**: 108. 1816.
Echinochloa zelayensis Schult. Mant. **2**: 269. 1824.
Echinochloa crusgalli zelayensis Hitchc. Contr. U. S. Nat. Herb. **22**: 147. 1920.

Culms erect, 30–100 cm. tall; panicles 10–20 cm. long, the spikes mostly appressed, sometimes spreading, more or less compound, 2–5 cm. long; spikelets about 4 mm. long, acuminate or short-awned. Resembling *E. crusgalli*, but the spikelets short-awned or awnless.

Moist places, especially along irrigating ditches, infrequent within our range, common in Mexico and southward in South America. Aug.–Sept. Type locality: Zelaya, Mexico.

2. Echinochloa colònum (L.) Link. Small Barnyard Grass. Fig. 245.

Panicum colonum L. Syst. Nat. ed. 10. **2**: 870. 1759.
Echinochloa colonum Link, Hort. Berol. **2**: 209. 1833.

Culms erect, spreading or prostrate, 30–60 cm. tall; leaves smooth; panicle of 5–10 dense racemes, 1.5–2.5 cm. long, rather distant, racemose along the axis; spikelets about 3 mm. long; glumes and sterile lemma pubescent, mucronate-pointed only.

Introduced from the Old World into the warmer parts of America. Occasional in moist places in southern California. Type locality: Jamaica.

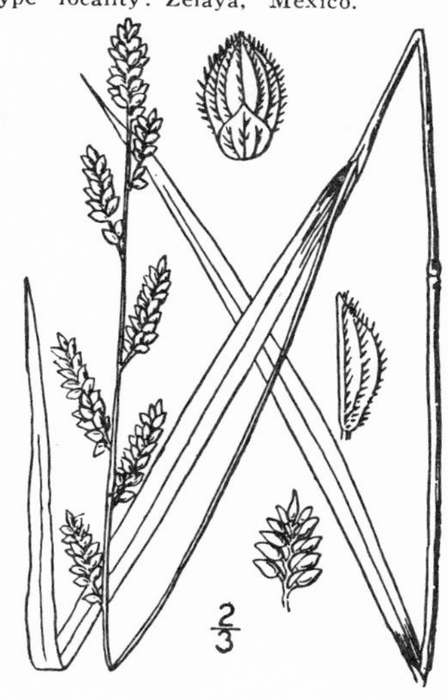

10. CHAETÓCHLOA Scribn. U. S. Dept. Agr. Div. Agrost. Bull. 4: 38. 1897.

[SETARIA Beauv. Ess. Agrost. 51. 1812, not Ach. 1798, nor Michx. 1803.]

Spikelets subtended by 1 to several bristles (sterile branchlets), falling free from the bristles, awnless; first glume broad, usually less than half the length of the spikelet, 3–5-nerved; second glume and sterile lemma equal or the former shorter, several-nerved; fertile lemma coriaceous-indurate, smooth or rugose. Annuals or perennials with narrow terminal panicles, these dense and spikelike or somewhat loose and open. [Greek, bristle-grass.]

Species about 65, in the tropical and warm temperate regions of both hemispheres. Type species, *Panicum viride*.

Plants annual.
  Bristles backwardly barbed. — 1. *C. verticillata.*
  Bristles forwardly barbed.
    Bristles below each spikelet 5 or more. — 2. *C. lutescens.*
    Bristles below each spikelet 1–3. — 3. *C. viridis.*
Plants perennial by short rhizomes or knotty crown. — 4. *C. geniculata.*

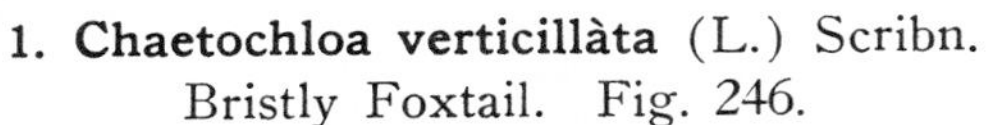

1. Chaetochloa verticillàta (L.) Scribn. Bristly Foxtail. Fig. 246.

Panicum verticillatum L. Sp. Pl. ed. 2. 82. 1762.
Setaria verticillata Beauv. Ess. Agrost. 51, 178. 1812.
Chaetochloa verticillata Scribn. U. S. Dept. Agr. Div. Agrost. Bull 4: 39. 1897.

Annual; culms 30–60 cm. tall; blades scabrous; panicle green, 5–10 cm. long, usually interrupted at base and somewhat lobed, tapering above; bristles downwardly barbed, 3–6 mm. long; spikelets 2–2.5 mm. long, the first glume one-third as long as the second, the latter equaling the sterile lemma; fertile lemma obscurely transverse-rugose.

A weed, southern California (Upland, *Johnston*); common in eastern United States. Introduced from Europe. Type locality, European.

2. Chaetochloa lutéscens (Weigel) Stuntz. Yellow Foxtail. Fig. 247.

Panicum lutescens Weigel, Obs. Bot. 20. 1772.
Chaetochloa lutescens Stuntz, U. S. Dept. Agr. Bur. Pl. Ind. Inv. Seeds 31: 36, 86. 1914.
Panicum glaucum of authors not L. 1753; *Chaetochloa glauca* of American authors.

Annual; culms branching at the base, compressed, erect or ascending, 30–60 cm. tall; blades flat, with a spiral twist; panicle dense, oblong, 2–8 cm. long; bristles 5 or more, 4–8 mm. long, tawny-yellow; spikelets 3 mm. long; fruit undulate-rugose.

A weed in fields and waste places. Infrequent in California and Oregon; common in the Eastern States. Introduced from Europe, the type locality. Aug.–Oct.

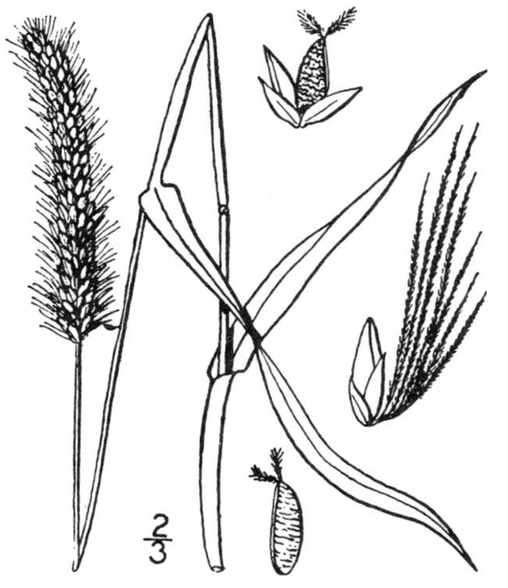

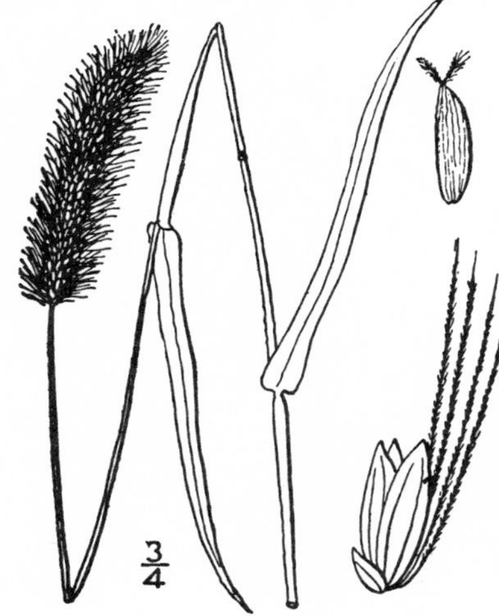

3. Chaetochloa víridis (L.) Scribn. Green Foxtail. Fig. 248.

Panicum viride L. Syst. Nat. ed. 10. 2: 870. 1759.
Setaria viridis Beauv. Ess. Agrost. 51, 178. *pl. 13. f. 3.* 1812.
Chaetochloa viridis Scribn. U. S. Dept. Agr. Div. Agrost. Bull. 4: 39. 1897.

Annual; culms 30–60 cm. tall; blades flat, not twisted; panicle oblong-ovate, 2–5 cm. long; bristles 1–3, slender, 6–12 mm. long, green or purple; spikelets 2 mm. long; fruit faintly wrinkled.

A weed in fields and waste places. Rare within our range, California (Rialto, *Parish*; Los Angeles, *Davidson*), Washington; common in the Eastern States. Introduced from Europe, the type locality. June–Aug.

The closely allied *Chaetochloa italica* (L.) Scribn. (*Setaria italica* Roem. & Schult., *S. californica* Kellogg, Proc. Calif. Acad. 1: ed. 2, 26. 1873) is cultivated under the name of millet in the Central States and is occasionally found as a stray or escape along railroads ("Sacramento River," Kellogg). The spike is larger, usually lobed, the bristles purple or yellow.

4. **Chaetochloa geniculàta** (Lam.) Millsp. & Chase.

Perennial Foxtail. Fig. 249.

Panicum geniculatum Lam. Encycl. **4**: 727. 1798.
Setaria gracilis H. B. K. Nov. Gen. & Sp. **1**: 109. 1816.
Panicum imberbe Poir. in Lam. Encycl. Suppl. **4**: 272. 1816
Chaetochloa imberbis Scribn. U. S. Dept. Agr Div. Agrost. Bull. **4**: 39. 1897.
Chaetochloa gracilis Scribn. & Merr. U. S. Dept. Agr. Div. Agrost. Bull. **21**: 15. *f. 4.* 1900.
Chaetochloa geniculata Millsp. & Chase, Field Mus. Bot. **3**: 37. 1903.

Perennial; culms erect, 60–150 cm. tall; blades elongate, narrow, 2–4 mm. wide, flat or folded; panicle slender, linear, 8–10 cm. long, 4–8 mm. thick; bristles 5–8, twice as long as spikelet, pale or tawny; spikelets 2–2.5 mm. long; fruit undulate-rugose. The blades are usually without hairs.

Open ground, dry or moist soil, southern United States to Argentina. June–Sept. Type locality: West Indies.

Chaetochloa nigriróstris (Nees) Skeels, an African species resembling *C. lutescens* but with brown bristles and dark tips to the fruit, has been found on ballast near Portland, Oregon (Albina and Linnton, *Suksdorf*).

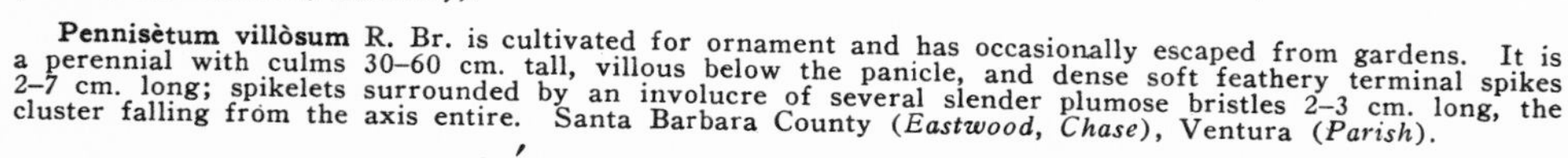

Pennisètum villòsum R. Br. is cultivated for ornament and has occasionally escaped from gardens. It is a perennial with culms 30–60 cm. tall, villous below the panicle, and dense soft feathery terminal spikes 2–7 cm. long; spikelets surrounded by an involucre of several slender plumose bristles 2–3 cm. long, the cluster falling from the axis entire. Santa Barbara County (*Eastwood, Chase*), Ventura (*Parish*).

11. **CÉNCHRUS** L. Sp. Pl. 1049. 1753.

Spikelets few in a cluster, surrounded and enclosed by a spiny bur made up of several coalescing bristles (sterile branchlets), the bur globular, the peduncle short and thick, articulate at base, the spines retrorsely barbed. Annual or sometimes perennial, low branching grasses, with flat blades and racemes of burs terminating the culms and branches, the burs readily deciduous. [Ancient Greek name for some grass.]

Species about 25 in the warmer parts of both hemispheres, but chiefly in America. Type species. *Cenchrus echinatus* L.

1. **Cenchrus paucifiòrus** Benth.

Sand-bur, Bur-grass. Fig. 250.

Cenchrus pauciflorus Benth. Bot. Voy. Sulph. 56. 1840.
Cenchrus tribuloides of authors, not L., and *Cenchrus carolinianus* of authors, not Walt.

Annual; culms much branched at base, ascending or spreading, flattened; sheaths loose; blades firm, flat or folded; burs about 8 mm. thick, 6–12 in short exserted racemes.

Sandy soil and desiccated pools, rare; California (San Bernardino, Gilroy, Stanford University), Oregon (The Willows, Linnton). Common eastward and south to Central America. July–Sept. Type locality: Lower California.

12. **HOMALOCÉNCHRUS** Mieg. Act. Helv. Phys. Math. **4**: 307. 1760.

[Leersia Swartz, Prodr. Veg. Ind. Occ. 21. 1788, not Hedw. 1782.]

Spikelets 1-flowered, strongly compressed laterally, articulate with the pedicel; glumes wanting, lemma chartaceous, broad, oblong, boat-shaped, usually 5-nerved, the lateral pair of nerves close to the margins, these and the keel often hispid-ciliate, the intermediate nerves sometimes faint; palea as long as the lemma, much narrower, usually 3-nerved, the keel usually hispid-ciliate, the lateral nerves close to the margins, the margins firmly held by the margins of the lemma; stamens 6 or fewer. Perennial grasses, usually with creeping rhizomes, flat scabrous blades, and open panicles, the spikelets nearly sessile along one side of the branchlets. [Greek, like-Cenchrus.]

Species 10, tropical America and temperate North America, one species also in Europe and Asia. Type species, *Phalaris oryzoides* L.

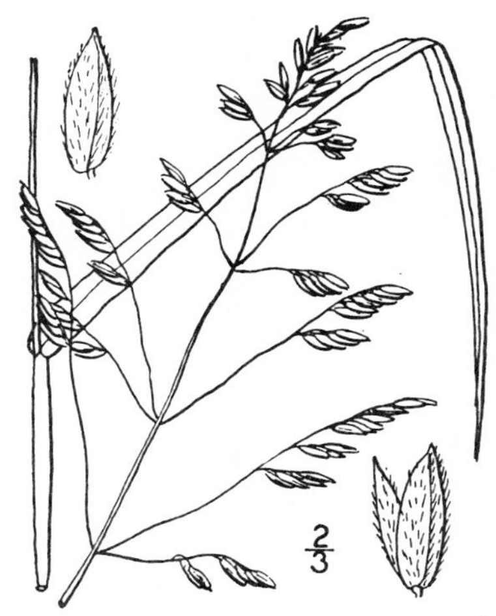

1. Homalocenchrus oryzoìdes (L.) Poll.

Rice Cut-grass. Fig. 251.

Phalaris oryzoides L. Sp. Pl. 55. 1753.
Homalocenchrus oryzoides Poll. Hist. Pl. Palat. **1**: 52. 1776.
Leersia oryzoides Swartz, Prodr. Veg. Ind. Occ. 21. 1788.

Culms erect from a decumbent base, much branched, bearing slender rhizomes; sheaths glabrous; ligule short; blades flat, very rough, 8–20 cm. long, 4–10 mm. wide; panicle diffusely branched, lax, 12–20 cm. long, the branches naked at base, finally drooping; spikelets 4–5 mm. long, narrowly oblong; lemma hispid, strongly bristly-ciliate on the keel.

Marshes and wet places, often forming zones around ponds; Transition Zone; eastern Washington and Oregon, eastward to the Atlantic Coast and northern Eurasia; also at San Bernardino, California (*Parish*). Aug.–Oct. Type locality, European.

13. PHÁLARIS L. Sp. Pl. 54. 1753.

Spikelets of one terminal perfect floret and 2 sterile lemmas, laterally compressed, disarticulating above the glumes, arranged in usually dense spikelike panicles; glumes 2, equal, boat-shaped, often winged on the keel; sterile lemmas reduced to 2 small scales (rarely only 1) remaining attached to the perfect floret; fertile lemma coriaceous, shorter than the glumes, inclosing the faintly 2-nerved palea. Annual or perennial erect grasses with flat blades. [An ancient Greek name for a grass.]

Species about 20, in temperate regions of Europe and America. Type species, *Phalaris canariensis* L.

- Spikelets in groups of 7, 1 fertile surrounded by 6 sterile. — 1. *P. paradoxa.*
- Spikelets single, all alike.
  - Plants perennial.
    - Rhizomes absent; panicle dense, ovate or oblong. — 2. *P. californica.*
    - Rhizomes present; panicle spreading during anthesis. — 3. *P. arundinacea.*
  - Plants annual.
    - Glumes broadly winged; panicle ovate or short-oblong.
      - Sterile lemma solitary; fertile lemma 3 mm. long. — 4. *P. minor.*
      - Sterile lemmas in pairs; fertile lemma 4–6 mm. long.
        - Sterile lemma .6 mm. long. — 5. *P. brachystachys.*
        - Sterile lemma half as long as fertile. — 6. *P. canariensis.*
    - Glumes wingless or nearly so; panicles oblong or linear, dense.
      - Glumes acuminate; fertile lemma turgid, the acuminate apex smooth. — 7. *P. lemmoni.*
      - Glumes acute; fertile lemma less turgid, villous to the acute apex.
        - Panicle 2–5 cm. long; sterile lemmas one-third as long as fertile. — 8. *P. caroliniana.*
        - Panicle 5–12 cm. long; sterile lemma half as long as fertile. — 9. *P. angusta.*

1. Phalaris paradóxa L.

Paradox Canary-grass. Fig. 252.

Phalaris paradoxa L. Sp. Pl. ed. 2. **2**: 1665. 1763.

Annual; culms cespitose, more or less spreading at base, 30–60 cm. tall; panicle dense, oblong, narrowed at base, 2–5 cm. long, often inclosed at base in the uppermost enlarged sheath; spikelets finally falling from the axis in groups of 7, the central fertile, nearly sessile, the others sterile, slender-pediceled; glumes of sterile spikelets narrow, with faint lateral nerves, the keel prominently winged above, the wing extending into a more or less well-marked tooth, the apex of the glume narrowed into an acuminate point or awn, the glumes of the 4 outer sterile spikelets in the lower part of the panicle more or less deformed; glumes of fertile central spikelet lanceolate, 6–8 mm. long including awn, the lateral nerves prominent, the wing on the keel more tooth-like, the apex of the glume narrowed into an awn about 2 mm. long; fertile lemma smooth and shining, 3 mm. long, the sterile lemmas obsolete.

Occasional in grain fields; California (Richmond, *Congdon*). Introduced from Europe. Type locality, the Orient.

Phalaris paradoxa praemórsa (Lám.) Coss. & Dur. Expl. Alg. **2**: 25. 1854. *Phalaris praemorsa* Lam. Fl. Franç. **3**: 566. 1778. Sterile spikelets short-pediceled, the 4 outer much reduced, the apex deformed or variously incurved; fertile spikelet somewhat indurate, several-nerved at base, acuminate, the wing fin-like in appearance.

Fields and waste places; rare, California (Berkeley Hills, *Davy;* San Diego, *Brandegee*), and Oregon (Linnton, *Nelson*). Introduced from Europe. Type locality: France.

2. **Phalaris califórnica** Hook. & Arn.
California Canary-grass. Fig. 253.

Phalaris californica Hook. & Arn. Bot. Beechey Voy. 161. 1841.
Phalaris amethystina of Thurber and others, not Trin.

Perennial; culms erect or somewhat geniculate at base; blades flat, rather lax, 6–12 mm. wide; panicle ovoid or oblong, 2–5 cm. long, 2–2.5 cm. thick, often purplish-tinged; glumes about 6–7 mm. long, narrow, gradually narrowed from below the middle to an acute apex, smooth or slightly scabrous on the keel, the lateral nerves somewhat nearer the margin than the keel; fertile lemma ovate-lanceolate, about 4 mm. long, rather sparsely villous, often exposing the palea, the sterile lemmas about half as long.

Ravines and open moist ground in the Transition Zone; in the Coast Ranges, Oregon to San Luis Obispo County, California. June. Type locality: west central California.

3. **Phalaris arundinàcea** L.
Reed Canary-grass. Fig. 254.

Phalaris arundinacea L. Sp. Pl. 55. 1753.

Perennial with creeping rhizomes; culms erect, 60–150 cm. tall; panicle 7–16 cm. long, narrow, the branches spreading during anthesis, the lower as much as 5 cm. long; glumes narrow, 4 mm. long, abruptly narrowed to an acute apex, the keel scabrous, not winged, the lateral nerves about midway between margin and keel; fertile lemma lanceolate, 3 mm. long, shining, sparsely villous; sterile lemmas villous, 1 mm. long.

Swamps and moist places, Transition Zone; infrequent in California (Ager, Warner Mountains, *Bouldin*); common in eastern Washington and eastward through the northern part of North America; also in northern Eurasia. June–Sept. A form with variegated leaves is cultivated under the name of ribbon-grass. Type locality, European.

4. **Phalaris mìnor** Retz.
Mediterranean Canary-grass. Fig. 255.

Phalaris minor Retz. Obs. Bot. 3: 8. 1783.

Annual; culms erect, 30–100 cm. tall; panicle ovate-oblong to oblong, 2–5 cm. long; glumes oblong, 4–6 mm. long, strongly winged on the keel as in *P. canariensis,* the green stripe less conspicuous, the wing scabrous on margin and more or less toothed; fertile lemma ovate, acute, villous but less so than *P. canariensis,* about 3 mm. long, the sterile lemma solitary, about 1 mm. long.

Introduced along the coast of California and at Linnton, Oregon (*Nelson*). A native of the Mediterranean region. August. Type locality, European.

5. **Phalaris brachystàchys** Link.
Short-spiked Canary-grass. Fig. 256.

Phalaris brachystachys Link, Neu. Journ. Bot. Schrad. 1^3: 134. 1806.

Annual; culms 30–60 cm. tall; panicle ovate, about 2.5 cm. long; glumes about 6 mm. long, similar to those of *P. canariensis;* fertile lemma 4–5 mm. long, densely short-villous; sterile lemmas short, brown, ovate, equal, about 7 mm. long.

Introduced in California (Nelson, Butte County, *Heller*), and on ballast at Linnton, Oregon (*Nelson*). A native of the Mediterranean region. Type locality, European.

6. **Phalaris canariénsis** L.
Canary Grass. Fig. 257.

Phalaris canariensis L. Sp. Pl. 54. 1753.

Annual; culms erect, 30–60 cm. tall; panicle ovate to oblong-ovate, 1.5–3 cm. long, pale with green markings; glumes 7–8 mm. long, widened above, smooth or sparsely villous, the keel prominently winged above, the wing entire or somewhat sinuous, the keel on each side at base of the white wing marked by a green stripe, the lateral nerves approaching the margin; fertile lemma elliptic, acute, densely short-villous, 5–6 mm. long; sterile lemmas about half as long as fertile.

Introduced occasionally, from Washington to California and eastward. A native of the Mediterranean region. June–Aug. Type locality: Canary Islands.

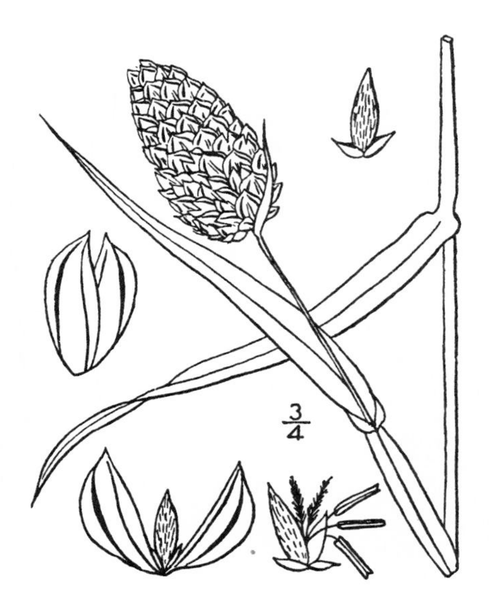

7. **Phalaris lémmoni** Vasey.
Lemmon's Canary-grass. Fig. 258.

Phalaris lemmoni Vasey, Contr. U. S. Nat. Herb. **3**: 42. 1892.

Annual; culms erect, 30–90 cm. tall; panicle dense, 5–10 cm. long; glumes about 5 mm. long, narrow, acuminate, the lateral nerves about midway between margin and keel; fertile lemma ovate-lanceolate, acuminate, dark-colored at maturity, villous except the acuminate tip, 3.5 mm. long; sterile lemmas less than one-third as long.

Central and southern California near the coast, Transition Zone. Apr.–May. Type locality: Santa Cruz.

8. Phalaris caroliniàna Walt.
Carolina Canary-grass. Fig. 259.

Phalaris caroliniana Walt. Fl. Carol. 74. 1788.

Annual; culms erect, 30–60 cm. tall; panicle oblong, 2.5–5 cm. long; glumes 5–6 mm. long, oblong, rather abruptly narrowed to an acute apex, the keel scabrous and narrowly winged above from below the middle, the lateral nerves about midway between keel and margin; fertile lemma ovate, acute, densely villous, about 4 mm. long, the close-appressed sterile lemmas about one-third as long.

A native of the southeastern States; introduced in Oregon (Grants Pass) and California. Mar.–May. Type locality: South Carolina.

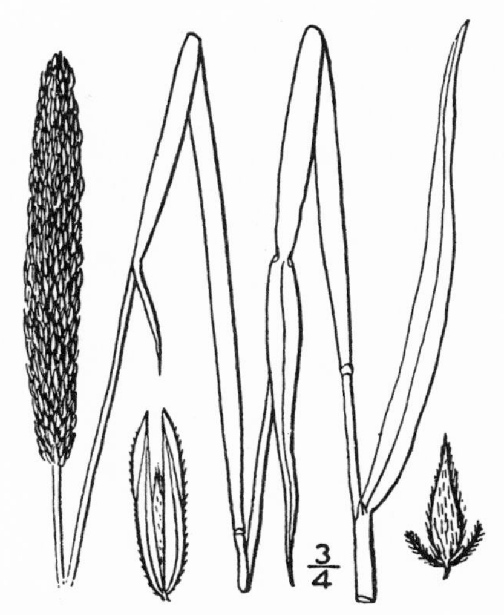

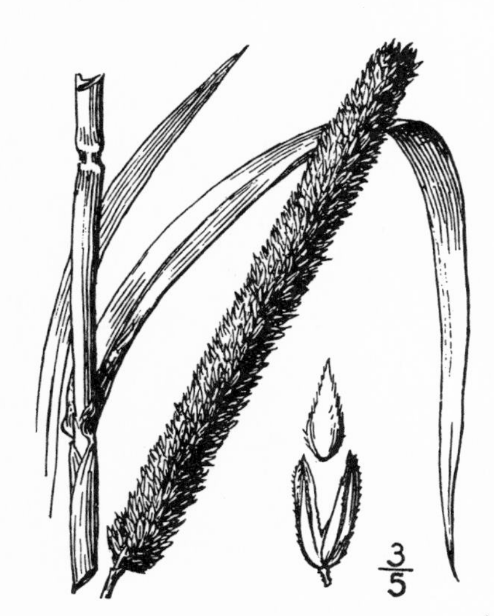

9. Phalaris angústa Nees.
Narrow Canary-grass. Fig. 260.

Phalaris angusta Nees; Trin. Gram. Icon. 1: *pl. 78.* 1827.
Phalaris intermedia angusta Chapm. Fl. South. U. S. 568. 1860.

Annual; 1–1.5 meters tall, smooth; blades flat, 6–8 mm. wide; panicle dense, linear-oblong, 5–12 cm. long, about 8 mm. thick; glumes about 4 mm. long, narrow, rounded at apex to a mucronate tip, scabrous on keel, nerves, and more or less on the back, especially near the apex, lateral nerves near the margin; fertile lemma ovate-lanceolate, acute, villous, 3 mm. long, sterile lemmas about half as long.

Open ground at low altitudes, California from San Francisco southward, and east to Louisiana; also in South America. Apr.–June. Type locality: Uruguay.

14. ANTHOXÁNTHUM L. Sp. Pl. 28. 1753.

Spikelets with 1 terminal perfect floret and 2 lateral sterile lemmas, the rachilla disarticulating above the glumes; unequal, acute or mucronate; sterile lemmas shorter than the glumes, empty, awned from the back; fertile lemma shorter than the sterile ones, awnless; palea 1-nerved, rounded on the back, inclosed in the lemma. Sweet-smelling annual or perennial grasses with flat blades and spikelike panicles. [Greek, flower-yellow.]

Species about 4, Europe and Asia, 2 introduced in the United States. Type species, *Anthoxanthum odoratum* L.

1. Anthoxanthum odoràtum L.

Sweet Vernal Grass. Fig. 261.

Anthoxanthum odoratum L. Sp. Pl. 28. 1753.

Perennial; culms slender, erect, 20–60 cm. tall, panicle 2–7 cm. long, pointed; spikelets brownish green, 8–10 mm. long; glumes sparsely pilose; first sterile lemma short-awned below the apex, the second bearing a strong, bent, scarcely exserted awn near its base.

Occasionally cultivated in the United States as a meadow grass and escaped or introduced in the cooler parts of the country. Washington to California. May–June. Type locality, European.

15. HIERÓCHLOË R. Br. Prodr. 208. 1810.

[SAVASTANA Schrank, 1789. TORRESIA Ruiz & Pav. 1798.]

Spikelets with one terminal perfect floret and 2 lateral staminate florets, disarticulating above the glumes; glumes equal, broad, thin and papery, smooth, acute; sterile lemmas 2, about as long as the glumes, mostly somewhat appressed-hispid, sometimes awned from between two lobes; fertile lemma somewhat indurate, about as long as the others, smooth or nearly so, awnless; palea 1–3-nerved, rounded on the back. Perennial low erect sweet-smelling grasses with small panicles of bronze-colored spikelets. [Name Greek, meaning holy grass.]

Species about 17 in the arctic and cool regions of the world. Type species, *Hierochloë odorata* (L.) Wahl.

Blades numerous in a basal cluster, usually 8–14 mm. wide, the uppermost leaf with a short blade; florets sparsely villous. 1. *H. macrophylla.*
Blades scattered, usually less than 5 mm. wide, the uppermost leaf reduced to a sheath; florets densely villous. 2. *H. odorata.*

1. Hierochloë macrophýlla Thurb. California Vanilla Grass. Fig. 262.

Hierochloë macrophylla Thurb.; Boland. Trans. Calif. Agr. Soc. 1864–65: 132. 1866.
Savastana macrophylla Beal, Grasses N. Am. **2**: 187. 1896.
Torresia macrophylla Hitchc. Am. Journ. Bot. **2**: 300. 1915.

Culms few, erect, 60–90 cm. tall; sheaths scabrous; blades crowded toward base, flat, rather stiffly upright, scabrous above, glaucous beneath, acuminate-pointed, 6–14 mm. wide; panicle somewhat open, 7–12 cm. long, the lower branches spreading or drooping, 2.5–5 cm. long; glumes 4 mm. long.

In woods in the redwood belt from Oregon southward to Monterey; also at Bingen, Washington (*Suksdorf*). May–July. Type locality: Marin County, California.

2. Hierochloë odoràta (L.) Wahl. Vanilla Grass. Fig. 263.

Holcus odoratus L. Sp. Pl. 1048. 1753.
Hierochloë borealis Roem. & Schult. Syst. Veg. **2**: 513. 1817.
Hierochloë odorata Wahl. Fl. Ups. 32. 1820.
Savastana odorata Scribn. Mem. Torrey Club **5**: 34. 1894.
Torresia odorata Hitchc. Am. Journ. Bot. **2**: 301. 1915.

Culms 30–60 cm. tall; panicle pyramidal, 4–8 cm. long, usually rather compact; spikelets 5 mm. long; staminate lemmas hispid-ciliate on the margins, awnless.

Moist meadows, Cascade Mountains, Washington and Oregon, eastward across the continent; also in Eurasia. Apr.–July. Type locality, European.

16. ARÍSTIDA L. Sp. Pl. 82. 1753.

Spikelets 1-flowered, the rachilla disarticulating obliquely above the glumes; glumes narrow, acute, acuminate or somewhat awned; lemma indurate, narrow, terete, convolute, tipped below by the pointed, usually minutely bearded callus, terminating above in a usually trifid awn. Annual or perennial, mostly low grasses, with narrow frequently convolute blades and narrow or sometimes open panicles. [Latin, *arista,* an awn.]

Species about 150 in the warmer regions of the world. Type species, *Aristida adscensionis* L.

Neck of fruit jointed at base. 1. *A. californica.*
Neck of fruit not jointed at base.
Lateral awns wanting or reduced to mere points. 2. *A. schiedeana.*
Lateral awns evident.
Panicle diffuse, open, the main branches usually horizontally spreading or divaricate; glumes about equal; plants perennial. 3. *A. divaricata.*
Panicle narrow, somewhat open, the branches not divaricate.
Plants annual.
Glumes 20–30 mm. long, nearly equal; awns 4–7 cm. long. 4. *A. oligantha.*
Glumes not over 10 mm. long, the lower about half as long as the upper; awns about 1 cm. long. 5. *A. adscensionis.*
Plants perennial.
Glumes about equal. 6. *A. parishii.*
Glumes unequal, the first half to two-thirds as long as the second.
Neck of fruit slender, straight or slightly twisted, about as long as the body; panicle narrow, strict. 7. *A. reverchoni.*
Neck of fruit short or not differentiated from the body.
Pedicels slender, more or less recurved. 8. *A. purpurea.*
Pedicels stiffly erect.
Panicle elongate, many-flowered; collar of leaf bearing a ciliate line. 9. *A. wrightii.*
Panicle short, few-flowered; no ciliate line on collar of leaf. 10. *A. fendleriana.*

1. **Aristida califórnica** Thurb.

California Aristida. Fig. 264.

Aristida californica Thurb. in S. Wats. Bot. Calif. **2**: 289. 1880.

Aristida jonesii Vasey, Contr. U. S. Nat. Herb. **3**: 48. 1892, as synonym.

Aristida californica fugitiva Vasey, Contr. U. S. Nat. Herb. **3**: 49. 1892.

Perennial; culms cespitose, wiry, much branched above base, 15–30 cm. tall, crisp-pubescent; blades short, involute, sharp-pointed, 1–4 cm. long; panicles numerous, loose, 2–5 cm. long, the few branches few-flowered; glumes smooth or the first slightly scabrous near apex, 1-nerved, awnless, unequal, the first 8 mm. long, the second about twice as long; lemma 6 mm. long, smooth except the short-pubescent callus, 2 mm. long, the narrowed apex articulated with the slender, spirally twisted 18-mm.-long neck of the awns; awns equal, spreading, about 2.5 cm. long.

Deserts of southern California, Arizona, and northern Mexico. Apr.–May. Type locality: Colorado Desert, California.

2. **Aristida schiedeàna** Trin. & Rupr.

Schiede's Aristida. Fig. 265.

Aristida schiedeana Trin. & Rupr. Mém. Acad. St. Pétersb. VI. Sci. Nat. **5**[1]: 120. 1842.

Aristida orcuttiana Vasey, Bull. Torrey Club **13**: 27. 1886.

Perennial; culms erect, scaberulous, 30–60 cm. tall; sheaths scaberulous, villous at the throat; blades involute in drying, the persistent basal ones flat and flexuous, 10–20 cm. long, .5–2 mm. wide; panicle somewhat open, 10–30 cm. long, the lower branches slender, ascending or spreading, 4–7 cm. long; glumes somewhat unequal, acuminate or short-awned, the first longer, about 1 cm. long; neck of fruit twisted, 5–8 mm. long, the central awn spreading, 7–10 mm. long, the lateral awns minute or obsolete; fruit, including neck, 15–18 mm. long.

Deserts and rocks, San Diego (*Orcutt*) and Arizona to southern Mexico. September. Type locality: Jalapa; of *A. orcuttiana*, Lower California.

3. **Aristida divaricàta** Humb. & Bonpl.

Divaricate Aristida. Fig. 266.

Aristida divaricata Humb. & Bonpl. in Willd. Enum. Pl. 99. 1809.

Aristida palmeri Vasey, Bull. Torrey Club **10**: 42. 1883.

Aristida humboldtiana minor Vasey, Contr. U. S. Nat. Herb. **3**: 47. 1892.

Perennial; culms cespitose, erect, 30–60 cm. tall; blades involute, as much as 15 cm. long; panicle usually more than half the length of the entire plant; branches distant, mostly in pairs, divaricately spreading, spikelet-bearing toward the ends; glumes nearly equal, 10–12 mm. long. 1-nerved, short-awned, the first scabrous on the keel; lemma about 10 mm. long, scabrous toward the straight or twisted neck; awns about equal, 12–20 mm. long, somewhat spreading.

Deserts and rocky hills, Bakersfield and southward, east to Texas, south to Oaxaca. May–Aug. Type locality: Mexico.

4. Aristida oligántha Michx.

Few-flowered Aristida. Fig. 267.

Aristida oligantha Michx. Fl. Bor. Am. 1: 41. 1803.

Annual; culms erect, branched at base and at all the nodes, 30–60 cm. high, often woolly at the very base; blades 2 mm. wide or less, usually involute, as much as 15 cm. long, sparingly pilose at base, the prophyllum often conspicuous at base of branches; panicles narrow, loosely few-flowered, bearing a few scattered large appressed short-pediceled spikelets; glumes about 2.5 cm. long, slightly unequal, long-awned from a bifid apex, the first strongly 7-nerved; lemma a little shorter than the glumes, the gradually narrowed neck scaberulous, the callus rather minutely pubescent; awns about equal, widely spreading, 5–8 cm. long.

Open ground, central Oregon to the interior valley of California; common in the southeastern States. July–Sept. Type locality: Illinois.

5. Aristida adscensiònis L.

Six-weeks Grass. Fig. 268.

Aristida adscensionis L. Sp. Pl. 82. 1753.
Aristida bromoides H. B. K. Nov. Gen. & Sp. 1: 122. 1816.

Annual; culms much-branched at the base, 10–30 cm. tall, erect or often spreading or prostrate; blades 2.5–5 cm. long, narrow, usually involute; panicle narrow, rather dense, 5–8 cm. long, the branches short, fascicled; glumes unequal, smooth except the keel of the first, 1-nerved, the first 5–6 mm. long, acutish, the second 7–8 mm. long, obtuse or slightly mucronate; lemma 8–10 cm. long, smooth except upper portion of keel, the callus with a dense tuft of short hairs, the apex scarcely narrowed; awns equal, finally spreading, about 10 mm. long, or the lateral sometimes shorter.

Open ground, southern California (San Luis Obispo and southward) to New Mexico and south to South America. Feb.–June. Type locality: Island of Ascension.

6. Aristida paríshii Hitchc.

Parish's Aristida. Fig. 269.

Aristida parishii Hitchc. in Jepson, Fl. Calif. 1: 101. 1912.

Perennial; culms tufted, 30–60 cm. tall, smooth; sheaths smooth, ciliate at the throat; blades ascending, firm, flat or more or less involute, scabrous on the upper surface, smooth below or scabrous toward the tip, 1–2 mm. wide, 15–30 cm. long; panicle narrow, about 15 cm. long, the branches rather stout, ascending or appressed, the lower 2–5 cm. long; glumes somewhat unequal, short-awned, smooth or scabrous on the keel, 1-nerved or the first 3-nerved, the second a little longer, 12 mm. long; lemma a little shorter than the second glume, very scabrous on the upper half, the neck rather stout, not twisted, the awns ascending, the central about 2 cm. long, and the lateral a little shorter.

Dry ground, southern California. Apr.–July. Type locality: Agua Caliente, Riverside County.

7. Aristida revérchoni Vasey.
Reverchon's Aristida. Fig. 270.

Aristida reverchoni Vasey, Bull. Torrey Club **13**: 52. 1886.

Perennial; culms densely cespitose, erect, 30–60 cm. tall; blades involute, more or less flexuous, as much as 15 cm. long; panicle narrow, 10–15 cm. long, the branches short, appressed; glumes unequal, awnless, smooth, or the first scabrous on the upper part of the keel, the first about 6 mm. long, the second about 10 mm. long; lemma 10–12 mm. long, smooth except the minutely pubescent callus, narrowed above but not twisted; awns equal, about 2 cm. long.

Deserts and plains, Mojave Desert, east to Texas. May. Type locality: Texas.

8. Aristida purpùrea Nutt.
Purple Aristida. Fig. 271.

Aristida purpurea Nutt. Trans. Am. Phil. Soc. II. **5**: 145. 1837.

Perennial; culms erect, about 60 cm. tall; blades flat or involute, 5–12 cm. long; panicles 10–15 cm. long, rather loose; branches and pedicels slender, more or less recurved; glumes unequal, smooth, short-awned, 1-nerved, the first 6 mm. long, the second about twice as long; lemma 12 mm. long, purple, strongly scabrous in lines, the apex somewhat narrowed, flattened and slightly twisted; awns equal, about 3 cm. long.

Plains and deserts, southern California to Texas and northern Mexico. Apr.–July. Type locality: Texas.

9. Aristida wrìghtii Nash.
Wright's Aristida. Fig. 272.

Aristida wrightii Nash in Small, Fl. Southeast. U. S. 116. 1903.

Culms erect, 30–60 cm. tall, smooth; sheaths smooth, slightly villous at the throat, the collar bearing a ciliate line; blades involute, scattered; panicle narrow, 15–25 cm. long, the short stiff branches appressed or ascending; glumes unequal, the first about 10 mm. long, the second about 15 mm. long; lemma papillose-scaberulous, about 1 cm. long; awns about equal, spreading, 15–25 mm. long.

Deserts, plains, and rocky hills, Colorado Desert (*Parish*), east to Texas and south to central Mexico. May. Type locality: Texas.

10. Aristida fendleriàna Steud. Fendler's Aristida. Fig. 273.

Aristida fendleriana Steud. Syn. Pl. Glum. 1: 420. 1854.

Perennial; culms densely cespitose, erect, usually less than 30 cm. tall; blades crowded at base of culms, forming a curly tuft, involute, arcuate, sharp-pointed, pilose at base, usually 3.2–5 cm. long, sometimes longer; panicle narrow, 5–10 cm. long, bearing a few, mostly short-pediceled loosely arranged, more or less appressed spikelets; glumes unequal, smooth, awnless, 1-nerved, the first about 6 mm. long, the second 10–12 mm. long, scaberulous and slightly narrowed above; callus minutely pubescent, 1 mm. long; awns equal, about 2–3 cm. long, ascending.

Deserts and plains, southern California (San Bernardino Mountains, *Parish*) to Texas. June. Type locality: New Mexico.

17. STÌPA L. Sp. Pl. 78. 1753.

Spikelets 1-flowered, articulate above the glumes, the articulation oblique, leaving a bearded sharp-pointed callus attached to the base of the floret; glumes membranaceous, often papery, acute, acuminate, or even aristate, usually long and narrow; lemma narrow, terete, firm or indurate, strongly convolute, terminating in a usually bent and twisted prominent persistent awn; palea inclosed in the convolute lemma. Perennial grasses with usually convolute blades and narrow panicles. [Greek, tow, referring to the feathery awns of some species.]

Species about 100 in the temperate regions of the world, especially on plains and steppes. Type species, *Stipa pennata* L.

- Awn plumose.
  - Awn with 1 bend, very plumose below the bend. — 1. *S. speciosa.*
  - Awn with 2 bends, plumose to second bend.
    - Ligule 3–6 mm. long. — 2. *S. thurberiana.*
    - Ligule very short.
      - Sheaths pubescent. — 3. *S. elmeri.*
      - Sheaths glabrous.
        - Blades involute, mostly basal. — 4. *S. occidentalis.*
        - Blades flat, tardily involute, scattered. — 5. *S. californica.*
- Awn more or less scabrous, but not plumose.
  - Lemma clothed with copious hairs, 4 mm. long.
    - Awn with 2 bends; plants 1 meter or more tall. — 6. *S. coronata.*
    - Awn with usually 1 bend; plant 30–50 cm. tall. — 7. *S. parishii.*
  - Lemma more or less hairy, the hairs not over 1 mm. long.
    - Panicle open, the branches naked below, more or less spreading.
      - Upper glume 5-nerved; lower glume 3- or 5-nerved. — 8. *S. comata.*
      - Upper and lower glumes 3-nerved.
        - Ligule evident; terminal segment of awn mostly 4 cm. or more long. — 9. *S. pulchra.*
        - Ligule very short; terminal segment of awn mostly less than 2 cm. long. — 10. *S. lepida.*
    - Panicle narrow, the branches erect.
      - Sheaths pubescent; culms pubescent below the nodes. — 11. *S. williamsii.*
      - Sheaths glabrous.
        - Glumes 15–20 mm. long; blades as much as 7 mm. wide. — 12. *S. stillmanii.*
        - Glumes 10 mm. or less long.
          - Sheaths hairy at the throat. — 13. *S. vaseyi.*
          - Sheaths not hairy at the throat.
            - Glumes broad, usually 5-nerved; fruit fusiform. — 14. *S. lemmoni.*
            - Glumes narrow, 3-nerved.
              - Blades slender, involute; lemma scarcely 5 mm. long. — 15. *S. lettermani.*
              - Blades flat; lemma usually more than 5 mm. long.
                - Lemma about 7 mm. long; culms usually 1–1.5 meters tall. — 16. *S. nelsoni.*
                - Lemma 5–6 mm. long; culms usually less than 1 meter tall. — 17. *S. minor.*

1. Stipa speciòsa Trin. & Rupr.
Desert Stipa. Fig. 274.

Stipa speciosa Trin. & Rupr. Mém. Acad. St. Pétersb. VI. Sci. Nat. 5[1]: 45. 1842.
Stipa chrysophylla Desv. in Gay, Fl. Chil. 6: 278. *pl.* 76. *f.* 2. 1853.

Culms numerous, cespitose, 30–60 cm. tall; sheaths smooth, or lower pubescent or even felty at the very base, the throat densely short-villous; ligule short; blades elongate, involute-filiform, mostly basal, more or less deciduous from the outer and older persistent sheaths; panicle narrow, dense, 10–15 cm. long, not much exceeding the leaves, white or tawny, feathery from the plumose awns; glumes smooth, 14–16 mm. long, 3-nerved, long-acuminate, papery; lemma 7–9 mm. long, narrow, densely short-pubescent, the callus sharp and smooth below; awn with one sharp bend, the first section 1.5–2 cm. long, densely long-pilose on the lower half or two-thirds, the hairs 6–8 mm. long, the remaining portion of the awn scabrous, the second section about 2.5 cm. long.

Deserts and arid hills of southern California, as far north as Mono Lake and San Luis Obispo; especially characteristic of the Colorado and Mojave Deserts; east to Colorado and south into Lower California; also in Chile. May–June. Type locality: Chile.

2. Stipa thurberiàna Piper.
Thurber's Stipa. Fig. 275.

Stipa thurberiana Piper, U. S. Dept. Agr. Div. Agrost. Circ. 27: 10. 1900.

Culms 15–50 cm. tall; sheaths smooth or somewhat scabrous, mostly basal; ligule, about 3–4 mm. long, acute; blades involute, scabrous; panicle 5–10 cm. long, often subtended by an enlarged sheath; glumes about 12 mm. long, acuminate, 3-nerved; lemma 7 mm. long, appressed-pilose, the callus acute; awn about 3 mm. long, indistinctly twice-geniculate, short-pilose to the second bend.

Open dry woods, Canadian Zone; central California to eastern Washington. June–July. Type locality: Washington.

3. Stipa élmeri Piper & Brodie.
Elmer's Stipa. Fig. 276.

Stipa viridula pubescens Vasey, Contr. U. S. Nat. Herb. 3: 50. 1892.
Stipa elmeri Piper & Brodie, U. S. Dept. Agr. Div. Agrost. Bull. 11: 46. 1898.

Culms 60–100 cm. tall, more or less puberulent, especially at the nodes; sheaths pubescent; ligule very short; blades flat or becoming involute, pubescent on the upper surface, or those of the innovations also on the lower surface; panicle narrow, 15–35 cm. long, rather loose; glumes 3-nerved, gradually acuminate, thin, papery, 12–14 mm. long, the first a little the longer; lemma about 7 mm. long, appressed-pubescent, the callus 1 mm. long, glabrous at the point; awn distinctly twice-geniculate, first section 8–10 mm. long, the second section somewhat shorter, both plumose, third section about 1.5–2 cm. long, scabrous.

Open ground in the mountains, Canadian Zone; Washington to southern California and Nevada. June–July. Type locality: Empire City, Nevada.

4. Stipa occidentàlis Thurb.

Western Stipa. Fig. 277.

Stipa occidentalis Thurb.; S. Wats. in King, Geol. Expl. 40th Par. **5**: 380. 1871.

Stipa stricta Vasey, Bull. Torrey Club **10**: 42. 1883.

Stipa stricta sparsiflora Vasey, Contr. U. S. Nat. Herb. **3**: 51. 1892.

Stipa oregonensis Scribn. U. S. Dept. Agr. Div. Agrost. Bull. **17**: 130. *f. 426.* 1899.

Stipa occidentalis montana Merr. & Davy, Univ. Calif. Publ. Bot. **1**: 62. 1902.

Culms slender, cespitose, 30–60 cm. tall; sheaths smooth; ligule 1 mm. long; blades narrow, involute; panicle narrow, 10–20 cm. long; glumes 8–10 mm. long, acuminate, 3-nerved, smooth; lemma 6 mm. long, long-pilose, the callus sharp; awn about 2.5 cm. long, twice-geniculate, pilose to the second bend or throughout, the first section 6–8 mm. long.

Open dry woods, Canadian Zone; Washington to California and Wyoming. June–Aug. Type locality: Yosemite Trail, California.

5. Stipa Califórnica Merr. & Davy.

California Stipa. Fig. 278.

Stipa californica Merr. & Davy, Univ. Calif. Publ. Bot. **1**: 61. 1902.

Culms 60–150 cm. tall, smooth, or the nodes pubescent; sheaths smooth, villous at the throat; ligule very short; blades flat, becoming involute, especially at the long slender point; panicle narrow; usually 30–40 cm. long; branches fascicled, short, appressed, or some of the lower as much as 12 cm. long; glumes thin and papery, equal, about 10 mm. long, smooth, 3-nerved, the lateral nerves rather indistinct; lemma about 6 mm. long, appressed-villous; awn twice-geniculate, the first section 6–10 mm. long, closely twisted, villous, the second section shorter, 4–6 mm. long, twisted, villous, the third section 10–16 mm. long, straight, scabrous only and lighter in color.

Meadows and open woods, Canadian Zone; Oregon southward through the Sierra Nevada Mountains to the San Jacinto Mountains. June–Aug. Type locality: San Jacinto Mountains.

6. Stipa coronàta Thurb.

Giant Stipa. Fig. 279.

Stipa coronata Thurb. in S. Wats. Bot. Calif. **2**: 287. 1880.

Culms stout, 1–2 meters tall, as much as 6 mm. thick at base, smooth or pubescent below the nodes; sheaths smooth, the margin and throat villous; ligule about 2 mm. long, ciliate-margined; blades very long, flat, with a slender involute point; panicle narrow, dense, stout, purplish, 30–40 cm. long; glumes gradually acuminate, 3-nerved, smooth except the scabrous keel of the first, unequal, the first about 2 cm. long, the second 2–4 mm. shorter; lemma about 8 mm. long, densely villous with long appressed hairs; awn twice-geniculate, the first section about 10 mm. long, twisted, scabrous but not villous, the second section similar but shorter, the third section about as long as the other two, straight.

Open ground, Arid Transition Zone of Coast Ranges; Monterey and southward into Lower California. Apr.–June. Type locality: southern California.

7. **Stipa paríshii** Vasey.

Parish's Stipa. Fig. 280.

Stipa parishii Vasey, Bot. Gaz. 7: 33. 1882.

Culms stout, 30–60 cm. tall; sheaths smooth, villous at the throat; ligule short, ciliate; blades firm, flat, with a slender involute point, very scabrous above, about 4 mm. wide; panicle 15–20 cm. long, narrow, dense, purple-tinged; glumes smooth, 3-nerved, long-acuminate, unequal, the first 1.5 cm. long, the second shorter; lemma 7 mm. long, densely long-villous, especially above; awn about 2.5 cm. long, once-geniculate, twisted below, straight above, nearly smooth.

Open ground, Arid Transition Zone; southern California and western Nevada. May–Aug. Type locality: San Bernardino Mountains.

8. **Stipa comàta** Trin. & Rupr.

Needle and Thread Grass. Fig. 281.

Stipa comata Trin. & Rupr. Mém. Acad. St. Pétersb. VI. Sci. Nat. 5[1]: 75. 1842.

Culms 60–120 cm. tall, smooth; sheaths smooth; ligule 4–6 mm. long; blades becoming involute, elongate; panicle loose, open, 15–25 cm. long; branches slender, ascending, or, in anthesis, spreading, the lower 8–10 cm. long, bearing usually 2 spikelets toward the extremities; glumes nearly 2.5 cm. long, gradually narrowed into an awn, smooth, 5-nerved, thin, papery; lemma 10–12 mm. long, rather sparsely appressed-villous; callus 3 mm. long; awn very long, the first section 2–4 cm. long, closely twisted, appressed-villous but becoming nearly smooth, the second like the first but shorter, the third section as long or longer than the other two, more or less flexuous but not twisted, scabrous, very slender.

Dry open ground, Upper Sonoran Zone; Washington to eastern California, east to the Great Plains. June–July. Type locality: Saskatchewan.

Stipa comata intermèdia Scribn. Bot. Gaz. 11: 171. 1886. *Stipa tweedyi* Scribn. U. S. Dept. Agr. Div. Agrost. Bull. 11: 47. 1898. Differs from the typical form in the shorter straight third section of the awn. California (Fallen Leaf Lake, *Eastwood*; Campito Mountain, *Jepson*), and eastward to Wyoming. Type locality: Yellowstone Park.

Stipa comata intónsa Piper, Contr. U. S. Nat. Herb. 11: 109. 1906. Differs from the typical form in having pubescent leaves. Near Rockland, Washington, the type locality.

9. **Stipa pùlchra** Hitchc.

Nodding Stipa. Fig. 282.

Stipa pulchra Hitchc. Am. Journ. Bot. 2: 301. 1915.
Stipa setigera of Californian botanies, not Presl.

Culms 60–100 cm. tall; blades long and narrow, flat or involute; ligule about 1 mm. long; panicle about 15 cm. long, loose, the branches spreading, slender, some of the lower 2.5–5 cm. long; glumes narrow, long-acuminate, purplish, 3-nerved, unequal, the first about 2 cm. long, the second 2–4 mm. shorter; lemma 8 mm. long, sparingly pilose, the callus sharp; awn 4–6 mm. long, short-pubescent to the second bend, the first section 1.5–2 cm. long, the second shorter, the third slender and flexuous.

Open ground, Arid Transition Zone; mostly in the Coast Ranges of central California, south into Lower California. Mar.–May. Type locality: Sonoma County, California.

10. **Stipa lépida** Hitchc.

Small-flowered Stipa. Fig. 283.

Stipa lepida Hitchc. Am. Journ. Bot. **2**: 303. 1915.
Stipa eminens of Californian botanies, not Cav.

Culms slender, puberulent below the nodes, 60–100 cm. tall; sheaths smooth, sparingly villous at throat; ligule very short; blades flat, narrow, 2–4 mm. wide, pubescent on upper surface near base; panicle rather loose and open, usually 15–20 cm. long, but sometimes more than 30 cm. long, the branches distant, slender; glumes 3-nerved, smooth, unequal, acuminate, the first 6–10 mm. long, the second about 2 mm. shorter; lemma about 6 mm. long, sparingly villous, nearly glabrous toward the hairy-tufted apex; awn indistinctly twice-geniculate, about 2.5–4 cm. long, scabrous but not villous.

Open ground, Arid Transition Zone; Coast Ranges from central California to Lower California. Apr.–June. Type locality: Santa Ynez Forest Reserve.

Stipa lepida andersòni (Vasey) Hitchc. Am. Journ. Bot. **2**: 303. 1915. *Stipa eminens andersoni* Vasey, Contr. U. S. Nat. Herb. **3**: 54. 1892. Differs from the typical form chiefly in the slender involute blades. This form is, on the average, a smaller plant, the culms being shorter, the panicles narrower and few-flowered, the spikelets usually smaller.

Range about the same as for *S. lepida* but extending north to Mount Shasta (*Jepson*). Type locality: Santa Cruz.

11. **Stipa williámsii** Scribn.

Williams's Stipa. Fig. 284.

Stipa williamsii Scribn. U. S. Dept. Agr. Div. Agrost. Bull. **11**: 45. *pl. 4.* 1898.

Culms erect, 60–100 cm. tall; puberulent, especially around the nodes; sheaths puberulent; ligule very short; blades more or less puberulent, 1–3 mm. wide; panicle narrow, 15–20 cm. long, the branches appressed; glumes thin, nearly equal, about 1 cm. long; lemma about 7 mm. long, the awn 3 cm. long.

Open ground, Arid Transition Zone; eastern Washington and eastern Oregon to Wyoming. May–Aug. Type locality: Wyoming.

12. **Stipa stillmánii** Boland.

Stillman's Stipa. Fig. 285.

Stipa stillmanii Boland. Proc. Calif. Acad. **4**: 169. 1872.

Culms stout, 60–100 cm. tall; sheaths smooth, puberulent at the throat and collar; ligule very short; blades scattered, folded or involute, firm, the uppermost filiform; panicle narrow, dense or interrupted at base, the branches short, fascicled; glumes equal, papery, minutely scabrous, acuminate into a scabrous awn-point, 14–16 mm. long, the first 3-nerved, the second 5-nerved; lemma 9 mm. long, short-pilose, bearing 2 slender teeth at the apex, the callus short; awn about 2.5 cm. long, once- or indistinctly twice-geniculate, scabrous.

Only known from the type collection by Bolander (Blue Cañon, Placer County, California).

13. **Stipa vàseyi** Scribn. Sleepy-grass. Fig. 286.

Stipa vaseyi Scribn. U. S. Dept. Agr. Div. Agrost. Bull. **11**: 46. 1898.

Stipa viridula robusta Vasey, Contr. U. S. Nat. Herb. **1**: 56. 1890.

Culms 60–120 cm. high; sheaths somewhat hairy at the throat; blades elongate, involute; panicle about 30 cm. long, dense, the branches and branchlets numerous, many-flowered; glumes narrow, acuminate, scabrous, the first a little longer, rather strongly 5–7-nerved, 10 mm. long; lemma about 6 mm. long, appressed-pilose, the callus short, pilose; awn twice-geniculate, about 2.5 cm. long, minutely puberulent.

Plains and open woods, Arid Transition Zone; from Arizona to Colorado and south to northern Mexico. A specimen from San Nicolas Island (*Trask*) is referred to this species. April. Type locality: Texas.

14. **Stipa lémmoni** (Vasey) Scribn. Lemmon's Stipa. Fig. 287.

Stipa pringlei lemmoni Vasey, Contr. U. S. Nat. Herb. **3**: 55. 1892.

Stipa lemmoni Scribn. U. S. Dept. Agr. Div. Agrost. Circ. **30**: 3. 1901.

Culms 60–100 cm. tall, sometimes pubescent below the nodes; sheaths smooth; ligule about 1 mm. long; blades usually flat, pubescent on upper surface; panicle narrow, the branches 2.5–5 cm. long, appressed; glumes nearly equal, rather broad, scarious, acuminate, 3–5-nerved, 12 mm. long; lemma 7 mm. long, rather thinly appressed-pilose, the callus short; awn about 2.5 cm. long, twice-geniculate, appressed-pilose to the second bend.

Open woods, Arid Transition Zone; Washington, south in the Coast Ranges and Sierra Nevada Mountains to Tehachapi. June–July. Type locality: Plumas County, California.

Stipa lemmoni jònesii Scribn. U. S. Dept. Agr. Div. Agrost. Circ. **30**: 4. 1901. Differs in the more slender, firm, involute blades, and smaller spikelets; glumes about 8 mm. long; lemma about 6 mm. long, the awn 2 cm. long, tending to be incurved, the pubescence shorter. A scarcely distinct variety. Range about the same as for *S. lemmoni*. Type locality: Emigrant Gap, California.

15. **Stipa lettermáni** Vasey. Letterman's Stipa. Fig. 288.

Stipa lettermani Vasey, Bull. Torrey Club **13**: 53. 1886.

Stipa viridula lettermani Vasey, Contr. U. S. Nat. Herb. **3**: 50. 1892.

Stipa pinctorum Jones, Proc. Calif. Acad. II. **5**: 724. 1895.

Culms cespitose, slender, 30–40 cm. tall; sheaths smooth; ligule very short; blades crowded at base of plant, short, slender, involute; panicle narrow, 7–20 cm. long; glumes narrow, acuminate, 3-nerved, about 8 mm. long; lemma narrow, 5 mm. long, pilose; awn very slender, 1–1.5 cm. long, nearly smooth, twice-geniculate, the first section short, about 3 mm. long.

Dry soil, Hudsonian and Alpine Zones; Oregon and south in the Sierra Nevada and San Bernardino Mountains, and on high peaks east to Colorado. July–Aug. Type locality: Idaho.

16. Stipa nélsoni Scribn.

Nelson's Stipa. Fig. 289.

Stipa nelsoni Scribn. U. S. Dept. Agr. Div. Agrost. Bull. **11**: 46. 1898.

Culms 1–1.5 meters tall, smooth; blades flat, 15–30 cm. long, 3–5 mm. wide; panicle narrow, 20–25 cm. long, rather densely flowered; glumes acuminate, nearly equal, about 12 mm. long; lemma about 7 mm. long; awn 4–5 cm. long.

Open ground, Arid Transition Zone; eastern Washington and Oregon to Wyoming. June–Aug. Type locality: Wyoming.

17. Stipa mìnor (Vasey) Scribn.

Subalpine Stipa. Fig. 290.

Stipa viridula minor Vasey, Contr. U. S. Nat. Herb. **3**: 50. 1892.

Stipa minor Scribn. U. S. Dept. Agr. Div. Agrost. Bull. **11**: 46. 1898.

Culms few in a cluster, 50–80 cm. tall; sheaths smooth; ligule very short; blades flat or becoming involute, narrow, as much as 30 cm. long; panicle narrow, 15–20 cm. long; glumes 6 mm. long, 3-nerved, slightly scabrous on the keels; lemma narrow, pilose, 5 mm. long; awn about 12 mm. long, nearly smooth, twice-geniculate, the first section 3 mm. long.

Open ground, Arid Transition and Canadian Zones; Washington, to the high Sierra Nevada Mountains of central California, and east to Wyoming and Colorado. July–Aug. Type locality: Colorado.

Stipa littoràlis Phil. and **Nassélla chilénsis** Desv., Chilean perennials, have been found on ballast at Linnton, Oregon (*Nelson*).

18. ORYZÓPSIS Michx. Fl. Bor. Am. **1**: 51. *pl. 9.* 1803.

Spikelets 1-flowered, disarticulating above the glumes; glumes about equal, obtuse or acuminate; lemma indurate, usually about as long as the glumes, broad, oval or oblong, nearly terete, usually pubescent, with a short blunt oblique callus, and a short deciduous sometimes bent and twisted awn; palea inclosed by the edges of the lemma. Perennial mostly low grasses with flat or often involute blades, and terminal narrow or open panicles. [Greek, Oryza-like.]

Species about 20 in the north temperate regions of both hemispheres. Type species, *Oryzopsis asperifolia* Michx.

Lemma smooth.
  Blades flat, 5 mm. wide or more; spikelets numerous, 3 mm. long. — 1. *O. miliacea.*
  Blades involute, less than 2 mm. wide; spikelets few, 5 mm. long. — 2. *O. hendersoni.*
Lemma pubescent.
  Pubescence short and appressed.
    Branches of panicle stiffly erect; awn about 5 mm. long, sometimes less; blades stiffly erect. — 3. *O. exigua.*
    Branches of panicle slender, ascending, somewhat flexuous; awn more than 5 mm. long; blades slender and flexuous. — 4. *O. kingii.*
  Pubescence long and silky.
    Branches of panicle and pedicels divaricately spreading. — 5. *O. hymenoides.*
    Branches of panicle and pedicels erect or ascending.
      Awn 6 mm. long; culms usually less than 30 cm. tall. — 6. *O. webberi.*
      Awn 12 mm. long; culms 30–60 cm. tall. — 7. *O. bloomeri*

1. Oryzopsis miliàcea (L.) Benth. & Hook.
Millet Mountain-rice. Fig. 291.

Agrostis miliacea L. Sp. Pl. 61. 1753.
Oryzopsis miliacea Benth. & Hook.; Aschers. & Schweinf. Mém. Inst. Égypte 2: 169. 1887.

Culms erect from a decumbent base, 60–100 cm. tall; sheaths smooth; ligule about 2 mm. long; blades flat, 8–10 mm. wide; panicle as much as 30 cm. long, loose, the branches spreading; glumes 3 mm. long, smooth, equal; lemma smooth, 2 mm. long, the deciduous straight awn about 4 mm. long.

Introduced in a few localities in California (Cahto, Santa Barbara, Los Angeles). Apr.–May. Type locality, European.

2. Oryzopsis hendersòni Vasey.
Henderson's Mountain-rice. Fig. 292.

Oryzopsis hendersoni Vasey, Contr. U. S. Nat. Herb. 1: 267. 1893.

Culms densely cespitose, scabrous, 10–40 cm. tall; sheaths scabrous, the basal pale, glabrescent; ligule very short; blades mostly basal, involute, scabrous, firm, about 1 mm. wide when unrolled, mostly less than 10 cm. long, the culm blades single, 4–5 cm. long; panicle rather few-flowered, 5–10 cm. long, the branches distant, stiff, scabrous, appressed or ascending, spikelet-bearing toward the end, the lower as much as 8 cm. long; glumes abruptly acute, 5 mm. long; lemma about as long as the glumes, glabrous, the awn very deciduous, nearly straight, 6–10 mm. long.

Only known from the type locality, "in Washington," and from the Ochoco National Forest, Oregon, at 4800 feet altitude.

3. Oryzopsis exígua Thurb.
Little Mountain-rice. Fig. 293.

Oryzopsis exigua Thurb. in Wilkes U. S. Expl. Exped. 17: 481. 1874.

Culms densely cespitose, scabrous, stiffly erect, 15–30 cm. tall; sheaths smooth or somewhat scabrous; ligule 2–3 mm. long; blades involute-filiform, stiffly erect, scabrous, 5–10 cm. long, the culm blades about 2, the upper mostly 3–4 cm. long; panicle narrow, 3–6 cm. long, the branches appressed, the lower 1-2 cm. long; glumes abruptly acute, 5 mm. long; lemma appressed-pilose, about as long as the glumes, the awn about 5 mm. long, untwisted, once-geniculate.

Dry open ground or open woods, Hudsonian and Alpine Zones; high mountains of Washington (Mount Adams) and Oregon, to Wyoming. June–Aug. Type locality: Cascade Mountains, Oregon.

4. Oryzopsis kíngii (Boland.) Beal.
King's Mountain-rice. Fig. 294.

Stipa kingii Boland. Proc. Calif. Acad. **4**: 170. 1872.
Oryzopsis kingii Beal, Grasses N. Am. **2**: 229. 1896.

Culms tufted, slender, 20-40 cm. tall; blades numerous at the base of the plant, involute, capillary; ligule about 1 mm. long; panicle narrow, loose, the short slender branches appressed or ascending, few-flowered; glumes broad, papery, nerveless, obtuse, purple at base, unequal, the first about 3.5 mm. long, the second a little longer; lemma elliptic, 3 mm. long, rather sparingly appressed-pubescent, the callus short; awn more or less sickle-shaped, bent in a wide curve or indistinctly geniculate below the middle, not twisted, minutely pubescent, not readily deciduous, about 12 mm. long.

Dry open ground, Hudsonian and Alpine Zones; high central Sierra Nevada Mountains. July–Aug. Type locality: Mount Dana.

5. Oryzopsis hymenoìdes (Roem. & Schult.) Ricker. Indian Mountain-rice. Fig. 295.

Stipa membranacea Pursh, Fl. Am. Sept. **2**: 728. 1814, not L. 1753.
Stipa hymenoides Roem. & Schult. Syst. Veg. **2**: 339. 1817.
Eriocoma cuspidata Nutt. Gen. Pl. **1**: 40. 1818.
Oryzopsis cuspidata Benth.; Vasey, Grasses U. S. 23. 1883.
Oryzopsis membranacea Vasey, U. S. Dept. Agr. Div. Bot. Bull. **12**: 10. *pl. 10.* 1891.
Eriocoma membranacea Beal, Grasses N. Am. **2**: 232. 1896.
Oryzopsis hymenoides Ricker; Piper, Contr. U. S. Nat. Herb. **11**: 109. 1906.

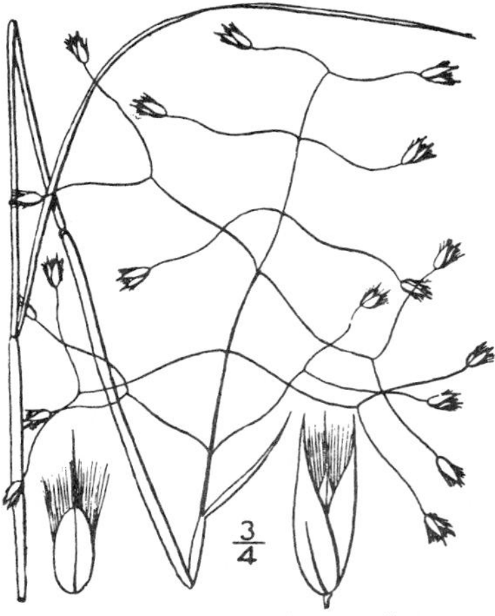

Culms cespitose, 30–60 cm. tall; sheaths smooth or minutely scabrous; ligule about 6 mm. long, acute; blades slender, elongate, nearly as long as the culms; panicle diffuse, 7–15 cm. long, the slender branches in pairs, the branchlets dichotomous, all divaricately spreading, the ultimate pedicels capillary, flexuous, enlarged below the spikelets; glumes equal, about 6 mm. long, puberulent, papery, ovate, 3-nervėd, abruptly narrowed into an awn-like point; lemma fusiform, turgid, about 3 mm. long, nearly black at maturity, densely long-pilose with hairs 3 mm. long; awn when present about 4 mm. long, straight, readily deciduous.

Deserts and plains, in the Upper Sonoran Zone; eastern Washington and Oregon, southern and eastern California, east to the Great Plains. Apr.–July. Type locality: "On the banks of the Missouri."

6. Oryzopsis wébberi (Thurb.) Benth.
Webber's Mountain-rice. Fig. 296.

Eriocoma webberi Thurb. in S. Wats. Bot. Calif. **2**: 283. 1880.

Oryzopsis webberi Benth.; Vasey, Grasses U. S. 23. 1883.

Culms cespitose, erect, 15–30 cm. tall; blades involute, filiform, scabrous; panicle narrow, 2.5–5 cm. long, the branches appressed; glumes equal, narrow, obscurely 5-nerved, minutely scaberulous, acuminate, about 8 mm. long; lemma narrow, 6 mm. long, densely long-pilose, the awn about 6 mm. long, straight or bent, not twisted.

Deserts and plains, Transition Zone; Lassen and Plumas Counties, California, to Colorado. August. Type locality: Sierra Valley, California.

7. Oryzopsis bloòmeri (Boland.) Ricker.
Bloomer's Mountain-rice. Fig. 297.

Stipa bloomeri Boland. Proc. Calif. Acad. **4**: 168. 1872.
Oryzopsis bloomeri Ricker; Piper, Contr. U. S. Nat. Herb. **11**: 109. 1906.

Culms tufted, 30–60 cm. tall, glabrous; sheaths glabrous; ligule about 1 mm. long; blades crowded at the base, involute, narrow, firm; panicle 7–15 cm. long, the branches slender, rather stiffly ascending, the longer 5–7 cm. long, bearing spikelets from about the middle; glumes comparatively broad, indistinctly 3–5-nerved, smooth, rather abruptly acuminate, equal, 8–10 mm. long; lemma elliptic, 5 mm. long, densely long-villous; awn about 12 mm. long, tardily deciduous, once-geniculate, the first section about 6 mm. long, slightly twisted, appressed-villous, indistinctly bent or flexuous, the second section straight, minutely scabrous.

Dry ground, Arid Transition Zone; eastern Washington to California (Moulton, Mount Diablo, Lancaster), east to Manitoba and New Mexico. June–July. Type locality: near Mono Lake, California.

19. MUHLENBÉRGIA Gmel. Syst. Nat. **2**: 171. 1791.

Spikelets 1-flowered, the rachilla disarticulating above the glumes; glumes usually shorter than the lemma, obtuse to acuminate or awned, the first sometimes small or rarely obsolete; lemma membranaceous, 3–5-nerved, with a very short usually minutely pilose callus, the apex acute, sometimes bidentate, extending into a straight or flexuous awn, or sometimes only mucronate. Perennial or rarely annual low or moderately tall grasses, tufted or bearing rhizomes, the culms simple or much-branched, the inflorescence a narrow or open panicle. [Named in honor of Rev. Dr. Henry Muhlenberg, a distinguished American botanist, 1753–1815.]

Species about 80, mostly in Mexico and southwestern United States, a few in the eastern part of the Old World. Type species, *Muhlenbergia schreberi* Gmel.

Plants annual (see also no. 9); lemma bearing a capillary awn. — 1. *M. microsperma.*
Plants perennial.
- Plants producing creeping scaly rhizomes.
  - Hairs at base of floret copious, as long as body of lemma. — 2. *M. andina.*
  - Hairs at base of floret inconspicuous, not more than half as long as lemma.
    - Blades short and narrow, usually less than 2 mm. wide, often involute; lemma awnless or mucronate.
      - Culms erect or decumbent at base; glumes 1 mm. long. — 3. *M. squarrosa.*
      - Culms widely spreading or creeping; glumes 1.5 mm. long. — 4. *M. repens.*
    - Blades rather long, more than 2 mm. wide (except in *M. lemmoni*), flat; lemma sometimes awned.
      - Glumes awned, the awn much exceeding the awnless lemma; panicle compact, like an interrupted spike. — 5. *M. racemosa.*
      - Glumes acuminate or short-awned, not much longer than the body of the lemma; panicles mostly slender.
        - Sheaths scabrous; blades firm; plants branched from the base; awn 2–3 mm. long.
          - Blades about 2 mm. wide. — 6. *M. lemmoni.*
          - Blades 3–5 mm. wide. — 7. *M. californica.*
        - Sheaths glabrous; blades thin; plants branched above the base; awn more than 3 mm. long. — 8. *M. foliosa ambigua.*
- Plants without creeping scaly rhizomes.
  - Lemma mucronate or short-awned.
    - Culms delicate, the base often appearing to be annual; blades mostly less than 2 cm. long; glumes entire. — 9. *M. filiformis.*
    - Culms moderately stout, from a firm base; blades mostly more than 3 cm. long; second glume truncate, toothed. — 10. *M. jonesii.*
  - Lemma distinctly awned.
    - Panicles diffuse; plants widely spreading, much-branched, fragile, wiry, the base knotty. — 11. *M. porteri.*
    - Panicles narrow; plants cespitose, erect from a short decumbent base. — 12. *M. montana.*

1. Muhlenbergia microspérma (DC.) Kunth. Annual Muhlenbergia. Fig. 298.

Trichochloa microsperma DC. Cat. Hort. Monsp. 151. 1813.
Podosaemum debile H. B. K. Nov. Gen. & Sp. 1: 128. 1816.
Muhlenbergia debilis Kunth, Rév. Gram. 1: 63. 1829.
Muhlenbergia microsperma Kunth, Rév. Gram. 1: 64. 1829.
Muhlenbergia purpurea Nutt. Journ. Acad. Phila. II. 1: 186. 1848.

Annual, often purple; culms spreading, 15–35 cm. tall, scaberulous, especially below the nodes; sheaths smooth or scaberulous; ligule 1 mm. long; blades 2.5–5 cm. long, 1 mm. wide, flat, scabrous; panicles narrow, loose, 2–7 cm. long; glumes ovate, obtuse or emarginate, 1-nerved, unequal, the second the longer, 1 mm. long; lemma narrow, acuminate, 3-nerved, 3 mm. long, appressed-pubescent on margins and callus; awn terminal, capillary, 10–14 mm. long. Cleistogamous spikelets are developed at the base of the lower sheaths. These are solitary or few in a fascicle in each axil, each spikelet included in an indurate thickened, tightly rolled narrowly conical reduced sheath, which readily disarticulates from the plant at maturity. The glumes are wanting and awn of the lemma reduced, but the grain is larger than that of the spikelets in the terminal inflorescence, being about the same length (2 mm.) but much thicker.

Open ground, Lower Sonoran Zone. Mar.–May. Type locality: Mexico.

2. Muhlenbergia andìna (Nutt.) Hitchc. Hairy Muhlenbergia. Fig. 299.

Calamagrostis andina Nutt. Journ. Acad. Phila. II. 1: 187. 1848.
Vaseya comata Thurb. Proc. Acad. Phila. **1863**: 79. 1863.
Muhlenbergia comata Thurb.; Benth. Journ. Linn. Soc. Bot. **19**: 83. 1881.
Muhlenbergia andina Hitchc. U. S. Dept. Agr. Bull. **772**: 145. 1920.

Perennial, with numerous scaly rhizomes; culms erect or sometimes spreading, smooth below, scabrous above, pubescent about the nodes, 50–100 cm. tall; sheaths smooth or slightly scabrous, keeled; ligule 1 mm. long, membranaceous, short-ciliate; blades flat, 2–6 mm. wide, scabrous; panicles narrow, spikelike, usually more or less lobed or interrupted, often purple-tinged, 7–15 cm. long; glumes narrow, acuminate, 1-nerved, smooth, ciliate-scabrous on the keels, 3–4 mm. long; lemma 3 mm. long, gradually narrowed into a capillary awn 4–8 mm. long, the hairs at base of floret copious, 2–3 mm. long.

Open ground, Arid Transition Zone; Cascade and Sierra Nevada Mountains from Washington to California, east to Wyoming. July–Aug. Type locality: "Upper California on the Colorado of the West."

3. Muhlenbergia squarròsa (Trin.) Rydb. Short-leaved Muhlenbergia. Fig. 300.

Vilfa squarrosa Trin. Mém. Acad. St. Pétersb. VI. Sci. Nat. 4¹: 100. 1840.
Vilfa richardsonis Trin. Mém. Acad. St. Pétersb. VI. Sci. Nat. 4¹: 103. 1840.
Vilfa depauperata Torr.; Hook. Fl. Bor. Am. **2**: 257. *pl. 236.* 1840, not *Muhlenbergia depauperata* Scribn.
Sporobolus depauperatus Scribn. Bull. Torrey Club **9**: 103. 1882.
Sporobolus richardsonis Merr. Rhodora **4**: 46. 1902.
Muhlenbergia squarrosa Rydb. Bull. Torrey Club **36**: 531. 1909.

Perennial from numerous hard creeping rhizomes; culms wiry, erect or decumbent at base, from 5 cm. to as much as 60 cm. in height; blades flat or usually involute, 1–5 cm. long; panicle narrow, interrupted, or sometimes rather close and spikelike, 2–15 cm. long; glumes ovate, 1 mm. long; lemma lanceolate, acute, mucronate, 2 mm. long.

Dry or moist open ground, Upper Sonoran and Arid Transition Zones; eastern Washington to California, east to Montana and south to central Mexico. June–Aug. Type locality: Menzies Island, Columbia River.

4. Muhlenbergia rèpens (Presl.) Hitchc.
Aparejo Grass, Creeping Muhlenbergia.
Fig. 301.

Sporobolus repens Presl, Rel. Haenk. 1: 241. 1830.
Muhlenbergia repens Hitchc. in Jepson Fl. Calif. 1: 111. 1912.

Perennial from woody creeping rhizomes; culms slender, wiry, freely branching, widely spreading or creeping, 15–40 cm. long; flower-bearing branches ascending; blades numerous, narrow, involute, arcuate, 1–3 cm. long; panicles narrow, interrupted, few-flowered, 1–2.5 cm. long; glumes ovate, acute, 1.5 mm. long, smooth; lemma exceeding the glumes, about 2 mm. long, smooth or sparsely pubescent, acute or mucronate.

Deserts of San Bernardino (Upland) and Inyo Counties, to Arizona and northern Mexico. January. Type locality: Mexico.

5. Muhlenbergia racemòsa (Michx.) B. S. P.
Wild Timothy, Satin-grass. Fig. 302.

Agrostis racemosa Michx. Fl. Bor. Am. 1: 53. 1803.
Muhlenbergia glomerata Trin. Gram. Unifl. 191. *pl. 5. f. 10.* 1824.
Muhlenbergia racemosa B. S. P. Prel. Cat. N. Y. 67. 1888.

Perennial from stout creeping scaly rhizomes; culms erect or reclining 50–100 cm. tall, or even more, slightly roughened below the nodes, simple or sparingly branched; sheaths smooth, keeled; ligule about 1 mm. long, slightly ciliate; blades flat, mostly appressed, scabrous, 5–10 cm. long, 2–5 mm. wide; panicle narrow, condensed or lobed, 3–10 cm. long, made up of several oblong dense clusters; glumes narrow, aristate, about equal, 4–6 mm. long, exceeding the acute lemma.

Moist meadows and low ground, Transition Zone; Washington to northern California, east across the continent, south to New Mexico. July–Sept. Type locality: along the Mississippi.

6. Muhlenbergia lémmoni Scribn.
Lemmon's Muhlenbergia. Fig. 303.

Muhlenbergia lemmoni Scribn. Contr. U. S. Nat. Herb. 1: 56. 1890.

Perennial, from a creeping branching woody rhizome; culms slender, wiry, erect or ascending, 30–60 cm. tall; blades flat or somewhat involute, 1–2 mm. wide; panicles narrow, interrupted, the branches short; glumes narrow, gradually acuminate, including the awn about 3 mm. long; lemma 3 mm. long, acuminate into an awn as much as 6 mm. long, the callus hairs rather sparse, about half as long as the body of the lemma.

Deserts, from southern California (Jamacha) to Texas and northern Mexico. July. Type locality: Arizona.

7. Muhlenbergia califórnica Vasey.
California Muhlenbergia. Fig. 304.

Muhlenbergia glomerata brevifolia Vasey, Bot. Gaz. **7**: 92. 1882.

Muhlenbergia sylvatica californica Vasey, Bot. Gaz. **7**: 93. 1882.

Muhlenbergia californica Vasey, Bull. Torrey Club **13**: 53. 1886.

Muhlenbergia parishii Vasey, Bull. Torrey Club **13**: 53. 1886.

Perennial, the base more or less creeping and rhizomatous; culms erect, somewhat woody below, smooth, puberulent about nodes, 30–60 cm. tall; sheaths scaberulous, keeled; ligule scarcely 1 mm. long; blades flat, 4-6 mm. wide, scabrous, usually short; panicles narrow, spikelike or interrupted, 7–15 cm. long; glumes narrow, acuminate or awn-pointed, 3–4 mm. long, scabrous on the keels; lemma about 3 mm. long, scabrous, the callus hairs rather sparse, about half as long as the lemma; awn 2 mm. long or less.

Borders of streams and irrigation ditches, confined to southern California. Apr.–Aug. Type locality: San Bernardino Mountains.

8. Muhlenbergia foliòsa ambígua (Torr.) Scribn.
Leafy Muhlenbergia. Fig. 305.

Muhlenbergia ambigua Torr. in Nicoll. Rep. Miss. 164. 1843.
Muhlenbergia foliosa ambigua Scribn. Rhodora **9**: 20. 1907.

Perennial from creeping scaly rhizomes; culms erect, more or less branched above, minutely scabrous below the nodes, 60–100 cm. tall; sheaths smooth or slightly scabrous, keeled; blades 5–15 cm. long, 2–5 mm. wide, scabrous, appressed; panicles narrow, 10–15 cm. long, interrupted, spikelike, the branches 1–3 cm. long, densely flowered; glumes narrow, acuminate or aristate, about 3 mm. long; lemma a little shorter, bearing a slender awn about 5 mm. long.

Moist soil, Arid Transition Zone; Washington (Spokane County) and Oregon (Coos River, Harper Ranch) east to Minnesota. August. Type locality: Minnesota.

9. Muhlenbergia filifórmis (Thurb.) Rydb.
Slender Muhlenbergia. Fig. 306.

Vilfa depauperata filiformis Thurb.; S. Wats. in King, Geol. Expl. 40th Par. **5**: 376. 1871.

Vilfa gracillima Thurb. in S. Wats. Bot. Calif. **2**: 268. 1880, not *Muhlenbergia gracillima* Torr. 1856.

Sporobolus filiformis Rydb. Contr. U. S. Nat. Herb. **3**: 189. 1895.

Muhlenbergia filiformis Rydb. Bull. Torrey Club **32**: 600. 1905.

Perennial or sometimes apparently annual, rather soft and lax, spreading from a cluster of fibrous roots or with decumbent creeping, apparently perennial bases; culms capillary, a few centimeters to as much as 30 cm. tall, often depauperate; blades flat, usually less than 3 cm. long; panicles narrow, interrupted, few-flowered, usually less than 3 cm. long; glumes ovate, 1 mm. long; lemma lanceolate, acute, mucronate, 2 mm. long, minutely pubescent, scaberulous at tip.

Open woods and mountain meadows, Hudsonian Zone; Washington to California, east to Colorado. July–Sept. Type locality: Yosemite Valley.

10. Muhlenbergia jònesii (Vasey) Hitchc.
Jones's Muhlenbergia. Fig. 307.

Sporobolus jonesii Vasey, Bot. Gaz. **6**: 297. 1881.
Muhlenbergia jonesii Hitchc. in Jepson, Fl. Calif. **1**: 111. 1912.

Perennial; culms cespitose, erect, slender, about 30 cm. tall; blades mostly basal, involute, flexuous, scabrous; panicles narrow, loose, 5–7 cm. long; glumes equal, obtuse, toothed at apex, a little more than 1 mm. long; lemma 4 mm. long, acuminate, awn-pointed.

Only known from northeastern California. July–Aug. Type locality: Soda Springs.

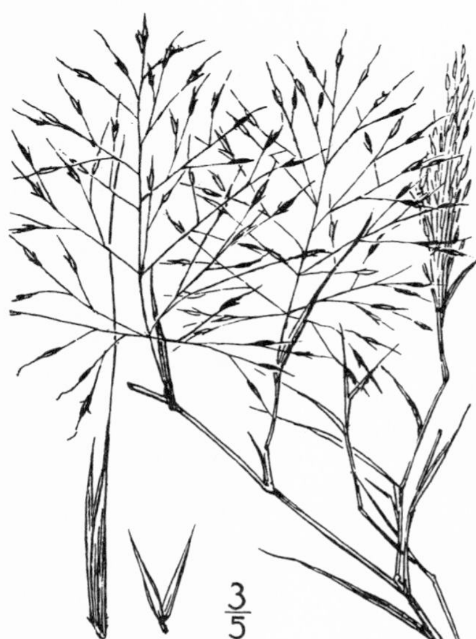

11. Muhlenbergia pòrteri Scribn.
Mesquite-grass, Porter's Muhlenbergia.
Fig. 308.

Muhlenbergia texana Thurb. in Port. & Coult. Syn. Fl. Colo. 144. 1874, not Buckl. 1863.
Muhlenbergia porteri Scribn.; Beal, Grasses N. Am. **2**: 259. 1896.

Perennial; culms woody or persistent at base, numerous, wiry, widely spreading or ascending through bushes, scaberulous, more or less branched from all the nodes, 30–100 cm. tall or more; sheaths smooth, spreading away from the branches, the prophyllum conspicuous; blades small, flat, 2–5 cm. long, early deciduous from the sheath; panicles 5–10 cm. long, open, the slender branches and branchlets brittle, widely spreading, bearing rather few long-pediceled spikelets; glumes narrow, acuminate, slightly unequal, the second longer, about 2 mm. long; lemma purple, acuminate, minutely pilose, 3–4 mm. long, the awn about 6 mm. long.

Rocky deserts, southern California (San Felipe) to Texas and northern Mexico. June. Type locality: Texas.

12. Muhlenbergia montàna (Nutt.) Hitchc.
Mountain Muhlenbergia. Fig. 309.

Calycodon montanum Nutt. Journ. Acad. Phila. II. **1**: 186. 1848.
Muhlenbergia trifida Hack. Repert. Nov. Sp. Fedde **8**: 518. 1910.
Muhlenbergia montana Hitchc. U. S. Dept. Agr. Bull. **772**: 145. 1920.
Muhlenbergia gracilis of authors, not (H. B. K.) Trin.

Perennial; culms densely cespitose, erect from a short decumbent rhizomatous base, smooth or scabrous above, 15–40 cm. tall; sheaths smooth or scabrous; ligule 4–6 mm. long; blades crowded at base, involute, scabrous, sharp-pointed; panicles narrow, loose, 5–10 cm. long; glumes broad, oblong, sparsely pubescent, 2 mm. long, obtuse or more or less erose at apex, the second 3-toothed; lemma 3 mm. long, sparsely pubescent at base and margins, gradually narrowed into a slender, more or less flexuous awn 1.5–2 cm. long.

Dry ground, in the Arid Transition Zone; middle Sierra Nevada Mountains (Yosemite Valley, Mount Tallac), east to Wyoming, south to Oaxaca. June–July. Type locality: Santa Fe, New Mexico.

20. CRÝPSIS Ait. Hort. Kew. 1: 48. 1789.

Spikelets 1-flowered, the rachilla disarticulating below the glumes, not prolonged behind the palea; glumes about equal, narrow, acute; lemma broad, thin, awnless; palea similar to the lemma, about as long, 2-nerved, readily splitting between the nerves; fruit a utricle, the seed free from the thin pericarp. A spreading annual, with capitate inflorescences in the axils of broad bracts, these being enlarged sheaths with short rigid blades. [Greek, hiding, from the partially concealed inflorescence.]

Species 1, in the Mediterranean region, sparingly introduced into the United States. Type species, *Schoenus aculeatus* L.

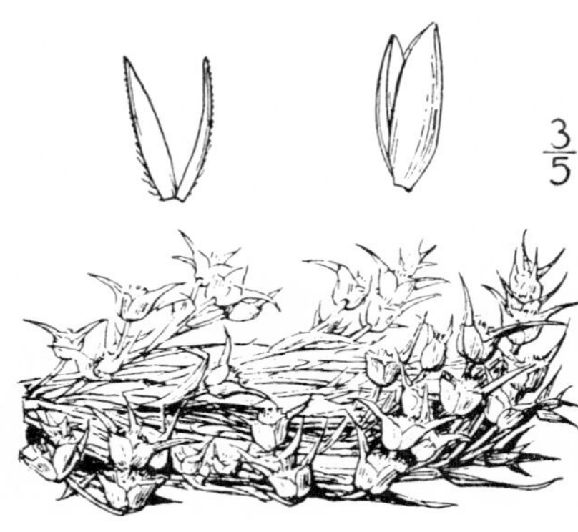

1. Crypsis aculeàta (L.) Ait.

Sharp-leaved Crypsis. Fig. 310.

Schoenus aculeatus L. Sp. Pl. 42. 1753.
Crypsis aculeata Ait. Hort. Kew. 1: 48. 1789.

Plants prostrate, the mats 30 cm. in diameter, or often depauperate, 1–2 cm. wide; glumes about 3 mm. long, minutely hispid, about equal in length, the first narrower; lemma about as long as the glumes, scabrous on the keel.

In overflowed land of the interior valley (Colusa, Chico, Norman, Stockton). September. Type locality, European.

Heleóchloa alopecuroìdes Host has been found on ballast near Portland (*Suksdorf*). It is a low spreading perennial with oblong dense spikelike panicles, the subtending leaves with inflated sheaths and reduced blades.

21. PHLEÙM L. Sp. Pl. 59. 1753.

Spikelets 1-flowered, laterally compressed, in dense cylindrical spikelike panicles; rachilla disarticulating above the glumes; glumes equal, membranaceous, keeled, abruptly mucronate or awned; lemma shorter than the glumes, hyaline, broadly truncate, 3–5-nerved; palea narrow, nearly as long as the lemma. Annual or perennial grasses with erect culms and flat blades. [Greek, a kind of reed.]

About 10 species in the temperate region of both hemispheres. Type species, *Phleum pratense* L.

Heads cylindrical, several times longer than wide. 1. *P. pratense.*
Heads ovoid or oblong, one and one-half to two times as long as wide. 2. *P. alpinum.*

1. Phleum praténse L.

Timothy. Fig. 311.

Phleum pratense L. Sp. Pl. 59. 1753.

Culms 50–100 cm. tall, from a swollen or bulb-like base, forming large clumps; panicles long-cylindrical, 3–10 cm. long; awn of glumes 1 mm. long.

Commonly escaped from cultivation, along roadsides and in fields and waste places throughout our range except in the deserts and alpine regions. May–June. Type locality, European.

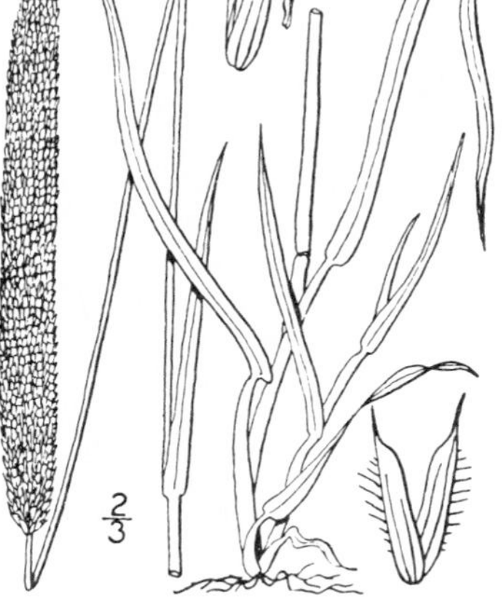

2. Phleum alpìnum L.

Mountain Timothy. Fig. 312.

Phleum alpinum L. Sp. Pl. 59. 1753.

Culms 20–50 cm. tall, from a decumbent, somewhat creeping base; panicles ellipsoid or short-cylindric; awn of glumes 2 mm. long, giving the head a bristly appearance.

Common in mountain meadows, in bogs and wet places throughout our range, Alpine and Hudsonian Zones; in the cooler regions of North America and Eurasia, extending south in the mountains to Mexico and South America. July–Aug. Type locality, European.

Phleum graècum Boiss. & Heldr. and **P. bellárdi** Willd., annual species introduced from Europe, have been found on ballast near Portland (*Suksdorf*).

22. ALOPECÙRUS L. Sp. Pl. 60. 1753.

Spikelets 1-flowered, disarticulating below the glumes, falling entire, strongly compressed laterally, the rachilla not produced; glumes equal, awnless, usually united at base, ciliate on the keel; lemma about as long as the glumes, 5-nerved, obtuse, connate at base, bearing from below the middle a slender dorsal awn, this included or exserted 2 or 3 times the length of the spikelet; palea wanting. Low or moderately tall perennial grasses with flat blades and soft dense spikelike panicles. [Greek, foxtail.]

Species about 25 in temperate regions of the northern hemisphere. Type species, *Alopecurus pratensis* L.

Awn included or exserted only about 1 mm. 1. *A. aequalis.*
Awn exserted the length of the glumes or more. 2. *A. saccatus.*

1. Alopecurus aequàlis Sobol.
Little Meadow-foxtail. Fig. 313.

Alopecurus aequalis Sobol. Fl. Petrop. 16. 1799.
Alopecurus aristulatus Michx. Fl. Bor. Am. **1**: 43. 1803.
Alopecurus fulvus J. E. Smith, Engl. Bot. *pl. 1467.* 1805.
Alopecurus geniculatus aristulatus Torr. Fl. North. & Mid. U. S. **2**: 97. 1823.
Alopecurus geniculatus fulvus Schrad. Linnaea **12**: 424. 1838.
Alopecurus howellii merrimani Beal, Grasses N. Am. **2**: 278. 1896.
Alopecurus howellii merriami Beal; Macoun in Jordan, Fur Seals North Pacif. **3**: 573. 1899.

Culms erect or spreading, 15–60 cm. tall; panicles narrow-cylindric, 2–7 cm. long, about 4 mm. wide; glumes 2 mm. long; awn of lemma short, scarcely exserted.

In water and wet places, in the Transition Zone; throughout the mountains of our region and the cooler parts of North America. May–July. Type locality: Russia.

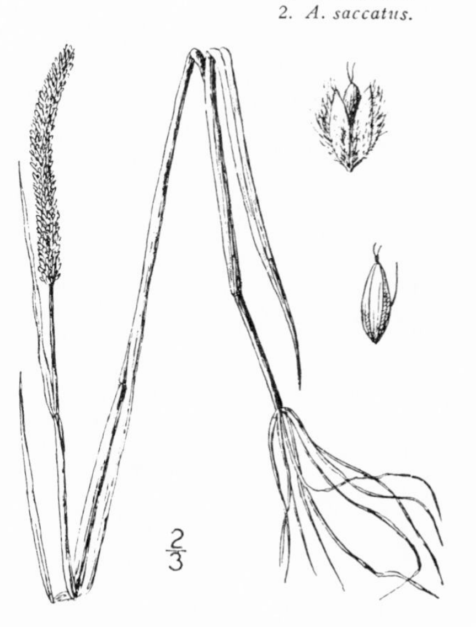

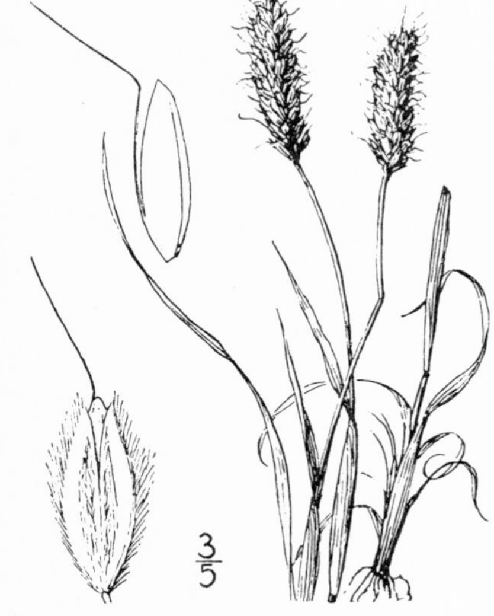

2. Alopecurus saccàtus Vasey.
Pacific Meadow-foxtail. Fig. 314.

Alopecurus saccatus Vasey, Bot. Gaz. **6**: 290. 1881.
Alopecurus howellii Vasey, Bull. Torrey Club **15**: 12. 1888.
Alopecurus californicus Vasey, Bull. Torrey Club **15**: 13. 1888.
Alopecurus pallescens Piper, Fl. Palouse 18. 1901.

Culms 15–60 cm. tall; sheaths inflated; panicles oblong, 2–5 cm. long, about 6 mm. wide; glumes 3–4 mm. long; lemma sparsely pilose on the sides, the awn exserted, about 3–5 mm.

Meadows and wet places, in the Transition Zone, California to Washington and Idaho. May–June. Type locality: eastern Oregon.

Alopecurus geniculàtus L., with more decumbent base and spikelets with awns protruding 2–3 mm., is known on Pacific Coast only from Falcon Valley, Washington, and Portland, Oregon.

Alopecurus praténsis L., meadow foxtail, with silky heads 5–10 cm. long and 6–8 mm. thick, has been found escaped from cultivation, at Corvallis, Oregon.

Alopecurus agréstis L., with more slender nearly glabrous heads, has been found at Linnton, Oregon (*Nelson*).

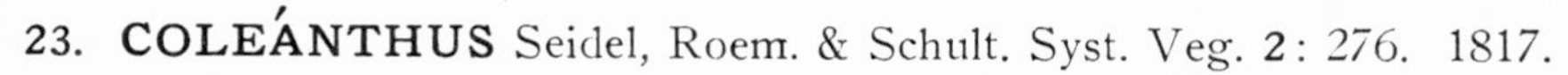

23. COLEÁNTHUS Seidel, Roem. & Schult. Syst. Veg. 2: 276. 1817.

[SCHMIDTIA Tratt. Fl. Oesterr. Kaiserth. **1**: 12. *pl. 10.* 1816, not Moench, 1802.]

Spikelets 1-flowered; glumes wanting; lemma ovate, hyaline, terminating in a short awn; palea broad, 2-keeled. A low annual with short flat blades and small few-flowered panicles. [Greek, sheath-flower, referring to the inflated sheaths.]

Species 1, northern Eurasia, introduced in America. Type species, *Schmidtia subtilis* Tratt.

1. Coleanthus súbtilis (Tratt.) Seidl.

Moss-grass. Fig. 315.

Schmidtia subtilis Tratt. Fl. Austr. 1: 12. 1816.
Coleanthus subtilis Seidel; Roem. & Schult. Syst. Veg. 2: 276. 1817.

Culms tufted, mostly 1–3 cm. tall; panicles terminal and axillary; sheaths inflated; spikelets about 1 mm. long.

In mud along the lower Columbia River. October. Type locality, European.

24. SPORÓBOLUS R. Br. Prodr. Fl. Nov. Holl. 169. 1810.

Spikelets 1-flowered, the rachilla disarticulating above the glumes, not produced beyond the palea; glumes usually unequal, the second often as long as the spikelet; lemma membranaceous, awnless; palea usually prominent and as long as the lemma or longer; seed free from the pericarp. Annual or perennial grasses with small spikelets in open or contracted panicles. [Greek, seed-throwing.]

Species about 95 in the warm regions of both hemispheres. Type species, *Agrostis indica* L.

Plants annual.
- Panicles spike-like, often inclosed in the sheaths. — 1. *S. neglectus.*
- Panicles diffuse, the pedicels capillary. — 2. *S. confusus.*

Plants perennial.
- Plants producing creeping rhizomes; blades short, flat; panicle diffuse. — 3. *S. asperifolius.*
- Plants without rhizomes.
  - Second glume about half as long as the spikelet, the first a little shorter; spikelets 5–6 mm. long. — 4. *S. asper.*
  - Second glume as long as the spikelet, the first glume shorter.
    - Panicles diffuse, the pedicels capillary. — 5. *S. airoides.*
    - Panicles contracted, open or spikelike, the pedicels short.
      - Panicles, or the exserted portion, somewhat open, the branches naked below. — 6. *S. cryptandrus.*
      - Panicles spikelike, the branches floriferous from the base. — 7. *S. contractus.*

1. Sporobolus negléctus Nash.

Small Rush-grass. Fig. 316.

Sporobolus neglectus Nash, Bull. Torrey Club 22: 464. 1895.
Sporobolus vaginaeflorus neglectus Scribn. U. S. Dept. Agr. Div. Agrost. Bull. 17: ed. 2. 170. *f. 466.* 1901.

Annual; culms spreading or erect, 10–30 cm. tall, smooth; sheaths smooth, often inflated; ligule a very short ciliate ring; blades flat, glabrous, scabrous on the margin, 2–8 cm. long, 1–2 mm. wide; panicles spike-like, 2–3 cm. long, scattered along the stem, partially or wholly concealed in the sheaths; spikelets about 2 mm. long; glumes and lemma nearly equal.

Open sterile ground, eastern Washington to Massachusetts and Tennessee. July. Type locality: northeastern United States.

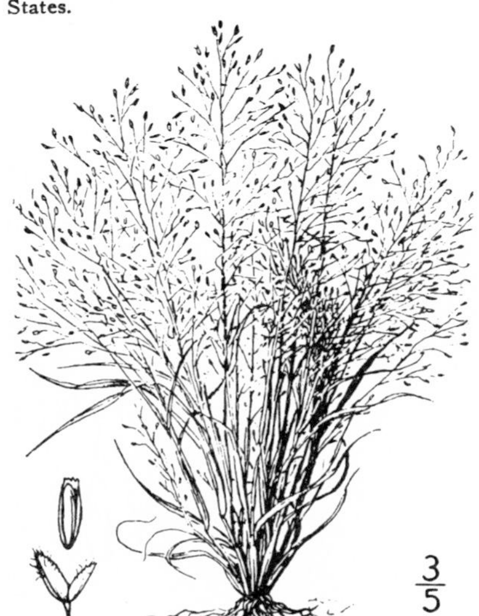

2. Sporobolus confùsus Vasey.

Tufted Annual Dropseed. Fig. 317.

Sporobolus confusus Vasey, Bull. Torrey Club 15: 293. 1888.

Annual; culms slender, 15–20 cm. tall, often depauperate; blades mostly less than 3 cm. long; panicles oblong, diffuse, often more than half the length of the entire plant, the branches capillary, spreading, 2.5–3.5 cm. long; spikelets 1–1.5 mm. long, the glumes about half as long, subequal, obtuse, sparsely pilose.

Open sandy or gravelly, usually moist ground, mostly near streams or lakes in the mountains, Upper Sonoran Zone; Washington and Montana and to Texas and Mexico. July–Sept. Type locality: Mexico.

3. **Sporobolus asperifòlius** Nees & Mey.
Rough-leaved Dropseed. Fig. 318.

Vilfa asperifolius Nees & Mey.; Trin. Mém. Acad. St. Pétersb. VI. Sci. Nat. 4[1]: 95. 1840.

Sporobolus asperifolius Nees & Mey. Nov. Act. Acad. Caes. Leop. Carol. 19: 141. 1843.

Perennial from creeping rhizomes; culms 30–60 cm. tall, ascending from a creeping or decumbent base; sheaths smooth, keeled; blades flat, 2.5–5 cm. long, about 2 mm. wide, scabrous; panicles diffuse, tardily exserted from the uppermost sheath, oval, 10–15 cm. long; spikelet 1.5 mm. long, the glumes slightly unequal, a little shorter than the spikelet.

Meadows and wet places, Upper Sonoran Zone; British Columbia to California, east to North Dakota and south into Mexico; also in South America. July–Aug. Type locality: Chile.

4. **Sporobolus ásper** (Michx.) Kunth.
Long-leaved Rush-grass. Fig. 319.

Agrostis aspera Michx. Fl. Bor. Am. 1: 52. 1803.
Agrostis composita Poir. in Lam. Encycl. Suppl. 1: 254. 1810.
Agrostis longifolia Torr. Fl. North. & Mid. U. S. 1: 90. 1823.
Sporobolus asper Kunth, Rév. Gram. 1: 68. 1829.
Sporobolus longifolius Wood, Class-book, 775. 1861.
Sporobolus compositus Merr. U. S. Dept. Agr. Div. Agrost. Circ. 35: 6. 1901.

Culms erect, glabrous, 40–150 cm. tall, mostly single; blades 10–40 cm. long, 2–4 mm. wide, attenuate to a long slender involute point; panicle spike-like, usually partly included in the uppermost sheath, 7–20 cm. long; spikelets 5–6 mm. long; glumes unequal, shorter than the lemma; lemma and palea equal, obtuse.

Dry soil, Washington (Kittitas County), east to Maine and south to Texas. September. Type locality: Illinois.

5. **Sporobolus airoìdes** Torr.
Alkali Saccaton, Hair-grass Dropseed. Fig. 320.

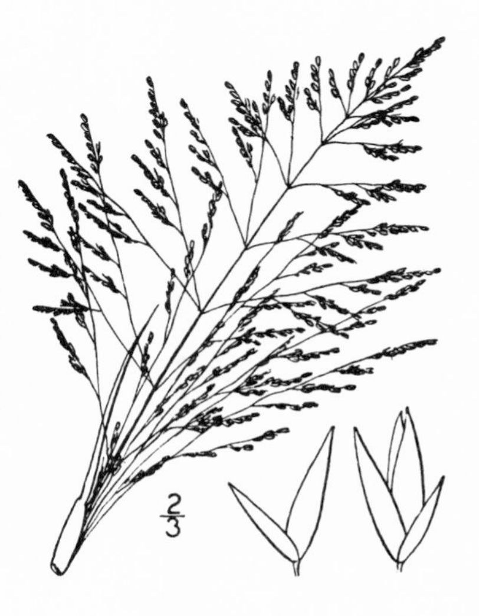

Agrostis airoides Torr. Ann. Lyc. N. Y. 1: 151. 1824.
Sporobolus airoides Torr. Pac. R. Rep. 7[3]: 21. 1856.

Perennial; culms densely cespitose, forming large tussocks, smooth, stout, spreading at base, 30–80 cm. tall; sheaths smooth, sparsely pilose at the throat; blades involute, elongate, the upper short; panicles diffuse, finally about half the length of the entire plant; spikelets 1.5–2 mm. long, obtuse; glumes unequal, the first oval, half as long as spikelet, the second as long as spikelet.

Bottom lands and valleys, often in saline or alkaline soil in the Upper Sonoran Zone; Washington (Okanogan River, near Oroville), south to California and Mexico, east to South Dakota. April–July (sometimes flowering in autumn). Type locality: Colorado.

6. Sporobolus cryptándrus (Torr.) A. Gray.
Sand Dropseed. Fig. 321.

Agrostis cryptandrus Torr. Ann. Lyc. N. Y. **1**: 151. 1824.
Sporobolus cryptandrus A. Gray, Man. 576. 1848.

Perennial; culms erect, usually branched at base, 50–100 cm. tall; sheaths smooth, densely pilose at the throat; blades flat, glabrous beneath, scabrous above and on the cartilaginous margin, 7–15 cm. long, 2–5 mm. wide, slender-pointed; panicles terminal and axillary, hidden in the sheaths or the former more or less exserted, the exserted portion spreading; spikelets about 2 mm. long; glumes unequal, the first one-third as long as the second, this and the lemma and palea about equal.

Open dry ground in the Upper Sonoran Zone; Washington to Maine, and south to Texas. June–Sept. Type locality: "on the Canadian River."

7. Sporobolus contráctus Hitchc.
Narrow-spiked Dropseed. Fig. 322.

Sporobolus cryptandrus strictus Scribn. Bull. Torrey Club **9**: 103. 1882.
Sporobolus strictus Merr. U. S. Dept. Agr. Div. Agrost. Circ. **32**: 6. 1901, not Franch. 1893.
Sporobolus contractus Hitchc. Am. Journ. Bot. **2**: 303. 1915.

Perennial; culms erect, smooth, 60–120 cm. tall; sheaths smooth; villous at the throat and on the collar; ligule a very short ciliate fringe; blades flat, 10–15 cm. long, 2–5 mm. wide, tapering to a fine involute point; panicles terminating the culms and the erect branches, narrow and spikelike, the terminal somewhat interrupted, 15–25 cm. long, all more or less concealed in the sheaths; spikelets about 2.5 mm. long; first glume one-third as long as the spikelet; second glume, lemma, and palea about equal, acute.

Dry or sandy soil, west end of Salton Sea (*Parish*), eastward to Utah and Texas. October. Type locality: Camp Lowell, Arizona.

Sporobolus elongàtus R. Br. Prodr. Fl. Nov. Holl. 170. 1810. With spike-like panicle, and with both glumes much shorter than the spikelet, has been collected on ballast near Portland (*Suksdorf*).

Sporobolus flexuòsus (Thurb.) Rydb., allied to *S. cryptandrus* but with flexuous panicle branches, has been collected in the southeastern part of the Mojave Desert (*Parish*).

25. EPICÁMPES Presl, Rel. Haenk. 1: 235. *pl. 39.* 1830.

Spikelets 1-flowered, the rachilla articulate above the glumes, not prolonged behind the palea; glumes about equal; lemma equaling or longer than the glumes, 3-nerved. Tall perennial grasses with open, narrow, or spikelike panicles. [Greek, curved.]

Species 15, northern South America to Mexico. Type species, *Epicampes strictus* Presl.

1. Epicampes rìgens Benth.
Deer-grass. Fig. 323.

Epicampes rigens Benth. Journ. Linn. Soc. Bot. **19**: 88. 1881.

Culms erect, 1–1.5 meters tall; sheaths smooth or slightly scabrous, covering the nodes; ligule truncate, 1–2 mm. long; blades scabrous, elongate, involute, tapering into a long slender point; panicle spikelike, slender, 30 cm. long or more; glumes 2–3 mm. long, oblong, obtuse or somewhat erose, puberulent, convex, scarcely keeled, striate; lemma slightly exceeding the glumes, scaberulous, sparsely pilose at base, 3-nerved toward the narrowed summit, awnless.

Dry or open ground, hillsides, gullies and open forest, Sonoran Zone; southern California to New Mexico and Mexico. July–Sept. Type locality: southern California.

26. POLYPÒGON Desf. Fl. Atlant. 1: 66. 1798.

Spikelets 1-flowered, the pedicel disarticulating a short distance below the glumes, leaving a short pointed callus attached, the rachilla not prolonged behind the palea; glumes equal, entire or 2-lobed, awned from the tip or from between the lobes, the awn slender, straight; lemma much shorter than the glumes, hyaline, bearing in our species a slender straight awn shorter than the awns of the glumes (awnless in *P. maritimus*). Annual or perennial usually decumbent grasses with flat blades and dense bristly spikelike panicles. [Greek, many beards.]

Species about 10 in the temperate regions of the world. Type species, *Alopecurus monspeliensis* L.

Plant perennial; awn of glumes 1–3 mm. long. — 1. *P. lutosus.*
Plant annual; awn of glumes 6–10 mm. long.
 Glumes slightly lobed, the lobes not ciliate. — 2. *P. monspeliensis.*
 Glumes prominently lobed, the lobes ciliate-fringed. — 3. *P. maritimus.*

1. Polypogon lutòsus (Poir.) Hitchc.
Beard-grass. Fig. 324.

Agrostis littoralis With. Bot. Arr. Veg. Brit. ed. 3. 2: 129. *pl. 23.* 1796, not Lam. 1791.
Polypogon littoralis J. E. Smith, Comp. Fl. Brit. 13. 1800.
Agrostis lutosa Poir. in Lam. Encycl. Suppl. 1: 249. 1810.
Polypogon lutosus Hitchc. U. S. Dept. Agr. Bull. 772: 138. 1920.

Perennial; culms geniculate at base, 30–80 cm. tall; sheaths scabrous; ligule 2–4 mm. long or the uppermost longer; panicles oblong, 5–15 cm. long, more or less interrupted or lobed; glumes equal, scabrous on back and keel, 2–3 mm. long, terminated by an awn as long; lemma smooth and shining, 1 mm. long, minutely toothed at the truncate apex; awn about as long as the glumes.

Introduced from Europe, from Vancouver Island to New Mexico. In California, in waste places, especially along irrigating ditches at moderate altitudes. June–Aug. Type locality, European.

2. Polypogon monspeliénsis (L.) Desf.
Annual Beard-grass, Rabbit's Foot. Fig. 325.

Alopecurus monspeliensis L. Sp. Pl. 61. 1753.
Polypogon monspeliensis Desf. Fl. Atlant. 1: 67. 1798.

Annual; culms erect or decumbent at base, scabrous below panicle, depauperate or as much as 1 meter tall; sheaths smooth, the ligule large; panicles dense and spike-like, 2–15 cm. long, 1–2 cm. wide, tawny-yellow when mature; glumes hispidulous, 2 mm. long, terminating in an awn 6–8 mm. long, or rarely longer; lemma as in *P. lutosus.*

Introduced from Europe; common in waste places and along irrigating ditches at moderate altitudes, Alaska to Mexico; occasional in the Atlantic States. June–Aug. Type locality, European.

3. Polypogon marítimus Willd.
Maritime Beard-grass. Fig. 326.

Polypogon maritimus Willd. Neue Schrift. Ges. Naturf. Freund. Berlin Mag. 3: 443. 1801.

Annual; culms 20–30 cm. tall, upright or spreading; sheaths smooth; ligule as much as 6 mm. long; blades usually less than 5 cm. long, 2–4 mm. wide; panicle cylindric, scarcely lobed, 5 cm. long or less, 1 cm. thick; glumes lanceolate, about 3 mm. long, villous, deeply 2-lobed, the lobes ciliate-fringed, the awn as much as 7 mm. long.

New York Falls, Amador County, California (*Hansen*). Introduced from Europe, the type locality. June.

27. CÍNNA L. Sp. Pl. 5. 1753.

Spikelets 1-flowered, disarticulating below the glumes, the rachilla forming a stipe below the floret and produced behind the palea as a minute bristle; glumes equal, 1-nerved; lemma similar to the glumes, nearly as long, 3-nerved, short-awned just below the tip; palea apparently 1-nerved, 1-keeled. Tall perennial grasses with flat blades and paniculate inflorescence. [A Greek name for a kind of grass.]

Species 3, North America and northern Eurasia, two in the United States and one in Mexico and Central America. Type species, *Cinna arundinacea* L.

1. Cinna latifòlia (Trevir.) Griseb.

Slender Wood or Sweet Reed-grass. Fig. 327.

Agrostis latifolia Trevir.; Goepp. Beschr. Bot. Gaertn. in Breslau 82. 1830.

Cinna pendula Trin. Mém. Acad. St. Pétersb. VI. Sci. Nat. 4[1]: 280. 1841.

Cinna latifolia Griseb. in Ledeb. Fl. Ross. 4: 435. 1853.

Cinna arundinacea pendula A. Gray, Man. ed. 2. 545. 1856.

Cinna bolanderi Scribn. Proc. Acad. Phila. 1884: 290. 1884.

Culms .5–1.5 meters tall; blades 10–14 mm. wide; panicle 15–30 cm. long, the flexuous capillary branches spreading or drooping; glumes about equal, scabrous, 4 mm. long; lemma about equaling the glumes; palea 2-nerved, the nerves close together.

In moist places in woods and along streams, Transition and Canadian Zones; in the cooler regions of North America and Eurasia, extending south in the mountains to Sequoia National Park, to Colorado and to North Carolina. July–Aug. Type locality, European.

28. AGRÓSTIS L. Sp. Pl. 61. 1753.

Spikelets 1-flowered, disarticulating above the glumes, the rachilla usually not prolonged; glumes equal or nearly so, acute, acuminate, or sometimes awn-pointed, carinate, usually scabrous on the keel and sometimes on the back; lemma obtuse, usually shorter than the glumes, thinner in texture than these, awnless or dorsally awned, often hairy at base; palea usually shorter than the lemma, rarely 2-keeled, usually small and nerveless or obsolete. Annual or usually perennial delicate or rather tall grasses with glabrous culms, flat or sometimes involute, scabrous blades and open or contracted panicles of small spikelets. [A Greek name for a forage grass.]

Species about 100 in the temperate and cold regions of the world, especially in the northern hemisphere. Type species, *Agrostis stolonifera* L.

Lemma pubescent. — 1. *A. retrofracta.*
Lemma glabrous.
  Rachilla prolonged behind the palea.
    Spikelets 3 mm. long, usually purple. — 2. *A. aequivalvis.*
    Spikelets 2 mm. long, usually pale. — 3. *A. thurberiana.*
  Rachilla not prolonged.
    Palea evident, 2-nerved.
      Panicle contracted and lobed or verticillate; glumes scabrous on keel and back. — 4. *A. verticillata.*
      Panicle open or if contracted not lobed or verticillate; glumes smooth on the back, more or less scabrous on the keel.
        Plant tufted; a dwarf alpine species. — 5. *A. humilis.*
        Plant with rhizomes or stolons; tall species of lower altitudes.
          Panicles spreading; culms erect or somewhat decumbent at base, with rhizomes but with no conspicuous stolons. — 6. *A. palustris.*
          Panicles narrow, contracted; culms long-decumbent or producing from base conspicuous stolons. — 7. *A. maritima.*
    Palea wanting, or a small nerveless scale.
      Lemma provided with a slender awn about 5 mm. long; a delicate annual. — 8. *A. exigua.*
      Lemma awnless or short-awned; perennials.
        Plants spreading by rhizomes (see also *A. lepida*).
          Tuft of hairs at base of lemma 1–2 mm. long. — 9. *A. hallii.*

Tuft of hairs minute or wanting.
Panicle spikelike. 10. *A. pallens.*
Panicle open. 11. *A. diegoensis.*
Plants tufted, not producing rhizomes or only short ones.
Panicle narrow, usually a part of the lower branches spikelet-bearing from the base.
Panicle strict, the branches short and appressed; plant low and cespitose. 12. *A. breviculmis.*
Panicle narrow or somewhat open, not strict.
Panicle short, 2–4 cm. long; a dwarf plant of high altitudes; lemma awnless. 13. *A. rossae.*
Panicle usually elongate; a usually taller plant of low or intermediate altitudes; lemma awned or awnless. 14. *A. exarata.*
Panicle open, sometimes diffusely spreading; usually no short branches in the lower whorls of branches.
Awn attached near base of lemma; panicle diffuse. 15. *A. howellii.*
Awn, if present, attached at or above the middle of the lemma.
Panicle very diffuse, the branches capillary, spikelet-bearing toward the tips. 16. *A. hiemalis.*
Panicle open but not diffuse.
Lemma awned. 17. *A. longiligula.*
Lemma awnless.
Plants producing short rhizomes; leaves mostly basal. 18. *A. lepida.*
Plants not producing rhizomes.
Plants delicate, 10–30 cm. tall. 19. *A. idahoensis.*
Plants slender, usually over 40 cm. tall. 20. *A. oregonensis.*

1. Agrostis retrofrácta Willd.

Hairy-flowered Bent-grass. Fig. 328.

Agrostis retrofracta Willd. Enum. Pl. **1**: 94. 1809.
Agrostis forsteri Roem. & Schult. Syst. Veg. **2**: 359. 1817.
Calamagrostis forsteri Steud. Nom. Bot. ed. 2. **1**: 250. 1840.
Calamagrostis retrofracta Link; Steud. Nom. Bot. ed. 2. **1**: 251. 1840.

Perennial; culms tufted, erect or decumbent at base, 20–60 cm. tall; sheaths smooth; ligule of culm leaves 3–5 mm. long; blades flat, scabrous, 1–2 mm. wide; panicle diffuse, 15–30 cm. long, the branches in distant whorls, capillary, reflexed at maturity, divided above the middle; glumes acuminate, 3 mm. long; lemma shorter than the glumes, thin, pubescent, short-bearded on the callus, bearing about the middle of the back a slender geniculate and twisted awn exserted about the length of the glumes; palea narrow, nearly as long as the lemma; rudiment slender, pilose, as long as the lemma.

Introduced 15 miles south of Stockton, California (*Ball*). A common Australasian species. Type locality: Australia.

2. Agrostis aequiválvis Trin.

Northern Bent-grass. Fig. 329.

Agrostis canina aequivalvis Trin. in Bong. Mém. Acad. St. Pétersb. VI. Math. Phys. Nat. **2**: 171. 1832.

Agrostis aequivalvis Trin. Mém. Acad. St. Pétersb. VI. Sci. Nat. **4**[1]: 362. 1841.

Deyeuxia aequivalvis Benth. Journ. Linn. Soc. Bot. **19**: 91. 1881.

Culms tufted, slender, smooth, 30–60 cm. tall; ligule 2 mm. long; blades narrow, upright, those of the culm 1 or 2, appressed, 4–6 cm. long, 1 mm. wide, somewhat scabrous; panicle open, 5–15 cm. long, the branches slender, somewhat scabrous; spikelets about 3 mm. long, usually purplish; glumes equal, sharp-pointed, minutely scabrous below the tip of the keel; lemma obtuse, about as long as the glumes; palea nearly as long as the glumes; rudiment minutely pubescent, one-fifth to one-half the length of the spikelet.

Wet meadows and moist places in the Hudsonian Zone; Alaska to Oregon, rare within our limits. July. Type locality: Sitka.

3. Agrostis thurberiàna Hitchc.
Thurber's Bent-grass. Fig. 330.

Agrostis thurberiana Hitchc. U. S. Dept. Agr. Bur. Pl. Ind. Bull. **68**: 23. *pl. 1. f. 1.* 1905.

Culms slender, in small tufts, erect, 20–40 cm. tall; panicle rather narrow, lax, more or less drooping, 5–7 cm. long; leaves somewhat crowded at base, the blades about 2 mm. wide; spikelets green or pale, rarely purple, 2 mm. long; lemma nearly as long as glumes, the palea about two-thirds as long; rachilla prolonged behind the palea as a minutely hairy pedicel, .3 mm. long.

Bogs and moist places in the Hudsonian Zone; British Columbia to Montana, and south to Sequoia National Park. July–Aug. Type locality: Skamania County, Washington.

4. Agrostis verticillàta Vill.
Water Bent-grass. Fig. 331.

Agrostis stolonifera L. Sp. Pl. 62. 1753, in part, but not as to the Swedish plant.
Agrostis verticillata Vill. Prosp. Pl. Dauph. 16. 1779.

Culms usually decumbent at base, sometimes with long creeping and rooting stolons; panicle contracted, lobed or verticillate, especially at base, 3–10 cm. long, light green or rarely purplish, the branches spikelet-bearing from the base; glumes equal, obtuse, scabrous on back and keel, 2 mm. long; lemma 1 mm. long, awnless, truncate and toothed at apex; palea nearly as long as the lemma. Resembles in habit *Polypogon lutosus*, which differs in having awned glumes and lemma.

Moist ground, especially along irrigation ditches, in the Sonoran Zone; Central California to Texas and Mexico; also at Linnton, Oregon (*Nelson*); introduced from Europe. May–June. Type locality, European.

5. Agrostis hùmilis Vasey.
Mountain Bent-grass. Fig. 332.

Agrostis humilis Vasey, Bull. Torrey Club **10**: 21. 1883.

Culms slender, tufted, erect, 5–30 cm. tall; leaves mostly basal, the blades flat, thin, 2–10 cm. long, as much as 2 mm. wide; panicle 2–8 cm. long, rather open or in dwarf plants narrow; spikelets usually purple, 1.5–2 mm. long; glumes equal, rather broad, abruptly pointed, smooth or slightly scabrous on the upper part of the keel; lemma awnless, nearly as long as the glumes; palea usually about two-thirds as long as the lemma.

Wet places in mountain meadows in the Alpine Zone; British Columbia to Oregon and Colorado. July–Aug. Type locality: Mount Adams.

6. Agrostis palùstris Huds.

Redtop. Fig. 333.

Agrostis palustris Huds. Fl. Angl. 27. 1762.
Agrostis alba of authors. *A. alba* L. is doubtful.

Culms erect or somewhat decumbent at base, 30–100 cm. tall; ligule mostly 2-5 mm. long, scabrous; panicle loose but not diffuse, 5–30 cm. long, the lower branches in whorls; glumes acute, 2–3 mm. long, scabrous on keel but not on back; lemma a little shorter than the glumes, obtuse, rarely awned on back; palea half to two-thirds as long as lemma.

Cultivated as a meadow grass and frequently escaped along roadsides and in waste places, at low and intermediate altitudes throughout the United States; apparently not native in America. July–Aug. Type locality: England.

7. Agrostis marítima Lam.

Maritime Bent-grass. Fig. 334.

Agrostis maritima Lam. Encycl. 1: 61. 1783.
Agrostis alba maritima Meyer, Chloris Hanov. 656. 1836.
Agrostis depressa Vasey, Bull. Torrey Club 13: 54. 1886.
Agrostis exarata stolonifera Vasey, Bull. Torrey Club 13: 54. 1886.

Differs from *Agrostis palustris* chiefly in habit; culms decumbent at base, often extensively creeping, rooting at the nodes, and with short spreading blades; flowering culms ascending, the blades longer; panicles narrow, the short branches ascending.

Along the coast of Europe and North America; Pacific Coast from British Columbia south to Sonoma County, California. Upon the moist sand dunes the stolons are conspicuous. July–Aug. Type locality, European.

8. Agrostis exígua Thurb.

Sierra Bent-grass. Fig. 335.

Agrostis exigua Thurb. in S. Wats. Bot. Calif. 2: 275. 1880.

Annual; culms delicate, 3–10 cm. tall, branching from the base; blades 5–20 mm. long, sub-involute, scabrous; panicle half the length of plant, finally open; glumes 1.5 mm. long, scaberulous; lemma equaling the glumes, scaberulous toward the summit, bearing below the tip a delicate bent awn 4 times as long; palea wanting.

Only known from two collections: "Foothills of the Sierras" (*Bolander*) (the type), and Howell Mountain, Napa County (*Tracy*). May.

9. **Agrostis hàllii** Vasey.

Hall's Bent-grass. Fig. 336.

Agrostis hallii Vasey, Contr. U. S. Nat. Herb. **3**: 74. 1892.
Agrostis davyi Scribn. U. S. Dept. Agr. Div. Agrost. Circ. **30**: 3. 1901.
Agrostis occidentalis Scribn. & Merr. Bull. Torrey Club **29**: 466. 1902.

Culms erect, stout, 60–90 cm. tall, bearing rhizomes; ligule usually conspicuous; panicle 10–12 cm. long, narrow, open; glumes about 4 mm. long; lemma awnless, 3 mm. long, with a tuft of hairs at base about half as long; palea wanting.

Mostly in woods near the coast from Oregon to Santa Barbara. June–July. Type locality: Oregon.

Agrostis hallii prínglei (Scribn.) Hitchc. U. S. Dept. Agr. Bur. Pl. Ind. Bull. **68**: 33. *pl. 12.* 1905. (*Agrostis pringlei* Scribn. U. S. Dept. Agr. Div. Agrost. Bull. **7**: 156. *f. 138.* 1897). Differs from *A. hallii* in the narrower and more compact panicles, narrower and more involute blades and the more stramineous appearance of the whole foliage.

Near the coast, Mendocino County, California, the type locality. June–July.

10. **Agrostis pállens** Trin.

Seashore Bent-grass. Fig. 337.

Agrostis pallens Trin. Mém. Acad. St. Pétersb. VI. Sci. Nat. **4**[1]: 328. 1841.
Agrostis exarata littoralis Vasey, Bull. Torrey Club **13**: 54. 1886.

Culms 20–40 cm. tall, erect, from creeping rhizomes; panicle contracted, almost spikelike, 5–10 cm. long; glumes 2.5–3 mm. long; lemma a little shorter than the glumes, awnless, the hairs at base minute; palea wanting.

Sandy soil near the coast, Washington to Marin County, California. June–Aug. Type locality: "Amer. borealis (*Hooker*)."

11. **Agrostis diegoénsis** Vasey.

Leafy Bent-grass. Fig. 338.

Agrostis foliosa Vasey, Bull. Torrey Club **13**: 55. 1886, not Roem. & Schult. 1817.
Agrostis diegoensis Vasey, Bull. Torrey Club **13**: 55. 1886.
Agrostis pallens foliosa Hitchc. U. S. Dept. Agr. Bur. Pl. Ind. Bull. **68**: 34. *pl. 14. f. 1.* 1905.

Differs from *A. pallens* in the taller culms, and more open panicles, the branches rather stiffly ascending; lemma awnless or with a straight or rarely a bent awn.

Meadows and open woods, in the Transition Zone; British Columbia to southern California. July–Aug. Type locality: San Diego.

12. Agrostis brevicùlmis Hitchc.

Dwarf Bent-grass. Fig. 339.

Trichodium nanum Presl, Rel. Haenk. 1: 243. 1830.
Agrostis nana Kunth, Rev. Gram. 1: Suppl. XVIII. 1830, not Delarbre. 1800.
Agrostis breviculmis Hitchc. U. S. Dept. Agr. Bur. Pl. Ind. Bull. **68**: 36. *pl. 18.* 1905.

Culms cespitose, erect, 10–15 cm. tall; panicle strict and narrow, 2–3 cm. long; glumes 3 mm. long, acute; lemma a little more than half as long as glumes, awnless or bearing a very short awn near the middle of the back; palea a minute nerveless scale.

Open ground and clay cliffs, Fort Bragg, California; also in Peru, the type locality.

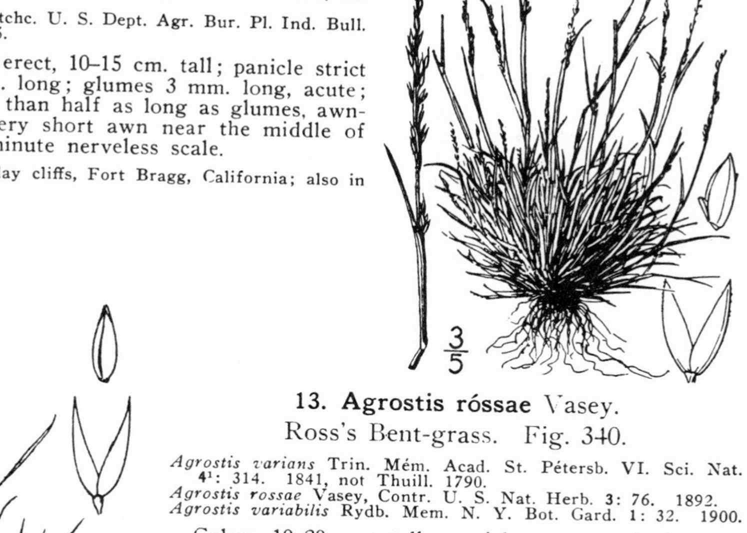

13. Agrostis róssae Vasey.

Ross's Bent-grass. Fig. 340.

Agrostis varians Trin. Mém. Acad. St. Pétersb. VI. Sci. Nat. **4**[1]: 314. 1841, not Thuill. 1790.
Agrostis rossae Vasey, Contr. U. S. Nat. Herb. **3**: 76. 1892.
Agrostis variabilis Rydb. Mem. N. Y. Bot. Gard. **1**: 32. 1900.

Culms 10–20 cm. tall; panicle contracted, 2–6 cm. long, the branches appressed; spikelets green or purple, 2 mm. long; lemma 1.5 mm. long, awnless; palea minute. Differs from *A. exarata* chiefly in the size, possibly an alpine form of that species; glumes not scabrous on the back as usual in *A. exarata*.

Rocky creeks and mountain slopes of the Alpine Zone; Washington to Sequoia National Park and east to Wyoming. Aug.–Sept. Type locality: Yellowstone Park.

14. Agrostis exaràta Trin.

Western Bent-grass. Fig. 341.

Agrostis exarata Trin. Gram. Unifl. 207. 1824.
Vilfa glomerata Presl, Rel. Haenk. **1**: 239. 1830.
Agrostis glomerata Kunth, Rev. Gram. **1**: Suppl. XVII. 1830.
Agrostis grandis Trin. Mém. Acad. St. Pétersb. VI. Sci. Nat. **4**[1]: 316. 1841.
Agrostis asperifolia Trin. Mém. Acad. St. Pétersb. VI. Sci. Nat. **4**[1]: 317. 1841.
Agrostis scouleri Trin. Mém. Acad. St. Pétersb. VI. Sci. Nat. **4**[1]: 329. 1841.
Agrostis californica Trin. Mém. Acad. St. Pétersb. VI. Sci. Nat. **4**[1]: 359. 1841.
Agrostis microphylla Steud. Syn. Pl. Glum. **1**: 164. 1854.
Polypogon alopecuroides Buckl. Proc. Acad. Phila. **1862**: 88. 1863.
Agrostis albicans Buckl. Proc. Acad. Phila. **1862**: 91. 1863.
Agrostis densiflora Vasey, Contr. U. S. Nat. Herb. **3**: 72. 1892.
Agrostis densiflora arenaria Vasey, Contr. U. S. Nat. Herb. **3**: 58, 72. 1892.
Agrostis microphylla major Vasey, Contr. U. S. Nat. Herb. **3**: 58, 72. 1892.
Agrostis inflata Scribn.; Canad. Record. Sci. **6**: 152. 1894.
Agrostis ampla Hitchc. U. S. Dept. Agr. Bur. Pl. Ind. Bull. **68**: 38. *pl. 20.* 1905.

Culms 20–120 cm. tall, smooth or scabrous below the panicle; sheaths smooth or somewhat scabrous; ligule prominent; blades flat, 1–8 mm. wide; panicle narrow, somewhat open, or close and spikelike, sometimes interrupted, 1–25 cm. long; glumes acute or awn-pointed, nearly equal, 2.5–3.5 mm. long, scabrous on the keel and often also on the back; lemma about 2 mm. long; awnless or bearing from about the middle of the back a straight or bent exserted awn; palea minute, less than .5 mm.

Moist or rather dry open ground, from the seacoast up to the Hudsonian Zone; Alaska to Mexico, east to the Rocky Mountains. June–Aug. Type locality: Alaska.

A. glomeràta is a seacoast form with dense panicles, inflated sheaths, and awned or awnless lemma; *A. microphylla* is a form with dense, often interrupted panicles and awned lemma. *A. ampla* has a somewhat open panicle and awned lemma.

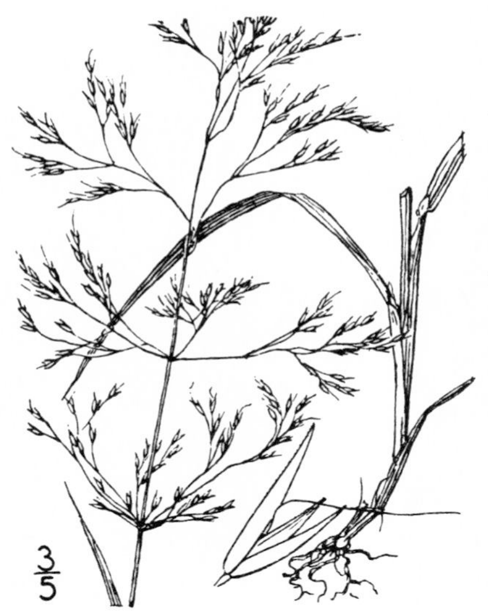

15. Agrostis howéllii Scribn.

Howell's Bent-grass. Fig. 342.

Agrostis howellii Scribn. Contr. U. S. Nat. Herb. **3**: 76. 1892.

Culms erect or decumbent at base, 40–60 cm. tall; blades lax, as much as 30 cm. long, 3–5 mm. wide; panicle loose and spreading, scabrous, 10–30 cm. long; spikelets pale, clustered toward the ends of the branches; glumes unequal, acuminate, rather narrow and firm in texture, somewhat scabrous on the keel, the first 3 mm. long, the second a little shorter lemma acute, 2.5 mm. long, bearing from near the base an exserted bent awn about 6 mm. long; palea none.

Only known from Oregon (Hood River, Bridal Veil). July.

16. Agrostis hiemàlis (Walt.) B. S. P.

Rough Hair-grass, Tickle-grass. Fig. 343.

Cornucopiae hyemalis Walt. Fl. Carol. 73. 1788.
Agrostis scabra Willd. Sp. Pl. **1**: 370. 1797.
Agrostis scabriuscula Buckl. Proc. Acad. Phila. **1862**: 90. 1863.
Agrostis hyemalis B. S. P. Prel. Cat. N. Y. 68. 1888.

Culms slender, tufted or scattered, 20–80 cm. tall, but usually delicate; leaves usually mostly basal, the blades narrow or almost setaceous; panicle very diffuse and open, as much as 30 cm. long, the branches scabrous, long and capillary, bearing spikelets near the extremities; glumes 1.5–2 mm. long, acute or acuminate; lemma two-thirds to three-fourths as long as glumes, awnless or rarely awned, palea wanting.

Meadows and moist places in the Transition and Canadian Zones; Washington to California and nearly throughout North America. July–Sept. Type locality: South Carolina.

Agrostis hiemalis geminàta (Trin.) Hitchc. U. S. Dept. Agr. Bur. Pl. Ind. Bull. **68**: 44. *pl. 28. f. 1.* 1905. (*Agrostis geminata* Trin. Gram. Unifl. 207. 1824.) Differs from *A. hiemalis* in having a smaller, less diffuse panicle, with divaricate branches; culms usually less than 30 cm. high; lemma awnless within our range.

Alpine Zone; Alaska to California. Type locality: Alaska.

17. Agrostis longilígula Hitchc.

Pacific Bent-grass. Fig. 344.

Agrostis longiligula Hitchc. U. S. Dept. Agr. Bur. Pl. Ind. Bull. **68**: 54. 1905.

Culms erect, about 60 cm. tall; ligule 5–6 mm. long; blades 10–15 cm. long, 3–4 mm. wide, scabrous; panicle narrow, but loosely flowered, bronze-purple, 10–15 cm. long, the branches very scabrous; glumes 4 mm. long; lemma 2.5 mm. long, bearing at the middle a bent exserted awn; palea minute.

Bogs and marshes near the coast, Alaska to Mendocino County, California. July–Sept. Type locality: Fort Bragg, California.

18. **Agrostis lépida** Hitchc.
Sequoia Bent-grass. Fig. 345.

Agrostis lepida Hitchc. in Jepson, Fl. Calif. **1**: 121. 1912.

Culms tufted, 30–40 cm. tall, erect, producing numerous short rhizomes; ligule long, especially on the innovations, these as much as 4 mm. long, narrow, pointed; blades mostly basal, firm, erect, flat or folded, the upper culm leaf below the middle of the culm, the blade short, 3 cm. long or less; panicle purple, erect, 10–15 cm. long, the branches in verticils stiff and straight, becoming divaricately spreading, the lowermost 2–5 cm. long, spikelet-bearing from about the middle, some short branches intermixed; glumes 3 mm. long, smooth or nearly so; lemma 2 mm. long, awnless; palea wanting or a minute nerveless scale.

Meadows and open woods Sequoia National Park, at about 10,000 feet altitude. Aug.–Sept. Type locality: Siberian Pass.

19. **Agrostis idahoénsis** Nash.
Idaho Bent-grass. Fig. 346.

Agrostis tenuis Vasey, Bull. Torrey Club **10**: 21. 1883, not Sibth. 1794.
Agrostis idahoensis Nash, Bull. Torrey Club **24**: 42. 1897.
Agrostis tenuiculmis Nash in Rydb. Mem. N. Y. Bot. Gard. **1**: 32. 1900.
Agrostis tenuiculmis erecta Nash in Rydb. Mem. N. Y. Bot. Gard. **1**: 32. 1900.

Culms slender, 10–30 cm. tall; panicle loosely spreading, 5–10 cm. long, the branches capillary, minutely scabrous; spikelets 1.5 mm. long; lemma 1 mm. long, awnless; palea minute. Differs from *A. hiemalis* in the narrow panicle with shorter branches.

Mountain meadows, Hudsonian and Canadian Zones; Washington to southern California, east to Colorado. July–Sept. Type locality: Idaho.

20. **Agrostis oregonénsis** Vasey.
Oregon Bent-grass. Fig. 347.

Agrostis oregonensis Vasey, Bull. Torrey Club **13**: 55. 1886.
Agrostis attenuata Vasey, Bot. Gaz. **11**: 337. 1886.
Agrostis hallii californica Vasey, Contr. U. S. Nat. Herb. **3**: 74. 1892.

Culms 60–90 cm. tall; panicle oblong, 10–30 cm. long, open, the branches verticillate, rather stiff and ascending, the lower whorls numerous, the longer 5–10 cm. long, branching above the middle; glumes 2.5–3 mm. long; lemma 1.5 mm. long, awnless; palea about .5 mm. long.

Marshes, bogs, and wet meadows, Transition Zone; British Columbia to California. July–Aug. Type locality: Oregon. This species has been incorrectly referred to *A. shiedeana* Trin.

Agrostis spica-vénti L. (*Apera spica-venti* Beauv.), an annual with many-flowered panicle, the lemma with a long slender subterminal awn, and the rachilla prolonged, has been found on ballast at Portland.

29. GASTRÍDIUM Beauv. Ess. Agrost. 21. *pl 6. f. 6.* 1812.

Spikelets 1-flowered, the rachilla disarticulating above the glumes, prolonged behind the palea as a minute bristle; glumes unequal, somewhat enlarged or swollen at the base; lemma much shorter than the glumes, hyaline, broad, truncate, awned or awnless; palea about as long as the lemma. Annual grasses with flat blades and pale shining spikelike panicles. [Greek, belly, referring to the saccate glumes.]

Species 2, in the Mediterranean region, one introduced into the United States. Type species, *Milium lendigerum* L.

1. Gastridium ventricòsum (Gouan) Schinz & Thell.

Nit-grass. Fig. 348.

Agrostis ventricosa Gouan, Hort. Monsp. 39. *pl. 1. f. 2.* 1762.
Milium lendigerum L. Sp. Pl. ed. 2. 91. 1762.
Gastridium australe Beauv. Ess. Agrost. 21, 164. 1812.
Gastridium lendigerum Desv. Obs. Angers. 48. 1918.
Gastridium ventricosum Schinz & Thell. Mitt. Bot. Mus. Univ. Zurich **58**: 39. 1913.

Culms about 30 cm. tall, smooth; panicle 5–8 cm. long, dense and spikelike; glumes 3 mm. long, gradually narrowed into an awn-point; lemma much shorter than the glumes, globular, pubescent, the awn 5 mm. long, geniculate.

Open ground and waste places, Transition Zone; mostly in the Coast Ranges, at moderate altitudes. Oregon to California and Texas; introduced from Europe. May–July. Type locality, European.

30. CALAMAGRÓSTIS Ádans. Fam. Pl. 2: 31. 530. 1763.

Spikelets 1-flowered, the rachilla disarticulating above the glumes, usually prolonged behind the palea as a short usually hairy bristle; glumes about equal, acute or acuminate; lemma shorter than the glumes, usually more delicate than these, glabrous in our native species, surrounded at base with a tuft of hairs, these usually copious, often as long as the lemma, awned from the back, usually below the middle, the awn delicate and straight, or stouter and exserted, bent and sometimes twisted; palea shorter than the lemma. Perennial, usually moderately tall or robust grasses with small spikelets in open or usually narrow sometimes spikelike panicles. [Greek, reed-agrostis.]

Species about 100 in the cool and temperate regions of both hemispheres. Type species, *Arundo calamagrostis* L.

Awn longer than the glumes, geniculate.
  Panicle open, the branches spreading, naked below.
    Blades scattered, broad, flat; plant mostly over 1 meter tall. — 1. *C. bolanderi.*
    Blades mostly basal, narrow, mostly not over 2 mm. wide, often involute.
      Awn about 1 cm. long, much longer than the glumes; blades nearly or quite as long as the flowering culms. — 2. *C. howellii.*
      Awn only a little exceeding the glumes; blades much shorter than the culms, capillary. — 3. *C. breweri.*
  Panicle compact, the branches appressed, floriferous from the base.
    Blades scattered, broad, flat, 6–10 mm. wide. — 4. *C. tweedyi.*
    Blades mostly basal, firm, narrow, becoming involute.
      Glumes about 1 cm. long, gradually long-acuminate; awn about 1 cm. long above the bend. — 5. *C. foliosa.*
      Glumes about 5 mm. long, abruptly acute or acuminate; awn usually less than 5 mm. above the bend.
        Marcescent lower sheaths prominent, extending up the culm for several centimeters; panicle compact. — 6. *C. purpurascens.*
        Marcescent sheaths inconspicuous; panicle rather loose, interrupted below. — 7. *C. vaseyi.*
Awn included or scarcely longer than the glumes, straight or geniculate.
  Awn geniculate, protruding sidewise from the glumes; callus hairs sparse, shorter than the lemma.
    Sheaths glabrous on the collar.
      Panicle loose, the branches spreading or ascending; plants tall, in large bunches; blades as much as 1 cm. wide. — 8. *C. nutkaensis.*
      Panicle compact.
        Culms stout, mostly over 1 meter tall. — 9. *C. densa.*
        Culms slender, mostly less than 1 meter tall. — 10. *C. koelerioides.*
    Sheaths pubescent on the collar. — 11. *C. rubescens.*
  Awn straight, included; callus hairs usually not much shorter than the glumes.
    Panicle rather loose and open.
      Callus hairs about half as long as the lemma. — 12. *C. californica.*
      Callus hairs about as long as the lemma.
        Spikelets about 2 mm. long. — 13. *C. macouniana.*
        Spikelets 3–5 mm. long. — 14. *C. canadensis.*
    Panicle contracted or spikelike.
      Blades short, as much as 5 mm. wide, nearly smooth. — 15. *C. crassiglumis.*
      Blades elongate, firm and rather rigid, scabrous. — 16. *C. inexpansa.*

1. **Calamagrostis bolánderi** Thurb.

Bolander's Reed-grass. Fig. 349.

Calamagrostis bolanderi Thurb. in S. Wats. Bot. Calif. **2**: 280. 1880.

Deyeuxia bolanderi Vasey, Grasses U. S. 28. 1883.

Culms 1–1.5 meters tall; sheaths scabrous; blades flat, scattered, nearly smooth; panicle open, 10–20 cm. long, the branches verticillate, spreading, naked below, the longer 5–10 cm. long; glumes 3–4 mm. long, purple, scabrous, acute; lemma very scabrous, about as long as glumes, the awn from near the base, geniculate, exserted, about 2 mm. long above the bend; rudiment pilose, 2 mm. long.

Bogs and moist ground, prairie or open woods, near the coast, Mendocino and Humboldt Counties, California. July–Aug. Type locality: Noyo River.

2. **Calamagrostis howéllii** Vasey.

Howell's Reed-grass. Fig. 350.

Calamagrostis howellii Vasey, Bot. Gaz. **6**: 271. 1881.

Culms cespitose, rather slender, 30–60 cm. tall; sheaths smooth or slightly scabrous; ligule 2–8 mm. long; blades slender, scabrous on the upper surface, flat or soon involute, especially toward the tip, about as long as the culms, the 2 cauline shorter, about 1 mm. wide; panicle pyramidal, 5–15 cm. long, rather open, the lower branches in whorls, ascending, naked below, 3–5 cm. long; spikelets pale or tinged with purple; glumes acuminate, 7–8 mm. long; lemma acuminate, a little shorter than the glumes, the callus hairs about 3 mm. long, the awn attached about 2 mm. above the base, geniculate, the exserted portion about 7 mm. long.

On perpendicular cliffs of the Columbia Gap. June–July. Type locality: "Oregon."

3. **Calamagrostis brèweri** Thurb.

Brewer's Reed-grass. Fig. 351.

Calamagrostis breweri Thurb. in S. Wats. Bot. Calif. **2**: 280. 1880.

Deyeuxia breweri Vasey, Grasses U. S. 28. 1883.

Calamagrostis lemmoni Kearn. U. S. Dept. Agr. Div. Agrost. Bull. **11**: 16. 1898.

Culms slender, erect, cespitose, 15–30 cm. tall; leaves mostly basal, usually involute-filiform; panicle ovate, open, purple, 3–8 cm. long, the lower branches slender, spreading, few-flowered, 1–2 cm. long; glumes 3–4 mm. long, 1-nerved, smooth, acute; lemma nearly as long as glumes, cuspidate-toothed, the awn borne near the base, geniculate, exserted, twisted below, about 2 mm. long above the bend; rudiment long-pilose, 2 mm. long.

Mountain meadows in the Hudsonian Zone of the high Sierra Nevada, California. Aug.–Sept. Type locality: Carson Pass.

4. **Calamagrostis tweèdyi** Scribn.
Tweedy's Reed-grass. Fig. 352.

Deyeuxia tweedyi Scribn. Bull. Torrey Club **10**: 64. 1883.
Calamagrostis tweedyi Scribn. Contr. U. S. Nat. Herb. **3**: 83. 1892.

Culms 1–1.5 meters tall, smooth, bearing short rhizomes; sheaths smooth, the lower becoming fibrose; blades flat, somewhat scabrous, the cauline 5–15 cm. long, as much as 1 cm. wide, those of the innovations narrower and longer; panicle oblong, rather compact, or interrupted below, about 10 cm. long; glumes abruptly acuminate, purplish tinged, about 7 mm. long; lemma about as long as the glumes, the awn exserted about 5 mm., the callus hairs scant, scarcely 1 mm. long; rudiment pilose, 3 mm. long.

Only known from the Cascade Mountains, Washington, the type locality.

5. **Calamagrostis foliòsa** Kearn.
Leafy Reed-grass. Fig. 353.

Calamagrostis sylvatica longifolia Vasey, Contr. U. S. Nat. Herb. **3**: 83. 1892.
Calamagrostis foliosa Kearn. U. S. Dept. Agr. Div. Agrost. Bull. **11**: 17. 1898.

Culms cespitose, erect, 30–60 cm. tall; leaves mostly basal, numerous, the blades involute, firm, smooth, nearly as long as culm; panicle dense, spikelike, 5–8 cm. long; glumes about 1 cm. long, acuminate; lemma 5–7 mm. long, acuminate, 4-nerved, the nerves ending in setaceous teeth; awn from near base, geniculate, about 8 mm. long above the bend; rudiment pilose, nearly as long as lemma; callus hairs numerous, 3 mm. long.

Humboldt and Mendocino Counties, California. August. Type locality: Humboldt County.

6. **Calamagrostis purpuráscens** R. Br.
Purple Reed-grass. Fig. 354.

Calamagrostis purpurascens R. Br. in Richards. Bot. App. Frankl. Journ. 731. 1823.
Calamagrostis sylvatica purpurascens Vasey, Contr. U. S. Nat. Herb. **3**: 83. 1892.

Culms cespitose, erect, 40–60 cm. tall; sheaths scabrous; blades flat or more or less involute, scabrous; panicle dense, spikelike, 5–12 cm. tall, pale or sometimes purple; glumes 6–8 mm. long, scabrous; lemma nearly as long as glumes, 4-nerved, 4-awned at apex, the dorsal awn from near base, finally geniculate, exserted about 2 mm.

Rocks and cliffs, Arctic regions to California and Colorado, Hudsonian and Canadian Zones. August. Type locality: British America.

7. Calamagrostis vàseyi Beal.
Vasey's Reed-grass. 355.

Calamagrostis vaseyi Beal, Grasses N. Am. **2**: 344. 1896.

Culms cespitose, smooth, 30–40 cm. tall; sheaths smooth or slightly scabrous; ligule 2–3 mm. long; blades flat, becoming involute at the slender point, scabrous on the upper surface, 1.5–2 mm. wide, 5–10 cm. long or the uppermost shorter; panicle narrow, long-exserted, spikelike or somewhat interrupted, 5–10 cm. long, purplish; glumes equal, acuminate, 6 mm. long; lemma about 5 mm. long, the awn attached near the base, bent, exserted about 5 mm., the callus hairs nearly half as long as the lemma; rudiment about 2 mm. long, long-villous.

Dry rocks, Arctic Zone; Washington and Oregon. July–Aug. Cascade Mountains of Washington.

8. Calamagrostis nutkaénsis (Presl) Steud.
Pacific Reed-grass. Fig. 356.

Deyeuxia nutkaensis Presl, Rel. Haenk. **1**: 250. 1830.
Calamagrostis aleutica Trin. in Bong. Mém. Acad. St. Pétersb. VI. Math. Phys. Nat. **2**: 171. 1832.
Calamagrostis nutkaensis Steud. Syn. Pl. Glum. **1**: 190. 1854.
Deyeuxia aleutica Munro; Hook. f. Trans. Linn. Soc. Bot. **23**: 345. 1862.
Calamagrostis albicans Buckl. Proc. Acad. Phila. **1862**: 92. 1863.
Calamagrostis pallida Nutt.; A. Gray, Proc. Acad. Phila. **1862**: 334. 1863.
Deyeuxia breviaristata Vasey, Bull. Torrey Club **15**: 48. 1888.

Culms stout, 1–1.5 meters tall; blades flat, becoming involute, elongate, gradually narrowed into a long point; panicle narrow, rather loose, 15–30 cm. long, the branches rather stiffly ascending; glumes 5–6 mm. long, acuminate; lemma 4 mm. long, indistinctly nerved, the callus hairs half as long; awn rather stout, attached below the middle, slightly geniculate, extending to summit of lemma.

Along the coast in moist soil or wet wooded hills, from Alaska to central California. July–Aug. Type locality: Nootka Sound.

9. Calamagrostis dénsa Vasey.
Dense Reed-grass. Fig. 357.

Calamagrostis densa Vasey, Bot. Gaz. **16**: 147. 1891.
Calamagrostis koelerioides densa Beal, Grasses N. Am. **2**: 345. 1896.
Calamagrostis vilfaeformis Kearn. U. S. Dept. Agr. Div. Agrost. Bull. **11**: 20. 1898.

Culms rather stout, smooth or scabrous just below the panicle, mostly over 1 meter tall; sheaths scabrous; ligule 3–5 mm. long; blades flat, scabrous, 15–25 cm. long, the uppermost shorter, 3–8 mm. wide; panicle spikelike, dense, pale, 10–15 cm. long; glumes acuminate, about 6 mm. long; lemma about 4 mm. long, the awn bent, about as long as the lemma, scarcely exserted at the side, the callus hairs scant, about 1 mm. long.

Dry hills, among shrubs, mountains east of San Diego, California. July. Type locality: Julian.

10. Calamagrostis koelerioìdes Vasey. Tufted Pine-grass. Fig. 358.

Calamagrostis koelerioides Vasey, Bot. Gaz. **16**: 147. 1891.

Culms erect, cespitose, rather slender, smooth, or scabrous just below the panicle, 40–80 cm. tall; sheaths smooth; ligule 2–3 mm. long, obtuse; blades flat, scabrous on both surfaces, 5–15 cm. long, 2–3 mm. wide; panicle spikelike, mostly purplish, 5–10 cm. long; glumes equal, acuminate, about 5 mm. long; lemma a little shorter than the glumes, the awn bent, shorter than the glumes but exserted at one side, the callus hairs rather scant, one-fourth to one-third as long as the lemma.

Dry hills and banks, Arid Transition Zone; Oregon to southern California. July–Aug. Type locality: Julian, California.

11. Calamagrostis rubéscens Buckl. Pine Grass. Fig. 359.

Calamagrostis rubescens Buckl. Proc. Acad. Phila. **1862**: 92. 1863.
Deyeuxia rubescens Vasey, Grasses U. S. 28. 1883.
Deyeuxia cusickii Vasey, Bot. Gaz. **10**: 224. 1885.
Deyeuxia suksdorfii Scribn. Bull. Torrey Club **15**: 9. *pl. 76.* 1888.
Calamagrostis aleutica angusta Vasey, Contr. U. S. Nat. Herb. **3**: 80. 1892.
Calamagrostis cusickii Vasey, Contr. U. S. Nat. Herb. **3**: 81. 1892.
Calamagrostis suksdorfii Scribn. in Vasey, Contr. U. S. Nat. Herb. **3**: 82. 1892.
Calamagrostis angusta Kearn. U. S. Dept. Agr. Div. Agrost. Bull. **11**: 21. 1898.
Calamagrostis subflexuosa Kearn. U. S. Dept. Agr. Div. Agrost. Bull. **11**: 22. 1898.
Calamagrostis fasciculata Kearn. U. S. Dept. Agr. Div. Agrost. Bull. **11**: 23. *f. 1.* 1898.
Calamagrostis suksdorfii luxurians Kearn. U. S. Dept. Agr. Div. Agrost. Bull. **11**: 24. 1898.

Culms slender, 60–100 cm. tall, from creeping rhizomes; sheaths smooth, but pubescent on the collar; blades flat or somewhat involute; panicle narrow, spikelike, pale or purple, 7–15 cm. long; glumes 4–5 mm. long, narrow, acuminate; lemma pale and thin, about as long as glumes, smooth, scarcely nerved, the callus hairs about one-third as long; awn attached near base, geniculate, exserted at side of glumes, the terminal portion about 1 mm. long.

Open pine woods, prairies, and banks in the Arid Transition Zone; British Columbia to central California, and east to Manitoba and Colorado. July–Aug. Type locality: Oregon.

12. Calamagrostis califórnica Kearn. California Reed-grass. Fig. 360.

Calamagrostis californica Kearn. U. S. Dept. Agr. Div. Agrost. Bull. **11**: 37. 1898.

Culms about 1 meter tall; blades flat, firm, rather rigid, 4 mm. wide, those of the innovations involute; panicle 20 cm. long, narrow, loose; glumes 3–4 mm. long, scabrous, acuminate; lemma shorter than the glumes, strongly nerved; callus hairs abundant, half as long as the lemma; awn delicate, straight, attached below the middle, extending to tip of lemma.

A little-known species resembling *C. canadensis* but having more rigid, firm blades, and callus hairs only half as long as lemma. The only specimens known are those of the type collection, from "Sierra Nevada Mountains," *Lemmon 444* in 1875.

13. Calamagrostis macouniàna Vasey.
Macoun's Reed-grass. Fig. 361.

Deyeuxia macouniana Vasey, Bot. Gaz. **10**: 297. 1885.
Calamagrostis macouniana Vasey, Contr. U. S. Nat. Herb. **3**: 81. 1892.

Culms slender, about 1 meter tall; sheaths smooth; ligule 2–3 mm. long; blades slender, flat, narrowed to a slender involute point, scabrous, 15–20 cm. long, or the uppermost shorter, about 3 mm. wide; panicle narrow, rather open, 10–15 cm. long; glumes broad, abruptly acuminate, 2 mm. long; lemma a little shorter than the glumes, bifid at apex, the lobes obtuse, the awn straight or a little bent, extending somewhat beyond the summit of the lemma, the callus hairs about as long as the lemma.

Marshes and wet places in the Arid Transition Zone; Alberta to Oregon, and east to the Rocky Mountains. July. Type locality: Assiniboia.

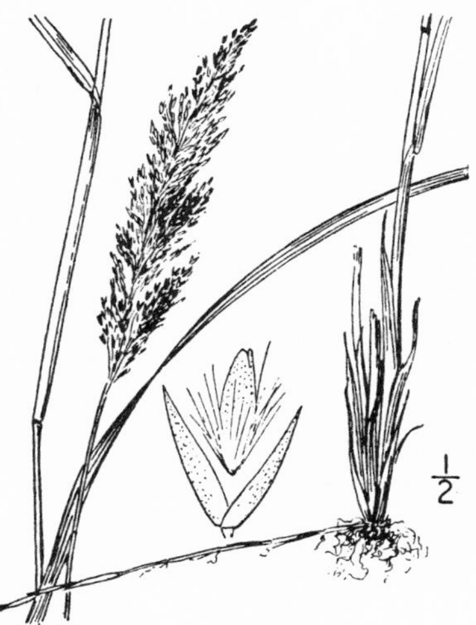

14. Calamagrostis canadénsis (Michx.) Beauv.
Blue-joint. Fig. 362.

Arundo canadensis Michx. Fl. Bor. Am. **1**: 73. 1803.
Calamagrostis canadensis Beauv. Ess. Agrost. 15, 157. 1812.
Calamagrostis oregonensis Buckl. Proc. Acad. Phila. **1862**: 92. 1863.
Calamagrostis columbiensis Nutt.; A. Gray, Proc. Acad. Phila. **1862**: 334. 1863.
Calamagrostis pallida Vasey & Scribn. Contr. U. S. Nat. Herb. **3**: 79. 1892.
Calamagrostis lactea Beal, Grasses N. Am. **2**: 346. 1896.
Deyeuxia lactea Suksdorf; Beal, Grasses N. Am. **2**: 346. 1896.
Calamagrostis blanda Beal, Grasses N. Am. **2**: 349. 1896.
Calamagrostis canadensis acuminata Vasey, U. S. Dept. Agr. Div. Agrost. Bull. **5**: 26. 1897.
Calamagrostis langsdorffii lactea Kearn. U. S. Dept. Agr. Div. Agrost. Bull. **11**: 28. 1898.
Calamagrostis inexpansa cuprea Kearn. U. S. Dept. Agr. Div. Agrost Bull. **11**: 37. 1898.

Culms 60–120 cm. tall, from creeping rhizomes; blades scattered, flat, rather lax, scabrous, 4–8 mm. wide; panicle narrow but loose, rather open, especially at base; glumes 3–4 mm. long, smooth, scabrous on keel, acuminate; lemma nearly as long as glumes, smooth, narrowed toward summit; callus hairs abundant, about as long as the lemma; awn delicate, straight, attached just below the middle and extending to or slightly beyond its tip; rudiment delicate, sparsely long-pilose.

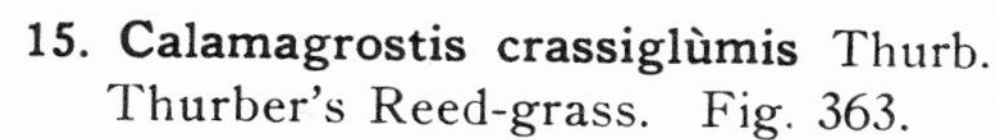

Marshes, wet places, open woods, and meadows, Transition to Hudsonian Zones; Alaska to Oregon and south in the Sierra Nevada Mountains to Mount Whitney, east to the Atlantic States. July–Sept. Type locality: Canada.

15. Calamagrostis crassiglùmis Thurb.
Thurber's Reed-grass. Fig. 363.

Calamagrostis crassiglumis Thurb. in S. Wats. Bot. Calif. **2**: 281. 1880.
Deyeuxia crassiglumis Vasey, Grasses U. S. 28. 1883.
Calamagrostis neglecta crassiglumis Beal, Grasses N. Am. **2**: 353. 1896.

Culms rather stout, 15–40 cm. tall; blades flat, or somewhat involute, smooth, firm, about 4 mm. wide; panicle narrow, spikelike, 2–5 cm. long; glumes 4 mm. long, ovate, rather abruptly acuminate, purple, scaberulous, firm or almost indurate; lemma about as long as glumes, broad, obtuse; callus hairs abundant, about 3 mm. long; awn attached at middle of back, straight, about as long as lemma; rudiment 1 mm. long, the white hairs reaching to apex of lemma.

Swampy soil, Transition Zone; Vancouver Island to Washington; also in Mendocino County, California, the type locality. July.

16. Calamagrostis inexpánsa A. Gray.

Narrow-spiked Reed-grass. Fig. 364.

Calamagrostis inexpansa A. Gray, Gram. & Cyp. 1: no. 20. 1834.
Calamagrostis stricta robusta Vasey in Rothr.; Wheeler, Rep. U. S. Surv. 100th Merid. 6: 285. 1878.
Calamagrostis hyperborea Lange, Fl. Dan. 17[50]: *pl. 2942. f. 1.* 1880.
Deyeuxia neglecta americana Vasey; Macoun, Cat. Can. Pl. 4: 206. 1888.
Calamagrostis americana Scribn. U. S. Dept. Agr. Div. Agrost. Bull. 5: 27. 1897.
Calamagrostis inexpansa barbulata Kearn. U. S. Dept. Agr. Div. Agrost. Bull. 11: 37. 1898.
Calamagrostis hyperborea stenodes Kearn. U. S. Dept. Agr. Div. Agrost. Bull. 11: 39. 1898.
Calamagrostis hyperborea elongata Kearn. U. S. Dept. Agr. Div. Agrost. Bull. 11: 40. 1898.
Calamagrostis hyperborea americana Kearn. U. S. Dept. Agr. Div. Agrost. Bull. 11: 41. 1898.

Culms often scabrous below the panicle, 50–120 cm. tall, producing stout rhizomes; sheaths smooth, or somewhat scabrous, the outer basal ones numerous, marcescent, persistent; blades loosely involute, scabrous, 2–4 mm. wide; panicle narrow, more or less spikelike, 5–15 cm. long; glumes 3 mm. long, scaberulous; lemma as long as glumes, scabrous, the callus hairs one-half to three-fourths as long; awn attached about the middle, straight, about as long as glumes; rudiment .5 mm. long, some of the hairs reaching to tip of lemma.

Meadows and marshes throughout British America, southward through Washington and Oregon, and in mountain meadows of the high Sierra Nevada of California to Sequoia National Park. June-Aug. Type locality: New York.

31. AMMÓPHILA Host, Icon. Gram. Austr. 4: 24. *pl. 41.* 1809.

Spikelets 1-flowered, the rachilla disarticulating above the glumes, produced beyond the palea as a short bristle hairy above; glumes about equal, chartaceous; lemma similar to the glumes, a little shorter, slightly emarginate and mucronate between the teeth, the callus bearing a tuft of short hairs; palea nearly as long as the lemma. Coarse erect perennials with stout creeping rhizomes, long tough involute blades and pale dense spikelike panicles. [Greek, sand-loving.]

Species 3, on the sandy seacoast of Europe and northern North America. Type species, ***Arundo arenaria*** L.

1. Ammophila arenària (L.) Link.

Beach Grass. Fig. 365.

Arundo arenaria L. Sp. Pl. 82. 1753.
Ammophila arenaria Link, Hort. Berol. 1: 105. 1827.
Ammophila arundinacea Host, Icon. Gram. Austr. 4: 24. *pl. 41.* 1809.

Culms stout, 60–100 cm. tall; blades elongate, gradually narrowed into an involute point, the ligule 1.5–3 cm. long; panicle 10–30 cm. long, dense; glumes as much as 10 mm. long, scabrous.

Sands of the seacoast; introduced on the Pacific Coast, where it has been used as a sandbinder, at Linnton and Coos Bay, Oregon, and southward to San Francisco. June–Aug. Type locality, European.

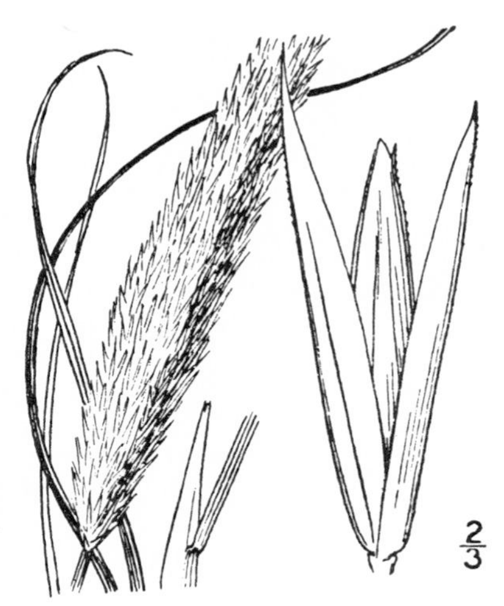

32. LAGÙRUS L. Sp. Pl. 81. 1753; Gen. Pl. ed. 5. 34. 1754.

Spikelets 1-flowered, the rachilla disarticulating above the glumes, pilose under the floret, produced beyond the palea as a bristle; glumes equal, thin, 1-nerved, villous, gradually tapering into a plumose aristiform point; lemma shorter than the glumes, thin, glabrous, narrowed above gradually into 2 slender naked awns about as long as the awns of the glumes, bearing on the back above the middle a slender exserted somewhat geniculate dorsal awn; palea narrow, thin, the 2 keels ending in short awns. An annual grass with dense ovoid or oblong woolly heads. [Greek, hare-tail.]

Species 1, in the Mediterranean region, and introduced sparingly in California. Type species, ***Lagurus ovatus*** L.

1. Lagurus ovàtus L.

Hare's Tail. Fig. 366.

Lagurus ovatus L. Sp. Pl. 81. 1753.

Culms branching at the base, 10–30 cm. tall, slender, pubescent; sheaths and blades pubescent, the sheaths somewhat inflated, the blades flat and lax; panicle 2–3 cm. long and nearly as thick, pale and downy, bristling with dark awns; glumes very narrow, 10 mm. long.

Cultivated for ornament and sparingly escaped; has been found at Pacific Grove, San Francisco, and Berkeley. May–June. Type locality, European.

33. HOLCUS L. Sp. Pl. 1047. 1753.

Spikelets 2-flowered, the pedicel disarticulating below the glumes, the rachilla curved and somewhat elongate below the first floret, not prolonged above the second floret; glumes about equal, longer than the 2 florets; first floret perfect, the lemma awnless; second floret staminate, the lemma bearing on the back a short awn. Perennials with flat blades and contracted panicles. [An old Latin name for a grass, taken from Pliny.]

Species about 8, Europe and Africa, 2 introduced into the United States. Type species, *Holcus lanatus* L.

Rhizomes wanting. 1. *H. lanatus.*
Rhizomes present. 2. *H. mollis.*

1. Holcus lanàtus L.

Velvet Grass. Fig. 367.

Holcus lanatus L. Sp. Pl. 1048. 1753.
Notholcus lanatus Nash; Jepson Fl. Calif. 1: 126. 1912.

Plant grayish, velvety-pubescent; culms erect, 30–60 cm. tall; panicles 5–10 cm. long, narrow, contracted, sometimes almost spikelike, purple-tinged; spikelets 4 mm. long; glumes villous, hirsute on the nerves, the second broader than the first, 3-nerved; lemmas ciliate at the apex; awn of the second floret hook-like.

A native of Europe, occasionally cultivated as a meadow grass in the United States, and abundantly introduced or escaped on the Pacific Coast, especially in the Coast Ranges. June–July. Type locality, European.

2. Holcus móllis L.

Creeping Velvet-grass. Fig. 368.

Holcus mollis L. Syst. Nat. ed. 10. 2: 1305. 1759.
Notholcus mollis Hitchc. Am. Journ. Bot. 2: 304. 1915.

Culms glabrous, 50–100 cm. tall; sheaths glabrous; ligule 2–4 mm.; blades flat, villous or velvety, 5–10 cm. long, 4–6 mm. wide, the uppermost smaller; panicle ovate or oblong, rather loose, 6–10 cm. long; glumes 4–5 mm. long, glabrous; awn of the second floret geniculate, exserted, about 3 mm. long.

Collected at Eureka, California (*Tracy*). Type locality, European. (A recently introduced species, that may become common.)

34. AÌRA L. Sp. Pl. 63. 1753; Gen. Pl. ed. 5. 31. 1754.

Spikelets 2-flowered, disarticulating above the glumes, the hairy rachilla prolonged behind the upper floret, this sometimes bearing a reduced floret; glumes about equal, acute or acutish, membranaceous; lemmas thin, truncate and 2- to 4-toothed at the summit, bearing a slender awn from or below the middle, the awn straight, bent, or twisted, sometimes absent from the lower floret. Low or rather tall annuals or usually perennials with shining pale or purplish spikelets in narrow or open panicles. [A Greek name for a grass.]

Species about 35, in the temperate and cool regions of the world. Type species, *Aira caespitosa* L.

Plants annual. — 1. *A. danthonioides.*
Plants perennial.
  Panicle elongate, narrow, loose, the distant branches appressed; blades filiform, lax. — 2. *A. elongata.*
  Panicle open or contracted but not elongate and loose.
    Blades thin, flat; glumes longer than the florets. — 3. *A. atropurpurea.*
    Blades firm, narrow, becoming folded or involute; glumes about as long as or shorter than the upper floret.
      Panicle narrow, compact, erect. — 4. *A. holciformis.*
      Panicle open, usually drooping. — 5. *A. caespitosa.*

1. Aira danthonioìdes Trin.

Annual Hair-grass. Fig. 369.

Aira danthonioides Trin. Mém. Acad. St. Pétersb. VI. Math. Phys. Nat. 1: 57. 1830.
Deschampsia calycina Presl. Rel. Haenk. 1: 251. 1830.
Deschampsia danthonioides Munro; Benth. Pl. Hartw. 342. 1857.
Deschampsia gracilis Vasey, Bot. Gaz. 10: 224. 1885.

Annual; culms slender, erect, 15–60 cm. tall; blades few, short and narrow; panicle open, 7–15 cm. long, the branches capillary, stiffly ascending, naked below, bearing a few spikelets toward the ends; glumes 4–8 mm. long, acuminate, smooth except the keel, longer than the florets; lemmas smooth and shining, somewhat indurate, 2–3 mm. long, the base of the florets and the rachilla pilose, the awns geniculate, 4–6 mm. long.

Open ground, from Alaska to Mexico, throughout our region except in the higher mountains. June–Aug. Type locality: western North America.

2. Aira elongàta Hook.

Slender Hair-grass. Fig. 370.

Aira elongata Hook. Fl. Bor. Am. 2: 243. *pl. 228.* 1840.
Deschampsia elongata Munro; Benth. Pl. Hartw. 342. 1857.
Deschampsia elongata ciliata Vasey; Beal, Grasses N. Am. 2: 371. 1896.
Deschampsia elongata tenuis Vasey; Beal, Grasses N. Am. 2: 372. 1896.

Perennial; culms slender, erect, 30–120 cm. tall; blades flat, narrow, the basal clusters usually capillary; panicle narrow, as much as 30 cm. long, the branches slender, appressed; glumes 4–6 mm. long; lemmas 2 mm. long, similar to those of *A. danthonioides,* the awns shorter.

Open ground in the Transition Zone; Alaska to Montana and south to southern California and Arizona. May–July. Type locality: islands of the Columbia River.

3. **Aira atropurpùrea** Wahl.

Mountain Hair-grass. Fig. 371.

Aira atropurpurea Wahl. Fl. Lapp. 37. 1812.
Aira latifolia Hook. Fl. Bor. Am. 2: 243. *pl. 227.* 1840.
Deschampsia atropurpurea Scheele, Flora 27: 56. 1844.
Deschampsia atropurpurea latifolia Scribn. in Macoun, Cat. Can. Pl. 2: 209. 1888.

Perennial; culms erect, smooth, purplish at base, 40–80 cm. tall; blades flat, rather soft, ascending or appressed, the apex acute or abruptly narrowed, 5–10 cm. long, 4–6 mm. wide; panicle loose and open, 5–10 cm. long, the branches few, slender, drooping, naked below; glumes about 5 mm. long, broad, purplish; lemmas about 2 mm. long, the awn of the first straight, included, of the second geniculate, exserted.

Woods and wet meadows, Hudsonian Zone; Alaska to New England, south in the mountains to Oregon and Colorado; also in northern Eurasia. July–Aug. Type locality, European.

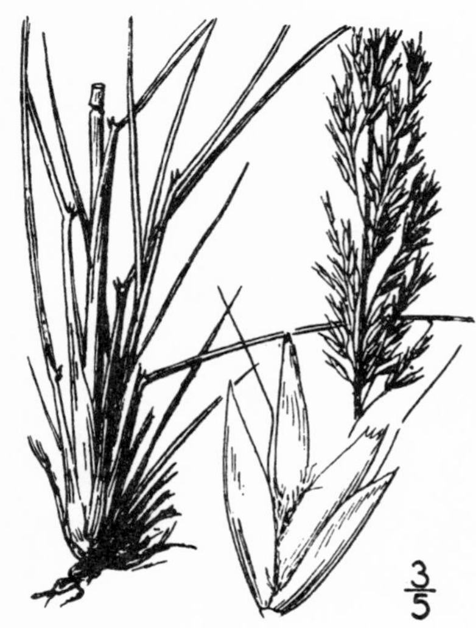

4. **Aira holcifórmis** (Presl) Steud.

California Hair-grass. Fig. 372.

Deschampsia holciformis Presl, Rel. Haenk. 1: 251. 1830.
Aira holciformis Steud. Syn. Pl. Glum. 1: 221. 1854.

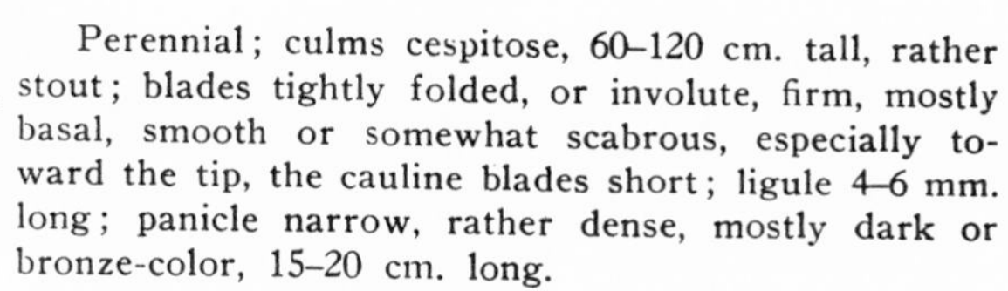

Perennial; culms cespitose, 60–120 cm. tall, rather stout; blades tightly folded, or involute, firm, mostly basal, smooth or somewhat scabrous, especially toward the tip, the cauline blades short; ligule 4–6 mm. long; panicle narrow, rather dense, mostly dark or bronze-color, 15–20 cm. long.

Marshes, bogs, and moist places near the coast, Oregon to Monterey County, California. May–July. Type locality: Monterey.

5. **Aira caespitòsa** L.

Tufted Hair-grass. Fig. 373.

Aira caespitosa L. Sp. Pl. 64. 1753.
Deschampsia caespitosa Beauv. Ess. Agrost. 91, 160. *pl. 18. f. 3.* 1812.
Deschampsia caespitosa confinis Vasey; Beal, Grasses N. Am. 2: 369. 1896.

Perennial; culms in dense tufts, erect, 60–120 cm. tall; sheaths smooth; blades flat or folded, scabrous above; panicle loose, drooping, 10–20 cm. long, the slender scabrous branches spikelet-bearing toward the ends; spikelets 4 mm. long, the florets distant, the rachilla half the length of the lower sessile floret; lemmas smooth, erose-truncate; awn from near the base, but little longer than the lemma, straight, articulate at the base, deciduous.

Bogs and wet places, Transition to Hudsonian Zones; of northern America and Eurasia, south in the mountains to southern California. July–Sept. Type locality, European.

35. ÁSPRIS Adans.

Spikelets 2-flowered, disarticulating above the glumes, the rachilla not prolonged; glumes about equal, acute, membranaceous or subscarious; lemmas firm, rounded on the back, tapering into 2 slender teeth, bearing on the back below the middle a slender geniculate twisted usually exserted awn, this rarely wanting in the lower floret or reduced; callus bearing a very short tuft of hairs. Delicate annuals with open or contracted panicles of small spikelets. [A Greek name for a grass.]

Species 5, warmer parts of Europe. Type species, *Aira praecox* L.

Panicle dense, spikelike. ..... 1. *A. praecox.*
Panicle open.
  Lower floret awnless or nearly so. ..... 2. *A. capillaris.*
  Lower floret with awn as long as that from the upper floret. ..... 3. *A. caryophyllea.*

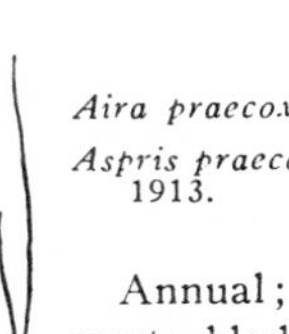

1. Aspris praècox (L.) Nash. Little Hair-grass. Fig. 374.

Aira praecox L. Sp. Pl. 65. 1753.
Aspris praecox Nash in Britt. & Brown, Illustr. Fl. ed. 2. **1**: 215. 1913.

Annual; culms tufted, up to 20 cm. tall, usually erect; blades setaceous; panicle narrow, dense, 1–3 cm. long; spikelets yellowish, shining, 3.5–4 mm. long; lemmas bidentate at apex, bearing a geniculate awn 2–4 mm. long from below the middle, the awn of the lower floret shorter than that of the upper.

Sandy open ground, Washington and Oregon (Seaside); also in the Atlantic States. June–July. Type locality, European.

2. Aspris capilláris (Host) Hitchc. Diffuse Hair-grass. Fig. 375.

Aira capillaris Host, Icon. Gram. Austr. **4**: 20. *pl. 35.* 1809.
Aspris capillaris Hitchc. U. S. Dept. Agr. Bull. **772**: 116. 1920.

Similar to *A. caryophyllea;* panicle more diffuse; spikelets scattered at the ends of the branches, 2.5 mm. long; lemma of lower floret awnless or with a minute awn just below the apex; the teeth short; lemma of upper floret bearing a geniculate awn 3 mm. long from below the middle, the teeth setaceous.

Open ground, sparingly introduced in California (Redwood Creek, *Kenyon*). May. Type locality, European.

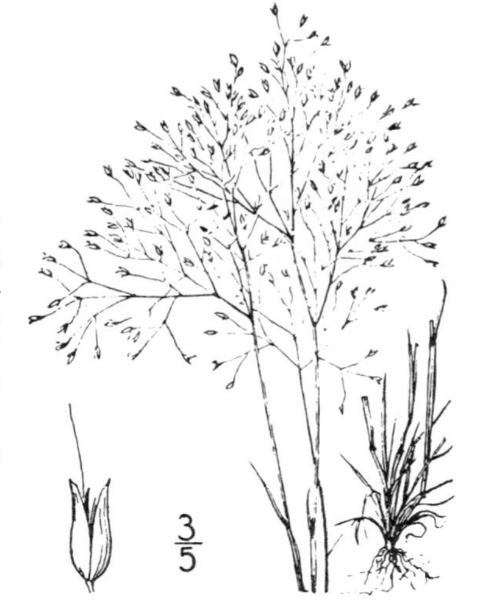

3. Aspris caryophýllea (L.) Nash. Silvery Hair-grass. Fig. 376.

Aira caryophyllea L. Sp. Pl. 66. 1753.
Aspris caryophyllea Nash in Britt. & Brown, Illustr. Fl. ed. 2. **1**: 214. 1913.

Annual; culms solitary or few, slender, erect, 10–30 cm. tall; blades short, setaceous; panicle open, the silvery shining spikelets 3 mm. long, clustered toward the ends of the spreading capillary branches; lemma of both florets with a geniculate awn 4 mm. long from below the middle, the teeth of the apex setaceous.

Open dry ground, Vancouver Island to southern California; also in the eastern States. May–June. Type locality, European.

36. TRISÈTUM Pers. Syn. Pl. 1: 97. 1805.

Spikelets 2-flowered, sometimes 3–5-flowered, the usually villous rachilla disarticulating above the glumes and between the florets, prolonged behind the upper floret; glumes somewhat unequal, acute, awnless, the second usually longer than the first floret; lemmas usually short-bearded at the base, 2-cleft at the apex, the teeth often awned, a straight and included, but usually bent and exserted awn from the back below the apex. Tufted perennials with flat blades and open or usually contracted or spikelike panicles. [Latin, three bristles.]

Species about 65 in the arctic and temperate regions of the world. Type species, *Trisetum flavescens* L.

Awn included or wanting. — 1. *T. wolfii.*
Awn exserted.
  Panicle loose and open, the branches naked at base. — 2. *T. cernuum.*
  Panicle narrow or compact.
    Panicle narrow but loose; blades flat, usually 5–10 mm. wide. — 3. *T. canescens.*
    Panicle spikelike.
      Sheaths pubescent. — 4. *T. spicatum.*
      Sheaths glabrous. — 5. *T. sesquiflorum.*

1. Trisetum wòlfii Vasey.
Beardless Trisetum. Fig. 377.

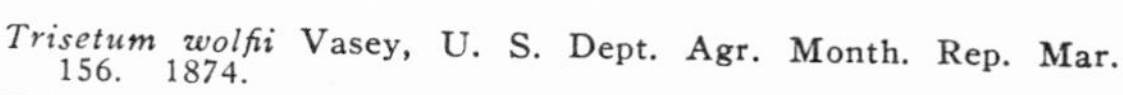

Trisetum wolfii Vasey, U. S. Dept. Agr. Month. Rep. Mar. 156. 1874.
Trisetum subspicatum muticum Boland.; S. Wats. Bot. Calif. 2: 296. 1880.
Trisetum brandegei Scribn. Bull. Torrey Club 10: 64. 1883.
Trisetum muticum Scribn. U. S. Dept. Agr. Div. Agrost Bull. 11: 50. *f. 10.* 1898.
Trisetum wolfii muticum Scribn. Rhodora 8: 88. 1906.

Culms smooth, erect, 30–60 cm. tall; sheaths smooth or sparsely retrorse-pilose; blades flat, erect, 2–4 mm. wide, scabrous or more or less pilose; panicle narrow, usually spikelike, 5–10 cm. long; glumes about 6 mm. long, scabrous on the keel, subequal, the first 1-nerved or obscurely 3-nerved, the second 3-nerved; lemmas scaberulous, the lower 4 mm. long, the awn reduced to a bristle scarcely reaching the tip, or on the upper lemma obsolete, the teeth acute, not aristate.

Mountain meadows, Washington (Spangle) to Colorado, and south in the Sierra Nevada to Sequoia National Park. July–Aug. Type locality: Twin Lakes, Colorado.

3/5

2. Trisetum cérnuum Trin.
Nodding Trisetum. Fig. 378.

Trisetum cernuum Trin. Mém. Acad. St. Pétersb. VI. Math. Phys. Nat. 1: 61. (Jan.) 1830.
Avena nutkaensis Presl, Rel. Haenk. 1: 254. 1830.
Trisetum sandbergii Beal. Grasses N. Am. 2: 378. 1896.
Trisetum nutkaense Scribn. & Merr.; Davy, Univ. Calif. Publ. Bot. 63. 1902.

Culms rather lax, 60–120 cm. tall; sheaths smooth; blades thin, flat, lax, scabrous, 6–12 mm. wide; panicle open, lax or drooping, 15–30 cm. long, the branches verticillate, slender, flexuous, spikelet-bearing toward the ends; spikelet 6–12 mm. long, with usually 3 distant florets, the first longer than the second glume; glumes very unequal, the first narrow, acuminate, 1-nerved, 1 mm. long, the second broad, 3-nerved, 3–4 mm. long; lemmas with setaceous teeth, the awns about as in *T. canescens.*

Moist woods in the Transition Zone; Alaska to Montana and south in western California to Mendocino County. July–Aug. Type locality: Sitka.

3. Trisetum canéscens Buckl.
Tall Trisetum. Fig. 379.

Trisetum canescens Buckl. Proc. Acad. Phila. **1862**: 100. 1863.

Culms erect, or decumbent at base, 60–120 cm. tall; sheaths more or less retrorse-pilose, at least the lower, and often also canescent; blades flat, scabrous or canescent; panicle narrow, loose, sometimes interrupted and spikelike, 10–20 cm. long; glumes smooth, except the keel, strongly unequal, the first narrow, acuminate, 1-nerved, the second broad, acute, 3-nerved, longer than the first, 5–7 mm. long; lemmas firm, scaberulous, the upper exceeding the glumes, the teeth aristate; awns geniculate, spreading, exserted, more or less twisted below, attached one-third below the apex, usually about 12 mm. long.

Mountain meadows, moist ravines, and along streams, Transition Zone; British Columbia to Montana and south in California to Santa Cruz and Tulare Counties. July–Aug. Type locality: Columbia Plains of Oregon.

4. Trisetum spicàtum (L.) Richt.
Downy Oat-grass. Fig. 380.

Aira spicata L. Sp. Pl. 64. 1753.
Aira subspicata L. Syst. Nat. ed. 10. **2**: 873. 1759.
Avena mollis Michx. Fl. Bor. Am. **1**: 72. 1803.
Trisetum subspicatum Beauv. Ess. Agrost. 88. 1812.
Trisetum subspicatum molle A. Gray, Man. ed. 2. 572. 1856.
Trisetum spicatum Richt. Pl. Eur. **1**: 59. 1890.

Culms erect, rather stout, 15–50 cm. tall, smooth or puberulent; sheaths and usually the blades puberulent; panicle dense and spikelike, pale or often dark-purple, 5–15 cm. long; spikelets 4–6 mm. long; glumes somewhat unequal in length, smooth except the keels, the first narrow, acuminate, 1-nerved, the second broader, 3-nerved, acute; lemmas scaberulous, 5 mm. long, the first longer than the glumes, the teeth setaceous; awns geniculate, exserted.

A characteristic grass of high altitudes, mostly in the Alpine Zone; arctic regions of the northern hemisphere, southward in the higher mountains to the southern hemisphere. Found on high mountains up to the limit of vegetation. July–Aug. Type locality, European.

5. Trisetum sesquiflòrum Trin.
Congdon's Trisetum. Fig. 381.

Trisetum sesquiflorum Trin. Mém. Acad. St. Pétersb. VI. Sci. Nat. **2**[1]: 14. 1836.
Trisetum congdoni Scribn. & Merr. Bull. Torrey Club **29**: 470. 1902.

Resembling *T. spicatum*, but differing in having smooth sheaths and blades, the latter usually flat but sometimes involute, and in having wider panicles and larger spikelets, 8 mm. long.

Meadows and slopes of the Alpine Zone; Alaska to California in the Sierra Nevada to Sequoia National Park. July–Aug. Type locality: Kamchatka.

Trisetum flavéscens (L.) Beauv., a European species, occasionally introduced in the eastern States, has been found at Blue Lake, California (*Tracy*). It differs from *T. canescens* in the relatively broader and more open yellow panicle.

37. SPHENÓPHOLIS Scribn. Rhodora 8: 142. 1906.

Spikelets 2- or 3-flowered, the pedicel disarticulating below the glumes, the rachilla produced beyond the upper floret; glumes unlike in shape, the first narrow, acute, 1-nerved, the second broadly obovate, 3–5-nerved, somewhat coriaceous; lemmas firm, scarcely nerved, awnless, the first a little shorter or a little longer than the second glume. Perennials with usually flat blades and narrow panicles. [Greek, wedge-scale.]

Species 4, United States, extending into Mexico and the West Indies. Type species, *Aira obtusata* Michx.

Panicle dense, erect, somewhat lobate; second glume obovate. 1. *S. obtusata lobata.*
Panicle lax, usually nodding; second glume oblanceolate. 2. *S. pallens.*

1. Sphenopholis obtusàta lobàta (Trin.) Scribn. Early Bunch-grass, Prairie-grass. Fig. 382.

Trisetum lobatum Trin. Mém. Acad. St. Pétersb. VI. Math. Phys. Nat. 1: 66. 1830.
Sphenopholis obtusata lobata Scribn. Rhodora 8: 144. 1906.

Culms erect, 30–60 cm. tall; sheaths and blades scabrous; panicle narrow and compact, often spikelike, more or less interrupted or lobed, especially near base, 5–10 cm. long; glumes subequal, the second subcucullate, the broad chartaceous margins smooth and shining.

Prairies and open woods throughout the United States. Frequent in Washington and Oregon in the Upper Sonoran Zone; rare in California. Apr.–July. Type locality: North America.

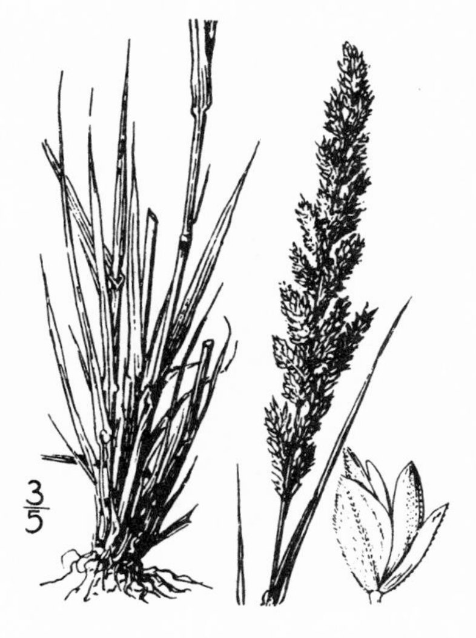

2. Sphenopholis pállens (Spreng.) Scribn.

Shining Prairie-grass. Fig. 383.

Aira pallens Spreng. Mant. Fl. Hal. 33. 1807.
Eatonia pallens Scribn. & Merr. U. S. Dept. Agr. Div. Agrost. Circ. 27: 7. 1900.
Sphenopholis pallens Scribn. Rhodora 8: 145. 1906.

Culms 30–100 cm. tall, usually slender; sheaths glabrous; blades 5–20 cm. long, 4–6 mm. wide, scabrous on the nerves; panicle lax, nodding, 8–20 cm. long; spikelets 3–4 mm. long, oblong-lanceolate; glumes unequal, scabrous on the keels, the first linear, the second broadly oblanceolate.

Moist open ground, in the Arid Transition Zone; eastern Washington and eastward to New England. June–July. Type locality: Pennsylvania.

38. KOELÈRIA Pers. Syn. Pl. 1: 97. 1805.

Spikelets 2–4-flowered, the rachilla disarticulating above the glumes and prolonged beyond the perfect florets, sometimes bearing a reduced or sterile floret at the tip; glumes usually about equal in length but unequal in shape, the lower narrow and sometimes shorter, 1-nerved, the upper somewhat broader above the middle, wider than the lower, 3–5-nerved; lemmas somewhat scarious and shining, the lowermost a little longer than the glume, obscurely 5-nerved, acute or short-awned. Annual or perennial slender low or moderately tall grasses with narrow blades and spikelike panicles. [Named for Professor G. L. Koeler, an early writer on grasses.]

Species about 20 in the temperate regions of both hemispheres. Type species, *Aira cristata* L.

Plants annual; culms glabrous below panicle. 1. *K. phleoides.*
Plants perennial; culm puberulent below panicle. 2. *K. cristata.*

1. Koeleria phleoìdes (Vill.) Pers.
Bristly Koeleria. Fig. 384.

Festuca phleoides Vill. Fl. Delph. 7. 1785.
Koeleria phleoides Pers. Syn. Pl. 1: 97. 1805.

Annual; culms 15–30 cm. tall, smooth throughout; sheaths and blades sparsely pilose; panicle close and spike-like, 2–7 cm. long, obtuse; spikelets 2–4 mm. long, glumes acute, the first narrower; lemmas short-awned from a bifid apex; glumes and lemmas in the typical form papillose-hirsute on the back, but commonly papillose only.

Introduced from Europe at several points in California (Lassen Peak, Butte County, Lathrop), and found on ballast near Portland, Oregon (typical form). May. Type locality, European.

2. Koeleria cristàta (L.) Pers.
Koeler's-grass. Fig. 385.

Aira cristata L. Sp. Pl. 63. 1753.
Koeleria cristata Pers. Syn. Pl. 1: 97. 1805.
Koeleria cristata pubescens Vasey; Davy in Jepson, Fl. W. Mid. Calif. 61. 1901.
Koeleria cristata pinetorum Abrams, Fl. Los Ang. 46. 1904.

Perennial; culms erect, 30–60 cm. tall, glabrous below, puberulent below panicle; sheaths pubescent, at least the lower; blades mostly basal, rather short; panicle compact, spikelike, pointed, 5–10 cm. long, often interrupted at base; spikelets 4–5 mm. long; glumes and lemmas scabrous.

Prairies in the Transition Zone; British Columbia to California and eastward. June–July. Type locality, European.

Koeleria cristata longifòlia Vasey; Davy in Jepson, Fl. W. Mid. Calif. 61. 1901. Differs in being taller and in having longer blades, the basal as much as 30 cm. long, and larger and looser panicles. Type locality: Santa Cruz, California.

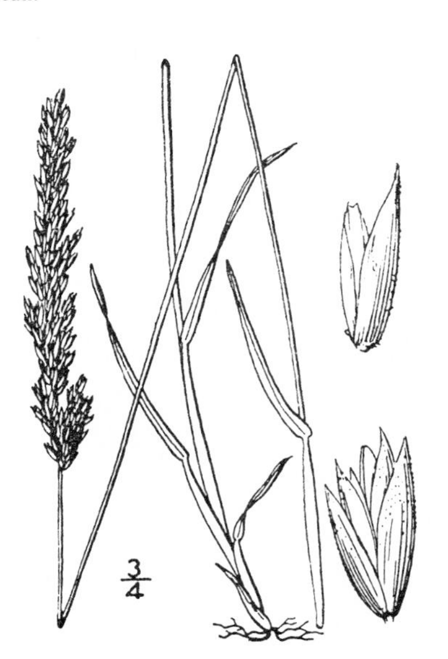

39. AVÈNA L. Sp. Pl. 79. 1753.

Spikelets two- to several-flowered, the rachilla bearded, disarticulating above the glumes and between the florets; glumes about equal, membranaceous or papery, several-nerved, longer than the lower floret, usually longer than the upper floret; lemmas indurate except toward the summit, 5–9-nerved, bidentate at the apex, bearing a dorsal bent and twisted awn (this straight or reduced in *A. sativa*). Annual or perennial low or rather tall grasses with narrow or open, usually rather few-flowered panicles of usually large spikelets. [The ancient Latin name.]

Species about 55 in the temperate regions of the world, but only a few in the western hemisphere; 7 species in the United States, only 2 being native. Type species, *Avena sativa* L.

Lemmas glabrous or nearly so.
  Spikelets usually 2-flowered; awn usually wanting, or if present weakly geniculate. 2. *A. sativa.*
  Spikelets usually 3-flowered; awn present, strongly geniculate. 1a. *A. fatua glabrata.*
Lemmas pubescent with long, usually brown hairs.
  Teeth of lemmas acute, not awned. 1. *A. fatua.*
  Teeth of lemmas awned. 3. *A. barbata.*

1. **Avena fátua** L.

Wild Oat. Fig. 386.

Avena fatua L. Sp. Pl. 80. 1753.

Culms 30–60 cm. tall, erect, stout; panicle loose and open, the slender branches usually horizontally spreading; spikelets usually 3-flowered; glumes about 2.5 cm. long; rachilla and lower part of the shining lemma clothed with long stiff brownish hairs; florets readily falling from the glumes; lemma nerved above, about 2 cm. long, the teeth acuminate but not awned; awn stout, geniculate, red-brown, twisted below, about 3 cm. long.

A common weed on the Pacific Coast; introduced from Europe. Apr.–June. Type locality, European.

Avena fatua glabràta Peterm. Fl. Bien. 13. 1841. (*A. fatua glabrescens* Coss. Fl. Alg. 113. 1867.) Differs in having nearly or quite glabrous lemmas. Introduced from Europe, in situations similar to those of the species.

2. **Avena satìva** L.

Cultivated Oat. Fig. 387.

Avena sativa L. Sp. Pl. 79. 1753.

Similar to *A. fatua;* florets not readily separating from the glumes; spikelets usually 2-flowered; lemma glabrous; awn straight, often wanting.

Commonly cultivated and occasionally escaped. Apr.–June. Type locality, European.

3. **Avena barbàta** Brot.

Slender Wild Oat. Fig. 388.

Avena barbata Brot. Fl. Lusit. 1: 108. 1804.

Similar to *A. fatua;* spikelets somewhat smaller, mostly 2-flowered, the pedicels curved and capillary; lemma clothed with stiff red hairs, the teeth acuminate and ending in fine awns 4 mm. long.

A native of Europe, introduced on the Pacific Coast; a common weed in fields and waste places. March–June. Type locality, European.

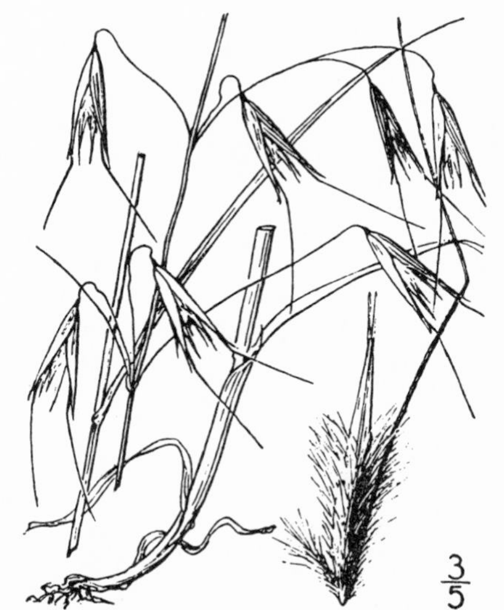

40. **ARRHENÁTHERUM** Beauv. Ess. Agrost. 55, 152. *pl. 11, f. 5.* 1812.

Spikelets 2-flowered, the lower floret staminate, the upper perfect, the rachilla disarticulating above the glumes and produced beyond the florets; glumes rather broad and papery, the first 1-nerved, the second a little longer than the first and about as long as the spikelet, 3-nerved; lemmas 5-nerved, hairy on the callus, the lower bearing near the base a twisted geniculate, exserted awn, the upper bearing a short straight slender awn just below the tip. Perennial rather tall grasses, with flat blades and rather dense panicles. [Greek, masculine-awn, in reference to the awned staminate flower.]

Species about 6 in the temperate regions of Eurasia; one species introduced into the United States. Type species, *Arrhenatherum avenaceum* Beauv.

1. **Arrhenatherum elàtius** (L.) Mert. & Koch. Tall Oat-grass. Fig. 389.

Avena elatior L. Sp. Pl. 79. 1753

Arrhenatherum avenaceum Beauv. Ess. Agrost. 55, 152. *pl. 11. f. 5.* 1812.

Arrhenatherum elatius Mert. & Koch, Deutsch. Fl. 1: 546. 1823.

Perennial; culms erect, smooth, 1–1.5 meters tall; blades flat, scabrous on both surfaces, 5–10 mm. wide; panicle pale or purplish, shining, narrow, 15–30 cm. long, the short branches verticillate, usually spikelet-bearing from the base; spikelets 7–8 mm. long; glumes minutely scabrous, the second about equaling the florets; lemmas scabrous, the awn of the staminate floret about twice the length of its lemma.

A native of Europe, cultivated occasionally as a meadow grass; escaped along roadsides in a few localities in our range. June–July. Type locality, European.

41. **DANTHÒNIA** Lam. & DC. Fl. Franç. ed. 3, 3: 32. 1805.

Spikelets several-flowered, the rachilla readily disarticulating above the glumes and between the florets; glumes about equal, broad and papery, acute, mostly exceeding the uppermost floret; lemmas rounded on the back, obscurely several-nerved, bearing at base a strong callus, the apex bifid, the lobes acute, usually extending into a slender awn, a stout awn arising from between the lobes, the awn flat, twisted, geniculate, exserted. Tufted low or moderately tall perennials, with few-flowered open or spikelike panicles of rather large spikelets. [Named for Étienne Danthoine, a French botanist.]

Species about 100 in the temperate regions of both hemispheres. Type species, *Avena spicata* L.

Panicle narrow and spikelike.
  Lemmas glabrous on the back. — 1. *D. intermedia.*
  Lemmas sparsely villous on the back. — 2. *D. thermale.*
Panicle open, the few branches spreading at anthesis, or reduced to a single spikelet.
  Spikelets solitary or rarely 2. — 3. *D. unispicata.*
  Spikelets 2 to several.
    Sheaths pubescent. — 4. *D. americana.*
    Sheaths glabrous. — 5. *D. californica.*

1. **Danthonia intermèdia** Vasey. Mountain Wild Oat-grass. Fig. 390.

Danthonia intermedia Vasey, Bull. Torrey Club **10**: 52. 1883.

Merathrepta intermedia Piper, Contr. U. S. Nat. Herb. **11**: 122. 1906.

Culms 15–40 cm. tall; sheaths smooth; blades becoming involute, more or less pilose; panicle narrow, compact, often 1-sided, 2.5–5 cm. long, the pedicels short and appressed; glumes about 12 mm. long; lemmas similar to those of *D. americana,* the teeth more gradually acuminate, the awns shorter, the dorsal awn flat, tightly twisted below, slightly twisted above; callus long, and pilose.

Mountain meadows of the Alpine Zone; British Columbia to California and Colorado. July–Aug. Type locality: Mount Albert, Lower Canada.

Danthonia intermedia cusíckii Williams, U. S. Dept. Agr. Div. Agrost. Circ. **30**: 7. 1901. *Merathrepta intermedia cusickii* Piper, Contr. U. S. Nat. Herb. **11**: 122. 1906. Differs in having larger and glabrous blades, usually stouter and taller culms, and rather larger spikelets.

British Columbia to Oregon and Wyoming. Type locality: eastern Oregon.

2. Danthcnia thermàle Scribn.
Northern Wild Oat-grass. Fig. 391.

Danthonia spicata pinetorum Piper, Erythea **7**: 103. 1899.
Danthonia thermale Scribn. U. S. Dept. Agr. Div. Agrost. Circ. **30**: 5. 1901.
Merathrepta pinetorum Piper, Contr. U. S. Nat. Herb. **11**: 122. 1906.

Culms tufted, erect, stouter than in *D. intermedia,* 30–60 cm. tall; blades flat, more or less pilose, 2–4 mm. wide, the upper much reduced; inflorescence as in *D. intermedia;* spikelets about 1 cm. long; lemma sparsely villous on the back, the callus short and scarcely pilose.

Moist slopes in the Transition Zone; British Columbia to Wyoming, south to Oregon. July–Aug. Type locality: Yellowstone Park.

3. Danthonia unispicàta Munro.
Few-flowered Wild Oat-grass. Fig. 392.

Danthonia unispicata Munro; Thurb. in S. Wats. Bot. Calif. **2**: 294. 1880.
Danthonia californica unispicata Munro; S. Wats. Bot. Calif. **2**: 294. 1880.
Merathrepta unispicata Piper, Contr. U. S. Nat. Herb. **11**: 123. 1906.

Culms short, 15–20 cm. tall, about as long as the numerous basal leaves; sheaths and blades pilose; panicle reduced to a single spikelet, (often an abortive spikelet, rarely a perfect one below) the pedicel about 12 mm. long, flexuous, pubescent above; spikelets about as in *D. americana,* the lemma more gradually acuminate into awns.

Rocky hills in the Arid Transition Zone; Washington to California and Wyoming. May–July. Type locality: California.

4. Danthonia americàna Scribn.
American Wild Oat-grass. Fig. 393.

Danthonia grandiflora Phil. Anal. Univ. Chile **48**: 568. 1873, not Hochst. 1851.
Danthonia americana Scribn. U. S. Dept. Agr. Div. Agrost. Circ. **30**: 5. 1901.
Merathrepta americana Piper, Contr. U. S. Nat. Herb. **11**: 123. 1906.

Culms 30–60 cm. tall, smooth, tending to disarticulate at the nodes; sheaths pilose; blades short, flat, or those of the innovations involute; panicle bearing 2–5 spikelets, the pedicels usually about 12 mm. long, spreading or somewhat reflexed; glumes 12–18 mm. long, smooth, acuminate, about 7-nerved; lemmas 5–7 mm. long, smooth and convex on the back, pilose at base and margins, broad, abruptly contracted into 2 teeth with awns 2–6 mm. long, the dorsal awn from between these teeth, geniculate, flat and twisted below, straight and divergent above, exserted. This and the other species have at the base of the sheaths cleistogenes (hidden spikelets) with 1 or 2 larger florets than those of the normal spikelets.

Wet meadows and moist places in rocks in the Transition Zone; British Columbia to Wyoming, south to the San Bernardino Mountains; also in Chile. June–July. Type locality: Chile.

5. Danthonia califórnica Boland.

California Wild Oat-grass. Fig. 394.

Danthonia californica Boland. Proc. Calif. Acad. **2**: 182. 1863.
Merathrepta californica Piper, Contr. U. S. Nat. Herb. **11**: 122. 1906.

Resembling *D. americana;* culms 60–90 cm. tall; sheaths smooth or somewhat pilose at the throat; blades scabrous above, longer, especially those of the less numerous innovations; teeth of lemma more gradually acuminate.

Dry hills in the Transition Zone; British Columbia south to San Luis Obispo, east to Colorado, rare in the Sierra Nevada. June–July. Type locality: Oakland.

42. CỲNODON Rich.; Pers. Syn. 1: 85. 1805.

Spikelets 1-flowered, awnless, sessile in 2 rows along one side of a slender continuous rachis, the rachilla disarticulating above the glumes, and prolonged behind the palea, sometimes bearing a minute bract; glumes narrow, acuminate, 1-nerved, about equal, shorter than the floret; lemma strongly compressed, pubescent on the keel, firm in texture, 3-nerved, the lateral nerves close to the margins. Perennial usually low grasses with creeping stolons or rhizomes, short blades and several slender spikes digitate at the summit of the upright flowering stems. [Name Greek, meaning dog and tooth.]

Species 6, one, *C. dactylon,* being widely distributed in the warm regions of the globe. Type species, *Panicum dactylon* L.

1. Cynodon dáctylon (L.) Pers.

Bermuda Grass. Fig. 395.

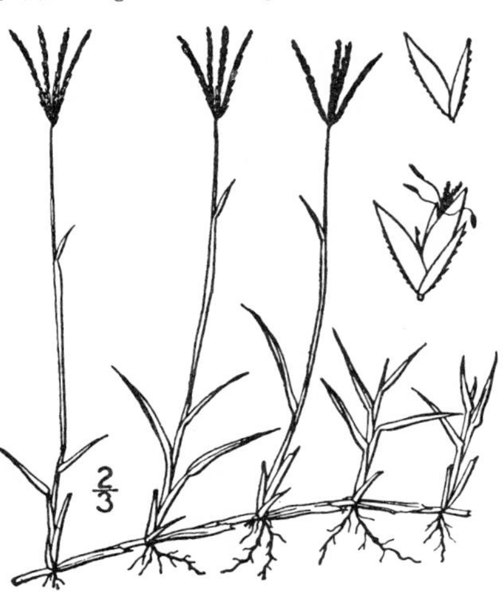

Panicum dactylon L. Sp. Pl. 58. 1753.
Cynodon dactylon Pers. Syn. Pl. **1**: 85. 1805.
Capriola dactylon Kuntze, Rev. Gen. Pl. **2**: 764. 1891.

Culms flattened, wiry, glabrous; ligule a conspicuous ring of white hairs; spikes 4 or 5, 2.5–5 cm. long; spikelets imbricated, 2 mm. wide, the lemma longer than the glumes.

A native of the warmer parts of the Old World, now widely cultivated in the western hemisphere from Virginia to Argentina. Not uncommon in California, especially along irrigating ditches; occasional in Oregon. Abundantly established in the southern part of the United States. June–Aug. Type locality, European.

43. SPARTÌNA Schreb.; Gmel. Syst. Veg. 1: 123. 1791.

Spikelets 1-flowered, much flattened laterally, sessile and usually closely imbricated in 2 rows on one side of a continuous rachis, the pedicel disarticulating below the glumes, the rachilla not produced beyond the floret; glumes keeled, 1-nerved, acute or short-awned, the first shorter, the second often exceeding the lemma; lemma firm, keeled, the lateral nerves obscure, narrowed to a rather obtuse point; palea 2-nerved, keeled and flattened, the keel between or at one side of the nerves. Stout erect often tall perennials with usually extensively creeping firm scaly rhizomes, long tough blades, and 2 to many appressed or sometimes spreading spikes scattered along the main axis. [Greek, a cord.]

Species about 14, all North American except 2 or 3 along the coast of Europe, Africa, and South America. Type species, *Spartina schreberi* Gmel.

Spikes closely approximate, forming a cylindric inflorescence; blades smooth. — 1. *S. foliosa.*
Spikes distinct, appressed or spreading; blades scabrous on margin or surface.
  Second glume awned; spikes numerous, ascending or spreading. — 2. *S. michauxiana.*
  Second glume acute but not awned; spikes few, appressed. — 3. *S. gracilis.*

1. Spartina foliòsa Trin.

California Cord-grass. Fig. 396.

Spartina foliosa Trin. Mém. Acad. St. Pétersb. VI. Sci. Nat. 4[1]: 114. 1840.

Culms stout, as much as 1 cm. thick at base, usually rooting from the lower nodes, 30–120 cm. tall, somewhat spongy in texture; blades 8–12 mm. broad at the flat base, gradually narrowed to a long involute tip, smooth on surface and margin; inflorescence dense, spikelike, about 15 cm. long; spikes approximate, numerous, close-appressed, 3–5 cm. long; spikelets indurate, very flat, about 12 mm. long; glumes ciliate on keel, acute but not awned, the first narrow, about ⅔ as long as second, smooth, the second sparingly hispidulous and striate-nerved; lemma hispidulous on sides, smooth on keel, a little shorter than the second glume; palea thin, longer than the lemma.

Salt marshes along the coast from San Francisco Bay southward. July–Nov. Type locality: California.

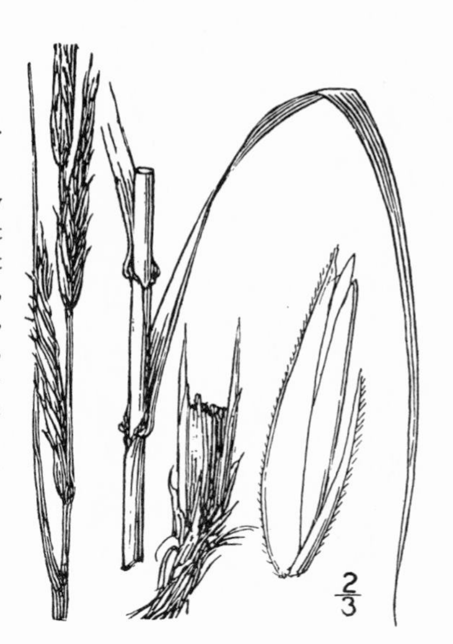

2. Spartina michauxiàna Hitchc.

Cord-grass. Fig. 397.

Spartina michauxiana Hitchc. Contr. U. S. Nat. Herb. 12: 153. 1908.

Spartina cynosuroides of authors not L.

Culms 1–2 meters tall; blades 60–120 cm. long, as much as 15 mm. wide, tapering to a slender point, flat but soon involute in drying, smooth except the margins; spikes usually numerous (5–20), scattered, ascending or spreading, usually 5–10 cm. long; glumes serrulate-hispid on the keel, the first acuminate and equaling the floret, the second tapering into an awn 7 mm. long; lemma 7–9 mm. long, glabrous except the serrulate-scabrous midnerve, this abruptly terminating below the emarginate or 2-toothed apex.

Banks of rivers and lakes, Nova Scotia to Oklahoma and eastern Washington. July–Sept. Type locality: Illinois.

3. Spartina grácilis Trin.

Alkali Cord-grass. Fig. 398.

Spartina gracilis Trin. Mém. Acad. St. Pétersb. VI Sci. Nat. 4[1]: 110. 1840.

Culms 60–100 cm. tall; blades flat, becoming involute, 15–20 cm. long, very scabrous above; spikes few, 4–8, closely appressed to the axis, 1–1.5 cm. long; spikelets about 6 mm. long; glumes smooth, except the ciliate keel, 1-nerved, acute but not awned, the first about half as long as the second; lemma about as long as second glume, ciliate on keel; palea as long as lemma, obtuse.

Alkaline meadows, Great Plains to eastern Washington, extending southward east of the Sierra Nevada. June–Aug. Type locality: North America, probably Saskatchewan.

44. CHLÒRIS Swartz, Prod. Veg. Ind. Occ. 25. 1788.

Spikelets 1-flowered, sessile in 2 rows along one side of a continuous rachis, the rachilla disarticulating above the glumes, produced beyond the perfect floret and bearing 1 to several reduced florets consisting of empty lemmas, these often truncate and, if more than one, the smaller ones enclosed in the lower, forming a usually club-shaped rudiment; glumes somewhat unequal, the first shorter, narrow, acute; lemma compressed, usually broad, 1–5 nerved, often villous on the callus, and villous or long-ciliate on the keel or marginal nerves, awned from between the short teeth of a bifid apex, the sterile lemmas awned or awnless. Perennial or sometimes annual tufted grasses with flat blades and 2 to several often showy and feathery spikes aggregate at or near the summit of the culms. [Named for Chloris, the goddess of flowers.]

Species about 60 in the warmer countries of the world. Type species, *Agrostis cruciata* L.

1. Chloris virgàta Swartz. Silky Chloris. Fig. 399.

Chloris virgata Swartz, Fl. Ind. Occ. 1: 203. 1797.
Chloris elegans H. B. K. Nov. Gen. & Sp. 1: 166. *pl. 49.* 1816.

Annual; culms erect or spreading, 30–100 cm. tall, smooth; sheaths smooth, much compressed, especially the basal, the uppermost often inflated around the base of the inflorescence; spikes several, feathery, suberect, 6–12, pale or dark-colored, 2–7 cm. long; spikelets imbricated; glumes 1-nerved, the second about 3 mm. long, awn-pointed; lemma somewhat fusiform, about 2 mm. long, 3-nerved, short-pilose at base and along the lower half of the keel, long-pilose on the margins near the apex, with a slender straight awn about 10 mm. long from just below the apex; rudiment reaching about to the tip of the fertile floret, truncate, the awn somewhat shorter.

Fields and waste places, Colorado Desert to Texas and tropical America. Dec.–Feb. Type locality: Island of Antigua.

Chloris verticillàta Nutt. Perennial; culms 15–35 cm. tall; sheaths compressed; blades 2.5–8 cm. long, 2–4 mm. wide; spikes slender, stiff, divergent, 5–10 cm. long, in 1 or 2 whorl; spikelets 3 mm. long, the awns of the lemmas about 5 mm. long.
Open dry ground, Missouri to Colorado and Texas; introduced in California (Berkeley). Type locality: Arkansas.

Chloris radiàta (L.) Swartz, with paler, more slender racemes than in *C. virgata,* has been collected on ballast at Linnton, Oregon (*Nelson*).

45. BOUTELOÙA Lag. Var. Cienc. 2⁴: 134. 1805.

Spikelets 1-flowered, with the rudiments of 1 or more florets above, sessile in 2 rows along one side of the rachis; glumes unequal, 1-nerved, acuminate or short-awned, the first shorter and narrower; lemma as long as the second glume or a little longer, 3-nerved, the nerves extending into short or usually rather long awns; the internerves usually extending into teeth; palea 2-nerved, sometimes 2-awned; rudiment various, usually 3-awned, a second rudimentary floret sometimes present. Perennial or sometimes annual, low or moderately tall grasses with 2 to several or many spikes, or sometimes solitary, along a main axis, the spikelets 2 to many in each spike, pectinate or more loosely arranged and appressed, the rachis of the spike usually produced beyond the insertion of the spikelets as a bristle or point. [Named for the brothers Boutelou, Spanish gardeners.]

Species 38, all American and chiefly North American. Type species, *Bouteloua racemosa* Lag.

Spikes containing 1–3 spikelets, numerous along a main axis.
  Plants perennial. — 1. *B. curtipendula.*
  Plants annual. — 2. *B. aristidoides.*
Spikes usually few, containing numerous spikelets.
  Plants annual.
    Awns about 3 mm. long; spikes 2–4. — 3. *B. arenosa.*
    Awns barely protruding; spikes 4–6 or more. — 4. *B. barbata.*
  Plants perennial.
    Spikes several.
      Spikes narrow, strictly 1-sided; spikelets numerous. — 5. *B. rothrockii.*
      Spikes broad, loose, irregularly 1-sided; spikelets few. — 6. *B. radicosa.*
    Spikes usually 1–3.
      Rachis not prominently produced; glumes sparsely hairy. — 7. *B. gracilis.*
      Rachis produced beyond the spikelets as a naked point; glumes prominently papillose-hispid. — 8. *B. hirsuta.*

1. **Bouteloua curtipéndula** (Michx.) Torr.
Side-oats Grama. Fig. 400.

Chloris curtipendula Michx. Fl. Bor. Am. 1: 59. 1803.
Bouteloua racemosa Lag. Var. Cienc. 2[4]: 141. 1805.
Bouteloua curtipendula Torr. in Emory, Mil. Reconn. 154. 1848.

Perennial; culms erect, 30–120 cm. high; spikes numerous on an elongated rachis, 5–15 mm. long. reflexed, mostly turned to one side, falling entire from the main axis; glumes narrow, acuminate, scabrous on keel and somewhat so on the back, the second about 5 mm. long; lemma as long as second glume, ovate-lanceolate, 3-nerved, scabrous toward tip, 3-toothed, the palea about as long; rudiment commonly as long as lemma, 4-lobed, 3-awned between the lobes, the lateral lobes and awns shorter, or the whole sometimes reduced to 1–3 awns.

Plains and rocky hills, Montana and Ontario, south to Mexico, extending into southern California (Santa Rosa Mountain). July–Aug. Type locality: Illinois.

2. **Bouteloua aristidoìdes** (H. B. K.) Griseb.
Needle Grama. Fig. 401.

Dinebra aristidoides H. B. K. Nov. Gen. & Sp. 1: 171. 1816.
Bouteloua aristidoides Griseb. Fl. Brit. W. Ind. 537. 1864.

Annual; culms spreading, slender, 15–40 cm. tall; spikes several, slender, about 1 cm. long, falling entire, the 1–3 spikelets distant, appressed to the rachis, the latter ending in a slender naked point; glumes narrow, acuminate, the first half as long as the second; lemma narrowly lanceolate, 3-nerved, the nerves pilose, the lateral ending in awned teeth as long as the central acuminate point; rudiment consisting of a pilose pedicel and 3 awns longer than the spikelet.

Open ground, deserts, and foothills, southern California to western Texas and south into South America. Type locality: Mexico.

3. **Bouteloua arenòsa** Vasey.
Sand Grama. Fig. 402.

Bouteloua arenosa Vasey, U. S. Dept. Agr. Div. Bot. Bull. 12[1]: 34. 1890.

Annual; culms spreading or prostrate, about 15 cm. long; spikes 2–4, many-flowered, about 1 cm. long; persistent on the main axis; glumes persistent, the florets falling, 1-nerved, the first 2 mm., the second 3 mm. long; lemma a little shorter than the second glume, pilose below, 4-lobed, the lateral lobes short, 3-awned from between the lobes, the awns about 3 mm. long; palea 4-toothed, 2-awned; rudiment 1 mm. long, triangular-truncate, pilose at base, 4-lobed, with 3 long awns between the lobes.

Loose sandy soil, deserts of northern Mexico, extending sparingly into the adjoining United States. Cargo Muchacho, Colorado Desert (*Orcutt*). Type locality: Guaymas, Sonora.

4. Bouteloua barbàta Lag.
Annual Grama. Fig. 403.

Bouteloua barbata Lag. Var. Cienc. 2[4]: 141. 1805.
Chondrosium polystachya Benth. Bot. Voy. Sulph. 56. 1844.
Bouteloua polystachya Torr. U. S. Rep. Expl. Miss. Pacif. 5: 366. *pl. 10.* 1857.

Annual; culms spreading or prostrate, 15–30 cm. tall; spikes usually 4–6, about 1–2 cm. long; persistent on the main axis; spikelets numerous, imbricate; glumes persistent, unequal, scabrous on keel and somewhat so on back, awn-pointed from a toothed apex, the second twice as long as the first, 3 mm. long; lemma pilose below, 3-awned, the awns about 2 mm. long, the lemma lobed between the awns; rudiment pilose at base, 2-lobed, 3-awned, enclosing an orbiculaar scale.

Deserts, Utah to southern California (The Needles, *Jones*) and south into Mexico. Type locality: Mexico.

5. Bouteloua rothróckii Vasey.
Rothrock's Grama. Fig. 404.

Bouteloua rothrockii Vasey, Contr. U. S. Nat. Herb. 1: 268. 1893.

Perennial; culms erect or spreading, 30–60 cm. tall; spikes several, usually 4–6, 1–2.5 cm. long; persistent on the main axis, spikelets numerous, imbricate; glumes persistent, unequal, scabrous on keel and back, cuspidate and 2-toothed at apex, the second 2.5 mm. long, about twice as long as the first; lemma pilose below, 4-lobed, 3-awned, the awns equal, 2.5 mm. long; rudiment pilose at base, consisting of 2 short truncate lobes, 3 equal awns about 1 mm. long and an included orbicular scale.

Mesas and foothills, Utah to southern California (Jamacha, *Canby*) and Mexico. Type locality: Arizona.

6. Bouteloua radicòsa Griffiths.
Purple Grama. Fig. 405.

Dinebra bromoides H. B. K. Nov. Gen. & Sp. 1: 172. *pl. 51.* 1816, not *Bouteloua bromoides* Lag.
Atheropogon radicosus Fourn. Mex. Pl. 2: 140. 1886.
Bouteloua radicosa Griffiths, Contr. U. S. Nat. Herb. 14: 411. *pl. 81.* 1912.

Perennial in dense tufts; culms 15–60 cm. tall, erect; blades mostly basal, flat, sparsely ciliate toward the base; spikes several to many, 2–2.5 cm. long, irregularly 1-sided, falling entire; glumes somewhat unequal, rather broad, the second about 6 mm. long; lemmas smooth, bearing 3 short awns; rudiment lanceolate, with 3 long awns.

Upper foothills and mountains, southern California (*Orcutt*) and New Mexico to Mexico. Type locality: Mexico.

7. Bouteloua grácilis (H. B. K.) Lag. Blue Grama. Fig. 406.

Chondrosium gracile H. B. K. Nov. Gen. & Sp. 1: 176. *pl.* 58. 1816.

Atheropogon oligostachyum Nutt. Gen. Pl. 1: 78. 1818.
Bouteloua gracilis Lag.; Steud. Nom. Bot. ed. 2. 1: 219. 1840.
Bouteloua oligostachya Torr.; Gray, Man. ed. 2. 553. 1856.

Perennial; culms erect, 15–45 cm. tall; sheaths and blades glabrous; spikes 1–3, 2–5 cm. long, usually a little curved, persistent on the main axis, the rachis not produced; spikelets 5–6 mm. long, densely crowded, pectinate; glumes persistent, narrow, the first about half as long as the second, the latter sparsely papillose-pilose on the keel; lemma pilose, 3-cleft, the lateral divisions awned, the terminal 2-toothed, awned between the teeth; rudiment 3-awned, pilose at base, a second rudimentary scale above.

Plains and hills, Manitoba to Mexico, extending into southern California. Type locality: Mexico.

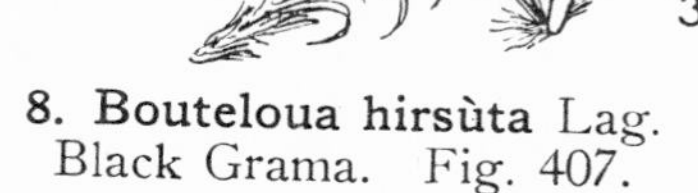

8. Bouteloua hirsùta Lag. Black Grama. Fig. 407.

Bouteloua hirsuta Lag. Var. Cienc. 2[4]: 141. 1805.

Perennial; culms erect, 20–40 cm. tall; sheaths smooth; blades sparsely papillose-hairy, especially on the margins; spikes 1–4, 2–5 cm. long, persistent, the rachis produced into a prominent point beyond the uppermost spikelets; glumes persistent, the first narrow, setaceous, the second acuminate, twice as long as first and equaling the floret; conspicuously tuberculate-hirsute on the back; lemma pubescent, 3-cleft; rudiment of 2 obtuse lobes and 3 equal awns, not pilose at base.

Mesas and dry hills, British Columbia to South Dakota, south to Mexico, extending into southern California (Jamacha). Type locality: Mexico.

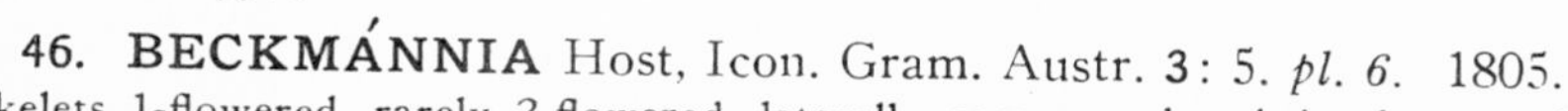

46. BECKMÁNNIA Host, Icon. Gram. Austr. 3: 5. *pl.* 6. 1805.

Spikelets 1-flowered, rarely 2-flowered, laterally compressed, subcircular, nearly sessile and closely imbricate in 2 rows along one side of a slender rachis, the pedicel disarticulating below the glumes, the spikelet falling entire; glumes equal, inflated, obovate, 3-nerved, rounded above but the apex apiculate; lemma narrow, 5-nerved, acuminate, about as long as the glumes; palea 2-nerved, nearly as long as the lemma. An erect rather stout annual with flat blades and numerous short pedunculate spikes in a narrow more or less interrupted panicle. [Named for Johann Beckmann, teacher of natural history at St. Petersburg.]

Species 1, in the cooler parts of America and Eurasia. Type species, *Phalaris erucaeformis* L.

1. Beckmannia erucaefórmis (L.) Host. Slough-grass. Fig. 408.

Phalaris erucaeformis L. Sp. Pl. 55. 1753.

Beckmannia erucaeformis Host, Icon. Gram. Austr. 3: 5. 1805.

Beckmannia erucaeformis uniflora Scribn.; Wats. & Coult. in A. Gray, Man. ed. 6. 628. 1890.

Plants light green; culms 30–100 cm. tall, branching at the base; panicle 10–25 cm. long; spikes 1–2 cm. long, erect or ascending; spikelets nearly circular, 3 mm. long; glumes with a deep keel, transversely wrinkled, the acuminate apex of the lemma protruding.

Swamps and ditches, cooler parts of the northern hemisphere, extending south through Washington and Oregon to San Francisco Bay. June–July. Type locality, European.

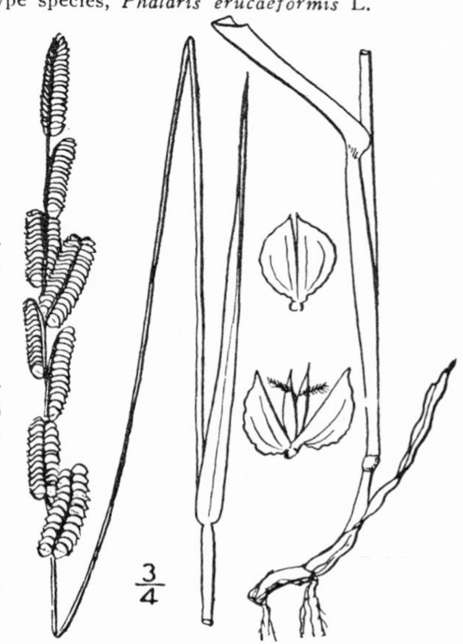

47. ELEUSÌNE Gaertn. Fruct. & Sem. 1: *7. pl. 1.* 1788.

Spikelets few- to several-flowered, compressed, sessile and closely imbricate in 2 rows along one side of a rather broad rachis, the latter not prolonged beyond the spikelets; glumes persistent, unequal, rather broad, acute 1-nerved, shorter than the first lemma; lemmas acute, with 3 strong green nerves close together, forming a keel, the uppermost somewhat reduced; seed dark brown, transversely minutely ridged. Annual grasses with 2 to several rather stout spikes digitate at the summit of the culms, sometimes with 1 or 2 a short distance below, or rarely with a single terminal spike. [Greek, Eleusine, the town where Ceres, the goddess of harvests, was worshipped.]

Species about 6 in the warm regions of the eastern hemisphere, one a common introduced weed in America. Type species, *Cynosurus indicus* L.

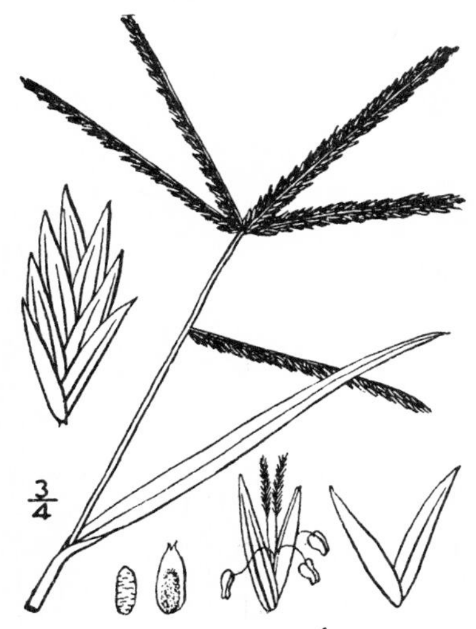

1. Eleusine índica (L.) Gaertn. Goose-grass. Fig. 409.

Cynosurus indicus L. Sp. Pl. 72. 1753.
Eleusine indica Gaertn. Fruct. & Sem. 1: 8. 1788.

Culms flattened, decumbent at base or prostrate-spreading; sheaths loose, overlapping, compressed; spikes 2–10, 2–8 cm. long; spikelets appressed, 3–5-flowered, about 5 mm. long.

A common roadside weed in the warmer parts of America, introduced from the Old World. Rare on the Pacific Coast. Linnton, Oregon (*Nelson*), Los Angeles (*Braunton*). July–Aug. Type locality: India.

Eleusine tristàchya Lam., with shorter, broader spikes, has been found at Linnton (*Nelson*).

48. LEPTÓCHLOA Beauv. Ess. Agrost. 71. *pl. 15. f. 1.* 1812.

Spikelets two- to several-flowered, sessile or short-pediceled, approximate or somewhat distant along one side of a slender rachis; glumes persistent, unequal or nearly equal, awnless or mucronate, 1-nerved, usually shorter than the first lemma; lemmas 3-nerved, the nerves sometimes pubescent, obtuse or acute, sometimes 2-toothed and mucronate or short-awned from between the teeth. Our species annual, with flat blades and numerous spikes or racemes, scattered along a main axis forming a long or sometimes short panicle. [Greek, slender grass.]

Species probably 20 in the warmer parts of the world. Type species, *Cynosurus virgatus* L.

Glumes longer than first lemma; sheaths papillose-hispid. — 1. *L. filiformis.*
Glumes shorter than first lemma; sheaths smooth.
  Lemmas awned. — 2. *L. fascicularis.*
  Lemmas awnless. — 3. *L. uninervia.*

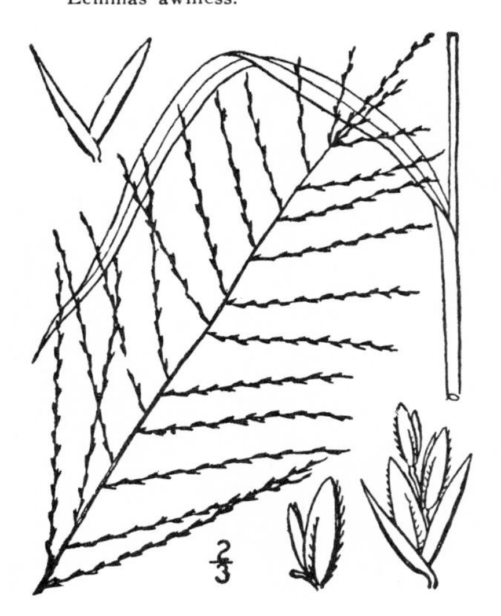

1. Leptochloa filifórmis (Lam.) Beauv. Sprangle-top, Feather-grass. Fig. 410.

Festuca filiformis Lam. Tabl. Encycl. 1: 191. 1791.
Eleusine mucronata Michx. Fl. Bor. Am. 1: 65. 1803.
Leptochloa filiformis Beauv. Ess. Agrost. 71, 166. 1812.
Leptochloa mucronata Kunth, Rév. Gram. 1: 91. 1829.

Culms 3–100 cm. tall, often depauperati; sheaths sparsely papillose-hairy; spikes numerous, spreading, 3–10 cm. long, slender, usually purple, the spikelets rather distant, about 3 mm. long; glumes more or less mucronate, nearly equaling the 3 or 4 small awnless florets.

Open ground, fields, and moist depressions, Imperial County (Colorado River); common in tropical America. Sept.–Dec. Type locality: tropical America.

2. Leptochloa fasciculáris (Lam.) A. Gray.
Loose-flowered Sprangle-top. Clustered Salt-grass. Fig. 411.

Festuca fascicularis Lam. Tabl. Encycl. 1: 189. 1791.
Leptochloa fascicularis A. Gray, Man. 588. 1848.

Culms erect or spreading, 30–60 cm. tall; sheaths smooth; blades erect, as long or longer than the culms; spikes numerous, 7–12 cm. long; spikelets short-pediceled, 7–11-flowered, the florets much longer than the lanceolate glumes; lemmas hairy-margined toward the base, short-awned from the toothed apex.

Ditches and moist, especially alkaline places, Maryland to Florida and the West Indies, west to the Great Basin, rare in California; Fresno County (*Griffiths*), Kern County (*Raymond*). June–Aug. Type locality: tropical America.

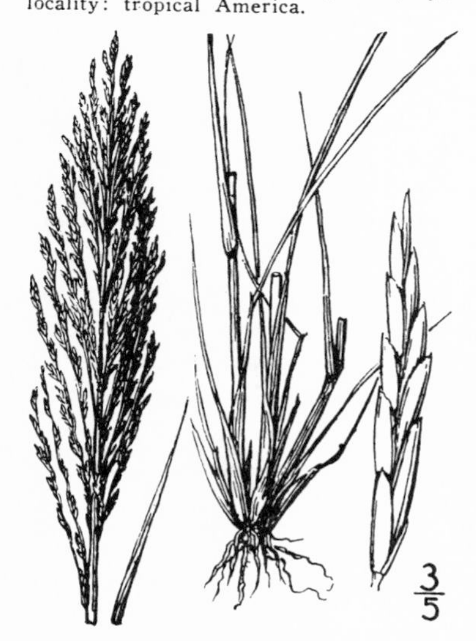

3. Leptochloa uninérvia (Presl.) Hitchc. & Chase.
Dense-flowered Sprangle-top. Fig. 412.

Megastachya uninervia Presl, Rel. Haenk. 1: 283. 1830.
Leptochloa imbricata Thurb. in S. Wats. Bot. Calif. 2: 293. 1880.
Leptochloa uninervia Hitchc. & Chase, Contr. U. S. Nat. Herb. 18: 383. 1917.

Resembles *L. fascicularis;* usually strictly erect, the panicle more oblong in outline, with shorter denser flowered spikes; glumes broader and more obtuse; lemmas apiculate but not awned.

Ditches and moist places, San Bernardino Mountains, southward to Mexico and east to Louisiana. May–Oct. Type locality: Mexico.

49. MONANTHÓCHLOË Engelm. Trans. Acad. St. Louis 1: 436. 1859.

Plants dioecious; spikelets 3–5-flowered, the uppermost rudimentary, the rachilla disarticulating tardily in the pistillate spikelets; glumes wanting; lemmas rounded on the back, convolute, narrowed above, 3-nerved, membranaceous; palea narrow, 2-nerved, in the pistillate spikelets convolute around the pistil, the rudimentary uppermost floret inclosed between the keels of the next floret below. A spreading wiry perennial, with clustered short subulate leaves, the spikelets at the ends of the short branches only a little longer than the leaves. [Greek, one-flowered grass.]

Species 1, muddy shores of the ocean in tropical and subtropical America. Type species, *Monanthochloë littoralis* Engelm.

1. Monanthochloë littorális Engelm.
Salt Cedar. Fig. 413.

Monanthochloë littoralis Engelm. Trans. Acad. St. Louis 1: 437. *pl. 13, 14.* 1859.

A low, extensively creeping stoloniferous perennial with wiry culms and rigid crowded leaves, the blades 5–10 cm. long, squarrose; spikelets almost hidden among the upper leaves.

Salt marshes and mucky or gravelly tidal flats along the coast of tropical seas in the western hemisphere, extending as far north on the Pacific Coast as Santa Barbara. May–June. Type locality: Texas.

50. ORCÚTTIA Vasey, Bull. Torrey Club 13: 219. *pl. 60.* 1886.

Spikelets several-flowered, the rachilla disarticulating above the glumes and between the florets, the upper florets reduced; glumes nearly equal, shorter than the lemmas, broad, irregularly 2–5-toothed, many-nerved, the nerves extending into the teeth; lemmas similar to the glumes, prominently 13–15-nerved, the broad summit with 5 long teeth or with numerous short teeth; palea broad, 2-nerved, as long as the lemma. Low cespitose annuals with short blades and terminal racemes, the upper spikelets aggregate, the lower more or less remote. [Named for C. R. Orcutt, a botanist of San Diego.]

Species 2, California and Lower California. Type species, *Orcuttia californica* Vasey.

Lemmas toothed; blades pilose on the upper surface; glumes and lemmas sparsely long-pilose and somewhat papillose. 1. *O. greenei.*
Lemmas several-awned; blades and spikelets not pilose. 2. *O. californica.*

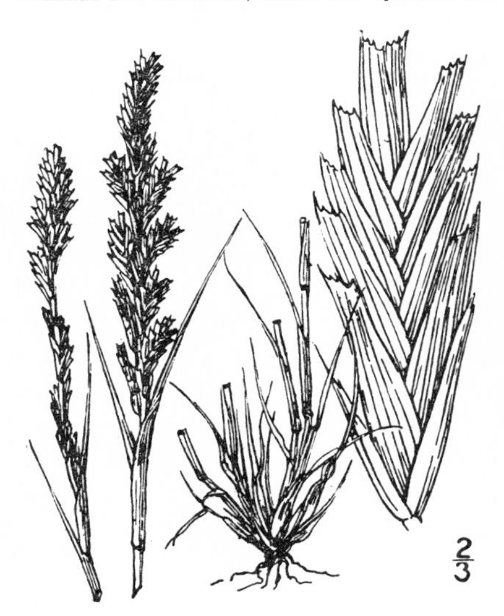

1. Orcuttia greènei Vasey.

Awnless Orcutt Grass. Fig. 414.

Orcuttia greenei Vasey, Bot. Gaz. 16: 146. 1891.

Culms 15–20 cm. tall, scabrous or appressed-pilose, especially at the nodes; sheaths finely papillose, shorter than the internodes, the ligule very short; blades 2–3 cm. long, pilose on the upper surface, inrolled; raceme 3–7 cm. long, pale; glumes and lemmas sparsely long-pilose and more or less papillose, the glumes 4 mm. long, the lemmas 6 mm. long, the obtuse or truncate tip several-toothed, the teeth mucronate but not awned.

Only known from the type collection, "Moist plains of the upper Sacramento, near Chico, California, June, 1890, by Prof. E. L. Greene."

2. Orcuttia califórnica Vasey.

Orcutt Grass. Fig. 415.

Orcuttia californica Vasey, Bull. Torrey Club 13: 219. *pl. 60.* 1886.

Culms 5–10 cm. tall, erect, smooth; sheaths and blades smooth, the latter 1–2 cm. long, mostly basal; raceme 2–6 cm. long, the spikelets distant below, contiguous above; spikelets appressed, about 1 cm. long; glumes 4–5 mm. long; lemmas 5 mm. long, strongly nerved, 5-toothed, the teeth acuminate, awn-tipped, rigid.

Only known from two collections, the type collection from San Quentin Bay, Lower California (*Orcutt*) and Goose Valley, Shasta County (*Eastwood*). June–July.

51. ARÚNDO L. Sp. Pl. 81. 1753.

Spikelets several-flowered, the rachilla glabrous, disarticulating above the glumes and between the florets; glumes somewhat unequal, membranaceous, 3-nerved, narrow, tapering into a slender point, about as long as the spikelet; lemmas thin, 3-nerved, densely long-pilose, gradually narrowed at the summit, the nerves ending in slender teeth, the middle one longer and extending into a straight awn. Tall perennial reeds with broad blades and large plume-like terminal panicles. [An ancient Latin name.]

Species about 6 in the warmer parts of the Old World one introduced in America. Type species, *Arundo donax* L.

1. Arundo dònax L.

Giant Reed. Fig. 416.

Arundo donax L. Sp. Pl. 81. 1753.

Culms stout, as much as 7 meters tall, and 2 cm. in diameter at base, from rough knotty branching rhizomes; blades numerous, broad, flat, 5–7 cm. wide on the main stem, smaller on the branches, the base cordate and more or less hairy-tufted; panicle large, 30–60 cm. long; spikelets about 12 mm. long.

A native of the Old World, frequently cultivated for ornament in tropical America. Rather common in gardens in the southern United States, and escaped along irrigating ditches from Texas to central and southern California. The only California specimen in the National Herbarium is from the Alameda marshes (*Davy*). September. Type locality, European.

52. PHRAGMÌTES Adans. Fam. Pl. 2: 34, 559. 1763.

Spikelets several-flowered, the rachilla disarticulating above the glumes and between the florets, clothed with long silky hairs, the lowest floret staminate or neuter; glumes 3-nerved or the upper 5-nerved, unequal, lanceolate, acute, the upper shorter than the florets, the lower about half as long as the upper; lemmas narrow, long-acuminate, glabrous, 3-nerved, the florets successively smaller, the summits about equal; palea much shorter than the lemma. Perennial reeds with broad flat blades and large terminal panicles. [Greek, growing in hedges.]

Species 3, one in Asia, one in Argentina, and one cosmopolitan. Type species, *Arundo phragmites* L.

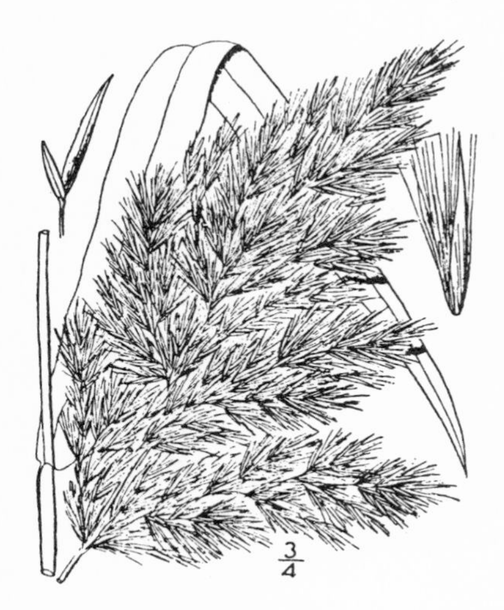

1. Phragmites phragmìtes (L.) Karst.

Common Reed. Fig. 417.

Arundo phragmites L. Sp. Pl. 81. 1753.
Arundo vulgaris Lam. Fl. Franç. 3: 615. 1778.
Phràgmites communis Trin. Fund. Agrost. 134. 1820.
Phragmites phragmites Karst. Deutsch. Fl. 379. 1880.
Phragmites vulgaris B. S. P. Prel. Cat. N. Y. 69. 1888.

Culms as much as 4 meters high, from long creeping rhizomes, these sometimes appearing on the surface of the ground as leafy stolons as much as 10 meters long; blades as much as 5 cm. wide, flat, the base somewhat narrowed, not hairy; panicle 15–40 cm. long; spikelets 12–14 mm. long.

Fresh-water swamps, marshes, and around springs, through the temperate regions of the world; here and there from Washington to California. July–Nov. Type locality, European.

53. TRIÒDIA R. Br. Prodr. Fl. Nov. Holl. 1: 182. 1810.

Spikelets several-flowered, the rachilla disarticulating above the glumes and between the florets; glumes membranaceous, often papery, nearly equal in length, the first sometimes narrower, 1-nerved, or the second rarely 3–5-nerved, acute or acuminate, shorter or longer than the first lemma; lemmas broad, rounded on the back, the apex narrow or broad, minutely emarginate or toothed, or deeply and obtusely lobed, 3-nerved, the nerves distant, parallel or somewhat converging, disappearing before reaching the margin or extending into mucros beyond the margin or extending into the lobes, the midnerve excurrent as a mucro, or short awn, or as a moderately long awn between the lobes, the lower portion of all the nerves pubescent or villous, or the lateral villous nearly the whole length, or in one of our species glabrous except a slight pubescence at base; palea broad, 2-nerved, the nerves near the margin, sometimes villous. Erect tufted perennials, rarely rhizomatous or stoloniferous, the blades usually flat, the inflorescence an open or contracted panicle, or a cluster of few-flowered spikes interspersed with leaves. [Greek, three-tooth.]

Species about 25, mostly in America. Type species, *Triodia pungens* R. Br.

Inflorescence a naked narrow panicle. 1. *T. muticus.*
Inflorescence a leafy head or umbel. 2. *T. pulchellus.*

1. Triodia mùtica (Torr.) Scribn.

Awnless Triodia. Fig. 418.

Tricuspis muticus Torr. U. S. Rep. Expl. Miss. Pacif. 4: 156. 1857.
Triodia mutica Scribn. Bull. Torrey Club 10: 30. 1883.
Tridens muticus Nash in Small, Fl. Southeast. U. S. 143. 1903.

Culms erect, 30–60 cm. high; blades involute, scabrous; panicle narrow, 7–15 cm. long, exserted, the branches short and appressed; spikelets terete, narrow, 8–10 mm. long; glumes about 4 mm. long, 1-nerved, shorter than the spikelets; lemmas pilose on nerves, obtuse, about 4 mm. long, entire or slightly emarginate, awnless.

Dry slopes and gravelly banks, central Sierra Nevada (Silver Mountain, *Brewer*), east to Colorado and Texas. July–Aug. Type locality: New Mexico.

2. Triodia pulchélla H. B. K.

Low Triodia. Fig. 419.

Triodia pulchella H. B. K. Nov. Gen. & Sp. 1: 155. *pl.* 47. 1816.
Tricuspis pulchellus Torr. U. S. Rep. Expl. Miss. Pacif. 4: 156. 1857.
Tridens pulchellus Hitchc. in Jepson, Fl. Calif. 1: 141. 1912.

Low and tufted, usually not over 15 cm. high; culms slender, scabrous or puberulous, consisting of 1 long internode, bearing at the top a fascicle of leaves, the fascicle finally bending over to the ground, taking root and producing other culms, the fascicles also producing the inflorescence; sheaths striate, papery-margined, pilose at base; blades involute, short, scabrous, sharp-pointed, striate; panicles much reduced, usually not exceeding the blades of the fascicle, consisting of 1–5 nearly sessile spikelets; glumes subequal, broad, acuminate, awn-pointed, 1-nerved, 6–8 mm. long, about as long as the spikelet; lemmas 4 mm. long, long-pilose below, cleft about halfway, the awn about as long as or a little longer than the obtuse lobes.

Mesas and rocky hills in the Mojave and Colorado deserts, east to Utah and Texas, and south into Mexico. February–May. Type locality: Mexico.

54. DISSANTHÈLIUM Trin. Linnaea 10: 305. 1836.

Spikelets mostly 2-flowered, the florets distant, the rachilla slender, disarticulating above the glumes and between the florets; glumes nearly equal, acuminate, much longer than the lower floret, mostly longer than all the florets, the first 1-nerved, the second 3-nerved; lemmas strongly compressed, oval or elliptic, acute, awnless, 3-nerved, the lateral nerves near the margin; palea somewhat shorter than the lemma. Annual or perennial grasses with flat blades and narrow panicles. [Greek, double-floret.]

Species 2, one in Mexico and South America, the other in California. Type species, *Dissanthelium supinum* Trin.

1. **Dissanthelium califórnicum** (Nutt.) Benth.

California Dissanthelium. Fig. 420.

Stenochloa californica Nutt. Journ. Acad. Phila. II. 1: 189. 1848.
Dissanthelium californicum Benth. in Hook. Icon. Pl. III. 4: 56. *pl. 1375.* 1881.

Culms 60–100 cm. tall, smooth; leaves smooth; ligule membranaceous, 2–6 mm. long; blades flat, lax; panicle narrow, loose, 15–20 cm. long, the lower clusters of branches rather remote; glumes somewhat unequal, the second about 4 mm. long; lemmas about 2 mm. long, minutely villous, especially below.

Known only from California (Tassajara Hot Springs, *Elmer;* San Clemente Island, *Trask;* Santa Catalina Island, *Gambel*). Type locality: Santa Catalina Island.

55. **ERAGRÓSTIS** Host, Icon. Gram. Austr. 4: 14. *pl. 24.* 1809.

Spikelets few- to many-flowered, the florets usually closely imbricate, the rachilla disarticulating above the glumes and between the florets, or continuous, the lemmas deciduous, the paleas persistent; glumes somewhat unequal, shorter than the first lemma, acute or acuminate, 1-nerved; lemmas acute or acuminate, keeled or rounded on the back, 3-nerved, the nerves usually prominent; palea 2-nerved, the keels sometimes ciliate. Annual or perennial grasses of various habit, the inflorescence an open or contracted panicle. [Greek, love-grass.]

Species more than 100, tropical and temperate regions of the world. Type species, *Briza eragrostis* L.

Plants perennial. — 1. *E. secundiflora.*
Plants annual.
  Spikelets unisexual; plants dioecious or polygamous. — 2. *E. hypnoides.*
  Spikelets perfect.
    Plants with minute glands on the foliage, panicle branches or keels of lemmas.
      Panicle branches appressed, spikelets about 1.5 mm. wide, not glandular. — 8. *E. lutescens.*
      Panicle branches more or less spreading; spikelets 2 mm. or more wide; lemmas usually with one or two glands on the keel.
        Spikelets about 2 mm. wide. — 7. *E. eragrostis.*
        Spikelets 3–4 mm. wide. — 3. *E. cilianensis.*
    Plants not glandular.
      Panicle pilose in the lower axils; pedicels appressed. — 4. *E. caroliniana.*
      Panicle glabrous in the axils; pedicels ascending, somewhat flexuous.
        Pedicels and branchlets noticeably flexuous; spikelets linear, about 1 mm. wide; panicle large and open. — 5. *E. orcuttiana.*
        Pedicels and branchlets not flexuous or only slightly so; spikelets more than 1 mm. wide; panicle large or contracted. — 6. *E. mexicana.*

1. **Eragrostis secundiflòra** Presl.

Sand Eragrostis. Fig. 421.

Eragrostis secundiflora Presl. Rel. Haenk. 1: 276. 1830.
Poa interrupta Nutt. Trans. Am. Phil. Soc. II. 5: 146. 1837, not Lam. 1791.
Poa oxylepis Torr. in Marcy, Expl. Red Riv. 301. *pl. 19.* 1853.
Eragrostis oxylepis Torr. U. S. Rep. Expl. Miss. Pacif. 4: 156. 1857.

Perennial; culms erect or decumbent at base, stiff, 30–60 cm. tall; sheaths pilose at the throat; panicles narrow, the branches ascending, compactly flowered, approximate or more or less remote; spikelets many-flowered, the florets closely imbricate, usually tinged with red; glumes 1-nerved, the second 2 mm. long; lemmas prominently 3-nerved, scabrous on keel, broad at base, the acuminate apex somewhat divergent.

Sandy prairies, Kansas to Florida and Mexico and west to southern California (San Diego, *Orcutt*). August. Type locality: Mexico.

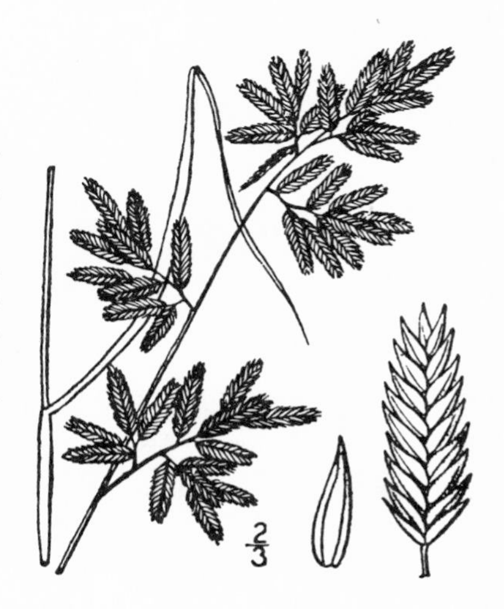

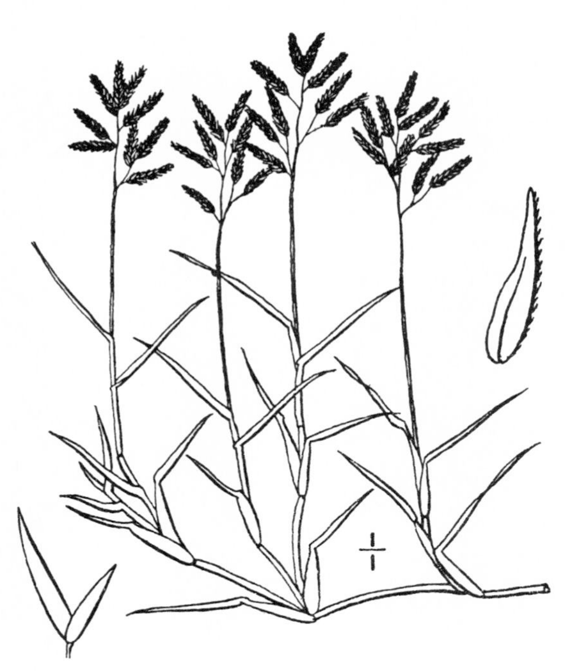

2. Eragrostis hypnoìdes (Lam.) B. S. P.

Creeping Eragrostis. Fig. 422.

Poa hypnoides Lam. Tabl. Encycl. **1**: 185. 1791.
Eragrostis reptans Nees, Agrost. Bras. 514. 1829.
Eragrostis hypnoides B. S. P. Prel. Cat. N. Y. 69. 1888.

Annual, extensively creeping, more or less dioecious; culms slender, 20–40 cm. long, with short, erect or ascending, panicle-bearing branches, 5–10 cm. tall; blades 1–3 cm. long; panicles mostly simple, of rather few lanceolate-oblong spikelets, the fertile inflorescence tending to be capitate; spikelets 10–35-flowered, 5–14 mm. long.

Sandbars and wet shores of rivers and lakes, throughout the United States and south to Argentine. Aug.–Sept. Type locality: tropical America.

3. Eragrostis cilianènsis (All.) Link.

Stink-grass, Candy-grass. Fig. 423.

Briza eragrostis L. Sp. Pl. 70. 1753, not *Poa eragrostis* L. (*Eragrostis eragrostis* Karst).
Poa ciliarensis All. Fl. Pedem. **2**: 246. 1785.
Poa megastachya Koel. Descr. Gram. 181. 1802.
Eragrostis major Host, Icon. Gram. Austr. **4**: 14. *pl. 24.* 1809.
Eragrostis megastachya Link, Hort. Berol. **1**: 187. 1827.
Eragrostis poaeoides megastachya A. Gray, Man. ed. 2. 563. 1856.
Eragrostis minor megastachya Davy in Jepson, Fl. West Mid. Calif. 60. 1901.
Eragrostis cilianensis Link; Vign. Lut. Malpighia **18**: 386. 1904.

Annual, strong-scented when fresh; culms erect or ascending from a decumbent base, rather flaccid, freely branching, 20–60 cm. tall; blades 4–15 cm. long, 3–6 mm. wide; panicles greenish lead-color, 4–15 cm. long, rather densely flowered; spikelets 5–14 mm. long, 3 mm. wide, 10–40-flowered, the florets closely imbricate; pedicels and keels of the acute glumes and lemmas sparingly glandular; lemmas thin, the lateral nerves prominent.

Fields, roadsides, and waste places; a common weed throughout the United States and Mexico, introduced from Europe; infrequent in California and Oregon (Eugene, *Nelson*). June–Aug. Type locality, European.

Several species of Eragrostis have been cultivated at Experiment Station grounds on the Pacific Coast to test their value as forage plants. The two species described below appear to have spread in the vicinity of the testing plots, and others, especially *E. chloromelas* Steud., *E. curvula* (Schrad.) Nees, and *E. plana* Nees, all of South Africa, are likely to spread from their original plantings.

Eragrostis abyssìnica (Jacq.) Link, Hort. Berol. **1**: 192. 1827. (*Poa abyssinica* Jacq. Misc. **2**: 364. 1781.) Annual, as much as 1 meter tall, smooth throughout; panicle large and spreading, glabrous in the axils; spikelets 5–8 mm. long, several-flowered, the florets turgid; lemmas 2–3 mm. long, acute. Kern County (*Leckenby*). Becoming frequent in Arizona. A native of Abyssinia and cultivated in central Africa as a cereal under the name of Teff.

Eragrostis suavèolens Becker in Clous. Beitr. Pflanz. Russ. Reich. **8**: 266. 1851. Annual, branching at the base, geniculate and rooting at the lower nodes; glabrous throughout or with a few long hairs at the mouth of the sheath; blades often elongate, 2–8 mm. wide; panicle one-third to half the entire height of the plant, diffuse, many-flowered, the branchlets and pedicels bearing minute glands; spikelets 5–10 mm. long. Plant sweet-scented when fresh. Kern County, California (*Leckenby*). A native of southeastern Europe.

4. Eragrostis caroliniàna (Spreng.) Scribn.
Carolina Eragrostis. Fig. 424.

Poa caroliniana Spreng. Mant. Fl. Hal. 33. 1807.
Eragrostis purshii Schrad. Linnaea **12**: 451. 1838.
Eragrostis caroliniana Scribn. Mem. Torrey Club **5**: 49. 1894.
Eragrostis pilosa of Jepson's Flora, not Beauv.

Annual; culms erect, decumbent at base or prostrate-spreading, 15–40 cm. tall, diffusely branched at base; sheaths sparingly pilose at summit; blades 2–10 cm. long, 2–3 mm. wide; panicle diffuse, 7–20 cm. long, the lower axils sparingly bearded; spikelets 5–18-flowered, becoming linear, 4–8 mm. long, about 1.5 mm. wide, equaling or shorter than the pedicels; glumes 1-nerved, smooth except the keel, the first 1 mm., the second 1.5 mm. long; lemmas smooth, slightly scabrous on keel, the lower about 1.5 mm. long, .5 mm. wide, the lateral nerves distinct but not prominent.

Fields, waste places, and open ground, Maine to Minnesota and south to Mexico. In California, San Luis Obispo and Tulare Counties southward; also Humboldt County (*Tracy*). July–Aug. Type locality: North Carolina.

5. Eragrostis orcuttiàna Vasey.
Orcutt's Eragrostis. Fig. 425.

Eragrostis orcuttiana Vasey, Contr. U. S. Nat. Herb. 1: 269. 1893.

Resembling *E. mexicana;* differing in the more slender, usually arcuate spikelets on shorter flexuous pedicels; spikelets 5–6 mm. long, .7–1 mm. wide; glumes short, the first 1 mm., the second 1.5 mm. long; lemmas scarcely .5 mm. wide.

Fields and waste places, known only from California (except on ballast at Linnton, Oregon). July–Oct. Type locality: San Diego.

6. Eragrostis mexicàna (Lag.) Link.
Mexican Eragrostis. Fig. 426.

Poa mexicana Lag. Gen. & Sp. Nov. 3. 1816.
Eragrostis mexicana Link, Hort. Berol. 1: 190. 1827.

Annual; culms erect or spreading, 30–60 cm. tall; sheaths hairy at the throat; blades often elongate; panicle large and diffuse, glabrous in the axils, 15–30 cm. long; spikelets 4–6 mm. long, 1.5 mm. wide, mostly 6–12-flowered, the pedicels slender, flexuous, mostly longer than the spikelet; glumes acuminate, the second about 2 mm. long; lemmas smooth, the lower 2 mm. long, .7 mm. wide, the lateral nerves not prominent.

A weed in fields and waste places, New Mexico and southern California (Los Angeles, Riverside) to Mexico. Sept.–Oct. Type locality: Mexico.

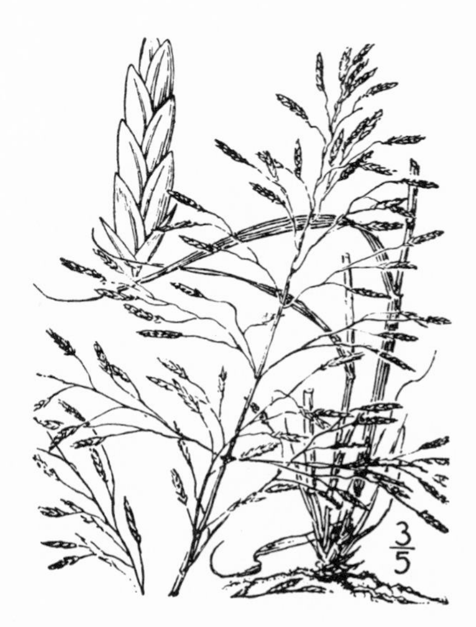

7. Eragrostis eragróstis (L.) Karst.
Low Eragrostis. Fig. 427.

Poa eragrostis L. Sp. Pl. 68. 1753.
Eragrostis poaeoides Beauv. Ess. Agrost. 162. 1812.
Eragrostis minor Host, Fl. Austr. 1: 135. 1827.
Eragrostis eragrostis Karst. Deutsch. Fl. 389. 1880.

Annual; culms tufted, mostly less than 30 cm. tall, glabrous; sheaths smooth, somewhat pilose at the throat; blades 2–6 cm. long, 1 mm. wide, thin, the margins and the sheaths beset with glandular depressions; panicles 5–10 cm. long, the branches spreading; spikelets 8–15-flowered, 6–10 mm. long, 2 mm. wide; lemmas about 1.5 mm. long.

In waste places, introduced from Europe, infrequent in eastern United States, rare in California (Clinton, Amador County, *Hansen*). October. Type locality, European.

8. Eragrostis lutéscens Scribn.
Viscid Eragrostis. Fig. 428.

Eragrostis lutescens Scribn. U. S. Dept. Agr. Div. Agrost. Circ. 9: 7. 1899.

Annual; culms slender, glabrous, cespitose, 10–20 cm. tall or often depauperate; blades flat, narrow, pale green or yellowish, 2–5 cm. long, 2–3 mm. wide; panicles narrow, rather densely flowered, 4–7 cm. long, the branches appressed; spikelets narrowly oblong, 2–4 mm. long, 1.5 mm. wide, about 8–12-flowered; glumes unequal, the first 1 mm. long, the second a little longer; lemmas obtuse, 2 mm. long, distinctly nerved; blades, sheaths, and main panicle branches beset with glandular depressions.

Sandy soil, Idaho to Washington (Almota) and Oregon (The Dalles). Type locality: Almota, Washington.

A specimen collected at Jamacha by Canby in 1894 was referred in the Flora of California (Jepson, Fl. Calif. 1: 142. 1912) to *Eragrostis lugens* Nees, Agrost. Bras. 505. 1829. It is a perennial, 50 cm. tall, with a loose spreading panicle of long-pedicelled linear spikelets, the branches pilose in the axils. It may be the same as forms that have been referred to *E. lugens* but, having glabrous blades, cannot be that species as described by Nees.

Eragrostis cyperoìdes (Thunb.) Beauv., a South African species with spikelets in lateral fascicles on the stout axis, has been found on ballast at Linnton, Oregon (*Nelson*).

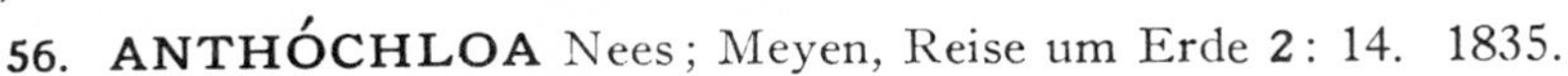

56. ANTHÓCHLOA Nees; Meyen, Reise um Erde 2: 14. 1835.

Spikelets few-flowered, subsessile on a simple axis and imbricate, the rachilla disarticulating above the glumes and between the florets; glumes (in our species) wanting; lemmas thin-membranaceous, flabelliform, petal-like, many-nerved; palea narrower than the lemma, hyaline. Low annuals or perennials with close spikes. [Greek, flower-grass.]

Species 3, 2 in the Andes, 1 in California. Type species, *Anthochloa lepidula* Nees.

1. Anthochloa colusàna (Davy) Scribn.
Colusa Grass. Fig. 429.

Stapfia colusana Davy, Erythea 6: 110. *pl. 3.* 1898.
Neostapfia colusana Davy, Erythea 7: 43. 1899.
Davyella Hack. Oesterr. Bot. Zeitschr. 49: 133. 1899.
Anthochloa colusana Scribn. U. S. Dept. Agr. Div. Agrost. Bull. 17: 221. *f. 517.* 1899.

Annual; culms ascending from a decumbent base, 7–30 cm. long; leaves overlapping, pale green, scarious between the nerves, loosely folded around the culm but not differentiated into sheath and blade, about 12 mm. wide at the middle, tapering to each end, 5–10 cm. long, keeled on the back above, plicate, minutely ciliate, with raised glands on the margins and nerves; panicles pale green, cylindric, at first partially included, never much exserted, 3–7 cm. long, 8–12 mm. wide, the upper portion of the axis bearing, instead of spikelets, lanceolate-linear empty bracts 8 mm. long; spikelets subsessile, usually 5-flowered, 6–7 mm. long, imbricate; glumes wanting; lemmas flabellate, very broad, many-nerved, 5 mm. long, ciliolate-fringed.

Only known from the type collection, "near Princeton, Colusa County, California, bordering rain-pools on the hard uncultivated alkali 'goose-lands,' beside the stage road to Norman" (*Davy*).

57. MÉLICA L. Sp. Pl. 66. 1753.

Spikelets 2 to several-flowered, the rachilla disarticulating above the glumes and prolonged beyond the perfect florets and bearing at the apex 2 or 3 gradually smaller empty lemmas, convolute together or the upper inclosed in the lower; glumes somewhat unequal, thin, often papery, scarious-margined, obtuse or acute, 3–5-nerved, sometimes nearly as long as the lower floret; lemmas convex, several-nerved, membranaceous or rather firm, scarious-margined, sometimes conspicuously so, awnless or sometimes awned from between the teeth of the bifid apex. Moderately tall perennials, with the base of the culm often swollen into a corm, closed sheaths, usually flat blades, narrow or sometimes open, usually simple panicles. [An Italian name for sorghum, from the Greek, *mel,* honey.]

Species about 60 in the cooler parts of the world. Type species, *Melica nutans* L.

Spikelets narrow; glumes usually narrow, scarious-margined; sterile lemmas similar to the fertile, the latter acute or awned. (Section Bromelica.)
  Lemmas long-awned from a bifid apex.
    Branches of panicle few, distant, spreading, naked on the lower half. — 1. *M. smithii.*
    Branches of panicle short, appressed, spikelet-bearing from near the base. — 2. *M. aristata.*
  Lemmas awnless or nearly so.
    Culms not bulbous at base; lemmas mucronate. — 3. *M. harfordii.*
    Culms bulbous at base; lemmas acute or acuminate.
      Lemmas acuminate, usually pilose; panicle narrow, the branches short, usually appressed, sometimes spreading. — 4. *M. subulata.*
      Lemmas narrowed to an obtuse or obscurely bifid apex, not pilose; panicle broad, the branches long and spreading. — 5. *M. geyeri.*
Spikelets broad; glumes broad and papery; sterile lemmas small and convolute, more or less hidden in the upper fertile lemmas.
  Culms bulbous at base.
    Pedicels capillary, flexuous or recurved. — 6. *M. spectabilis.*
    Pedicels stouter, appressed.
      Panicle narrow; branches short, erect. — 7. *M. bella.*
      Panicle open; branches spreading.
        First glume about 3 mm. long. — 8. *M. fugax.*
        First glume about 8 mm. long. — 9. *M. inflata.*
  Culms not distinctly bulbous at base (somewhat bulbous in *M. bulbosa*).
    Spikelets large, reflexed. — 10. *M. stricta.*
    Spikelets smaller, not reflexed.
      Fertile florets 3 or 4 in each spikelet; spikelets 10–12 mm. long.
        Spikelets silvery white; glumes about as long as spikelet; plant tall and somewhat woody. — 11. *M. frutescens.*
        Spikelets tawny or purplish; glumes shorter than spikelet; plant lower, herbaceous. — 12. *M. bulbosa.*
      Fertile florets 1 or 2 in each spikelet; spikelets 4–6 mm. long.
        Fertile lemmas pubescent. — 13. *M. torreyana.*
        Fertile lemmas glabrous. — 14. *M. imperfecta.*

1. Melica smíthii (Porter) Vasey. Smith's Melica. Fig. 430.

Avena smithii Porter; A. Gray, Man. ed. 5. 640. 1867.
Melica smithii Vasey, Bull. Torrey Club **15**: 294. 1888.
Melica retrofracta Suksdorf, Deutsch. Bot. Monatschr. **19**: 92. 1901.
Bromelica smithii Farw. Rhodora **21**: 77. 1919.

Culms slender, 60–120 cm. tall; sheaths retrorsely scabrous; blades flat, lax, scabrous, 10–20 cm. long, 6–12 mm. wide; panicle 12–25 cm. long, the branches solitary, distant, spreading, naked below; spikelets 3–6-flowered, 18–20 mm. long, sometimes purplish; glumes acute; lemmas glabrous, about 10 mm. long, awned, the awn 3–5 mm. long.

Moist woodlands, Washington and Oregon to northern Michigan. July. Type locality: Lake Superior region.

2. Melica aristàta Thurb. Awned Melica. Fig. 431.

Melica aristata Thurb.; Boland. Proc. Calif. Acad. **4**: 103. 1870.
Bromelica aristata Farw. Rhodora **21**: 77. 1919.

Culms erect or decumbent below, not bulbous at at base, smooth, 60–100 cm. tall; sheaths scabrous or pubescent; blades flat, more or less pubescent; panicle narrow, the branches short and appressed; glumes narrow, 5-nerved, 10–12 mm. long; lemmas 5-nerved, scabrous, bifid at apex, awned, the awn 6–10 mm. long.

Dry woods, slopes, and meadows, Washington, and southward in the Sierra Nevada to Fresno County. June–Aug. Type locality: Clark's (now Wawona), California.

3. **Melica harfòrdii** Boland.
Harford's Melica. Fig. 432.

Melica harfordii Boland. Proc. Calif. Acad. **4**: 102. 1870.
Bromelica harfordii Farw. Rhodora **21**: 78. 1919.

Culms 60–120 cm. tall, decumbent below, smooth, not bulbous at base; sheaths smooth or villous; blades scabrous, firm; panicle narrow, the branches appressed; glumes narrow, about 6 mm. long, obtuse; lemmas 7-nerved, pilose on lower part of margin, the apex emarginate, mucronate or short-awned; awn less than 2 mm. long.

Open dry woods and slopes, British Columbia south in the Coast Ranges to Monterey County. June–July. Type locality: Santa Cruz road near Lexington, California.

Melica harfordii minor Vasey, Bull. Torrey Club **15**: 48. 1888. (*Melica harfordii tenuior* Piper, Contr. U. S. Nat. Herb. 11: 127. 1906. *Bromelica harfordii minor* Farw. Rhodora **21**: 78. 1919.) Culms more slender, mostly less than 50 cm. tall; blades short and narrow, about 1 mm. wide, mostly involute. Washington and Oregon (Siskiyou Mountains, the type locality; McKenzie River). July.

4. **Melica subulàta** (Griseb.) Scribn.
Alaska Onion-grass. Fig 433.

Bromus subulatus Griseb. in Ledeb. Fl. Ross. **4**: 358. 1853.
Festuca cepacea Phil. Linnaea **33**: 297. 1865.
Melica acuminata Boland. Proc. Calif. Acad. **4**: 104. 1870.
Melica subulata Scribn. Proc. Acad. Phila. **1885**: 47. 1885.
Melica cepacea Scribn. U. S. Dept. Agr. Div. Agrost. Circ. **30**: 8. 1901.
Bromelica subulata Farw. Rhodora **21**: 78. 1919.

Culms 60–120 cm. tall, bulbous at base; panicle narrow, the branches appressed or sometimes spreading; spikelets narrow, 2–2.5 cm. long, loosely several-flowered; glumes narrow, obscurely nerved, the second about 8 mm. long, shorter than the lower lemma; lemmas prominently 7-nerved, gradually narrowed to an acuminate point, awnless, the keel and marginal nerves pilose-ciliate.

Meadows, banks, and shady slopes, Alaska to Lake Tahoe and Mount Tamalpais, east to Wyoming, May–July. Type locality: Unalaska.

5. **Melica geỳeri** Munro.
Geyer's Onion-grass. Fig. 434.

Melica geyeri Munro; Boland. Proc. Calif. Acad. **4**: 103. 1870.
Melica bromoides Boland.; A. Gray, Proc. Am. Acad. **8**: 409. 1872.
Bromelica geyeri Farw. Rhodora **21**: 78. 1919.

Culms 1–1.5 meters tall, bulbous at base; sheaths glabrous or sometimes pubescent; blades scabrous, flat; panicle open, the lower branches slender, spreading, bearing a few spikelets above the middle; spikelets narrow, 12–20 mm. long; glumes broad, smooth, papery, the second about 6 mm. long; lemmas scaberulous, obtuse, awnless.

Wooded ravines and along streams, Oregon to central California; also in British Columbia. May–July. Type locality: Russian River Valley.

6. Melica spectábilis Scribn.

Showy Onion-grass. Fig. 435.

Melica spectabilis Scribn. Proc. Acad. Phila. 1885: 45. 1885.

Culms 30–100 cm. tall, bulbous at base; panicle narrow, the branches appressed; sheaths pubescent; spikelets purple-tinged, 10–15 mm. long, 4 or 5-flowered, the pedicels slender, curved; glumes broad and papery, shorter than the lower lemma; lemmas strongly 7-nerved, obtuse, awnless.

Rocky or open woods and thickets, in the Arid Transition Zone; Washington to Montana and Colorado, south to northwestern California (Sherwood, *Hitchcock*). May–July. Type locality: Colorado.

7. Melica bélla Piper.

Onion-grass. Fig. 436.

Melica bella Piper, U. S. Dept. Agr. Div. Agrost. Circ. 27: 10. 1900.

Culms 30–60 cm. tall, bulbous at base; sheaths and blades glabrous or scabrous; panicle narrow, the branches short and appressed; spikelets 10–15 mm. long, papery with age; glumes broad, the second 8 mm. long; lemmas obscurely nerved, obtuse or slightly emarginate, awnless.

Rocky woods and hills, Arid Transition Zone; Washington to central California, east to Montana and Washington. May–Aug. Type locality: "Upper Platte."

Melica bella intónsa Piper, Contr. U. S. Nat. Herb. 11: 128. 1906. Leaves softly pubescent. Washington to central California and Nevada. Type locality: Wenas, Washington.

8. Melica fùgax Boland.

Small Onion-grass. Fig. 437.

Melica fugax Boland. Proc. Calif. Acad. 4: 104. 1870.
Melica fugax madophylla Piper, Contr. U. S. Nat. Herb. 11: 128. 1906.

Culms 30–60 cm. tall, bulbous at base; sheaths retrorsely scabrous; blades scabrous, usually pubescent beneath (glabrous in var. *madophylla*); panicle narrow, open, the few lower branches 2–5 cm. long, stiffly spreading, few-flowered; spikelet about 6 mm. long, 2- or 3-flowered, usually purple-tinged; glumes broad, papery, the second nearly as long as the lower lemma; lemmas obscurely nerved, obtuse or emarginate, awnless.

Dry hills in the Arid Transition Zone; Washington to central California (Lake Tahoe) and east to Nevada. June–Aug. Type locality: Donner Lake, California.

9. **Melica inflàta** (Boland.) Vasey.

Sierra Onion-grass. Fig. 438.

Melica poaeoides inflata Boland. Proc. Calif. Acad. **4**: 101. 1870.
Melica inflata Vasey, Contr. U. S. Nat. Herb. **1**: 269. 1893.

Culms 60–100 cm. tall, bulbous at base; panicle more or less open, the few branches long, spreading at least in anthesis; spikelets several-flowered, about 12 mm. long, broad, pale green; glumes broad, shorter than the lemmas, scabrous on the strong nerves; lemmas scabrous, strongly nerved.

Wet meadows; known only from California (Mount Shasta; Yosemite National Park). June. Type locality: Yosemite Valley.

10. **Melica stricta** Boland.

Nodding Melica. Fig. 439.

Melica stricta Boland. Proc. Calif. Acad. **3**: 4. 1863.

Culms 15–40 cm. tall, the base somewhat thickened but not bulbous; panicle narrow, few-flowered, nearly simple, usually 1 or 2 branches below; spikelets large, about 12 mm. long, reflexed on rather delicate pedicels; glumes nearly as long as the spikelet, longer than the lower lemma; lemma scabrous, obtuse, awnless.

Rocky slopes and banks in the Arid Transition Zone; Oregon to San Bernardino Mountains, and east to Utah. June–July. Type locality: Yosemite Valley.

11. **Melica frutéscens** Scribn.

Tall Melica. Fig. 440.

Melica frutescens Scribn. Proc. Acad. Phila. **1885**: 45. *pl. 1. f. 15, 16.* 1885.

Culms 1–2 meters tall, rather woody below, not bulbous at base; blades short, especially on the branches and innovations; panicle silvery-shining, narrow, the branches short and appressed; spikelets about 12 mm. long; glumes about as long as the spikelet, prominently 5-nerved; lemmas acute, entire, awnless, 7-nerved.

Hills and cañons, Lower Sonoran Zone; southern California to Lower California. Apr.–May. Type locality: San Diego.

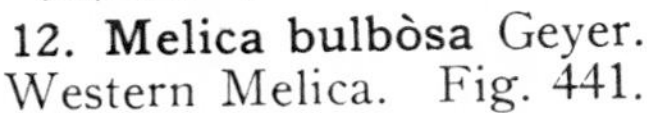

12. **Melica bulbòsa** Geyer.

Western Melica. Fig. 441.

Melica bulbosa Geyer; Thurb. in S. Wats. Bot. Calif. **2**: 304. 1880.
Melica californica Scribn. Proc. Acad. Phila. **1885**: 46. 1885.
Melica longiligula Scribn. & Kearn. U. S. Dept. Agr. Div. Agrost. Bull. **17**: 225. *f. 521.* 1899.

Culms 60–120 cm. tall, the base usually decumbent and often more or less bulbous or corm-like; lower sheaths on the older culms persistent, brown and split into numerous fibers; panicle narrow, rather densely flowered, 10–20 cm. long, tawny or purplish, not silvery-shining; spikelets 10–12 mm. long, papery, 3–4-flowered; second glume about 7 mm. long; lemmas rather prominently 7-nerved, awnless.

Mountain meadows and rocky woods, Oregon to Nevada and south to southern California. May–June. Type locality: Santa Ynez.

13. **Melica torreyàna** Scribn.

Torrey's Melica. Fig. 442.

Melica torreyana Scribn. Proc. Acad. Phila. **1885**: 43. 1885.

Culms from a loose and decumbent base, 30–100 cm. tall, not bulbous; blades flat, lax; panicle narrow, rather loose, the branches more or less fascicled, appressed or ascending, the lower fascicles distant; spikelets 4–6 mm. long, with 1 or 2 perfect florets and a rudiment; glumes strongly nerved, nearly as long as spikelet; lemmas pubescent; rudiment long-pediceled, obovoid-truncate, divergent.

Thickets and banks at low altitudes, in the Arid Transition Zone; central California, especially in the Bay region. Apr.–June. Type locality: New Almaden.

14. **Melica imperfécta** Trin.

Small-flowered Melica. Fig. 443.

Melica imperfecta Trin. Mém. Acad. St. Pétersb. VI. Sci. Nat. 2[1]: 59. 1836.
Melica colpodioides Nees, Ann. Nat. Hist. **1**: 283. 1838.
Melica panicoides Nutt. Journ. Acad. Phila. II. **1**: 188. 1848.
Melica poaeoides Nutt. Journ. Acad. Phila. II. **1**: 188. 1848.
Poa thurberiana Vasey, U. S. Dept. Agr. Div. Bot. Bull. **13**[2]: *pl. 84.* 1893 (as to plant figured).
Melica parishii Vasey; Beal, Grasses N. Am. **2**: 500. 1896.

Culms erect, 30–100 cm. tall; blades narrow, usually not over 2 mm. wide; panicle narrow, 5–30 cm. in length, the branches more or less fascicled, long and short together; spikelets 4–6 mm. long, purple-tinged, usually with 1 perfect floret and a rudiment; glumes indistinctly nerved; lemma a little longer than the glumes, smooth, indistinctly nerved, obtuse; rudiment oblong, short-stipitate, appressed to the palea.

Dry open woods and rocky hillsides, Arid Transition and Upper Sonoran Zones; Coast Ranges from the Bay region south into Lower California. Mar.–June. Type locality: California.

Melica imperfecta refrácta Thurb. in S. Wats. Bot. Calif. **2**: 303. 1880. Lower branches of panicle spreading or reflexed; blades pubescent. Southern California. Type locality: San Bernardino.

Melica imperfecta flexuòsa Boland. Proc. Calif. Acad. **4**: 101. 1870. Lower branches of panicle spreading or reflexed; blades glabrous. Central California to Lower California. Type locality: "Mariposa to Clark's."

Melica imperfecta mìnor Scribn. Proc. Acad. Phila. **1885**: 42. 1885. Southern California and Lower California. Blades glabrous, very narrow; plant low, less than 30 cm. tall, a scarcely distinct variety. Type locality: San Bernardino Mountains.

58. **PLEUROPÒGON** R. Br. Suppl. App. Parry's Voy. 189. *pl. D.* 1823.

Spikelets several- to many-flowered, linear, the rachilla disarticulating above the glumes and between the florets; glumes unequal, membranaceous or subhyaline, scarious at the somewhat lacerate tip, the first 1-nerved, the second indistinctly 3-nerved; lemmas membranaceous, 7-nerved, with a round indurate callus at base, the apex entire or 2-toothed, the midnerve extending into a short mucro or into an awn; palea 2-keeled, the keels winged. Soft annuals or perennials with flat blades and loose racemes of rather large spikelets. [Greek, side-beard.]

Species 3, 1 in the arctic region, 2 on the Pacific Coast of the United States. Type species, *Pleuropogon sabinii* R. Br.

Lemmas about 6 mm. long, scabrous and strongly nerved; spikelets not refracted; culm usually not over 60 cm. tall. 1. *P. californicus.*
Lemmas about 8 mm. long, smooth or slightly scabrous; spikelets often refracted; culms usually over 1 meter tall. 2. *P. refractus.*

1. Pleuropogon califórnicus (Nees) Benth.

California Pleuropogon. Fig. 444.

Lophochlaena californica Nees, Ann. Nat. Hist. **1**: 283. 1838.
Pleuropogon californicus Benth.; Vasey, Grasses U. S. 40. 1883.

Annual; culms 30–60 cm. tall; blades short, abruptly narrowed at apex; racemes 15–20 cm. long; spikelets distant, about 2.5 cm. long, erect or somewhat spreading, short-pediceled; glumes obtuse, erose at apex, the first 4 mm., the second 6 mm. long; scabrous, toothed and scarious at apex, the nerves prominent; the awn variable; usually 6–12 mm. long, sometimes wanting; wings of palea prominent, cleft, forming a tooth about the middle.

Wet meadows and marshy ground, Transition Zone; Mendocino County to the San Francisco Bay region. Apr.–June. Type locality: California.

2. Pleuropogon refráctus (A. Gray) Benth.

Nodding Pleuropogon. Fig. 445.

Lophochlaena refracta A. Gray, Proc. Am. Acad. **8**: 409. 1872.
Pleuropogon refractus Benth.; Vasey, Grasses U. S. 40. 1883.

Perennial; culms 1–1.5 meters tall; spikelets about as in *P. californicus,* spreading, or often reflexed; lemmas 8 mm. long, only minutely scabrous, the nerves less prominent; awn variable, as much as 12 mm. long or nearly wanting; palea narrow, keeled to about the middle, scarcely toothed.

Bogs, wet meadows, and mountain streams, Transition Zone; west of the Cascades, Washington to Mendocino County, California. May–Aug. Type locality: Oregon.

59. DISTÍCHLIS Raf. Journ. de Phys. 89: 104. 1819.

Spikelets dioecious, several- to many-flowered, the rachilla of the pistillate spikelets disarticulating above the glumes and between the florets; glumes unequal, broad, acute, keeled, mostly 3-nerved, the lateral nerves sometimes faint, or obscured by striations and intermediate nerves; lemmas closely imbricate, firm, the pistillate coriaceous, the margins bowed out near base, acute or acutish, 3-nerved, with several intermediate nerves or striations; palea as long as the lemma or shorter, the pistillate coriaceous, inclosing the grain. Low perennials with extensively creeping scaly rhizomes, erect rather rigid culms, and short dense rather few-flowered panicles. [Greek, two-ranked.]

Species about 6 in salt marshes of the coast and interior in America. Type species, *Uniola spicata* L.

1. Distichlis spicàta (L.) Greene.

Salt Grass. Fig. 446.

Uniola spicata L. Sp. Pl. 71. 1753.
Distichlis maritima Raf. Journ. de Phys. **89**: 104. 1819.
Uniola stricta Torr. Ann. Lyc. N. Y. **1**: 155. 1824.
Distichlis spicata Greene, Bull. Calif. Acad. **2**: 415. 1887.
Distichlis dentata Rydb. Bull. Torrey Club **36**: 536. 1909.

Plants forming extensive patches of tough sod, pale or glaucous; culms 10–60 cm. tall· sheaths overlapping; ligule short; blades often conspicuously distichous, rigidly ascending, long-pilose at the very base; panicle narrow, 2–7 cm. long; spikelets 8–16 mm. long, the florets closely imbricate; pistillate spikelets more turgid than the staminate.

Salt marshes and alkaline soil, at low altitudes, Nova Scotia to British Columbia, south to Mexico. Apr.–July. Type locality: north Atlantic Coast of America.

60. BRÌZA L. Sp. Pl. 70. 1753.

Spikelets several-flowered, broad, often cordate, the florets crowded and spreading horizontally, the rachilla glabrous, disarticulating above the glumes and between the florets, the uppermost floret reduced; glumes about equal, broad, papery-chartaceous with scarious margins; lemmas papery, broad and scarious-margined, semicordate at base, several-nerved, the nerves often obscure, the apex in our species obtuse or acutish; palea much shorter than the lemma. Low annuals or perennials, with erect culms, flat blades, and usually open showy panicles, the pedicels in our species being capillary, allowing the spikelets to vibrate in the wind. [Greek for a kind of grain.]

Species about 20, the two found in the United States introduced from Europe. Type species, *Briza media* L.

Panicle drooping; spikelets 10 mm. broad. 1. *B. maxima.*
Panicle erect; spikelets not over 4 mm. broad. 2. *B. minor.*

1. Briza máxima L.

Large Quaking-grass. Fig. 447.

Briza maxima L. Sp. Pl. 70. 1753.

Annual; culms erect or decumbent at base, 30–60 cm. tall; panicle drooping, few-flowered; spikelets ovate, 12 mm. long or more, 10 mm. broad, the pedicels slender, drooping; glumes and lemmas usually purple- or brown-margined.

A native of Europe; sometimes cultivated in gardens for ornament; sparingly escaped in California (Big Sur, *Davy*). Type locality, European.

2. Briza mìnor L.

Small Quaking-grass. Fig. 448.

Briza minor L. Sp. Pl. 70. 1753.

Annual; culms erect, 10–40 cm. tall; panicle erect, pyramidal, many-flowered, the main branches stiffly ascending, the capillary branchlets spreading; spikelets triangular-ovate, 3 mm. long.

Naturalized from Europe, rather common from central California to British Columbia; also in the Eastern States. May–July. Type locality, European.

61. DACTYLIS L. Sp. Pl. 71. 1753.

Spikelets few-flowered, compressed, nearly sessile in dense 1-sided fascicles, these arranged at the ends of the few branches of a panicle, the rachilla finally disarticulating between the spikelets; glumes 2, unequal, carinate, acute, hispid-ciliate on the keel; lemmas compressed-keeled, mucronate, 5-nerved, ciliate on the keel. Perennials with flat blades and fascicled spikelets. [Greek, a finger.]

Species 2 or 3, one cultivated and naturalized in the United States. Type species, *Dactylis glomerata* L.

1. Dactylis glomeràta L.

Orchard-grass. Fig. 449.

Dactylis glomerata L. Sp. Pl. 71. 1753.

Plants in large tussocks; culms erect, 60–120 cm. tall; blades broadly linear; panicle 7–15 cm. long, the few stiff branches naked below, contracted after flowering; spikelets crowded in dense 1-sided clusters at the ends of the branches.

A native of Europe, commonly cultivated in the cooler regions of the United States as a meadow grass. Escaped along roadsides and in waste places, especially at intermediate altitudes, throughout our area. June–Aug. Type locality, European.

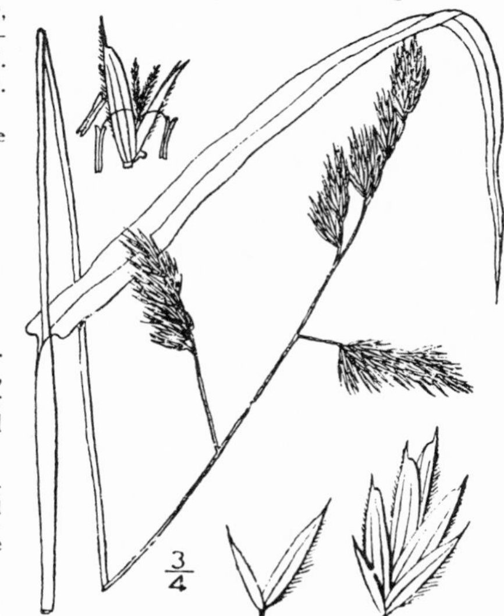

62. CYNOSÙRUS L. Sp. Pl. 72. 1753.

Spikelets of two kinds, sterile and fertile intermixed in unilateral fascicles, these arranged in a dense spike-like panicle; sterile spikelets consisting of 2 glumes and several narrow acuminate 1-nerved lemmas on a continuous rachilla; fertile spikelets 2–3-flowered, the glumes narrow, the lemmas broader, rounded on the back, bearing a slender short awn, the rachilla disarticulating above the glumes. [Greek, dog-tail.]

Species 4, in the Mediterranean region, 1 occasionally cultivated in the United States, and sparingly escaped into waste places. Type species, *Cynosurus cristatus* L.

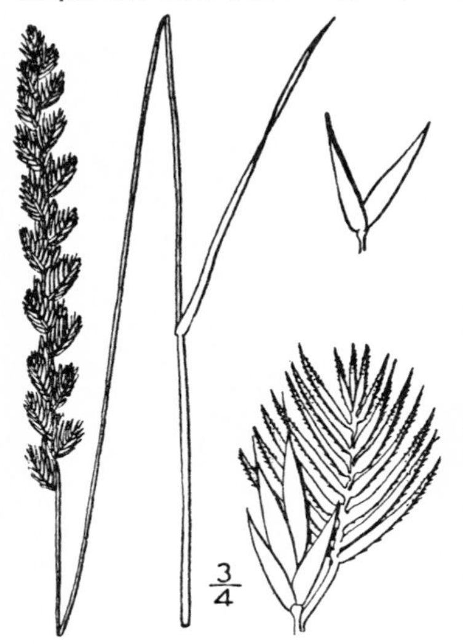

1. Cynosurus cristàtus L.

Crested Dog's-tail Grass. Fig. 450.

Cynosurus cristatus L. Sp. Pl. 72. 1753.

Perennial; culms slender, erect, 30–80 cm. tall; blades short, narrow, erect, glabrous or scabrous toward apex; panicles spike-like, 3–10 cm. long about 5 mm. thick.

Sparingly occurring in fields and waste places in the eastern States, and in Oregon (Portland, Salem). June. Type locality, European.

Cynosurus echinàtus L., a low grass with ovate heads 1–3 cm. long, and bristly-awned spikelets, has been found at Sausalito, California (*Eastwood*), and Eugene, Oregon (*Bradshaw*); also on Vancouver Island.

63. LAMÁRCKIA Moench, Meth. Pl. 201. 1794.

[ACHYRODES Boehmer in Ludw. Def. Gen. Pl. 420. 1760.]

Spikelets of 2 forms, in fascicles, the terminal one of each fascicle fertile, the others sterile; fertile spikelet with 1 perfect floret, the rachilla produced beyond the floret as a slender stipe, bearing a small awned empty lemma or reduced wholly to an awn; glumes narrow, acuminate or short-awned, 1-nerved; lemma broader, raised on a slender stipe; scarcely nerved, bearing just below the apex a slender straight awn; sterile spikelets linear, 1–3 in each fascicle, consisting of 2 glumes similar to those of the fertile spikelet, and numerous distichously imbricate, obtuse awnless empty lemmas. A low erect annual with flat blades and oblong 1-sided compact panicles of crowded fasciculate spikelets, the fertile being hidden, except the awns, by the numerous sterile ones. [Name in honor of Antoine Pierre Monnet, Chevalier de La Marck, French naturalist.]

Species 1, a native of southern Europe, naturalized in southern California. Type species, *Cynosurus aureus* L.

1. Lamarckia aùrea (L.) Moench.

Golden-top. Fig. 451.

Cynosurus aureus L. Sp. Pl. 73. 1753.
Lamarckia aurea Moench, Meth. Pl. 201. 1794.
Achyrodes aureum Kuntze, Rev. Gen. Pl. **2**: 758. 1891.

Culms erect, or decumbent at base, 10–40 cm. tall; leaves smooth; ligule prominent, decurrent as a broad, scarious margin; panicle dense, 2–7 cm. long, 1–2 cm. wide, shining, golden-yellow or purplish, the branches close, short, erect; pedicels fascicled, somewhat clavate, pubescent, spreading at right angles or drooping, the fascicle with a tuft of long whitish hairs at the base; fertile spikelet about 2 mm. long, the sterile 6–8 mm. long; glumes narrow, hyaline, 2 mm. long; lemmas awned from below the apex.

A native of the Mediterranean region, abundantly naturalized in southern California, rarer northward to Santa Clara County; also in northern Mexico. Mar.–May. Type locality, European.

64. FLUMÍNEA Fries, Summ. Veg. Scand. 247. 1846.

[SCOLOCHLOA Link, Hort. Berol. 1: 136. 1827, not Mert. & Koch, 1823.]

Spikelets 3–4-flowered, the rachilla disarticulating above the glumes and between the florets; glumes somewhat unequal, somewhat scarious and lacerate at summit, the first 3-nerved, the second 5-nerved, about as long as the first lemma; lemmas firm, rounded on the back, villous on the callus, 7-nerved, the nerves rather faint and unequal, extending into the scarious lacerate apex; palea narrow, flat, about as long as the lemma. Tall perennials with succulent rhizomes, flat blades, and spreading panicles. [Latin, belonging to rivers.]

Species 2, one in Saghalin, the other in northern Eurasia and northern North America, extending south to Iowa. Type species, *Festuca borealis* Mert. & Koch.

1. Fluminea festucàcea (Willd.) Hitchc. Fluminea. Fig. 452.

Arundo festucacea Willd. Enum. Pl. 1: 126. 1809.
Festuca borealis Mert. & Koch, Deutschl. Fl. 1: 664. 1823.
Scolochloa festucacea Link, Hort. Berol. 1: 137. 1827.
Fluminea arundinacea Fries, Summ. Veg. Scand. 1: 247. 1846.
Graphephorum festucaceum A. Gray, Ann. Bot. Soc. Can. 1: 57. 1861.
Fluminea festucacea Hitchc. U. S. Dept. Agr. Bull. 772: 38. 1920.

Culms erect, smooth, 1–1.5 meters tall; blades smooth, with scabrous margins, 15–30 cm. long, 4–8 mm. wide; panicle 15–30 cm. long, open, the branches ascending, several at the lower nodes, scabrous, naked below, the longer 7–10 cm. long; spikelets 6–8 mm. long; glumes acute; lemmas scabrous, 7-nerved.

Wet places, Manitoba to British Columbia, south to Iowa, and extending into eastern Oregon; also in northern Europe. July. Type locality, European.

65. PÒA L. Sp. Pl. 67. 1753.

Spikelets 2 to several-flowered, the rachilla disarticulating above the glumes and between the florets, the uppermost floret reduced or rudimentary; glumes acute, keeled, somewhat unequal, the first 1-nerved, the second usually 2-nerved; lemmas somewhat keeled, obtuse or acutish, awnless, membranaceous, often somewhat scarious at the tip, 5-nerved, the nerves sometimes pilose. Annual or usually perennial species of low or rather tall grasses with spikelets in open or contracted panicles, the blades, at least when flat, ending in a navicular point. [Greek, grass, a fodder.]

Species probably over 200 in the temperate and cool regions of the world. Type species, *Poa pratensis* L.

Plants annual.
  Lemmas villous on the nerves below.
    Panicle pyramidal, open; sheaths smooth. — 1. *P. annua.*
    Panicle narrow, contracted; sheaths scabrous. — 2. *P. bigelovii.*
  Lemmas not villous on the keel and nerves.
    Sheaths usually rough; lemmas pubescent on the back. — 3. *P. howellii.*
    Sheaths smooth; lemmas smooth. — 4. *P. bolanderi.*
Plants perennial.
  Creeping rhizomes present.
    Culms conspicuously flattened. — 5. *P. compressa.*
    Culms terete or slightly flattened.
      Glumes about 8 mm. long; spikelet 1–1.5 cm. long. — 6. *P. macrantha.*
      Glumes and spikelets smaller.
        Blades involute; first two species plants of seacoast dunes.
          Lemmas pilose on the keel; panicle a dense ovoid head; plants dioecious. — 7. *P. douglasii.*
          Lemmas not pilose; panicle small but rather open.
            Glumes about 3 mm. long; web sparse. — 8. *P. confinis.*
            Glumes about 5 mm. long; web copious. — 9. *P. piperi.*
        Blades flat or folded; plants of the interior.
          Lemmas without cobwebby hairs at base.
            Panicle almost spikelike; glumes 2 mm. long. — 10. *P. atropurpurea.*
            Panicle open; glumes 4 mm. long. — 11. *P. nervosa.*
          Lemmas with a tuft of cobwebby hairs at base.
            Lemmas not pilose on the keel or nerves. — 12. *P. kelloggii.*
            Lemmas pilose on keel and marginal nerves.
              Lemmas 3 mm. long. — 13. *P. pratensis.*
              Lemmas 5 mm. long. — 14. *P. rhizomata.*
  Creeping rhizomes wanting.
    Lemmas villous on the nerves or provided with a tuft of cobwebby hairs at base.
      Lemmas with a tuft of cobwebby hairs at base.
        Keel of lemma slightly villous, the lateral nerves glabrous. — 15. *P. trivialis.*

Keel and lateral nerves villous.
Sheaths retrorsely scaberulous; culms 1–1.5 meters tall. 17. *P. occidentalis.*
Sheaths glabrous or very slightly scaberulous.
Plants of the Alpine Zone, low and tufted; web scant. 16. *P. leptocoma.*
Plants of the Transition Zone, mostly 50–100 cm. tall; web abundant. 18. *P. palustris.*

Lemma with no tuft of cobwebby hairs at base.
Blades short and flat, 2–5 mm. wide; an alpine plant. 19. *P. alpina.*
Blades long and slender, mostly folded or involute.
Blades soft and lax, flat or loosely involute.
Panicles long-exserted; culms stiffly erect; plants of alpine regions. 20. *P. rupicola.*
Panicles short-exserted; culms somewhat decumbent at base; plants of the seacoast. 21. *P. pachypholis.*
Blades firm and stiff, mostly basal, the panicles long-exserted; plants of deserts.
Ligule 5–7 mm. long, acute. 22. *P. longiligula.*
Ligule short, rounded or truncate. 23. *P. fendleriana.*

Lemma not villous on nerves nor with tuft of cobwebby hairs at base, sometimes pubescent on lower part of back.
Lemma minutely pubescent on the lower half or third, the hairs somewhat curled.
Sheaths scaberulous; panicle usually narrow and rather long, the branches short and appressed. 24. *P. scabrella.*
Sheaths smooth.
Culms decumbent and loose at base; plants usually lax; panicle open, the branches slender, spreading. 25. *P. gracillima.*
Culms erect from a densely tufted base; panicle narrow, the branches appressed except during anthesis.
Plants slender, averaging less than 30 cm. tall; blades in a short basal cluster, usually folded. 26. *P. sandbergii.*
Plants stout, averaging over 50 cm. tall; blades not usually conspicuously basal, usually flat. 27. *P. canbyi.*
Lemmas glabrous or scabrous.
Panicle narrow, mostly over 10 cm. long, the branches appressed, except in anthesis; plants robust, usually over 50 cm. tall.
Sheaths scabrous; ligule long. 28. *P. nevadensis.*
Sheaths smooth; ligule short.
Blades flat. 29. *P. ampla.*
Blades involute. 30. *P. brachyglossa.*
Panicle short, compact or open, less than 10 cm. long.
Blades of the culm flat, those of the lower culm 2–3 mm. wide; alpine grasses.
Panicle rather loose, short-exserted, mostly green; culms low, in tufts from firm branched caudices. 31. *P. curtifolia.*
Panicle compact, spikelike, long-exserted, mostly purple; culms 20–40 cm. tall, slender, mostly solitary or few in a tuft. 32. *P. epilis.*
Blades narrow or filiform, usually folded or involute.
Panicle oblong, compact and spikelike, 2–6 cm. long, 1–2 cm. thick; plants of seacoast cliffs. 33. *P. unilateralis.*
Panicle open, or if narrow, few-flowered; plants of the interior.
Panicle short and open, few-flowered, the capillary branches bearing 1–2 spikelets; culms 10–30 cm. tall. 34. *P. vaseyochloa.*
Panicle narrow and few-flowered or somewhat open and many-flowered.
Panicle somewhat open; blades filiform, scabrous; plants of the Arid Transition Zone. 35. *P. cusickii.*
Panicle narrow, the branches short and appressed; blades short and narrow, rather firm, glabrous; plants of the alpine regions.
Glumes 4–5 mm. long; panicle shining, usually pale. 36. *P. pringlei.*
Glumes 3 mm. long; panicle usually purple.
Blades more than 5 cm. long; panicle long-exserted. 37. *P. leibergii.*
Blades less than 5 cm. long; panicle short-exserted. 38. *P. lettermani.*

1. Poa ánnua L.
Annual Bluegrass. Fig. 453.

Poa annua L. Sp. Pl. 68. 1753.
Poa infirma H. B. K. Nov. Gen. & Sp. 1: 158. 1816.
Poa aestivalis Presl, Rel. Haenk. 1: 272. 1830.

Annual, tufted; bright green, glossy; culms flattened, decumbent at base, sometimes rooting at the lower nodes; sheaths loose; blades soft and lax; panicle pyramidal, open, 3–7 cm. long; spikelets crowded, 3–6 flowered, about 4 mm. long; lemma not webbed at base, distinctly 5-nerved, the nerves pilose on lower half.

Open ground, along roadsides and in waste places, Alaska to Mexico, east to the Atlantic Ocean; introduced from Europe. Jan.–June. Type locality, European.

2. **Poa bigelòvii** Vasey & Scribn.
Bigelow's Bluegrass. Fig. 454.

Poa annua stricta Vasey, Bull. Torrey Club **10**: 31. 1883.
Poa bigelovii Vasey & Scribn.; Vasey, Descr. Cat. Grasses U. S. 81. 1885.

Annual; culms erect, 15–35 cm. tall; panicle narrow, 7–15 cm. long, the branches short, appressed; spikelets ovate, about 6 mm. long; glumes acuminate, 3-nerved, 4 mm. long; lemmas 4 mm. long, webbed at base, copiously pilose on the lower part of the lateral nerves and keel, villous on lower portion of back between.

Open ground in the Lower Sonoran Zone; southern California to western Texas, south into Mexico. Apr.–May. Type locality: Rillito River, Arizona.

3. **Poa howéllii** Vasey & Scribn.
Howell's Bluegrass. Fig. 455.

Poa howellii Vasey & Scribn. U. S. Dept. Agr. Div. Bot. Bull. **13**[2]: *pl. 78.* 1893.
Poa howellii microsperma Vasey, Contr. U. S. Nat. Herb. **1**: 273. 1893.

Annual; culms 30–60 cm. tall; sheaths retrorsely scabrous or sometimes smooth; panicle including one-third to half the plant, open, the branches in rather distant fascicles, spreading, scabrous, naked below, some short branches intermixed; spikelets 3–4 mm. long, usually 3- or 4-flowered; glumes narrow, acuminate, the first 1.5 mm. long, 1-nerved or rarely 3-nerved, the second 2 mm. long, 3-nerved; lemmas webbed at base, 2 mm. long, ovate, pubescent over the lower half or two-thirds, the nerves all rather distinct.

Rocky banks and shaded slopes, in the Humid Transition Zone; Vancouver Island to southern California, mostly in the Coast Ranges. Apr.–Aug. Type locality: Portland, Oregon.

4. **Poa bolánderi** Vasey.
Bolander's Bluegrass. Fig. 456.

Poa bolanderi Vasey, Bot. Gaz. **7**: 32. 1882.
Poa howellii chandleri Davy, Univ. Calif. Publ. Bot. **1**: 60. 1902.
Poa bolanderi chandleri Piper, Contr. U. S. Nat. Herb. **11**: 132. 1906.

Annual; culms erect, 15–60 cm. tall; sheaths smooth; panicle open, about half the length of the entire plant, the branches few and distant, smooth, stiffly spreading or somewhat reflexed, naked below; spikelets usually 2- or 3-flowered; glumes broad, the first 1-nerved, 2 mm. long, the second 3-nerved, 3 mm. long; lemma scantily webbed at base, smooth, scabrous on the keel, acute, the marginal nerves rather indistinct, the intermediate nerves obsolete.

Open ground or open woods, in the Canadian Zone; Blue Mountains of Washington, through the Sierra Nevada to the high mountains of southern California, east to Utah. June–Aug. Type locality: Yosemite National Park.

5. **Poa compréssa** L.

Canada Bluegrass. Fig. 457.

Poa compressa L. Sp. Pl. 69. 1753.

Perennial from creeping rhizomes; culms not tufted, geniculate-ascending, flattened, wiry, bluish green, 15–45 cm. tall; panicle narrow, 3–7 cm. long, the usually short branches in pairs, spikelet-bearing to the base; spikelets crowded, subsessile, 3–6-flowered, 4–6 mm. long; glumes about 2 mm. long, 3-nerved; lemmas firm, obscurely nerved, 2–2.5 mm. long, sparingly webbed at base, short-pubescent below on keel and marginal nerves.

Open ground, open woods, meadows, and waste places throughout the United States; introduced from Europe. June–Aug. Type locality: Europe.

6. **Poa macrántha** Vasey.

Seashore Bluegrass. Fig. 458.

Poa macrantha Vasey, Bull. Torrey Club **15**: 11. 1888.

Perennial, from extensively creeping rhizomes; culms erect from a decumbent base, 15–40 cm. tall, the sterile shoots widely spreading; sheaths smooth, tawny and papery; blades smooth, involute, more or less curved or flexuous; panicle narrow, contracted, sometimes dense and spike-like, 5–12 cm. long, pale or tawny; spikelets large, about 12 mm. long, about 5-flowered; glumes smooth, 3-nerved, or the second indistinctly 5-nerved, about 8 mm. long; lemmas 8 mm. long; not webbed at base, short-pilose on the keel and marginal nerves below, slightly scabrous on the keel above and sparingly on the back near margins; palea ciliate on keels.

Sand dunes along the coast, Washington to Eureka, California. June–July. Type locality: mouth of Columbia River.

7. **Poa douglásii** Nees.

Douglas's Bluegrass. Fig. 459.

Poa douglasii Nees, Ann. Nat. Hist. 1: 284. 1838.

Perennial from extensively creeping rhizomes; culms ascending from a decumbent base, usually less than 30 cm. tall; sheaths smooth, tawny and papery; blades involute, some of these usually exceeding the culm; panicles dioecious, ovoid or oblong, dense and spike-like, 2–5 cm. long, nearly 1 cm. wide, pale or tawny; spikelets 6–10 mm. long, about 5-flowered; glumes broad, 3-nerved, smooth, scabrous on upper part of keel, nearly equal, 4–6 mm. long; lemmas 6 mm. long, slightly webbed at base, villous on the lower part of keel and marginal nerves, scabrous on keel above, 1–3 pairs of indistinct intermediate nerves; palea ciliate on keels.

Sand dunes near the coast, Point Arena to Monterey. Apr.–July. Type locality: "California."

8. Poa confìnis Vasey.

Dune Bluegrass. Fig. 460.

Poa confinis Vasey, U. S. Dept. Agr. Div. Bot. Bull. 13[2]: *pl. 75.* 1893.

Perennial, from creeping rhizomes; culms low, often geniculate or ascending at base, usually less than 15 cm. tall; sheaths and the numerous involute blades smooth; panicle narrow, contracted, 1–3 cm. long, tawny; spikelets about 4 mm. long; glumes unequal, the second 3 mm. long; lemmas 3 mm. long, scaberulous, sparsely webbed at base, the nerves faint.

Sand dunes and sandy meadows near the coast, southern Alaska to Mendocino County, California. June–July. Type locality: Tillamook Bay, Oregon.

9. Poa pìperi Hitchc. n. sp. Piper's Bluegrass. Fig. 461.

Perennial, pale green; culms loosely tufted with numerous innovations, from slender creeping rhizomes, smooth, terete, 40–60 cm. tall; sheaths of the culm smooth, of the innovations pubescent or puberulous, especially around the throat, becoming glabrous; ligule of the upper culm leaf about 1.5 mm. long, scabrous, of the innovation leaves very short, less than 0.5 mm.; blades narrow, involute, the margins meeting, pubescent on the upper surface, glabrous beneath, rather strongly nerved, those of the culm about 2, the upper at about the middle, 2–4 cm. long, those of the innovations 10–15 cm. long, firm and rather stiff; panicle open, 5–8 cm. long, the lower branches mostly in twos, ascending, 2–3 cm. long; spikelets 3–5-flowered, about 1 cm. long; glumes about 5 mm. long, indistinctly 3-nerved, the first a little narrower than the second; lemmas keeled, acute, about 5 mm. long, 5-nerved, the intermediate nerves distinct, glabrous on the back, minutely scabrous on the lateral nerves and on the upper part of the midnerve, the tuft of cobwebby hairs at base copious; palea a little shorter than the lemma, strongly ciliate on the nerves.

"Mountains 8 miles southwest of Waldo, Oregon, on dry mountain side under yellow pine, June 14, 1904." Collected by C. V. Piper (no. 6496). Only known from the type collection.

10. Poa atropurpùrea Scribn.

San Bernardino Bluegrass. Fig. 462.

Poa atropurpurea Scribn. U. S. Dept. Agr. Div. Agrost. Bull. 11: 53. *pl. 10.* 1898.

Perennial, from creeping rhizomes; culms 30–40 cm. tall, slender; sheaths smooth; blades mostly basal, folded or involute, firm, smooth on under surface, the uppermost culm-leaf below the middle; panicle narrow, contracted, almost spike-like, purple-tinged, 3–5 cm. long; spikelets 3–4 mm. long, turgid; glumes broad, less than 2 mm. long; lemmas a little over 2 mm. long, broad, smooth, not webbed, the nerves faint.

Only known from the San Bernardino Mountains (*Parish*).

11. **Poa nervòsa** (Hook.) Vasey. Hooker's Bluegrass. Fig. 463.

Festuca nervosa Hook. Fl. Bor. Am. **2**: 251. *pl. 232.* 1840.
Poa nervosa Vasey, U. S. Dept. Agr. Div. Bot. Bull. **13^{2}**: *pl. 81.* 1893.
Poa olneyae Piper, Erythea **7**: 101. 1899.

Perennial, from creeping rhizomes; culms 30–60 cm. tall; sheaths smooth or the lower scaberulous, rarely retrorsely puberulent; ligule 1–2 mm. long; blades mostly flat, rather thin, smooth beneath, scaberulous above, as much as 4 mm. wide; panicle open, the slender branches flower-bearing only at the extremities, lemmas typically strongly nerved, smooth or nearly so, sometimes scaberulous, rarely minutely pubescent.

Moist rocks and banks, Humid Transition Zone; British Columbia to California. May–Aug. Type locality: Nootka Sound.

12. **Poa kellóggii** Vasey. Kellogg's Bluegrass. Fig. 464.

Poa kelloggii Vasey, U. S. Dept. Agr. Div. Bot. Bull. **13^{2}**: *pl. 79.* 1893.

Perennial, from creeping rhizomes; culms 30–60 cm. tall, smooth; sheaths smooth, mostly basal; blades flat or folded, scabrous on upper surface; panicle pyramidal, open, 7–15 cm. long, the branches mostly in ones or twos, slender, spreading or reflexed, bearing a few spikelets toward the extremities; spikelets rather loosely flowered, 4–6 mm. long; glumes 2 and 3 mm. long; lemmas acute or almost cuspidate, 3–4 mm. long, smooth, rather obscurely nerved, conspicuously webbed at base.

Coast Ranges in the Transition Zone; Oregon (Corvallis) to Santa Cruz County, California. May–June. Type locality: Mendocino County.

13. **Poa praténsis** L. Kentucky Bluegrass. Fig. 465.

Poa pratensis L. Sp. Pl. 67. 1753.

Perennial, from creeping rhizomes; culms tufted, 30–100 cm. tall, terete or slightly flattened; sheaths smooth, compressed; ligule about 2 mm. long; blades soft, flat or folded, the basal often elongated; panicle pyramidal, open, the slender branches in remote fascicles of 3–5, ascending or spreading, naked at base, some of them short; spikelets crowded, 3–5-flowered, 4–5 mm. long; lemmas 3 mm. long, copiously webbed at base, silky-pubescent on the keel and marginal nerves, the intermediate nerves prominent.

Open woods, banks, open ground, widely distributed except in the deserts; extending throughout the northern part of North America and Eurasia; escaped from cultivation. May–Aug. Type locality, European.

14. **Poa rhizomàta** Hitchc.

Timber Bluegrass. Fig. 466.

Poa rhizomata Hitchc. in Jepson, Fl. Calif. 1: 155. 1912.

Perennial, from creeping rhizomes; culms erect, 30–60 cm. tall, smooth; sheaths smooth, the lower loose and papery; ligule 2–3 mm. long; blades flat or folded, 1–2 mm. wide, 3–7 cm. long, the culm blades about 2, the upper erect, about 2 cm. long; panicle long-exserted, oblong, contracted, 2–5 cm. long, the branches short, slender, mostly in twos, ascending, few-flowered; spikelets about 6 mm. long, 3–5-flowered; glumes unequal, rather broad, acute, scabrous on the keels, the first 1-nerved, 3 mm. long, the second 3-nerved, 4 mm. long; lemmas 5 mm. long, acutish, copiously webbed at base, short-pilose on keel below, and sparingly so on lower part of marginal nerves, the intermediate nerves faint, sparingly scabrous between the nerves; palea ciliate on the keels.

Damp, shady woods, in the Transition Zone; Oregon, northern California (Siskiyou County), and Idaho. Apr.–June. Type locality: Oro Fino, Cailfornia.

15. **Poa triviàlis** L.

Rough-stalked Meadow Grass. Fig. 467.

Poa trivialis L. Sp. Pl. 67. 1753.

Perennial, without rhizomes, culms erect from a decumbent base, often rather lax, scabrous below the panicle, 30–100 cm. tall; sheaths retrorsely scaberulous; ligule 4–6 mm. long; blades flat, scabrous, 7–15 cm. long, 2–4 mm. wide; panicle oblong, 6–15 cm. long, the lower branches about 5 in a whorl; spikelets usually 2–3-flowered, about 3 mm. long; lemma 2.5–3 mm. long, glabrous except the slightly pubescent keel, the web at base conspicuous, the nerves prominent.

Moist places, through the northern United States; introduced from Europe; on the west coast from Alaska to Humboldt County, California (*Tracy*). May–June. Type locality, European.

16. **Poa leptocòma** Trin.

Bog Bluegrass. Fig. 468.

Poa leptocoma Trin. Mém. Acad. St. Pétersb. VI. Math. Phys. Nat. 1: 374. 1830.

Poa laxiflora Buckl. Proc. Acad. Phila. **1862**: 96. 1863.

Poa paucispicula Scribn. & Merr. Contr. U. S. Nat. Herb. **13**: 69. *pl. 15.* 1910.

Perennial, without rhizomes, culms loosely tufted, erect from a slightly decumbent base, smooth, rather lax, 10–50 cm. tall; sheaths smooth or slightly scaberulous; blades flat, lax, scabrous, 1–3 mm. wide; panicle loose and open, 5–10 cm. long, the slender branches naked below, spreading, or the lower finally reflexed, mostly in twos; spikelets mostly 2–3-flowered; lemmas about 3 mm. long, villous on the keel and less so on the lateral nerves, the web at base inconspicuous, the intermediate nerves indistinct.

Damp, often shady places, boggy or springy soil, in the Hudsonian Zone; Alaska to California (Mount Dana). July–Aug. Type locality: Sitka.

17. Poa occidentàlis Vasey.

Tall Western Bluegrass. Fig. 469.

Poa occidentalis Vasey, Contr. U. S. Nat. Herb. 1: 274. 1893.
Poa leptocoma elatior Scribn. & Merr. Contr. U. S. Nat. Herb. 13: 71. 1910.

Perennial, without rhizomes; culms rather stout, erect, scabrous, 1–1.5 meters tall; sheaths somewhat keeled, retrorsely scabrous; ligule triangular, 6–8 mm. long; blades flat, scabrous, 10–20 cm. long, 3–6 mm. wide; panicle open, 15–30 cm. long, the branches in distant whorls of threes to fives, slender, spreading or reflexed, the lower as much as 10 cm. long, spikelet-bearing toward the end; spikelets 3–6-flowered; glumes acuminate, 4 mm. long; lemmas about 5 mm. long, conspicuously webbed at base, villous on lower part of keel, slightly villous on lateral nerves.

Open woods, Alaska, Oregon (Sauvies Island, *Howell*), California (Hot Springs, Sequoia National Forest, *Jardine*), Colorado, Utah, and New Mexico. May. Type locality: Las Vegas, New Mexico.

18. Poa palústris L.

Fowl Meadow Grass. Fig. 470.

Poa palustris L. Syst. Veg. ed. 10. 2: 874. 1759.
Poa serotina Ehrh. Beitr. Naturk. 6: 83. 1791 (nom. nud.) and later authors.
Poa triflora Gilib. Exerc. Phyt. 2: 531. 1792.

Perennial, without rhizomes; culms loosely tufted, smooth, decumbent at the flattened purplish base, 30-150 cm. tall; sheaths smooth; ligule 3–5 mm. long; blades soft, 8–15 cm. long, 2–4 mm. wide; panicle pyramidal or oblong, nodding, yellowish green or purplish, 10–30 cm. long, the branches in rather distant fascicles, naked below; spikelets 2–4-flowered, about 4 mm. long; lemmas 2.5–3 mm. long, villous on the keel and marginal nerves, the web copious, the intermediate nerves faint.

Meadows and moist open ground, in the Transition Zone; Alaska to California; Sierra Valley (*Lemmon*), east to Nova Scotia. June–Aug. Type locality, European.

19. Poa alpìna L.

Alpine Bluegrass. Fig. 471.

Poa alpina L. Sp. Pl. 67. 1753.

Perennial, from a rather thick vertical caudex, with no creeping rhizomes; culms erect, comparatively stout, smooth, 10–30 cm. tall; sheaths smooth; blades short and flat, smooth, 2–5 mm. wide, the uppermost often very short; panicle ovoid, rather compact, 1–4 cm. long; spikelets broad, purple or purplish-tinged; lemmas 3–4 mm. long, strongly villous on the keel and lateral nerves, and pubescent below between the nerves.

Arctic regions of the northern hemisphere, extending to alpine meadows of the high mountains of Washington and Oregon (Wallowa Mountains, *Cusick*). July–Aug. Type locality, European.

Poa stenántha Trin. has been found at Crater Lake (*Clemens*). It has flat lax blades, ligule 3–5 mm. long, open nodding panicle of rather large spikelets, the lemma pubescent on the nerves but without cobwebby hairs at base. Alaska to Montana.

20. **Poa rupícola** Nash.
Timberline Bluegrass. Fig. 472.

Poa rupestris Vasey, Bull. Torrey Club **14**: 94. 1887, not Roth 1821.
Poa rupicola Nash in Rydb. Mem. N. Y. Bot. Gard. **1**: 49. 1900.

Perennial; culms tufted, erect, rather stiff, smooth, 10–20 cm. tall; sheaths smooth; ligule about 1 mm. long; blades erect, flat or involute, smooth, 1–5 cm. long, .5–1 mm. wide; panicle narrow, purplish, 2–4 cm. long, the branches ascending or appressed, the lower as much as 1 cm. long; spikelets 5 mm. long, about 3-flowered; lemmas 3–4 mm. long, villous below on the keel and marginal nerves, the intermediate nerves indistinct, no web at base.

Rocky slopes in the Arctic Zone; Oregon (Wallowa Mountains, *Cusick*), California (Mono Pass, *Eastwood*), and in the Rocky Mountains from Montana to Colorado. August. Type locality: Rocky Mountains, no definite locality.

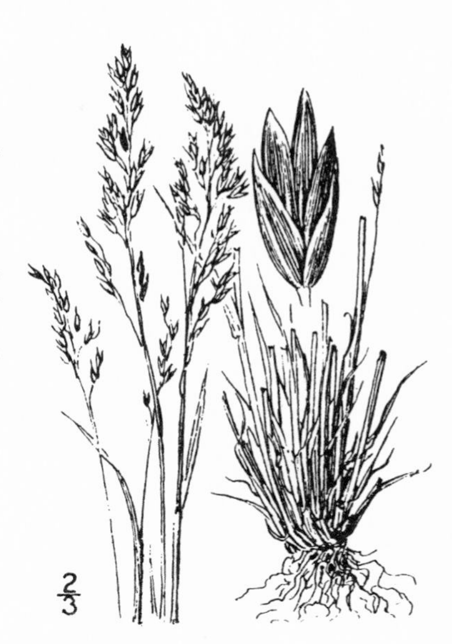

21. **Poa pachýpholis** Piper.
Seacliff Bluegrass. Fig. 473.

Poa pachypholis Piper, Proc. Biol. Soc. Washington **18**: 146. 1905.

Perennial, without rhizomes; culms densely tufted, wholly glabrous below the panicles, 15–30 cm. tall; basal leaves numerous, the sheaths long-persisting; ligule 2 mm. long; blades erect, pale or glaucous, flat or loosely involute, 4–10 cm. long, 1–2 mm. wide, the culm leaves usually 3; panicle dense, oblong, 2–5 cm. long, the lower branches mostly in twos; spikelets 3–5-flowered, pale; lemmas 4 mm. long, villous on the keel and lateral nerves, no web at base.

Only known from the type collection from ocean cliffs at Ilwaco, Washington. June.

22. **Poa longilígula** Scribn. & Williams.
Long-liguled Mutton-grass. Fig. 474.

Poa longiligula Scribn. & Williams, U. S. Dept. Agr. Div. Agrost. Circ. **9**: 3. 1899.

Tufted perennial, similar to *P. fendleriana;* culms smooth, 30–60 cm. tall; sheaths and blades smooth; ligule 5–7 mm. long, or on the innovations somewhat shorter; panicle looser and often longer; spikelets as in *P. fendleriana.*

Cañons and dry mountain slopes, Oregon (Stein's Mountain), California (San Bernardino Mountains), east to Montana and New Mexico. June. Type locality: Silver Reef, Utah.

23. Poa fendleriàna (Steud.) Vasey.
Mutton Grass. Fig. 475.

Eragrostis fendleriana Steud. Syn. Pl. Glum. **1**: 278. 1854.

Atropis californica Munro; Thurb. in S. Wats. Bot. Calif. **2**: 309. 1880.

Poa californica Scribn. Bull. Torrey Club **10**: 31. 1883, not Steud. 1854.

Poa fendleriana Vasey; U. S. Dept. Agr. Div. Bot. Bull. **13**²: *pl. 74.* 1893.

Tufted perennial; culms erect, smooth, scabrous below panicle, 30–50 cm. tall; sheaths somewhat scabrous; ligule less than 1 mm. long; blades mostly basal, involute or folded, scabrous, firm; panicle long-exserted, narrow, contracted, 2–7 cm. long; glumes broad, 3 mm. long, the first 1-nerved; lemmas 4 mm. long, long-pilose on the lower portion of keel and marginal nerves, the intermediate nerves obscure.

Mesas and rocky hills, in the Upper Sonoran Zone; southern California (Bear Valley, *Parish;* Owens Valley, *Jones;* Panamint, *Hall & Chandler*), east to Wyoming and New Mexico. May–June. Type locality: New Mexico.

24. Poa scabrélla (Thurb.) Benth.
Malpais Bluegrass. Fig. 476.

Poa tenuifolia Buckl. Proc. Acad. Phila. **1862**: 96. 1863, not A. Rich. 1851.

Atropis scabrella Thurb. in S. Wats. Bot. Calif. **2**: 310. 1880.

Poa scabrella Benth.; Vasey, Grasses U. S. 42. 1883.

Poa orcuttiana Vasey, West Am. Sci. **3**: 165. 1887.

Poa buckleyana Nash, Bull. Torrey Club **22**: 465. 1895.

Poa capillaris Scribn. U. S. Dept. Agr. Div. Agrost. Bull. **11**: 51. *f. 11.* 1898, not L. 1753.

Poa nudata Scribn. U. S. Dept. Agr. Div. Agrost. Circ. **9**: 1. 1899.

Poa acutiglumis Scribn. U. S. Dept. Agr. Div. Agrost. Circ. **9**: 4. 1899.

Poa limosa Scribn. & Williams, U. S. Dept. Agr. Div. Agrost. Circ. **9**: 5. 1899.

Tufted perennial; culms erect, 50–100 cm. tall, usually scabrous, at least below panicle; sheaths scabrous; ligule 3–5 mm. long; blades mostly basal, flat, narrow, usually about 1 mm. wide, lax, more or less scabrous; panicle narrow, usually contracted, sometimes rather open at base, 5–12 cm. long; spikelets narrow, 6–10 mm. long; glumes scabrous, 3 mm. long; lemmas 4 mm. long, puberulent or scabrous on back, and more or less crisp-pubescent at base.

Meadows, open woods, rocks, and hills, Transition Zone; Oregon to northern Mexico, east to Nevada and Arizona. May–July. Type locality: Oakland, California.

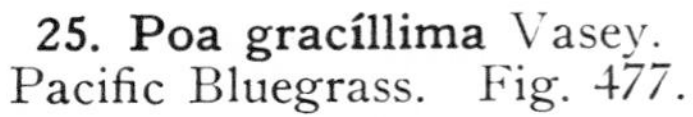

25. Poa gracíllima Vasey.
Pacific Bluegrass. Fig. 477.

Sporobolus bolanderi Vasey, Bot. Gaz. **11**: 337. 1886, not *Poa bolanderi* Vasey. 1882.

Poa gracillima Vasey, Contr. U. S. Nat. Herb. **1**: 272. 1893.

Poa saxatilis Scribn. & Williams, U. S. Dept. Agr. Div. Agrost. Circ. **9**: 1. 1899.

Poa tenerrima Scribn. U. S. Dept. Agr. Div. Agrost. Circ. **9**: 4. 1899.

Poa invaginata Scribn. & Williams, U. S. Dept. Agr. Div. Agrost. Circ. **9**: 6. 1899.

Poa multnomae Piper, Bull. Torrey Club **32**: 435. 1905.

Poa alcea Piper, Bull. Torrey Club **32**: 436. 1905.

Tufted perennial; culms 30–60 cm. tall, erect from usually a decumbent base; sheaths smooth; ligule 2–5 mm. long, or shorter on the innovations; blades flat or folded, lax, smooth, mostly basal; panicle pyramidal, loose, rather open, 5–10 cm. long, the branches in whorls, the lower 2 to 6, 2–7 cm. long, slender, spreading or sometimes reflexed, naked below; spikelets 4–6 mm. long; glumes smooth, the second 3–4 mm. long; lemmas about as long as second glume, minutely scabrous, crisp-pubescent near base, especially on the nerves.

Cliffs and rocks, Transition and Arctic Zones; Alaska to the southern Sierra Nevada. June–Aug. Type locality: Mount Paddo (*Adams*).

26. **Poa sandbérgii** Vasey.
Sandberg's Bluegrass. Fig. 478.

Poa sandbergii Vasey, Contr. U. S. Nat. Herb. 1: 276. 1893.
Poa incurva Scribn. & Williams, U. S. Dept. Agr. Div. Agrost Circ. 9: 6. 1899.

Similar to *P. scabrella;* differing in being smooth, averaging lower and more slender, the panicle smaller, the blades short and soft, often involute. A very variable species, sometimes in small tufts of one or two culms, sometimes in large tussocks 30 cm. in diameter. The blades of the basal tuft are usually curly at maturity.

Plains, dry woods, and rocky slopes, in the Arid Transition and Canadian Zones; British Columbia to the mountains of southern California, east to Wyoming. June–Aug. Type locality: Lewiston, Idaho.

27. **Poa cánbyi** (Scribn.) Piper.
Canby's Bluegrass. Fig. 479.

Glyceria canbyi Scribn. Bull. Torrey Club 10: 77. 1883.
Atropis canbyi Beal, Grasses N. Am. 2: 580. 1896.
Poa leckenbyi Scribn. U. S. Dept. Agr. Div. Agrost. Circ. 9: 2. 1899.
Poa wyomingensis Scribn. Proc. Davenport Acad. Sci. 7: 242. 1899.
Poa canbyi Piper, Contr. U. S. Nat. Herb. 11: 132. 1906.

Perennial; culms tufted, erect, smooth, 50–120 cm. tall; sheaths smooth or slightly scabrous; ligule 2–5 mm. long; blades scabrous above, usually flat, 1–2 mm. wide, those of the culm usually 2, the basal rather long; panicle narrow, compact or somewhat loose, 10–15 cm. long, the branches short and appressed; spikelets pale or purplish, 3–5-flowered; lemmas rounded on the back, faintly nerved, pubescent on the lower part.

Sandy or dry ground, valleys and cliffs, in the Arid Transition Zone; eastern Washington and Oregon. May–June. Type locality: Cascade Mountains, Washington.

28. **Poa nevadénsis** Vasey.
Nevada Bluegrass. Fig. 480.

Atropis pauciflora Thurb. in S. Wats. Bot. Calif. 2: 310. 1880, not *Poa pauciflora* Roem. & Schult. 1817.
Poa nevadensis Vasey; Scribn. Bull. Torrey Club 10: 66. 1883.
Panicularia thurberiana Kuntze, Rev. Gen. Pl. 2: 783. 1891.
Poa thurberiana Vasey, U. S. Dept. Agr. Div. Bot. Bull. 13[2]: *pl. 84.* 1893 (as to name).

Tufted perennial; culms 50–100 cm. tall, smooth; sheaths scabrous; ligule 4 mm. long, decurrent; blades scabrous, usually firm, involute; panicle narrow, 10–15 cm. long; spikelets 6–8 mm. long, narrow; glumes narrow, the second 3 mm. long; lemmas smooth or scaberulous, not pilose nor webbed, 3 mm. long.

Plains and meadows, in the Upper Sonoran Zone; eastern Washington (Bingen, *Suksdorf*) to southern California, east to Colorado. July–Aug. Type locality: "California," according to type specimen.

29. **Poa ámpla** Merr.
Merrill's Bluegrass. Fig. 481.

Poa ampla Merr. Rhodora **4**: 145. 1902.
Poa laeviculmis Williams, Bot. Gaz. **36**: 55. 1903.

Perennial; culms tufted, erect, smooth, 80–120 cm. tall; sheaths smooth; ligule short and rounded; blades flat, 1–3 mm. long; panicle narrow, 10–15 cm. long, the branches appressed; spikelets 4–7-flowered; lemmas rounded on the back, smooth or minutely scabrous, 4–5 mm. long.

Meadows and moist open ground, in the Arid Transition Zone; British Columbia to Oregon and Idaho. May–July. Type locality: Steptoe, Washington.

30. **Poa brachyglóssa** Piper.
Alkali Bluegrass. Fig. 482.

Poa brachyglossa Piper, Proc. Biol. Soc. Washington **18**: 145. 1905.

Perennial; culms tufted, erect, smooth, 50–100 cm. tall; sheaths smooth; ligules of the innovations very short, of the culm leaves 1–2 mm. long; blades rather stiff, folded or involute, 5–20 cm. long; panicle narrow, 10–20 cm. long, the branches ascending or appressed; spikelets 7–10 mm. long, 3–6-flowered; glumes 4–5 mm. long; lemmas rounded on the back, 4 mm. long.

Alkaline meadows, in the Arid Transition and Upper Sonoran Zones; eastern Washington to eastern California (Mount Lola, Dixie Valley), east to Idaho and Nevada. June–July. Type locality: Douglas County, Washington.

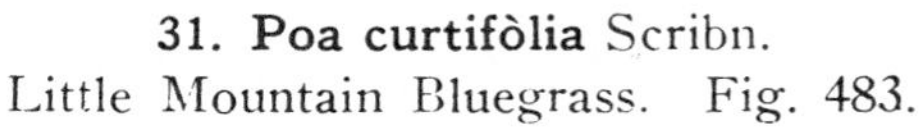

31. **Poa curtifòlia** Scribn.
Little Mountain Bluegrass. Fig. 483.

Poa curtifolia Scribn. U. S. Dept. Agr. Div. Agrost. Circ. **16**: 3. 1899.

Perennial; culms several in a tuft from firm branched caudices, 10–20 cm. tall; sheaths smooth; ligule long, the uppermost as much as 5 mm. long; blades short and broad, flat, spreading, those of the lower culm 1.5–2 cm. long, about 2 mm. wide, those of the innovations narrower; panicle narrow, rather loose, short-exserted, mostly green, 3–6 cm. long; spikelets about 3-flowered; glumes equal, smooth and shining, 5 mm. long; lemmas smooth, about 5 mm. long.

Rocky slopes, in the Arctic Zone; Washington (Mount Stuart, *Elmer;* Yakima region, *Tweedy*). August. Type locality: Mount Stuart. A little-known species which may prove to be the same as the next.

32. Poa épilis Scribn.

Mountain Bluegrass. Fig. 484.

Poa purpurascens Vasey, Bot. Gaz. 6: 297. 1881, not Spreng. 1819.

Poa epilis Scribn. U. S. Dept. Agr. Div. Agrost. Circ. 9: 5. 1899.

Poa paddensis Williams, U. S. Dept. Agr. Div. Agrost. Bull. 17, ed. 2: 261. *f. 557.* 1901.

Perennial; culms erect, smooth, solitary or few in a tuft, 20–40 cm. tall; sheaths smooth; ligule of the culm leaves about 3 mm. long; blades of the culm about 3, flat, 3–6 cm. long, 2–3 mm. wide, of the innovations narrower, longer and usually folded, all smooth; panicle usually compact and spike-like, 2–6 cm. long, usually purple, long-exserted; glumes about 5 mm. long; lemmas about 6 mm. long, smooth.

Mountain meadows in the Alpine Zone; British Columbia to the Mount Whitney region, California, east to Montana and Colorado. July–Aug. Type locality: Colorado.

33. Poa unilateràlis Scribn.

San Francisco Bluegrass. Fig. 485.

Poa unilateralis Scribn.; Vasey, U. S. Dept. Agr. Div. Bot. Bull. 13²: *pl. 85.* 1893.

Tufted perennial; culms 10–40 cm. tall; sheaths smooth, tawny and papery; blades flat or folded, shorter than the culms; panicle oblong, dense and spike-like, or somewhat interrupted below, 2–6 cm. long; spikelets 6–8 mm. long, perfect; glumes broad, acute, smooth, indistinctly scabrous on keel near apex, the first 1-nerved or indistinctly 3-nerved, the second 3-nerved; lemmas 4 mm. long, not webbed at base nor pilose, scabrous on base of marginal nerves and apex of keel, the intermediate nerves faint; palea ciliate on keels.

Cliffs, bluffs, and rocky meadows near the seashore, Humboldt County to Monterey, California. May. Type locality: San Francisco.

34. Poa vaseyòchloa Scribn.

Vasey's Bluegrass. Fig. 486.

Poa pulchella Vasey, Bot. Gaz. 7: 32. 1882, not Salisb. 1796.

Poa vaseyochloa Scribn. U. S. Dept. Agr. Div. Agrost. Circ. 9: 1. 1899.

Perennial; culms tufted, erect, slender, smooth, 10–20 cm. tall; sheaths smooth; blades smooth or somewhat scabrous, slender, flat or involute, those of the innovations mostly filiform, mostly less than 5 cm. long, the 2 culm blades a little wider; panicle ovate, 2–4 cm. long, the slender branches spreading, bearing 1–2 spikelets; spikelets purple, 3–6-flowered; glumes 2–3 mm. long, rather broad; lemmas smooth, or minutely scabrous, 3 mm. long.

Rocks and hills, Cascade Mountains of Washington and Oregon in the region of the Columbia River, and in the Wallowa Mountains. Mar.–May. Type locality: Klickitat County, Washington.

35. **Poa cusíckii** Vasey.
Cusick's Bluegrass. Fig. 487.

Poa cusickii Vasey, Contr. U. S. Nat. Herb. **1**: 271. 1893.
Poa filifolia Vasey, Contr. U. S. Nat. Herb. **1**: 271. 1893, not Schur, 1866.
Poa idahoensis Beal, Grasses N. Am. **2**: 539. 1896.
Poa scabrifolia Heller, Bull. Torrey Club **24**: 310. 1897.
Poa spillmani Piper, Erythea **7**: 102. 1899.
Poa capillarifolia Scribn. & Williams, U. S. Dept. Agr. Div. Agrost. Circ. **9**: 1. 1899.
Poa cottoni Piper, Proc. Biol. Soc. Washington **18**: 146. 1905.

Perennial; culms tufted, erect, smooth, or scabrous below the panicle, 25–50 cm. tall; sheaths smooth; blades filiform, scabrous; panicle somewhat open, 3–8 cm. long; spikelets 7–9 mm. long; lemmas smooth, 4–5 mm. long.

Open woods and rocky slopes, in the Arid Transition Zone; eastern Washington to Oregon and Idaho. May–June. Type locality: Oregon.

36. **Poa prínglei** Scribn.
Pringle's Bluegrass. Fig. 488.

Poa pringlei Scribn. Bull. Torrey Club **10**: 31. 1883.
Poa argentea Howell, Bull. Torrey Club **15**: 11. 1888.
Poa suksdorfii Vasey; Beal, Grasses N. Am. **2**: 574. 1896.

Tufted perennial; culms 10–20 cm. tall, the base sometimes decumbent and rhizome-like; sheaths smooth, loose and papery; blades mostly basal, involute, usually not over 2–5 cm. long, smooth, the uppermost culm blade at or below the middle; panicle narrow, contracted, few- to several-flowered, short- or long-exserted; spikelets 6–8 mm. long, about 3-flowered; glumes equal, broad, 3-nerved, 4–5 mm. long; lemmas 5–6 mm. long, smooth or scabrous, not webbed nor pilose.

Rocks and sand, in the Alpine Zone; summits of mountains, Washington, south through the Sierra Nevada to the Mount Whitney region of California. July–Aug. Type locality: headwaters of the Sacramento River.

37. **Poa leibérgii** Scribn.
Leiberg's Bluegrass. Fig. 489.

Poa leibergii Scribn. U. S. Dept. Agr. Div. Agrost. Bull. **8**: 6. *pl. 2.* 1897.
Poa hanseni Scribn. U. S. Dept. Agr. Div. Agrost. Bull. **11**: 53. *pl. 9.* 1898.
Poa pringlei hanseni Smiley, Univ. Calif. Publ. **9**: 104. 1921.

Tufted perennial; culms 7–20 cm. tall; sheaths smooth; ligule about 2 mm. long; blades mostly basal, firm, involute, smooth, short; panicle narrow, 2–5 cm. long, usually purple, the branches short, appressed; spikelets 4–6 mm. long; glumes about 3 mm. long; lemmas 3 mm. long, smooth or scaberulous.

Alpine meadows and sterile gravelly alpine flats in the Hudsonian Zone; Malheur County, Oregon, through the Sierra Nevada to the Mount Whitney region, California. June–July. Type locality: Malheur County, Oregon.

38. **Poa lettermáni** Vasey. Letterman's Bluegrass. Fig. 490.

Poa lettermani Vasey, Contr. U. S. Nat. Herb. 1: 273. 1893.

Plants rather soft and lax; culms numerous in small dense tufts, smooth, 5–10 cm. tall; sheaths smooth; blades mostly flat, 1–3 cm. long, .5–1 mm. wide; panicle purplish, small and narrow, .5–1.5 cm. long, the branches short and appressed; glumes and lemmas about 3 mm. long, broad, the lemmas broad and toothed at the summit.

Alpine region of Mount Rainier and Mount Adams; also in Colorado. August. Type locality: Grays Peak, Colorado.

The viviparous form of *Poa bulbosa* L. has been found at Bingen and Walla Walla. The spikelets are replaced by dark-purple bulblets.

66. **GLYCÈRIA** R. Br. Prodr. Fl. Nov. Holl. 179. 1810.

Spikelets few- to many-flowered, subterete or slightly compressed, the rachilla disarticulating above the glumes and between the florets; glumes unequal, short, obtuse or acute, usually scarious, mostly 1-nerved; lemmas broad, convex on the back, firm, usually obtuse, awnless, scarious at apex, 5–9-nerved, the nerves usually prominent. Usually tall aquatic or marsh grasses, with flat blades and open or contracted panicles. [Name Greek, meaning sweet.]

Species about 35 in the temperate regions of both hemispheres. Type species, *Glyceria fluitans* (L.) R. Br.

Spikelets linear, usually as much as 1 cm. long; panicles erect.
  Lemmas smooth between the slightly scabrous nerves, thin, 3.5–4 mm. long. — 1. *G. borealis.*
  Lemmas scabrous between the usually distinctly scabrous nerves, firm, 3–6 mm. long.
    Lemmas about 3 mm. long, broadly rounded at summit. — 2. *G. leptostachya.*
    Lemmas 4–6 mm. long, narrowed toward the summit. — 3. *G. fluitans.*
Spikelets ovate or oblong, usually not over 5 mm. long; panicles nodding.
  Lemmas with 5 prominent nerves. — 4. *G. pauciflora.*
  Lemmas with 7 prominent nerves.
    Glumes short and rounded, the lower about 1 mm. long.
      Blades 1–3 mm. wide; culms usually not over 1 meter tall, firm, not succulent. — 5. *G. nervata.*
      Blades mostly 6–10 mm. wide; culms mostly 1–3 meters tall, succulent. — 6. *G. elata.*
    Glumes oblong, the lower about 2 mm. long, whitish. — 7. *G. grandis.*

1. **Glyceria boreàlis** (Nash) Batch. Northern Manna-grass. Fig. 491.

Panicularia borealis Nash, Bull. Torrey Club **24**: 348. 1897.

Glyceria borealis Batchelder, Proc. Manchester Inst. 1: 74. 1900.

Culms 30–100 cm. tall, erect from a more or less decumbent and rooting base; sheaths smooth or slightly scabrous, keeled; blades flat or usually folded, scabrous above, erect, 3–4 mm. wide; panicle long and narrow, the branches and slender pedicels appressed; spikelets narrow, nearly terete, 12–18 mm. long, 1 mm. wide, pale, not purple-tinged; glumes 1-nerved, the first 1.5 mm., the second 3 mm. long; lemmas oblong, 4 mm. long, 7-nerved, smooth or indistinctly scabrous on the nerves.

In shallow water, in the Canadian Zone; New England to central California and northward. June–Aug. Type locality: Maine.

2. Glyceria leptostáchya Buckl.

Davy's Manna-grass. Fig. 492.

Glyceria leptostachya Buckl. Proc. Acad. Phila. **1862**: 95. 1863.
Panicularia davyi Merr. Rhodora **4**: 145. 1902.
Panicularia leptostachya Piper; Piper & Beattie, Fl. Northw. Coast 59. 1915, not Maclosk. 1904.

Culms 1–1.5 meters tall; sheaths smooth; blades minutely and sparsely scabrous above, about 4 mm. wide; panicle long and narrow; spikelets 12–18 mm. long, about 1.5 mm. wide; lemmas oblong, truncate, more or less purple-tinged, about 3 mm. long, prominently 7-nerved, distinctly scabrous on and between the nerves.

In shallow water, in the Transition Zone; Washington to central California (Guerneville, *Davy*; Sonoma, *Heller*). May–July. Type locality: Guerneville, California.

3. Glyceria flùitans (L.) R. Br.

Water Manna-grass. Fig. 493.

Festuca fluitans L. Sp. Pl. **1**: 75. 1753.
Glyceria fluitans R. Br. Prodr. Fl. Nov. Holl. **1**: 179. 1810.
Panicularia fluitans Kuntze, Rev. Gen. Pl. **2**: 782. 1891.
Panicularia occidentalis Piper in Piper & Beattie, Fl. Northw. Coast 59. 1915.

Culms ascending from a decumbent rooting base, rather thick and succulent, 1–1.5 meters tall; sheaths smooth; blades 3–10 mm. wide, scabrous above; panicle long and narrow; glumes very unequal, obtuse, the second 3 mm. long; lemmas purple-tinged, broad, obtuse, 5 mm. long, prominently 7-nerved, with an additional short pair near the margin, very scabrous on the nerves and somewhat so between them; palea about as long as lemma.

In shallow water, in Transition Zone; northern parts of Eurasia and America; British Columbia to Mendocino County, California. June–Aug. Type locality, European.

4. Glyceria paucifl̀ora Presl.

Few-flowered Manna-grass. Fig. 494.

Glyceria pauciflora Presl, Rel. Haenk. **1**: 257. 1830.
Panicularia pauciflora Kuntze, Rev. Gen. Pl. **2**: 783. 1891.
Panicularia holmii Beal, Torreya **1**: 43. 1901.
Panicularia multifolia Elmer, Bot. Gaz. **36**: 54. 1903.
Panicularia flaccida Elmer, Bot. Gaz. **36**: 55. 1903.
Glyceria erecta Hitchc. in Jepson, Fl. Calif. **1**: 161. 1912.
Panicularia erecta Hitchc. Am. Journ. Bot. **2**: 309. 1915.

Culms 30–120 cm. tall, from a decumbent rooting base, with creeping rhizomes; sheaths smooth or scabrous; blades scattered, 4–12 mm. wide, scabrous; panicle pyramidal, nodding, 10–20 cm. long, open, the branches spreading, naked below, rather densely flowered toward the ends; spikelets about 4 mm. long, oblong, about 5-flowered; glumes short, broad, obtuse, 1–1.5 mm. long; lemmas 2 mm. long, oblong, rounded and somewhat erose at summit, prominently 5-nerved, very scabrous on the nerves and somewhat so between.

Swamps, shallow water, and wet meadows, in the Transition Zone; Montana to Colorado, west to Alaska and central California. June–Aug. Type locality: Nootka Sound.

The form described as *Panicularia erecta* has short culms, narrow blades, and narrow panicles with ascending branches. Springy places in mountain meadows in the Sierra Nevada to southern Oregon. Type locality: Sunrise Creek, Yosemite National Park.

5. Glyceria nervàta (Willd.) Trin.

Fowl Manna-grass. Fig. 495.

Poa nervata Willd. Sp. Pl. 1: 389. 1797.
Glyceria nervata Trin. Mém. Acad. St. Pétersb. VI. Math. Phys. Nat. 1: 365. 1830.
Panicularia nervata Kuntze, Rev. Gen. Pl. 1: 783. 1891.

Culms erect, smooth, 30–100 cm. tall; sheaths smooth or retrorsely scabrous; blades erect or somewhat spreading, smooth or somewhat scabrous; 5–20 cm. long, 1–3 mm. wide, panicle ovate, open, 7–20 cm. long, the branches naked below, ascending and somewhat drooping; spikelets oval, 2–3 mm. long, mostly 4–6-flowered, often purplish; glumes very short, the second about 1 mm. long, the first shorter; lemmas broad and obtuse above, prominently 7-nerved, slightly scarious at apex, about 2 mm. long; palea large, about as long as the lemma.

Marshes and wet places in the Transition Zone; Newfoundland to Florida, and west to eastern Washington and Oregon. June–Aug. Type locality: North America.

6. Glyceria elàta (Nash) Hitchc.

Tall Manna-grass. Fig. 496.

Panicularia elata Nash in Rydb. Mem. N. Y. Bot. Gard. 1: 54. 1900.
Glyceria latifolia Cotton, Bull. Torrey Club 29: 573. 1902.
Panicularia nervata elata Piper, Contr. U. S. Nat. Herb. 11: 140. 1906.
Glyceria elata Hitchc. in Jepson, Fl. Calif. 1: 162. 1912.

Culms erect, smooth, succulent, 1–2 meters tall; sheaths scabrous; blades flat, usually 6–10 mm. or sometimes only 4 mm. wide, scabrous; panicle large and diffuse, becoming oblong, 15–30 cm. long, the branches naked below, the lower usually reflexed at maturity; spikelets 3–5 mm. long, oblong or ovate-oblong, usually 6–8-flowered; glumes broad, obtuse, much shorter than the lower lemmas, nerveless, the first about 1 mm. long; lemmas firm, obovoid, obtuse, or acutish, prominently 7-nerved, the apex distinctly scarious.

Wet meadows, springs, and shady moist soil in woods, in the Coast Ranges to the Bay region, in the Sierra Nevada, and the high southern mountains, north to British Columbia and east to Idaho. June–Aug. Type locality: Montana.

This may be only a variety of *P. nervata*, but in our area it appears distinct.

7. Glyceria grándis S. Wats.

Reed Manna-grass. Fig. 497.

Poa aquatica americana Torr. Fl. North & Mid. U. S. 1: 108. 1823.
Glyceria grandis S. Wats. in A. Gray, Man. ed. 6. 667. 1890.
Panicularia americana MacM. Met. Minn. Val. 81. 1892.
Panicularia grandis Nash in Britt. & Brown, Illustr. Fl. ed. 2. 1: 265. 1913.

Culms erect, rather stout, glabrous, 1–2 meters tall; sheaths smooth or slightly scabrous; blades flat, glabrous beneath, scabrous above, 15–35 cm. long, 6–15 mm. wide; panicle 20–40 cm. long, open, the branches spreading, 10–20 cm. long; spikelets oblong, 4–7-flowered, mostly purplish, 4–6 mm. long; glumes whitish, the first about 2 mm. long, the second a little longer; lemmas oblong, narrowed to an obtuse apex.

Marshes and wet soil, in the Transition Zone; Nova Scotia to Alaska, south to Pennsylvania, Colorado and eastern Oregon. June–Aug. Type locality: Massachusetts.

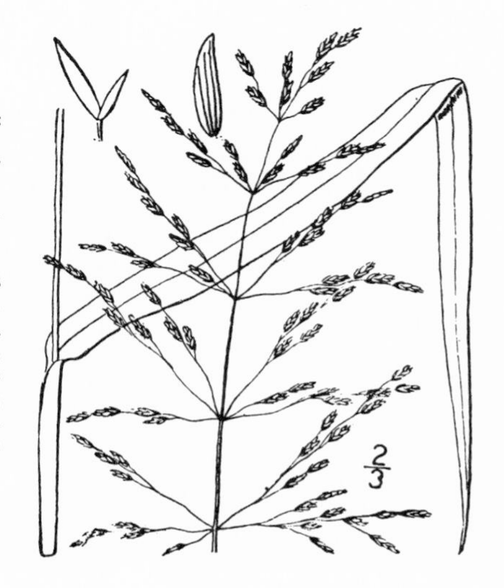

67. PUCCINÉLLIA Parl. Fl. Ital. 1: 366. 1848.

Spikelets several-flowered, usually terete or only slightly flattened, the rachilla disarticulating above the glumes and between the florets; glumes shorter than the first lemma, unequal, obtuse or acute, rather firm, often scarious at the tip, the first 1-nerved or sometimes 3-nerved, the second 3-nerved; lemmas usually firm, rounded on the back, obtuse or acute, rarely acuminate, usually scarious and often erose at the tip, glabrous or puberulent toward the base, 5-nerved, the nerves obscure or indistinct, rarely rather prominent; palea about as long as the lemma or somewhat shorter. Annual or usually perennial low pale smooth cespitose grasses with narrow or open panicles. [Named for Professor Benedetto Puccinelli, an Italian botanist.]

Species about 25, mostly along coasts and on interior alkali soil of the cool and arctic regions of the northern hemisphere. Type species, *Poa distans* L.

Panicle open, the branches spreading or reflexed.
  Lemmas short and broad, truncate; panicle branches finally reflexed. 1. *P. distans.*
  Lemmas narrow, narrowed to an acute or obtusish point; panicle branches spreading, usually not reflexed.
    Leaves mostly in a short basal cluster; panicle usually less than 10 cm. long; lemmas smooth. 2. *P. lemmoni.*
    Leaves scattered; panicle usually more than 10 cm. long; lemmas pubescent on lower part. 3. *P. nuttalliana.*
Panicle narrow, the branches ascending or appressed.
  Plant annual; panicles strict. 4. *P. simplex.*
  Plant perennial; panicles narrow but not strict.
    Lemma entire at apex; plants mostly less than 30 cm. tall; panicle not much exceeding the leaves, the branches smooth or nearly so. 5. *P. paupercula alaskana.*
    Lemma ciliolate at apex; plants mostly more than 30 cm. tall; panicles large, overtopping the foliage, the lower branches as much as 7 cm. long, strongly scabrous. 6. *P. nutkaensis.*

1. Puccinellia dístans (L.) Parl.
European Alkali Grass. Fig. 498.

Poa distans L. Mant. Pl. 1: 32. 1767.
Glyceria distans Wahl. Fl. Ups. 36. 1820.
Puccinellia distans Parl. Fl. Ital. 1: 367. 1848.

Culms tufted, spreading or erect, rather thick and somewhat succulent, 20–30 cm. tall; blades 2–10 cm. long; panicle ovoid, open, 5–10 cm. long, the branches finally reflexed; spikelets 4–5 mm. long, 4–6-flowered; glumes slightly unequal, the first about 1 mm. long; lemmas short and broad, about 2 mm. long, the apex truncate or very obtuse.

Wet soil, introduced in a few localities in Washington (Wenas, Yakima City, Rockland), Oregon (Portland, The Dalles, Klamath Falls), and California (San Diego); a native of Europe, introduced in the eastern States. June. Type locality, European.

2. Puccinellia lémmoni (Vasey) Scribn.
Lemmon's Alkali Grass. Fig. 499.

Poa lemmoni Vasey, Bot. Gaz. 3: 13. 1878.
Puccinellia lemmoni Scribn. U. S. Dept. Agr. Div. Agrost. Bull. 17: *276. f. 572.* 1899.
Puccinellia rubida Elmer, Bot. Gaz. 36: 56. 1903.

Culms slender, 15–40 cm. tall; blades short, filiform, mostly basal, smooth, involute; panicle 5–10 cm. long, becoming open, the branches spreading; spikelets 5–6 mm. long; glumes 1-nerved, the first 2 mm., the second 3 mm. long; lemmas 3 mm. long, smooth.

Alkaline soil, eastern Oregon to Nevada and northern California (Sierra Valley, *Bolander*). June–Aug. Type locality: Sierra County, California.

3. Puccinellia nuttalliàna (Schult.) Hitchc.
Nuttall's Alkali Grass. Fig. 500.

Poa airoides Nutt. Gen. Pl. **1**: 68. 1818, not Koeler, 1802.
Poa nuttalliana Schult. Mant. **2**: 303. 1824.
Puccinellia airoides Wats. & Coult. in Gray, Man. ed. 6. 668. 1890.
Puccinellia nuttalliana Hitchc. in Jepson, Fl. Calif. **1**: 162. 1912.
Puccinellia cusickii Weatherby, Rhodora **18**: 182. 1916.

Culms tufted, erect, 40–60 cm. tall; sheaths and involute blades smooth; panicle open, 15–20 cm. long, the branches spreading, naked below; spikelets terete, about 6 mm. long, usually pale; glumes acutish, the first 1-nerved, 1 mm. long, the second 3-nerved, 2 mm. long; lemmas about 3 mm. long, sparsely pubescent at base.

Alkaline soil, in the Transition and Upper Sonoran Zones; Dakotas to California and Texas. June–Aug. Type locality: Mandan, North Dakota.

4. Puccinellia símplex Scribn.
Little Alkali Grass. Fig. 501.

Puccinellia simplex Scribn. U. S. Dept. Agr. Div. Agrost. Circ. **16**: 1. *f. 1.* 1899.

Apparently annual; culms 7–20 cm. tall; blades narrow, soft, flat, scattered; panicle narrow, about half the length of the entire plant, the branches few, short and appressed; spikelets 6–8 mm. long, appressed; glumes strongly 3-nerved, 1 and 2 mm. long; lemmas 2.5 mm. long, tapering from below the middle to the acute apex, pubescent on lower half.

Alkaline soil, California (Woodland, *Blankinship;* Livermore Pass, *Davy;* Rabbit Springs, *Parish;* Tulare County, *Congdon*). Mar.–Apr. Type locality: Woodland.

5. Puccinellia paupércula alaskàna. (Scribn. & Merr.) Fern. & Weatherby.
Alaska Alkali Grass. Fig. 502.

Puccinellia alaskana Scribn. & Merr. Contr. U. S. Nat. Herb. **13**: 78. 1910.
Puccinellia paupercula alaskana Fern. & Weath. Rhodora **18**: 18. *f. 68–72.* 1916.

Culms erect, tufted, about 30 cm. tall; blades involute, erect, smooth; panicles narrow, 5–7 cm. long, the branches about 2–3 cm. long, appressed; spikelets about 6 mm. long; glumes 3-nerved, the first 2 mm., the second 3 mm. long; lemmas 3 mm. long, sparingly pubescent at base, especially on the lower part of the rather prominent marginal nerves.

Saline soil along the coast, Alaska to Mendocino County, California, and Newfoundland to Connecticut. Whatcom, *Suksdorf;* Umpqua River, *Howell;* Fort Bragg, *Davy.* June–July. Type locality: Alaska.

This species was referred to *P. angustata* (R. Br.) Rand & Redf. (Hitchc. in Jepson, Fl. Calif. **1**: 163. 1912).

6. Puccinellia nutkaénsis (Presl). Fern. & Weath.

Pacific Alkali Grass. Fig. 503.

Poa nutkaensis Presl, Rel. Haenk. **1**: 272. 1830.
Puccinellia nutkaensis Fern. & Weath. Rhodora **18**: 22. *f. 49–53.* 1916.

Culms in small tufts, rather stout, 40–60 cm. tall; leaves scattered, smooth, the blades loosely involute, more or less spreading; panicle narrow, 10–15 cm. long, the branches appressed, the lower as much as 7 cm. long; spikelets about 8 mm. long; glumes nearly equal, 3-nerved, narrow, about 3 mm. long; lemmas 4 mm. long, smooth.

Saline soil near the coast, British Columbia to central California (Point Reyes, *Davy;* San Mateo County, *Jaffa*). June–Aug. Type locality: Nootka Sound, Vancouver Island.

This species was referred to *P. festucaeformis* (Host) Parl. (Hitchc. in Jepson, Fl. Calif. **1**: 163. 1912).

68. FESTÙCA L. Sp. Pl. 73. 1753.

Spikelets few to several-flowered, the rachilla disarticulating above the glumes and between the florets; glumes narrow, acute, unequal, the first sometimes very small; lemmas rounded on the back, membranaceous or somewhat indurate, 5-nerved, the nerves often obscure, acute or rarely obtuse, awned from the tip or rarely from between the minute teeth of the bifid apex. Annual or perennial low or moderately tall grasses of various habit, the spikelets in narrow or open panicles. [Latin, an old name for a grass.]

Species about 100 in the temperate and cool regions of the world. Type species, *Festuca ovina* L.

Plants annual.
  Spikelets densely 5–13-flowered; lemmas without a scarious margin. — 1. *F. octoflora.*
  Spikelets loosely 1–5-flowered; lemmas with a narrow scarious margin.
    Panicle long and narrow, the branches appressed.
      First glume about two-thirds as long as the second. — 2. *F. bromoides.*
      First glume not more than half as long as the second.
        Lemmas ciliate. — 3. *F. megalura.*
        Lemmas not ciliate. — 4. *F. myuros.*
    Panicle rather short, the principal branches and sometimes the pedicels divergent or even reflexed.
      Spikelets glabrous.
        Pedicels of nearly all the spikelets finally reflexed, notably those of the upper part of the main axis; branches of panicle reflexed; spikelets mostly 1- or 2-flowered. — 5. *F. reflexa.*
        Pedicels of spikelets appressed; lower branches of the panicle usually finally reflexed; spikelets usually 3–5-flowered. — 6. *F. pacifica.*
      Spikelets pubescent, the pubescence on glumes or lemmas or both.
        Pedicels of spikelets appressed or slightly spreading; lower branches of panicle usually spreading or reflexed.
          Lemmas pubescent.
            Lemmas hirsute; glumes glabrous or pubescent; lower branches of panicle spreading or reflexed. — 7. *F. grayi.*
            Lemmas woolly-pubescent; glumes glabrous; panicle nearly simple, the branches scarcely spreading. — 8. *F. arida.*
          Lemmas glabrous; glumes pubescent. — 9. *F. confusa.*
        Pedicels of spikelets and the branches of the panicle all finally spreading or reflexed.
          Glumes glabrous; lemmas pubescent. — 10. *F. microstachys.*
          Glumes pubescent; lemmas glabrous. — 11. *F. tracyi.*
          Glumes pubescent; lemmas pubescent. — 12. *F. eastwoodae.*
Plants perennial.
  Rhizomes usually present; blades broad and flat, firm; panicle narrow; spikelets awnless. — 13. *F. confinis.*
  Rhizomes wanting (base of culms decumbent in *F. rubra*).
    Blades flat, rather soft and lax.
      Lemmas awnless (or rarely short-awned), indurate or firm, not keeled. — 14. *F. elatior.*
      Lemmas awned, membranaceous, more or less keeled.
        Florets long-stipitate, the rachilla appearing to be jointed a short distance below the florets. — 15. *F. subuliflora.*
        Florets not stipitate.
          Lemma indistinctly nerved; awn terminal; blades broad and thin. — 16. *F. subulata.*
          Lemma distinctly 5-nerved; awn from between 2 minute teeth; blades narrower, rather firm. — 17. *F. elmeri.*
    Blades narrow or involute, usually not lax.
      Collar and mouth of sheath villous. — 18. *F. californica.*
      Collar and mouth of sheath not villous.
        Culms scabrous, densely cespitose, rather stout, 50–100 cm. tall. — 19. *F. scabrella.*
        Culms smooth.
          Culms slender and decumbent at the usually red and fibrillose base; awn of lemma shorter than the body; blades smooth. — 20. *F. rubra.*

Culms erect, or at least not with decumbent rhizome-like base.
 Lemmas awnless or nearly so. 21. *F. viridula.*
 Lemmas awned.
  Awn of lemma longer than the body; blades soft, sulcate. 22. *F. occidentalis.*
  Awn of lemma shorter than the body; culms closely tufted.
   Blades scabrous, usually elongate. 23. *F. idahoensis.*
   Blades smooth.
    Plants about 1 meter tall or more. 24. *F. howellii.*
    Plants low, usually less than 15 cm.
     Blades hard, involute, shining, not angled. 25. *F. supina.*
     Blades soft, angled in drying, the tissue soft between the angles. 26. *F. brachyphylla.*

1. **Festuca octoflòra** Walt.
Slender Fescue. Fig. 504.

Festuca octoflora Walt. Fl. Carol. 81. 1788.
Festuca tenella Willd. Sp. Pl. 1: 419. 1797.

Culms slender, erect, usually 15–30 cm. tall; blades narrow, involute; panicle narrow, the branches short, appressed; spikelets 6–8 mm. long, densely 5–13-flowered; glumes subulate-lanceolate, the first 1-nerved, 3 mm. long, the upper 3-nerved, 4 mm. long; lemmas firm, convex, lanceolate, glabrous or scabrous, the margins not scarious, 4–5 mm. long, attenuate into a scabrous awn 2–4 mm. long.

Open ground, in the Upper Sonoran and Tansition Zones; throughout most of the United States, extending into eastern Washington and Oregon, and into southeastern California. Apr.–July. Type locality: South Carolina.

Festuca octoflora hirtélla Piper, Contr. U. S. Nat. Herb. **10**: 12. 1906. Differs in being usually in low spreading tufts; foliage usually pubescent; lemmas hirtellous or pubescent. More frequent than the species, growing in more arid ground; Arizona and northern Mexico to California, San Luis Obispo County, to the Mojave Desert; also on Mount Tamalpais (*Chase*) and in northern Inyo County (*Heller*). Type locality: Santa Catalina Mountains, Arizona.

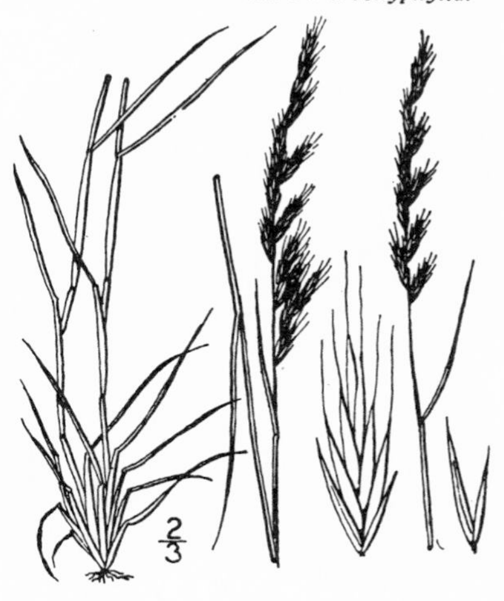

2. **Festuca bromoìdes** L.
Six-weeks Fescue. Fig. 505.

Festuca bromoides L. Sp. Pl. 75. 1753.

Similar to *F. megalura;* culms 10–30 cm. tall; panicle dense, 5–10 cm. long; glumes unequal, the first 4 mm. long, the second 6–7 mm. long; lemma 7–8 mm. long, the awn 10–12 mm. long.

Dry hills and meadows, Santa Barbara and San Bernardino Counties to Vancouver Island; introduced from Europe. Apr.–June. Type locality, European.

3. **Festuca megalùra** Nutt.
Western Six-weeks Fescue. Fig. 506.

Festuca megalura Nutt. Journ. Acad. Phila. II. 1: 188. 1848.

Culms 20–60 cm. tall; sheaths and blades smooth; panicle narrow, 7–20 cm. long, the branches appressed; spikelets 4- or 5-flowered; glumes glabrous, very unequal, the first about 2 mm. long or less, the second 4–5 mm. long; lemmas linear-lanceolate, scabrous above, ciliate on the upper half, attenuate into an awn about twice its length. The cilia on the lemmas, by which this species is distinguished from *F. myuros,* are sometimes hidden by the incurved edges of the lemma at maturity.

Cultivated or open ground, sandy soil and waste places, in the Transition Zone; mostly in the Coast Ranges; extending from British Columbia to Idaho and Lower California. Apr.–June. Type locality: Santa Barbara.

4. Festuca myùros L.
Rattail Fescue. Fig. 507.

Festuca myuros L. Sp. Pl. 74. 1753.

Culms solitary or in small tufts, usually 20–60 cm. tall; sheaths and blades smooth, the sheaths overlapping, the leaves mostly involute; panicle as in *F. megalura,* commonly somewhat shorter; first glume a little shorter than in that species; lemmas scabrous above, not ciliate.

Open ground, introduced from Europe into eastern States; rare on the Pacific Coast. Washington (Vancouver, *Suksdorf*), Oregon (Portland, *Chase*), California (San Francisco, Wilkes Exped.; San Diego, *Brandegee;* Santa Catalina Island, *Brandegee*). May–June. Type locality, European.

5. Festuca refléxa Buckl.
Few-flowered Fescue. Fig. 508.

Festuca reflexa Buckl. Proc. Acad. Phila. **1862**: 98. 1863.
Festuca microstachys pauciflora Scribn.; Beal, Grasses N. Am. **2**: 586. 1896.

Culms 20–40 cm. tall; sheaths smooth or pubescent; blades narrowly linear, flat or loosely involute; panicle 5–12 cm. long, the solitary rays and the spikelets all at length divaricate; spikelets 1–3-flowered, 5–7 mm. long; glumes glabrous, the first 2–4 mm. long, the second 4–5 mm. long; lemmas glabrous or somewhat scabrous, 5–6 mm. long, attenuate into a scabrous awn, usually 5–8 mm. long.

Mesas, rocky slopes, and wooded hills in the Transition Zone; Washington to California and Utah. Common in California, rare in Washington and Oregon. May–June. Type locality: "Upper California."

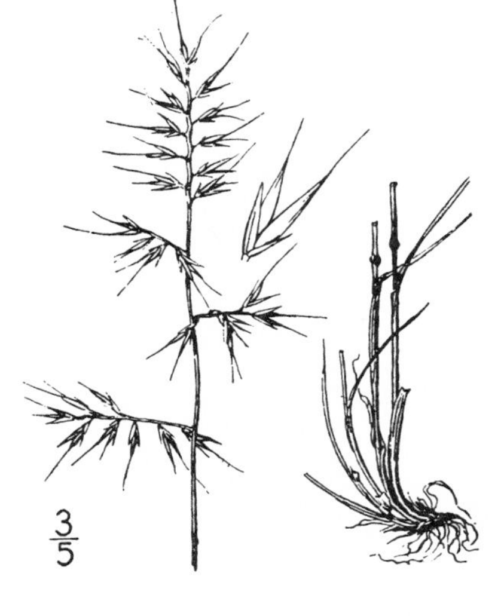

6. Festuca pacífica Piper.
Pacific Fescue. Fig. 509.

Festuca pacifica Piper, Contr. U. S. Nat. Herb. **10**: 12. 1906.

Culms slender, erect, 30–60 cm. tall; blades soft, glabrous, loosely involute; panicle 5–12 cm. long, the lower branches solitary, divaricate; spikelets 3–6-flowered; glumes glabrous, the first subulate-lanceolate, 1-nerved, 4 mm. long, the second lanceolate-acuminate, 3-nerved, 5 mm. long; lemmas lanceolate, scabrous, except in the lowermost floret (this smooth) 6–7 mm. long, attenuate into a scabrous awn 10–15 mm. long; glumes averaging longer and the second more distinctly nerved than in *F. reflexa.*

Open ground, mountain slopes, and open woods, in the Upper Sonoran or Transition Zone; British Columbia to Lower California and Arizona; common throughout California except the great valley and the desert region. May–July. Type locality: Pullman, Washington.

7. **Festuca gràyi** (Abrams) Piper. Gray's Fescue. Fig. 510.

Festuca microstachys ciliata Gray; Beal, Grasses N. Am. **2**: 585. 1896.
Festuca microstachys grayi Abrams, Fl. Los Ang. 52. 1904.
Festuca grayi Piper, Contr. U. S. Nat. Herb. **10**: 14. *pl. 3.* 1906.

Habit of *F. pacifica* but somewhat stouter; sheaths and sometimes blades pubescent; glumes glabrous or more or less villous; lemmas pubescent or puberulent, or sometimes villous.

Open ground and rocky slopes, in the Transition Zone; Oregon to California and Arizona. May–June. Type locality: Grants Pass, Oregon.

8. **Festuca árida** Elmer. Desert Fescue. Fig. 511.

Festuca arida Elmer, Bot. Gaz. **36**: 52. 1903.

Culms erect or spreading, mostly less than 15 cm. tall; sheaths glabrous; blades glabrous, loosely involute, mostly less than 4 cm. long; panicle narrow, 2–5 cm. long, the branches appressed or the lowermost somewhat spreading; glumes about equal, glabrous, 5–6 mm. long; lemmas densely woolly, about 5 mm. long, the awn 5–10 mm. long.

Sandy or dry ground, in the Transition Zone; eastern Washington (Coulee City, North Yakima), western Nevada and eastern California (Truckee, Jess Valley to Blue Lake, Doyle Station). May–July. Type locality: North Yakima.

9. **Festuca confùsa** Piper. Hairy-leaved Fescue. Fig. 512.

Festuca confusa Piper, Contr. U. S. Nat. Herb. **10**: 13. *pl. 1.* 1906.

Habit of *F. pacifica* but commonly lower and more slender; sheaths and blades pubescent; axis and branches of the panicle ciliate on the angles; spikelets 2–3 flowered; glumes hirsute; lemma attenuate, the awn often somewhat flexuous.

Dry hillsides, in the Arid Transition Zone; Washington to middle California. May–June. Type locality: west Klickitat County, Washington.

10. Festuca microstàchys Nutt.

Nuttall's Fescue. Fig. 513.

Festuca microstachys Nutt. Journ. Acad. Phila. II. 1: 187. 1848.

Culms erect, glabrous, 10–30 cm. tall; sheaths smooth or sparsely pubescent; ligule nearly obsolete; blades flat or loosely involute; panicle erect, 4–10 cm. long, often nearly simple, the spikelets 1–3-flowered, finally reflexed, rather distant; glumes glabrous; lemmas pubescent.

Open ground in the Transition Zone; California; rare (Pasadena, Costella, Napa City). Apr.–May. Type locality: Los Angeles.

11. Festuca tràcyi Hitchc. *n. sp.*

Tracy's Fescue. Fig. 514.

Annual; culms erect, smooth, 30–60 cm. tall; sheaths pubescent; ligule very short; blades glabrous, narrow, loosely involute, 3–6 cm. long; panicle 4–8 cm. long, a simple raceme or compound, the lower as much as 3 cm. long, widely spreading; spikelets mostly 1–3-flowered, the pedicels divaricately spreading or reflexed, about 0.5 mm. long; glumes rather sparsely hispid-villous, slightly unequal in length, the first narrow, 1-nerved, 2–3 mm. long, the second 3-nerved; lemmas glabrous, about 4 mm. long, the awns 4–7 mm. long. The species has the aspect of *F. eastwoodae* and *F. reflexa* but differs in having pubescent glumes and glabrous lemmas.

Type specimen collected at head of Moore's Creek, 3–4 miles east of Angwin's, Howell Mountain, Napa County, California, May 15, 1902, by J. P. Tracy (no. 1479). No other collection is known.

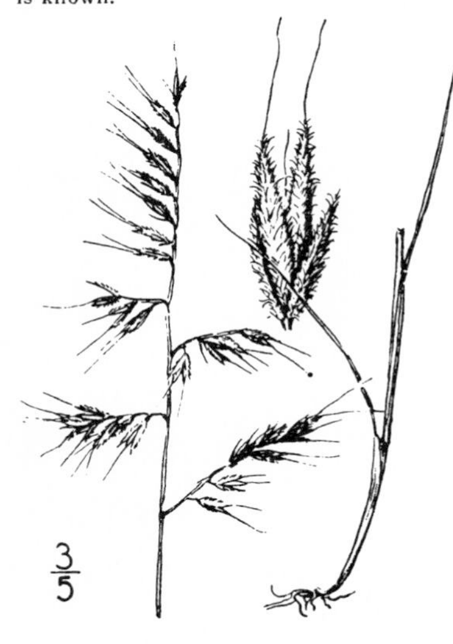

12. Festuca èastwoodae Piper.

Eastwood's Fescue. Fig. 515.

Festuca eastwoodae Piper, Contr. U. S. Nat. Herb. 10: 16. 1906.

Culms erect, glabrous, as much as 30 cm. tall; sheaths glabrous or puberulent, the lower overlapping; ligule minute; blades soft, loosely involute, attenuate, puberulent or glabrous; panicle 10 cm. long, axis and branches pubescent; spikelets finally divaricate; glume and lemmas densely pubescent; awns 3–6 mm. long.

Open pine forests, rare. Type specimen from Santa Lucia Mountains, California (*Eastwood*). The only other specimen known is from Volcano, California, the collector unknown.

13. Festuca confìnis Vasey.
Spiked Fescue. Fig. 516.

Poa kingii S. Wats. in King, Expl. 40th Par. **5**: 387. 1871, not *F. kingiana* Steud., 1854.
Festuca confinis Vasey, Bull. Torrey Club **11**: 126. 1884.
Festuca kingii Cassidy, Colo. Agr. Exp. Sta. Bull. **12**: 36. 1890.
Festuca watsoni Nash in Britton Man. 148. 1901.
Hesperochloa kingii Rydb. Bull. Torrey Club **39**: 106. 1912.
Wasatchia kingii Jones, Contr. West Bot. **14**: 16. 1912.

Culms stout, erect, glabrous, 40–100 cm. tall; rhizomes usually present; sheaths smooth, striate; blades firm, flat or loosely involute, coarsely striate, 3–6 mm. wide; panicle narrow, erect, 7-20 cm. long, the branches short and appressed, floriferous nearly to base; glumes broadly lanceolate, subscarious, nearly smooth, the first 3–4 mm. long, the second a half longer; lemmas ovate, acuminate, convex, faintly nerved, scabrous all over the back, 5–8 mm. long.

Dry meadows and hills, in the Transition Zone; Oregon to southern California, east to Montana and Colorado. June–Aug. Type locality: Colorado.

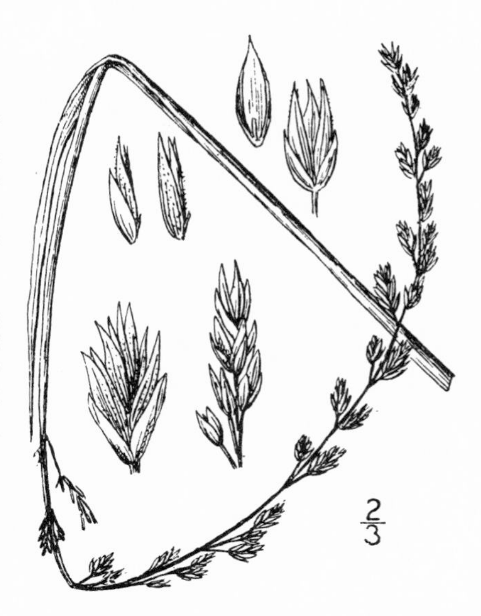

14. Festuca elàtior L.
Meadow Fescue. Fig. 517.

Festuca elatior L. Sp. Pl. 75. 1753.

Culms smooth, 50–120 cm. tall; sheaths smooth; blades flat, 4–8 mm. wide; scabrous above; panicle erect, or nodding at summit, 10–20 cm. long, contracted after flowering, much-branched or nearly simple, the branches spikelet-bearing nearly to base; spikelets usually 6–8-flowered, 8–12 mm. long; glumes 3 and 4 mm. long, lanceolate; lemmas oblong-lanceolate, coriaceous, 5–7 mm. long, the scarious apex acutish, rarely short-awned.

Meadows and roadsides. A native of Europe, cultivated in the United States under the name of Meadow Fescue, and escaped into fields and waste places throughout the cooler portion of America. June–July. Type locality, European.

15. Festuca subuliflòra Scribn.
Coast Range Fescue. Fig. 518.

Festuca subuliflora Scribn. in Macoun, Cat. Can. Pl. **5**: 396. 1890.
Festuca ambigua Vasey, Contr. U. S. Nat. Herb. **1**: 277. 1893.
Festuca denticulata Beal, Grasses N. Am. **2**: 589. 1896.

Culms rather slender, glabrous, 60–100 cm. tall; sheaths sparsely hispidulous; blades flat, rather soft, hirsutulous above, 3–6 mm. wide; panicle loose, open, somewhat drooping, 10–20 cm. long, the branches slender, mostly solitary, naked below the middle; spikelets loosely 3–4-flowered; glumes subulate, glabrous, 1-nerved, 3 and 4 mm. long; lemmas lanceolate, scabrous toward the apex, keeled above, 6–8 mm. long, tipped with a more or less flexuous awn 10–15 mm. long, abruptly contracted at base into a hispidulous tubular structure including the rachilla, the latter apparently disarticulating half way between the florets.

Moist places and borders of rich woods, in the Humid Transition Zone; Vancouver Island to northern California in the Coast Ranges. June–July. Type locality: Vancouver Island.

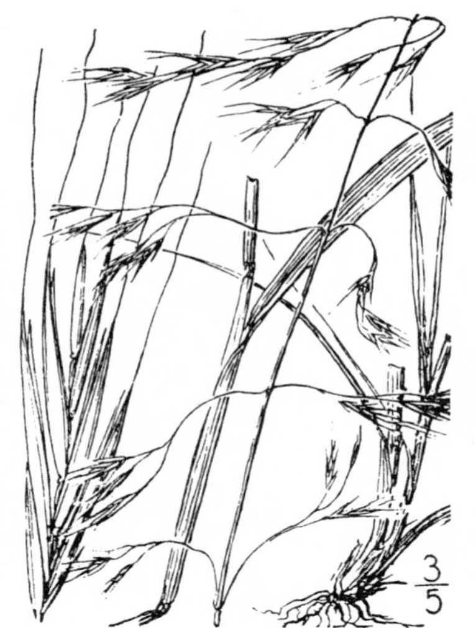

16. **Festuca subulàta** Trin.
Nodding Fescue. Fig. 519.

Festuca subulata Trin. in Bong. Mém. Acad. St. Pétersb. VI. Math. Phys. Nat. **2**: 173. 1832.
Festuca jonesii Vasey, Contr. U. S. Nat. Herb. **1**: 278. 1893.

Culms scaberulous, 40–120 cm. tall; sheaths nearly smooth; blades flat, thin, 3–10 mm. wide, auriculate at base, usually scabrous on both surfaces, lax and spreading; panicle very loose, somewhat drooping, 15–40 cm. long, the branches mostly in pairs, naked below; spikelets 3–5-flowered; glumes subulate; lemmas membranaceous, narrowly lanceolate, 3-nerved, somewhat keeled, attenuate into a scabrous awn 6–20 mm. long.

Moist rocky woods and shady banks, from the Humid Transition to the Hudsonian Zone; Alaska to California and Wyoming. June–Aug. Type locality: Sitka.

17. **Festuca élmeri** Scribn. & Merr.
Elmer's Fescue. Fig. 520.

Festuca elmeri Scribn. & Merr. Bull. Torrey Club **29**: 468. 1902.

Culms slender, 40–100 cm. tall, glabrous, sheaths nearly smooth; blades flat, scabrous or pubescent above, 2–4 mm. wide; panicle 10–20 cm. long, loose, open, the branches mostly in pairs, smooth or nearly so, naked below; spikelets 3 or 4-flowered; glumes lanceolate, glabrous, the first 2 mm., the second 3–4 mm. long; lemmas lanceolate, membranaceous, minutely hispidulous, 6 mm. long, cleft at the apex and bearing between the short teeth a scabrous awn 2–8 mm. long.

Wooded hillsides, in the Transition Zone; Oregon to California, mostly in the Coast Ranges. May–June. Type locality: Stanford University.

Festuca elmeri luxùrians Piper, Contr. U. S. Nat. Herb. **10**: 38. 1906. *F. jonesii conferta* Hack.; Beal, Grasses N. Am. **2**: 593. 1896. Panicle rather close; spikelets 5- or 6-flowered. Moist groves, San Francisco Bay region. Type locality: San Jose.

18. **Festuca califórnica** Vasey.
California Fescue. Fig. 521.

Bromus kalmii aristulatus Torr. U. S. Rep. Expl. Miss. Pacific **4**: 157. 1856.
Festuca californica Vasey, Contr. U. S. Nat. Herb. **1**: 277. 1893.
Festuca aristulata Shear; Piper, Contr. U. S. Nat. Herb. **10**: 32. 1906.

Culms tufted, stout, coarse, usually 1–1.5 meters tall, scabrous; sheaths somewhat scabrous, the lower persistent, the collar and auricles pilose; blades flat or becoming involute when dry, hard and firm, scabrous, the lower much elongated; panicle large, usually loose and open, the branches few, long and slender, naked below, bearing a few large spikelets toward the ends; spikelets compressed, about 5-flowered; glumes oblong-lanceolate, firm, smooth, except the scabrous keel, 6–8 mm. long; lemmas 8–10 mm. long, lanceolate, convex, firm, scabrous, acuminate or short-awned.

Meadows, shady banks, and borders of woods, Transition Zone; in the Coast Ranges from Oregon to Monterey. Apr.–June. Type locality: San Francisco.

Festuca californica paríshii (Piper) Hitchc. *Festuca aristulata parishii* Piper, Contr. U. S. Nat. Herb. **10**: 33. 1906. *Festuca parishii* Hitchc. in Jepson, Fl. Calif. **1**: 169. 1912. Culms lower, about 40–60 cm. tall; sheaths puberulent; blades 15–25 cm. long, closely involute, smooth below or nearly so; panicle 10–12 cm. long; awn of lemma 3–4 mm. long. Only known from the San Bernardino Mountains (*Parish*).

19. **Festuca scabrélla** Torr.

Buffalo Bunch-grass. Fig. 522.

Festuca scabrella Torr.; Hook. Fl. Bor. Am. **2**: 252. 1840.
Melica hallii Vasey, Bot. Gaz. **6**: 296. 1881.
Festuca scabrella major Vasey, Contr. U. S. Nat. Herb. **1**: 278. 1893.
Festuca campestris Rydb. Mem. N. Y. Bot. Gard. **1**: 57. 1900.
Festuca hallii Piper, Contr. U. S. Nat. Herb. **10**: 31. 1906.

Culms densely tufted, erect, smooth or scabrous, 30–100 cm. tall; upper sheaths scabrous, the lower smooth, enlarged at base; blades firm, involute, pale or glaucous, 10–30 cm. long, sharp-pointed, usually scabrous; panicle narrow, 3–15 cm. long, the rays solitary or in pairs, scabrous, ascending or appressed, spikelet-bearing near the end; spikelets oblong, 8–12 mm. long, 4–6-flowered; glumes unequal, smooth, or scabrous near the apex, the first lanceolate, 1-nerved, 7–8 mm. long, the second ovate-lanceolate, 3-nerved, 8–9 mm. long; lemmas firm, keeled above, scaberulous, 8–10 mm. long, acute, or short-awned.

Rocky cliffs and dry woods, in the Arid Transition Zone; British Columbia to Oregon and Colorado. May–July. Type locality: Rocky Mountains.

20. **Festuca rùbra** L.

Red Fescue. Fig. 523.

Festuca rubra L. Sp. Pl. 74. 1753.
Festuca vallicola Rydb. Mem. N. Y. Bot. Gard. **1**: 57. 1900.

Culms erect from a decumbent or somewhat creeping base, smooth, 40–100 cm. tall; sheaths smooth, the lowermost usually purple; blades smooth, soft, usually folded or involute; panicle 3–20 cm. long, usually contracted and narrow, the rays mostly erect; spikelets 4–6-flowered, pale green or glaucous, often purple-tinged; lemmas 5–7 mm. long, smooth, or scabrous toward apex, bearing a scabrous awn usually about half as long.

Meadows, hills, and marshes, in the cooler parts of the northern hemisphere, extending south in the Coast Ranges to Monterey and in the Sierra Nevada to the San Bernardino Mountains. In the sand of the seacoast the plants are often glaucous. June–Aug. Type locality, European.

Festuca rubra subvillòsa Mert. & Koch in Röhling, Deutsch. Fl. ed. 3. **1**: 654. 1823. *F. kitaibeliana* Schult. Mant. **2**: 398. 1824. *F. rubra pubescens* Vasey; Beal, Grasses N. Am. **2**: 607. 1896. Lemmas villous. Alaska to Greenland and northern Europe. Introduced along the Columbia River (Portland; west Klickitat County).

21. **Festuca virídula** Vasey.

Mountain Bunch-grass. Fig. 524.

Festuca viridula Vasey, U. S. Dept. Agr. Div. Bot. Bull. **13**^{2}: *pl. 93.* 1893.

Culms rather loosely tufted, erect, smooth, 60–100 cm. high; sheaths smooth; blades erect, 2 mm. wide or less, soft, scaberulous above, often more or less involute; panicle loose and open, 10–15 cm. long, the branches ascending; spikelets 3–6-flowered; glumes membranaceous, smooth, about 3 mm. long; lemmas firm, membranaceous, keeled toward the apex, acute or somewhat mucronate, 6–7 mm. long.

Subalpine meadows in the Hudsonian Zone; Washington to central California. July–Aug. Type locality: California.

22. Festuca occidentàlis Hook.

Western Fescue. Fig. 525.

Festuca occidentalis Hook. Fl. Bor. Am. **2**: 249. 1840.
Festuca ovina polyphylla Vasey; Beal, Grasses N. Am. **2**: 597. 1896.

Culms densely tufted, slender, erect, shining, 40–80 cm. tall; sheaths smooth; blades numerous, mostly basal, filiform-involute, bright green, soft, 3–20 cm. long; panicle loose, subsecund, 7–20 cm. long, often drooping above, the rays solitary or the lowest in pairs; spikelets loosely 3–5-flowered, 6–10 mm. long, mostly on slender pedicels, pale green; lemmas rather thin, 5–6 mm. long, scaberulous toward the apex, attenuate into a slender awn about as long.

Dry, rocky wooded slopes and banks, in the Transition Zone; British Columbia to central California, east to Wyoming and northern Michigan. June–Aug. Type locality: mouth of the Columbia River.

23. Festuca idahoénsis Elmer.

Blue Bunch-grass. Fig. 526.

Festuca ovina ingrata Hack.; Beal, Grasses N. Am. **2**: 598. 1896.
Festuca ovina columbiana Beal, Grasses N. Am. **2**: 599. 1896.
Festuca ovina oregona Hack.; Beal, Grasses N. Am. **2**: 599. 1896.
Festuca idahoensis Elmer, Bot. Gaz. **36**: 53. 1903.
Festuca ingrata Rydb. Bull. Torrey Club **32**: 608. 1905.

Culms densely tufted, smooth or somewhat scabrous above, 30–100 cm. tall; blades numerous, mostly basal, rather stiff and firm, more or less flexuous, scabrous, 15–30 cm. long, sometimes shorter; panicle narrow, 10–20 cm. long, the branches appressed or ascending, very scabrous; spikelets about as in *F. rubra*, the lemmas firmer, the awn 2–4 mm. long.

Open woods and rocky slopes, in the Arid Transition Zone; British Columbia to central California, east to Alberta and Colorado. May–July. Type locality: Shoshone County, Idaho.

24. Festuca howéllii Hack.

Howell's Fescue. Fig. 527.

Festuca howellii Hack.; Beal, Grasses N. Am. **2**: 591. 1896.

Culms densely tufted, erect, 60–120 cm. tall; sheaths smooth; ligule very short; blades loosely involute, shining, 8–15 cm. long, about 4 mm. wide; panicle open, 8–10 cm. long, the lower branches in pairs; spikelets 4 or 5-flowered, 8–12 mm. long; glumes lanceolate, glabrous or nearly so, the first 2–3 mm., the second 4 mm. long; lemma membranaceous, linear-lanceolate, strongly 5-nerved, appressed-hispidulous, 6 mm. long, the awn about 2 mm. long.

The type collection is from Deer Creek, Josephine County, Oregon. Certain California specimens may belong to this species (Sherwood Valley, *Davy, Hitchcock;* Mount Hood, Sonoma County, *Heller*).

25. **Festuca supìna** Schur.

Small Sheep Fescue. Fig. 528.

Festuca supina Schur, Enum. Pl. Trans. 784. 1866.
Festuca ovina supina Hack. Bot. Centralbl. **8**: 405. 1881.

Culms erect, densely tufted, 7–15 cm. tall; blades numerous, usually less than half the length of the culms, involute, smooth, firm and hard, scarcely angled in drying; inflorescence about as in *F. brachyphylla,* the lemmas firmer, narrower, involute and more scabrous, the florets looser and more numerous and the awn longer.

In the Alpine Zone; extending south on the high peaks of the Sierra Nevada and San Bernardino Mountains. July–Aug. Type locality, European.

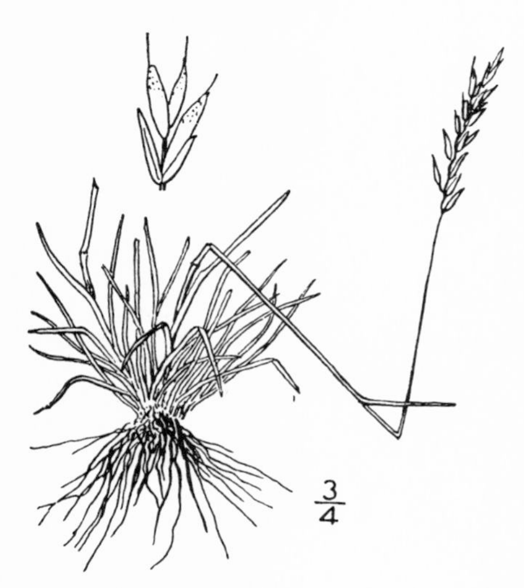

26. **Festuca brachyphýlla** Schult.

Alpine Fescue. Fig. 529.

Festuca brevifolia R. Br. Suppl. App. Parry's Voy. 289. 1824, not Muhl. 1817.
Festuca brachyphylla Schult. Mant. **3**: 646. 1827.
Festuca ovina brevifolia S. Wats. in King, Geol. Expl. 40th Par. **5**: 389. 1871.
Festuca ovina brachyphylla Piper, Contr. U. S. Nat. Herb. **10**: 27. 1906.

Culms erect, tufted, 10–15 cm. high; blades about half as long as the culms, filiform, soft, angled in drying, the tissue soft between the angles; panicle short and narrow, 2–5 cm. long, few-flowered; glumes and lemmas broad, rather soft; awn of the lemma about 1 mm. long.

Alpine Zone; extending south in Rocky Mountains to Arizona; in the Blue Mountains of Oregon; also on Mount Dana, California. July–Aug. Type locality: Melville Island, arctic America.

Sclerópoa rígida (L.) Griseb. Spic. Fl. Rum. **2**: 431. 1844. *Poa rigida* L. Amoen. Acad. **4**: 265. 1759. This European species is introduced at a few localities in the United States and has been found at Salem, Oregon (*Nelson*). It is a low annual with narrow panicles or racemes of spikelets resembling those of *Puccinellia,* 1-nerved glumes, and obscurely nerved lemmas convex on the back.

69. **BRÒMUS** L. Sp. Pl. 76. 1753.

Spikelets several to many-flowered, the rachilla disarticulating above the glumes and between the florets; glumes unequal, acute, the first 1–3-nerved, the second usually 3–5-nerved; lemmas convex on the back or keeled, 5–9 nerved, 2-toothed at apex, awnless or usually awned from between the teeth; palea usually shorter than the lemma. Annual or perennial, low or moderately tall grasses with closed sheaths, flat blades and open or contracted panicles of large spikelets. [Greek, an ancient name for the oat.]

Species about 100 in the temperate regions of the world. Type species, *Bromus secalinus* L.

Spikelets strongly flattened, the lemmas distinctly compressed-keeled; plants mostly perennial.
  Lemmas acuminate, mucronate, the awn, if present, usually not more than 2 mm. long; blades scabrous; annual. — 1. *B. unioloides.*
  Lemmas distinctly awned.
    Blades canescent and densely pilose, narrow or involute. — 2. *B. subvelutinus.*
    Blades not canescent, glabrous or somewhat pilose.
      Panicle narrow, the branches short and erect; spikelets subsessile or short-pediceled, fascicled or solitary along the main axis and the few short branches; a somewhat succulent beach plant. — 3. *B. maritimus.*
      Panicle open, the branches spreading or drooping (sometimes reduced in *B. marginatus*); spikelets not subsessile, usually rather long-pediceled, not fascicled.
        Branches of the very open panicle, long, slender, drooping, bearing 1 or 2 large spikelets at the end, the naked portion as much as 10 or 15 cm. long in the lowermost; sheaths smooth; spikelets smooth or scabrous. — 4. *B. sitchensis.*
        Branches of panicle not greatly elongate.
          Lemmas smooth or scabrous; sheaths smooth. — 5. *B. polyanthus.*
          Lemmas pubescent, at least below.
            Awn less than 7 mm. long. — 6. *B. marginatus.*
            Awn more than 7 mm. long. — 7. *B. carinatus.*

Spikelets not distinctly flattened and keeled.
Plants perennial.
Creeping rhizomes present; lemmas glabrous, awnless or mucronate; panicle somewhat open, the branches ascending. 8. *B. inermis.*
Creeping rhizomes absent (base of culm decumbent in *B. laevipes*).
Panicle narrow, the branches erect.
Lemma glabrous or evenly scabrous. 9. *B. erectus.*
Lemma appressed-pubescent on the margins and on the lower part. 10. *B. suksdorfii.*
Panicle open, the branches spreading or drooping.
Lemmas pubescent along the margin and on the lower part of the back, the upper dorsal part glabrous.
First glume 3-nerved; plant pale or glaucous. 11. *B. laevipes.*
First glume 1-nerved, or only faintly 3-nerved at base.
Ligule prominent, 3–5 mm. long; lemmas narrow, the awn usually more than 5 mm. long. 12. *B. vulgaris.*
Ligule inconspicuous, about 1 mm. long; lemmas broad, the awn 3–5 mm. long. 13. *B. richardsoni.*
Lemmas pubescent rather evenly over the back.
Branches of the panicle short, stiffly spreading; blades short, mostly on the lower part of the plant. 14. *B. orcuttianus.*
Branches of panicle lax or drooping; blades elongate, scattered.
Panicles small, nodding or drooping, usually not over 10 cm. long; spikelets densely and conspicuously pubescent; blades narrow, 2–4 mm. wide. 15. *B. porteri.*
Panicles larger, erect, the branches more or less drooping; blades mostly wide and lax.
Blades pilose above, scabrous or smooth beneath; ligule 3–4 mm. long; panicle large and open, the slender branches long and drooping. 16. *B. pacificus.*
Blades pubescent on both surfaces, but not pilose; ligule short. 17. *B. grandis.*
Plants annual.
Panicle contracted, usually dense, the branches erect or ascending.
Awn 5–8 mm. long.
Lemmas pubescent. 18. *B. hordeaceus.*
Lemmas glabrous. 19. *B. racemosus.*
Awn 15–20 mm. long.
Culms pubescent below panicle. 20. *B. rubens.*
Culms glabrous. 21. *B. madritensis.*
Panicle open, the branches spreading.
Awn short or wanting; lemmas broad, obtuse, inflated. 22. *B. brizaeformis.*
Awn well developed.
Awn twisted and bent. 23. *B. trinii.*
Awn not twisted and bent (sometimes curved or divaricate).
Sheaths smooth. 24. *B. secalinus.*
Sheaths pubescent.
Lemma narrow, gradually acuminate; awn 1–5 cm. long.
Second glume usually less than 1 cm. long; pedicels capillary, flexuous. 25. *B. tectorum.*
Second glume more than 1 cm. long; pedicels sometimes flexuous but not capillary; spikelets scabrous.
Awn about 2 cm. long; first glume about 8 mm. long. 26. *B. sterilis.*
Awn 3–5 cm. long; first glume about 15 mm. long. 27. *B. rigidus.*
Lemma broad, abruptly narrowed above.
Branches of panicle rather stiffly spreading or drooping, not flexuous; awn straight. 28. *B. commutatus.*
Branches slender and flexuous.
Lemmas pubescent; awn straight. 29. *B. arenarius.*
Lemmas glabrous; awn becoming divergent. 30. *B. japonicus.*

1. **Bromus unioloìdes** H. B. K.
Rescue Grass. Fig. 530.

Bromus unioloides H. B. K. Nov. Gen. & Sp. **1**: 151. 1816.
Ceratochloa haenkeanus Presl, Rel. Haenk. **1**: 285. 1830.
Bromus haenkeanus Kunth, Rev. Gram. **1**: Suppl. XXXII. 1830.
Bromus schraderi Kunth, Enum. Pl. **1**: 416. 1833.
Ceratochloa breviaristata Hook. Fl. Bor. Am. **2**: 253. *pl. 234.* 1840.
Bromus unioloides haenkeanus Shear, U. S. Dept. Agr. Div. Agrost. Bull. **23**: 52. *f. 31.* 1900.

Annual; culms 60–100 cm. tall; sheaths pilose; blades narrow, very scabrous; panicle open or sometimes reduced and narrow; spikelets about 2.5 cm. long, 5–9 mm. broad; glumes smooth, the first 5-nerved, 7–10 mm. long; the second 7-nerved, 10–12 mm. long; lemmas acute, subcoriaceous, glabrous or scabrous, 12–16 mm. long; awn 2 mm. long or less; palea half to three-fourths as long as the lemma.

Native country not certainly known, but probably the Andes; now distributed from Chile to southern United States. Introduced in southern and central California. Apr.–June.

2. Bromus subvelùtinus Shear.

Narrow-leaved Brome-grass. Fig. 531.

Bromus subvelutinus Shear, U. S. Dept. Agr. Div. Agrost. Bull. **23**: 52. *f. 32.* 1900.

Perennial; culms 30–60 cm. tall; sheaths canescent; blades narrow, rather rigid, becoming involute, canescent and also pilose; panicle 5–10 cm. long, narrow, erect, the branches short, erect; spikelets 2.5 cm. long; glumes puberulent, the first 3–5-nerved, 8–10 mm. long; the second 7-nerved, 10–12 mm. long; lemmas appressed-puberulent, 12–14 mm. long; awn 3–4 mm. long.

Dry wooded hills and meadows, in the Transition Zone; eastern Washington to California and Wyoming. May–July. Type locality: Reno, Nevada.

3. Bromus marítimus (Piper) Hitchc.

Seaside Brome-grass. Fig. 532.

Bromus marginatus maritimus Piper, Proc. Biol. Soc. Wash. **18**: 148. 1905.
Bromus maritimus Hitchc. in Jepson, Fl. Calif. **1**: 177. 1912.

Perennial; culms robust, mostly not over 60 cm. tall, more or less geniculate at base and with numerous leafy basal shoots; sheaths smooth or minutely scaberulous; ligule 2–3 mm. long; blades 12–30 cm. long, mostly 6–8 mm. wide, scabrous; panicles mostly 10–20 cm. long, strict, the branches short and erect; spikelets 3–4 cm. long.

Near the coast, from Sonoma County to Monterey County, California. May–July. Type locality: Point Reyes.

4. Bromus sitchénsis Trin.

Alaska Brome-grass. Fig. 533.

Bromus sitchensis Trin. in Bong. Mém. Acad. St. Pétersb. VI. Math. Phys. Nat. **2**: 173. 1832.

Perennial; culms smooth, stout, 120–180 cm. tall; sheaths smooth; ligule 3–5 mm. long; blades smooth beneath, sparsely short-pilose above, 20–40 cm. long, 7–12 mm. wide; panicle large, lax, drooping, 25–35 cm. long, the lower branches 2–4, long, spreading or drooping, bearing usually 1–3 spikelets on long slender pedicels; spikelets 2.5–3 cm. long, 6–8 mm. wide, strongly compressed; first glume 3-nerved, 8–9 mm. long, the second 5–7-nerved, 10–12 mm. long; lemmas 7-nerved, smooth or scabrous, 12–14 mm. long, the straight awn 6–9 mm. long.

Woods and banks, in the Humid Transition Zone; near the coast, Alaska to Washington. July–Aug. Type locality: Sitka.

5. Bromus polyánthus Scribn.
Great Basin Brome-grass. Fig. 534.

Bromus multiflorus Scribn. U. S. Dept. Agr. Div. Agrost. Bull. **13**: 46. 1898, not Weig. 1772.
Bromus polyanthus Scribn. U. S. Dept. Agr. Div. Agrost. Bull. **23**: 56. *f. 34.* 1900.

Perennial; culms stout, 60–100 cm. tall; sheaths smooth; ligule about 2 mm. long; blades scabrous; panicle erect, the branches erect or slightly spreading; spikelets 3–3.5 cm. long, compressed, 7–11-flowered; glumes smooth or somewhat scabrous; lemmas smooth or scabrous, 13–15 mm. long, the awn 4–6 mm. long.

Woods and banks, in the Arid Transition Zone; eastern Washington and Oregon to Montana and Colorado. July–Aug. Type locality: Wyoming.

6. Bromus marginàtus Nees.
Large Mountain Brome-grass. Fig. 535.

Bromus marginatus Nees; Steud. Syn. Pl. Glum. **1**: 322. 1854.
Bromus breviaristatus Buckl. Proc. Acad. Phila. **1862**: 98. 1863, not *Ceratochloa breviaristata* Hook. 1840.

Short-lived perennial; culms rather stout, 60–120 cm. tall; sheaths pilose; blades broad, flat, more or less pilose; panicle erect, rather narrow, 10–20 cm. long, the lower branches erect or somewhat spreading; spikelets 2.5–3.5 cm. long, 5–7 mm. wide, 7 or 8-flowered; glumes broad, scabrous, or scabrous-pubescent, the first subacute, 3–5-nerved, 7–9 mm. long, the second obtuse, 5–7-nerved, 9–11 mm. long; lemmas subcoriaceous, coarsely pubescent, ovate-lanceolate, acute, 11–14 mm. long; awn 4–7 mm. long.

Open ground, open woods, roadsides, and waste places, in the Transition Zone; British Columbia to southern California, and Alberta. June–Aug. Type locality: Columbia River.

Bromus marginatus seminùdus Shear, U. S. Dept. Agr. Div. Agrost. Bull. **23**: 55. 1900. Sheaths glabrous; plant often tall and stout with large spreading panicle. Woods or near streams, Transition and Canadian Zones; mostly from 3000–9000 feet altitude, British Columbia to the San Jacinto Mountains, east to Colorado. Type locality: Wallowa Lake, Oregon.

Bromus marginatus làtior Shear, U. S. Dept. Agr. Div. Agrost. Bull. **23**: 55. 1900. A tall stout form with pubescent sheaths and large panicles, the lower branches spreading, as much as 20 cm. long. With the species. Type locality: Walla Walla, Washington.

7. Bromus carinàtus Hook. & Arn.
California Brome-grass. Fig. 536.

Bromus carinatus Hook. & Arn. Bot. Beechey Voy. 403. 1841.

Perennial or annual; culms 60–100 cm. tall; sheaths pilose; blades narrow, flat, more or less pilose; panicle pyramidal, rather lax, the lower branches spreading or drooping; spikelets about 2.5 cm. long, 5 mm. wide, 5–9-flowered; glumes lanceolate, acute, glabrous or slightly scabrous-pubescent, the first 3-nerved, 7–9 mm. long, the second 5-nerved, 9–11 mm. long; lemmas lanceolate, puberulent or short-pubescent, 13–16 mm. long; awn 7–10 mm. long.

Open ground, open woods, roadsides, and waste places, Upper Sonoran and Transition Zones; British Columbia to southern California and Idaho. June–Aug. Type locality: California.

Bromus carinatus califórnicus Shear, U. S. Dept. Agr. Div. Agrost. Bull. **23**: 60. 1900; *B. carinatus densus* Shear. Sheaths smooth; lemmas scabrous instead of pubescent. Common in the southern Coast Ranges of California. Type locality: California.

Bromus carinatus hookeriànus (Thurb.) Shear, U. S. Dept. Agr. Div. Agrost. Bull. **23**: 60. *f. 38.* 1900; *Ceratochloa grandiflora* Hook. Fl. Bor. Am. **2**: 253. *pl. 235.* 1840, not

Weig. 1772; *Bromus virens* Buckl. Proc. Acad. Phila. **1862**: 98. 1863, not Nees 1829; *Bromus hookerianus* Thurb. in Wilkes, U. S. Expl. Exped. **17**: 493. 1874. A robust plant with large spreading panicle, the spikelets 3–4 cm. long 5–7 mm. wide, the lemmas scabrous. Washington to California and Idaho. Type locality: Willamette River, Oregon.

Bromus carinatus lineàris Shear, U. S. Dept. Agr. Div. Agrost. Bull. **23**: 61. *f. 39.* 1900. A very narrow-leaved form, with narrow panicle. A depauperate form found only in a few localities in California (Berkeley Hills, *Davy;* Mount Lyell, *Hitchcock*) and Oregon (Eight Dollar Mountain, *Piper;* Siskiyou, *Hitchcock*).

Bromus marginatus, B. carinatus, and their varieties form a polymorphous series difficult to segregate into definite nomenclatural groups.

8. Bromus inérmis Leyss.

Smooth Brome. Hungarian Brome. Fig. 537.

Bromus inermis Leyss. Fl. Hal. 16. 1761.

Perennial, with creeping rhizomes; culms erect, smooth, 50–100 cm. tall; sheaths smooth; ligule 1.5–2 mm. long; blades flat, smooth or nearly so, 5–10 mm. wide; panicle oblong, somewhat spreading, 10–20 cm. long; spikelets narrow, 2–2.5 cm. long, terete; glumes smooth, the first 4–5 mm. long, the second 6–8 mm. long; lemmas emarginate, 9–12 mm. long, glabrous, awnless or with a short awn as much as 2 mm. long.

Introduced from Europe and cultivated as a forage grass. Occasionally escaped along roads and in waste places, Washington to northern California and eastward in the northern States. June–Aug. Type locality, European.

9. Bromus eréctus Huds.

English Brome-grass. Fig. 538.

Bromus erectus Huds. Fl. Angl. 39. 1762.
Bromus macounii Vasey, Bull. Torrey Club **15**: 48. 1888.

Perennial, cespitose; culms erect, glabrous, 60–90 cm. tall; sheaths sparsely pilose or glabrous; ligule about 1.5 mm. long; blades sparsely pubescent; panicle narrow, 10–20 cm. long,. the branches ascending or erect; spikelets 5–10-flowered; glumes acuminate, the first 6–8 mm., the second 8–10 mm. long; lemmas glabrous or evenly scabrous-pubescent over the back, 10–12 mm. long, the awn 5–6 mm. long.

Introduced in a few localities in the eastern States. It has been collected at Steptoe, Washington (*G. R. Vasey*). Type locality, European.

10. Bromus suksdórfii Vasey.

Suksdorf's Brome-grass. Fig. 539.

Bromus suksdorfii Vasey, Bot. Gaz. **10**: 223. 1885.

Culms 60–100 cm. tall; sheaths and blades smooth, scattered; panicle narrow, erect, rather dense, 7–12 cm. long, the branches erect or ascending; spikelets about 2.4 cm. long, longer than the pedicels; glumes glabrous, the first 1-nerved, 8–10 mm. long, the second 3-nerved, 10–12 mm. long; lemmas 12–14 mm. long, appressed-pubescent near the margin and on the lower part of the midnerve.

Rocky woods and slopes, in the Hudsonian Zone; Washington to the southern Sierra Nevada. July–Aug. Type locality: Mount Adams, Washington.

11. Bromus laèvipes Shear.

Woodland Brome-grass. Fig. 540.

Bromus laevipes Shear, U. S. Dept. Agr. Div. Agrost. Bull. **23**: 45. *f. 25.* 1900.

Culms 50–100 cm. tall, the base often decumbent and rooting; sheaths and blades glabrous; panicle broad, lax, drooping, 15–20 cm. long, the branches slender, drooping; glumes smooth, the first 3-nerved, 6–8 mm. long, the second 5-nerved, 10–12 mm. long; lemmas obtuse, 7-nerved, 12–14 mm. long, densely pubescent on the margin nearly to the apex and on the back at the base; awn 3–5 mm. long.

Moist woods and shady banks, southern Washington to southern California. June–Aug. Type locality: west Klickitat County, Washington.

12. Bromus vulgáris (Hook.) Shear.

Narrow-flowered Brome-grass. Fig. 541.

Bromus purgans vulgaris Hook. Fl. Bor. Am. **2**: 252. 1840.
Bromus ciliatus pauciflorus Vasey; Beal, Grasses N. Am. **2**: 619. 1896.
Bromus vulgaris Shear, U. S. Dept. Agr. Div. Agrost. Bull. **23**: 43. *f. 24.* 1900.

Culms 80–120 cm. tall, the nodes pubescent; sheaths pilose; ligule prominent, 3–5 mm. long; blades scattered, more or less pilose; panicle open, 10–15 cm. long, the branches slender and drooping; spikelets slender, about 2.5 cm. long; glumes narrow, sparsely pubescent, the first 1-nerved, 5–8 mm. long, acute, the second 3-nerved, broader and longer than the first, obtuse or acutish; lemmas 8–10 mm. long, sparsely pubescent over the back, pubescent or ciliate near the margins or nearly glabrous; awn 6–8 mm. long.

Rocky woods and shady ravines, in the Transition Zone; British Columbia to California and Montana. June–Aug. Type locality: Columbia River.

Bromus vulgaris exímius Shear, U. S. Dept. Agr. Div. Agrost. Bull. **23**: 44. 1900. Plant more robust, the sheaths glabrous, the lemma nearly glabrous. Washington to Mendocino County, California. Type locality: Wallowa Lake, Oregon.

Bromus vulgaris robústus Shear, U. S. Dept. Agr. Div. Agrost. Bull. **23**: 44. 1900. Robust like the preceding variety but with pilose sheaths; panicle large. British Columbia to Oregon. Type locality: Seaside, Oregon.

13. Bromus richardsòni Link.

Richardson's Brome-grass. Fig. 542.

Bromus richardsoni Link, Hort. Berol. **2**: 281. 1833.
Bromus purgans longispicatus Hook. Fl. Bor. Am. **2**: 252. 1840.
Bromus ciliatus scariosus Scribn. U. S. Dept. Agr. Div. Agrost. Bull. **13**: 46. 1898.

Plant pale green; culms erect, smooth, 60–130 cm. tall; sheaths smooth or sparsely pilose; ligule short and truncate, 1–2 mm. long; blades mostly less than 5 mm. wide; panicle drooping; 10–20 cm. long; spikelets 2–3 cm. long; glumes smooth, the first 1-nerved, 8–10 mm. long, the second 3-nerved, 9–12 mm. long; lemma 12–15 mm. long, silky-villous between the outer nerves and the margins nearly to the apex.

Open woods and rocky slopes, in the Canadian Zone; Alaska to Arizona and Nebraska, extending into Washington and eastern Oregon. June–Aug. Type locality: western North America.

The form found in our area belongs to the variety *B. richardsoni pállidus* (Hook.) Shear, U. S. Dept. Agr. Div. Agrost. Bull. **23**: 34. 1900 (*B. purgans pallidus* Hook. Fl. Bor. Am. **2**: 252. 1840), which has smaller panicles than the species, narrower blades and more densely silky lemmas.

14. Bromus orcuttiànus Vasey.
Orcutt's Brome-grass. Fig. 543.

Bromus orcuttianus Vasey, Bot. Gaz. **10**: 223. 1885.
Bromus brachyphyllus Merr. Rhodora **4**: 146. 1902.

Culms erect, leafy below, nearly naked above, 80–120 cm. tall, pubescent at and below the nodes; sheaths pilose or more or less velvety; blades glabrous, rather short and erect; panicle narrow-pyramidal, erect, 10–15 cm. long, the branches few, divaricate and rather rigid in fruit; spikelets about 2 cm. long, subterete, on short stout pedicels; glumes narrow, smooth or scabrous, the first acute, 6–8 mm. long, 1-nerved, or sometimes with a faint lateral pair, the second broader, obtuse, 8–10 mm. long, 3-nerved; lemmas 10–12 mm. long, narrow, scabrous, or scabrous-pubescent over the back, the awn 5–7 mm. long.

Open woods, in the Canadian Zone; Washington to California. July–Aug. Type locality: San Diego, California.

Bromus orcuttianus hállii Hitchc. in Jepson, Fl. Calif. **1**: 175. 1912. Blades soft-pubescent on both surfaces; glumes and lemmas pubescent. Dry, mostly wooded ridges and slopes, 5000–9000 feet, California. Type locality: San Jacinto Mountains.

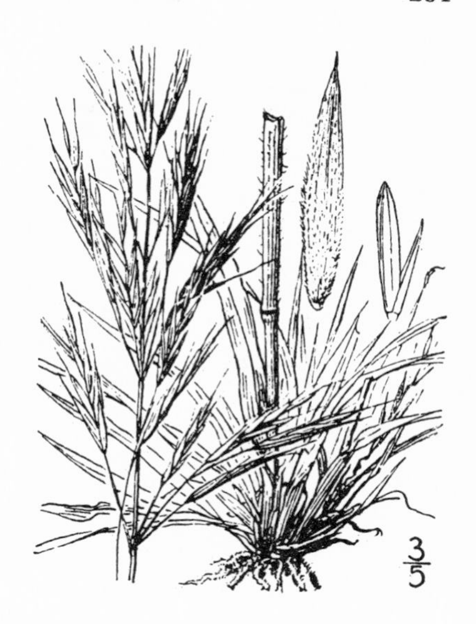

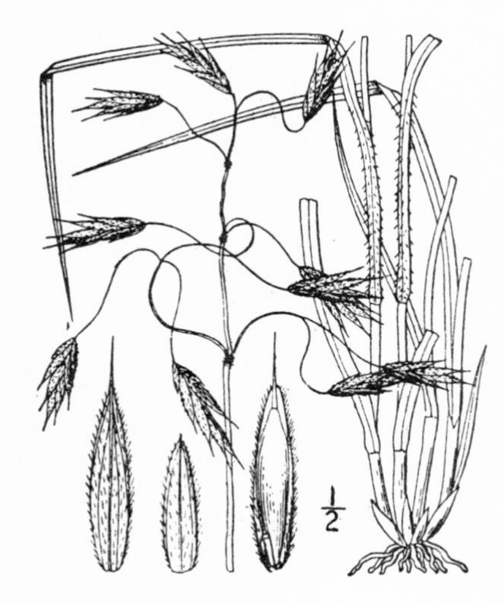

15. Bromus pòrteri (Coulter) Nash.
Porter's Brome-grass. Fig. 544.

Bromus kalmii porteri Coulter, Man. Rocky Mount. 425. 1885.
Bromus porteri Nash, Bull. Torrey Club **22**: 512. 1895.
Bromus kalmii occidentalis Vasey; Beal, Grasses N. Am. **2**: 624. 1896.
Bromus ciliatus montanus Vasey; Beal, Grasses N. Am. **2**: 619. 1896.

Culms slender, erect, pubescent at the nodes, 30–60 cm. tall; sheaths sparsely pilose or glabrous; ligule short, about 1 mm. long; blades scabrous, 3–5 mm. wide; panicle nodding, rather few-flowered, about 10 cm. long; spikelets 2–2.5 cm. long; lemmas about 12 mm. long, evenly pubescent over the back, the awn 2–4 mm. long.

Open woods, in the Transition and Canadian Zones; Idaho to Arizona, east to Colorado; Ontario, California (Johnston). June–Aug. Type locality: Colorado.

16. Bromus pacíficus Shear.
Pacific Brome-grass. Fig. 545.

Bromus pacificus Shear, U. S. Dept. Agr. Div. Agrost. Bull. **23**: 38. *f. 21.* 1900.
Bromus magnificus Elmer, Bot. Gaz. **36**: 53. 1903.

Culms stout, erect, pubescent at the nodes, 1–1.5 meters tall; sheaths sparsely retrorse-pilose; ligule prominent, 3–4 mm. long; blades sparsely pilose above, 8–10 mm. wide; panicle large, drooping, very open, 10–20 cm. long, the branches slender, drooping; spikelets 2–2.5 cm. long, coarsely pubescent throughout; lemmas 11–12 mm. long, the pubescence somewhat dense on the margin, the awn 4–6 mm. long.

Moist thickets near the shore, southern Alaska to western Oregon. August. Type locality: Seaside, Oregon.

17. Bromus grándis (Shear) Hitchc.

Tall Brome-grass. Fig. 546.

Bromus orcuttianus grandis Shear, U. S. Dept. Agr. Div. Agrost. Bull. **23**: 43. 1900.
Bromus porteri assimilis Davy, Univ. Calif. Publ. Bot. **1**: 55. 1902.
Bromus grandis Hitchc. in Jepson, Fl. Calif. **1**: 175. 1912.

Culms 1–1.5 m. tall; sheaths pubescent; blades pubescent, elongated, spreading, rather lax; panicle broad, open, the branches slender and drooping, naked below, the lower usually in pairs, as much as 15 cm. long; first glume usually distinctly 3-nerved; lemmas 12–15 mm. long, densely pubescent over the back; awn 5–7 mm. long.

Dry hills at moderate altitudes, in the Transition Zone; Monterey and Madera Counties, California, south to San Diego. June–July. Type locality: San Diego.

18. Bromus hordeàceus L.

Soft Cheat. Fig. 547.

Bromus hordeaceus L. Sp. Pl. 77. 1753.

Culms 20–80 cm. tall; sheaths retrorsely softly pilose-pubescent; blades usually pubescent; panicle contracted, erect, 5–10 cm. long, or, in depauperate plants, reduced to a few spikelets; glumes broad, obtuse, coarsely pilose or scabrous-pubescent, the first 3–5-nerved, 4–6 mm. long, the second 5–7-nerved, 7–8 mm. long; lemmas broad, obtuse, 7-nerved, coarsely pilose or scabrous-pubescent, rather deeply bidentate, 8–9 mm. long, the margin and apex hyaline; awn rather stout, 6–9 mm. long, palea about three-fourths as long as lemma.

A weed in waste places and cultivated soil, abundant on the Pacific Coast, occasional in the eastern States, introduced from Europe. June–July. Type locality, European.

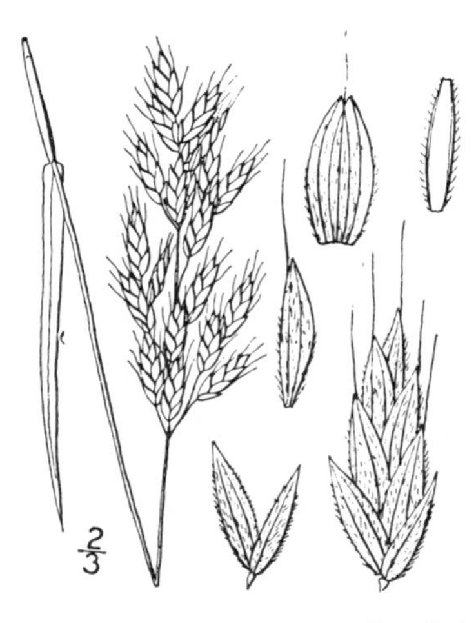

Bromus scopàrius L. Cent. Pl. **1**: 6. 1755. This has been collected in California at Mariposa (*Congdon*) and Santa Barbara (*Somes*). It differs from *B. hordeaceus* in having flexuous, somewhat divaricate or spreading awns. Introduced from Europe.

19. Bromus racemòsus L.

Smooth-flowered Soft Cheat. Fig. 548.

Bromus racemosus L. Sp. Pl. ed. 2. 114. 1762.
Bromus hordeaceus leptostachys Beck, Fl. Niederosterr. 109. 1890.
Bromus hordeaceus glabrescens Shear, U. S. Dept. Agr. Div. Agrost. Bull. **23**: 20. 1900.

Resembles *B. hordeaceus,* but the panicle often more open; lemmas glabrous or scabrous, not pubescent.

A weed in waste places, introduced from Europe. June–July. Type locality, European.

20. **Bromus rùbens** L.

Foxtail Brome-grass. Fig. 549.

Bromus rubens L. Cent. Pl. 1: 5. 1755.

Culms 15–40 cm. tall, puberulent below the panicle; sheaths and blades pubescent; panicle erect, compact, ovoid, usually purplish, 3–7 cm. long; spikelets 7–11-flowered, about 2.5 cm. long; glumes narrow, acuminate, pubescent or sometimes smooth, the first 1-nerved, 7–9 mm. long, the upper 3-nerved, 10–12 mm. long; lemmas lanceolate, acute, 5-nerved, pubescent or smooth, 12–16 mm. long, the apex deeply cleft into 2 long-acuminate hyaline teeth, 4–5 mm. long; awn straight, 18–22 mm. long.

Dry hills and in waste or cultivated ground, in the Arid Transition Zone; Washington to southern California; common in middle and southern California; introduced from southern Europe. Apr.–June. Type locality, European.

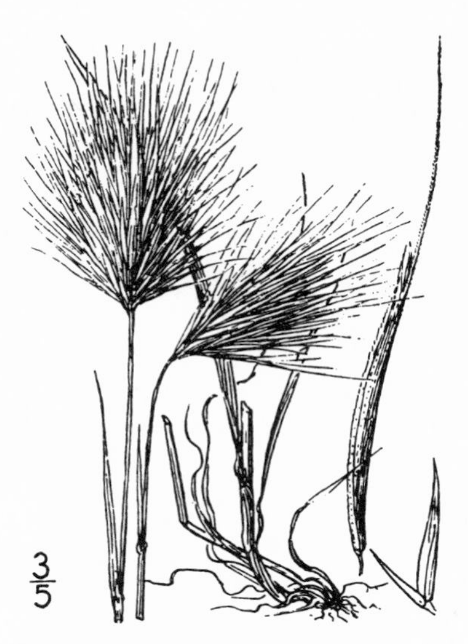

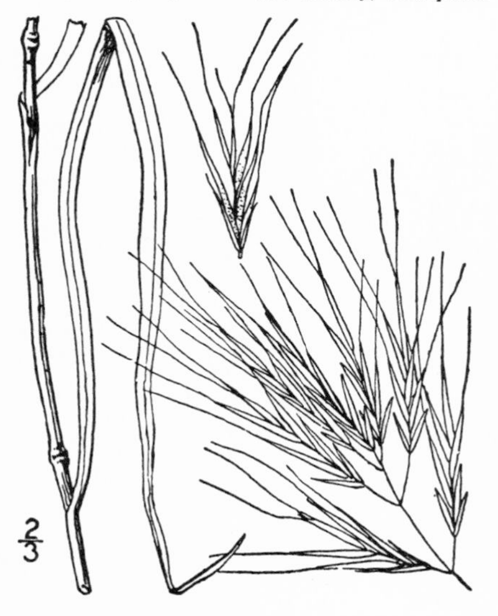

21. **Bromus madriténsis** L.

Spanish Brome-grass. Fig. 550.

Bromus madritensis L. Cent. Pl. 1: 5. 1755.

Culms tufted, 30–60 cm. tall; sheaths smooth or the lower slightly pubescent; blades puberulent or nearly smooth; panicle erect, 5–10 cm. long, oblong-ovoid in outline, contracted and rather dense; glumes narrow, lanceolate, acuminate, the first 1-nerved, 9–12 mm. long, the second 3-nerved, 14–16 mm. long; lemmas narrow, linear-lanceolate, 14–18 mm. long, usually glabrous or merely scabrous, somewhat curved outward when old, distinctly 3- or faintly 5–7-nerved, with 2 acute hyaline teeth, 2–3 mm. long; awn rather stout, tapering, somewhat curved, 16–22 mm. long.

Open ground and waste places, in the Transition Zone; Oregon to Washington; introduced from Europe. May–June. Type locality, European.

22. **Bromus brizaefórmis** Fisch. & Mey.

Rattlesnake Grass. Fig. 551.

Bromus brizaeformis Fisch. & Mey. Ind. Sem. Hort. Petrop. 3: 30. 1837.

Culms 30–60 cm. tall; sheaths and blades pilose-pubescent; panicle 5–15 cm. long, lax, secund, nodding; spikelets oblong-ovate, laterally much compressed, 1.5–2.5 cm. long, about 10 mm. wide; glumes broad, obtuse, smooth or minutely scabrous, the first 3–5-nerved, about half the length of the broader second, the second 5–9-nerved, 6–8 mm. long; lemma 10 mm. long, very broad, obtuse, smooth, with a broad scarious margin, awnless or nearly so.

Sandy fields and waste ground; occasional in the western States, Washington to California, rare in the eastern States, introduced from Europe.

23. Bromus trínii Desv.

Chilian Brome-grass. Fig. 552.

Bromus trinii Desv. in Gay, Fl. Chil. **6**: 441. 1853.
Bromus trinii pallidiflorus Desv. in Gay, Fl. Chil. **6**: 441. 1853.
Trisetum barbatum Steud. Syn. Pl. Glum. **1**: 229. 1854.

Culms erect, 30–60 cm. tall; sheaths pilose or nearly smooth; blades usually pilose; panicle narrow, 10–20 cm. long, rather dense; spikelets narrow, 5–7-flowered, 1.5–2 cm. long; glumes lanceolate, acuminate, smooth, the first mostly 1-nerved, 8–10 mm. long, the second broader, mostly 3-nerved, 12–16 mm. long; lemmas coarsely and sparsely pubescent, 5-nerved, 12–14 mm. long, acuminate, with 2 narrow teeth 2 mm. long; awn 1.5–2 cm. long, twisted below, bent below the middle and strongly divaricate when old.

Dry plains and rocky or wooded slopes, in the Upper and Lower Sonoran Zones; California to Colorado and Mexico; Chile. Mar.–May. Type locality: Chile.

Bromus trinii excélsus Shear, U. S. Dept. Agr. Div. Agrost. Bull. **23**: 25. 1900. A little known form from the Panamint Mountains (*Coville & Funston*). It differs in having larger spikelets, 7-nerved lemmas, and a scarcely twisted or bent awn.

24. Bromus secálinus L.

Chess, Cheat. Fig. 553.

Bromus secalinus L. Sp. Pl. 76. 1753.

Culms erect, 30–60 cm. tall; sheaths smooth; panicle pyramidal, drooping, 7–12 cm. long, open, the lower branches 3–5, unequal; spikelets ovoid-lanceolate, becoming somewhat laterally compressed and turgid in fruit; 1–2 cm. long, 6–8 mm. wide; glumes smooth, obtuse, the first 3–5-nerved, 4–6 mm. long, the second 7-nerved, 6–7 mm. long; lemmas 7-nerved, 6–8 mm. long, elliptic, obtuse, smooth or scabrous, the margin strongly involute in fruit, shortly bidentate at apex, the undulate awn usually 3–5 mm. long; palea about as long as lemma. In fruit the turgid florets are somewhat distant so that, viewing the spikelet sidewise, the light passes through the small openings at the base of each floret.

A weed in grain fields and waste places, more or less throughout the United States; introduced from Europe. June–July. Type locality, European.

25. Bromus tectòrum L.

Downy Brome-grass. Fig. 554.

Bromus tectorum L. Sp. Pl. 77. 1753.

Culms 30–60 cm. tall, smooth, slender; sheaths and blades pubescent; panicle broad, rather dense, drooping, 5–15 cm. long, the branches slender; spikelets nodding, linear, becoming cuneiform in flower, 12–20 mm. long; glumes narrow, acute, villous, the first 1-nerved, 4–6 mm. long, the second 3-nerved, 8–10 mm. long; lemmas lanceolate, acute, villous or pilose, 5-nerved, 10–12 mm. long, bidentate at apex; awn straight, 12–14 mm. long.

Along roadsides, banks, and waste places; introduced from Europe; common on the Pacific Coast in Washington and Oregon, occasional elsewhere. June–July. Type locality, European.

Bromus tectorum nùdus Klett. & Richter, Fl. Leipz. 109, 1830. Differs in having glabrous spikelets. Oregon and California.

26. **Bromus stérilis** L.

Sterile Brome-grass. Fig. 555.

Bromus sterilis L. Sp. Pl. 77. 1753.

Culms erect, smooth, 50–100 cm. tall; sheaths pubescent; blades pubescent; panicle lax and open, drooping, 10–20 cm. long; spikelets 2.5–3.5 cm. long, 6–10-flowered; glumes lanceolate-subulate, smooth or scabrous, the first 1-nerved, 7–9 mm. long, the second 3-nerved, 11–13 mm. long; lemmas linear-lanceolate, 5–7-nerved, 17–20 mm. long, scabrous or scabrous-pubescent, deeply bidentate, the teeth 2 mm. long, the awn 2–3 cm. long.

Introduced at a few localities from Washington to California. May–June. Type locality, European.

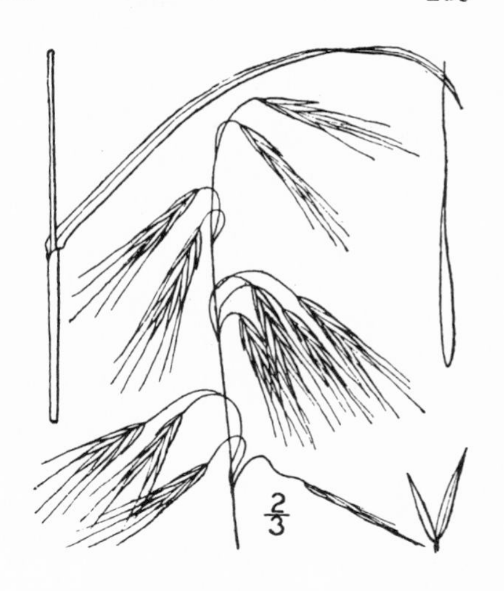

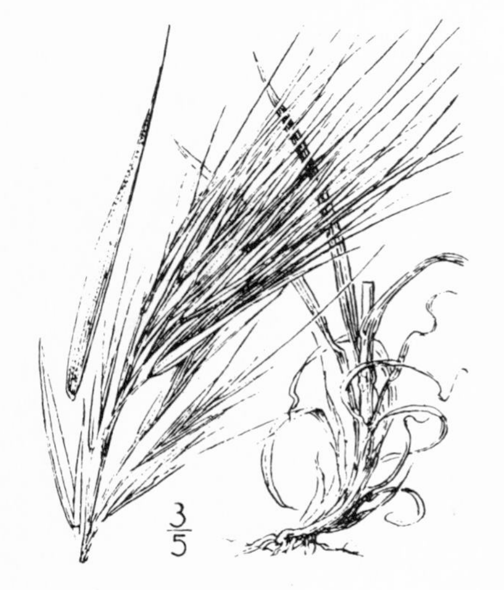

27. **Bromus rígidus** Roth.

Ripgut-grass. Fig. 556.

Bromus villosus Forsk. Fl. Aegypt. Arab. 23. 1775, not Scop. 1772.

Bromus rigidus Roth, Roem. & Ust. Mag. Bot. **10**: 21. 1790.

Bromus maximus Desf. Fl. Atlant. 1: 95. *pl. 26.* 1798, not Gilib. 1792.

Culms 40–70 cm. tall; sheaths and blades pilose; panicle open, rather few-flowered, 7–12 cm. long, the lower branches 1–2 cm. long; spikelets usually 5–7-flowered, 3–4 cm. long; glumes smooth, narrow, acuminate, the first 1.5–2 cm. long, 1-nerved, the second 2.5–3 cm. long, 3-nerved; lemmas 5-nerved, 2.5–3 cm. long, scabrous or puberulent, 2-toothed, the teeth 3–4 mm. long; awn stout, 3.5–5 cm. long.

A weed in open ground and waste places, introduced from the Mediterranean region; occasional in Washington and Oregon, common from San Francisco southward. Injurious to stock. Apr.–June. Type locality, European.

Bromus rigidus gussònei (Parl.) Coss. & Dur. Expl. Sci. Alger. **2**: 159. 1867. *B. maximus gussonei* Parl. Fl. Ital. 1: 407. 1848. Differs in having a more open panicle, the lower branches as much as 10–12 cm. long. Washington to California and Arizona. More common than the species in middle and northern California. Type locality, European.

28. **Bromus commutàtus** Schrad.

Downy-sheathed Cheat. Fig. 557.

Bromus pratensis Ehrh.; Hoffm. Deutschl. Fl. ed. 2. **2**: 52. 1800, not Lam. 1785.

Bromus commutatus Schrad. Fl. Germ. 353. 1806.

Resembling *B. secalinus;* sheaths pilose with short retrorse hairs; lemmas with an obtuse angle on the margin just above the middle, the margin not as strongly inrolled in fruit as in *B. secalinus*, the awn straight and rather longer. In fruit the less turgid florets are imbricate, leaving no spaces at the base of the florets as in *B. secalinus*.

A weed in fields and waste places, Washington to California and Montana, and more sparingly in the eastern States. May–July. Type locality, European.

Bromus commutatus apricòrum Simonkai, Enum. Fl. Transs. 583. 1886. Lemmas pubescent. Oregon. Introduced from Europe.

29. Bromus arenàrius Labill.

Australian Brome-grass. Fig. 558.

Bromus arenarius Labill. Nov. Holl. Pl. 1: 23. *pl. 28.* 1804.

Culms 15–40 cm. tall; sheaths and blades pilose; panicle pyramidal, open, nodding, the spreading branches and slender pedicels sinuously curved; glumes densely pilose, acute, scarious-margined, the first narrower, 3-nerved, 8 mm. long, the second 7-nerved, 10 mm. long; lemmas densely pilose, 7-nerved, 10 mm. long, 2-toothed at apex; awn straight, 10–16 mm. long.

Sandy roadsides, gravelly or sterile hills, Oregon to southern California; introduced from Australia. Apr.–June. Type locality, Australian.

30. Bromus japónicus Thunb.

Japanese Brome-grass. Fig. 559.

Bromus japonicus Thunb. Fl. Japon. 52. 1784.
Bromus patulus Mert. & Koch, Deutschl. Fl. 1: 685. 1823.

Culms erect or geniculate at base, 40–70 cm. tall; sheaths and blades soft-pubescent; panicle 12–20 cm. long, broadly pyramidal, diffuse, somewhat drooping, the lower branches 3–5, slender; glumes rather broad, the first narrower, acute, 3-nerved, 4–6 mm. long, the second obtuse, 5-nerved, 6–8 mm. long; lemmas broad, obtuse, smooth, 9-nerved, the marginal pair faint, 7–9 mm. long, the hyaline margin obtusely angled above the middle, the apex emarginate; awns 8–10 mm. long, somewhat twisted and strongly divaricate at maturity, those of the lower florets shorter than the upper; palea conspicuously shorter than the lemma.

A weed in waste places, Washington to California and Kansas, occasional in the eastern States, introduced from Europe and the Orient. May–July. Type locality: Japan.

70. LÒLIUM L. Sp. Pl. 83. 1753.

Spikelets several-flowered, solitary, sessile, placed edgewise to the continuous rachis, one edge fitting to the alternate concavities, the rachilla disarticulating above the glumes and between the florets; first glum wanting (except on the terminal spikelet), the second outward, strongly 3–5-nerved, equaling or exceeding the second floret; lemmas rounded on the back, 5–7-nerved, obtuse, acute or awned. Annuals or perennials with flat blades and simple terminal flat spikes. [An ancient Latin name.]

Species about 8, in Eurasia, 4 of these being introduced in the United States. Type species, *Lolium perenne* L.

Glume shorter than the spikelet.
  Lemmas awned. — 1. *L. multiflorum.*
  Lemmas nearly or quite awnless. — 2. *L. perenne.*
Glume as long or longer than the spikelet; annuals.
  Spikelets conspicuous. — 3. *L. temulentum.*
  Spikelets inconspicuous, the florets hidden behind the appressed glumes. — 4. *L. subulatum.*

1. **Lolium multiflòrum** Lam.

Italian or Australian Rye-grass. Fig. 560.

Lolium multiflorum Lam. Fl. Franç. **3**: 621. 1778.
Lolium italicum A. Br. Flora **17**: 241. 1834.
Lolium perenne italicum Parn. Grasses Scotl. **1**[1]: 142. *pl. 65.* 1842.
Lolium perenne multiflorum Parn. Grass. Brit. 302. *pl. 140.* 1845.

Short-lived perennial; culms 30–60 cm. tall, erect or often decumbent at base, often rough below the spike and on the convex portion of the rachis; spike as much as 30 cm. long; spikelets as much as 2.5 cm. long, twice as long as glume, 10–20-flowered; lemmas 7–8 mm. long, at least the upper awned.

Roadsides and waste places, mostly in the Coast Ranges; introduced from Europe, common on the Pacific Coast and frequent in the eastern States. Frequently cultivated for lawns and as a meadow or pasture grass. June–Aug. Type locality, European.

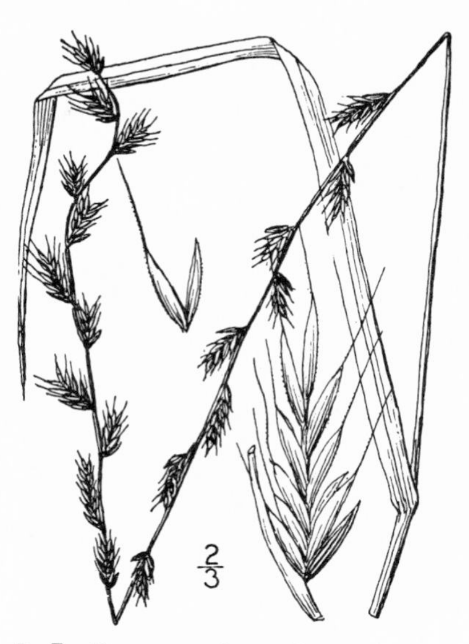

2. **Lolium perénne** L.

Perennial or English Rye-grass. Fig. 561.

Lolium perenne L. Sp. Pl. 83. 1753.

Resembling *L. multiflorum,* but usually more delicate, with narrower blades and smaller spikes; culm and convex surface of rachis smooth; spikelets usually 8–10-flowered, but not much exceeding the glume; lemmas smaller, awnless.

Roadsides and waste places, throughout the cooler and moister portion of the United States. Introduced from Europe. Sometimes cultivated as a lawn or pasture grass. Rare on the Pacific Coast. Type locality, European.

Lolium perenne cristàtum (Pers.) Doell, with ovate spikes, the spikelets horizontally spreading, has been collected at Eola, Oregon (*Nelson*).

Lolium stríctum Presl, an annual with rather stiff culms and spikes, the glume shorter than the spikelet, has been collected at West Berkeley (*Pendleton*) and near Portland (on ballast, *Suksdorf*).

3. **Lolium temuléntum** L.

Darnel. Fig. 562.

Lolium temulentum L. Sp. Pl. 83. 1753.

Annual; culms 60–90 cm. tall; spike stout and strict, 15–20 cm. long; glume about 2.5 cm. long, as long or longer than the 5–7-flowered spikelet, firm, pointed; lemmas as much as 8 mm. long, obtuse, awned; awn as much as 8 mm. long.

Fields and waste places, rather common throughout the State and northward along the Pacific Coast, rare in the eastern States; introduced from Europe. Apr.–June. Type locality, European.

Lolium temulentum arvénse (With.) Bab. Man. Brit. Bot. 377. 1843; *L. arvense* With. Arr. Brit. Pl. ed. 3. **2**: 168. 1796. Differs in having awnless spikelets. Less common than the species, introduced from Europe.

4. Lolium subulàtum Vis.

Narrow-spiked Rye-grass. Fig. 563.

Lolium subulatum Vis. Fl. Dalm. 1: 90. *pl. 3.* 1842.

Annual; culms bushy-branched at base, stiffly spreading or prostrate; sheaths and blades smooth; spike stout, rigid, often curved; spikelets partially sunken in the excavations of the rachis, the florets partially hidden by the appressed obtuse, strongly nerved glume; lemmas 5 mm. long.

Introduced from Europe; rare. Oregon (Portland, on ballast, *Sheldon*). July. Type locality, European.

71. LEPTÙRUS R. Br. Prodr. Fl. Nov. Holl. 207. 1810.

[MONERMA Beauv. Ess. Agrost. 116. *pl. 20. f. 10.* 1812.]

Spikelets 1-flowered, imbedded in the hard cylindric articulate rachis, placed edgewise thereto, the first glume wanting, except on the terminal spikelet, the second glume closing the cavity of the rachis and flush with the surface, indurate, nerved, acuminate, longer than the joint of the rachis; lemma lying next the rachis, hyaline, shorter than the glume, 3-nerved; palea hyaline, 2-nerved, a little shorter than the lemma. Low annuals or perennials with hard cylindric spikes. [Greek, slender-tail.]

Species 3, all from the eastern hemisphere, one introduced in California. Type species, *Rottboellia repens* Forst.

1. Lepturus cylíndricus (Willd.) Trin.

Thin Tail. Fig. 564.

Rottboellia cylindrica Willd. Sp. Pl. 1: 464. 1797.
Lepturus cylindricus Trin. Fund. Agrost. 123. 1820.
Monerma cylindrica Coss. & Dur. Expl. Algér. 2: 214. 1867.

Culms bushy-branched, spreading or prostrate, 10–30 cm. tall; spike cylindric, curved, narrowed upward; glume 6 mm. long, acuminate; lemmas 5 mm. long, pointed, scarious; axis disarticulating at maturity, the spikelets remaining attached to the joints.

Salt marshes, San Francisco Bay, south to San Diego; introduced from Europe. June. Type locality, European.

72. PHOLIÙRUS Trin. Fund. Agrost. 131. 1820.

Spikelets 1 or 2-flowered, imbedded in the articulate rachis; glumes 2, placed in front of the floret and inclosing it, coriaceous, 5-nerved, acute, unsymmetric, appearing like halves of a single split glume; lemma lying next to the axis, smaller than the glumes, hyaline, keeled, scarcely more than 1-nerved; palea a little shorter than the lemma, hyaline, 2-nerved. Low annuals with cylindric spikes. [Greek, scaly-tail.]

Species 4, in the eastern hemisphere, one introduced in the United States. Type species, *Rottboellia pannonica* Host.

1. Pholiurus incúrvus (L.) Schinz & Thell.

Sickle Grass. Fig. 565.

Aegilops incurva L. Sp. Pl. 1051. 1753.
Aegilops incurvata L. Sp. Pl. ed. 2. **2**: 1490. 1763.
Lepturus incurvatus Trin. Fund. Agrost. 123. 1820.
Lepturus incurvus Druce List Brit. Pl. 85. 1908.
Pholiurus incurvatus Hitch. U. S. Dept. Agr. Bull. **772**: 106. 1920.
Pholiurus incurvus Schinz & Thell. Viertel. Naturf. Gesellsch. Züruch **66**: 265. 1921.

Culms tufted, decumbent at base, 10–20 cm. tall; blades short and narrow; spike 7–10 cm. long, cylindric, curved; spikelets 7 mm. long, pointed.

Mud flats and salt marshes, Marin County, California, to San Diego; ballast near Portland, Oregon; adventive on ballast on the Atlantic Coast. Apr.–June. Type locality, European.

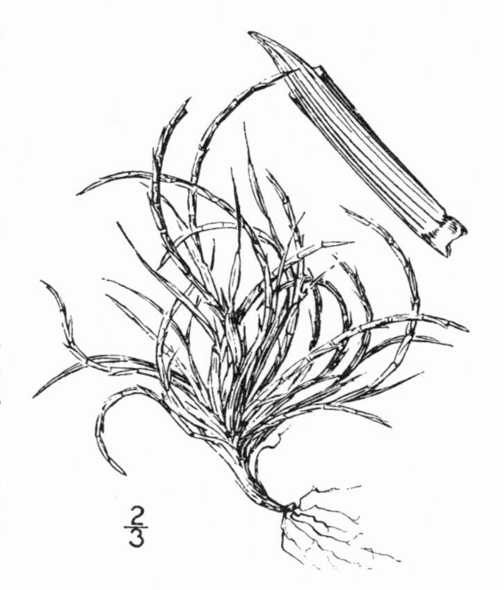

73. SCRIBNÈRIA Hack. Bot. Gaz. 11: 105. 1886.

Spikelets 1-flowered, solitary, sessile, placed with the floret lateral to the continuous rachis, the rachilla disarticulating above the glumes, prolonged behind the floret as a minute stipe; glumes equal, narrow, firm, acute, keeled on the outer nerves, the first 2-nerved; lemma shorter than the glumes, membranous, obscurely nerved, the apex minutely bifid, awned between the teeth, the callus hairy; palea about as long as the lemma. A low slender annual with short narrow blades and slender spikes. [Named for Prof. F. Lamson-Scribner, the eminent American agrostologist.]

Species 1, Pacific Coast. Type species, *Lepturus bolanderi* Thurb.

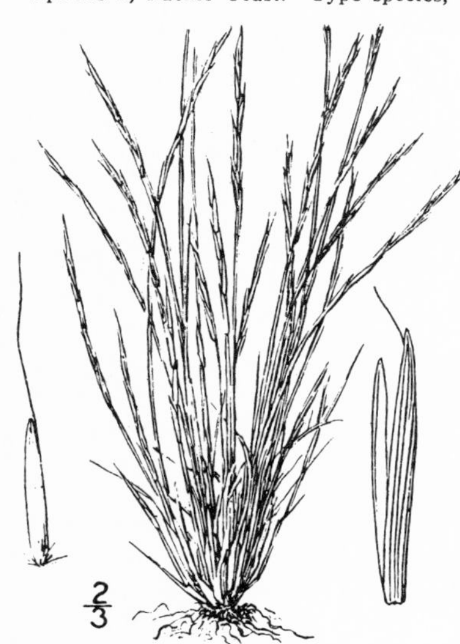

1. Scribneria bolánderi (Thurb.) Hack.

Scribner Grass. Fig. 566.

Lepturus bolanderi Thurb. Proc. Am. Acad. **7**: 401. 1868.
Scribneria bolanderi Hack. Bot. Gaz. **11**: 105. *pl. 5.* 1886.

Culms 7–30 cm. tall, tufted, erect or ascending; spike slender, about 1 mm. thick, the joints 4–6 mm. long; awn of the lemma not much exceeding the glumes.

Sandy or sterile ground, in the mountains, Washington (west Klickitat County, *Suksdorf*) to California (middle Sierra Nevada); rare. May–June. Type locality: Russian River Valley, California.

74. AGROPỲRON Gaertn. Nov. Comm. Acad. Sci. Petrop. 14: 539. *pl. 19. f. 4.* 1770.

Spikelets several-flowered, solitary (rarely in pairs), sessile, placed flatwise at each joint of a continuous (rarely disarticulating) rachis, the rachilla disarticulating above the glume and between the florets; glumes 2, equal, firm, several-nerved, usually shorter than the first lemma, acute or awned, rarely obtuse or notched; lemmas convex on the back, rather firm, 5 to 7-nerved, usually acute or awned from the apex; palea shorter than the lemma. Perennials or sometimes annuals, often with creeping rhizomes, with usually erect culms, and green or purplish usually erect spikes. [Greek, field-wheat.]

Species about 60 in the temperate regions of both hemispheres. Type species, *Agropyron triticeum* Gaertn.

Creeping rhizomes present.
Glumes and lemmas not awn-pointed.
Lemmas pubescent. 1. *A. dasystachyum.*
Lemmas glabrous or minutely pubescent. 2. *A. riparium.*
Glumes and usually lemmas awn-pointed.
Blades thin, flat, usually sparsely pilose above; rhizomes yellowish; lemmas glabrous. 3. *A. repens.*
Blades firm, not pilose, usually becoming involute, the nerves prominent on the upper surface; rhizomes pale.
Spikelets rather distant; spike mostly more than 15 cm. long; lemmas pubescent, rather thin and not strongly pointed. 4. *A. elmeri.*
Spikelets crowded; spike mostly less than 15 cm. long; lemmas glabrous (pubescent in *molle*), firm and strongly pointed or short-awned. 5. *A. smithii.*
Creeping rhizomes wanting.
Lemmas awnless or short-awned.
Nodes appressed pubescent. 6. *A. parishii.*
Nodes glabrous.
Glumes narrow, blunt or abruptly pointed; blades narrow, usually involute. 7. *A. inerme.*
Glumes broad, acute or acuminate; blades flat, mostly 3–6 mm. wide.
Spikes slender, the spikelets more loosely arranged, green. 8. *A. tenerum.*
Spikes relatively short and thick, the spikelets closely overlapping, usually violet-tinged. 9. *A. violaceum.*
Lemmas long-awned.
Awn straight or nearly so; blades flat.
Spikelets distant, scarcely reaching the one above on the opposite side. 10. *A. laeve.*
Spikelets overlapping. 11. *A. caninum.*
Awn strongly divergent.
Glumes acute or awn-pointed. 12. *A. spicatum.*
Glumes distinctly awned.
Rachis disarticulating at maturity; glumes slender, 2-nerved. 13. *A. saxicola.*
Rachis not disarticulating at maturity; glumes broad, 3–5-nerved. 14. *A. pringlei.*

1. **Agropyron dasystàchyum** (Hook.) Vasey. Downy Wheat-grass. Fig. 567.

Triticum repens dasystachyum Hook. Fl. Bor. Am. **2**: 254. 1840.
Triticum repens subvillosum Hook. Fl. Bor. Am. **2**: 254. 1840.
Agropyron dasystachyum Vasey, Grasses U. S. 45. 1883.
Agropyron dasystachyum subvillosum Scribn. & Smith, U. S. Dept. Agr. Div. Agrost. Bull. **4**: 33. 1897.
Agropyron lanceolatum Scribn. & Smith, U. S. Dept. Agr. Div. Agrost. Bull. **4**: 34. 1897.
Agropyron subvillosum E. Nels. Bot. Gaz. **38**: 378. 1904.

Culms 40–100 cm. tall, from creeping rhizomes; blades narrow, mostly involute, scabrous; spike erect, 5–12 cm. long; spikelets few-flowered, 1–1.5 cm. long; glumes lanceolate, the first narrow, 3–5-nerved, 8 mm. long, the second broader, 5–7-nerved, 9 mm. long; lemmas 8–10 mm. long, villous-pubescent or only scabrous-pubescent, acute or awn-pointed.

Dry or sandy soil, in the Upper Sonoran Zone; eastern Washington to northeastern California (Lassen County), east to Colorado. June–July. Type locality: on the Saskatchewan.

2. **Agropyron ripàrium** Scribn. & Smith. Montana Wheat-grass. Fig. 568.

Agropyron riparium Scribn. & Smith, U. S. Dept. Agr. Div. Agrost. Bull. **4**: 35. 1897.

Plant pale, often glaucous; culms erect, 50–100 cm. tall, from slender pale rhizomes; sheaths smooth; blades flat or involute, rather stiff, 1–3 mm. wide, narrowed to a sharp point, scabrous above, glabrous beneath; spikes slender, 5–15 cm. long; spikelets rather distant, often not reaching the one above on the same side, 10–15 mm. long, glabrous; glumes 4–7 mm. long, rather abruptly acute; lemmas acutish or rather obtuse, the lower 7–10 mm. long.

Dry sandy or gravelly soil, in the Transition Zone; eastern Washington and Oregon to North Dakota, south to Nevada and Colorado. July–Aug. Type locality: Montana.

3. Agropyron rèpens (L.) Beauv.
Quack Grass. Fig. 569.

Triticum repens L. Sp. Pl. 86. 1753.
Agropyron repens Beauv. Ess. Agrost. 102, 146. 1812.

Culms 60–120 cm. tall, from a bright yellow-green creeping scaly rhizome; blades thin, flat, usually sparsely pilose above; spike 5–15 cm. long; spikelets about 5-flowered, 10–14 mm. long; glumes 8–10 mm. long, acuminate or awn-pointed, strongly nerved; lemmas 10 mm. long, glabrous or scabrous, strongly nerved, pointed or awned.

A common and troublesome weed in the eastern United States, rare on the Pacific Coast; introduced from Europe; Washington, Oregon (Garibaldi, *Hitchcock;* Menomonei, *Suksdorf*), California (San Francisco, *Bolander*). June–Aug. Type locality, European.

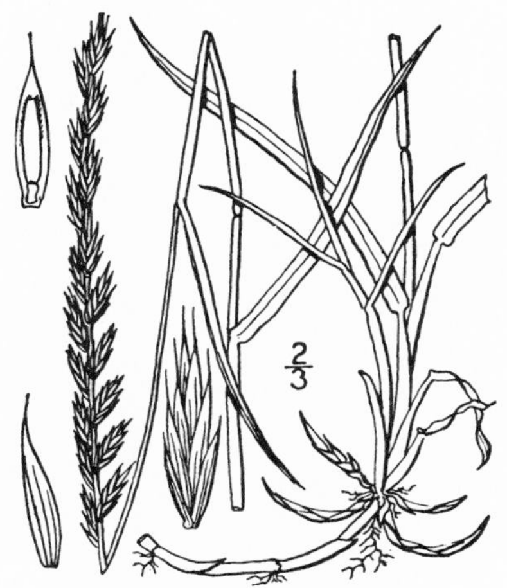

4. Agropyron élmeri Scribn.
Elmer's Wheat-grass. Fig. 570.

Agropyron elmeri Scribn. U. S. Dept. Agr. Div. Agrost. Bull. **11**: 54. *pl. 12.* 1898.

Culms erect, smooth, 60–120 cm. tall, with numerous pale creeping rhizomes; sheaths smooth; blades flat, becoming involute, smooth beneath, scabrous or scabrous-puberulent above, rather stiff and firm, 15–30 cm. long, the upper shorter, 2–8 mm. wide; spike 15–30 cm. long, erect or nodding, the spikelets usually not reaching the one above on the same side; spikelets about 7-flowered, pale, usually 2–2.5 mm. long, sometimes 2 at a node; glumes narrow, firm, acuminate, scabrous or puberulent, about 1 cm. long; lemmas 12–15 mm. long, villous or puberulent, sometimes only toward the base.

Sandy soil, in the Upper Sonoran Zone; eastern Washington and Oregon. June–July. Type locality: Wawawai, Washington.

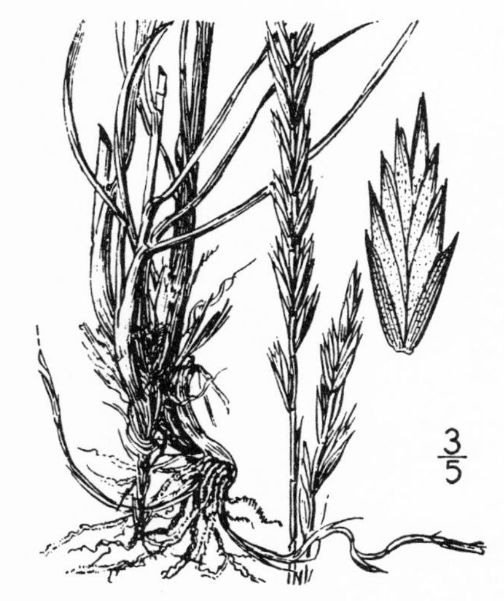

5. Agropyron smíthii Rydb.
Western Wheat-grass. Fig. 571.

Agropyron glaucum occidentale Scribn. Trans. Kans. Acad. **9**: 119. 1885.
Agropyron smithii Rydb. Mem. N. Y. Bot. Gard. **1**: 64. Feb. 1900.
Agropyron occidentale Scribn. U. S. Dept. Agr. Div. Agrost. Circ. **27**: 9. Dec. 1900.

Plant usually glaucous, from gray or tawny creeping scaly rhizomes; culms 30–150 cm. tall, rigid; blades bluish green, scabrous, firm, striate, becoming involute; spikelets 7–13-flowered, somewhat distant, glabrous or nearly so, acute, compressed, divergent, sometimes in pairs; glumes acuminate, half to two-thirds as long as the spikelet, the nerves usually faint; lemmas mucronate or awn-pointed, hard, faintly nerved.

Dry, especially alkaline soil, in the Transition and Upper Sonoran Zones; eastern Washington to northeastern California (Smoke Creek, *Griffiths and Hunter*), east to Michigan and Kansas. June–Aug. Type locality: "Upper Missouri."

Agropyron smithii mólle (Scribn. & Smith) Jones, Contr. West. Bot. **14**: 18. 1912.

Agropyron spicatum molle Scribn. & Smith U. S. Dept. Agr. Div. Agrost. Bull. **4**: 33. 1897.

Agropyron occidentale molle Scribn. U. S. Dept. Agr. Div. Agrost. Circ. **27**: 9. 1900.

Agropyron molle Rydb. Mem. N. Y. Bot. Gard. **1**: 65. 1900. Differs from *A. smithii* chiefly in having pubescent lemmas. Eastern Washington to Alberta, south to Arizona. Type locality: not definitely indicated.

6. Agropyron paríshii Scribn. & Smith.

Parish's Wheat-grass. Fig. 572.

Agropyron parishii Scribn. & Smith, U. S. Dept. Agr. Div. Agrost. Bull. **4**: 28. 1897.

Culms 1–1.5 meters tall, without rhizomes, the nodes pubescent; blades flat; spike narrow, as much as 30 cm. long; spikelets narrow, distant, mostly shorter than the internodes of the rachis, about 2 cm. long; glumes about 1.5 cm. long, several-nerved, acute, more than half as long as spikelet; lemmas smooth, faintly nerved, short-awned or awn-pointed.

Only known from California, Pico Blanco, Monterey County (*Davy*), San Bernardino Mountains (*Parish*).

7. Agropyron inérme (Scribn. & Smith) Rydb.

Great Basin Wheat-grass. Fig. 573.

Agropyron divergens inerme Scribn. & Smith, U. S. Dept. Agr. Div. Agrost. Bull. **4**: 27. 1897.

Agropyron spicatum inerme Heller, Cat. N. Am. Pl. ed. 2. 3. 1900.

Agropyron inerme Rydb. Bull. Torrey Club **36**: 539. 1909.

Culms tufted, without rhizomes, erect, slender, glabrous, 50–100 cm. tall; sheaths smooth; blades slender, soon involute, pubescent on upper surface, scabrous below, especially toward the tip; spike slender erect, 15–20 cm. long, the spikelets about as long as the rachis joints; spikelets 1–1.5 cm. long; glumes acutish but not long-acuminate, 5–8 mm. long; lemmas smooth, scarcely nerved, obtusish, 8–10 mm. long.

Dry plains and hills, in the Arid Transition Zone; British Columbia to eastern Oregon, east to Utah. June–July. Type locality: Idaho.

8. Agropyron ténerum Vasey.

Slender Wheat-grass. Fig. 574.

Agropyron tenerum Vasey, Bot. Gaz. **10**: 258. 1885.

Agropyron violaceum major Vasey, Contr. U. S. Nat. Herb. **1**: 280. 1893.

Agropyron tenerum longifolium Scribn. & Smith, U. S. Dept. Agr. Div. Agrost. Bull. **4**: 30. 1897.

Agropyron pseudorepens Scribn. & Smith, U. S. Dept. Agr. Div. Agrost. Bull. **4**: 34. 1897.

Agropyron caninum tenerum Pease & Moore, Rhodora **12**: 71. 1910.

Culms erect, tufted, 60–120 cm. tall, without rhizomes; blades narrow, flat or involute; spike cylindrical, slender, erect, 10–15 cm. long; glumes firm, nearly as long as the spikelet, gradually tapering into an awned point; lemmas short-awned.

Open woods, rocky slopes, and upland plains, in the Transition and Upper Sonoran Zones; Alaska to southern California, east to Labrador and Colorado. June–Aug. Type locality: Fort Garland, Colorado.

9. **Agropyron violàceum** (Hornem.) Lange.

Alpine Wheat-grass. Fig. 575.

Triticum violaceum Hornem. Fl. Dan. *pl. 2044.* 1832.
Agropyron violaceum Lange, Consp. Fl. Groenland **3**: 155. 1880.
Agropyron caninum hornemanni Pease & Moore, Rhodora **12**: 73. 1910.

Differs from *A. tenerum* in having lower culms, usually less than 50 cm. tall, and a shorter denser spike, mostly 5–10 cm. long, usually tinged with violet, the spikelets closely overlapping.

Alpine meadows and rocky summits, in the Alpine Zone; Washington (Mount Baker, *Hitchcock*), Oregon (Imnaha National Forest, *Sampson*), alpine regions, southward in the mountains to Colorado and the high summits of New York and New England. August. Type locality, European. This species has been referred to *Agropyron biflorum* (Brign.) Roem. & Schult.

Agropyron violaceum andìnum Scribn. & Smith, U. S. Dept. Agr. Div. Agrost. Bull. **4**: 30. 1897; *A. brevifolium* Scribn. U. S. Dept. Agr. Div. Agrost. Bull. **11**: 55. *pl. 13.* 1898. *A. caninum andinum* Pease & Moore, Rhodora **12**: 75. 1910. Differs in having short-awned lemmas. Alaska to Washington (North Fork of Bridge Creek, *Elmer*), Oregon (Summit Wallowa Mountains, *Cusick*), and the mountains of Colorado. Type locality: Colorado.

3/5

10. **Agropyron laève** (Scribn. & Smith). Hitchc.

California Wheat-grass. Fig. 576.

Agropyron parishii laeve Scribn. & Smith, U. S. Dept. Agr. Div. Agrost. Bull. **4**: 28. 1897.
Agropyron laeve Hitchc. in Jepson, Fl. Calif. **1**: 181. 1912.

Culms 1–1.5 meters tall, without rhizomes; blades flat; spike as much as 30 cm. long; spikelets distant, usually shorter than the internodes of the rachis; glumes obtuse, several-nerved, about 10 mm. long; lemmas about 10 mm. long, long-awned; awn 1.5–2.5 cm. long.

Only known from California, Clinton (*Hansen*), Dunlap to Millwood (*Griffiths*), San Jose (*Norton*), Lassen National Forest, Cuyamaca Mountains (*Palmer*), the latter being the type locality. June–July.

11. **Agropyron canìnum** (L.) Beauv.

Bearded Wheat-grass. Fig. 577.

Triticum caninum L. Sp. Pl. 86. 1753.
Agropyron caninum Beauv. Ess. Agrost. 102, 146. 1812.

Culms erect, 30–90 cm. tall, without rhizomes; blades flat, rather lax, 2–6 mm. wide, scabrous; spike more or less nodding at apex, rather dense, 7–15 cm. long; spikelets 12–14 mm. long; glumes pointed or awned; lemmas 3–5-nerved, awn straight, or somewhat spreading, once or twice the length of the lemma.

Dry hillsides and mountain meadows, Transition to Hudsonian Zones; Alaska to California, east to New England. June–Aug. Type locality, European.

12. Agropyron spicàtum (Pursh). Scribn. & Smith.

Blue Bunch Wheat-grass. Fig. 578.

Festuca spicata Pursh, Fl. Am. Sept. **1**: 83. 1814.
Agropyron divergens Nees; Steud. Syn. Pl. Glum. **1**: 347. 1854.
Agropyron divergens tenuispicum Scribn. & Smith, U. S. Dept. Agr. Div. Agrost. Bull. **4**: 27. 1897.
Agropyron vaseyi Scribn. & Smith, U. S. Dept. Agr. Div. Agrost. Bull. **4**: 27. 1897.
Agropyron spicatum Scribn. & Smith, U. S. Dept. Agr. Div. Agrost. Bull. **4**: 33. 1897.

Culms erect, tufted, smooth, 60–120 cm. tall; sheaths smooth; blades rather short; spike 10–20 cm. long, the spikelets distant, often shorter than the internodes of the rachis; glumes acute, not awned, 8–10 mm. long; lemmas smooth, 12–15 mm. long, the awn 1–2 cm. long, becoming divergent.

Plains and dry hills in the Arid Transition and Upper Sonoran Zones; Yukon Territory to eastern Washington and northeastern California, east to Colorado. June–July. Type locality: Idaho.

Agropyron spicatum pubescens Elmer, Bot. Gaz. **36**: 52. 1903. *A. spicatum puberulentum* Piper, Contr. U. S. Nat. Herb. **11**: 147. 1906. Culms and foliage puberulent. Mount Stewart, Kittitas County, Washington, the type locality.

13. Agropyron saxícola (Scribn. & Smith) Piper.

Rock Wheat-grass. Fig. 579.

Elymus saxicolus Scribn. & Smith, U. S. Dept. Agr. Div. Agrost. Bull. **11**: 56. *pl. 15.* 1898.
Sitanion flexuosum Piper, Erythea **7**: 99. 1899.
Agropyron flexuosum Piper, Proc. Biol. Soc. Washington **18**: 149. 1905.
Agropyron saxicola Piper, Contr. U. S. Nat. Herb. **11**: 148. 1906.

Culms 60–100 cm. tall, slender, without rhizomes; sheaths smooth; blades short, flat or loosely involute, puberulent or glabrous; spike 7–10 cm. long, flexuous, long-exserted, the rachis disarticulating; spikelets sometimes in pairs; glumes subulate or narrowly lanceolate, mostly 2-nerved, narrowed into a slender spreading awn 1.5–2.5 cm. long; lemmas 8 mm. long, smooth and rounded below, 5-nerved and somewhat scabrous above, tipped with a slender spreading awn about 2.5 cm. long.

Mountain slopes in the Arid Transition Zone; eastern Washington to northeastern California, east to Idaho. July–Aug. Type locality: Okanogan County, Washington.

14. Agropyron prínglei (Scribn. & Smith) Hitchc.

Pringle's Wheat-grass. Fig. 580.

Agropyron gmelini pringlei Scribn. & Smith, U. S. Dept. Agr. Div. Agrost. Bull. **4**: 31. 1897.
Agropyron pringlei Hitchc. in Jepson, Fl. Calif. **1**: 183. 1912.

Culms 30–40 cm. tall; blades usually flat, short; spikes 5–10 cm. long, not disarticulating, the florets falling; glumes lanceolate, 3–5-nerved, ending in a short straight awn; lemmas ending in stout horizontally spreading awns about 2 cm. long.

Gravelly slides and rocky slopes, in the Boreal Zone; Sierra Nevada Mountains. August. Type locality: above Summit Valley.

Agropyron glaùcum (Desf.) Roem. & Schult., **A. júnceum** (L.) Beauv. and **A. púngens** (Pers.) Roem. & Schult. have been found on ballast at Linnton, Oregon (*Nelson*).

75. HÓRDEUM L. Sp. Pl. 84. 1753.

Spikelets 1-flowered, 3 (sometimes 2) together at each node of the articulate rachis (continuous in *H. vulgare*), the middle one sessile or subsessile, the lateral ones pediceled; rachilla disarticulating above the glumes and, in the central spikelet, prolonged behind the palea as a bristle, this sometimes bearing a rudimentary floret; lateral spikelets usually imperfect, sometimes reduced to bristles; glumes narrow, often subulate and awned, rigid, standing in front of the spikelet; lemmas rounded on the back, 5-nerved, usually obscurely so, tapering into a usually long awn. Annual or perennial low or moderately tall grasses with flat blades and dense terminal cylindrical spikes. [An ancient Latin name for barley.]

Species about 20 in the temperate regions of both hemispheres. Type species, *Hordeum vulgare* L.

Plants annual.
  Glumes, or some of them, ciliate. — 1. *H. murinum.*
  Glumes not ciliate.
    Glumes of the fertile spikelet dilated above the middle. — 2. *H. pusillum.*
    Glumes not dilated.
      Glumes glabrous or nearly so. — 3. *H. gussoneanum.*
      Glumes very scabrous. — 4. *H. depressum.*
Plants perennial.
  Awns 2–5 cm. long.
    Awns 4–5 cm. long. — 5. *H. jubatum.*
    Awns 2–3 cm. long. — 6. *H. caespitosum.*
  Awns not over 1 cm. long.
    Spikes rather stout, the central florets rather prominent on each side of the spike; culms 50–100 cm. tall. — 7. *H. boreale.*
    Spikes slender, the awns obscuring the fertile central florets; culms mostly less than 50 cm. tall. — 8. *H. nodosum.*

1. Hordeum murìnum L.

Wall Barley. Fig. 581.

Hordeum murinum L. Sp. Pl. 85. 1753.

Annual; culms bushy-branched, spreading; sheaths and blades smooth; spike 5–7 cm. long, often partially enclosed by the uppermost inflated sheath; glumes of the central spikelet narrowly spindle-form, 3-nerved, long-ciliate on both margins, the nerves scabrous; awn about 2.5 cm. long; glumes of the lateral spikelets unlike, the inner similar to the central, the outer setaceous, not ciliate; lemmas all broad, 8–10 mm. long, the awns somewhat exceeding those of the glumes.

Fields, waste places, and open ground, introduced from Europe. British Columbia to Lower California, east to Idaho and New Mexico, rare in the eastern States. Apr.–July. Type locality, European. Also called Wild Barley and Foxtail.

2/3

2. Hordeum pusíllum Nutt.

Little Barley. Fig. 582.

Hordeum pusillum Nutt. Gen. Pl. 1: 87. 1818.

Annual; culms 10–35 cm. tall; blades erect, flat; spike erect, 2–7 cm. long, 10–14 mm. wide; lateral pair of spikelets abortive, the first glume of each and both glumes of the fertile spikelet dilated above the base, attenuate into a slender awn 8–15 mm. long; glumes very scabrous; lemma of central spikelet awned, of lateral spikelets awn-pointed.

Plains and open, especially alkaline ground; southern California (San Diego, *Baker, Orcutt;* Santa Catalina Island, *Trask*); east to Ohio. Apr.–June. Type locality: plains of the Missouri.

3. Hordeum gussoneànum Parl.
Mediterranean Barley. Fig. 583.

Hordeum gussoneanum Parl. Fl. Palerm. **1**: 246. 1845.
Hordeum maritimum gussoneanum Richt. Pl. Eur. **1**: 131.

Annual; culms numerous, spreading or geniculate at base, 15–40 cm. tall; sheaths and flat blades, especially the lower, more or less pubescent; spike erect, oblong, 1.5–3 cm. long, 6–10 mm. wide, rounded at base; glumes setaceous, glabrous or minutely scabrous, about 12 mm. long; lemma of lateral spikelets small, narrowed into an awn about 3 cm. long; lemma of central spikelet 5 mm. long, the awn somewhat longer than the glumes.

Fields and waste places, introduced from Europe. British Columbia to southern California, east to Idaho. May–Aug. Type locality, European.

4. Hordeum depréssum (Scribn. & Smith) Rydb.
Low Barley. Fig. 584.

Hordeum nodosum depressum Scribn. & Smith, U. S. Dept. Agr. Div. Agrost. Bull. **4**: 24. 1897.
Hordeum depressum Rydb. Bull. Torrey Club **36**: 539. 1909.

Annual; culms erect or geniculate below, smooth, 10–30 cm. tall; sheaths smooth or pubescent, the lower papery; blades flat, pubescent on both surfaces, 2–10 cm. long, 2–3 mm. wide; spike 2–4 cm. long; glumes 1.5–2 cm. long, subulate, scabrous; lemma of the lateral spikelets awnless; central spikelets about as in *H. nodosum,* but the awns longer.

Open ground, Washington (Adams County, *Cotton;* Columbus, *Suksdorf*), Oregon (Lexington, *Leiberg*), California (Yolo County, *Heller*), Idaho and Arizona, rare. Apr.–May. Type locality: near Lexington, Morrow County, Oregon.

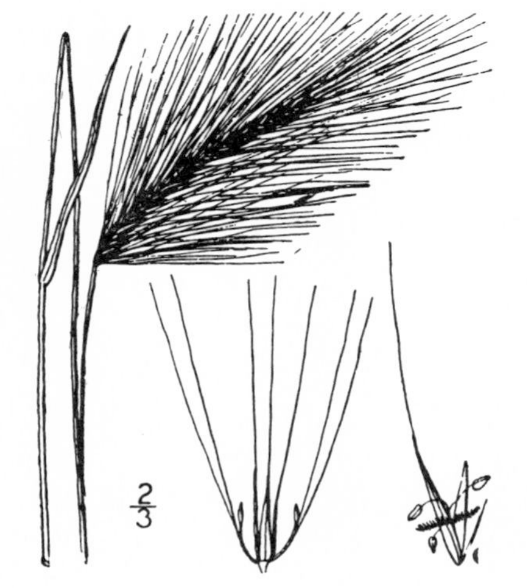

5. Hordeum jubàtum L.
Squirrel Tail. Fig. 585.

Hordeum jubatum L. Sp. Pl. 85. 1753.

Perennial; culms erect, or decumbent at base, 30–60 cm. tall; blades 5 mm. wide, scabrous; spike nodding, 5-10 cm. long, about 2 cm. wide, soft; spikelets lateral, each reduced to 1–3 spreading awns; glumes of perfect spikelets awn-like, 2.5–6 cm. long, spreading; lemma 6–8 mm. long with an awn as long as the glumes.

Open ground, fields and waste places, in the Arid Transition and Upper Sonoran Zones; Alaska to California, east to Ontario and Kansas, and sparingly further east. June–Aug. Type locality: Canada.

6. **Hordeum caespitòsum** Scribn.

Tufted Barley. Fig. 586.

Hordeum caespitosum Scribn. Proc. Davenport Acad. **7**: 245. 1899.

Differs from *H. jubatum* chiefly in the shorter awns, about 2 cm. long.

Plains and alkali meadows, in the Arid Transition Zone; eastern Washington and Oregon to South Dakota. June–July. Type locality: Edgemont, South Dakota.

7. **Hordeum boreàle** Scribn. & Smith.

Northern Barley. Fig. 587.

Hordeum boreale Scribn. & Smith, U. S. Dept. Agr. Div. Agrost. Bull. **4**: 24. 1897.

Perennial; culms erect, smooth, 60–100 cm. tall; sheaths smooth; blades 5–10 cm. long, 4–6 mm. wide; spikes 5–8 cm. long; glumes setaceous, scabrous 1.5–2 cm. long; floret of central spikelet sessile, 1 cm. long, smooth below, the awn about 1 cm. long; floret of lateral spikelets pedicellate, as large as the central or smaller.

Along the coast, Alaska to Oregon (Gearhart, *Chase*) and California (Bucksport, *Tracy;* Fort Bragg, *Hitchcock*). July. Type locality: Atka Island, Aleutian Islands.

8. **Hordeum nodòsum** L.

Meadow Barley. Fig. 588.

Hordeum nodosum L. Sp. Pl. ed. 2. **1**: 126. 1762.

Perennial; culms erect or sometimes spreading, 10–50 cm. tall; spike slender, 2–8 cm. long; glumes all setaceous, 8–15 mm. long; floret of lateral spikelets much reduced, similar to *H. pusillum* but with setaceous glumes.

Open ground and meadows, in the Transition Zone; Alaska to California, east to Indiana. May–July. Type locality, European.

76. **ÉLYMUS** L. Sp. Pl. 83. 1753.

Spikelets 2–6-flowered, sessile in pairs (rarely in groups of 3 or more), at each node of a continuous rachis, the rachilla disarticulating above the glumes and between the florets; glumes equal, usually rigid, sometimes indurate below, narrow, sometimes subulate, one- to several-nerved, acute to aristate, somewhat unsymmetric, often placed in front of the spikelets; lemmas rounded on the back or nearly terete, obscurely 5-nerved, acute or usually awned from the tip. Erect usually rather tall grasses with flat or rarely convolute blades, and terminal spikes, the spikelets usually crowded, sometimes somewhat distant. [Greek, an ancient name for a kind of millet.]

Species about 45 in the temperate regions of the northern hemisphere. Type species, *Elymus sibiricus* L.

Plants annual. — 1. *E. caput-medusae.*
Plants perennial.
  Creeping slender rhizomes present.
    Glumes lanceolate.
      Glumes about 2 cm. long, lemmas distinctly nerved. — 2. *E. mollis.*
      Glumes 1–1.5 cm. long; lemmas faintly nerved, or distinctly so only at apex. — 3. *E. vancouverensis*
    Glumes subulate or very narrow.
      Lemmas glabrous. — 4. *E. triticoides.*
      Lemmas pubescent.
        Pubescence short. — 5. *E. arenicola.*
        Pubescence woolly. — 6. *E. flavescens.*
  Creeping rhizomes wanting (short and thick in *E. condensatus*).
    Glumes subulate.
      Plant cinereus-pubescent. — 7. *E. cinereus.*
      Plant glabrous or scabrous, or rarely pubescent but not cinereus. — 8. *E. condensatus.*
    Glumes lanceolate or narrower, nerved, not subulate.
      Glumes and lemmas acuminate or awn-pointed.
        Sheaths pubescent. — 12a. *E. glaucus jepsoni.*
        Sheaths glabrous. — 9. *E. virescens.*
      Glumes and lemmas awned.
        Lemmas sparsely ciliate above; glumes several-nerved; spikes interrupted. — 10. *E. hirsutus.*
        Lemmas glabrous or variously pubescent but not ciliate above.
          Glumes not indurate at base.
            Spike slender and dense, about 5 mm. thick; blades narrow, mostly 2–4 mm. wide; glumes narrow, faintly nerved. — 11. *E. macounii.*
            Spike stouter, about 8–10 mm. thick, if slender then not dense; blades usually 5–10 mm. wide; glumes broader, distinctly nerved. — 12. *E. glaucus.*
          Glumes indurate and usually divergent at base.
            Awn short and straight; glumes bowed out, the induration at base prominent, usually yellow; spikes erect. — 13. *E. virginicus submuticus.*
            Awn long and spreading; glumes divergent but not much bowed out at base, the induration less conspicuous; spike nodding. — 14. *E. canadensis.*

1. Elymus caput-medùsae L.

Medusa Head. Fig. 589.

Elymus caput-medusae L. Sp. Pl. 84. 1753.

Annual; culms branched at base, erect or decumbent at base, slender, 20–60 cm. tall; blades narrow and short; spike 2–5 cm. long, long-awned; glumes awl-shaped, smooth, indurate below, narrowed into a slender awn 1–2.5 cm. long; lemmas lanceolate, 3-nerved, 6 mm. long, flat, very scabrous, gradually narrowed into a flat awn, 5–10 cm. long.

Open ground, Washington to California, infrequent; introduced from Europe. June–July. Type locality, European.

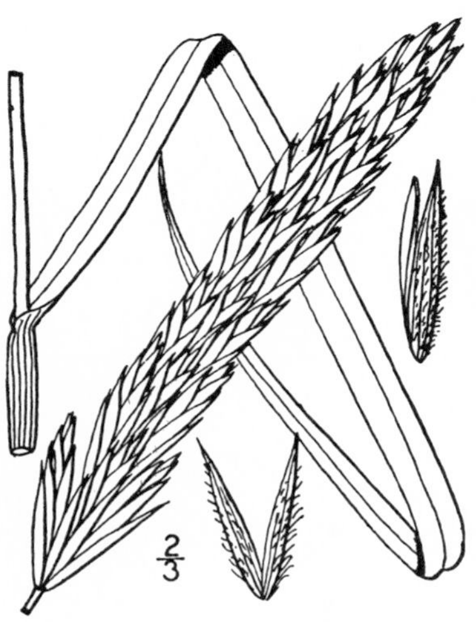

2. Elymus móllis Trin.

Sea Lyme-grass. Fig. 590.

Elymus mollis Trin. in Spreng. Neu. Entd. 2: 72. 1821.
Elymus dives Presl, Rel. Haenk. 1: 265. 1830.
Elymus arenarius villosus E. Mey. Pl. Labrad. 20. 1830.
Elymus ampliculmis Provancher, Fl. Canad. 2: 706. 1862.
Elymus capitatus Scribn. U. S. Dept. Agr. Div. Agrost. Bull. 11: 55. *pl. 14.* 1898.
Elymus villosissimus Scribn. U. S. Dept. Agr. Div. Agrost. Bull. 17: 326. *f. 622.* 1899.
Elymus arenarius mollis Koidzumi, Journ. Coll. Sci. Tokyo 27: 24. 1910.
Elymus arenarius compositus St. John, Rhodora 17: 102. 1915.

Culms stout, smooth, or pubescent above, glaucous, 60–120 cm. tall from creeping rhizomes; sheaths and blades smooth or the latter scabrous above; spike erect, dense, 7–25 cm. long; glumes lanceolate, flat, many-nerved, scabrous or pubescent, 12–25 cm. long, acuminate, awnless, about as long as the spikelet; lemmas about as long as glumes, scabrous or felty-pubescent, acuminate or mucronate.

Sand dunes along the coast, Alaska to Santa Cruz, California; also Labrador to Maine, and along the Great Lakes. June–Aug. Type locality: Kamchatka.

3. Elymus vancouverénsis Vasey.
Vancouver Ryé-grass. Fig. 591.

Elymus vancouverensis Vasey, Bull. Torrey Club **15**: 48. 1888.

Perennial, with abundant creeping rhizomes; culms erect, 80–150 cm. tall, smooth, or puberulent below the spike; sheaths smooth; blades flat, gradually narrowed to a slender involute point, scabrous above, glabrous beneath, 5–8 mm. wide; spike erect, 10–20 cm. long; glumes narrowly lanceolate, firm, gradually acuminate, 1–1.5 cm. long, sparsely long-villous, especially toward apex; lemmas firm 1–1.5 cm. long, narrowed to a short awn.

Along the seacoast, Vancouver Island to Eureka, California. July–Aug. Type locality: Vancouver Island.

4. Elymus triticoìdes Buckl.
Alkali Rye-grass. Fig. 592.

Elymus triticoides Buckl. Proc. Acad. Phila. **1862**: 99. 1863.

Elymus condensatus triticoides Thurb. in S. Wats. Bot. Calif. **2**: 326. 1880.

Elymus orcuttianus Vasey, Bot. Gaz. **10**: 258. 1885.

Agropyron arenicola Davy in Jepson, Fl. West. Mid. Calif. 76. 1901.

Culms usually glaucous, 60–120 cm. tall, usually in large masses, from extensively creeping, scaly rhizomes; sheaths smooth or scabrous; blades narrow, mostly 2–6 mm. wide, flat or soon involute; spike erect, slender, sometimes branched; glumes subulate, 10–14 mm. long, lemmas 6–10 mm. long glabrous, brownish or tawny, short-pointed.

Moist bottom land and alkaline soil, in the Upper Sonoran Zone; Washington to California, east to Colorado and Arizona. June–July. Type locality: Rocky Mountains.

Elymus triticoides pubéscens Hitchc. in Jepson, Fl. Calif. 1: 186. 1912. Sheaths and involute blades hirsute-pubescent. Only known from the type collection, Griffin, California (*Elmer*).

5. Elymus arenícola Scribn. & Smith.
Sand Rye-grass. Fig. 593.

Elymus arenicola Scribn. & Smith, U. S. Dept. Agr. Div. Agrost. Circ. **9**: 7. 1899.

Perennial, with creeping rhizomes; culms erect, smooth, 60–120 cm. tall; sheaths smooth; blades firm, scabrous above, glabrous beneath, 2–4 mm. wide; spike slender, nodding, 10–25 cm. long, the spikelets often solitary, the lower rather distant; spikelets about 2 cm. long, several-flowered; glumes about 1 cm. long, glabrous; lemmas short pubescent at least toward base, about 1 cm. long.

Sandy river bottoms and alkali meadows in the Upper Sonoran Zone; eastern Washington and Oregon to Idaho. May–July. Type locality: Suferts, Oregon.

6. Elymus flavéscens Scribn. & Smith. Yellow Rye-grass. Fig. 594.

Elymus flavescens Scribn. & Smith, U. S. Dept. Agr. Div. Agrost. Bull. **8**: 8. *f. 1.* 1897.

Perennial with creeping rhizomes; culms erect, smooth, 50–100 cm. tall; sheaths smooth; blades firm, glabrous beneath, scabrous above, 2–5 mm. wide; spike 10–25 cm. long, sometimes branching; spikelets 2–3 cm. long, approximate or somewhat distant; glumes very narrow or subulate, 1–1.5 cm. long; lemmas densely silky villous, the hairs long, yellowish or brownish.

Sand dunes, in the Upper Sonoran Zone; eastern Washington, Oregon, and east to Idaho. June–July. Type locality: Columbus, Washington.

7. Elymus cinèreus Scribn. & Merr. Gray Rye-grass. Fig. 595.

Elymus cinereus Scribn. & Merr. Bull. Torrey Club **29**: 467. 1902.

Culms erect, stout, puberulent, 1–3 meters tall; sheaths and blades cinereus-pubescent, the latter with an indurate point; spike erect, 15–20 cm. long, dense, interrupted below; spikelets 1.5–2 cm. long; glumes subulate, about 12 mm. long, scabrous-pubescent; lemmas scabrous-pubescent, especially the apex, faintly nerved, obtuse, mucronate or with a short awn-point.

Meadows and hills, in the Arid Transition Zone; southern California (San Bernardino, *Parish;* Lone Pine, *Jepson;* Lancaster, *Elmer;* without locality, *Davy*) and Nevada. June–July. Type locality: Pahrump Valley, Nevada.

8. Elymus condensàtus Presl. Giant Rye-grass. Fig. 596.

Elymus condensatus Presl, Rel. Haenk. **1**: 265. 1830.

Culms in large tufts, stout, 1–3 meters tall, producing stout knotty rhizomes; sheaths smooth; blades flat, as much as 2 cm. wide; spike erect, usually dense, as much as 30 cm. long, sometimes branched; glumes narrowly lanceolate or subulate, awn-pointed, usually only 1-nerved, or nerveless, about as long as the first lemma; lemma awnless or mucronate.

Dry plains and hillsides and along gullies and ditches, in the Arid Transition and Upper Sonoran Zones; British Columbia to southern California, east to Nebraska. June–Aug. Type locality: Monterey, California.

Elymus condensatus pùbens Piper, Erythea **7**: 101. 1899. Sheaths and blades pubescent. Yakima City, Washington, the type locality, and Santa Barbara (*Hitchcock*).

9. Elymus viréscens Piper.

Pacific Rye-grass. Fig. 597.

Elymus virescens Piper, Erythea **7**: 101. 1899.
Elymus howellii Scribn. & Merr. Contr. U. S. Nat. Herb. **13**: 88. 1910.
Elymus strigatus St. John, Rhodora **17**: 102. 1915.

Culms tufted smooth, 30–120 cm. tall; sheaths smooth; blades flat, smooth or scabrous, 5–20 cm. long, 5–10 mm. wide; spike 5–15 cm. long, rather dense; spikelets few-flowered; glumes firm, lanceolate, acute or acuminate, strongly nerved, the nerves often scabrous, 10–15 mm. long; lemmas firm, rounded on the back, scarcely nerved, awnless or short-awned.

Damp soil, in woods or open ground, in the Humid Transition Zone; southern Alaska to Mendocino County, California. May–Aug. Type locality: Olympic Mountains.

10. Elymus hirsùtus Presl.

Northern Rye-grass. Fig. 598.

Elymus hirsutus Presl, Rel. Haenk. **1**: 264. 1830.
Elymus ciliatus Scribn. U. S. Dept. Agr. Div. Agrost. Bull. **11**: 57. *pl. 16.* 1898, not Muhl. 1817.
Elymus borealis Scribn. U. S. Dept. Agr. Div. Agrost. Circ. **27**: 9. 1900.

Culms tufted erect, or geniculate at base, smooth, 1-1.5 meters tall; sheaths smooth; blades flat, thin, lax, rather sparsely pilose above, scabrous beneath, 15–20 cm. long, 5–12 mm. wide; spike loosely flowered, nodding, 10–15 cm. long; glumes lanceolate, strongly nerved, scabrous, extending into a short awn or into an awn as long as the body; lemmas rather thin, nerved, ciliate-margined, awned, the awn as much as 2 cm. long.

Moist woods in the Hudsonian Zone; southern Alaska to Oregon (Gearhart, *Chase*). June–Aug. Type locality: Nootka Sound.

11. Elymus macoùnii Vasey.

Macoun's Rye-grass. Fig. 599.

Elymus macounii Vasey, Bull. Torrey Club **13**: 119. 1886.

Culms tufted, erect, slender, smooth, 50-100 cm. tall; sheaths smooth; blades usually scabrous on both surfaces, erect and rather firm, 10–20 cm. long, 2–5 mm. wide; spike slender, erect or somewhat nodding, 4–10 cm. long, about 5 mm. thick; glumes narrowly lanceolate, scabrous, about 1 cm. long, extending into a short awn; lemmas smooth and rounded below, scabrous toward the apex, extending into a slender straight awn 1–1.5 cm. long.

Meadows and open ground, in the Arid Transition and Upper Sonoran Zones; eastern Washington, Oregon, and California (Klamath, *Copeland;* Donner Creek, *Hitchcock*), north to Alberta, east to Minnesota and New Mexico. June–Aug. Type locality: Great Plains of British America.

12. Elymus glaùcus Buckl.

Western Rye-grass. Fig. 600.

Elymus glaucus Buckl. Proc. Acad. Phila. **1862**: 99. 1863.
Elymus nitidus Vasey, Bull. Torrey Club **13**: 120. 1886.
Elymus americanus Vasey and Scribn.; Macoun, Cat. Can. Pl. **4**: 245. 1888.
Elymus glaucus tenuis Vasey; Contr. U. S. Nat. Herb. **1**: 280. 1893.
Elymus glaucus breviaristatus Davy in Jepson, Fl. West. Mid. Calif. 79. 1901.
Elymus glaucus maximus Davy in Jepson, Fl. West. Mid. Calif. 79. 1901.
Elymus angustifolius Davy in Jepson, Fl. West. Mid. Calif. 80. 1901.
Elymus angustifolius caespitosus Davy in Jepson, Fl. West. Mid. Calif. 81. 1901.

Culms erect, 60–120 cm. tall, without rhizomes; sheaths smooth or scabrous; blades flat, as much as 1 cm. wide, scabrous on both surfaces, sometimes narrow and more or less involute; spike erect, usually dense, long-exserted, 5–15 cm. long, rarely longer; glumes about as long as the spikelet, lanceolate, 8–15 mm. long, acuminate or awn-pointed, with 2–4 scabrous nerves; lemmas awned, the awn 1–2 times as long as the body.

Open woods, copses, and dry hillsides in the Transition Zone; Alaska to the mountains of southern California, east to Michigan. June–Aug. Type locality: Columbia River.

Elymus glaucus jépsoni Davy, in Jepson, Fl. West. Mid. Calif. 79. 1901. *E. hispidulus* Davy, loc. cit.; *E. pubescens* Davy, op. cit. 78. *E. divergens* Davy, op. cit. 80. *E. velutinus* Scribn. & Merr. Bull. Torrey Club **29**: 466. 1902. *E. parishii* Davy & Merr. Univ. Calif. Publ. Bot. **1**: 58. 1902. Distinguished by the more or less pubescent sheaths and blades. Lemmas sometimes only awn-pointed. California.

Elymus glaucus aristàtus (Merr.) Hitchc. *E. aristatus* Merr. Rhodora **4**: 147. 1902. Glumes narrow, almost subulate; spike densely flowered. Mountains of Oregon and Washington.

13. Elymus virgínicus submùticus Hook.

Awnless Rye-grass. Fig. 601.

Elymus virginicus submuticus Hook. Fl. Bor. Am. **2**: 255. 1840.
Elymus curvatus Piper, Bull. Torrey Club **30**: 233. 1903.

Culms erect or geniculate at base, about 1 meter tall, glabrous; sheaths glabrous; blades flat, scabrous on both sides, 15–20 cm. long, 5–10 mm. wide; spike erect, often included at base in the upper sheath, 10–12 cm. long; glumes rigid, lanceolate, thick, indurate and bowed out at base, 3-5-nerved, scabrous, acuminate, 13–17 mm. long, often curved or twisted; spikelets mostly 3-flowered; lemmas sparsely hispid above the middle, 8-10 mm. long, awn-pointed, the awn 1–2 mm. long.

Box Cañon, Pend Oreille River, Washington, east to Ontario and Kansas. August. Type locality: Cumberland House Fort, Saskatchewan River.

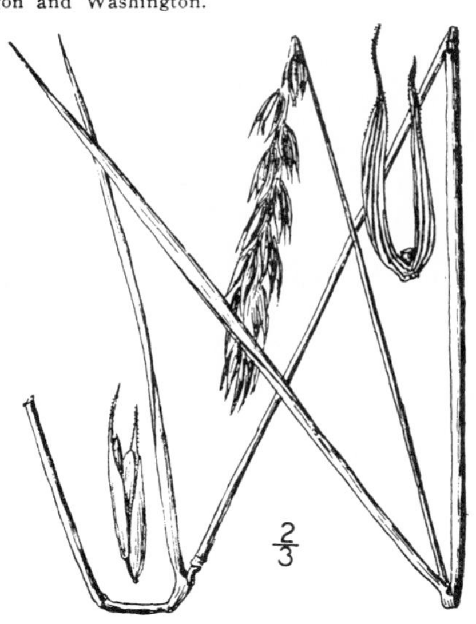

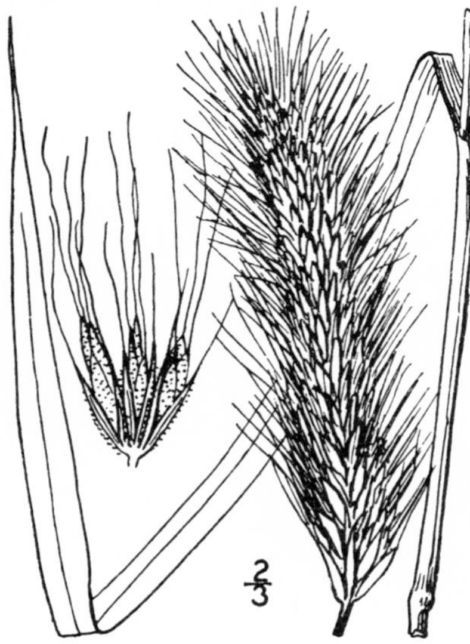

14. Elymus canadénsis L.

Canada Rye-grass. Fig. 602.

Elymus canadensis L. Sp. Pl. 83. 1753.
Sitanion brodiei Piper, Erythea **7**: 100. 1899.

Culms tufted, erect, smooth, rather stout, 1–1.5 meters tall, sometimes lower; sheaths smooth or scabrous; blades flat, firm, scabrous above, smooth or scabrous below, 15–30 cm. long, 5–15 mm. wide; spike thick, nodding, usually interrupted below, 10–20 cm. long, bristly with the numerous divergent awns; glumes narrow, somewhat indurate but not definitely bowed out at base, about 3-nerved, scabrous, narrowed into an awn; lemmas, scabrous-pubescent, about 1 cm. long, narrowed into an awn 2–3 cm. long, divergent when dry.

Moist places, copses, edges of woods, and rocky shaded banks in the Arid Transition and Upper Sonoran Zones; eastern Washington to northern California (Sonoma County, *Samuels*), east to Nova Scotia and Florida. June–Aug. Type locality: Canada.

77. SITÀNION Raf. Journ. de Phys. **89**: 103. 1819.

Spikelets 2 to few-flowered, sessile, usually 2 at each node of a disarticulating rachis, the rachis breaking at base of each joint and remaining attached to spikelets above; glumes narrow or setaceous, 1–3-nerved, the nerves prominent, extending into 1 to several awns, these (when more than one) irregular in size, sometimes mere lateral appendages of the long central awn, sometimes equal, the glume being bifid; lemmas firm, convex on the back, nearly terete, the apex slightly 2-toothed, 5-nerved, the nerves obscure, the central nerve extending into a long slender finally spreading awn, sometimes one or more of the lateral nerves also extending into short awns; palea firm, nearly as long as the body of the lemma, the 2 keels serrulate. Low or rather tall erect cespitose perennials with bristly spikes. [Greek, a kind of food.]

Species about 6 in the dry regions of the western United States. Type species, *Sitanion elymoides* Raf.

Glumes, or some of them, 3-nerved, lanceolate, entire or sometimes bifid.
- Plant glabrous. — 1. *S. hanseni.*
- Plant pubescent. — 2. *S. anomalum.*

Glumes subulate or very narrow, 2-nerved, entire or cleft into 2–several lobes.
- Glumes cleft into 3–several lobes. — 3. *S. jubatum.*
- Glumes entire or 2-cleft. — 4. *S. hystrix.*

1. **Sitanion hánseni** (Scribn.) J. G. Smith. Hansen's Squirrel-tail. Fig. 603.

Elymus hanseni Scribn. U. S. Dept. Agr. Div. Agrost. Bull. **11**: 56. *f. 12.* 1898.
Sitanion planifolium J. G. Smith, U. S. Dept. Agr. Div. Agrost. Bull. **18**: 19. 1899.
Sitanion hanseni J. G. Smith, U. S. Dept. Agr. Div. Agrost. Bull. **18**: 20. 1899.
Sitanion leckenbyi Piper, Erythea **7**: 100. 1899.
Sitanion rubescens Piper, Bull. Torrey Club **30**: 234. 1903.
Elymus leckenbyi Piper, Contr. U. S. Nat. Herb. **11**: 151. 1906.

Culms smooth and usually glaucous, 50–100 cm. tall; sheaths glabrous; blades flat, scabrous, 2–8 mm. wide; spike somewhat nodding, 10–20 cm. long; glumes narrowly lanceolate, sometimes bifid, 2–3-nerved, long-awned; lemmas smooth except toward the scabrous tip, about 8 mm. long, the awn about 4 cm. long, becoming somewhat divergent.

Open woods and rocky slopes in the Arid Transition and Upper Sonoran Zones; eastern Washington to middle California, east to Utah. June–Aug. Type locality: Amador County, California.

2. **Sitanion anómalum** J. G. Smith. Slender Squirrel-tail. Fig. 604.

Sitanion anomalum J. G. Smith, U. S. Dept. Agr. Div. Agrost. Bull. **18**: 20. *pl. 4.* 1899.

Differs from *S. hanseni* in having more or less pubescence on the culms, sheaths, and blades, and in having usually narrower blades, especially on the numerous innovations; awns commonly erect or narrowly ascending.

Dry woods and rocky slopes, in the Arid Transition Zone; Washington (Wawawai, *Piper*) to San Luis Obispo County, California. June–July. Type locality: Allen, California.

3. Sitanion jubàtum J. G. Smith.
Big Squirrel-tail. Fig. 605.

Polyanthrix hystrix Nees, Ann. Nat. Hist. **1**: 284. 1838, not *Sitanion hystrix* J. G. Smith.
Sitanion jubatum J. G. Smith, U. S. Dept. Agr. Div. Agrost. Bull. **18**: 10. 1899.
Sitanion villosum J. G. Smith, U. S. Dept. Agr. Div. Agrost. Bull. **18**: 11. *pl. 1.* 1899.
Sitanion multisetum J. G. Smith, U. S. Dept. Agr. Div. Agrost. Bull. **18**: 11. 1899.
Sitanion polyanthrix J. G. Smith, U. S. Dept. Agr. Div. Agrost. Bull. **18**: 12. 1899.
Sitanion breviaristatum J. G. Smith, U. S. Dept. Agr. Div. Agrost. Bull. **18**: 12. 1899.
Sitanion strictum Elmer, Bot. Gaz. **36**: 59. 1903.

Culms erect, 30–60 cm. tall, rarely taller; sheaths smooth, scabrous or villous-pubescent; blades flat, often becoming involute, smooth or usually more or less pubescent, at least on upper surface, usually not over 3 mm. wide; spike erect, dense, 3–7 cm. long, thick and bushy from the numerous long awns; glumes split into 2 or usually 3 or more lobes or divisions, each extending into a long awn; lemmas mostly 8–10 mm. long, smooth, or scabrous toward apex, the awns and those of the glumes 3–10 cm. long, rarely shorter.

Rocky or brushy hillsides and open dry woods and plains, in the Arid Transition and Upper Sonoran Zones; eastern Washington to southern California, east to Idaho. May–July. Type locality: Waitsburg, Washington.

4. Sitanion hýstrix (Nutt.) J. G. Smith.
Bottle-brush Squirrel-tail. Fig. 606.

Aegilops hystrix Nutt. Gen. Pl. **1**: 86. 1818.
Sitanion elymoides Raf. Journ. de Phys. **89**: 103. 1819.
Elymus elymoides Swezey, Nebr. Fl. Pl. 15. 1891.
Sitanion minus J. G. Smith, U. S. Dept. Agr. Div. Agrost. Bull. **18**: 12. 1899.
Sitanion rigidum J. G. Smith, U. S. Dept. Agr. Div. Agrost. Bull. **18**: 13. 1899.
Sitanion californicum J. G. Smith, U. S. Dept. Agr. Div. Agrost. Bull. **18**: 13. 1899.
Sitanion glabrum J. G. Smith, U. S. Dept. Agr. Div. Agrost. Bull. **18**: 14. 1899.
Sitanion cinereum J. G. Smith, U. S. Dept. Agr. Div. Agrost. Bull. **18**: 14. 1899.
Sitanion hystrix J. G. Smith, U. S. Dept. Agr. Div. Agrost. Bull. **18**: 15. *pl. 2.* 1899.
Sitanion montanum J. G. Smith, U. S. Dept. Agr. Div. Agrost. Bull. **18**: 16. 1899.
Sitanion strigosum J. G. Smith, U. S. Dept. Agr. Div. Agrost. Bull. **18**: 17. 1899.
Sitanion velutinum Piper, Bull. Torrey Club **30**: 233. 1903.
Sitanion rigidum californicum Smiley, Univ. Cal. Publ. **9**: 99. 1921.

Culms tufted erect and rather stiff, 10–50 cm. tall; sheaths glabrous, puberulent or softly pubescent; blades flat or involute, glabrous or puberulent below, usually puberulent above, sometimes softly pubescent all over, rather stiffly ascending or somewhat spreading, 5–20 cm. long, or the uppermost shorter, 1–3 mm. wide, rarely as much as 5 mm. wide; spike erect, 2–7 cm., rarely 10 cm. long; glumes very narrow, almost subulate, 1–2-nerved, extending into a slender scabrous awn, sometimes bifid at the middle, or bearing a bristle, or awn along one margin; lemmas convex, smooth or scabrous, sometimes glaucous, extending into a slender awn, the awns of glumes and lemmas, 2–7 cm. long; rachis readily breaking up at maturity. Variable in aspect, as to length of blades and awns, and as to amount of pubescence. At high altitudes the plants are dwarf.

Dry hills, plains, open woods, and rocky slopes from the Arid Transition to the Alpine Zone; eastern Washington to southern California, east to western Kansas. July–Aug. Type locality: arid plains of the Missouri.

78. HÝSTRIX Moench, Meth. Pl. 294. 1794.

[GYMNOSTICHUM Schreb. Beschr. Gräs. **3**: 127. *pl. 47.* 1810.]

Spikelets 2–4-flowered, sessile, 1–3 at each node of a continuous flattened rachis, horizontally divergent at maturity; glumes reduced to short or minute awns, the first usually obsolete, both often wanting in the upper spikelets; lemmas convex, rigid, tapering into long awns, 5-nerved, the nerves obscure except toward the tip; palea about as long as the body of the lemma. Erect perennials with flat blades and bristly loosely flowered spikes. [Greek, porcupine.]

Species 4 in temperate regions. Type species, *Elymus hystrix* L.

1. Hystrix califórnica (Boland.) Kuntze.
California Bottle-brush. Fig. 607.

Gymnostichum californicum Boland.; Thurb. in S. Wats. Bot. Calif. 2: 327. 1880.

Hystrix californica Kuntze, Rev. Gen. Pl. 778. 1891.
Asprella californica Beal, Grasses N. Am. 2: 657. 1896.

Culms stout, 1–2 meters tall; sheaths hispid or the upper smooth; blades flat, as much as 2 cm. wide, scabrous; spike stout, dense and somewhat nodding above, more or less interrupted below, 12–25 cm. long; spikelets mostly in pairs, 1–3-flowered, on short callus-like pedicels; glumes wanting; lemmas 12–15 mm. long, 5-nerved above, the nerves, especially the marginal ciliate-hispid with short stiff hairs; awn stout, straight, rough, about 2 cm. long.

Woods and shaded banks, near the coast, Marin County to Santa Cruz County, California. June–Aug. Type locality: San Francisco.

Family 11. CYPERÀCEAE.*

Sedge Family.

Grass-like or rush-like herbs. Stems (culms) slender, solid (rarely hollow), triangular, quadrangular, terete or flattened. Roots fibrous (many species perennial by long rootstocks). Leaves narrow, with closed sheaths. Flowers perfect or imperfect, arranged in spikelets, 1 (rarely 2) in the axil of each scale (glume, bract), the spikelets solitary or clustered, one- to many-flowered. Scales 2-ranked or spirally imbricated, persistent or deciduous. Perianth hypogynous, composed of bristles, or interior scales, rarely calyx-like, or entirely wanting. Stamens 1–3, rarely more. Filaments slender or filiform. Anthers 2-celled. Ovary 1-celled. Ovule 1, anatropous, erect. Style 2–3-cleft or rarely simple or minutely 2-toothed. Fruit a lenticular, plano-convex, or trigonous achene. Endosperm mealy. Embryo minute.

About 65 genera and 3000 species, of very wide geographic distribution.

Fertile flowers perfect.
  Basal empty scales of the spikelets none, or not more than 2 (except in *Eriophorum*).
    Scales of the spikelets 2-ranked; bristles none. — 1. *Cyperus.*
    Scales of the spikelets spirally imbricated.
      Base of the style persistent as a tubercle on the achene.
        Spikelet 1; culms leafless; bristles usually present. — 2. *Eleocharis.*
        Spikelets mostly several or numerous; culms leaf-bearing; bristles none. — 3. *Stenophyllus.*
      Base of the style not persistent as a tubercle.
        Flowers without any inner scales.
          Base of the style swollen; bristles none. — 4. *Fimbristylis.*
          Base of the style not swollen; bristles usually present.
            Bristles many, much elongated. — 5. *Eriophorum.*
            Bristles few, short. — 6. *Scirpus.*
        Flowers with a minute inner scale. — 7. *Hemicarpha.*
  Basal empty scales of the spikelets 3 or more.
    Style 2-cleft.
      Spikelets breaking up into 1-fruited joints; scales 2-ranked. — 8. *Dulichium.*
      Rachis of the spikelets not jointed; scales spirally imbricated. — 9. *Rynchospora.*
    Style 3-cleft.
      Bristles none. — 10. *Mariscus.*
      Bristles present. — 11. *Schoenus.*
All the flowers imperfect.
  Pistillate flower partly enwrapped by a scale. — 12. *Kobresia.*
  Pistillate flower enclosed by a perigynium. — 13. *Carex.*

1. CYPÈRUS L. Sp. Pl. 44. 1753.

Annual or perennial sedges. Culms in our species simple, triangular, leafy near the base, and with 1 or more leaves at the summit forming an involucre to the simple or compound, umbellate or capitate inflorescence. Rays of the umbel sheathed at the base, usually very unequal, one or more of the heads or spikes commonly sessile. Spikelets flat or subterete, the scales falling away from the rachis as they mature, or persistent and the spikelets falling away from the axis of the head or spike with the scales attached. Scales concave, condupli-

* Text contributed, except for the genus *Carex*, by Nathaniel Lord Britton.

cate or keeled, 2-ranked, all flower-bearing or the lower ones empty. Flowers perfect. Perianth none. Stamens 1–3. Style 2–3-cleft, deciduous from the summit of the achene. [Ancient Greek name for these sedges.]

About 600 species, of wide distribution in tropical and temperate regions. Type species, *Cyperus esculentus* L.

Style-branches 2; achene lenticular.
  Achene laterally flattened.
    Spikelets 2–2.5 mm. wide.
      Spikelets in a single capitate cluster. — 1. *C. melanostachyus.*
      Spikelets in umbeled clusters. — 2. *C. rivularis.*
    Spikelets about 5 mm. wide. — 3. *C. unioloides.*
  Achene dorsally flattened. — 4. *C. laevigatus.*
Style-branches 3; achene trigonous.
  Spikelets flattened, not breaking up at the nodes.
    Rachis of the spikelet persistent, the scales falling away from it.
      Rachis-wings very narrow, or none.
        Annuals with fibrous roots.
          Spikelets capitate.
            Scales awned. — 5. *C. inflexus.*
            Scales acute. — 6. *C. acuminatus.*
          Spikelets spicate. — 7. *C. parishii.*
        Perennials, with rootstocks.
          Spikelets subcapitate; base of the culm tuber-like; stamens 3. — 8. *C. bushii.*
          Spikelets digitate; base of the culm not tuber-like; stamen 1. — 9. *C. vegetus.*
      Rachis distinctly winged.
        Perennial by tuberiferous rootstocks; rachis-wings persistent.
          Scales straw-color; achene obovoid. — 10. *C. esculentus.*
          Scales purple-brown; achene linear-oblong. — 11. *C. rotundus.*
        Annual; rachis-wings separating. — 12. *C. erythrorrhizos.*
    Spikelets falling away from the 2 lower empty scales.
      Spikes loose, not densely cylindric, with two to several achenes.
        Spikelets distinctly flattened. — 13. *C. strigosus.*
        Spikelets little flattened. — 14. *C. hanseni.*
      Spikes densely cylindric; spikelets about 3 mm. long, with only 1 achene. — 15. *C. regiomontanus.*
  Spikelets subterete, breaking up at the nodes into 1-fruited joints. — 16. *C. ferax.*

1. **Cyperus melanostàchys** H. B. K.

Brown Cyperus. Fig. 608.

Cyperus melanostachys H. B. K. Nov. Gen. **1**: 207. 1815.
Cyperus diandrus capitatus Britton, Bull. Torrey Club **13**: 205. 1886.

Rootstocks short; roots fibrous. Culms glabrous, slender, trigonous, tufted, 1–6 dm. high. Leaves glabrous, shorter than the culm or equaling it, 1–3 mm. wide, those of the involucre about 3, sometimes much elongated; inflorescence a capitate cluster of spikelets, rarely with one umbel-ray bearing a smaller cluster; spikelets lanceolate, acute or acutish, 5–12 mm. long, about 2.5 mm. wide; scales ovate, acutish or obtuse, brown or green-brown; achene oblong, pointed, its superficial cells quadrate.

Wet grounds of the Sonoran Zones; California northward to Butte County; Arizona to Texas and Mexico; Colombia. Referred by Watson to *C. diandrus castaneus* Torr. Type locality: Colombia.

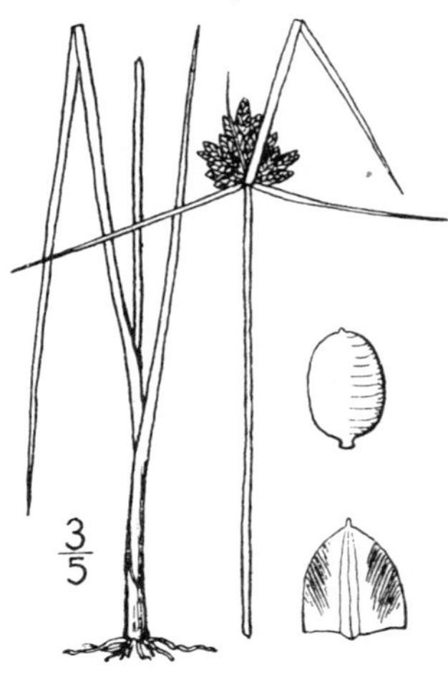

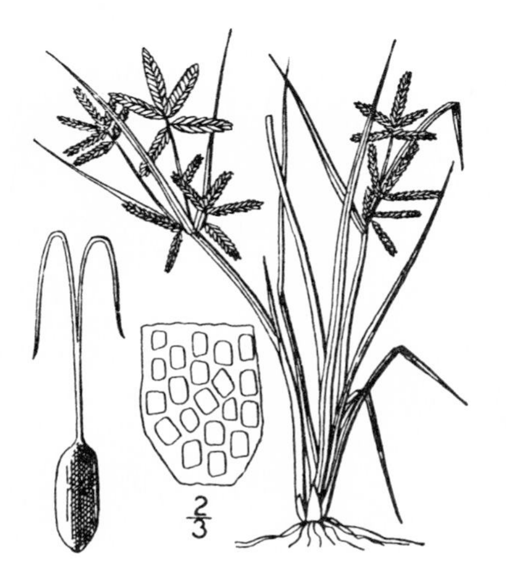

2. **Cyperus rivulàris** Kunth.

Shining Cyperus. Fig. 609.

Cyperus rivularis Kunth, Enum. **2**: 6. 1837.

Culms slender, tufted, 1–4 dm. high. Umbel usually simple; spikelets linear or linear-oblong, acutish, 8–20 mm. long; scales green or dark brown, or with brown margins, appressed, firm, subcoriaceous, shining, obtuse; stamens mostly 3; style 2-cleft, scarcely exserted; achene oblong or oblong-obvate, lenticular, somewhat pointed, dull, its superficial cells quadrate.

Wet soil, Transition or Canadian Zone; Sisson, California; Michigan to Kansas, Ontario, Maine, and North Carolina. Type locality: Georgia.

3. **Cyperus unioloìdes** R. Br.

Uniola-like Cyperus. Fig. 610.

Cyperus unioloides R. Br. Prodr. 1: 216. 1810.

Perennial by slender rootstocks; culms rather slender, 5–8 dm. tall. Leaves mostly shorter than the culm, 2–4 mm. wide, those of the involucre 3–5, one of them usually much exceeding the inflorescence; rays of the umbel 2–8, about 10 cm. long or less, sometimes very short and the inflorescence almost glomerate; spikelets short-spicate, lanceolate to ovate-lanceolate, many-flowered, 12–18 mm. long, about 5 mm. wide, flat, acute; scales yellow, shining, ovate or ovate-lanceolate, acute, appressed; achene obovate, short-pointed, nearly black, its superficial cells subquadrate.

Cienega, Upper Sonoran Zone, Los Angeles County, California; also Mexico, Jamaica, Cuba, Hispaniola, continental tropical America, South Africa, tropical Asia, and Australia. Type locality: presumably Australia.

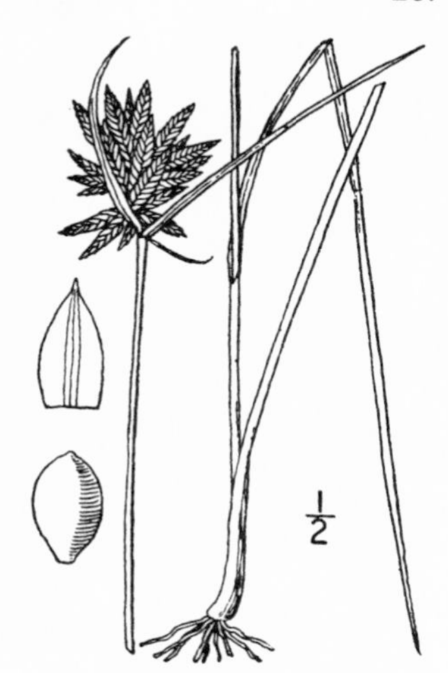

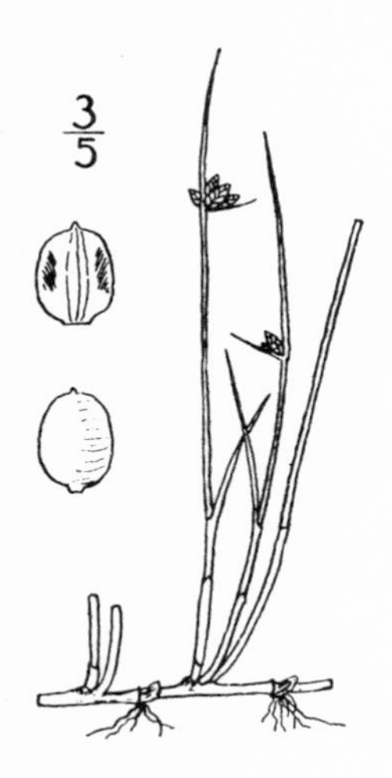

4. **Cyperus laevigàtus** L.

Smooth Cyperus. Fig. 611.

Cyperus laevigatus L. Mant. 1: 179. 1771.

Perennial by horizontal rootstocks; culms glabrous, erect, 1–6 dm. high, trigonous above. Basal leaves sheathing, 2–4 mm. wide, sometimes as long as the culm, usually much shorter, sometimes reduced to sheaths, those of the involucre 1 or 2, when solitary erect, usually much exceeding the spikelets, spikelets one to many, sessile, appearing lateral, 6–12 mm. long, 2–3 mm. wide, compressed, linear-oblong, many-flowered; scales ovate, obtuse, brown or nearly white; achene plano-convex, brown, elliptic to obovate, dorsally flattened, 1.5–2 mm. long.

Moist or wet soil, Sonoran Zones; California, north to Santa Clara and Inyo Counties; also Mexico, tropical America and Old World tropics. Type locality: Cape of Good Hope.

5. **Cyperus infléxus** Muhl.

Awned Cyperus. Fig. 612.

Cyperus inflexus Muhl. Gram. 16. 1817.

Annual; culms slender or almost filiform, tufted, about equaled by the leaves. Leaves 2 mm. wide or less, those of the involucre 2–3, exceeding the umbel; umbel sessile, capitate, or 1–3-rayed; spikelets linear-oblong, 6–10-flowered, 4–6 mm. long; scales light brown, lanceolate, rather firm, strongly several-nerved, tapering into a long recurved awn, falling from the rachis at maturity; stamen 1; style 3-cleft; rachis winged, the wings persistent; achene 3-angled, brown, dull, narrowly obovoid or oblong, obtuse, mucronulate.

In wet, sandy soil, Sonoran and Transition Zones; southern California to British Columbia, east across the continent to Florida and New Brunswick. Included by authors in *C. aristatus* Rottb. Type locality: Pennsylvania.

6. Cyperus acuminàtus Torr. & Hook.

Short-pointed Cyperus. Fig. 613.

Cyperus acuminatus Torr. & Hook. Ann. Lyc. N. Y. **3**: 435. 1836.

Annual; culms very slender, tufted, 7–40 cm. tall. Leaves light green, usually less than 2 mm. wide, those of the involucre much elongated; umbel 1–4-rayed, simple; rays short; spikelets flat, ovate-oblong, obtuse, 4–8 mm. long, many-flowered, densely capitate; scales oblong; pale green, 3-nerved, coarsely cellular, conduplicate, with a short, sharp, more or less recurved tip; stamen 1; style 3-cleft; achene sharply 3-angled, gray, oblong, narrowed at each end, about one-half as long as the scale.

In moist soil, Upper Sonoran Zone; Tulare County, California, to Washington; Texas to Georgia, South Dakota and Illinois. Type locality: St. Louis, Missouri.

7. Cyperus paríshii Britton.

Parish's Cyperus. Fig. 614.

Cyperus parishii Britton; Parish, Bull. S. Calif. Acad. Sci. **3**: 52. 1904.

Annual with fibrous roots, or perhaps of longer duration; culms tufted, 1–2.5 dm. high. Leaves 2–5 mm. wide, mostly shorter than the culm, those of the involucre 5 or fewer, the longer ones exceeding the inflorescence; umbel simple or somewhat compound, usually several-rayed, the rays 0.5–5 cm. long; spikelets several or many, rather densely short-spicate, linear, acute, 12–20 mm. long, about 2 mm. wide; rachis wingless; scales oblong-lanceolate, purple-green, obtuse, about 2 mm. long, several-nerved; achene trigonous, oblong or obovoid-oblong, nearly black, mucronulate.

Wet meadows, Sonoran Zones; San Bernardino County, California, to Arizona and New Mexico. Related to *C. sphacelatus* Rottb. of tropical America. Type locality: vicinity of San Bernardino, California.

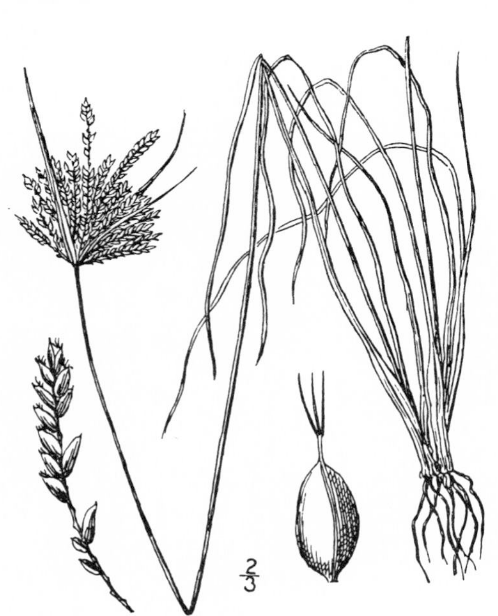

8. Cyperus bùshii Britton.

Bush's Cyperus. Fig. 615.

Cyperus bushii Britton, Man. 1044. 1901.

Perennial by small tuber-like corms. Leaves 3–4 mm. wide, smooth; scapes smooth, 3–6 dm. high, longer than the leaves; longer involucral leaves much exceeding the umbel; umbel capitate, or with 1–5 rays; spikelets loosely capitate, flat, linear, acute, 8–16 mm. long; scales firm, shining, oblong, mucronate, strongly about 11-nerved; achene oblong, 3-angled, nearly twice as long as thick, apiculate, two-thirds as long as the scale.

Moist soil along streams, Transition Zone; Oregon and Washington, east to Minnesota, Colorado and Texas. Has been referred to *C. Houghtoni* Torr. and to *C. Schweinitzii* Torr., and confused with *C. filiculmis* Vahl. Type locality: Arkansas, Oklahoma.

9. **Cyperus végetus** Willd.

Tall Cyperus. Fig. 616.

Cyperus vegetus Willd. Sp. Pl. 1: 283. 1797.
Cyperus serrulatus S. Wats. Proc. Am. Acad. 17: 382. 1882.

Perennial by rootstocks. Culms sometimes thickened at the base, smooth, stout, 3–10 dm. tall; basal leaves as long as the culm or shorter, 10 mm. wide or less, more or less transversely lineolate, those of the involucre several, similar to the basal ones, the longer much exceeding the compound umbel; umbel-rays 2–10 cm. long; spikelets several to many, flat, digitate, 10–18 mm. long, about 3 mm. wide; scales ovate or ovate-lanceolate, yellow or yellowish white, the keel sometimes serrulate; stamen 1; achene oblong.

Moist soil, Upper Sonoran Zone; central and south central California, north to Butte County; San Luis Potosi to Mexico; western South America. The California plant has been referred to *C. virens* Michx. Type locality not cited.

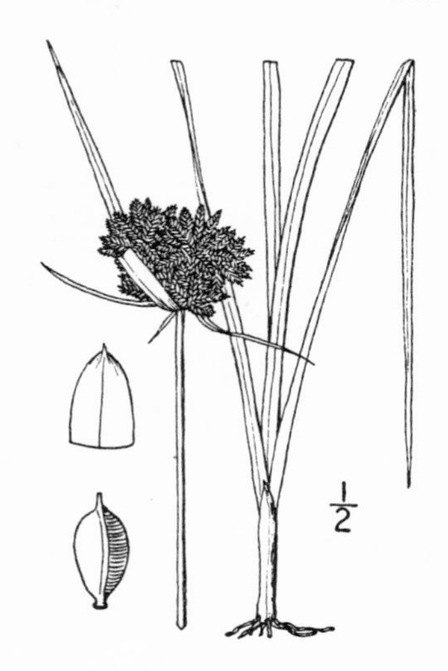

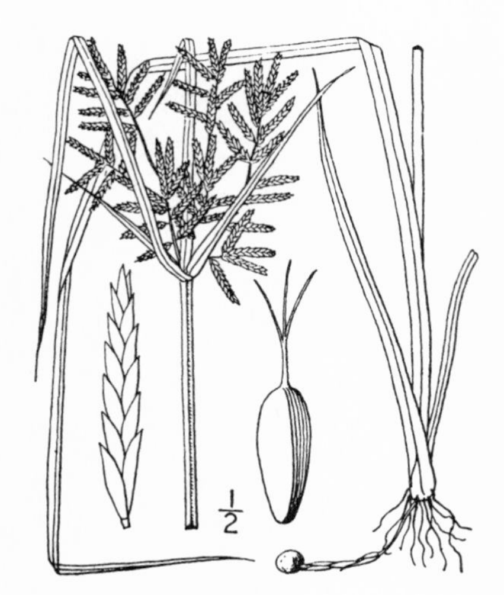

10. **Cyperus esculéntus** L.

Yellow Nut-grass. Fig. 617.

Cyperus esculentus L. Sp. Pl. 45. 1753.
Cyperus phymatodes Muhl. Gram. 23. 1817.
Cyperus hermanni Buckley, Proc. Acad. Phila. 1862: 10. 1862.
C. esculentus macrostachyus Boeckl. Linnaea 36: 291. 1869.
C. phymatodes hermanni S. Wats. Bot. Calif. 2: 215. 1880.
C. esculentus hermanni Britton, Bull. Torrey Club 13: 211. 1886.

Perennial by scaly tuber-bearing rootstocks; culm usually stout, 0.3–0.8 m. tall, commonly shorter than the leaves. Leaves light green, 4–8 mm. wide, the midvein prominent, those of the involucre 3–6, the longer much exceeding the inflorescence; umbel 4–10-rayed, often compound; spikelets numerous in loose spikes, straw-color or yellowish brown, flat, spreading, 1–2.5 cm. long, about 3 mm. wide, many-flowered, scales ovate-oblong, subacute, 3–5-nerved; rachis narrowly winged; stamens 3; style 3-cleft; achene obovoid, obtuse, 3-angled.

Fields and banks, Canadian to Sonoran Zone; southern California to Washington; Alaska; eastward across the continent to Florida and New Brunswick; tropical America and widely distributed in the Old World. Consists of many races. Type locality: Montpellier, France.

11. **Cyperus rotúndus** L.

Nut-grass. Fig. 618.

Cyperus rotundus L. Sp. Pl. 45. 1753.

Perennial by scaly tuber-bearing rootstocks, culm rather stout, 1.5–5 dm. high, usually longer than the leaves. Leaves 3–6 mm. wide, those of the involucre 3–5, the longer equaling or exceeding the inflorescence; umbel compound or nearly simple, 3–8-rayed, the longer rays 5–12 cm. long; spikelets linear, closely clustered, few in each cluster, acute, 8–20 mm. long, 2–3 mm. wide; scales dark purple-brown or with green margins and center, ovate, acute, closely appressed when mature, about 3-nerved on the keel; stamens 3; style 3-cleft, its branches exserted; achene 3-angled, about one-half as long as the scale.

Local in southern California, becoming a weed, apparently recently naturalized. Widely distributed in the eastern United States and in temperate and tropical regions of both the Old World and the New. Type locality: India.

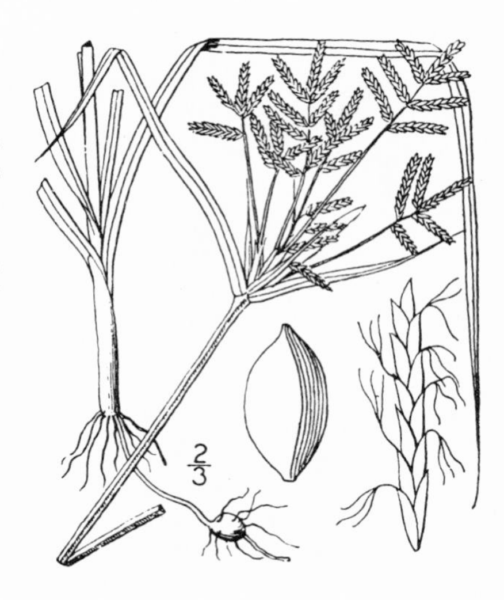

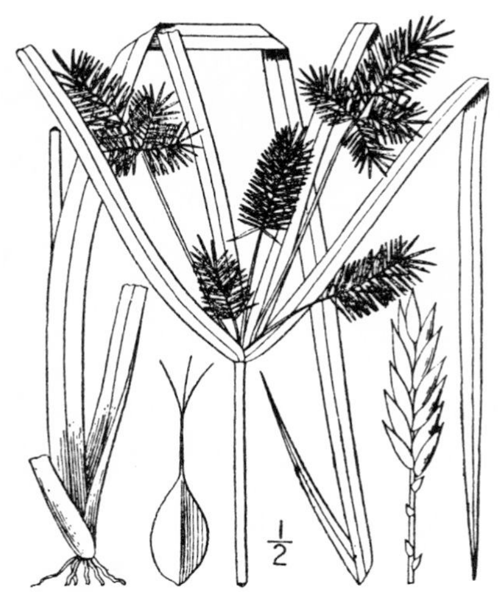

12. Cyperus erythrorrhìzos Muhl.

Red-rooted Cyperus. Fig. 619.

Cyperus erythrorrhizos Muhl. Gram. 20. 1817.
Cyperus occidentalis Torr. Ann. Lyc. N. Y. 3: 259. 1836.
Cyperus cupreus Presl, Rel. Haenk. 1: 172. 1828.

Annual; culms tufted, stout or slender, 7–50 cm. tall. Leaves 3–8 mm. wide, rough-margined, those of the involucre 3–7, some of them 3–5 times as long as the inflorescence; umbel mostly compound; spikelets linear, subacute, 6–25 mm. long, less than 2 mm. wide, compressed, many-flowered, clustered in oblong, nearly or quite sessile spikes; scales bright chestnut-brown, oblong-lanceolate, mucronulate, appressed, separating from the rachis at maturity, the membranous wings of the rachis separating as a pair of hyaline interior scales; stamens 3; style 3-cleft; achene sharply 3-angled, oblong, pointed at both ends, pale, one-half as long as the scale.

Wet or moist soil, Sonoran Zones; southern California to southern Washington; eastward at lower elevations across the continent to Florida and Massachusetts. Either a large race of this species, or a related species occurs along the Colorado River near the boundary. Type locality: Pennsylvania.

13. Cyperus strigòsus L.

Straw-colored Cyperus. Fig. 620.

Cyperus strigosus L. Sp. Pl. 47. 1753.

Perennial by basal tuber-like corms; culms rather stout, 0.3–0.9 m. tall. Leaves rough-margined, 4–6 mm. wide, the longer ones of the involucre much exceeding the umbel; umbel several-rayed, some of the primary rays often 10–15 cm. long, their sheaths terminating in 2 bristles; involucels setaceous; heads oblong or ovoid; spikelets flat, linear, 8–19 mm. long, 2 mm. wide or less, 7–15-flowered, separating from the axis at maturity; scales straw-colored, oblong-lanceolate, subacute, strongly several-nerved; stamens 3; style 3-cleft; achene linear-oblong, 3-angled, acute, about one-third as long as the scale.

Moist soil, Upper Sonoran Zone; Butte, Mariposa and Amador Counties, California; near springs, west Klickitat County, Oregon; Texas to Minnesota, Florida and Maine. Type locality: Virginia.

14. Cyperus hánseni Britton, sp. nov.

Hansen's Cyperus. Fig. 621.

Perennial by short rootstocks; culms rather stout, smooth, sharply trigonous, 3–5 dm. tall. Leaves 4–6 mm. wide, the basal ones usually shorter than the culm, those of the involucre 1–5 dm. long; umbel compound, the rays few, 7 cm. long or less; spikes cylindric, dense, 2–4 cm. long; spikelets yellowish brown, linear, spreading, 6–9 mm. long, about 1 mm. thick, bearing 2–4 achenes, falling away above the lower empty scales; fertile scales narrowly oblong; young achene narrowly linear.

Moist soil, Upper Sonoran Zone; Amador and Shasta Counties and in the valley of the Sacramento River, California. Type collected at Jackson, Amador County (*George Hansen*, No. 821, in 1904).

Referred by S. Watson to *Cyperus stenolepis* Torr., and included by C. B. Clarke in *Mariscus jacquini* H. B. K. [*Cyperus hermaphroditus* (Jacq.) Standley.]

15. Cyperus regiomontànus Britton.

Monterey Cyperus. Fig. 622.

Cyperus regiomontanus Britton, Rose, Contr. U. S. Nat. Herb. 1: 362. 1895.
Mariscus haenkei Presl, Rel. Haenk. 1: 181. 1828, not *Cyperus haenkeanus* Kunth.

Perennial, with slightly thickened culm-bases; culms smooth, 4 dm. tall or less. Leaves 1.5–3 mm. wide, the basal ones as long as the culm or shorter, those of the involucre spreading or reflexed, 1.5 dm. long or less; spikes 1–4, sessile or very short-stalked, cylindric, 1–2 cm. long, 6–7 mm. thick, obtuse; spikelets linear-oblong, brown, about 3 mm. long, usually with but 1 achene; fertile scale striate; achene linear, 2 mm. long, 0.5 mm. wide.

Mexico; Costa Rica. Mr. C. B. Clarke connected the Mexican and Central American plant with Presl's *Mariscus haenkei*, of which he had an authentic co-type at Kew in 1893. Type locality: Monterey, California (according to Presl). The plant has not since been collected in California and the type locality may be erroneous. It is here illustrated from Dr. E. Palmer's No. 1080 from Colima, Mexico.

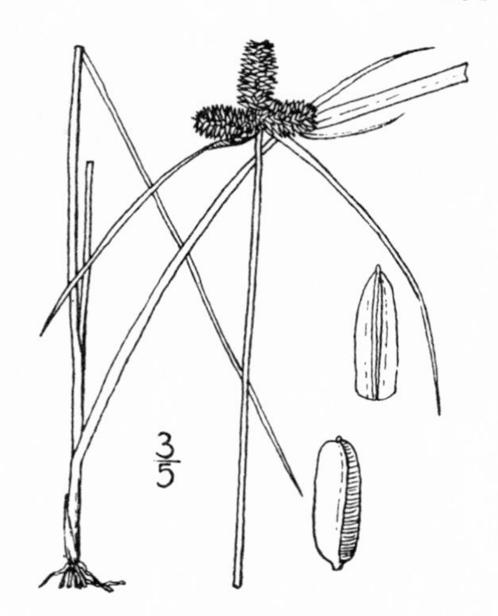

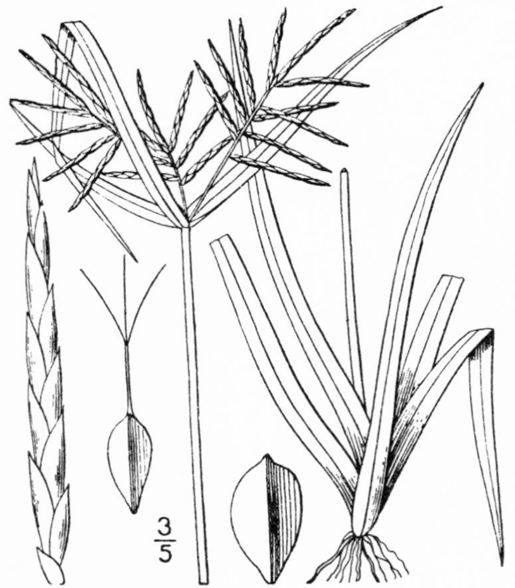

16. Cyperus fèrax L. C. Rich.

Coarse Cyperus. Fig. 623.

Cyperus ferax L. C. Rich. Act. Soc. Hist. Nat. Paris 1: 106. 1792.
Cyperus longispicatus Norton, Trans. St. Louis Acad. 12: 37. 1902.

Annual, with smooth-margined leaves, those of the involucre sometimes but little exceeding the inflorescence. Umbel often compact, the rays mostly short; spikelets linear, subterete, 10–20-flowered, 16–25 mm. long, about 2 mm. thick, falling away from the axis at maturity; scales ovate-oblong, appressed, imbricated, obtuse, rather firm, green and 7–9-nerved on the back, yellowish on the sides; stamens 3; style 3-cleft; rachis broadly winged; achene 3-angled, narrowly obovoid, obtuse.

In wet soil, Sonoran Zones; southern California north to Los Angeles and San Bernardino Counties; Arizona; Texas; Florida; extending northward along the Atlantic Coast to New England; tropical America; India. Type locality: Cayenne.

The record of *Cyperus pubescens* Presl, from Monterey, California, is an error in locality.

2. ELEÓCHARIS R. Br. Prodr. Fl. Nov. Holl. 1: 224. 1810.

Annual or perennial sedges. Culms simple, triangular, quadrangular, terete, flattened or grooved, the leaves reduced to sheaths or the lowest very rarely blade-bearing. Spikelets solitary, terminal, erect, several- to many-flowered, not subtended by an involucre. Scales concave, spirally imbricated all around. Perianth of 1–12 bristles, usually retrorsely barbed, wanting in some species. Stamens 2–3. Style 2-cleft and achene lenticular or biconvex, or 3-cleft and achene 3-angled, but sometimes with very obtuse angles and appearing turgid. Base of the style persistent on the summit of the achene, forming a terminal tubercle. [Greek, referring to the growth of most of the species in marshy ground.]

About 140 species, widely distributed. Besides the following, some 28 others occur farther east in North America. Type species, *Scirpus palustris* L.

Style-branches mostly 2; achene lenticular or plano-convex.
 Fibrous-rooted annuals.
  Achene black.
   Bristles brown; scales pale; achene 1 mm. long. — 1. *E. caribaea.*
   Bristles pale; scales purple-brown; achene 0.5 mm. long. — 2. *E. atropurpurea.*
  Achene brown.
   Spikelet ovoid to oblong.
    Tubercle narrower than the top of the achene. — 3. *E. ovata.*
    Tubercle about as broad as the top of the achene. — 4. *E. obtusa.*
   Spikelet oblong-cylindric. — 5. *E. engelmanni.*
 Perennial by rootstocks.
  Scales brown to purple-brown. — 6. *E. palustris.*
  Scales pale green to straw-color. — 7. *E. macrostachya.*
Style-branches 3; achene trigonous.
 Tubercle manifestly distinct from the achene.
  Achene ribbed and cancellate. — 8. *E. acicularis*

Achene smooth or merely papillose or reticulated.
  Fibrous-rooted annual. — 9. *E. disciformis.*
  Perennial by rootstocks or stolons.
    Tubercle depressed, nearly flat; achene smooth. — 10. *E. bolanderi.*
    Tubercle conic or subulate.
      Rootstock very stout, with scarious scales 2–3 cm. long. — 11. *E. decumbens.*
      Rootstocks slender, nearly naked.
        Spikelet nearly linear; tubercle conic-subulate. — 12. *E. parishii.*
        Spikelet oblong; tubercle conic.
          Achene finely rigid-reticulate or nearly smooth. — 13. *E. montana.*
          Achene papillose. — 14. *E. acuminata.*
Tubercle continuous with the achene. — 15. *E. rostellata.*

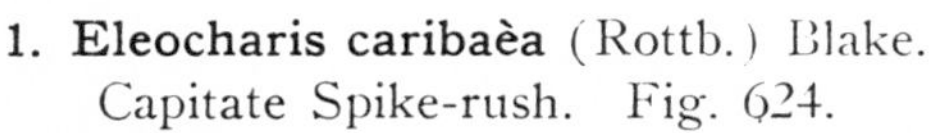

1. **Eleocharis caribaèa** (Rottb.) Blake. Capitate Spike-rush. Fig. 624.

Scirpus caribaeus Rottb. Descr. 24. 1772.
Scirpus capitatus L. Sp. Pl. 48. 1753.
Eleocharis caribaea Blake, Rhodora **20**: 24. 1918.

Annual; roots fibrous; culms densely tufted, nearly terete, almost filiform, 5–25 cm. tall. Upper sheath 1-toothed; spikelet ovoid, obtuse, much thicker than the culm, 3–5 mm. long, 2–3 mm. thick, many-flowered; scales broadly ovate, obtuse, firm, pale or dark brown with a greenish midvein, narrowly scarious-margined, persistent; stamens mostly 2; style 2-cleft; bristles 5–8, slender, downwardly hispid, as long as the achene; achene obovate, jet black, smooth, shining, nearly 1 mm. long; tubercle depressed, apiculate, constricted at the base, very much shorter than the achene.

Moist grounds, Sonoran Zones; Riverside and San Bernardino Counties, California; east to Indiana, Maryland and Florida; West Indies; Mexico; tropical continental America and Old World tropics. Type locality: St. Croix. The specific name *capitatus* has long been erroneously applied to this species.

2. **Eleocharis atropurpùrea** (Retz.) Kunth. Purple Spike-rush. Fig. 625.

Scirpus atropurpureus Retz. Obs. **5**: 14. 1789.
Eleocharis atropurpurea Kunth, Enum. **2**: 151. 1837.

Annual; roots fibrous; culms tufted, very slender, 2–9 cm. high. Upper sheath 1-toothed; spikelet ovoid, many-flowered, subacute, 3–4 mm. long, 2 mm. in diameter or less; scales minute, ovate-oblong, persistent, purple-brown with green midvein and very narrow scarious margins; stamens 2 or 3; style 2–3-cleft; bristles 2–4, fragile, white, minutely downwardly hispid, about as long as the achene; achene jet black, shining, 0.5 mm. long, smooth, lenticular; tubercle conic, minute, depressed but rather acute, constricted at the base.

Moist grounds, Lake Chelan, Washington; Nebraska to Texas, Iowa, and Florida. Europe and Asia. Type locality: India. Recorded by Piper as *E. capitata.*

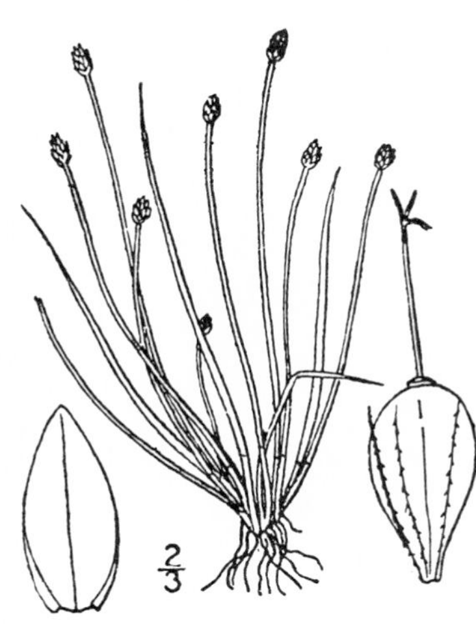

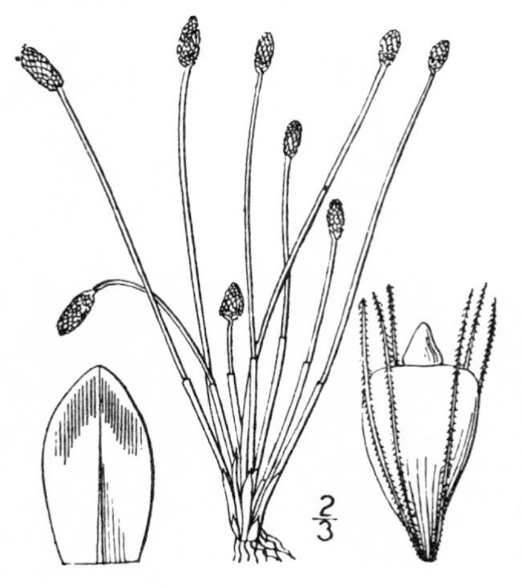

3. **Eleocharis ovàta** (Roth) R. & S. Ovoid Spike-rush. Fig. 626.

Scirpus ovatus Roth, Catal. Bot. **1**: 5. 1797.
Eleocharis ovata R. & S. Syst. **2**: 152. 1817.

Annual, with fibrous roots; culms tufted, slender to filiform, nearly terete, mostly erect, 0.5–4 dm. long. Upper sheath 1-toothed; spikelet ovoid to oblong, many-flowered, obtuse, 4–10 mm. long; scales oblong-orbicular, obtuse, brown; bristles 6–8, sometimes fewer, or wanting, usually longer than the achene; stamens 2 or 3; style mostly 2-cleft; achene pale brown, smooth, shining, lenticular, obovate-oblong, about 1 mm. long; tubercle deltoid, acute, its base narrower than the top of the achene.

Oregon (*E. Hall*), the specimen formerly referred by Watson to *E. olivacea* Torr. and by me to *E. capitata.* Michigan to New Brunswick and Connecticut. Europe. Type locality: Germany.

4. **Eleocharis obtùsa** (Willd.) Schultes. Blunt Spike-rush. Fig. 627.

Scirpus obtusus Willd. Enum. 76. 1809.
Eleocharis obtusa Schultes, Mant. **2**: 89. 1824.
E. ovata gigantea Clarke; Britton, Journ. N. Y. Micros. Soc. **5**: 103. Hyponym. 1889.
E. obtusa gigantea Fernald, Proc. Am. Acad. **34**: 493. 1899.

Annual with fibrous roots. Culms slender, tufted, 1–5 dm. long; upper sheath 1-toothed; spikelet ovoid to ovoid-oblong, obtuse, 2–13 mm. long, densely many-flowered; scales obovate to nearly orbicular, brown, obtuse, with scarious margins; style 2-cleft; achene compressed, obovate, pale brown, smooth, shining, about 1 mm. long; tubercle deltoid, acute, about one-third as long as the achene and nearly or quite as broad as its top; bristles usually longer than the achene.

Muddy places, Transition Zone; Siskiyou County, California, to British Columbia; Texas to Florida, Minnesota and Cape Breton Island. Type locality: North America. The species was formerly included in *E. ovata*, and as such recorded by me from Oregon.

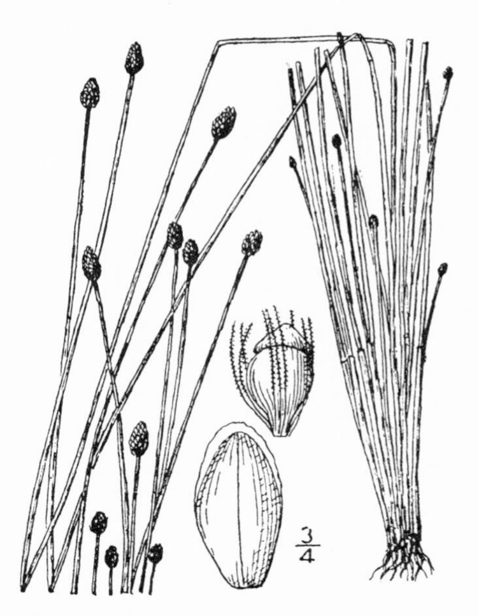

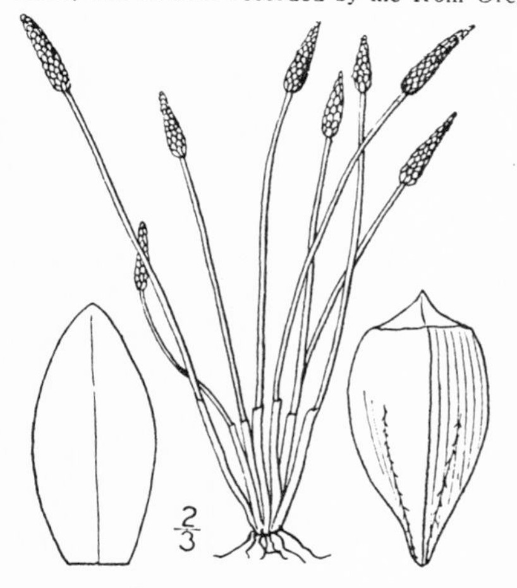

5. **Eleocharis engelmánni** Steud. Engelmann's Spike-rush. Fig. 628.

Eleocharis Engelmanni Steud. Syn. Pl. Cyp. 79. 1855.
Eleocharis monticola Fernald, Proc. Am. Acad. **34**: 496. 1899.

Annual, similar to the preceding species. Upper sheath obliquely truncate or 1-toothed; spikelet oblong-cylindric or ovoid-cylindric, obtuse or subacute, 4–16 mm. long, 2–3 mm. in diameter, many-flowered; scales pale brown with a green midvein and narrow scarious margins, ovate, obtuse, deciduous; style 2-cleft or sometimes 3-cleft; bristles about 6, not longer than the achene; achene broadly obovate, brown, smooth, lenticular; tubercle broad, low, covering the top of the achene, less than one-fourth its length.

Muddy places, Transition Zone; Washington to Mariposa County, California, east to Idaho, Manitoba, South Dakota, Texas, Indiana, Massachusetts, and New Jersey. Type locality: St. Louis, Missouri. Referred by S. Watson to *E. obtusa*.

6. **Eleocharis palústris** (L.) R. & S. Creeping Spike-rush. Fig. 629.

Scirpus palustris L. Sp. Pl. 47. 1753.
Eleocharis palustris R. & S. Syst. **2**: 151. 1817.

Perennial by horizontal rootstocks; culms terete or somewhat compressed, striate, 0.3–1.6 m. tall. Basal sheaths brown, rarely bearing a short blade, the upper one obliquely truncate; spikelet ovoid-cylindric, 6–25 mm. long, 3–4 mm. in diameter, many-flowered, thicker than the culm; scales ovate-oblong or ovate-lanceolate, purplish brown with scarious margin and a green midvein, or pale green all over; bristles usually 4, slender, retrorsely barbed, longer than the achene and tubercle, or sometimes wanting; stamens 2–3; style mostly 2-cleft; achene lenticular, smooth, yellow, over 1 mm. long; tubercle conic-triangular, constricted at the base, flattened, one-fourth to one-half as long as the achene.

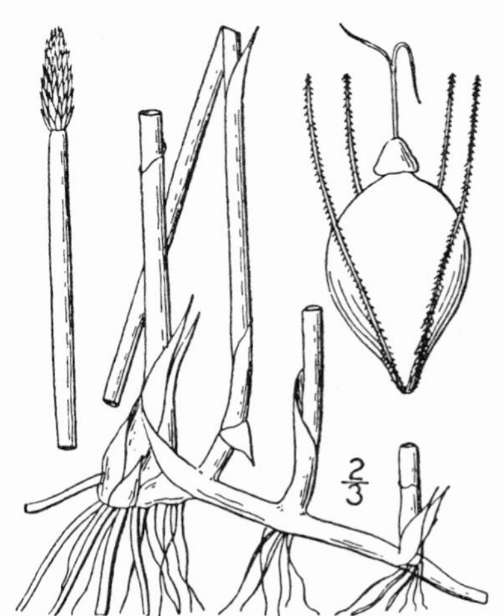

Wet grounds and in water, Sonoran to Canadian Zones; southern California to Alaska, east to Labrador and Florida. Europe and Asia. Type locality: Europe. Consists of many races, the culms slender to stout, the tubercle narrow or quite broad.

7. Eleocharis macrostàchya Britton.

Pale Spike-rush. Fig. 630.

E. macrostachya Britton; Small, Fl. S. E. U. S. 184, 1327. 1903.

Perennial by rootstocks, pale green. Culms tufted, slender or stout, sometimes twisted, 1.3 m. high or less; spikelet lanceolate-cylindric, about 2.5 cm. long or less, acute, many-flowered: scales oblong-ovate to oblong-lanceolate, acute, light green to straw-color, with a somewhat darker midvein; bristles as long as the achene and tubercle, or shorter, sometimes very short; style 2-cleft; achene obovate, lenticular, 1.5 mm. long, brown, the cap-like tubercle small, yellow.

Moist soil, Upper Sonoran Zone; Amador County, California, to Nevada, Missouri, Louisiana, and Jalisco. Type locality: Indian Territory [Oklahoma]. Perhaps better regarded as a race of *E. palustris*.

8. Eleocharis aciculàris (L.) R. & S.

Needle Spike-rush. Fig. 631.

Scirpus acicularis L. Sp. Pl. 48. 1753.
Eleocharis acicularis R. & S. Syst. **2**: 154. 1817.
E. acicularis minima Torr.; Britton, Journ. N. Y. Micros. Soc. **5**: 104. 1889.
E. acicularis radicans Britton, loc. cit. 105. 1889. Probably not *Scirpus radicans* Poir.
E. acicularis bella Piper, Fl. Palouse 35. 1901.

Perennial by filiform stolons or rootstocks; culms tufted, finely filiform or setaceous, obscurely 4-angled and grooved, weak, 1–20 cm. long. Sheath truncate; spikelet compressed, narrowly ovate or linear-oblong, acute, 3–10-flowered, 2–6 mm. long, 1 mm. wide; scales oblong, obtuse or the upper subacute, thin, pale green, deciduous, many of them commonly sterile; bristles 3–4, fragile, fugacious, shorter than the achene; stamens 3; style 3-cleft; achene obovoid-oblong, pale, obscurely 3-angled with a rib on each angle and 6–9 lower intermediate ribs connected by fine ridges; tubercle conic, acute, one-fourth as long as the achene.

Moist grounds, Sonoran to Canadian Zones; southern California to British Columbia, east to Missouri, New Jersey, and Newfoundland. Europe and Asia. Type locality: Europe. Consists of many races, differing in size of the plant, of the spikelet and of the achene.

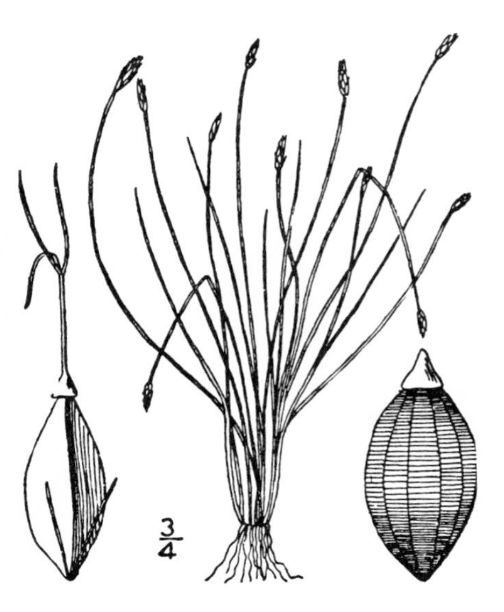

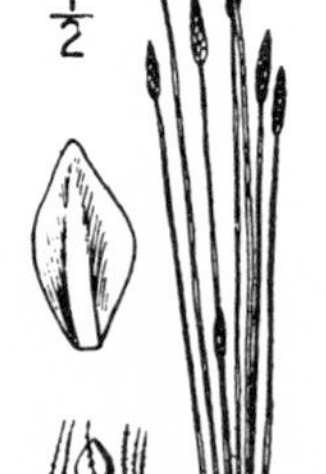

9. Eleocharis discifórmis Parish.

San Jacinto Spike-rush. Fig. 632.

Eleocharis disciformis Parish, Bull. S. Calif. Acad. Sci. **3**: 81. 1904.

Annual with fibrous roots and filiform tufted striate culms 10–15 cm. long; upper sheath nearly truncate, 1-toothed. Spikelet linear-lanceolate, acute, 10–15-flowered, 5–10 mm. long, about 1.5 mm. thick; scales ovate, obtuse, about 2 mm. long, brown with a pale midvein; bristles a little longer than the achene and tubercle; achene ovoid, 1 mm. long; tubercle short or depressed, disciform, and abruptly tipped, about one-sixth as long as the achene.

Known only from the type locality, eastern base of the the San Jacinto Mountains, Lower Sonoran Zone, San Diego County, California.

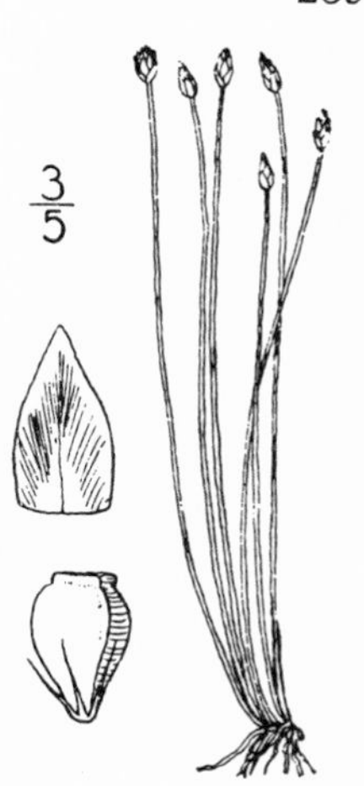

10. Eleocharis bolánderi A. Gray.

Bolander's Spike-rush. Fig. 633.

Eleocharis bolanderi A. Gray, Proc. Am. Acad. **7**: 392. 1868.

Perennial by short rootstocks; culms slender, angular-striate, 1–3 dm. high, the upper sheath truncate, 1-toothed. Spikelet ovoid-oblong, many-flowered, about 8 mm. long; scales dark brown, acute; bristles much shorter than the achene or wanting; style 3-cleft; achene trigonous, obovoid, smooth, white to yellow, much shorter than the scale, truncate, capped by the broad, nearly flat tubercle.

Stream banks and moist places, Transition Zone; Sequoia Grove, Mariposa County; wet adobe flat near Forestdale, Modoc County, California. Type locality: banks of stream near Clark's, Mariposa County, California.

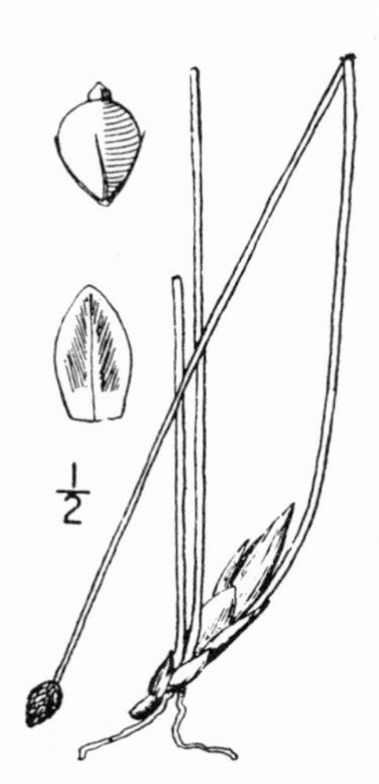

11. Eleocharis decúmbens Clarke.

Decumbent Spike-rush. Fig. 634.

Eleocharis decumbens Clarke, Kew Bull. Add. Ser. **8**: 23. 1908.

Perennial by stout rootstocks; culms stout, subterete, 5–6 dm. long. Basal scales lanceolate, obtuse, brown, 2–3 cm. long, loose; spikelet ellipsoid, obtuse, 5–7 mm. long, about 4 mm. thick; scales brown, ovate, obtuse; style 3-cleft, its base squamose; achene trigonous, yellow, papillose, obovoid-oblong, 1 mm. long; tubercle conic, its base narrower than the top of the achene; bristles as long as the achene and tubercle or shorter.

Wet meadows, Transition and Canadian Zones; Siskiyou County and Yosemite National Park, California. Type locality: Mount Shasta, California, at 7,500 feet.

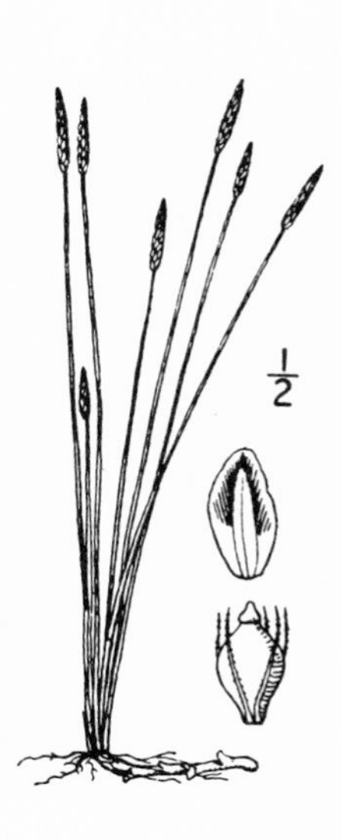

12. Eleocharis paríshii Britton.

Parish's Spike-rush. Fig. 635.

Eleocharis parishii Britton, Journ. N. Y. Micros. Soc. **5**: 110. 1889.

Perennial by slender rootstocks; culms slender, tufted, 1–2 dm. tall, the upper sheath truncate, 1-toothed. Spikelet linear-lanceolate, acute, 1–1.5 cm. long, 2 mm. thick; scales ovate, pale or chestnut-brown, subcarinate; style 3-cleft; achene trigonous, ellipsoid, yellow-brown, smooth, 1 mm. long, narrowed at each end; tubercle short-subulate, acute, about one-fourth as long as the achene; bristles about as long as the achene.

Wet soil, Sonoran Zones; California, from Kern and San Bernardino Counties, southward; Nevada. Type locality: Agua Caliente, San Diego County, California.

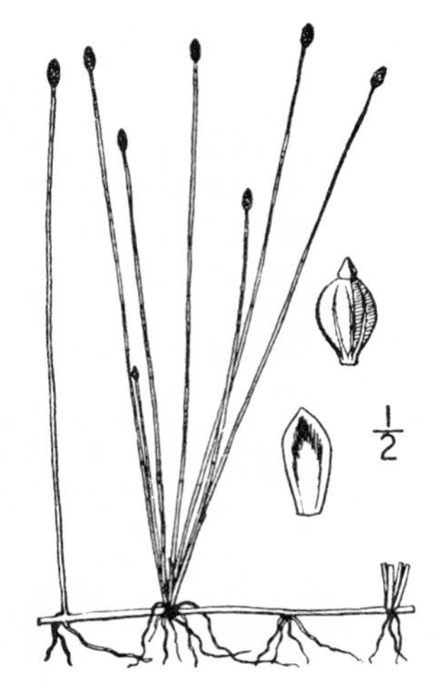

13. Eleocharis montàna (H. B. K.) R. & S.

Dombey's Spike-rush. Fig. 636.

Scirpus montanus H. B. K. Nov. Gen. **1**: 206. 1815.
Eleocharis montana R. & S. Syst. **2**: 153. 1817.
Eleocharis dombeyana Kunth, Enum. **2**: 145. 1837.
Eleocharis arenicola Torr.; Engelm. & Gray, Bost. Journ. Nat. Hist. **5**: 237. 1847.

Perennial by rootstocks; culms slender, 1–4 dm. high, the upper sheath truncate, usually 1-toothed. Spikelet oblong to lanceolate-oblong, obtuse or acutish, 5–12 mm. long, 2–3 mm. thick, densely many-flowered; scales ovate, brown to yellowish, obtuse or the upper acute; bristles mostly as long as the achene or somewhat longer; style 3-cleft; achene obovoid or oblong-obovoid, trigonous, yellow-brown, 1 mm. long, finely reticulated or nearly smooth; tubercle conic, much shorter than the achene.

Wet soil, Sonoran Zones; Butte County, California, to Lower California, east to New Mexico, San Luis Potosi, Louisiana and South Carolina and Florida; Mexico to northern South America. Type locality: Mount Quindiu, Colombia. I follow C. B. Clarke in reducing *E. arenicola* to *E. montana*, but now with some hesitation.

14. Eleocharis acuminàta (Muhl.) Nees.

Flat-stemmed Spike-rush. Fig. 637.

Scirpus acuminatus Muhl. Gram. 27. 1817.
Eleocharis acuminata Nees, Linnaea **9**: 204. 1835.

Perennial by stout rootstocks, similar to the preceding species but stouter; culms flattened, striate, slender but rather stiff, tufted, 0.2–0.5 m. tall. Upper sheath truncate, sometimes slightly 1-toothed; spikelet ovoid or oblong, obtuse, thicker than the culm, many-flowered, 6–12 mm. long; scales oblong or ovate-lanceolate, acute or the lower obtusish, purple-brown with a greenish midvein and hyaline white margins, deciduous; bristles 1–5, shorter than or equaling the achene, fugacious, or wanting; stamens 3; style 3-cleft, exserted; achene obovoid, obtusely 3-angled, light yellowish brown, papillose, much longer than the depressed-conic acute tubercle.

Wet grounds, Transition and Boreal Zones; Washington (?) and British Columbia; Montana to Ontario, New York, Louisiana and Georgia. Type locality: not cited.

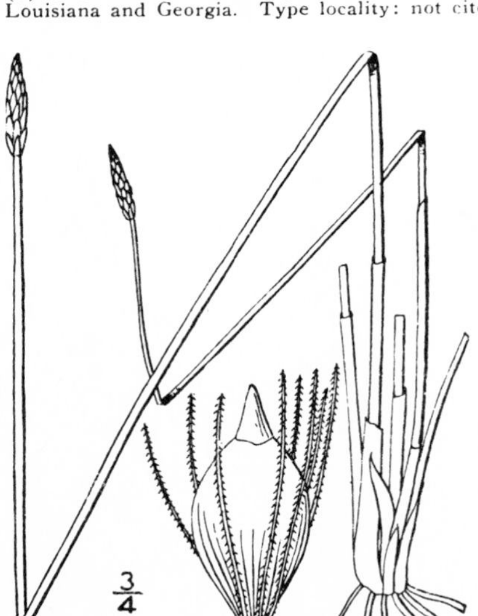

15. Eleocharis rostellàta Torr.

Beaked Spike-rush. Fig. 638.

Scirpus rostellatus Torr. Ann. Lyc. N. Y. **3**: 318. 1836.
Eleocharis rostellata Torr. Fl. N. Y. **2**: 347. 1843.
E. rostellata occidentalis S. Wats. Bot. Calif. **2**: 222. 1880.

Perennial by a short caudex; culms slender, wiry, the fertile erect or ascending, the sterile reclining and rooting at the summit, grooved, 0.3–1.5 m. long. Upper sheath truncate; spikelet oblong, narrowed at both ends, thicker than the culm, 10–20-flowered, 6–12 mm. long, about 2 mm. in diameter; scales ovate, obtuse or the upper acute, green with a somewhat darker midvein; bristles 4–8, retrorsely barbed, longer than the achene and tubercle; stamens 3; style 3-cleft; achene oblong-obovoid, obtusely 3-angled, its surface finely reticulated; tubercle conic-subulate, about one-half as long as the achene or shorter, capping its summit, partly or entirely falling away at maturity.

Wet grounds, Transition and Sonoran Zones; southern California to British Columbia, east across the continent to Montana, Michigan, New Hampshire, and Florida; Bermuda; Cuba. Type locality: Penn Yan, Yates County, New York.

3. STENOPHÝLLUS Raf. Neog. 4. 1825.

Mostly annual sedges, with slender erect culms, leafy below, the leaves narrowly linear or filiform, with ciliate or pubescent sheaths. Spikelets umbellate, capitate or solitary, subtended by a one- to several-leaved involucre, their scales spirally imbricated all around, mostly deciduous. Flowers perfect. Perianth none. Stamens 2 or 3. Style 2–3-cleft, glabrous, its base much swollen and persistent as a tubercle on the achene as in *Eleocharis*. Achene 3-angled, turgid or lenticular. [Greek, referring to the narrow leaves.]

A genus of some 20 species, natives of temperate and warm regions. Besides the following, 5 others occur in the southeastern United States. Type species, *Scirpus stenophyllus* Ell.

1. Stenophyllus capillàris (L.) Britton.

Hair-like Stenophyllus. Fig. 639.

Scirpus capillaris L. Sp. Pl. 49. 1753.
Fimbristylis capillaris A. Gray, Man. ed. 1, 530. 1848.
Stenophyllus capillaris Britton, Bull. Torrey Club **21**: 30. 1894.

Annual; roots fibrous; culms filiform, densely tufted, erect, grooved, smooth, 5–25 cm. tall. Leaves roughish, much shorter than the culm, their sheaths more or less pubescent with long hairs; involucral leaves 1–3, setaceous; spikelets narrowly oblong, somewhat 4-sided, 5–8 mm. long, less than 2 mm. thick, several in a terminal umbel, or in depauperate forms solitary; scales oblong, obtuse or emarginate, puberulent, dark brown with a green keel; stamens 2; style 3-cleft; achene yellow-brown, narrowed at the base, very obtuse or truncate at the summit, 0.5 mm. long, transversely wrinkled; tubercle minute, depressed.

Grassy and rocky situations, Transition Zone; southern California to Oregon; Arizona to Florida, Minnesota and Maine; Cuba; Central and South America. Type locality: Virginia.

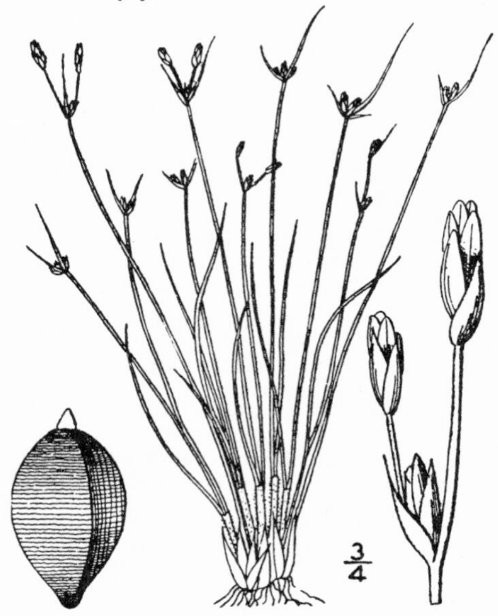

4. FIMBRÍSTYLIS Vahl, Enum. 2: 285. 1806.

Annual or perennial sedges. Culms leafy below. Spikelets umbellate or capitate, terete, several- to many-flowered, subtended by a one- to many-leaved involucre, their scales spirally imbricated all around, mostly deciduous, all fertile. Perianth none. Stamens 1–3. Style 2–3-cleft, pubescent or glabrous, its base much enlarged, falling away from the summit of the achene at maturity. Achene lenticular, biconvex, or 3-angled, reticulated, cancellate, or longitudinally ribbed or striate in our species. [Greek, in allusion to the fringed style of some species.]

A large genus, the species widely distributed in temperate and tropical regions. Besides the following, some 8 others occur in the southern and eastern parts of North America. Type species, *Fimbristylis acuminata* Vahl.

Style-branches 2; achene lenticular.
- Spikelets in a terminal head; scales acuminate; plant diminutive. — 1. *F. vahlii.*
- Spikelets umbeled; plant tall. — 2. *F. thermalis.*

Style-branches 3; achene trigonous. — 3. *F. miliacea.*

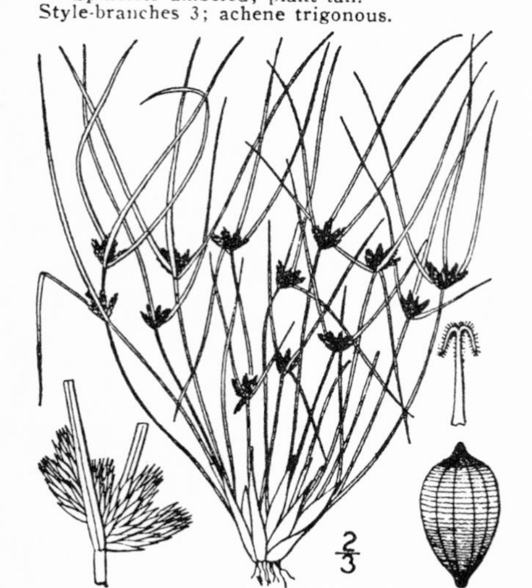

1. Fimbristylis vàhlii (Lam.) Link.

Vahl's Fimbristylis. Fig. 640.

Scirpus vahlii Lam. Tabl. Encycl. **1**: 139. 1791.
Fimbristylis Vahlii Link, Hort. Berol. **1**: 287. 1827.
Fimbristylis congesta Torr. Ann. Lyc. N. Y. **3**: 345. 1836.

Annual; culms very slender, densely tufted, compressed, striate, 2–10 cm. high, longer than or equaling the leaves. Leaves setaceous or almost filiform, rough, those of the involucre 3–5, erect, much exceeding the simple capitate cluster of 3–8 spikelets; spikelets oblong-cylindric, obtuse, 4–8 mm. long, about 1 mm. thick, many-flowered; scales lanceolate, pale greenish brown, acuminate; stamen 1; style 2-cleft, glabrous below; achene minute, biconvex, yellowish white, cancellate.

Shore of Clear Lake, Tulare County; sand-bar, Fort Yuma, California; Texas to Florida, Missouri and North Carolina. Central and South America. Type locality: cited as Spain, presumably in error.

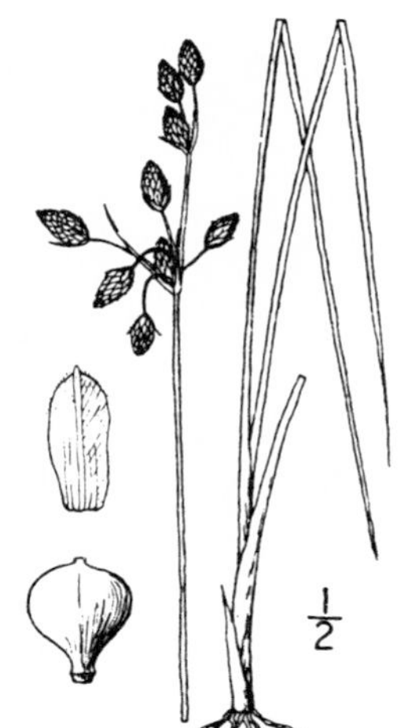

2. Fimbristylis thermàlis S. Wats.

Hot Spring Fimbristylis. Fig. 641.

Fimbristylis thermalis S. Wats. Bot. King's Exped. 360. 1871.

Perennial by short rootstocks. Culms slender, erect, roughish, 3–5 dm. long. Leaves more or less pubescent, flat, 1–2.5 mm. wide, shorter than the culm, sometimes involute above, those of the involucre few, short and narrow; umbel simple or somewhat compound; spikelets few or several, oblong to subcylindric, 8–18 mm. long, many-flowered; scales dull, puberulent, mucronate or blunt; style-branches 2; achene lenticular, broadly obovate, longitudinally finely striate, about 1.5 mm. long.

Hot Springs, Sonoran Zones, Inyo, Kern, and San Bernardino Counties, California; Nevada. Type locality: Ruby Valley, Nevada.

3. Fimbristylis miliàcea (Thunb.) Vahl.

Grass-like Fimbristylis. Fig. 642.

Scirpus miliaceus Thunb. Fl. Jap. 37. 1784.
Fimbristylis miliacea Vahl, Enum. 2: 287. 1806.

Annual with fibrous roots; culms weak, tufted, 7 dm. long or less, angled above. Leaves soft, 1.5–4 mm. wide, the basal ones usually about half as long as the culm, long-attenuate at the apex, those of the involucre similar, but much shorter and narrower, shorter than the loose, large, decompound umbel; spikelets subglobose, 2–4 mm. long, their lower scales early deciduous; scales ovate, the midvein broad; style-branches 3; style villous; achene obovoid, trigonous, brownish, minutely tuberculate, less than 1 mm. long.

San Francisco, collected by A. Wood in 1868; I have seen the specimen preserved in the Gray Herbarium; Florida; Cuba; Trinidad; continental tropical America; Old World tropics. Type locality: East Indies. *Fimbristylis junciformis* Kunth and *F. cyperoides* R. Br. are erroneously recorded from Monterey, California.

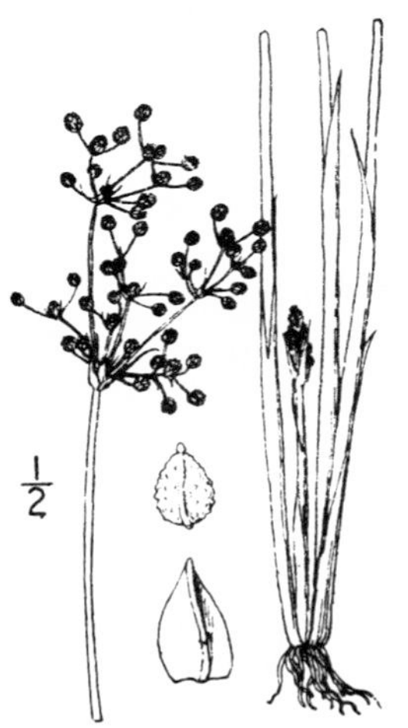

5. ERIÓPHORUM L. Sp. Pl. 52. 1753.

Bog sedges, perennial by rootstocks, the culms erect, triangular or nearly terete, the leaves linear, or 1 or 2 of the upper ones reduced to bladeless sheaths. Spikelets terminal, solitary, capitate or umbeled, subtended by a one- to several-leaved involucre, or naked. Scales spirally imbricated, usually all fertile. Flowers perfect. Perianth of 6 or numerous filiform smooth soft bristles, which are white or brown, straight or crisped, and exserted much beyond the scales at maturity. Stamens 1–3. Style 3-cleft. Achene 3-angled, oblong, ellipsoid or obovoid. [Greek, signifying wool-bearing, referring to the soft bristles.]

About 15 species, in the northern hemisphere. Besides the following, 7 others occur in northern and eastern North America. Type species, *Eriophorum vaginatum* L.

Spikelet solitary; no involucre. — 1. *E. chamissonis.*
Spikelets several, umbeled; involucre 1–4-leaved.
  Leaves triangular-channeled. — 2. *E. gracile.*
  Leaves flat, at least below the middle. — 3. *E. angustifolium.*

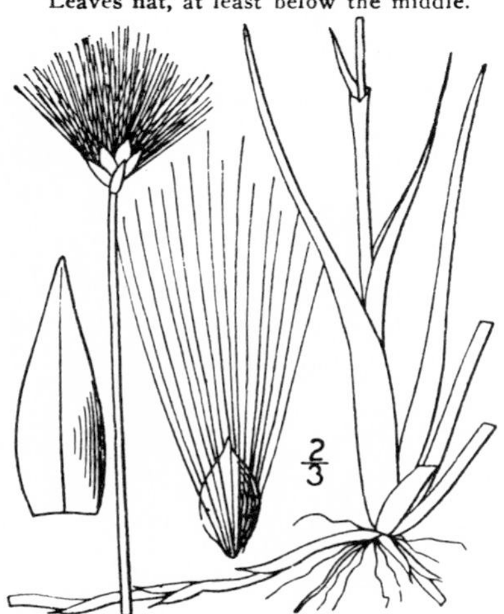

1. Eriophorum chamissònis C. A. Meyer.

Russet Cotton-grass. Fig. 643.

Eriophorum chamissonis C. A. Meyer; Ledeb. Fl. Alt. 1: 70. 1829.
Eriophorum russeolum Fries, Novit. Mant. 3: 67. 1842.

Stoloniferous; culms triangular, erect, smooth, longer than the leaves, 1–8 dm. high. Upper sheath inflated, bladeless, mucronate, or rarely with a short subulate blade usually borne below the middle of the culm; leaves filiform, triangular-channeled, mucronate, 2–10 cm. long; spikelet erect; involucre none; scales ovate-lanceolate, acuminate, thin, purplish brown with nearly white margins; bristles numerous, bright reddish brown, or sometimes white, 3–5 times as long as the scale; achene oblong, narrowed at each end, apiculate.

In bogs, Transition and Canadian Zones; Oregon, north to Alaska, east to Montana, Ontario, New Brunswick and Newfoundland. Europe and Asia. Type locality: Unalaska.

2. **Eriophorum grácile** Koch.

Slender Cotton-grass. Fig. 644.

Eriophorum gracile Koch; Roth, Catal. Bot. **2**: 259. 1800.

Culms slender, smooth, nearly terete, spreading or reclining, 6 dm. long or less. Leaves triangular-channeled, the basal ones mostly wanting at flowering time, those of the culm 2 or 3, the upper one with a blade shorter than its sheath, 3 cm. long or less; involucral leaf about 1 cm. long; spikelets 2–4, rarely 6, the slender peduncles pubescent, mostly less than 2 cm. long; scales ovate, gray to nearly black, acutish, the midvein prominent; achene obovoid-oblong, about 2 mm. long; bristles bright white, 1–2 cm. long.

In bogs, Transition and Canadian Zones; Calaveras County, California, north to British Columbia, east to Colorado, Iowa, Pennsylvania, New Jersey, and Quebec. Europe and Asia. Type locality: not cited.

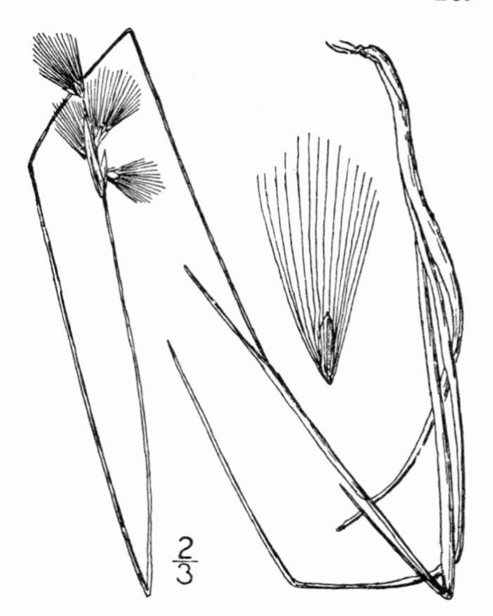

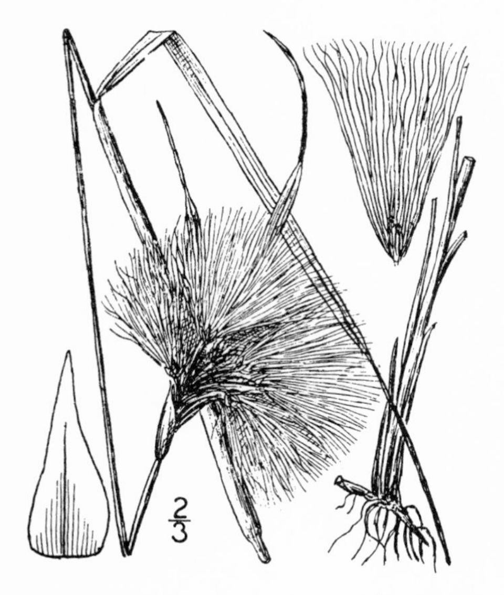

3. **Eriophorum angustifòlium** Roth.

Tall Cotton-grass. Fig. 645.

Eriophorum angustifolium Roth, Tent. **1**: 24. 1788.

Culm stiff, smooth, obtusely triangular above, 0.4–0.6 m. tall, all the sheaths blade-bearing. Leaves flat, 3–8 mm. wide, tapering to a channeled rigid tip, those of the involucre 2–4, the longer commonly equaling or exceeding the inflorescence, 3–8 mm. wide; spikelets 2–12, drooping, in a terminal umbel; rays filiform, smooth; scales ovate-lanceolate, acute or acuminate, purple-green or brown; bristles numerous, bright white, about 2.5 cm. long, 4–5 times as long as the scale; achene obovoid, obtuse, light brown.

In bogs, Canadian to Arctic Zones, base of Mount Rainier, north to Alaska, east to Colorado, Illinois, Maine, and Newfoundland. Europe and Asia. Type locality: Germany. Included by authors in *E. polystachyum* L., an Old World species.

6. **SCÍRPUS** L. Sp. Pl. 47. 1753.

Annual or perennial very small or very large sedges, with leafy culms or the leaves reduced to basal sheaths. Spikelets terete or somewhat flattened, solitary, capitate, spicate or umbellate, subtended by a one- to several-leaved involucre or the involucre wanting in some species. Scales spirally imbricated all around, usually all fertile, the 1–3 lower sometimes empty. Flowers perfect. Perianth of 1–6, slender or rigid, short or elongated, barbed, pubescent or smooth bristles, or none in some species. Stamens 2–3. Style 2–3-cleft, not swollen at the base, wholly deciduous from the achene or its base persistent as a subulate tip. [Latin name of the bulrush.]

About 150 species of wide geographic distribution. Besides the following some 20 others occur in the eastern United States. Type species, *Scirpus lacustris* L.

Bristles not longer or but little longer than the scales, or wanting.
 Spikelet solitary, terminal (rarely 2 or 3 in No. 2).
  Fibrous-rooted low annuals.
   No involucral bract; bristles 6. — 1. *S. nanus.*
   Involucral bract one; bristles none.
    Scales of the spikelet keeled. — 2. *S. carinatus.*
    Scales of the spikelet not keeled. — 3. *S. cernuus.*
  Perennials.
   None of the sheaths leaf-bearing; no involucral bract. — 4. *S. pauciflorus.*
   One or more of the sheaths leaf-bearing; an involucral bract present or wanting.
    Involucral bracts shorter than the spikelet or none; terrestrial species.
     Perianth-bristles present. — 5. *S. caespitosus.*
     Perianth-bristles none. — 6. *S. pumilus.*
    Involucral bract longer than the spikelet; aquatic species. — 7. *S. subterminalis*

Spikelets normally more than one, if only one, appearing as if lateral.
  Involucre of one bract, rarely 2.
    Spikelets few, 1–12, appearing lateral.
      Culm sharply 3-angled.
        Spikelet much shorter than the bract; scales awned. — 8. *S. americanus.*
        Spikelet not much shorter than the bract; scales mucronulate. — 9. *S. olneyi.*
      Culm not sharply 3-angled. — 10. *S. nevadensis.*
    Spikelets several or numerous, umbeled.
      Perianth-bristles downwardly barbed.
        Achenes lenticular.
          Achenes nearly as long as the scales. — 11. *S. validus.*
          Achenes much shorter than the scales. — 12. *S. acutus.*
        Achenes trigonous. — 13. *S. heterochaetus*
      Perianth-bristles short-plumose below. — 14. *S. californicus.*
  Involucre of 2 leaves or more.
    Spikelets several or few, large, capitate or subumbellate.
      Achene lenticular or plano-convex.
        Scales red-brown; achene cuneate at base. — 15. *S. pacificus.*
        Scales pale brown or whitish; achene narrowed at base. — 16. *S. paludosus.*
      Achene trigonous. — 17. *S. fluviatilis.*
    Spikelets numerous, small, in umbeled heads or compound umbels.
      Perianth-bristles about as long as the achene or little longer, mostly downwardly barbed.
        Style-branches 2; achene plano-convex. — 18. *S. microcarpus.*
        Style-branches 3; achene trigonous.
          Perianth-bristles stiff, downwardly barbed. — 19. *S. pallidus.*
          Perianth-bristles weak, flexuous, a little roughened. — 20. *S. congdoni.*
      Perianth-bristles much longer than the achene, smooth, entangled. — 21. *S. lineatus.*
Bristles much longer than the scales; spikelets densely capitate. — 22. *S. criniger.*

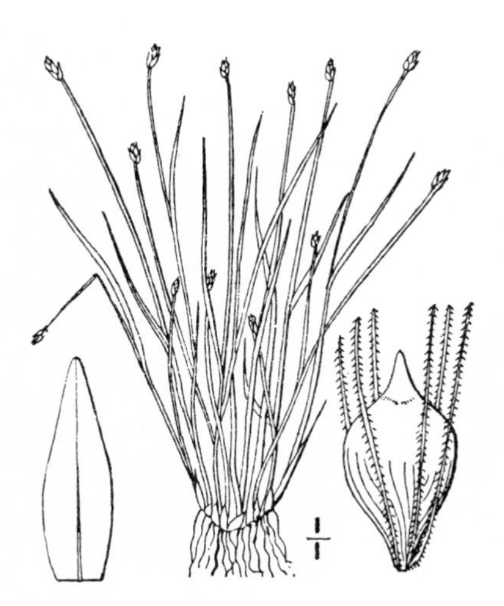

1. Scirpus nànus Spreng.

Dwarf Club-rush. Fig. 646.

Scirpus nanus Spreng. Pug. 1: 4. 1815.
Eleocharis pygmaea Torr. Ann. Lyc. N. Y. 3: 313. 1836.

Annual; roots fibrous; culms filiform, flattened, grooved, tufted, erect or ascending, 2–5 cm. high, bearing a scarious bladeless sheath near the base. Spikelet solitary, terminal, ovoid-oblong, rather acute, 3–8-flowered, 2–3 mm. long, not subtended by a bract; scales ovate or lanceolate, pale green, the lower obtuse, the upper subacute; bristles about 6, downwardly barbed, longer than the achene; stamens 3; style 3-cleft; achene oblong, 3-angled, pale, pointed at each end, smooth.

Muddy places in salt marshes, coasts of Washington and British Columbia; ?Cucamonga, California; about salt springs in Michigan and New York; Atlantic Coast from Cape Breton Island to Florida; Gulf Coast, Florida to Texas; Cuba; coasts of Europe. Type locality: Germany.

2. Scirpus carinàtus (H. & A.) A. Gray.

Keeled Club-rush. Fig. 647.

Isolepis carinata H. & A.; Torr. Ann. Lyc. N. Y. 3: 349. 1836.
Isolepis koilepis Steud. Syn. Pl. Cyp. 318. 1855.
Scirpus carinatus A. Gray, Proc. Am. Acad. 7: 392. 1868.

Annual, fibrous-rooted; culms slender or nearly filiform, tufted, smooth, 2.5–12 cm. high, trigonous with slightly channeled sides, the upper sheath bearing a linear-filiform leaf-blade 2 cm. long or less; spikelet usually solitary, rarely 2 or 3, ovoid, 5–10-flowered, 4–7 mm. long, subtended by a filiform bract 2 cm. long or less; scales broadly ovate, greenish brown, sharply keeled, acute, or the lower one long-tipped; bristles none; achene sharply trigonous, smooth, brown, about 1 mm. long.

Moist, grassy situations, Upper Sonoran Zone; near the coast, Mendocino to Monterey Counties, California; Texas to Missouri, Alabama and Tennessee. Type locality: near New Orleans, Louisiana.

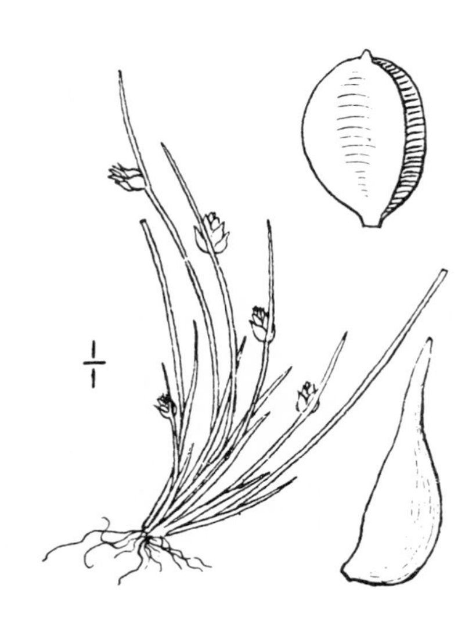

3. **Scirpus cernùus** Vahl.
Low Club-rush. Fig. 648.

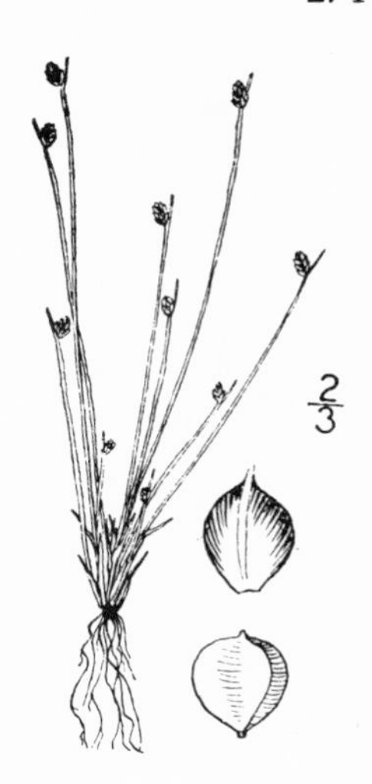

Scirpus cernuus Vahl, Enum. **2**: 245. 1806.
Scirpus riparius Spreng. Syst. **1**: 208. 1825.
Isolepis pygmaea Kunth, Enum. **2**: 191. 1837.
Scirpus leptocaulis Torr. Bot. Whipple Exp. 97. 1857.
Scirpus pygmaeus A. Gray, Proc. Am. Acad. **7**: 392. 1868.
Isolepis pygmaea californica Torr. Bot. Wilkes Exp. 476. 1874.

Annual, with fibrous roots; culms nearly filiform, tufted, 5–25 cm. long, the upper sheath bearing a filiform leaf 2–20 mm. long, or merely acuminate. Spikelet solitary, ovoid, 2–5 mm. long, few-flowered, subtended by a bract 2–20 mm. long; scales broadly ovate, pale to dark brown, concave, scarcely keeled, obtuse or acutish; bristles none; style 3-cleft; nutlet sharply trigonous, smooth, brown, somewhat less than 2 mm. long.

Moist or wet ground, Sonoran and Transition Zones; Lower California to British Columbia; western South America; Australia. Widely distributed in the Old World. Type locality: Lusitania.

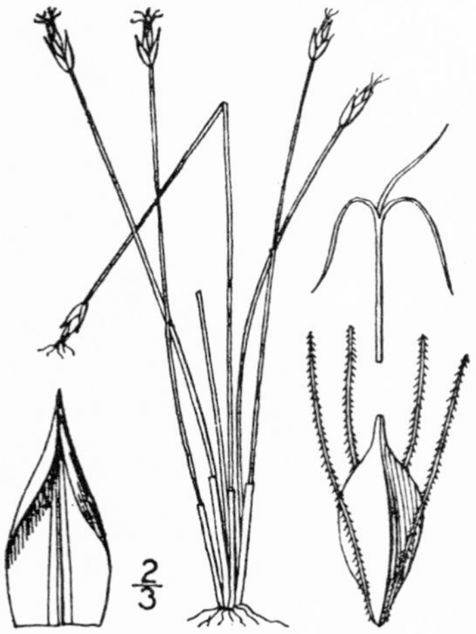

4. **Scirpus pauciflòrus** Lightf.
Few-flowered Club-rush. Fig. 649.

Scirpus pauciflorus Lightf. Fl. Scot. 1078. 1777.
Eleocharis pauciflora Link, Hort. Berol. **1**: 284. 1827.

Perennial by filiform rootstocks; culms very slender, little tufted, 3-angled, grooved, leafless, 7–25 cm. tall, the upper sheath truncate. Spikelet terminal, solitary, not subtended by an involucral bract, oblong, compressed, 4–10-flowered, 4–6 mm. long, nearly 2 mm. wide; scales brown with lighter margins and midvein, lanceolate, acuminate; bristles 2–6, hispid, as long as the achene or longer, or wanting; stamens 3; style 3-cleft; achene obovoid-oblong, gray, rather abruptly beaked, its surface finely reticulated.

Wet soil, Canadian Zone; San Bernardino County, California, to British Columbia, east to Colorado, Ontario, New York, Maine, and Anticosti. Europe. Type locality: Highlands of Scotland.

5. **Scirpus caespitòsus** L.
Tufted Club-rush. Fig. 650.

Scirpus caespitosus L. Sp. Pl. 48. 1753.
Eleocharis caespitosa Nees, Linnaea **9**: 294. 1835.

Perennial; culms smooth, terete, densely tufted, light green, erect or ascending, almost filiform, wiry, 10–40 cm. long. Basal sheaths numerous, membranous, imbricated, acuminate, the upper one bearing a short very narrow blade; spikelet solitary, terminal, few-flowered, ovoid-oblong, about 4 mm. long, subtended by a subulate involucral leaf of about its own length; scales yellowish brown, ovate, obtuse or subacute, deciduous; bristles 6, smooth, longer than the achene; stamens 3; style 3-cleft; achene oblong, smooth, 3-angled, brown, acute.

In bogs and on moist rocks, Arctic Zone; Washington to Alaska, east to Colorado, Minnesota, New York, Maine, and Greenland and on the southern Alleghenies. Europe and Asia. Type locality: Europe.

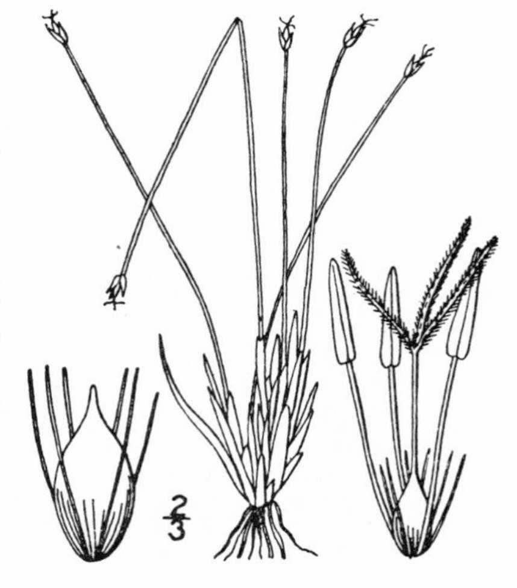

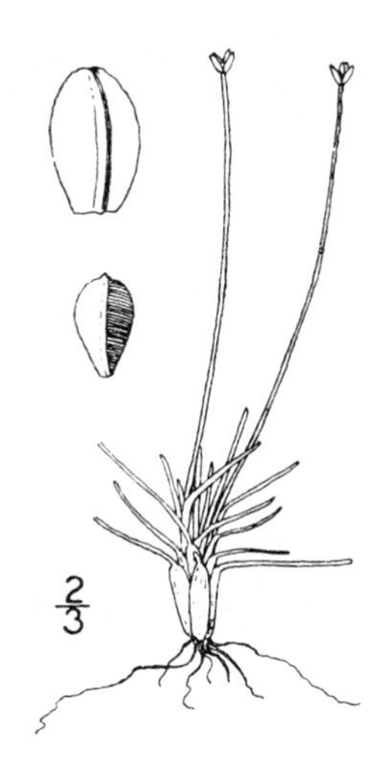

6. Scirpus pùmilus Vahl.

Low Club-rush. Fig. 651.

Scirpus pumilus Vahl, Enum. **2**: 243. 1806.
Isolepis pumila R. & S. Syst. **2**: 106. 1817.

Perennial, pale green, densely tufted, the slender culms 5–12 cm. high, the upper 1 to 3 sheaths bearing linear blades 3 cm. long or less; spikelet solitary, terminal, few-flowered, about 3 mm. long, bractless, or rarely the lower scale tipped by a prolongation 2–3 mm. long; scales subcoriaceous, ovate or ovate-oblong, obtuse or acutish, brown, indistinctly nerved, about 2 mm. long; achene oblong or oblong-obovoid, trigonous, brown or nearly black, 1–1.5 mm. long; bristles none.

Hocket Meadow, Canadian Zone; southern Sierra Navada, Tulare County, California; mountains of British Columbia. Europe. Type locality: Switzerland. American specimens have somewhat narrower and longer achenes than European.

7. Scirpus subterminàlis Torr.

Water Club-rush. Fig. 652.

Scirpus subterminalis Torr. Fl. U. S. **1**: 47. 1824.

Perennial(?), aquatic, culms slender, terete, nodulose, 0.3–1 m. long. Leaves slender, channeled, 15–50 cm. long, 0.5–1 mm. wide; spikelet solitary, terminal, oblong-cylindric, narrowed at each end, several-flowered, 6–10 mm. long, subtended by a subulate erect involucral leaf 1–3 cm. long, thus appearing lateral; scales ovate-lanceolate, acute, membranous, light brown with a green midvein; bristles about 6, downwardly barbed, as long as the achene or shorter; stamens 3; style 3-cleft to about the middle; achene obovoid, 3-angled, dark brown, smooth, rather more than 2 mm. long, obtuse, abruptly beaked by the slender base of the style.

In ponds and streams or sometimes on their borders, Canadian and Transition Zones; Washington and British Columbia, east to Idaho, Michigan, South Carolina, and Newfoundland. Type locality: near Deerfield, Massachusetts.

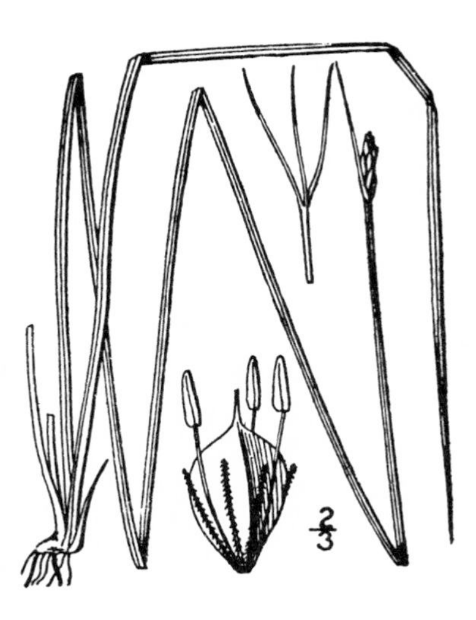

8. Scirpus americànus Pers.

Three-square. Fig. 653.

Scirpus americanus Pers. Syn. **1**: 68. 1805.
Scirpus pungens Vahl, Enum. **2**: 255. 1806.

Perennial by long rootstocks; culms sharply triangular, erect, stiff, 0.3–1.1 m. tall. Leaves 1–3, narrowly linear, keeled, shorter than the culm; spikelets oblong-ovoid, acute, 8–12 mm. long, capitate in a cluster of 1–7, appearing as if lateral; involucral leaf slender, 3–10 cm. long; scales broadly ovate, brown, often emarginate or sharply 2-cleft at the apex, the midvein extended into a subulate awn sometimes 2 mm. long, the margins scarious; bristles 2–6, downwardly barbed, shorter than or equaling the achene; stamens 3; achene obovate, plano-convex, smooth, dark brown, mucronate.

Wet grounds, Sonoran to Canadian Zones; southern California to British Columbia, east to Florida and Newfoundland. Bermuda. Hispaniola. South America. Europe. Type locality: Carolina.

9. **Scirpus ólneyi** A. Gray.
Olney's Bulrush. Fig. 654.

Scirpus olneyi A. Gray, Bost. Journ. Nat. Hist. 5: 238. 1845.

Similar to the preceding species; culms stout, sharply 3-angled with concave sides, 0.5–2 m. tall. Leaves 1–3, 2–13 cm. long, or sheaths sometimes leafless; spikelets capitate in a dense cluster of 5–12, oblong or ovoid-oblong, obtuse, 5–8 mm. long, the involucral leaf short, stout, erect, 1–3 cm. long; scales oval or orbicular, dark brown with a green midvein, emarginate or mucronulate, glabrous; bristles usually 6, slightly shorter than or equaling the achene, downwardly barbed; stamens 2–3; style 2-cleft; achene obovate, plano-convex, brown, mucronate.

Marshes, Sonoran and Transition Zones; Lower California to Oregon, mostly near the coast; Arkansas; Michigan; Atlantic and Gulf Coasts from New Hampshire to Texas and Mexico; Bermuda; Cuba; Hispaniola; Porto Rico. Type locality: Providence, Rhode Island.

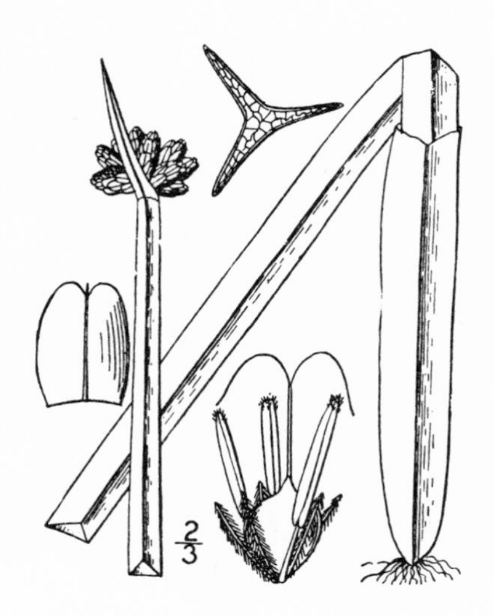

10. **Scirpus nevadénsis** S. Wats.
Nevada Club-rush. Fig. 655.

Scirpus nevadensis S. Wats. Bot. King's Exped. 360. 1871.

Perennial by rather stout rootstocks; culms sometimes clustered, slender, erect, subterete, 1.5–4 dm. high. Leaves convolute, narrowly linear, about 2 mm. wide, shorter than the culm, their margins sometimes roughish; involucral bract 1 (rarely 2), similar to the leaves, 1–7 cm. long; spikelets 2–8 (rarely 1), capitate, sessile, 8–16 mm. long; scales ovate, chestnut-brown, shining, obtuse or acutish; stamens 3; style 2-cleft; achene plano-convex, obovate, shining, abruptly short-tipped or rounded; bristles very short.

Wet grounds, mostly alkaline, Upper Sonoran and Transition Zones; Inyo County, California, north to Washington, east to Wyoming and Saskatchewan. Type locality: Soda Lake, Carson County, Nevada.

11. **Scirpus válidus** Vahl.
American Great Bulrush, Tule. Fig. 656.

Scirpus validus Vahl, Enum. 2: 268. 1806.

Perennial by stout rootstocks, culm stout, terete, smooth, erect, 1–3 m. tall, sheathed below, the upper sheath occasionally extended into a short blade. Involucral leaf solitary, erect, shorter than the umbel, appearing as if continuing the culm; umbel compound, appearing lateral, its primary rays slender, spreading, 1–7 cm. long; spikelets oblong-conic, sessile or some of them peduncled in capitate clusters of 1–5, obtuse or acute, 5–12 mm. long, 3–4 mm. in diameter; scales ovate to suborbicular, slightly pubescent, with a rather strong midvein which is sometimes excurrent into a short tip; bristles 4–6, downwardly barbed, equaling or longer than the achene; stamens 3; style 2-cleft; achene plano-convex, obovate, nearly as long as the scale, gray to brown, abruptly mucronate, a little more than 3 mm. wide.

Ponds and marshes, Sonoran and Transition Zones; southern California to British Columbia, east to Florida and Newfoundland; West Indies. Type locality: Caribbean Islands. Referred by authors to *S. lacustris* L.

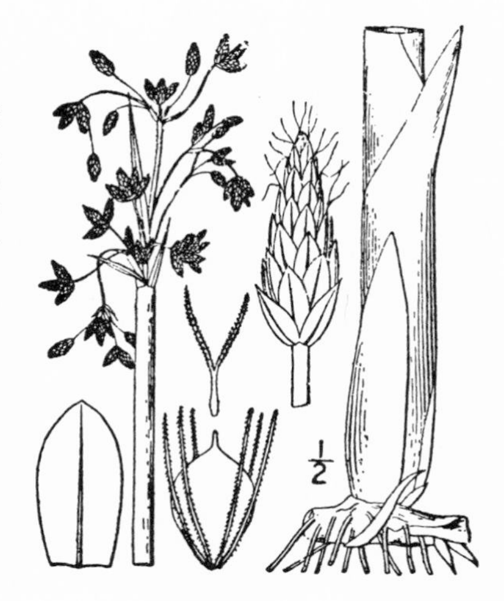

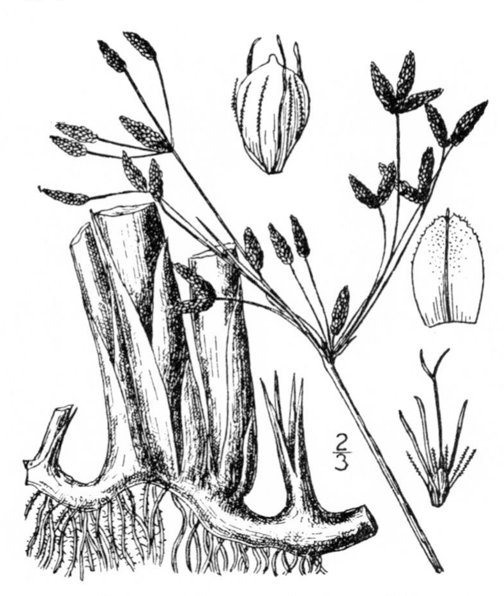

12. Scirpus acùtus Muhl.

Viscid Bulrush, or Tule. Fig. 657.

Scirpus acutus Muhl.; Bigel. Fl. Bost. 15. 1814.
Scirpus lacustris occidentalis S. Wats. Bot. Calif. **2**: 218. 1880.
Scirpus occidentalis Chase, Rhodora **6**: 68. 1904.

Similar to *S. validus*, tall, the culms firmer in texture, the margins of the basal sheaths becoming fibrillose. Involucral leaf shorter than the compound umbél; primary rays rather stiff; bracts viscid at the tip; spikelets clustered in twos to sevens, or solitary, oblong-cylindric, 2 cm. long or less, about 4 mm. thick, acute or bluntish; scales ovate, short-awned, viscid above; style 2-cleft; achene biconvex, obovate, dull, nearly 2 mm. wide, much shorter than the scale.

Ponds and marshes, Sonoran and Transition Zones; southern California to British Columbia, east to Utah, Missouri, New York, and Newfoundland. Type locality: San Diego County, California.

13. Scirpus heterochaètus Chase.

Pale Great Bulrush. Fig. 658.

Scirpus heterochaetus Chase, Rhodora **6**: 70. 1904.

Perennial by rather stout rootstocks; culms slender, sheathed below, 2 m. high or less. Involucral leaf much shorter than the compound umbel; primary rays slender, 1 dm. long or less; bracts acuminate, glabrous; spikelets solitary, ovoid to ellipsoid, acutish; 8–15 mm. long, about 5 mm. thick; scales ovate, glabrous, often erose-margined; style 3-cleft; bristles 2–4, unequal, as long as or shorter than the achene; achene about 2 mm. wide, trigonous, obovate, yellowish, shorter than the scale.

Swan Lake, Klamath County, Oregon; Idaho to Nebraska, New York, Massachusetts, and Vermont. Type locality: Havana, New York.

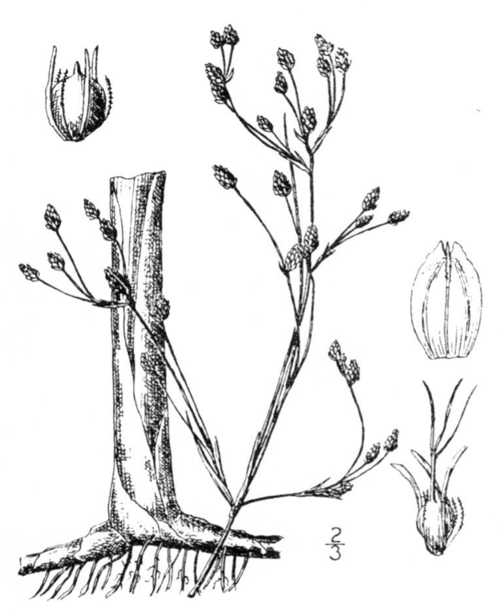

14. Scirpus califórnicus (C. A. Meyer) Britton.

California Bulrush, or Tule. Fig. 659.

Elytrospermum californicum C. A. Meyer, Mém. Acad. St. Petersb. Sav. Etrang. **1**: 201. *pl. 2.* 1830.
Scirpus tatora Kunth, Enum. **2**: 166. 1837.
Scirpus riparius Presl, Rel. Haenk. **1**: 193. 1830.
Scirpus californicus Britton, Trans. N. Y. Acad Sci. **11**: 80. 1892.

Perennial, tall, up to 2 m. high, the leaves reduced to basal sheaths, the involucral bract short and subulate. Umbel compound; spikelets 6–10 mm. long, oblong, acute, stalked, or some of them sessile; scales ovate, brown, the midvein excurrent; bristles shorter than the achene or as long, short-plumose below, barbed above; stamens 2 or 3, style 2-cleft; achene obovate, plano-convex, nearly white, short-pointed, its surface cellular-reticulated.

Swamps and lagoons, Sonoran Zones; Contra Costa County to southern California; east to Louisiana and Florida, south to Guatemala, Peru, and Argentina. Erroneously recorded from New York. Type locality: California.

15. Scirpus pacíficus Britton.
Pacific Coast Bulrush. Fig. 660.

Scirpus pacificus Britton; Parish, Bull. S. Calif. Acad. Sci. **4**: 8. 1905.
Scirpus robustus compactus Davy; Jepson, Fl. W. Mid. Calif. 87. 1901.

Perennial by rootstocks, culm stout, roughish above, trigonous, 3–6 dm. tall. Leaves 10 mm. wide or less, the longer often equaling the culm, those of the involucre 2–5, at least one of them usually longer than the inflorescence; rays of the umbel 1–3, stout, 7 cm. long or less, or wanting and the spikelets densely clustered; spikelets ovoid, obtuse, 1–2 cm. long, about 6 mm. thick; scales brown, ovate, puberulent, tipped with a recurved awn 2–4 mm. long, bristles shorter than the achene; stamens 3; achene orbicular-obovate, cuneate at the base, smooth, somewhat shining, about 3 mm. long.

Saline marshes, along the coast, Sonoran and Transition Zones; southern California to Washington. Recorded by Jepson and by Abrams as *S. robustus* Pursh, of the Atlantic and Gulf Coasts, which it much resembles. Type locality: Long Beach, California.

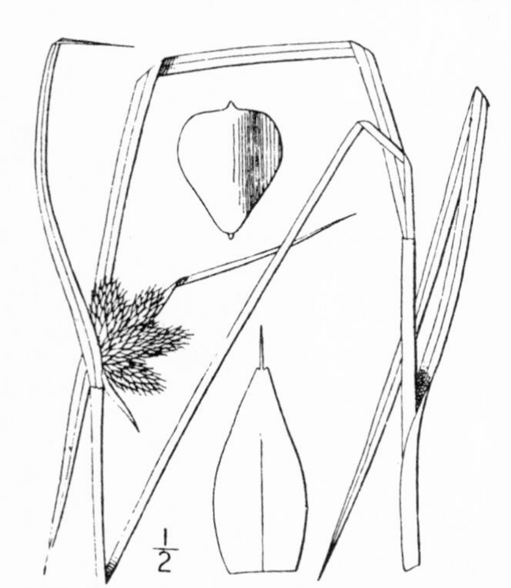

16. Scirpus paludòsus A. Nelson.
Prairie Bulrush. Fig. 661.

Scirpus campestris Britton, in Britton & Brown, Ill. Fl. ed. 1, 1: 267. 1896, not Roth.
Scirpus paludosus A. Nelson, Bull. Torrey Club **26**: 5. 1899.
Scirpus brittonianus Piper, Contr. U. S. Nat. Herb. **11**: 157. 1906.
S. robustus paludosus Fernald, Rhodora **2**: 241. 1900.

Perennial by slender rootstocks, culm slender, smooth, sharply triangular, 3–7 dm. tall. Leaves usually pale green, smooth, shorter than or overtopping the culm, 2–4 mm. wide, those of the involucre 2 or 3, the longer much exceeding the inflorescence; spikelets 3–10 in a dense terminal simple head, oblong-cylindric, mostly acute, 1.5–2.5 cm. long, 5–8 mm. in diameter; scales ovate, membranous, puberulent or glabrous, pale to brown, 2-toothed at the apex, the midvein excurrent into an ascending or spreading awn about 2 mm. long; bristles 1–3, much shorter than the achene or none; style 2-cleft; achene lenticular, obovate or oblong-ovate, mucronulate, yellow-brown.

Saline and alkaline marshes, Sonoran and Transition Zones; Lower California and southern California to British Columbia, east to Texas, Manitoba, Minnesota and on the Atlantic Coast from New Jersey to Quebec. About salt springs in the middle States. Type locality: Granger, Wyoming.

17. Scirpus fluviátilis (Torr.) A. Gray.
River Bulrush. Fig. 662.

Scirpus maritimus fluviatilis Torr. Ann. Lyc. N. Y. **3**: 324. 1836.
Scirpus fluviatilis A. Gray, Man. 527. 1848.

Perennial by rootstocks; culm stout, smooth, triangular with nearly flat sides, 0.9–2 m. tall. Leaves 8–16 mm. wide, smooth, attenuate to a very long tip, those of the involucre 3–5, erect or spreading, often 20 cm. long; spikelets in a terminal umbel, solitary or 2–3 together at the ends of its long spreading or drooping rays, or the central spikelets sessile, oblong-cylindric, acute, 1.6–2.5 cm. long, about 7 mm. in diameter; scales ovate, scarious, puberulent, the midvein excurrent into a curved awn 3–4 mm. long; bristles 6, downwardly barbed, about as long as the achene; style 3-cleft; achene sharply 3-angled, obovoid, rather dull, short-pointed, 4 mm. long.

Tule land and in sand along Calavaras River, near Stockton, and in Sutter County, California; North Dakota to Kansas, Quebec, and New Jersey. Type locality: western New York.

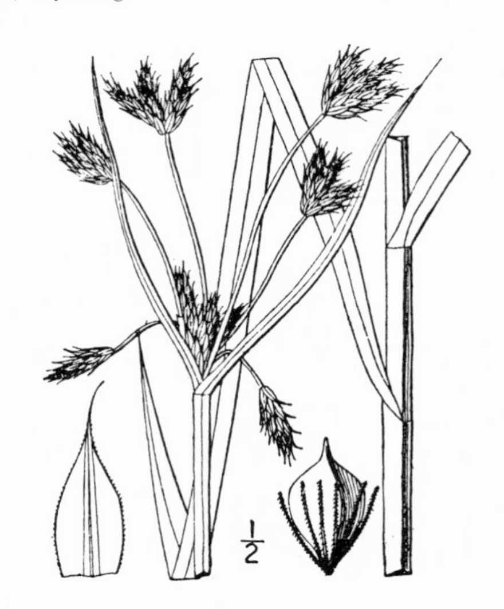

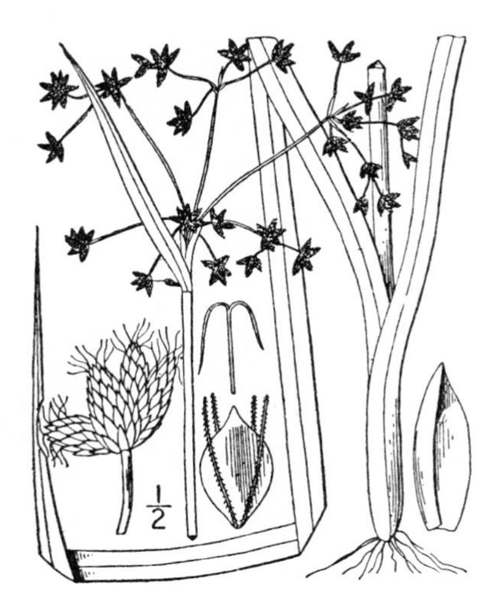

18. Scirpus microcàrpus Presl.
Small-fruited Bulrush. Fig. 663.

Scirpus microcarpus Presl, Rel. Haenk. **1**: 195. 1828.
Scirpus lenticularis Torr. Ann. Lyc. N. Y. **3**: 328. 1836.
Scirpus sylvaticus digynus Boeckl. Linnaea **36**: 727. 1870.

Perennial; culms 0.9–1.6 m. tall, often stout, overtopped by the leaves. Longer leaves of the involucre usually exceeding the inflorescence; spikelets ovoid-oblong, acute, 3–4 mm. long, in capitate clusters at the ends of the usually spreading raylets; scales brown with a green midvein, bristles 4, barbed downwardly nearly or quite to the base, somewhat longer than the achene; stamens 2; style 2-cleft; achene oblong-obovate, nearly white, plano-convex or with a low ridge on the back, pointed.

In swamps and wet woods, Upper Sonoran and Transition Zones; southern California to Alaska, east to Nevada, Colorado, Minnesota, Connecticut, and Newfoundland. Type locality: Nootka Sound, Vancouver.

19. Scirpus pállidus (Britton) Fernald.
Pale Bulrush. Fig. 664.

Scirpus atrovirens pallidus Britton, Trans. N. Y. Acad. Sci. **9**: 14. 1889.
Scirpus pallidus Fernald, Rhodora **8**: 162. 1906.

Perennial, with short rootstocks; culms stout, triangular, 0.9–1.3 m. high. Leaves pale, elongated, 6–15 mm. wide, somewhat nodulose, umbel usually compound; spikelets oblong, numerous in dense capitate clusters 6–8 mm. in diameter; scales pale, ovate, acute, tipped with an awn about one-half as long as the body; bristles 6, downwardly barbed, about as long as the oblong, trigonous achene.

In brooks and swamps, Transition Zone; Oregon and Washington, east to New Mexico, Texas, Kansas, and Manitoba. Reported from Minnesota. Type locality: Indian Territory [Oklahoma]. Recorded from Washington by Piper, and by Watson from Oregon as *S. atrovirens.*

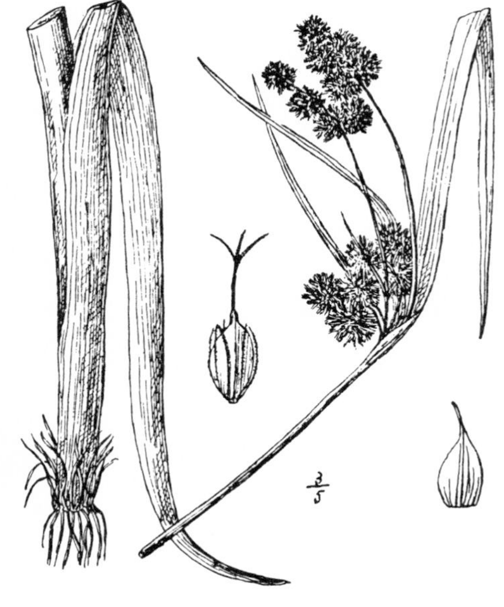

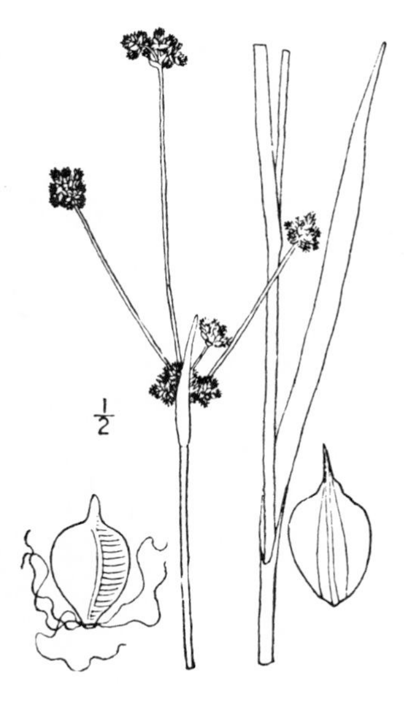

20. Scirpus cóngdoni Britton.
Congdon's Bulrush. Fig. 665.

Scirpus congdoni Britton, Torreya **18**: 36. 1918.

Perennial by short stout rootstocks; culms slender, erect, smooth, somewhat trigonous, 4–5 dm. high, leaf-bearing below the middle. Leaves linear, smooth, acuminate, 5–6 mm. wide, the lower ones 1.5–2 dm. long, the upper shorter; umbel more or less compound, its principal rays 3–6, very slender, 8 cm. long or less; spikelets capitate at the ends of the rays and raylets, several-flowered, the heads 7–10 mm. in diameter; scales brown, ovate to ovate-lanceolate, abruptly acuminate; style 3-cleft; achene pale, trigonous, about 2 mm. long, short-beaked; bristles filiform, curled, slightly roughened, rather longer than the achene.

Madera, Plumas, and Fresno Counties in the Transition Zone, California. Type locality: Upper San Joaquin, Madera County, California. Referred by Watson to *S. atrovirens.*

21. **Scirpus lineàtus** Michx. Reddish Bulrush. Fig. 666.

Scirpus lineatus Michx. Fl. Bor. Am. 1: 32. 1803.

Perennial by stout rootstocks; culms triangular, erect, 0.3–1 m. high, leafy. Leaves 4–8 mm. wide, not exceeding the inflorescence, light green, flat, rough-margined; umbels decompound, the rays very slender, becoming pendulous; spikelets mostly solitary at the ends of the raylets, oblong, obtuse, 6–10 mm. long, about 2 mm. in diameter; scales ovate or oblong, reddish-brown with a green midvein; bristles 6, weak, smooth, entangled, much longer than the achene, equaling the scales or longer; stamens 3; style 3-cleft; achene oblong or oblong-obovoid, pale brown, narrowed at both ends, 3-angled, short-beaked.

In swamps and wet meadows, Transition Zone; Jackson and Josephine Counties, Oregon; Texas to Georgia, Ontario and New Hampshire. Type locality: Carolina.

22. **Scirpus crìniger** A. Gray Fringed Bulrush. Fig. 667.

Scirpus criniger A. Gray, Proc. Am. Acad. 7: 392. 1867.

Perennial by rootstocks; culms slender, roughish above, trigonous, erect, 7 dm. high or less. Leaves linear or linear-lanceolate, flat, rough-margined, 3–6 mm. wide, the basal ones sometimes as long as the culm, those of the culm few, distant, short; spikelets 5–10, oblong, 1 cm. long or less, in a dense terminal head subtended by a few involucral bracts 4–8 mm. long; scales membranous, oblong; bristles about 6, upwardly barbed, much exserted, 3–4 times as long as the achene; filaments elongated; style 3-cleft; achene oblong-obovate, trigonous, short-tipped, 2–3 mm. long.

Upland marshes, Canadian Zone, Mendocino and Eldorado Counties, California, to Josephine County, Oregon. Type locality: Red Mountain, Humbóldt County, California.

8. **DULÍCHIUM** L. C. Richard; Pers. Syn. 1: 65. 1805.

A tall perennial sedge with terete hollow-jointed culms, leafy to the top, the lower leaves reduced to sheaths. Spikes axillary, peduncled, simple or compound. Spikelets 2-ranked, flat, linear, falling away from the axis at maturity(?), many-flowered. Scales 2-ranked, carinate, conduplicate, decurrent on the joint below. Flowers perfect. Perianth of 6–9 retrorsely barbed bristles. Stamens 3. Style 2-cleft at the summit, persistent as a beak on the summit of the achene. Achene linear-oblong. [Name said to be from *dulcichimum*, a Latin name for some sedge.]

A monotypic genus of eastern North America.

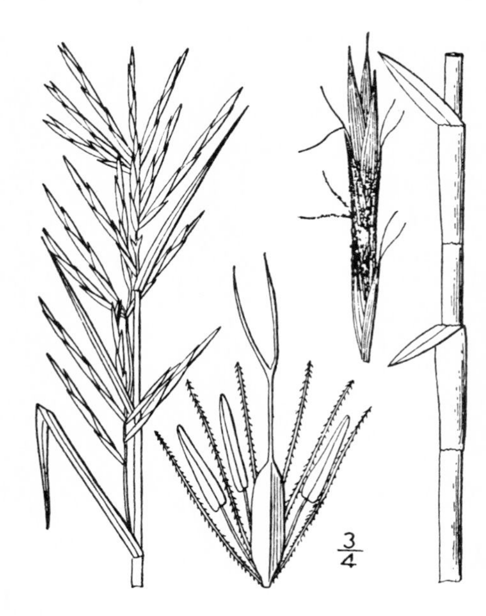

1. Dulichium arundinàceum (L.) Britton.
Dulichium. Fig. 668.

Cyperus arundinaceus L. Sp. Pl. 44. 1753.
Cyperus spathaceus L. Syst. ed. 12, **2**: 275. 1767.
Dulichium spathaceum Pers. Syn. **1**: 65. 1805.
Dulichium arundinaceum Britton, Bull. Torrey Club **21**: 29. 1894.

Culm stout, 0.3–1 m. tall, erect. Leaves numerous, flat, 2–8 cm. long, 4–8 mm. wide, spreading or ascending, the lower sheaths bladeless, brown toward their summits; peduncles 4–25 mm. long; spikelets narrowly linear, spreading, 1–2.5 cm. long, about 2 mm. wide, 6–12-flowered; scales lanceolate, acuminate, strongly several-nerved, appressed, brownish; bristles of the perianth rigid, longer than the achene; style long-exserted, persistent.

Swamps, Transition Zone; Humboldt County, California, to Oregon, British Columbia, east to Montana, Nebraska, Texas, Florida, and Nova Scotia. Type locality: Virginia.

7. HEMICÀRPHA Nees & Arn. Edinb. New Phil. Journ. **17**: 263. 1834.

Low tufted mostly annual sedges, with erect or spreading, almost filiform culms and leaves, and terete small terminal capitate or solitary spikelets subtended by a 1–3-leaved involucre. Scales spirally imbricated, deciduous, all subtending perfect flowers. Perianth of a single hyaline sepal (bract?) between the flower and the rachis of the spikelet; bristles none. Stamens 1–3. Style 2-cleft, deciduous, not swollen at the base. Achene oblong, turgid or lenticular. [Greek, in allusion to the single sepal.]

About 5 species, natives of temperate and tropical regions. Type species, *Hemicarpha isolepis* Nees.

Scales short-tipped or mucronate. — 1. *H. micrantha.*
Scales awned.
  Scales rhombic-obovate, abruptly awned. — 2. *H. aristulata.*
  Scales lanceolate to ovate-lanceolate, narrowed above into the awn. — 3. *H. occidentalis.*

1. Hemicarpha micrántha (Vahl) Pax.
Common Hemicarpha. Fig. 669.

Scirpus micranthus Vahl, Enum. **2**: 254. 1806.
Hemicarpha subsquarrosa Nees in Mart. Fl. Bras. **21**: 61. 1842.
Hemicarpha micrantha Pax in E. & P. Nat. Pflf. **22**: 105. 1887.

Annual; glabrous; culms compressed, grooved, 2–10 cm. long, mostly longer than the setaceous smooth leaves. Spikelets ovoid, many-flowered, obtuse, about 2 mm. long; involucral leaves usually much exceeding the spikelets; scales brown, obovate, with a short blunt spreading or recurved tip; stamen 1; achene obovate-oblong, obtuse, mucronulate, little compressed, light brown.

Moist soil, Transition Zone; Tuolumne County, California, north to Washington; east to Florida and New Hampshire; West Indies; Mexico; continental tropical America. Type locality: (?) South America.

2. Hemicarpha aristulàta (Coville) Smyth.

Awned Hemicarpha. Fig. 670.

Hemicarpha micrantha aristulata Coville, Bull. Torrey Club 21: 36. 1894.

Hemicarpha aristulata Smyth, Trans. Kans. Acad. Sci. 16: 163. 1899.

Hemicarpha intermedia Piper; Piper & Beattie, Fl. Palouse Reg. 36. 1901.

Annual, similar to the preceding species; culms 2 dm. high or less, longer than the setaceous leaves. Leaves of the involucre 1–3, sometimes 2 cm. long; spikelets ovoid, 4–8 mm. long; scales rhombic-obovate, brown, rather abruptly contracted into a subulate, spreading or somewhat recurved awn about as long as the body; style short; achene oblong or narrowly obovate, black.

Moist soil, Tehama County, California; Almota, Washington; Texas to Kansas, Colorado and Wyoming. Type locality: Texas.

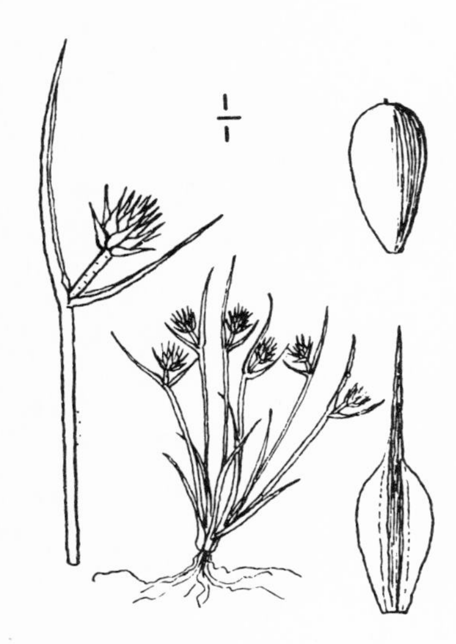

3. Hemicarpha occidentàlis A. Gray.

Western Hemicarpha. Fig. 671.

Hemicarpha occidentalis A. Gray, Proc. Am. Acad. 7: 391. 1867.

Scirpus occidentalis Clarke, Kew Bull. Add. Ser. 8: 30. 1908, not Chase 1904.

Annual, diminutive, pale green; culms filiform, tufted, 2–4 cm. high. Leaves filiform, as long as the culm or shorter, those of the involucre similar to the lower ones, much exceeding the spikelets; spikelets subglobose, about 2 mm. in diameter, several- to many-flowered; scales lanceolate to ovate-lanceolate, tapering into a spreading awn nearly as long as the body.

Yosemite Valley, California; Falcon Valley, Washington. Type locality: Yosemite Valley, California. *Cyperus inflexus* Muhl. has been mistaken for this plant.

9. RYNCHÓSPORA Vahl, Enum. 2: 229. 1806.

Leafy sedges, mostly perennial by rootstocks, with erect 3-angled or terete culms, narrow, flat or involute leaves, and ovoid-oblong or fusiform, variously clustered spikelets. Scales thin, 1-nerved, imbricated all around, usually mucronate by the excurrent midvein, the lower empty. Upper flowers imperfect, the lower perfect. Perianth of 1–20 (mostly 6) upwardly or downwardly barbed or scabrous bristles, wanting in some species. Stamens commonly 3. Style 2-cleft, 2-toothed, or rarely entire. Achene lenticular or swollen, not 3-angled, smooth or transversely wrinkled, capped by the persistent base of the style (tubercle), or in some species by the whole style. [Greek, referring to the beak-like tubercle.]

About 200 species, widely distributed, most abundant in warm regions. Besides the following, some 48 other species occur in the southern and eastern United States. Type species, *Rynchospora aurea* Vahl.

Spikelets nearly white; bristles 9–15. 1. *R. alba.*
Spikelets brown; bristles about 6. 2. *R. capitellata.*

1. Rynchospora álba (L.) Vahl.
White Beaked-rush. Fig. 672.

Schoenus albus L. Sp. Pl. 44. 1753.
Rynchospora alba Vahl, Enum. 2: 236. 1806.

Pale green; rootstocks short; culms slender or filiform, glabrous, 1.5–5 dm. tall. Leaves bristle-like, 0.5–1 mm. wide, shorter than the culm; spikelets in 1–4 dense corymbose clusters, narrowly oblong, acute at both ends, 4–6 mm. long; scales ovate or ovate-lanceolate, nearly white, acute; bristles 9–15, downwardly barbed, slender, about as long as the achene and tubercle; achene obovate-oblong, smooth, pale brown, lenticular; tubercle triangular-subulate, flat, one-half as long as the achene.

In bogs, Transition and Canadian Zones; Yosemite Valley, California, north to Alaska, east to Idaho, Minnesota, Kentucky, Florida, and Newfoundland. Europe and Asia. Type locality: Europe.

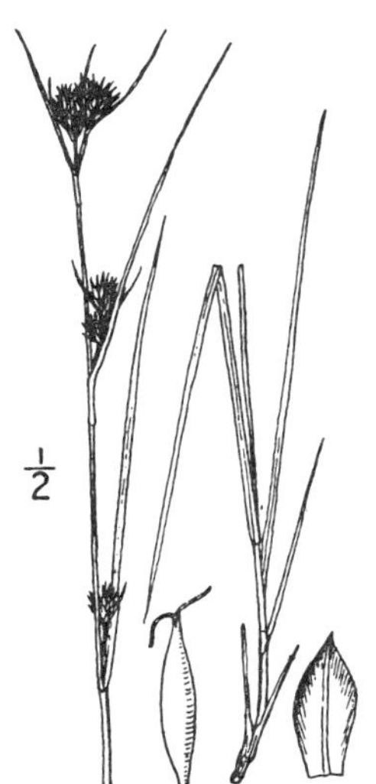

2. Rynchospora capitellàta (Michx.) Vahl.
Brownish Beaked-rush. Fig. 673.

Schoenus capitellatus Michx. Fl. Bor. Am. 1: 36. 1803.
Rynchospora capitellata Vahl, Enum. 2: 235. 1805.

Rootstocks slender; culms trigonous, smooth, slender, erect, 2–10 dm. high. Leaves flat, 1–3 mm. wide, shorter than the culm; spikelets glomerate in few or several axillary dense clusters, oblong, narrowed at both ends, 3–4 mm. long; scales brown, lanceolate; bristles about 6, downwardly barbed or rarely smooth, about as long as the achene and tubercle; achene obovate, brown, smooth; tubercle subulate.

Sphagnum bog near Brookings, Oregon (*Morton E. Peck*); Michigan to Arkansas, Màine, and Georgia. The Oregon specimen is provisionally referred to this species, but the specimens collected are young, and achenes have not been seen.

10. MARÍSCUS (Hall.) Zinn, Cat. Hort. Goett. 79. 1757.

Perennial leafy sedges, the spikelets oblong or fusifòrm, few-flowered, variously clustered. Scales imbricated all around, the lower empty, the middle ones mostly subtending imperfect flowers, the upper usually fertile. Perianth none. Stamens 2 or sometimes 3. Style 2–3-cleft, deciduous from the summit of the achene, its branches sometimes 2–3-parted. Achene ovoid to globose, smooth or longitudinally striate. Tubercle none. [Latin, referring to the swamp or marsh habitat of some species.]

About 40 species, natives of tropical and temperate regions. Besides the following, two others occur in the eastern United States. Type species, *Schoenus mariscus* L.

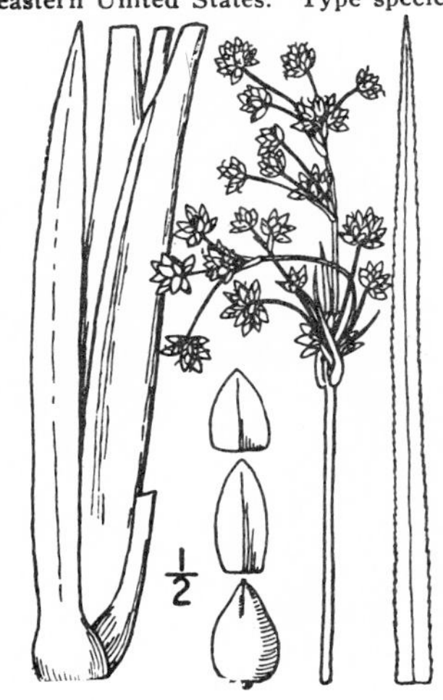

1. Mariscus califórnicus (S. Wats.) Britton.
Saw-grass. Fig. 674.

Cladium mariscus californicum S. Wats. Bot. Calif. 2: 224. 1880.

Culm stout, 1–2 m. high, obtusely 3-angled. Leaves very long, glabrous, 7–10 mm. wide, the margins spinulose-serrulate; umbels numerous, decompound, forming a large panicle; spikelets glomerate at the ends of the raylets, narrowly ovoid, acute, about 5 mm. long; scales acute, striate; uppermost scale subtending a perfect flower; stamens 2; achene ovoid, pointed, smooth, shining, about 2 mm. long.

Swamps, Sonoran Zones; San Bernardino and Inyo Counties and recorded from near San Gabriel, Los Angeles County; Nevada; Arizona; northern Mexico. Type locality: near San Gabriel, California.

11. SCHOÈNUS L. Sp. Pl. 42. 1753.

Perennial sedges, with stiff, tufted culms, basal, narrow or semiterete leaves, the one- to few-flowered spikelets clustered, capitate or paniculate, the clusters subtended by 1 bract or few. Scales imbricated in 2 series, the lower ones empty, the upper subtending flowers. Perianth of 3–6, scabrous or plumose bristles. Stamens 3. Style slender, scarcely enlarged at the base, 3-cleft, deciduous. Achene trigonous; tubercle none. [Greek, a rush.]

Sixty species or more, mostly natives of the Old World, the following typical.

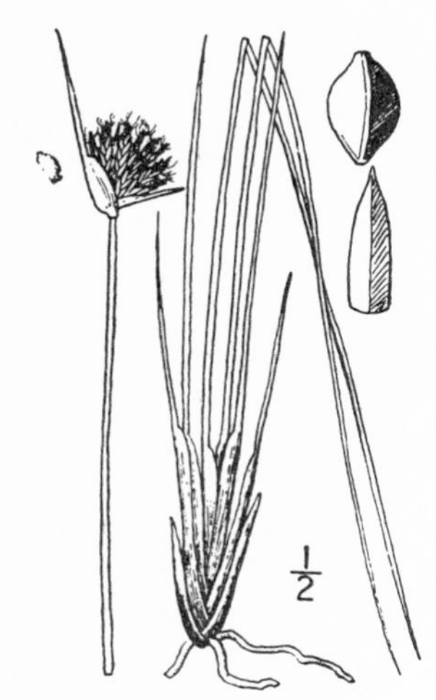

1. Schoenus nìgricans L.
Black Sedge. Fig. 675.

Schoenus nigricans L. Sp. Pl. 43. 1753.

Glabrous; culms 2–7 dm. tall, mostly longer than the leaves. Leaves similar to the culm, stiff, sharp-pointed, semiterete, 0.5–1.5 mm. thick, their bases dark brown or nearly black, shining; involucre of 2 bracts, one of them elongated, sometimes 8 cm. long; spikelets about 1 cm. long, in a dense, terminal, capitate cluster, 5–8-flowered, compressed, their scales dark chestnut-brown, or nearly black, lanceolate, carinate, acuminate, somewhat shining; perianth-bristles 6, plumose; achene ellipsoid, white, shining, about 2 mm. long; shorter than the bristles.

About Hot Springs, San Bernardino County, California; Florida; Bahamas; Cuba; Europe. Type locality: Europe.

12. KOBRESIA Willd. Sp. Pl. 4: 205. 1805.

Slender arctic and mountain sedges, with erect culms leafy below, and few-flowered spike lets variously clustered. Scales of the spikelets 1-flowered, the lower usually pistillate, and the upper staminate. Stamens 3. Perianth-bristles or perigynium wanting. Ovary oblong, narrowed into a short style; stigmas 2–3, linear. Achene obtusely angled, sessile. [In honor of Von Kobres, a naturalist of Augsburg.]

About 30 species, widely distributed in arctic and mountainous regions. Type species, *Kobresia scirpina* Willd.

1. Kobresia bellàrdi (All.) Degland.
Bellard's Kobresia. Fig. 676.

Carex bellardi All. Fl. Ped. 2: 264. *pl. 92, f. 2.* 1785.
Kobresia scirpina Willd. Sp. Pl. 4: 205. 1805.
Kobresia bellardi Degland, in Loisel. Fl. Gall. 2: 626. 1807.

Culms very slender, 1–4.5 dm. tall, longer than the very narrow leaves. Old sheaths fibrillose, brown; margins of the leaves more or less revolute; spike subtended by a short bract, or bractless, densely flowered or sometimes interrupted below, 1.5–3 cm. long, 3–4 mm. in diameter; achenes rather less than 2 mm. long, 1 mm. thick, appressed.

Alpine ridge of the Wallowa Mountains, Oregon; British Columbia to Alaska, east to Greenland, south on the Rocky Mountains to New Mexico. Europe and Asia. Type locality: Europe.

13. CÁREX* L. Sp. Pl. 972. 1753.

Grass-like sedges, perennial by rootstocks. Culms mostly triangular, often strongly phyllopodic or aphyllopodic. Leaves three-ranked, the upper (bracts) elongate or short, and subtending the spikes of flowers, or wanting. Plants monoecious or sometimes dioecious. Spikes one to many, either wholly pistillate, wholly staminate, androgynous or gynaecandrous, sessile or peduncled, the base of the peduncle often with a perigynium-like or spathe-like organ (cladoprophyllum) surrounding it. Flowers solitary in the axils of scales (glumes). Perianth none. Staminate flowers of three (or rarely two) stamens, the filaments filiform. Pistillate flowers of a single pistil, with a style and two or three stigmas. Style either (1) jointed with the apiculate-tipped or rounded achene and withering and at length deciduous; or (2) continuous with achene, persistent, indurated and not withering. Achene triangular, lenticular or plano-convex, completely surrounded by the perigynium or rarely rupturing it in ripening. Rhacheola occasionally developed. [The classical Latin name, probably derived from the Greek, to cut, referring to the sharp leaf-blades of some species].

Species more than 1000, all continents, but least developed in the tropics. Type species, *Carex pulicaris* L.

Spike one.
  Stigmas two; achenes lenticular.
    Perigynia many-striate. — 1. NARDINAE.
    Perigynia nerveless ventrally. — 5. CAPITATAE.
  Stigmas three; achenes triangular.
    Perigynia strongly inflated, sessile, not becoming reflexed; pistillate scales persistent. — 2. INFLATAE.
    Perigynia not inflated.
      Pistillate scales deciduous; perigynia stipitate, at least the lower reflexed at maturity.
        Perigynia dark-tinged; style jointed with achene. — 3. ATHROCHLAENAE.
        Perigynia straw-color; style continuous with achene. — 39. PAUCIFLORAE.
      Pistillate scales persistent; perigynia not reflexed at maturity.
        Perigynia rounded and beakless at apex, many-nerved. — 19. POLYTRICHOIDEAE.
        Perigynia not rounded at apex, beakless or beaked.
          Perigynia lanceolate. — 6. CIRCINATAE.
          Perigynia not lanceolate.
            Perigynia finely striate. — 1. NARDINAE.
            Perigynia 2-keeled, otherwise nerveless.
              Culms dioecious or nearly so. — 22. SCIRPINAE.
              Spikes androgynous.
                Perigynia pubescent or puberulent, 2.5–4 mm. long. — 21. FILIFOLIAE.
                Perigynia glabrous, 5–7 mm. long. — 20. FIRMICULMES.
Spikes more than one.
  Stigmas two; achenes lenticular.
    Lateral spikes sessile, short (rarely somewhat elongated), the terminal one androgynous or gynaecandrous.
      Perigynia not white puncticulate.
        Rootstocks long-creeping, the culms arising singly or few together.
          Spikes densely aggregated into a globose or ovoid head, appearing like one spike. — 6. FOETIDAE.
          Spikes, at least the lower, distinct.
            Perigynia plano-convex, sharp-edged, the beak obliquely cut, bidentulate in age. — 7. DIVISAE.
            Perigynia much flattened, wing-margined at least above, the beak deeply bidentate. — 8. ARENARIAE.
        Cespitose or rootstocks short-creeping.
          Spikes androgynous.
            Perigynia with body abruptly contracted into the beak.
              Spikes few (usually ten or less); perigynia green or tinged with reddish brown. — 10. MUHLENBERGIANAE.
              Spikes numerous; perigynia yellowish or brownish.
                Perigynia yellowish; opaque part of leaf-sheaths usually transversely rugulose. — 11. MULTIFLORAE.
                Perigynia brownish; opaque part of leaf-sheaths not transversely rugulose. — 12. PANICULATAE.
            Perigynia with body tapering into the beak. — 13. STENORHYNCHAE.
          Spikes gynaecandrous.
            Perigynia at most thin-edged.
              Perigynia spreading or ascending at maturity. — 14. STELLULATAE.
              Perigynia appressed. — 15. DEWEYANAE.
            Perigynia narrowly to broadly wing-margined. — 16. OVALES.
      Perigynia white puncticulate. — 17. CANESCENTES.
    Lateral spikes elongated, peduncled or sessile; terminal spike staminate, or if rarely gynaecandrous, the lateral spikes peduncled.
      Lowest bract long-sheathing, pistillate spikes not very many-flowered. — 26. BICOLORES.
      Lowest bract sheathless or, if more or less sheathing, the pistillate spikes very many-flowered.
        Achenes not constricted in the middle. — 35. ACUTAE.
        Achenes constricted in the middle; scales sharp-pointed, three-nerved. — 36. CRYPTOCARPAE.
  Stigmas three; achenes triangular.
    Lateral spikes very numerous, sessile, short; perigynia plano-convex, strongly lacerate-winged. — 9. MACROCEPHALAE.
    Lateral spikes not very numerous, usually not sessile; perigynia not lacerate-winged.
      Perigynia pubescent or at least puberulent.
        Pistillate spikes with from few to about 25 perigynia.
          Bracts of non-basal pistillate spikes sheathless or very nearly so. — 23. MONTANAE.
          Bracts of non-basal pistillate spikes sheathing.
            Bracts more or less purplish-tinged, the blades absent or rudimentary. — 24. DIGITATAE.

* Text (except zonal distribution) contributed by and illustrations (with a few exceptions) prepared under the direction of Mr. KENNETH K. MACKENZIE.

Bracts with blades well developed.
Terminal spike on many of the culms wholly pistillate; the culms largely dioecious. 22. SCIRPINAE.
Terminal spike staminate; culms never dioecious. 25. TRIQUETRAE.
Pistillate spikes very many-flowered. 37. HIRTAE.
Perigynia glabrous.
Style jointed with achene, at length withering and deciduous.
Lowest bract strongly sheathing.
Rootstocks long-creeping; scales purplish-tinged. 27. PANICEAE.
Rootstocks not long-creeping; scales greenish to reddish brown or blackish tinged.
Scales greenish or light reddish-brown-tinged.
Perigynia with beak not bidentate with stiff teeth.
Pistillate spikes erect on stiff peduncles. 28. LAXIFLORAE.
Pistillate spikes slender on slender peduncles, the lower drooping. 29. DEBILES.
Perigynia with beak bidentate with stiff teeth. 38. EXTENSAE.
Scales dark reddish-brown- to blackish-tinged. 30. FRIGIDAE.
Lowest bract sheathless or very short-sheathing.
Pistillate spikes narrowly cylindric, elongate. 32. ANOMALAE.
Pistillate spikes ovoid, oblong or linear.
Freely long stoloniferous, not cespitose (perigynia glaucous green; pistillate spikes drooping). 33. LIMOSAE.
Cespitose. 34. ATRATAE.
Style continuous with achene, indurated and persistent.
Perigynia nerveless except for marginal nerves. 31. TRACHYCHLAENAE.
Perigynia several- to many-nerved.
Perigynia or leaves or both pubescent. 37. HIRTAE.
Perigynia and leaves not pubescent.
Perigynia coarsely ribbed. 40. PHYSOCARPAE.
Perigynia finely and closely ribbed. 41. PSEUDO-CYPEREAE.

1. NARDINAE.

Represented by one species in our range. 1. *C. hepburnii.*

2. INFLATAE.

Scales 1-nerved; perigynia ovoid, inflated, tapering at apex; staminate part of spike scarcely conspicuous; achenes 1.25 mm. long. 2. *C. engelmannii.*
Scales 3-nerved; perigynia broadly ovoid, strongly inflated, very abruptly short-beaked; staminate part of spike conspicuous; achenes 2 mm. long. 3. *C. breweri.*

3. ATHROCHLAENAE.

Densely cespitose; leaf-blades involute, 1 mm. wide or less; staminate flowers few; perigynia erect until full maturity. 4. *C. pyrenaica.*
Short stoloniferous; leaf-blades flat, 1.5 mm. wide or more; staminate flowers conspicuous; perigynia early spreading or deflexed. 5. *C. nigricans.*

4. CIRCINATAE.

Represented by one species in our range. 6. *C. circinata.*

5. CAPITATAE.

Represented by one species. 7. *C. capitata.*

6. FOETIDAE.

Represented by one species in our range. 8. *C. vernacula.*

7. DIVISAE.

Rootstocks slender, light brownish; culms obtusely triangular, normally smooth; leaf-blades narrowly involute.
Perigynia long-beaked; heads dioecious or nearly so. 9. *C. douglasii.*
Perigynia short-beaked; heads androgynous. 10. *C. stenophylla.*
Rootstocks stout; culms acutely triangular, normally rough above.
Perigynia chestnut, thick, the beak about one-fifth as long as the body. 11. *C. simulata.*
Perigynia blackish in age, the beak one-third to one-half as long as the body.
Scales very dark chestnut-brown, shining; perigynia 3.5–4.5 mm. long, polished, scarcely hyaline at apex. 12. *C. pansa.*
Scales lighter-colored, dull; perigynia 3–4 mm. long, dull, strongly hyaline at orifice. 13. *C. praegracilis.*

8. ARENARIAE.

Represented by one species in our area. 14. *C. siccata.*

9. MACROCEPHALAE.

Represented by one species. 15. *C. macrocephala.*

10. MUHLENBERGIANAE.

Densely cespitose; head ovoid; capitate; body of perigynium serrulate to the middle. 16. *C. hoodii.*
Rootstocks elongate; head linear, interrupted; body of perigynium usually serrulate on upper margin only.
Scales obtuse- to short-cuspidate, half as long as the spreading perigynia. 17. *C. vallicola.*
Scales strongly cuspidate, concealing the appressed perigynia. 18. *C. tumulicola.*

11. MULTIFLORAE.

Scales strongly hyaline-margined; sheaths little or not at all cross-rugulose.
Perigynia sharp-margined above, brownish black at maturity. 19. *C. alma.*
Perigynia narrowly thin-winged to base, straw-color at maturity. 20. *C. stenoptera.*
Scales not or but little hyaline-margined; sheaths normally more or less cross-rugulose; perigynia straw-color, yellowish or tawny at maturity, sharp-margined to base.
Perigynia 3–4.5 mm. long; pistillate scales (except lowest) acute to short-awned; beak of perigynium shorter than body.
Ligule conspicuous, as long as wide; scales brownish-tinged; perigynia strongly nerved ventrally. 21. *C. densa.*

Ligule very short; scales reddish-brown-tinged.
Perigynia flat and nerveless or nearly so ventrally, the body sparingly serrulate above, contracted into beak. 22. *C. vicaria.*
Perigynia low-convex and strongly nerved ventrally, the body strongly serrulate above, abruptly contracted into the beak. 23. *C. brevi-ligulata.*
Perigynia 2.25–3.25 mm. long; pistillate scales strongly awned; beak of perigynium about length of body.
Leaf-blades shorter than culm, little roughened. 24. *C. dudleyi.*
Leaf-blades exceeding culms, very rough. 25. *C. vulpinoidea.*

12. Paniculatae.

Leaf-blades 1–2.5 mm. wide; head little interrupted; perigynia 2–2.75 mm. long, shining, not concealed by the scales; sheaths not copper-tinged at mouth. 26. *C. diandra.*
Leaf-blades 2.5–6 mm. wide; head interrupted, compound; perigynia 3–4 mm. long, dull, concealed by the scales; sheaths copper-tinged at mouth. 27. *C. cusickii.*

13. Stenorhynchae.

Perigynia 3–4 mm. long, the beak much shorter than the body; scales strongly dark-tinged.
Leaves clustered at base; sheaths not green and white-mottled dorsally; culms slender; perigynia spongy at base. 28. *C. jonesii.*
Leaves not clustered at base; sheaths green and white-mottled dorsally; culms thick at base, strongly aphyllopodic.
Sheaths not cross-rugulose ventrally, concave or truncate and not prolonged at mouth; perigynia strongly spongy at base, sharply bidentate; stigmas long; culms weak. 29. *C. nervina.*

Sheaths cross-rugulose ventrally, convex and prolonged at mouth; perigynia little spongy at base, very shallowly bidentate; stigmas short; culms not weak. 30. *C. neurophora.*
Perigynia 4–6 mm. long, the beak about equaling or longer than the body; scales not dark-tinged. 31. *C. stipata.*

14. Stellulatae.

Spikes in a small (6–10 mm. long) densely capitate brownish-black head. 32. *C. illota.*
Spikes more or less widely separate, not brownish-black.
Beak of perigynium with few weak serrulations, the body broadest near middle. 33. *C. laeviculmis.*
Beak of perigynium strongly serrulate, the body broadest near base.
Beak of perigynium bluntly bidentate, one-fourth to one-third length of body, the ventral false suture obsolete or inconspicuous; scale obtuse, half length of body of perigynium. 34. *C. interior.*
Beak of perigynium sharply bidentate, the ventral false suture conspicuous.
Culms obtusely triangular; beak of perigynium chestnut-brown-tipped; pistillate scales obtuse or obtusish, chestnut-brown-tinged with broad shining margins and apex, rounded on back and not sharply keeled, the midvein obscure at apex.
Spikes widely separate, the terminal long-clavate; perigynia 3.5–4 mm. long. 35. *C. ormantha.*
Spikes approximate, the terminal short-clavate; perigynia 3.75–4.5 mm. long. 36. *C. phyllomanica.*
Culms sharply triangular; beak of perigynium reddish-brown-tipped; pistillate scales obtusish to cuspidate, yellowish-brown-tinged, the margins and apex narrow, opaque or dull whitish, the back keeled with the sharp midvein which is prominent to apex.
Perigynia lanceolate, the beak more than half length of body obscurely serrulate; leaf-blades 0.75–2 mm. wide. 37. *C. angustior.*
Perigynia ovate, the beak much shorter than the body strongly serrulate; leaf-blades 1.5–2.5 mm. wide. 38. *C. cephalantha.*

15. Deweyanae.

Perigynia shallowly bidentate, 3.5–4 mm. long, the beak about one-third the length of the body. 39. *C. leptopoda.*
Perigynia deeply bidentate, 4–4.5 mm. long, the beak about one-half the length of the body. 40. *C. bolanderi.*

16. Ovales.

Lower bracts leaflet-like, much exceeding head.
Beak of perigynium hyaline at orifice, obliquely cleft, bidentulate; lowest bract not appearing like a continuation of the culm. 41. *C. athrostachya.*
Beak of perigynium ferruginous at apex, bidentate; lowest bract appearing like a continuation of the culm. 42. *C. unilateralis.*
Lower bracts not leaflet-like, from much shorter than to slightly exceeding head.
Sheaths green-striate opposite blades, except at mouth; perigynia and scales light green. 43. *C. feta.*
Sheaths white-hyaline opposite blades.
Sheaths strongly prolonged upward at mouth opposite blade in a quickly ruptured very membranaceous appendage; perigynia and scales greenish. 44. *C. fracta.*
Upper sheaths (at least) concave or truncate at mouth opposite blades, not quickly ruptured.
I. Beak of perigynium flat and serrulate to the often strongly bidentate tip.
Perigynia subulate to lanceolate, at least two and one-half times as long as wide.
Perigynia subulate, the margin at the base almost obsolete. 45. *C. crawfordii.*
Perigynia lanceolate, the margin conspicuous to the base. 46. *C. scoparia.*
Perigynia lanceolate-ovate or broader, at most twice as long as wide.
Perigynia narrowly to broadly ovate, 3–4 mm. long.
Perigynia brownish; spikes closely aggregated, rounded at base.
Culms rough above; perigynia with beak flattened to tip. 47. *C. bebbii.*
Culms smooth; perigynia with beak subterete towards tip. 67. *C. subfusca.*
Perigynia green; spikes contiguous to widely separate, usually clavate at base. 48. *C. tenera.*
Perigynia averaging 4–8.5 mm. long; or if shorter with orbicular bodies.
Scales little if at all reddish-brown-tinged. 49. *C. brevior.*
Scales strongly reddish-brown- or chestnut-brown-tinged.
Perigynia membranaceous, flattened, concave-convex, with wide thin margins conspicuously crinkled dorsally; scales with light slender midvein. 50. *C. straminiformis.*
Perigynia subcoriaceous, plano-convex and thick with narrower margins not crinkled dorsally; scales with light 3-nerved center. 51. *C. multicostata.*
II. Beak of perigynium terete towards apex; the upper 1–3 mm. smooth or nearly so (except in *C. subfusca* and *C. preslii*).
Perigynia lanceolate, 5.5–8.5 mm. long, three to five times as long as wide.
Scales about length and width of perigynia and largely concealing them. 52. *C. petasata.*

Scales shorter and narrower than perigynia, the latter therefore conspicuous in the spikes.
Spikes about two or three, not capitate; many scales little more than half length of perigynia at maturity. 53. *C. davyi.*
Spikes six to twelve, capitate; scales equaling bodies of perigynia. 54. *C. specifica.*

Perigynia ovate or broader, or if lanceolate shorter than 5.5 mm. in length.
A. Perigynia appressed, nearly or entirely covered by scales, the beaks not conspicuous in the spikes.
Culms slender; head flexuous or moniliform.
Scales light reddish-brown-tinged. 55. *C. praticola.*
Scales chestnut-brown-tinged. 56. *C. piperi.*
Culms and head stiff; spikes approximate.
Culms 1–6 dm. high, the leaves not bunched near the base; blades 2–3 mm. wide, flat; beak of perigynium not hyaline at orifice. 57. *C. tracyi.*
Culms 1–3 dm. high, the leaves bunched near the base; blades 1.5–2 mm. wide, more or less involute; beak of perigynium hyaline at orifice.
Perigynium oblong-ovate, rather sharply margined. 58. *C. phaeocephala.*
Perigynium linear-lanceolate, very narrowly margined, boat-shaped. 59. *C. leporinella.*

B. Upper part of perigynia conspicuous in the spikes, not covered by the scales.
(1) Perigynia with thinnish submembranaceous walls.
* Perigynia thin, save where distended by achene.
Perigynia 3.5–5 mm. long; culms slender.
Perigynia lance-ovate, very narrowly margined, spreading. 60. *C. microptera.*
Perigynia ovate, strongly margined, appressed. 61. *C. festivella.*
Perigynia 4.5–6 mm. long; culms low, ascending or decumbent. 62. *C. nubicola.*
** Perigynia plano-convex, abruptly short-beaked with conspicuous raised nerves on inner face at maturity.
Spikes densely capitate; perigynia conspicuously hyaline-tipped. 63. *C. abrupta.*
Spikes more or less strongly separate; perigynia reddish-tipped. 64. *C. mariposana.*

(2) Perigynia plano-convex with thick firm walls; nerveless or inconspicuously nerved on inner face (except in *C. Harfordii* and *C. Montereyensis*).
Perigynia very small, 2.5–3.5 mm. long.
Margins of perigynia entire (or very obscurely subserrulate). 65. *C. integra.*
Margins of perigynia strongly serrulate.
Culms slender; leaf-blades deep green, 1–2.5 mm. wide; perigynia loosely appressed or spreading-ascending. 66. *C. teneraeformis.*
Culms stiff; leaf-blades light green, 2–3.5 mm. wide; perigynia closely appressed. 67. *C. subfusca.*
Perigynia 3.5 mm. or more in length.
Perigynia beaks at tip and scales brownish or blackish-tinged. 68. *C. pachystachya.*
Perigynia beaks at tip and scales reddish- or reddish-brown-tinged.
Perigynia beaks and scales little reddish-brown-tinged; lower bracts at least strongly amplectant. 69. *C. amplectens.*
Perigynia beaks and scales strongly reddish-brown-tinged; lower bracts not strongly amplectant.
Perigynia strongly nerved ventrally.
Sterile shoots not conspicuous; lower bladeless sheaths short; culms slender; scales mostly acute, reddish brown. 70. *C. harfordii.*
Sterile shoots numerous, elongate; lower bladeless sheaths very long; culms very slender; scales cuspidate or short-awned, yellowish brown. 71. *C. montereyensis.*
Perigynia nerveless or obscurely nerved ventrally.
Perigynia obscurely nerved dorsally.
Culms slender; lower spikes more or less strongly separate; scales not conspicuously hyaline-margined; perigynia 3.5–4 mm. long. 72. *C. preslii.*
Culms stiff; spikes densely capitate; scales conspicuously white-hyaline-margined; perigynia 5 mm. long. 73. *C. pachycarpa.*
Perigynia with prominent raised nerves dorsally.
Spikes strongly capitate; blades averaging 2.5–3 mm. wide; culms 3.5–12 dm. high. 74. *C. sub-bracteata.*
Spikes not capitate, the head slender; blades averaging 1.5–2 mm. wide; culms 1–6 dm. high.
Culms slender; leaves not clustered, the blades 0.5–2 dm. long; spikes with 10–20 perigynia. 75. *C. gracilior.*
Culms stiff; leaves clustered, the blades short; spikes with few perigynia. 76. *C. paucifructus.*

17. Canescentes

Spikes androgynous; pergynia unequally bi-convex. 77. *C. disperma.*
Spikes gynaecandrous; perigynia plano-convex.
Perigynia broadest near middle; beak short, smooth or moderately serrulate.
Plant glaucous; leaf-blades 2–4 mm. wide; spikes many-flowered; perigynia appressed-ascending, scarcely beaked, with emarginate or entire orifice. 78. *C. canescens.*
Plant not glaucous; leaf-blades 1–2.5 mm. wide; spikes fewer-flowered; perigynia loosely spreading, distinctly beaked with minutely bidentate orifice. 79. *C. brunnescens.*
Perigynia ovate, broadest near base; beak conspicuous, strongly serrulate. 80. *C. arcta.*

18. Phyllostachyae.

Represented by one species in our area. 81. *C. saximontana.*

19. Polytrichoideae.

Represented by one species. 82. *C. leptalea.*

20. Firmiculmes.

Not stoloniferous; culms smooth, terete; leaf-blades 1.5 mm. wide; bracts very long-awned. 83. *C. multicaulis.*
Stoloniferous; culms very rough, triangular; leaf-blades 2–3.5 mm. wide; bracts not very long-awned. 84. *C. geyeri.*

21. Filifoliae.

Perigynia essentially beakless, 2 mm. long, tipped by the long persistent exserted style; pistillate scales with narrow dull white margins. 85. *C. exserta.*
Perigynia distinctly beaked, 2.5–3.5 mm. long; pistillate scales with very broad bright white hyaline margins. 86. *C. filifolia.*

22. Scirpinae.

Culms phyllopodic with 6–10 well-developed leaves; perigynia blackish-tinged at least at apex.
 Perigynia sparsely pubescent towards apex only.
  Perigynium body broadly oval, 3 mm. long, very abruptly beaked. 87. *C. gigas.*
  Perigynium body lanceolate, 4 mm. long, tapering into the beak. 88. *C. scabriuscula.*
 Perigynia strongly pubescent, obovoid-triangular. 89. *C. pseudo-scirpoidea.*
Culms aphyllopodic with 3–5 well-developed leaves; perigynia lighter in color.
 Perigynia lanceolate, flattish, 4 mm. long. 90. *C. stenochlaena.*
 Perigynia broader, triangular, 3 mm. long or less. 91. *C. scirpoidea.*

23. Montanae.

Basal spikes not developed. 92. *C. inops.*
Basal spikes present.
 Perigynia finely many-ribbed as well as strongly 2-keeled.
  Scales purplish-tinged, obtuse to cuspidate; body of perigynium globose; staminate spikes many-flowered; basal pistillate spikes on elongated very slender peduncles. 93. *C. globosa.*
  Scales reddish-brown-tinged, cuspidate or long-awned; body of perigynium oval; staminate spikes few-flowered; basal pistillate spikes on short erect peduncles. 94. *C. brainerdii.*
 Perigynia strongly 2-keeled, otherwise ribless.
  Bract of lowest non-basal pistillate spike leaflet-like, exceeding culm; if at all colored, purplish-brown-tinged at base.
   Perigynia 2.5–3 mm. long, the beak 0.25–0.75 mm. long, shallowly bidentate. 95. *C. brevipes.*
   Perigynia 3–4.5 mm. long, the beak longer, bidentate. 96. *C. rossii.*
  Bract of lowest non-basal pistillate spike squamiform and shorter than culm; or, if longer, auriculate and strongly reddish-brown-tinged at base. 97. *C. brevicaulis.*

24. Digitatae.

Represented by one species in our range. 98. *C. concinnoides.*

25. Triquetrae.

Represented by one species in our range. 99. *C. triquetra.*

26. Bicolores.

Perigynia short-tapering at apex, straw-colored, 2.5–3.75 mm. long. 100. *C. salinaeformis.*

Perigynia rounded or truncate at apex, orange-colored or white-pulverulent, smaller.
 Mature perigynia whitish, ellipsoid, not fleshy or translucent; scales appressed. 101. *C. hassei.*
 Mature perigynia orange or brownish, broader, fleshy, translucent; scales spreading. 102. *C. aurea.*

27. Paniceae.

Perigynia beakless or nearly so; bract sheaths short; plant glaucous; leaf-blades narrow, involute. 103. *C. livida.*
Perigynia strongly beaked; bract sheaths long; plant not glaucous; leaf-blades broad, flat. 104. *C. californica.*

28. Laxiflorae.

Represented by one species in our range. 105. *C. hendersonii.*

29. Debiles.

Lowest bract sheathless or very short-sheathing; spikes approximate; perigynia glabrous, strongly ribbed or nerved.
 Perigynia ovoid, 3.5–4.25 mm. long, abruptly short-beaked, the sides several-nerved. 106. *C. flaccifolia.*
 Perigynia ovoid-lanceolate, 4–5 mm. long, tapering into the beak, the sides strongly ribbed. 107. *C. whitneyi.*
Lowest bract long-sheathing; lower spikes more or less strongly separate; perigynia faintly nerved or nerveless.
 Perigynia and leaf-blades hairy; spikes oblong or short-oblong, the upper approximate.
  Perigynia 4–5 mm. long, round-tapering at base, finely many-nerved. 108. *C. gynodynama.*
  Perigynia 3.5–4 mm. long, tapering at base, 2-keeled, otherwise nerveless. 109. *C. hirtissima.*
 Perigynia glabrous or puberulent above; leaf-blades very sparsely hairy; pistillate spikes linear, widely separate. 110. *C. mendocinensis.*

30. Frigidae.

Perigynia compressed-triangular or slightly flattened, the beak bidentulate; scales obtusish, the midvein not prominent at apex.
 Perigynia somewhat compressed-triangular, 3.5 mm. long or less; spikes widely separate, the staminate one usually strongly overtopping the uppermost pistillate one; lower bracts with conspicuous blades. 111. *C. lemmonii.*
 Perigynia compressed-triangular, longer; uppermost pistillate spikes bunched, usually little exceeded by the staminate one; blades of bracts usually less developed.
  Pistillate spikes oblong; scales reddish-brown. 112. *C. luzulina.*
  Pistillate spikes linear-oblong; scales dark-tinged. 113. *C. ablata.*
Perigynia strongly flattened, the beak bidentate; scales sharp-pointed with midvein prominent to apex.
 Perigynia glabrous; scales smooth; bract sheaths strongly enlarged upwards; leaf-blades very leathery. 114. *C. luzulaefolia.*
 Perigynia sparsely hairy; scales more or less hairy; bract sheaths scarcely enlarged upwards; leaf-blades not leathery. 115. *C. fissuricola.*

31. Trachychlaenae.

Represented by one species in our area. 116. *C. spissa.*

32. Anomalae.

Represented by one species in our area. 117. *C. amplifolia.*

33. Limosae.

Represented by one species in our area. 118. *C. limosa.*

34. Atratae.

Terminal spike staminate or sometimes with perigynia in the middle.
Basal sheaths not filamentose.
Culms few-leaved, purplish-tinged at base, the lower culm leaves much reduced.
Pistillate scales with midvein excurrent as a long, slender awn; perigynia finely nerved. 119. *C. macrochaeta.*
Pistillate scales with prominent excurrent midvein; perigynia nerveless or nearly so on inner face. 120. *C. spectabilis.*
Culms many-leaved, clothed at base with dried-up leaves of previous year, brownish-tinged at base.
Perigynia flattened. 121. *C. tolmiei.*
Perigynia round in cross-section, many-nerved. 122. *C. raynoldsii.*
Basal sheaths filamentose. 123. *C. bifida.*
Terminal spike gynaecandrous, i. e., the terminal flowers pistillate.
Culms aphyllopodic, strongly purplish-red at base, the lower sheaths filamentose. 124. *C. buxbaumii.*
Culms phyllopodic.
Spikes 3–5, not oblong-cylindric; perigynia 2.5–5 mm. long, nerveless or obscurely nerved on face, dull green to brownish black, the walls not papery.
Perigynia not papillate-roughened.
Spikes contiguous, sessile or nearly so, forming a dense head; scales lanceolate, strongly exceeding perigynia; culms stiff, erect. 125. *C. helleri.*
Lower spike or spikes peduncled, usually distant, erect or nodding; scales wider, shorter than or about equaling perigynia; culms more slender.
Scales with midvein largely obsolete; mature perigynia 3.5–4.5 mm. long, wider on either side than achene, the latter on stipe of nearly its own length; sheaths not purplish-tinged ventrally. 126. *C. epapillosa.*
Scales with prominent midvein; mature perigynia 3.5 mm. long, narrower on either side than achene, the latter much longer than its stipe; sheaths normally purplish-tinged ventrally. 127. *C. heteroneura.*
Perigynia papillate-roughened, especially on upper margins.
Perigynia slightly inflated and sub-triangular, not strongly compressed at maturity. 128. *C. atrosquama.*
Perigynia strongly compressed.
Spikes contiguous, sessile or short-peduncled, forming a dense head; culms stiff, erect, scales black with conspicuous white-hyaline upper margins. 129. *C. albo-nigra.*
Lower spike or spikes strongly peduncled, usually distant, erect or nodding; culms not stiff. 130. *C. viridior.*
Spikes 6–10, oblong-cylindric; perigynia 5 mm. long, lightly 3-nerved, light-green, the walls papery; scales much shorter than perigynia. 131. *C. mertensii.*

35. Acutae.

Flowering culms arising from the center of previous year's tuft of leaves and surrounded at base with dried-up leaves of previous year.
Lower sheaths of flowering culms not breaking and becoming filamentose.
Strongly stoloniferous, the culms arising one to few together, low; lowest bract normally much shorter than inflorescence; scales with obsolete or slender midvein.
Dried first-year leaf-blades at base of fertile culm stiff, rigid and conspicuous, concealing the culms; fertile culm leaves all blade-bearing, the lower sheaths not purplish or hispidulous dorsally. 132. *C. scopulorum.*
Dried first-year leaf-blades at base of fertile culms much desiccated, not stiff, rigid or conspicuous, and not concealing the culms; lowest fertile culm leaves (of season's growth) not blade-bearing, the lower sheaths purplish and more or less strongly hispidulous dorsally.
Lower bladeless sheaths of fertile culms (of season's growth) inconspicuous and largely hidden by the old dead leaves; lower sheaths sparingly hispidulous; culms sharply triangular and rough above.
Perigynia coriaceous, spreading, olive-green; scales very short, largely hidden by perigynia. 133. *C. accedens.*
Perigynia membranaceous, straw-color or blackish-tinged; scales conspicuous.
Perigynia plano-convex or slightly bi-convex, appressed-ascending. 134. *C. gymnoclada.*
Perigynia mostly deeply concave ventrally, convex dorsally, curved outwardly and spreading. 135. *C. campylocarpa.*
Lower bladeless sheaths of fertile culm (of season's growth) elongated and conspicuous; lower sheaths strongly rough-hispidulous; culms narrowly wing-angled and very serrulate. 136. *C. prionophylla.*
Culms taller, less stiff, in larger clumps; lowest bract usually equaling or exceeding inflorescence; scales with slender midvein or broader light-colored center.
Perigynia strongly nerved ventrally, the nerves raised.
Perigynia coriaceous, sessile or nearly so, the beak bidentate; plants strongly stoloniferous. 137. *C. nebraskensis.*
Perigynia membranaceous, more or less slenderly stipitate, the beak entire; plants cespitose.
Perigynia substipitate, sub-orbicular, minutely papillate-roughened; scales deciduous. 138. *C. paucicostata.*
Perigynia strongly stipitate, ovate.
Perigynia yellowish-green, ribbed, papillate-roughened; scales deciduous. 139. *C. hindsii.*
Perigynia light-green or in age glaucous green, nerved, very minutely granular; scales long persistent. 140. *C. kelloggii.*
Perigynia nerveless ventrally or with obscure impressed nerves.
Perigynia turgid and scales divaricate at maturity. 141. *C. aperta.*
Perigynia not turgid; scales appressed.
Perigynia ovate-orbicular, olive-green, scarcely 2 mm. long. 142. *C. interrupta.*
Perigynia oblong-ovate or lanceolate, greenish straw-color, longer.
Sheaths colored ventrally at mouth; lower pistillate spikes subcernuous on long peduncles; scales in age whitened at tip. 143. *C. sitchensis.*
Sheaths not colored ventrally at mouth; lower pistillate spikes not nodding; scales not whitened at tip. 144. *C. aquatilis.*
Lower sheaths of season's growth of flowering culms breaking and becoming filamentose.
Beak of perigynium bidentate, hispidulous between teeth; scales strongly rough awned. 145. *C. barbarae.*

Beak of perigynium entire or emarginate, not hispidulous between teeth; scales not rough awned.
Lower culm sheaths (of the year's growth) strongly yellowish-brown tinged, sharply keeled; culms stout; leaf-blades 6–12 mm. wide. 146. *C. schottii.*
Lower culm sheaths (of the year's growth) purplish-tinged, not sharply keeled; culms more slender; leaf-blades narrower. 147. *C. senta.*
Some or all of the flowering culms arising laterally and not enveloped at base by previous year's tuft of leaves; lower sheaths filamentose.
Cespitose or with short ascending stolons; lowest bract from very short to about length of inflorescence; foliage light green; lower sheaths strongly filamentose, smooth or nearly so dorsally.
Perigynia plano-convex, broadly ovate to oblong-lanceolate, 3–4 mm. long. 148. *C. nudata*
Perigynia unequalling bi-convex, ovoid-orbicular, scarcely 2.5 mm. long. 149. *C. suborbiculata.*
Loosely cespitose with long slender horizontal stolons; lowest bract exceeding inflorescence; foliage deep-green; lower sheaths little filamentose, hispidulous dorsally.
Perigynia sub-orbicular or broadly obovate. 150. *C. eurycarpa.*
Perigynia oblanceolate or narrowly obovate. 151. *C. oxycarpa.*

36. Cryptocarpae.

Perigynia dull, straw-color or light-brown, slightly granular; lower sheaths of sterile shoot not filamentose. 152. *C. lyngbyei.*
Perigynia shining, brown, smooth; lower sheaths of sterile shoots strongly filamentose. 153. *C. obnupta.*

37. Hirtae.

Perigynia hairy, the teeth 1.5 mm. long or less.
Beak of perigynium obliquely cut, shallowly bidentate at maturity; foliage pubescent; staminate scales long ciliate. 154. *C. yosemitana.*
Beak of perigynium deeply bidentate; staminate scales at most erose.
Foliage not pubescent; teeth of perigynium beak short.
Lowest bract strongly sheathing; fertile culms phyllopodic with many leaves, the sheaths not breaking and becoming filamentose. 155. *C. oregonensis.*
Lowest bract not sheathing; fertile culms aphyllopodic with few leaves, the sheaths breaking and becoming filamentose. 156. *C. lanuginosa.*
Leaf-blades flat, more than 2 mm. wide.
Leaf-blades involute, 2 mm. wide or less. 157. *C. lasiocarpa.*
Sheaths and under surface of leaf-blades hairy; teeth of perigynium beak conspicuous. 158. *C. sheldonii.*
Perigynia glabrous, the teeth 1.5 mm. long or more; sheaths and leaf-blades hairy. 159. *C. atherodes.*

38. Extensae.

Represented by one species in our range. 160. *C. viridula.*

39. Pauciflorae.

Represented by one species in our range. 161. *C. pauciflora.*

40. Physocarpae.

Lower perigynia not reflexed; bracts moderately exceeding the spikes.
Perigynia ascending; lower sheaths more or less strongly filamentose; rootstocks short-creeping with short ascending stolons; leaves sparingly nodulose.
Perigynia 6–8 mm. long, contracted into the beak. 162. *C. vesicaria.*
Perigynia 8–10 mm. long, tapering into the beak. 163. *C. exsiccata.*
Perigynia spreading at maturity; lower sheaths not filamentose; densely cespitose, sending out long horizontal stolons; leaves strongly nodulose. 164. *C. rostrata.*
Lower perigynia reflexed; bracts many times exceeding the spikes. 165. *C. retrorsa.*

41. Pseudo-cypereae.

Perigynia suborbicular in cross-section, more or less inflated; teeth of perigynium beak erect, 0.5–1 mm. long. 166. *C. hystricina.*
Perigynia obtusely triangular, scarcely inflated, closely enveloping the achene; teeth of perigynium beak 1.5–2 mm. long, recurved or spreading. 167. *C. comosa.*

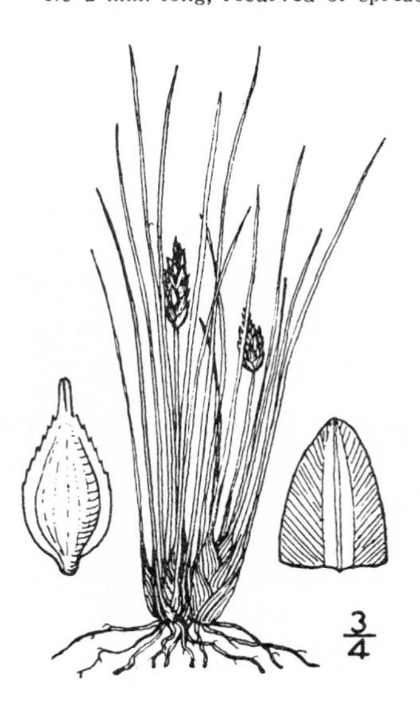

1. **Carex hépburnii** Boott.
Hepburn's Sedge. Fig. 677.

Carex hepburnii Boott in Hook. Fl. Bor. Am. **2**: 209. *pl. 207.* 1840.
Carex nardina hepburnii (Boott) Kukenth. in Engler Pflanzenreich **4**[20]: 70. 1909.

Very densely cespitose, the culms 2–15 cm. high. Leaf-blades filiform; sheaths strongly hyaline-margined above, abruptly contracted into the blades; spike solitary, bractless, linear-oblong, 5–12 mm. long, androgynous, densely flowered with 1–10 perigynia; scales ovate, acutish, the midvein conspicuous; perigynia 3 mm. long, ascending, biconvex, elliptic-ovate, not inflated, the walls thin, glabrous, striate, stipitate, the beak hyaline tipped, in age bidentulate, obliquely cut and with a closed suture dorsally; achenes usually triangular, or sometimes lenticular; stigmas 3 or sometimes 2.

Dry alpine slopes, Hudsonian and Arctic-Alpine Zones; Alberta and Colorado to British Columbia and Washington. Type locality: Rocky Mountains.

2. Carex engelmànnii Bailey.
Engelmann's Sedge. Fig. 678.

Carex engelmannii Bailey, Proc. Am. Acad. **22**: 132. 1886.
Carex paddoensis Suksdorf, Allg. Bot. Zeitschr. **12**: 43. 1906.

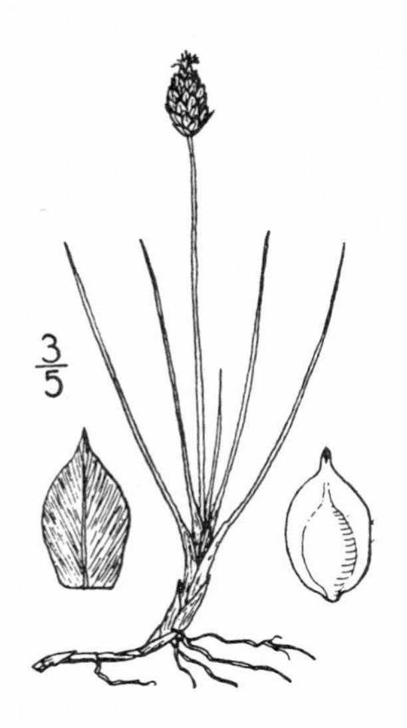

Rootstocks slender, tough, elongate, the culms 5–20 cm. high, smooth, brownish-tinged at base. Leaf-blades filiform, 3–15 cm. long, 0.3–0.5 mm. wide, smooth or nearly so, the sheaths loose, many-striate; spike solitary, bractless, ovoid, 10–15 mm. long, 6–10 mm. wide, androgynous, densely flowered, the lower three-quarters pistillate with 15–40 ascending perigynia; scales 1-nerved, acute to cuspidate, all except lower shorter than perigynia; perigynia inflated, 4.5–5 mm. long, 2.25 mm. wide, sessile, rounded at base, the walls very thin, glabrous, nerveless, tapering at apex into a minute smooth beak, 0.5 mm. long, which is hyaline tipped, and obliquely cut, in age bidentulate; achenes triangular, 1.25 mm. long; stigmas 3.

Isolated stations on high mountain summits, Hudsonian and Arctic-Alpine Zones; Colorado, Wyoming, Washington, and California (Tulare County). Type locality: Colorado Springs, Colorado.

3. Carex brèweri Boott.
Brewer's Sedge. Fig. 679.

Carex breweri Boott. Ill. Carex **4**: 142. *pl. 455.* 1867.

Rootstocks tough, elongate, the culms 1–2.5 dm. high, smooth, brownish-tinged at base. Leaf-blades filiform, but rigid, 5–10 cm. long, less than 1 mm. wide, the sheaths loose, many-striate and scarious; spike solitary, bractless, ovoid, 10–15 mm. long, 6–10 mm. wide, androgynous, densely flowered, the upper third staminate; scales ovate, 3-nerved, short acuminate, narrower and shorter than perigynia; perigynia strongly inflated, broadly ovoid, 5 mm. long, 3.5 mm. wide, sessile, rounded at base, the walls very thin, glabrous, nerveless, abruptly beaked, the beak smooth, 0.5–1 mm. long, obliquely cut, at length bidentulate; achenes triangular, 2 mm. long; stigmas 3.

High alpine peaks, Arctic-Alpine Zone; Washington and Oregon and south in California on the higher summits of the Sierra Nevada to Mount Whitney. Type locality: Mount Shasta, California.

4. Carex pyrenaìca Wahl.
Pyrenaen Sedge. Fig. 680.

Carex pyrenaica Wahl. in Vet. Akad. Handl. Stockholm **24**: 139. 1803.
Callistachys pyrenaica Heuff. Flora **27**: 528. 1844.

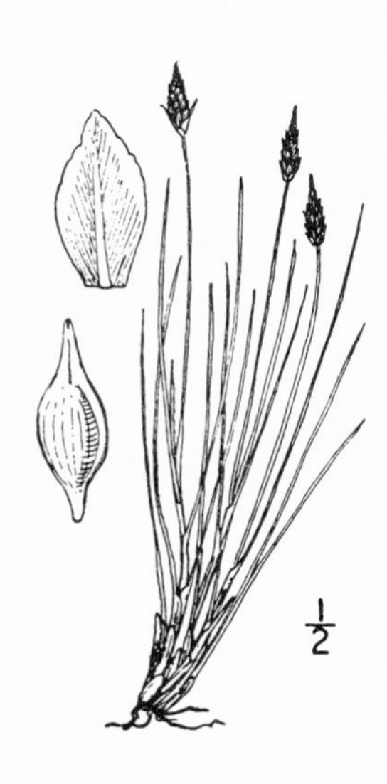

Densely cespitose, the culms 3–20 cm. high, wiry, slender, smooth, longer or shorter than the leaves, brownish at base. Leaves 2–3 to a fertile culm, the blades 2–10 cm. long, 0.25–1 mm. wide, involute, attenuate at apex; spike solitary, androgynous, bractless, narrow, 5–20 mm. long, 3–5 mm. wide, many-flowered, the upper part with a few staminate flowers, the lower with 10–many perigynia; scales ovate, obtusish, shorter than perigynia, chesnut tinged, with narrow hyaline margins, at length deciduous; perigynia 3–4 mm. long, scarcely 1 mm. wide, compressed orbicular in cross-section, brownish, nerveless, membranaceous, glabrous, rounded and strongly stipitate at base, at full maturity spreading or deflexed, tapering into a smooth beak 0.5 mm. long with a dorsal suture and with obliquely cut hyaline orifice becoming bidentulate; achenes triangular and stigmas 3.

Alpine localities, Hudsonian and Arctic-Alpine Zones; Mackenzie and Alaska south to Oregon and Colorado; widely distributed in Eurasia. Type locality: Europe.

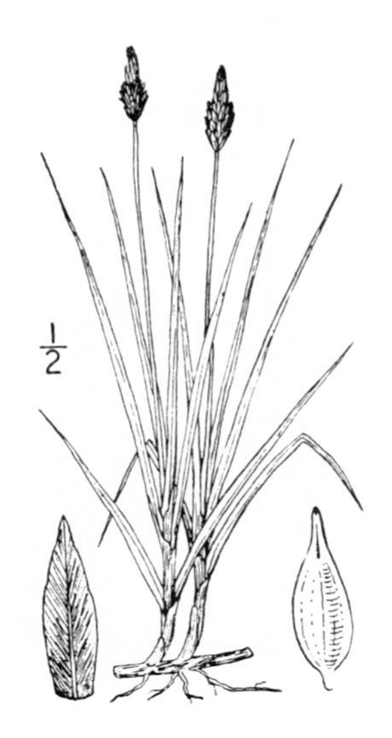

5. Carex nìgricans C. A. Meyer.

Blackish Sedge. Fig. 681.

Carex nigricans C. A. Meyer, Mem. Acad. St.-Petersb. 1: 211. *pl.* 7. 1831.

Rootstocks stout, lignescent, creeping, the culms 5–20 cm. high, stiff, firm, smooth, exceeding leaves, brownish tinged at base. Leaves 4–9 to a fertile culm, the blades 4–10 cm. long, 1.5–3 mm. wide, flat or channeled at base, attenuate at apex; spike solitary, androgynous, bractless, narrow, 8–15 mm. long, 6–9 mm. wide, densely many-flowered, the upper half staminate, the lower with 10–25 perigynia; scales ovate obtuse to acutish, half length of perigynia, dark brown tinged with hyaline margins, soon falling; perigynia 4 mm. long, 1 mm. wide, compressed orbicular and obscurely triangular in cross-section, brownish, nerveless, membranaceous, glabrous, rounded and strongly stipitate at base, at maturity deflexed, tapering into a long smooth beak with a dorsal suture and with an obliquely cut hyaline orifice becoming bidentulate; achenes usually triangular and stigmas 3.

Hudsonian and Arctic-Alpine Zones; Colorado to Tulare County, California, northward to Alberta and Alaska. Type locality: Unalaska.

6. Carex circinàta C. A. Meyer.

Coiled Sedge. Fig. 682.

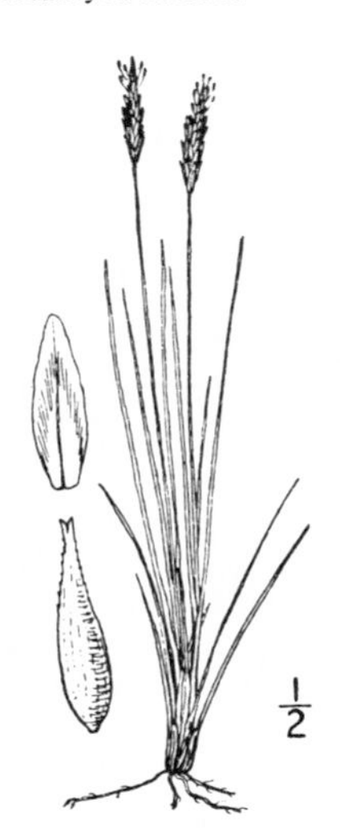

Carex circinata C. A. Meyer, Mem. Acad. St. Petersb. 1: 209. *pl.* 6. 1831.

Cespitose from lignescent slightly creeping rootstocks, the culms 5–15 cm. high, slender, smooth or nearly so, the many-striate lower sheaths conspicuous. Leaves several, densely clustered, the blades involute filiform, 0.5 mm. wide, 2–10 cm. long; spike solitary, androgynous, bractless, narrow, 1.5–2.5 cm. long, 3–4 mm. wide, the upper fourth staminate, the lower with 8–15 perigynia; scales (except lowest) broadly obovate, very obtuse, shorter than perigynia, reddish brown with hyaline margins; perigynia erect-ascending, 5–6 mm. long, 1.5 mm. wide, spindle-shaped, obscurely triangular in cross-section, light straw-colored, lightly nerved, glabrous, membranaceous, attenuate at base, contracted into a minutely serrulate bidentulate beak 1 mm. long, not sutured dorsally, but with an obliquely cut ferruginous orifice; achenes triangular; stigmas 3.

Along the coast, Arctic-Alpine Zone; Unalaska to northwestern Washington (Olympic Mountains). Local. Type locality: Unalaska.

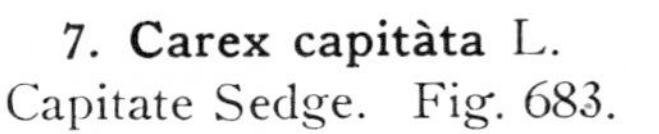

7. Carex capitàta L.

Capitate Sedge. Fig. 683.

Carex capitata L. Syst. Nat. (ed. 10) 1261. 1759.

Rootstocks slightly elongate, the culms cespitose, 1–3.5 dm. high, erect, roughish above, the basal sheaths purplish, sparingly filamentose. Leaf-blades filiform, about 0.5 mm. wide, rigid, stiff, shorter or longer than the culm; spike solitary, ovoid, bractless, densely flowered, androgynous, orbicular or oblong-orbicular, 4–8 mm. long, 3–6 mm. wide, with 6–25 ascending perigynia below and a short but conspicuous cone of staminate flowers above; scales ovate-orbicular, obtuse, shorter and narrower than perigynia, chestnut-brown with broad hyaline margins; perigynia plano-convex, sharp-edged, not inflated, 2–2.5 mm. long and rather narrower, pale green, smooth, nerveless, substipitate, rounded at base, the walls thin, abruptly beaked, the beak smooth, slender, dark-colored, less than 1 mm. long, with closed dorsal suture, conspicuously hyaline tipped, in age bidentulate; achenes lenticular; stigmas 2.

Arctic-Alpine Zone; Greenland to Alaska, and also occurring very locally on mountain summits southward; New Hampshire; Alberta; Nevada; Tulare County, California; Mexico. Type locality: Europe.

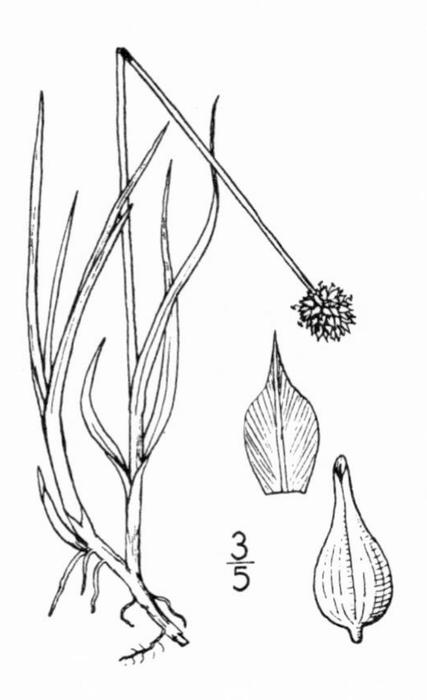

8. **Carex vernácula** Bailey.

Vernacular or Native Sedge. Fig. 684.

Carex foetida W. Boott in S. Wats. Bot. Calif. **2**: 232. 1880. Not All.
Carex vernacula Bailey, Bull. Torrey Club **20**: 417. 1893.
Carex foetida vernacula (Bailey) Kükenth. in Engler Pflanzenreich **4**[20]: 115. 1909.

Rootstocks creeping, lignescent, dark-colored, the culms in small clumps, 1–2 dm. high, smooth, usually exceeding the leaves. Leaf-blades 5–12 cm. long, 2–4 mm. wide, stiff; spikes few to several, androgynous, aggregated into a very dense nearly orbicular head, about 1 cm. in diameter, the spikes not distinguishable and the staminate flowers inconspicuous; scales ovate, rather wider and longer or shorter than perigynia, brown, sharp-pointed; perigynia ovoid, 3.5–4 mm. long, 1.5 mm. wide, spreading, plano-convex, membranaceous, scarcely inflated, not margined, rounded and short stipitate at base, more or less nerved, tapering into the smooth obliquely cut, somewhat bidentate beak, one-third length of body, with closed dorsal suture; achenes lenticular; stigmas 2.

Alpine slopes, Hudsonian and Arctic-Alpine Zones; Washington south to Mount Whitney, California, and east to Wyoming and Colorado. Type locality: western United States.

9. **Carex doùglasii** Boott.

Douglas's Sedge. Fig. 685.

Carex douglasii Boott in Hook. Fl. Bor. Am. **2**: 213. *pl. 214.* 1840.
Carex nuttallii Dewey, Am. Journ. Sci. **43**: 92. *pl. 2, f. 97.* 1842.
Carex meckii Dewey, Am. Journ. Sci. II. **24**: 48. 1857.
Carex irrasa Bailey, Bot. Gaz. **25**: 271. 1898.

Rootstocks slender but tough, brownish, the culms 6–30 cm. high, smooth, obtusely triangular, light brownish at base. Leaf-blades 1–2.5 mm. wide, flat or channelled at base; heads dioecious or nearly so; spikes few–many, densely aggregated; lower 1–several bracts developed; staminate spikes linear-elliptic, 8–15 mm. long, 2.5–4 mm. wide, the scales straw-colored or brownish, pointed; pistillate spikes wider, the scales ovate to lanceolate, concealing the perigynia, yellowish brown with broad hyaline margins and lighter center; perigynia appressed-ascending, lanceolate, 4 mm. long, 1.75 mm. wide, plano-convex, coriaceous, sharp-edged, lightly nerved ventrally, strongly nerved dorsally, rounded and short stipitate at base, tapering into the strongly serrulate beak nearly 2 mm. long, its apex hyaline, obliquely cut and with closed suture dorsally, in age bidentate; achenes lenticular; styles elongate; stigmas 2.

Dry or alkaline soil, Upper Sonoran and Arid Transition Zones; Manitoba to New Mexico, westward to California and British Columbia. Type locality: Northwest Coast.

10. **Carex stenophýlla** Wahl.

Involute-leaved Sedge. Fig. 686.

Carex stenophylla Wahl. in Vet. Acad. Handl. Stockholm **24**: 142. 1803.
Carex eleocharis Bailey, Mem. Torrey Club **1**: 6. 1889.

Rootstocks long-creeping, slender, brownish, the culms arising one–few together, 5–20 cm. high, slender, obtusely triangular, normally smooth, light brownish at base, exceeding leaves. Leaf-blades 1.5 mm. wide at the base, involute above; spikes few, densely aggregated into a head 7–15 mm. long; lower bracts little developed, shorter than head; scales slightly exceeding perigynia, ovate, acute, brownish with broad white hyaline margins; perigynia few to a spike, ascending, coriaceous, plano-convex, 3 mm. long, 1.75 mm. wide, blackish at maturity, the body ovate, shining, more or less strongly nerved at least dorsally, rounded at base, sharp edged but not margined, contracted into a short serrulate beak which is obliquely cut, and with closed suture dorsally, the apex at length bidentulate; achenes lenticular; styles short; stigmas 2.

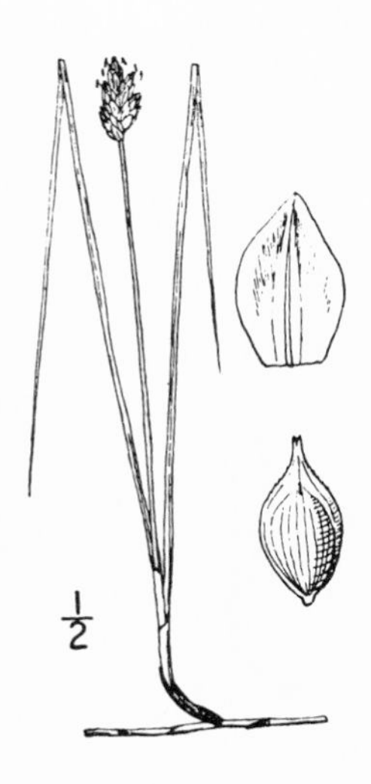

Dry soil, Canadian Zone; Manitoba to Yukon, south to Iowa, New Mexico, Utah, and eastern Oregon. Type locality: Europe.

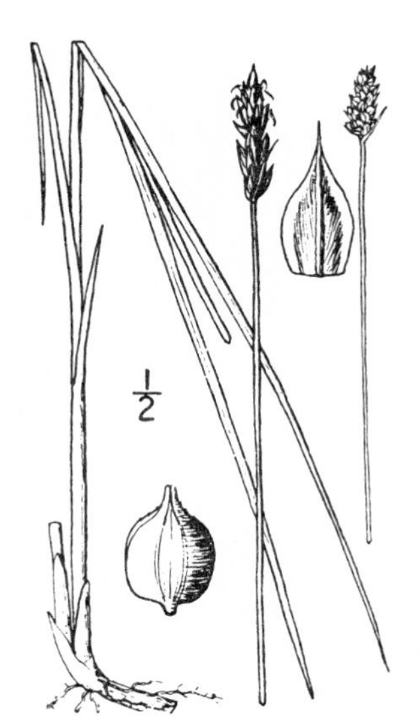

11. **Carex simulàta** Mackenzie.
Short-beaked Sedge. Fig. 687.

Carex gayana Boott Ill. Car. **3**: 126. *pl. 411.* 1862. Not Desv.
Carex simulata Mackenzie Bull. Torrey Club **34**: 604. 1908.

Rootstocks slender, light brown, long-creeping, the culms arising one–few together, 3–5 dm. high, roughened on the angles above, much exceeding the leaves. Leaf-blades 2–4 mm. wide, flat; head linear-oblong or oblong-ovoid, 12–25 mm. long, 6–10 mm. wide, the 5–15 spikes densely aggregated, pistillate, staminate or androgynous; bracts shorter than head; scales concealing the perigynia, cuspidate, the midvein sharp, brown with hyaline margins, not dull; perigynia ascending, plano-convex, smooth, shining, coriaceous, broadly ovate, 1.8–2.25 mm. long, 1.5 mm. wide, round truncate and little spongy at base, sharp edged but not margined, nerveless ventrally, nerved dorsally, serrulate above, abruptly beaked, the beak 0.25 mm. long, obliquely cut and with closed dorsal suture, the apex at length bidentulate; achenes lenticular; stigmas 2.

Wet soil, Transition and Canadian Zones; Montana to New Mexico westward to Washington and the Sierra Nevada of California. Type locality: Chug Creek, Albany County, Wyoming.

12. **Carex pánsa** Bailey.
Sand Dune Sedge. Fig. 688.

Carex pansa Bailey, Bot. Gaz. **13**: 82. 1888.

Rootstocks long-creeping, stout, dark-colored, the culms arising one–few together, 1.5–3 dm. high, sharply triangular, roughened above, exceeding leaves. Leaf-blades 1–3 mm. wide, flattened or channelled; spikes closely aggregated in an oblong or oblong-ovoid head, 1.5–2.5 cm. long, the spikes lanceolate-ovoid, 7–8 mm. long, 5 mm. wide; lower bracts developed but shorter than head; scales concealing perigynia, very dark chesnut brown with conspicuous white hyaline margins, shining; perigynia several–many, appressed, plano-convex, smooth, coriaceous, shining, blackish in age, 3.5–4.5 mm. long, 1.5–2 mm. wide, the body ovate, nerveless ventrally, nerved dorsally, round tapering at base, sharp edged, but not wing-margined, tapering into the beak ⅓–½ length of body, the beak 1 mm. long, serrulate, obliquely cut and with closed suture dorsally, the apex little hyaline at length bidentulate; achenes lenticular; stigmas 2.

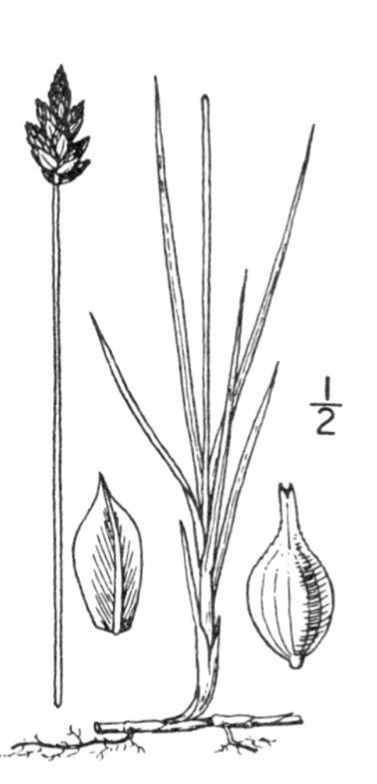

Drifting sands along the seacoast, Humid Transition Zone; Monterey County, California, to southwestern Washington. Type locality: Clatsop, Oregon.

13. **Carex praegrácilis** W. Boott.
Clustered Field Sedge. Fig. 689.

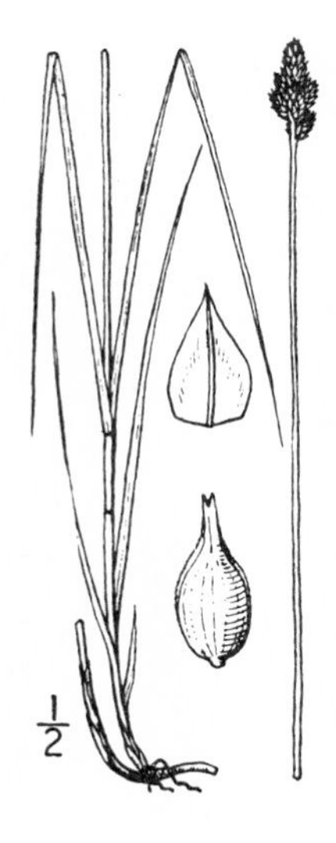

Carex marcida Boott in Hook. Fl. Bor. Am. **2**: 212. *pl. 213.* 1840. Not J. F. Gmel. 1791.
Carex praegracilis W. Boott, Bot. Gaz. **9**: 87. 1884.
Carex usta Bailey, Mem. Torrey Club **1**: 20. 1889.
Carex alterna C. B. Clarke, Kew Bull. Misc. Inf. Add. Ser. **8**: 69. 1908.
Carex siccata obscurior Kükenth. in Engler Pflanzenreich 4^{20}: 133. 1909
Carex camporum Mackenzie, Bull. Torrey Club **37**: 244. 1910.

Rootstocks long-creeping, stout, dark-colored, the culm arising 1–few together, 2–5 dm. high, sharply triangular and roughened above and normally exceeding leaves. Leaf-blades 1.5–3 mm. wide, flattened or canaliculate; spikes closely aggregated in a linear-oblong to oblong-ovoid head, 1–5 cm. long, 6–12 mm. wide, the spikes normally androgynous with 4–10 perigynia; lower bracts developed, but exceeded by head; scales nearly concealing perigynia, ovate-lanceolate, acute to cuspidate, dull light-brownish with hyaline margins; perigynia appressed-ascending, plano-convex, smooth, coriaceous, dull, blackish in age, 3–4 mm. long, 1.5 mm. wide, the body broadly ovate, nerved dorsally, nerveless or nearly so ventrally, round tapering at base, sharp-edged but not wing-margined, tapering into a beak ⅓–½ length of body, the beak about 1 mm. long, serrulate, obliquely cut and prominently hyaline at orifice and with closed suture dorsally, the apex at length bidentulate; achenes lenticular, stigmas 2.

Meadows, widely distributed and variable, Upper Sonoran and Transition Zones; Manitoba, Iowa and Kansas to Yukon, British Columbia, California, and central Mexico. Type locality: San Diego, California.

14. Carex siccàta Dewey.
Dry Sedge. Fig. 690.

Carex siccata Dewey, Am. Journ. Sci. **10**: 278. *pl. F, f. 18.* 1826.

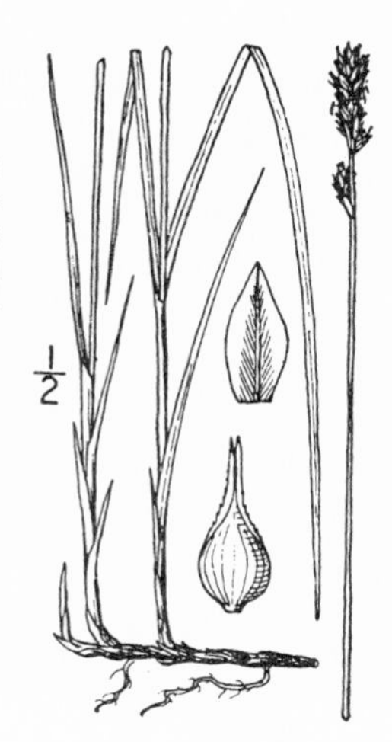

Rootstocks long-creeping, light brown, slender but tough, the culms arising singly or in small clumps, 2–9 dm. high, rough above, exceeding leaves, stiff, dark tinged at base. Leaf-blades 2–3 mm. wide, flat or channelled; spikes 6–12, closely aggregated except the lower 1–3 in a linear-oblong head 2–3.5 cm. long, 5–10 mm wide, some or all the spikes gynaecandrous or staminate and some pistillate or androgynous; lower bracts developed but shorter than head; scales shorter than perigynia, lance-ovate, acute, fulvous, green-keeled with very broad white hyaline margins; perigynia appressed-ascending, plano-convex, sharp- or wing-margined at least above, 5–6 mm. long, 2 mm. wide, light green, much flattened, coriaceous, many-nerved on both faces, spongy and short stipitate at base, serrulate on the margins above, contracted into a sharply bidentate beak nearly as long as the body; achenes lenticular; stigmas 2.

Dry soil. Transition, Canadian and Hudsonian Zones; Maine to New Jersey, west to Washington (Kittitas County) and south in the mountains to Arizona. Erroneously recorded from California. Type locality: Westfield, Massachusetts.

15. Carex macrocéphala Willd. Large-headed Sedge. Fig. 691.

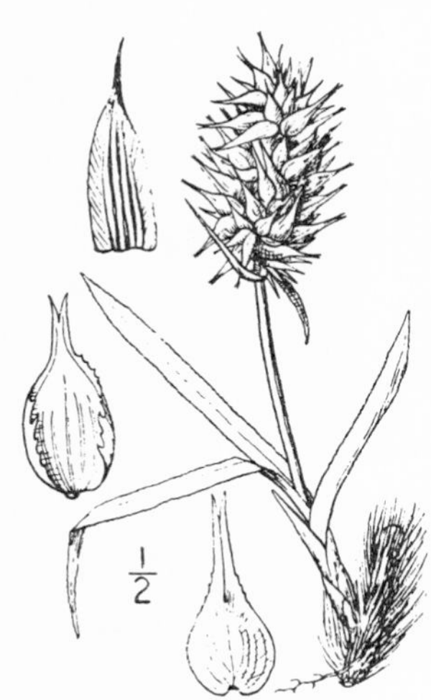

Carex macrocephala Willd. Spreng. Syst. **3**: 808. 1826.
Carex anthericoides Presl. Reliq. Haenk. **1**: 204. 1828.

Culms few together from elongated rootstocks, long-fibrillose at base, 1.5–3.5 dm. high, stout, often strongly roughened on the angles. Leaf-blades 4–8 mm. wide, the margins minutely but sharply serrulate; heads usually dioecious, the pistillate very large, 4–6 cm. long, 2.5–5 cm. thick, ovoid-orbicular to oblong, the spikes very numerous, several–many flowered, closely aggregated and scarcely distinguishable, 1.5 cm. long, 6–9 mm. wide; bracts from little to very strongly developed; staminate heads about 4 cm. long and 1 cm. wide; pistillate scales ovate, cuspidate or acuminate, exceeded by perigynia, striate, brownish with green center and hyaline margins; perigynia plano–convex, at length spreading, 10–15 mm. long, coriaceous, the body 4–6 mm. wide, 6–8 mm. long, rounded and somewhat spongy at base, tapering into a beak of about its own length, strongly many-nerved on both faces, and with an irregularly cut membranaceous raised wing margin, the beak sharply bidentate, smooth above; achenes obtusely triangular, thick, short-oblong, 4 mm. long, constricted in middle; style slightly enlarged at base, slender, jointed with achene; stigmas 3, very long.

Sands along the seacoast, Humid Transition and Canadian Zones; Oregon to Alaska, and on the Asiatic coast southward to northern China and Japan. Type locality: Asia.

16. Carex hoòdii Boott.
Hood's Sedge. Fig. 692.

Carex hoodii Boott in Hook. Fl. Bor. Am. **2**: 211. *pl. 211.* 1840.
Carex muricata confixa Bailey, Bot. Gaz. **10**: 203. 1885.
Carex hoodii nervosa Bailey, Bull. Torrey Club **1**: 14. 1889.

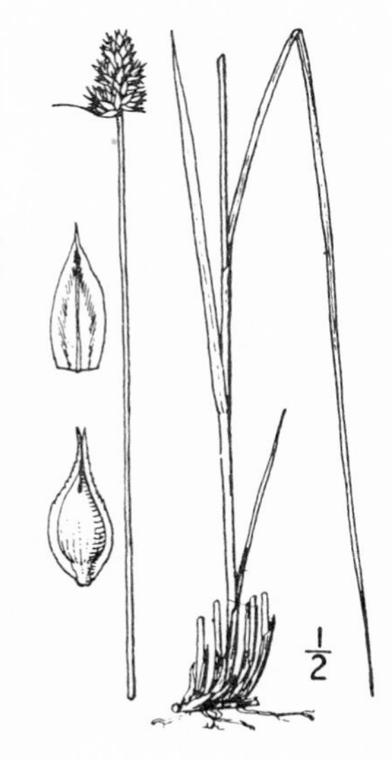

Very densely cespitose, the culms 3–6 dm. high, slender, but rather stiff, rough above, brownish at base. Leaf-blades 1.5–3.5 mm. wide, flat, the sheaths thin at mouth, not prolonged beyond base of blade; head capitate, orbicular or oblong-ovoid, 1–2 cm. long, 8–15 mm. wide, the spikes several, androgynous with 5–10 ascending perigynia; lower bracts more or less developed; scales about length of perigynia, ovate, sharp-pointed, chestnut-brown with lighter keel and broad hyaline margins; perigynia plano-convex, lance-ovate, green, more or less brownish-tinged, membranaceous, smooth, 4–5 mm. long, 1.75–2 mm. wide, obsoletely nerved ventrally towards the base, green-margined above, serrulate to the middle, somewhat spongy at base, contracted into a sharply bidentate beak half the length of the body; achenes lenticular; stigmas 2.

Mountain meadows and slopes, Arid Transition and Canadian Zones; Alberta to Colorado, westward to California (Sierra Nevada) and north to British Columbia. Type locality: Columbia River.

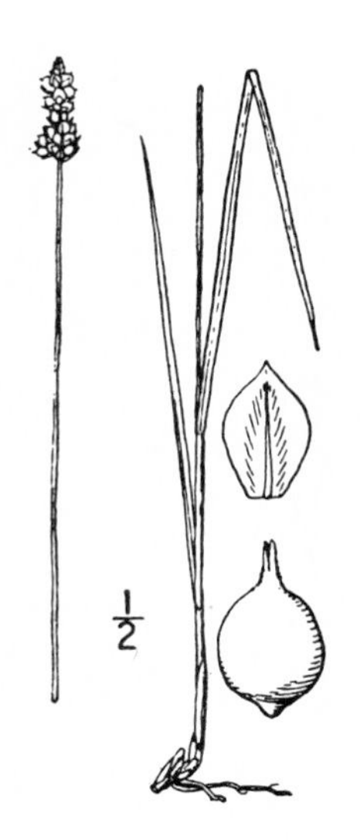

17. Carex vallícola Dewey.
Valley Sedge. Fig. 693.

Carex vallicola Dewey, Am. Journ. Sci. II. **32**: 40. 1861.
Carex brevisquama Mackenzie, Bull. Torrey Club **34**: 152. 1907.

Rootstocks short-creeping, scaly, the culms slender, 2.5–6 dm. high, slender, roughened on the angles, much exceeding the leaves. Leaf-blades narrow, about 1 mm. wide; spikes androgynous, closely aggregated into a dense terminal head 15–20 mm. long and about 7 mm. wide, the individual spikes poorly defined with 2–10, ascending or at maturity spreading perigynia; bracts little developed; scales broadly ovate, strongly exceeded at maturity by perigynia, acuminate or short cuspidate, brownish straw-color with hyaline margins; perigynia plano-convex, light green, 3.5 mm. long, 2 mm. wide, the body ovate, smooth, polished, nerveless, round-tapering at base, margined above, abruptly narrowed into the short, minutely serrulate bidentate beak about 1 mm. long; achenes lenticular; stigmas 2.

Dry slopes, Transition Zone; Black Hills of South Dakota and Wyoming to Oregon. Type locality: Jackson's Hole, Snake River.

18. Carex tumulícola Mackenzie.
Foot-hill Sedge. Fig. 694.

Carex muricata gracilis W. Boott in S. Wats. Bot. Calif. **2**: 232. 1880, not Boott.
Carex tumulicola Mackenzie, Bull. Torrey Club **34**: 154. 1907.

Short-creeping from tough rootstocks, the culms 4.5–8 dm. high, rough above, exceeding the leaves, brownish tinged at base. Leaf-blades 1.5–2.5 mm. wide; head 2–5 cm. long, slender and often rather flexuous, the spikes 5–10, androgynous, the upper aggregated, the lower separate, with 10 or fewer appressed-ascending perigynia; bracts especially the lower strongly developed and long cuspidate; scales largely concealing perigynia, brownish straw color with hyaline margins and green midrib, acuminate to cuspidate; perigynia plano-convex, lanceolate, glabrous, 4–5 mm. long, 1.5 mm. wide, prominently narrow-margined, nerved dorsally, more or less strongly nerved ventrally, round-tapering and substipitate at base, contracted into a serrulate bidentate beak one-third to one-half length of body; achenes lenticular; stigmas 2.

Dry soil, Upper Sonoran and Arid Transition Zones; Washington south through Oregon and the coastal counties of California to Monterey County. Type locality: Lake Temescal, Alameda County, California.

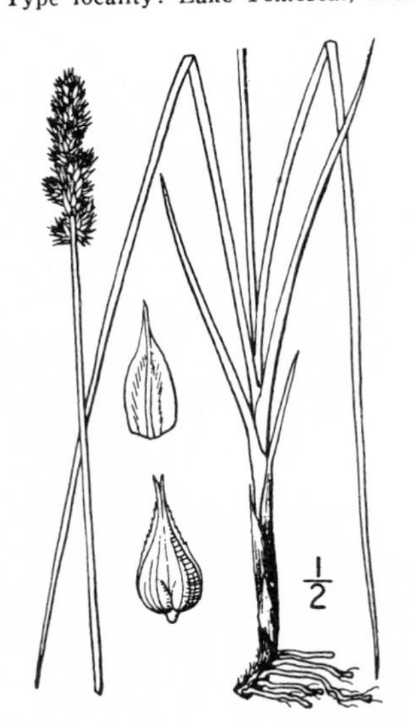

19. Carex álma Bailey. Sturdy Sedge. Fig. 695

Carex alma Bailey, Mem. Torrey Club **1**: 50. 1889.
Carex vitrea Holm, Am. Journ. Sci. IV. **17**: 302–3. *fig. 5–7.* 1904.

Cespitose, the culms 3–12 dm. high, strict, sharply triangular and rough on the angles. Leaf-blades 3–6 mm. wide, the sheaths very thin at mouth, exceeding the base of the blade, normally not cross-rugulose ventrally; head 2.5–10 cm. long, more or less strongly decompound, the clusters from closely aggregated to strongly separate, the individual spikes androgynous, hardly recognizable, the perigynia few, spreading; lower bracts more or less developed; scales ovate, about equalling perigynia, sharp pointed to obtusish, straw-colored or brownish, strongly hyaline-margined; perigynia plano-convex, 3.5 mm. long, 1.8 mm. wide, smooth, shining, in age brownish-black, broadly ovate from a rounded base, narrowly green-margined above, serrulate from middle, lightly few-nerved on both faces, tapering into the serrulate bidentate beak half length of body; achenes lenticular; stigmas 2.

Along streams in southern California, Upper Sonoran Zone; Monterey and Tulare Counties southward, eastward to southern Nevada and Arizona. Type locality: San Bernardino County, California.

20. **Carex stenóptera** Mackenzie. Narrow-winged Sedge. Fig. 696.

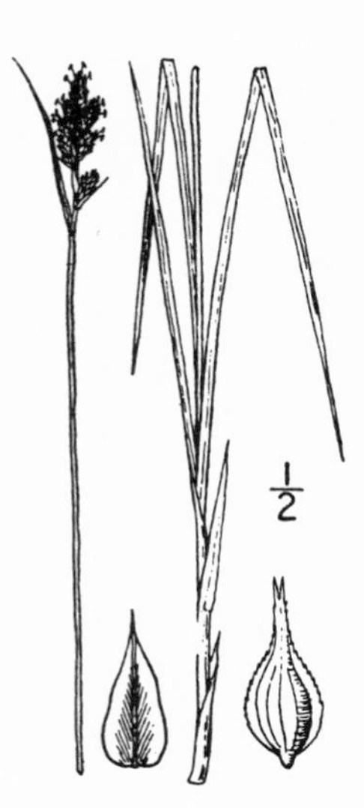

Carex stenoptera Mackenzie, Erythea **8**: 28. 1922.

Rootstocks thick, dark, short-creeping, the culms 2.5–4 dm. high, smooth; leaf-blades 2–3 mm. wide, channelled, the sheaths tight, not cross-rugulose, obscurely red-dotted; head decompound, 2.5–5 cm. long, 6–10 mm. thick, with numerous androgynous spikes 5–8 mm. long, 4–6 mm. wide, with several–many perigynia; bracts inconspicuous; scales ovate-lanceolate, obtusish to short cuspidate, about length of perigynia, brownish with broad green center and conspicuous hyaline margins; perigynia ovate-lanceolate, plano-convex, 3–3.5 mm. long, 1.5 mm. wide, narrowly thin-winged to base, stramineous, nerved dorsally, nerveless ventrally or nearly so, round tapering at base, tapering or contracted into a serrulate bidentate beak more than half length of the body; achenes lenticular; style slender, jointed with achene; stigmas 2.

San Antonio Mountains, southern California. Type locality: Ice House Cañon, San Antonio Mountains.

21. **Carex dénsa** Bailey. Dense Sedge. Fig. 697.

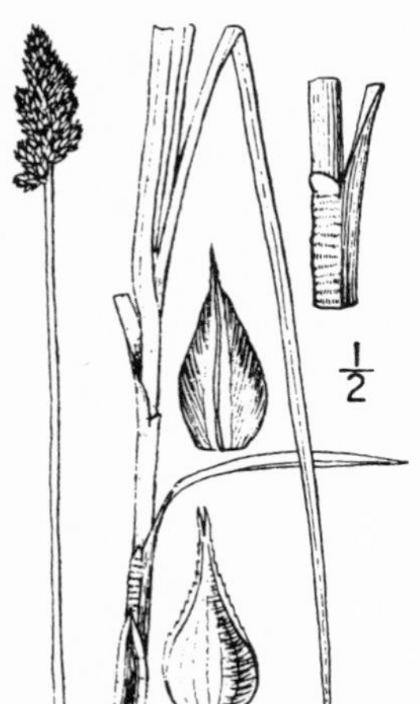

Carex brongniartii densa Bailey, Proc. Am. Acad. **22**: 137. 1886.
Carex densa Bailey, Mem. Torrey Club **1**: 50. 1889.
Carex chrysoleuca Holm, Am. Journ. Sci. IV. **17**: 302. 1904.

Culms 3–6 dm. high, sharply triangular, stiff, smooth or roughened immediately beneath head, from exceeding to shorter than the leaves. Leaf-blades 3–6 mm. wide, the sheaths loose, conspicuously septate dorsally, and thin, hyaline and more or less cross-rugulose ventrally, prolonged and convex at the mouth; head 2–5 cm. long, 1.5 cm. wide, dense, decompound, the clusters closely aggregated, the individual spikes hardly recognizable, the perigynia few, appressed-ascending; bracts inconspicuous, except one or two lower ones; scales ovate, shorter than perigynia, dark chestnut-brown with green midvein, acute to cuspidate; perigynia 4–4.5 mm. long, 1.5 mm. wide, unequally slightly bi-convex, smooth, ovate or ovate-lanceolate from a round tapering substipitate base, straw-colored or at length brownish, strongly several nerved on both faces, narrowly green-margined, the upper half of body serrulate, more or less abruptly beaked, the beak more than half length of body; achenes lenticular; stigmas 2.

In dry soil, Upper Sonoran Zone; west of the higher ranges of the Sierra Nevada from Santa Clara and Mariposa Counties, California, northward into Oregon. Type locality: Mark West Creek and Napa, California.

22. **Carex vicària** Bailey. Western Fox Sedge. Fig. 698.

Carex vicaria Bailey, Mem. Torrey Club **1**: 49. 1889.
Carex vulpinoidea vicaria (Bailey) Kükenth. in Engler Pflanzenreich **4[20]**: 148. 1909.

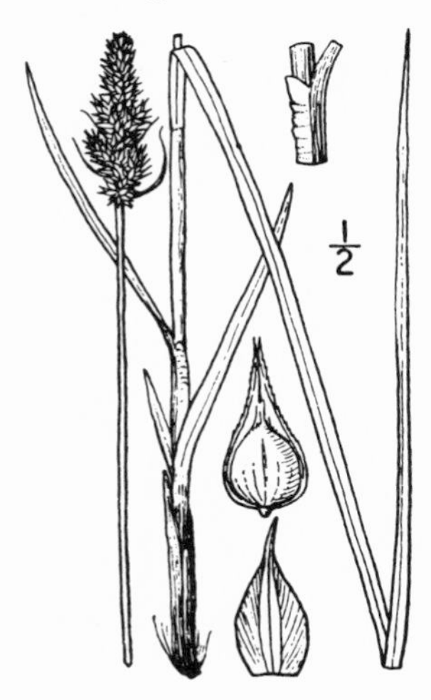

Culms 3–6 dm. high, sharply triangular, strongly roughened on angles above, exceeding the leaves. Leaves with blades 3–4.5 mm. wide, the sheaths tight, not conspicuously septate dorsally; and thin, hyaline and more or less cross-rugulose ventrally, the upper short-prolonged and convex at mouth; head 1.5–3 cm. long, about 12 mm. wide, decompound, the clusters closely aggregated, or the lower slightly separate, the individual spikes hardly recognizable, the perigynia in each few, spreading at maturity; bracts inconspicuous, except one or two lower ones; scales ovate, acute to cuspidate; shorter than perigynia, reddish-brown tinged with green midvein; perigynia 3–3.5 mm. long, 1.5 mm. wide, plano-convex, smooth, ovate from a rounded base, substipitate, reddish-tinged or yellowish-brown with green margin, few-nerved dorsally, nerveless ventrally, the body sparingly serrulate above, contracted into the serrulate, bidentate beak half length of body; achenes lenticular; stigmas 2.

Marshes and swales, Humid Transition and Canadian Zones; Washington to northern California. Type locality: Oregon.

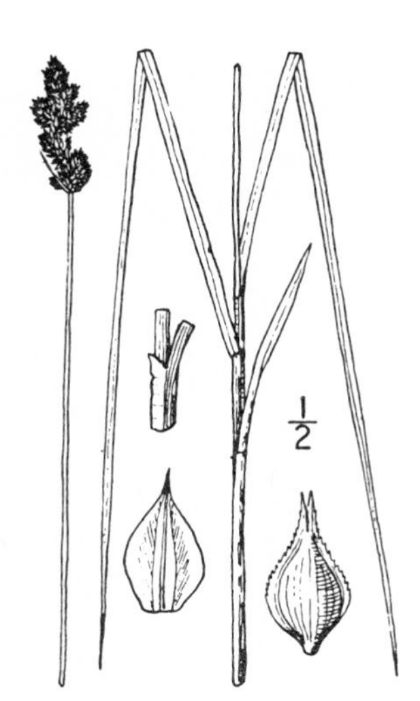

23. Carex brevi-ligulàta Mackenzie.
Short-liguled Sedge. Fig. 699.

Carex vicaria costata Bailey, Mem. Torrey Club, 1: 49. 1889.
Carex brevi-ligulata Mackenzie, Erythea 8: 92. 1922.

Culms 3–6 dm. high, sharply triangular, strongly roughened on angles above, exceeding the leaves. Leaves with blades 3–4.5 mm. wide, the sheaths tight, not conspicuously septate dorsally; and thin, hyaline and more or less cross-rugulose ventrally, short prolonged at mouth; head 1.5–3.5 cm. long, 8–12 mm. wide, decompound, the clusters closely aggregate, the individual spikes hardly recognizable, the perigynia 6–15, spreading at maturity; bracts usually not conspicuous; scales ovate, acute to cuspidate, shorter than perigynia, light reddish-brown with green center; perigynia 3.25–3.75 mm. long, 1.5–2 mm. wide, thick, plano-convex, smooth, ovate, round-tapering at base, substipitate, yellowish brown with green margin, several ribbed dorsally, strongly fewer ribbed ventrally, strongly serrulate from middle, abruptly contracted into the serrulate bidentate beak, half length of body; achenes lenticular, stigmas 2.

Marshes and swales, Transition Zone; northern California and Oregon. Type locality: Grant's Pass, Oregon.

24. Carex dúdleyi Mackenzie.
Dudley's Sedge. Fig. 700.

Carex dudleyi Mackenzie, Erythea 8: 30. 1922.

Culms 3–6 dm. high, roughened on angles and sharply triangular beneath head, exceeding leaves. Leaf-blades flat, 4–7 mm. wide, little roughened towards apex, the sheaths tight, inconspicuously septate dorsally, white hyaline and scarcely if at all cross-rugulose ventrally, short prolonged and convex at mouth; head 2–3.5 cm. long, 9–12 mm. wide, decompound, the spikes all very closely aggregated and hardly recognizable, the perigynia in each few, spreading at maturity; bracts setaceous, at least the lower conspicuous; scales ovate-lanceolate, mostly shorter than perigynia, yellowish-green, strongly cuspidate or awned; perigynia 2.5–3 mm. long, 1.25–1.5 mm. wide, plano-convex, smooth, narrowly ovate from a rounded base, sub-stipitate, brownish yellow with green margin, few nerved dorsally, obscurely nerved ventrally, the body not serrulate, contracted into the serrulate bidentate beak of its own length; achenes lenticular; stigmas 2.

In the Coast Ranges of California, Upper Sonoran Zone; Monterey County to Lake County; apparently local. Type locality: Tassajara Hot Springs, Monterey County, California.

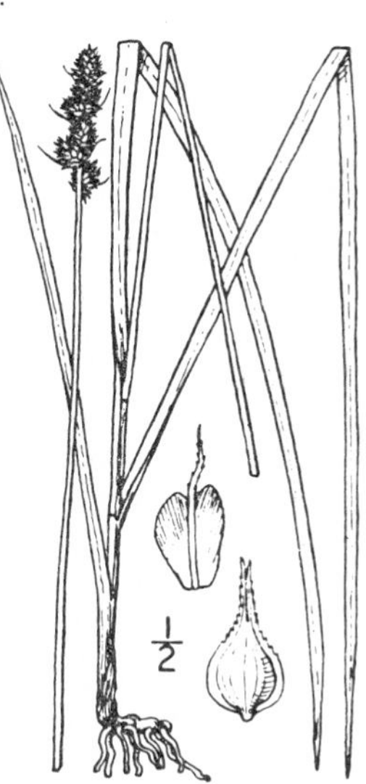

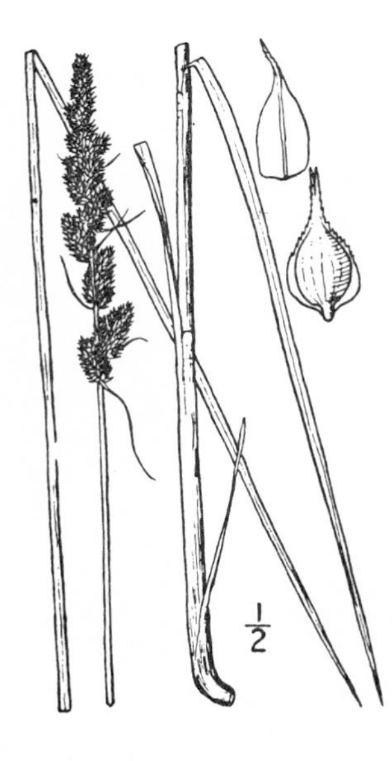

25. Carex vulpinoìdea Michx.
Fox Sedge. Fig. 701.

Carex vulpinoidea Michx. Fl. Bor. Am. 2: 169. 1803.
Carex microsperma Wahl. Vet. Acad. Handl. Stockholm 24: 144. 1803.
Carex multiflora Muhl. in Willd. Sp. Pl. 4: 243. 1805.
Carex vulpinaeformis Tuckerm. Enum. Meth. 9. 1843.

Cespitose, the rootstocks woody, short, the culms 3–9 dm. high, sharply triangular, very rough above. Leaf-blades flat, 2–5 mm. wide, very rough above, many exceeding the culms, the sheaths tight, inconspicuously septate dorsally, cross-rugulose ventrally, more or less thickened at the mouth and little if at all prolonged beyond the base of the blade; head 5–12.5 cm. long, decompound, the spikes very numerous, small, androgynous, closely aggregated and the upper scarcely distinguishable; bracts bristle-like, the lower conspicuous; scales ovate lanceolate, strongly awned, about as long as upper and longer than lower perigynia, ferruginous or straw-colored with green keel; perigynia ovate, plano-convex, 2–2.5 mm. long, 1.2 mm. wide, greenish yellow, membranaceous, rounded at base, several nerved dorsally, 1–3 nerved or nearly nerveless ventrally, ascending or spreading at maturity, contracted into a serrulate bidentate beak about length of the body; achenes lenticular; style slightly thickened at base, jointed with the achene; stigmas 2.

Swampy places, Arid Transition Zone; New Brunswick to British Columbia south to Florida, Texas, Colorado, and Oregon. Type locality: Canada and New England.

26. **Carex diándra** Schrank.

Lesser Panicled Sedge. Fig. 702.

Carex diandra Schrank, Acta. Acad. Mogunt. 49. 1782.
Carex teretiuscula Good. Trans. Linn. Soc. **2**: 163. *pl. 19, f. 3.* 1794.
Carex bernardina Parish, Bull. South. Calif. Acad. **5**: 24. *pl. 21.* 1906.

Loosely cespitose, the culms slender, 3–7 dm. high, sharply triangular, the sides convex. Leaf-blades 1.25 mm. wide, canaliculate at base; sheaths not copper-colored at the mouth, strongly red-dotted ventrally; head 2.5–5 cm. long, somewhat compound, the spikes numerous, small, androgynous or pistillate, the lower more or less separated; bracts not conspicuous; scales nearly equaling perigynia, acute, brownish with lighter midvein and hyaline margins; perigynia 2-2.75 mm. long, strongly convex dorsally, low convex ventrally, dark chestnut, sharp-edged but not margined, shining, smooth, hard, nerveless on inner face except at base, nerved on outer face, spongy, round-truncate and stipitate at base, spreading at maturity, contracted into a flat bidentate beak nearly length of body; achenes lenticular; stigmas 2.

Wet meadows, Transition and Boreal Zones; Newfoundland to Yukon south to New Jersey, Colorado, and very locally in the higher mountains to southern California. Type locality: southern Bavaria, Germany.

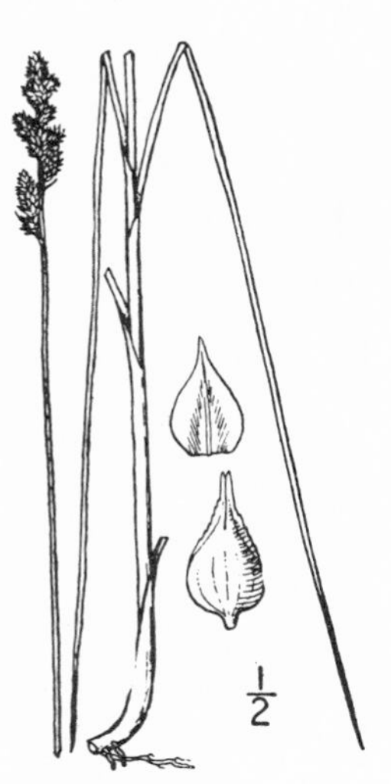

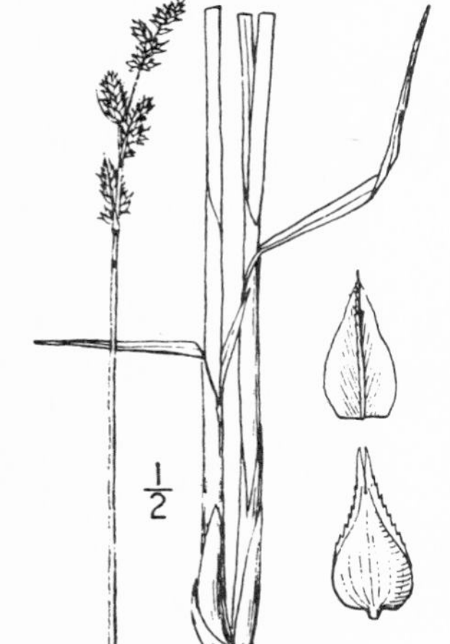

27. **Carex cùsickii** Mackenzie.

Cusick's Sedge. Fig. 703.

Carex teretiuscula ampla Bailey Mem. Torrey Club **1**: 53. 1889.
Carex cusickii Mackenzie in Piper & Beattie Flora Northwest Coast 72. 1915.

More densely cespitose, the culms sharply triangular, stout, 7–12 dm. high, rough above, much exceeding the leaves. Leaf-blades 2.5–6 mm. wide, flat with slightly revolute margins, the sheaths strongly red-dotted; head 4–8 cm. long, 1–2 cm. wide, decompound, the lower branches separated, the spikes numerous, small, ovoid, 3–6 mm. long, 2.5–4 mm. wide, androgynous or pistillate; bracts inconspicuous; scales covering the perigynia, chestnut tinged with lighter sharply rough-keeled midvein and hyaline margins; perigynia 3–4 mm. long, 1.5 mm. wide, dull brownish-black, very thick, strongly convex and nerved dorsally, slightly convex and lightly nerved at base ventrally, spreading in age, truncate and short stipitate at base, abruptly beaked, the beak setulose-serrulate, shallowly bidentate, about length of body; achenes lenticular; stigmas 2.

Wet meadows, Humid Transition and Canadian Zones; Montana to British Columbia south to San Francisco, California; rare southward and there confined to the Coast Ranges. Type locality: head of Burn River, eastern Oregon.

28. **Carex jònesii** Bailey.

Jones' Sedge. Fig. 704.

Carex jonesii Bailey, Mem. Torrey Club **1**: 16. 1889.
Carex nervina jonesii Kükenth. in Engler Pflanzenreich **4**[20]: 167. 1909.

Cespitose from somewhat elongated woody rootstocks, the culms triangular, 2–4 dm. high, slender, rough above, exceeding the leaves. Leaves clustered near base, the blades 1–2 mm. wide; sheath not green and white mottled dorsally and not cross-rugulose, ventrally truncate at mouth; spikes androgynous, in a dense ovoid head 8–12 mm. long, 8–10 mm. wide, the larger with about 5–10 ascending perigynia, the staminate flowers often inconspicuous; bracts inconspicuous; scales exceeding the perigynia, ovate, dark brown with inconspicuous midvein and hyaline margins; perigynia ovate-lanceolate, 3–4 mm. long, 1.5–1.75 mm. wide, plano-convex, rounded short-stipitate and but little spongy at base, strongly many nerved dorsally, more or less strongly many nerved ventrally at maturity, slightly margined above, tapering into a very slightly serrulate shallowly bidentate beak one-third length of body; achenes lenticular; stigmas 2, short.

High mountains, Canadian Zone; Montana to Wyoming, westward to Washington and California, where confined to the Sierra Nevada and San Bernardino Mountains. Type locality: Soda Springs, Nevada County, California.

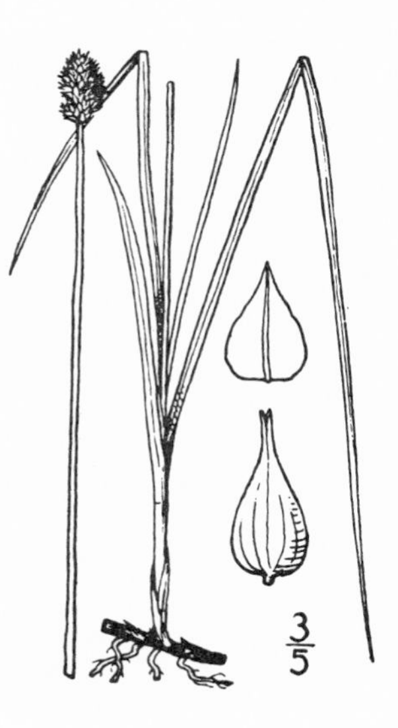

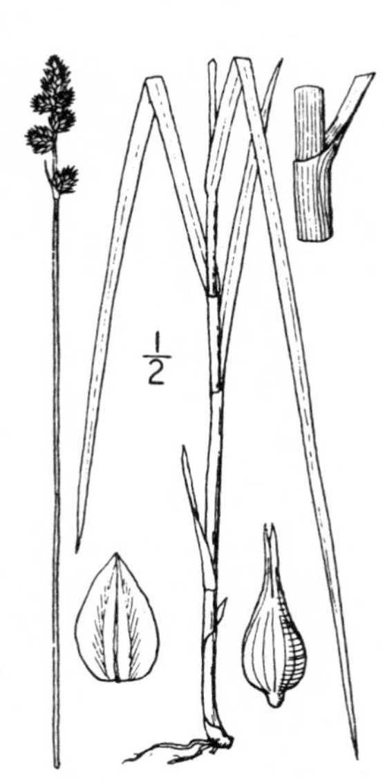

29. **Carex nervìna** Bailey.

Sierra Nerved Sedge. Fig. 705.

Carex nervina Bailey, Bot. Gaz. **10**: 203. *pl. 3, f. 68.* 1885.

Cespitose from somewhat elongated stout rootstocks, the culms 3–9 dm. high, strongly aphyllopodic, flattened in drying, weak, roughened above, about equaling the leaves. Well-developed leaves on lower fourth of culms, the blades 3.5–5 mm. wide; opaque part of sheath olive-tinged, not cross-rugulose, truncate or concave at the mouth, green and white-mottled dorsally, spikes androgynous in a dense ovoid or oblong head 1.3–3 cm. long, 8–10 mm. wide, the larger with about 6–12 ascending or at length spreading perigynia, the staminate flowers not conspicuous; bracts little developed; scales ovate, exceeded by the perigynia, brownish with green center and hyaline margins; perigynia plano-convex, lanceolate, or ovate-lanceolate, 3.5–4 mm. long, 1.5–1.75 mm. wide, rounded short-stipitate and strongly spongy at base, strongly many-nerved on both faces, sharp angled above, tapering into a slightly serrulate sharply bidentate beak 1 mm. long, the teeth erect; achenes lenticular; stigmas 2, long.

Mountains, Transition and Canadian Zones; southern Oregon and California, where found in the Sierra Nevada as far south as Tulare County. Type locality: Summit Camp, California.

30. **Carex neuróphora** Mackenzie. n. sp.

Alpine Nerved Sedge. Fig. 706.

Carex nervina Mackenzie in Rydberg Fl. Rocky Mts. (in part) 123. 1918, not Bailey.

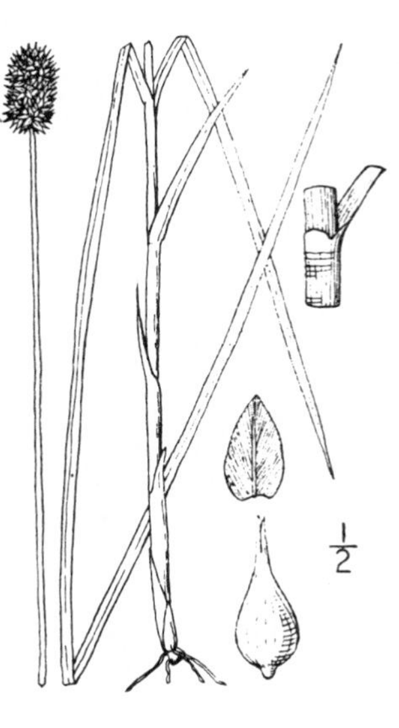

Cespitose from slightly elongated rootstocks, the culms 3–7 dm. high, strongly aphyllopodic, 3.5 mm. thick at base, little roughened above, exceeding the leaves. Well developed leaves on lower fourth of culms, the blades about 3.5 mm. wide, the sheaths green and white mottled dorsally, cross-rugulose ventrally, thickish and convex and short-prolonged at mouth; spikes androgynous in a dense ovoid or oblong head 1.5–2.5 cm. long, 8–12 mm. wide, the larger with several–many at length widely spreading perigynia, the staminate flowers inconspicuous; bracts not developed; scales ovate, brownish with lighter midrib, about half length of perigynia; perigynia plano-convex, lanceolate, 3.5–4 mm. long, 1.5 mm. wide, rounded, short stipitate and somewhat spongy at base, several–many nerved on both faces, sharp-edged above, tapering into a slightly serrulate bidentate beak half length of body, the teeth very short; achenes lenticular; stigmas 2.

High mountains, Hudsonian Zone; Wyoming to Washington and Oregon. Type: Stevens Pass, Cascade Mountains, Washington, *Sandberg and Leiberg, 773.*

31. **Carex stipàta** Muhl.

Awl-fruited Sedge. Fig. 707.

Carex stipata Muhl. in Willd. Sp. Pl. **4**: 233. 1805.

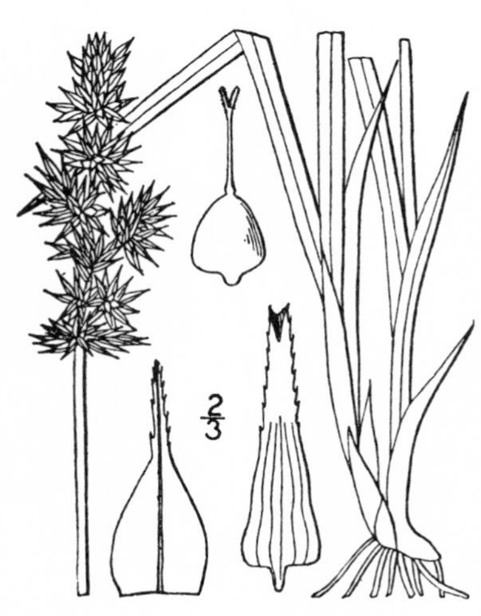

Cespitose, the rootstocks short, stout, the culms 3–10 dm. high, rather weak, sharply triangular, strongly serrulate above, mostly exceeding leaves. Leaves with blades 4–8 mm. wide, flat, flaccid, the sheaths strongly septate dorsally, the opaque part thin, quickly broken, cross-rugulose, prolonged above base of blade; head 3–10 cm. long, yellowish brown, the spikes numerous, androgynous, the lower often separate; bracts inconspicuous; scales ovate-triangular, about length of body of perigynia, light brownish with light midvein and hyaline margins; perigynia lanceolate, plano-convex, 4–5 mm. long, 1.5 mm. wide, strongly nerved dorsally, less strongly ventrally, yellowish green or at length brownish, round cordate, spongy and stipitate at base, narrowly margined ventrally, tapering into a serrulate bidentate beak nearly equalling or longer than the body; achenes lenticular; stigmas 2.

Swamps and wet meadows, Transition Zone; Newfoundland to North Carolina west to middle California and southern Alaska. Type locality: Pennsylvania.

32. **Carex illòta** Bailey.
Small-headed Sedge. Fig 708.

Carex bonplandii minor Boott, Proc. Acad. Phila. 77. 1863.
Carex bonplandii angustifolia Boott, in S. Wats. Bot. Calif. **2**: 233. 1880.
Carex illota Bailey, Mem. Torrey Club **1**: 15. 1889.

Cespitose with short-prolonged rootstocks, the culms 1–2.5 dm. high, slender but strict, roughened above. Leaf-blades short, flat, 1.5–3 mm. wide; spikes 3–5, gynaecandrous or pistillate, forming a dense capitate brownish black head 6–10 mm. long and nearly as wide; bracts not developed; scales broadly ovate, obtuse, shorter than perigynia, brownish black with light midvein and scarcely hyaline margins; perigynia ovate, plano-convex, 3 mm. long, membranaceous, smooth, shining, nerved on both faces, rounded and spongy at base, brownish black, at length spreading, the beak one-third the length of the body, smooth or nearly so, emarginate, hyaline at orifice; achenes lenticular; stigmas 2.

High mountains, Canadian Zone; Wyoming and Colorado west to Washington and the Sierra Nevada of California. Type locality: Colorado.

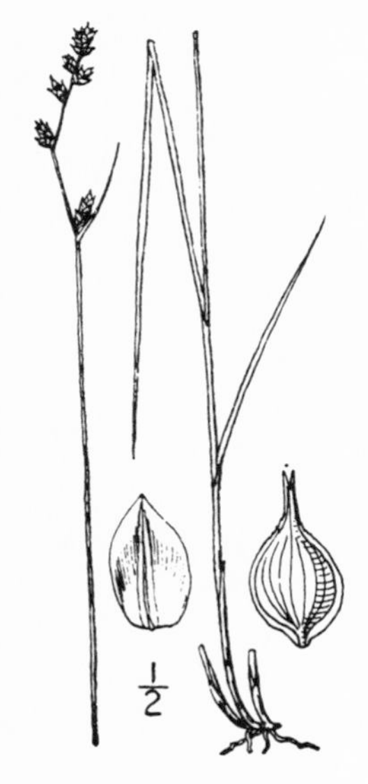

33. **Carex laevicúlmis** Meinsh.
Smooth-stemmed Sedge. Fig. 709.

Carex bolanderi sparsiflora Olney Proc. Am. Acad. **7**: 407. 1872.
Carex deweyana sparsiflora Bailey, Bot. Gaz. **13**: 87. 1888.
Carex laeviculmis Meinsh. Bot. Centralbl. **55**: 195. 1893.

Cespitose from slender short-elongated rootstocks, the culms 3–6 dm. high, weak, light brownish at base, roughened above. Leaf blades 1.5–2 mm. wide, light green, flat, soft; spikes 3–8, gynaecandrous, widely separate or upper approximate, suborbicular, 3–10 mm. long, 3–6 mm. wide, with 3–10 appressed or at length spreading perigynia, the beaks spreading; uppermost spike long clavate at base; lowest bract more or less developed; scales ovate, about length of body of perigynia, hyaline, with conspicuous green midvein; perigynia oblong-ovate, plano-convex or concave-convex, 2.5–4 mm. long, 1.5 mm. wide, broadest near middle, green or brownish-green, thin-walled, few nerved dorsally, lightly nerved ventrally, rounded and substipitate at base, tapering into a sparingly subserrulate beak one-quarter to one-third length of body, the apex entire or bidentulate, obliquely cut dorsally; achenes lenticular; stigmas 2.

Wet shaded places, Transition and Canadian Zones; Idaho to Alaska, south to El Dorado County, California; also in eastern Siberia. Type locality: Kamtschatka.

34. **Carex intèrior** Bailey.
Inland Sedge. Fig. 710.

Carex scirpoides Schk. in Willd. Sp. Pl. **4**: 237. 1803 (in small part), not *C. scirpoidea* Michx. 1805.
Carex interior Bailey, Bull. Torrey Club **20**: 426. 1893.

Densely cespitose, the culms 2–3.5 dm. high, slender and wiry, somewhat roughened beneath head. Leaf-blades 1–2 mm. wide, flat or somewhat canaliculate; head 1–2 cm. long, the 3–4 spikes approximate, the lateral pistillate, suborbicular, 4 mm. long, with 3–10 widely spreading perigynia, the upper long tapering and staminate at base; bracts little developed; scales half the length of the perigynia, ovate-orbicular, very obtuse, brownish, hyaline-margined all around, the center lighter-colored, the midvein not sharply defined and not reaching the tip; perigynia plano-convex, ovoid, plump, 2.5 mm. long, 1.5 mm. wide, straw-color or light brownish, rounded and spongy at base, very narrowly sharp-margined, nerved dorsally, nerveless or obscurely nerved at base ventrally, sparingly serrulate on upper margins, abruptly beaked, the beak one-third length of body or less, its teeth very short, the ventral suture inconspicuous, the dorsal more developed; achenes lenticular; stigmas 2.

Boggy meadows, Hudsonian Zone; Maine to Pennsylvania, westward to British Columbia and south to Arizona, northern Mexico and northern California. Type locality: Penn Yan, New York.

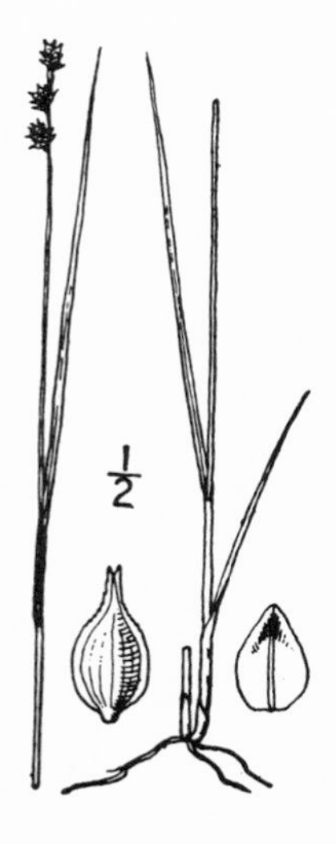

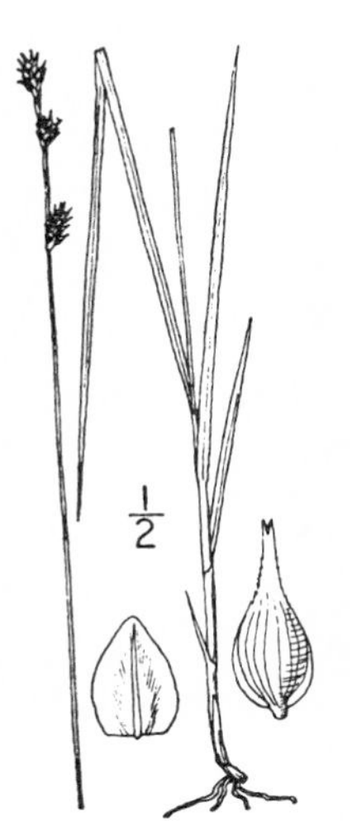

35. **Carex ormántha** (Fernald) Mackenzie.
Western Stellate Sedge. Fig. 711.

Carex echinata ormantha Fernald, Proc. Am. Acad. **37**: 483. *pl. 4, f. 89.* 1902.
Carex stellulata ormantha Fernald, Rhodora **4**: 222. 1902.
Carex ormantha Mackenzie, Erythea **8**: 35. 1922.

Densely cespitose from very short creeping slender rootstocks, the culms 1.5–4 dm. high, obtusely triangular, slender but rather stiff, smooth. Leaf-blades slightly canaliculate, 1.5–2 mm. wide; head 2–6 cm. long, the 3–4 spikes widely separate, the terminal gynaecandrous, long clavate at base, the lateral suborbicular, 6–8 mm. wide, 3–5 mm. long, with 2–12 widely radiating perigynia; lowest bract more or less developed; scales ovate, obtuse, nearly length of body of perigynia, chestnut brown with lighter center and broad shining hyaline margins and apex, not sharply keeled, the midvein obscure at apex; perigynia plano-convex, lance-ovate, 3.5–4 mm. long, 1.5 mm. wide, soon brownish yellowish, round truncate at base, conspicuously many striate on both faces, the margins scarcely elevated, not serrulate, tapering into a serrulate beak more than half length of body, the dorsal suture inconspicuous, the ventral prominent, the teeth triangular, short, erect, chestnut-brown; achenes lenticular; stigmas 2.

Boggy places, Transition Zone; north from the San Bernardino range through the Sierra Nevada to northern California and Oregon. Type locality: Strawberry Creek, El Dorado County, California.

36. **Carex phyllománica** W. Boott.
Coastal Stellate Sedge. Fig. 712.

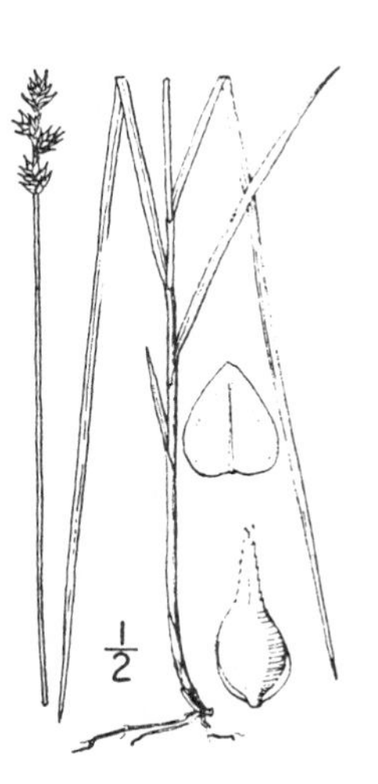

Carex phyllomanica W. Boott, in S. Wats. Bot. Calif. **2**: 233. 1880.
Carex sterilis W. Boott in S. Wats. Bot. Calif. **2**: 236. 1880; not. Willd.

Densely cespitose from slender creeping rootstocks, the culms 2.5–6 dm. high, obtusely triangular (sharply above), smooth or nearly so. Leaf-blades flat, 1.75–2.75 mm. wide; head 1.5–3.5 cm. long, the 3–4 spikes approximate, the terminal gynaecandrous, clavate at base, the lateral suborbicular, 7 mm. wide, with 8–15 widely spreading perigynia; lowest bract somewhat developed; scales ovate, obtuse, two-thirds length of body of perigynia, chestnut brown with lighter center and hyaline margins, not sharply keeled, the midvein obscure at apex; perigynia ovate-lanceolate, plano-convex, 3.75–4.5 mm. long, 1.5 mm. wide, round-truncate at base, conspicuously striate on both faces, the margins scarcely raised, entire, tapering into a serrulate beak scarcely half length of body, the apex hyaline and chestnut tinged, bidentate, the teeth short, erect, the ventral suture conspicuous; achenes lenticular; stigmas 2.

Swampy places near the coast, Canadian Zone; northern California to Alaska. Type locality: Mendocino City, California.

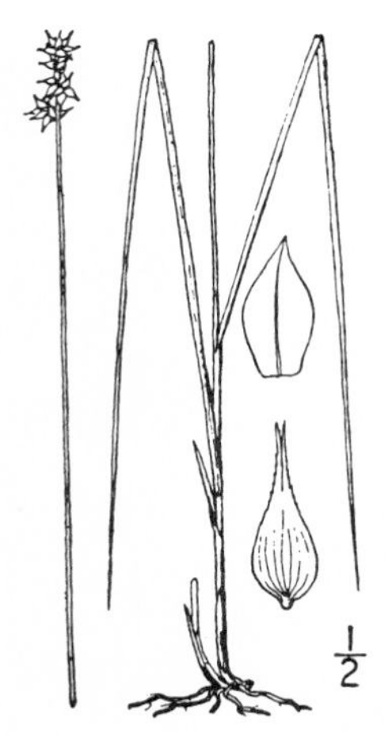

37. **Carex angústior** Mackenzie.
Narrower Sedge. Fig. 713.

Carex stellulata angustata Carey, in A. Gray Man. 544. 1848.
Carex echinata angustata Bailey, Mem. Torrey Club **1**: 59. 1889.
Carex angustior Mackenzie in Rydb. Flora Rocky Mts. 124. 1917.

Densely cespitose, the culms 1.5–3 dm. high, slender, sharply triangular, roughened above. Leaf-blades flat or canaliculate, 0.75–2 mm. wide; spikes 3–5, approximate or little separate forming a head 1–2 cm. long, the terminal gynaecandrous, the lateral pistillate, 5–15 flowered, suborbicular or short-oblong, 4–6 mm. long; lowest bract more or less developed; scales ovate, short cuspidate, equaling body of perigynia, yellowish with sharply defined green midvein extending to apex and ill-defined opaque margins; perigynia lanceolate, plano-convex, 3 mm. long, 1.25 mm. wide, yellowish, fragile, nerveless or obscurely impressed nerved ventrally, more or less nerved dorsally, rounded at base, the margins slightly raised, not serrulate, tapering into a sparingly serrulate beak more than half length of body, the apex bidentate, reddish brown tipped, the ventral suture prominent, achenes lenticular; stigmas 2.

In boggy places, Transition and Boreal Zones; Newfoundland to North Carolina, west to Washington and the Sierra Nevada of California. Type locality: Fairfield, New York.

38. Carex cephalántha (Bailey) Bicknell.
Larger Stellate Sedge. Fig. 714.

Carex echinata cephalantha Bailey, Mem. Torrey Club **1**: 58. 1889.
Carex cephalantha Bicknell, Bull. Torrey Club **35**: 493. 1908.

Densely cespitose, the culms 2.5–7.5 dm. high, stiff, sharply triangular, somewhat roughened above. Leaf-blades flat or somewhat canaliculate, 1.5–2.25 mm. wide; spikes 3–7, more or less strongly separate, forming a head 2.5–5 cm. long, gynaecandrous, 5–25-flowered, subglobose or short-oblong, 4–10 mm. long; lowest bracts developed; scales ovate, acute or somewhat cuspidate, equaling body of perigynia, yellowish or chestnut-brown tinged with sharply keeled green midvein extending to apex and ill-defined opaque margins; perigynia ovate, plano-convex, 3.5–4 mm. long, 1.5–2 mm. wide, subcoriaceous, strongly many nerved dorsally, finely nerved ventrally, round-truncate at base, serrulate above, contracted into a serrulate beak half length of body, the apex strongly bidentate, reddish brown tipped; achenes lenticular; stigmas 2.

Sphagnum bogs, Humid Transition and Canadian Zones; Newfoundland to Maryland, west to Ontario; and on the Pacific Coast from Vancouver to Washington. Type locality: eastern Pennsylvania.

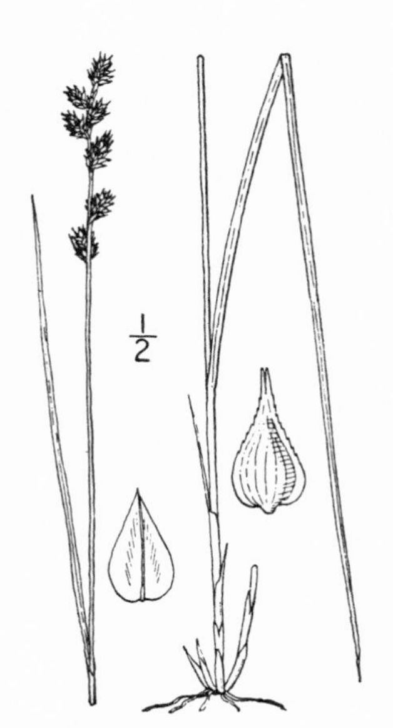

39. Carex leptópoda Mackenzie.
Shorter-scaled Sedge. Fig. 715.

Carex deweyana W. Boott, in S. Wats. Bot. Calif. **2**: 236. 1880; not Schw.
Carex leptopoda Mackenzie in Rydb. Flora Rocky Mts. 124. 1917.

Rootstocks slender, elongate, the culms slender, erect, 3–7.5 dm. high, little brownish tinged at base, roughened beneath head, exceeding leaves. Leaf-blades 2.5–5 mm. wide, flat, smooth, pale-green; spikes 4–7, approximate except lower 1–3, gynaecandrous, ovoid-oblong or linear-oblong, with 6–18 appressed perigynia; the lower bracts usually shorter than the head; scales shorter than body of perigynia, not reddish-brown tinged, mostly cuspidate; perigynia ovate-lanceolate, plano-convex, light green, 3.5–4 mm. long, 1.5 mm. wide, rounded, substipitate and somewhat spongy at base, obscurely nerved on outer face, nerved at base only on inner face, tapering into the serrulate shallowly bidentate beak, about one-third length of body; achenes lenticular; stigmas 2.

Damp woods, Transition Zone; Idaho to British Columbia south to Santa Cruz and Tulare Counties, California. Type locality: Elk Rock, near Oswego, Clackamas County, Oregon.

40. Carex bolánderi Olney.
Bolander's Sedge. Fig. 716.

Carex bolanderi Olney, Proc. Am. Acad. **7**: 393. 1868.
Carex deweyana bolanderi W. Boott in S. Wats. Bot. Calif. **2**: 236. 1880.

Rootstocks slender, elongate, the culms slender but strict, erect, 4–9 dm. high, brownish tinged at the base, little roughened beneath the head, exceeding the leaves. Leaf-blades 2.5–5 mm. wide, flat, smooth, pale green; spikes 4–8, linear-oblong or linear, the lower separate, the 8–30 perigynia appressed; lower bracts usually shorter than the head; scales covering body of perigynia, usually reddish brown tinged, mostly acute or mucronate; perigynia lanceolate, plano-convex, light green, 4–4.5 mm. long, 1.5 mm. wide, rounded, substipitate and spongy at base, nerved dorsally and at base on the inner face, rather abruptly tapering into the serrulate deeply bidentate beak half length of body; achenes lenticular; stigmas 2.

Woods, Transition Zone; British Columbia to California east to New Mexico, Utah, and western Montana. Type locality: Yosemite Valley, California.

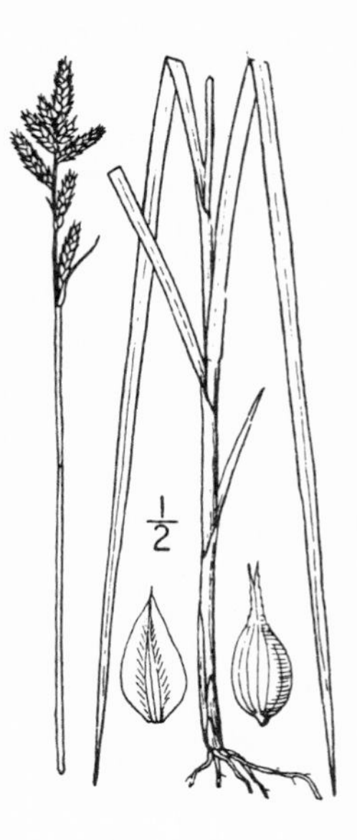

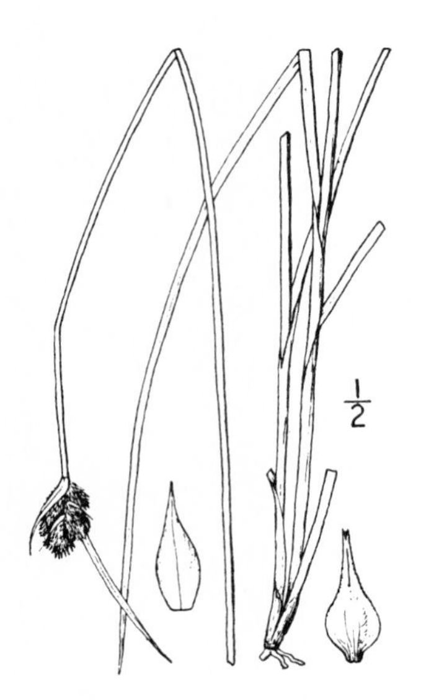

41. Carex athrostáchya Olney.
Slender-beaked Sedge. Fig. 717.

Carex athrostachya Olney, Proc. Am. Acad. **7**: 393. 1868.
Carex tenuirostris Olney, Am. Nat. **8**: 214. 1874.
Carex macloviana pachystachya f. involucrata Kükenth. in Engler Pflanzenreich 4[20]: 197. 1909.

Densely cespitose, the rootstock very short, the culms 1–9 dm. high, strict, slender, sharply triangular, smooth or somewhat roughened above. Leaf-blades 1.5–2.5 mm. wide; spikes 4–20, gynaecandrous, ovoid, 5 mm. long, 3–4 mm. wide, densely aggregated into an ovoid head, 1–3 cm. long; lowest 2–3 bracts elongated and much exceeding head, dilated and strongly brownish hyaline margined at base, spreading; scales ovate or lanceolate-ovate, shorter than the perigynia, acute or short cuspidate, brownish with hyaline margins; perigynia ascending, ovate-lanceolate, thin, 3–4 mm. long, 1.5 mm. wide, more or less nerved ventrally, rounded and substipitate at base, wing margined, serrulate above, tapering into a beak 1 mm. long, the tip slender, terete, brownish, hyaline at apex, obliquely cut dorsally; achenes lenticular; style slender, jointed with the achene; stigmas 2.

Meadows and copses, Arid Transition Zone; Saskatchewan and Yukon south to Colorado and California, extending as far south (along the Sierra Nevada) as the San Bernardino Mountains. Type locality: Yosemite Valley, California.

42. Carex unilaterális Mackenzie.
One-sided Sedge. Fig. 718.

Carex athrostachya Kükenth. in Engler Pflanzenreich 4[20]: 193. 1909; not Olney.
Carex unilateralis Mackenzie, Erythea **8**: 43. 1922.

Cespitose from very short creeping rootstocks, the culms erect, slender, obtusely triangular, slightly roughened above. Leaf-blades 2–4 mm. wide, flat, little roughened; spikes 6–20, gynaecandrous, ovoid, 5 mm. long, 3–4 mm. wide, densely aggregated into an ovoid head, 1–3 cm. long; lowest 2–3 bracts elongated and much exceeding head, dilated and strongly brownish hyaline margined at base, the lowermost erect appearing like a continuation of the culm, the head unilateral; scales ovate, strongly cuspidate, nearly equaling perigynia, reddish with lighter center and narrow hyaline margins; perigynia appressed, ovate-lanceolate, thin, 3–4 mm. long, 1.5 mm. wide, nerveless or nearly so ventrally, rounded and substipitate at base, wing margined, serrulate above, tapering into a short beak, the beak flat, serrulate nearly to apex, reddish tipped, not hyaline, at length bidentate; achenes lenticular; style slender, jointed with the achene; stigmas 2.

Wet meadows and copses, Transition and Canadian Zones; northwestern California to British Columbia. Type locality: Alton, Humboldt County, California.

43. Carex fèta Bailey.
Green-sheathed Sedge. Fig. 719.

Carex straminea mixta Bailey, Proc. Am. Acad. **22**: 151. 1886.
Carex feta Bailey, Bull. Torrey Club **20**: 417. 1893.
Carex feta multa Bailey, Bot. Gaz. **21**: 8. 1896.

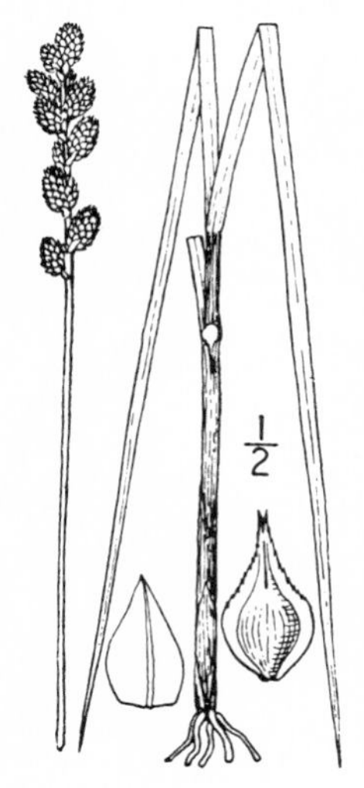

Densely cespitose, the culms 5–12 dm. high, obtusely triangular, smooth, the lower nodes exposed. Leaf-blades 2.5–4 mm. wide, flat, the sheaths green striate ventrally and hyaline only at mouth; head 2–8 cm. long, the spikes 5–15, gynaecandrous, greenish, aggregated or more or less separate, short oblong or suborbicular, 6–10 mm. long, 4–6 mm. wide, short clavate at base, rounded at apex, the 20–40 perigynia appressed or spreading in age; bracts inconspicuous; scales ovate, shorter than perigynia, greenish straw colored or light brownish tinged, obtusish or acutish; perigynia ovate, plano-convex, thickish, greenish, 3–3.5 mm. long, 1.75–2 mm. wide, nearly nerveless ventrally, rounded to a substipitate base, narrowly margined, serrulate above, contracted into a flat, serrulate, reddish tipped beak less than half length of body, obliquely cut dorsally and minutely bidentate; achenes lenticular; style slender, jointed to achene; stigmas 2.

Wet meadows, Humid Transition Zone; British Columbia to southern California; a characteristic species of the Pacific Coast States. Type locality: Cloverdale, Sonoma County, California.

44. Carex frácta Mackenzie. Fragile-sheathed Sedge. Fig. 720.

Carex specifica f. *brevi-fructus* Kükenth. Engler Pflanzenr. 4[20]: 199. 1909.
Carex specifica Mackenzie Bull. Torrey Club 43: 602. 1917; not Bailey.
Carex fracta Mackenzie, Erythea 8: 38. 1922.

Densely cespitose, the culms 5–12 dm. high, obtusely triangular, smooth or nearly so, the lower nodes not exposed. Leaf-blades 3–5 mm. wide, flat, the sheaths very thin ventrally, strongly prolonged at mouth above insertion of blade, fragile and soon breaking; head 2.5–7.5 cm. long, the spikes 7–15, greenish, aggregated or the lower slightly separate, short oblong or obovoid, 8–12 mm. long, 5–7 mm. wide, rounded or short clavate at base, rounded at apex, the 15–30 perigynia appressed or ascending in age; bracts inconspicuous; scales lance-ovate, acuminate or cuspidate, shorter than perigynia, greenish with hyaline margins; perigynia lance-ovate, 3–4.5 mm. long, thickish over achene, greenish, strongly nerved ventrally, round-tapering to a stipitate base, narrowly margined, serrulate above, tapering into the flat, serrulate bidentate beak nearly length of body; achenes lenticular; style slender, jointed to achene; stigmas 2.

Washington to southern California. Type locality: Mount Shasta, California.

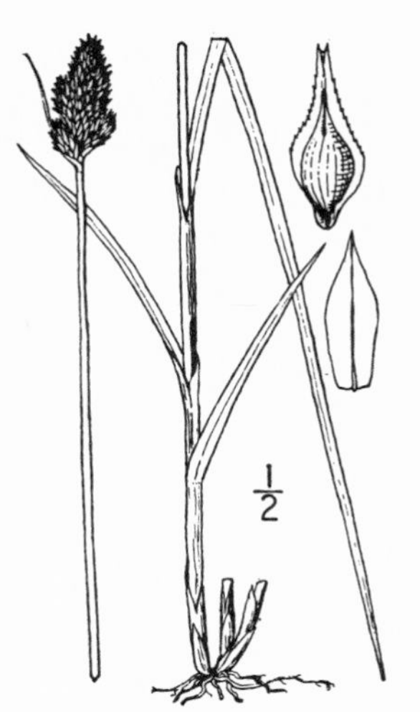

45. Carex cràwfordii Fernald. Crawford's Sedge. Fig. 721.

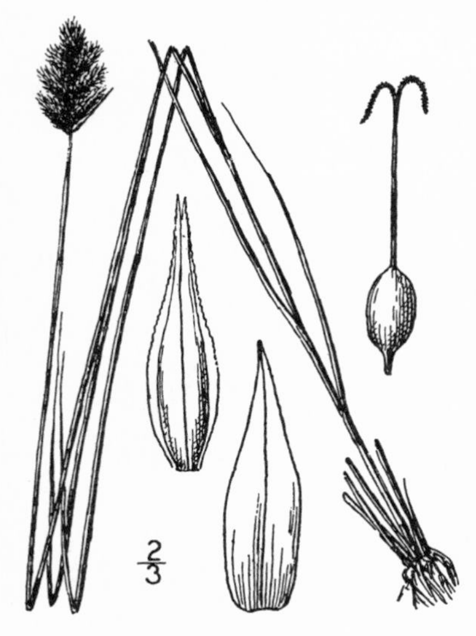

Carex scoparia minor Boott, Ill. Car. 3: 116. *pl. 369.* 1862.
Carex crawfordii Fernald, Proc. Am. Acad. 37: 469. 1902.
Carex crawfordii vigens Fernald, Proc. Am. Acad. 37: 470. 1902.

Densely cespitose, the culms slender, 2–3.5 dm. high, acutely triangular, roughened above. Leaf-blades 1–3 mm. wide; leaves of sterile shoots not very numerous, the blades erect or ascending; spikes 3–12, rather closely aggregated into an ovoid or linear-oblong head 12–26 mm. long, 4–8 mm. wide, the spikes light brownish, oblong to suborbicular, 10–18 mm. long, 6–10 mm. wide, gynaecandrous, many-flowered tapering at base, bluntly tapering at apex; lower 1–2 bracts developed but not conspicuous; scales ovate, acute or short acuminate, thin, shorter than perigynia, brown with hyaline margins; perigynia subulate, thin but distended over achene, the margin at base nearly obsolete, 4 mm. long, about 1 mm. wide at base, erect-ascending with erect beak, obscurely nerved on both faces, rounded at base, tapering into the serrulate bidentate beak half the length of the body; achenes lenticular, jointed with the slender style; stigmas 2.

Open places, Transition Zone; Newfoundland to British Columbia south to New Jersey, Michigan, and Washington. Type locality: base of White Mountains, New Hampshire.

46. Carex scopària Schk. Pointed Broom Sedge. Fig. 722.

Carex scoparia Schk.; Willd. Sp. Pl. 4: 230. 1805.
Carex lagopodioides scoparia Boeckl. Linnaea 39: 114. 1875.

Cespitose, the culms 1.5–8 dm. high, slender, erect, roughened above. Leaf-blades 1.5–3 mm. wide; leaves of sterile shoots not very numerous, the blades erect or ascending; spikes 3–10, closely aggregated or scattered, 6–16 mm. long, 4–6 mm. wide, gynaecandrous, brownish or straw-colored, narrowed at base, narrowed or sometimes truncate at apex, densely many flowered; lower bracts inconspicuous; scales ovate, shorter than the perigynia, thin, acute or acuminate, straw-colored or light brownish with narrow hyaline margins; perigynia lanceolate, very thin, narrowly wing-margined, 4–6.5 mm. long, 1.2–1.9 mm. wide, erect-ascending with appressed beak, nerved on both faces, round-tapering at base, tapering into the serrulate bidentate beak, half the length of the body; achenes lenticular, jointed with the slender style; stigmas 2.

Moist soil, Arid Transition Zone; Newfoundland to British Columbia south to North Carolina, Missouri, Colorado, and Oregon. Type locality: Boreal America.

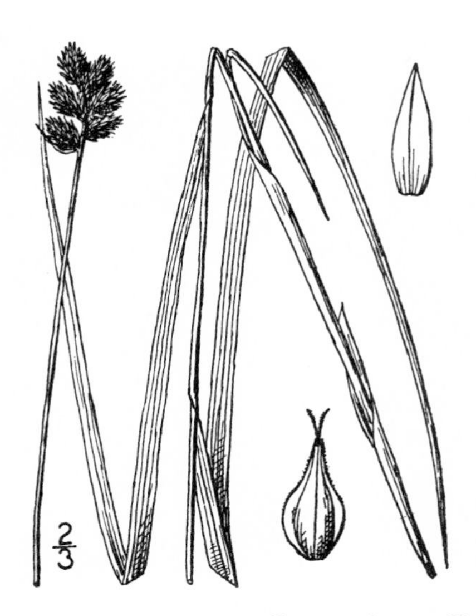

47. Carex bébbii Olney.
Bebb's Sedge. Fig. 723.

Carex bebbii Olney; Bailey Bot. Gaz. **10**: 379. 1885.
Carex tribuloides bebbii Bailey, Mem. Torrey Club **1**: 55. 1889.

Densely cespitose, the culms erect, slender, 2–8 dm. high, sharply triangular and roughened above. Leaf-blades 2–4.5 mm. wide; head oblong or linear-oblong, 14–28 mm. long, 8–12 mm. thick, the spikes 5–10, aggregated, gynaecandrous, brownish tinged, densely many-flowered, subglobose to broadly ovoid, 4–9 mm. long, 3–6 mm. wide, blunt; bracts inconspicuous; scales oblong-ovate, acute or short acuminate, shorter than perigynia, brownish; perigynia narrowly ovate, plano-convex, ascending, 3–4 mm. long, 1.5–2 mm. wide, distended over achene, wing-margined to the rounded base, obscurely nerved, tapering into a beak less than half length of body, the beak bidentate, flattened and serrulate nearly to tip; achenes lenticular; style slender, jointed with the achene; stigmas 2.

Low grounds, Arid Transition Zone; Newfoundland to British Columbia south to New Jersey, Montana, and Washington. Type locality: Illinois.

48. Carex ténera Dewey. Slender Sedge. Fig. 724.

Carex tenera Dewey, Am. Journ. Sci. **8**: 97. 1824.
Carex festucacea tenera Carey in A. Gray Man. 545. 1848.
Carex straminea tenera Boott, Ill. Car. **3**: 120. *pl. 384.* 1862.
Carex tenera erecta Olney; Kükenth., in Engler Pflanzenreich **4**[20]: 205 (as synonym). 1909.

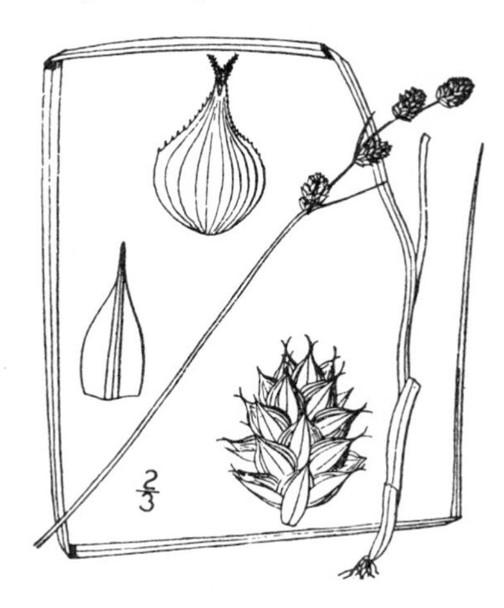

Densely cespitose, the culms slender, 3–7.5 dm. high, sharply triangular, rough above. Leaf-blades 1.5–2.5 mm. wide; head 2.5–5 cm. long, more or less strongly moniliform, the spikes 4–8, gynaecandrous, ovoid, 6–10 mm. long, 4.5–6 mm. wide, ovoid, rounded at apex, rounded or the terminal one short clavate at base, the 10–20 perigynia appressed; bracts inconspicuous; scales ovate, acute, shorter than perigynia, hyaline and tawny tinged with green midvein; perigynia ovate, plano-convex, thickish, 3.5 mm. long, nearly 2 mm. wide, green or in age straw-colored, nerved on both faces, rounded and sessile at base, contracted into a beak half length of body, the beak tawny tipped, flat, strongly serrulate and obliquely cut dorsally; achenes lenticular; style slender, jointed with the achene; stigmas 2.

Dry soil, Upper Sonoran Zone; Maine to Washington and Oregon, south to New Jersey, Illinois, and New Mexico; rare in our range. Type locality: probably western Massachusetts.

49. Carex brévior (Dewey) Mackenzie. Shorter-beaked Sedge. Fig. 725.

Carex straminea brevior Dewey, Am. Journ. Sci. **11**: 158. 1826.
Carex festucacea Schk. *brevior* Fernald, Proc. Am. Acad. **37**: 477. 1902.
Carex brevior Mackenzie, Bull. Torrey Club **42**: 605. 1915.

Cespitose, the culms erect, rather stiff, 3–10 dm. high, sharply triangular, more or less roughened above. Leaf-blades flat, 1.5–3 mm. wide; head 1.5–5 cm. long, 7–15 mm. wide, the spikes 3–10, gynaecandrous, aggregated or at times moniliform, subglobose, ovoid or oblong, 7–15 mm. long, 5–9 mm. wide, blunt at apex and from rounded to clavate at base, 8–30-flowered; lowest bract more or less strongly developed; scales ovate, obtuse to short acuminate, shorter than perigynia, brownish with green midvein and hyaline margins; perigynia broadly ovate to suborbicular, plano-convex 4–5.5 mm. long, 2.5–3.5 mm. wide, the walls thick, truncate, or rounded at base, strongly nerved dorsally, nerveless or nearly so ventrally, abruptly narrowed into the flat, strongly serrulate bidentate beak about 1 mm. long; achenes lenticular, sessile, 1.5–2 mm. long; style slender, jointed with the achene, stigmas 2.

Open sunny places, Transition Zone; New Brunswick to British Columbia, south to Tennessee, Texas, New Mexico, and Washington; apparently rare in our area. Type locality: probably western Massachusetts.

50. Carex straminifórmis Bailey.

Mount Shasta Sedge. Fig. 726.

Carex straminea congesta Olney, Proc. Am. Acad. 7: 393. 1868.
Carex straminiformis Bailey, Mem. Torrey Club 1: 24. 1889.

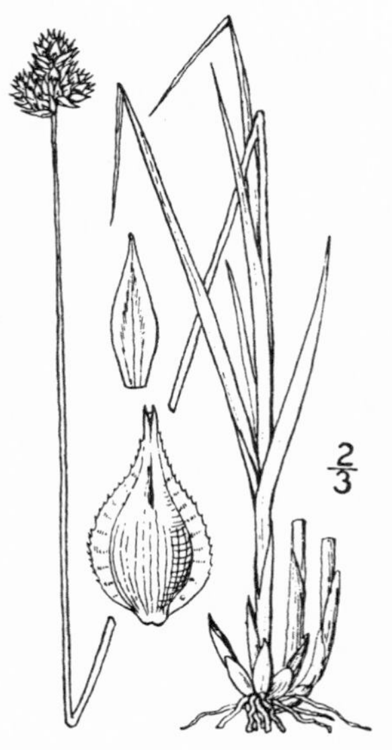

Densely cespitose, the culms 2.5–4 dm. high, strict but slender, slightly roughened beneath head. Leaf-blades 2–3 mm. wide; head 1.5–2.5 cm. long containing 3–6 closely aggregated suborbicular spikes 6–9 mm. long and nearly as wide with many spreading-ascending perigynia; bracts inconspicuous; scales ovate-lanceolate, acute, shorter than perigynia, reddish-brown with lighter midvein and hyaline margins; perigynia broadly ovate, 5 mm. long, 3 mm. wide, serrulate to middle, strongly winged to the rounded base, thin, lightly nerved dorsally, nerveless or nearly so ventrally, the margins conspicuously wrinkled dorsally, abruptly beaked, the beak flat, bidentate, one-third length of body; achenes lenticular; style slender, jointed with achene; stigmas 2.

High mountain summits, Canadian to Arctic-Alpine Zones; Washington through Oregon and northern California to western Nevada, and Mount Whitney, California. Type locality: Mount Shasta, California.

51. Carex multicostàta Mackenzie.

Many-ribbed Sedge. Fig. 727.

Carex multicostata Mackenzie, Bull. Torrey Club 43: 604. 1917.

Cespitose, the culms rather stout, 3–9 dm. high, many striate, bluntly triangular, slightly roughened beneath head. Leaf-blades 2.5–6 dm. wide; spikes about 10, closely aggregated, oblong or oblong-ovoid, 8–16 mm. long, 6–8 mm. wide, gynaecandrous, with 20–30 appressed perigynia; lower bracts little developed; scales ovate, obtuse to acute, reddish brown with lighter midvein and conspicuous hyaline margins; perigynia ovate, plano-convex, thick, the margins not wrinkled, green or in age straw-colored, 3.5–4.5 mm. long, 2–2.5 mm. wide, winged to the rounded base, serrulate to below middle, conspicuously many nerved dorsally and several-nerved ventrally, abruptly contracted into a broad, flat, bidentate non-hyaline reddish brown tipped beak, winged and serrulate to the tip; achenes lenticular, substipitate; stigmas 2.

Mountains of southern California, Transition and Canadian Zones. Type locality: Bear Valley Dam, San Bernardino County, California.

52. Carex petasàta Dewey.

Liddon's Sedge. Fig. 728.

Carex petasata Dewey, Am. Journ. Sci. 29: 246. *pl. W, f. 72.* 1836.
Carex liddonii Boott in Hook. Fl. Bor. Am. 2: 214. *pl. 215.* 1840.

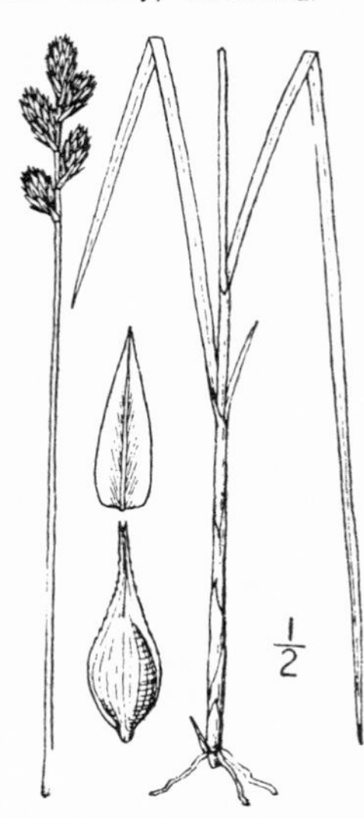

Cespitose, the culms 3–8 dm. high, obtusish, smooth or nearly so. Leaf-blades flat, 2–3 mm. wide; head erect, 2–4 cm. long, 1–1.5 cm. wide, the spikes 3–6, gynaecandrous, aggregated, oblong-ovoid, 8–16 mm. long (or the terminal longer and long clavate), 6–9 mm. wide, many-flowered; bracts little developed; scales ovate, acute, slightly longer and wider than mature perigynia, reddish brown with lighter center and broad white hyaline margins; perigynia narrowly ovate-lanceolate, appressed, 6–7 mm. long, 2.25 mm. wide, narrowly winged to the base, flat but strongly dilated over the achene, strongly many-nerved on both faces, round-tapering at base, serrulate winged above, tapering into the shallowly bidentate beak scarcely 2 mm. long; achenes lenticular, jointed with the slender style; stigmas 2.

Meadows and open woods, Transition and Canadian Zones; Saskatchewan to Colorado, west to Nevada, eastern Oregon and Washington. Type locality: Rocky Mountains.

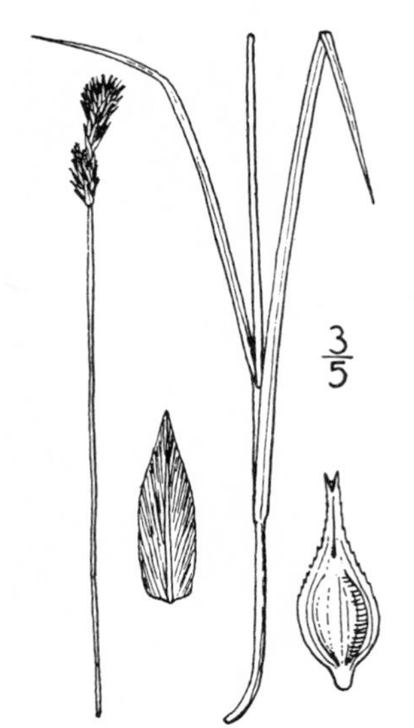

53. **Carex dàvyi** Mackenzie.
Davy's Sedge. Fig. 729.

Carex siccata W. Boott in S. Wats. Bot. Calif. **2**: 230. 1880; not Dewey.
Carex davyi Mackenzie, Bull. Torrey Club **43**: 606. 1917.

Densely cespitose, the culms 2.5–3.5 dm. high, smooth, erect, slender, much exceeding the leaves. Leaf-blades 1.5–2.5 dm. wide; head about 2.5 cm. long, the spikes usually three, approximate, oblong-obovoid, 12–18 mm. long, 6–8 mm. wide, with 10–15 appressed perigynia; bracts not developed; scales oblong-ovate, very obtuse, about half length of perigynia, chestnut with lighter center and hyaline margins; perigynia lanceolate, thin, green or in age straw-colored, 7.5 mm. long, 2 mm. wide, strongly many striate dorsally, less so ventrally, narrowly margined from base, contracted into a substipitate base, serrulate from middle of achene, tapering into the sharply bidentate beak, obliquely cut dorsally; the apex and dorsal suture reddish-tinged; achenes lenticular, jointed to the slender style; stigmas 2.

In the Sierra Nevada of California known from Placer to Tulare Counties, Transition and Canadian Zones. Type locality: Truckee River, Placer County, California.

54. **Carex specífica** Bailey.
Narrow-fruited Sedge. Fig. 730.

Carex scoparia fulva W. Boott in S. Wats. Bot. Calif. **2**: 237. 1880.
Carex specifica Bailey, Mem. Torrey Club **1**: 21. 1889.
Carex lancifructus Mackenzie, Bull. Torrey Club **43**: 607. 1917.

Culms 2.5–4.5 dm. high, smooth or nearly so, erect, stiff, much exceeding leaves. Leaf-blades 2–3.5 mm. wide, flat or canaliculate; head globose, 1.5–2 cm. long and nearly as wide, the spikes 6–10, gynaecandrous, oblong-ovoid, 6–9 mm. long, 4 mm. wide, tapering at each end, with 8–15 appressed perigynia; bracts little developed; scales lance-ovate, acute, exceeded by perigynia, reddish brown with lighter midvein and narrow hyaline margins; perigynia lanceolate, thin, plano-convex, straw-colored, 6 mm. long, 1.5 mm. wide, finely several to many nerved on both faces, contracted to a substipitate base, narrowly margined (serrulate above) to base, tapering into a bidentate beak one-third length of body, the beak reddish-tipped and obliquely cut dorsally; achenes lenticular, jointed with the slender style; stigmas 2.

In the Sierra Nevada of California from El Dorado to Tulare Counties; Transition and Canadian Zones. Type locality: Silver Valley, Alpine County, California.

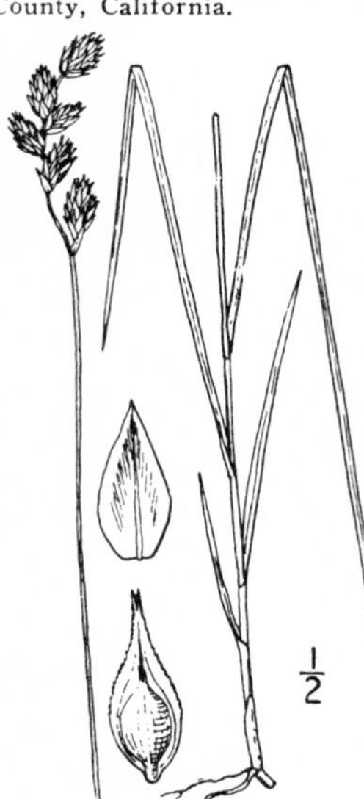

55. **Carex pratícola** Rydb.
Meadow Sedge. Fig. 731.

Carex adusta minor Boott in Hook. Fl. Bor. Am. **2**: 215. 1840.
Carex pratensis Drejer, Revis. Car. Bor. 24. 1841; not Hose 1797.
Carex praticola Rydb. Mem. N. Y. Bot. Garden **1**: 84. 1900.

Cespitose, the rootstocks short, the culms slender, often nodding, 2.5–6 dm. high, roughened beneath the head, much exceeding the leaves. Leaf-blades 1–2 mm. wide, flat; spikes 2–6, gynaecandrous, elliptic, 6–16 mm. long, 5 mm. wide, the upper contiguous, the lower remote, in a more or less moniliform head; bracts except lowest not developed; scales ovate, covering perigynia, obtuse, shining, brownish tinged with green midvein and hyaline margins; perigynia ovate-lanceolate, appressed, 4.5–6.5 mm. long, narrowly winged, round-tapering at the base, lightly nerved dorsally, nearly nerveless ventrally, pale green, membranaceous, tapering to a beak one-third length of body, terete and smooth above, the apex hyaline, bidentulate, obliquely cut and fissured dorsally; achenes lenticular; style slender, articulated to achene; stigmas 2.

Greenland to Alaska, south to northern Maine, Michigan, Colorado, and northern California; Transition to Boreal Zones. Type locality: Godthaab, Greenland.

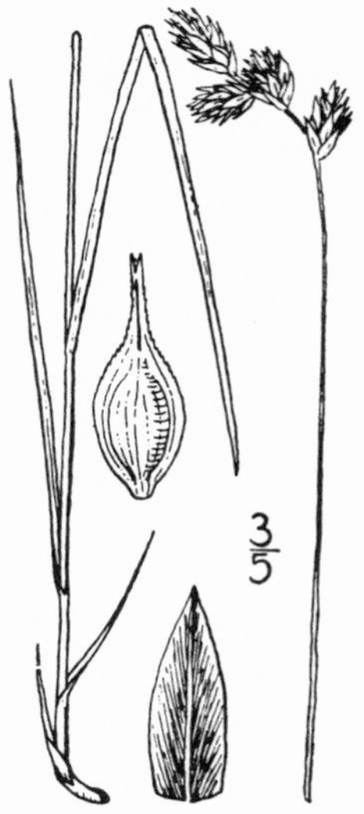

56. **Carex pìperi** Mackenzie.
Piper's Sedge. Fig. 732.

Carex pratensis furva Bailey in Macoun, Catal. Canad. Pl. **5**: 377. 1890.
Carex furva Piper, Contr. U. S. Nat. Herb. **11**: 166. 1906; not Webb 1838.
Carex piperi Mackenzie; Piper & Beattie Flora Northwest Coast 75. 1915.

Cespitose, the culms 3–8 dm. high, slightly roughened above. Leaf-blades 2–3.5 mm. wide, flat, on lower third of culm, but not bunched: head 2–4.5 cm. long, erect, the spikes 3–9, gynaecandrous, aggregated or slightly separate, ovoid-oblong, 10–18 mm. long, 5–6 mm. wide, short tapering at base, slightly pointed at apex, many-flowered; bracts little developed; scales ovate, acutish, slightly longer and wider than mature perigynia, brown with lighter center and conspicuous hyaline margins; perigynia ovate or lanceolate-ovate, erect appressed, 4–5 mm. long, 1.5–1.75 mm. wide, thinly plano-convex, narrowly winged to the base, strongly many nerved on outer surface, nearly nerveless on inner surface, round tapering at base, tapering into the shallowly bidentate beak 1 mm. long, the apex hyaline; achenes lenticular, jointed with the slender style; stigmas 2.

Damp meadows, Canadian Zone; Alberta to Wyoming, west to Oregon and British Columbia. Type locality: Vancouver Island.

57. **Carex tràcyi** Mackenzie.
Tracy's Sedge. Fig. 733.

Carex tracyi Mackenzie, Erythea **8**: 41. 1922.

Cespitose, the culms 2–8 dm. high, leafy nearly to middle, strict, roughened on the angles above, exceeding the leaves. Leaf-blades 2–4 mm. wide, flat, light green; head stiff, narrow, 1.5–4 cm. long, the spikes 4–7, gynaecandrous, aggregated or the lower a little separate, ovoid, or short-oblong, 7–15 mm. long, 5–9 mm. wide, obtuse at apex, the perigynia numerous, appressed-ascending; lower bracts little developed; scales ovate, acute, covering perigynia, brownish red with lighter midvein and broad hyaline margins; perigynia ovate, 4–4.5 mm. long, 2.25 mm. wide, plano-convex, membranaceous, at length brownish tinged, winged to the rounded base, serrulate to middle, strongly many nerved dorsally, lightly several nerved ventrally, abruptly beaked, the beak nearly length of body, bidentate, tawny tipped; achenes lenticular; stigmas 2.

Wet meadows and low swampy grounds, Canadian Zone; northern California to British Columbia. Type locality: Bald Mountain, Humboldt County, California.

58. **Carex phaeocéphala** Piper.
Mountain Hare Sedge. Fig. 734.

Carex leporina americana Olney, Proc. Am. Acad. **8**: 407. 1872, name only; ex. Bailey Proc. Am. Acad. **22**: 152. 1886.
Carex phaeocephala Piper, Contr. U. S. Nat. Herb. **11**: 172. 1906.

In large stools from densely matted rootstocks, the culms stiff, 1–3 dm. high, exceeding leaves, more or less roughened beneath head. Leaves bunched at the base, the blades more or less involute, 1.5–2 mm. wide; spikes 2–5 (rarely 7), gynaecandrous, aggregated, 6–12 mm. long, 5–8 mm. wide, forming an erect head, 12–25 mm. long; lowest bract occasionally developed; scales ovate, acute, covering perigynia, dark brownish with strongly hyaline margins and lighter midvein; perigynia ascending, oblong-ovate, plano-convex, 4.5 mm. long, 1.8 mm. wide, membranaceous, at length brownish, round tapering at the base, strongly nerved dorsally, obscurely nerved or nerveless ventrally, contracted into a minutely bidentate, serrulate beak about 1 mm. long, hyaline at the orifice; achenes lenticular; stigmas 2.

High mountain summits, Arctic-Alpine Zone; Alberta to Colorado, west to Alaska and British Columbia, and south to California, where confined to the higher peaks of the Sierra Nevada. Type locality: Oregon.

59. Carex leporinélla Mackenzie.
Sierra Hare Sedge. Fig. 735.

Carex leporinella Mackenzie, Bull. Torrey Club **43**: 605. 1917.

Very densely cespitose from short-creeping rootstocks, the culms 1.5–3 dm. high, smooth, exceeding leaves. Leaves bunched near the base, the blades 1.5–2 mm. wide, more or less involute; spikes 3–6, gynaecandrous, forming a head 1.5–3 cm. long, the spikes narrowly oblong-oval, 6–15 mm. long, 3–5 mm. wide, short clavate at base, the perigynia 8–20, appressed; lowest bract occasionally somewhat developed; scales ovate, covering perigynia, acute, reddish-brown with lighter midvein and hyaline margins; perigynia linear-lanceolate, boat-shaped, 4 mm. long, scarcely 1 mm. wide, very narrowly margined, serrulate above middle, finely striate dorsally, few nerved ventrally, tapering at base and at apex into the short (1 mm. long) beak, which is terete and smooth above, hyaline at apex and obliquely cut dorsally; achenes lenticular; style slender, articulated to achene; stigmas 2.

Summits of high mountains, Hudsonian and Arctic-Alpine Zones; Washington to California, where confined to the Sierra Nevada. Type locality: Pyramid Peak, El-dorado County, California.

60. Carex micróptera Mackenzie.
Small-winged Sedge. Fig. 736.

Carex microptera Mackenzie, Muhlenbergia **5**: 56. 1909.

Cespitose, the rootstocks short, stout, the culms 5–10 dm. high, smooth. Leaf-blades flat, 2–3.5 mm. wide; head ovoid or sub-orbicular, 12–18 mm. long, 10–16 mm. wide, the spikes 5–10, ovoid, closely aggregated, 5–8 mm. long, 4–6 mm. wide, gynaecandrous, 15–30-flowered; bracts little developed; scales ovate-lanceolate, acute, shorter than perigynia, brown with lighter midvein; perigynia lanceolate, 3.5–4 mm. long, 1–1.5 mm. wide, plano-convex, narrowly margined, ascending with ascending or somewhat spreading tips, light brownish or straw-colored, lightly nerved, rounded at base, tapering into the serrulate beak, one-third to one-half the length of the body, obliquely cut dorsally and becoming shallowly bidentate, the tip more or less hyaline; achenes lenticular, jointed with the slender style; stigmas 2.

Mountains, Transition and Canadian Zones; Alberta to Colorado, west to eastern Oregon and Washington. Type locality: Deeth, Elko County, Nevada.

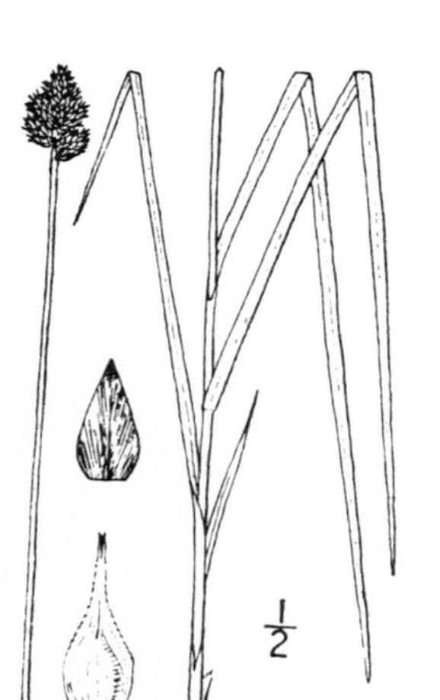

61. Carex festivélla Mackenzie.
Mountain Meadow Sedge. Fig. 737.

Carex festiva Bailey, Proc. Am. Acad. **22**: 153 (in greater part). 1886; not Dewey.

Carex festivella Mackenzie, Bull. Torrey Club **42**: 609. 1915.

Cespitose, the culms slender, mostly annual, 3–10 dm. high, smooth or roughened beneath the head. Leaf-blades 2–6 mm. wide; spikes 5–20, densely aggregated into a suborbicular to oblong-ovoid head 12–25 mm. long, 10–18 mm wide, the spikes ovoid, 5–12 mm. long, 4–8 mm. wide; bracts inconspicuous; scales ovate, obtuse to acutish, shorter than perigynia, dark-chestnut to brownish black with light midvein; perigynia 15–30 to a spike, appressed, 3.75–5 mm. long, 1.5–2 mm. wide, ovate, flat except where distended by achene, straw-colored or light brownish at maturity, winged, lightly nerved ventrally, tapering into a serrulate shallowly bidentate beak one-third length of whole, the beak terete and more or less hyaline tipped at apex, fissured and obliquely cut dorsally; achenes lenticular; style slender, jointed with achene; stigmas 2.

Meadows and mountain sides, Transition Zone; Alberta to New Mexico, west to Arizona, the Sierra Nevada of California, eastern Oregon and British Columbia. Type locality: Albany County, Wyoming.

62. **Carex nubícola** Mackenzie.
Cloud Sedge. Fig. 738.

Carex haydeniana Olney in Bot. King Explor. 366. 1871; not *C. haydenii* Dewey 1854.
Carex festiva haydeniana, W. Boott in S. Wats. Bot. Calif. **2**: 234. 1880.
Carex festiva decumbens Holm, Am. Journ. Sci. IV. **16**: 20, 26. 1903.
Carex nubicola Mackenzie, Bull. Torrey Club **36**: 480. 1909.

Densely cespitose, the culms 1–3.5 dm. high, erect or decumbent, smooth or nearly so. Leaf-blades 2–3 mm. wide; head ovoid or globose, 12–18 mm. long, 9–18 mm. wide, containing 4–7 densely aggregated, gynaecandrous, ovoid or subglobose spikes 5–9 mm. long, 4.5–8 mm. wide, each with 15–35 ascending perigynia with spreading beaks; bracts inconspicuous; scales ovate, acute, much narrower and shorter than perigynia, blackish with lighter center and hyaline margins; perigynia ovate, very flat and strongly winged, 4.5–5.5 mm. long, 2–2.75 mm. wide, rather weakly nerved, dark-tinged, rounded at base, abruptly contracted into a serrulate bidentulate beak one-half length of body; achenes lenticular, jointed with the slender style; stigmas 2.

Summits of high mountains, Hudsonian and Arctic-Alpine Zones; Alberta and Oregon south to Colorado, Utah, and California, where recorded only from Mount Dana. Type locality: Pagosa Peak, Colorado.

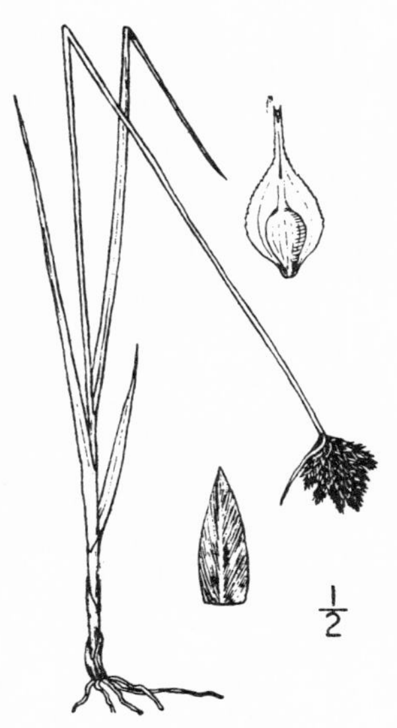

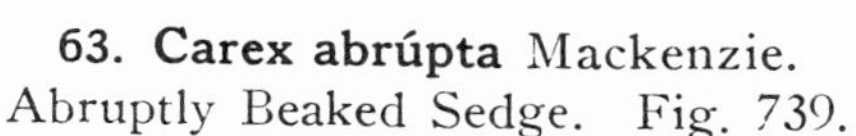

63. **Carex abrúpta** Mackenzie.
Abruptly Beaked Sedge. Fig. 739.

Carex abrupta Mackenzie, Bull. Torrey Club **43**: 618. 1917.

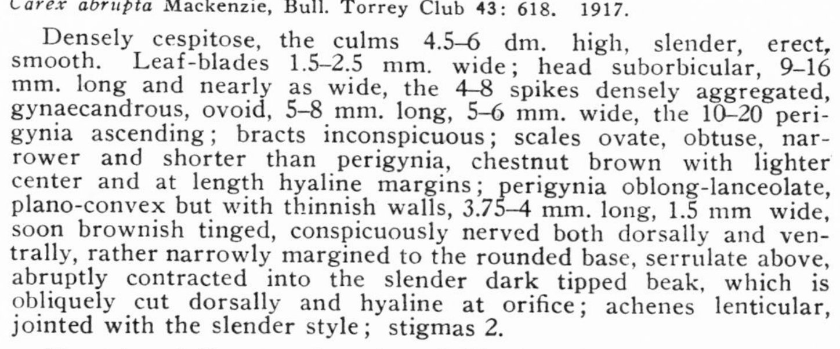

Densely cespitose, the culms 4.5–6 dm. high, slender, erect, smooth. Leaf-blades 1.5–2.5 mm. wide; head suborbicular, 9–16 mm. long and nearly as wide, the 4–8 spikes densely aggregated, gynaecandrous, ovoid, 5–8 mm. long, 5–6 mm. wide, the 10–20 perigynia ascending; bracts inconspicuous; scales ovate, obtuse, narrower and shorter than perigynia, chestnut brown with lighter center and at length hyaline margins; perigynia oblong-lanceolate, plano-convex but with thinnish walls, 3.75–4 mm. long, 1.5 mm wide, soon brownish tinged, conspicuously nerved both dorsally and ventrally, rather narrowly margined to the rounded base, serrulate above, abruptly contracted into the slender dark tipped beak, which is obliquely cut dorsally and hyaline at orifice; achenes lenticular, jointed with the slender style; stigmas 2.

Mountains of Oregon and northern California and throughout the Sierra Nevada, and in the San Bernardino and San Jacinto Ranges, Transition and Canadian Zones. Type locality: West Branch of North Fork of Feather River, near Stirling, Butte County, California.

64. **Carex mariposàna** Bailey.
Mariposa Sedge. Fig. 740.

Carex mariposana Bailey, Bull. Torrey Club **43**: 619. 1917.

Densely cespitose, the culms 2.5–6 dm. high, slender, smooth. Leaf-blades 2–3 mm. wide; head oblong or ovoid, 2–3.5 cm. long, the spikes 4–12, gynaecandrous, the upper approximate, the lower 1–3 slightly separate, ovoid or oblong-ovoid, 8–12 mm. long, 4.5–7 mm. wide, the 10–20 perigynia closely appressed with erect tips; bracts inconspicuous; scales ovate, acute, narrower and shorter than perigynia, reddish with lighter center; perigynia narrowly ovate, plano-convex at maturity, the walls membranaceous, 5 mm. long, 1.75 mm. wide, deep green or in age greenish straw-colored and contrasting strongly with scales, strongly nerved on both faces, narrowly winged to the round-tapering base, serrulate to middle, tapering into the slender beak one-fourth length of the body, the beak light reddish tipped and obliquely cut dorsally; achenes lenticular, jointed to the slender style; stigmas 2.

In the California Mountains, Transition to Canadian Zones; Shasta to Tulare Counties in the Sierra Nevada and in the higher southern mountains. Type locality: Tuolumne Meadows, Yosemite National Park.

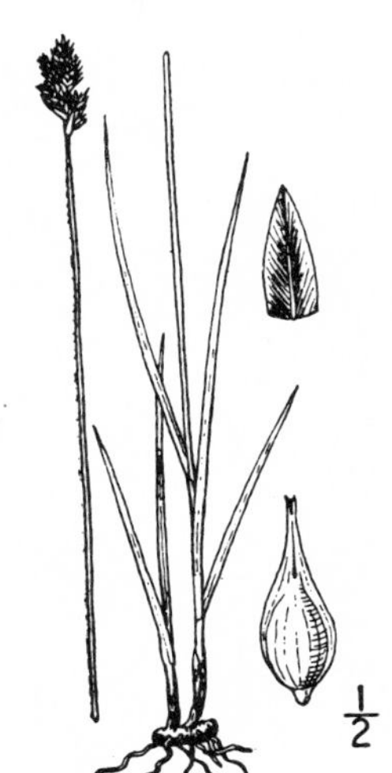

65. Carex intégra Mackenzie.
Smooth-beaked Sedge. Fig. 741.

Carex integra Mackenzie, Bull. Torrey Club **43**: 608. 1917.

Very densely cespitose, the culms 1.5–3.5 dm. high, slender but erect, smooth. Leaf-blades 1–2 mm. wide; head 1–2 cm. long, the spikes 4–8, gynaecandrous, densely aggregated, obovoid or oblong-obovoid, 4–8 mm. long, 3.5–5 mm. wide, the 10–20 perigynia appressed or appressed-ascending; lowest bract somewhat developed; scales ovate, acute or short cuspidate, shorter than perigynia, dark chestnut with prominent midvein and in age hyaline margins; perigynia lanceolate, thickish, plano-convex, very small, 2.25–2.75 mm. long, 0.75–1 mm. wide, brownish tinged, lightly few nerved dorsally, nerveless or very obscurely nerved ventrally, narrowly margined to the round-tapering base, more or less contracted into the slender beak half length of body or more, the margins smooth, chestnut-brown tipped with white hyaline orifice and obliquely cut dorsally; achenes lenticular jointed with the slender style; stigmas 2.

In the Sierra Nevada of California, known from Tulare to Siskiyou Counties, and extending north in the Cascade Mountains of Oregon, Transition Zone. Type locality: Summit, Placer County, California.

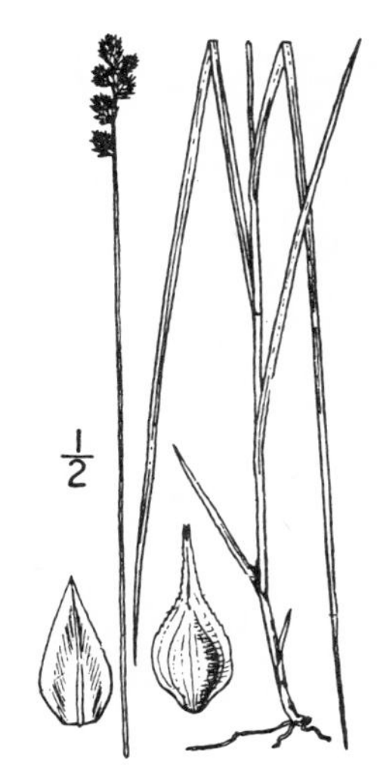

66. Carex teneraefórmis Mackenzie.
Sierra Slender Sedge. Fig. 742.

Carex teneraeformis Mackenzie, Bull. Torrey Club **43**: 609. 1917.

Cespitose, the culms 3–4.5 dm. high, very slender, smooth or nearly so. Leaf-blades deep green, 1–2.5 mm. wide; head 1.5–2.5 cm. long, the spikes 5–8, gynaecandrous, readily distinguishable and more or less separate, 3.5–6 mm. long, 3.5–4.5 mm. wide, the 6–12 perigynia loosely appressed or spreading-ascending; lowest bract somewhat developed; scales ovate, acute, shorter than perigynia, light brown with lighter midvein and inconspicuous hyaline margins; perigynia ovate, plano-convex, green, 3.25 mm. long, 1.25 mm. wide, thickish, winged to base, the walls submembranaceous, conspicuously several nerved on outer face, nerveless or nearly so on inner, rounded at base, serrulate above, tapering into a slender white hyaline-tipped beak 1 mm. long, obliquely cut dorsally, the tip terete and scarcely serrulate; achenes lenticular, jointed with slender style; stigmas 2.

In the Sierra Nevada of California, Transition Zone; Butte to Tulare Counties; also on Mount Sanhedrin, Lake County. Type locality: Jonesville, Butte County, California.

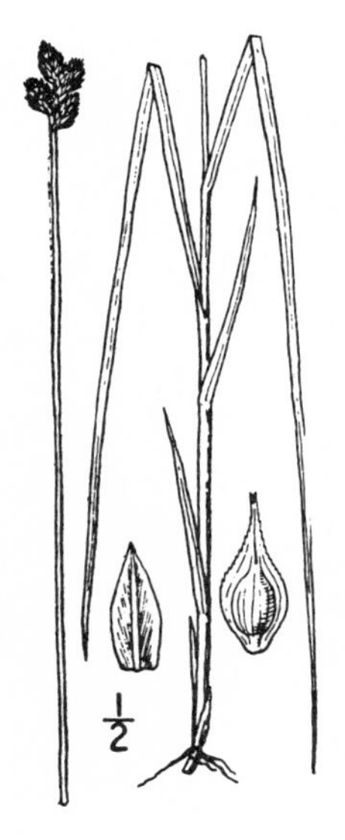

67. Carex subfúsca W. Boott.
Rusty Sedge. Fig. 743.

Carex subfusca W. Boott in S. Wats. Bot. Calif. **2**: 234. 1880.
Carex macloviana subfusca (W. Boott) Kükenth. in Engler Pflanzenreich **4**[20]: 197. 1909.

Cespitose, the culms 2–6.5 dm. high, slender but firm, smooth or very nearly so. Leaf-blades 1.5–3 mm. wide; head oblong or ovoid, 1–2 cm. long, the spikes 4–8, gynaecandrous, well-defined but closely aggregated, ovoid or oblong, 4–10 mm. long, 2.5–4.5 mm. wide. rounded at apex, rounded or somewhat tapering at base, the 8–15 perigynia appressed-ascending; bracts little developed; scales shorter than perigynia, ovate, acute, brownish with lighter midvein and inconspicuous hyaline margins; perigynia ovate, 3–3.5 mm. long, 1–1.5 mm. wide, plano-convex, thickish, winged to the rounded base, serrulate above, nerved ventrally, contracted into a beak half the length of the body or more, somewhat flattened nearly to apex and obliquely cut dorsally; achenes lenticular; style slender, jointed with achenes; stigmas 2.

Mountains and foothills, Transition and Canadian Zones; Oregon to southern California, and eastward to Arizona. Type locality: Lake Tahoe.

68. **Carex pachystáchya** Cham.

Thick-headed Sedge. Fig. 744.

Carex pachystachya Cham.; Steud. Synop. Cyper. 197. 1855.
Carex festiva pachystachya Bailey, Mem. Torrey Club 1: 51. 1889.
Carex olympica Mackenzie, Bull. Torrey Club 43: 610. 1917.

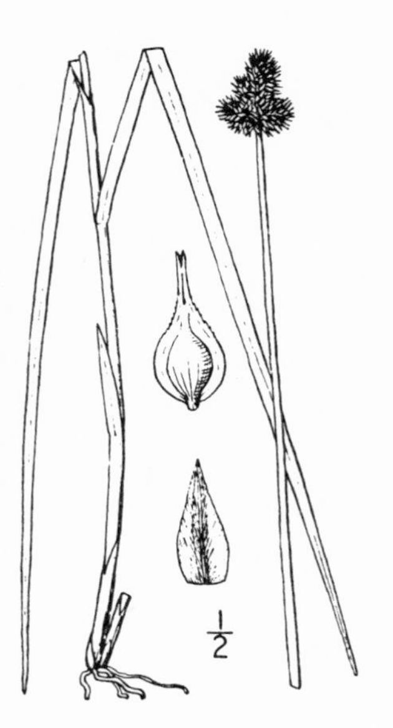

Densely cespitose, the short rootstocks conspicuously fibrillose, the culms erect, slender, 1.5–3 dm. high, smooth. Leaf-blades 1–4 mm. wide; head 12–24 mm. long, the spikes 3–8, gynaecandrous, densely aggregated or the lower more or less separated, ovoid or suborbicular, 4–8 mm. long, 4–6 mm. wide, the 8–30 perigynia ascending or in age more or less spreading; bracts inconspicuous; scales ovate, shorter than the perigynia, chestnut-brown with lighter midvein and slightly hyaline margins; perigynia ovate, strongly plano-convex, thickish, 3.5–5 mm. long, 1.5–2 mm. wide, greenish or in age olive-brown, rounded at base, nerveless or nearly so ventrally, narrowly winged, the body little serrulate, contracted into a beak about half length of body, obliquely cut dorsally, and terete and nearly smooth at apex; achenes lenticular, jointed with the slender style; stigmas 2.

A very variable species extending from the Aleutian Islands south to Colorado and northern California; abundant in parts of Washington and Oregon; Tansition and Canadian Zones. Type locality: Unalaska.

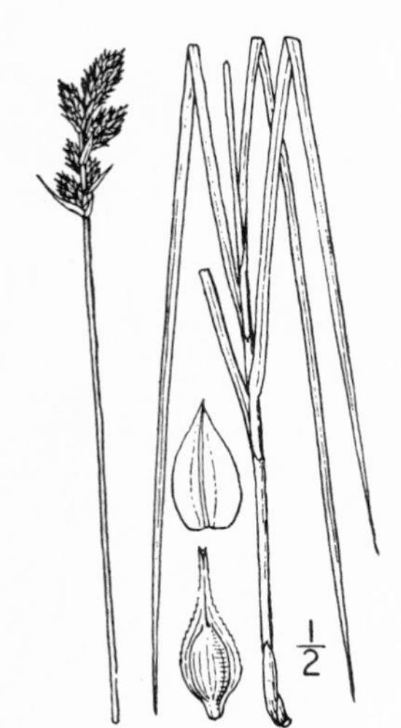

69. **Carex ampléctens** Mackenzie.

Clasping-bracted Sedge. Fig. 745.

Carex amplectens Mackenzie, Bull. Torrey Club 43: 611. 1917.

Cespitose, the culms stiff, 5–8 dm. high, slightly roughened above, the sterile shoots well-developed. Leaf-blades 2.5–4 mm. wide, the sheaths rather loose; head 2.5–3.5 cm. long, the 6–12 spikes approximate or the lower slightly separate, ovoid or oblong-ovoid, 7–15 mm. long, 4–5 mm. wide, short clavate and sparingly staminate at base, the 15–35 perigynia closely appressed; bracts conspicuous, dilated at base, closely appressed to spikes, the lower 3 or 4 usually prolonged and from nearly equalling to exceeding head; scales ovate, acute or short-cuspidate, shorter than perigynia, greenish and slightly tawny tinged, the midvein prominent; perigynia ovate, 3.5–4 mm. long, 1.75 mm. wide, light green, rounded and contracted at base, margined, several nerved on both faces, strongly serrulate, contracted into a beak one-third to one-half length of body, the tip slightly tawny-tinged and obliquely cut dorsally; achenes lenticular, jointed to the slender style; stigmas 2.

In the Sierra Nevada of California, Transition Zone; Shasta County to Tulare County. Type locality: Lover's Leap, Eldorado County, California.

70. **Carex hàrfordii** Mackenzie.

Harford's Sedge. Fig. 746.

Carex festiva stricta Bailey, Mem. Torrey Club 1: 51 (at least in part). 1889.
Carex harfordii Mackenzie, Bull. Torrey Club 43: 615. 1917.

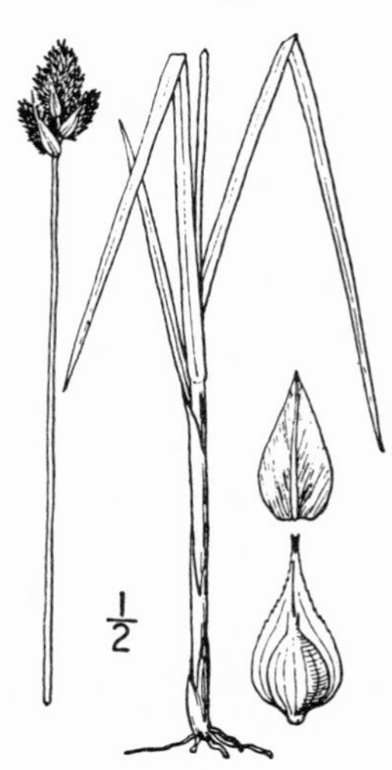

Cespitose, the rootstocks thick, short; the culms 2.5–8 dm. high, erect, stiff, smooth. Leaf-blades 2.5–4.5 mm. wide; head 1.5–2.5 cm. long, globose to oblong-ovoid, the spikes 10–20, gynaecandrous, closely aggregated, ovoid, 6–10 mm. long, 4–6 mm. wide, the perigynia 10–30, appressed-ascending or in age spreading; lower one or more bracts conspicuous; scales ovate, acute or short cuspidate, reddish brown with sharply defined lighter midvein and very narrow hyaline margins; perigynia narrowly ovate, thick and strongly plano-convex at maturity, 3.5–4.25 mm. long, 1.5 mm. wide, several nerved ventrally, strongly finely nerved dorsally, round-tapering to a substipitate base, narrowly margined (serrulate above) from base, tapering into a slender beak one-fourth the length of the body, the beak chestnut-tinged, obliquely cut dorsally and hyaline at orifice; achenes lenticular, jointed with the slender style; stigmas 2.

Coastal counties of California, Humid Transition Zone; Humboldt Bay to San Francisco Bay. Type locality: California; locality not indicated.

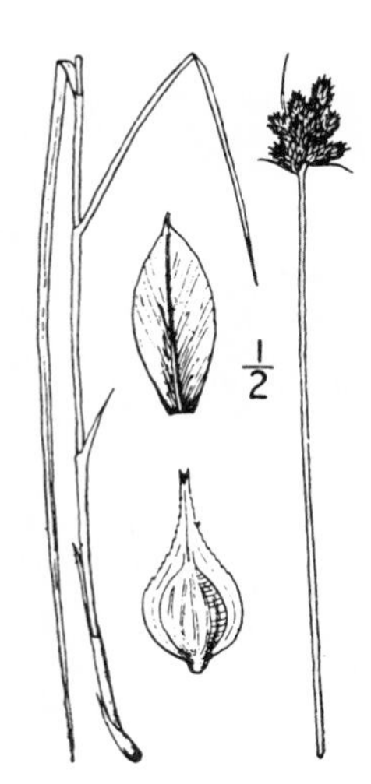

71. Carex montereyénsis Mackenzie.
Monterey Sedge. Fig. 747

Carex montereyensis Mackenzie, Erythea **8**: 92. 1922.

Cespitose, the rootstock very short, thick, the culms 8–10 dm. high, erect, very slender, smooth; sterile shoots numerous, elongated; leaves 2.5–3 mm. wide; the basal bladeless sheaths very conspicuous; head 1.5–2.5 cm. long, ovoid or short-oblong, the spikes 8–12, gynaecandrous, closely aggregated, ovoid or oblong-ovoid, 6–9 mm. long, 4.5–6 mm. wide, the perigynia 15–35, ascending or spreading-ascending; lower one or more bracts conspicuous; scales ovate-lanceolate, cuspidate or short awned, yellowish brown with sharply defined lighter midvein and very narrow hyaline margins; perigynia ovate, plano-convex, 3.25 mm. long, 1.5 mm. wide, several–many nerved dorsally, prominently few-nerved ventrally, rounded and short stipitate at base, narrowly margined (serrulate above) from base, contracted into a slender beak half the length of the body, the beak yellowish brown tinged, obliquely cut dorsally and hyaline at orifice; achenes lenticular, jointed with the slender style; stigmas 2.

Pine forests, Transition Zone; Pacific Grove, Monterey County, California.

72. Carex préslii Steud.
Presl's Sedge. Fig. 748.

Carex preslii Steud. Synops. Cyper. 242. 1855.
Carex multimoda Bailey, Bot. Gaz. **21**: 5. 1896.

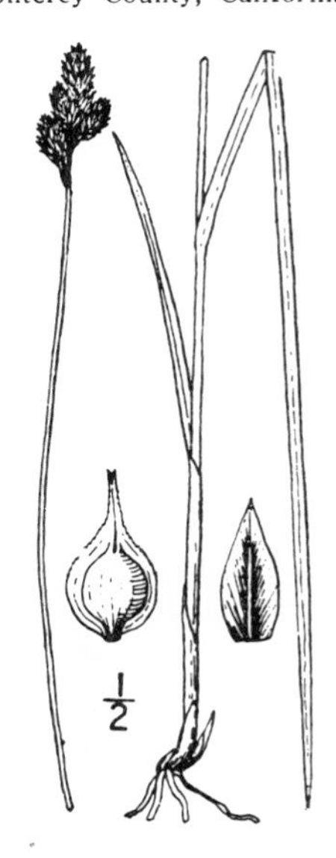

Densely cespitose, the culms erect, slender, 2.5–7 dm. high, light brownish at base, roughened above. Leaf-blades flat, 1.5–4 mm. wide; head 12–20 mm. long, 6–10 mm. wide, the spikes 3–8, gynaecandrous, aggregated or the lower somewhat separate, ovoid or suborbicular, 5–8 mm. long and slightly narrower, rounded at apex, the lateral rounded and the terminal short-tapering at base, 10–25-flowered; bracts little developed; scales ovate, sharp pointed, shorter than the perigynia, reddish brown with lighter midvein and narrow hyaline margins; perigynia ovate, with spreading tips, distended by the thick achene, 3.5 mm. long, 2 mm. wide, green or yellowish brown tinged in age, strongly winged, rounded at the base, nerveless ventrally or obscurely nerved, abruptly contracted into the yellowish brown tipped beak 1 mm. long, the erect teeth closely contiguous; achenes lenticular, jointed with the slender style; stigmas 2.

Mountains, Canadian and Hudsonian Zones; Montana west to British Columbia and Oregon. Type locality: Nootka Sound.

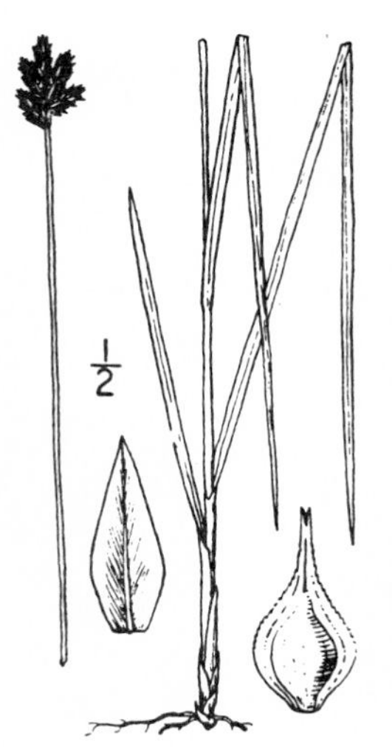

73. Carex pachycàrpa Mackenzie.
Thick-fruited Sedge. Fig. 749.

Carex adusta congesta W. Boott in S. Wats. Bot. Calif. **2**: 238. 1880.
Carex liddoni incerta Bailey, Bot. Gaz. **13**: 88. 1888.
Carex pachycarpa Mackenzie, Bull. Torrey Club **43**: 616. 1917.

Culms 3–6 dm. high, obtusely triangular, smooth. Leaf-blades 2.5–4 mm. wide; head 1.5–2.5 cm. long, globose or short ovoid, the spikes 5–8, gynaecandrous, closely aggregated, ovoid, 6–10 mm. long, 6–9 mm. wide, the 10–20 perigynia appressed, the beaks not spreading; bracts inconspicuous; scales ovate, acute or subcuspidate, shorter than perigynia, light reddish-brown with sharply defined lighter midvein and hyaline margins; perigynia ovate, plano-convex, thick, 5 mm. long, 2 mm. wide, many striate dorsally, finely many striate ventrally, narrowly margined, serrulate above, contracted into the slender bidentate beak half length of body, the apex slender, reddish-tinged, obliquely cut dorsally and hyaline at orifice; achenes lenticular, jointed with the slender style; stigmas 2.

Mountains of Oregon and northern California, extending south in the Sierra Nevada to Tulare County; Transition Zone. Type locality: Big Tree Road, Silver Valley, California.

74. **Carex sub-bracteàta** Mackenzie.
Small-bracted Sedge. Fig. 750.

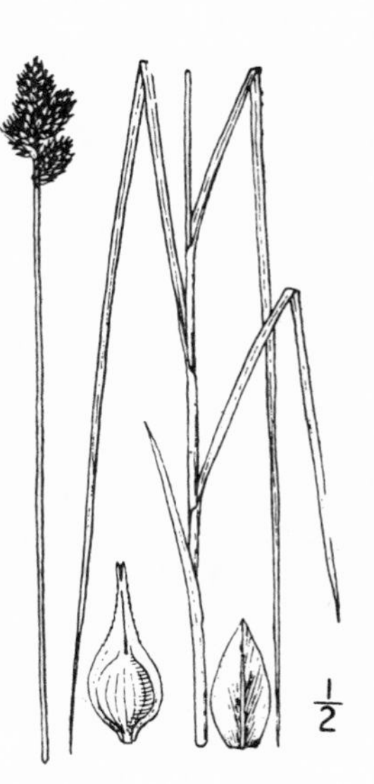

Carex sub-bracteata Mackenzie, Bull. Torrey Club **43**: 612. 1917.

Cespitose, the rootstocks short-creeping, tough, black fibrillose, the culms 3.5–12 dm. high, stoutish, smooth, obtusely triangular. Leaf-blades 2.5–4 mm. wide; head 1.5–2.5 cm. long, globose or ovoid, the spikes 5–10, gynaecandrous, closely aggregated, ovoid, 6–10 mm. long, 4–6 mm. wide, the perigynia 10–20, appressed or in age appressed-ascending; one or two of the lower bracts conspicuous, dilated at base; scales ovate, obtuse or acutish, shorter than perigynia, reddish-brown with lighter center and hyaline margins; perigynia narrowly ovate, thick, plano-convex, 3.5–4.5 mm. long, 1.5 mm. wide, dull green or soon yellowish brown, finely few-nerved dorsally, nerveless or essentially so ventrally, round-tapering at base, narrowly margined (serrulate above) from base, contracted into a beak one-third length of body, the beak chestnut tinged, obliquely cut dorsally and hyaline at apex; achenes lenticular, jointed with the slender style; stigmas 2.

In the Coast Ranges of California from Santa Barbara County north to Mendocino County; Upper Sonoran and Transition Zones. Type locality: Oakland, California.

75. **Carex gracílior** Mackenzie.
Slenderer Sedge. Fig. 751.

Carex gracilior Mackenzie, Bull. Torrey Club **43**: 614. 1917.

Cespitose from slender short-creeping black fibrillose rootstocks, the culms 3–6 dm. high, slender, smooth or nearly so on the angles. Leaves on lower third of culm, but not bunched, the blades 1–2 mm. wide, 0.5–2 dm. long; head narrow or oblong, 12–20 mm. long, the spikes 3–6, gynaecandrous, aggregated or lower 1–2 more or less separate, suborbicular, bluntish, 5–8 mm. long, 4–6 mm. wide, the 4–12 perigynia ascending or spreading-ascending, with conspicuous beaks; lower bract developed; scales ovate, obtusish or acutish, shorter than perigynia, chestnut brown with lighter midvein and narrow hyaline margins; perigynia narrowly ovate, thick, plano-convex, 3.5–4.5 mm. long, 1.5 mm. wide, several nerved dorsally, nerveless or nearly so ventrally, round-tapering at base, tapering into the slender beak half length of body, the apex chestnut brown and hyaline tipped, obliquely cut dorsally at mouth; achenes lenticular, jointed with the slender style; stigmas 2.

In the Coast Ranges of California from San Mateo County to Mendocino County; Transition Zone. Type locality: Cloverdale, Sonoma County, California.

76. **Carex paucifrúctus** Mackenzie.
Few-fruited Sedge. Fig. 752.

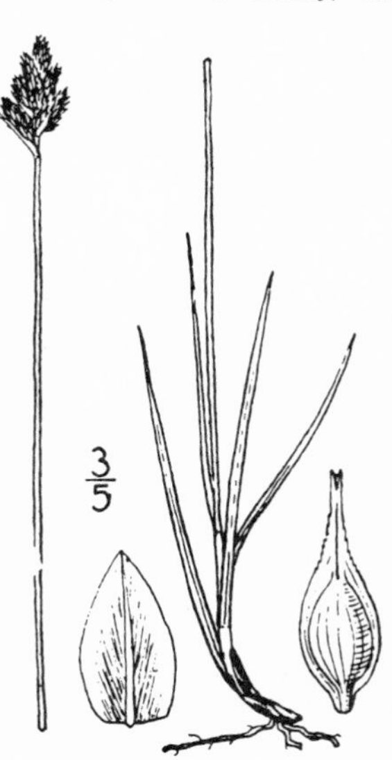

Carex paucifructus Mackenzie, Bull. Torrey Club **43**: 615. 1917.

Densely cespitose, the culms 1–2.5 dm. high, erect, stiff, smooth, sharply triangular. Leaves bunched above the base, the blades 1.5–3 mm. wide, 3–7 cm. long; head 1–2 cm. long, ovoid or oblong, the spikes 4–8, gynaecandrous, aggregated, ovoid, tapering at base, the basal staminate flowers rather conspicuous, round-tapering at apex, 6–9 mm. long, 4.5 mm. wide, the perigynia 6–12, appressed or appressed-ascending, the beaks not conspicuous; lowest bract much shorter than head; scales ovate, shorter than perigynia, chestnut with lighter midvein and conspicuous hyaline margins; perigynia ovate, thick, strongly plano-convex, 4 mm. long, 1.5 mm. wide, nerveless ventrally, finely many-nerved dorsally, margined to the round-tapering base, serrulate above, tapering into the beak one-third length of body, the tip slender, reddish tipped, obliquely cut dorsally; achenes lenticular, jointed with the slender style; stigmas 2.

Known only in the Sierra Nevada from Sierra and Eldorado Counties, California; Canadian Zone. Type locality: Devil's Basin, Eldorado County, California.

77. **Carex dispérma** Dewey. Soft-leaved Sedge. Fig. 753.

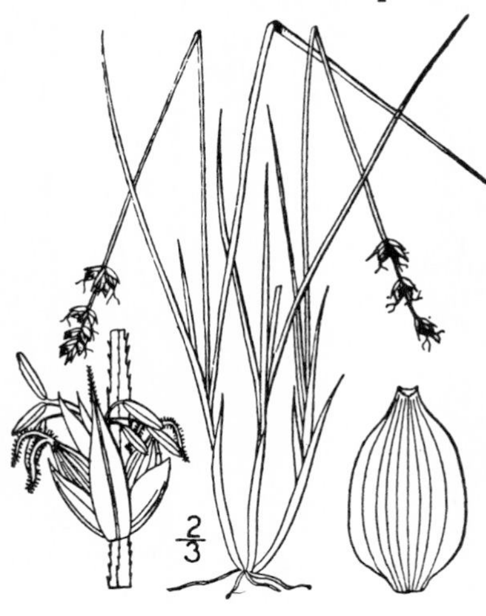

Carex tenella Schk. Riedgr. **1**: *23. pl. pp., f. 104.* 1801; not Thuill 1799.
Carex disperma Dewey, Am. Journ. Sci. **8**: 266. 1824.
Carex blytii Nylander, Spic. Fl. Fenn. **2**: 35. 1846.

In large clumps, the rootstocks sending out long slender stolens, the culms very weak 1.5–6 dm. high, rough above, exceeding leaves. Leaf-blades 1–1.5 mm. wide, flat, flaccid, deep green; spikes distant or the upper aggregated, in a terminal inflorescence 1.5–2.5 cm. long, each with 1–5 ascending perigynia below and 1–2 staminate flowers above; bracts wanting or the lower slightly developed; scales ovate triangular, shorter than perigynia, sharp-pointed, hyaline with green midvein; perigynia 2 mm. long, 1.5 mm. wide, ovoid-elliptic, flattened suborbicular in cross-section, smooth, light green, not margined, finely nerved, rounded and slightly stipitate at base, rounded and abruptly beaked at apex, the minute beak smooth, 0.25 mm. long, hyaline at orifice; achenes lenticular; style slender, jointed with achene; stigmas 2.

Boggy woods, Transition to Hudsonian Zones; Newfoundland to Alaska, south to New Jersey, Indiana, New Mexico, and the Sierra Nevada of California, where rare. Type locality: Massachusetts.

78. **Carex canéscens** L. Silvery or Hoary Sedge. Fig. 754.

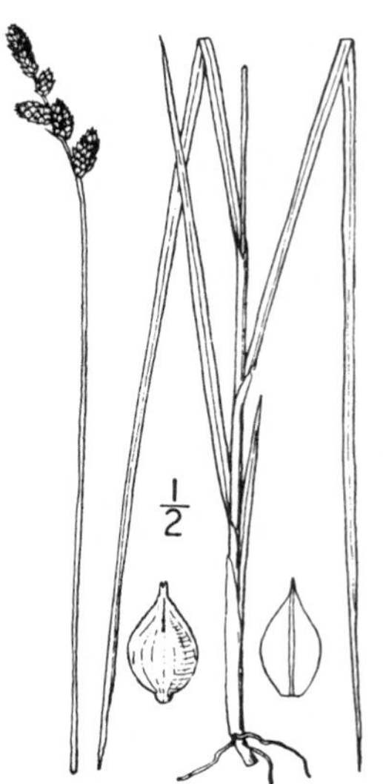

Carex canescens L. Sp. Pl. 974. 1753.
Carex curta Good. Trans. Linn. Soc. **2**: 145. 1794.
Carex lagopina W. Boott in S. Wats. Bot. Calif. **2**: 233. 1880; not Wahl.
Carex canescens dubia Bailey, Bot. Gaz. **9**: 119. 1884.

Cespitose in large clumps, the culms erect, slender, 2.5–8 dm. high, roughened immediately beneath head, mostly exceeding the leaves. Leaf-blades glaucous, flat, 2–4 mm. wide; spikes 4–9, the lower remote (sometimes but little so), 3–12 mm. long, 3–5 mm. wide, closely 10–30-flowered; scales ovate, sharp-pointed, shorter than perigynia, hyaline with green keel; perigynia appressed-ascending, 1.8–2.8 mm. long, 1–1.8 mm. wide, membranaceous, pale green, white puncticulate, faintly few-nerved, rounded short stipitate and spongy at base, minutely beaked, the beak with margins minutely serrulate and orifice emarginate or entire; achenes lenticular, closely enveloped; style slender, jointed with achenes; stigmas 2.

Swamps and bogs, Canadian Zone; Labrador to Alaska, south to Virginia and the Sierra Nevada of California, where rare. Widely distributed in Eurasia and recorded from South America and Australia. Type locality: Europe.

79. **Carex brunnéscens** (Pers.) Poir. Brownish Sedge. Fig. 755.

Carex canescens alpicola Wahl. Vet. Akad. Handl. Stockholm **24**: 147. 1803.
Carex curta brunnescens Pers. Syn. **2**: 539. 1807.
Carex brunnescens Poir. Encycl. Suppl. **3**: 286. 1813.
Carex vitilis Fries. Mant. **3**: 137. 1842.
Carex sphaerostachya Dew. Am. Journ. Sci. **49**: 44. *pl. EE, f. 110.* 1845.

Cespitose in large clumps, the culms slender, roughish above 2–5 dm. high, roughened above, mostly exceeding the leaves. Leaf-blades rather dark green, not glaucous, 1–2.5 mm. wide; spikes 4–8, the lower remote, the upper approximate, gynaecandrous, subglobose or short oblong, 4–10 mm. long, 4–10-flowered; lowest bract usually developed; scales ovate, shorter than the perigynia, thin, hyaline with green midvein and brownish tinged; perigynia ascending-spreading, 2 mm. long, oval-ovate, membranaceous, plano-convex, green at length brownish, glabrous, densely white puncticulate, obsoletely nerved, rounded short-stipitate and spongy at base, abruptly beaked, the beak serrulate, bidentate, one-fourth length of body; achenes lenticular, closely enveloped; style slender, jointed with the achene; stigmas 2.

Wet places, banks and open woods, Arctic-Alpine Zone; Labrador to Alaska, south in the mountains to North Carolina, Colorado, and Oregon; widely distributed in northern Eurasia. Type locality: Europe.

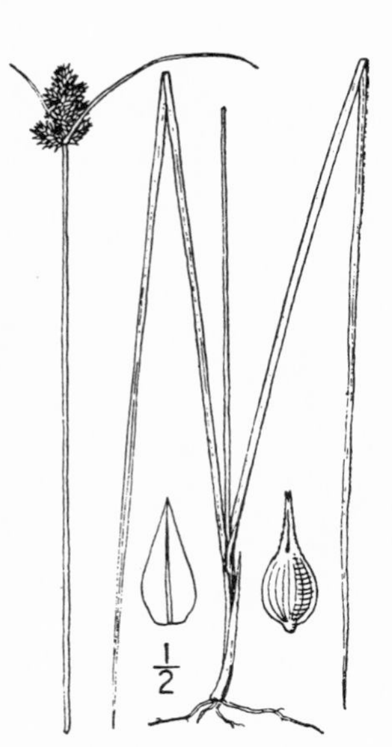

80. **Carex àrcta** Boott.

Northern Clustered Sedge. Fig. 756.

Carex canescens polystachya Boott in Richards. Arct. Exped. 2: 344. 1852.
Carex arcta Boott, Ill. Car. 4: 155. *pl. 497.* 1867.
Carex canescens oregana Bailey, Mem. Torrey Club 1: 75. 1889.

Cespitose, the culms slender, erect, 1.5–8 dm. high, very rough above, usually strongly exceeded by the leaves. Leaf-blades flat, 2–4 mm. wide, glaucous or light-green; spikes 5–15, 5–10 mm. long, 4–6 mm. wide, many-flowered, aggregated into a head 1.5–3 cm. long, 7–12 mm. wide; lower one or two bracts developed; scales ovate, obtusish to short cuspidate, shorter than perigynia, hyaline with green midvein, more or less brownish tinged; perigynia ovate, plano-convex, 2–3 mm. long, nearly 1.25 mm. wide, ascending or somewhat spreading, sharp-edged but not winged, many nerved dorsally, lightly nerved at base ventrally, white puncticulate, rounded and short stipitate at base, tapering into the short, serrulate, shallowly bidentate beak, which is obliquely cut and fissured on the dorsal side; achenes lenticular; style slender, jointed with achene; stigmas 2.

Swamps and wet woods, Transition and Canadian Zones; New Brunswick to British Columbia, south to New York, Montana, and northwestern California. Type locality: Canada.

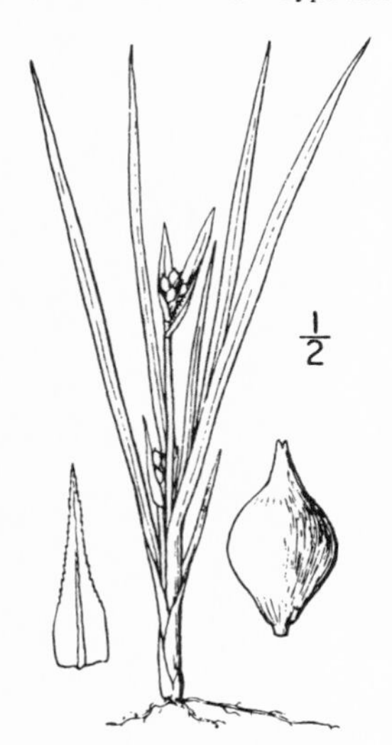

81. **Carex saximontàna** Mackenzie.

Rocky Mountain Sedge. Fig. 757.

Carex saximontana Mackenzie, Bull. Torrey Bot. Club 33: 439. 1906.

Cespitose, the culms up to 2.5 dm. long, triangular, weak, strongly winged above. Leaves nearly basal, strongly exceeding culms, strongly glaucous, the blades 1.5–3 dm. long, 3–5 mm. wide; spikes 2–4, androgynous, the rachis zigzag, dilated, one terminal, the others basal on slender often much elongated peduncles, the staminate flowers with small tight scales, the pistillate flowers 2–5, the lower subtended by enlarged, leaf-like, saccate scales, the upper by scales which are ovate lanceolate and shorter than the perigynia; perigynia 4 mm. long, the body oblong-orbicular, 2.5 mm. long, 2 mm. thick, nearly round in cross-section, 2-keeled but otherwise nearly nerveless, tapering to a stipitate base, contracted into the short, hyaline, slightly toothed, flattened, triangular beak scarcely 1 mm. long; achenes triangular, stipitate, closely enveloped, the apex rounded, constricted at the base, the sides convex; style jointed with achene, slender, soon withering; stigmas 3, short.

Woods and thickets, Transition Zone; Colorado and western Nebraska to eastern Oregon. Type locality: Fort Collins, Colorado.

82. **Carex leptàlea** Wahl.

Bristle-stalked Sedge. Fig. 758.

Carex leptalea Wahl. Vet. Akad. Handl. Stockholm 139. 1803.
Carex microstachya Michx. Fl. Bor. Am. 2: 169. 1803; not Ehrh. 1788.
Carex polytrichoides Muhl. in Willd. Sp. Pl. 4: 213. 1805.

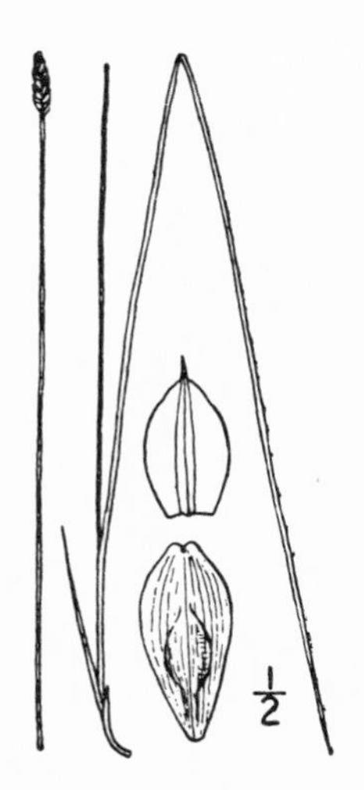

Densely tufted from slender rootstocks, the culms very slender, 2–6 dm. high, obscurely triangular, smooth or slightly roughened, mostly exceeding leaves, the leaf-blades 0.5–1.25 mm. wide, flat or channelled. Spike solitary, linear, androgynous, bractless, 4–15 mm. long, 2–3 mm. wide, the staminate part varying from inconspicuous to occupying nearly the whole spike; pistillate scales (except lowest) ovate, very obtuse to short-pointed, half length of perigynia, reddish-brown tinged with hyaline margins and green center; perigynia 1–10, oval-elliptic, round-triangular, or somewhat flattened in cross-section, more or less strongly overlapping, 2.5–4.25 mm. long, 1–1.5 mm. wide, membranaceous, finely many striate, substipitate, rounded and beakless at apex; achenes triangular; style slender, flexuous; stigmas 3, short.

Bogs and wet meadows very widely distributed, Humid Transition and Canadian Zones; Labrador to Alaska, south to Florida, Texas, and Humboldt County, California. Uncommon in our range. Type locality: Pennsylvania.

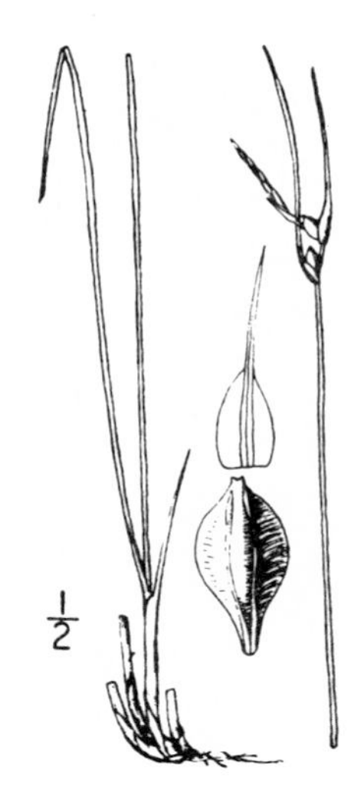

83. Carex multicaùlis Bailey.
Many-stemmed Sedge. Fig. 759.

Carex geyeri Boott, Ill. Car. **1**: 42. *pl. 105* (in part). 1858.
Carex multicaulis Bailey, Bot. Gaz. **9**: 118. 1884

Rootstocks lignescent. Culms 2–4 dm. high, deep green, smooth, obtusely triangular; leaves with well-developed blades 1–2 to a culm, the blades 1.5 mm. wide, very rough above, flat or somewhat involute; inflorescence of 2–several perigynia, the lower in the axils of long-cuspidate scales, the upper in the axils of short-cuspidate scales which are enlarged and white hyaline at base; and a terminal staminate part 7–25 mm. long, 1.5–2 mm. wide, with very obtuse broadly white hyaline margined scales; perigynia appressed-ascending, oblong-obovoid, triangular, 5–7 mm. long, the sides concave and 2.5 mm. wide, pale green, 2-keeled and finely and obscurely nerved, subcoriaceous, tapering to the stipitate base, abruptly beaked, the beak minute, entire or nearly so, minutely denticulate; achenes triangular, closely enveloped; stigmas 3.

Dry soil, Transition Zone; southern Oregon through northern and middle California and to southern California along the Sierra Nevada. Type locality: Yosemite Valley.

84. Carex gèyeri Boott.
Geyer's Sedge. Fig. 760.

Carex geyeri Boott, Trans. Linn. Soc. **20**: 118. 1846.

Rootstocks thick, lignescent, elongated. Culms up to 3.5 dm. high, sharply triangular, very rough on the angles; leaves with well-developed blades usually two to a culm, the blades erect, thick, flat, 2–3.5 mm. wide, very rough on the margins, developing after flowering; inflorescence of two or three perigynia, the lower in the axils of short-awned scales, the upper in the axils of obtusish or acutish scales, which are straw-colored with hyaline margins and largely conceal the perigynia; and a terminal staminate part 5–10 mm. long, 1.5–3 mm. wide, with oblong-ovate, striate, obtusish, straw-colored scales; perigynia appressed-ascending, oblong, triangular, 6 mm. long, straw-colored, smooth, shining, the sides 2.5 mm. wide, 2-keeled but otherwise nerveless, tapering to the short stipitate base, rounded at apex and abruptly minutely beaked, the beak entirely or nearly so, minutely denticulate around the base; achenes triangular, closely enveloped; stigmas 3.

Dry mountain sides and open woods, Arid Transition Zone; Alberta and Colorado to Washington and northern California. Type locality: Rocky Mountains.

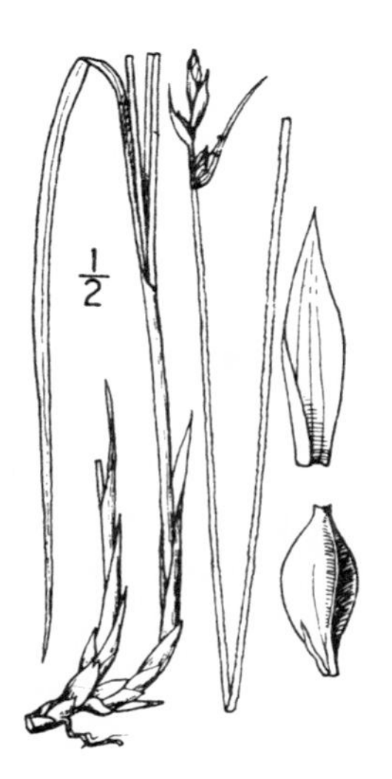

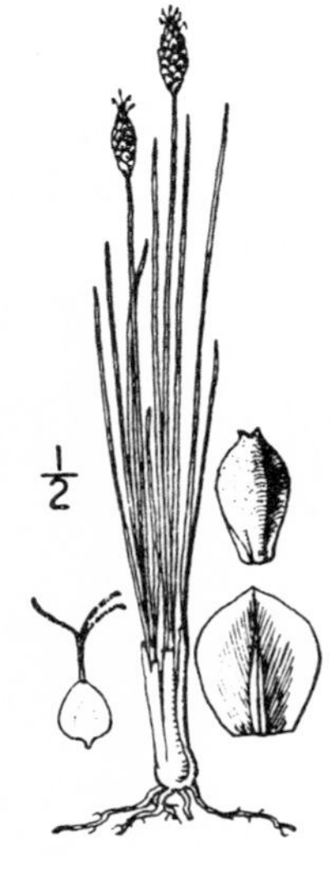

85. Carex exsérta Mackenzie.
Short-grass Sedge. Fig. 761.

Carex filifolia W. Boott in S. Wats. Bot. Calif. **2**: 229. 1880; not Nutt.
Carex filifolia erostrata Kükenth. in Engler Pflanzenreich **4**[20]: 86. 1909.
Carex exserta Mackenzie, Bull. Torrey Club **42**: 620. 1915.

Densely cespitose, the culms very slender and wiry, 5–25 cm. high, obtusely triangular, smooth, the basal sheaths filamentose. Leaf-blades acicular, 0.25–0.5 mm. wide; spike androgynous, bractless, 7–15 mm. long, the terminal staminate part often more than half the whole, the pistillate part up to 6 mm. wide with 2–12 ascending perigynia; pistillate scales orbicular-ovate, obtuse, dull reddish brown with hyaline margins; perigynia obovoid, triangular, 2-ribbed but otherwise nerveless, 2.5 mm. long, papillose-puberulent, rounded at base and apex, essentially beakless, hyaline and obliquely cut at mouth; achenes triangular; style black, exserted; stigmas 3; rhacheola conspicuous.

In dry places at elevations between 1500 and 3500 meters from southern Oregon, south in the Sierra Nevada of California to Tulare County and in the San Bernardino Mountains; Transition Zone. Type locality: Echo Lake, Eldorado County, California.

86. **Carex filifòlia** Nutt.
Thread-leaved Sedge. Fig. 762.

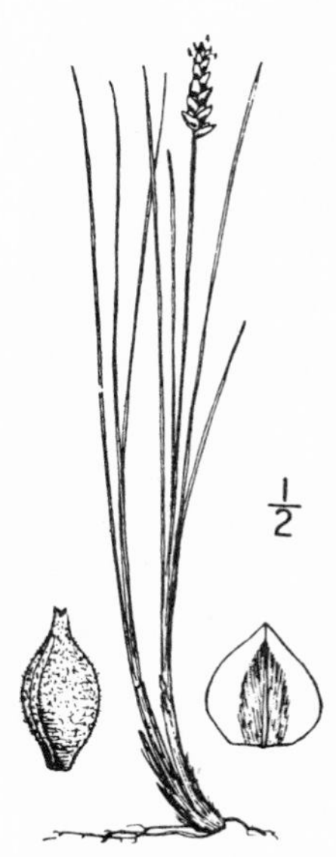

Carex filifolia Nutt. Gen. N. Am. Pl. **2**: 204. 1818.
Kobresia globularis Dewey, Am. Journ. Sci. **29**: 253. 1836.
Uncinia breviseta Torr. Ann. Lyc. N. Y. **3**: 428. 1836.

Densely cespitose, the culms slender and wiry, 8–30 cm. high, obtusely triangular, smooth, the basal sheaths usually strongly filamentose. Leaf-blades acicular, 0.25–0.5 mm. wide; spike androgynous, bractless, 1–2 cm. long, 3–4 mm. wide, the upper half staminate, the lower with 5–10 erect-ascending perigynia; pistillate scales very broadly obovate, very obtuse or mucronulate, concealing perigynia, brownish with very conspicuous shining white hyaline margins; perigynia obovoid-orbicular, obtusely triangular, 2-ribbed but otherwise nerveless, 3 mm. long, 2 mm. wide, puberulent, round-tapering at base, abruptly narrowed into a stoutish obliquely cut beak, 0.4 mm. long, which is hyaline at orifice; achenes triangular, closely enveloped; stigmas 3; rhacheola usually developed.

Plains and ridges, Arid Transition Zone; Saskatchewan to Yukon, south to Texas, New Mexico, and eastern Oregon. Type locality: dry plains and hills of the Missouri River.

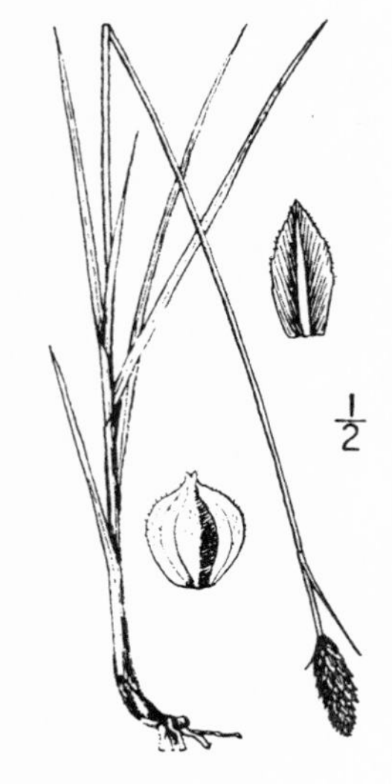

87. **Carex gìgas** (Holm) Mackenzie.
Siskiyou Sedge. Fig. 763.

Carex scirpoidea gigas Holm, Am. Journ. Sci. IV. **18**: 20. 1904.
Carex gigas Mackenzie, Bull. Torrey Club **35**: 268. 1908.

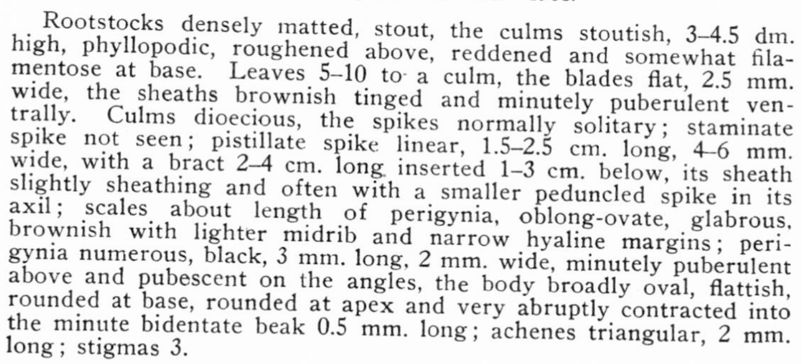

Rootstocks densely matted, stout, the culms stoutish, 3–4.5 dm. high, phyllopodic, roughened above, reddened and somewhat filamentose at base. Leaves 5–10 to a culm, the blades flat, 2.5 mm. wide, the sheaths brownish tinged and minutely puberulent ventrally. Culms dioecious, the spikes normally solitary; staminate spike not seen; pistillate spike linear, 1.5–2.5 cm. long, 4–6 mm. wide, with a bract 2–4 cm. long inserted 1–3 cm. below, its sheath slightly sheathing and often with a smaller peduncled spike in its axil; scales about length of perigynia, oblong-ovate, glabrous, brownish with lighter midrib and narrow hyaline margins; perigynia numerous, black, 3 mm. long, 2 mm. wide, minutely puberulent above and pubescent on the angles, the body broadly oval, flattish, rounded at base, rounded at apex and very abruptly contracted into the minute bidentate beak 0.5 mm. long; achenes triangular, 2 mm. long; stigmas 3.

Mountains of Siskiyou County, California, reported at elevations of 6800–8000 feet; Transition Zone. Type locality: Mount Eddy, California.

88. **Carex scabriúscula** Mackenzie.
Cascade Sedge. Fig. 764.

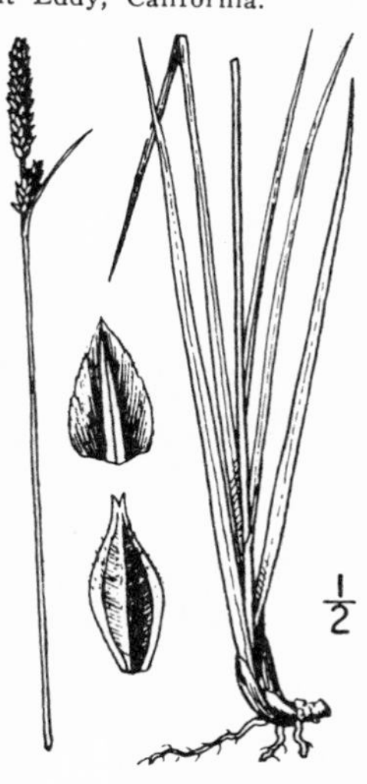

Carex scabriuscula Mackenzie, Bull. Torrey Club **35**: 268. 1908.

Rootstocks short-creeping stout, the culms phyllopodic, stoutish, about 3 dm. high, roughened above, somewhat reddened at base. Leaves 5–10 to a culm, the blades flat, 2–3 mm. wide, brownish tinged and minutely puberulent ventrally. Culms dioecious, the spikes normally solitary; staminate spike not seen; pistillate spike linear, 1.5–4 cm. long, 4–8 mm. wide with a bract 2–4 cm. long inserted 1–3 cm. below, often with a smaller peduncled spike in its axils; scales shorter than perigynia, oblong-ovate, glabrous, brownish-black with lighter midvein and narrow hyaline margins; perigynia numerous, black, 4–4.5 mm. long, 1.5 mm. wide, minutely puberulent and pubescent on the angles above, the body lanceolate, flattish, rounded at base, tapering at apex into the minute entire beak 0.5 mm. long; achenes triangular, 2 mm. long; stigmas 3.

Known only from the Cascade Mountains, Oregon; Transition Zone. Type: (*Cusick 2849.*)

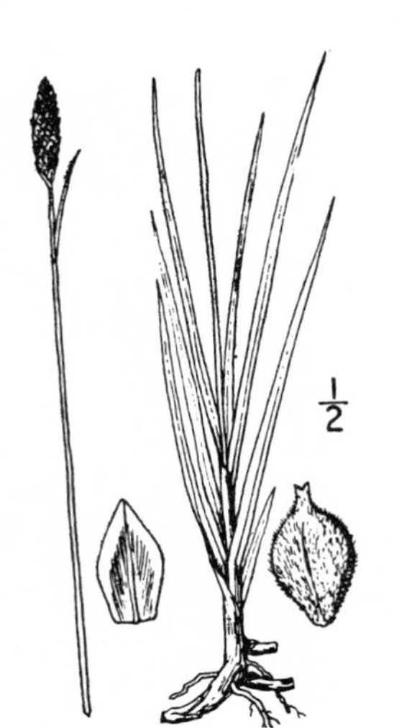

89. **Carex pseudo-scirpoìdea** Rydb. Western Single-spiked Sedge. Fig. 765.

Carex pseudo-scirpoidea Rydb. Mem. N. Y. Bot. Garden 1: 78. 1900.

Rootstocks stout, long-creeping, the culms phyllopodic, 1–4 dm. high, little roughened above, reddened at base. Leaves 5–10 to a culm, the blades flat, 2–3 mm. wide, the sheaths reddish-brown tinged and puberulent ventrally; culms dioecious, the spikes solitary, 12–36 mm. long, 2–5 mm. wide, often with a bract 1–2.5 cm. long inserted a short distance below the head; scales covering perigynia, broadly ovate, brownish black with conspicuous hyaline margins; perigynia numerous, appressed-ascending, obovoid, obtusely triangular in cross-section, 2.5 mm. long, 1.25 mm. wide, strongly pubescent all over, 2-keeled, tapering at base, very abruptly contracted into the minutely bidentate beak about 0.4 mm. long; achenes triangular, 1.5 mm. long; stigmas 3.

Mountain sides, Canadian and Hudsonian Zones; Montana and Colorado, west to Utah and Washington. Type locality: Montana.

90. **Carex stenochlaèna** (Holm) Mackenzie. Alaskan Single-spiked Sedge. Fig. 766.

Carex scirpoidea stenochlaena Holm, Am. Journ. Sci. 18: 20. 1904.
Carex stenochlaena Mackenzie, Bull. Torrey Club 35: 269. 1908.

Rootstocks densely matted, stout, the culms stoutish, 2.5–4 dm. high, aphyllopodic, strongly roughened, reddened and somewhat filamentose at base. Leaves 3–6 to a culm, the blades flat, 2–2.5 mm. wide, the sheaths brownish tinged and puberulent ventrally; culms dioecious, the spike solitary; staminate spike 2 cm. long, 3–5 mm. wide; pistillate spikes linear, 1.5–3 cm. long, 4–7 mm. wide with an elongated bract 0.5–5 cm. long inserted 0.5–3 cm. below; scales somewhat shorter than perigynia, oblong-ovate, puberulent and slightly ciliate, black with lighter midvein and narrow hyaline margin; perigynia numerous, black (at least towards apex), 4 mm. long, 1.5 mm. wide, appressed-pubescent, the body lanceolate, flattish, rounded at base, tapering at apex into the nearly entire beak, 0.5 mm. long; achenes triangular, 2 mm. long; stigmas 3.

Mountains, Transition and Canadian Zones; Alaska to Alberta, Idaho and Washington. Type locality: Juneau, Alaska.

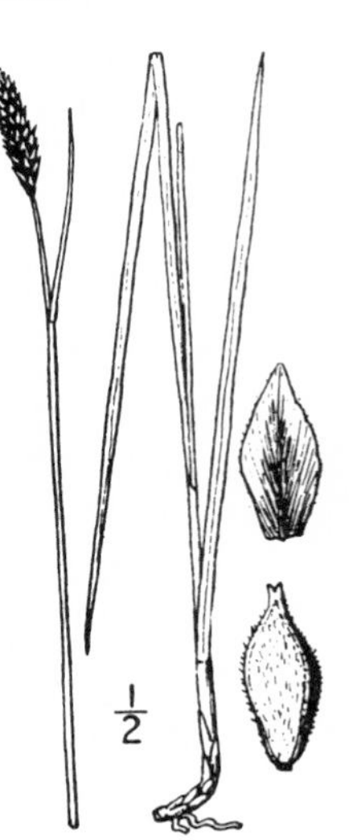

91. **Carex scirpoìdea** Michx. Canadian Single-spiked Sedge. Fig. 767.

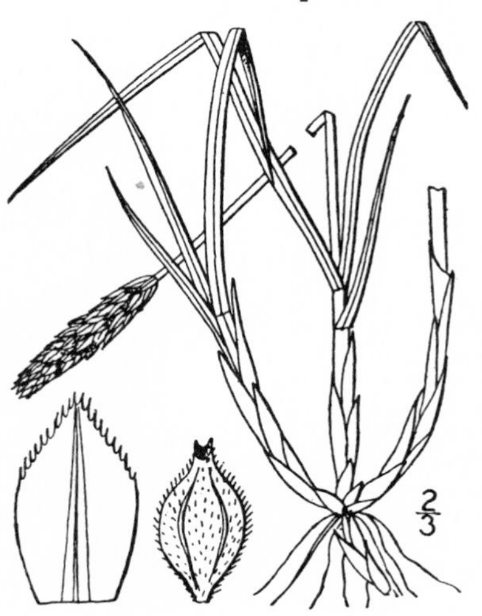

Carex scirpoidea Michx. Fl. Bor. Am. 2: 171. 1803.
Carex wormskjoldiana Hornem. Fl. Dan. *pl. 1528.* 1816.
Carex michauxii Schw. Ann. Lyc. N. Y. 1: 64. 1824.

Rootstocks creeping, stout, the culms aphyllopodic, 2–4 dm. high, roughened above, reddened at base. Leaves 2–4 to a culm, the blades flat, 1–2 mm. wide, brownish tinged and minutely puberulent ventrally; culms dioecious, the spike solitary; staminate spike somewhat smaller than the pistillate; pistillate spike linear 1.5–3 cm. long, 2.5–5 mm. wide with a bract some 3–10 mm. long at base; scales shorter than perigynia, oblong obovate, puberulent, obtusish, chocolate brown with lighter midrib and narrow hyaline margins; perigynia numerous, erect-appressed, 3 mm. long, 1 mm. wide, oblong-elliptic, compressed orbicular in cross-section, 2-keeled, obscurely triangular, whitish pubescent, rounded at base, rounded at apex and abruptly narrowed into a nearly entire beak about 0.25 mm. long; achenes triangular, 1.5 mm. long; stigmas 3.

Along streams, Arctic-Alpine Zone; Greenland to Alaska, south to New Hampshire, New York, Michigan, Montana, and Washington. Type locality: Hudson Bay.

92. **Carex ínops** Bailey.

Long-stoloned Sedge. Fig. 768.

Carex inops Bailey, Proc. Am. Acad. **22**: 126. 1886.
Carex pennsylvanica vespertina Bailey, Mem. Torrey Club **1**: 74. 1889.
Carex verecunda Holm, Am. Journ. Sci. IV. **16**: 461. 1903.
Carex vespertina Howell, Fl. N. W. Am. **1**: 705. 1903.

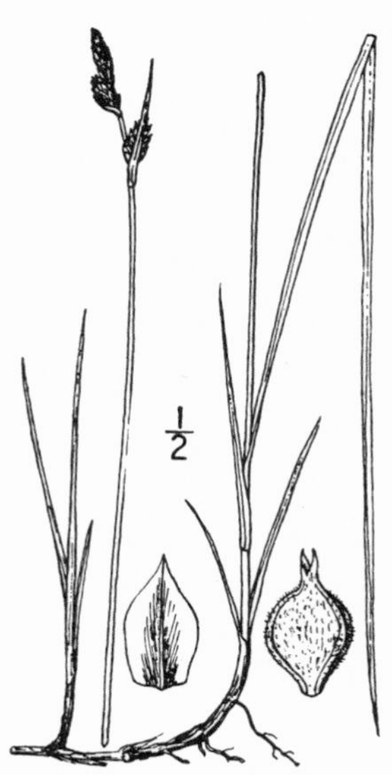

Cespitose and strongly stoloniferous, the culms 2–3.5 dm. high, very slender, roughened above, reddened and fibrillose at base. Leaves clustered towards base, the blades 1.5–2.5 mm. wide, very rough above; staminate spike solitary, 1.5–2.5 cm. long, sessile or short-peduncled, the reddish-brown or purplish-brown scales with conspicuous white-hyaline margins; pistillate spikes 1–3, approximate or more or less separate, erect, sessile or short-peduncled, oblong or short oblong, 7–12 mm. long, 4–5 mm. wide, with 4–10 ascending perigynia; lowest bract leaflet like, reddish-tinged at base, sheathless; scales ovate, sharp-pointed, chestnut-brown or purplish-brown with conspicuous white hyaline margins; perigynia 3.5 mm. long, 2 mm. wide, nearly orbicular in cross-section, membranaceous, pubescent, 2-keeled but otherwise nerveless, strongly stipitate, abruptly beaked, the beak 0.75–1.5 mm. long, deeply bidentate, the teeth strongly hyaline in age; achenes triangular, closely enveloped; style short, jointed with achene; stigmas 3, long.

Dry soil, Transition and Canadian Zones; chiefly in the Cascade Mountains, British Columbia to Siskiyou County, California. Type locality: Mount Hood, Oregon.

93. **Carex globòsa** Boott.

Round-fruited Sedge. Fig. 769.

Carex globosa Boott, Proc. Linn. Soc. **1**: 259. 1845.
Carex umbellata globosa Kükenth. in Engler Pflanzenreich **4**[20]: 453. 1909.

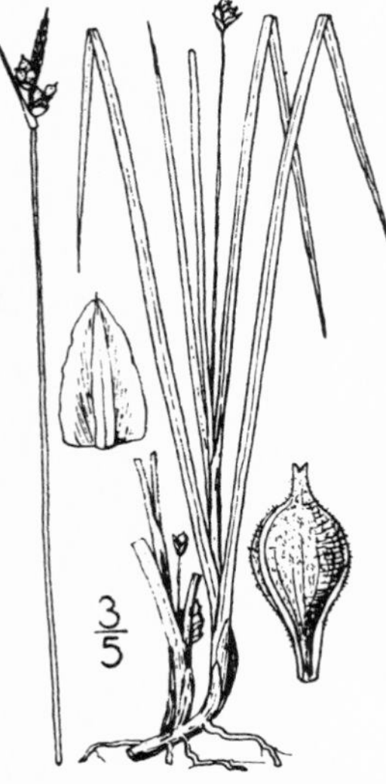

Stoloniferous, the rootstocks slender, elongate, the culms 1.5–3.5 dm. high, slender, roughened above. Leaf-blades 1.5–2.5 mm. wide, strongly roughened; staminate spike solitary, sessile or short-peduncled, 1–2 cm. long, many-flowered; pistillate spikes 2–3 (with additional basal ones on long capillary peduncles), approximate, sessile or short-peduncled, 5–10 mm. long, 4–8 mm. wide, with 4–10 ascending perigynia; lower bract leaflet-like, shorter than to exceeding inflorescence; scales ovate, obtuse to cuspidate, purplish-tinged; perigynia 5 mm. long, the body globose, 2.25 mm. wide, membranaceous, pubescent, strongly 2-keeled and finely many ribbed, abruptly narrowed to a prominent stipitate base and abruptly beaked, the beak 0.75–1.25 mm. long, strongly bidentate; achenes triangular, closely enveloped; style short, jointed with achene; stigmas 3, long.

Coastal counties of California and the islands off the coast from Santa Cruz Island and San Diego County northward to Sonoma County; Humid Transition Zone. Type locality: California.

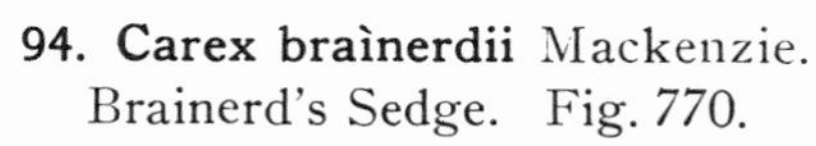

94. **Carex braìnerdii** Mackenzie.

Brainerd's Sedge. Fig. 770.

Carex brainerdii Mackenzie, Bull. Torrey Club **40**: 534. 1913.

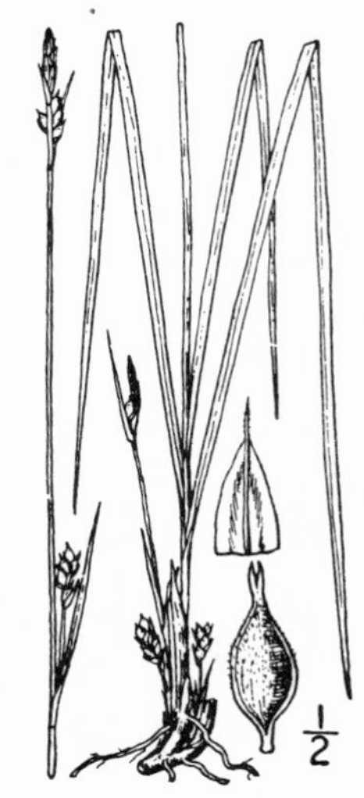

Rootstocks slender, elongate, the culms from very short to 15 cm. high, slender, very rough on the sharp angles. Leaf-blades 1.5–3 mm. wide, much roughened; staminate spike solitary, sessile or short-peduncled, 5–8 mm. long, few-flowered; pistillate spikes 4–6, 1–4-flowered, the upper 2 or 3 approximate, sessile or short-peduncled, the others basal on erect peduncles; lower bract of upper spikes exceeding inflorescence, chestnut-tinged, more or less strongly sheathing; scales ovate, reddish-brown tinged, cuspidate or long-awned; perigynia 4.5 mm. long, the body oval, 1.75 mm. wide, membranaceous, pubescent, strongly 2-keeled and finely many-ribbed, strongly stipitate, abruptly contracted into the serrulate, hyaline-tipped bidentate beak 1 mm. long; achenes triangular, closely enveloped; style short, jointed with achene; stigmas 3, long.

In the Sierra Nevada of California from Eldorado County north to Siskiyou County; and in southern Oregon; Transition Zone. Type locality: Slippery Ford, Eldorado County, California.

95. Carex brévipes W. Boott. Short Sedge. Fig. 771.

Carex brevipes W. Boott, in S. Wats. Bot. Calif. 2: 246. 1880.
Carex globosa brevipes, W. Boott, l.c. 485. 1880.
Carex deflexa boottii Bailey, Mem. Torrey Club 1: 43. 1889.
Carex rossii brevipes Kükenth. in Engler Pflanzenreich 4[20]: 452. 1909.

In dense clumps from short matted rootstocks, the culms from very short to 18 cm. high, slender, roughened above. Leaf-blades 1.5–2.5 mm. wide, roughened towards apex; staminate spike solitary, short-peduncled or sessile, 4–12 mm. long, 2.5 mm. wide, several–many-flowered; pistillate spikes 3–5, usually 10–20-flowered, the upper 1–2 approximate, sessile to strongly-peduncled, the others basal, long-peduncled; bract of lowest non-basal pistillate spike leaflet-like, exceeding culm, more or less purplish brown tinged; perigynia 2.5–3 mm. long, the body obscurely triangular, little longer than wide, membranaceous, pubescent, 2-keeled but otherwise nerveless, strongly stipitate, abruptly beaked, the beak 0.25–0.75 mm. long, minutely serrulate, shallowly bidentate; achenes triangular, closely enveloped; style short, jointed with achene; stigmas 3, long.

From Washington south in the mountains, especially in the Sierra Nevada, to southern California; Transition Zone. Closely related to the following species. Type locality: Lake Tahoe to Bear Valley, California.

96. Carex róssii Boott. Ross' Sedge. Fig. 772.

Carex rossii Boott, in Hook. Fl. Bor. Am. 2: 222. 1840.
Carex deflexa rossii Bailey, Mem. Torrey Club 1: 43. 1889.
Carex deflexa farwellii Brit. in Brit. & Brown Ill. Fl. 1: 334. 1896.
Carex farwellii Mackenzie, Bull. Torrey Club 37: 244. 1910.

Densely cespitose from stout rootstocks, the culms wiry, smooth or slightly roughened above, 5–25 cm. high. Leaf-blades 1–2.5 mm. wide, roughened towards apex; staminate spike solitary, sessile or nearly so, usually conspicuous, 3–10 mm. long, 1 mm. wide; pistillate spikes globose to short-oblong, 3–5 mm. long, 3–4 mm. wide, 2–12-flowered, the upper contiguous, the lower basal and long-peduncled; bract of lowest non-basal pistillate spike leaflet-like, exceeding culm, purplish brown tinged at base; scales ovate, sharp-pointed; perigynia 3–4.5 mm. long, the body 1.25 mm. wide, nearly globose in cross-section, membranaceous, pubescent, 2-keeled but otherwise nerveless, strongly stipitate, abruptly beaked, the beak 0.75–1.5 mm. long, deeply bidentate; achenes triangular, closely enveloped; style short, jointed with achene; stigmas 3, long.

Dry soil, Canadian Zone; Michigan to Yukon, south in the mountains to Colorado and California (Mariposa County). Type locality: northwest coast.

97. Carex brevicaùlis Mackenzie. Short-stemmed Sedge. Fig. 773.

Carex brevicaulis Mackenzie, Bull. Torrey Club 40: 547. 1913.

Stoloniferous, the culms 5–10 cm. high, slender, very rough on the angles. Leaf-blades 1.5–3.5 mm. wide, roughened above; staminate spike solitary, short-peduncled, few-flowered, 6–10 mm. long, 1.5–2 mm. wide; lateral spikes 2–4, 4–6 mm. long and nearly as wide, the upper 1–2 sessile and approximate, the others basal, slender peduncled; bract of lowest non-basal spike squamiform and shorter than culm, reddish-brown tinged and auriculate at base; scales ovate, acute to short cuspidate, reddish brown with lighter center and hyaline margins; perigynia about 4 mm. long, membranaceous, loosely short pubescent, 2-keeled but otherwise nerveless, stipitate, the body globose, 2.25 mm. wide, abruptly contracted into the slender, serrulate, rather shallowly bidentate beak 1 mm. long; achenes triangular, closely enveloped; style short, jointed with achene; stigmas 3, long.

Dry soil near the coast, Transition and Canadian Zones; British Columbia south to Santa Cruz County, California. Type locality: Yaquina Bay, Oregon.

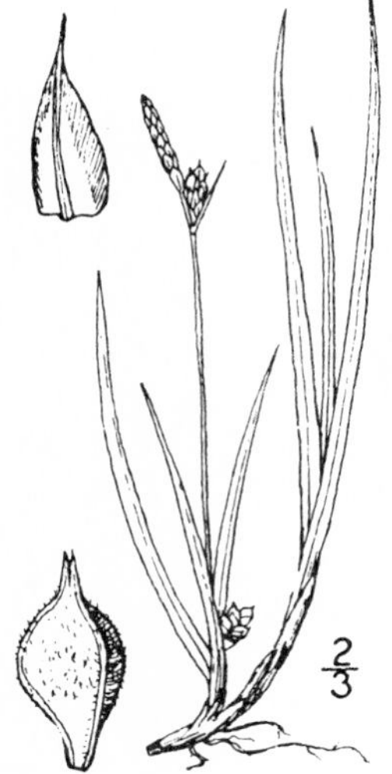

98. **Carex concinnoìdes** Mackenzie.
Northwestern Sedge. Fig. 774.

Carex richardsonii W. Boott, in S. Wats. Bot. Calif. **2**: 246. 1880; not R. Br.
Carex concinnoides Mackenzie, Bull. Torrey Club **33**: 440. 1906.

Strongly stoloniferous, the culms smooth, slender, 2.5 dm. high or less. Leaf-blades light green, 2–4 mm. wide; staminate spike solitary, nearly sessile, 8–22 mm. long; pistillate spikes one or two, approximate, rather closely 5–10-flowered, 5–10 mm. long, 4–5 mm. wide, sessile, or short-peduncled; bracts short-sheathing, more or less purplish tinged, the blades rudimentary; scales narrowly ovate, narrower and shorter than the perigynia, hyaline-margined, acute to acuminate; perigynia 2.5–3 mm. long, 1.5 mm. wide, membranaceous, appressed, loosely pubescent, tapering to a short substipitate base, the body oblong-elliptic, triangular, abruptly contracted into the short entire beak, 0.5 mm. long; achenes triangular, 2 mm. long, closely enveloped; style short, thickened, jointed with achene; stigmas 3, early deciduous.

Dry soil, Canadian Zone; British Columbia to Mendocino County, California, east to Alberta and Montana. Type locality: Columbia Falls, Montana.

99. **Carex triquètra** Boott.
Triangular-fruited Sedge. Fig. 775.

Carex triquetra Boott, Trans. Linn. Soc. **20**: 126. 1846.
Carex monticola Dewey, Bot. Mex. Bound. 229. 1858.

Cespitose, the culms 3–6 dm. high, sharply triangular, stiffish, smooth or nearly so, leafy towards base. Leaf-blades rigid, light green, 2.5–6 mm. wide, the sheaths cinnamon-brown tinged and purplish spotted ventrally; staminate spike solitary, 1–3 cm. long, short-peduncled; lateral spikes three or four, pistillate, or staminate at apex, erect, the upper little exsert-peduncled, approximate and often exceeding the staminate spike, the lower 1 or 2 often widely separate and long exsert-peduncled, 1–4.5 cm. long, 4–7 mm. wide, the 5–30 perigynia ascending, closely arranged in few rows; bracts long-sheathing, the lowest leaflet-like; scales ovate, short-cuspidate, brownish copper colored with broad 3-nerved center and hyaline margins; perigynia broadly obovoid, 4–4.5 mm. long, 2.75 mm. wide, softly short pubescent, light green, membranaceous, sharply triangular, several-nerved, short tapering at base, very abruptly beaked, the beak 0.3 mm. long, bidentate; achenes triangular; stigmas 3.

Dry hillsides below 2000 feet altitude, Lower and Upper Sonoran Zones; southwestern California from Santa Barbara County south and extending into the northern part of Lower California. Type locality: California, probably near Santa Barbara (*Nuttall*).

100. **Carex salinaefórmis** Mackenzie.
Deceiving Sedge. Fig. 776.

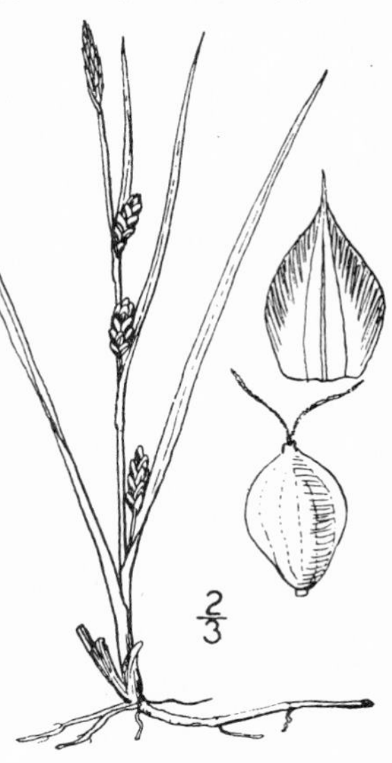

Carex salina minor W. Boott, in S. Wats. Bot. Calif. **2**: 242. 1880; not Boott.
Carex salinaeformis Mackenzie, Bull. Torrey Club **36**: 477. 1909.

From long-creeping, slender rootstocks, the culms 5–15 cm. high, phyllopodic, smooth, bluntly triangular, not reddened or fibrillose at base. Leaf-blades 2–3 mm. wide; staminate spike solitary, more or less peduncled, 8–12 mm. long, 2.5 mm. wide; pistillate spikes 3 or 4, erect, the upper approximate, short-peduncled, the lower widely separate, long-peduncled, 6–12 mm. long, 3–4 mm. wide, with 8–15 appressed-ascending perigynia; scales reddish brown, obtuse or cuspidate; perigynia oblong-ovoid, straw-colored, flattened sub-orbicular in cross-section, 2.5–3.75 mm. long, 1.75 mm. wide, nerved, glabrous, rounded at base, and beakless but short-tapering and slightly constricted at apex; achenes lenticular, apiculate, closely enveloped; style slender, short, jointed with achene; stigmas 2.

Only known from near the coast in Mendocino County, California, Humid Transition Zone. Type locality: Mendocino City.

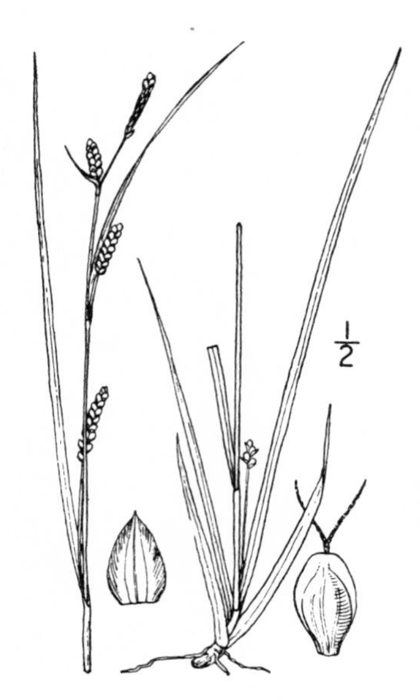

101. Carex hássei Bailey.
Hasse's Sedge. Fig. 777.

Carex aurea celsa Bailey, Mem. Torrey Club **1**: 75. 1889.
Carex hassei Bailey, Bot. Gaz. **21**: 5. 1896.
Carex celsa Piper, in Piper & Beattie Flora Northwest Coast 79. 1915; not Boott, 1862.

Rootstocks very slender, whitish, elongated, the culms 1.5–6 dm. high, phyllopodic, sharply triangular, usually much roughened above, light brownish at base. Leaf-blades 2–4 mm. wide; staminate spike solitary, more or less peduncled, 6–12 mm. long, 2–3 mm. wide, often pistillate at apex; pistillate spikes usually 3–5, the upper approximate and short-peduncled, the lower long-peduncled, linear-oblong, 8–20 mm. long, 3.5 mm. wide, with 6–20 ascending perigynia; bracts sheathing, the lowest exceeding the culm; scales ovate, acute, reddish-brown tinged; perigynia obovoid-ellipsoid, suborbicular in cross section, nerved, glabrous, at first straw-colored, soon white pulverulent, 2.5–3 mm. long, 2 mm. wide, substipitate or tapering at base, rounded or slightly pointed; achenes lenticular, apiculate, closely enveloped; style slender, short, jointed with achene, somewhat persistent and becoming short-exserted; stigmas 2.

River banks and wet rocks, Transition and Boreal Zones; Labrador to Yukon, south to Maine, Pennsylvania, Alberta, and in the mountains to southern California. Type locality: San Antonio Cañon, San Bernardino Mountains.

102. Carex aùrea Nutt.
Golden-fruited Sedge. Fig. 778.

Carex aurea Nutt. Gen. N. Am. Pl. **2**: 205. 1818.
Carex mutica R. Br. in Richardson App. Narr. Franklin Voy. 35. 1823.
Carex pyriformis Schw.; Dewey, Am. Journ. Sci. **9**: 69. 1825.

Rootstocks slender, whitish, elongated, the culms 0.3–4 dm. high, phyllopodic, smooth or somewhat roughened, triangular, light brownish at base. Leaf-blades 2–4 mm. wide; staminate spike solitary, sessile or short-peduncled, 3–10 mm. long, 1.5–3 mm. wide; pistillate spikes 3 to 5, the upper approximate and short-peduncled, the lower from little to strongly separate and often strongly peduncled, 4–20 mm. long, 3–5 mm. wide, with 4–20 ascending perigynia; bracts sheathing, exceeding culm; scales ovate, obtusish to short cuspidate, reddish brown tinged; perigynia broadly obovoid, brownish or at maturity golden yellow, nerved, glabrous, broadly oval in cross-section, 2–3 mm. long, 1.5–2 mm. wide, tapering at base, umbonate and beakless; achenes lenticular, apiculate, closely enveloped; style slender, short, jointed with achene, not exserted or persistent; stigmas 2.

Wet places, Canadian and Hudsonian Zones; Newfoundland to Yukon, south to Connecticut, Michigan, and, in the mountains, to New Mexico and southern California. Type locality: shores of Lake Michigan.

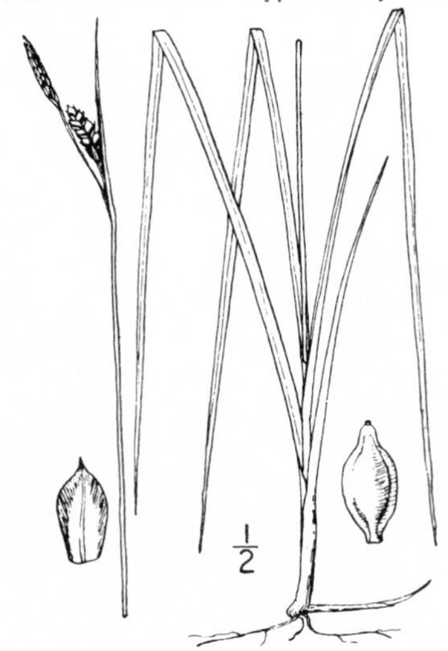

103. Carex lívida (Wahl.) Willd.
Livid Sedge. Fig. 779.

Carex limosa livida Wahl. Vet. Akad. Handl. Stockholm **24**: 162. 1803.
Carex livida Willd. Sp. Pl. **4**: 285. 1805.
Carex grayana Dewey, Am. Journ. Sci. **25**: 141. 1834.

Rootstocks very slender, elongated, the culms 1.5–5 dm. high, smooth, light brownish at base, phyllopodic. Leaf-blades glaucous green, more or less involute, 2 mm. wide or less; staminate spike solitary, short-peduncled, 1.5–2.5 cm. long; pistillate spikes one or two, approximate, erect, sessile or short-peduncled, 1–2 cm. long, 5 mm. wide, closely 5–15-flowered; bracts often exceeding culm, the sheaths short; scales ovate, chestnut or copper tinged with more or less hyaline margins; perigynia 3.75 mm. long, 1.75 mm. wide, membranaceous, the body ellipsoid, obscurely triangular in cross-section, slightly inflated, glaucous green, faintly nerved, narrowed and pointed but not beaked at apex, tapering at base, exceeding the scales; achenes triangular; style slender; stigmas 3.

Sphagnum bogs, Boreal Zone; Labrador to Alaska, south to New Jersey, Michigan, and northern California. In our range very local, being known only from Mendocino County, California. Type locality: Europe.

104. **Carex califórnica** Bailey.

California Sedge. Fig. 780.

Carex polymorpha W. Boott in S. Wats. Bot. Calif. **2**: 247. 1880; not Muhl.
Carex californica Bailey, Mem. Torrey Club **1**: 9. 1889.
Carex polymorpha californica Kükenth. in Engler Pflanzenreich **4**[20]: 515. 1909.

Rootstocks stout, elongated, the culms 2–4.5 dm. high, sharply triangular, smooth, reddish-purple at base, strongly aphyllopodic. Culm-leaves 2–4, the blades 1.5–5 mm. wide, flat with revolute margins; staminate spike 1.5–3.5 cm. long, strongly peduncled; pistillate spikes 2–4, erect, strongly separate, the upper short exsert-peduncled, the lower often nearly basal and long exsert-peduncled, linear-oblong, 1–3 cm. long, 3–5 mm. wide, with 7–20 appressed perigynia; bracts leaf-like, strongly exceeding spikes; scales ovate, obtusish, shorter than perigynia, purplish brown with lighter midvein; perigynia 3.5–4 mm. long, 2 mm. wide, membranaceous, yellowish green, the body ovoid, suborbicular in cross-section, somewhat inflated, several nerved, rounded at base, abruptly beaked, the beak 0.75 mm. long with oblique orifice; achenes triangular; style slender; stigmas 3.

Meadows and prairies, Transition Zone; northwestern California through western Oregon to southwestern Washington. Apparently rare and local. Type locality: Mendocino City, California.

105. **Carex hendersònii** Bailey.

Henderson's Sedge. Fig. 781.

Carex laxiflora plantaginea Boott; Olney Proc. Am. Acad. **8**: 407. 1872.
Carex hendersonii Bailey, Proc. Am. Acad. **22**: 115. 1886.

Rather loosely cespitose, the culms 4–8 dm. high, slender, sharply triangular, rough above. Sterile shoots developing conspicuous culms; culm blades 3–8 mm. wide, 5–25 cm. long, those of the sterile shoots 4–10 mm. wide, 2–5 dm. long; terminal spike staminate, more or less peduncled, 2–3 cm. long; pistillate spikes 2–4, erect, linear, 12–25 mm. long, 4–6 mm. wide, with 5–12 alternate ascending perigynia, the upper spikes approximate and little if at all exsert-peduncled, the lower widely separate and often long exsert-peduncled; bracts leaf-like, long sheathing; scales broadly obovate, mucronate, exceeded by perigynia, 3-nerved, green with hyaline margins, often reddish brown tinged; perigynia narrowly ovoid, 5–6 mm. long, 2 mm. wide, obtusely triangular, many-nerved, long-tapering and substipitate at base, tapering into a long straight, scarcely differentiated beak, obliquely cut at orifice; achenes triangular; style slender; stigmas 3.

Damp woods, Humid Transition and Canadian Zones; Coast Ranges from southwestern British Columbia to Sonoma County, California. Type locality: bogs at Portland, Oregon.

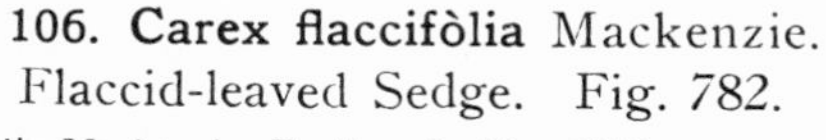

106. **Carex flaccifòlia** Mackenzie.

Flaccid-leaved Sedge. Fig. 782.

Carex flaccifolia Mackenzie, Erythea **8**: 92. 1922.

Rootstocks not seen, the culms 6–9 dm. high, sharply triangular, slender, weak, not roughened on the angles, leafy on the lower fifth, light brownish at base. Leaf-blades flaccid, light green, about 3 mm. wide, flat, pubescent, the sheaths strongly pubescent, white hyaline ventrally; staminate spike solitary, 1–2.5 cm. long, short-peduncled; lateral spikes mostly three, pistillate, erect, approximate or somewhat separate, the upper sessile, the lower short-peduncled, 1–2.5 cm. long, 5–7 mm. wide, the 8–25 perigynia ascending, loosely arranged in few rows; lowest bract leaflet-like, very short-sheathing; scales shorter than perigynia, ovate, cuspidate, hyaline with 3-nerved green center; perigynia broadly ovoid, 3.5–4.25 mm. long, 2.25 mm. wide, glabrous, light green, membranaceous, sharply triangular, several nerved on each side, short tapering at base, abruptly beaked, the beak 0.5 mm. long, bidentulate; achenes triangular; style jointed with achene; stigmas 3.

Southwestern California. Type collected by George B. Grant, May 1–2, 1902, in dry sunny plains, sheet 468192 in United States National Herbarium, without definite locality label.

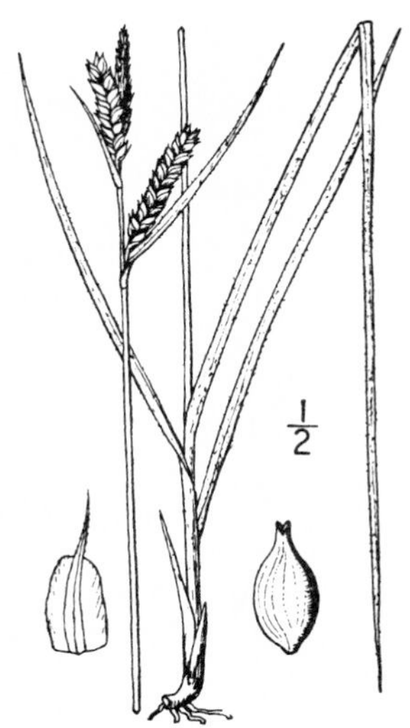

107. **Carex whítneyi** Olney.
Whitney's Sedge. Fig. 783.

Carex whitneyi Olney, Proc. Am. Acad. **7**: 394. 1868.
Carex pilosiuscula Boeckl. Flora **65**: 61. 1882.

Cespitose, the culms 2.5–10 dm. high. Foliage soft-pubescent, the sheaths loose, the blades 2.5–8 mm. wide; staminate spike solitary, short- or long-peduncled, 5–30 mm. long; pistillate spikes 2–4, approximate or the lower more or less separate, erect, sessile, or short-peduncled, suborbicular to linear-oblong, 7–30 cm. long, 5–7 mm. wide, closely 5–30-flowered; lowest bract short sheathing, about equalling culm; scales ovate, acute or short cuspidate, exceeding perigynia, green, 3-nerved, broadly hyaline margined; perigynia ascending, obliquely lanceolate-ovoid, sharply triangular, 4–5 mm. long, 2 mm. wide, pale green, smooth, puncticulate, about 5-ribbed on each side, round-tapering at base, tapering into a smooth beak 0.75 mm. long, the orifice minutely bidentate in age; achenes triangular; style jointed with achene, its base scarcely thickened; stigmas 3.

Southern Oregon and northern California, and south in the Sierra Nevada to Tulare County; Canadian and Hudsonian Zones. Type locality: Yosemite Valley.

108. **Carex gynodynáma** Olney.
Olney's Hairy Sedge. Fig. 784.

Carex gynodynama Olney, Proc. Am. Acad. **7**: 394. 1868.
Carex blakinshipii Fernald, Erythea **7**: 121. 1899.

Cespitose, the culms 2–7 dm. high, slender, sharply triangular, brownish at base. Leaves mostly clustered near base, the blades 3–9 mm. wide, sparsely and softly short pubescent; terminal spike staminate or with a few perigynia, sessile or short-peduncled, 1–2 cm. long; lateral spikes 2–4, oblong-cylindric, 1–3.5 cm. long, 6–9 mm. wide, erect, closely 20–40-flowered, the upper approximate and often overtopping the staminate spike, short-peduncled, the lower strongly separate, long-peduncled; bracts sheathing, the blades short; scales exceeded by perigynia, ovate-orbicular, short cuspidate or obtuse, reddish brown with lighter center and conspicuous hyaline margins; perigynia ascending, oblong-ovoid, 4–5 mm. long, 2 mm. wide, finely many nerved, strongly long hairy, membranaceous, triangular, round-tapering at base, rounded and abruptly beaked at apex, the beak 0.75 mm. long, at length bidentulate; achenes triangular; style jointed with achenes; stigmas 3.

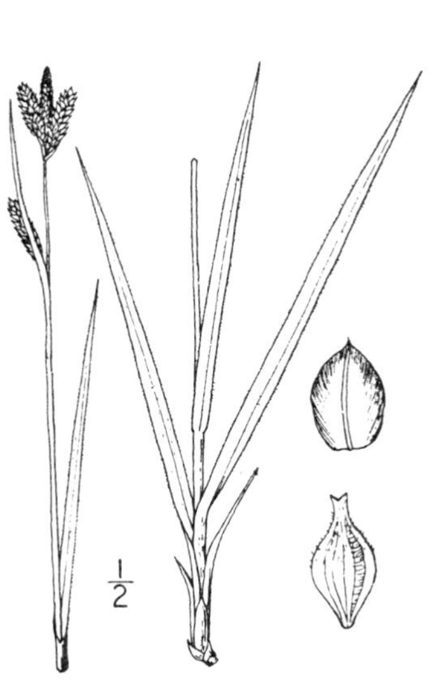

Moist places, Transition Zone; coast ranges from southern Oregon to San Mateo County, California. Type locality: Mendocino City, California.

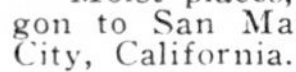

109. **Carex hirtíssima** W. Boott.
Hairy Sierra Sedge. Fig. 785.

Carex hirtissima W. Boott in S. Wats. Bot. Calif. **2**: 247. 1880.

Cespitose, the culms 3–5 dm. high, strictly erect, sparingly pubescent, not roughened on the angles, much exceeding the leaves. Leaf-blades flat, 3–4 mm. wide or up to 7 mm. on the sterile shoots, loosely hirsute on both surfaces; terminal spike staminate or gynaecandrous, from short to long peduncled, 1.5–2.5 cm. long, the scales oblong-obovate, glabrous, or slightly pubescent, strongly white margined; pistillate spikes 2 or 3, oblong, 1–2.5 cm. long, 5 mm. wide, more or less strongly separate, the lower on long-exserted peduncles, closely flowered with 20–30 ascending perigynia; lower bract leaf-like, sheathing, about equalling the culm; scales ovate or obovate, cuspidate or mucronate, shorter than perigynia, obscurely pubescent, the midrib green and the margins broad white hyaline; perigynia loosely pubescent, 3.5–4 mm. long, 1.75 mm. wide, sessile, the body obovoid, 2.5–3 mm. long, obtusely triangular, the sides very obscurely striate, abruptly short-beaked, the beak 1 mm. long, hyaline, very shallowly bidentate; achenes triangular, closely enveloped; style very short, jointed with achene; stigmas 3.

Rare and local in the central part of the Sierra Nevada, California; also Mount Sanhedrin, Lake County; Transition and Canadian Zones. Type locality: Summit Camp, Sierra Nevada.

110. **Carex mendocinénsis** Olney.
Mendocino Sedge. Fig. 786.

Carex cinnamomea Olney, Proc. Am. Acad. **7**: 396. 1868; not Boott 1846.
Carex mendocinensis Olney; W. Boott in S. Wats. Bot. Calif. **2**: 249. 1880.
Carex debiliformis Mackenzie, Bull. Torrey Club **37**: 244. 1910.

Cespitose from elongated rootstocks, the culms slender, 3–8 dm. high, much exceeding the leaves, reddish-purple at base. Leaves sparsely pubescent, the culm blades 1.75–3 mm. wide, 5–12 cm. long, those of the sterile shoots somewhat wider and 1.5–3.5 dm. long; terminal spike 2–3.5 cm. long, staminate or with a few perigynia; pistillate spikes 2 or 3, slender, erect, linear, 1.5–4 cm. long, 3.5–5 mm. wide, closely flowered above, more loosely towards the base, the 20–40 perigynia appressed-ascending; lowest bract about equalling culm, sheathing; scales somewhat exceeded by perigynia, ovate, obtuse or short cuspidate, cinnamon-brown with broad hyaline margins; perigynia 3.5–5 mm. long, 1.5 mm. wide, oblong-obovoid, somewhat flattened, triangular, membranaceous, lightly nerved, glabrous or minutely puberulent, somewhat tapering at base, abruptly beaked, the beak 0.5 mm. long, bidentate; achenes triangular; style jointed with achene; stigmas 3.

Along streams, Transition Zone; Coast Ranges of northwestern California and southwestern Oregon. Type locality: Mendocino City, California.

111. **Carex lémmoni** W. Boott. Lemmon's Sedge. Fig. 787.

Carex fulva hornschuchiana W. Boott in S. Wats. Bot. Calif. **2**: 250. 1800; not Boott.
Carex lemmoni W. Boott, Bot. Gaz. **9**: 93. 1884.
Carex abramsii Mackenzie, Bull. Torrey Club **36**: 482. 1909.

Rootstocks slender, more or less elongated, the culms slender, 4–8 dm. high, smooth, exceeding leaves, fibrillose at base. Leaves mostly clustered near base, the blades flat, 1.75–4 mm. wide, erect or ascending; staminate spike solitary, 6–25 mm. long, 2.5–4 mm. wide, sessile or short-peduncled; pistillate spikes 2–4, linear-oblong, 0.5–2 cm. long, 3–6 mm. wide, 5–30-flowered, often staminate at apex, the upper approximate, the lower separate and exsert-peduncled; bracts long sheathing; scales ovate, obtusish, exceeded by perigynia, reddish-brown with lighter center and broad hyaline margins; perigynia 3.5 mm. long or less, triangular, closely enveloping achene, puncticulate, glabrous, varying from green and slightly purplish-black tinged to purplish black, obscurely nerved, tapering at base, abruptly beaked, the beak 1 mm. long, sparingly ciliate-serrulate, bidentulate, slightly hyaline at mouth; achenes triangular, jointed with the slender style; stigmas 3.

Confined to Sierra Nevada of California from Tehama to Tulare Counties and the San Bernardino and San Jacinto Mountains of southern California; Transition Zone. Type locality: Sierra Nevada.

112. **Carex luzulìna** Olney.
Luzula-like Sedge. Fig. 788.

Carex luzulina Olney, Proc. Am. Acad. **7**: 395. 1868.
Carex cherokeensis W. Boott in S. Wats. Bot. Calif. **2**: 250. 1880; not Schw.
Carex albida Bailey, Mem. Torrey Club **1**: 9. 1889.

Culms densely cespitose, 1.5–7.5 dm. high. Leaves near the base, the blades soon spreading, stiff, 3–7 mm. wide; spikes 3–6, the uppermost staminate, the upper clustered, the lower widely separated on long-exserted peduncles, the lateral pistillate, oblong, 6–8 mm. wide, not compound at base; bracts long sheathing, the blades short; scales obtuse or acutish, exceeded by perigynia, reddish brown; perigynia 3.75–5 mm. long, the body compressed-triangular, closely enveloping the achene, not inflated or hispidulous, obsoletely nerved, contracted into the short, sparingly ciliate-serrulate, shallowly bidentate beak; achenes triangular, jointed with the slender style; stigmas 3.

This and the preceding and following species are very closely related. Mountain meadows and bogs, Transition and Canadian Zones; in the Coast Ranges of northwestern California and southwestern Oregon. Type locality: Mendocino City, California.

113. **Carex áblata** Bailey.
American Cold-loving Sedge. Fig. 789.

Carex frigida Olney in Bot. King's Explor. 371. 1871; not All.
Carex ablata Bailey, Bot. Gaz. **13**: 82. 1888.
Carex owyheensis A. Nelson, Bot. Gaz. **53**: 219. 1912.

Rootstocks somewhat elongated, the culms slender, smooth or nearly so, 2.5–6 dm. high, much exceeding leaves, the latter 4–9 to a fertile culm, the blades 3–4.5 mm. wide, flat. Spikes 3–7, the upper clustered and sessile or nearly so, the lower usually widely separate and on slender exserted peduncles, the terminal staminate or with a few perigynia, the lateral pistillate or staminate at apex, linear-oblong or linear-cylindric, 8–20 mm. long, 4–7 mm. wide; bracts shorter than inflorescence, long-sheathing; scales ovate, obtuse, strongly exceeded by perigynia, dark reddish brown or brownish black with lighter center and hyaline margins; perigynia lanceolate, greenish, compressed triangular, 3.5–4 mm. long, about 1.25 mm. wide, obscurely nerved, slightly ciliate serrulate, rounded at base, tapering into the minutely bidentate beak scarcely 1 mm. long; achenes triangular, jointed with the slender style; stigmas 3.

Mountain bogs and meadows, Canadian and Hudsonian Zones; Montana and Wyoming to British Columbia and the extreme northern part of California. Type locality: Mount Mark, Vancouver Island.

114. **Carex luzulaefòlia** W. Boott.
Luzula-leaved Sedge. Fig. 790.

Carex luzulaefolia W. Boott in S. Wats. Bot. Calif. **2**: 250 (in greater part). 1880.
Carex luzulaefolia strobilantha Holm, Am. Journ. Sci. **20**: 305. *f. 18.* 1905.
Carex pseudo-japonica C. B. Clarke, Kew Bull. Misc. Inf. Add. Ser. **8**: 81. 1908.

Culms 6–10 dm. high, the leaves mostly clustered at the base, the blades 5–15 mm. wide, mostly 1–3 dm. long, thick and leathery. Terminal spike staminate, more or less peduncled, usually about 1 cm. long, and with one or two sessile staminate spikes at its base; pistillate spikes 3 or 4, all or only the lower strongly exsert-peduncled, widely separate, the upper often equalling the staminate spikes, linear-oblong, 1.5–2.5 cm. long, 7.5 mm. wide, closely-flowered, the 20–50 perigynia appressed; bracts long sheathing, the sheaths enlarged upwards, the blades rudimentary or short; scales lanceolate, sharp-pointed, exceeding perigynia, glabrous, purplish black with conspicuous light colored midvein; perigynia 5–7.5 mm. long, 2.5 mm. wide, with oblong ovate strongly flattened body, glabrous or with a very few ciliae on the margins, strongly purplish black tinged, loosely enveloping achene, rounded at base, abruptly beaked, the beak slender, 1.5–2 mm. long, bidentate; achenes triangular, jointed with slender style; stigmas 3.

In the Sierra Nevada of California from Shasta County south to Tulare County; Hudsonian Zone. Type locality: above Ebetts Pass, near lake.

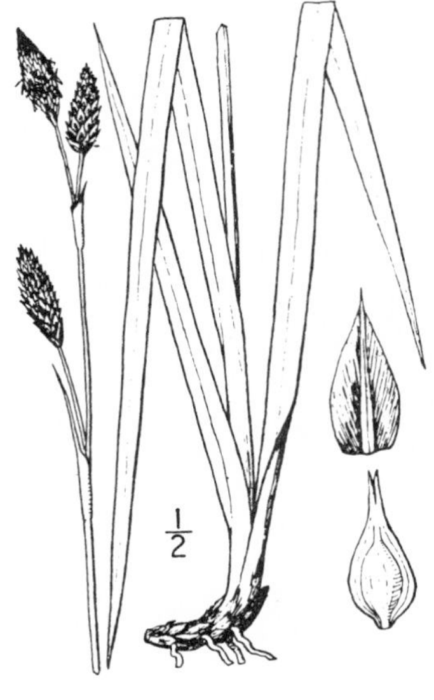

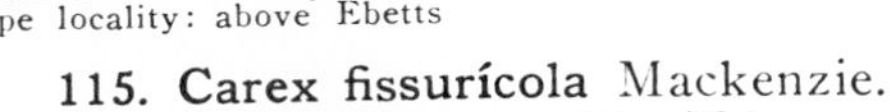

115. **Carex fissurícola** Mackenzie.
Canyon Sedge. Fig. 791.

Carex luzulaefolia W. Boott in S. Wats. Bot. Calif. **2**: 250 (in part). 1880.
Carex ablata luzuliformis Bailey, Bot. Gaz. **25**: 272. 1898.
Carex fissuricola Mackenzie, Muhlenbergia **5**: 53. 1909.

Culms 5–8 dm. high, the leaves mostly clustered at the base, the blades 3–6 mm. wide, 7–14 cm. long. Terminal spike staminate or often slightly pistillate, sessile or short-peduncled; lateral spikes 4 or 5, the upper contiguous and sessile or short-peduncled, the lower separate and strongly peduncled; bracts long sheathing, the sheaths scarcely enlarged upwards, the blades short or rudimentary; scales ovate, acute to cuspidate, exceeding perigynia, sparsely hairy when young, brown with lighter midrib conspicuous to apex; perigynia 5 mm. long, 2 mm. wide, the body narrowly ovate, much flattened, loosely enveloping the achene, sparsely hairy when young, remotely ciliate-serrulate on the margins, contracted into a shallowly bidentate beak; achenes triangular, jointed with the slender style; stigmas 3.

Mountain meadows, Transition Zone; western Nevada and the central part of the Sierra Nevada of California extending as far south as Tulare County. Type locality: south fork of Humboldt River, Elko County, Nevada.

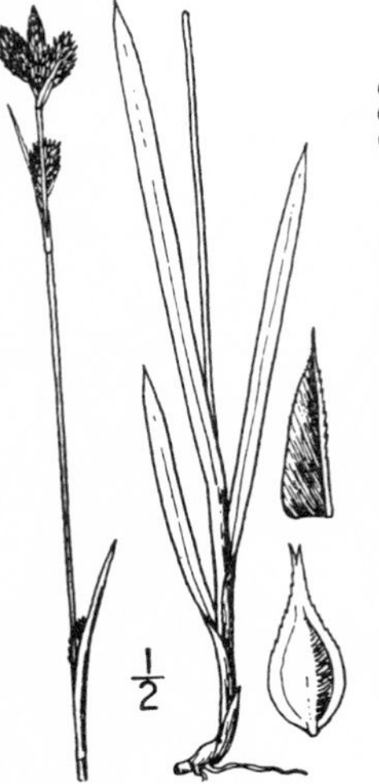

116. **Carex spíssa** Bailey.
San Diego Sedge. Fig. 792.

Carex hispida W. Boott Bot. Gaz. **9**: 89. 1884; not Willd.
Carex spissa Bailey, Proc. Am. Acad. **22**: 70. 1886.

Rootstocks stout, woody, the culms very stout, 10–18 dm. high, obtusely triangular, smooth. Leaves clustered above base, the blades glaucous-green, 7–14 mm. wide, flat with revolute strongly serrulate margins, the sheaths brownish tinged and strongly filamentose ventrally; staminate spikes 3–6, 4–10 cm. long; pistillate spikes 3–6, staminate at apex, approximate or the lower more or less separate, erect, sessile, or nearly so, linear-cylindric, 6–14 cm. long, 7–10 mm. wide, the 150–300 perigynia appressed-ascending or at length spreading; lower bracts leafy, exceeding inflorescence, the lowest short-sheathing; scales exceeding perigynia, narrowly ovate, serrulate-awned, light brownish with green center and hyaline margins; perigynia obovoid, flattened-triangular, 3–4.5 mm. long, light green, very obscurely nerved, membranaceous, glabrous, tapering at base, very abruptly beaked, the beak 0.5 mm. long, emarginate; achenes triangular, oblong-elliptic, short stipitate; style slender; stigmas 3.

Banks of streams at low altitudes, Upper and Lower Sonoran Zones; Los Angeles County, California, southward into Lower California and eastward to southern Arizona. Type locality: San Diego County, California.

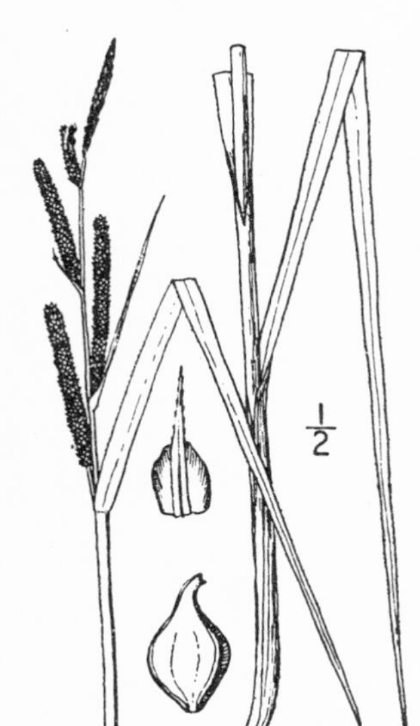

117. **Carex amplifòlia** Boott.
Ample-leaved Sedge. Fig. 793.

Carex amplifolia Boott in Hook. Fl. Bor. Am. **2**: 228. *pl. 226.* 1840.

Stoloniferous, with stout stolons, the culms 5–10 dm. high, sharply triangular, rough above, phyllopodic. Sheaths hispidulous, purplish-brown tinged; leaf-blades glaucous-green, sub-flaccid, 8–18 mm. wide; terminal spike staminate, 4–7 cm. long, 3–5 mm. wide; pistillate spikes four or five, the upper approximate, the lower more or less strongly separate, short-peduncled or nearly sessile, linear-cylindric, densely flowered, 3.5–8 cm. long, 6–7 mm. wide, with very many spreading perigynia; lower bracts leafy, the lowest short sheathing; scales exceeded by perigynia, oblong-ovate, acute to mucronate; perigynia 3 mm. long, obovoid, inflated triangular, brownish green, glabrous, nerveless except for keels, abruptly long-beaked, the beak often excurved, and with oblique hyaline orifice; achenes triangular, short stipitate; style slender jointed with achene; stigmas 3, short.

Wet soil, Transition Zone; Idaho to British Columbia and south to San Mateo and Tulare Counties, California. Type locality: Columbia River.

118. **Carex limòsa** L.
Shore Sedge. Fig. 794.

Carex limosa L. Sp. Pl. 977. 1753.

Strongly long stoloniferous, the culms rising obliquely, slender, few, 2–4.5 dm. high, sharply triangular, rough above, the basal sheaths sparingly filamentose. Leaves glaucous, shorter than the culm, the blades channelled, 3 mm. wide; terminal spike staminate, long-stalked; lateral spikes 1–3, pistillate, filiform stalked and drooping or the upper nearly erect, oblong, 10–26 mm. long, 5–8 mm. thick, closely 8–30-flowered; bracts leafy, very short sheathing, the auricles brownish; scales persistent, ovate, acute or short cuspidate, slightly shorter than perigynia, copper colored with 1–3 nerved green keel; perigynia glaucous-green, broadly ovate, appressed, coriaceous, 2.5–4 mm. long, 2 mm. wide, compressed triangular, densely puncticulate, glabrous, many nerved, rounded and stipitate at base, contracted into a very short beak, the orifice entire or emarginate; achenes triangular; style slender; stigmas 3.

In sphagnum bogs, Canadian Zone; Newfoundland to Alaska, south to New Jersey, Iowa, Montana, and southern Oregon; widely distributed in northern Eurasia. Type locality: Europe.

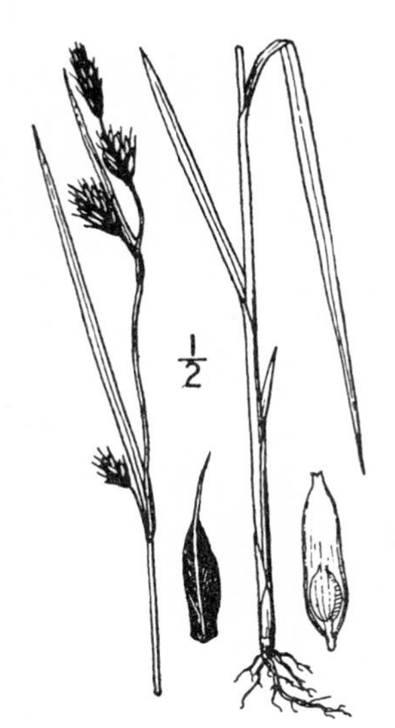

119. Carex macrochaèta C. A. Meyer.
Alaskan Long-awned Sedge. Fig. 795.

Carex macrochaeta C. A. Meyer, Mem. Acad. St. Petersb. **1**: 224. *pl. 13* 1831.
Carex excurrens Cham.; Steud. Syn. Cyper. 228. 1855.

Loosely cespitose, the rootstocks densely matted, tough, fibrous, the culms aphyllopodic, 2–6 dm. high, smooth or slightly roughened above, remotely leafy to the middle, the basal sheaths purplish tinged. Leaf-blades flat, 2.5–5 mm. wide; terminal spike staminate, peduncled, about 2 cm. long; lateral spikes 2–4, pistillate, more or less strongly separate, the upper short-peduncled and erect, the lower longer peduncled and erect to drooping, oblong or oblong-cylindric, 1–3 cm. long, 6–8 mm. wide, closely 15–40-flowered, the perigynia appressed; lowest bract leaflet-like, not sheathing; scales ovate-oblong, shorter than perigynia, black with whitish midrib excurrent as a slender awn 2–12 mm. long; perigynia obovoid- or oblong-oval, triangular-flattened, 4.5–6 mm. long, 2 mm. wide, smooth, sessile or nearly so, obscurely nerved, greenish straw-colored, dark purple spotted, abruptly very minutely tipped, the orifice entire or nearly so; achenes triangular, obovoid-oblong, nearly sessile; style slender; stigmas 3.

Abundant along the Alaskan coast, southward to Vancouver Island; Canadian Zone. Reported as found locally at Multnomah Falls, Oregon (*Piper*). It also is reported from the Pacific Coast of Siberia. Type locality: Unalaska.

120. Carex spectábilis Dewey.
Showy Sedge. Fig. 796.

Carex spectabilis Dewey, Am. Journ. Sci. **29**: 248. *pl. 10, f. 76.* 1836.
Carex nigella Boott in Hook. Fl. Bor. Am. **2**: 225. 1840.
Carex podocarpa W. Boott in S. Wats. Bot. Calif. **2**: 245. 1880; not R. Br.
Carex invisa Bailey, Proc. Am. Acad. **22**: 82. 1886
Carex tolmiei invisa Kükenth. in Engler Pflanzenreich **4**[20]: 412. 1909.

Culms purplish tinged at base, 2.5–5 dm. high from densely matted, tough, fibrillose rootstocks, the lower culm-leaves much reduced. Leaf-blades 2–3.5 dm. wide; terminal spike staminate, its scales with conspicuous more or less excurrent midvein; pistillate spikes 2–4, erect, oblong, 1–2 cm. long, 3.5–5 mm. wide, closely 15–30-flowered, not aggregated, the upper short-peduncled, the lower long-peduncled; bracts sheathless, the lowest about equalling the inflorescence; scales exceeded by perigynia, purplish-black with white usually excurrent midvein; perigynia ovate, 4 mm. long, 2 mm. wide, much flattened, sessile, rounded at base and apex, abruptly minutely beaked, the beak bidentulate; achenes triangular, short stipitate, jointed with the slender style; stigmas 3.

Mountains and meadows, Arctic-Alpine Zone; Alaska to Montana and south in the higher mountains to Tulare County, California. Type locality: "Arctic regions" [Rocky Mountains of British Columbia].

121. Carex tòlmiei Boott.
Tolmie's Sedge. Fig. 797.

Carex tolmiei Boott in Hook. Fl. Bor. Am. **2**: 224. 1840.
Carex microchaeta Holm, Am. Journ. Sci. IV. **17**: 305. 1904.
Carex paysonis Clokey, Am. Journ. Sci. V. **3**: 89. 1922.

Rootstocks woody, creeping, the culms phyllopodic, triangular, many-leaved, brownish at base, 1.5–5 dm. high. Leaf-blades 2.5–4 mm. wide; terminal spike staminate, sessile or short-peduncled, 8–20 mm. long, its scales purplish-black with white non-excurrent midvein; pistillate spikes 2–6 (8) erect, short oblong, 8–15 mm. long, 6–8 mm. wide, closely 20–40-flowered, approximate or the lower separate, sessile or short-peduncled; bracts sheathless, the lowest from shorter than to exceeding the culm; scales ovate, acute, shorter than the perigynia, purplish-black with white often obsolete usually non-excurrent midvein; perigynia oblong ovate, much flattened, 4 mm. long, 1.8 mm. wide, straw color and purplish tinged above, membranaceous, granular, lightly few-nerved, sessile, rounded at base and apex, abruptly minutely beaked, the beak 0.5 mm. long, bidentulate or nearly entire, purplish black; achenes triangular, short stipitate; style slender jointed with the achene; stigmas 3, slender.

Mountain meadows, Canadian and Hudsonian Zones; from Wyoming and Alberta to Idaho and Washington. Type locality: Columbia River.

122. **Carex raynóldsii** Dewey.

Raynolds's Sedge. Fig. 798.

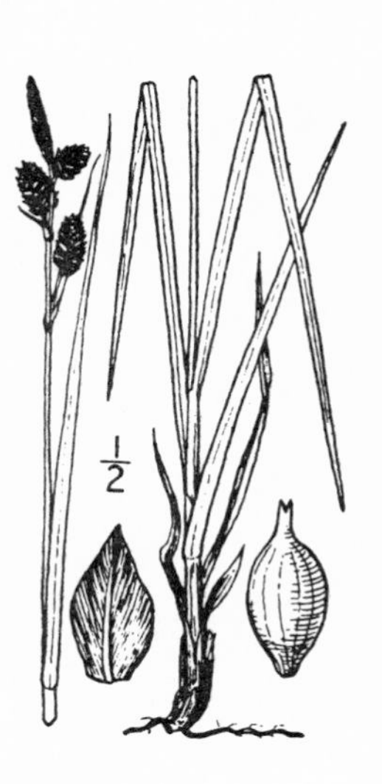

Carex raynoldsii Dewey, Am. Journ. Sci. II. **32**: 39. 1861.
Carex lyallii Boott, Ill. Car. **4**: 150. *pl. 483.* 1867.

Rootstocks stout, soloniferous, the culms 2–4 dm. high, stout, sharply angled, smooth or nearly so, little or not at all fibrillose, but purpish tinged at base, clothed at base with the dried up leaves of previous year. Leaf-blades 3–8 mm. wide, flat; terminal spike staminate, about 1.5 cm. long, 4 mm. wide, the lateral pistillate spikes 2 or 3, approximate or lowest separate, peduncled, erect, oblong, 1–2 cm. long, 7–8 mm. wide, closely 15–40-flowered; lowest bract about equalling culm, sheathless; scales ovate, cuspidate, nearly as wide as but exceeded by perigynia, blackish with lighter midvein; perigynia oblong-oval, round in cross-section, 4.5 mm. long, 2 mm. wide, densely puncticulate, prominently 2-keeled and slender nerved, membranaceous, greenish straw-colored, rounded and substipitate at base, the very short beak minutely bidentate; achenes triangular, broadly obovoid, 2.25 mm. long; style slender, straight, articulated to achene and in age deciduous; stigmas 3.

Mountain meadows and bogs, Canadian and Hudsonian Zones; Alberta and Colorado to Washington and northern California, extending south in the Sierra Nevada to Tulare County. Type locality: Pierre's Hole, Snake River Valley, altitude 6000 feet.

123. **Carex bífida** Boott.

Bifid Sedge. Fig. 799.

Carex bifida Boott; Olney, Proc. Am. Acad. **7**: 394. 1868.
Carex serratodens W. Boott in S. Wats. Bot. Calif. **2**: 245. 1880.
Carex aequa C. B. Clarke, Kew Bull. Misc. Inf. Add. Ser. **8**: 86. 1908.

Cespitose but stoloniferous, the culms sharply triangular, slender, smooth, 4–8 dm. high, the basal sheaths purplish-tinged and filamentose. Leaf-blades 1.75–3.5 mm. wide; terminal spike 1.5–3 cm. long, sessile or short-peduncled, staminate or with some perigynia in the middle; pistillate spikes 3–5, erect, the upper approximate and sessile, the lower more or less separate and short-peduncled, oblong, 8–18 mm. long, 6–8 mm. wide, with 20–40 at length spreading perigynia; lowest bract exceeding culm, purplish tinged at base, scarcely sheathing; scales ovate, more or less strongly exceeded by perigynia, acute or rough mucronate, reddish-brown with lighter center; perigynia narrowly ovate, flattened triangular, about 10-nerved, 3–4.5 mm. long, green, puncticulate, membranaceous, sessile, rounded at base, abruptly short-beaked, the beak 0.5–1 mm. long, bidentate, the teeth minute, rough and purplish tinged within; achenes triangular; style slender, straight, articulated to achene and in age deciduous; stigmas 3.

From Jackson County, Oregon, southward in California, mostly in the Coast Ranges, to San Luis Obispo and Kern Counties, Upper Sonoran Zone. Type locality: Salinas Valley, Monterey County, California.

124. **Carex buxbaùmii** Wahl.

Buxbaum's Sedge. Fig. 800.

Carex polygama Schk. Riedgr. **1**: 84. *pl. 10, f. 76.* 1801; not J. F. Gmel. 1791.
Carex subulata Schum. Fl. Saell. 1: 270. 1801; not J. F. Gmel. 1791.
Carex buxbaumii Wahl. Vet.-Akad. Handl. Stockholm **24**: 163. 1803.
Carex holmiana Mackenzie, Bull. Torrey Club **36**: 481. 1909.

Densely cespitose, but with long stolons, the culms 2–9 dm. high, aphyllopodic, sharply angled, rough above, slender but stiff, strongly reddish purple and filamentose at base. Leaf-blades 2–4 mm. wide, glaucous-green, long-pointed; spikes usually 3 or 4, erect, 8–40 mm. long, 8 mm. wide, with many perigynia, sessile or short-peduncled, the terminal gynaecandrous, the lateral pistillate; bracts sheathless, the lowest scarcely equalling the culm; scales oblong ovate, exceeding perigynia, awned, 1–3-nerved; perigynia 3–4 mm. long, sub-erect, obovoid, triangular, scarcely inflated, glaucous green, lightly many nerved, densely papillose, short stipitate, very short-beaked, the beak minutely bidentate; achenes triangular; style slender, straight, articulated to achene, and in age deciduous; stigmas 3.

Bogs, Canadian and Transition Zones; Greenland to Alaska, south to Georgia, Arkansas, Colorado, and California; widely distributed in northern Eurasia; rare and local in our range. Type locality: Europe.

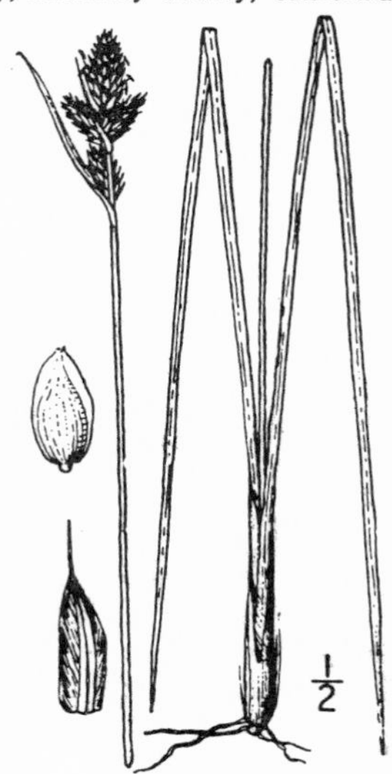

125. Carex hélleri Mackenzie. Heller's Sedge. Fig. 801.

Carex helleri Mackenzie, Erythea 8: 70. 1922.

Very densely cespitose, the culms 0.5–3 dm. high, slender but strict, sharply triangular, roughened above, purplish tinged at base. Leaf-blades 2–3 mm. wide, flat with slightly revolute margins; spikes 3–5, closely approximate, the terminal gynaecandrous, sessile or short-peduncled, the lateral pistillate, sessile or nearly so, oblong, 10–20 mm. long, 4.5–6 mm. wide, densely 25–50-flowered; lowest bract shorter than culm, sheathless; scales ovate-lanceolate or lanceolate, acuminate, strongly exceeding but much narrower than perigynia, purplish-black with light midvein; perigynia broadly oval to sub-orbicular, much flattened, 2.5–3.5 mm. long and nearly as wide, obscurely nerved, substipitate, puncticulate, not granular roughened, but sparingly minutely ciliate-serrulate when young, abruptly beaked, the beak 0.25 mm. long, bidentate; achenes triangular, short stipitate; style slender, straight, articulated to achene and in age deciduous; stigmas 3.

At high altitudes, Boreal Zone; central part of the Sierra Nevada of California extending as far south as Tulare County; also in the White Mountains of California, and the high mountains of western Nevada. Type locality: Mount Rose, Washoe County, Nevada.

126. Carex epapillòsa Mackenzie. Smooth-fruited Sedge. Fig. 802.

Carex epapillosa Mackenzie, in Rydberg Fl. Rocky Mts. 138. 1917.

Densely cespitose, but rootstocks more or less creeping, the culms 1.5–6 dm. high, slender, sharply triangular, smooth. Leaves all towards the base of the culms, the blades 3–8 mm. wide, nearly flat; spikes 3–6, approximate or the lowest a little separate, sessile or short-peduncled, oblong-obovoid, 1–2.5 cm. long, 6–10 mm. wide, closely flowered in many rows, the terminal one gynaecandrous; bracts sheathless; scales lance-ovate, sharp-pointed, somewhat narrower than but about equalling the perigynia, brownish black, the midvein more or less prominent; perigynia broadly oval or obovate, much flattened, 3.5–4.5 mm. long, 1.5–2 mm. wide, obscurely nerved, puncticulate but not granular roughened, the beak 0.5 mm. long, bidentate; achenes triangular, slender-stipitate, much narrower than perigynia; style slender, straight, articulated to achene and in age deciduous; stigmas 3.

Mountain meadows, Canadian to Arctic-Alpine Zones; Wyoming to Washington, Oregon and the Sierra Nevada of California. Type locality: Marysvale, Utah.

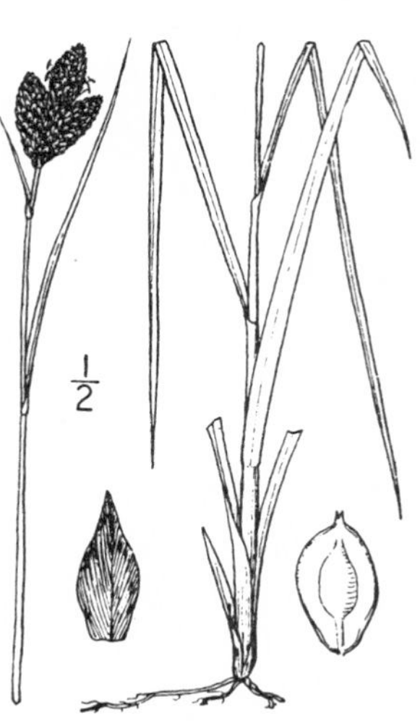

127. Carex heteroneùra W. Boott. Various-nerved Sedge. Fig. 803.

Carex atrata and var. *erecta* W. Boott in S. Wats. Bot. Calif. 2: 239. 1880, not L.
Carex heteroneura W. Boott, l.c. 240. 1880.
Carex quadrifida and var. *lenis* Bailey, Proc. Calif. Acad. II. 3: 104. 1891.
Carex quadrifida caeca Bailey, Bot. Gaz. 21: 8. 1896.

Densely cespitose, the rootstocks little if at all creeping, the culms 2.5–5 dm. high, slender, smooth or somewhat roughened above. Leaf-blades 2–4 mm. wide, the sheaths normally purplish-tinged ventrally; spikes about four, approximate or lower more or less strongly separate, the terminal gynaecandrous (rarely staminate), the lateral pistillate, the lower on peduncles of half their length, the upper sessile or short-peduncled, oblong, 1–2 cm. long, 5–7 mm. wide, closely flowered with 15–40 appressed-ascending perigynia; bracts sheathless, the lower shorter than to exceeding culm; scales acute, half width of perigynia, purplish-brown with conspicuous midvein; perigynia suborbicular, strongly flattened, 2.5–3.5 mm. long, 1.75–2.5 mm. wide, rounded at base and apex, puncticulate, but not granular roughened, the beak 0.25 mm. long, minutely bidentate; achenes triangular, short-stipitate, somewhat narrower than perigynia; style slender, straight, articulated to achene and in age deciduous; stigmas 3.

Sierra Nevada of California from Tulare County to Placer County; in the high mountains of southern California and in the mountains of western Nevada; Transition Zone. Type locality: Lake Tahoe to Bear Valley.

128. **Carex atrosquàma** Mackenzie.
Black-scaled Sedge. Fig. 804.

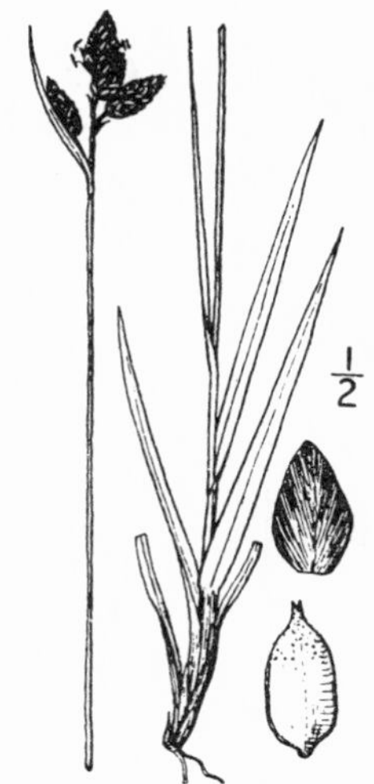

Carex atrosquama Mackenzie, Proc. Biol. Soc. Wash. **25**: 51. 1912.
Carex apoda Clokey Am. Journ. Sci. V. 3: 88. 1922.

Densely cespitose, the culms 2–4.5 dm. high, phyllopodic, slender, erect or the apex nodding, sharply triangular, purplish-brown tinged and slightly fibrillose at base. Leaf-blades deep green, 2.5–3.5 mm. wide; spikes 3 or 4, approximate or the lower slightly separate, the upper sessile or nearly so, the lower on erect peduncles half to twice their length, oblong, 6–12 mm. long, 5 mm. wide, densely 15–30-flowered, the terminal gynaecandrous, the lateral staminate; lowest bract slightly sheathing, shorter than to exceeding inflorescence; scales broadly ovate, obtuse or acutish, shorter than perigynia, black, the midvein obsolete; perigynia narrowly obovoid, 3.25 mm. long, 1.75 mm. wide, slightly inflated and subtriangular, olive green, membrănaceous, nerveless, puncticulate, granular roughened above, glabrous, round-tapering at base, abruptly beaked, the beak purplish black, 0.5 mm. long, shallowly bidentate; achenes triangular, short-stipitate; style slender, jointed with achene; stigmas 3.

Mountain meadows, Canadian Zone; Alberta and Montana to British Columbia and Washington. Type locality: Smoky River, Alberta.

129. **Carex albonìgra** Mackenzie.
Black and White-scaled Sedge. Fig. 805.

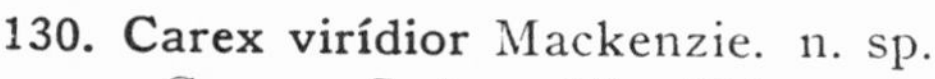

Carex albonigra Mackenzie, in Rydberg, Fl. Rocky Mts. 137. 1917.

Densely cespitose, the culms 1–2.5 dm. high, stiff, rigid and erect, sharply triangular, roughened above, purplish-brown tinged and but slightly fibrillose at base. Leaf-blades not glaucous, 3 mm. wide, flat with revolute margins; spikes mostly three, approximate or lowest slightly separate, the lowest short peduncled, the others sessile or nearly so, the lateral pistillate, narrowly oblong, 8–10 mm. long, 4 mm. wide, with about 8–15 appressed perigynia, the terminal gynaecandrous, longer, clavate at base; lowest bract about equalling inflorescence, sub-sheathing; scales broadly ovate, obtuse or acutish, nearly length of perigynia, black with conspicuous shining white margins, the midvein more or less obsolete; perigynia broadly ovate, much flattened, 3 mm. long, 2 mm. wide, blackish, membranaceous, nerveless, puncticulate, glabrous, minutely roughened on margins above, rounded at base, very abruptly contracted into a shallowly bidentate beak scarcely 0.5 mm. long; achenes triangular, short stipitate, much narrower than perigynia; style slender, jointed with achene; stigmas 3, short.

Mountain meadows at high altitudes, Transition and Candiana Zones; Montana to Washington, south to Colorado and Arizona. Type locality: Needle Mountain, Wyoming.

130. **Carex virídior** Mackenzie. n. sp.
Greener Sedge. Fig. 806.

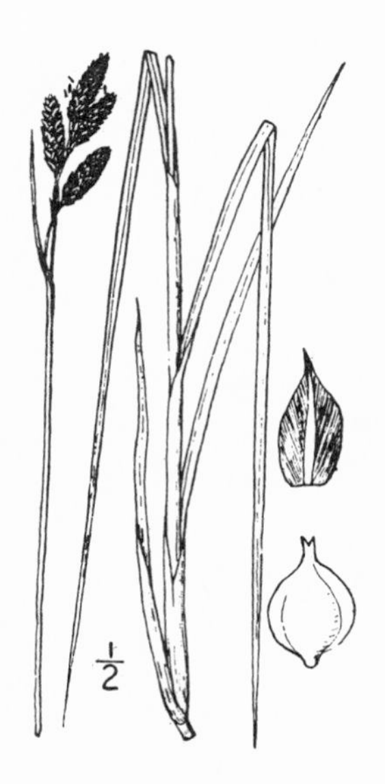

Carex atratiformis Mackenzie in Rydberg, Fl. Rocky Mts. 137 (in part). 1917; not Brit.

Rootstocks slender, short-creeping, the culms loosely cespitose, 4–7 dm. high, slender but erect, sharply triangular, smooth or little roughened, phyllopodic, purplish tinged at base. Leaf-blades 2.5–4 mm. wide, flat with revolute margins, green, not at all glaucous, the sheaths white hyaline ventrally; spikes 3 or 4, the terminal ovoid, gynaecandrous, clavate at base, the lateral pistillate (with a very few staminate basal flowers), oblong or linear-oblong, 1–2 cm. long, 4–5 mm. wide, closely 20–40-flowered, the upper approximate and short peduncled, the lower more or less strongly remote and on erect sharply triangular peduncles one-half to twice length of spikes; lowest bract purplish black tinged at base, sheathless, nearly equalling culm; scales exceeding perigynia, brownish black with obsolete midvein, obtuse or acutish; perigynia obovate or oblong-oval, 3.5 mm. long, 2 mm. wide, much flattened, rounded at base, nerveless, granular roughened above, yellowish brown tinged, membranaceous, abruptly beaked, the beak 0.25 mm. long, bidentulate; achenes triangular, one-third to one-half width of perigynia, slenderly long stipitate; style very slender, straight, jointed with the achene; stigmas 3, short.

Mountains from Alberta and Montana to Washington. Type: Sheep Mountain, Okanogan County, Washington, *Eggleston 3329*.

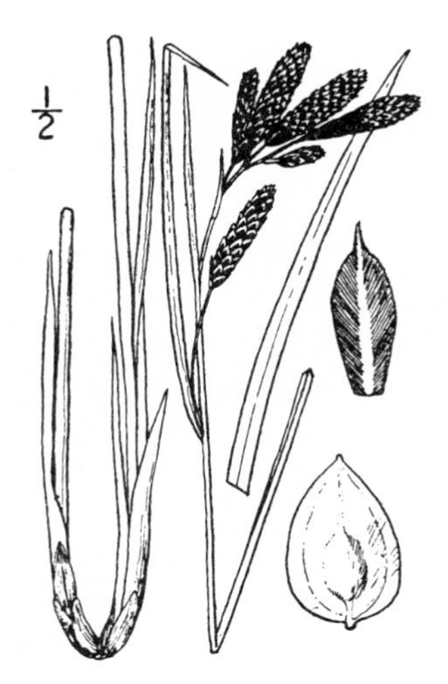

131. Carex merténsii Prescott.
Mertens's Sedge. Fig. 807.

Carex mertensii Prescott, Mem. Acad. St. Petersb. II. **6**: 168. 1833.
Carex columbiana Dewey, Am. Journ. Sci. **30**: 62. 1836.

Cespitose and short stoloniferous, the culms 3–10 dm. high, sharply triangular, rough. Leaf-blades flat, 4–7 mm. wide, the lower sheaths brownish purple tinged; spikes 6–10, gynaecandrous, the uppermost strongly staminate at base, the lateral sparingly, 1–4 cm. long, 7–10 mm. wide, the upper approximate, the lower more remote on capillary peduncles, the perigynia numerous, appressed-ascending; bracts not sheathing; scales lance-ovate, acute, much shorter than perigynia, brown with light midvein and margins; perigynia broadly ovate, much flattened, light green, purple spotted, 5 mm. long, 3 mm. wide, lightly 3-nerved, rounded at base, tapering at the apex, minutely beaked, the beak 0.25 mm. long, entire; achenes triangular, 2 mm. long, 0.8 mm. wide, strongly stipitate; style slender, straight, articulated to achene and in age deciduous; stigmas 3.

A very well marked and handsome species found from Alaska south in the mountains to Trinity County, California, and eastward to Montana; Hudsonian Zone. Type locality: Sitka, Alaska.

132. Carex scopulòrum Holm.
Holm's Rocky Mountain Sedge. Fig. 808.

Carex tolmiei subsessilis Bailey, Mem. Torrey Club **1**: 47 (in part). 1889.
Carex scopulorum Holm, Am. Journ. Sci. IV. **14**: 421–2. *f. 1–6.* 1902.

Loosely stoloniferous, the culms solitary or in small clumps, 1–4 dm. high, stiff, smooth, sharply triangular. Leaf-blades 3–7 mm. wide with revolute margins; terminal spike staminate or androgynous; lateral spikes 2–3, pistillate or androgynous, approximate, erect, sessile or stort-stalked, 1–2.5 cm. long, 6–7 mm. wide; bracts not sheathing, the lowest shorter than the inflorescence; scales obtuse, exceeded by perigynia, black, the midvein obsolete; perigynia 2.5–3.5 mm. long, soon turgid, membranaceous, papillose, spreading, greenish straw-colored, at length brownish, nerveless except the two marginal ribs, abruptly contracted into a short entire more or less strongly recurved beak; achenes lenticular; style slender, its base not enlarged; stigmas 2.

High mountains, Hudsonian and Arctic-Alpine Zones; Montana and Washington to Colorado and the Sierra Nevada of California (Tulare County). Type locality: Clear Creek Cañon, Colorado.

133. Carex áccedens Holm.
Mount Rainier Sedge. Fig. 809.

Carex stylosa virens Bailey, Proc. Am. Acad. **22**: 79. 1886.
Carex spreta Bailey, Mem. Torrey Club **1**: 6. 1889; not Steud. 1855.
Carex accedens Holm, Am. Journ. Sci. IV. **16**: 457. 1903.

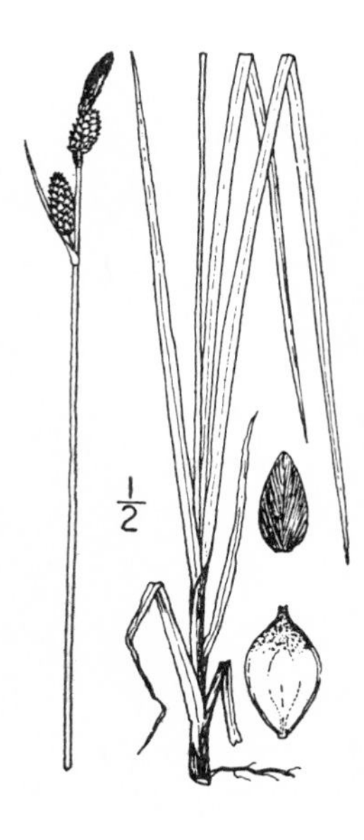

Loosely stoloniferous, the rootstocks stoutish, elongated, the stolons ascending, the culms phyllopodic, 2.5–4 dm. high, strict, very sharply triangular, roughened above, purplish tinged at base. Culms with old leaves clustered at base, the new 4–5 on lower fourth, 3–4 mm. wide, flat with revolute margins, elongated, light green, the lower sheaths short puberulent dorsally, white hyaline and purplish spotted ventrally, not filamentose; terminal spike sessile or nearly so, staminate or gynaecandrous, 18 mm. long; pistillate spikes 3–4, aggregated or lowest separate, sessile or short-peduncled, oblong, 7–20 mm. long, 6 mm. wide, closely flowered, the 20–40 perigynia spreading-ascending; bracts dark-auricled, the lowest much shorter than culm; scales lance-ovate, obtuse or acutish, exceeded by perigynia, purplish black with white midvein; perigynia broadly obovate or suborbicular, much flattened, 3 mm. long, 1.75–2.25 mm. wide, sessile, nerveless, but with thick margins, green not dark tinged, rather coriaceous, puncticulate, abruptly beaked, the beak 0.25 mm. long, entire; achenes lenticular; style jointed with achene, short protruding; stigmas 2.

Rare and local, Arctic-Alpine Zone; Mount Adams and Mount Rainier, Washington; reported from Sauvie's Island, Oregon, probably erroneously. Type locality: Sauvie's Island, Oregon.

134. **Carex gymnocl̀ada** Holm.

Sierra Alpine Sedge. Fig. 810.

Carex vulgaris alpina W. Boott in S. Wats. Bot. Calif. **2**: 240. 1880; not Boott.
Carex vulgaris bracteosa Bailey, Proc. Am. Acad. **22**: 81. 1886.
Carex gymnoclada Holm, Am. Journ. Sci. IV. **14**: 424. *f. 12–14.* 1902.
Carex brachypoda Holm, Am. Journ. Sci. IV. **20**: 302. *f. 4–6.* 1905.

Loosely strongly stoloniferous, the culms phyllopodic, stiff, sharply triangular and usually roughened above, exceeding leaves, brownish at base, 2.5–6 dm. high. Culms developing some short blades the first year, and in the flowering (second) year 2–4 erect blades, 2.5–4 mm. wide, flat, with revolute margins; terminal spike normally staminate, short-peduncled, 5–15 mm. long; lateral spikes 2–3, pistillate or staminate above, approximate, sessile or short-peduncled, oblong or linear-oblong, 5–15 mm. long, 4–6 mm. wide, the perigynia 8–30, spreading-ascending, closely packed; bracts sheathless, the lower normally shorter than the inflorescence; scales ovate or lanceolate, obtuse to acuminate, exceeded by perigynia, purplish-black, the midvein usually obsolete; perigynia obovoid, plano-convex, not turgid, 3 mm. long, 1.75 mm. wide, stramineous, often dark-tinged, nerveless except for the two marginal ribs, membranaceous, granular, rounded at base, abruptly minutely beaked, the beak straight, 0.25 mm. long; achenes lenticular; style slender; stigmas 2.

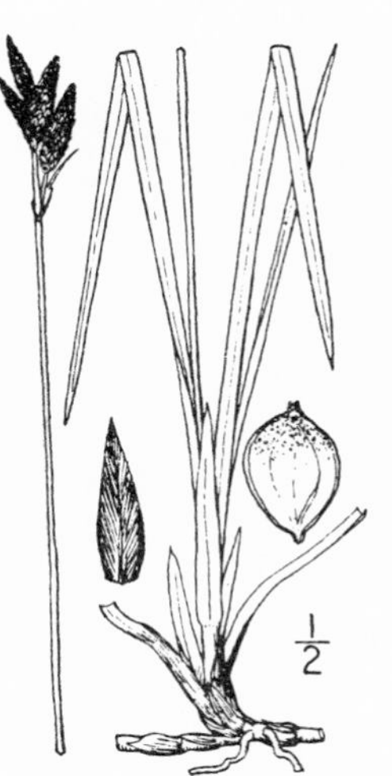

In the higher mountains, Arctic-Alpine Zone; Washington south to Tulare County, California. Type locality: bogs of Hurricane Creek, eastern Oregon.

135. **Carex campylocàrpa** Holm.

Crater Lake Sedge. Fig. 811.

Carex campylocarpa Holm, Am. Journ. Sci. IV. **20**: 304. *f. 13–15.* 1905.

Rootstock slender, ascending or horizontal, elongated, the culms 4–6 dm. high, rather slender, sharply triangular and very rough on the angles, exceeding leaves, phyllopodic, strongly purplish tinged at base. Leaf-blades erect, about 3 mm. wide, flat with revolute margins, the sheaths olive-tinged ventrally; terminal spike staminate, slightly clavate, 6–8 mm. long, sessile or short-peduncled; pistillate spikes mostly two, sessile or nearly so, the lower separate, short-oblong, 7–10 mm. long, 4.5–6 mm. wide, closely 10–25 flowered; lowest bract shorter than culm, purplish-black at base; scales ovate, obtuse, half width and half length of perigynia, purplish-black, the midvein obsolete or nearly so; perigynia ascending or spreading, ovoid, mostly convex-concave, 3.75 mm. long, 1.75 mm. wide, more or less turgid, dull green below, purplish mottled above, sub-membranaceous, nerveless, granular, prominently sparingly serrulate above, rounded and substipitate at base, tapering above, prominently short-beaked, the beak black, entire, the middle ones often excurved; achenes lenticular; style slender, jointed with achene; stigmas 2.

Alpine stream banks, Boreal Zone; Washington and Oregon. Possibly but a form of the last. Type locality: Crater Lake, National Park, Oregon.

136. **Carex prionophýlla** Holm.

Rough-sheathed Sedge. Fig. 812.

Carex prionophylla Holm, Am. Journ. Sci. IV. **14**: 423, 425. *f. 7–11.* 1902.

Rootstocks thick, the culms 5–9 dm. high, strongly aphyllopodic, narrowly wing-angled and very serrulate on the angles, strongly reddish purple at base. Leaves 3–5 to a culm, clustered well above the base, the blades flat, 4–5 mm. wide, very long and scabrous, the sheaths strongly hispidulous and often red-tinged; terminal spike staminate, short-peduncled, 18–25 mm. long; pistillate spikes 3 or 4, erect, sessile or short peduncled, approximate or lower separate, oblong or linear-cylindric, 1–2 cm. long, 3–4 mm. wide, closely flowered with 15–40 appressed perigynia; lowest bract shorter than culm, dark auricled; scales oblong-ovate, acutish or obtusish, half width but nearly length of perigynia, blackish throughout or rarely with lighter midvein; perigynia obovate, much flattened, 2.5 mm. long, 1.5 mm. wide, green, very obscurely nerved dorsally, round tapering and substipitate at base, abruptly rounded and minutely beaked at apex, the beak 0.25 mm. long, black, entire; achenes lenticular; style slender, jointed with achene; stigmas 2.

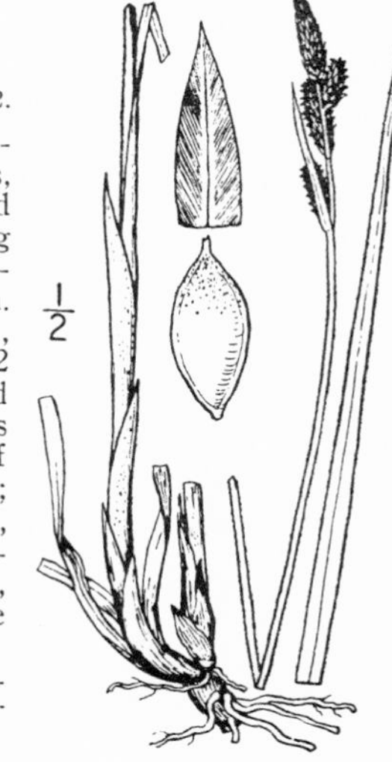

Mountain streams, Transition Zone; northern Idaho and eastern Washington. A well marked species. Type locality: between St. Joe and Clearwater Rivers, Idaho.

137. **Carex nebraskénsis** Dewey.

Nebraska Sedge. Fig. 813.

Carex jamesii Torr. Ann. Lyc. N. Y. **3**: 398. 1836; not Schw. 1824.
Carex nebraskensis Dewey, Am. Journ. Sci. II. **18**: 102. 1854.
Carex nebraskensis praevia Bailey, Mem. Torrey Club **1**: 49. 1889.
Carex nebraskensis ultriformis Bailey, Bot. Gaz. **21**: 8. 1896.
Carex jacintoensis Parish, Bull. South. Calif. Acad. Sci. **4**: 110. *pl. 16.* 1905.

Rootstocks creeping and stoloniferous, the culms 2.5–10 dm. high, stout, rigid, phyllopodic, roughened or smooth above. Leaf-blades 3–8 mm. wide, pale green, flat, the sheaths nodulose, white hyaline ventrally; staminate spikes 1–2, more or less peduncled, 1.5–3.5 cm. long, 3–6 mm. wide; pistillate spikes 2–5, oblong, sessile or short-peduncled, 1.5–6 cm. long, 6–9 mm. wide, contiguous or the lower separate with many ascending perigynia; lowest bract about equalling culm; scales lanceolate, obtusish to acuminate, blackish with light midvein; perigynia 3–3.5 mm. long, 2 mm. wide, compressed by-convex, ribbed, coriaceous, greenish straw color, rounded and sessile at the base, contracted at apex into the bidentate beak; achenes lenticular; style slender; stigmas 3.

Meadows and swamps, Arid Transition Zone; South Dakota and Kansas to New Mexico, California, and British Columbia. Type locality: Nebraska.

138. **Carex paucicostàta** Mackenzie.

Few-ribbed Sedge. Fig. 814.

Carex lenticularis W. Boott in S. Wats. Bot. Calif. **2**: 242 (in part). 1880; not Michx.
Carex interrupta impressa Bailey, Mem. Torrey Club **1**: 18. 1889.
Carex paucicostata Mackenzie, Erythea **8**: 74. 1922.

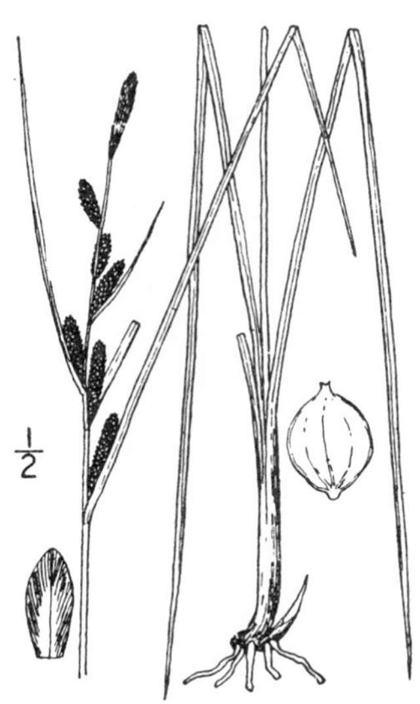

Cespitose with very short ascending stolons, the culms slender, phyllopodic, 2.5–4 dm. high, sharply triangular, smooth or little roughened. Leaf-blades light green, 2–4 mm. wide, flat or channelled at the base, the sheaths smooth, very thin and hyaline ventrally; terminal spike staminate, short-peduncled or nearly sessile; pistillate spikes 4–6, linear, 3.5–5 mm. wide, 1–4 (mostly 2–3), cm. long, the numerous perigynia appressed-ascending; lower bracts exceeding the culms; scales oblong, obtuse or acutish, somewhat shorter than the perigynia, falling early but after perigynia, dark-colored with lighter center and hyaline apex; perigynia very broadly ovate or obovate, or suborbicular, 2 mm. long, 1.5–2 mm. wide, glaucous green, papillate roughened, 3–5 ribbed ventrally, minutely and abruptly black apiculate tipped; achenes lenticular; style slender; stigmas 2.

Wet places, mostly around lakes, Transition and Canadian Zones; Sierra Nevada of California from Tulare County to El Dorado County. Type locality: Summit Camp, California.

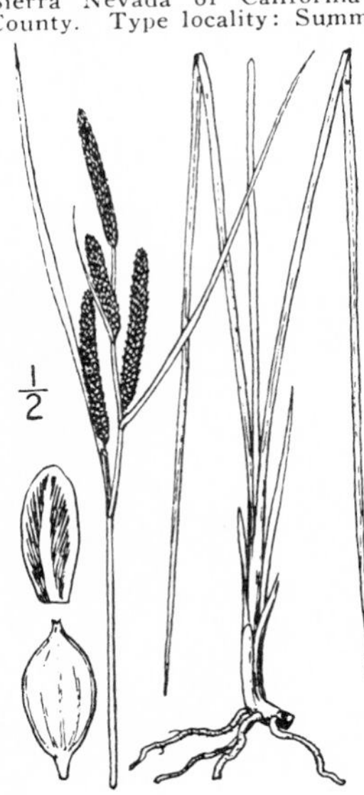

139. **Carex híndsii** C. B. Clarke.

Hinds's Sedge. Fig. 815.

Carex decidua Boott, Ill. Car. 1: 163 (as to N. Am. plant). 1858.
Carex hindsii C. B. Clarke, Kew Bull. Misc. Inf. Add. Ser. **8**: 70. 1908.

Cespitose with stout usually very short ascending stolons, the culms slender, 1–6 dm. high, phyllopodic, brownish at base, sharply triangular, smooth or little roughened. Leaf-blades 1.5–3 mm. wide, flat or channelled at the base, the sheaths smooth, very thin and hyaline ventrally; terminal spike mostly staminate, long-peduncled; pistillate spikes 4–6, linear or oblong-linear, 1.5–4.5 cm. long, 5–7 mm. wide, the numerous perigynia appressed-ascending; lower bracts exceeding culms; scales oblong, obtuse or acutish, much shorter than perigynia, early deciduous, blackish with lighter center and hyaline apex; perigynia ovate, 2.5–3.5 mm. long, 1.5–2 mm. wide, yellowish green, ribbed, papillate roughened, strongly stipitate, the upper part empty, minutely black apiculate-tipped; achenes lenticular; styles slender; stigmas 2.

Wet places along the coast, Canadian Zone; Aleutian Islands and Alaska (where abundant) southward locally to Del Norte County, California. Type locality: Columbia River.

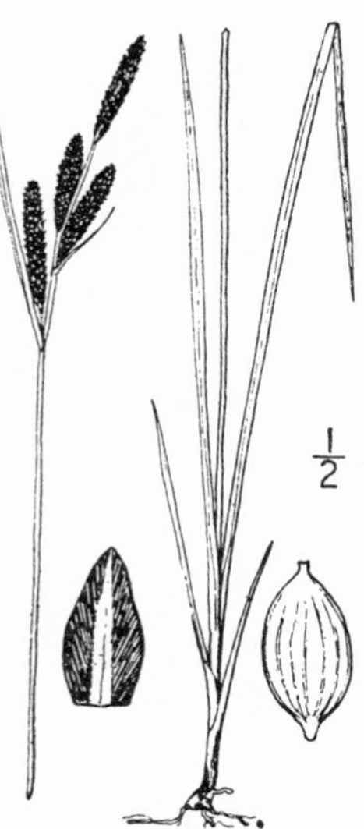

140. **Carex kellóggii** W. Boott.

Kellogg's Sedge. Fig. 816.

Carex acuta pallida Boott, Ill. Car. **4**: 166. *pl. 554.* 1867.
Carex kelloggii W. Boott in S. Wats. Bot. Calif. **2**: 240. 1880.
Carex vulgaris lipocarpa Holm, Am. Journ. Sci. IV. **17**: 308 (in part). 1904.
Carex limnaea Holm, Am. Journ. Sci. IV. **20**: 301. *f. 1–3.* 1905.
Carex hindsii brevigluma Kükenth. in Engler Pflanzenreich **4**^20^: 307. 1909.
Carex lenticularis paullifructus Kükenth. l.c. 308. 1909.

Cespitose with very short ascending stolons, the culms phyllopodic, slender, 3–7 dm. high, light brownish and somewhat fibrillose at base, somewhat roughened above. Leaf-blades light green, flat or somewhat channelled at base, 1.5–3 mm. wide, the sheaths smooth, very thin and white hyaline ventrally; staminate spike usually solitary, 12–36 mm. long, 3–4 mm. wide; pistillate spikes 3–5, sessile or nearly so, approximate or slightly separate, linear, 1.5–4 cm. long, 4–6 mm. wide, often attenuate at base, the numerous perigynia appressed-ascending; lowest bract exceeding culm; scales obtuse or acutish, exceeded by perigynia, dark with broad light colored center; perigynia 2.5 mm. long, 1.25 mm. wide, pale green, slenderly nerved, densely granular, strongly stipitate, abruptly minutely beaked, the beak dark-colored, entire; achenes lenticular; style slender; stigmas 2.

Wet places, Transition and Canadian Zones; Montana to Colorado, west to Alaska and California, where confined to the higher Sierra Nevada. Type locality: Alta, Sierra Nevada Mountains.

141. **Carex apérta** Boott.

Columbia Sedge. Fig. 817.

Carex aperta Boott in Hook. Fl. Bor. Am. **2**: 218. *pl. 219.* 1840.
Carex acuta prolixa Bailey, Proc. Am. Acad. **22**: 86. 1886; not Hornem.
Carex turgidula Bailey, Bot. Gaz. **25**: 271. 1898.
Carex bovina Howell, Fl. N. W. Am. **1**: 702. 1903.

Stoloniferous, the culms 5–10 dm. high, sharply triangular, rough above, dark-tinged at base. Leaf-blades flat with slightly revolute margins, 2.5–5 mm. wide, roughened towards apex; terminal spike staminate, peduncled, 2–3.5 cm. long, 3 mm. wide; lateral spikes 2–3, pistillate, oblong, or linear-oblong, sessile to slender-peduncled, approximate, 12–48 mm. long, 5 mm. wide, closely flowered (or loosely at base) with numerous, ascending perigynia; lowest bract about equalling culm; scales narrower than but mostly exceeding perigynia, lanceolate, sharp-pointed; perigynia obovoid, 3 mm. long, 1.5 mm. wide, somewhat inflated at maturity, greenish or straw-color, membranaceous, nerveless, rounded to a slightly stipitate base, abruptly minutely beaked, the beak bidentate, 0.5 mm. long; achenes lenticular; style slender; stigmas 2.

Low grounds, Transition Zone; British Columbia to northern Oregon, east to Montana. Type locality: Columbia River.

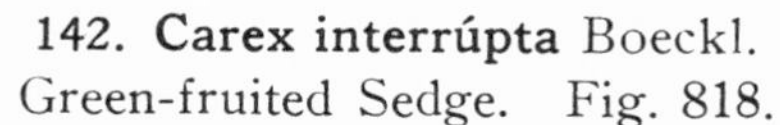

142. **Carex interrúpta** Boeckl.

Green-fruited Sedge. Fig. 818.

Carex angustata verticillata Boott in Hook. Fl. Bor. Am. **2**: 218. 1840.
Carex verticillata Boott, Ill. Car. **1**: 67. *pl. 183, f. 2.* 1858; not Zoll. & Mor. 1854–5.
Carex interrupta Boeckl. Linnaea **40**: 432. 1876.

Cespitose with short ascending stolons, the culms 3–6 dm. high, smooth, exceeding leaves. Leaf-blades thin, flat, 3–5 mm. wide, not glaucous, the sheaths smooth, not filamentose; terminal spike staminate, 1.5–3.5 cm. long; pistillate spikes 3–5, erect, more or less strongly separate, occasionally staminate at apex, oblong to linear, the upper 1–3 cm. long, densely flowered and sessile or nearly so, the lower 4–9 cm. long, loosely flowered at base, and strongly peduncled, the perigynia very numerous, ascending; lowest bract about equalling the culm; scales ovate or ovate-lanceolate, acute or obtusish, about length of perigynia, reddish purple with lighter center and narrow hyaline margins; perigynia suborbicular or short oval, olive green, very small, 1.5–1.75 (2) mm. long, 1–1.25 mm. wide, biconvex, 2-edged, nerveless, membranaceous, rounded at base and apex, very abruptly and very minutely short-beaked, the beak emarginate or bidentulate; achenes lenticular; style slender, jointed with achene; stigmas 2.

Low grounds along streams, Humid Transition Zone; Oregon and Washington. Type locality: Columbia River.

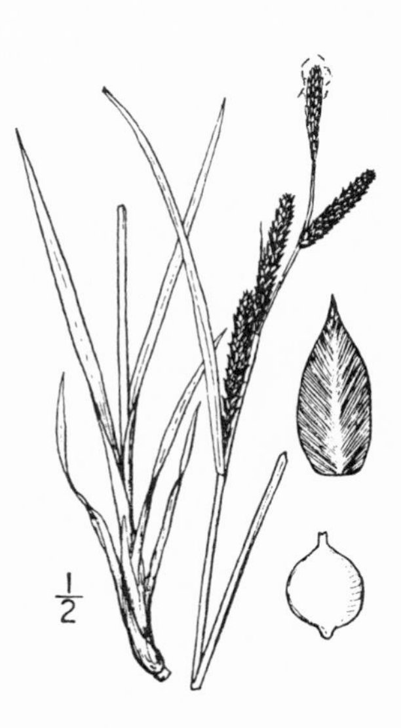

143. Carex sitchénsis Prescott. Sitka Sedge. Fig. 819.

Carex sitchensis Prescott, Mem. Acad. St. Petersb. II. **6**: 169. 1832.
Carex aquatilis W. Boott in S. Wats. Bot. Calif. **2**: 241. 1880; not Wahl.
Carex howellii Bailey, Mem. Torrey Club **1**: 45. 1889.
Carex dives Holm, Am. Journ. Sci. IV. **17**: 312. 1904.
Carex pachystoma Holm, Am. Journ. Sci. IV. **20**: 302. *f. 7–8.* 1905.

Densely cespitose from elongated rootstocks, the culms 6–12 dm. high, stout, sharply triangular, strongly reddened at base, phyllopodic, the sheaths smooth, not filamentose, the upper and lower leaves with well-developed blades, the middle one shorter. Leaf-blades flat or channelled at base, 2–9 mm. wide; staminate spikes 2–5, slender; pistillate spikes 3–5, strongly separate, on long slender peduncles, linear-cylindric, 3–9 cm. long, 4.5–6 mm. wide, very many flowered; bracts sheathless, the lower usually exceeding the culm; scales lanceolate, sharp-pointed, brownish with lighter center, the apex often noticeably hyaline in age; perigynia ovate or oval, 3–5.3 mm. long, 1.5–2 mm. wide, widest near middle, obscurely nerved, substipitate, puncticulate, greenish straw-colored, little red-dotted, apiculate; achenes lenticular; style slender; stigmas 2.

Near the coast, Humid Transition and Canadian Zones; Prince William Sound, Alaska, to Santa Cruz County, California. Very local southward. Type locality: Sitka, Alaska.

144. Carex aquátilis Wahl. Water Sedge. Fig. 820.

Carex aquatilis Wahl. in Vet. Akad. Nya Handl. Stockholm **24**: 165. 1803.
Carex stans Drejer, Revis. Caric. Bor. 40. 1841.
Carex variabilis Bailey, Mem. Torrey Club **1**: 18. 1889.

Rootstocks slender, sending forth long horizontal stolons, the culms cespitose, 2–7 dm. high, slender, sharply triangular above, reddened at the base. Leaf-blades 2–6 mm. wide; staminate spikes 1–2, slender; pistillate spikes 2–4, sessile or short-peduncled, not aggregated, linear, 1.5–6 cm. long, 4–6 mm. wide; lowest bract exceeding culm; scales oblong-obovate to lanceolate, obtuse or acutish, blackish to reddish brown, 1-nerved or with a lighter center; perigynia elliptic-obovate, 2.5 mm. long, 1.25 mm. wide, broadest below apex, nerveless but with a median ridge, substipitate, puncticulate, granular, red-dotted, minutely beaked, the beak entire; achenes lenticular; style slender; stigmas 2.

Swampy grounds, Transition and Canadian Zones; Quebec to Alaska south to New Mexico and northern California; widely distributed in northern Eurasia. Rare in our range. Type locality: Europe.

145. Carex bàrbarae Dewey. Santa Barbara Sedge. Fig. 821.

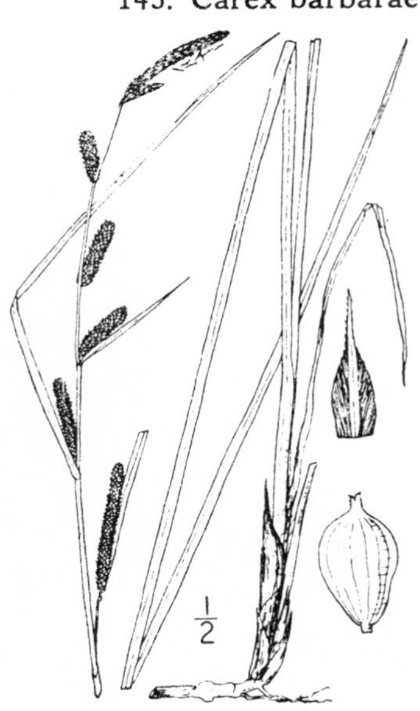

Carex barbarae Dewey, Bot. Mex. Bound. Surv. 231. 1858.
Carex laciniata Boott in Benth. Pl. Hartw. 341. 1857 (name only); and Ill. Carex **4**: 175. *pl. 594* (in part). 1867.
Carex lacunarum Holm, Am. Journ. Sci. IV. **17**: 303. *f. 12–13.* 1904.

Cespitose with long horizontal stolons, the culms 3–10 dm. high, phyllopodic, stout, sharply triangular, serrulate at least in inflorescence. Leaves 7–12, the sheaths brownish-puberulent, the middle more or less filamentose, the blades light green, thick, flat or channelled, 3.5–9 mm. wide, serrulate, the middle ones much reduced; staminate spikes one or two, narrowly linear; pistillate spikes 2–5, sessile or short-peduncled, oblong or linear-cylindric, 2.5–8 cm. long, 5–8 mm. wide, the very many perigynia ascending; lowest bract shorter than to exceeding the inflorescence; scales narrowly ovate, hispid mucronate, occasionally some merely acute, reddish-purple with lighter center; perigynia narrowly to broadly obovate, substipitate, 3–4.5 mm. long, 2–2.5 mm. wide, obscurely nerved on both faces, straw-colored at length brownish, puncticulate, often granular, slightly serrulate above, the beak 0.5 mm. long, sharply bidentate and hispidulous between the teeth; achenes lenticular; style slender; stigmas 2.

Southern Oregon south to San Bernardino, California, west of the higher Sierra Nevada; Upper Sonoran and Transition Zones. One of the most characteristic California species. Type locality: Santa Barbara, California.

146. Carex schóttii Dewey. Schott's Sedge. Fig. 822.

Carex schottii Dewey, Bot. Mex. Bound. Surv. 231. 1858.
Carex barbarae Parish, Bull. South. Calif. Acad. Sci. **4**: 108. *pl. 14.* 1905; not Dewey.

Culms stout, in large clumps, 10–15 dm. high, sharply triangular and rough above. Leaf-blades flat with revolute margins, 6–12 mm. wide, serrulate, the new lower sheaths strongly yellowish-brown tinged, sharply keeled and hispidulous dorsally, the ventral side very fragile, breaking and becoming filamentose; staminate spikes about 3, elongate linear, 8–14 cm. long, 4 mm. wide; pistillate spikes (usually staminate above) mostly three, sessile or nearly so, strongly separate, erect, elongate linear, 11–14 cm. long, 5–6 mm. wide, the perigynia appressed-ascending, closely packed in several ranks; lowest bracts mostly exceeding culms; scales narrowly lanceolate, acute or obtusish, usually exceeding perigynia, purplish black with broad 3-ribbed lighter center; perigynia obovate, plano-convex, 3 mm. long, 1.75 mm. wide, membranaceous, greenish straw colored, several nerved on both faces, round tapering and nearly sessile at base, abruptly beaked, the beak 0.25 mm. long submarginate; achenes lenticular; stigmas 2.

Stream banks, Upper Sonoran Zone; southern California from Santa Barbara and San Diego Counties eastward into the San Bernardino Mountains below 2500 feet. Type locality: Santa Barbara.

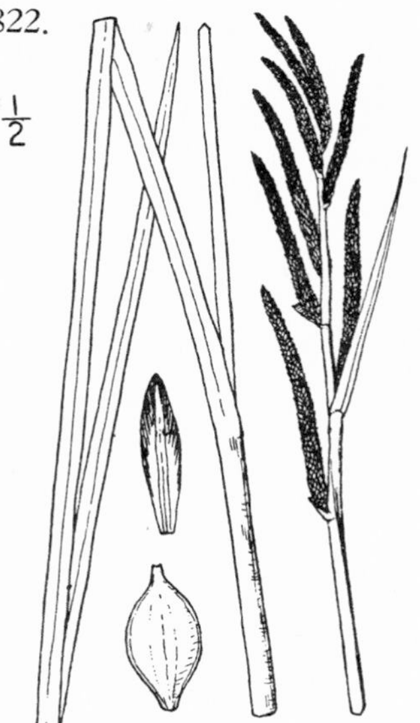

147. Carex sénta Boott. Rough Sedge. Fig. 823.

Carex senta Boott, Ill. Carex **4**: 174. 1867.
Carex auriculata Bailey, Mem. Torrey Club **1**: 19. 1889.
Carex austromontana Parish, Bull. South. Calif. Acad. Sci. **4**: 108. *f. 15.* 1905.
Carex bishallii C. B. Clarke, Kew Bull. Miss. Inf. Add. Ser. **8**: 70. 1908.
Carex nudata f. *firmior* and *sessiliflora* Kükenth. in Engler Pflanzenreich **4**[20]: 337. 1909.

Loosely cespitose and stoloniferous, the culms slender, 4–7 dm. high, sharply triangular, brownish at base. Leaves 6–12, the middle sheaths sparingly hispidulous dorsally and filamentose ventrally, the blades flat or slightly revolute, 3–5 mm. wide, the middle ones much reduced, the lower and upper 1–4 dm. long; terminal spike staminate, peduncled, 3–4.5 cm. long; pistillate spikes 1–3, sessile or short-peduncled, 2.5–5 cm. long, 5–9 mm. wide, the numerous peryginia ascending; lowest bract shorter than to exceeding inflorescence; scales oblong-ovate or lanceolate, obtuse to acuminate, purplish-black with lighter center; perigynia ovate, rounded and substipitate at base, conspicuously several nerved on both faces, green or straw-colored, puncticulate, round-tapering at apex, abruptly minutely beaked, the beak 0.25 mm. long, entire; achenes lenticular; stigmas 2.

Swamps, Upper Sonoran and Transition Zones; southern California north in the coastal counties to Alameda County, and in the Sierra Nevada to Amador County. Type locality: Santa Inez Mountains, California.

148. Carex nudàta W. Boott. Torrent Sedge. Fig. 824.

Carex nudata W. Boott in S. Wats. Bot. Calif. **2**: 241. 1880.
Carex hallii Bailey, Proc. Am. Acad. **22**: 82. 1881.
Carex acutina Bailey, Mem. Torrey Club **1**: 52. 1889.

Extremely densely cespitose, the rootstocks descending obliquely, the culms slender, aphyllopodic, 3–8 dm. high, sharply triangular, somewhat roughened above, strongly dark purplish at base. Leaf-blades light green, flat with revolute margins, 2.5–3.5 mm. wide, very smooth except on margins above and towards apex, the basal sheaths filamentose, rounded and hispidulous dorsally; staminate spike short-peduncled, 1.5–3 cm. long; lateral spikes 3–5, sessile or short-peduncled, 1–4 cm. long, 6–7 mm. wide, the numerous perigynia ascending; lowest bract shorter than culm, the upper ones very short, conspicuously bi-auriculate; scales ovate or oblong-ovate, much exceeded by perigynia, blackish with lighter midvein; perigynia lanceolate or ovate, compressed, 2.5–4 mm. long, 1.75 mm. wide, finely nerved on both faces, membranaceous, greenish straw colored or purplish-black tinged, smooth or slightly granular at apex, rounded and substipitate at base, the upper portion empty, abruptly beaked, the beak 0.25 mm. long, entire; achenes lenticular; stigmas 2.

Rocky stream beds, Transition Zone; western Oregon south to Santa Clara and Mariposa Counties, California. Type locality: San Francisco Bay.

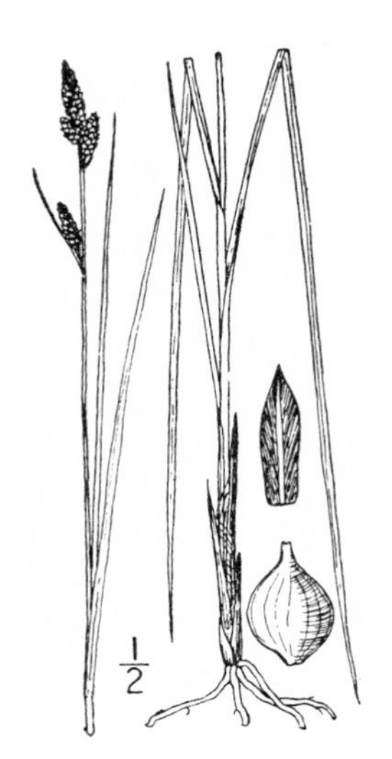

149. **Carex suborbiculàta** Mackenzie. n. sp.

Round-fruited Mountain Sedge. Fig. 825.

Densely cespitose; the culms erect, stiff but slender, 3–4.5 dm. high, sharply triangular and slightly roughened above, strongly aphyllopodic, reddish-tinged at base. Leaves 2–4 to a culm, the blades 1.25–2 mm. wide, with revolute margins, long attenuate, the lower sheaths filamentose; terminal spike staminate or pistillate below, short-peduncled, 1.5 cm. long, the scales oblong, very obtuse, purplish black with conspicuous light colored midrib; pistillate spikes 3 or 4, erect, linear, 1–2.5 cm. long, 3–4 mm. wide, the upper sessile and aggregated, the lower peduncled and separate, closely flowered with 15–40 spreading perigynia; lowest bract from much shorter than to equalling inflorescence, not dark-auricled; scales oblong, very obtuse, much narrower and shorter than the perigynia, purplish black with conspicuous light-colored midrib; perigynia biconvex and in age turgid, suborbicular to broadly obovoid-oval in cross-section, 2.5 mm. long, 1.5–1.75 mm. wide, whitish, purplish blotched, membranaceous, puncticulate, rather strongly several-nerved, substipitate, rounded at base, abruptly minutely beaked, the beak entire straight or somewhat bent; achenes lenticular; style slender, jointed with the achenes; stigmas 2.

Klickitat and Skamania Counties, Washington. Type: along mountain streams, Klickitat County, Washington (*Suksdorf 1315*).

150. **Carex eurycàrpa** Holm.

Well-fruited Sedge. Fig. 826.

Carex eurycarpa Holm, Am. Journ. Sci. IV. **20**: 303. 1905.

Loosely cespitose and long stoloniferous from creeping rootstocks, the culms aphyllopodic, 4–9 dm. high, slender, sharply triangular, roughened above. Leaves several to a culm, the blades flat with revolute margins, 3–4.5 mm. wide, roughened towards apex, the lower sheaths minutely hispidulous and rounded dorsally, olive tinged and sparingly filamentose ventrally; staminate spike more or less peduncled, usually 3–4 cm. long; lateral spikes 3–5, the upper often staminate above, sessile or short-peduncled, 2–4 cm. long, 4–6 mm. wide, the numerous perigynia appressed; lowest bract usually exceeding the culm; scales lanceolate, short acuminate, purplish brown with prominent light midvein; perigynia suborbicular or broadly obovate, 2.75 mm. long, 2 mm. wide, several nerved on both faces, puncticulate, minutely roughened, greenish or straw colored, rounded and substipitate at base, and rounded and abruptly minutely beaked at orifice, the beak 0.5 mm. long, emarginate; achenes lenticular; style slender; stigmas 2.

In boggy meadows, Transition Zone; Washington to northern California and extending south in the Sierra Nevada to El Dorado County. Type locality: Falcon Valley, Klickitat County, Washington.

151. **Carex oxycàrpa** Holm.

Sharp-fruited Sedge. Fig. 827.

Carex oxycarpa Holm, Am. Journ. Sci. IV. **20**: 303. 1905.
Carex eurycarpa oxycarpa Kükenth. in Engler Pflanzenreich 4[20]: 339. 1909.
Carex egregia Mackenzie, Bull. Torrey Club **42**: 414. 1915.

Loosely cespitose and stoloniferous from creeping rootstocks, the culms aphyllopodic, 4.5–9 dm. high, slender, sharply triangular, slightly roughened above. Leaves 3–4 to a culm, the blades flat with revolute margins, 2–3 mm. wide, roughened towards apex, the lower sheaths hispidulous and scarcely carinate dorsally and light brownish ventrally; terminal spike staminate, stalked; lateral spikes 4–5, the upper 1–2 staminate, the lower 2–4 pistillate or androgynous, sessile or short-peduncled, 2.5–4.5 cm. long, 5 mm. wide, the perigynia numerous, appressed-ascending; lowest bract exceeding inflorescence; scales lanceolate, acute, purplish black with lighter midvein; perigynia oblanceolate or narrowly obovate, 3.5 mm. long, 1.5 mm. wide, 3–5 striate on both faces, granular roughened, brownish, rounded and substipitate at base and rounded and abruptly minutely apiculate at apex, the orifice entire; achenes lenticular; style slender; stigmas 2.

Wet meadows, Transition Zone; from northern California to Washington and east to Idaho. Probably a form of *Carex eurycarpa* Holm Type locality: west Klickitat County, Washington.

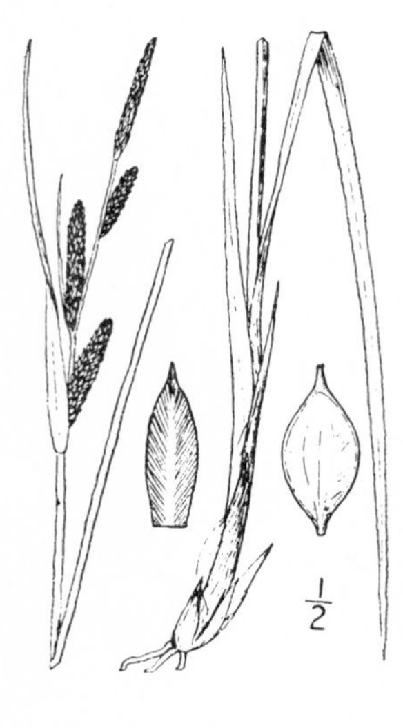

152. Carex lýngbyei Hornem.
Lyngbye's Sedge. Fig. 828.

Carex lyngbyei Hornem. Fl. Dan. Pl. 1888. 1827.
Carex cryptocarpa C. A. Meyer, Mem. Acad. St. Petersb. 1: 226. *pl. 14.* 1831.
Carex scouleri Torr. Ann. Lyc. N. Y. 3: 399. 1836.
Carex filipendula Drejer Rev. Crit. Carex 46. 1841.
Carex macounii A. Bennett in Macoun Cat. Canad. Pl. 4: 147. 1888.
Carex qualicumensis Bailey, Bull. Torrey Club 20: 428. 1893.

Strongly long stoloniferous, the culms varying from rather slender to very stout, 3–9 dm. high, the lowermost leaves (of first year's growth) with very long blades, the lower of the second year's growth with shorter blades than the upper. Leaf-blades flat, 2–12 mm. wide; uppermost spike staminate, long-peduncled; lateral spikes 2–6, the upper 1 or 2 often staminate or androgynous, the lower pistillate, dropping on slender smooth peduncles, densely many-flowered, linear or oblong, 2–8 cm. long, 5–10 mm. wide; bracts leaf-like, the lower exceeding the culm; scales lanceolate, acuminate, much exceeding perigynia, dark brown with straw-colored keels; perigynia oblong-oval, biconvex, 2.5–3 mm. long, more or less nerved, straw colored or at length brownish, coriaceous, puncticulate, rounded and slightly stipitate at base, abruptly minutely beaked, the beak entire; achenes lenticular; stigmas 2.

Along the coast, Canadian Zone; Humboldt County, California, to the Aleutian Islands; also known from eastern Asia, Greenland, Iceland, and northern Europe. Type locality: Faroe Islands, Europe.

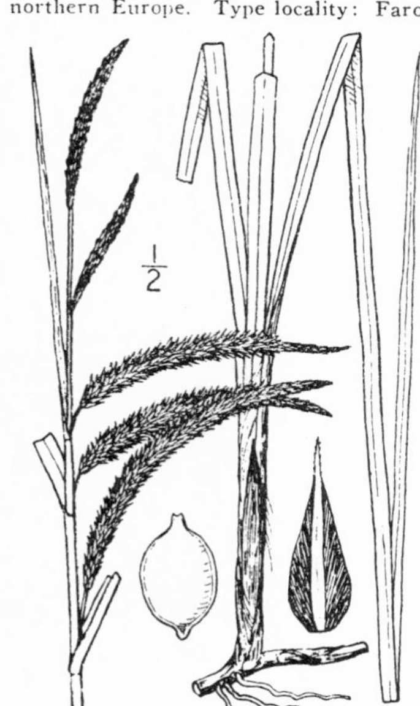

153. Carex obnúpta Bailey.
Slough Sedge. Fig. 829.

Carex sitchensis Boott in Hook. Fl. Bor. Am. 2: 220. *pl. 221.* 1840; not Prescott.
Carex aquatilis W. Boott in S. Wats. Bot. Calif. 2: 241. 1880; not Wahl.
Carex obnupta Bailey, Proc. Calif. Acad. Sci. II. 3: 104. 1891.
Carex magnifica Dewey; Holm, Am. Journ. Sci. IV. 17: 316. 1904.

With long stout stolons, the culms 5–15 dm. high, sharply triangular, roughened above. Leaves 5–10 clustered near the base, the blades 5–8 mm. wide, thick, much roughened above; staminate spikes 2–3, linear; pistillate spikes 2–4, oblong to linear-cylindric, 3–10 cm. long, 4–8 mm. wide, many-flowered, the upper sessile or nearly so, the lower more or less strongly peduncled; bracts exceeding culms; scales narrowly ovate, concealing perigynia, sharp-pointed, blackish; perigynia 3–3.5 mm. long, 1.5–2 mm. wide, coriaceous, abruptly minutely beaked, the beak entire or nearly so; achenes lenticular; style slender; stigmas 2.

In the coastal counties, Transition and Canadian Zones; Monterey Bay, California, north to Vancouver and British Columbia. Type locality: San Mateo County, California.

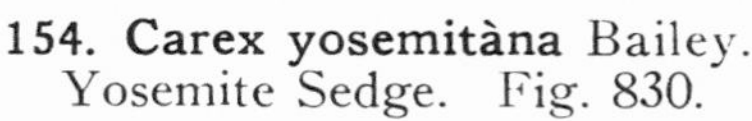

154. Carex yosemitàna Bailey.
Yosemite Sedge. Fig. 830.

Carex sartwelliana Olney, Proc. Am. Acad. 7: 396. 1868, not C. Sartwellii Dewey, 1842.
Carex yosemitana Bailey, Mem. Torrey Club 1: 8. 1889.
Carex congdoni Bailey, Bot. Gaz. 21: 6. 1896.

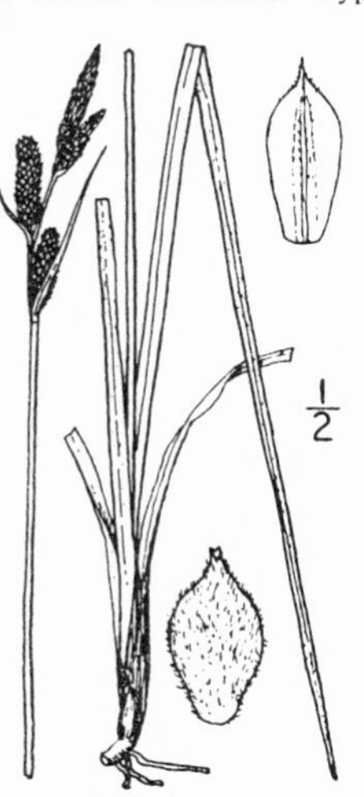

Cespitose from stout rootstocks, the culms 3–9 dm. high, sharply triangular, much exceeding the leaves. Foliage softly pubescent, the blades not rigid, flat with revolute margins, 3–7 mm. wide; terminal spike staminate, linear, 12–25 mm. long, more or less peduncled, occasionally with a few perigynia, the scales ciliate; pistillate spikes 3–4, more or less separate, sessile or slightly peduncled, erect, oblong-cylindric, 12–20 mm. long, 4.5–6 mm. wide, closely flowered with 40–200 appressed perigynia, often staminate at apex; scales lance-ovate, sharp-pointed, ciliate and pubescent, chestnut-brown with 3-nerved green center and hyaline margins; perigynia 3–3.5 mm. long, 1.25 mm. wide, obovoid or oblong obovoid, obscurely nerved, tapering at base, abruptly short-beaked, the beak 0.5 mm. long, obliquely cut, at length bidentulate; achenes triangular; style slender, jointed with achene; stigmas 3.

Central and southern Sierra Nevada and San Jacinto Mountains of California; Transition Zone. Type locality: Yosemite Valley.

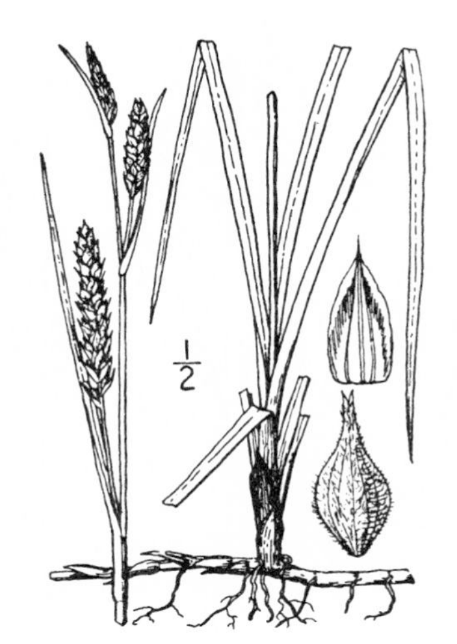

155. Carex oregonénsis Olney.

Oregon Sedge. Fig. 831.

Carex oregonensis Olney, Proc. Am. Acad. **8**: 407. 1872.
Carex halliana Bailey, Bot. Gaz. **9**: 117. 1884; not *Carex hallii* Olney 1871.

Rootstocks slender, woody, creeping, the culms 1.5–4 dm. high, rigid, smooth, sharply angled, the basal sheaths sparingly reddish purple. Leaves clustered towards base, the blades thick and rigid, glabrous, 3–5 mm. wide, canaliculate, mostly exceeding the culms; terminal 2–3 spikes staminate, linear, 8–16 mm. long; pistillate spikes 3–4, 1.5–4.5 cm. long, 4.5–6 mm. wide, approximate or the lower more or less separate, erect-appressed, closely flowered above or loosely below, the peduncles little exserted, the 20–40 perigynia appressed-ascending; lowest bract exceeding culm; scales ovate, acute to cuspidate, chestnut brown with hyaline margins and 3-nerved green center; perigynia ovoid, obtusely triangular, 4–5 mm. long, 2–2.5 mm. wide, densely short-pubescent, rounded at base, tapering into a bidentate beak one-third to one-half length of body, the teeth 0.5 mm. long; achenes triangular, jointed with the style; stigmas 3.

Mountain meadows, Boreal Zone; from southern Washington to Siskiyou County, California. Type locality: Oregon.

156. Carex lanuginòsa Michx.

Woolly Sedge. Fig. 832.

Carex lanuginosa Michx. Fl. Bor. Am. **2**: 175. 1803.
Carex pellita Muhl. in Willd. Sp. Pl. **4**: 302. 1805.
Carex watsoni Olney, Bot. King's Exped. 370. 1871.
Carex filiformis latifolia Boeckl. Linnaea **41**: 309. 1877.

Rootstocks stout, long-creeping, the culms stoutish, more or less reddened and filamentose at base, 6–9 dm. high, sharp angled and rough above. Leaf-blades flat, 2–4 mm. wide, rough; staminate spikes 1–3, 3–4 mm. wide, up to 3 cm. long, distant; pistillate spikes 1–3, oblong-cylindric, 1–5 cm. long, 5–7 mm. wide, sessile or short-peduncled, closely flowered with 25–50 perigynia; lower bracts mostly exceeding the culm; scales lanceolate, acuminate or aristate, narrower than and shorter or longer than the perigynia, reddish-brown tinged, the margins hyaline; perigynia ovoid, 2.5–3.5 mm. long, 2 mm. wide, densely pubescent, the nerves obscure, rounded at base, abruptly short-beaked, the beak bidentate, the teeth 0.5–1 mm. long; achenes triangular, jointed with style; stigmas 3.

Swampy places, Transition Zone; Nova Scotia to British Columbia, south to District of Columbia, Missouri, New Mexico, and southern California. Type locality: "ad lacus Mistassins," Canada.

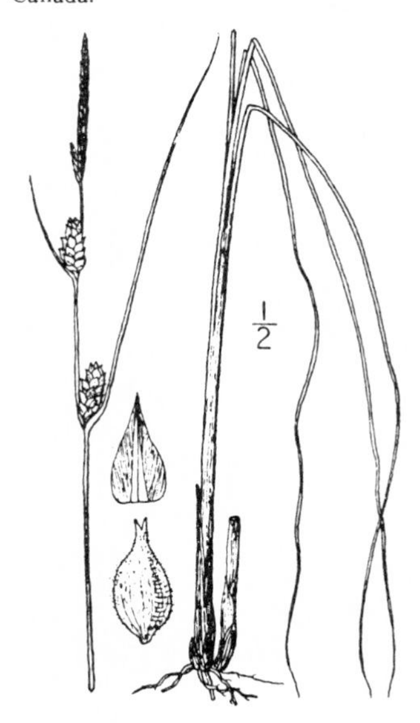

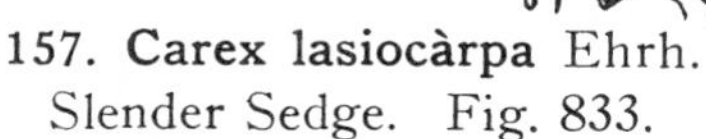

157. Carex lasiocàrpa Ehrh.

Slender Sedge. Fig. 833.

Carex lasiocarpa Ehrh. in Hannoev. Magaz. **9**: 132. 1784.
Carex splendida Willd. Prodr. Fl. Berol, 33, *pl. 1, f. 3.* 1787.
Carex filiformis Good. Trans. Linn. Soc. **2**: 172. *pl. 20, f. 5.* 1794; not L.

Cespitose, freely long stoloniferous, the stolons tough, slender, the culms 3–10 dm. high, slender but stiff, smooth or roughened above, obtusely angled, strongly reddened and filamentose at base. Leaves shorter than the culm, the blades very narrow and attenuate prolonged, involute, 2 mm. wide or less, rough on the inrolled margin; staminate spikes 1–3, usually 2, long-peduncled, 3–6 cm. long; pistillate spikes 1–3, cylindric, 1–5 cm. long, about 6 mm. thick, erect, sessile or the lower distant and short-peduncled; bracts leaf-like, the lowest about equalling the culm; scales ovate, shorter than or equalling the perigynia, acute or short-awned, green with hyaline margins and reddish brown tinged; perigynia oval-ovoid, ascending, 3–3.5 mm. long, 1.75 mm. thick, densely pubescent, green, coriaceous, contracted at base, tapering into the short-bidentate beak, the teeth 0.5–1 mm. long, achenes triangular; style slender, jointed with the achene; stigmas 3.

Sphagnum swamps, Transition Zone; Newfoundland to British Columbia, south to New Jersey, Iowa, Colorado, and Washington. Type locality: Europe.

158. **Carex sheldònii** Mackenzie.
Sheldon's Sedge. Fig. 834.

Carex sheldonii Mackenzie, Bull. Torrey Club **42**: 618. 1915.

Strongly stoloniferous, the culms 6–9 dm. high, very smooth below the spikes, neither bright colored nor fibrillose at the base. Leaves about four, the blades 5–6 mm. wide, 2–4 dm. long (or longer on the sterile shoots), sparingly short-pubescent as are the sheaths, the latter dark-tinged at the mouth, the basal breaking and slightly filamentose; staminate spikes 2–3, distant; pistillate spikes usually two, widely separate, sessile or short-peduncled, oblong-cylindric, 2–5 cm. long, 8–10 mm. wide, rather closely 25–60-flowered; lowest bract sheathing, exceeding inflorescence; scales ovate-lanceolate, sharp pointed, exceeded by the perigynia, brownish with several nerved green center and hyaline margins; perigynia lanceolate, obscurely triangular, 5–6 mm. long, 2 mm. wide, little inflated, short-pubescent, prominently about 15-nerved, tapering into a bidentate beak 2 mm. long, the teeth less than 1 mm. long; achenes triangular; style slender, straight, continuous with achene; stigmas 3.

In swamps, Transition Zone; Idaho to Oregon and northeastern California. Type locality: Clark's Creek, Oregon.

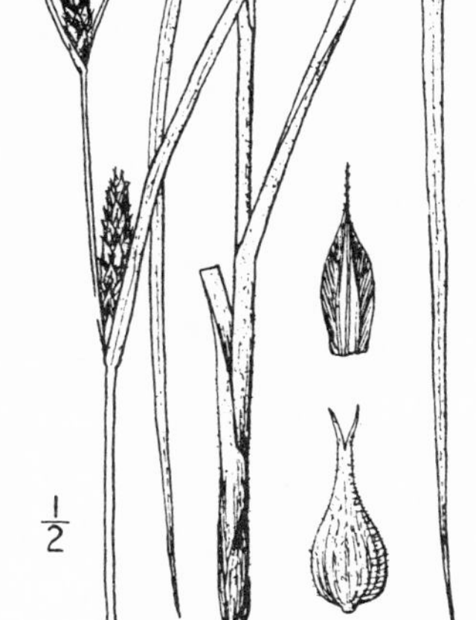

159. **Carex atheròdes** Spreng.
Awned Sedge. Fig. 835.

Carex aristata R. Br. Frank. Journ. 751. 1823; not Honck. 1792.
Carex atherodes Spreng. Syst. Veg. **3**: 828. 1826.
Carex trichocarpa aristata Bailey, Bot. Gaz. **10**: 294. 1885.

Rootstocks with stout stolons, the culms 6–12 dm. high, stout, sharp-angled, smooth or roughish above. Leaf-blades elongated, 4–12 mm. wide, flat, nodulose, more or less pubescent beneath, the sheaths soft pubescent, brownish or purple tinged, quickly broken; staminate spikes 2–6, slender, long-stalked; pistillate spikes 3–5, remote, cylindric, sessile or the lower short-stalked, loosely flowered at the base, densely above, 2–10 cm. long, 12–16 mm. wide; bracts leaf-like, the lower 1–2 exceeding the culm; scales oblong-lanceolate, the upper shorter than the perigynia, rough awned, 3-nerved; perigynia ascending, lanceolate, glabrous, conspicuously many ribbed, 8–12 mm. long, rounded and substipitate at base, tapering into the very deeply bidentate beak, the teeth widely spreading, 1–4 mm. long; achenes triangular, short stipitate; style slender, straight, continuous with achene; stigmas 3.

Swamps, Transition Zone; Ontario to Yukon, south to New York, Missouri, Kansas, Utah, and Oregon; also widely distributed in northern Eurasia. Type locality: Arctic America.

160. **Carex virídula** Michx.
Green Sedge. Fig. 836.

Carex viridula Michx. Fl. Bor. Am. **2**: 170. 1803.
Carex oederi Schw. & Torr. Ann. Lyc. N. Y. **1**: 334. 1825; not Retz.
Carex urbanii Boeckl. Bot. Jahrb. **7**: 280. 1886.
Carex flava recterostrata Bailey, Bot. Gaz. **13**: 84. 1888.

Densely cespitose, the culms 0.7–4 dm. high, smooth, bluntly triangular, not yellowish green. Leaf-blàdes 1.5–3 mm. wide, canaliculate, the sheaths not prolonged at the throat; terminal spike usually staminate, sessile or short-peduncled, 3–15 mm. long; pistillate spikes 2–16, oblong or globose-oblong, 4–12 mm. long, 4–7 mm. wide, aggregated and sessile or the lower separate and exsert-peduncled, with 15–30 spreading perigynia; bracts leaf-like, usually erect and the lower much exceeding culm, strongly sheathing; scales ovate, much shorter than the perigynia, obtuse or acutish, straw-color with greenish midvein and hyaline margins, often reddish brown tinged; perigynia 2–3 mm. long, obovoid, obtusely triangular, not inflated, many-nerved, tapering at base, abruptly beaked, the beak whitish tipped, minutely bidentate, scarcely half length of body; achenes triangular; style slender, jointed with achene; stigmas 3.

Lake and river banks, Transition and Canadian Zones; Newfoundland to Alaska, south to New Jersey, Indiana, Colorado, Utah, and northern California. Type locality: Canada.

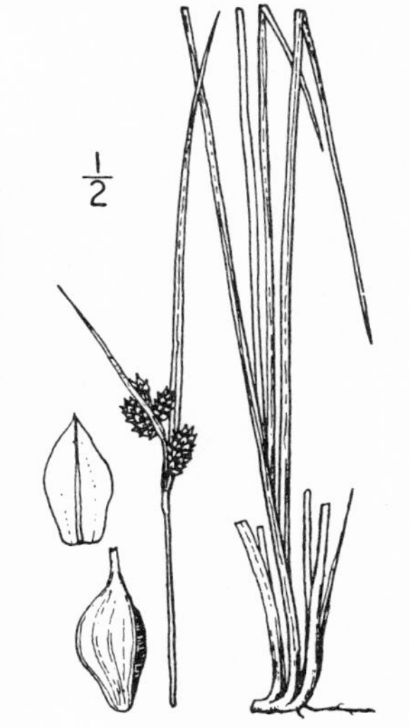

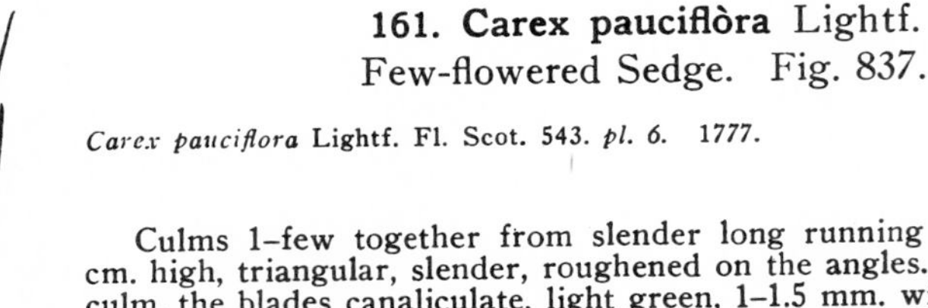

161. **Carex paucifl̀ora** Lightf.

Few-flowered Sedge. Fig. 837.

Carex pauciflora Lightf. Fl. Scot. 543. *pl. 6.* 1777.

Culms 1–few together from slender long running rootstocks, 10–25 cm. high, triangular, slender, roughened on the angles. Leaves 2–3 to a culm, the blades canaliculate, light green, 1–1.5 mm. wide; spike solitary androgynous, the staminate flowers 1–6, conspicuous; bracts none; scales lanceolate, acutish, the staminate closely appressed, the pistillate early falling; perigynia 1–6, reflexed, jointed to rachis and soon deciduous, 2–3 times length of scale, greenish straw color, linear-lanceolate, obscurely triangular, 6–8 mm. long, barely 1 mm. wide, obscurely several nerved, glabrous, rounded and substipitate at base, tapering into the slender entire beak with oblique orifice; achenes linear-oblong, 2 mm. long, continuous with the long slender style; stigmas 3.

Sphagnum bogs, Canadian Zone; Newfoundland to Alaska, south to Connecticut, Pennsylvania, Michigan, and Washington. Type locality: Scotland.

162. **Carex vesicària** L.

Inflated Sedge. Fig. 838.

Carex vesicaria L. Sp. Pl. 979. 1753.
Carex monile Tuckerm. Enum. Meth. 20. 1843.
Carex vesicaria obtusisquamis Bailey, Bot. Gaz. 9: 121. 1884.
Carex monile colorata Bailey, Mem. Torrey Club 1: 39. 1889.
Carex monile pacifica Bailey, Proc. Calif. Acad. Sci. II. 3: 105. 1891.

Rootstocks short-creeping and stoloniferous, the culms 3–9 dm. high, acutely angled and rough above, slender to stoutish, aphyllopodic and purplish tinged at base. Leaf-blades 3–6 mm. wide, the sheaths sparingly nodulose dorsally and usually somewhat breaking and filamentose ventrally; staminate spikes 2–4, linear, 2–4 cm. long, 2.5–4 mm. wide; pistillate spikes 1–3, sessile or short-peduncled, erect, oblong-cylindric, 2.5–7 cm. long, 6–15 mm. wide, many-flowered, more or less strongly separate; lower bracts exceeding culms; scales ovate or lanceolate, one-half to two-thirds length of perigynia, acute, acuminate or short awned; perigynia ovoid, round in cross section, ascending, 5–8 mm. long, yellowish green or darker-tinged, 8–10-nerved, the beak smooth 2 mm. long, the teeth erect, 0.5 mm. long; achenes triangular; style continuous with achene; stigmas 3.

Wet meadows and swamps, Transition Zone; Quebec to British Columbia, south to Pennsylvania, Ohio, and in California as far as Marin and Tulare Counties; widely distributed in Eurasia and found in northern Africa. Very variable. Type locality: Europe.

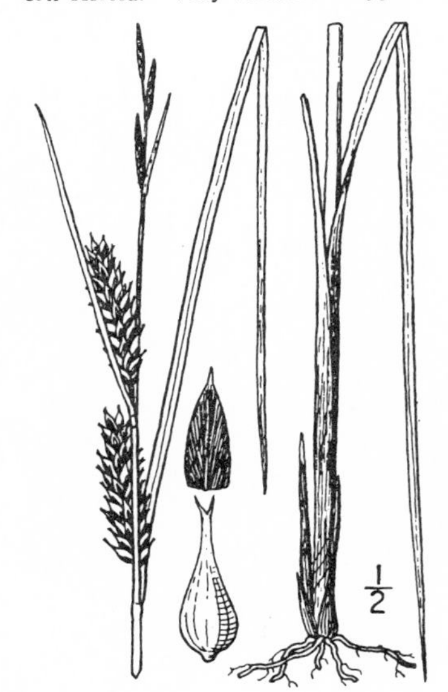

163. **Carex exsiccàta** Bailey.

Western Inflated Sedge. Fig. 839.

Carex vesicaria major Boott in Hook. Fl. Bor. Am. 2: 221. 1840.
Carex vesicaria lanceolata Olney, Proc. Am. Acad. 8: 407. 1872.
Carex vesicaria globosa Olney, l.c. 408. 1872.
Carex exsiccata Bailey, Mem. Torrey Club 1: 6. 1889.

Rootstocks stout, short-creeping, the culms stout, 3–10 dm. high, acutely triangular and rough above, aphyllopodic, more or less purplish tinged at base. Leaf-blades 3–6 mm. wide, the sheaths sparingly nodulose dorsally and usually somewhat breaking and filamentose ventrally; staminate spikes 2–4, 2–4.5 cm. long, narrow; pistillate spikes 1–3, sessile or short-peduncled, more or less strongly separate, erect, cylindric, 2–7 cm. long, 10–14 mm. wide, closely many-flowered; lower bracts exceeding culm; scales lanceolate ovate, narrower than and about half length of perigynia, sharp-pointed; perigynia lanceolate, ascending, 7–10 mm. long, little inflated, olive green, 8–10-ribbed, tapering to the beak, the beak 1.5–2 mm. long, smooth, the teeth erect, 0.5 mm. long; achenes triangular; style slender, continuous with the achene; stigmas 3.

Wet places, Transition Zone; southern Alaska to middle California, east to Montana; in California confined to the Coast Ranges from San Mateo County northward. Type locality: Columbia River.

164. **Carex rostràta** Stokes.
Beaked Sedge. Fig. 840.

Carex vesicaria β L. Sp. Pl. 979. 1753.
Carex rostrata Stokes in With. Arrang. Brit. Pl. ed. 2, **2**: 1059. 1787.
Carex ampullacea Good. Trans. Linn. Soc. **2**: 207. 1794.
Carex utriculata Boott in Hook. Fl. Bor. Am. **2**: 221. 1840.

Cespitose and stoloniferous, the culms stout, phyllopodic, 3–12 dm. high, obtusely triangular, rough above the lowest spike. Leaf-blades 2–12 mm. wide, the sheaths strongly nodulose dorsally and little if at all breaking and filamentose ventrally; staminate spikes 2–4, slender, 1–6 cm. long, 4 mm. wide; pistillate spikes 2–4, erect, remote, cylindric, densely many-flowered, sessile or short-peduncled, 5–15 cm. long, 6–20 mm. wide, the perigynia spreading-ascending or at maturity spreading; lower bracts exceeding culm; scales lanceolate, narrower than and from shorter to longer than the perigynia, acute or awned; perigynia ovoid, round in cross-section, 4–6 mm. long, 2 mm. wide, several-nerved, membranaceous, inflated, greenish straw color or darker tinged, abruptly beaked, the beak smooth, 1.5–2 mm. long, with erect or spreading teeth 1 mm. long or less; achenes triangular; style slender, continuous with achene; stigmas 3.

Swampy places, Transition and Canadian Zones; Labrador to Alaska, south to Delaware, New Mexico, and California; widely distributed in Eurasia. Type locality: Great Britain.

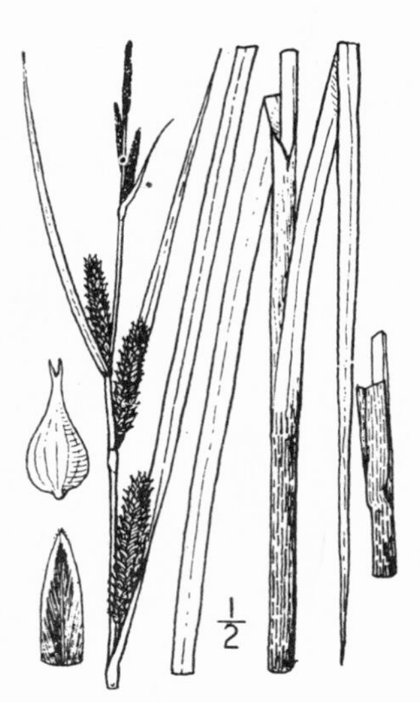

165. **Carex retrórsa** Schw.
Retrorse Sedge. Fig. 841.

Carex retrosa Schw. Ann. Lyc. N. Y. **1**: 71. 1824.
Carex hartii Dewey, Am. Journ. Sci. II. **41**: 226. 1866.

Cespitose from short rootstocks, stoloniferous, the culms stout, 5–10 dm. high, smooth or slightly rough, much exceeded by upper leaves and bracts. Leaf-blades 4–10 mm. wide, flat, thin, rough-margined, nodulose, not clustered at base, the sheaths brownish hyaline ventrally; terminal 1–3 spikes staminate, linear, more or less peduncled; pistillate spikes 3–8, cylindric, densely many-flowered, erect, aggregated and short-peduncled or the lower strongly separate and long-peduncled; bracts short-sheathing; scales lanceolate, acute or acuminate, about equalling body of perigynia, straw color with green midvein, often purplish brown tinged; perigynia spreading or the lower reflexed at maturity, yellowish green, inflated, 7–10 mm. long, 3 mm. wide, the body ovoid, suborbicular in cross section, coarsely about 10-nerved, rounded at base, membranaceous, rather abruptly tapering into the smooth bidentate beak 2–3.5 mm. long, the slender teeth erect, 0.5 mm. long; achenes triangular; style slender, very abruptly bent, continuous with achene; stigmas 3.

Swamps and wet meadows, Upper Sonoran and Transition Zones; Newfoundland to British Columbia, south to northern New Jersey, Iowa, and Oregon. Type locality: Massachusetts.

166. **Carex hystricìna** Muhl.
Porcupine Sedge. Fig. 842.

Carex hystricina Muhl. in Willd. Sp. Pl. **4**: 282. 1805.
Carex cooleyi Dewey, Am. Journ. Sci. **48**: 144. 1845.

Cespitose and stoloniferous, the culms 3–9 dm. high, reddish purple at the base, rough above. Leaf-blades 3–8 mm. wide, the basal sheaths often breaking and filamentose; staminate spike 1–5 cm. long, 2.5–4 mm. wide, slender-stalked, the scales rough-awned; pistillate spikes 1–4, approximate or strongly separate, densely many-flowered, oblong or oblong-cylindric, 1–6 cm. long, 10–14 mm. wide, the lower slender-stalked; lower bracts exceeding the culm; scales rough awned, narrower and mostly shorter than the perigynia, green, 3-nerved; perigynia narrowly ovoid, ascending or at length spreading, 5–6 mm. long, 1.5–2 mm. wide, finely 15–20-nerved, greenish straw color, rounded and short-stipitate at base, tapering into a smooth bidentate beak 2 mm. long, the slender teeth erect; achenes triangular; style slender, continuous with the achene; stigmas 3.

Swampy soil, Transition Zone; New Brunswick to Washington, south to Pennsylvania, Texas, Arizona, and Trinity County, California. Rare and local in our area. Type locality: Pennsylvania.

167. **Carex comòsa** Boott.
Bristly Sedge. Fig. 843.

Carex furcata Ell. Sketch Bot. **2**: 552. 1824; not Lappeyr 1813.
Carex comosa Boott, Trans. Linn. Soc. **20**: 117. 1846.
Carex pseudo-cyperus comosa Boott, Ill. Car. **4**: 141. 1867.
Carex pseudo-cyperus americana Hochst.; Bailey, Mem. Torrey Club **1**: 54. 1889.

Cespitose and not stoloniferous, the culms stout, 5–15 dm. high, very sharply angled, strongly roughened to smooth. Leaves very nodulose, the blades 6–14 mm. wide, flat with revolute margins, the basal sheaths not breaking nor filamentose; staminate spike 3–7 cm. long, 4–7 mm. wide, slender-stalked, the scales rough-awned; pistillate spikes 1–4, densely many-flowered, oblong-cylindric, 1–6 cm. long, 12–14 mm. wide, the upper erect and short-peduncled, the lower slender-stalked and at length nodding; lower bracts exceeding culm; scales narrow, very rough-awned, mostly shorter than the perigynia, greenish or brownish tinged; perigynia lanceolate, rigid, 5–7 mm. long, 1.5 mm. wide, greenish or brownish tinged, round-tapering and stipitate at base, closely many-ribbed, reflexed when mature, tapering into a smooth strongly bidentàte beak with recurved spreading awns 1.5–2 mm. long; achenes triangular, continuous with style; stigmas 3.

Swamps, Transition Zone; Nova Scotia to Minnesota, south to Florida and Louisiana, and very locally on the Pacific Coast from San Francisco Bay to Washington, and east to Idaho. A stray plant has been collected in the San Bernardino Valley. Type locality: Georgia and Carolina.

Family 12. **PHOENICÀCEAE.**

Palm Family.

Tall or low trees with unbranched cylindrical trunks, bearing a tuft of large leaves at the apex, clothed below by the persistent leaf-bases. Leaves plaited in the bud, fan-shaped or pinnate, long-petioled, with the stalks clasping at the base by broad tough fibrous sheaths. Flowers small, perfect or unisexual, in large compound axillary spadices, surrounded by large spathes. Perianth of 6 segments in two sets, usually firm in texture, the segments distinct or more or less united. Stamens mostly 6, sometimes 9–12; filaments dilated at base and more or less united; anthers introrse. Ovary of 3 more or less united or distinct carpels; styles 1–3; ovules 1 to each cell. Fruit a drupe or berry. Seeds often hollow; endosperm horny or cartilaginous, rarely channelled, with the small cylindrical embryo imbedded near the surface.

A large and important tropical and subtropical family comprising 150 genera and about 1000 species.

1. **WASHINGTÒNIA** H. Wendl. Bot. Zeit. **37**: 68. 1879.

Trees with tall cylindrical trunks, clothed sometimes almost to the ground with a dense thatch of dead drooping leaves. Leaves large, fan-shaped, on long petioles furnished on the margins with stout, straight or recurved, flattened spines, the leaf-divisions separating on the margins into slender light colored fibres. Spadix with long conspicuous spathes subtending the principal branches. Flowers perfect. Fruit a small rounded or ellipsoid black berry, with dry thin flesh. Seeds shallowly excavated on the raphal face. [Name in honor of George Washington.]

A genus of three species, peculiar to the arid regions of southwestern United States and northern Mexico. Type species, *Washingtonia filifera* (Linden) Wendl.

1. Washingtonia filífera (Linden) Wendl. California Fan Palm. Fig. 844.

Pritchardia filamentosa Drude, Bot. Zeit. **34**: 806. 1876, nomen nudum.
Pritchardia filifera Linden, Ill. Hort. **2**: 32, 105. 1877.
Washingtonia filifera Wendl. Bot. Zeit. **37**: 68. 1879.
Washingtonia filamentosa Kuntze, Rev. Gen. **2**: 737. 1891.
Neowashingtonia filamentosa Sudw. U. S. Div. Forest Bull. **14**: 105. 1897.
Neowashingtonia filifera Sudw. Forest Trees Pacif. Slope, 199. 1908.

A tall stately tree, often 20–40 m. high, with a trunk 6–10 dm. in diameter, clothed sometimes almost to the base with a dense thatch of dead leaves; wood light and soft. Leaves 1.5–2 m. long and 1–1.5 m. wide, pale green, sparsely tomentose on the folds; petioles nearly as long as the leaves, about 5 cm. broad, with sheaths about 3 dm. wide, and 4 dm. long; spadix often 3–4 m. long; spathes yellowish brown, those of the secondary branches laciniate; fruits about 1 cm. long.

Occurs in a few scattered groves mostly in cañons along the base of the desert mountains, from White Water to Carizo Creek, southern California. Especially fine specimens are in Palm Springs Cañon at the eastern base of the San Jacinto Mountains. Type locality: described from plants cultivated in Europe. The source of the seeds is not definitely known.

Washingtonia filífera robústa (Wendl.) Parish, Bot. Gaz. **44**: 420. 1907. *Washingtonia robusta* Wendl. Garten Zeit. **2**: 198. 1883. A robust form with stout trunks; petioles armed throughout. Those of the typical form unarmed near the base of the blade. Growing with the typical form on the western borders of the Colorado Desert. Frequently cultivated.

Washingtonia grácilis Parish, Bot. Gaz. **44**: 420. 1907. Trunk tall and slender; leaf blades destitute of filaments or nearly so. This species is not known in the wild state, but is frequently cultivated in California, usually under the name *W. robusta*. It is a much more rapid grower than *W. filifera*, and is known to reach a height of 30 m.

Family 13. ARÀCEAE.

Arum Family.

Herbs with mostly basal, simple or compound leaves, and spathaceous inflorescence, the spathe inclosing or subtending the spadix. Spadix densely flowered, the staminate flowers above, the pistillate below, or the plants dioecious, or with perfect flowers in some species. Perianth of 4–6 scale-like segments, or wanting. Stamens 4–10; filaments very short; anthers 2-celled, commonly with a thick truncate connective, the sacs opening by dorsal pores or slits. Ovary 1–several-celled; ovules 1–several in each cell; style short or wanting; stigma terminal, commonly minute and sessile. Fruit a berry or utricle. Seeds various. Endorsperm copious, sparse or none.

A large family containing about 105 genera and 900 species, mostly tropical, a few in the temperate zones.

1. LYSICHÌTON Schott, Oestr. Bot. Wochenbl. 7: 62. 1857.

Acaulescent swamp herb, with an ill-scented shiny sap, and large leaves arising from a thick horizontal rootstock. Spathe sheathing at base, usually expanded above into a broad colored lamina. Spadix at first enveloped by the spathe, becoming long-exserted upon a stout peduncle. Flowers perfect, crowded and covering the cylindrical spadix. Perianth 4-lobed. Stamens 4, opposite the perianth-lobes; filaments short, flat; anthers 2-celled, opening upward. Ovary conical, 2-celled; 2-ovuled; stigma depressed; ovules horizontal, orthotropus. [Greek, meaning loose and tunic, in allusion to the spathe.]

A monotypic genus peculiar to the Pacific region of North America and Asia.

1. Lysichiton camtschaténsis (L.) Schott.
Yellow Skunk Cabbage. Fig 845.

Dracontium camtschatense L. Sp. Pl. 968. 1753.
Lysichiton camtschatensis Schott, Oestr. Bot. Wochenbl. **7**: 62. 1857.
Arctodracon camtschaticum A. Gray, Mem. Am. Acad. II. **6**: 408. 1858.

Acaulescent ill-scented swamp-inhabitating herb, with thick peppery rootstock. Leaves erect or ascending, 3–15 dm. tall, oblong-lanceolate, acutish, narrowed below to a sessile base or a short petiole, rather thin, glabrous; spathe with a broad acute blade, 6–10 cm. long, yellowish; spadix 4–8 mm. long, greenish yellow, on a stout peduncle becoming 3–5 dm. long.

In swamps, Transition and Canadian Zones; Alaska to northern Idaho, and in the Coast Ranges to the Santa Cruz Mountains, California, also on the Pacific Coast of Asia to Japan. The peppery roots are reputed to have medicinal value, and are used for the treatment of such skin diseases as ringworm. Type locality: Asia.

Zantedéschia aethiòpica (L.) Spreng. (*Richardia aethiopica* Hort.) Common Calla. The common white calla of the gardens is extensively cultivated, and especially in central and southern California occasionally established in moist situations. Native of southern Africa.

Family 14. LEMNÀCEAE.

Duckweed Family.

Minute stemless and leafless perennial plants, floating free on the surface of fresh water. The plant body consisting of a disc-shaped, elongated or irregular thallus, loosely cellular, rootless or bearing 1 or more simple rootlets. Inflorescence composed of one or more naked monoecious flowers borne in a sac-like spathe on the edge or upper surface of the frond. Flowers consisting of a single stamen or a single flask-shaped ovary. Anther with 2 or 4 pollen-sacs; pollen-grains barbellate. Pistil narrowed to the funnel-shaped scar-like stigmatic apex, 1-celled; ovules 1–6, erect or inverted. Fruit a 1–6-seeded utricle; seeds large; embryo straight.

A family of 4 genera and about 26 mostly widely distributed species. The simplest and smallest of flowering plants, propagating by the proliferous growth of a new individual from a cleft in the edge or base of the frond, also by autumnal fronds in the form of minute bulblets which sink to the bottom of the water, but arise and vegetate in spring. The flowers are scarce, in some species hardly ever seen.

Rootlets present, 1 or more.
 Rootlets more than 1, fascicled. — 1. *Spirodela.*
 Rootlets 1. — 2. *Lemna.*
Rootlets none. — 3. *Wolffiella.*

1. SPIRODÈLA Schleid. Linnaea 13: 391. 1839.

Fronds disc-shaped, 7–12-nerved, the stipe attached to the frond back of and under the basal margin. Rootlets several, fascicled, these and the nerves of the frond with a single bundle of vascular tissue. Reproductive pouches 2, triangular, opening as clefts in either margin of the basal portion of the frond. Spathe sac-like. Spadix of 1 pistillate and 2 staminate flowers from the reproductive pouches. Filaments curving upward from the margin of the frond; anthers 2-celled, longitudinally dehiscent. Ovule 2, anatropous. Fruit rounded wing-margined. [Greek, referring to the cluster of rootlets.]

A genus of 3 species, two American and one Asiatic and Australian. Type species, *Lemna polyrhiza* L.

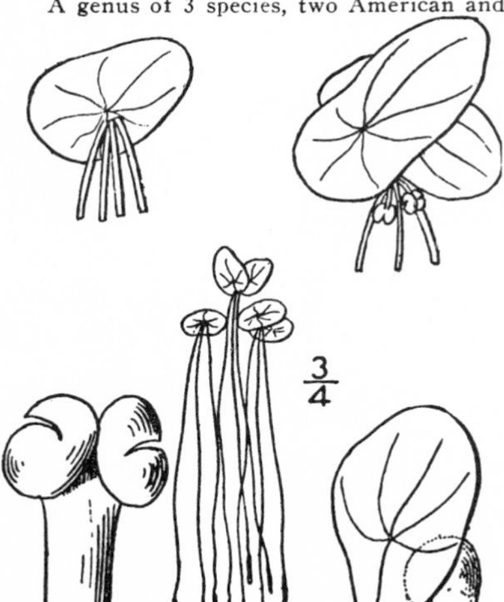

1. Spirodela polyrhìza (L.) Schleid.
Greater Duckweed. Fig. 846.

Lemna polyrhiza L. Sp. Pl. 970. 1753.
Spirodela polyrhiza Schleid. Linnaea **13**: 392. 1839.

Fronds solitary or united in colonies of 2–5, round-obovate, 3–6 mm. long, flat and dark green on the upper surface, slightly convex and purplish beneath, 5–11-nerved; rootlets 5–11; rootcap large, sharp-pointed; spathe a complete sac, opening at the upper end; seed slightly compressed, smooth.

In slow-running streams, ponds and shallow lakes, widely distributed in the Old World and in North and tropical America. On the Pacific Coast ranging from British Columbia to southern California, but not common in California. Type locality: Europe.

2. LÉMNA L. Sp. Pl. 970. 1753.

Thallus disc-shaped, stipitate or sessile, usually with a central nerve and with or without 2–4 lateral nerves. Rootlet solitary on each thallus, devoid of vascular tissue, the rootcap thin, blunt or pointed. Reproductive pouches 2, triangular, opening as clefts in either margin of the basal portion of the frond. Spadix of 1 pistillate and 2 staminate flowers. Spathe sac-like or open. Stamens with filaments curving upwards from the margin of the frond; anthers 2-celled, transversely dehiscent. Ovule 1–6, orthotropus, amphitropus or anatropus. Utricle ovoid more or less ribbed. Endosperm in 1 or 3 layers. [Greek, in allusion to the growth of these plants in swamps.]

A genus of about eight, mostly widely distributed species. Type species, *L. trisulca* L.

Fronds long-stipitate. — 1. *L. trisulca.*
Fronds short-stipitate, or sessile.
  Spathe open; thallus 1-nerved or nerveless.
    Frond thin, without papules; rootcap strongly curved, tapering. — 2. *L. cyclostasa.*
    Frond thick, with a row of papules along the nerve; rootcap little curved, cylindric. — 3. *L. minima.*
  Spathe sac-like; thallus more or less faintly 3–5-nerved.
    Frond green or purplish beneath; fruit not winged. — 4. *L. minor.*
    Frond pale beneath, usually strongly gibbous; fruit winged. — 5. *L. gibba.*

1. Lemna trisúlca L.
Ivy-leaved Duckweed. Fig. 847.

Lemna trisulca L. Sp. Pl. 970. 1753.

Fronds remaining connected, and forming chain-like colonies oblong to oblong-lanceolate, 6–10 mm. long, narrowed at base to a slender stipe, denticulate toward the apex, obscurely 3-nerved, with or without rootlets; spathe sac-like; seeds ovate, amphitropus, 12–15-ribbed.

Growing in pools and slow streams, widely distributed over North America and the Old World. On the Pacific Coast ranging from British Columbia to the San Bernardino Mountains, southern California. Type locality: Europe.

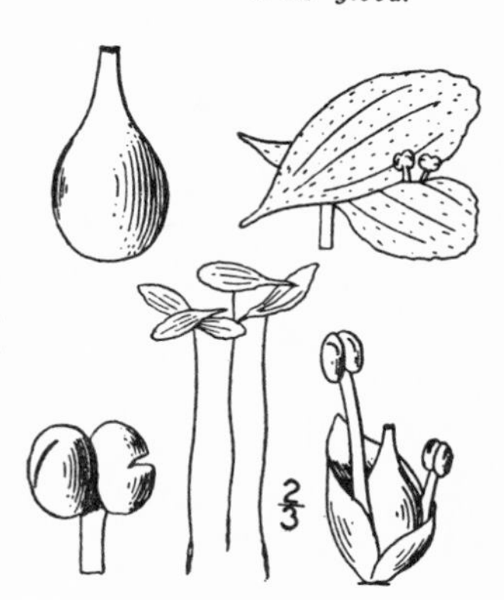

2. Lemna cyclóstasa (Ell.) Chev.
Valdivia Duckweed. Fig. 848.

Lemna minor cyclostasa Ell. Bot. S. C. and Ga. **2**: 518. 1824.
Lemna cyclostasa Chev. Fl. Paris **2**: 256. 1827.
Lemna valdiviana Philippi, Linnaea **33**: 239. 1864.

Fronds elliptic-oblong, 2.5–4 mm. long, rather thick, subfalcate and shortly stalked at the base, obscurely 1-nerved; rootcap curved, short-tapering; spathe broadly reniform; fruit ovoid-oblong, pointed by the long style, unsymmetrical; seed oblong, 12–29-ribbed, amphitropus.

Widely distributed over North and South America, but on the Pacific Coast apparently not found north of Lake County, California, and common only in southern California. Type locality: southeastern United States.

3. Lemna mínima Philippi.
Least Duckweed. Fig. 849.

Lemna minima Philippi, Linnaea **33**: 239. 1844.

Frond oblong to elliptic, 1.5–4 mm. long, 1-nerved or commonly nerveless, with a row of papules along the nerve on the upper surface, the lower surface flat or slightly convex; rootcap slightly curved or straight, cylindric and blunt; spathe open; ovary short, clavate; ovule 1, obliquely orthotropus; seed oblong, pointed, about 16-ribbed.

In pools and streams, Georgia to Florida and Kansas, also in Wyoming and California, where it ranges from the San Francisco Bay region, and the San Joaquin Valley to southern California. Type locality: Santiago, Chile.

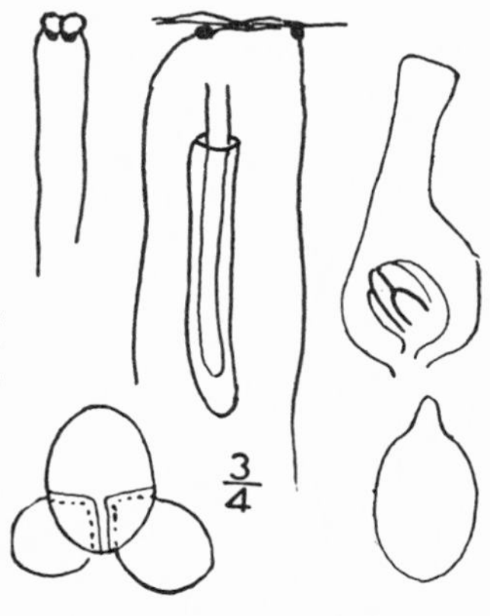

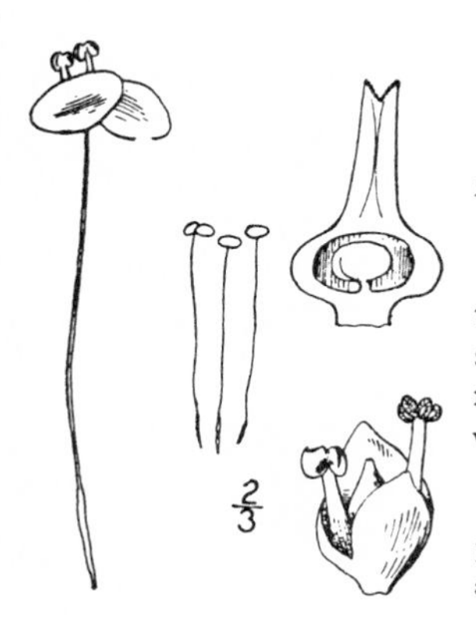

4. Lemna mìnor L.
Lesser Duckweed. Fig. 850.

Lemna minor L. Sp. Pl. 970. 1753.

Fronds rounded to obovate-oblong, 2–4 mm. long, symmetrical, thickish, convex and green or purplish on both surfaces, upper surface sometimes keeled and with a row of papulae along the midnerve, obscurely 3-nerved; fruit not winged; seeds solitary, with a prominent protruding hilum, 12–15-ribbed.

In ponds, lakes and stagnant waters, widely distributed throughout North America, except the extreme north; also in Europe, Asia, Africa and Australia. Type locality: Europe.

5. Lemna gíbba L.
Gibbous Duckweed. Fig. 851.

Lemna gibba L. Sp. Pl. 970. 1753.

Fronds rounded to obovate, 2–5 mm. long, unsymmetrical, thick, convex, slightly keeled and green on the upper surface, pale beneath and more or less gibbous, usually 3–5-nerved; fruit winged with rounded lobes on either side of the stigma; seeds thick, deeply and unequally ribbed.

In ponds and slow streams; Nebraska and Texas to California and Mexico; also in Europe and Asia. In California this is a common species, especially in the valleys and foothills. Type locality: Europe.

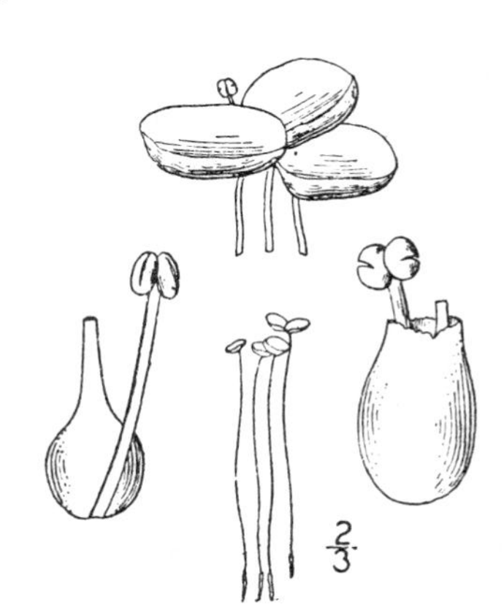

3. WOLFFIÉLLA Hegelm. in Engler Bot. Jahrb. **21**: 303. 1895.

Fronds rootless, thin, unsymmetrical, curved in the form of a segment of a band, the two ends submerged, abundantly punctuate on both surfaces with brown epidermal pigment cells. Reproductive pouch triangular, opening as a cleft in the basal margin of the frond. Flowers and fruit unknown. [Name a diminutive of *Wolffia*, a closely related genus.]

Seven species, mostly of tropical distribution. Besides the following, *W. floridana* (J. D. Smith) Thompson occurs in the southern states. Type species, *Wolffiella oblonga* (Philippi) Hegelm.

Fronds saber-shaped; stipe insertion at the lower angle of the two walls of the pouch. 1. *W. oblonga.*
Fronds tongue-shaped; stipe insertion on the margin of the lower wall of the pouch. 2. *W. lingulata.*

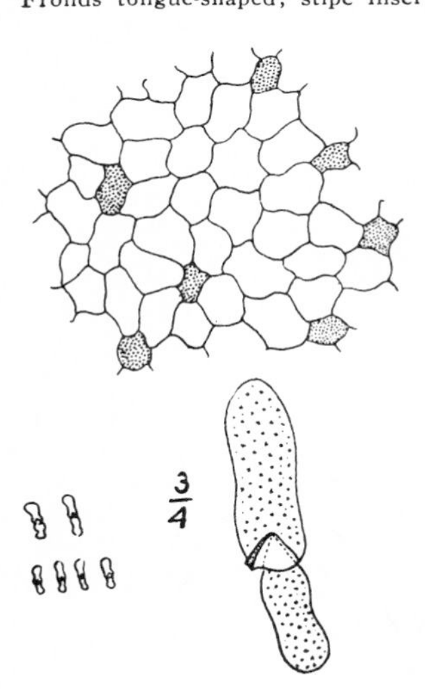

1. Wolffiella oblónga (Philippi) Hegelm.
Saber-shaped Wolffiella. Fig. 852.

Lemna oblonga Philippi, Linnaea **29**: 45. 1857.
Wolffia oblonga Hegelm. Monogr. Lemnac. 131. 1868.
Wolffiella oblonga Hegelm. in Egler Bot. Jahrb. **21**: 303. 1895.

Fronds solitary or in pairs, oblong or commonly tapering from the obliquely rounded base to the slightly narrower bluntly rounded apex; 0.5–1 mm. wide, 1.7–4.6 mm. long; stipe insertion at the lower angle of the two walls of the pouch.

A Mexican and South American species, which was collected in 1881 with *Spirodela polyrhiza* near San Bernardino, southern California, but has not been rediscovered, and is otherwise unknown in the United States. Type locality: Santiago, Chile.

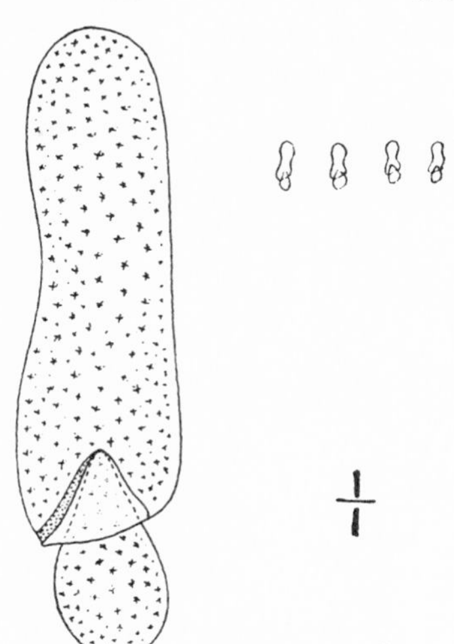

2. **Wolffiella lingulàta** Hegelm.
Tongue-shaped Wolffiella. Fig. 853.

Wolffia lingulata Hegelm. Monogr. Lemnac. 132. 1868.
Wolffiella lingulata Hegelm. in Engler Bot. Jahrb. **21**: 303. 1895.

Fronds solitary or rarely in pairs, ovate to oblong, tongue-shaped, slightly unsymmetrical, 1.7–3 mm. wide, 2.7–6.6 mm. long; stipe insertion on the margin of the wall of the pouch.

A Mexican species extending into California, where it has been collected in San Mateo, Monterey, Kern, San Bernardino and Orange Counties. Type locality: Mexico.

Family 15. **PONTEDERIÀCEAE.**
PICKEREL-WEED FAMILY.

Perennial aquatic or bog plants, the leaves petioled, with thick blades, or long and grass-like. Flowers perfect, regular, more or less irregular, solitary or spiked, subtended by leaf-like spathes. Perianth free from the ovary, petaloid, 6-parted. Stamens 3 or 6, inserted on the tube or the base of the perianth; mostly dissimilar or unequal; anthers 2-celled, introrse, usually linear-oblong. Ovary 3-celled, with axile placentae, or 1-celled with 3 parietal placentae. Style 1; stigma 3-lobed or 6-toothed. Ovules anatropus, numerous. Fruit a many-seeded capsule, or a 1-celled, 1-seeded utricle. Endosperm copious; embryo central, cylindric.

A family of about 5 genera and 25 species, inhabiting the temperate and tropical regions of America, Asia and Africa.

1. **HETERANTHÈRA** Ruiz & Pav. Prodr. Fl. Per. 9. 1794.
[SCHOLLERA Schreb., 1789, not Roth, 1788.]

Herbs with creeping, ascending or floating stems, the leaves petioled, with oval to reniform blades, or grass-like. Flowers 1–several in a spathe, small, petaloid. Lobes of the perianth nearly or quite equal, linear. Stamens 3, equal or unequal, inserted on the throat of the perianth. Ovary fusiform, entirely or incompletely 3-celled by the intrusion of the placentae; ovules numerous; stigma 3-lobed. Fruit an ovoid, many-seeded capsule, enclosed in the withered perianth-tube. Seeds ovoid, many-ribbed. [Greek, referring to the unequal anthers of some species.]

A genus of about 10 species, 2 in tropical Africa, the others American, four of which occur in the United States, Type species, *Heteranthera reniformis* Ruiz & Pav.

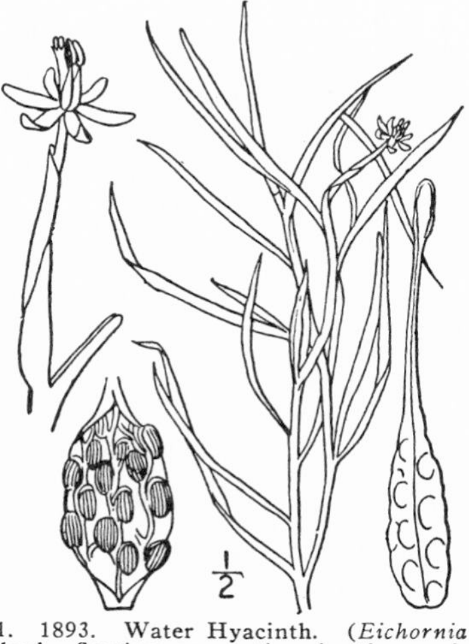

1. **Heteranthera dùbia** (Jacq.) MacM.
Water Star-grass. Fig. 854.

Commelina dubia Jacq. Obs. Bot. **3**: 9. *pl. 59.* 1768.
Leptanthus gramineus Michx. Fl. Bor. Am. **1**: 25. 1803.
Heteranthera graminea Vahl. Enum. **2**: 45. 1806.
Schollera graminea A. Gray, Man. 511. 1848.
Heteranthera dubia MacM. Met. Minn. 138. 1892.
Zosterella dubia Small, Fl. Lancaster Co. 68, 319. 1913.

Submerged aquatic, with slender forked stems, 6–10 dm. long, often rooting at the nodes. Leaves linear, elongated, sessile, acutish, finely parallel-veined, their sheaths thin, tipped with small acute stipule-like appendages; spathe terminal, bearing a single flower exposed on the surface of the water; perianth light yellow, with a long thread-like tube and very narrow segments; stamens equal; filaments dilated below; anthers sagittate, linear, 4 mm. long; stigma several-lobed; capsule 1-celled with 3 parietal placentae, many-seeded.

In still water, Quebec to Oregon, south to Cuba and Mexico. Only three collections have been made in the Pacific States: Latah Creek near Marshall Junction, Washington (*Suksdorf*); Sauvies Island, Willamette River (*Howell*), and Mendocino County, California (*Vasey*).

Piaròpus cràssipes (Mart.) Britton, Ann. N. Y. Acad. Sci. **7**: 241. 1893. Water Hyacinth. (*Eichornia crassipes* Solms. in DC. Monog. Phan. **4**: 527. 1883.) Aquatic herb, floating or rooting in the mud. Leaves clustered at the nodes, ovate to orbicular, 3–8 cm. wide, leathery; petioles elongated, wholly or partly inflated; flowers in a spadix terminating a simple scape 10–40 cm. high, blue and showy, irregular; perianth-segments 6, united into a tube below; stamens 6, irregularly adnate to the perianth, 3 included, 3 exserted; ovary 3-celled; ovules numerous. An attractive ornamental plant frequently grown in water gardens, naturalized in a few places in California; Yolo, Fresno and San Bernardino Counties. A troublesome plant in the tropics and the Gulf States, where it often covers slow-running streams, greatly impeding navigation. Native of tropical America.

Family 16. JUNCÀCEAE.*

Rush Family.

Perennial or sometimes annual, grass-like, usually tufted herbs, commonly growing in moist places. Inflorescence usually compound or decompound, paniculate, corymbose, or umbelloid, rarely reduced to a single flower, bearing its flowers singly, or loosely clustered, or aggregated into spikes or heads. Flowers small, regular, with or without bractlets (prophylla). Perianth 6-parted, the parts glumaceous. Stamens 3 or 6, rarely 4 or 5, the anthers adnate, introrse, 2-celled, dehiscing by a slit. Pistil superior, tricarpous, 1-celled or 3-celled, with 3–many ascending anatropous ovules, and 3 filiform stigmas. Fruit a loculicidal capsule. Seeds 3–many, small, cylindric to subglobose, with loose or close seed-coat, with or without caruncular or tail-like appendages.

Eight genera and about 300 species, widely distributed.

Leaf-sheaths open; capsule 1- or 3-celled, many-seeded; placentae parietal or axial. 1. *Juncus.*
Leaf-sheaths closed; capsule 1-celled, 3-seeded, its placenta basal. 2. *Luzula.*

1. JÚNCUS L. Sp. Pl. 325. 1753.

Usually perennial plants, principally of swamp habitat, with glabrous herbage, stems leaf-bearing or scapose, leaf-sheaths with free margins, and leaf-blades terete, gladiate, grass-like, or channelled. Inflorescence paniculate or corymbose, often unilateral, sometimes congested, bearing its flowers either singly and with 2 bractlets (prophylla), or in heads and without bractlets, but each in the axil of a bract; bractlets almost always entire; stamens 3 or 6; ovary 1-celled or by the intrusion of the placentae 3-celled, the placentae correspondingly parietal or axial; seeds several–many, usually distinctly reticulated or ribbed, often tailed. [Latin from jungo, to bind, in allusion to the use of these plants for withes.]

About 215 species, most abundant in the north temperate zone. The plants bloom in summer. Type species, *Juncus acutus* L.

Lowest leaf of the inflorescence terete, not conspicuously channelled, erect, exactly simulating a continuation of the stem, the inflorescence therefore appearing lateral.
  Flowers inserted singly on the branches of the inflorescence, each with a pair of bractlets at the base. I. Genuini.
  Flowers inserted in few-flowered paniculate heads, each in the axil of a bract but without bractlets. II. Thalassici.
Lowest leaf of the inflorescence not exactly simulating a continuation of the stem, or if so conspicuously channelled along the inner side; inflorescence usually appearing terminal.
  Leaf-blades transversely flattened (inserted with the flat surface facing the stem) not provided with septa.
    Flowers inserted singly on the branches of the inflorescence, each subtended by 2 bractlets. III. Poiophyllii.
    Flowers inserted in true heads on the branches of the inflorescence, in the axils of a single bract but without bractlets. IV. Graminifolii.
  Leaf-blades terete or gladiate, never transversely flattened, the septa usually conspicuous and complete (except in the species with gladiate leaves).
    Leaf-blades terete or somewhat compressed but not gladiate, the septa complete. V. Nodosi.
    Leaf-blades gladiate, the septa incomplete. VI. Ensifolii.

I. Genuini

Inflorescence usually 1–3-flowered, rarely 4–5-flowered; seeds with conspicuous white tails at either end.
  Blade of the uppermost basal leaf-sheath well developed; capsule acute. 1. *J. parryi.*
  Blade of the uppermost basal leaf-sheath reduced to a mere rudiment; capsule retuse. 2. *J. drummondi.*
Inflorescence of many flowers, rarely reduced to 3–5 in depauperate plants; seeds not tailed.
  Stamens 3, opposite the outer perianth-segments. 3. *J. effusus.*
  Stamens 6, opposite the inner and outer perianth-segments.
    Anthers about as long as the filaments.
      Stems numerous in dense tufts; inflorescence decompound; capsule globose-ovoid. 4. *J. patens.*
      Stems barely tufted; inflorescence simply umbelloid or nearly so; capsule oblong. 5. *J. filiformis.*
    Anthers much longer than the filaments.
      Leaf-blades usually well developed on the uppermost basal leaf-sheaths; stems compressed. 6. *J. mexicanus.*
      Leaf-blades wanting; stem terete.
        Stems smooth or in dried specimens rugulose with irregularly longitudinal ridges; woody tissue very thin, composed of 1 or 2 rows of vascular bundles and without fascicles of subepidermal strengthening cells.
          Perianth-segments 4–5 mm. long. 7. *J. balticus.*
          Perianth-segments 5–6 mm. long. 8. *J. lescurii.*
        Stems finely and evenly sulcate, the woody tissue thick, composed of 3–4 rows of vascular bundles and with well defined fascicles of subepidermal strengthening cells. 9. *J. textilis.*

---

* Text prepared with the assistance of Mr. Frederick V. Coville.

II. Thalassici

Perianth-segments acute and rigid at the apex; capsule oblong, acute, about equalling the perianth. 10. *J. cooperi.*
Perianth-segments broad and scarious at the apex; capsule obovoid, about twice the length of the perianth. 11. *J. acutus.*

III. Poiophyllii

Annuals; inflorescence exclusive of its leaves more than one-third the height of the plant.
Capsule oblong to ovoid, 3–4.5 mm. long, closely enclosed by the inner perianth-segments in fruit. 12. *J. bufonius.*
Capsule subglobose or short-ovoid, 2–3 mm. long; perianth-segments not closely embracing the fruiting capsule. 13. *J. sphaerocarpus.*
Perennials; inflorescence exclusive of its leaves less than one-third the height of the plant.
Stem-leaves 1 or 2, rarely wanting; perianth-segments obtuse at the apex. 14. *J. gerardi.*
Stem-leaves none; perianth-segments acute or acuminate.
Auricles of the leaf-sheaths cartilaginous, opaque and yellowish. 15. *J. dudleyi.*
Auricles of the leaf-sheaths membranous, transparent and usually white.
Capsule completely 3-celled.
Stem stout; bracts leaf-like; perianth-segments 3.5–4 mm. long, pungent at apex. 16. *J. brachyphyllus.*
Stem slender; bracts filiform; perianth-segments 5 mm. long, merely acute at apex. 17. *J. confusus.*
Capsule 1-celled, the septa extending only half way to the center.
Leaves over half the length of the stem; perianth-segments with brown scarious margins. 18. *J. occidentalis.*
Leaves less than half the length of the stem; perianth-segments with white scarious margins. 19. *J. tenuis.*

IV. Graminifolii

Perennials, only depauperate plants less than 10 cm. high; leaf-blades flat.
Junction of leaf-blade and sheath inconspicuous, the auricles rudimentary or wanting; perianth-segments minutely roughened on the back.
Perianth longer than the capsule.
Seeds not tailed.
Inner perianth-segments distinctly longer than the outer; flowers usually in many heads; montane species. 20. *J. orthophyllus.*
Inner perianth-segments about equalling the outer; flowers in 1 or 2 heads; maritime species. 21. *J. falcatus.*
Seeds tailed. 22. *J. regelii.*
Perianth equalling or shorter than the capsule.
Perianth-segments 3–4 mm. long, dark brown, except the well defined roughened portion. 23. *J. covillei.*
Perianth-segments 2–3 mm. long, pale brown, the roughened portion very small or obsolete. 24. *J. obtusatus.*
Junction of the leaf-sheath and blade well marked, and the auricles well developed (except in *J. macrophyllus*); perianth-segments smooth and usually shining.
Outer perianth-segments equalling or longer than the inner; capsule with a broad truncate or retuse apex and slender mucro. 25. *J. longistylis.*
Outer perianth-segments shorter than the inner; capsule tapering into a stout beak. 26. *J. macrophyllus.*
Annuals, less than 10 cm. and seldom more than 5 cm. high; leaf-blades filiform.
Inflorescence few-flowered or when reduced to 1 flower still retaining 2 bracts at its base.
Style 2 or 3 mm. long; anthers longer than the filaments. 27. *J. triformis.*
Style about 0.5 mm. long; anthers minute, much shorter than the filaments. 28. *J. brachystylus.*
Inflorescence 1-flowered, subtended by a single bract. 29. *J. uncialis.*

V. Nodosi

Capsule obovate to linear-lanceolate in outline, more than two-thirds the length of the perianth.
Stamens 6.
Epidermis of the leaves and stems not rugosely roughened.
Anthers shorter than the filaments.
Plants with two kinds of leaves, one normal, the other submersed and capillary. 30. *J. supiniformis.*
Plants without submersed capillary leaves.
Heads with few to several flowers, turbinate at maturity.
Perianth and fruiting capsule 5 mm. or more in length; capsule shorter than the segments. 31. *J. oreganus.*
Perianth and fruiting capsule less than 5 mm. long; capsule longer than the segments.
Branches of the inflorescence divaricate; capsule acute below the mucronation; seeds transversely lineolate in the reticulations. 32. *J. articulatus.*
Branches of the inflorescence erect; capsule obtuse below the mucronation; seeds smooth in the reticulations. 33. *J. richardsonianus.*
Heads many-flowered and at maturity spheroidal. 34. *J. mertensianus.*
Anthers equalling or longer than the filaments.
Perianth greenish white or sometimes tinged with red. 35. *J. chlorocephalus.*
Perianth dark or pale brown.
Body of the capsule obtuse or abruptly acute below the mucronation.
Perianth dark brown.
Capsule obtuse or truncate. 36. *J. badius.*
Capsule abruptly acute; heads often reduced to 1 or 2 and much congested. 37. *J. nevadensis.*
Perianth pale brown. 38. *J. columbianus.*
Body of the capsule tapering above into a sharp beak; perianth usually pale brown. 39. *J. dubius.*

Epidermis of the leaves and stems transversely rugosely roughened. 40. *J. rugulosus.*
Stamens 3.
Heads few, spherical, densely many-flowered. 41. *J. bolanderi.*
Heads many, turbinate, few-flowered. 42. *J. acuminatus*
Capsule subulate; stamens 6.
Leaf-blade abruptly divergent from its sheath; inner perianth-segments shorter than the outer 43. *J. torreyi.*
Leaf-blade erect; inner perianth-segments longer than the outer. 44. *J. nodosus.*

VI. Ensifolii

Anthers much longer than the filaments.
Perianth 4–5 mm. long; capsule acute below the mucronation. 45. *J. phaeocephalus.*
Perianth 3–4 mm. long; capsule obtuse below the mucronation. 46. *J. macrandrus.*
Anthers usually much shorter than the filaments, rarely equalling them.
Capsule tapering gradually into a slender beak. 47. *J. oxymeris.*
Capsule abruptly contracted into a short beak.
Stamens 6.
Heads commonly 25–100; perianth usually straw-colored; leaves 4–10 mm. wide. 48. *J. xiphioides.*
Heads commonly 2–12; perianth dark brown; leaves usually 2–4 mm. wide. 49. *J. saximontanus.*
Stamens 3. 50. *J. ensifolius.*

1. **Juncus párryi** Engelm.

Parry's Rush. Fig. 855.

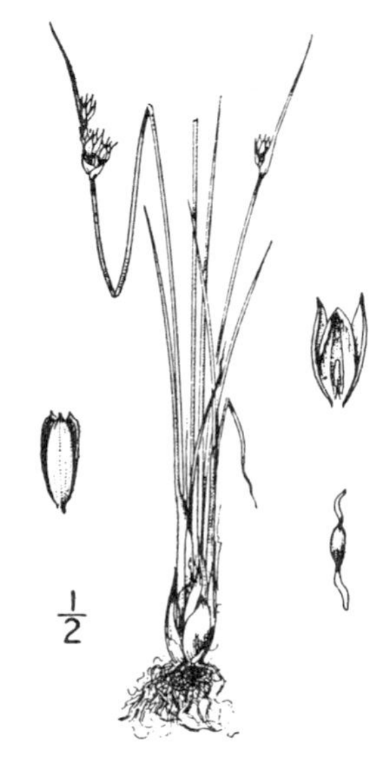

Juncus parryi Engelm. Trans. St. Louis Acad. **2**: 446. 1866.

Stems tufted, 10–30 cm. high, slender, from matted rootstocks. Only the uppermost basal leaf-sheath bearing a blade; blade sulcate at base, terete above, 3–6 or rarely 8 cm. long, scarcely 1 mm. thick; inflorescence 1–3-flowered, the flowers inserted singly and each with a pair of bractlets at the base; lowest leaf of inflorescence 1.5–6 cm. long; perianth 5–7 mm. long, tinged with brown, its segments lanceolate, acuminate, the inner a little shorter and acute, with broad scarious margins; stamens 6, scarcely half the length of the perianth; anthers much longer than the filaments; capsule oblong, triangular, acute, slightly exceeding the perianth; seeds ovate, 2 mm. long, tailed, finely striate.

Usually in dry grassy or stony slopes, Arctic to Canadian Zones; British Columbia to Montana south to Colorado and southern California. Type locality: Colorado.

2. **Juncus drúmmondi** Meyer.

Drummond's Rush. Fig. 856.

Juncus compressus subtriflorus Meyer, Linnaea 3: 368. 1828.
Juncus drummondi Meyer; Ledeb. Fl. Ross. **4**: 235. 1853.
Juncus subtriflorus Coville, Contr. Nat. Herb. **4**: 208. 1893.

Stems tufted, mostly 20–45 cm. high, from matted rootstocks. Basal leaf-sheaths all bladeless or with the mere rudiments of blades; inflorescence 1–3-flowered rarely 4–5-flowered, the flowers inserted singly and each with a pair of bractlets at the base; lowest leaf of the inflorescence mostly 2–3 cm. long; perianth 6 mm. long, its segments lanceolate, acute or acuminate, with broad brown margins, the inner equalling the outer, or nearly so; stamens 6, scarcely half the length of the segments; anthers longer than the filaments; capsule oblong, retuse at apex, equalling the segments; seeds ovate, 2 mm. long, caudate, very finely striate.

Moist alpine slopes, Arctic and Hudsonian Zones; Alaska south to the southern Sierra Nevada, California and New Mexico. Type locality: Rocky Mountains.

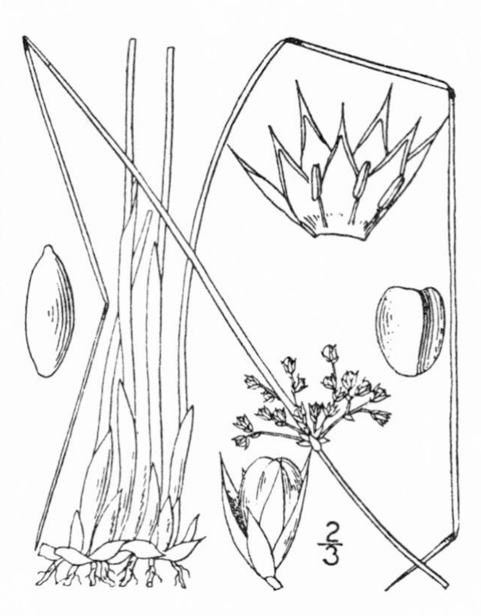

3. Juncus effùsus L.

Common Rush, Bog Rush, Soft Rush. Fig. 857.

Juncus effusus L. Sp. Pl. 326. 1753.

Stems densely tufted, from stout branching proliferous rootstocks, 5–15 dm. high, soft. Basal leaf-blades reduced to short filiform rudiments; inflorescence many-flowered, 2.5–10 cm. long; lowest leaf of the inflorescence appearing as a prolongation of the stem, 5–20 cm. long; perianth 2–3 mm. long, green, lanceolate, acuminate; stamens 3, the anthers shorter than the filaments; capsule obovoid, 3-celled, muticous, regularly dehiscent; seeds reticulate in about 16 longitudinal rows, the reticulations smooth, two or three times broader than long.

In swamps and moist places, Canadian to Upper Sonoran Zones; nearly throughout North America, also in Europe and Asia. Common throughout the Pacific States except the desert areas. Type locality: Europe.

Juncus effusus brùnneus Engelm. Trans. St. Louis Acad. 2: 491. 1868. (*Juncus effusus hesperius* Piper). Distinguished from the typical form by the usually more compact panicle and dark brown perianth and capsule. This is the common form along the coast, extending from British Columbia to southern California.

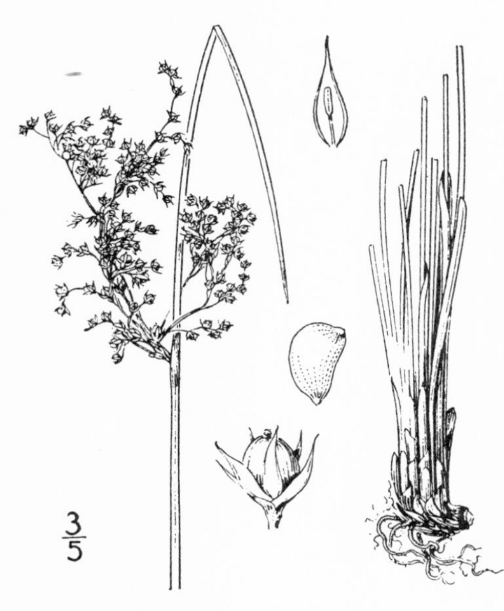

4. Juncus pàtens Meyer.

Spreading Rush. Fig. 858.

Juncus patens Meyer, Syn. Luzul. 28. 1823.

Stems more or less tufted, from stout branching rootstocks, 5–8 dm. high, 1.5–3 mm. thick above the leaf-sheaths. Basal leaf-sheaths with or without filiform rudiments of blades, the uppermost narrowed above, and often separating from the stem, thus suggesting blades in dried specimens; inflorescence open or somewhat compact, usually many-flowered, 2.5–7 cm. long; lowest leaf of the inflorescence appearing as a continuation of the stem, 2–15 cm. long; perianth pale or brownish, the segments lanceolate, acuminate, 3 mm. long; stamens 6, the anthers shorter than the filaments; capsule subglobose, shorter than the segments and widely spreading them, slightly 3-angled, very obtuse and apiculate at apex; seeds obliquely oblong, obscurely and irregularly reticulate.

Along streams and on moist hillsides, Transition and Upper Sonoran Zones; Rogue River region, southern Oregon, south in the Coast Ranges to Los Angeles County, California. Type locality: Monterey, California.

5. Juncus filifórmis L.

Thread Rush. Fig. 859.

Juncus filiformis L. Sp. Pl. 326. 1753.

Stems arising from creeping rootstocks, 1–6 cm. high, 1 mm. thick. Basal leaf-sheaths with the blades reduced to very short filiform rudiments; lowest leaf of inflorescence appearing as a continuation of the stem and usually as long or longer; inflorescence 15–25 mm. long, open, commonly about 8-flowered, rarely over 20-flowered; perianth 2.5–3.5 mm. long; segments narrowly lanceolate, acute or the inner obtuse, green with hyaline margins; stamens 6, about half as long as the perianth; anthers shorter than the filaments; capsule obovoid, green, barely pointed, about three-fourths as long as the perianth, 3-celled; seed obliquely oblong pointed at either end, with an irregularly wrinkled coat, seldom reticulate.

Moist ground, borders of lakes and streams, Canadian Zone; Alaska to Labrador, south to northern Oregon, Idaho, Utah, and Pennsylvania, also in Europe and Asia. Type locality: Europe.

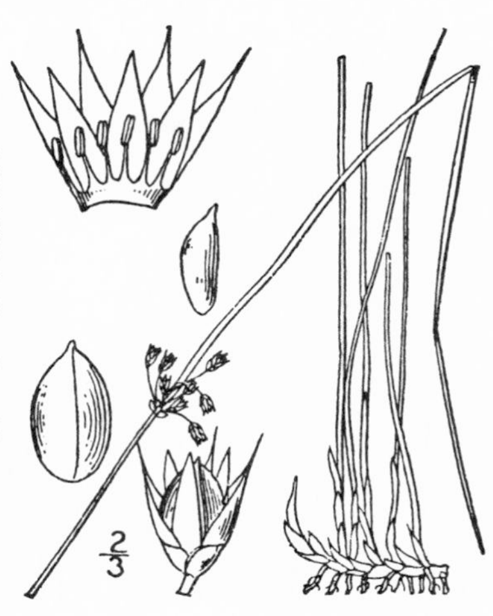

6. Juncus mexicànus Willd.
Mexican Rush. Fig. 860.

Juncus compressus H. B. K. Nov. Gen. & Sp. 235. 1815, not Jacq.
Juncus mexicanus Willd. in Roem. & Schult. Syst. 7: 178. 1829.
Juncus balticus mexicanus Kuntze, Rev. Gen. Pl. 3: 320. 1893.

Stems from creeping rootstocks, mostly 2–6 dm. high, seldom over 2–3 mm. thick, distinctly compressed, longitudinally wrinkled in dried specimens, but not striate. Basal leaf-sheaths brown or stramineous, resembling the stem, 10 cm. or more in length; lowest leaf of the inflorescence 3–15 cm. long; infloresence mostly loosely 5-many-flowered, 2–6 cm. high; perianth 5 mm. long, pale brown, the segments lanceolate, acuminate, with broad scarious margins, the inner about equalling the outer; stamens a little over half the length of the segments; anthers much longer than filaments; capsule equalling the perianth, narrowly ovoid, conspicuously mucronate, brown, 3-celled; seeds 0.7–0.8 mm. long, irregularly reticulate, the meshes longer than broad.

Moist ground especially in slightly saline soils, Upper and Lower Sonoran Zones; San Joaquin Valley, and Monterey County southward to Lower California and Mexico. Aphyllous forms are often confused with *J. balticus*. Type locality: Mexico.

7. Juncus bàlticus Willd.
Baltic Rush. Fig. 861.

Juncus balticus Willd. Berlin Mag. 3: 298. 1809.

Stems arising at intervals from creeping rootstocks, 2–8 dm. high, 2–3 mm. thick. Basal leaf-sheaths bladeless, the uppermost with or without a slender mucronation; lowest leaf of inflorescence 3–20 cm. long; panicle commonly 25–60 mm. high; perianth 4–5 mm. long, the segments lanceolate, acuminate, or the inner acute, nearly equal, usually purplish brown; stamens about two-thirds the length of the segments; anthers much longer than the filaments; capsule about equalling the perianth, narrowly ovoid, conspicuously mucronate, pale or dark brown; seeds 0.8–1 mm. long oblong to narrowly obovoid, oblique, about 40-striate.

Moist places. Boreal and Austral regions; Alaska to Labrador south to southern California, New Mexico, and Pennsylvania. Also Europe and Asia. A variable species segregated into several varieties by some authors. Type locality: Europe.

8. Juncus lescùrii Boland.
Salt Rush. Fig. 862.

Juncus lescurii Boland. Proc. Calif. Acad. 2: 179. 1862.
Juncus balticus pacificus Englem. Trans. St. Louis Acad. 2: 442. 1866.

Stems usually stout soft and smooth, 3–10 dm. high, 1.5–3 mm. thick, arising at intervals from stout creeping rootstocks. Basal leaf-sheaths bladeless; lowest leaf of inflorescence often as long or nearly as long as the stem, spiny tipped; panicle usually 3–4-rayed, the rays 2–3 cm. long, many-flowered near the summit, or much reduced forming a compact head-like inflorescence; perianth 5–6 mm. long, the segments lanceolate-acuminate, with the inner acute and a little shorter, dark brown with scarious margins; stamens about two-thirds the length of segments; anthers much longer than filaments; capsule oblong-ovoid, acute not mucronate, scarcely equalling the perianth, brown; seeds ovoid, 0.8 mm. long, finely reticulate, the reticulations much broader than long.

Borders of salt marshes, Transition and Sonoran Zones, Vancouver Island, south near coast to Monterey County, California. Type locality: salt marshes near San Francisco Bay.

9. **Juncus téxtilis** Buch.

Basket Rush. Fig. 863.

Juncus lescurii elatus Engelm. in S. Wats. Bot. Calif. **2**: 205. 1880.

Juncus textilis Buch. Abh. Nat. Ver. Bremen **17**: 336. 1903.

Stems stout, rigid, distinctly striate and pale, 10–20 dm. high, 4–5 mm. thick, arising from creeping rootstocks. Basal leaf-sheaths bladeless, the uppermost sometimes with a short slender mucronation; lowest leaf of inflorescence 5–15 cm. long, spinescent; panicle lax, 5–10 cm. high, many-flowered; perianth 4 mm. long, the segments lanceolate acute, the inner a little shorter, pale green with broad scarious margins somewhat tinged with pale brown; stamens half the length of the segments, the anthers much longer than the very short filaments; capsule oblong-ovoid, obtuse, short-beaked, purplish-brown, 3-celled; seeds oblong to narrowly obovoid, 0.8 mm. long, finely reticulate.

Borders of streams and springs, Upper Sonoran Zone; Los Angeles and San Bernardino Counties, and Santa Catalina Island. Type locality: San Gabriel Cañon, Los Angeles County.

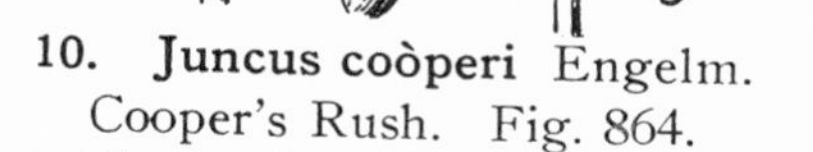

10. **Juncus coòperi** Engelm.

Cooper's Rush. Fig. 864.

Juncus cooperi Engelm. Trans. St. Louis Acad. **2**: 590. 1868.

Stems erect, rigid, 6–8 dm. high, densely tufted, arising from short stout much branched rootstocks. Basal leaf-sheaths with short stiff terete spinescent blades; lowest leaf of the inflorescence spinescent, 6–10 cm. long; panicle compound, with very unequal branches, the longer often 10 cm. long; flowers borne in the axils of scarious bracts, but with bractlets; perianth pale green or stramineous, 5–6 mm. long, the segments oblong-lanceolate, broadly scarious-margined, the outer prominently cuspidate and longer than the inner; stamens 6, nearly equalling the inner segments, anthers longer than the filaments, capsule narrowly oblong, 3-angled, acute, slightly exceeding the perianth, rigid-coriaceous, 3-celled; seeds with short broad appendages, reticulate, the reticulations not lineolate.

In alkaline places, Lower Sonoran Zone; Inyo to Imperial Counties, California and southern Nevada. Type locality: Camp Cody, California.

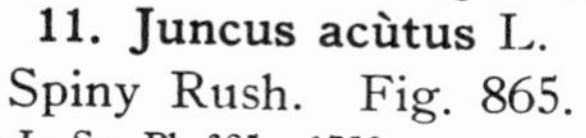

11. **Juncus acùtus** L.

Spiny Rush. Fig. 865.

Juncus acutus L. Sp. Pl. 325. 1753.

Juncus acutus sphaerocarpus Engelm. Rep. U. S. Geol. Surv. West 100 Merid. **6**: 376. 1878.

Stems growing in large tufts, 6–12 dm. high, stout, rigid, and spinescent. Basal leaf-sheaths usually bearing terete leaves resembling and nearly as long as the stems; lowest leaf of inflorescence 5–15 cm. long, stout and spinescent; panicle composed of very unequal branches, the longer 10–20 cm. long; flower clusters 2–4-flowered; perianth 3–4 mm. long, pale brown; the outer segments broadly lanceolate, obtuse with a broad scarious margin, the inner shorter, retuse at the scarious-margined apex; stamens nearly equalling the perianth; anthers longer than the filaments; capsule broadly obovate, or subglobose, apiculate, 5 mm. long; seeds acute at each end or slightly appendaged, finely reticulate.

Salt marshes, Upper and Lower Sonoran Zones; San Luis Obispo County to Lower California along the Coast; and western borders of the Colorado Desert, southern California. Also in slightly modified forms in South America and the Old World. Type locality: Europe.

12. **Juncus bufònius** L. Common Toad Rush. Fig. 866.

Juncus bufonius L. Sp. Pl. 328. 1753.
Juncus kelloggii Engelm. Trans. St. Louis Acad. **2**: 494. 1866.
Juncus congdoni S. Wats. Proc. Am. Acad. **22**: 480. 1887.

Annual, branching from the base, the stems erect, seldom exceeding 20 cm. in height. Leaves 1–3 on the stem below the inflorescence, the blade flat, 1 mm. wide or in small plants narrower and involute; inflorescence about half as high as the plants, with blade-bearing leaves at the lower nodes; flowers inserted singly on the branches, in one form fasciculate; perianth 4–5 mm. long, pale brown and scarious-margined; segments lanceolate, acuminate, the outer usually a little longer; stamens usually 6, rarely 3, seldom half as long as the segments; anthers shorter than the filaments; capsule oblong, obtuse, mucronate 3-celled, shorter than the segments; seeds broadly oblong, minutely reticulate with 30–40 longitudinal rows.

Dried up pools, border of streams; a cosmopolitan species occurring throughout North America, except the extreme north. The form with fasciculate flowers has been segregated as variety *halophilus* Fern. & Buch. Type locality: Europe.

13. **Juncus sphaerocàrpus** Nees. Round-fruited Toad Rush. Fig. 867.

Juncus sphaerocarpus Nees; Funk, Flora **1**: 521. 1818.

Annual, branching from the base, 5–20 cm. high, the branches very slender. Leaves 1–3 below the inflorescence, their blades flat or involute, about 1 mm. wide, often 5 cm. long; inflorescence usually over half the length of the stem; flowers inserted singly on the branches; perianth 3–4 mm. long, pale green and scarious margined; segments lanceolate acuminate, spreading in fruit; stamens 6; anthers shorter than the filaments; capsule subglobose to broadly ovoid, obtuse 3-celled, green, two-thirds as long as the segments; seeds oblong, minutely reticulate with 30–40 rows.

Borders of pools and streams, Transition and Sonoran Zones; Idaho and Oregon to southern California and Arizona. Also in Asia, North Africa, and southern Europe. Type locality: Europe.

14. **Juncus geràrdi** Lois. Black-grass Rush. Fig. 868.

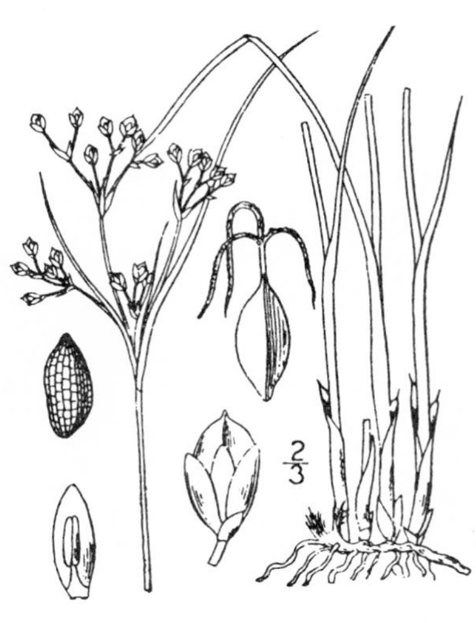

Juncus gerardi Lois. Journ. de Bot. **2**: 284. 1809.

Stems tufted from creeping rootstocks 2–4 dm. high. Basal leaf-sheaths loosely clasping, auriculate, their blades flat; stem leaves usually present, 1 or 2, similar to the basal; lowest leaf of the inflorescence often exceeding the panicle; panicle erect, strict or slightly spreading, 3–5 cm. high; perianth 2–3 mm. long, the segments oblong, obtuse, with green midrib and broad dark brown margins, straw-colored in age; stamens 6, nearly as long as the perianth; anthers much longer than the filaments; capsule one-fourth to one-half longer than the perianth, obovoid, mucronate, dark brown, shining, 3-celled; seeds dark brown, obovate, acute at base, obtuse and often depressed at summit, marked by 12–16 conspicuous ribs, the intervening spaces cross lined.

In salt marshes, Canadian and Transition Zones; Vancouver Island and adjacent mainland. Common on the Atlantic Coast. Type locality: France.

15. **Juncus dúdleyi** Wiegand.
Dudley's Rush. Fig. 869.

Juncus dudleyi Wiegand, Bull. Torrey Club **27**: 524. 1904.

Stems tufted, 3–12 dm. high, pale green, wiry, striate-grooved. Leaves basal, with blades about half the length of the stems, narrowly linear, flat or somewhat involute; panicle 25–50 mm. high, usually rather congested, few-flowered, considerably exceeded by its lowest leaf; perianth green or pale straw-colored, 4–5 mm. long, the segments lanceolate-subulate, acute, more or less spreading, firm, scarious-margined, nearly equal; stamens 6, about half as long as the perianth; anthers slightly shorter than the filaments; capsule ovoid-oval, about three-fourths the length of the segments, somewhat apiculate; seed oblong, apiculate at each end.

In damp soil and open places. Canadian and Transition Zones: eastern Washington to Maine, Missouri and Arizona. Type locality: Ithaca, New York.

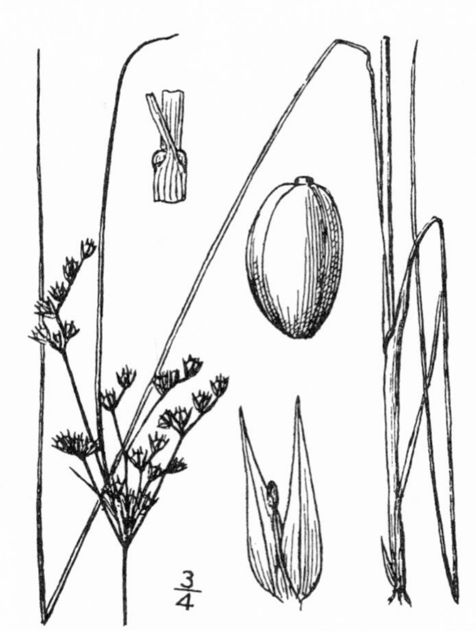

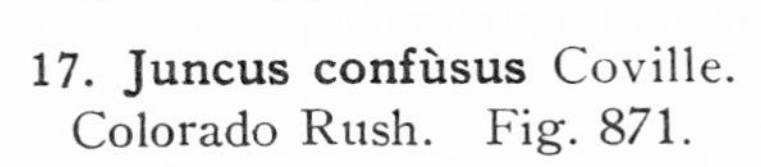

16. **Juncus brachyphýllus** Wiegand.
Short-leaved Rush. Fig. 870.

Juncus brachyphyllus Wiegand, Bull. Torrey Club. **27**: 519. 1900.

Stems tufted, erect and stiff, 4–5 dm. high, about 1.5 mm. in diameter, conspicuously grooved. Leaves basal, their blades 8–12 cm. long, spreading, flat, 1.5–2 mm. wide, rarely involute, auricles of the sheaths produced above the insertion and scarious; panicle 2–6 cm. long or sometimes capitate, densely many-flowered; bracts 2–9 cm. long, leaf-like; perianth pale straw-colored, 5 mm. long, the segments subulate, slightly unequal, the outer scarious-margined below the very acute pungent apex, the inner scarious to the top; stamens one-half the length of the perianth; anthers nearly as long as the filaments; capsule rather narrowly oblong, obtuse or retuse, equalling the perianths; 3-celled; seeds apiculate at each end, areolate.

Open moist grassy places, Transition Zone; eastern Washington and Oregon to Wyoming and New Mexico. Type locality: Upper Platte River.

17. **Juncus confùsus** Coville.
Colorado Rush. Fig. 871.

Juncus confusus Coville, Proc. Biol. Soc. Wash. **10**: 127. 1896.

Stems sparingly tufted, slender, erect, 4–5 dm. high, somewhat grooved. Leaves basal, two-thirds the length of the stems, the sheaths narrow, close, the blades almost filiform, thick but flat and usually involute; auricles produced above the insertion of the blade, scarious and whitish; inflorescence compact and capitate, 0.5–2 cm. long; bracts 2–7 cm. long, filiform; perianth 3.5–4 mm. long, pale; the segments straw-colored, subulate, merely acute, not pungent, with a fuscous stripe on each side, the margins all broadly scarious to the apex; anthers shorter than the filaments; capsule oblong, firm, a little shorter than the perianth, conspicuously triangular and retuse at the apex, completely 3-celled; seeds apiculate at both ends, coarsely and shallowly areolate.

Open moist grassy places, Transition Zone; eastern Washington to Montana, south to Wyoming and Colorado. Type locality: North Park, Colorado.

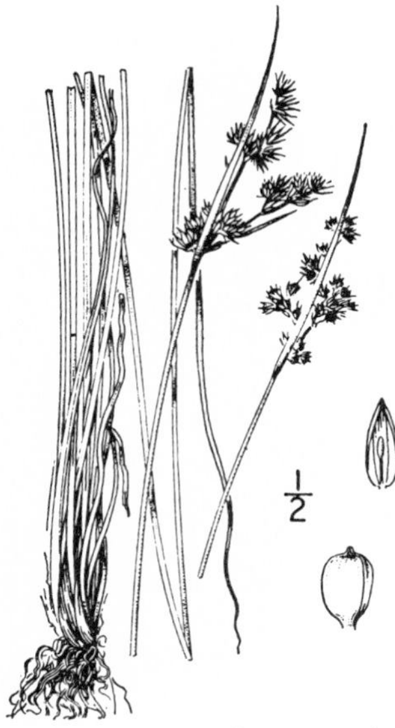

18. Juncus occidentàlis (Coville) Wiegand. Western Rush. Fig. 872.

Juncus tenuis congestus Engelm. Trans. St. Louis Acad. **2**: 450. 1866.
Juncus tenuis occidentalis Coville, Proc. Biol. Soc. Wash. **10**: 129. 1896.
Juncus occidentalis Wiegand, Bull. Torrey Club **27**: 521. 1900.

Stems sparingly tufted, stiff, erect, 3–6 dm. high, finely striate. Leaves densely tufted at the base, one-third to one-half the length of the stems; blades flat and flexuous, 1–1.5 mm. wide; sheaths loose and expanded, the auricles short, scarious; panicle open, many-flowered or congested, 1.5–3 cm. long, fuscous; bract 1, leaf-like, exceeding the panicle; perianth 4–5 mm. long; the segments erect or slightly spreading, lanceolate-subulate, fuscous with green midrib and brown scarious margin extending to the apex; stamens half the length of the perianth, anthers much shorter than the filaments; capsule oblong-oval, obtuse, becoming retuse when mature, three-fourths the length of the perianth, 1-celled, the placentae extending only half way to the middle; seed apiculate, areolate, reticulate, not striate.

Open moist grassy places, Transition and Upper Sonoran Zones; Multnomah County, Oregon, through the Coast Ranges and Sierra Nevada to central California. Type locality: San Francisco, California.

19. Juncus ténuis Willd. Slender Rush. Fig. 873.

Juncus tenuis Willd. Sp. Pl. **2**: 214. 1799.

Stems tufted, bright green, 2–6 dm. high, usually spreading, finely striate. Leaves one-half to nearly as long as the stem; blades flat, about 1 mm. wide, rarely slightly involute; sheaths short, loose and often expanded, their auricles scarious, large, often extending 15 mm. beyond the insertion of the blade; panicle often 1–7 cm. long, many-flowered, pale green; bracts 2 or 3, leaf-like, much exceeding the panicle; perianth 3–4.5 mm. long; segments lanceolate, very acute, green with white scarious margins, conspicuously spreading; stamens half the length of the perianth; anthers much shorter than the filament; capsule thin-walled, broadly ovoid, obscurely triangular above, 1-celled, the placentae not extending half way to the axis; seeds with transversely oblong reticulations.

Open grassy places, Boreal and Austral regions; Washington and Idaho to Newfoundland, south to Oregon, Arkansas, and Florida. In the Pacific States it is much less common but occurs on both sides of the Cascade Mountains, in Washington. Type locality: North America, locality not designated.

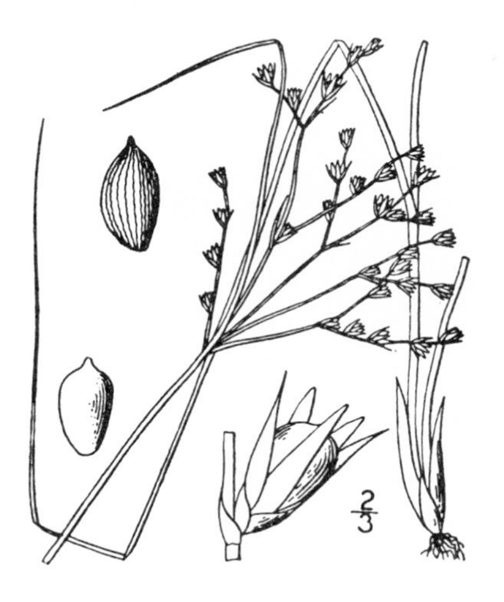

20. Juncus orthophýllus Coville. Straight-leaved Rush. Fig. 874.

Juncus longistylis latifolius Engelm. Trans. St. Louis Acad. **2**: 496. 1868.
Juncus latifolius Buch. in Engler Bot. Jahrb. **18**: 425. 1890, not Wulf, 1789.
Juncus orthophyllus Coville, Contr. Nat. Herb. **4**: 207. 1893.

Stems arising from creeping rootstocks, 2–4 dm. high, compressed, 1–1.5 mm. thick, pale green. Basal leaves grass-like, one-third to nearly as long as the stem, 3–5 mm. broad, many-nerved, usually spreading, without auricles; stem leaves none, or sometimes 1 (rarely 2) above the middle, 2–5 cm. long; inflorescence commonly of 2 (rarely 1) heads, these usually 8–10 flowered; perianth 6 mm. long; segments with a green midrib bordered by brown, and with a scarious fuscous margin, minutely roughened on the back, the inner distinctly longer than the outer, stamens about two-thirds the length of the perianth; anthers longer than the filaments; capsule oblong-ovoïd, obtuse, mucronate, scarcely equalling the perianth; seeds oblique, obovate, shortly apiculate.

Mountain meadows, Arid Transition and Canadian Zones; Cascade Mountains, Washington to San Bernardino Mountains, California, also Utah. Type locality: Alpine meadows, Yosemite Valley, California.

21. Juncus falcàtus Meyer.
Sickle-leafed Rush. Fig. 875.

Juncus falcatus Meyer, Syn. Luzul. 34. 1823.
Juncus menziesii R. Br.; Hook. Fl. Bor. Am. 2: 192. 1838.

Stems 10–20 cm. or rarely 30 cm. high, arising from slender stoloniferous rootstocks. Basal leaves flat and grass-like, the union of blade and sheath ill-defined, usually without a trace of auricle, falcate, two-thirds as long to longer than the stems, 1.5–3 mm. wide; stem leaves solitary, near the middle of the stem, 3–4 cm. long; heads solitary or sometimes 2 or 3, 5–25-flowered; lowest bracts 8–20 mm. long; perianth 5–6 mm. long; segments equal, dark brown or with a greenish midrib, finely roughened, the outer shortly acuminate, the inner obtuse; stamens half the length of the perianth; anthers much longer than the filaments; capsule obovoid, equalling or nearly equalling the perianth, dark brown, rounded or depressed at the apex; seeds oblong-ovate, 0.7 mm. long, obtuse, longitudinally reticulate.

In sands or borders of marshes along the coast; Humid Transition Zone; British Columbia to Monterey, California. Also in modified forms to Unalaska, Japan and Tasmania. Type locality: Monterey, California.

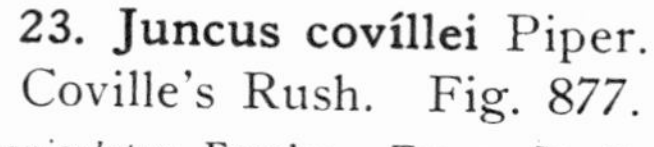

22. Juncus regèlii Buch.
Regel's Rush. Fig. 876.

Juncus regelii Buch. Engler Bot. Jahrb. 12: 414. 1890.

Stems erect, slender, 15–50 cm. high, 12 mm. in diameter, from stoloniferous rootstocks. Basal leaves grass-like, the junction of the blade and sheath inconspicuous, the auricles rudimentary or wanting; blades 10–15 cm. long, 3–4 mm. wide; stem leaves 1–3, usually with very minute auricles; heads 1–3 or rarely more, few–many-flowered, 8–15 mm. broad; perianth 4–5 mm. long, minutely roughened; the outer segments ovate-lanceolate, short acuminate, with very narrow scarious margins, the inner ovate, obtuse, with broad scarious margins; stamens 6, about two-thirds the length of the segments; anthers equalling the filaments; capsule ovoid, scarcely equalling the segments, obtuse and slightly depressed, mucronate, brown, 3-celled; seeds linear with a long white appendage at each end.

Wet places, Canadian Zone; Cascade Mountains, Washington to Montana and Utah. Type locality: Washington, but locality not designated.

23. Juncus covíllei Piper.
Coville's Rush. Fig. 877.

Juncus falcatus paniculatus Engelm. Trans. St. Louis Acad. 2: 495. 1868.
Juncus covillei Piper, Contr. Nat. Herb. 11: 182. 1906.

Stems from slender stoloniferous rootstocks, 8–30 cm. high, slender, pale green. Basal leaves grass-like, the junction of sheath and blade inconspicuous, without auricles, about two-thirds as long to longer than the stems, 1.5–3 mm. wide; stem leaves 1 or rarely 2, usually with a small auricle; heads paniculate, 2–6, rarely solitary, lowest bract foliaceous, 2–5 cm. long, or commonly the sheath brownish, broadly scarious-margined and bladeless or with a green foliaceous blade 10–25 mm. long; perianth 4 mm. long, minutely roughened; segments oblong-ovate, obtuse, about equal, the outer minutely mucronate, dark brown throughout or the midrib green; stamens scarcely two-thirds as long as the segments; anthers longer than the filaments; capsule oblong-ovoid, equalling or distinctly longer than the perianth, obtuse and slightly depressed, minutely mucronate, 3-celled; seeds reticulate, usually truncate at apex.

Wet places, about springs and lakes, Transition Zone; Whatcom and Challam Counties, Washington, south to Mendocino and Lake Counties, California. Type locality: Mendocino County, California.

24. **Juncus obtusàtus** Engelm.
Blunt-scaled Rush. Fig. 878.

Juncus obtusatus Engelm. Trans. St. Louis Acad. 2: 495. 1868.

Stems tufted, 10–25 cm. high, from creeping rootstocks. Basal leaves grass-like, the junction of the sheath and the blade inconspicuous, without auricles, usually about equalling the stems, 2–3 mm. wide; stem leaves 1, rarely 2, sometimes none; heads paniculate, mostly 4–8, sometimes reduced to 1 or 2, commonly 3–5-flowered; perianth 2–3 mm. long, pale brown, the roughened portion narrow and obscure; segments ovate, obtuse, about equal; capsule longer than the perianth, oblong-ovoid, depressed at the obtuse apex and mucronate, dark brown; seeds minutely reticulate.

Moist places along mountain streams, Transition Zone; central Sierra Nevada to the San Bernardino Mountains, California. Type locality: near the Mariposa Grove, California.

25. **Juncus longístylis** Torr.
Long-styled Rush. Fig. 879.

Juncus longistylis Torr. Bot. Mex. Bound. 223. 1859.

Stems erect, loosely tufted, 2–4 dm. high, rather stiff, slender, compressed. Basal leaves flat, grass-like, one-third to one-half the length of the stem; sheaths scarious-margined and with a well-developed auricle; blades 1.5–4 mm. wide, acute, striate; stem leaves 1–3, their blades commonly 5–7 cm. long; heads usually 2–8, 3–8-flowered, or reduced to a single larger one; perianth 5–6 mm. long, the segments lanceolate, accuminate, greenish-brown with hyaline margins; stamens 6, one-half to two-thirds as long as the perianth; anthers longer than the filaments; style 1 mm. long; capsule oblong, brown, angled above, obtuse or depressed at summit, mucronate, 3-celled; seeds oblong, white-tipped, 0.5 mm. long, 14–20-ribbed.

Moist meadows, Canadian and Transition Zones; British Columbia to western Ontario and Nebraska, south to Sierra Nevada, California, and New Mexico. Also in Newfoundland. Type locality: near the Copper Mines, New Mexico.

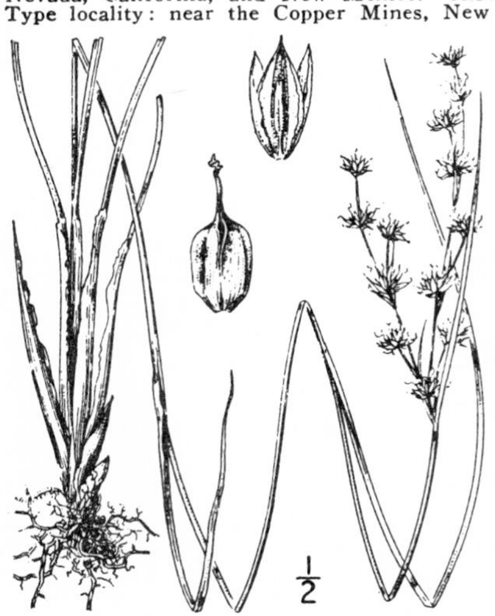

26. **Juncus macrophyllus** Coville.
Long-leaved Rush. Fig. 880.

Juncus canaliculatus Engelm. Bot. Gaz. 7: 6. 1882, not Liebm. 1850.
Juncus macrophyllus Coville. Univ. Calif. Pub. Bot. 1: 65. 1902.

Stems erect, 3–9 dm. high, rather stiff, compressed. Basal leaves equalling or about half the length of the stems; sheath scarious-margined and more or less distinctly auriculate; blades flat but rather thick and firm, 1.5–4 mm. wide, striate, long attenuate and pungent; stem leaves 1–3, their blades mostly 8–15 cm. long; inflorescense loosely paniculate; heads usually 12–25, 3–5-flowered; perianth green tinged with light brown, 5–6 mm. long, the segments ovate, acute or obtuse, hyaline-margined, the outer distinctly shorter than the inner; stamens 6, half the length of the segments; anthers reddish-brown, much longer than the filaments; capsule shorter than the perianth, tapering at the apex into a short beak; seeds 0.5 mm. long, obliquely obovate, about 20-ribbed, the reticulations lineolate.

Dry hillsides and marshes, Transition and Upper Sonoran Zones; southern Sierra Nevada, California, to northern Lower California and Arizona. Type locality: Mill Creek, San Bernardino Mountains, California.

27. **Juncus trifórmis** Engelm.
Yosemite Dwarf Rush. Fig. 881.

Juncus triformis Engelm. Trans St. Louis Acad. **2**: 492. 1868.
Juncus triformis stylosus Engelm. Trans. St. Louis Açad. **2**: 492. 1868.

Annual with very short stems from fibrous roots. Leaves 25 mm. long or less, filiform, channelled; peduncles several, scape-like, 5–10 cm. long; flowers usually 3–7, rarely solitary in a small terminal head; perianth 2–3 mm. long, brownish; segments narrowly lanceolate, acuminate, nearly equal; stamens 3, about three-fourths the length of the perianth; anthers slightly longer than the filaments; capsule nearly equalling the perianth, ovate, obtuse, apiculate; style 2–3 mm. long; seeds ovate, obtuse, faintly few-ribbed, the areolae cross-lined.

Moist places, Transition Zone; Sierra Nevada and San Bernardino Mountains, California. Type locality: Yosemite Valley, California.

28. **Juncus brachystỳlus** (Engelm.) Piper.
Short-styled Dwarf Rush. Fig. 882.

Juncus triformis brachystylus Engelm. Trans. St. Louis Acad. **2**: 492. 1868.
Juncus brachystylus Piper, Contr. Nat. Herb. **11**: 181 1906.

Annual, with very short stems, from fibrous roots. Leaves 1–2.5 cm. long, filiform, channelled; peduncles several, scape-like, 5–8 cm. long; flowers 3–7, rarely solitary, in a terminal head; stamens 3; anthers minute, much shorter than the filaments; styles very short, about 0.5 mm. long.

Moist places, Transition and Upper Sonoran Zones; Washington to Mendocino County, California. Type locality: Ukiah, California.

29. **Juncus unciàlis** Greene.
Inch-high Dwarf Rush. Fig. 883.

Juncus uncialis Greene, Pittonia **2**: 105. 1890.

Annual with very short stems, from fibrous roots. Leaves 3–8 mm. long, filiform, flat and somewhat fleshy; peduncles several, 2–3 mm. long; flowers solitary, terminal, subtended by a single bract; perianth hyaline, with purple or greenish midrib, 1.5 mm. long, the segments oblong-lanceolate, acute; stamens 3; capsule a little shorter than the perianth, ovoid, obtuse; seed ovate, apiculate, few-ribbed, the areolae cross-lined.

Moist places, Upper Sonoran and Transition Zones; Washington, south to Utah and central California. Type locality: Suisun marshes, Solano County, California.

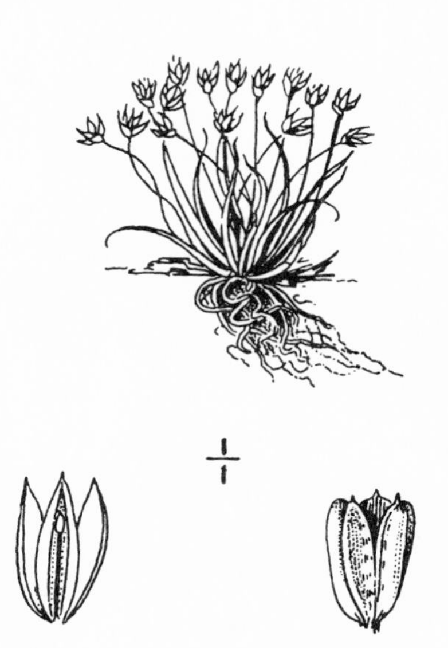

30. **Juncus supinifórmis** Engelm.
Hair-leaved Rush. Fig. 884.

Juncus supiniformis Engelm. Trans. St. Louis Acad. **2**: 461. 1868.

Stems tufted, the rhizomes much-branched and matted, clothed with numerous fibrous roots. Early leaves elongated and capillary, floating; stem-leaves terete, much longer than the stems, about 1 mm. in diameter, not conspicuously septate; inflorescence a simple panicle of 2–6, 3–9-flowered heads; perianth light brown, 4 mm. long; segments narrowly lanceolate, acute, 3–4-nerved; stamens 3; anthers shorter than the filaments; capsule narrowly oblong, about a third longer than the perianth, light brown, acute at the summit and rather stoutly beaked, the beak about 0.5 mm. long; seeds obovate, apiculate at each end.

In swamps, Humid Transition Zone; near the coast from Humboldt to Mendocino Counties, California. Type locality: near Mendocino City, California.

31. **Juncus oregànus** S. Wats.
Oregon Rush. Fig. 885.

Juncus oreganus S. Wats. Proc. Am. Acad. **23**: 267. 1888.
Juncus paucicapitatus Buch. Engler, Bot. Jahrb. **12**: 367. 1890.

Stems numerous, tufted, from slender matted rootstocks, 10–20 cm. high, very slender. Basal leaves early deciduous; stem leaves 2–4, the sheaths with conspicuous hyaline margins and auricles, the blades 25–60 cm. long, terete, conspicuously septate, panicle of usually 2–4 small few-flowered heads; perianth-segments nearly equal, narrowly lanceolate, acute, brown; stamens half as long as the perianth; anthers much shorter than the filaments; capsule linear-oblong, often twice as long as the perianth, acute and mucronate, dark brown; seeds with about 20 ribs, the areolae cross-lined.

In bogs near the coast, Humid Transition Zone; Ilwaco, Seattle, Sproat Lake, Vancouver Island. Type locality: Ilwaco, Washington.

32. **Juncus articulàtus** L.
Jointed Rush. Fig. 886.

Juncus articulatus L. Sp. Pl. 327. 1753.

Stems erect or spreading, 2–6 dm. high, tufted, from branching rootstocks. Basal blade-bearing leaves only 1 or 2, usually dying early, stem leaves with rather loose sheaths and conspicuously septate terete blades; inflorescence 5–10 cm. high, its branches divaricately spreading; heads hemispheric to top-shaped, 6–12-flowered; perianth 2–3 mm. long, the segments nearly equal, lanceolate acuminate, reddish brown with a green midrib, or green throughout; stamens 6, one-half to three-fourths as long as the perianth; anthers shorter than the filaments; capsule longer than the perianth, dark brown, 3-angled, sharply acute, tapering to a conspicuous tip, 1-celled; seeds oblong-obovoid, about 0.5 mm. long, reticulate in about 16–20 rows, the areolae finely cross-lined.

Moist ground about springs and marshes, Canadian and Humid Transition Zones; British Columbia to Clatsop County, Oregon, and Labrador to New York, and Michigan; also in Europe and Asia. Type locality: Europe.

33. **Juncus richardsoniànus** Schult.
Richardson's Rush. Fig. 887.

Juncus richardsonianus Schult. in Roem. & Schult. Syst. **7**: 201. 1829.

Juncus alpinus insignis Fries; Engelm. Trans. St. Louis Acad. **2**: 458. 1866.

Stems erect, 15–25 cm. high, in loose tufts, from creeping rootstocks. Stem leaves 1–3, usually borne below the middle, conspicuously septate; panicle 6–20 cm. high, loose, its branches strict or slightly spreading; heads 3–12-flowered; perianth 2–2.5 mm. long; inner segments shorter than the outer, obtuse, usually purplish toward the apex, the outer paler, obtuse, mucronate or acute; stamens 6, one-half to two-thirds as long as the perianth; anthers much shorter than the filaments; capsule ovoid-oblong, slightly exceeding the perianth, straw-colored or brown, broadly acute or obtuse with a short tip; seed ovoid to oblong, apiculate, acute or acuminate at the base, shallowly reticulate, the areolae smooth.

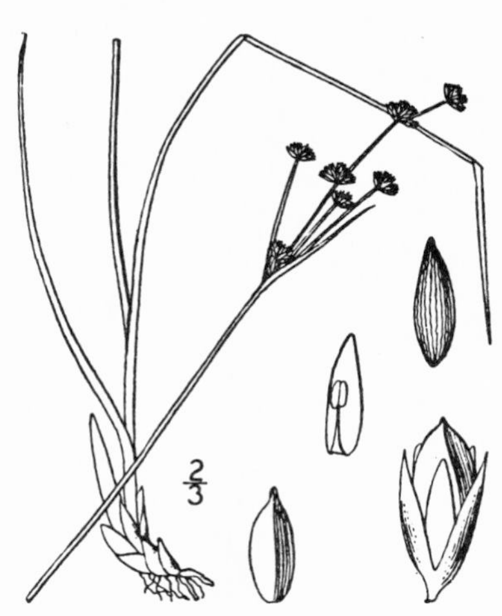

Shores of lakes and ponds, Canadian Zone; Alaska to Greenland, south to Washington, Idaho, Nebraska and Pennsylvania. In Washington it has been found in Chelan and Whatcom Counties. Type locality: "In sylvis Americae arcticae."

34. **Juncus mertensiànus** Bong.
Mertens' Rush. Fig. 888.

Juncus mertensianus Bong. Mem. Acad. St. Petersb. VI. **2**: 167. 1832.

Juncus mertensianus filifolius Suksdorf, Deutsch. Bot. Monatss. **19**: 92. 1901.

Stems slender and weak, 15–45 cm. high, from slender matted rootstocks. Leaves 2–3 on the stem, sheaths with auricles, blades very slender, somewhat compressed, equalling or shorter than the stems, 1 mm. wide, the septa often obscure; heads usually solitary, densely many-flowered, becoming spheroidal; perianth 4 mm. long, dark brown, the segments nearly equal, lanceolate, the outer acuminate, the inner acute; stamens nearly equalling the perianth; anthers much shorter than the filaments; capsule narrowly oval, obtuse or emarginate at apex, mucronate, equalling the perianth, dark brown; seeds ovate-lanceolate, the reticulations finely cross-lined.

Mountain meadows, Canadian to Arctic Zones; Sitka to southern California, east to Colorado. Type locality: Sitka, Alaska.

35. **Juncus chlorocéphalus** Engelm.
Green-headed Rush. Fig. 889.

Juncus chlorocephalus Engelm. Trans. St. Louis Acad. **2**: 485. 1868.

Stems tufted from slender matted rootstocks, very slender, 2–6 dm. high. Basal sheath bladeless, often rose-colored, stem leaves 3–4, the sheaths with broad scarious margins and prominent auricles, the blades somewhat compressed, 4–10 cm. long, scarcely 1 mm. in diameter, conspicuously septate; heads solitary or in few-headed panicles, few–many-flowered; perianth 4 mm. long, light green with broad scarious margins; the segments equal, oblong-lanceolate, obtuse; stamens three-fourths the length of the segments; anthers longer than the filaments; capsule oblong, obtuse, about two-thirds the length of the segments.

Mountain meadows, Canadian Zone; Sierra Nevada, California from Eldorado to Mariposa Counties. Also adjacent Nevada. Type locality: Sierra Nevada, California.

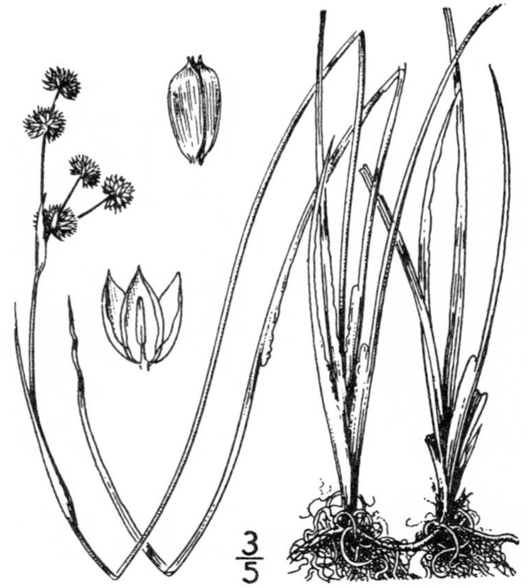

36. Juncus bàdius Suskdorf. Chestnut Rush. Fig. 890.

Juncus badius Suksdorf, Deutsche Bot. Monatss. **19**: 92. 1901.

Stems tufted from slender creeping rootstocks, 15–50 cm. high, very slender, slightly compressed. Leaves with distinctly auricled sheaths, the blades slightly compressed, completely septate, 5–15 cm. long, or sometimes longer, 1–1.5 mm. wide; heads solitary to several, mostly 7–12-flowered; lowest bract of inflorescence brownish or purplish, 8–10 mm. long or sometimes longer; perianth 3–3.5 mm. long, dark brown, the segments about equal, lanceolate, acuminate; anthers linear, longer than the filaments; capsule slightly exceeding the perianth, oblong-obovoid, rounded at the apex, dark brown; seeds apiculate at each end, areolate, the reticulations in about 15 rows.

Moist soils, Arid Transition Zone; eastern Washington and Idaho. Type locality: Falcon Valley, Klickitat County, Washington.

37. Juncus nevadénsis S. Wats. Sierra Rush. Fig. 891.

Juncus phaeocephalus gracilis Engelm. Trans. St. Louis Acad. **2**: 473. 1868.
Juncus nevadensis S. Wats. Proc. Am. Acad. **14**: 303. 1879.
Juncus suksdorfii Rydb. Bull. Torrey Club **26**: 541. 1899.

Stems somewhat tufted from slender creeping rootstocks, 15–60 cm. high, very slender, somewhat compressed. Leaves with distinctly auricled sheaths, the blades 5–15 cm. long or sometimes longer, 1–1.5 mm. wide, slightly compressed, completely but often obscurely septate; heads often 1–2 and much congested, or 4–10 and few-flowered; perianth 3.5 mm. long, dark brown, the segments about equal, lanceolate, acuminate; anthers linear, longer than the filaments; stigmas long-exserted; capsules scarcely equalling the perianth, oblong, abruptly acute and mucronate at apex, dark brown; seed minute, apiculate at each end, areolate, the reticulations in about 15 rows.

Mountain meadows, Canadian Zone; Washington and western Idaho to southern California. Type locality: Tuolumne Meadows, Sierra Nevada, California.

38. Juncus columbiànus Coville. Columbia Rush. Fig. 892.

Juncus columbianus Coville, Proc. Biol. Soc. Wash. **14**: 87. 1901.

Stems tufted from creeping rootstocks, 20–60 cm. high, slender, slightly compressed. Leaves with auricled sheaths, the blades 5–15 cm. long or longer, slightly compressed, 1–1.5 mm. wide, completely septate; heads several to numerous, 4–12-flowered; perianth pale brown, 3.5 mm. long, the segments narrowly lanceolate, acuminate; anthers longer than the filaments; capsule oblong, acute at the apex and prominently mucronate, shorter than the perianth; seeds with the reticulations in about 15 rows, areolate lineolate.

In wet meadows, Arid Transition Zone; eastern Washington to Montana south to Oregon. Type locality: in wet meadows near Pullman, Washington.

39. **Juncus dùbius** Engelm.

Mariposa Rush. Fig. 893.

Juncus dubius Engelm. Trans. St. Louis Acad. **2**: 459. 1868.

Stems stout, 5–12 dm. high, from stout creeping rootstocks. Leaf-sheaths with distinct auricles, the blades 1–4 dm. long, 1.5–3 mm. broad, compressed, conspicuously septate; panicle compound, diffuse, 7–25 cm. long; heads numerous, 6–20-flowered; perianth 3 mm. long, rather pale brown, the segments about equal, lanceolate, acuminate; anthers much longer than the filaments; capsule narrowly oblong, tapering above into a long rostrate beak; seeds abruptly apiculate at each end, reticulate, the reticulations transversely lineolate.

Mountain meadows and stream banks, Arid Transition Zone; Oregon through the Coast Ranges, the Cascade Mountains and Sierra Nevada to San Diego County, California. Type locality: Clark's Meadow, near the Mariposa Big Tree Grove.

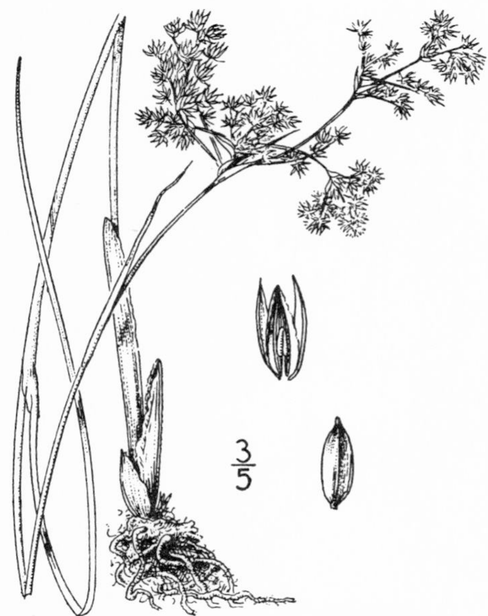

40. **Juncus rugulòsus** Engelm.

Wrinkled Rush. Fig. 894.

Juncus rugulosus Engelm. Bot. Gaz. **6**: 224. 1881.

Stems short, 5–12 dm. high, from stout creeping rootstocks, conspicuously rugose, as also the sheaths and blades. Basal sheaths bladeless, usually colored; stem leaves 2–4, their sheaths with a conspicuous auricle, the blades terete, 15–45 cm. long, 3–4 mm. broad, distinctly septate; panicle decompound and diffuse, 6–25 cm. long; heads numerous, 5–10-flowered; perianth 3 mm. long, light greenish brown and more or less tinged with purple, the segments narrowly lanceolate, acuminate; anthers about equalling the filaments; capsule narrowly oblong, tapering above into a beak; seeds reticulate, the reticulations transversely lineolate.

Stream banks and springs, Upper Sonoran Zone; Kern and San Luis Obispo Counties, south to San Diego County. Type locality: "running streamlet at the base of the San Bernardino Mountains."

41. **Juncus bolánderi** Engelm.

Bolander's Rush. Fig. 895.

Juncus bolanderi Engelm. Trans. St. Louis Acad. **2**: 470. 1868.
Juncus bolanderi riparius Jepson, Fl. Calif. **1**: 255. 1921.

Stems from creeping rootstocks, slender, 5–8 dm. high. Basal sheaths bladeless; stem leaves 3–4, their sheaths with conspicuous auricles; the blades slightly compressed, 10–20 cm. long, 1–1.5 mm. wide, distinctly septate; heads 2 or 3 or sometimes solitary, many-flowered, subglobose; perianth 3.5 mm. long, greenish brown, the segments narrowly lanceolate, setaceously acuminate; stamens 3; anthers much shorter than the filaments; capsule clavate-oblong, shorter than the segments, obtuse and apiculate at the apex; seeds minute, obovate, reticulate.

Stream banks and swamps, Humid Transition and Upper Sonoran Zones; Umpqua Valley, Oregon, to Mendocino County, California, also occurring locally in the lower Sacramento Valley and Orange County. Type locality: swamps near Mendocino City, California.

42. **Juncus acuminàtus** Michx.
Sharp-fruited Rush. Fig. 896.

Juncus acuminatus Michx. Fl. Bor. Am. **1**: 192. 1803.

Stems tufted, 3–10 dm. high, from short and inconspicuous rootstocks. Leaves 1–3 on a stem, the blades of the lower 10–20 cm. long, 1–2 mm. thick, the upper shorter; panicle 5–15 cm. long, with 5–50 heads or sometimes reduced to a single head, its branches usually spreading; heads top-shaped to subspheric, 3–20-flowered; perianth 2.5–3.5 mm. long, the segments lanceolate-subulate, nearly equal; stamens 3, about half or as long as the perianth; anthers shorter than the filaments; capsule ovate-lanceolate, broadly acute, mucronate, equalling the perianth, light brown; seeds oblong, apiculate at each end, reticulate with 16–20 longitudinal rows, the areolae transversely many-lined.

Stream banks and swamps, Transition Zone; British Columbia to Oregon, east to Maine and Georgia. Type locality: South Carolina.

43. **Juncus tórreyi** Coville.
Torrey's Rush. Fig. 897.

Juncus nodosus megacephalus Torr. Fl. N. Y. **2**: 326. 1843.
Juncus megacephalus Wood, Bot. ed. 2, 724. 1861, not Curtis, 1835.
Juncus torreyi Coville, Bull Torrey Club **22**: 303. 1895.

Stems arising singly from tuberiform thickenings on the slender rootstock, stout, 2–5 dm. high. Leaves 1–4 on the stem, the blades abruptly divergent from the stem, stout, terete, 2–5 mm. thick; panicle congested, consisting of 1–20 heads, exceeded by its lowest bract; heads 10–15 mm. in diameter; perianth 4–5 mm. long, the segments subulate, the outer longer than the inner; stamens 6, about half as long as the perianth; capsule subulate, 3-sided, 1-celled, the beak 1–1.5 mm. long, exceeding the perianth and holding the valves together throughout dehiscence; seed oblong, acute at both ends, reticulate in about 20 longitudinal rows, the areolae finely cross-lined.

Wet soil, Transition and Upper Sonoran Zones; Washington to southern California, east to Massachusetts, Texas and Alabama. In the Pacific States it is east of the Cascade-Sierra Nevada divide, except in southern California. Type locality: on the shores of Lake Ontario.

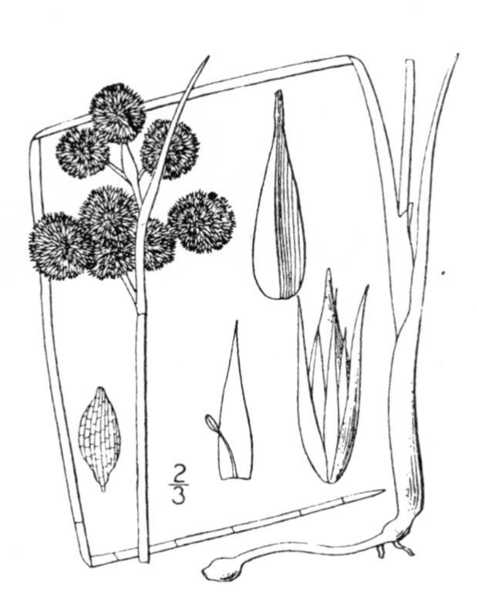

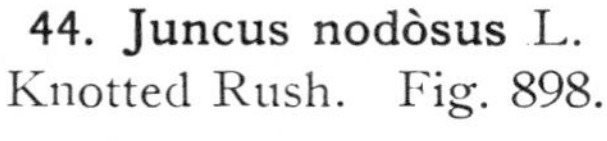

44. **Juncus nodòsus** L.
Knotted Rush. Fig. 898.

Juncus nodosus L. Sp. Pl. ed. 2, 466. 1762.

Stems 15–60 cm. high, arising singly from tuber-like thickenings of a slender, nearly scaleless rootstock. Basal leaves with long erect blades, stem leaves 2–4, their blades erect, the upper exceeding the inflorescence; panicle shorter than its lowest bract, seldom exceeding 5 or 6 cm., bearing 1–30 heads; heads spherical, several–many-flowered, 7–12 mm. in diameter; perianth 3.5 mm. long, usually reddish brown above; segments lanceolate-subulate, the outer shorter than the inner; stamens 6, about half as long as the perianth; anthers equalling the filaments; capsule lanceolate-subulate, 3-sided, 1-celled, exceeding the perianth; seed oblong, acute below, apiculate above, reticulate in 20–30 longitudinal rows, the areolae finely cross-lined.

Wet places, Transition Zone; British Columbia south, east of the Cascade Mountains, to Nevada, and east to Nova Scotia and Virginia. Type locality: eastern North America.

45. Juncus phaeocéphalus Engelm.

Brown-headed Rush. Fig. 899.

Juncus phaeocephalus Engelm. Trans. St. Louis Acad. 2: 484. 1868.
Juncus phaeocephalus glomeratus Engelm. Trans. St. Louis Acad. 2: 484. 1868.
Juncus phaeocephalus paniculatus Engelm. Trans. St. Louis Acad. 2: 484. 1868.

Stems flattened, 2-edged, 3–9 dm. high, from creeping rootstalks. Leaves flattened laterally, equitant, without auricles, the blades 8–20 cm. long, 2.5 mm. wide; inflorescence of 1–few large heads or paniculate with numerous few-flowered heads; perianth 4–5 mm. long, dark brown or purplish, the segments lanceolate, about equal, the outer acuminate, the inner acute; stamens 6, two-thirds the length of the perianth; anthers much longer than the filaments; capsule oblong, abruptly acute below the prominent beak; seeds reticulate, the reticulations in about 20 rows, the areolae cross-lined.

Wet places, Transition and Upper Sonoran Zones, Coastal region and Coast Range of California from Mendocino to Los Angeles and San Bernardino Counties. The variety *glomeratus* inhabits the sea coast from Mendocino to Monterey County and has few–many-flowered heads; the variety *paniculatus* is inland ranging from Napa County to southern California. Type locality: Monterey, California.

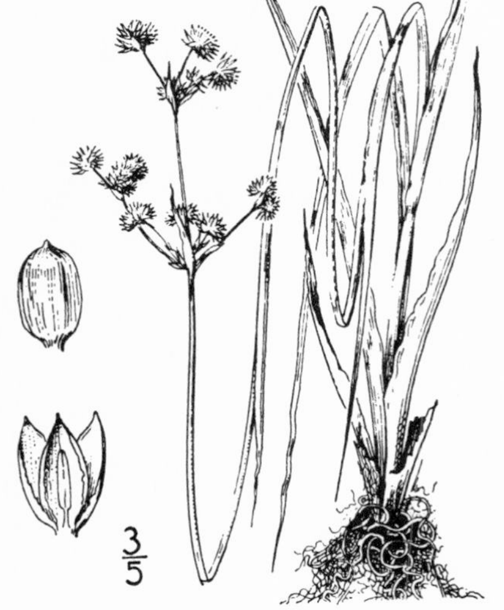

46. Juncus macrándrus Coville, n. sp.

Long-anthered Rush. Fig. 900.

Stems tufted from creeping rootstocks, flattened, 2-edged, 3–5 dm. high. Leaves flattened laterally, equitant, without auricles, the blades of the lower stem leaves usually about 10 cm. long, 2–3 mm. wide; inflorescence paniculate, 2–8 cm. long, the heads few to many, usually 3–5-flowered; perianth 3–4 mm. long, dark purplish brown; segments oblong-lanceolate, the outer acuminate, the inner acute; stamens 6, the anthers much longer than the filaments; the capsule broadly oblong, shorter than the perianth, obtuse below the beak; seeds reticulate, the areolae in about 20 rows, cross-lined.

Springs and wet meadows, Transition and Canadian Zones; northern Sierra Nevada to the San Jacinto Mountains, southern California. Type: Owens River Valley, Inyo County, California, *Coville & Funston 1774.*

47. Juncus oxymèris Engelm.

Pointed Rush. Fig. 901.

Juncus oxymeris Engelm. Trans. St. Louis Acad. 2: 483. 1868.

Stems compressed and 2-edged, 3–6 dm. high, arising from creeping rootstocks. Leaves flattened laterally and equitant; auricles usually obscure; blades 5–20 cm. long, 3–5 mm. wide; panicle usually ample, compound, with numerous few-flowered heads; perianth 3 mm. long, light green, with broad scarious margins, usually tinged with purple toward the acuminate and sharp-pointed tips; stamens 6, about two-thirds the length of the segments; anthers twice the length of the filaments; capsule tapering into a slender beak, longer than the perianth; seeds finely reticulate, the areolae in about 20 rows, smooth.

Stream banks and mountain meadows, Transition and Upper Sonoran Zones; western Washington to southern California. Type locality: Sacramento Valley.

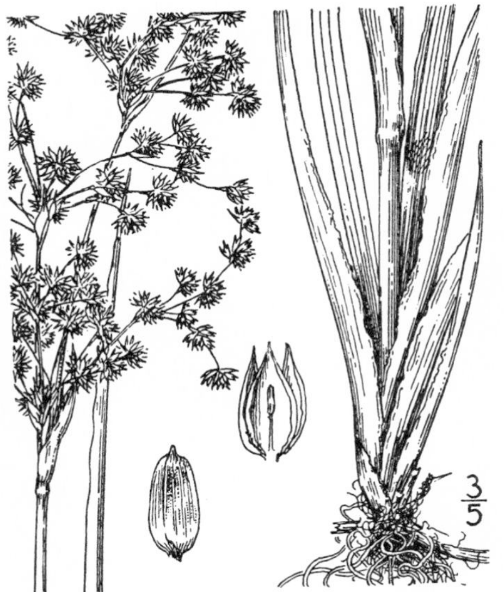

48. Juncus xiphioìdes Meyer. Iris-leaved Rush. Fig. 902.

Juncus xiphioides Meyer, Syn. Junc. 50. 1822.
Juncus xiphioides littoralis Engelm. Trans. St. Louis Acad. **2**: 481. 1868.
Juncus xiphioides auratus Engelm. Trans. St. Louis Acad. **2**: 481. 1868.

Stems 6–8 dm. high, compressed and acutely 2-edged, from a thick creeping rootstock. Leaves flattened laterally and equitant, the sheaths without auricles, the blades 10–40 cm. long or the upper shorter, 3–10 mm. wide; heads numerous in a compound panicle, 3–20-flowered; perianth brownish, 3 mm. long, the segments lanceolate, acuminate; stamens 6, half as long as the perianth; anthers shorter than or sometimes nearly equalling the filaments; capsule oblong, acute below the mucronation, slightly longer than the perianth; seeds ovate-lanceolate, reticulate, the areolae cross-lined.

Wet places, Transition and Upper Sonoran Zones; foothills of the northern Sierra Nevada and Coast Ranges of California to northern Lower California and Arizona. Type locality: Monterey, California.

49. Juncus saximontànus A. Nels. Rocky Mountain Rush. Fig. 903.

Juncus ensifolius major Hook, Fl. Bor. Am. **2**: 191. 1840.
Juncus xiphioides montanus Engelm. Trans. St. Louis Acad. **2**: 481. 1868.
Juncus saximontanus A. Nels. Bull. Torrey Club **29**: 401. 1902.

Stems 3–5 dm. high, compressed and 2-edged, from stout creeping rootstocks. Leaves flattened laterally and equitant, usually auricled, the blades 2–4 mm. wide; heads panicled, commonly 2–12, 3–10-flowered; perianth dark brown, 2.5–3 mm. long, the outer segments lanceolate, acuminate, the inner acute, shorter than the outer; stamens 6, about two-thirds the length of the perianth; anthers shorter than the filaments; capsule oblong, equalling the perianth, obtuse below the mucronation, dark brown; seeds reticulate, the reticulations in about 20 longitudinal rows, the areolae cross-lined.

Wet places, Transition and Canadian Zones; British Columbia to Oregon, east to Alberta and New Mexico. Type locality: Rocky Mountains.

50. Juncus ensifòlius Wiks. Three-stamened Rush. Fig. 904.

Juncus ensifolius Wiks. Kongl. Vet. Akad. Handl. **2**: 274. 1823.
Juncus xiphioides triandrus Engelm. Trans. St. Louis Acad. **2**: 482. 1868.

Stems compressed and 2-edged, 3–6 dm. high, arising from creeping rootstocks. Leaves flattened laterally and equitant, auricles wanting, blades 7–15 cm. long, 2–5 mm. wide; panicle of 2–7 rather large many-flowered heads, or compound, with numerous small heads, often only 3–6-flowered; perianth 3 mm. long, brown or dark brown, the segments equal, narrowly lanceolate, acuminate; stamens 3, about two-thirds the length of the perianth; anthers shorter than the filaments; capsule oblong, slightly exceeding the perianth, obtuse below the mucronation, dark brown; seeds reticulate, the reticulations in about 20 rows, areolae cross-lined.

Wet places, Transition and Canadian Zones; Alaska to the Sierra Nevada and northern Coast Ranges, California, east to Alberta and Utah. Type locality: Unalaska.

2. LÙZULA DC. Fl. Fr. 3: 158. 1805.

[JUNCOIDES Adans. Fam. Pl. 2: 47. 1763.]

Perennial plants, with herbage either glabrous or sparingly webbed, stems leaf-bearing, leaf-sheaths with united margins, and leaf-blades grass-like. Inflorescence umbelloid, paniculate, or corymbose, often congested; flowers always bracteolate, the bractlets usually lacerate or denticulate; stamens 6 in our species; ovary 1-celled, its 3 ovùles with basal insertion; seeds 3, indistinctly reticulate, sometimes carunculate at base or apex, but not distinctly tailed. [Name from *lucciola,* an old Italian name for the Glowworm.]

About 65 species, widely distributed, mostly flowering in spring. Type species, *Luzula lutea* (All.) DC.

Flowers in clusters of 2 or 3 or solitary in an open panicle.
  Cymes open; bracts entire or somewhat lacerate, not ciliate-fimbriate.
    Perianth 3–3.5 mm. long; leaves 10–12 mm. wide. — 1. *L. glabrata.*
    Perianth 1.5–2.5 mm. long; leaves 6–8 or rarely 10 mm. wide.
      Rays of the panicle drooping; leaves sparsely pilose at base.
        Perianth and capsule pale green; leaves thin, shining; seeds brown, ellipsoid. — 2. *L. parviflora.*
        Perianth and capsule dark brown; leaves thick, dull; seeds yellow, constricted at each end. — 3. *L. piperi.*
      Rays of the panicle divaricate; leaves not pilose. — 4. *L. divaricata.*
  Cymes congested; bracts ciliate-fimbriate. — 5. *L. subcongesta.*
Flowers congested into 1 to several spike-like or head-like clusters.
  Inflorescence nodding, usually of a single spike-like cluster; seeds without a strophiole-like base. — 6. *L. spicata.*
  Inflorescence erect or its individual branches rarely nodding; heads or spikes 2–12; seeds with a white strophiole-like base. — 7. *L. campestris.*

1. Luzula glabràta (Hoppe) Desv.
Smooth Wood-rush. Fig. 905.

Juncus glabratus Hoppe; Rostk. Mon. Junc. 27. 1801.
Luzula glabrata Desv. Journ. de Bot. 1: 145. 1808.
Luzula spadicea glabrata Meyer, Syn. Luzul. 8. 1823.
Juncoides glabratum Sheldon, Minn. Bot. Stud. Bull. 9: 63. 1894.

Stems stoloniferous, but slightly tufted, 2–5 dm. high. Stem leaves usually 4 or 5, blades 4–6 cm. long, 8–12 mm. wide, abruptly acute, flat, glabrous; basal leaves 10–20 cm. long; panicles decompound, erect, 6–10 cm. long, 5–7 mm. wide; bracts ciliate on the margins, the lowest 15–20 mm. long, brownish; rays often divaricate; flowers usually single, on rather stout divaricate pedicels; perianth 3–3.5 mm. long, lanceolate, acute, dark purplish brown; capsule ovoid, equalling or slightly exceeding the perianth, acute, almost black; seeds ellipsoid, dark brown.

Moist alpine woods, Hudsonian Zone; Alaska to Crater Lake, Oregon, east to Montana. Type locality: Europe.

2. Luzula parviflòra (Ehrh.) Desv.
Small-flowered Wood-rush. Fig. 906.

Juncus parviflorus Ehrh. Beitr. 6: 139. 1791.
Luzula parviflora Desv. Journ. de Bot. 1: 144. 1808.
Juncoides parviflorum Coville, Contr. Nat. Herb. 4: 209. 1893.

Stems stoloniferous, single or few in a tuft, erect 1–3 dm. high, 2–5-leaved. Leaves glabrous, their blades 3–10 mm. wide, tapering to a sharp or blunt apex; inflorescense a nodding decompound panicle, commonly 6–10 cm. long; lowest bract foliose, one-fourth to one-half the length of the panicle; flowers borne singly or sometimes 2 or 3 together, on very slender pedicels; bractlets ovate, entire or lacerate; perianth 2–2.5 mm. long, the segments lanceolate, acute, green or more or less tinged with brown; capsule ovoid, slightly exceeding the perianth, green or brownish; seeds ellipsoid, brown.

Moist woods and meadows; Transition to Hudsonian Zones; Alaska to Labrador, south to California, Minnesota and New York, also in Europe. In the Pacific States it reaches its southern limits in Humboldt County in the Coast Ranges, and Kern County in the Sierra Nevada. Type locality: Europe.

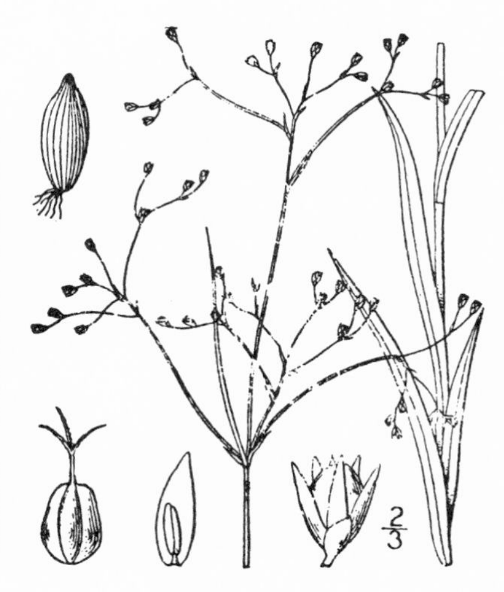

3. Luzula pìperi (Coville) M. E. Jones.
Piper's Wood-rush. Fig. 907.

Juncoides piperi Coville, Contr. Nat. Herb. **11**: 185. 1906.

Luzula piperi M. E. Jones, Bull. Univ. Montana Biol. Ser. **15**: 22. 1910.

Stems densely tufted from usually matted rootstocks, erect, 10–35 cm. high. Leaves mostly basal, linear-lanceolate, about one-fourth as long as the stem, 2–4 mm. wide, dull, sparsely pilose on the sheaths and margins; panicle decompound, 5–8 cm. long, nodding; lowest bract foliaceous, 8–15 mm. long; bractlets brown, lacerate; flowers borne singly or sometimes in clusters of 2 or 3; perianth 1.5–2 mm. long, dark purplish brown, the segments ovate, acuminate; capsule ovoid, exceeding the perianth, dark brown, the valves not apiculate; seeds buff or amber color, oblong-lanceolate, narrowed at each end.

Dry sandy or gravelly soils, Hudsonian Zone; Washington to Montana. Type locality: Bridge Creek, Okanogan County, Washington, "on dry sandy-gravelly moraines just below the glaciers."

4. Luzula divaricàta S. Wats.
Forked Wood-rush. Fig. 908.

Luzula divaricata S. Wats. Proc. Am. Acad. **14**: 302. 1879.

Juncoides divaricatum Coville, Contr. Nat. Herb. **4**: 209. 1893.

Stems tufted from matted rootstocks, 10–50 cm. high, 2–4-leaved. Basal leaves flat, linear-lanceolate, acuminate, 8–25 cm. long, 3–10 mm. wide; stem leaves 3–10 cm. long; panicle decompound, diffuse with divaricately spreading branches and pedicels, 8–15 cm. long, and nearly as wide; bractlets ovate, entire or sometimes sparsely lacerate; perianth 2–2.5 mm. long, tinged with brown, the segments lanceolate, acuminate; capsule ovoid, scarcely equalling the perianth, green, the valves apiculate; seeds brown, the basal end darker.

Gravelly soils in open woods or borders of meadows, Canadian Zone; Mt. St. Helens, Washington south to Mariposa County, California. Type locality: Sierra Nevada, California.

5. Luzula subcongésta
(S. Wats.) Jepson.
Donner Wood-rush. Fig. 909.

Luzula spadicea subcongesta S. Wats. Bot. Calif. 1: 202. 1876.

Juncoides subcongestum Coville, Muhlenbergia 1: 105. 1904.

Luzula subcongesta Jepson, Fl. Calif. 1: 258. 1921.

Stems tufted from a matted rootstock, 20–50 cm. high, 3–4-leaved. Basal leaves flat, linear-lanceolate, 8–20 cm. long, 4–8 mm. wide; stem-leaves 5–12 cm. long; cyme more or less congested into few–several rather dense head-like clusters terminating more or less elongated rays; bracts small, fimbriate; pedicels very short; perianth 1.5 mm. long, ovate, rather abruptly acuminate, dark brown with a hyaline apex; stamens half the length of the perianth; anthers about equalling the filaments; capsule ovoid, nearly as long as the perianth, apiculate, light brown; seeds oblong-ovate, oblique, brown, darker at apex.

Alpine slopes, Canadian and Hudsonian Zones; Sierra Nevada, California. Type locality: Donner Pass, California.

6. Luzula spicàta (L.) DC.

Spiked Wood-rush. Fig. 910.

Juncus spicatum L. Sp. Pl. 330. 1753.
Luzula spicata DC. Fl. Fr. **3**: 161. 1805.
Juncoides spicatum Kuntze, Rev. Gen. Pl. 725. 1891.

Stems closely tufted without rootstocks, erect, 10–40 cm. high, distantly 1–3-leaved, tapering into a filiform summit. Leaf-blades 1–3 mm. broad, often involute, tapering to a short apex, stiffly erect, sparingly webbed especially toward the base; inflorescence nodding, spike-like, often interrupted, commonly 12–25 mm. long, usually exceeded by the lowest involute foliose bract; bractlets ovate-lanceolate, acuminate, equalling the perianth, sparingly lacerate; perianth 2–2.5 mm. long, brown with hyaline margins, the segments lanceolate, aristate-acuminate; capsule broadly ovoid, bluntly acute, about two-thirds as long as the perianth; seeds narrowly and obliquely obovoid.

Alpine slopes, Hudsonian and Arctic Zones; Alaska to Labrador, south to the southern Sierra Nevada, California, Colorado and New York; also in Europe and Asia. Type locality: Europe.

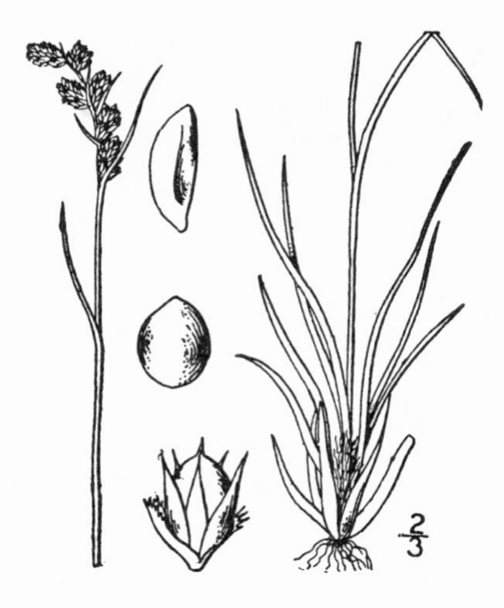

7. Luzula campéstris (L.) DC.

Common Wood-rush. Fig. 911.

Juncus campestre L. Sp. Pl. 329. 1753.
Luzula campestris DC. Fl. Fr. **3**: 161. 1805.
Luzula comosa Meyer, Syn. Luzul. 21. 1823.
Juncoides campestre Kuntze, Rev. Gen. Pl. 722. 1891.

Stems densely tufted, erect, 10–50 cm. high, 2–4-leaved. Leaf-blades flat, 2–7 mm. wide, tapering at the apex to a blunt almost gland-like point, sparingly webbed when young; inflorescense umbelloid, sometimes congested; its several branches unequal, each bearing an oblong to short-cylindric dense spike; lowest bract foliose often exceeding the inflorescence; bractlets ovate, acuminate, fimbriate at apex; perianth 2–3 mm. long, brown, the segments lanceolate-ovate, acuminate; capsule ovoid or broadly oblong; seed with an oblong body, about 1 mm. long, supported on a narrow white loosely cellular strophiole-like base about one-half as long.

In woodlands, Upper Sonoran to Arctic Zones; widely distributed almost throughout the United States and British America; also in Europe and Asia. Type locality: Europe. Blooming in early spring and consisting of several different races or varieties.

Family 17. MELANTHÀCEAE.

Bunch-flower Family.

Leafy-stemmed or scapose perennial herbs, with elongated or bulb-like rootstocks. Leaves alternate, sometimes all basal, broad or grass-like, parallel-veined, the veins often connected by transverse veinlets. Flowers perfect, polygamous, or dioecious, regular, in terminal spikes, racemes or panicles, or solitary. Perianth usually petaloid, of 6 separate or nearly separate, usually persistent segments. Stamens 6, borne on the base of the perianth-segments. Anthers small, 2-celled, oblong or ovate, or confluently 1-celled and cordate or reniform, mostly versatile and extrorsely or introrsely dehiscent. Ovary 3-celled, superior or rarely slightly inferior; ovules few or numerous in each cavity; anatropous or amphitropous. Styles 3, distinct or nearly so. Fruit a capsule, with septicidal dehiscence except in *Narthecium*. Seeds commonly tailed or appendaged; embryo small; endosperm usually copious.

About 40 genera and 145 species, widely distributed.

Anthers oblong or ovate, 2-celled.
  Anthers introrsely dehiscent.
    Flowers involucrate; capsule septicidal. — 1. *Tofieldia.*
    Flowers not involucrate; capsule loculicidal. — 2. *Narthecium.*
  Anthers extrorsely dehiscent. — 3. *Xerophyllum.*
Anthers cordate or reniform, confluently 1-celled.
  Plants from a coated bulb, glabrous; leaves grass-like.
    Flowers nodding; perianth-segments glandless. — 4. *Stenanthium.*
    Flowers erect; perianth-segments with a prominent gland. — 5. *Zygadenus.*
  Plants from a thick rootstock, pubescent; leaves broad. — 6. *Veratrum.*

1. TOFIÈLDIA Huds. Fl. Angl. ed. 2, 157. 1778.

Perennial herbs, with short erect or horizontal rootstocks, and slender erect stems, leafless above or nearly so. Leaves linear, somewhat 2-ranked and equitant, clustered at the base. Flowers small, green or white, in a close terminal centrifugal panicle or raceme. Pedicels bracted at base, solitary or clustered. Perianth-segments oblong or obovate, subequal, persistent, glandless, subtended by 3 scarious somewhat united bractlets. Stamens 6; filaments filiform; anthers ovate sometimes cordate, marginally dehiscent. Ovary sessile, 3-lobed at the summit; styles 3, short, recurved. Capsule 3-lobed, 3-beaked, septicidally dehiscent to the base, many-seeded. Seeds unappendaged or with tail-like appendages. [Name in honor of Mr. Tofield, an English correspondent of Hudson.]

A genus of 12 species, native of the North Temperate Zone, also the Andes of South America. Type species, *Tofieldia palustris* Huds.

1. Tofieldia occidentàlis S. Wats.
Western Tofieldia. Fig. 912.

Tofieldia occidentalis S. Wats. Proc. Am. Acad. **14**: 283. 1879.

Stems viscid with short-stalked or nearly sessile blackish glands, 2–6 dm. high, scape-like with 2–4 leaves toward the base. Basal leaves narrowly linear, mostly 8–20 cm. long and 3–4 mm. wide, acute or obtuse; inflorescence a narrow panicle or raceme, pedicels becoming 6–10 mm. long, commonly clustered in 3's; perianth light yellow, the outer segments narrowly obovate, 4–5 mm. long, the inner slightly narrower and longer; capsule 5–7 mm. long, tipped with the slender recurved styles; seeds 6–8 in each cell, angular-ovate, with a very loose spongy white testa and a slender appendage at the outer end about equalling the seed.

About cool springs and mountain meadows, Transition and Canadian Zones; southern Alaska to Mendocino County, and the southern Sierra Nevada, California. Type locality: near cold springs, Red Mountain, Mendocino County, California.

2. NARTHÈCIUM Juss. Gen. 47. 1789.

[ABAMA Adans. Fam. Pl. 2: 47, 511. 1763.]

Perennial herbs, with creeping rootstocks, fibrous roots, erect simple slender stems, and linear grass-like equitant basal leaves, those of the stem short and distant. Flowers in a terminal raceme, small, greenish yellow, perfect. Pedicels bracted at base and with or without a small bractlet. Perianth-segments persistent, linear or linear-lanceolate, obscurely 3–5-nerved, glandless. Stamens 6; filaments subulate, woolly; anthers linear-oblong, erect, introrse. Ovary sessile; style very short or none; stigma slightly 3-lobed. Capsule oblong, loculicidally dehiscent, many-seeded. Seeds linear, ascending from near the base, with a long bristle-like tail at each end. [Name an anagram of *Anthericum*.]

A genus of four closely allied species, the others in the Atlantic States, Europe, and eastern Asia, respectively. Type species, *Anthericum ossifragum* L.

1. Narthecium califórnicum Baker.
California Bog-asphodel. Fig. 913.

Narthecium ossifragum occidentale A. Gray, Proc. Am. Acad. 7: 391. 1867.
Narthecium californicum Baker, Journ. Linn. Soc. **15**: 351. 1876.
Abama californica Heller, Cat. N. Am. Pl. 3. 1898.
Abama occidentalis Heller, Muhlenbergia 1: 47. 1904.

Stems slender, 3–6 dm. high, glabrous. Basal leaves 10–25 cm. long, 3–4 mm. broad, usually 7-nerved, acute; stem leaves usually 3 or 4, remote, raceme loose, becoming 10–25 cm. long; pedicels 6–12 mm. long, perianth-segments, narrowly linear-lanceolate, 6–8 mm. long, yellow; filaments densely white-wooly to near the top, about two-thirds the length of the perianth-segments; capsule well exserted, tapering above to a slender beak, fully half as long as the body; seeds 10–15 in each cell, stramineous, 2 mm. long, or with the tails 9 mm. long.

About springs and bogs in mountain meadows, Transition Zone; Siskiyou Mountains, southern Oregon to Mendocino and Fresno Counties. Type locality: swamps, Red Mountain, Humboldt County, California.

3. XEROPHÝLLUM Michx. Fl. Bor. Am. 1: 210. 1803.

Tall perennial herbs, with thick short woody rootstocks, and stout simple leafy stems. Leaves narrowly linear, dry, rough-margined, numerous and longer toward the base. Flowers numerous, white, in a large dense terminal raceme, the lower first expanding. Perianth-segments withering-persistent, oblong or ovate, 5–7-nerved, spreading, glandless. Stamens 6, filaments subulate, glabrous; anthers oblong. Ovary sessile, 3-grooved; styles 3, filiform, reflexed or recurved, stigmatic along the inner side; ovules 2–4 in each cell. Capsule ovoid, 3-grooved, loculicidally and sometimes also septicidally dehiscent. Seeds 5, oblong, without or with only minute appendages. [Greek, in allusion to its xerophytic foliage.]

An American genus of 2 or 3 species. *Xerophyllum asphodeloides* (L.) Nutt., the type species, is restricted to the pine barrens of the Atlantic States.

1. Xerophyllum tènax (Pursh) Nutt.
Western Turkey-beard. Fig. 914.

Helonias tenax Pursh, Fl. Am. Sept. 1: 243, *pl.* 9. 1814.
Xerophyllum tenax Nutt. Gen. 1: 235. 1818.

Stems slender, 2–15 dm. high, with scattered ascending leaves dilated at base. Basal leaves, 5–8 dm. long, 3–6 mm. wide, rather rigid, flat above and somewhat carinate, pale green; raceme dense, becoming 3–6 dm. long, the lower bracts foliaceous and serrulate, the upper scarious and often upon the lower part of the pedicel; pedicels 3–5 cm. long, very slender, erect in fruit; flowers fragrant; perianth-segments 8–10 mm. long, scarcely equalling the stamens; styles 4 mm. long, exceeding the ovary; capsule about 6 mm. long, broadly ovate, acute, loculicidally 3-valved; seeds narrowly oblong, 4 mm. long.

Open dry ridges or borders of mountain meadows, Transition and Upper Sonoran Zones; Yellowstone Park to British Columbia, southward through the Pacific States to Plumas and Monterey Counties, California. Indian Basket Grass. Type locality: Collins Creek, Idaho.

Xerophyllum douglásii S. Wats. Proc. Am. Acad. **14**: 284. 1879. Distinguished by the somewhat smaller flowers and the slightly included stamens. A little known and doubtfully distinct species, collected by Douglas in western North America, but the locality not known.

4. STENÁNTHIUM Kunth, Enum. 4: 189. 1842.

Erect glabrous bulbous herbs, with few mainly basal narrow leaves, and racemose or paniculate perfect, greenish, brownish or purplish flowers. Perianth-segments subequal, withering-persistent, narrowly lanceolate, with reflexed tips, at length involute, without gland and distinct claw. Stamens 6, free, included; anthers reniform, confluently 1-celled. Ovary ovoid, superior; styles 3. Capsule ovoid-oblong, 3-beaked, septicidal to the base, wholly superior. Seeds about 4 in each cell, oblong, winged. [Greek, in allusion to the narrow perianth-segments.]

A genus of five species, four North American and one Asiatic (Sakhalin Island). Type species, *S. angustifolium* (Pursh) Kunth.

1. Stenanthium occidentàle A. Gray. Western Stenanthium. Fig. 915.

Stenanthium occidentale A. Gray, Proc. Am. Acad. 8: 405. 1872.
Stenanthella occidentalis Rydb. Bull. Torrey Club 27: 531. 1900.

Stems slender, erect, 3–6 dm. high, sparingly leafy. Leaves mainly basal, linear to narrowly oblanceolate, 15–30 cm. long, 6–20 mm. wide, acute, those on the stem narrower and shorter; raceme terminal, erect, simple or sometimes compound below; bracts scarious, lanceolate, 6–10 mm. long; pedicels slender, slightly exceeding the bracts; flowers drooping, narrowly campanulate, 12–15 mm. long, greenish- or brownish-purple; perianth-segments narrowly lanceolate, with recurved tips; capsule erect in fruit, 12–16 mm. long, attenuate into the elongated slender styles; seeds narrowly oblong, flat and winged, 3–4 mm. long.

Mossy stream banks, Canadian to Arctic Zones; British Columbia and Alberta to Montana and Washington south to Trinity Mountains, California. Type locality: Cascade Mountains, Oregon.

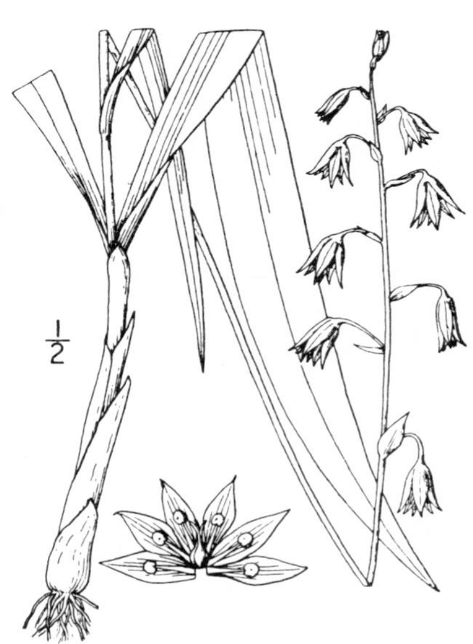

5. ZYGADÈNUS Michx. Fl. Bor. Am. 1: 213. 1803.

Glabrous or obscurely scabrous perennial herbs, with membranous-coated bulbs, leafy stems, and rather large greenish or yellowish white flowers in terminal racemes. Leaves linear, mainly basal. Flowers perfect or polygamous. Perianth withering-persistent, adnate to the lower part of the ovary or entirely free, its segments bearing 2 or (in ours) a single obcordate or obovate gland. Stamens distinct from the perianth-segments. Capsule 3-celled, the cavities dehiscent to the base. Seeds numerous. [Name Greek, meaning yoke and gland, in reference to the pair of glands in some species.]

About 18 species, distributed over the cooler parts of North America and northern Asia. Type species, *Zygadenus glaberrimus* Michx.

Ovary partly inferior; glands obcordate. — 1. *Z. elegans.*
Ovary wholly superior; glands obovate or semi-orbicular.
  Stamens half as long as the perianth-segments; flowers all perfect. — 2. *Z. fremontii.*
  Stamens about equalling or exceeding the perianth-segments; lower flowers staminate.
    Perianth-segments acute or acuminate; flowers mostly paniculate.
      Stamens shorter than the perianth-segments. — 3. *Z. exaltatus.*
      Stamens longer than the perianth-segments. — 4. *Z. paniculatus.*
    Perianth-segments obtuse; flowers normally racemose.
      Fruiting pedicels widely divergent.
        Perianth-segments about 4.5 mm. long. — 5. *Z. micranthus.*
        Perianth-segments about 7 mm. long. — 6. *Z. brevibracteatus.*
      Fruiting pedicels ascending.
        Petals long-clawed and subcordate at base; sepals short-clawed; margin of gland thin and ill-defined. — 7. *Z. intermedius.*
        Petals and sepals both long-clawed; margin of gland thick and well-defined. — 8. *Z. venenosus.*

1. **Zygadenus élegans** Pursh. Glaucous Zygadene. Fig. 916.

Zygadenus elegans Pursh, Fl. Am. Sept. 1: 241. 1814.
Zygadenus longus Greene, Pittonia 4: 241. 1901.
Anticlea elegans Rydb. Bull. Torrey Club 30: 273. 1903.
Anticlea longa Heller, Muhlenbergia 6: 12. 1910.

Plant glaucous, with a slender stem 2–10 dm. high; bulb long-ovoid, 3 cm. long, its coats membranous. Leaves 4–14 mm. wide, keeled, the lower 10–30 cm. long, the upper much shorter; bracts lanceolate, 1 cm. long or more, green or purplish; inflorescence a simple or compound raceme, open, its branches ascending; flowers greenish or yellowish white; perianth-segments about 10 mm. long, oval or obovate, obtuse, bearing a large obcordate gland just above the short claw; capsule oblong, 20–25 mm. long, exceeding the erect withered perianth-segments.

Moist places, mainly Canadian Zone; Alaska to Colorado, Utah, Nevada, and Oregon. In the Pacific States it is mostly in eastern Washington and northeastern Oregon, but occurs in the Olympic Mountains. Type locality: Big Blackfoot River, Montana.

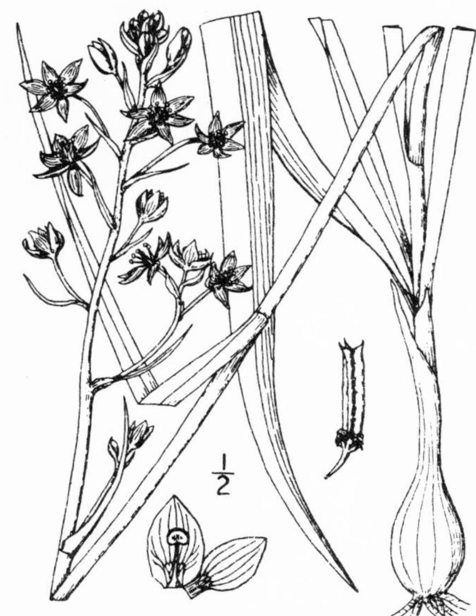

2. **Zygadenus fremóntii** Torr. Fremont's Zygadene. Star Lily. Fig. 917.

Anticlea fremontii Torr. Pacif. R. Rep. 4: 144. 1857.
Zygadenus douglasii Torr. Pacif. R. Rep. 7: 20. 1857.
Zygadenus fremontii Torr.; S. Wats. Bot. King Expl. 343. 1871.
Zygadenus speciosus Dougl.; Greene, Man. Bot. Bay Reg. 315. 1894.

Plants 5–8 dm. high; bulbs oblong-ovoid, 2–3 cm. long, with dark membranous coats. Basal leaves usually falcate, 3–5 dm. long, 1–2 cm. wide, the cauline few and much reduced; inflorescence a simple or more commonly compound raceme, often 30 cm. long; perianth-segments 10–14 mm. long, ovate, obtuse, sepals short-clawed, the petals long-clawed; gland greenish yellow, toothed on its upper margin; stamens about half as long as the perianth-segments; capsule 15–30 mm. long; seeds 10–20 in each cell, about 5 mm. long.

Hillsides, especially among bushes, Upper Sonoran Zone; California Coast Ranges, from Mendocino to San Diego Counties. Type locality: Santa Cruz, California.

Zygadenus fremontii minor (Greene) Jepson, Fl. W. Mid. Calif. 122. 1901. (*Zygadenus speciosus minor* Greene, Man. Bay Reg. 315. 1894.) Smaller than the species; inflorescence a few-flowered (usually 3–5) corymbose raceme. Open moist fields near the coast, Upper Sonoran and Transition Zones; San Francisco to Monterey.

3. **Zygadenus exaltàtus** Eastw. Giant Zygadene. Fig. 918.

Zygadenus exaltatus Eastw. Bot. Gaz. 41: 283. 1906.
Toxicoscordion exaltatum Heller, Muhlenbergia 6: 83. 1910.

Plants stout, 6–8 dm. tall; bulbs oblong-ovoid, 10 cm. long, 5 cm. thick. Basal leaves 6 dm. or more long, 2 cm. wide, the lower cauline sheathing at base, the upper not sheathing; inflorescence paniculate, 2–3 dm. long, the lower branches usually only staminate, perianth-segments 7–8 mm. long, the sepals sessile, elliptic, obtuse, the petals distinctly clawed; gland toothed on the upper margin; filaments dilated toward the base, 5 mm. long; capsule 2 cm. long.

Wooded hillsides, Upper Sonoran and Transition Zones; western slope of the Sierra Nevada, Butte to Tulare Counties, California. Type locality: Prindle's Ranch, above Mokelumne Hill, Calaveras County, California.

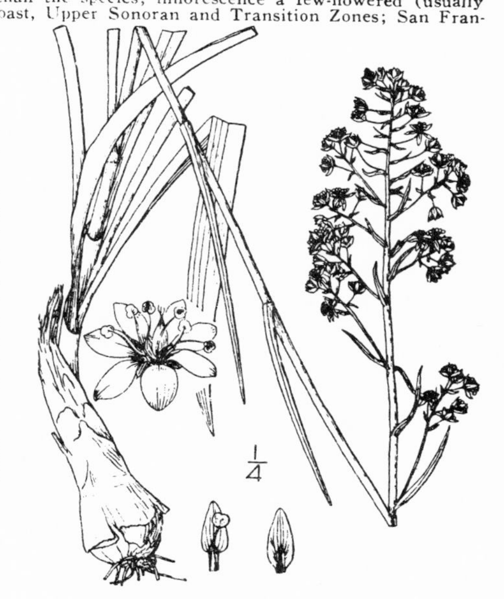

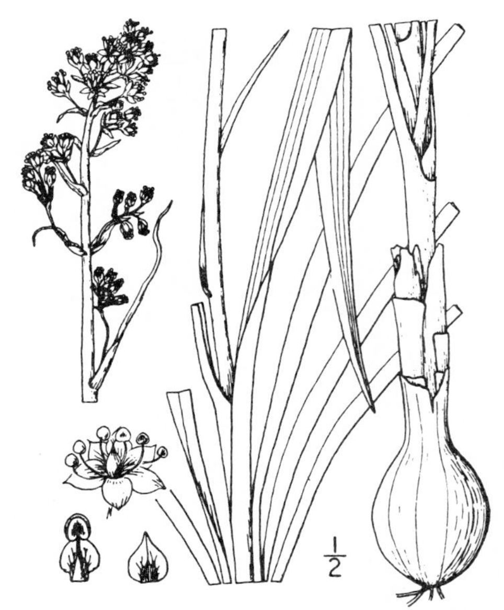

4. **Zygadenus paniculàtus** (Nutt.) S. Wats.
Panicled Zygadene. Fig. 919.

Helonias paniculatus Nutt. Journ. Acad. Philad. **7**: 57. 1834.
Zygadenus paniculatus S. Wats. King Expl. 344. 1871.
Toxicoscordion paniculatum Rydb. Bull. Torrey Club **30**: 272. 1903.

Plants 3–7 dm. tall, rather stout; bulb ovoid, about 4 cm. long. Basal leaves 3–5 dm. long, 8–12 mm. wide, the cauline sheathing at base; inflorescence rather a dense panicle, often 15 cm. long, the lower branches elongated, the others shorter and bearing several closely crowded flowers; bracts lax and scarious; perianth-segments 4 mm. long, acute or acuminate, deltoid, the sepals sessile, the petals subcordate and distinctly clawed; stamens exceeding the perianth-segments; anthers slightly over 1 mm. long; capsule ellipsoidal, 1 cm. long; seeds 4–5 mm. long.

Open grassy places or rocky slopes, mainly Transition Zone; Great Basin region, ranging from eastern Washington to Modoc County, California, eastward to Wyoming, Utah, and Nevada.

5. **Zygadenus micránthus** Eastw.
Small-flowered Zygadene. Fig. 920.

Zygadenus micranthus Eastw. Bull. Torrey Club **30**: 483. 1903.
Toxicoscordion micranthum Heller, Muhlenbergia **6**: 83. 1910.

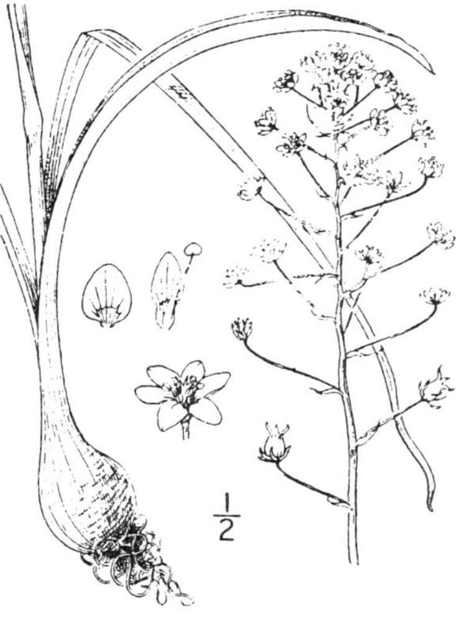

Plants slender, 1–5 dm. tall, scabrous; bulbs ovoid, 15 mm. thick. Basal leaves two-thirds to nearly as long as the stem, falcate, 4–7 mm. wide, scabrous; inflorescence a simple raceme or with a few branches at base, 10–25 cm. long; bracts long-attenuate; pedicels widely spreading and rigid, becoming 3–5 cm. long; perianth-segments 4–5 mm. long, the sepals ovate, obtuse, sessile, the petals oblong, narrowed at base to a short claw; gland toothed on the upper margin; stamens scarcely .equalling the perianth-segments; anthers 1.5 mm. long; capsule becoming 12 mm. long, slightly narrowed at the apex; seeds 4.5 mm. long.

Wooded slopes in dry situations, Upper Sonoran and Transition Zones; Siskiyou Mountains, southern Oregon, through the Coast Ranges east of the fog belt to Mendocino County, California. Type locality: road between Cahto and the Eel River, Mendocino County, California.

6. **Zygadenus brevibracteàtus** (Jones) Hall.
Desert Zygadene. Fig. 921.

Zygadenus fremonti brevibracteatus M. E. Jones, Contr. West. Bot. **12**: 78. 1908.
Zygadenus brevibracteatus Hall, Univ. Calif. Pub. Bot. **6**: 165. 1915.

Plants erect, 4–6 dm. tall, glabrous; bulbs ovoid, 2–2.5 cm. thick. Basal leaves 15–20 cm. long, 6–8 mm. wide, scabrous only on the margins; cauline leaves not sheathing at base; bracts scarious, 5 mm. long, abruptly short-acuminate; inflorescence a simple or compound open raceme, 10–25 cm. long; pedicles horizontally spreading, rigid, 1–3 cm. long; perianth-segments 7 mm. long, the sepals broadly elliptic, narrowed to a very short claw, the petals narrowly elliptic, narrowed to a distinct claw; stamens 5 mm. long; capsule oblong, 15 mm. long.

Gravelly or rocky situations, Lower Sonoran Zone; Mojave and Colorado Deserts, southern California. Type locality: Victor, California, on the Mojave Desert, altitude about 800 meters.

7. **Zygadenus intermèdius** Rydb.

Rocky Mountain Zygadene. Fig. 922.

Zygadenus intermedius Rydb. Bull. Torrey Club **27**: 535. 1900.

Toxicoscordion intermedium Rydb. Bull. Torrey Club **30**: 272. 1903.

Plants rather stout, 3–6 dm. tall, glabrous; bulb elongated ovoid, 3 cm. long, 1.5 cm. thick. Basal leaves 2–3 dm. long, 5–9 mm. wide, scabrous on the midrib and margins; cauline leaves prominently sheathed at base; inflorescence a simple raceme or with one or two basal branches, becoming 10–20 cm. long; bracts membranous, long-attenuate; pedicels slender, erect in fruit, 1–2 cm. long; perianth-segments pale yellow, 6–8 mm. long, obtuse, the sepals narrowed or rounded at the base into a short claw, the petals oblong, subcordate at base and with a claw 1 mm. long; gland with the upper margin toothed, thin and ill-defined, capsule cylindric, 12 mm. long; seeds 4.5 mm. long.

Wet places, Transition and Boreal Zones; Montana and Utah to western Idaho and eastern Washington. Type locality: Nez Perces County, Idaho.

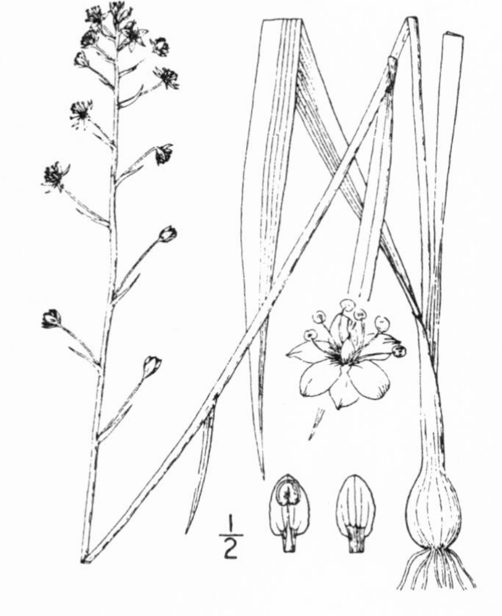

8. **Zygadenus venenòsus** S. Wats.

Deadly Zygadene or Death Camass. Fig. 923.

Zygadenus venenosus S. Wats. Proc. Am. Acad. **14**: 279. 1879.

Toxicoscordion venenosum Rydb. Bull. Torrey Club **30**: 272. 1903.

Toxicoscordion arenicola Heller, Muhlenbergia **2**: 182. 1906.

Plants 2.5–6 dm. tall, glabrous; bulbs broadly ovoid, 20 mm. thick or less. Basal leaves about two-thirds the length of the stem, 12 mm. wide or commonly less, faintly scabrous on the margins; cauline leaves with only the lowest scariously sheathed; inflorescence 10–20 cm. long, usually a simple raceme, occasionally with one or more basal branches; bracts membranous, setaceously attenuate, about 1 cm. long; pedicels ascending or erect in fruit, about 2 cm. long; perianth-segments narrowly ovate, subcordate at base, all distinctly and about equally clawed; gland with a thick, well-defined upper margin; capsule cylindric 12–15 mm. long.

Usually in wet meadows, but also on rocky slopes, Upper Sonoran to Canadian Zones; British Columbia and Idaho to Utah, Nevada and California, where it extends in both the Sierra Nevada and Coast Ranges to the Cuyamaca Mountains. Type locality: Monterey County, California.

7. **VERÀTRUM** L. Sp. Pl. 1044. 1753.

Tall perennial herbs, with thick short poisonous rootstocks, the leaves mostly broad, clasping, strongly veined and plaited, the stem and inflorescence pubescent. Flowers greenish, white or purple, rather large, polygamous or monoecious, on short stout pedicels in large terminal panicles. Perianth-segments 6, glandless or nearly so, not clawed, sometimes adnate to the base of the ovary. Stamens opposite the perianth-segments and free from them, short, mostly curved. Anthers cordate, their sacs confluent ovoid, styles 3, persistent. Capsule 3-lobed, 3-celled, the cavities several-seeded. [Ancient name of Hellebore.]

About 12 species, natives of the northern hemisphere. Besides the following, 3 other species occur in the eastern United States. Type species, *Veratrum album* L.

Ovary glabrous or nearly so.
- Perianth-segments serrulate or entire.
  - Panicle drooping; flowers green. — 1. *V. viride.*
  - Panicle erect; flowers white.
    - Terminal branch of panicle much elongated; perianth-segments narrow (2–3 mm.). — 2. *V. caudatum.*
    - Terminal branch of panicle rather short; perianth-segments broader. — 3. *V. californicum.*
- Perianth-segments fimbriate. — 4. *V. fimbriatum.*

Ovary densely woolly; perianth-segments erose or slightly fringed. — 5. *V. insolitum.*

1. Veratrum víride Ait.

Green False Hellebore. Fig 924.

Veratrum viride Ait. Hort. Kew. **3**: 422. 1789.
Veratrum lobelianum eschscholtzianum Roem. & Sch. Syst. **7**: 1555. 1829.
Veratrum eschscholtzii A. Gray, Ann. Lyc. N. Y. **4**: 119. 1837.
Veratrum eschscholtzianum Rydb.; Heller, Muhlenbergia **1**: 120. 1905.

Stems stout, 5–20 dm. tall, very leafy; rootstock erect, 5–8 cm. long, 2.5–5 cm. thick, with numerous fibrous-fleshy roots. Leaves acute, the lower broadly oval or elliptic, 15–30 cm. long, 7–15 cm. wide, sessile or short-petioled, sheathing, the upper successively narrower and acuminate; panicle 2–6 dm. long, the lower branches spreading and more or less drooping; pedicels 2–6 mm. long, mostly shorter than the bracts; flowers yellowish green; perianth-segments oblong or oblanceolate, acute, 8–10 mm. long, woolly-pubescent, ciliolate-serrulate; stamens about half the length of the perianth-segments; capsule 20–25 mm. long, 8–12 mm. thick, many-seeded; seeds 4–10 mm. long.

Swamps and wet woods, Hudsonian Zone; Alaska to Mount Hood, Oregon, and Montana, eastward to the Atlantic Coast. Type locality: North America.

2. Veratrum caudàtum Heller.

Tailed False Hellebore. Fig. 925.

Veratrum caudatum Heller, Bull. Torrey Club **26**: 588. 1899.

Stems tall, 2–2.5 m. high, leafy, short woolly-pubescent. Leaves glabrous, ciliate on the margins, elliptic or elliptic-lanceolate, 3–4 dm. long, 1.5–2 dm. wide, gradually becoming smaller and narrower above; panicle 4 dm. or more in length, branched below, the main rachis prolonged above into a tail-like extension 2 dm. long or more; bracts lanceolate acuminate, extending to about the middle of the perianth; pedicels 3–5 mm. long; perianth-segments linear-lanceolate, 12–15 mm. long, 2–3 mm. wide, acuminate, serrulate toward the apex, white with a green and pubescent base; stamens 2–3 mm. long.

Wet meadows, Humid Transition Zone; western Washington. Type locality: Montesano, Chehalis County.

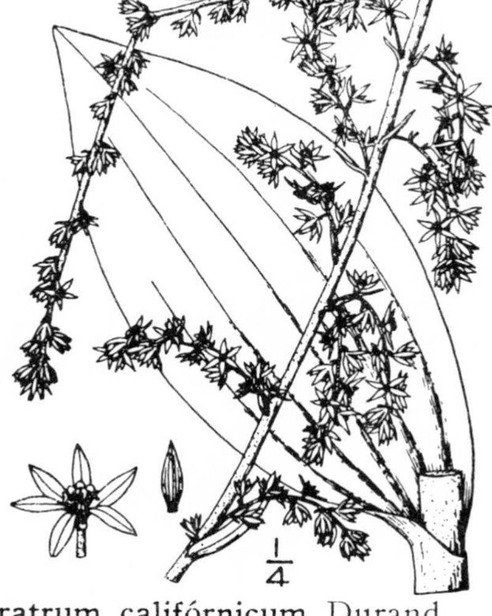

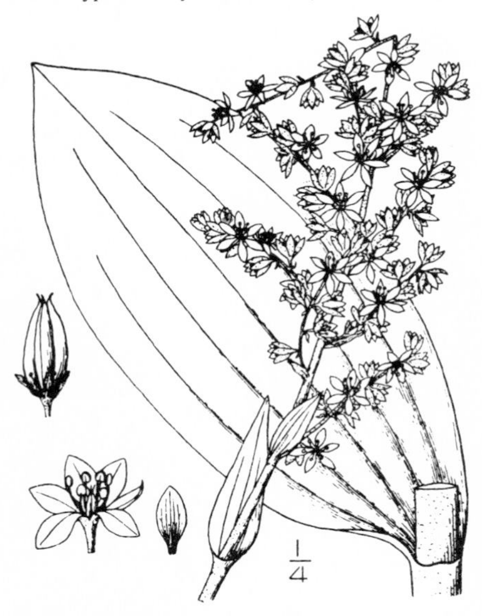

3. Veratrum califórnicum Durand.

California False Hellebore. Fig. 926.

Veratrum californicum Durand, Journ. Philad. Acad. II. **3**: 103. 1855.

Stems stout, 1–2 m. tall, tomentose above, leafy. Leaves sparsely and inconspicuously pubescent, ciliate on the margin, the lower broadly oval to ovate, acute or obtuse, 2–5 dm. long, 1–2 dm. wide, the upper becoming narrower, finely lanceolate and acuminate; inflorescence tomentose, 2–5 dm. long, the branches of the panicle spreading mostly simple, the lower usually sterile; bracts membranous, ovate-lanceolate, equalling or exceeding the pedicels, these stout rarely over 3 mm. long; perianth-segments ovate, mostly obtuse, 12–15 mm. long, 5–8 mm. wide, white with greenish veins and a green V-shaped spot at base; stamens 6–8 mm. long; capsule 25–35 mm. long; seeds 12–15 in each cavity, broadly winged, 10–12 mm. long.

Wet meadows, Arid Transition Zone; Washington to western Montana, Utah and northern Lower California. Type locality: near Nevada City, Nevada County, California. This is one of the characteristic plants of the mountain meadows of the Sierra Nevada and Cascade Mountains.

4. **Veratrum fimbriàtum** A. Gray. Fringed False Hellebore. Fig. 927.

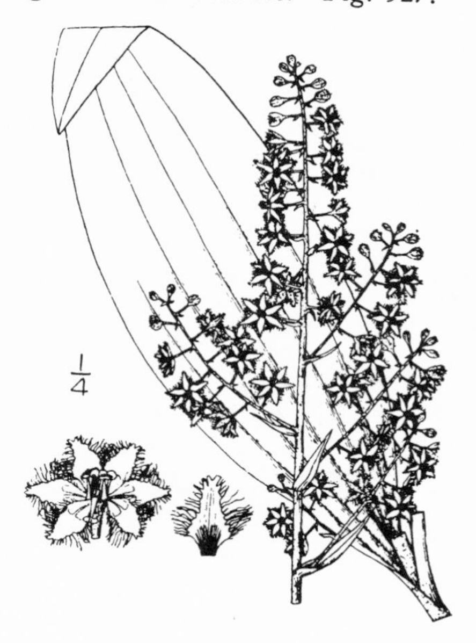

Veratrum fimbriatum A. Gray, Proc. Am. Acad. 7: 391. 1868.

Stems 1–2 m. tall. Leaves lanceolate, acute or acuminate, narrowed to the base, 2–5 dm. long, sparsely and inconspicuously pubescent; inflorescence densely tomentose, usually 3–5 dm. long, the branches spreading, simple or the lowest sometimes branched; bracts ovate to ovate lanceolate shorter than the pedicels; pedicels divergent, stout, 6–12 mm. long; perianth-segments rhombic-ovate, 6–10 mm. long, irregularly fimbriate to near the broad base, with 2 oblong subglandular spots extending from the base to the middle of the segment and separated by a narrow furrow; stamens 4 mm. long; styles slender: capsule obovate, depressed and somewhat emarginate at the apex, 8 mm. long; seeds 5–7 in each cavity, 6 mm. long, scarcely margined.

Frequent in wet meadows, Humid Transition Zone; along the coast of northern Sonoma and Mendocino Counties, California. Type locality: plains west of the redwood belt in Mendocino County.

5. **Veratrum insolìtum** Jepson. Siskiyou False Hellebore. Fig. 928.

Veratrum insolitum Jepson, Fl. Calif. 1: 266. 1921.

Stems 1–1.5 m. high. Leaves 15–20 cm. long, elliptic or the uppermost lanceolate; panicle 30–50 cm. long, lanate-tomentose; perianth-segments 6–8 mm long, white, obovate, irregularly ciliate or erose, or the inner shallowly fimbriate; ovary densely wholly.

Usually in thickets, Humid Transition Zone; Illinois River, southwestern Oregon to Del Norte County, California. Type locality: Adams Station to Shelly Creek, California.

Family 18. **LILIÀCEAE.**

Lily Family.

Scapose or leafy-stemmed herbs from bulbs, corms, or rarely with rootstocks, or sometimes with a woody caudex and arboreal, the leaves various. Flowers solitary or clustered, regular and mostly perfect. Perianth of 6 separate or nearly separate segments, or these more or less united into a tube. Stamens 6, hypogynous, or borne on the perianth; anthers 2-celled, introrse, or sometimes extrorse. Ovary superior, 3-celled; ovules few or numerous in each cavity, anatropous or amphitropous; styles united; stigma 3-lobed or capitate. Fruit a loculicidal (septicidal in *Calochortus*) capsule. Embryo in copious endosperm.

About 120 genera and 1250 species, widely distributed.

Herbaceous plants, arising from bulbs or corms.
Perianth-segments all similar, petaloid; capsule loculicidal.
Flowers in umbels terminating a scape.
Plants alliaceous (with the odor and taste of onions). 1. *Allium.*
Plants not alliaceous.
Filaments separate, not united to form a corona.
Perianth 6-parted to near the base, the tube scarcely evident; segments closely 2–3-nerved.
Filaments subulate or dilated, without appendages; pedicels not jointed; flowers greenish-white. 2. *Muilla.*
Filaments surrounded at base by cup-shaped winged appendages; pedicels jointed; flowers yellow. 3. *Bloomeria.*
Perianth-tube evident; campanulate to tubular; segments 1-nerved.
Anthers versatile, 6.
Perianth-segments rotate-spreading, the tube narrow; flowers yellow. 4. *Calliprora.*
Perianth-segments little or not at all spreading.
Perianth-tube open campanulate; flowers white. 5. *Hesperoscordum.*
Perianth-tube funnel-form or tubular and more or less inflated; flowers blue. 6. *Triteleia.*
Anthers basifixed, 3 except in *Dichelostemma capitata.*
Capsule sessile; flowers blue or purple, rarely red.
Perianth-tube funnel-form, not at all inflated, the segments widely spreading. 7. *Brodiaea.*
Perianth-tube tubular and more or less inflated, the segments erect or but little spreading. 8. *Dichelostemma.*
Capsule stipitate; perianth-tube red, the lobes green. 9. *Brevoortia.*
Filaments united to form a tubular corona, with erect bifid segments between the anthers. 10. *Androstephium.*
Flowers not umbellate, or if so, not terminating an evident scape.
Perianth united below into a more or less evident tube.
Plants acaulescent, from a short rootstock; flowers arising from subterranean pedicels. 11. *Leucocrinum.*
Plants caulescent from bulbs; flowers in simple or compound racemes.
Perianth funnel-form; stems leafy. 12. *Hesperocallis.*
Perianth salver-form; leaves basal. 13. *Odontostomum.*
Perianth-segments separate, not united below into a tube.
Bracts scarious; perianth persistent.
Perianth scarious, not twisted over the fruit. 14. *Schoenolirion.*
Perianth not scarious, often twisted over the fruit.
Flowers scattered in a widely branching panicle. 15. *Chlorogalum.*
Flowers in a simple raceme terminating a scape. 16. *Camassia.*
Bracts, if any, foliaceous; perianth deciduous (except *Lloydia*).
Plants from scaly bulbs, the scales appearing as bulblets in many species of *Fritillaria.*
Anthers versatile; perianth-segments oblanceolate with a linear nectariferous groove. 17. *Lilium.*
Anthers fixed near the base, slightly or not at all versatile; nectary a shallow pit. 18. *Fritillaria.*
Plants from tunicated bulbs; anthers strictly basifixed.
Leaves 2, broad, basal or nearly so; flowers nodding. 19. *Erythronium.*
Leaves several, grass-like; flowers not nodding. 20. *Lloydia.*
Perianth-segments unlike, the inner petaloid, the outer smaller and often sepal-like; capsule septicidal. 21. *Calochortus.*
Plants with a woody caudex, arborescent or appearing acaulescent.
Flowers large, perfect; perianth-segments many-nerved; ovules numerous in each cell.
Style stout; stigma 6-notched, openly perforate. 22. *Yucca.*
Style filiform; stigma capitate, long-papillate. 23. *Hesperoyucca.*
Flowers polygamo-dioecious; perianth-segments 1-nerved; ovules 2 or 3 in each cell. 24. *Nolina.*

1. ÁLLIUM L. Sp. Pl. 294. 1753.

Scapes from a tunicated bulb or a coated corm, with mostly narrowly linear basal leaves. Herbage with the characteristic odor and taste of onions. Flowers in a terminal simple umbel, subtended by 2–4 membranous, separate or united bracts. Pedicels slender, not jointed. Perianth persistent, its segments distinct or united at the base. Stamens inserted on the bases of the perianth-segments; filaments filiform or dilated, sometimes toothed. Style filiform, jointed. Capsule obovate-globose, obtusely 3-lobed, often crested, loculicidally dehiscent. Seeds obovoid, wrinkled, black. [Latin for garlic.]

A genus of about 300 species, widely distributed; especially developed in California, where the majority of the North American species are to be found. Type species, *Allium ampeloprasum* L.

Leaves hollow, terete; pedicels much shorter than the flowers. 1. *A. sibiricum.*
Leaves solid, flat or channelled, or when terete solitary.
Bulbs with fibrous coats. 5. *A geyeri.*
Bulbs with membranous coats.
Bulbs oblong, cespitose, with a more or less developed rhizome, the coats without reticulations.
Stamens exserted.
Umbels nodding; perianth-segments acutish. 2. *A. cernuum.*
Umbels erect; perianth-segments long-acuminate. 3. *A. validum.*
Stamens included. 4. *A. haematochiton.*
Bulbs ovoid or subglobose, mostly solitary, not rhizomateous, the center usually solid and therefore a coated corm.
Outer bulb-coats without distinct reticulations (with fine transversely oblong reticulations in *tribracteatum.*

Leaves flat or channelled, 2 or more, very rarely reduced to 1.
Ovary obscurely or not at all crested; scapes terete, or somewhat 2-edged.
Leaves shorter than or little exceeding the scapes.
Perianth-segments lanceolate, acuminate. 6. *A. douglasii.*
Perianth-segments ovate-oblong. 7. *A madidum.*
Leaves much exceeding the very short scapes.
Bracts 3, or sometimes 2; bulb-coats with fine hexagonal reticulations. 8. *A. tribracteatum.*
Bracts 2; bulb-coats without reticulations.
Perianth-segments obtuse. 9. *A. parvum.*
Perianth-segments acuminate. 10. *A. macrum.*
Ovary distinctly crested; scapes compressed and 2-edged or 2-winged (nearly terete in *A. cusickii*); leaves broad and falcate.
Bracts 2.
Ovary with 3 crests, the edges of the crests forming 2 ridges on the back of each ovary-cell.
Bracts ovate-lanceolate, acuminate.
Perianth-segments narrowly attenuate from the base, spreading above. 11. *A. falcifolium.*
Perianth-segments ovate-lanceolate, acute, the tips not spreading. 12. *A. breweri.*
Bracts round-ovate, obtuse, mucronate. 13. *A. watsoni.*
Ovary 6-crested.
Stamens about equalling the narrow perianth-segments. 14. *A. anceps.*
Stamens well-included.
Ovary with prominent crests; stamens half the length of the segments. 15. *A. pleianthum.*
Ovary with low dorsal crests; stamens three-fourths the length of the segments.
Scapes scarcely compressed. 16. *C. cusickii.*
Scapes distinctly compressed.
Scapes with the winged margins entire. 17. *A. tolmiei.*
Scapes with the winged margins crenulate. 18. *A. crenulatum.*
Bracts 3–5; perianth-segments attenuate. 19. *A. platycaule.*
Leaves solitary; ovary 6-crested.
Stigma entire or obscurely lobed.
Crests of the ovary low and entire; perianth-segments obtuse; leaves flat.
Bracts obtuse; anthers purple. 20. *A. burlewii.*
Bracts acuminate; anthers yellow. 21. *A. obtusum.*
Crests of the ovary prominent, more or less toothed or lacerate; leaves terete.
Perianth-segments 10–12 mm. long, little exceeding the stamens. 22. *A. atrorubens.*
Perianth-segments 13–17 mm. long, twice the length of the stamens. 23. *A. parishii.*
Stigma 3-lobed, the lobes linear.
Stamens included; bracts 2, rarely 3 in *A. parryi.*
Stamens half the length of the segments. 24. *A. fimbriatum.*
Stamens nearly equalling the segments. 25. *A. parryi.*
Stamens exserted; bracts usually 4. 26. *A. sanbornii.*
Outer bulb-coats with evident reticulations.
Reticulations of coats oblong or quadrate, not contorted.
Walls of the reticulations similar.
Reticulations prominent, the meshes forming distinct pits; ovary 3-crested.
Inner perianth-segments serrate. 27. *A. acuminatum.*
Inner perianth-segments entire. 28. *A. lacunosum.*
Reticulations fine, the compressed transversely oblong meshes not forming pits; ovary with 6 crests. 29. *A. nevii.*
Walls of the reticulations dissimilar, the vertical thickened and sinuous.
Ovary prominently 6-crested.
Perianth-segments broadly ovate-lanceolate, acute, thin and spreading. 30. *A. campanulatum.*
Perianth-segments lanceolate or narrowly ovate-lanceolate, acuminate, not thin and spreading.
Crests of the ovary widely spreading. 31. *A. austinae.*
Crests of the ovary thin, erect. 32. *A. bisceptrum.*
Ovary with low broad obscure crests. 33. *A. lemmoni.*
Reticulations of bulb-coats much contorted.
Reticulations not in serrated rows, the walls irregularly contorted or wavy.
Ovary obscurely ridged near the summit. 34. *A. fibrillum.*
Ovary prominently 6-crested. 35. *A. nevadense.*
Reticulations of transverse linear V-shaped meshes, appearing in serrated rows.
Bracts scarcely 1 cm. long, rounded at the apex, mucronate.
Ovary without crests; perianth-segments not becoming lax. 36. *A. hickmani.*
Ovary distinctly 6-crested; perianth-segments becoming lax. 37. *A. amplectens.*
Bracts 15–25 mm. long, acuminate.
Ovary with 6 central crests at the base of the style; perianth-segments erect or spreading only near the tip.
Perianth-segments broadly oblong-ovate, obtuse. 38. *A. dichlamydeum.*
Perianth-segments ovate-lanceolate, acute or acuminate.
Umbels congested and subcapitate; perianth-segments 8 mm. long. 39. *A. serratum.*
Umbels open, the pedicels widely spreading.
Outer and inner perianth-segments dissimilar, the outer broader and more spreading.
Inner perianth-segments crisped. 40. *A. crispum.*
Inner perianth-segments not crisped. 41. *A. peninsulare.*
Outer and inner perianth-segments similar in shape, the inner conspicuously serrate. 42. *A. bolanderi.*
Ovary not crested; perianth open-campanulate; style much longer than the capsule.
Scapes arising directly from the bulbs.
Perianth becoming hyaline. 43. *A. hyalinum.*
Perianth not becoming hyaline. 44. *A praecox.*
Scapes arising from lateral off-shoots of the coated corm. 45. *A. unifolium.*

1. **Allium sibíricum** L. Wild Chives. Fig. 929.

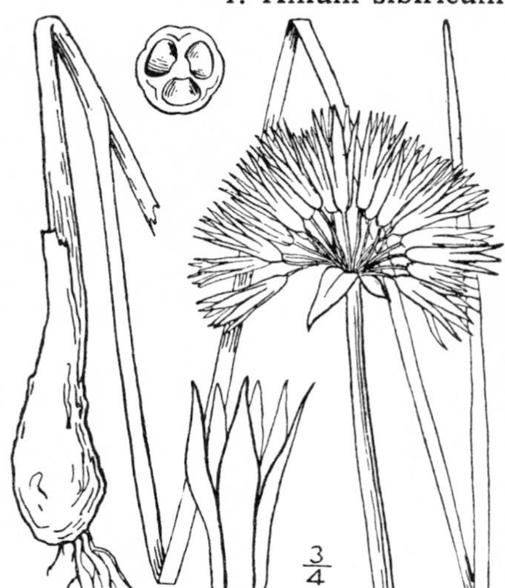

Allium sibiricum L. Mant. 2: 562. 1771.

Bulbs narrowly ovoid, clustered, 25 mm. long or less, their coats membranous, usually white, without reticulations. Scape rather stout, 2–6 dm. high, bearing below the middle 1 or 2 elongated linear terete hollow leaves 1–2 mm. in diameter, or the leaves all basal; bracts 2, ovate, acute, veiny; umbel many-flowered, capitate, the pedicels 2–6 mm. long; perianth-segments 8–12 mm. long, lanceolate, acuminate, pale rose-color; stamens about half as long as the perianth; filaments subulate; ovules 2 in each cavity; capsule obtusely 3-lobed, half as long as the perianth.

In moist soil, Arid Transition Zone; Alaska to Newfoundland, New York, Wyoming and Oregon. In the Pacific States it is found in Oregon and Washington, east of the Cascade Mountains. Type locality: Siberia.

2. **Allium cérnuum** Roth. Nodding Onion. Fig. 930.

Allium cernuum Roth; Roem. Arch. 1[3]: 40. 1789.
Allium recurvatum Rydb. Mem. N. Y. Bot. Gard. 1: 94. 1900.

Bulbs narrowly ovoid with a long neck, 3–6 cm. long, usually clustered on a short rootstock, the coats membranous without distinct reticulations. Scape slender, slightly rigid, 3–6 dm. high; leaves linear, channelled or nearly flat, 2–4 mm. wide, mostly shorter than the scape; bracts 2, short, deciduous; umbel many-flowered, nodding in flower; pedicels filiform 15–30 mm. long; flowers white or rose, campanulate; perianth-segments ovate-oblong, acute or obtusish, 4–6 mm. long; stamens longer than the perianth; filaments nearly filiform; ovules 2 in each cavity; capsule 3-lobed, each valve bearing 2 short processes near the summit.

On rocky slopes and dry hillsides, Transition Zone; British Columbia to northern Oregon, and in the Rocky Mountains to Arizona, east to New York and Virginia. Type locality: not indicated.

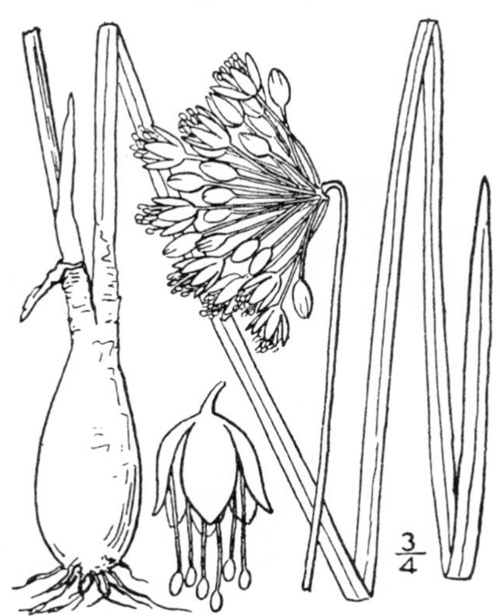

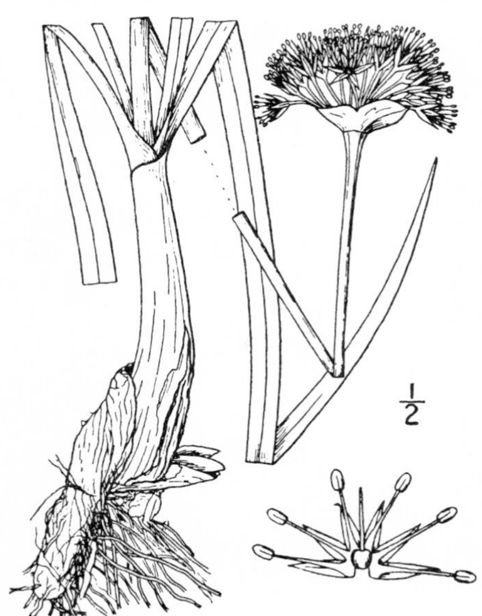

3. **Allium válidum** S. Wats. Swamp Onion. Fig. 931.

Allium validum S. Wats. Bot. King Expl. 350. 1871.

Bulb oblong-ovoid, 3–5 cm. long, solitary or clustered, arising obliquely from a horizontal rootstock, the membranous coats white or sometimes tinged with red, marked with fine vertical reticulations. Scape 5–7 dm. high, rather stout, slightly compressed and 2-edged; leaves several, flat or slightly carinate, nearly equalling the scape, 5–15 mm. wide; bracts 2–4, united at the base, acute or short acuminate; umbels many-flowered, rather dense, the pedicels 10–15 mm. long; perianth-segments rose to nearly white, lanceolate, narrowly acuminate, 6–8 mm. long; stamens about equalling or usually longer than the perianth; filaments subulate; style 7–8 mm. long; capsule subglobose, 5 mm. long, not crested.

Alpine meadows of the Pacific region, Canadian and Hudsonian Zones; Mount Rainier to the southern Sierra Nevada and in the Coast Ranges to Lake County, California, also western Nevada. Type locality: Mono Pass, California.

4. **Allium haematochìton** S. Wats. Red-skinned Onion. Fig. 932.

Allium haematochiton S. Wats. Proc. Am. Acad. **14**: 227. 1879.
Allium marvinii Davidson, Bull. S. Calif. Acad. **20**: 49. 1921.

Bulb narrowly oblong-ovoid, 2–3 cm. long, with a long neck, solitary or usually clustered, the membranous coats deep reddish purple. Scape 10–40 cm. high, slightly 2-edged; leaves several, flat and rather thick, about equalling the scape, 1–4 mm. wide; bracts 2–4, connate, short, obtuse; umbel erect; pedicels 8–30, slender, 15 mm. long or less; perianth-segments 6–8 mm. long, ovate-lanceolate, acute, deep purple or rose with a dark midrib; stamens and style included; filaments dilated at base; ovary truncate with very short rounded crests; capsule obcordate, 4 mm. long.

On dry hillsides or along water courses, Upper Sonoran Zone; San Luis Obispo to northern Lower California. Type locality: San Luis Obispo, California.

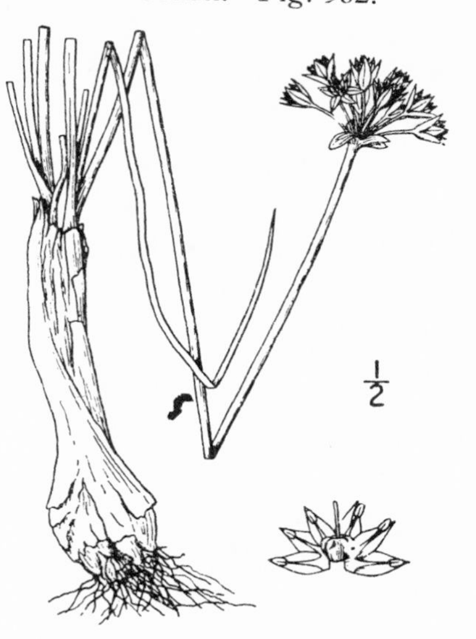

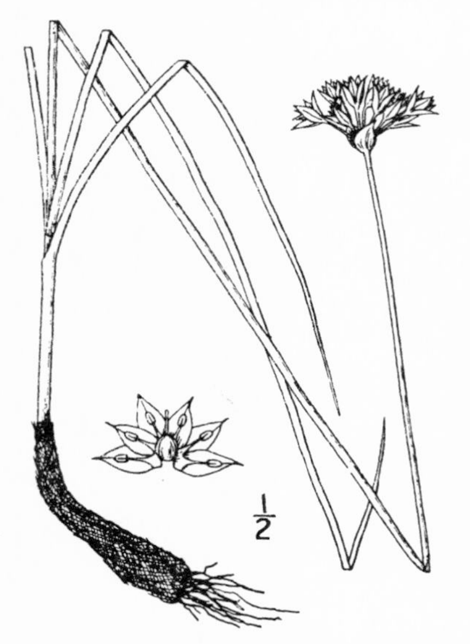

5. **Allium geỳeri** S. Wats. Geyer's Onion. Fig. 933.

Allium geyeri S. Wats. Proc. Am. Acad. **14**: 227. 1879.
Allium geyeri tenerum M. E. Jones, Contr. West. Bot. **10**: 28. 1902.

Bulbs ovoid or oblong-ovoid, 2–4 cm. long, the coats fibrous. Scape 20–35 cm. high, slightly 2-edged; leaves several, about 2 mm. wide, shorter than the scapes; umbels erect, 2–25-flowered; bracts 2, united on one side, ovate, short-acuminate; pedicels 10–25 mm. long; perianth-segments 8–9 mm. long, ovate, acute or short-acuminate, rose with a deeper colored midrib, strongly nerved and rigid in fruit; stamens about two-thirds the length of the perianth; filaments slightly dilated below; styles 5 mm. long; capsule obovoid, 4 mm. long, obscurely crested.

Water courses and mountain meadows, Arid Transition Zone; British Columbia to the Blue Mountains, eastern Oregon, east to Wyoming. Type locality: "on stony banks of the Kooskooskia River," Idaho. The "omoir" of the Nez Perce Indians.

6. **Allium douglásii** Hook. Douglas's Onion. Fig. 934.

Allium douglasii Hook. Fl. Bor. Am. **2**: 184. 1838.
Allium hendersonii Robins. & Seaton, Bot. Gaz. **18**: 237. 1893.

Bulb ovoid, 15–20 mm. long, the coats without reticulations. Scape rather stout, 20–25 cm. high, rather stout, terete; leaves 2, much shorter than the scape, 6–8 mm. wide, falcate and long-attenuate; bracts 2, ovate, acuminate; umbels many-flowered; pedicels slender, 12–18 mm. long; perianth-segments ovate-lanceolate, acuminate, 8–9 mm. long, rose, strongly carinate; stamens about equalling or slightly exceeding the perianth; filaments slightly dilated below; style slender about 7 mm. long; ovary not crested; capsule narrowly obovoid, 4 mm. long.

In springy places, Arid Transition Zone; eastern Washington and adjacent Idaho. Type locality: near Kettle Falls, Stevens County, Washington.

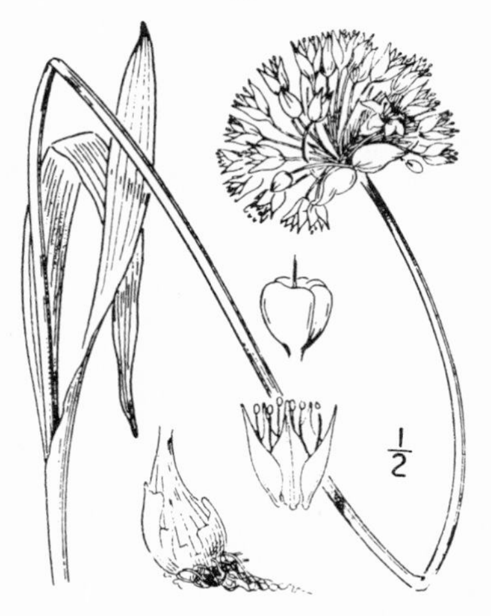

Allium scillioìdes Dougl.; S. Wats. Proc. Am. Acad. **14**: 229. 1879. An obscure species described from fragmentary specimens collected by Douglas at "Priest's Rapids," Columbia River. The original description is as follows: "Perianth-segments oblong-lanceolate, obtuse, 3 lines (6 mm.) long, a half longer than the stamens; ovary not at all crested."

7. **Allium mádidum** S. Wats. Swamp Onion. Fig. 935.

Allium madidum S. Wats. Proc. Am. Acad. **14**: 228. 1879.

Bulb ovoid, 10–15 mm. long, the outer coats without definite reticulations, the inner white. Scape rather stout, slightly angled, 10–20 cm. high; leaves 2, thick and channelled, 7–12 cm. long, 3–6 mm. wide; bracts 2, ovate to lanceolate, acute 8–10 mm. long; umbel usually many-flowered; pedicels 8–12 mm. long; perianth-segments white or pale rose, ovate-oblong, acute, 8 mm. long; stamens almost equalling the segments; style exceeding the segments; ovary with 2 fleshy ridges at the summit of each cell.

In wet places, Arid Transition and Canadian Zones; Blue Mountains, eastern Oregon. Type locality: Union County, Oregon, "in small streams or wet places on high ground."

8. **Allium tribracteàtum** Torr. Three-bracted Onion. Fig. 936.

Allium tribracteatum Torr. Pacif. R. Rep. **4**: 148. 1857.
Allium ambiguum M. E. Jones, Contr. West. Bot. **8**: 18. 1902.

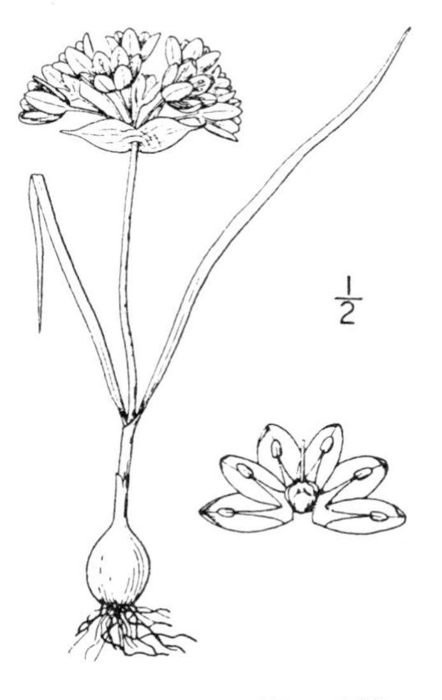

Bulb ovoid, 15–20 mm. long, the coats thin, with distinct transversely oblong reticulations. Scapes 3–6 cm. high; leaves about twice as long as the scape, 2–3 mm. wide; bracts 2 or 3, abruptly acuminate; umbels many-flowered; pedicels 4–6 mm. long; perianth-segments pale rose with a prominent dark greenish midvein, 6 mm. long, narrowly oblong-lanceolate, acutish, not gibbous at base; stamens two-thirds to nearly as long as the perianth; anthers purple; capsule crested with three pairs of low dorsal ridges.

Dry rocky ridges, Arid Transition and Canadian Zones; southern Oregon, south in the Sierra Nevada to Mono County. Type locality: Duffield's Ranch, Sierra Nevada, California.

9. **Allium párvum** Kell. Dwarf Onion. Fig. 937.

Allium parvum Kell. Proc. Calif. Acad. **3**: 54. 1863.
Allium tribracteatum andersoni S. Wats. Bot. King Expl. 353. 1871.

Bulb ovoid, about 15 mm. long, the coats without definite reticulations. Scape 3–5 cm. high, one-half to two-thirds under ground, terete; leaves 2, about twice as long as the scape, 2–3 mm. wide; bracts 2, round-ovate, very obtuse or rounded at apex, with or without a very short acumination; umbel 8–20-flowered; pedicels 5–8 mm. long; perianth rose-purple, turbinate at base, not at all gibbous, the segments with a prominent dark midvein, narrowly oblong-lanceolate, obtuse, 6–8 mm. long, erect; stamens about two-thirds the length of the perianth-segments, the filaments slightly dilated below; capsule obscurely crested.

Stony soil, Arid Transition Zone; Sierra Nevada, from Lassen Peak, Shasta County, to the Kern River, also in the Washoe Mountains, western Nevada. Type locality: Washoe, Nevada.

10. **Allium mácrum** S. Wats. Rock Onion. Fig. 938.

Allium macrum S. Wats. Proc. Am. Acad. **14**: 233. 1879.

Bulb ovoid, about 15 mm. long, the coats without definite reticulations. Scapes 2–5 cm. high, terete or slightly angled; leaves 2, much longer than the scape, 2–3 mm. wide, slightly falcate; bracts 2, ovate, abruptly acute; umbel spreading, many-flowered; pedicels slender, 5–10 mm. long; perianth-segments white or pale rose, veined with purple, 4–6 mm. long, narrowly lanceolate, acuminate, scarcely longer than the stamens and style; filaments slightly dilated below; capsule-cells bordered above by a thick obtuse ridge.

Barren rocky soil, Hudsonian Zone; Blue Mountains of eastern Washington and Oregon. Type locality: Union County, Oregon.

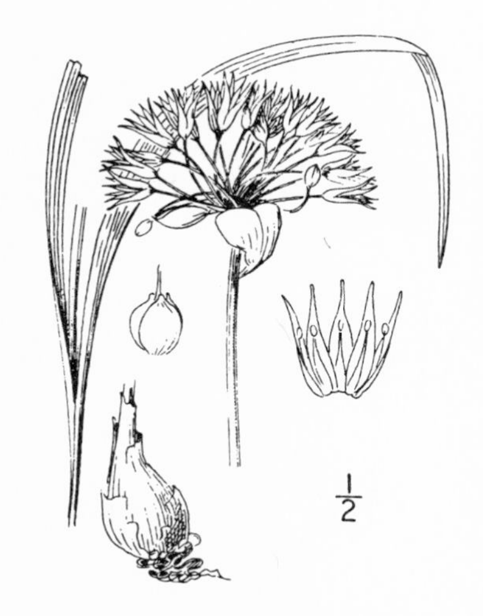

11. **Allium falcifòlium** Hook. & Arn. Scythe-leaved Onion. Fig. 939.

Allium falcifolium Hook. & Arn. Bot. Beechey 400. 1841.

Bulbs ovoid, about 15 mm. long, the outer coats reddish brown, without reticulations. Scape 5–10 cm. high, flattened and distinctly 2-edged, 2.5–5 mm. wide; leaves 2, flat and falcate, thick, much longer than the scape, 6–8 mm. wide; bracts 2, ovate-lanceolate, acuminate, 18–20 mm. long, 12–14-nerved; umbels rather crowded, 10–30-flowered; pedicels 6–12 mm. long; perianth deep rose, 10–14 mm. long, the segments narrowly lanceolate, attenuate and spreading above, minutely glandular-serrulate; stamens scarcely over half the length of the perianth-segments; stigma entire; capsule acute, with 3 central narrow entire crests.

Usually on serpentine outcrops, Upper Sonoran and Transition Zones; Coast Ranges from the Siskiyou Mountains, Oregon to Sonoma County, California. Type locality: California.

12. **Allium brèweri** S. Wats. Brewer's Onion. Fig. 940.

Allium breweri S. Wats. Proc. Am. Acad. **14**: 233. 1879.
Allium falcifolium breweri M. E. Jones, Contr. West. Bot. **10**: 83. 1902.

Bulb ovoid, 15–20 mm. long, the outer coats without reticulations, light brown, the inner reddish. Scape 6–8 cm. long, usually only 25–35 mm. above the ground, flattened and 2-winged; leaves 2, basal, flat, thick, much exceeding the scape, 4–7 mm. wide, falcate; umbles densely flowered; pedicels stout, 15 mm. long; perianth-segments deep rose-purple, oblong-lanceolate, obtuse or acutish, becoming acuminate in age, entire or sometimes minutely glandular-serrate, 10–12 mm. long; stamens two-thirds the length of the segments; capsule with 3 broadly triangular, sometimes lobed apical crests.

Heavy or rocky soil, usually of basalt or serpentine origin, Upper Sonoran Zone; Coast Ranges of central California. Type locality: summit of Mount Diablo.

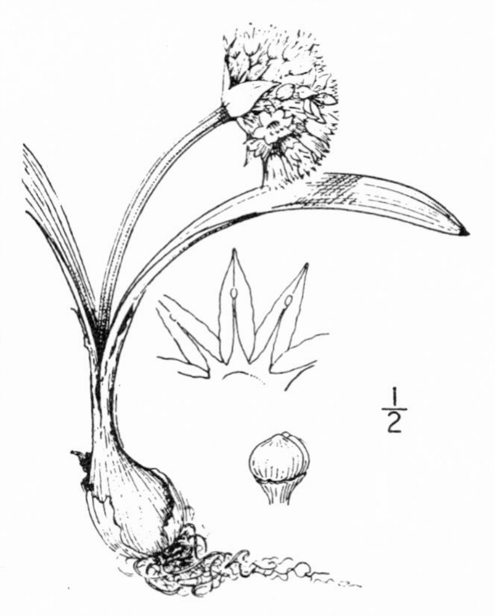

13. Allium wátsoni Howell. Watson's Onion. Fig. 941.

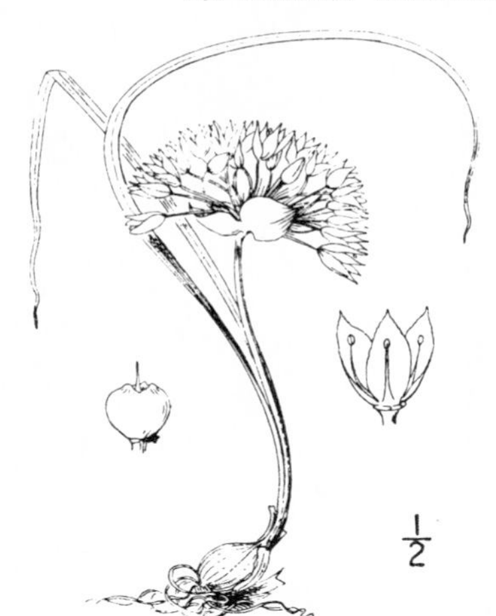

Allium watsoni Howell, Fl. N. W. Am. 1: 642. 1902.

Bulb ovoid, about 15 mm. long, the outer coats reddish brown, without reticulations, the inner white. Scape 3–7 cm. high, slightly compressed and narrowly 2-winged, 1.5–2 mm. wide; leaves 2, much exceeding the scape, 2–3 mm. wide; bracts 2, round-ovate, connate at base, obtuse at apex and abruptly short-acuminate or merely mucronate, 8–12 mm. long; umbels 10–20-flowered; pedicels 6–12 mm. long; perianth bright rose-purple, the segments 8–10 mm. long, ovate-lanceolate, broadly acuminate, erect, saccate at base, midvein prominent, margins more or less distinctly serrulate; stamens about two-thirds the length of the perianth-segments, the filaments dilated below; style 2–3 mm. long; stigma entire; ovary with 3 central entire crests.

Rocky slopes, Transition and Canadian Zones; Coast Ranges of central Oregon to the Siskiyou Mountains, northern California. Type locality: "coast mountains of middle Oregon."

14. Allium ánceps Kell. Kellogg's Onion. Fig. 942.

Allium anceps Kell. Proc. Calif. Acad. 2: 109. 1861.

Bulbs ovoid, about 15 mm. long, outer coats gray-brown, the inner very pale brown; reticulations not evident. Scape flattened, about 1 dm. high; leaves 2, usually much exceeding the scape, 4–5 mm. wide; bracts 2, connate below, broadly ovate, obtuse and mucronate, 7–9-nerved, rose-purple; umbel rather compact, 15–25-flowered, the pedicles erect or ascending, slender, 8–12 mm. long, enlarged at apex; perianth pale rose with a purple midvein, 8 mm. long, not gibbous at base, the segments narrowly linear-lanceolate, acuminate; stamens equalling or slightly exceeding the perianth-segments, the filaments flattened, dilated at base; anthers whitish, 1.5 mm. long; style 6 mm. long, the stigma obscurely 3-lobed; crests of the ovary 6, broad.

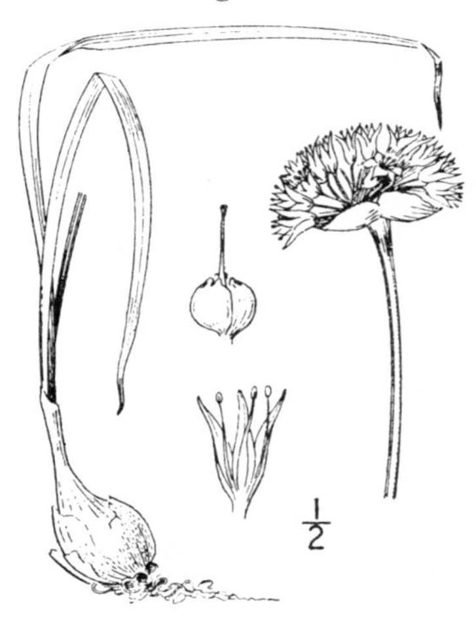

Foothills along the eastern base of the Sierra Nevada; frequent in the vicinity of Reno, Nevada; Upper Sonoran Zone. Type locality: Washoe, Nevada.

15. Allium pleiánthum S. Wats. Many-flowered Onion. Fig. 943.

Allium pleianthum S. Wats. Proc. Am. Acad. 14: 233. 1879.

Bulb ovoid, 2 cm. long, the outer coats grayish brown, without evident reticulations, the inner with rather conspicuous longitudinal veins. Scape 10–15 mm. high, flattened and 2-winged; leaves 2, basal, flat and falcate, longer than the scape; bracts 2, ovate-lanceolate, acuminate, 15 mm. long; umbels rather open; pedicels 15-20 mm. long; perianth 8–10 mm. long, gibbous at base, the segments lanceolate, acuminate; stamens scarcely over half the length of the segments, equalling the style; stigma entire; ovary and capsule prominently 6-crested.

Rocky ridges, Arid Transition Zone; Idaho and eastern Oregon. Type locality: Blue Mountains, Oregon.

16. Allium cusíckii S. Wats.
Cusick's Onion. Fig. 944.

Allium cusickii S. Wats. Proc. Am. Acad. **14**: 228. 1879.
Allium anceps aberrens M. E. Jones, Contr. West. Bot. **10**: 10. 1902.

Bulb ovoid, 15 mm. long and nearly as broad, the outer bulb-coats without evident reticulations, the inner usually whitish. Scape 8–10 cm. high, slightly compressed and angled, about 1.5 mm. thick; leaves flat, somewhat falcate, longer than the scape, 3 mm. wide; umbels many-flowered open; pedicels 12–16 mm. long, lanceolate, broadly acuminate; stamens and style about three-fourths the length of the segments; capsule with 6 small nipple-like dorsal crests.

On bare slopes in well drained soil, Arid Transition Zone; Blue Mountains, eastern Oregon, to western Idaho. Type locality: Union County, Oregon.

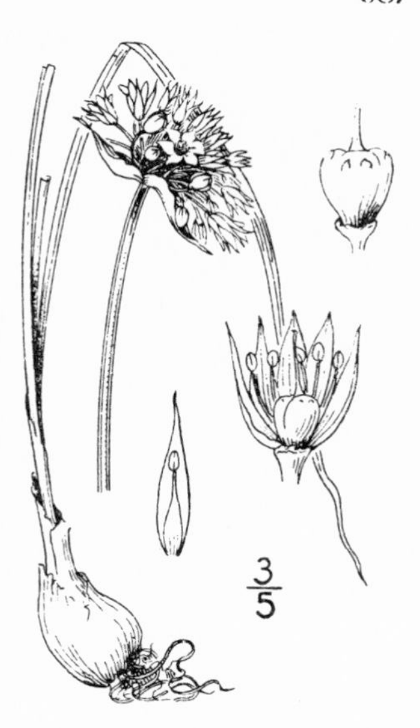

17. Allium tòlmiei Baker.
Tolmie's Onion. Fig. 945.

Allium douglasii var. *β* Hook. Fl. Bor. Am. **2**: 185. 1840.
Allium tolmiei Baker, Bot. Mag. II. **32**: under *pl. 6227*. 1876.
Allium platyphyllum Tidestrom, Torreya **16**: 242. 1916.

Bulb ovoid, the coats without reticulations. Scape 8–10 cm. high, compressed and 2-winged, 2.5–4 mm. wide above, narrower toward the base; leaves 2, exceeding the scapes, flat, but rather thick, more or less falcate, 5–12 mm. wide; bracts ovate-acuminate, 12–18 mm. long; umbels many-flowered; pedicels rather stout, 12–18 mm. long; perianth light rose-purple, the segments lanceolate, acute, 8 mm. long; stamens three-fourths the length of the segments; capsule with 6 low rather broad dorsal crests.

On rocky, especially volcanic ridges, Arid Transition Zone; eastern Oregon and southern Idaho to Utah. Type locality: "Snake Country."

18. Allium crenulàtum Wiegand.
Olympic Onion. Fig. 946.

Allium vancouverense Macoun, Cat. Can. Pl. **2**: 37. 1888, *nomen nudum*.
Allium crenulatum Wiegand, Bull. Torrey Club **26**: 135, *pl. 355*. 1899.

Bulb narrowly ovoid, 15 mm. long, propagating by fission, the outer bulb-coats gray, without reticulations, the inner white. Scape 5–8 cm. high, flattened, the winged margins more or less crenate; leaves 2, basal, flat, longer than the scape, 3–6 mm. wide; bracts 2, ovate-lanceolate, acuminate, 12–15 mm. long; umbels 12–18-flowered; pedicels rather stout, 10–15 mm. long; perianth-segments oblong-lanceolate, acutish, 8 mm. long, deep rose-purple; stamens about three-fourths the length of the segments; capsule with 6 low dorsal crests.

Stony ground, Canadian Zone; western British Columbia and Vancouver Island to the Olympic Mountains, Washington. Type locality: "loose gravel near the summit of the Olympic Mountains in the vicinity of the headwaters of the Quilcene River."

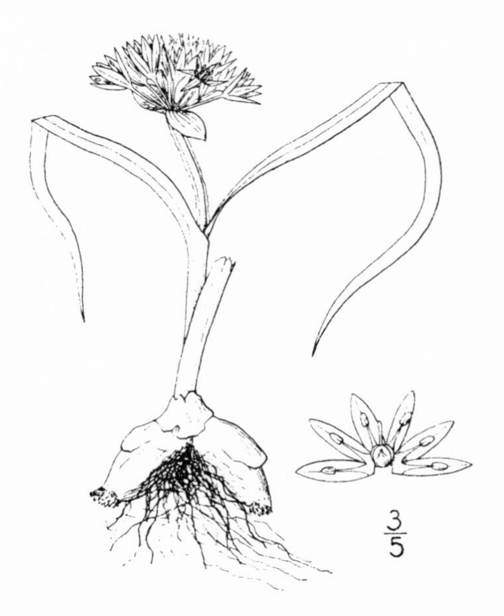

19. **Allium platycaùle** S. Wats. Broad-stemmed Onion. Fig. 947.

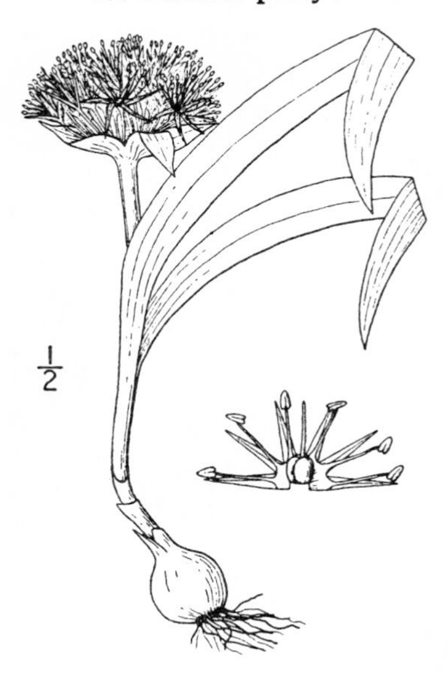

Allium platycaule S. Wats. Proc. Am. Acad. **14**: 234. 1879.

Bulb ovoid, 20–25 mm. long, the outer coats dark gray, without reticulations, the inner white. Scape 10–15 cm. high, flattened, 3–5 mm. wide; leaves usually 2, exceeding the scape, falcate, 10–20 mm. wide; bracts 3 or 4, connate at base, ovate-lanceolate, long-acuminate, 20 mm. long; umbels compact, 30–90-flowered, the pedicels 15–20 mm. long; perianth rose, 10–14 mm. long, not saccate at base, the segments very narrow, long-acuminate from the base; stamens equalling the perianth-segments; anthers purple, 1.5 mm. long; style 8–10 mm. long; ovary distinctly 3-lobed, without crests.

Sub-alpine meadows, Canadian Zone; Warner Mountains, Modoc County, California, and the Sierra Nevada from Plumas to Placer Counties. Type locality: high mountain valley of Plumas County.

20. **Allium burlèwii** Davidson. Burlew's Onion. Fig. 948.

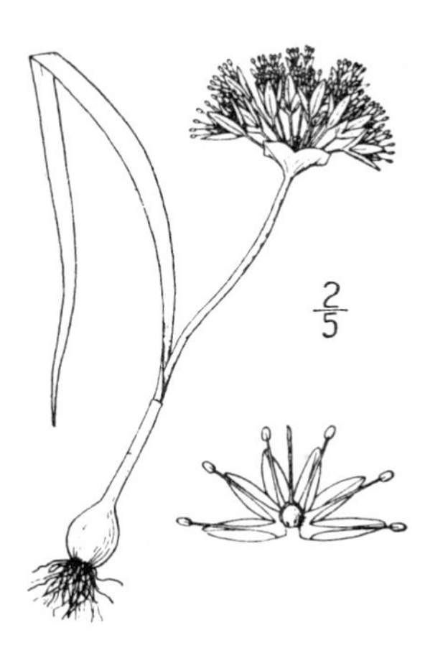

Allium burlewii Davidson, Bull. S. Calif. Acad. **15**: 17. 1916.

Bulb ovoid, about 15 mm. long, the outer coats gray-brown, without evident reticulations, or with obscure rectangular meshes. Scape 2–7 cm. high, somewhat flattened and 2-edged; leaf solitary, flat, well exceeding the scape, 3–6 mm. wide, bracts 3, ovate, abruptly short-acuminate or merely mucronate; umbels many-flowered; pedicels slender, 15–25 mm. long; perianth not saccate at base, the segments narrowly oblong-lanceolate, obtuse, 8–10 mm. long, rose purple with a conspicuous dark midvein; stamens equalling or slightly exceeding the segments; anthers purple; capsule with 6 low crests.

Granitic gravel, Canadian and Transition Zones; southern Sierra Nevada (Kern County) and Mount Pinos to the San Jacinto Mountains, California. Type locality: Mount San Antonio, San Gabriel Mountains.

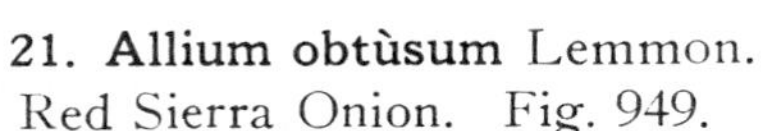

21. **Allium obtùsum** Lemmon. Red Sierra Onion. Fig. 949.

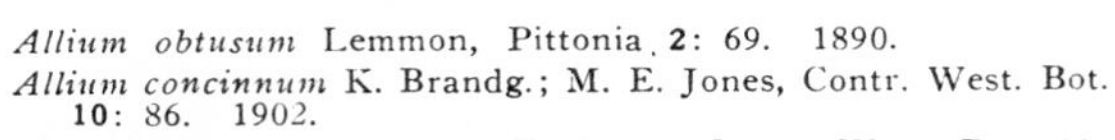

Allium obtusum Lemmon, Pittonia **2**: 69. 1890.

Allium concinnum K. Brandg.; M. E. Jones, Contr. West. Bot. **10**: 86. 1902.

Allium parvum brucae M. E. Jones, Contr. West. Bot. **10**: 12, *fig. 16.* 1902.

Bulb ovoid, 12–18 mm. long, the outer coats reddish brown, with fine rectangular reticulations. Scape 4–8 cm. high, terete or slightly compressed above; leaf solitary, flat, well exceeding the scape, 5–8 mm. broad; bracts 3, broadly ovate, acuminate, 15 mm. long; umbels many-flowered; pedicels 10-15 mm. long; perianth-segments dark rose-purple, oblong-ovate, obtuse, 8–10 mm. long; stamens about two-thirds the length of the segments; anthers yellow; capsule with 6 low crests.

In granitic or clay soils, Canadian and Arid Transition Zones; Sierra Nevada, California, from Plumas County to the Tehachapi Pass; also San Jacinto Mountains. Type locality: "sub-alpine region of Gold Lake, Plumas County."

22. **Allium atroruùbens** S. Wats. Dark Red Onion. Fig 950.

Allium atrorubens S. Wats. Bot. King Expl. 352. 1871.
A. decipiens M. E. Jones, Contr. West. Bot. **10**: 16. 1902.
A. inyonis M. E. Jones, Contr. West. Bot. **10**: 86. 1902.

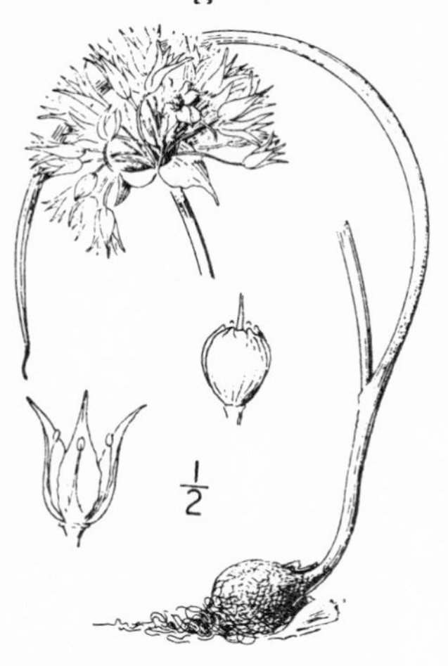

Bulb ovoid, about 15 mm. long, propagating by bulblets either sessile at the base of the bulb, or on short slender rhizomes, the coats reddish brown, without reticulations. Scape 6–12 cm. high, terete; leaf solitary, longer than the scape; umbel open, 20–25-flowered, the pedicels stout, 10–15 mm. long; perianth reddish purple, 10–12 mm. long, the segments spreading, lanceolate, acuminate; stamens very slender, about two-thirds the length of the segments; stigma entire; crests 6, prominent, 1.5 mm. high, acute, laciniately toothed.

On dry hillsides, Upper Sonoran Zone; northwestern Nevada to the eastern base of the Sierra Nevada, south to Owens Valley, Inyo County, California. Type locality: Unionville, Nevada.

23. **Allium paríshii** S. Wats. Parish's Onion. Fig. 951.

Allium parishii S. Wats. Proc. Am. Acad. **17**: 380. 1882.
Allium piersoni Jepson, Fl. Calif. **1**: 274. 1921.
Allium monticola Davidson, Bull. S. Calif. Acad. **20**: 51. 1921.

Bulb ovoid, with numerous reddish brown coats, without reticulations. Scape stout, 10–15 cm. high; leaf solitary, sheathing to above the ground, longer than the scape; bracts 2 or 3, broadly ovate, acuminate, 15–20 mm. long, strongly nerved; umbels few–many-flowered, the pedicels stout, 6–12 mm. long; perianth rose-purple, 13–17 mm. long, the segments lanceolate, acuminate; stamens half the length of the perianth-segments, the filaments lanceolate; stigma slightly lobed; ovary prominently 6-crested, the crests acutish, irregularly toothed.

Rocky soil, Upper Sonoran and Arid Transition Zones; San Gabriel and San Bernardino Mountains, especially on the desert slopes, extending to 9000 feet altitude. Type locality: Cushenberry Springs, desert slope of the San Bernardino Mountains, California.

24. **Allium fimbriàtum** S. Wats. Fringed Onion. Fig. 952.

Allium fimbriatum S. Wats. Proc. Am. Acad. **14**: 232. 1879.

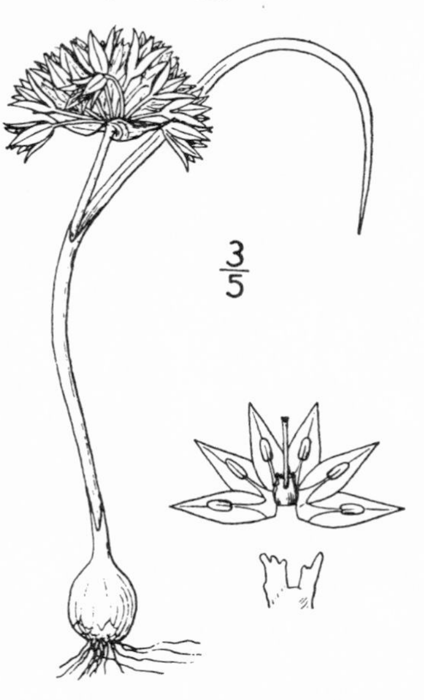

Bulb ovoid, 15 mm. long, the coats reddish brown, without reticulations. Scape 5–10 cm. long, terete, rather stout; leaf solitary, about equalling or exceeding the scape, 2–3 mm. wide, usually becoming revolute; bracts 2 or 3, 10–12 mm. long, setaceous-acuminate; umbel loose, the pedicels 12–20 mm. long, rather stout; perianth rose-purple, 8–10 mm. long, the outer segments lanceolate, acute or acuminate, the inner slightly narrower; stamens about half as long as the perianth-segments; style included, 4–5 mm. long, the stigma with 3 linear lobes; crests 6, prominent, 2 mm. high, fimbriate.

Usually in rocky soil, Upper and Lower Sonoran Zones; Santa Clara County south on the inner Coast Ranges and the desert slopes of the southern California mountains to the Mexican boundary. Type locality: Mojave River, southern California.

25. Allium párryi S. Wats. Parry's Onion. Fig. 953.

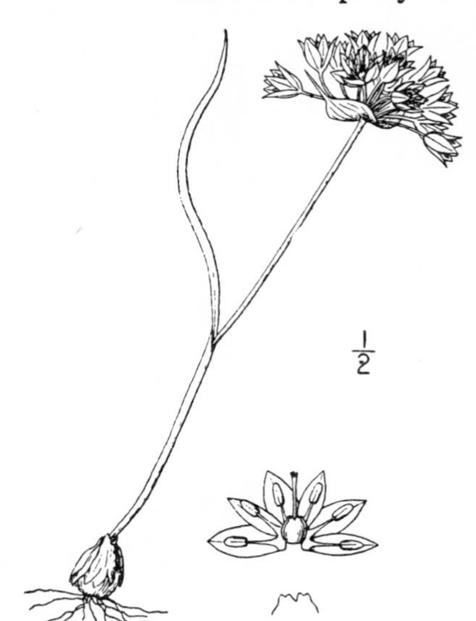

Allium parryi S. Wats. Proc. Am. Acad. **14**: 231. 1871.
Allium kessleri Davids. Bull. S. Calif. Acad. **20**: 49. 1921.

Bulb round-ovoid, scarcely over 1 cm. long, the coats reddish, without reticulations. Scape 7–20 cm. high, terete, rather slender; leaf solitary, the sheath extending above the ground, exceeding the scape, 2–3.5 mm. wide, becoming revolute; bracts 2 or 3, ovate, 10–12 mm. long, abruptly setaceously-acuminate; umbel loose, the pedicels 12–30, 8–15 mm. long, slender; perianth rose-purple, the outer segments lanceolate, acuminate, 6–8 mm. long, erect or the tips slightly spreading, the inner scarcely narrower; stamens equalling to three-fourths the length of the perianth-segments; stigma with 3 linear lobes; capsule with 6 prominent erect crests often 2 mm. high and emarginate or erose.

Usually in adobe soil, Upper Sonoran and Arid Transition Zones; Santa Clara and Kern Counties, California, south, especially on the desert slopes to the Mexican boundary. Type locality: San Bernardino Mountains, California.

26. Allium sanbórnii Wood.
Sanborn's Onion. Fig. 954.

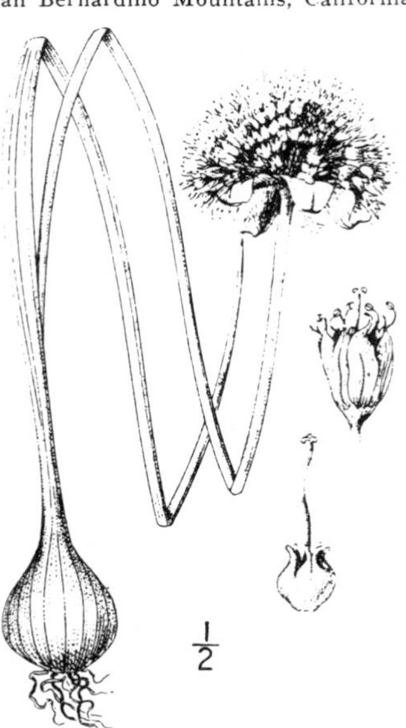

Allium sanbornii Wood, Proc. Acad. Philad. **1868**: 171. 1868.
Allium intactum Jepson, Fl. Calif. 1: 273. 1922.

Bulb ovoid, 15–20 mm. long, the outer coats usually reddish tinged, without reticulation, marked with a few longitudinal ribs and with obscure venation between the ribs. Scapes 2–4 dm. high, rather slender, terete; leaves solitary, much shorter than the scapes, 2–3 mm. wide, channelled or nearly terete; bracts 4, lanceolate, acuminate, about 15 mm. long; umbels rather crowded, 40–100-flowered, the pedicels filiform, about 15 mm. long; perianth pale rose, 6 mm. long, the segments ovate, acute; stamens slightly exceeding perianth-segments, the filaments filiform, slightly dilated at base; style 3 mm. long; stigma 3-lobed; crests of ovary prominent, entire or erose.

On shady hillsides, Arid Transition Zone; western slopes of the Sierra Nevada from Shasta to Mariposa Counties, California. Type locality: near Foster's Bar, Yuba County.

27. Allium acuminàtum Hook. Hooker's Onion. Fig. 955.

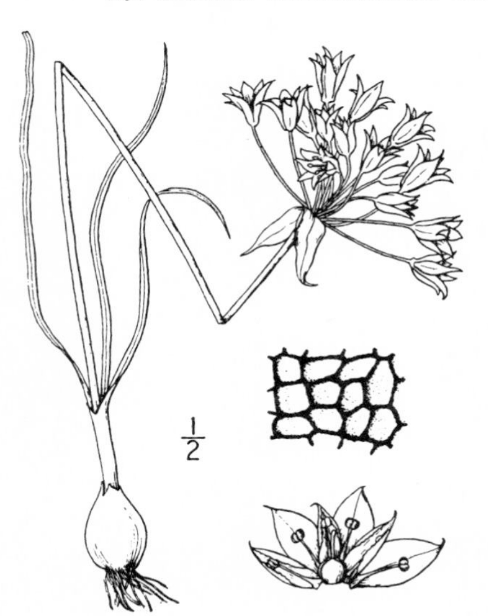

Allium acuminatum Hook. Fl. Bor. Am. **2**: 184. 1840.
Allium murrayanum Regel, Gartenfl. **1873**: 260. 1873.

Bulb ovoid, about 15 mm. long, the outer coats gray, the inner white; reticulations of distinct quadrate or hexagonal pitted meshes, the walls thick, not at all sinuous. Scape 1–3 dm. high, terete; leaves 2 or more, shorter than the scape, 2–3 mm. wide; bracts 2, ovate-lanceolate, long-acuminate, 20–25 mm. long; umbels open, 7–25-flowered, the pedicels 10–20 mm. long; perianth deep rose-purple, 10–15 mm. long, the outer segments lanceolate long-acuminate with recurved tips, entire, the inner shorter and serrulate; stamens two-thirds the length of the perianth-segments; crests of ovary 3, very small, becoming obsolete.

Heavy soils, mainly Transition Zone; British Columbia to the Siskiyou Mountains, California, east to northern Idaho and Salt Lake. Type locality: Nootka Sound.

Allium acuminatum cuspidàtum Fernald, Zoe **4**: 380. 1894. Small plant with more slender scapes; perianth 4–10 mm. long, the segments oblong, abruptly cuspidate; stamens nearly equalling the perianth-segments. Dry plains of the Great Basin region, extending from eastern Washington to Utah. Type locality: Wawawai, Washington.

28. Allium lacunòsum S. Wats.
Pitted Onion. Fig. 956.

Allium lacunosum S. Wats. Proc. Am. Acad. **14**: 231. 1879.
Allium davisiae M. E. Jones, Contr. West. Bot. **12**: 78. 1908.

Bulbs ovoid, the coats light colored, thick, distinctly pitted by the quadrate or transversely oblong reticulations, the walls of the reticulations very minutely sinuous. Scape 7–15 cm. high, terete; leaves about equalling the scape; bracts 2, broadly ovate, acuminate 10–15 mm. long; umbel open, 5–20-flowered; pedicels 6–10 mm. long; perianth pale rose, 6–8 mm. long, the segments oblong-lanceolate, acuminate; stamens nearly as long as the perianth-segments, the filaments dilated at base; ovary cells with an obtuse thickened ridge toward the summit on each side.

Rocky ridges, especially in serpentine, mainly Upper Sonoran Zone; California Coast Ranges from Marin to San Diego Counties. Type locality: Mariposa Peak, Santa Clara County, California.

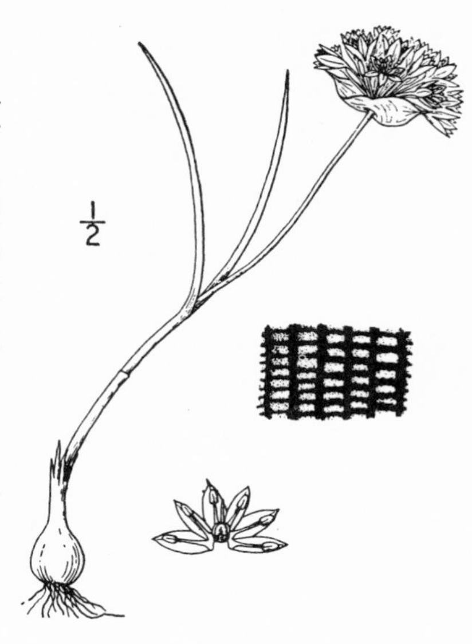

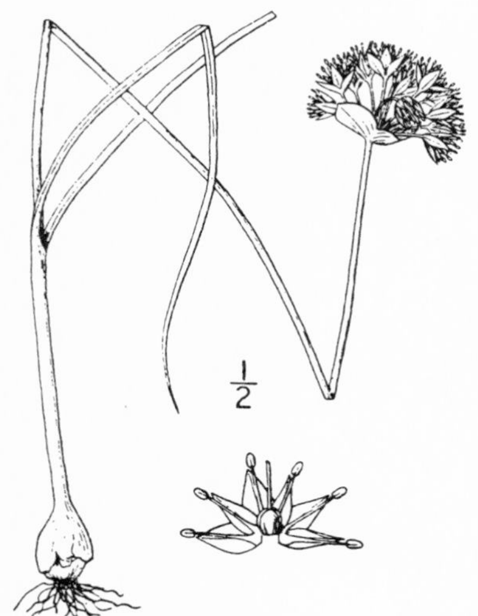

29. Allium névii S. Wats.
Nevius's Onion. Fig. 957.

Allium nevii S. Wats. Proc. Am. Acad. **14**: 231. 1879.

Bulbs ovoid, 10–12 mm. long, outer coats purplish brown, with fine compressed transversely oblong reticulations, the outlines of the meshes not sinuous. Scape slender, 15–25 cm. high, terete or slightly compressed and 2-edged; leaves 2, usually shorter than the scape; pedicels slender, 8–12 mm. long; perianth-segments light rose color, lanceolate, acuminate, 6 mm. long; stamens equalling the segments; filaments dilated below; capsule with 6 low crests near the summit.

Heavy soils, Arid Transition Zone; eastern Washington to western Idaho and eastern Oregon. Type locality: Hood River, Oregon.

30. Allium campanulàtum S. Wats. Sierra Onion. Fig. 958.

Allium campanulatum S. Wats. Proc. Am. Acad. **14**: 231. 1879.
Allium bidwelliae S. Wats. Proc. Am. Acad. **14**: 231. 1879.

Bulb ovoid, 15–20 mm. long, the outer coats brown, the inner white or often tinged with red; reticulations of minute meshes, the transverse walls inconspicuous, the vertical prominent, extending from the base to the apex of the bulb and very sinuous. Scape 2–3 dm. high, terete; leaves 2, shorter than the scape, flat, 3–4 mm. wide; bracts 2, ovate, abruptly acuminate, 12–15 mm. long; umbel open, often 30–40-flowered, the pedicels slender, 15–20 mm. long; perianth pale rose, 6–8 mm. long, the segments thin spreading, broadly ovate-lanceolate, acute or abruptly short-acuminate; stamens about three-fourths the length of the perianth-segments, the filaments slender, dilated at base; ovary with 6 prominent central crests.

Open coniferous forests, Arid Transition and Canadian Zones; Klamath Mountains south in the Sierra Nevada, and the higher altitudes of the inner Coast Ranges to the Mount Pinos region, Ventura County, California. Type locality: Mariposa County, California.

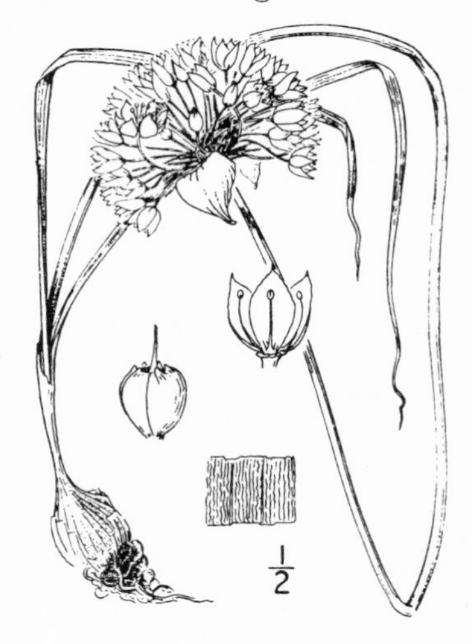

31. **Allium aùstinae** M. E. Jones. Austin's Onion. Fig. 959.

Allium austinae M. E. Jones, Contr. West. Bot. **10**: 85. 1902.

Bulb ovoid, about 10 mm. long, propagating by bulblets borne on the ends of slender rhizomes sometimes several centimeters long, the outer bulb coats reddish brown, the inner brownish white; reticulations with thickened approximate very sinuous vertical walls, as thick as the meshes. Scapes slender, 1–2 dm. high; leaves 2, shorter than or about equalling the scape, 2 mm. wide; bracts 2, ovate-lanceolate, acuminate, 10 mm. long, 3-nerved; umbels open, 15–35-flowered, the pedicels 10–15 mm. long, slender; perianth rose-purple, 6–7 mm. long, the segments lanceolate, acuminate, widely spreading; stamens about two-thirds the length of the perianth-segments; crests 6, prominent, spreading.

In gravelly soil, at high altitudes, Arid Transition and (mainly) Canadian Zones; Sierra Nevada, California, extending from Plumas County to the Tehachapi Mountains. Type locality: "Summit and Castle Peak (Mount Stanford)," Sierra Nevada.

32. **Allium biscéptrum** S. Wats. Patis Onion. Fig. 960.

Allium bisceptrum S. Wats. King. Expl. 351. 1871.

Bulb round-ovoid, about 15 mm. long, the outer coats gray, the inner white; reticulations of fine transversely oblong meshes, the walls, especially the vertical ones, very sinuous. Scapes 1–2 dm. high, often in pairs, terete or often somewhat 2-edged; umbels open, 15–40-flowered, the pedicels filiform, 15–20 mm. long; perianth rose to nearly white, 6–8 mm. long, the segments ovate-lanceolate, acute or acuminate; stamens a little shorter than the perianth-segments, the filaments narrowly deltoid at base; crests of the ovary thin, conspicuous.

Dry or rocky soil, mainly Arid Transition Zone; eastern slopes of the northern Sierra Nevada to the San Bernardino Mountains and eastwrrd to Utah. Type locality: Downieville, California.

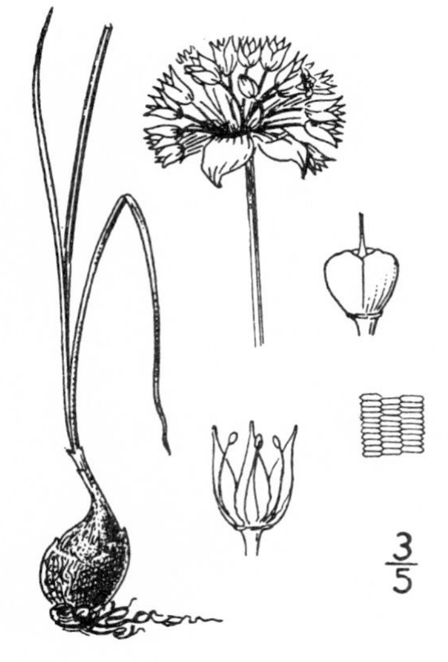

33. **Allium lémmoni** S. Wats. Lemmon's Onion. Fig. 961.

Allium lemmoni S. Wats. Proc. Am. Acad. **14**: 234. 1879.

Bulb ovoid, 15–20 mm. long, the outer coats gray, the reticulations of fine transversely oblong reticulations, arranged in rather definite vertical rows; propagation by division. Scape 12–15 cm. high, about 1.5 mm. wide, scarcely flattened, narrowly 2-edged; leaves 2, about equalling the scape, 4 mm. wide; bracts 2, broadly ovate, abruptly short-acuminate, 15 mm. long, 9-nerved; umbel open, many-flowered, the pedicels slender, 10–15 mm. long; perianth pale rose, 7–8 mm. long, the segments ovate-lanceolate, acuminate, gibbous at base; stamens about equalling the perianth-segments; crests obscure, broad and low.

Open places in coniferous forests, Transition Zone; northern Sierra Nevada in Placer and Sierra Counties. Type locality: Sierra County, California.

34. **Allium fibríllum** Jones. Jones's Onion. Fig. 962.

Allium collinum Dougl.; S. Wats. Proc. Am. Acad. **14**: 228. 1879, not Guss. 1842.
Allium fibrillum M. E. Jones, Contr. West. Bot. **10**: 24. 1902.

Bulb broadly ovoid, without bulblets, propagating by fission, with several thin coats, the reticulations very narrow and much contorted. Scape slender, about 10 cm. high; leaves flat, somewhat falcate, about equalling or exceeding the scapes, 2–4 mm. wide; umbels many-flowered, open; pedicels filiform, 8–12 mm. long; perianth-segments white or pale rose, ovate-lanceolate, acute, 8 mm. long; stamens and style one-half the length of the segments; ovary very obscurely rigid toward the summit.

Heavy soils, Arid Transition Zone; western Idaho to the Blue Mountains, Oregon. Type locality: Cuddy Mountains, Idaho.

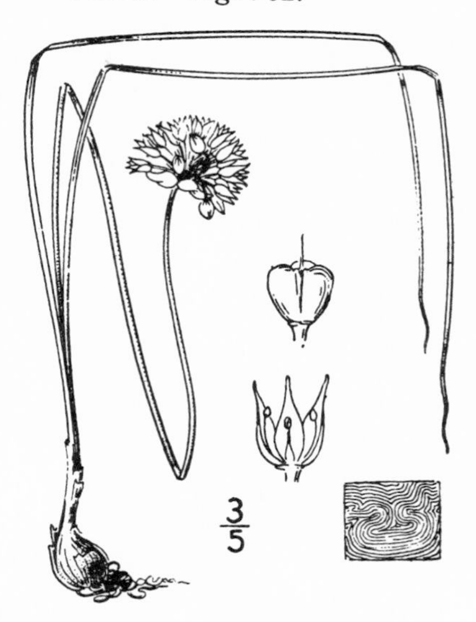

35. **Allium nevadénse** S. Wats. Nevada Onion. Fig. 963.

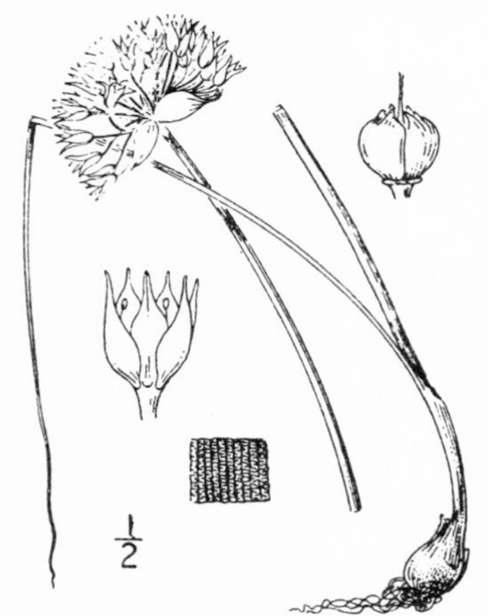

Allium nevadense S. Wats. Bot. King. Expl. 351. 1871.

Bulb ovoid, about 15 mm. long, the inner bulb coats white, the outer gray-brown with much contorted transverse reticulations. Scape slender, 15–20 cm. high; leaves solitary or rarely 2, mostly shorter than the scape; bracts ovate, 12–15 mm. long, abruptly acuminate; umbels open, 15–30-flowered, the pedicels very slender, 10–15 mm. long; perianth white or pale rose, 7–8 mm. long, the segments oblong-ovate, acuminate; stamens two-thirds the length of the perianth-segments, deltoid at the base; stigma entire; capsule prominently 6-crested, the crests about 2 mm. high, acutish or obliquely truncate, entire or nearly so.

A common species of the Great Basin region, Arid Transition and Upper Sonoran Zones; Salt Lake to northern Utah west to the eastern slopes of the Sierra Nevada, California. Type locality: dry foothills and ridges from the Trinity to the East Humboldt Mountains, at 5000 to 7000 feet altitude, Nevada.

36. **Allium híckmani** Eastw. Hickman's Onion. Fig. 964.

Allium hickmani Eastw. Bull. Torrey Club **30**: 483. 1903.
Allium hyalinum hickmani Jepson, Fl. Calif. **1**: 276. 1921.

Bulb round-ovoid, scarcely 1 cm. long, the outer coats gray, the inner white, the reticulations transversely undulate-serrate. Scapes 10–15 cm. long, slender, terete; leaves sheathing underground, filiform, equalling or exceeding the scapes; bracts round-ovate, very obtuse or rounded at the apex, and mucronate, less than 1 cm. long; umbels rather compact, 8–22-flowered; pedicels filiform, 1 cm. or usually less in length; perianth white or tinged with pink, the segments similar, ovate, 5 mm long, acuminate; stamens nearly as long as the perianth-segments, the filaments dilated at base; ovary without crests.

In moist grassy situations, Transition and Upper Sonoran Zones; Monterey Peninsula and near Jolon, in the Santa Lucia Mountains. Type locality: trail between Monterey and Pacific Grove, California.

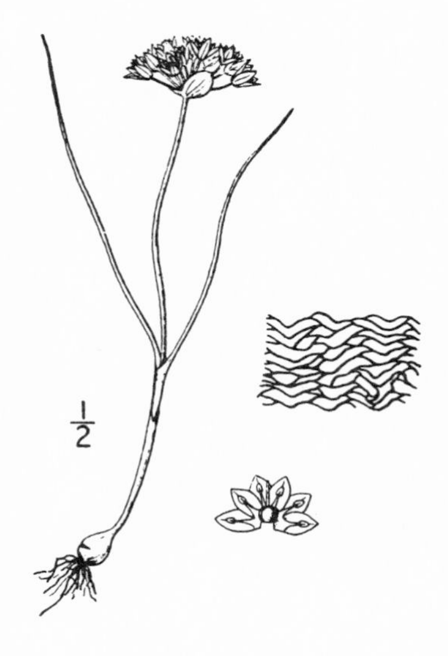

37. **Allium ampléctens** Torr. Narrow-leaved Onion. Fig. 965.

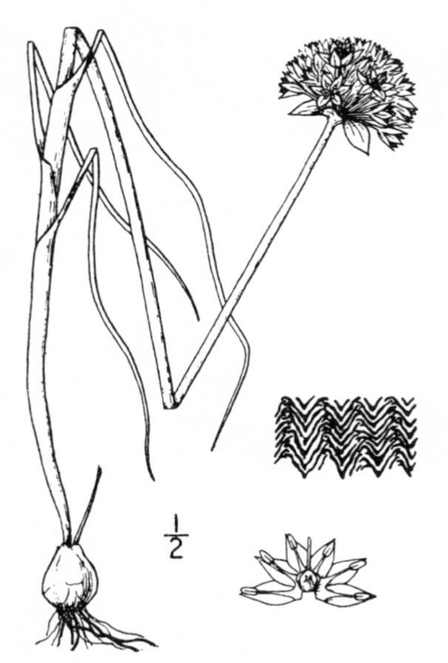

Allium amplectens Torr. Pacif. R. Rep. **4**: 148. 1856.
Allium attenuifolium Kell. Proc. Calif. Acad. **2**: 110. 1863.
Allium occidentale A. Gray, Proc. Am. Acad. **7**: 390. 1867.
Allium acuminatum gracile Wood, Proc. Acad. Philad. **1868**: 171. 1868.
Allium monospermum Jepson; Greene, Man. Bay-Region 321. 1894.

Bulb ovoid, 15 mm. long, the coats reddish, the reticulations of transverse linear V-shaped meshes often appearing in perpendicular rows, or sometimes more irregular. Scapes 15–45 cm. high, slender, terete; leaves 2–4, shorter than the scape, narrow, becoming convolute-filiform above the sheathing base; bracts 2, scarcely 1 cm. long, broadly ovate, rounded at the apex and mucronate; umbel dense, almost capitate; pedicels 6–15 mm. long; perianth-segments white or tinged with pink, oblong-lanceolate, acuminate, 6–8 mm. long, becoming lax in age; stamens scarcely shorter than the perianth-segments; capsule distinctly 6-crested.

Open fields and hillsides, mainly Upper Sonoran Zone; Klickitat County, Washington, to San Diego County, California. Type locality: "hillsides, Sonoma, California."

38. **Allium dichlamýdeum** Greene. Coastal Onion. Fig. 966.

Allium dichlamydeum Greene, Pittonia **1**: 166. 1888.

Bulb round ovoid, scarcely 1.5 cm. long, the coats gray-brown, with close conspicuous transverse serrate reticulations, becoming fissile when dry. Scape 2–3 dm. high, terete; bracts 2, ovate-lanceolate, about 15 mm. long; umbels open, 6–20-flowered; pedicels stout, spreading, 15–20 mm. long; perianth deep rose-purple, 10 mm. long, the outer segments broadly oblong-ovate, abruptly acute, somewhat spreading, the inner segments narrowly oblong, erect; stamens two-thirds the length of the perianth segments, the filaments deltiod and united at base; capsule with narrow central crests at the base of the style.

Open fields and hillsides, Transition and Upper Sonoran Zones; outer Coast Ranges of central California, from Mendocino to Monterey Counties. Type locality: seaward slopes of the Mission Hills, and at the Marine Hospital, San Francisco.

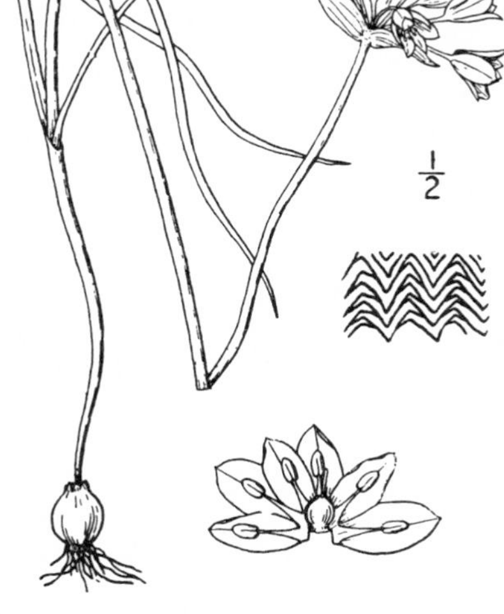

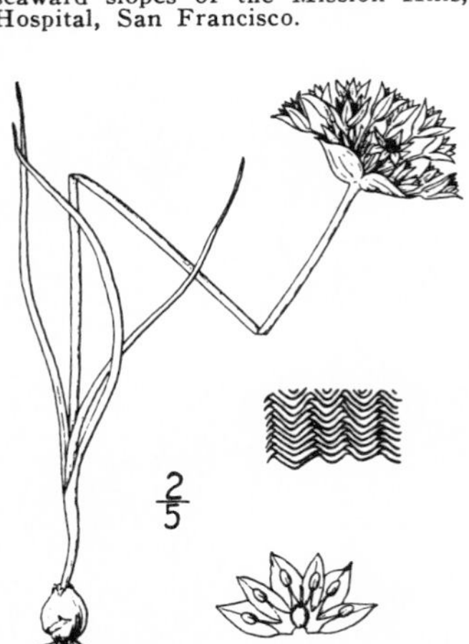

39. **Allium serràtum** S. Wats. Serrated Onion. Fig. 967.

Allium serratum S. Wats. Bot. King Expl. 487. 1871.

Bulb round-ovoid, scarcely over 1 cm. long, the coats gray-brown, distinctly marked with close transverse serrate reticulations, readily fissile when dry. Scape 2–3 dm. high, terete; leaves 2–4, shorter than the scape, 1.5–3 mm. wide; bracts 2, ovate-lanceolate, 10–15 mm. long, acuminate; umbels congested and subcapitate, many-flowered; pedicels very slender, 8–15 mm. long; perianth rose-pink, 8 mm. long, the outer segments ovate-lanceolate, the acuminate apex usually slightly spreading, the inner segments narrower; stamens scarcely two-thirds the length of the segments, the broadly dilated bases united; capsule with narrow central crests at the base of the style.

Usually in heavy soil on serpentine outcrops, valleys and foothills of the inner Coast Ranges of central California, from Solano to Santa Clara County. Upper Sonoran Zone. Type locality: California. Collected by Douglas.

40. Allium críspum Greene. Crinkled Onion. Fig. 968.

Allium crispum Greene, Pittonia 1: 166. 1888.
Allium peninsulare crispum Jepson, Fl. Calif. 1: 278. 1921.

Bulb round-ovoid, 1 cm. long, the coats gray-brown, distinctly marked with close transverse serrated reticulations, fissile when dry. Scape 10–20 cm. high, terete; leaves 2, a little shorter than the scape, scarcely 2 mm. wide; bracts ovate, about 1 cm. long, acute or abruptly short-acuminate; umbels open, 15–40-flowered; pedicels stout, spreading, 12–20 mm. long; perianth bright rose-purple, 9–11 mm. long, the outer segments ovate-oblong, slightly spreading and acute or somewhat acuminate at apex, entire; the inner segments narrower, canaliculate, with distinctly crisped margins; stamens scarcely two-thirds the length of the segments, dilated at base and slightly united; capsule with rather small crests central at the base of the style.

Heavy soil of open fields and hillsides, Upper Sonoran Zone; inner Coast Ranges, San Benito County to San Luis Obispo County, also foothills of the southern Sierra Nevada in Kern County. Type locality: near Paso Robles, California.

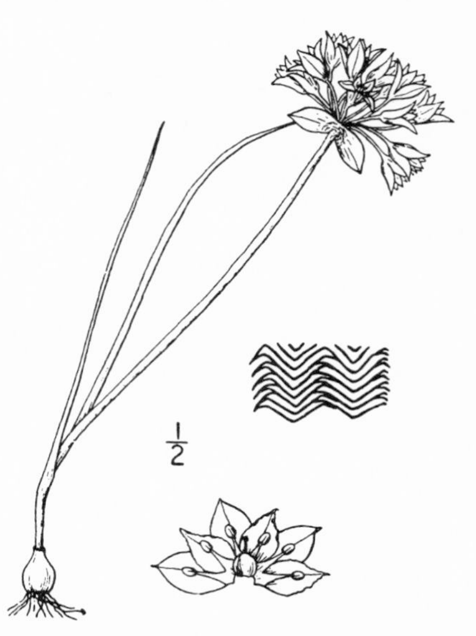

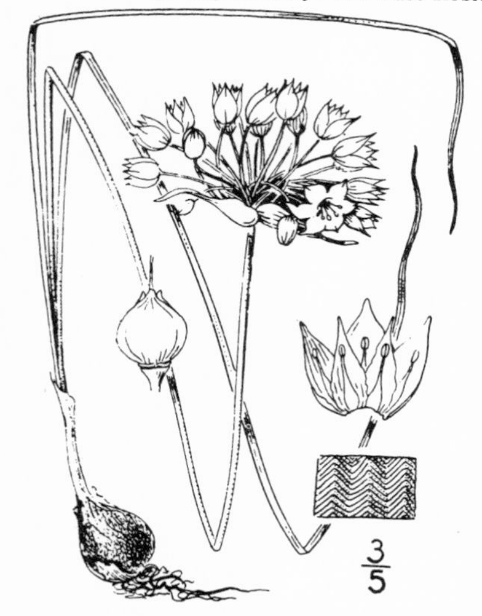

41. Allium peninsulàre Lemmon. Peninsular Onion. Fig. 969.

Allium peninsulare Lemmon, Pittonia 1: 165. 1888.

Bulb round-ovoid, about 15 mm. long, the coats gray-brown, the outer marked with close conspicuous transverse serrate reticulations, when dry readily fissile along the reticulations. Scape 2–3 dm. high, terete; leaves 2–4, usually about equalling the scape, flat, 1–3 mm. wide; bracts ovate-lanceolate, 15–20 mm. long, acuminate; umbels open, 8–20-flowered; pedicels widely spreading, stout, 2–3 cm. long; perianth deep rose-purple, 10–13 mm. long, the outer segments ovate to ovate-lanceolate, acuminate and spreading at apex, the inner narrower, erect, entire, finely serrate or crisped; stamens scarcely two-thirds the length of the segments, the filaments deltoid at the base and united; capsule with narrow central crests at the base of the style.

Dry open fields and hillsides, Upper Sonoran Zone; Shasta County to northern Lower California. Type locality: Las Cruces Cañon, east of Ensenada, Lower California.

42. Allium bolánderi S. Wats. Bolander's Onion. Fig. 970.

Allium bolanderi S. Wats. Proc. Am. Acad 14: 229. 1879.

Bulb round-ovoid, about 1 cm. long, the outer coats gray, rather faintly marked with close transverse serrate reticulations, fissile when dry, the inner coats white, the central part solid, propagating by short lateral offshoots. Scape 10–20 cm. long, slender, terete; leaves usually 2, narrowly linear, shorter than the scapes, 2 mm. wide; bracts ovate-lanceolate, 12–15 mm. long, acuminate; umbels open, 8–25-flowered; pedicels slender, 10–15 mm. long; perianth rose pink to white, 9–11 mm. long, the outer segments ovate-lanceolate, long-acuminate, nearly straight or slightly spreading at the apex, the inner segments narrower and a little shorter, distinctly serrulate on the margin; stamens about half the length of the perianth-segments, the filaments but little dilated at base; capsule obscurely crested.

Heavy, usually clay soil, Siskiyou Mountains, southern Oregon to Mendocino County, California. Humid Transition Zone. Type locality: Humboldt County, California.

Allium bolanderi stenánthum (Drew) Jepson, Fl. Calif. 1: 278. 1922. (*Allium stenanthum* Drew, Bull. Torrey Club 16: 152. 1889.) A slightly taller form with white or only slightly pinkish flowers and narrower perianth-segments. Open pine forest, Transition Zone; eastern Humboldt County, California. Type locality: Pilot Ridge, California.

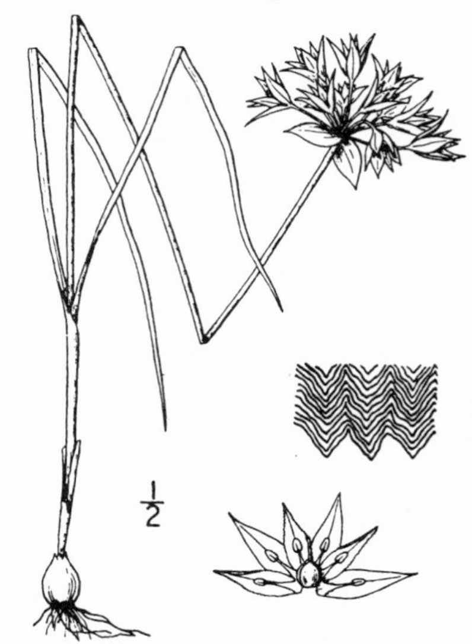

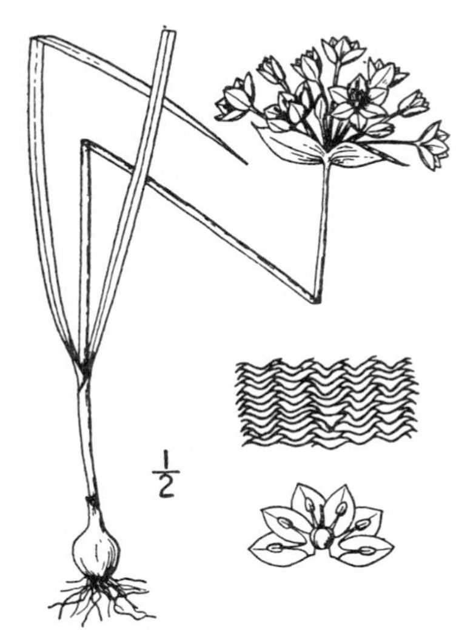

43. **Allium hyalìnum** Curran. Paper-flowered Onion. Fig. 971.

Allium hyalinum Curran, Bull. Calif. Acad. 1: 155. 1885.

Bulbs ovoid, 10 mm. high, the coats thin, gray, the reticulations forming narrowly linear V-shaped meshes appearing in perpendicular rows. Scape slender, terete, 15–30 mm. high; leaves 2 or sometimes solitary, flat, 1–5 mm. wide, shorter or nearly equalling the scape; bracts ovate-lanceolate, 15–20 mm. long, acuminate; umbel open, 6–10-flowered; pedicels 20–25 mm. long, widely spreading; perianth-segments white or tinged with pink, ovate-lanceolate, acute, 7–9 mm. long; stamens two-thirds the length of the perianth-segments, the filaments dilated at base and united; capsule not crested.

On shaded cañon slopes, foothills of the Sierra Nevada, from El Dorado to Kern Counties, California. Upper Sonoran Zone. Type locality: Salmon Falls, El Dorado County.

44. **Allium praècox** Brandg. Early Onion. Fig. 972.

Allium praecox Brandg. Zoe 5: 228. 1908.
Allium hyalinum praecox Jepson, Fl. Calif. 1: 276. 1921.

Bulb ovoid, about 10 mm. broad, the outer coats gray-brown, with conspicuous serrated reticulations. Scape 3–5 dm. high, rather stout; leaves 2–4, from near the base, flat, shorter than the scape, 2–4 mm. wide; bracts 2, 20–25 mm. long, acuminate; umbels open, 10–20-flowered; pedicels 2–3 cm. long; perianth-segments white with rose-purple midvein, ovate-acuminate, becoming 9–10 mm. long, the inner equalling the outer in length, but narrower; stamens two-thirds the length of the segments; capsule without crests.

On shaded slopes, Upper and Lower Sonoran Zones; San Bernardino County, California, to northern Lower California, also on Santa Cruz Island. Type locality: northern slopes of cañons about San Diego.

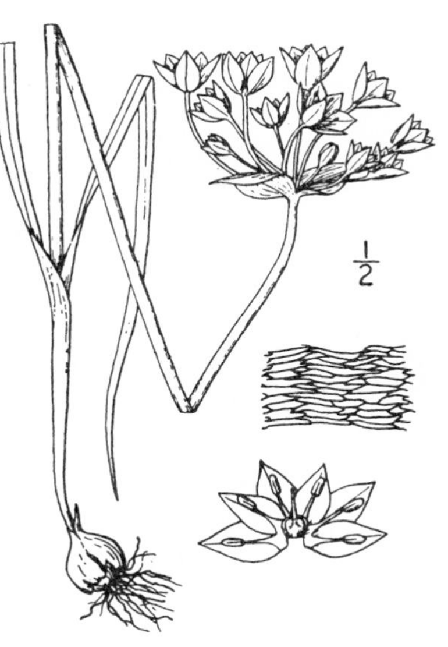

45. **Allium unifòlium** Kell. One-leaved Onion. Fig. 973.

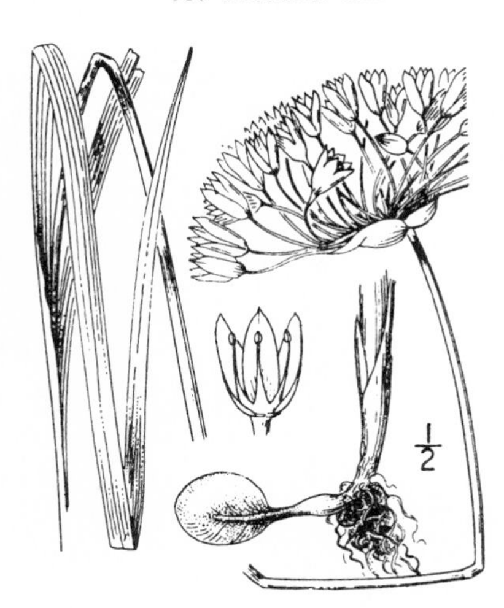

Allium unifolium Kell. Proc. Calif. Acad. 2: 112. 1861.
Allium unifolium lacteum Greene, Pittonia 2: 55. 1890.

Bulb ovoid, 12–15 mm. long, from lateral bulblets produced on stout rhizomes, the outer coats with linear transverse undulating reticulations. Scape stout, terete, 25–40 cm. high; leaves 2–4, flat or somewhat carinate, shorter than the scape, 2–8 mm. wide; bracts 2, 2–3 cm. long, ovate-lanceolate, acuminate; umbels many-flowered; pedicels about 3 cm. long; perianth-segments spreading, bright rose, 10–14 mm. long, oblong-lanceolate, acute or somewhat acuminate; stamens scarcely two-thirds the length of the segments, the filaments deltoid below and united at base; ovary with 3 low flat ridges, disappearing in age; capsule not crested.

Usually in adobe soil, Upper Sonoran Zone; California Coast Ranges from Mendocino County to northern Lower California. Type locality: vicinity of Oakland, California. The name is ill-advised, as the leaves are rarely solitary.

2. MUÍLLA S. Wats. Proc. Am. Acad. **14**: 235. 1879.

Scape from a fibrous corm and bearing an umbel subtended by several small scarious bracts. Leaves mostly few, very narrow, nearly terete. Pedicels not jointed. Perianth subrotate, persistent, of 6 nearly equal slightly united oblong-lanceolate segments, greenish or yellowish white with a dark 2-nerved midrib. Stamens inserted near the base; filaments filiform, slightly thickened toward the base or petaloid; anthers versatile. Ovules 8–10 in each cell; style clavate, persistent and at length splitting. Capsule globose, scarcely lobed, loculicidal. Seeds compressed and angled. [Anagram of *Allium.*]

A California genus of 2 species. Type species, *Muilla maritima* (Torr.) S. Wats.

Filaments more or less dilated at base, but not overlapping and crown-like; anthers lurid purple.
 Leaves very sparsely scabrous on the margins; anthers over 1 mm. long. 1. *M. maritima.*
 Leaves retrorsely scabrous on the margins and striae; anthers about .75 mm. long. 2. *M. serotina.*
Filaments hyaline-petaloid, their broad margins overlapping forming a cylindrical crown; anthers yellow. 3. *M. coronata.*

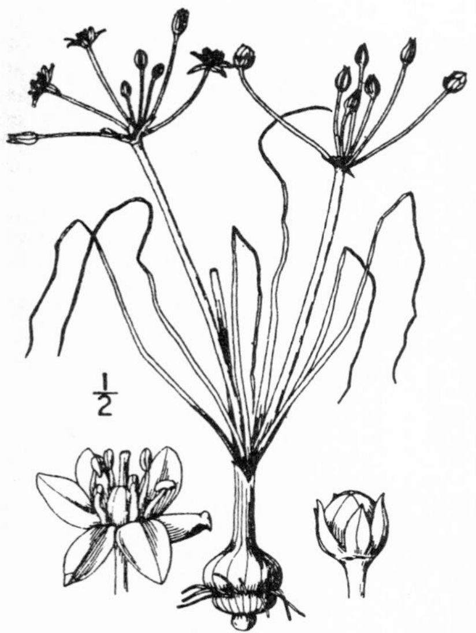

1. Muilla marítima (Torr.) S. Wats.
Common Muilla. Fig. 974.

Hesperoscordum(?) *maritimum* Torr. Pacif. R. Rep. **4**: 148. 1857.
Allium maritimum Torr. Benth. Pl. Hartw. 339. 1857.
Muilla maritima S. Wats. Bot. King. Expl. 354. 1871.
Nothoscordum maritimum Hook. f. Bot. Mag. **27**: under *pl. 5896.* 1871.
Muilla maritima S. Wats. Proc. Am. Acad. **14**: 235. 1879.
Bloomeria maritima Macbr. Contr. Gray Herb. II. **56**: 8. 1918.

Corm 12–18 mm. broad, usually 4–5 cm. under the ground. Leaves several, semi-terete, 10–20 cm. long, 2 mm. or less wide, sparsely scabrous on the margins; scapes 10–25 cm. long; bracts 4–6, lanceolate, acuminate, mostly 3-nerved; umbels 5–25-flowered, the pedicels slender, usually 15–25 mm. long; perianth-segments greenish white, with a prominent dark 2-nerved midrib, 4–5 mm. long, somewhat sacculate at the apex, the outer oblong, slightly revolute, the inner oval, plane; filaments slightly dilated at base; anthers lurid purple, slightly over 1 mm. long; style 1.5 mm. long; capsule subglobose, about 6 mm. high; seeds about 2 mm. long.

Subsaline soils, or stony, especially serpentine, ridges, Upper Sonoran and Transition Zones; Marin County and Sacramento Valley to San Diego County. Type locality: Point Reyes.

Muilla ténuis Congdon, Zoe **5**: 135. 1901. A small slender form with leaves scarcely 1 mm. wide, and pedicels filiform; perianth-segments about 3 mm. long; anthers less than 1 mm. long; seeds scarcely 1 mm. long. Described from specimens collected at Raymond, Madera County. Similar forms occur along the coast in San Diego County.

2. Muilla serotìna Greene.
Rough Muilla. Fig. 975.

Muilla serotina Greene, Erythea **1**: 152. 1893.
Bloomeria maritima serotina Macbr. Contr. Gray Herb. II. **56**: 8. 1918.

Scapes 3–5 dm. high, glabrous. Leaves 3–4 dm. long, nearly terete, the upper surface plane, the lower convex and sharply 7-striate, the strae and margins retrorsely scabrous; umbel 40–70-flowered; pedicels often 4 cm. long; perianth rotate, about 12 mm. broad; greenish white; outer segments oblong-linear, the inner oblong with pit-like glands; filaments stout, subulate, little compressed; anthers lurid purple, .75 mm. long.

Dry gravelly mesas and hillsides, Upper Sonoran Zone; Tehachapi Mountains, southern California, to northern Lower California. Type locality: mountains in the interior of southern California.

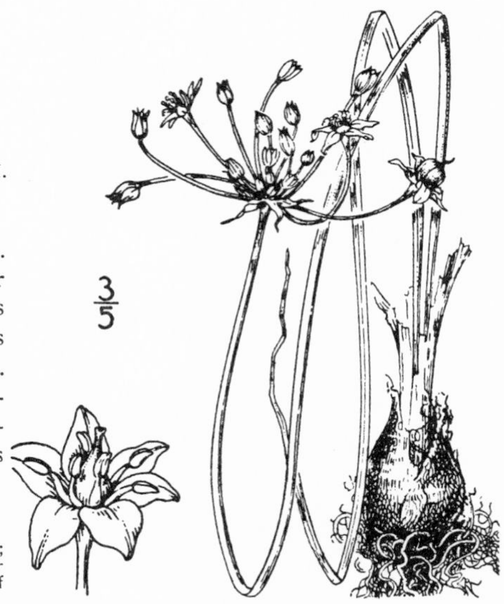

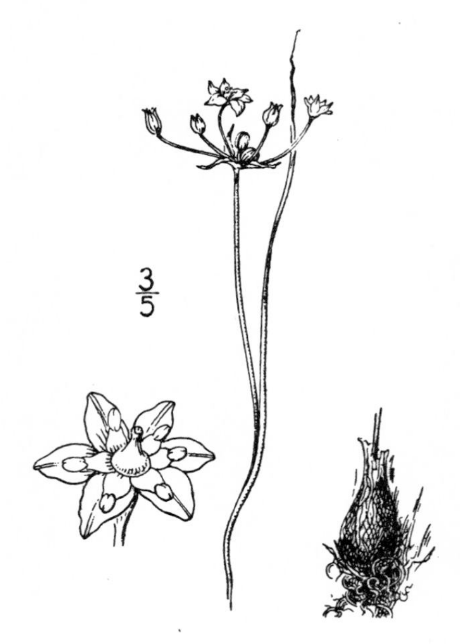

3. Muilla coronàta Greene.
Crowned Muilla. Fig. 976.

Muilla coronata Greene, Pittonia **1**: 165. 1888.

Corm 12–20 mm. in diameter, usually 2–3 cm. under the ground. Leaves 2 or 3, semi-terete, twice the length of the scapes, retrorsely scabrous on the margins; scapes very slender, 5–10 cm. high; umbels 2–4-bracted, 3–10-flowered; perianth rotate, its segments 3 or 4 mm. long, green on the outer surface, pale blue or whitish within; filaments conspicuously dilated, hyaline, petaloid, their broad margins overlapping, obtuse to deeply retuse at summit; anthers yellow.

A little known species of the Mojave Desert, originally collected at Lancaster, Los Angeles County.

3. BLOOMÈRIA Kell. Proc. Calif. Acad. 2: 11. 1859.

Scape from a fibrous coated corm, with linear carinate basal leaves and many yellow flowers in a terminal umbel, subtended by membranous bracts. Pedicels jointed at the summit. Perianth persistent, of 6 nearly equal distinct linear-oblong somewhat spreading segments. Stamens 6, inserted on the base of the segments and a little shorter; filaments filiform with a somewhat cup-shaped winged and often bicuspidate appendage surrounding the base; anthers oblong, attached near the base but versatile. Ovules several in each cell; style filiform-clavate, persistent and splitting with the capsule. Capsule subglobose, membranous, obtusely 3-lobed, loculicidally dehiscent. Seeds subovoid, angular and wrinkled, black. [Name in honor of H. G. Bloomer, a pioneer California botanist.]

A California genus of 2 species. Type species, *Bloomeria aurea* Kell.

Leaf solitary, 6–12 mm. wide; filament-appendages papillose, bicuspidate at apex. 1. *B. crocea.*
Leaves several, about 2 mm. wide; appendages of the filaments smooth, obtuse at apex. 2. *B. clevelandi.*

1. Bloomeria cròcea (Torr.) Coville. Common Golden Stars. Fig. 977.

Allium crocea Torr. Bot. Mex. Bound. 218. 1859.
Bloomeria aurea Kell. Proc. Calif. Acad. **2**: 11. 1859.
Nothoscordum aureum Hook. f. Bot. Mag. **27**: *pl. 5896.* 1871.

Bulb about 15 cm. in diameter, becoming densely covered with brownish fibers; scape scabrous, 2–5 dm. high; leaf solitary, equalling or exceeding the scape, 6–12 mm. broad; bracts narrowly lanceolate; pedicels numerous, 3–6 cm. long; perianth nearly rotate in bloom; segments 8–12 mm. long; appendages about 2 mm. long, bicuspidate, minutely papillose; style about 5 mm. long, much longer than the ovary; capsule about 5 mm. long.

Coast Ranges from Monterey County to San Diego and northern Lower California, also on the Channel Islands. Lower Sonoran to Transition Zones. Type locality: "summit of the mountains east of San Diego, California."

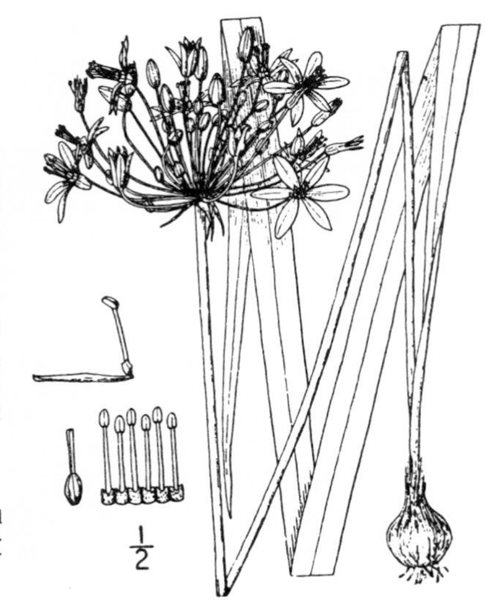

2. **Bloomeria clèvelandi** S. Wats. Cleveland's Golden Star. Fig. 978.

Bloomeria clevelandi S. Wats. Proc. Am. Acad. **20**: 376. 1885.

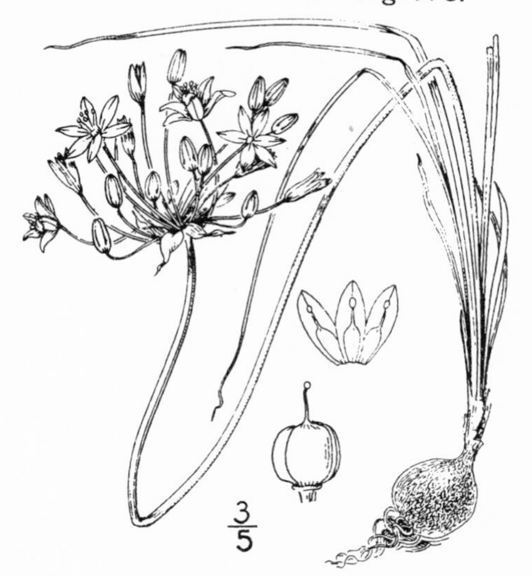

Bulbs about 15 mm. in diameter, densely covered with brownish fibres. Scape 10–30 cm. long, scabrous; leaves several, 6–15 cm. long, 2 mm. or less wide; bracts 5–10 mm. long; umbels 8–20-flowered; pedicels slender, spreading 25–35 mm. long; perianth-segments 6–8 mm. long; yellow with a brown midrib; appendages smooth, obtuse at the apex; style 1.5 mm. long, shorter than the ovary.

Dry mesas and hillsides in the vicinity of San Diego; Lower Sonoran Zone. Type locality: on the mesas near San Diego.

4. CALLIPRÒRA Lindl. Bot. Reg. 19: under *pl. 1590.* 1833.

Scapes slender erect, arising from a depressed fibrous-coated corm. Leaves basal, few, linear, flat. Flowers in an open bracted umbel, yellow with dark brown midveins. Perianth-tube short, turbinate, not at all saccate, the segments rotate-spreading, about equalling the tube. Stamens 6, all anther-bearing, inserted in 1 row on the throat, those opposite the outer segments shorter; filaments with broad lateral appendages extending their entire length and usually prolonged above in acute teeth; anthers linear-oblong to oval, versatile. Ovary ovoid, longer than the short stipe. Capsul ovoid, narrowed to the short style; seeds several in each cell. [Greek meaning pretty face.]

A California genus with only the following species. Type species, *Calliprora ixioides* (Ait. f.) Greene.

Anthers linear-oblong, 2–5 mm. long, creamy-white. — 1. *C. ixioides.*
Anthers oval, 1–1.5 mm. long, creamy-white or blue. — 2. *C. scabra.*

1. **Calliprora ixioìdes** (Ait. f.) Greene. Common Calliprora or Pretty Face. Fig. 979.

Ornithogalum ixioides Ait. f. Hort. Kew. ed 2, **2**: 257. 1811.
Calliprora lutea Lindl. Bot. Reg. **19**: *pl. 1590.* 1833.
Calliprora ixioides Greene, Man. Bay Reg. 319. 1894.

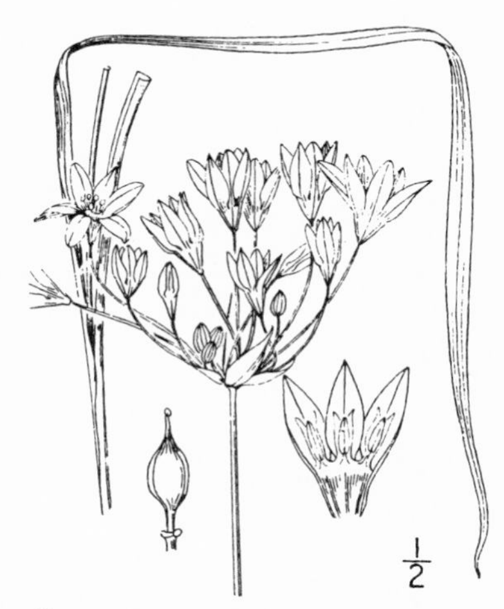

Corm round-ovoid, 15–20 mm. in diameter, seated 3–10 cm. Scape 10–45 cm. long, glabrous or sparsely and minutely scabrous; leaves exceeding the scape, 5–10 mm. wide, glabrous or very sparsely scabrous; bracts lanceolate, 10 mm. long; umbel open, 3–35-flowered; rays 20–40 mm. long; perianth yellow with broad brownish purple midveins, 15–20 mm. long; tube turbinate; segments ovate-lanceolate, nearly twice as long as the tube; filaments winged their whole length, at the apex produced into an acute tooth about equalling the anther; anthers creamy-white, narrowly oblong, 2.5 mm. long; style 3–4 mm. long; capsule, 6 mm. long.

Open hillsides and valleys usually in heavy soil, Upper Sonoran and Transition Zones; California Coast Ranges, Mendocino County, to Santa Barbara. Type locality: California. Collected by Menzies, 1796, probably at Monterey.

Calliprora ixioides lùgens (Greene) Abrams. (*Triteleia lugens* Greene, Bull. Calif. Acad. **2**: 142. 1886. *Calliprora lugens* Greene, Man. Bay. Reg. 319. 1894. *Brodiaea lugens* Baker, Gard. Chron. III, **20**: 459. 1896. *Hookera ixioides lugens* Jepson, Fl. West. Middle Calif. 117. 1901. *Brodiaea ixioides lugens* Jepson, Fl. West. Middle Calif. ed. 2, 101. 1911.) Differs from the typical form in having the apex of the appendages rounded and not prolonged into free acute teeth, the tube of the corolla is also brownish purple. Vaca Mountains, in the north Coast Ranges of California.

2. **Calliprora scàbra** Greene. Sierra Calliprora. Fig. 980.

Calliprora scabra Greene, Erythea **3**: 126. 1895.
Brodiaea scabra Baker, Gard. Chron. III. **20**: 459. 1896.

Corm round-ovoid, about 10 mm. in diameter, seated 4–10 cm. Scape 20–40 cm. long, minutely scabrous; leaves 1–several, shorter than or exceeding the scapes, 5–10 mm. wide; bracts 10–15 mm. long; umbels 5–30-flowered; pedicels 2–8 cm. long; perianth yellow with broad brownish purple midveins, 12–18 mm. long; tube turbinate, 3–4 mm. long; segments spreading, oblong-lanceolate, 10–12 mm. long; filaments with winged appendages extending their whole length and produced at apex into slender acute teeth often exceeding the anthers; anthers creamy white, oval, 1.5 mm. long; style 3–4 mm. long; capsule 6 mm. long; seeds several in each cavity.

Open hillsides of the Upper Sonoran and Transition Zones; western slopes of the Sierra Nevada, Butte County, to the Tehachapi Mountains, California. Type locality: Placerville, California.

Calliprora scabra analìna Greene, Erythea **3**: 126. 1895. (*Calliprora analina* Heller, Muhlenbergia **2**: 14. 1905.) Distinguished from typical *scabra* by the blue anthers, and less scabrous herbage. Resembling *ixioides* in herbage, but clearly distinguished by the small oval anthers of the *scabra* type. Transition and Canadian Zones; southern Oregon south through the Sierra Nevada to Tuolumne County, California. Type locality: Middle Sierra Nevada, California.

5. HESPEROSCÓRDUM Lindl. Bot. Reg. **15**: under *pl. 1293*. 1829.

Scapes erect, arising from a fibrous-coated corm. Leaves few, all basal, linear, thin and flat. Flowers in an open bracteate umbel, white or lilac. Perianth-tube campanulate, the segments equalling the tube, not spreading. Stamens 6, arranged in one row; filaments scarcely dilated or usually deltoid-dilated and united at base; anthers versatile. Capsule subglobose, stipitate, beaked by the persistent style; seeds several in each cell. [Greek meaning western onion.]

A genus of two species, native of western North America. Type species, *Hesperoscordum hyacinthinum* Lindl.

Filaments deltoid-dilated at base. 1. *H. hyacinthinum.*
Filaments not dilated at base. 2. *H. lilacinum.*

1. **Hesperoscordum hyacínthinum** Lindl. Wild Hyacinth. Fig. 981.

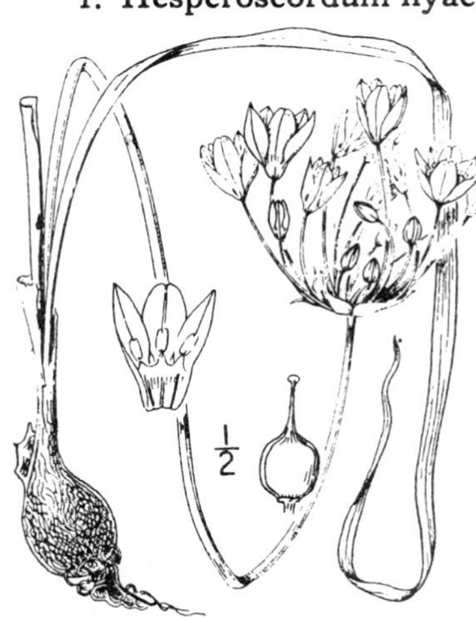

Hesperoscordum hyacinthinum Lindl. Bot. Reg. **15**: under *pl. 1293.* 1829.
Hesperoscordum lacteum Lindl. Bot. Reg. **19**: *pl. 1639.* 1833.
Hesperoscordon lewisii Hook. Fl. Bor. Am. **2**: 185. *pl. 198.* 1839.
Veatchia crystallina Kell. Proc. Calif. Acad. **2**: 11. 1859.
Allium tilingi Regel. Act. Hort. Petrop. III. **2**: 124. 1875.
Brodiaea lactea S. Wats. Proc. Am. Acad. **14**: 238. 1879.
Brodiaea hyacinthina Baker, Gard. Chron. III. **20**: 459. 1896.

Scapes 3–5 dm. high, sparsely scabrous or smooth. Leaves usually 2, equalling or shorter than the scapes, 6–12 mm. wide; umbels open, few–many-flowered; bracts hyaline, narrowly lanceolate; pedicels 2–4 cm. long; perianth open-campanulate, 10–15 mm. long, cleft below the middle, white or lilac with green midveins; segments mostly oblong-ovate, obtuse; filaments equal, about 2 mm. long, deltoid at base and slightly united; anthers yellow or purple, oblong, 1 mm. long; capsule about 5 mm. in diameter.

Moist heavy soil of the Upper Sonoran and Transition Zones; British Columbia and Idaho south through the Pacific States to San Luis Obispo and Tulare Counties, California. Type locality: "native of the plains of the Missouri and of the northwest of America."

2. Hesperoscordum lilacìnum (Greene) Heller in herb.

Lilac-flowered Wild Hyacinth. Fig. 982.

Triteleia lilacina Greene, Bull Calif. Acad. **2**: 143. 1886.
Brodiaea lilacina Baker, Gard. Chron. III. **20**: 459. 1896.

Scapes very slender, scabrous at least below, 25–35 cm. high. Leaves 2, equalling or shorter than the scape, 3–6 mm. wide; umbel open, 10–15-flowered; bracts linear-lanceolate; pedicels slender, 25–40 mm. long; perianth cleft below the middle, pale lilac, with green midribs, 10–12 mm. long; the segments oblong-ovate; filaments slender, 2 mm. long, not dilated below; anthers narrowly oblong, purple.

A little known Californian species, and possibly only a variant of *hyacinthinum*, originally collected in Amador County (*Mrs. Brandegee*) and since in Butte County (*Heller*), where it is said to be common "on stony volcanic uplands in red clay soil."

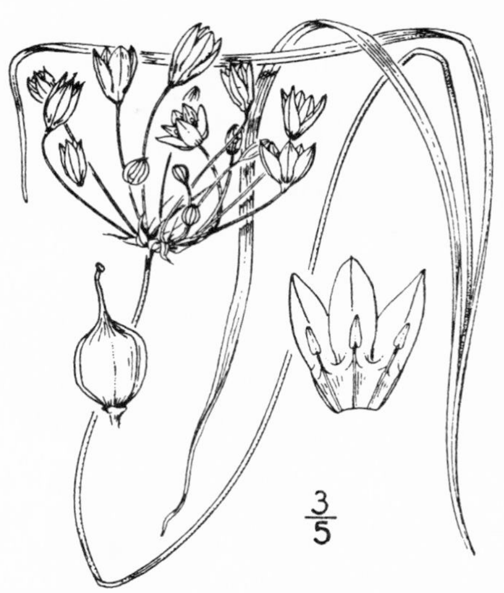

6. TRITELEÌA Dougl.; Lindl. Bot. Reg. **15**: under *pl. 1293*. 1830.

Scapes slender, erect, arising from a fibrous-coated, depressed corm. Leaves few, all basal, linear, thin and flat. Flowers in a close or open bracteate umbel, blue or purple. Perianth-tube funnel-form or subsaccate and broadly tubular; the segments erect or but little spreading, shorter than or equalling the tube. Stamens 6, all anther-bearing, arranged in 1 or 2 rows; filaments dilated below or for their whole length and produced above the anther attachment as lateral teeth; anthers erect, more or less versatile. Ovary on a slender stipe. Capsule beaked by the persistent style; seeds black, angled, several in each cavity. [Greek, alluding to the perfect ternary arrangement of the parts of the flower.]

A genus of western North America, with only the following species. Type species, *Triteleia grandiflora* Lindl.

Perianth tubular and more or less inflated; filaments dilated.
  Filaments without appendages. — 1. *T. grandiflora.*
  Filaments of inner row with lateral winged appendages extending the entire length of the filament.
    Perianth-segments much shorter than the tube. — 2. *T. bicolor.*
    Perianth-segments about equalling the tube. — 3. *T. howellii.*
Perianth funnel-form; filaments slender, not dilated or appendaged.
  Stamens in 2 distinct rows.
    Flowers blue or deep violet-purple.
      Perianth-tube attenuate into a slender clavate base; stipe longer than the ovary. — 4. *T. laxa.*
      Perianth-tube little or not at all attenuate at base; stipe shorter than the ovary.
        Pedicels short, forming a subcapitate umbel. — 5. *T. modesta.*
        Pedicels elongated, often 10 cm. long or more. — 6. *T. peduncularis.*
    Flowers yellow. — 7. *T. crocea.*
  Stamens in 1 row.
    Flowers blue; filaments longer than the anthers. — 8. *T. bridgesii.*
    Flowers yellow with brownish purple median band.
      Anthers blue; filaments stout. — 9. *T. hendersoni.*
      Anthers yellow; filaments filiform. — 10. *T. gracilis.*

1. Triteleia grandiflòra Lindl.

Large-flowered Triteleia. Fig. 983.

Triteleia grandiflora Lindl. Bot Reg. **15**. under *pl. 1293*. 1829.
Brodiaea douglasii S. Wats. Proc. Am. Acad. **14**: 237. 1879.
Hookera douglasii Piper, Contr. Nat. Herb. **11**: 190. 1906.

Leaves 2–several, linear, carinate, 30–50 cm. long, 5–10 mm. wide, glabrous. Scape erect, 40–70 cm. long, glabrous; umbel at length open, 6–20-flowered; pedicels 20–30 mm. long; perianth broadly tubular, 15–25 mm. long, blue, the lobes about equalling the subsaccate tube, oblong-ovate, spreading; stamens in 2 rows, the inner with the filaments dilated below, about equalling the anthers; anthers pale blue, 2 mm. long; capsule oblong-ovoid, 6–7 mm. long, beaked by a slightly shorter style, the stipe 3–5 mm. long.

Open hillsides and meadows in sandy or heavy soil, mainly Arid Transition Zone; east of the Cascade Mountains, ranging from British Columbia to Oregon, Montana, Wyoming, and Utah. Type locality: "Northwest America." Collected by Douglas.

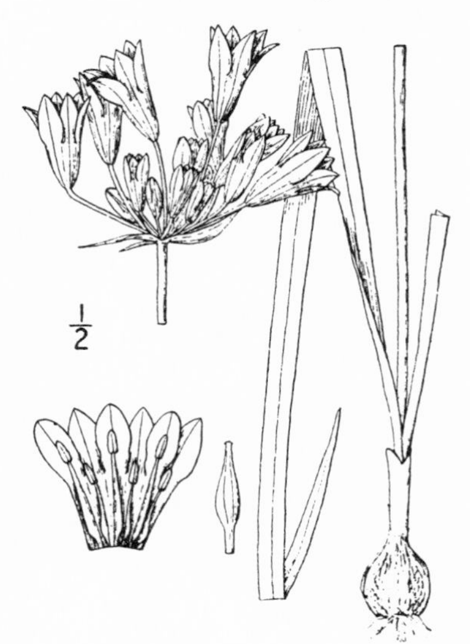

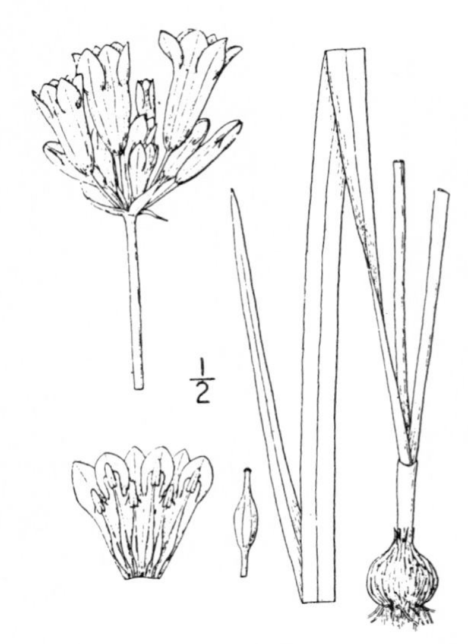

2. Triteleia bìcolor (Suksdorf) Abrams.

Bicolored Triteleia. Fig. 984.

Brodiaea bicolor Suksdorf, West. Am. Sci. **14**: 2. 1902.
Hookera bicolor Piper, Contr. Nat. Herb. 11: 190. 1906.

Leaves several, carinate, about two-thirds the length of the scape, 5–10 mm. wide. Scape stout, erect, 40–70 cm. long; umbels at length open, few-many-flowered; pedicels stout, erect, 5–20 mm. long; perianth with a blue tube and whitish or pale lobes, 15–25 mm. long, tube broadly tubular, subsaccate, the lobes about half the length of the tube, broadly ovate; stamens 6, the outer on the throat, subsessile, the very short filament forming a wing on the throat, the inner row on laterally-winged filaments, lobed at the apex; anthers 4 mm. long; capsule ovoid, 8–10 mm. long, beaked by the short style, on a stipe of about equal length.

Low ground, Upper Sonoran Zone; eastern Washington. Type locality: Falcon Valley, Klickitat County.

3. Triteleia howéllii (S. Wats.) Abrams.

Howell's Triteleia. Fig. 985.

Brodiaea howellii S. Wats. Proc. Am. Acad. **14**: 301. 1879.
Hookera howellii Piper, Contr. Nat. Herb. **11**: 190. 1906.

Leaves about two-thirds the length of the scape, 4–8 mm. wide, glabrous. Scapes slender, 40–60 cm. long; umbel 6–15-flowered, open; pedicels 10–25 mm. long, perianth deep violet-purple, 15–20 mm. long; the tube turbinate-campanulate, slightly saccate above; the lobes ovate, a little shorter than the tube; stamens in 2 rows, the filaments of the outer short and deltoid, those of the inner broadly winged the whole length, the wing truncate or rounded at apex; anther 4 mm. long; capsule oblong, 6 mm. long, attenuate into the short style, the stipe about 4 mm. long.

Dry hillsides and moist meadow-lands, mainly Transition Zone; Puget Sound region of western Washington to Ellensburg, eastern Washington, south to the Hood River, Oregon. Type locality: Klickitat County, Washington.

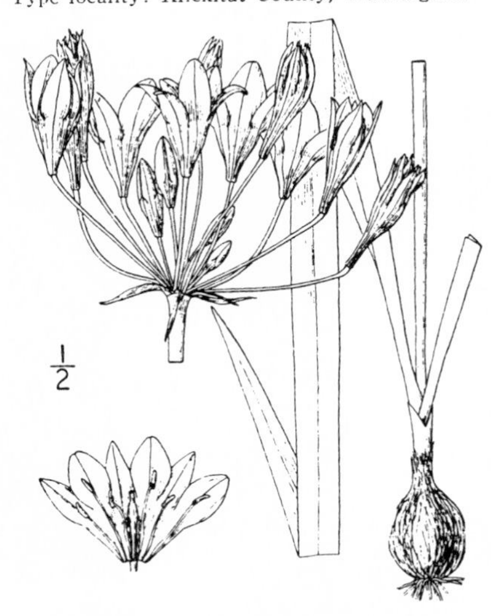

4. Triteleia láxa Benth.

Common Triteleia or Triplet Lily. Fig. 986.

Triteleia laxa Benth. Trans. Hort. Soc. 1: 413. *pl. 15.* 1835.
Seubertia laxa Kunth, Enum. Pl. **4**: 475. 1842.
Brodiaea laxa S. Wats. Proc. Am. Acad. **14**: 237. 1879.
Triteleia candida Greene, Bull. Calif. Acad. **2**: 139. 1886.
Hookera laxa Jepson, Fl. W. Mid. Calif. 117. 1901.

Leaves about two-thirds the length of the scape, 5–20 mm. wide, carinate. Scape slender or stout, erect, 25–50 cm. long; umbels open, 6–30-flowered; pedicels 15–80 mm. long; perianth violet-purple, rarely white, 25–40 mm. long; the tube funnel-form, attenuate at base, the segments oblong-ovate, a little shorter than the tube; stamens in 2 rows, all anther-bearing; the free portion of the filament slender, not dilated, below coalescent with the perianth-tube and reappearing near the base as low longitudinal crests; anthers oblong-lanceolate, 4 mm. long, erect, deeply 2-lobed at base; ovary purplish; capsule oblong-ovoid, 8–10 mm. long, shorter than the slender stipe.

Open hillsides and valleys, especially in adobe soils, Transition and Upper Sonoran Zones: Siskiyou and Humboldt Counties, California, south through the Coast Ranges and foothills of the Sierra Nevada to Los Angeles, but most common in the central Coast Ranges. Type locality: California. Collected by Douglas, probably near Monterey.

Triteleia angustiflòra Heller, Bull. So. Calif. Acad. **2**: 66. 1903. Scarcely more than a small-flowered form of *Triteleia laxa* Benth., distinguished by its smaller (20–25 mm. long) and narrowly funnel-form perianth. Type locality: Tiburon Peninsula, on the Bay Road, Marin County, California.

5. Triteleia modésta (Hall) Abrams.
Hall's Triteleia. Fig. 987.

Brodiaea modesta Hall, Univ. Calif. Pub. Bot. **6**: 166. 1915.

Scape erect, 15–20 cm. long, slender, glabrous; leaves about equalling the scape, flat, 1–4 mm. broad; bracts acuminate, 6 mm. long; umbel subcapitate, 4–8-flowered; pedicels 4–10 mm. long; perianth light blue, narrowly flunnel-form, attenuate at base, 15 mm. long; segments equalling the tube, the outer lanceolate, acute, the inner a little shorter and obtuse; stamens in 2 rows, the lower attached on the throat, the upper on the inner segments; filaments 2 mm. long, slender; anthers blue, 1.5–2 mm. long, linear-oblong; ovary ovoid-oblong, 6 mm. long including the short stipe; style 1.5 mm. long.

Known only from the type locality: Castle Lake, Siskiyou County, California, at an altitude of about 6,500 feet, Canadian Zone.

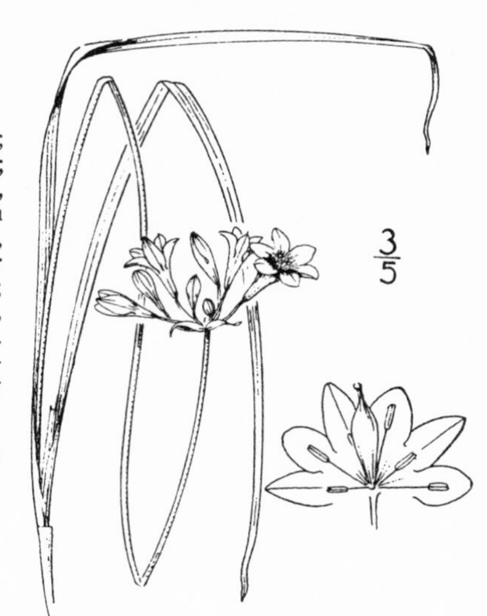

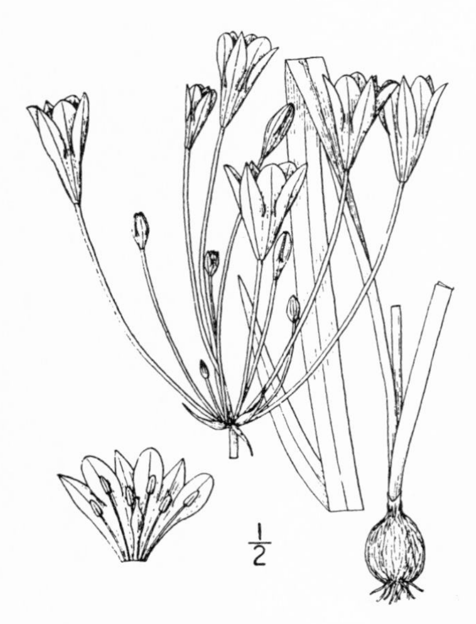

6. Triteleia pedunculàris Lindl.
Long-rayed Triteleia. Fig. 988.

Triteleia peduncularis Lindl. Bot. Reg. **20**: under *pl. 1685.* 1834.
Brodiaea peduncularis S. Wats. Proc. Am. Acad. **14**: 237. 1879.
Hookera peduncularis Jepson, Fl. W. Mid. Calif. 117. 1901.

Scape erect, 2–8 dm. long, glabrous; leaves usually 2 or 3, often longer than the scapes, 5–10 mm. wide; umbel open, 5–17-flowered; pedicels slender 5–20 cm. long; perianth 15–25 mm. long; violet-purple, the tube turbinate not long-attenuate at base; segments equalling or exceeding the tube, the outer lanceolate, acute, the inner oval, oblong; stamens in 2 rows, all anther-bearing, the outer sessile, the inner with short slender filaments; anthers oblong-lanceolate, 3 mm. long, creamy-white; ovary yellow; capsule broadly oblong, 5–6 mm. long, about equalling the stipe.

Moist places, Upper Sonoran Zone; northern Coast Ranges of California. Type locality: "California."

7. Triteleia cròcea (Wood) Greene.
Yellow Triteleia. Fig. 989.

Seubertia crocea Wood, Proc. Acad. Philad. **1868**: 172. 1868.
Brodiaea crocea S. Wats. Proc. Am. Acad. **14**: 238. 1879.
Triteleia crocea Greene, Bull. Calif. Acad. **2**: 141. 1886.

Scapes rather slender, 8–25 cm. long, sparsely and minutely retrorsely scabrous; leaves shorter than or exceeding the scapes, 3–5 mm. wide, very minutely scabrous; umbel 5–15-flowered rather close; pedicels slender, 8–40 mm. long; perianth 15–18 mm. long, yellow with a prominent greenish brown or purplish midvein; tube turbinate, not distinctly attenuate at base; segments ovate-lanceolate, a little longer than the tube; stamens in 2 rows, the lower subsessile, the upper with short slightly dilated filaments; anthers scarcely 2 mm. long; ovary pubescent on the angles; capsule broadly oblong-ovoid, 6 mm. long, on a stipe of about equal length.

Open woods, Transition Zone; Siskiyou and Shasta Counties, California. Type locality: Yreka, California.

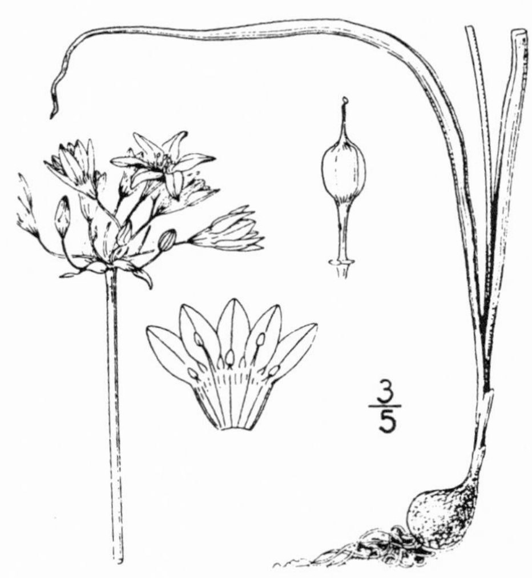

8. **Triteleia bridgèsii** (S. Wats.) Greene. Bridges' Triteleia. Fig. 990.

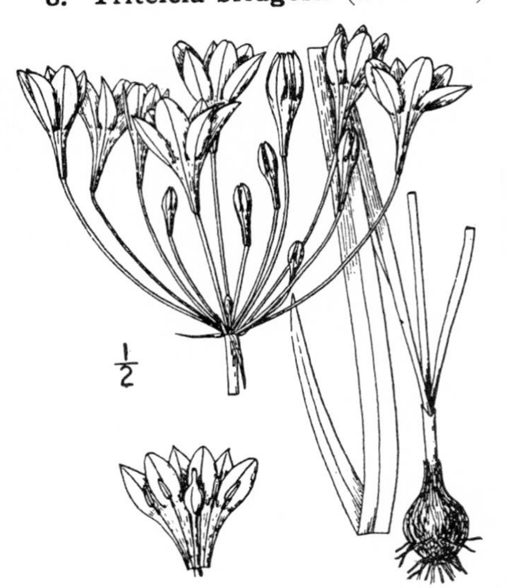

Brodiaea bridgesii S. Wats. Proc. Am. Acad. **14**: 237. 1879.

Triteleia bridgesii Greene, Bull. Calif. Acad. **2**: 141. 1886.

Scape slender, 15–45 cm. long, scabrous. Leaves shorter than or exceeding the scape, 3–10 mm. wide; umbel open, 5–20-flowered; pedicels slender, 1–5 cm. long; perianth violet-purple, funnel-form, 25–35 mm. long; the tube becoming very slender and long-attenuate below; the segments ovate-lanceolate, 8–10 mm. long; stamens inserted on the throat in 1 row, filaments 2–3 mm. long, dilated downward; anthers linear, 3–4 mm. long; capsule ovoid, 6 mm. long, scarcely half the length of the slender stipe.

Heavy soil in open woods or chaparral, Upper Sonoran and Transition Zones; Coast Ranges, from Curry County, Oregon, southward to Lake County, California, and in the foothills of the Sierra Nevada to Mariposa County. Type locality: Central California, probably in the foothills of the Sierra Nevada.

9. **Triteleia hendersòni** (S. Wats.) Greene. Henderson's Triteleia. Fig. 991.

Brodiaea hendersoni S. Wats. Proc. Am. Acad. **23**: 266. 1888.

Triteleia hendersoni Greene, Pittonia **1**: 164. 1888.

Scape rather slender, 20–50 cm. long, minutely scabrous. Leaves shorter than or exceeding the scapes, 5–10 mm. wide, glabrous; umbels open, 6–15-flowered; pedicels slender, 2–3 cm. long; perianth salmon-color with broad dark purplish midveins, 15–20 mm. long; the tube turbinate, short-attenuate at base; segments a little longer than the tube, oblong-lanceolate; stamens in 1 row, inserted on the throat; filaments 4–5 mm. long, slender, slightly dilated at base; anthers blue, 1.5 mm. long; capsule broadly ovoid, abruptly beaked by the style, 6 mm. long, shorter than the stipe; seeds 4–5 in each cavity.

Cañon slopes, and rocky hillsides, Transition Zone; Siskiyou Mountains of southern Oregon and northern California. Type locality: near Ashland, Jackson County, Oregon.

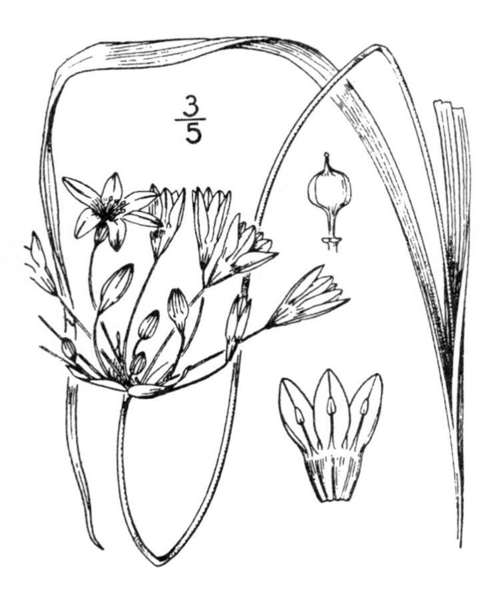

10. **Triteleia grácilis** (S. Wats.) Greene. Slender Triteleia. Fig. 992.

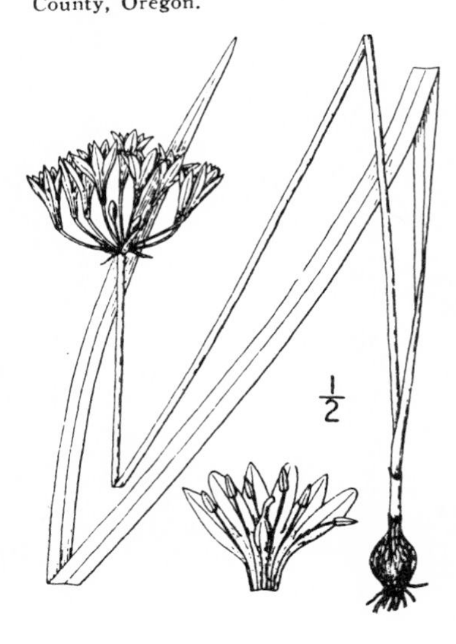

Calliprora aurantea Kell. Proc. Calif. Acad. **2**: 20. 1859.

Brodiaea gracilis S. Wats. Proc. Am. Acad. **14**: 238. 1879.

Triteleia gracilis Greene, Bull. Calif. Acad. **2**: 141. 1886.

Scape 10–30 cm. long, glabrous or minutely scabrous. Leaves usually exceeding the scape, mostly solitary, 4–8 mm. wide; bracts narrowly lanceolate, 10 mm. long; umbels open, 6–30-flowered; pedicels slender, 10–25 mm. long; perianth yellow with brown-purple midveins, 10–15 mm. long; tube narrowly turbinate, short-attenuate at base; segments lanceolate, a little longer than the tube; stamens inserted in 1 row on the throat; filaments filiform, 2–3 mm. long; anthers creamy-white, 1.5 mm. long; capsule ovate-oblong, 5 mm. long, attenuate into the short style; stipe slender, about equalling the capsule; seeds 2 in each cavity.

Borders of mountain meadows and open forests. Transition and Canadian Zones; Sierra Nevada from Plumas to Tuolumne Counties, California. Type locality: Spanish Peak near Gold Lake, Plumas County.

7. BRODIAÈA Smith, Trans. Linn. Soc. 10: 2. 1811.

[HOOKERA Salisb. Parad. Lond. *pl. 98.* 1808.]

Scapes erect, straight from a fibrous-coated corm, with few linear leaves and a solitary umbel subtended by several membranous bracts. Perianth-tube thick or becoming membranous turbinate, segments equalling the tube, spreading at the tip. Anther-bearing stamens 3, inserted on the throat opposite the inner segments, the outer set of stamens reduced to staminodia; anthers basifixed. Ovary and capsule sessile or subsessile. Seeds angled, black, 2–8 in each cavity. [Name in honor of James Brodie, Scotch botanist.]

A genus of 10 species restricted to western North America and mainly California. Type species, *Brodiaea grandiflora* Smith.

Staminodia prominent, oblong or lanceolate, emarginate to acute.
  Filaments not appendaged.
    Staminodia obtuse or emarginate; perianth violet-purple.
      Perianth funnel-form, the tube turbinate.
        Scapes usually 2 dm. or more high; filaments as thick as broad and not winged. — 1. *B. coronaria.*
        Scapes very short, mainly subterranean; filaments flat and winged. — 2. *B. terrestris.*
      Perianth with rotate segments, the tube not turbinate.
        Perianth-tube scarcely constricted over the ovary, becoming membranous and splitting away from the capsule. — 3. *B. californica.*
        Perianth-tube strongly constricted over the ovary, not becoming membranous. — 4. *B. minor.*
    Staminodia acute; perianth rose-purple. — 5. *B. rosea.*
  Filaments with dorsal appendages behind the anthers. — 6. *B. stellaris.*
Staminodia wanting or when present small, scale-like and acuminate.
  Anthers on slender filaments about 3 mm. long. — 7. *B. orcuttii.*
  Anthers sessile. — 8. *B. filifolia.*

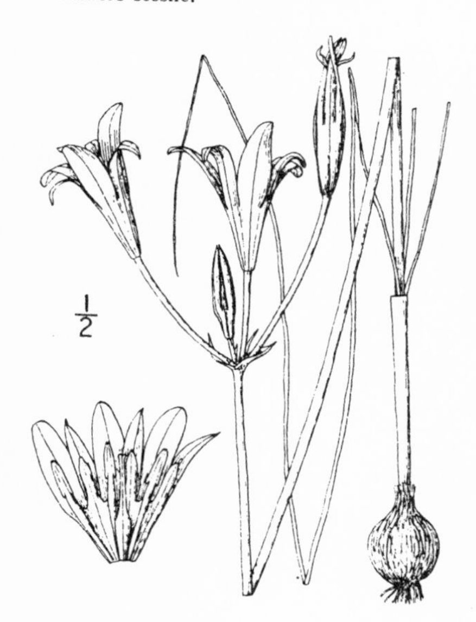

1. Brodiaea coronària (Salisb.) Engler.

Harvest Brodiaea. Fig. 993.

Hookera coronaria Salisb. Parad. Lond. *pl. 98.* 1808.
Brodiaea grandiflora Smith. Trans. Linn. Soc. 10: 2. 1811.
Brodiaea coronaria Engler, Notizb. 2: 317. 1899.

Corm sub-globose, 18–25 mm. broad, seated 10–15 cm. Leaves usually about equalling the scape, 2 mm. wide, thick or somewhat terete; scapes stout, 2–5 dm. high; umbels open, 2–11-flowered; pedicels stout, unequal, 3–8 cm. long; perianth violet-purple, 25–40 mm. long; the tube funnel-form, about half the length of the narrowly oblong, thick and persistent, segments; filaments 4–5 mm. long, slender, not laterally winged; anthers 8–10 mm. long, sagittate at base, entire or nearly so at apex; staminodia white, shorter than or equalling the stamens, oblong-lanceolate, mostly acute; capsule sessile oblong, narrower at base, attenuate at apex into the stout persistent style; seeds 6–8 in each cavity.

Usually in heavy soils, and the most common species, Transition and Upper Sonoran Zones: western British Columbia to the Mexican Boundary, west of the Cascades and the Sierra Nevada. Type locality: "in California."

2. Brodiaea terréstris Kell.

Dwarf Brodiaea. Fig. 994.

Brodiaea terrestris Kell. Proc. Calif. Acad. 2: 6. 1859.
Brodiaea torreyi Wood, Proc. Acad. Philad. 1868: 172.
Hookera terrestris Britten, Journ. Bot. 24: 51. 1886.

Leaves 2–3, much exceeding the scapes, channelled, 2–4 mm. wide. Scapes 2–6 cm. long, scarcely rising above the surface of the ground or wholly subterranean; umbels open, 1–10-flowered; pedicels slender unequal, 3–9 cm. long; perianth violet-purple, 15–20 mm. long, the tube broadly funnel-form; segments oblong-oval, about a third longer than the tube, rotate; filaments about 2 mm. long, flat and laterally winged; anthers 3 mm. long; staminodia yellow, exceeding the anthers, the margins revolute, emarginate at apex; capsule narrowed at base, attenuate at apex into the beak-like style-base.

In clay, serpentine or sandy soils, Upper Sonoran and Humid Transition Zones: Mendocino County to southern California. Type locality: vicinity of San Francisco.

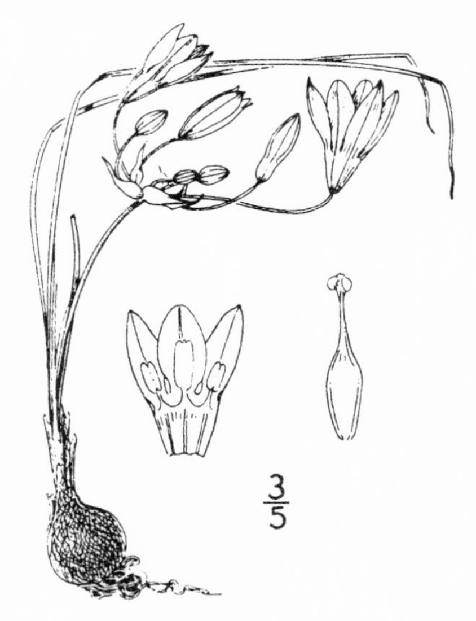

3. **Brodiaea califórnica** Lindl. California Brodiaea. Fig. 995.

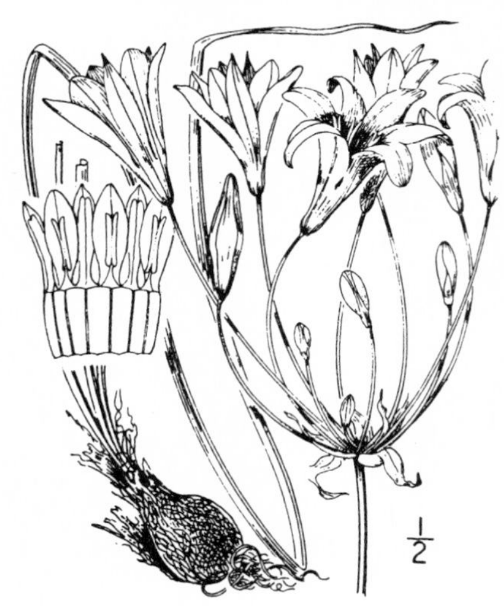

Brodiaea colifornica Lindl. Journ. Hort. Soc. **4**: 84. 1849.
Brodiaea grandiflora major Benth.; S. Wats. Bot. Calif. **2** 153. 1880.
Hookera californica Greene, Bull. Calif. Acad. **2**: 136. 1886.
Hookera leptandra Greene, Pittonia **1**: 74. 1887.
Hookera synandra Heller, Bull. S. Calif. Acad. **2**: 65. 1903.

Leaves thick and channelled conduplicate, 3 mm. wide. Scape stout, 30–50 cm. high, scabrous; umbels open, 3–13-flowered; pedicels stout, unequal, 2–9 cm. long; perianth violet-purple, 30–40 mm. long, the tube narrowly funnel-form, thin and finally splitting away from the capsule, scarcely one-third the length of the oblong spreading segments, outer segments acute, the inner rounded at apex; filaments not flattened, 5–6 mm. long; anthers oblong, 8 mm. long; staminodia white, oblong, the margins slightly involute, usually a little exceeding the anthers; 2 mm. wide, blunt or retuse; capsule obovoid, sessile, abruptly attenuated at apex; seeds 6–8 in each cavity.

Dry gravelly hillsides, Upper Sonoran and Transition Zones; Sierra Nevada foothills, from Shasta to Nevada Counties, California. Type locality: Sacramento Valley.

4. **Brodiaea mìnor** (Benth.) S. Wats. Small Brodiaea. Fig. 996.

Brodiaea grandiflora minor Benth. Pl. Hartw. 340. 1857.
Brodiaea minor S. Wats. Bot. Calif. **2**: 153. 1880.
Hookera minor Britten, Journ. Bot. **24**: 51. 1886.
Brodiaea purdyi Eastw. Proc. Calif. Acad. II. **6**: 427. *pl. 58.* 1896.
Hookera purdyi Heller, Muhlenbergia, **6**: 83. 1910.

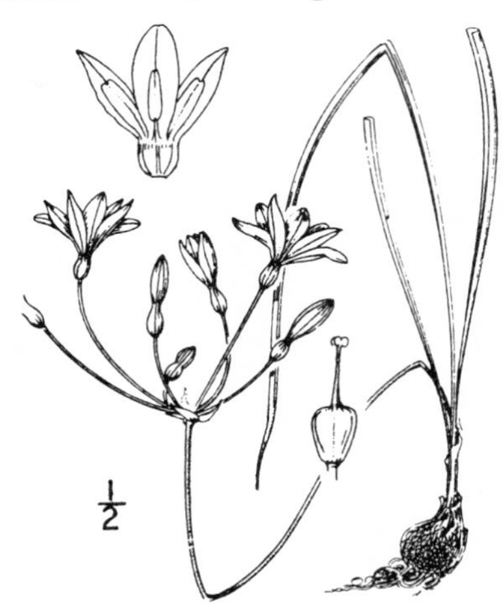

Leaves 2–3 mm. wide, channelled. Scape slender, 1–2 dm. high; umbels open, 2–9-flowered; pedicels 2–4 cm. long; perianth violet-purple, 20–25 mm. long, the tube scarcely half the length of the segments, constricted at the throat above the ovary, the segments linear-oblong, widely spreading and recurved; anthers subsessile, scarcely 4 mm. long; staminodia white, erect, 13 mm. long, 3 mm. wide, dentate or emarginate at apex; capsules sessile, round-obovoid, 4–5 mm. long, rounded at apex, closely enclosed by the constricted persistent perianth-tube; seeds 2 in each cavity.

Foothills of the Sierra Nevada in Butte County, Upper Sonoran Zone. Type locality: foothills of the Sierra Nevada, probably in Butte County, California.

5. **Brodiaea ròsea** (Greene) Baker. Rose-flowered Brodiaea. Fig. 997.

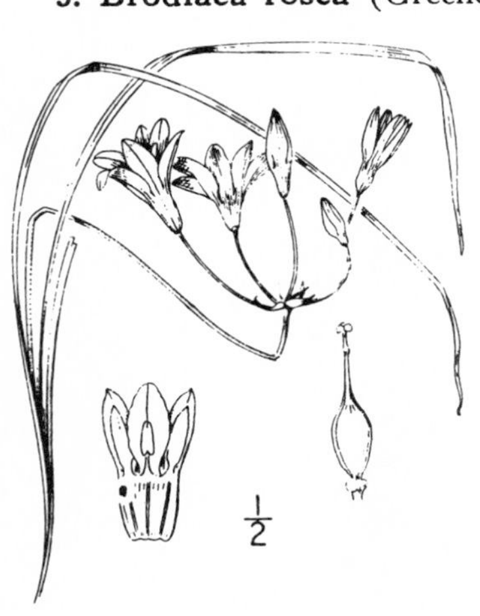

Hookera rosea Greene, Bull. Calif. Acad. **2**: 137. 1886.
Brodiaea rosea Baker, Gard. Chron. III. **20**: 214. 1896.

Corm round-ovoid, seated 4–6 cm. Leaves 2–3, usually exceeding the scape, thick and subterete, 1.5–2.5 mm. wide; scapes slender, 8–15 cm. high; bracts 6–7 mm. long; umbels open, 2–8-flowered; pedicels slender, 1–2 cm. long; perianth 15–20 mm. long, rose-red, the tube broadly funnel-form; segments but little longer than the tube, oblong oval, campanulate-spreading; filaments deltoid-dilated, 1.5 mm. long; anthers oblong, 4 mm. long, sagittate at base, cleft at apex; staminodia lanceolate, acute and entire at apex, slightly involute; capsule narrowed to a subsessile base.

Moist gravelly creek bottoms, Upper Sonoran Zone; Lake County, California. Type locality: Hough's Springs, Lake County, California.

6. **Brodiaea stellàris** S. Wats. Star-flowered Brodiaea. Fig. 998.

Brodiaea stellaris S. Wats. Proc. Am. Acad. 17: 381. 1882.
Hookera stellaris Greene, Bull. Calif. Acad. 2: 137. 1886.

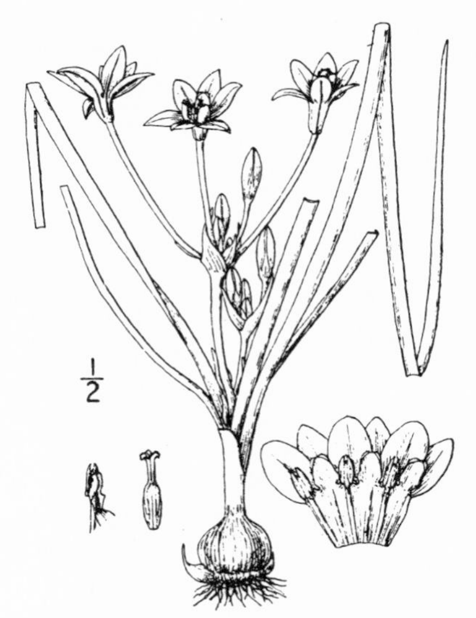

Corms round-ovoid, 12–16 mm. broad, seated 3–4 cm. Leaves 3–7, usually much exceeding the scapes, semi-terete, scarcely 2 mm. wide; scapes rather stout, glabrous, 8–15 cm. long; bracts ovate, 5–6 mm. long; umbels open, 3–8-flowered; pedicels unequal, 3–7 cm. long; perianth 15–20 mm. long, violet-purple, the tube funnel-form, greenish; segments a third longer than the tube, oblong-oval, obtuse, spreading; stamens 3; filaments very short, with 2 dorsal appendages appearing as wings back of the anthers; anthers 3 mm. long, sagittate; staminodia white, deeply emarginate, exceeding the anthers; capsule narrowed to a subsessile base.

Moist soils, Upper Sonoran Zone; Humboldt and Mendocino Counties, California. Type locality: mountain sides near Ukiah, California.

7. **Brodiaea orcúttii** (Greene) Baker. Orcutt's Brodiaea. Fig. 999.

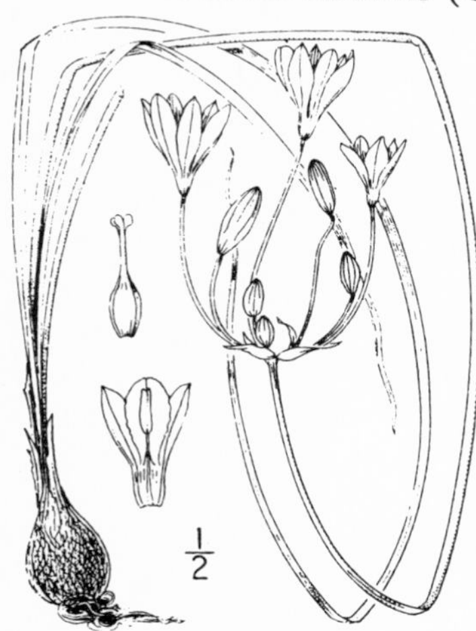

Hookera orcuttii Greene, Bull. Calif. Acad. 2: 138. 1886.
Brodiaea orcuttii Baker, Gard. Chron. III. 20: 214. 1896.
Hookera multipedunculata Abrams, Bull. Torrey Club 32: 537. 1905.

Corm subglobose, 15–20 mm. broad, seated 6–10 cm. Leaves often longer than the scape, usually 2, flat or channelled, 4–5 mm. broad, glabrous; scape rather stout, 1–4 dm. high, one to several from each corm; bracts ovate, acuminate, 6–8 mm. long; umbels open, 6–14-flowered; pedicels unequal, 2–5 cm. long; perianth violet-purple, 18–20 mm. long, the short tube about half the length of the oblong spreading segments, outer segments cuspidate; stamens 3; filaments slender, 3 mm. long; anthers 4 mm. long; staminodia wanting; capsule oblong-ovoid, sessile, 5 mm. long.

Heavy adobe soil of the Upper Sonoran and Transition Zones; southern California, on the mountains and mesas of San Diego County. Type locality: near San Diego.

8. **Brodiaea filifòlia** S. Wats. Thread-leaved Brodiaea. Fig. 1000.

Brodiaea filifolia S. Wats. Proc. Am. Acad. 17: 381. 1882.
Hookera filifolia Greene, Bull. Calif. Acad. 2: 138. 1886.

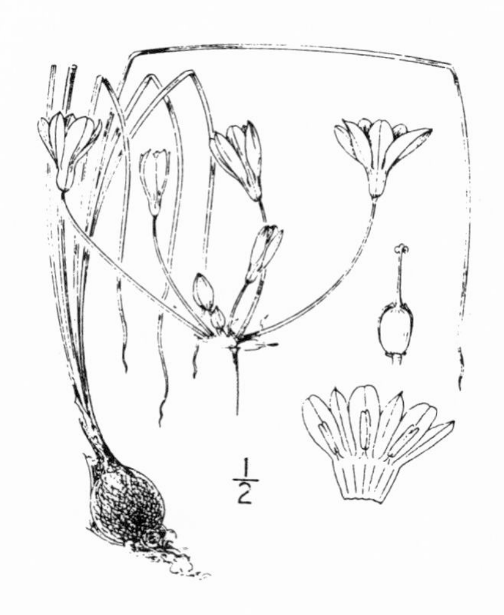

Leaves several about equalling or shorter than the scapes, mostly 1–2 mm. wide. Scapes 20–40 cm. high, smooth; umbel 5–8-flowered, open; pedicels 25–60 mm. long, slender; perianth violet-purple, usually tinged with rose; 12–18 mm. long, the tube rather narrow, shorter than the broadly oblong segments, becoming membranous and splitting from the capsule, the outer segments mucronate; anthers sessile, 4 mm. long; staminodia triangular, about 2 mm. long; capsule oblong-obovate, sessile, 6 mm. long; seeds 2 in each cavity.

Heavy soils of the Upper Sonoran Zone; San Bernardino Valley, southern California. Type locality: Arrowhead Hot Springs, near San Bernardino, California.

8. DICHELÓSTEMMA Kunth. Enum. Pl. 4: 469. 1843.

Scape tortuous or twining from a depressed fibrous-coated corm. Leaves usually 2, fleshy, linear. Umbel subtended by 3 or more thin spathaceous bracts. Perianth-tube thin, more or less inflated and angular or saccate, about equalled by the segments. Stamens 6, the inner with a free lanceolate appendage on each side, sterile in most species, the outer naked; anthers basifixed. Ovary sessile; ovules 3–8 in each cell; style persistent, with short divergent stigmas. Capsule ovate to oblong, more or less attenuate above. Seeds angled, black. [Greek meaning a bifid corona, referring to the appendages.]

A genus of 5 species peculiar to western America and mostly Californian. Type species, *Dichelostemma congestum* (Smith) Kunth.

Stamens with the outer set reduced to staminodia, the inner anther-bearing.
  Filaments winged with emarginate appendages; flowers rose-red, umbellate. 1. *D. californicum.*
  Filaments naked; flowers violet-purple, capitate.
    Staminodia deeply cleft. 2. *D. pulchellum.*
    Staminodia obtuse. 3. *D. multiflorum.*
Stamens all anther-bearing, the outer with naked filaments, the inner with winged appendages. 4. *D. capitatum.*

1. Dichelostemma califórnicum (Torr.) Wood. Twining Saitas or Brodiaea. Fig. 1001.

Stropholirion californicum Torr. Pacif. R. Rep. 4: 149. *pl. 21.* 1857.
Rupalleya volubilis Morière, Bull. Soc. Linn. Norm. 8: 317. 1864.
Dichelostemma californicum Wood, Proc. Acad. Philad. 1868: 173. 1868.
Brodiaea volubilis Baker, Journ. Linn. Soc. 11: 377. 1871.

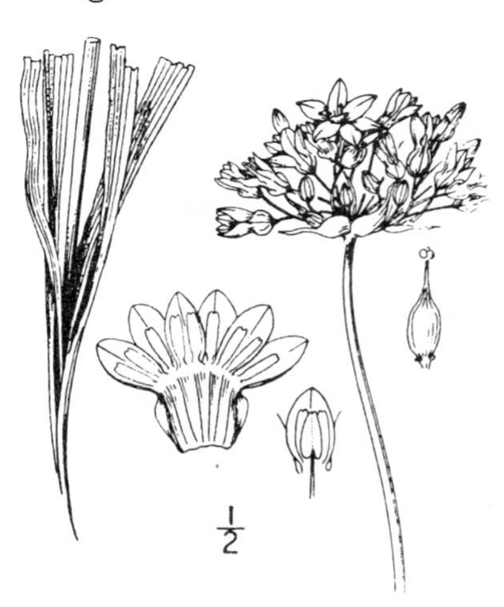

Leaves several, 30 cm. long or more, 8–12 mm. wide, carinate. Scape erect and more or less contorted, 5–10 dm. high, or often twining and 2–3 m. long; umbels dense, 15–30-flowered; pedicels filiform, 15–25 mm. long; perianth rose-red or pinkish, 12–16 mm. long, the tube 6–8 mm. long and nearly as wide, 6-angled, the angles produced into sacs above the middle, the segments equalling the tube, oblong, rotate, their tips recurved; stamens 3, opposite the inner perianth-segments; filaments short-winged with emarginate appendages; staminodia 3, opposite outer perianth-segments, ligulate, emarginate; capsule ovate, accuminate, subsessile.

On open grassy slopes, or in thickets, Upper Sonoran Zone; Coast Ranges of northern California and in the foothills of the Sierra Nevada to Kern County. Type locality: "Knight's Ferry, Stanislaus River," California.

2. Dichelostemma pulchéllum (Salisb.) Heller. Northern Saitas or Ookow. Fig. 1002.

Hookera pulchella Salisb. Parad. Lond. under *pl. 98.* 1806.
Brodiaea congesta Smith, Trans. Linn. Soc. 10: 2. 1811.
Dichelostemma congestum Kunth, Enum. Pl. 4: 470. 1843.
Hookera congesta Jepson, Fl. W. Mid. Calif. 116. 1901.
Dichelostemma pulchellum Heller, Muhlenbergia, 1: 132. 1906.

Leaves 2 or more, subterete, usually nearly equalling the scape, 6–12 mm. wide. Scape more or less flexuose, 5–10 dm. high; umbels 6–20-flowered, congested into short racemose head-like clusters; pedicels 2–8 mm. long; perianth deep violet-purple, 14–18 mm. long, the tube about equalling the segments, slightly constricted at the throat, the segments oblong-lanceolate, obtuse, spreading; anthers 3, sessile; staminodia deeply cleft, longer than the anthers, purple, spreading with the segments; capsule sessile, 10 mm. long.

Open hillsides, Upper Sonoran and Transition Zones; western Washington southward to the western slopes of the Sierra Nevada to Mariposa County and in the Coast Ranges to Santa Clara County, California. Type locality: collected by Menzies somewhere on the Pacific Coast, but definite locality not known.

3. Dichelostemma multiflòrum (Benth.) Heller. Many-flowered Saitas or Brodiaea. Fig. 1003.

Brodiaea multiflora Benth. Pl. Hartw. 339. 1857.
Brodiaea parviflora Torr. & Gray, Pacif. R. Rep. **2**: 125. 1857.
Dichelostemma multiflorum Heller, Muhlenbergia **2**: 15. 1905.

Leaves 3 or more, about equalling or sometimes much exceeding the scape. Scape 3–7 dm. high, slightly scabrous; umbel forming a more or less congested, few–many-flowered head; bracts ovate-lanceolate, acute; pedicels 5–15 mm. long; perianth deep violet-purple, 12–20 mm. long, the segments oblong, about equalling the tube; anthers very nearly sessile; staminodia equalling the anthers, entire, obtuse; capsule sessile, broadly ovoid, pointed, about 8 mm. long; seeds 3–4 in each cavity.

Grassy hillsides, and mountain meadows, Upper Sonoran and Transition Zones; southern Oregon southward through the Coast Ranges to Santa Clara County, the Sacramento Valley, and the Sierra Nevada to Mariposa County, California. Type locality: "Mendelin Pass of the Sierra Nevada."

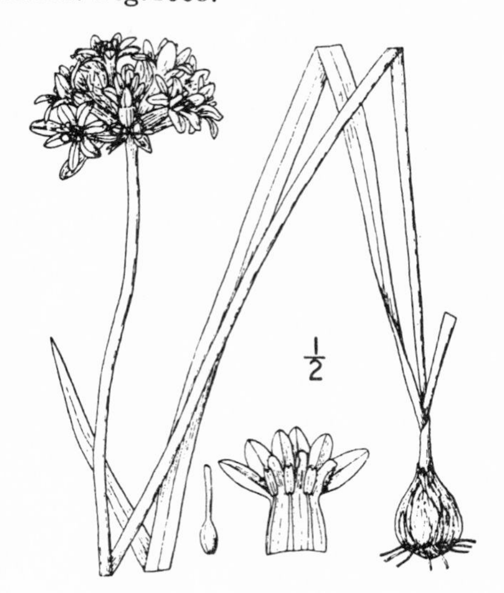

4. Dichelostemma capitàtum (Benth) Wood. Common Saitas or Brodiaea. Fig. 1004.

Brodiaea capitata Benth. Pl. Hartw. 339. 1857.
Dichelostemma capitatum Wood, Proc. Acad. Philad. **1868**: 173. 1868.
Brodiaea insularis Greene, Bull. Calif. Acad. **2**: 134. 1886.
Dipterostemon capitatum Rybd. Bull. Torrey Club **39**: 111. 1912.

Leaves usually shorter than the scapes, 6–15 mm. wide; umbels subcapitate, 2–20-flowered; pedicels 4–15 mm. long; perianth deep violet-purple, rarely reddish-purple, or white, broadly funnel-form, 12–20 mm. long, the segments equalling or somewhat exceeding the tube, oblong-ovate, obtuse; stamens 6, the inner anthers nearly sessile, 4 mm. long, a little shorter and concealed by the erect oblong-lanceolate appendages, the outer anthers about 2 mm. long, on short naked filaments broadly dilated at base; capsule sessile, broadly ovoid, 6 mm. long, backed by the slender style; seeds several in each cavity.

Dry open ridges and grassy fields, Sonoran and Transition Zones; southern Oregon to northern Lower California. A very common species, showing considerable variation in the size of the flowers. Plants from the desert slopes of southern California approach very closely the *Dichelostemma pauciflora* (Torr.) Standley, of Arizona and New Mexico, but are not specifically distinct from the common form.

9. BREVOÒRTIA Wood, Proc. Acad. Philad. **1867**: 81. 1867.

Scapes erect from a fibrous-coated corm, with a few linear basal leaves. Infloresense a terminal several-flowered umbel, with jointed pedicels and 2 or more scarious or colored bracts. Perianth rather broadly tubular, shortly saccate at the truncate base, slightly constricted at the throat, the segments short, more or less spreading or reflexed, faintly 1-nerved. Stamens 3, on the throat opposite the inner perianth-segments, alternate with 3 very broad truncate corona-like staminodia; filaments very short, naked; anthers basifixed, narrowly oblong, emarginate at each end. Style elongated, persistent with a slightly 7-lobed stigma. Ovary stipitate; ovules 4–6 in each cavity, capsule triangular-ovate, acuminate, loculicidal. Seeds angled, black. [Named for J. Carson Brevoort, a patron of natural science.]

A monotypic Californian genus. Type species, *Brevoortia ida-maia* Wood.

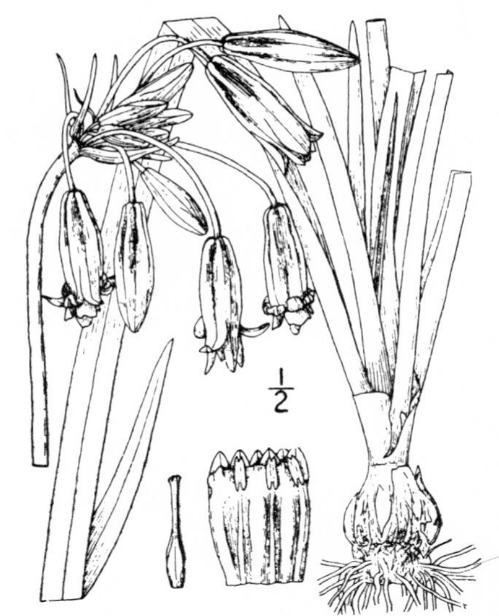

1. Brevoortia ida-maìa Wood.
Fire-cracker Flower. Fig. 1005.

Brevoortia ida-maia Wood. Proc. Acad. Philad. **1867**: 82. 1867.
Brodiaea coccinea A. Gray, Proc. Am. Acad. **7**: 389. 1867.
Brevoortia coccinea S. Wats. Proc. Am. Acad. **14**: 239. 1879.

Leaves mostly 3, linear, 30–45 cm. long, 4–8 mm. wide, carinate. Scape rather slender, 3–10 dm. long; umbels 5–15-flowered; pedicels 15–35 mm. long; flowers mostly nodding, 25–35 mm. long; perianth-tube scarlet, broadly tubular, 20–30 mm. long; perianth-segments 5–6 mm. long, ovate, obtuse, erect or often reflexed, chrome-green; anthers 5 mm. long, erect, creamy-white; staminodia 2–3 mm. long, pale yellow; capsule ovate-oblong, 10 mm. long, attenuate with the persistent style, narrowed at base to the 5–6 mm. long stipe.

Grassy hillsides in open forests, Humid Transition and Upper Sonoran Zones; Siskiyou Mountains, southern Oregon, to Marin and Contra Costa Counties, California. Type locality: Trinity Mountains, Shasta County, near the old "stage-road from Shasta City to Yreka."

Brevoortia venústa Greene (Pittonia **2**: 230. 1892). According to Carl Purdy this is a hybrid between *Brevoortia ida-maia* and *Dichelostemma congestum*. Distinguished by the rose-purple perianth, constricted under the segments, and the pinkish staminodia. Near Orr's Springs, Mendocino County, California.

10. ANDROSTÉPHIUM Torr. Bot. Mex. Bound. 218. 1859.

Scapose herbs, with fibrous-coated corm. Leaves basal, narrowly linear and channelled. Bracts several. Flowers in few-flowered umbels on short inarticulate pedicels. Perianth violet-purple, funnel-form, 6-cleft to the middle or below, the segments narrowly oblong, spreading, 1-nerved. Stamens 6, in one row on the throat, the filaments united to form a tubular corona with erect bifid segments between the anthers; anthers versatile, introrse. Ovary ovate, sessile, with elongated persistent style. Capsule subglobose, 3-angled. Seeds several in each cavity, large, black. [Name Greek, meaning stamen crown, in allusion to the united filaments.]

A genus of two species native of southwestern United States. Type species, *Androstephium violaceum* Torr.

1. Androstephium breviflòrum S. Wats.
Small-flowered Androstephium. Fig. 1006.

Androstephium breviflorum S. Wats. Am. Nat. **7**: 303. 1873.

Scape rather stout, 1–3 dm. high, scabrous especially below. Leaves several, equalling or exceeding the scape, 2 mm. wide, scabrous; bracts lanceolate, scarious; umbels 3–12-flowered; pedicels 15–30 mm. long; perianth light violet-purple, 15–20 mm. long; the segments much longer than the tube; lobes of the crown shorter than the anthers; anthers 3 mm. long; capsule 15–18 mm. in diameter, somewhat deeply 3-lobed; seeds usually 6 in each cavity, 6–8 mm. long.

Lower Sonoran Zone; vicinity of the Needles, southern California eastward to southern Utah and southwestern Colorado. Type locality: southern Utah.

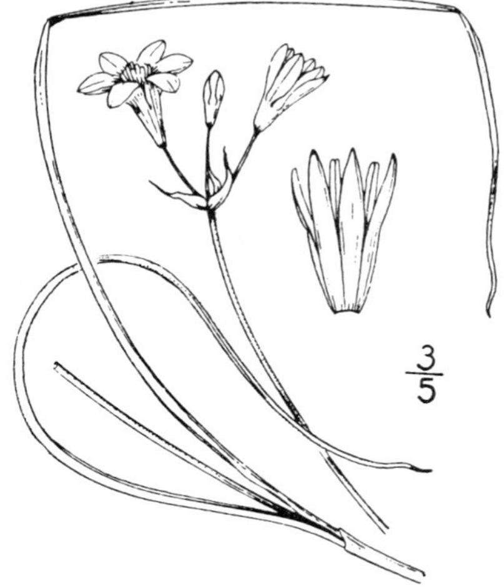

11. LEUCOCRÌNUM Nutt.; A. Gray, Ann. Lyc. N. Y. 4: 110. 1837.

Acaulescent herb, with a deep-seated short rhizome, fleshy roots and narrowly linear leaves surrounded at base by scarious bracts. Flowers in central sessile umbels, showy and fragrant, the pedicels with sheathing bracts. Perianth salver-form with a slender elongated tube and narrowly lanceolate several-nerved segments, persistent. Stamens 6; filaments filiform, inserted below the throat; anthers linear, attached near the base, introrse. Ovary sessile ovate-oblong; style persistent, filiform, tubular, the orifice somewhat expanded and slightly 3-lobed; ovules several in each cavity. Capsule triangular-obovate, subcoriaceous, loculicidally dehiscent. Seeds obovate, strongly angled, with a dull black seed coat. [Name Greek, meaning white lily.]

A monotypic genus of western north America.

1. **Leucocrinum montànum** Nutt. Mountain Lily. Fig. 1007.

Leucocrinum montanum Nutt.; A. Gray, Ann. Lyc. N. Y. **4**: 110. 1848.

Leaves 6–15 or more, linear, 10–20 cm. long, 2–6 mm. wide, flat and thick, obtuse, the underground portion narrower, 5–7 cm. long, sheathed by broad scarious, acutish bracts; flowers 4–8, white, on pedicels 15–35 mm. long from the summit of the rhizome; perianth-tube, very slender, 5–7 cm. long; perianth-segments linear-lanceolate, 20–25 mm. long; stamens scarcely half the length of the perianth-segments; anthers 4–6 mm. long; styles slender, about equalling the stamens; capsule somewhat wrinkled, truncate at apex, 6–8 mm. long; seeds 4–6 in each cavity.

In low moist ground, Arid Transition Zone; Oregon and northern California to Colorado and New Mexico. Type locality: plains of the upper Platte River.

12. **HESPEROCÁLLIS** A. Gray, Proc. Am. Acad. 7: 390. 1867.

Stout erect plants, with a tunicated bulb, mainly basal linear leaves and large funnel-form flowers in an elongated simple raceme, subtended by large scarious bracts. Pedicels jointed at apex. Perianth of 6 spatulate segments, united below the middle into a tube, closely 5–7-nerved. Stamens 6, inserted on the throat; filaments filiform, equal, shorter than the perianth-segments; anthers versatile, linear. Ovary sessile, oblong; ovules numerous; style filiform, equalling the perianth-segments, persistent; stigma depressed capitate, slightly lobed. Capsule subglobose, deeply 3-lobed, loculicidally dehiscent, the valves rather thick. Seeds horizontal, flattened, with a dull black seed coat. [Greek, meaning western beauty.]

A monotypic genus of the desert regions of southern California and Arizona.

1. **Hesperocallis undulàta** A. Gray. Desert Lily. Fig. 1008.

Hesperocallis undulata A. Gray, Proc. Am. Acad. 7: 391. 1867.

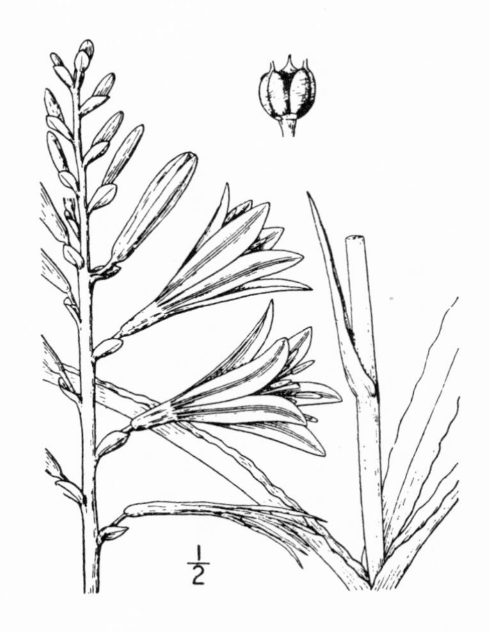

Bulb ovoid; stem erect, stout, simple or sometimes with 1 or 2 branches at the base of the inflorescence; basal leaves linear-lanceolate, 20–40 cm. long, 10–15 mm. wide, glabrous, undulate on the margins, the 2 or 3 stem leaves reduced and attenuated above; raceme 10–30 cm. long, elongating with age and many-flowered, but only a few flowering at one time, the flowers soon withering and deciduous by means of the jointed pedicel; bracts broadly ovate, 10–15 mm. long, abruptly short-acuminate; pedicels about 10 mm. long; perianth-segments 30–40 mm. long, 3–8 mm. broad, somewhat spreading, the tube scarcely half the length of the segments; anthers 7 mm. long; capsule 15 mm. long, acute by the persistent style-base, the valves transversely ridged; seeds about 5 mm. broad.

Dry desert slopes usually in sandy soil, Lower Sonoran Zone; Arizona west to the Imperial Valley, southern California. Type locality: Jessup Rapids, Arizona.

13. ODONTÓSTOMUM Torr. Pacif. R. Rep. 4: 150. *pl. 24.* 1857.

Erect, glabrous herb with fibrous-coated corm, narrow mainly basal leaves and paniculate flowers subtended by scarious bracts. Perianth tubular with a spreading or reflexed 6-parted limb, white or yellowish, tardily deciduous, the segments oblong, about equalling the tube, the outer 5–7-nerved, the inner 3–5-nerved. Stamens 6, inserted on the throat, alternating with as many very short linear staminodia; filaments very short, narrowly subulate; anthers basifixed, dehiscent at the summit. Ovary globose, sessile; ovules 2 in each cavity, ascending; style filiform, slightly 3-cleft, deciduous. Capsule triangular-obovate, 3-lobed, loculicidally dehiscent. Seeds solitary, obovate, with a close thin dark brown slightly rugose testa. [Greek *odous,* tooth, and *stoma,* mouth, referring to the staminodia on the throat.]

A monotypic genus peculiar to California.

1. Odontostomum hártwegi Torr.
Hartweg's Odontostomum. Fig. 1009.

Odontostomum hartwegi Torr. Pacif. R. Rep. 4: 150. 1857.

Corm deep-seated, subglobose, about 25 mm. in diameter. Stem 20–45 cm. high, simple below or usually paniculate from near the base; leaves basal or near the base, linear-lanceolate, 8–12 cm. long, 5–7 mm. wide, the lower stem leaves similar, the upper becoming reduced to bracts subtending the panicle branches; bracts subulate about equalling the pedicels; pedicels slender, 3–5 mm. long, with a small bractlet near the middle; flowers rather numerous, the tube cylindric, 5–6 mm. long, 12-nerved, the segments about equalling the tube, at length reflexed; capsule about 4 mm. in diameter.

Dry usually adobe hillsides, Upper Sonoran Zone; vicinity of Redding south to Mariposa County in the Sierra foothills and to Napa Valley in the Coast Range. A rather rare plant, but sometimes locally abundant. Type locality: Sacramento Valley, probably in Butte County.

14. SCHOENOLÍRION Durand, Journ. Acad. Philad. II. 3: 103. 1855.

Stem naked or sparingly leafy, from a coated bulb. Leaves linear, flat. Inflorescence a densely many-flowered sparingly panicled raceme, with small scarious bracts. Perianth white or greenish, lax and becoming scarious, persistent, divided to the base into 6 distinct, oblong, closely 3-nerved segments. Stamens 6, adnate to the base of the segments; anthers versatile, linear-oblong. Ovary ovate, very shortly stipitate; style short, persistent; ovules 2 in each cavity. Capsule ovoid, broken by the slender persistent style, loculicidally dehiscent. Seeds 2 in each cavity, oblong, black. [Greek, meaning rush and lily.]

A genus of two species, native of the Pacific Coast. Type species, *Schoenolirion album* Durand.

Perianth-segments 3 mm. long. 1. *S. album.*
Perianth-segments 8–10 mm. long. 2. *S. bracteosum.*

1. Schoenolirion album Durand.
White-flowered Schoenolirion. Fig. 1010.

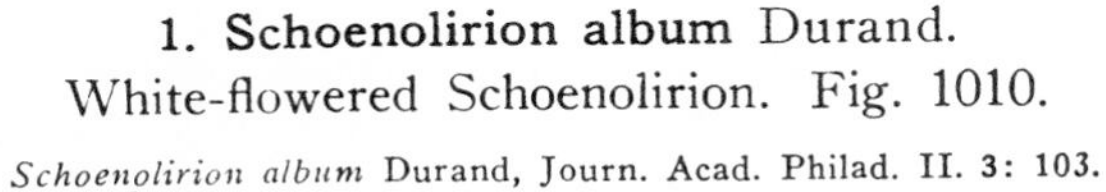

Schoenolirion album Durand, Journ. Acad. Philad. II. 3: 103. 1855.
Hastingsia alba S. Wats. Proc. Am. Acad. 14: 242. 1879.

Bulbs membranously coated or the outer coats somewhat fibrous, ovoid-oblong, 12–18 mm. broad. Stem and branches erect, often stout, 4–9 dm. high, glabrous; leaves several, 3–5 dm. long, 4–12 mm. wide; raceme simple or sparingly branched, densely many-flowered; bracts narrowly accuminate, 2–3 mm. long; pedicels about 1 mm. long, the lowermost sometimes longer; perianth-segments white tinged with green or pink, oblong, obtuse, 5 mm. long, prominently 3-nerved; stamens about equalling the segments; anthers about 1 mm. long; capsule broadly ovoid, 6 mm. long, very short stipitate; seeds oblong, 4 mm. long.

Stream banks and mountain meadows, Transition Zone; Siskiyou Mountains, southern Oregon, to Lake and Nevada Counties, California. Type locality: Nevada City, California.

2. Schoenolirion bracteòsum (S. Wats.) Jepson. Bracted Schoenolirion. Fig. 1011.

Hastingsia bracteosa S. Wats. Proc. Am. Acad. **20**: 377. 1885.
Schoenolirion bracteosum Jepson, Fl. Calif. **1**: 268. 1922.

Bulb oblong-obovid, 20–25 mm. in diameter. Stem stout, about 6 dm. high; leaves ? 5 dm. long, 2–6 mm. wide; raceme elongated, sparingly branched below; bracts filiform-attenuate, 7–11 mm. long; pedicels 2–3 mm. long; perianth-segments narrowly linear-lanceolate, 10–12 mm. long, white, prominently nerved; stamens about half the length of the perianth-segments; style stout, nearly as long as the ovary; capsule unknown.

Mountain meadows and springs, Transition Zone; Siskiyou Mountains, southern Oregon, and Del Norte County, California. Type locality: "Eight Dollar Mountain," near Waldo, Oregon.

15. CHLORÓGALUM Kunth, Enum. 4: 681. 1843.

[Laothoe Raf. Fl. Tellur. **3**: 53. 1836.]

Stems from a fibrous or membranously coated bulb, tall, almost leafless, paniculately branched above, the branches loosely racemose. Basal leaves tufted, long-linear, the stem leaves much reduced. Bracts small and scarious. Pedicels jointed at the summit. Perianth white or purplish, persistent and at length twisted over the ovary, its segments distinct, ligulate, spreading, with 3 closely approximate nerves. Stamens 6, inserted on the base of the perianth-segments; anthers versatile. Style long-filiform, slightly 3-cleft. Capsule broadly turbinate, 3-valved, loculicidal. Seeds 1 or 2 in each cell, obovate, with a thin somewhat rugose testa. [Name Greek, meaning green and juice.]

A California genus of four species. Type species, *Chlorogalum pomeridianum* (DC.) Kunth.

Bulb with thick fibrous coats; pedicels usually 10 mm. or more in length. — 1. *C. pomeridianum.*
Bulb with membranous coats; pedicels usually 5 mm. or less in length.
  Leaves undulate; flowers not white.
    Leaves glabrous; bracts attenuated. — 2. *C. purpureum.*
    Leaves more or less scabrous; bracts triangular. — 3. *C. parviflorum.*
  Leaves narrowly linear, not undulate, becoming revolute; flowers white. — 4. *C. angustifolium.*

1. Chlorogalum pomeridiànum (DC.) Kunth. Common Soap Plant or Amole. Fig. 1012.

Scilla pomeridiana DC. Cat. Hort. Monsp. 143. 1813.
Laothoe pomeridiana Raf. Fl. Tellur. **3**: 53. 1836.
Ornithogalum divaricatum Lindl. Bot. Reg. **28**: *pl. 28.* 1841.
Chorogalum pomeridianum Kunth, Enum. Pl. **4**: 682. 1843.
Laothoe divaricata Greene, Leaflets **1**: 91. 1904.

Bulbs oblong-ovate, often 10 cm. long, densely coated with coarse brown fibres. Stem 4–12 dm. high, glabrous; basal leaves broadly linear, 15–45 cm. long, 8–20 mm. wide, glabrous, carinate, the margins strongly undulate, the stem leaves only 1 or 2, much reduced and attenuated; panicle often 7 dm. long, the branches ascending or widely spreading; pedicels 10–20 mm. long; perianth-segments narrowly ligulate, 15–20 mm. long, scarcely 2 mm. wide, widely spreading from the base and more or less recurved, white with a purple midrib formed by the 3 approximate nerves; stamens about two-thirds the length of the perianth-segments; capsule subglobose, 6 mm. long.

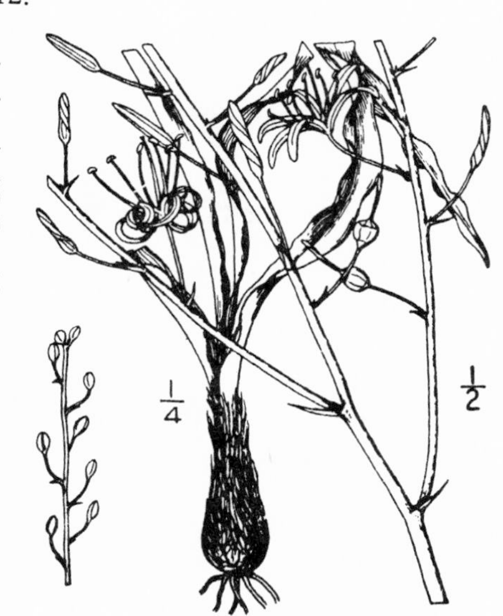

Open valleys and foothills, especially in stony ground, Upper Sonoran Zone; upper Sacramento Valley and Mendocino County to southern California. Type locality: not given. The bulbs are sometimes used as a substitute for soap. The flowers, which bloom in midsummer, after the leaves have begun to dry up, open only toward evening.

2. Chlorogalum purpùreum Brandg.
Purple Amole. Fig. 1013.

Chlorogalum purpureum Brandg. Zoe **4**: 159. 1893.
Laothoe purpurea Greene, Leaflets **1**: 91. 1904.

Bulb ovoid, 2–3 cm. in diameter, 4–5 cm. long, membranously coated. Stem and panicle 30–45 cm. high, slender; basal leaves narrowly linear, 3–4 mm. wide, about 10 cm. long, the margins undulate; stem leaves 1 or 2 attenuated and bract-like; bracts much attenuated, 2–5 mm. long; pedicels 3–5 mm. long, the lower sometimes longer; flowers not vespertine; perianth-segments, oblong, 5–7 mm. long, bluish purple, spreading from above the base; stamens nearly as long as the segments, the filaments filiform purple; style exceeding the segments, curved; ovary sessile; ovules 1 in each cell.

Known only from the Santa Lucia Mountains, California, Upper Sonoran Zone. Type locality: Santa Lucia Mountains.

3. Chlorogalum parviflòrum S. Wats.
Small-flowered Amole. Fig. 1014.

Chlorogalum parviflorum S. Wats. Proc. Am. Acad. **14**: 243. 1879.

Bulb about 3 cm. in diameter, 5–6 cm. long, with dark brown membranous coats. Stem with the panicle 6–10 dm. high; basal leaves linear, 2 dm. or less in length, 5–8 mm. wide, more or less scabrous, the margins undulate; stem leaves 1–3, much reduced and attenuated; panicle 30–50 cm. long, the branches more or less divaricate; bracts triangular, with or without a short acumination, seldom over 3–4 mm. in length; pedicels usually much shorter than the flowers (2–5 mm.); perianth-segments oblong-oblanceolate, 7–8 mm. long, white or pinkish above, purple beneath on the midrib; stamens nearly equalling the segments; style exceeding the segments; ovary broad and obtuse; capsule 3–4 mm. broad.

Dry hillsides and mesas, Upper and Lower Sonoran Zones; San Diego County from the vicinity of Oceanside and Ramona south to northern Lower California. Type locality: Cajon Valley, San Diego County, California.

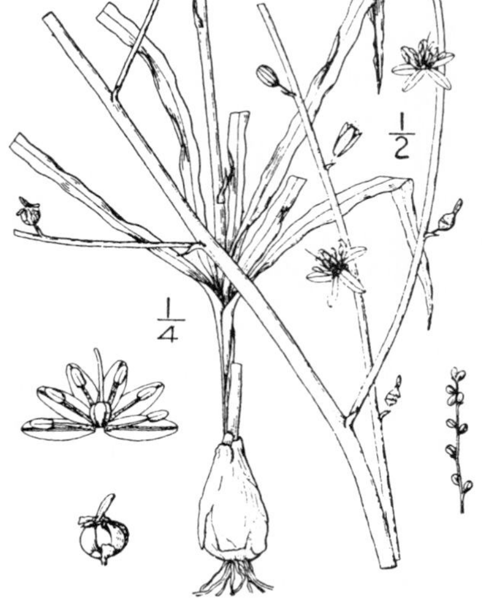

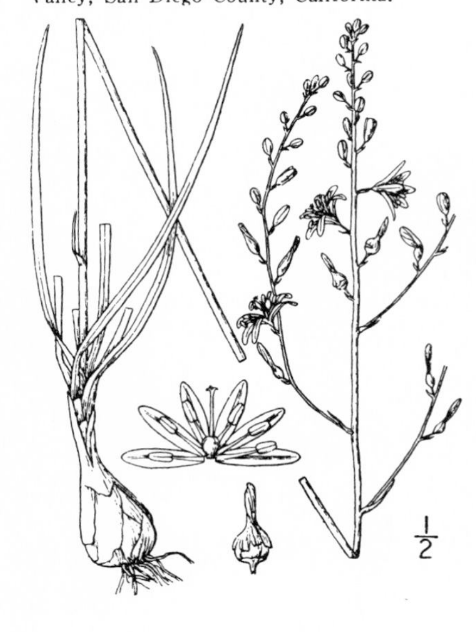

4. Chlorogalum angustifòlium S. Wats.
Narrow-leaved Amole. Fig. 1015.

Chlorogalum angustifolium Kellogg, Proc. Calif Acad. **2**: 104. 1861.
Laothoe angustifolia Greene, Leaflets **1**: 91. 1904.

Bulb ovoid, 15–25 mm. in diameter, tapering at apex, with dark brown, close-fitting membranous coats. Stem slender, 20–45 cm. high; basal leaves grass-like, linear, 10–20 cm. long, 2–5 mm. wide, glabrous, more or less revolute, the margins not undulate; stem leaves 1 or 2, much reduced and subulate; panicle 10–25 cm. long, the branches spreading; pedicels 2–4 mm. long, or slightly longer in fruit; perianth-segments oblong-oblanceolate, 10–12 mm. long, white with a yellowish-green midrib, spreading above the base; stamens and style included; ovary short-stipitate; ovules 2 in each cell; capsule about 4 mm. in diameter.

Dry open hillsides and mesas, Upper Sonoran Zone; Redding south through the foothills to Tuolume and Mendocino Counties, California. Type locality: described from cultivated specimens said to have been collected at Shasta, California.

16. CAMÁSSIA Lindl. Bot. Reg. 18: *pl. 1486.* 1832.

[QUAMASIA Raf. Am. Month. Mag. 2: 265. 1818.]

Herbaceous plants with slender scapes, linear basal leaves, membranous coated bulbs, and rather showy flowers in a simple terminal raceme. Pedicels jointed at the base of the flower, in the axils of scarious bracts. Perianth-segments 6, distinct, persistent, 3–7-nerved. Stamens 6, on the base of the segments; anthers versatile. Style filiform, its base persistent; stigma 3-lobed. Capsule ovate, 3-angled, loculicidal. Seeds several in each cavity, black and shiny. **[From quamash or camass, the Indian name.]**

A North American genus of about 5 species. Type species, *Camassia esculenta* Lindl.

Perianth-segments not connivent and twisted together over the ovary after anthesis; fruiting pedicels erect.
- Leaves 10–20 mm. wide; scapes usually solitary; bulbs edible; wet meadow plant. 1. *C. quamash.*
- Leaves often 35 mm. wide; bulbs nauseous; plants growing on open hillsides. 2. *C. cusickii.*

Perianth-segments connivent and twisted together over the ovary after anthesis; fruiting pedicels spreading.
- Perianth-segments often 20–25 mm. long; capsule oblong, 15–20 mm. long; seeds several in each cell. 3. *C. leichtlinii.*
- Perianth-segments 12–15 mm. long; capsule sub-globose, 6–8 mm. long; seeds 2–3 in each cell. 4. *C. howellii.*

1. Camassia quàmash (Pursh) Greene. Common Camass. Fig. 1016.

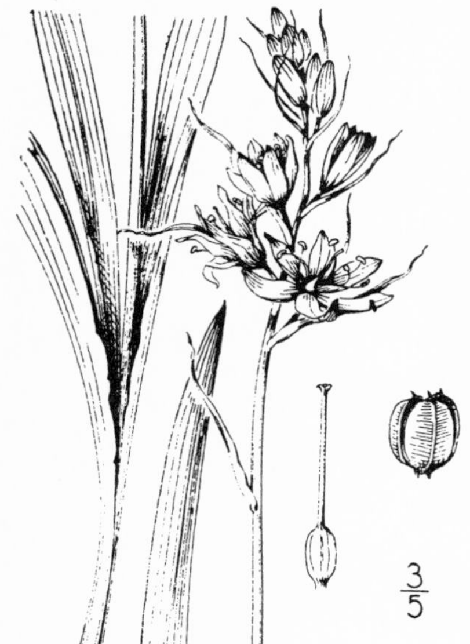

Phalangium quamash Pursh, Fl. Am. Sept. 1: 226. 1814.
Camassia esculenta Lindl. Bot. Reg. 18: *pl. 1486.* 1832.
Camassia quamash Greene, Man. Bay Reg. 313. 1894.
Quamasia quamash Coville, Proc. Biol. Soc. Wash. 11: 64. 1897.
Quamasia walpolei Piper, Proc. Biol. Soc. Wash. 29: 81. 1916.

Bulb broadly ovoid or globose, 15–30 mm. broad, its coats usually black. Scape often rather stout, 2–6 cm. high, sometimes with 1 or 2 short linear scarious leaves; basal leaves linear, shorter than the scape, paler on the upper surface; raceme few to many-flowered, 10–20 cm. long; pedicels curved upward near the base becoming erect in fruit, 10–15 mm. long, shorter than the bracts; perianth 10–25 mm. long, dark bluish purple, the segments twisting separately after anthesis; capsule oblong-ovoid, the cells several-seeded.

Mountain meadows, mainly Canadian Zone; British Columbia to Montana, Utah and the Coast Ranges of northern California. Type locality: Weippe, Idaho.

2. Camassia cusíckii S. Wats. Cusick's Camass. Fig. 1017.

Camassia cusickii S. Wats. Proc. Am. Acad. 22: 479. 1887.
Quamasia cusickii Coville, Proc. Biol. Soc. Wash. 11: 64. 1897.

Bulbs clustered, ovoid, 3–5 cm. broad, inedible, pungent and nauseous. Scapes stout, few to several in a cluster 60–80 cm. high; linear, shorter than the scapes, deeply keeled, glaucous above; 25–35 mm. broad; raceme often 30 cm. long, rather densely many-flowered; perianth pale blue; perianth-segments irregular, 5 of them ascending, 1 deflexed, 15–20 cm. long, mostly 5-nerved; fruiting pedicels erect, about 15 mm. long, scarcely equalling the bracts; capsule oblong, obtuse.

Open hillsides, Canadian Zone; Powder River Mountains, northeastern Oregon, where it was originally collected by Cusick at "4,000 to 6,000 feet altitude."

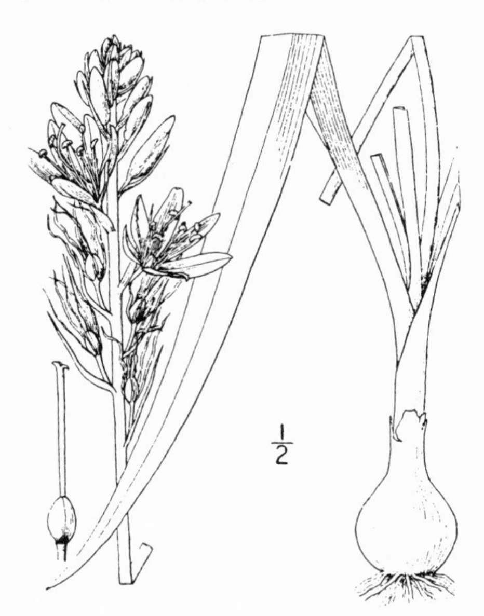

3. **Camassia leichtlínii** (Baker) S. Wats. Leichtlin's Camass. Fig. 1018.

Chlorogalum leichtlinii Baker, Gard. Chron. II. **1**: 689. 1874.
Camassia leichtlinii S. Wats. Proc. Am. Acad. **20**: 376. 1885.
Quamasia leichtlinii Coville, Proc. Biol. Soc. Wash. **11**: 63. 1897.
Camassia suksdorfii Greene, Bot. Gaz. **34**: 307. 1902.
Quamasia azurea Heller, Bull. Torrey Club **26**: 547. 1899.

Bulb broadly ovoid, 15–30 mm. broad, edible. Scape rather stout, 2–6 dm. high, with or without 1 or 2 scarious bracts; basal leaves linear, shorter than the scape, keeled, paler on the upper surface; raceme many-flowered, 10–20 cm. long; bracts shorter than the spreading pedicels; perianth blue or creamy white, regular, the segments 20–35 mm. long, 3-nerved, 5-nerved or sometimes 7-nerved, or alternately 3 and 5, or 5 and 7-nerved, connivent and twisted together over the ovary after anthesis; capsule oblong, 15 mm. long; seeds several in each cavity.

In meadows, Transition and Canadian Zones: Vancouver Island to Utah, the southern Sierra Nevada and northern Coast Ranges, California. Type locality: probably on the Umqua River, Oregon, where the typical white-flowered form occurs.

4. **Camassia howéllii** S. Wats. Howell's Camass. Fig. 1019.

Camassia howellii S. Wats. Proc. Am. Acad. **25**: 135. 1890.
Quamasia howellii Coville, Proc. Biol. Soc. Wash. **11**: 65. 1897.

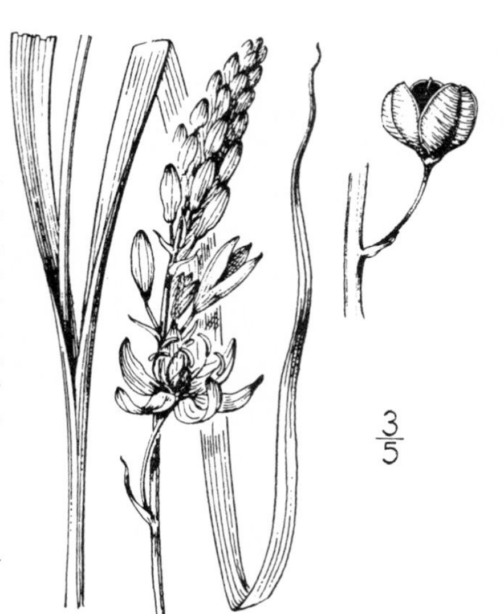

Bulbs ovoid, about 15 mm. broad. Scape rather slender, 3–5 dm. high, bearing 1 or 2 bract-like leaves above the middle; basal leaves several, linear, shorter than the scape, keeled; racemes often 20–25 cm., many-flowered; perianth pale blue, the segments 12–15 mm. long, 3-nerved or rarely 4–5-nerved; fruiting pedicels spreading, 15–25 mm. long, twice longer than the bracts; capsules broadly ovoid or subglobose, very obtuse, 6–8 mm. long, the cells 2–3-seeded.

In meadows, Transition Zone; southern Oregon. Type locality: Grants Pass, Oregon.

17. **LÍLIUM** L. Sp. Pl. 302. 1753.

Tall bulbous herbs, with simple leafy stems and large erect or drooping flowers. Perianth deciduous, funnel-form or campanulate, of 6 distinct spreading or recurved segments, each with a nectar-bearing groove at its base within. Stamens 6, mostly shorter than the perianth, slightly attached to the segments; filaments filiform or subulate; anthers linear or oblong, versatile. Ovules numerous; style long, somewhat clavate above; stigma 3-lobed. Capsule oblong or obovoid, loculicidally dehiscent. Seeds numerous, flat, packed in 2 rows in each cell. [Latin, from the Greek name of the lily.]

A genus of about 50 species widely distributed over the North Temperate Zone. Besides the following there are about 10 species in the Rocky Mountains and eastern United States. Type species, *Lilium candidum* L.

Perianth-segments narrowing gradually to a distinct claw, spotless or finely dotted; flowers funnel-form bulb-scales not jointed.
  Flowers white or pale lilac, often turning purple.
    Flowers horizontal, 7–9 cm. long. 1. *L. washingtonianum.*
    Flowers erect or ascending, 3.5–5 cm. long. 2. *L. rubescens.*
  Flowers pale lemon-yellow, horizontal. 3. *L. parryi.*
Perianth-segments oblanceolate, without a claw, orange-yellow or reddish, mostly conspicuously spotted.
  Flowers erect or ascending, funnel-form, only the upper third spreading or recurved.
    Flowers reddish; bulbs conical or ovate, the scales not jointed.
      Perianth-segments slightly spreading at top, not recurved, reddish purple, spotted with brown-purple; leaves mostly verticillate. 4. *L. bolanderi.*
      Perianth-segments reddish orange, spotted with purple, the tips recurved; leaves mostly scattered. 5. *L. maritimum.*
    Flowers bright orange-yellow, spotted with purple; bulbs on branching rhizomes, the scales jointed. 6. *L. parvum.*
  Flowers nodding, the segments revolute to below the middle.
    Bulbs on branching rhizomes, oblique and mat-like, the scales jointed.
      Anthers 5–6 mm. long; capsule subspherical; bulb-scales 3-jointed. 7. *L. occidentalis.*
      Anthers 10–14 mm. long; capsule narrowly oblong; bulb-scales jointed near the base. 8. *L. pardalinum.*
    Bulbs ovoid, the scales not jointed.
      Anthers 4–6 mm. long; herbage glabrous. 9. *L. columbianum.*
      Anthers 12–15 mm. long; herbage scabrous. 10. *L. humboldtii.*

1. Lilium washingtonianum Kell.
Washington Lily. Fig. 1020.

Lilium washingtonianum Kell. Proc. Calif. Acad. **2**: 13. 1863.

Bulbs 15–20 cm. long, somewhat rhizomatous and oblique, the imbricated scales 25–50 mm. long, not jointed. Stem 6–15 dm. high, glabrous or slightly scabrous; leaves in several whorls of 6–12, oblanceolate, acute or acutish, 4–12 cm. long, 10–35 mm. wide, more or less undulate; inflorescence a 2–20-flowered thyrsoid raceme; flowers horizontally declinate, very fragrant, pure white, becoming purplish, or often sparingly and finely dotted, funnel-form; perianth-segments 7–9 cm. long, 8–15 mm. wide, the upper third spreading; anthers 8–12 mm. long; capsule ovate-oblong, 25 mm. or more in length, truncate at apex, obtusely 6-angled or sometimes slightly winged.

Growing in chaparral, in comparatively dry situations, mainly Arid Transition Zone; Columbia River, Oregon, southward through the Cascades and Sierra Nevada to the southern end of the range. Type locality: Sierra Nevada, California. Also known as Shasta Lily.

2. Lilium rubéscens S. Wats.
Lilac Lily. Fig. 1021.

Lilium washingtonianum purpureum Masters, Gard. Chron. II. **2**: 322. *f. 67.* 1874.

Lilium rubescens S. Wats. Proc. Am. Acad. **14**: 256. 1879.

Bulbs about 5 cm. in diameter, somewhat rhizomatous and oblique, the thick broadly lanceolate scales about 25 mm. long. Stem usually stout, 5–20 dm. high, smooth; leaves oblanceolate, 4–8 cm. long, or sometimes longer, 12–25 mm. wide, acute or acutish, glabrous, glaucous beneath, undulate, scattered below, the upper in 3–7 whorls; flowers 3–8, erect or ascending, pale lilac or nearly white, becoming rose-purple, finely dotted with brown, funnel-form, the upper third revolute, 35–50 mm. long; perianth-segments about 1 cm. broad, acutish; anthers 5–6 mm. long; capsule 25–30 mm. long, about 10 mm. wide, narrowed at base, slightly winged.

On wooded slopes, Transition Zone; Siskiyou Mountains, southern Oregon to Marin and Lake Counties, California. Type locality: Marin County, California.

Lilium kellóggii Purdy. (Garden **59**: 330. 1901.) Perianth-segments pink, dotted with maroon, revolute to the base. Closely related to *Lilium rubescens* and probably best considered as a color form. Wet places, Del Norte County to Red Mountain, northern Mendocino County. Type locality: Kneeland Prairie.

3. **Lilium párryi** S. Wats. Parry's Lily. Fig. 1022.

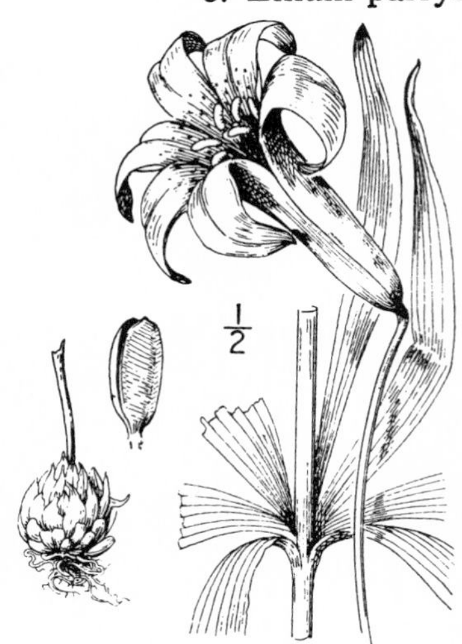

Lilium parryi S. Wats.; Parry, Davenport Proc. Acad. **2**: 189. *pl. 5, 6.* 1878.

Bulb small, about 25 mm. in diameter, somewhat rhizomatous, with numerous jointed scales, 12–20 mm. long. Stem slender, 6–12 dm. high, glabrous; leaves usually scattered, linear-oblanceolate, 10–15 cm. long, 6–10 mm. wide, mostly acuminate; flowers 1 or 2, rarely more, horizontal, funnel-form, pale lemon yellow and minutely dotted, very fragrant; pedicels stout, erect, usually 8–10 cm. long; perianth-segments 6–10 cm. long, 8–11 mm. wide, acuminate, the upper third spreading or the tips finally recurved; anthers oblong, brownish, 6–8 mm. long; capsule 4–5 cm. long, 12 mm. wide, narrowly oblong, acutish, the lobes not winged nor angled.

Wet places about mountain springs and meadows, Transition and Canadian Zones; San Gabriel River, San Gabriel Mountains, southward through the San Bernardino and San Jacinto Mountains, southern California. Type locality: San Bernardino Mountains, "at Ring Brothers' Potato Ranch."

4. **Lilium bolánderi** S. Wats. Bolander's Lily. Fig. 1023.

Lilium bolanderi S. Wats. Proc. Am. Acad. **20**: 377. 1885.

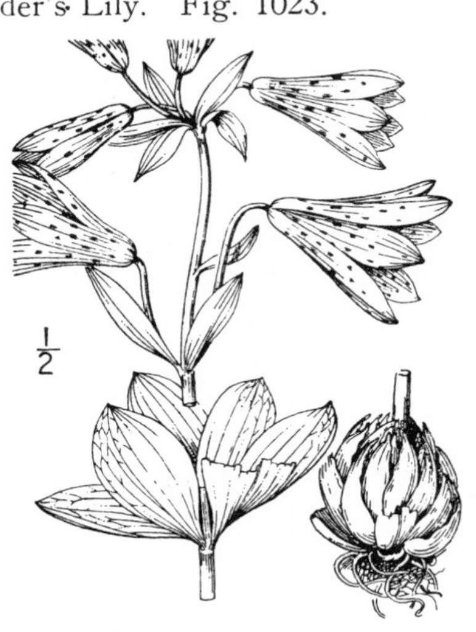

Bulb ovate, 3–4 cm. broad; scales numerous, lanceolate, closely appressed, 25–30 mm. long, not jointed. Stems stout, 3–10 dm. high; leaves mostly in whorls, obovate to oblanceolate, acutish, 35–45 mm. long, glaucous; inflorescence racemose or umbellate, 2–3-flowered to many-flowered; pedicels 3–4 cm. long; flowers horizontal or somewhat nodding, 30–35 mm. long, funnel-form, the upper third somewhat spreading, with the tips little or not at all recurved, reddish-purple, spotted with dark purple; perianth-segments about 8 mm. wide; anthers 5–6 mm. long; ovary 10 mm. long.

Wet places about springs and meadows, Humid Transition and Canadian Zones; Siskiyou Mountains, southern Oregon to Humboldt County, California. Type locality: in the Red Hills, Humboldt County, California.

5. **Lilium marítimum** Kell. Coast Lily. .Fig. 1024.

Lilium màrtimum Kell. Proc. Calif. Acad. **6**: 140. 1875.

Bulb conical 25–35 mm. in diameter, the scales close-appressed, 15–25 mm. long, not jointed. Stems rather slender, 3–6 dm. high; leaves usually scattered, very rarely somewhat verticillate, oblanceolate to linear, 3–8 cm. long, 7–15 mm. wide, obtuse; flowers 1–5, on long peduncles, horizontal, deep reddish-orange, spotted within with purple, funnel-form, the upper third somewhat recurved; perianth-segments 35 mm. long, 6–8 mm. wide, lanceolate, acute; anthers 3 mm. long; ovary 10–12 mm. long; capsule 3–4 cm. long, oblong, narrowed below to a short stout stipe, obtuse at apex, the angles narrowly winged.

Growing in bleak fog-swept meadows near the coast in central California, Humid Transition Zone; Ten Mile River, Mendocino County, south to Marin County. Type locality: low meadows along the coast "in the vicinity of San Francisco (probably Marin County)."

6. **Lilium pàrvum** Kell.
Small Tiger Lily. Fig. 1025.

Lilium parvum Kell. Proc. Calif. Acad. **2**: 179. *f. 52.* 1862.
Lilium canadense parvum Baker, Journn. Linn. Soc. **14**: 241. 1875.

Bulbs small, oblique, rhizomatous, the scales short, thick, 12–20 mm. long, 3-jointed. Stem slender, 4–15 dm. high; leaves scattered or sometimes with 1–3 whorls, broadly lanceolate to linear, 5–12 cm. long, 5–30 mm. broad, acute or acuminate; inflorescence few–many-flowered; flowers erect or ascending, scattered or somewhat verticillate, funnel-form; segment with the upper third recurved and more or less revolute, yellow or orange within and usually spotted with purple, reddish above, narrowly oblanceolate, about 35 mm. long, pubescent toward the apex; stamens 20–25 mm. long; anthers 3 mm. long; style about equalling the stamens; ovary 8 mm. long; capsule subspherical, 12–16 mm. long, truncate at apex.

Mountain springs and stream banks, mainly Canadian Zone; Cascade Mountains of southern Oregon, southward through the Sierra Nevada. Type locality: not given.

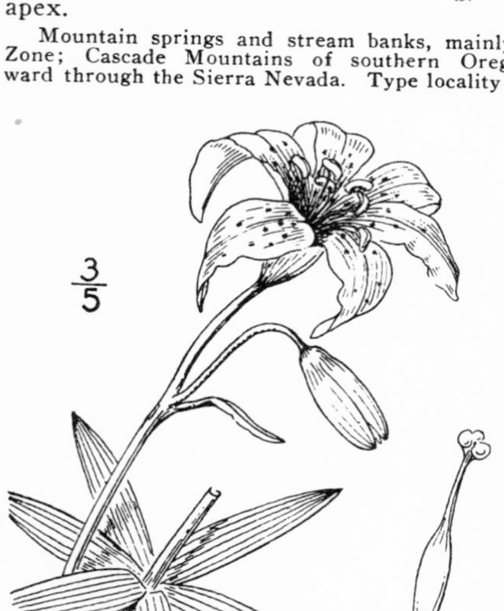

7. **Lilium occidentàlis** Purdy.
Western Lily. Fig. 1026.

Lilium occidentalis Purdy, Erythea **5**: 103. 1897.

Bulbs small, on a branching rhizome; scales 12–18 mm. long, 3-jointed; stems slender, 4–15 dm. high; leaves lanceolate to linear, scattered or usually with 1–3 whorls in the middle, 6–12 cm. long, 5–25 mm. wide, acute or acuminate, pale green; inflorescence few–many-flowered; pedicels 6–12 cm. long, spreading or ascending, recurved above the middle, scattered or somewhat verticillate; flowers nodding; perianth-segments revolute to below the middle, lanceolate, 35–50 mm. long, dark red or usually orange within and usually spotted with purple, reddish or orange above; stamens 25–30 mm. long; anthers 5–6 mm. long; style shorter than the stamens; ovary 9–12 mm. long; capsule subspherical, about 15 mm. long, truncate at apex.

Springs and stream banks, Transition Zone; southern Oregon to northern Humboldt County near the coast.

8. **Lilium pardalìnum** Kell.
Leopard Lily. Fig. 1027.

Lilium pardalinum Kell. Proc. Calif. Acad. **2**: 12. 1859.
Lilium pardalinum angustifolium Kell. Proc. Calif. Acad. **2**: 12. 1859.
Lilium californicum Lindl. Florist. *pl. 33.* 1873.

Rhizomes thick and branching, forming mat-like masses of roundish oblate bulbs; bulb-scales jointed near the base, 15–20 mm. long. Stem stout, 10–25 dm. high, glabrous; leaves usually in 3 or 4 whorls of 9–15, scattered above and below, or sometimes all scatte.ed, linear to lanceolate, acute or acuminate, 6–25 mm. wide; flowers racemose, or the lower whorled, nodding, strongly revolute to below the middle, bright orange-red with a light orange center, conspicuously spotted with large purple spots on the lower half; perianth-segments 5–8 cm. long, 12–18 mm. wide; stamens and style about two-thirds the length of the segments; anthers red, 10–15 mm. long; ovary 20–25 mm. long; capsule narrowly oblong, about 3 cm. long, acutely angled, umbilicate at the apex.

Springs and bogs, Transition and Upper Sonoran Zones; Jackson County, Oregon, southward through the Coast Ranges and lower elevations of the Sierra Nevada to Monterey and Kern Counties, California. Type locality: vicinity of San Francisco.

9. **Lilium columbiànum** Hanson. Columbia Lily. Fig. 1028.

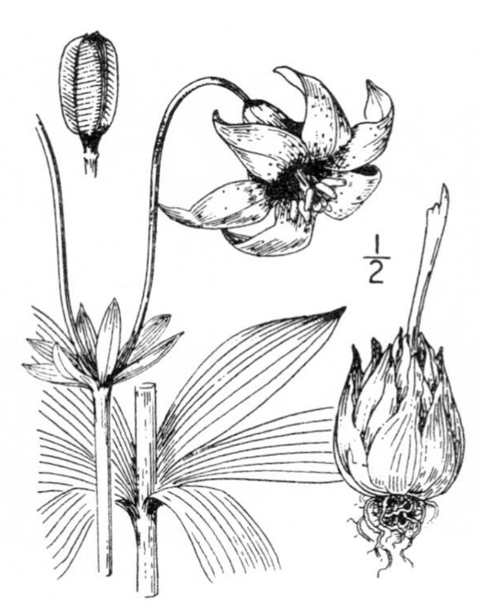

Lilium canadense parviflorum Hook. Fl. Bor. Am. **2**: 181. 1840.
Lilium canadense minus Wood, Proc. Acad. Philad. 166. 1868.
Lilium canadense walkeri Wood, Proc. Acad. Philad. **1868**: 166. 1868.
Lilium columbianum Hanson; Baker, Journ. Linn. Soc. **14**: 243. 1874.
Lilium lucidum Kell. Proc. Calif. Acad. **6**: 144. 1875.
Lilium parviflorum Holzinger, Contr. U. S. Nat. Herb. **3**: 253. 1895.
Lilium bakeri Purdy, Erythea **5**: 104. 1897.
Lilium purdyi Waugh. Bot. Gaz. **27**: 356. 1899.

Bulb small, ovoid, 4–5 cm. in diameter; the whitish scales lanceolate, acute, closely appressed, not jointed. Stems slender, 6–12 cm. high, glabrous; leaves in whorls of 5–9 or sometimes more, the upper and lower scattered, narrowly to broadly oblanceolate, or lanceolate, acute or acuminate, 5–10 cm. long, 12–30 mm. wide, glabrous; inflorescense many-flowered, racemose or whorled below; pedicels slender, 10–15 cm. long; flowers recurved, bright reddish orange, thickly spotted with purple; perianth-segments strongly revolute to below the middle, 4–6 cm. long, 8–12 mm. wide; stamens a little longer than the style; anthers 4–6 mm. long, yellow; ovary about 15 mm. long; capsule 3–4 mm. long, oblong, acutely 6-angled.

In open woods and meadows, moist situations, Transition Zone; southern British Columbia south through Washington and Oregon to Humboldt and Sierra Counties, California, east to western Idaho. Type locality: "Oregon."

10. **Lilium humbòldtii** Roezl. & Leichtl. Humboldt's Lily. Fig. 1029.

Lilium canadense puberulum Torr. Pacif. R. Rep. **4**: 146. 1856.
Lilium humboldtii Roezl. & Leichtl. Duch. Journ. Soc. Hort. Par. II. **4**: 216. 1871.
Lilium bloomerianum Kell. Proc. Calif. Acad. **4**: 160. 1871.
Lilium bloomerianum ocellatum Kell. Proc. Calif. Acad. **5**: 88. *pl. 4.* 1873.

Bulbs ovoid, 5–15 cm. in diameter, white or purplish, bitter; scales very fleshy, ovate-lanceolate, 4–7 cm. long, not jointed. Stems stout, 10–25 dm. high, purplish, more or less puberulent; leaves usually in 4–6 whorls of 10–20 each, 9–12 cm. long, 15–25 mm. wide, oblanceolate, more or less undulate, acute, bright green, scabrous or pubescent on the margins and beneath; inflorescence few-many-flowered, racemose; bracts often ovate; pedicels stout, usually widely spreading, 10–15 cm. long; flowers nodding, bright reddish orange, conspicuously spotted with purple; perianth-segments 7–9 cm. long, 12–25 mm. wide, strongly revolute to the abruptly narrowed short claw, papillose-ridged toward the base; stamens about 5 cm. long, equalling the style; anthers 12–15 mm. long; ovary about 20 mm. long; capsule obovoid, long; acutely 6-angled.

In dry open situations of the foothills and lower altitudes of the mountains, Upper Sonoran and Arid Transition Zones; western slopes of the Sierra Nevada from the head of the Sacramento Valley southward, at least to Amador County (not reported in the southern Sierra), also in southern California from Ventura to San Diego County, and on the channel islands. Type locality: Sierra Nevada. The southern California form often develops bulblets in the leaf-axils.

18. FRITILLÀRIA [Tourn.] L. Sp. Pl. 303. 1753.

Stems erect from scaly bulbs, with thick fleshy scales. Leaves scattered or verticillate, mostly narrow and sessile. Flowers solitary or racemose, leafy-bracted, often dull-colored, nodding. Perianth campanulate or funnel-form, deciduous, of 6 distinct equal oblong-oblanceolate concave segments, more or less blotched or tinged with purple or yellow or white and with a smooth nectariferous pit near the base. Stamens 6, inserted on the base of the segments; filaments slender; anthers oblong, basifixed little or not at all versatile, extrorse, dehiscing laterally. Ovules many; style slender, united to the middle or throughout, deciduous. Capsule membranous, ovate or oblong, 6-angled or 6-winged, loculicidally 3-valved. Seeds flat, in 2 rows in each cell, brownish. [Latin, from *fritillus,* a dice-box or chess-board, in allusion to the form or to the checkered markings of the perianth in some species.]

About 50 species, natives of the North Temperate Zone. Type species, *Fritillaria pyrenaica* L.

Styles distinct above; stigmas linear.
  Capsule rather obtusely angled, not winged; bulbs with few to several large fleshly scales.
    Flowers white or whitish, at least not scarlet or dark purple.
      Segments not mottled.
        Flowers creamy-white, fragrant. 1. *F. liliacea.*
        Flowers greenish white, very ill-scented. 2. *F. agrestis.*
      Segments mottled with purple. 3. *F. purdyi.*
    Flowers scarlet or dark purple.
      Flowers scarlet, yellowish within. 4. *F. recurva.*
      Flowers dark brownish purple throughout. 5. *F. biflora.*
  Capsule acutely angled and winged.
    Bulbs with numerous rice-bulblets.
      Leaves approximate below the flowers; flowers campanulate. 6. *F. camtschatcensis.*
      Leaves in 1–3 whorls, near the center of the stem remote from the flowers; flowers bowl-shaped.
        Leaves broadly lanceolate; flowers deep purple, often mottled with greenish yellow. 7. *F. lanceolata.*
        Leaves narrowly lanceolate to linear; flowers green mottled with bronze-purple.
          Herbage glaucous; flowers mostly 20–30 mm. long. 8. *F. mutica.*
          Herbage pale green, not glaucous; flowers mostly 15 mm. long. 9. *F. multiflora.*
    Bulbs with few large fleshy scales; herbage very glaucous.
      Leaves ovate to lanceolate; flowers usually only 1 or 2. 10. *F. glauca.*
      Leaves narrowly linear; flowers only 1 or usually several.
        Capsule wings without horn-like processes; ovary obtuse. 11. *F. atropurpurea.*
        Capsule wings with horn-like processes at the base and apex; ovary truncate. 12. *F. pinetorum.*
Styles connate to the 3-lobed stigmas.
  Flowers usually several, not yellow.
    Perianth reddish purple, glabrous, 25–35 mm. long. 13. *F. pluriflora.*
    Perianth purple, with a tuft of hairs near the apex of the segments, 20 mm. long. 14. *F. brandegei.*
  Flower usually solitary, nodding, yellow. 15. *F. pudica.*

1. **Fritillaria liliàcea** Lindl. Fragrant Fritillary. Fig. 1030.

Fritillaria liliacea Lindl. Bot. Reg. **20**: under *pl. 1663.* 1834.
Liliorhiza lanceolata Kell. Proc. Calif. Acad. **2**: 46. *fig. 1.* 1860.

Bulbs 15–20 mm. in diameter, of few thick scales, 6–8 mm. long. Stem slender, 2–4 dm. high; leaves 5–20, usually approximate or verticillate near the base, the lower oblanceolate, 5–8 cm. long, 6–15 mm. wide, the upper few, becoming linear; flowers 1–5, campanulate, fragrant; perianth-segments, oblong-ovate to obovate, 12–18 mm. long, creamy-white, more or less tinged with green, with a greenish purple gland at base; stamens about half the length of the perianth-segments; anthers 3 mm. long; styles exceeding the stamens, divided to below the middle; stigmas slender; capsule stipitate, truncate at each end, 10–15 mm. broad and as long, obtusely angled.

Open grassy slopes, usually on outcrops of basalt, Upper Sonoran Zone; Sonoma to Santa Clara Counties, California. Type locality: California; collected by Douglas probably in the vicinity of San Francisco.

2. **Fritillaria agréstis** Greene. Ill-scented Fritillary, Stink Bells. Fig. 1031.

Fritillaria biflora agrestis Greene, Man. Bay Reg. 311. 1894.
Fritillaria agrestis Greene, Erythea **3**: 67. 1895.
Fritillaria succulenta Elmer, Bot. Gaz. **41**: 311.' 1906.

Bulbs with several large fleshy scales, usually seated 15–20 cm. Stems leafy at base, naked or nearly so above, 2–4 dm. high; leaves 5–12 approximate near the base, mostly oblanceolate, 5–7 cm. long, 8–15 mm. wide, thick and more or less succulent; racemes 1–5-flowered; pedicels slender, 4 cm. or less in length; flowers creamy-white tinged with green and purplish at base, campanulate, 15 mm. long, with a strong disagreeable fishy odor; perianth-segments oblong-ovate; stamens 10 mm. long; styles distinct above, somewhat exceeding the stamens; capsule obtusely angled, abruptly rounded at base to a short stipe.

Low ground in usually adobe soil, Upper Sonoran Zone; upper Sacramento Valley south to southern San Joaquin Valley and in the Inner Coast Range valleys from Contra Costa to San Benito and Monterey Counties. Type locality: "grain-fields among the valleys of the Mt. Diablo Range, California."

3. **Fritillaria púrdyi** Eastw.

Purdy's Fritillary. Fig. 1032.

Fritillaria purdyi Eastw. Bull. Torrey Club **29**: 75. *pl. 6.* 1902.

Bulb with 1 or 2 large fleshy scales, seated 10–15 cm. Stem stout, 2–4 dm. long; basal leaves approximate, opposite, in 2–3 pairs, thick, pale green, 6–9 cm. long, 15–30 mm. wide, obtuse, the cauline 3 or 4, alternate about half as wide; flowers 1 or 2, the lower on a slender erect pedicel, recurved at top, campanulate white mottled with purple lines and spots; perianth-segments 20–25 mm. long, the oûter obovate, the inner oblong-lanceolate; stamens half the length of the segments; styles surpassing the stamens, connate to the middle; capsule obtusely angled, abruptly tapering at base, 15 mm. long.

Upper Sonoran and Transition Zones; Humboldt and Mendocino Counties, California. Type locality: Kneeland, Humboldt County.

4. **Fritillaria recúrva** Benth.

Scarlet Fritillary. Fig. 1033.

Fritillaria recurva Benth. Pl. Hartw. 340. 1857.

Bulb 20–25 mm. in diameter, of numerous thick scales 6–8 mm. long or less. Stems rather stout, 4–7 dm. high; leaves usually 8–10, mostly in two whorls near the middle of the stem, linear-lanceolate, 5–8 cm. long, 3–10 mm. wide; flowers 1–9, on erect or ascending pedicels, campanulate-funnelform, 20–30 mm. long; perianth-segments scarlet, slightly tinged with purple without, yellow spotted with scarlet within, oblanceolate, acutish, recurved at tip; nectary obscure; stamens nearly equalling the segments; anthers 3 mm long; style very slender, equalling the stamens, united to near the apex, the free portion 2 or 3 mm. long.

Dry hillsides in open chaparral or woods, Upper Sonoran and Transition Zones; southern Oregon to Mendocino County in the Coast Ranges, and to Placer County in the Sierra Nevada, California. Type locality: "in montibus Sacramento," the Sierrà Nevada probably in Butte or Placer County, California.

Fritillaria recurva coccínea Greene, Pittonia **2**: 230. 1892. Distinguished from the typical form in perianth-segments not recurved at apex, and the flowers usually more brilliantly scarlet. Originally collected on Mt. Hood, Sonoma County, California. Considered by some (*Fritillaria coccinea* Greene, Pittonia **2**: 250. 1892) as a distinct species, and including with it all the Coast Range plants; but, as plants of Mendocino County sometimes have recurved segments, and plants in the Sierra Nevada sometimes straight petals equally bright scarlet, the distinction does not hold geographically.

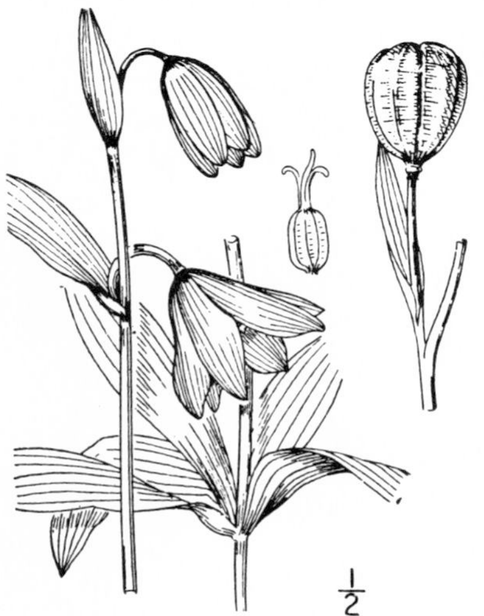

5. **Fritillaria bìflora** Lindl.

Black Fritillary. Fig. 1034.

Fritillaria biflora Lindl. Bot. Reg. **20**: under *pl. 1663.* 1834.
Fritillaria grayana Reichenb. & Baker, Journ. Bot. **16**: 263. 1878.

Bulb 15–20 mm. in diameter, of a few thick fleshy ovate scales, 6–10 mm. long. Stem stout, 15–45 cm. high; leaves 3–7, mostly approximate or somewhat verticillate at base, oblong-oblanceolate, 5–10 cm. long, 10–25 mm. wide, the cauline smaller, lanceolate; flowers 1–4, campanulate, 25–35 mm. long, dark brownish purple tinged with green; perianth-segments spreading, oblong-lanceolate; stamens 8–10 mm. long; anthers 3 mm. long, mucronate; styles distinct above, exceeding the stamens; stigmas linear; capsule broadly obovoid, 12–18 mm. long, obtusely 6-angled.

Usually in stony adobe soil, Upper Sonoran Zone; San Luis Obispo to Riverside and San Diego Counties. Type locality: California; collected by Douglas, exact station not known.

6. **Fritillaria camtschatcénsis** (L.) Ker-Gawl.

Kamchatka Fritillary. Fig. 1035.

Lilium camtschatcense L. Sp. Pl. **1**: 303. 1753.

Fritillaria camtschatcensis Ker-Gawl. Bot. Mag. **30**: under *pl. 1216*. 1809.

Fritillaria lunnellii A. Nels. Proc. Biol. Soc. Wash. **20**: 35. 1907.

Bulbs of several large fleshy scales, subtended by numerous rice bulblets. Stems stout, 3–5 dm. high; leaves usually in 2 or 3 whorls, approximate near the summit, lanceolate to oblong-lanceolate, 5–7 cm. long; flowers usually 2, approximate and short-pedicelled; campanulate, 25–35 mm. long, deep-purple tinged with yellowish green without; perianth-segments oblong-oblanceolate, with a very dark oblong gland at base; stamens 12 mm. long; styles well exceeding the stamens, distinct to the middle; ovary 10 mm. long; capsule obtusely angled.

Moist open woods, Canadian Zone; Aleutian Islands southward along the coast to western Oregon, also on the Asiatic coast to Kamchatka.

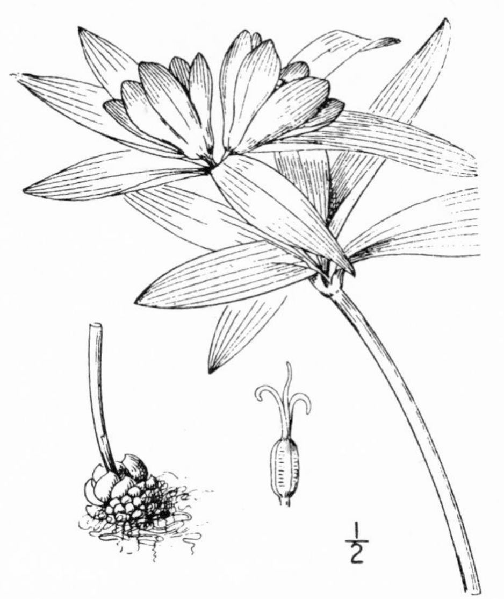

7. **Fritillaria lanceolàta** Pursh.

Purple Rice-bulbed Fritillary or Mission Bells. Fig. 1036.

Fritillaria lanceolata Pursh, Fl. Am. Sept. **1**: 230. 1814.

Bulb with numerous rice-grain bulblets. Stem 4–6 dm. high; leaves 6–9, in 2 or 3 whorls, or scattered, ovate-lanceolate, 5–10 cm. long; flowers 1–4, racemose, deeply bowl-shaped, dark-purple and usually more or less mottled with greenish yellow; perianth-segments ovate to oblong, 25–35 mm. long, deeply concave, with a large ovate lanceolate, usually dark purple gland sharply defined; stamens half the length of the segments; anthers 3 mm. long; styles exceeding the stamens, distinct above; capsule broadly ovoid, truncate at each end, 2 cm. high, broadly winged on the angles.

Shaded slopes and along streams, Humid Transition Zone; British Columbia to northern Idaho, and western Washington and south near the coast to northern California. In Butte County, California, there is a smaller flowered form with the perianth-segments more attenuated, and the styles dark purple.

8. **Fritillaria mùtica** Lindl. Mission Bells. Fig. 1037.

Fritillaria mutica Lindl. Bot. Reg. **20**: under *pl. 1663*. 1834.

Fritillaria lanceolata floribunda Benth. Pl. Hartw. 338. 1857.

Bulb subtended by numerous rice-grain bulblets. Stem 4–8 dm. high, glaucous; leaves in 2 or 3 central whorls, and a few scattering, linear-lanceolate, 3–9 cm. long, early radical leaf, long-petioled, oval, often 8 cm. broad; raceme 1–10-flowered, elongated; flowers greenish yellow blotched with bronze-purple, bowl-shaped; perianth-segments oblong-lanceolate, mostly 25–35 mm. long, concave, with a prominent oblong greenish purple gland; stamens a little over half the length of the segments; styles slightly surpassing the stamens, connate to the middle; capsule truncate at each end, 15–20 mm. long and as broad, conspicuously winged.

Open wooded slopes, Upper Sonoran Zone; California Coast Ranges, mostly east of the fog belt, Mendocino County to Santa Barbara. Type locality: California.

Fritillaria mutica grácilis (S. Wats.) Jepson, Fl. West. Mid. Calif. 108. 1901. (*Fritillaria lanceolata gracilis* S. Wats.) A smaller form with the perianth-segments 15 mm. long, narrower and somewhat attenuated at apex. Marin and Napa Counties, California.

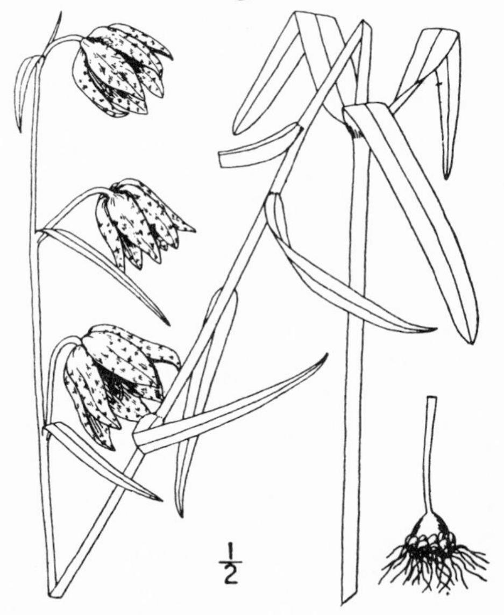

9. **Fritillaria multiflòra** Kell. Many-flowered Fritillary. Fig. 1038.

Fritillaria multiflora Kell. Proc. Calif. Acad. 1: 57. 1855.
Fritillaria parviflora Torr. Pacif. R. Rep. 4: 146. 1857., not Mart. 1838.
Fritillaria viridia Kell. Proc. Calif. Acad. 2: 9. 1863.
Fritillaria micrantha Heller, Muhlenbergia 6: 83. 1910.

Bulb with the solid centers subtended by numerous rice-grain bulblets. Stems light green, but not glaucous, slender, 3–5 dm. high; racemes 3–20-flowered; pedicels slender, slightly recurved; flowers broadly campanulate, greenish yellow, more or less blotched with brownish purple, 15 mm. long; perianth-segments oblanceolate, concave, with shallow nectaries; stamens a little over half the length of the segments; styles exceeding the stamens, distinct above the middle; capsule prominently 6-winged.

On shaded cañon slopes or borders of springs, Arid Transition Zone; Sierra Nevada from Siskiyou to Mariposa County, also in the San Carlos Mountains, San Benito County. Type locality: probably Placerville, California.

10. **Fritillaria glaùca** Greene. Siskiyou Fritillary. Fig. 1039.

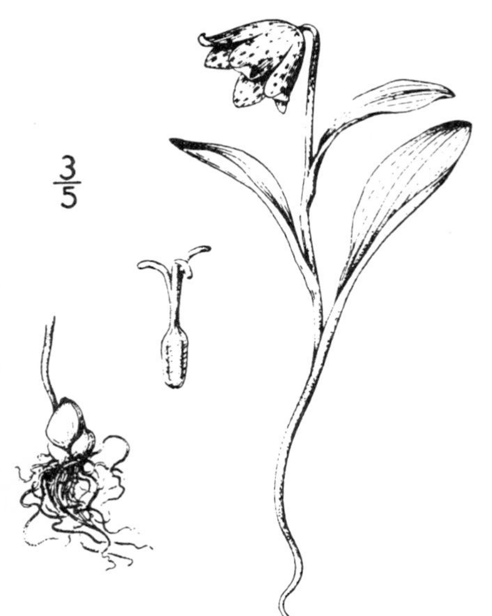

Fritillaria glauca Greene, Erythea 1: 153. 1893.

Bulb small, with few large fleshy scales. Stem 8–15 cm. high; leaves 2–4, scattered, oblong-lanceolate, glaucous; flowers 1–3, purple marked with greenish yellow; perianth-segments oblong-lanceolate, 15–20 mm. long, with large oblong nectary; stamens 12 mm. long; styles distinct to the middle; capsule broadly winged.

Barren slopes, Arid Transition Zone; Siskiyou Mountains, southern Oregon. Type locality: near Waldo, Oregon.

11. **Fritillaria atropurpùrea** Nutt. Purple Fritillary. Fig. 1040.

Fritillaria atropurpurea Nutt. Journ. Acad. Philad. 7: 54. 1834.

Bulb 10–15 mm. in diameter, with several large fleshy scales. Stem 1–5 dm. high, naked below the middle; leaves 6–20, scattered or the upper verticillate, narrowly linear, 3–8 cm. long; flowers 1–6, dull purple, more or less blotched with yellowish green; perianth-segments 8–20 mm. long, elliptic to linear, with obscure nectary; stamens about two-thirds the length of the segments; styles distinct almost to the base; capsule broadly obovoid, about 15 mm. long, acutely angled.

Gravelly slopes, Canadian and Arid Transition Zones; Blue Mountains, Oregon to the southern Sierra Nevada and Lake County, California, east to North Dakota and Utah. Type locality: on the borders of Flathead River, Montana.

12. **Fritillaria pinetòrum** Davidson. Davidson's Fritillary. Fig. 1041.

Fritillaria pinetorum Davidson, Muhlenbergia **4**: 67. 1908.
Fritilliaria atropurpurea pinetorum Johnston, Plant World **22**: 84. 1919.

Plants glaucous throughout, 10–30 cm. high. Leaves few to many, scattered, narrowly linear, 5–15 cm. long; flowers erect or scarcely nodding, 1–3, dull purple mottled with greenish yellow; perianth-segments elliptic, 12–15 mm. long, with obscure nectary; styles distinct nearly to the base; capsule 12–15 mm. long and as broad, the winged angles with horn-like processes at the summit.

Open pine forests, Arid Transition Zone; southern Sierra Nevada, especially on the desert slopes to Mt. Pinos and the San Bernardino Mountains. Type locality: Mt. Cummings, Tehachapi Mountains.

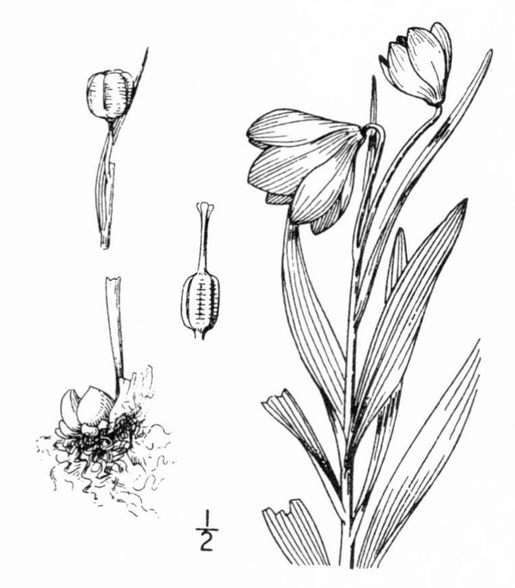

13. **Fritillaria pluriflòra** Torr. Wine-colored Fritillary. Fig. 1042.

Fritillaria pluriflora Torr.; Benth. Pl. Hartw. 338. 1857.

Bulbs of large fleshy scales, seated 10–15 cm. Stems stout, 2–4 dm. high; leaves 6–20, scattered from near the base to the summit, sometimes somewhat verticillate, narrowly lanceolate, 6–9 cm. long; flowers 4–12, nodding on slender elongated pedicels, uniformly reddish purple, not at all blotched, campanulate, 20–35 mm. long; perianth-segments, oblanceolate, with obscure nectaries; stamens half the length of the segments; styles connate exceeding the stamens; stigmas shortly 3-lobed; capsule obtusely angled.

Adobe soil on the plains and foothills, Upper Sonoran Zone; Sacramento Valley. Butte and Yolo Counties. Type locality: Feather River, California. Locally known as Adobe Lily.

14. **Fritillaria brandègei** Eastw. Brandegee's Fritillary. Fig. 1043.

Fritillaria brandegei Eastw. Bull. Torrey Club **30**: 484. 1903.

Bulbs unknown. Stems apparently tall, stout; leaves in whorls of 5–9, lanceolate, 7–9 cm. long, 1–2 cm. wide; flowers about 7, in rather stout recurved pedicels, purplish, campanulate, 2 cm. long; perianth-segments oblong-lanceolate, becoming involute and spreading, somewhat revolute from the base, with a tuft of silky hairs at apex; nectaries obscure; filaments acuminate, 2 mm. broad, 6 mm. long; styles exceeding the stamens, 1 cm. long connate; stigma capitate, scarcely 3-lobed; ovary 6-winged, truncate, 6 mm. long; capsule unknown.

Open pine forests, Transition Zone; southern Sierra Nevada. Type locality: Coburn's Mills, Tulare County, California.

15. Fritillaria pùdica (Pursh) Spreng. Yellow Fritillary, Yellow Bell. Fig. 1044.

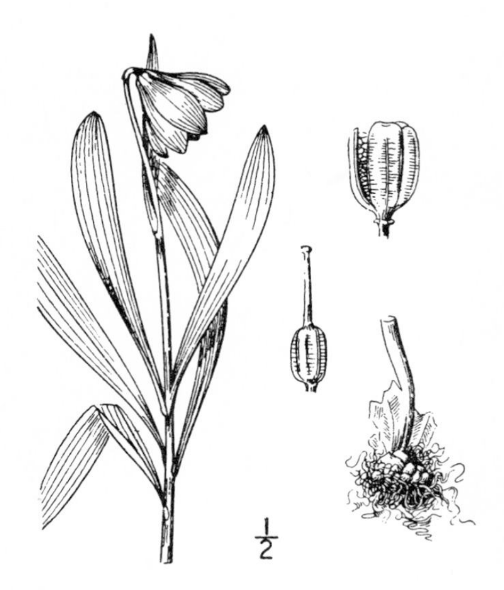

Lilium pudicum Pursh, Fl. Am. Sept. **1**: 228. *pl. 8.* 1814.
Fritillaria pudica Spreng. Syst. **2**: 64. 1825.
Ochrocodon pudicus Rydb. Fl. Rocky Mts. 164. 1917.

Bulb of 2 or 3 large fleshy scales and numerous smaller ones. Stems 8–20 cm. high, slender; leaves 3–8, scattered or somewhat verticillate, linear to narrowly oblanceolate, 5–10 cm. long; flowers 1 or 2, nodding, yellow or orange tinged with purple, campanulate, 10–25 mm. long; nectaries obscure; stamens nearly equalling the segments; styles connate; stigma shallowly 3-lobed; capsule oblong-subglobose, 15–20 mm. long, obtusely lobed.

Grassy slopes, Arid Transition Zone; British Columbia and Montana, south to New Mexico, eastern Oregon and Siskiyou and Plumas Counties, California, and Carson Valley, Nevada. Type locality: Clearwater River, Idaho. Several segregates have been proposed (Gandg. Bull. Soc. Bot. France **66**: 291. 1920), but they are based upon superficial differences.

19. ERYTHRÒNIUM L. Sp. Pl. 305. 7153.

Low herbaceous plants, from deep-seated membranous-coated corms, arising in succession on a short rhizome, or in some species propagating by offshoots, the stem simple, bearing two broad or narrow unequal leaves, usually below the middle, the leaves thus appearing basal. Flowers large, nodding, bractless, solitary or in few–several-flowered racemes or umbels. Many plants are flowerless and 1-leaved. Perianth-segments separate, lanceolate to oblanceolate, deciduous, with a nectariferous groove, and sometimes with two short processes at the base. Stamens 6, hypogynous, shorter than the perianth; anthers linear-oblong, not versatile. Ovary sessile, 3-celled; ovules numerous or several in each cavity; style filiform or thickened above; 3-lobed or entire. Capsule obovoid or oblong, somewhat 3-angled, loculicidal. Seeds compressed, or somewhat angled and swollen. [Greek, in allusion to the red flowers of some species.]

A genus of about 15 species, all but one North American. Type species, *Erythronium dens-canis* L.

Corms propagating from slender offshoots produced from the base; flowers on elongated scape-like pedicels, arising from a sessile umbel. 1. *E. multiscapoideum.*
Corms arising in succession from short rhizomes; propagation only by seed; flowers not in a sessile umbel.
  Stigma 3-lobed, the lobes finally recurved.
    Flowers bright yellow; the leaves not mottled.
      Anthers purple. 2. *E. grandiflorum.*
      Anthers cream white. 3. *E. parviflorum.*
    Flowers not bright yellow, usually cream, with yellow center, often fading pink.
      Leaves not mottled; filaments filiform. 4. *E. montanum.*
      Leaves mottled.
        Filaments not dilated below, stoutish. 5. *E. californicum.*
        Filaments dilated below.
          Capsule obtuse or retuse at apex; perianth-segments cream-color with yellow base and often with a greenish brown spot above. 6. *E. giganteum.*
          Capsule acutish at apex; perianth-segments cream-color usually tinged with purple. 7. *E. revolutum.*
  Stigmas entire or obscurely notched; style clavate.
    Inner perianth-segments appendaged.
      Perianth-segments 25 mm. or more in length.
        Perianth light yellow with orange base. 8. *E. citrinum.*
        Perianth pale purple with a dark purple base. 9. *E. hendersonii.*
      Perianth-segments less than 25 mm. long. 10. *E. purpurascens.*
    Inner perianth-segments not appendaged. 11. *E. howellii.*

1. Erythronium multiscapoìdeum (Kell.) Nelson & Kennedy.
Sierra Fawn Lily. Fig. 1045.

Fritillaria multiscapoidea Kell. Proc. Calif. Acad. 1: 46. 1855.
Erythronium hartwegii S. Wats. Proc. Am. Acad. **14**: 261. 1879.
Erythronium multiscapoideum Nelson & Kennedy, Muhlenbergia 3: 137. 1908.

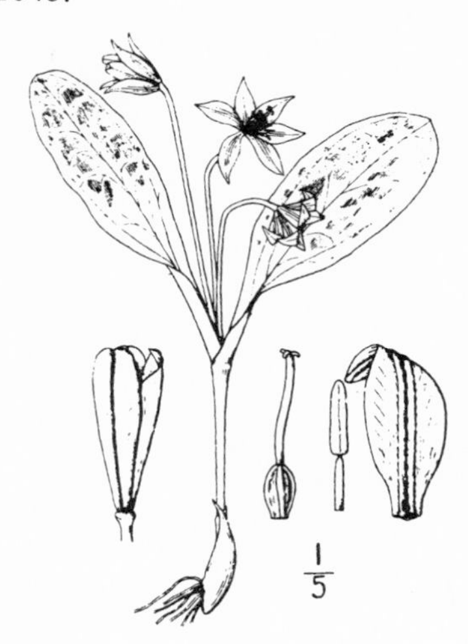

Corm ovate-oblong, 12–18 mm. long, offshoots produced at the base of the corm on slender stolons. Leaves oblanceolate, mottled; flowers solitary or 2 or 3 in a cluster arising from a sessile umbel, each on an elongated scape-like pedicel, cream-yellow above orange toward the base, fading to light pink; perianth-segments spreading, becoming recurved in age, lanceolate, 25–45 mm. long; the inner with a small rounded appendage at the base of the medial folds and a smaller one on the edge; filaments about 4 mm. long, stout, not dilated below, white, the outer abruptly narrowed at the apex; anthers cream-yellow, 10 mm. long; style 10–12 mm. long; stigma 3-lobed, the lobes recurved; ovary ovate-oblong, 4 mm. long; capsule oblong-obovoid, 15 mm. long, slightly winged.

Open woods and moist rocky situations, Upper Sonoran and Transition Zones; foothills of the Sierra Nevada from Placer to Mariposa Counties. Type locality: Placerville, Eldorado County, California. *Fritillaria multiscapoidea* Kell., heretofore referred to *E. purpurascens* S. Wats., undoubtedly is this species.

2. Erythronium grandiflòrum Pursh. Yellow Fawn Lily. Fig. 1046.

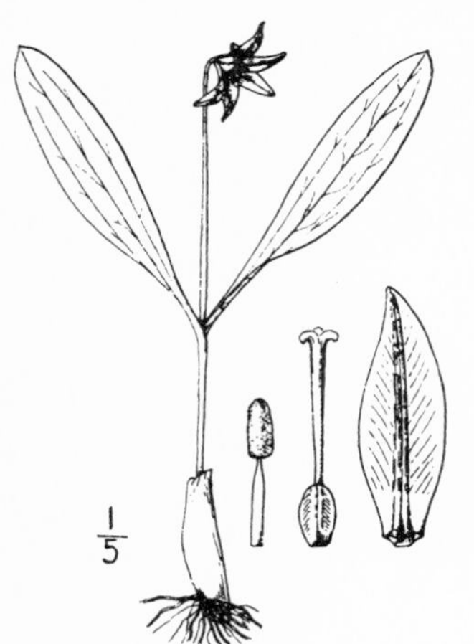

Erythronium grandiflorum Pursh, Fl. Sept. 1: 231. 1814.
Erythronium grandiflorum minus Hook. Fl. Bor. Am. **2**: 182. 1840.
Erythronium grandiflorum candidum Piper, Piper & Beattie, Fl. Southeast Wash. 61. 1914.

Leaves 7–15 cm. long, oblong-lanceolate, acute or acutish, narrowed to a broad and usually short petiole, not mottled; scape elongated, bearing a solitary flower or often a raceme of 2–6 flowers; perianth-segments bright yellow lanceolate, acuminate, strongly recurved, 3–5 cm. long; filaments about 12 mm. long, dilated below the middle; anthers reddish purple, 4–6 mm. long after dehiscence; ovary narrowly oblong, narrowed to a short stipe; stigma 3-lobed, the lobes at length recurved; capsule narrowly oblong-oblanceolate, gradually narrowed to a short stipe.

Open forests and grassy hillsides, Arid Transition Zone; British Columbia to Oregon, east of the Cascade Mountains, east to Montana and Utah. A white-flowered form (var. *candidum*) sometimes occurs. Type locality: "on the Kooskooskee," near the present town of Kamiah, Idaho.

3. Erythronium parviflòrum (S. Wats.) Goodding.
Small-flowered Fawn Lily. Fig. 1047.

Erythronium grandiflorum parviflorum S. Wats. Proc. Am. Acad. **26**: 129. 1891.
Erythronium parviflorum Goodding, Bot. Gaz. **33**: 67. 1902.

Corm narrowly oblong, 3–5 cm. long, arising in succession from a short rhizome. Leaves oblong, gradually narrowed at both ends, 10–15 cm. long, 3–4 cm. wide, not mottled; scape slender, 10–20 cm. long; flowers rarely more than one, terminating an abruptly curved peduncle; perianth-segments bright yellow with a very pale greenish base, broadly lanceolate-acuminate, 2–3 cm. long, 1 cm. wide, strongly reflexed; filaments slender, slightly dilated below, 1–1.5 cm. long; anthers cream color; style clavate, 3-lobed; capsule broadly oblong to oval, 2–3 cm. long.

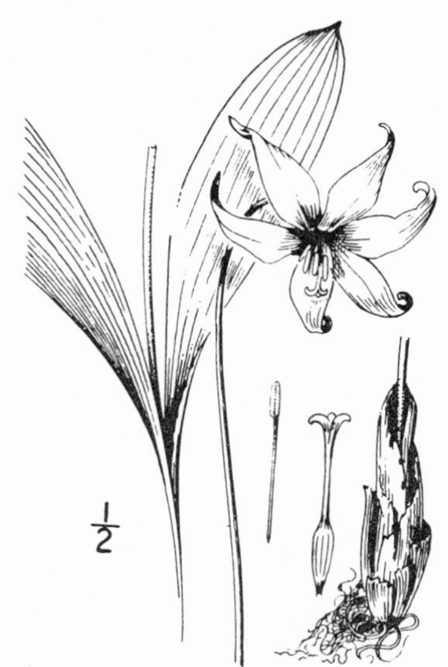

In subalpine meadows, Canadian and Hudsonian Zones; British Columbia, south through the Olympic and Cascade Mountains to northern California, extending eastward to Wyoming and Colorado. Type locality: Telephone Mines, Wyoming.

4. Erythronium montànum S. Wats. Alpine Fawn Lily. Fig. 1048.

Erythronium montanum S. Wats. Proc. Am. Acad. **26**: 130. 1891.

Corm narrowly oblong, 3–4 cm. long, arising in succession from the rhizome, which is often 4–5 cm. long and composed of a double row of imbricated thickened scales. Leaves not mottled, oblong-ovate, 7–12 cm. long, 2.5–4 cm. wide, acute or acutish at apex, abruptly narrowed at base; petiole 3–5 cm. long; scape rather stout, 25–45 cm. long, 1–several-flowered; perianth-segments white with an orange base, often drying pinkish, lanceolate-acuminate 35–40 mm. long, 10–13 mm. wide, somewhat recurved; filaments slender, not dilated below, 10–12 mm. long; anthers white, 3–4 mm. long; style slender, 12 mm. long, 3-lobed; capsule oblong-obovate, rounded at the apex, about 3 cm. long.

Alpine meadows, Hudsonian Zone; Washington and Oregon. Type locality: Mt. Hood, Oregon.

5. Erythronium califórnicum Purdy. California Fawn Lily. Fig. 1049.

Erythronium californicum Purdy, Flora and Sylva **2**: 253. 1904.

Leaves strongly mottled, oblong, 10–15 cm. long, 25–40 mm. wide, obtuse at apex, narrowed to a short (20–25 mm.) winged petiole; scape stout, 20–35 cm. long, 1-flowered or commonly 2–several-flowered; perianth-segment creamy-white, orange at base, broadly lanceolate, 30–40 mm. long, 10–14 mm. wide, appendaged at base; filaments rather stout, but not dilated below, about 8 mm. long; anthers cream color, 3–4 mm. long; style 10 mm. long, somewhat clavate, 3-lobed; capsule narrowly obovoid, 3–4 cm. long, 8–10 mm. wide, rounded at apex.

Open woods, Upper Sonoran and Transition Zones; California Coast Ranges from Mendocino to Lake and Sonoma Counties. Locally known as "Easter Lily." Type locality: Sonoma to Trinity County, California.

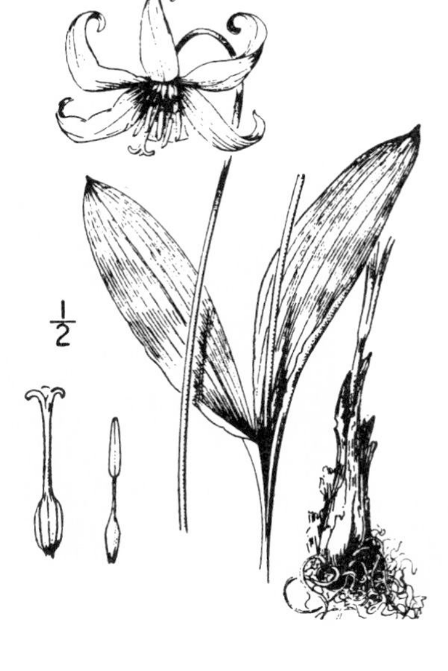

6. Erythronium gigánteum Lindl. Giant Fawn Lily. Fig. 1050.

Erythronium giganteum Lindl. Bot. Reg. under *pl. 1786*. 1835.
Erythronium grandiflorum albiflorum Hook. Fl. Bor. Am. **2**: 182. 1840.

Leaves mottled, oblong-lanceolate, 10–15 cm. long, 25–45 mm. wide; scape 25–40 mm. long, 1–6-flowered; perianth-segments creamy-white, marked with orange at base and sometimes with a band of greenish brown, lanceolate, acuminate, 30–50 mm. long, 8–14 mm. wide, strongly revolute, the inner appendaged at base; filaments about 10 mm. long, strongly dilated to near the middle; anthers cream-yellow, 7–9 mm. long; style usually longer than the stamens, slightly clavate, deeply 3-lobed, the lobes reflexed; capsule oblong-ovoid, about 3 cm. long, very obtuse or retuse at apex.

Stony ridges, Transition Zone; Puget Sound, Washington, to the Rogue River, southern Oregon. Type locality: "northwest America."

7. **Erythronium revolùtum** Smith. Coast Fawn Lily. Fig. 1051.

Erythronium revolutum Smith, Rees Cyclop. **13**: no. 3.
Erythronium grandiflorum smithii Hook. Fl. Bor. Am. **2**: 182. 1840.
Erythronium revolutum bolanderi S. Wats. Proc. Am. Acad. **26**: 129. 1891.
Erythronium smithii Orcutt, West. Am. Sci. **7**: 129. 1891.
Erythronium johnsonii Bolander, Erythea **3**: 127. 1895.
Erythronium revolutum johnsoni Purdy, Bailey Cyclop. Hort. 548. 1900.

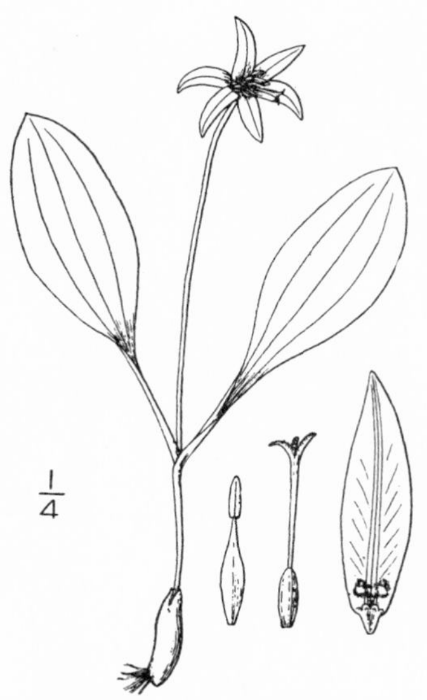

Leaves mottled, oblong-lanceolate, attenuate at base to a narrowly winged petiole; scape 10–30 cm. high, 1–2-flowered, rarely 3–4-flowered; perianth-segments narrowly lanceolate, 6–7 mm. broad, strongly recurved, creamy white, yellow at base, often turning purplish with age; filaments broadly dilated below; anthers white; style distinctly 3-lobed, the lobes recurved; capsule abruptly acutish at apex.

Coastal region, Canadian and Humid Transition Zones; Vancouver Island to Sonoma County, California. Type locality: Vancouver Island.

8. **Erythronium citrìnum** S. Wats. Lemon-colored Fawn Lily. Fig. 1052.

Erythronium citrinum S. Wats. Proc. Am. Acad. **22**: 480. 1887.

Leaves mottled, broadly lanceolate, obtuse, and short-apiculate, narrowed to a short winged petiole; scape rather stout, 10–20 cm. high, 1–9-flowered; perianth-segments strongly recurved, lemon-yellow with orange base, the tips becoming pink with age, the inner auriculate; filaments cream color, slender, the inner set a little stouter; anthers cream color, 5 mm. long; style slightly clavate; stigma obscurely 3-lobed.

Open woods, Transition Zone; a local species known only from the Siskiyou Mountains, southern Oregon, and Tuolumne County, California. Type locality: Deer Creek Mountains, Josephine County, Oregon.

9. **Erythronium hendersònii** S. Wats. Henderson's Fawn Lily. Fig. 1053.

Erythronium hendersonii S. Wats. Proc. Am. Acad. **22**: 479. 1887.

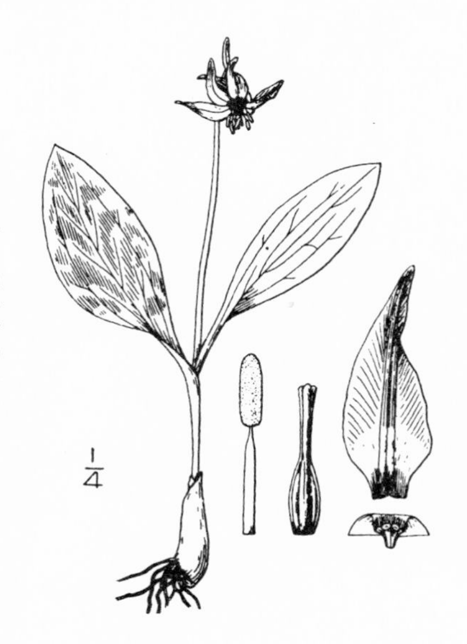

Corm oblong, 3–5 cm. long, arising in succession from a short rhizome. Leaves mottled, oblong-lanceolate, 10–15 cm. long, 30–45 mm. wide, gradually or rather abruptly narrowed to a winged petiole, 25–35 mm. long; scape 10–25 cm. high; perianth-segments pale purple, with a dark purple base surrounded by a tinge of yellow, strongly recurved, 25–35 mm. long, the inner with fleshy subsaccate auricles above the claw; filaments purple, very slender and attenuate; anthers brownish, 5 mm. long; style narrowly clavate, scarcely equalling the stamens; stigma obscurely 3-lobed.

Grassy fields and open woods, Transition Zone; southern Oregon, west of the Cascade Mountains. Type locality: Deer Creek Mountains, Josephine County.

10. **Erythronium purpuráscens** S. Wats. Purple Fawn Lily. Fig. 1054.

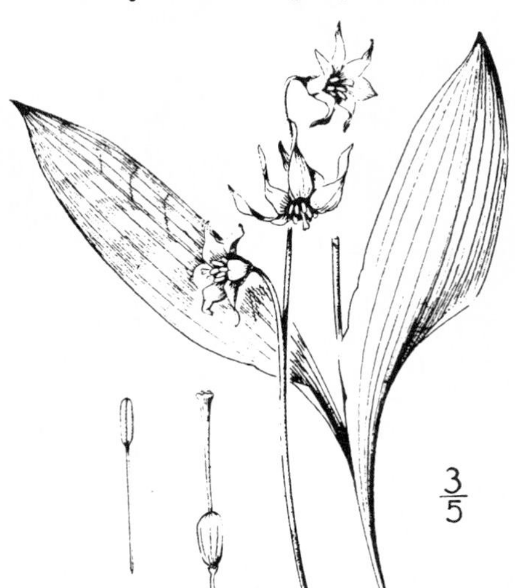

Erythronium purpurascens S. Wats. Proc. Am. Acad. **12**: 277. 1877.

Leaves narrowly lanceolate to oblong-lanceolate, 10–15 cm. long, undulate, not mottled, narrowed to the broadly winged petiole; scape 10–25 cm. high, bearing a single flower or generally a 2–8-flowered raceme; pedicels very unequal, 5–40 mm. long; perianth-segments 12–15 mm. long, lanceolate, spreading, light yellow tinged with purple, deep orange at base; filaments very slender; anthers yellow, 3 mm. long; styles clavate; stigma obscurely 3-lobed.

Moist granitic soils, Canadian Zone; Sierra Nevada, from Plumas to Nevada Counties, California. Type locality: near Downieville, Sierra County.

11. **Erythronium howéllii** S. Wats. Howell's Fawn Lily. Fig. 1055.

Erythronium howellii S. Wats. Proc. Am. Acad. **22**: 480. 1887.

Corm narrowly oblong, 3–4 cm. long. Leaves mottled, oblong-lanceolate or generally lanceolate, 8–15 cm. long, 15–40 mm. wide, attenuate at base to a narrowly winged petiole; scape 1–4-flowered, slender; perianth-segments pale yellow with an orange base, tinged with pink in age, without auriculate appendages; filaments very slender; anthers white, 4 mm. long; style clavate; stigma obscurely 3-lobed.

Open woods, Transition Zone; Siskiyou Mountains, southern Oregon, south to Trinity Mountains, California. Type locality: Waldo, Josephine County, Oregon.

1/2

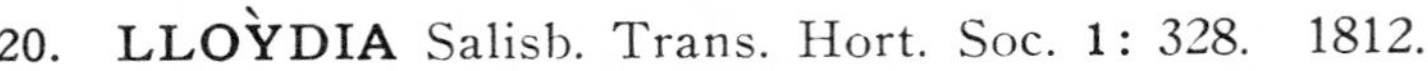

20. **LLOỲDIA** Salisb. Trans. Hort. Soc. **1**: 328. 1812.

Dwarf-caulescent herb with tunicated bulb arising from a creeping rootstock. Leaves narrow grass-like. Flowers white, solitary or in a terminal raceme. Perianth-segments with a transverse fold-like gland near the base. Stamens 6, separate; filaments subulate; anthers basifixed, dehiscent by marginal slits. Ovary triangular, 3-celled; ovules numerous in 2 rows in each cavity, anatropous; style persistent; stigma 3-lobed. Capsule loculicidal at apex. [Name in honor of George Lloyd.]

About 12 species, mainly European and Asiatic, only the following in America. Type species, *Lloydia serotina* (L.) S. Wats.

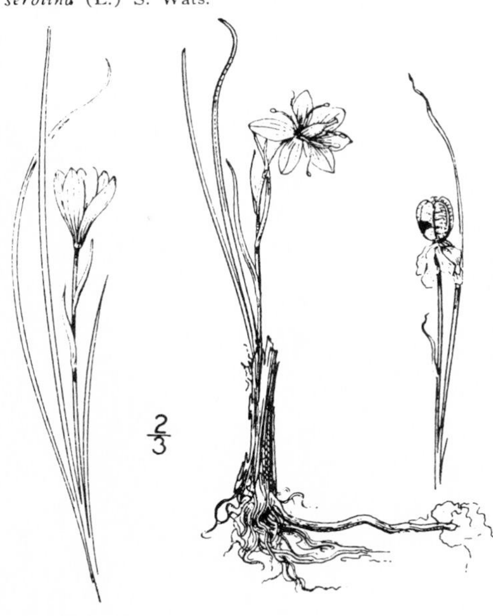

1. **Lloydia serotìna** (L.) S. Wats. Alp Lily. Fig. 1056.

Anthericum serotinum L. Sp. Pl. ed. 2. **1**: 444. 1762.

Lloydia alpina Salisb. Trans. Hort. Soc. **1**: 328. 1872.

Lloydia serotina S. Wats.; Hort. Brit. ed. 2. 527. 1830.

Bulb oblong, covered with grayish fibrous coats, ending in a creeping rootstock. Stem slender erect, 5–15 cm. high; leaves several, 5–10 cm. long, 1–2 mm. wide; flowers usually solitary, broadly turbinate; perianth-segments oblanceolate, about 1 cm. long, obtuse, creamy white, purple-veined and tinged with rose on the back; stamens about two-thirds the length of the perianth-segments; filaments slender; anthers 1 mm. long; style 2 mm. long; capsule obovoid, 8 mm. long.

Wet rocks, Arctic-Alpine Zone; Unalaska to Oregon, Nevada and New Mexico. Type locality: Europe.

21. CALOCHÓRTUS Pursh, Fl. Am. Sept. 1: 240. 1814.

Stems usually flexuous and branching from membranous or rarely fibrous coated corms, with few linear-lanceolate leaves, those of the stems alternate, clasping. Flowers few, showy, terminal on the branches or umbellately fascicled. Perianth deciduous, of 6 distinct more or less concave segments, the 3 outer sepal-like, narrow, the 3 inner petaloid, mostly broadly cuneate-obovate, usually with a conspicuous glandular pit near base. Stamens 6, inserted on base of segments; anthers linear to oblong, basifixed. Ovules many; stigmas sessile, recurved, persistent. Capsule elliptic to oblong. [Greek, signifying beautiful grass.]

About 50 species, natives of western North America and Mexico. Type species, *Calochortus elegans* Pursh.

Capsule winged on the angles.
Flowers nodding, globose, the petals being so strongly arched that the tips often cross each other (sometimes more spreading in *amoenus*).
Flowers white or purple.
Flowers white or faintly tinged with rose-purple. 1. *C. albus.*
Flowers purple. 2. *C. amoenus.*
Flowers yellow.
Petals hairy within to near the apex. 3. *C. pulchellus.*
Petals naked within or very sparsely hairy on each side of the gland. 4. *C. amabilis.*
Flowers open-campanulate, usually erect.
Capsule nodding in fruit.
Petals ciliate, and the inner surface usually covered with hairs.
Flowers bright yellow. 5. *C. monophyllus.*
Flowers creamy white or blue, often tinged with purple.
Anthers oblong, obtuse, not at all apiculate. 6. *C. caeruleus.*
Anthers acute to acuminate, or apiculate.
Sepals without a pit at base.
Flowers white more or less tinged with purple; anthers acuminate.
Gland with a scale.
Capsule suborbicular, rounded at both ends.
Gland scale half as wide as the petal or more, deeply laciniate into very narrow segments. 7. *C. elegans.*
Gland scale one-third as wide as the petal, but little incised. 8. *C. nanus.*
Capsule oblong-elliptic, acutish at both ends.
Petals less than 20 mm. long; umbels short-peduncled (2–3 cm.). 9. *C. maweanus.*
Petals more than 20 mm. long; umbels on peduncles 7–10 cm. long. 10. *C. purdyi.*
Gland without a scale. 11. *C. tolmiei.*
Flowers pale yellow; anthers conspicuously apiculate. 12. *C. apiculatus.*
Sepals with a distinct pit near the base. 13. *C. lobbii.*
Petals not ciliate, naked except for a few hairs around the gland.
Stems not producing bulblets near the base.
Anthers linear-lanceolate, 4–5 mm. long. 14. *C. nudus.*
Anthers oblong, 3 mm. long. 15. *C. umbellatus.*
Stems producing bulblets near the base. 16. *C. uniflorus.*
Capsule erect in fruit.
Flowers nodding on very slender pedicels.
Petals ovate-lanceolate, acute, strongly ciliate. 17. *C. lyalli.*
Petals obovate, rounded or obtuse, not ciliate. 18. *C. shastensis.*
Flowers strictly erect, on stoutish pedicels.
Stems normally producing 1 or more bulblets near the base. 19. *C. longebarbatus.*
Stems not producing bulblets near the base.
Petals without a central dark purple spot.
Petals long-hairy.
Petals long-ciliate, with a large lunate purple band above the gland. 20. *C. nitidus.*
Petals rarely ciliate, barred with yellow below. 21. *C. greenei.*
Petals with short crisped hairs. 22. *C. howellii.*
Petals with a rounded central dark purple spot. 23. *C. eurycarpus.*
Capsule not winged.
Capsule obtuse at apex. 24. *C. catalinae.*
Capsule attenuate at apex.
Petals not ciliate, hairy only below the middle (except *splendens*); corms membranously coated.
Gland depressed, surrounded by a laciniate membrane or more or less connected scales.
Flowers yellow or red.
Petals yellow.
Hairs of the petals clavate. 25. *C. clavatus.*
Hairs of the petals not clavate. 26. *C. concolor.*
Petals red. 27. *C. kennedyi.*
Flowers white or lilac.
Sepals long-acuminate, usually exceeding the petals. 28. *C. macrocarpus.*
Sepals acute or short-acuminate, shorter than the petals. 29. *C. nuttallii.*
Gland not depressed, without membrane or scales.
Stems not producing bulblets near the base.
Petals lilac, without a spot at base.
Hairs of gland stellate-branched at apex.
Anthers 4–6 mm. long; petals hairy only on the lower third. 30. *C. davidsonianus.*
Anthers 8–10 mm. long; petals hairy to the middle. 31. *C. splendens.*
Hairs of the gland linear, simple. 32. *C. palmeri.*
Petals white with reddish brown spot at base. 33. *C. dunnii.*
Stems normally producing bulblets near the base.
Petals never oculate or pencilled.
Hairs of gland linear, simple.
Anthers not sagittate. 34. *C. paludicola.*
Anthers sagittate. 35. *C. leichtlinii.*
Hairs of gland club-shaped. 36. *C. invenustus.*
Petals normally oculate or penciled.
Flowers yellow, with or without central spot, penciled with brown. 37. *C. luteus.*
Flowers purple, oculate. 38. *C. venustus.*
Petals normally ciliate on the margin, hairy to near the apex; corms coarsely fibrous-coated.
Petals purple, sometimes penciled with brown, the margins usually not ciliate. 39. *C. plummerae.*
Petals yellow, blotched or penciled with brown.
Petals fan-shaped, truncate, much longer than the stamens. 40. *C. weedii.*
Petals ovate-lanceolate, shorter than the stamens. 41. *C. obispoensis.*

1. **Calochortus álbus** (Benth.) Dougl. White Fairy Lantern. Fig. 1057.

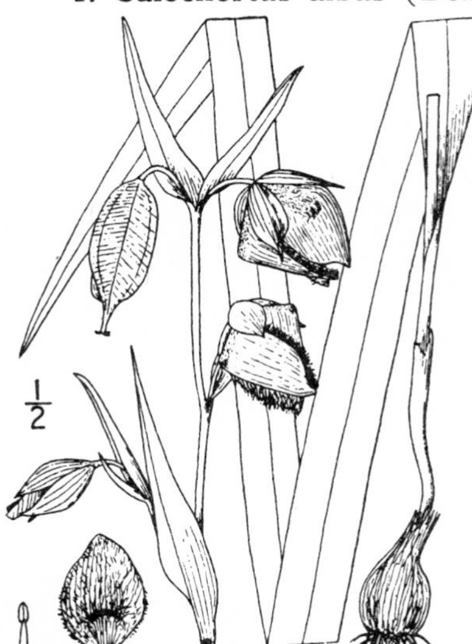

Cyclobothra alba Benth. Trans. Hort. Soc. **1**: 413. *pl. 14, fig. 3.* 1835.

Calochortus albus Dougl.; Benth. Maund & Hensl., Bot. **2**: *pl. 98.* 1838.

Cyclobothra paniculata Lindl. Bot. Reg. **20**: under *pl. 1662.* 1834.

Stems stout, branching, glaucous. Basal leaves lanceolate, acuminate, 3–5 dm. long, 10–30 mm. wide; bracts foliaceous, lanceolate acuminate, 5–15 mm. long; flowers subglobose, nodding on slender pedicels; sepals shorter than the petals, ovate, more or less acuminate, greenish white, often tinged with reddish purple; petals white, purplish at base, ovate-orbicular, obtuse or acutish, 20–30 mm. long, incurved and strongly arched, clothed above the gland with long yellowish hairs; gland lunate, shallow, with 4 transverse upwardly imbricate scales, fringed with close short yellow or white glandular hairs; anthers oblong, mucronate, 4 mm. long; capsule 20–40 mm. long, 15–25 mm. broad, more or less prominently winged.

Open woods and shaded slopes of the Upper Sonoran and Transition Zones; Mendocino and Butte Counties, southward through the Coast Ranges and Sierra Nevada to San Diego County. Type locality: California, exact station not known.

Calochortus albus rubèllus Greene. (Erythea 1: 152. 1893.) Distinguished from the typical form by the rose-tinged perianth-segments. Coast Ranges from the Santa Cruz to the Santa Lucia Mountains, California. Originally collected at Pacific Grove.

2. **Calochortus amoènus** Greene. Purple Fairy Lantern. Fig. 1058.

Calochortus amoenus Greene, Pittonia **2**: 71. 1890.

Calochortus albus amoenus Purdy, in Bailey Cyclop. Hort. **2**: 632. 1914.

Stem flexuous, branching, 15–40 cm. high, glaucous. Basal leaf linear-lanceolate, usually surpassing the stem, 8–12 mm. wide; bracts foliaceous, lanceolate, acuminate; flowers nodding on slender pedicels, rose-purple with maroon glands; sepals ovate, short-acuminate, two-thirds the length of the petals; petals broadly ovate, sparsely clothed above the gland to well above the middle with long hairs, long-ciliate on the margins to above the middle; gland nearly basal, deep maroon, clothed with very short hairs; anthers creamy white, scarcely 3 mm. long; capsule broadly elliptic, 20 mm. long.

Western slopes of the Sierra Nevada in the Upper Sonoran and lower edge of the Transition Zones; Fresno and Tulare Counties, California. Type locality: "mountains east of Visalia, California."

3. **Calochortus pulchéllus** (Benth.) Wood. Mt. Diablo Fairy Lantern. Fig. 1059.

Cyclobothra pulchella Benth. Trans. Hort. Soc. **1**: 412. *pl. 14, fig. 1.* 1835.

Calochortus pulchellus Wood, Proc. Acad. Philad. **1868**: 168. 1868.

Stem stout, branching 25–50 cm. high. Basal leaf linear-lanceolate, acuminate, 8–25 mm. wide; bracts foliaceous, lanceolate, 4–7 cm. long; flowers 2–3 terminating each branch, nodding, on slender pedicels; sepals ovate-acuminate, shorter than or about equalling the petals, yellow tinged with brown exteriorly, petals bright yellow, 20–30 mm. long, round-ovate, ciliate on the margin with short thick hairs, sparsely clothed above the gland with yellowish hairs; gland deep, appearing on the back of the petal as a prominent sometimes brownish knob, bordered with elongated stiff light colored hairs; anthers oblong, 5 mm. long, mucronate.

A local species only known on Mount Diablo, Contra Costa County, California. Upper Sonoran Zone.

4. Calochortus amàbilis Purdy.
Golden Fairy Lantern. Fig. 1060.

Calochortus pulchellus maculosus S. Wats.; Purdy, Zoe **1**: 245. 1890.
Calochortus amabilis Purdy, Proc. Calif. Acad. III. **2**: 119. 1901.
Calochortus pulchellus amabilis Jepson, Fl. West. Middle Calif. 113. 1901.

Stem flexuous, branching, 2–5 dm. high, glaucous. Sepals 20–30 mm. long, ovate, abruptly short-acuminate, yellow tinged with green and brown; petals clear golden yellow or the gland marked exteriorly by a brown spot, scarcely surpassing the sepals, ciliate on the margin with short thickish hairs, naked above the gland but often very sparsely hairy on either side of the gland; gland lunate, deeply sunken, appearing as a knob exteriorly, covered by a dense fringe of appressed light yellowish hairs growing from the upper margin, and crossing each other over the pit; anthers oblong, 4 mm long.

Open woods and grassy slopes of the Upper Sonoran Zone; California Coast Ranges, Humboldt and Trinity Counties south to Sonoma and Solano Counties. Possibly only a variety of the preceding. Type locality: not given.

5. Calochortus monophýllus (Lindl.) Lem.
Yellow Star Tulip or Cat's Ear. Fig. 1061.

Cyclobothra monophyllus Lindl. Journ. Hort. Soc. **5**: 81. 1849.
Calochortus monophyllus Lem. Fl. Serres **5**: 430. *b.* 1849.
Calochortus benthami Baker, Journ. Linn. Soc. **14**: 304. 1875.

Stems flexous, simple or sparingly branched, 10–25 cm. high. Flowers 1–6, weakly erect; sepals ovate, 16–20 mm. long, abruptly acuminate, mucronate, greenish yellow; petals yellow, about equalling the sepals, oval to obovate, rounded at apex, narrowed at base to a brownish claw, inner surface rather densely covered with yellow hairs; gland lunate, situated just above the claw, bordered above with short yellow often clavate hairs; anthers lanceolate, acute, 3 mm. long; capsule nodding 12–18 mm long and nearly as broad, broadly winged.

Wooded slopes, of the Upper Sonoran and Lower Transition Zones; western slopes of the Sierra Nevada from Butte to Mariposa Counties, California. Type locality: Bear Valley, Nevada County.

6. Calochortus caerùleus (Kell.) S. Wats.
Blue Star Tulip or Cat's Ear. Fig. 1062.

Cyclobothrya caerulea Kell. Proc. Calif. Acad. **2**: 4. 1858.
Calochortus caeruleus S. Wats. Proc. Am. Acad. **14**: 263. 1879.

Stem mostly simple, slender and flexuous, 7–15 cm. high. Flowers 2–4, umbellate, open-campanulate, erect, blue; sepals usually equalling or surpassing the petals, oblong-lanceolate, more or less acuminate; petals rhomboidal, rounded or broadly triangular at apex, narrowed at base to a rather slender claw, the upper surface densely covered with long slender bluish hairs above the gland, naked below; gland with a transverse fringed scale; anthers oblong, obtuse, 3–4 mm. long; capsule broadly oblong-elliptic, rounded at both ends, 10-15 mm. long and nearly as broad.

Wooded slopes of the Arid Transition Zone; western slopes of the Sierra Nevada, Plumas to Placer Counties, California. Type locality: "above Forest City, not far from the region of perpetual snow," Sierra Nevada, California.

7. **Calochortus élegans** Pursh.

Elegant Star Tulip or Cat's Ear. Fig. 1063.

Calochortus elegans Pursh, Fl. Am. Sept. 1: 240. 1814.
Cyclobothrya elegans Benth.; Lindl. Bot. Reg. **20**: under *pl. 1662.* 1834.

Stem very slender, flexuous, 1–2 dm. high. Basal leaf surpassing the stem, 2–9 mm. wide; bracts linear, attenuated, 15–30 mm. long; flowers 1–3, open-campanulate, erect or somewhat drooping, on slender pedicels 2–5 cm. long; sepals ovate-lanceolate, acuminate, 12–16 mm. long, greenish white tinged with purple at base; petals white tinged with purple, obovate, obtuse, 12–20 mm. long, densely hairy on the inner surface and ciliate on the margins; scale narrow, ascending, deeply lacerate; anthers 5 mm. long, acuminate; capsule nodding, elliptic, 2 cm. long, 12–15 mm. wide.

Shaded slopes, of the Arid Transition and Canadian Zones; eastern slopes of the Cascade Mountains, Washington to Montana and Utah. Type locality: "on the headwaters of the Kooskooky," Idaho.

8. **Calochortus nànus** (Wood) Piper.

Dwarf Star Tulip. Fig. 1064.

Calochortus elegans nanus Wood, Proc. Acad. Philad. **1868**: 168. 1868.
Calochortus nanus Piper, Bull. Torrey Club **33**: 537. 1906.

Stems erect slender, 10–15 cm. high, from ovoid, deep-seated bulbs. Basal leaf surpassing the stem, 2–7 mm. wide; bracts linear-lanceolate; flowers 1 to several, open-campanulate, erect or somewhat drooping, on slender pedicels about 25 mm. long; sepals oblong-ovate, short-acuminate, 12–15 mm. long, tinged with violet-purple; petals broadly obovate 15–18 mm. long, faintly tinged with violet, hairy over the entire inner surface, ciliate on the margin; scale about one-third as broad as the petal, nearly semi-circular, but little incised; anthers acuminate, equalling the flattened filaments; capsule nodding, orbicular at maturity, 15 mm. long.

Open coniferous forests, Transition and Canadian Zones; Siskiyou Mountains of northern California and perhaps southern Oregon. Type locality: mountains west of Yreka.

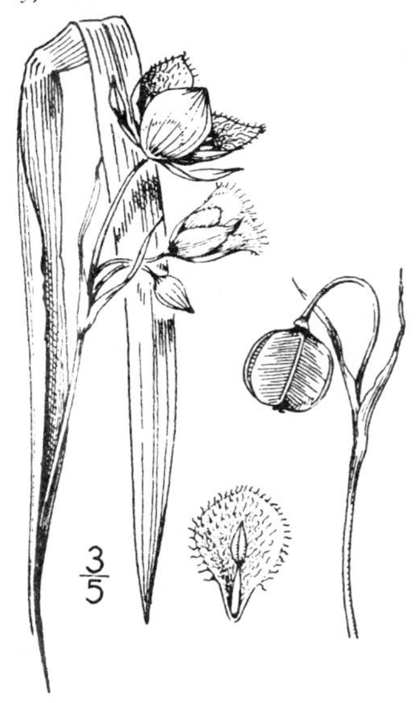

9. **Calochortus maweànus** Leicht.

Hairy Star Tulip. Fig. 1065.

Calochortus maweanus Leicht.; Baker, Journ. Linn. Soc. **14**: 305. 1875.

Stem flexuous, usually simple, 8–20 cm high. Flowers open-campanulate, erect, white or more or less tinged with purplish blue; petals 12–20 mm. long, nearly orbicular, obovate or somewhat rhomboidal, broadly or abruptly acute, the upper surface covered with long slender white or bluish, spreading hairs above the gland, naked below; gland covered above with a narrow transverse scale, the entire surface thickly bearded with long erect white or bluish hairs; anthers lanceolate, acuminate, 4–5 mm. long; capsule oblong-elliptic, 25 mm. long, 10–12 mm. thick, narrowly winged.

Moist wooded slopes, Upper Sonoran and Transition Zones; Josephine and Curry Counties, Oregon, to Santa Cruz and Butte Counties, California. Type locality: California.

10. **Calochortus púrdyi** Eastw. Purdy's Star Tulip. Fig. 1066.

Calochortus purdyi Eastw. Proc. Calif. Acad. III. 1: 137. *pl. 11.* 1898.

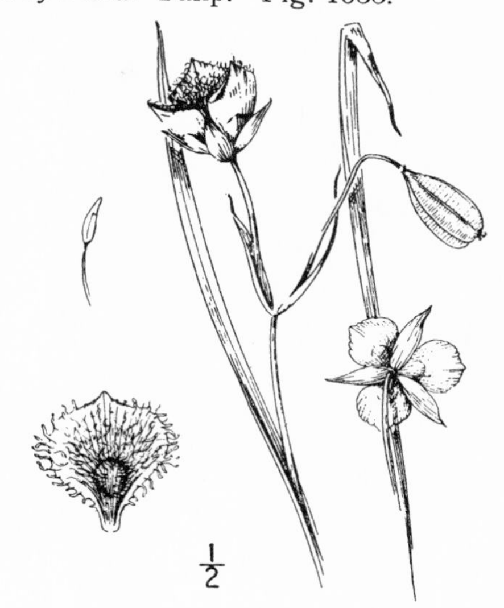

Stem rather stout, 2–3 dm. high, glaucous, branching, not bulbiferous at base. Flowers 2 to several, broadly open-campanulate; sepals elliptic to narrowly ovate, abruptly acuminate, tinged with purple, 12–15 mm. long; petals 15–20 mm. long, broadly obovate-cuneate, rounded or abruptly acute at apex, somewhat arched, creamy white, tinged with purple, bearded over the entire inner surface with long hairs, these white on the upper half of the petal and purple on the lower; gland conspicuous, semi-circular, covered by a densely hairy narrow scale; anthers lanceolate, abruptly acuminate, creamy white or purple, 4–5 mm. long, shorter than the filaments; capsule 25–30 mm. long, broadly elliptic.

Grassy meadows and hillsides, Humid Transition Zone; east of the Coast Ranges, Seattle to Grants Pass, Oregon. This species is very closely related to *Calochortus tolmiei*, from which it is distinguished by the presence of a scale over the gland. Type locality: Willamette Valley, in the foothills on dry gravelly soil.

11. **Calochortus tòlmiei** Hook. & Arn. Tolmie's Star Tulip. Fig. 1067.

Calochortus tolmiei Hook. & Arn. Bot. Beechey 398. 1841.

Stem erect or ascending, slender and flexuous, usually branched, 15–45 cm. high. Flowers 2 to several, open-campanulate; sepals oblong-lanceolate, 12–18 mm. long, tinged with purple; petals creamy white tinged with purple, cuneate-obovate, 15–20 mm. long, bearded over the entire inner surface with long hairs, these on the lower part of petal purple, on the upper white; the gland without a scale, but the upper circular edge with a dense fringe of reflexed hairs; anthers lanceolate, acute, 5 mm. long; capsule elliptic, acute at both ends, 2–3 cm. long, nodding.

On dry grassy plains of the Transition Zone; Willamette Valley, Oregon. Type locality: "on the banks of the Wahlamet (Willamette) River."

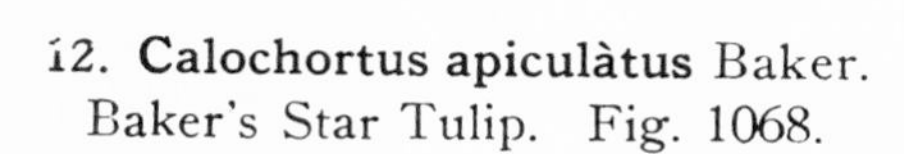

12. **Calochortus apiculàtus** Baker. Baker's Star Tulip. Fig. 1068.

Calochortus apiculatus Baker, Journ. Linn. Soc. **14**: 305. 1875.

Stem 2–4 dm. high, flexuous. Flowers 1–9, umbellate, on slender elongated pedicels, erect, campanulate; sepals lanceolate, 15–20 mm. long, acute, greenish white; petals pale yellow, 20–25 mm. long, obovate, obtuse, sparsely hairy on the inner surface; gland without scale; anthers 6–8 mm. long, acuminate, equalling the filaments; capsules nodding, 20–30 mm. long, oblong.

Moist woods of the Canadian Zone; British Columbia to Idaho and eastern Washington. Type locality: probably close to where Washington, British Columbia and Idaho meet.

13. **Calochortus lóbbii** (Baker) Purdy. Lobb's Star Tulip. Fig. 1069.

Calochortus elegans lobbii Baker, Journ. Linn. Soc. **14**: 305. 1875.

Calochortus lobbii Purdy, Proc. Calif. Acad. III. **2**: 122. 1901.

Calochortus subalpinus Piper, Contr. Nat. Herb. **11**: 195. 1906.

Stems flexuous, erect, 15–20 cm. high. Sepals lanceolate-ovate, 15–25 mm. long, strongly arched at base forming a shallow pit marked by a purple spot inside; petals creamy white, purplish at base, obovate or rhombic-orbicular, 2–3 cm. long, slightly hairy over the inner surface above the gland, to near the apex; gland minutely puberulent, covered above by a narrow entire scale, this covered with long retrorse hairs; anthers 7–8 mm. long, long-beaked, sagittate at base, equalling the broadly winged filaments; capsule nodding, 2–3 cm. long, narrowly elliptic, acutish at both ends.

Subalpine slopes and meadows, Hudsonian Zone; Mount Adams, Washington, to the Three Sisters, Oregon. Type locality: Mount Jefferson, Oregon.

14. **Calochortus nùdus** S. Wats. Naked Star Tulip. Fig. 1070.

Calochortus elegans subclavatus Baker, Journ. Linn. Soc. **14**: 305. 1875.

Calochortus nudus S. Wats. Proc. Am. Acad. **14**: 263. 1879.

Stem very slender, rather weak, 5–10 cm. high. Flowers 1–7, in a single umbel, open-campanulate, erect; sepals narrowly oblong-ovate, acute, 10–12 mm. long; petals greenish white or lilac, 12–16 mm. long, obovate, rounded or obtuse, denticulate, glabrous on the inner surface, or sometimes with a few hairs at the ends of the narrow closely appressed scale covering the upper margin of the gland; anthers oblong, 4–6 mm. long, blue, a third shorter than the subulate filaments; capsule oblong, acute at both ends, 15–20 mm. long.

Moist grassy places, Transition and (mainly) Canadian Zones; Sierra Nevada, California, from Plumas to Tulare Counties. Type locality: Yosemite Valley, California.

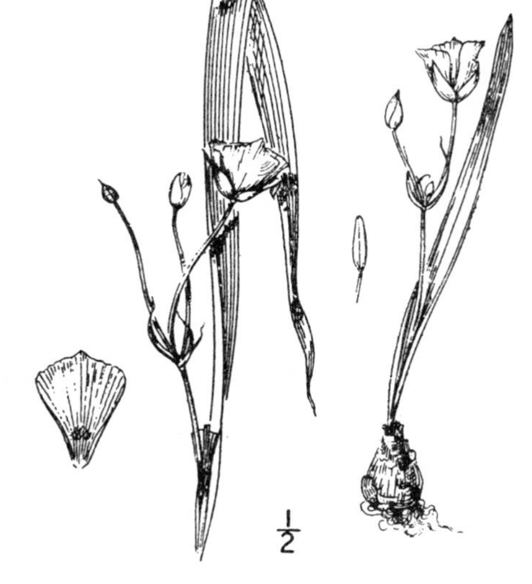

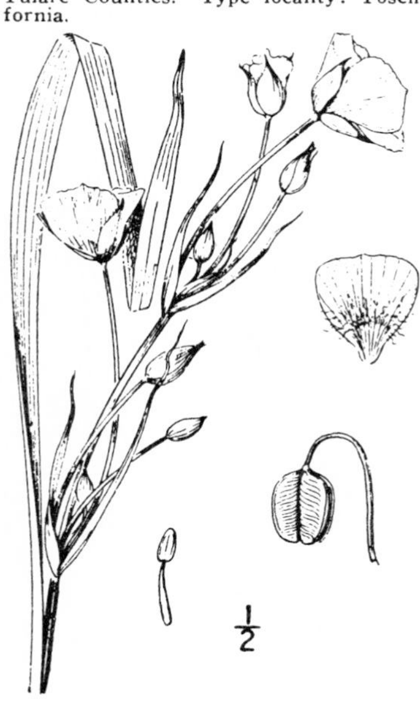

15. **Calochortus umbellàtus** Wood. Oakland Star Tulip. Fig. 1071.

Calochortus umbellatus Wood, Proc. Acad. Philad. **1868**: 168. 1868.

Calochortus collinus Lemmon, Erythea **3**: 49. 1895.

Stem 8-30 cm. high, flexuous. Umbel usually 5–10-flowered; sepals ovate to oblong-lanceolate; greenish white rarely tinged with lilac, 12–17 mm. long; petals white, rarely tinged with lilac, broadly cuneate-obovate, 15–20 mm. long, truncate-rounded, denticulate, naked except for a few hairs on either side and just above the gland; scale triangular, appressed to the upper portion of the gland, fimbriate on the margin; filaments twice as long as the anthers, slender; anthers oblong, obtuse, 3 mm. long, white; capsule nodding, broadly elliptic, rounded at both ends, 10–12 mm. long and nearly as broad.

On dry, wooded hillsides, Upper Sonoran Zone; California Coast Ranges, Lake and Mendocino to Marin and Alameda Counties. Type locality: Oakland, California.

16. **Calochortus uniflòrus** Hook. & Arn.
Large-flowered Star Tulip. Fig. 1072.

Calochortus uniflorus Hook. & Arn. Bot. Beechey. 398. *pl. 94.* 1841.

Calochortus lilacinus Kell. Proc. Calif. Acad. **2**: 5. 1858.

Stem 10–25 cm. high, flexuous, usually branched, bulbiferous near the base. Umbels 1–3, and usually 1–4-flowered, sepals ovate to oblong-lanceolate, 12–16 mm. long, greenish lilac; petals lilac often with a purple spot on each side of the gland, cuneate, nearly truncate, denticulate, 20–25 mm. long, naked above, sparingly hairy just above the gland; gland shallow, not pitted, with a narrow appressed triangular scale; filaments slender, twice the length of the anthers; anthers oblong, 4 mm. long, obtuse, lilac; capsule elliptic, rounded at both ends, 10–15 mm. long, nodding.

Wet meadows, Upper Sonoran and Humid Transition Zones; Coast Ranges, from the Siskiyou Mountains, near Grants Pass, Oregon, to Monterey, California. Type locality: Monterey, California.

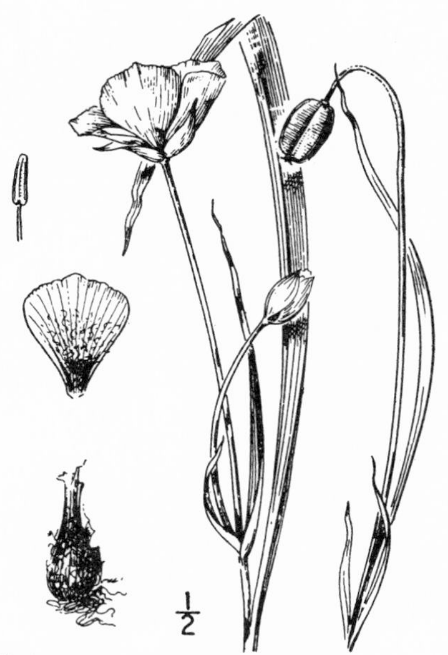

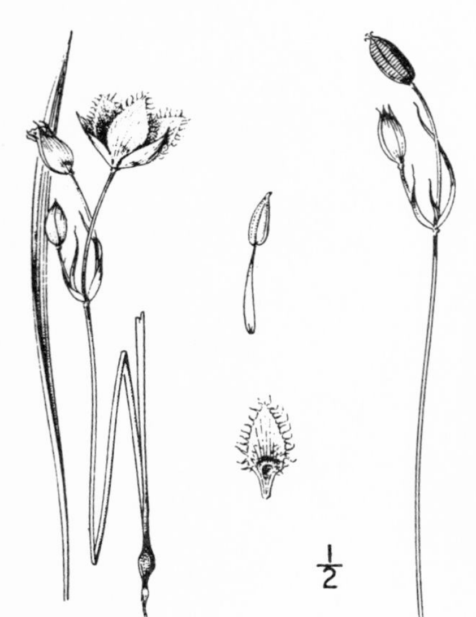

17. **Calochortus lyállii** Baker.
Lyall's Star Tulip. Fig. 1073.

Calochortus lyallii Baker, Journ. Linn. Soc. **14**: 305. 1875.

Calochortus ciliatus Robins. & Seaton, Bot. Gaz. **18**: 238. 1893.

Stem flexuous, 15–20 cm. high. Flowers 2–8, erect, broadly campanulate; sepals greenish white, lanceolate, acuminate, 8–15 mm. long; petals white tinged with bluish purple, scarcely longer than the sepals, narrowly ovate, acute, conspicuously ciliate on the margins, glabrous except the yellow doubly fringed lunate scale; stamens half as long as petals; anthers 4–5 mm. long, oblong, sagittate, apiculate; capsule erect in front, elliptic, 14–18 mm. long, acutely 3-winged.

Grassy slopes among pines, of the Hudsonian Zone; Wenatchee Mountains, eastern Washington. Type locality: "Columbia brittanica ad apicem montis alt 5,800 pedes inter fluv. Columbia et Yakima."

18. **Calochortus shasténsis** Purdy.
Shasta Star Tulip. Fig. 1074.

Calochortus shastensis Purdy, Proc. Calif. Acad. III. **2**: 125. 1901.

Calochortus nudus shastensis Jepson, Fl. Calif. **1**: 301. 1921.

Stem slender, erect, 10–25 cm. high. Sepals ovate, acute or acuminate 10–12 mm. long, green without, spotted with purple near the base within; petals white or lilac, broadly fan-shaped, 15 mm. long, broadly obtuse or somewhat truncate at apex, denticulate, naked or sometimes with a few hairs above the gland; scale ascending, fringed; filaments slender, 6 mm. long; anthers linear-oblong, 4 mm. long, obtuse at the apex, slightly sagittate at base, creamy white or lilac; capsule, strictly erect, elliptic, acutish at both ends, 15–20 mm. long.

Open moist meadows, in the Transition Zone; Siskiyou and Shasta Counties, California. Type locality: open moist meadows in the vicinity of Sisson, Shasta County, California.

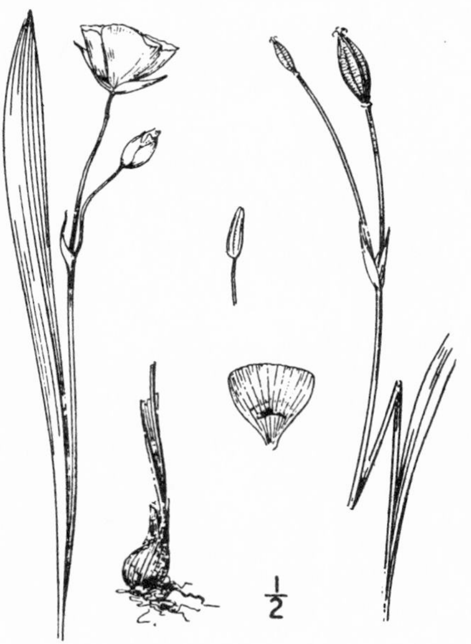

19. **Calochortus longebarbàtus** S. Wats. Long-haired Star Tulip. Fig. 1075.

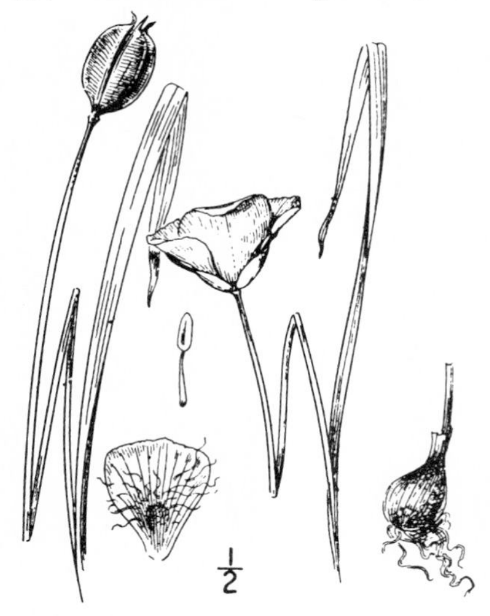

Calochortus longebarbatus S. Wats. Proc. Am. Acad. **17**: 381. 1882.

Stem slender, about 25–35 cm. high, bulbiferous near the base. Flowers 1 or 2, campanulate, erect; sepals lanceolate, 15–20 mm. long; petals lilac, yellowish at base, obovate, 20–25 mm. long, rounded at the summit and erose-denticulate, nearly naked except at gland, this bordered with long flexuous hairs; anthers narrowly oblong, obtuse, half the length of the slender filaments, 3–4 mm. long; capsule 20–25 mm. long, broadly elliptic, abruptly pointed at both ends, conspicuously winged.

In low grassy ground, Upper Sonoran and Arid Transition Zones; Klickitat County, Washington, and adjacent Oregon. Type locality: "Falcon Valley, Klickitat County," Washington.

20. **Calochortus nítidus** Dougl. Brilliant Mariposa. Fig. 1076.

Calochortus nitidus Dougl. Trans. Hort. Soc. **7**: 277. *pl. 9.* 1830.
Calochortus pavonaceus Fernald, Bot. Gaz. **19**: 335. 1894.

Stem erect, stiff, 3–4 dm. high. Umbel 1–4-flowered; sepals purplish, ovate-lanceolate, petals cuneate-obovate, 30–35 mm. long, lavender to purple, with a lunate dark purple band immediately above the gland, long-villous over the lower third, long-ciliate on the margin to the denticulate summit; gland small, orbicular, covered with yellow hairs; filaments 10 mm. long, dilated below; anthers oblong, obtuse 6–8 mm. long; capsule broadly elliptic, 20 mm. long, broadly winged, abruptly rounded to the short beak.

Grassy hillsides and meadows, of the Transition Zone; eastern Washington to Montana and northeastern Oregon. Type locality: "on the chain of the Columbia, from the confluence of the Spokane River upwards."

21. **Calochortus greènei** S. Wats. Greene's Mariposa. Fig. 1077.

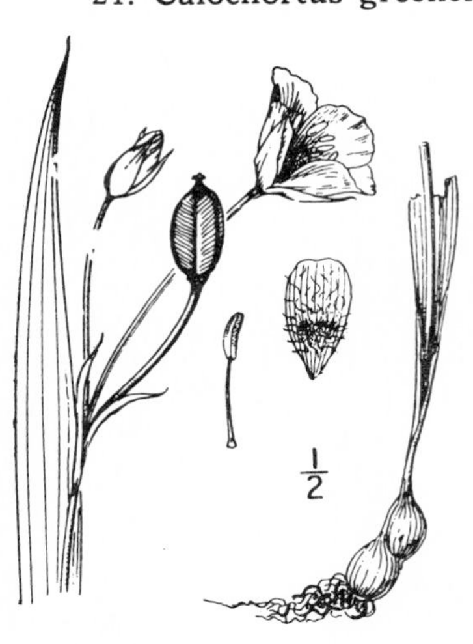

Calochortus greenei S. Wats. Proc. Am. Acad. **14**: 264. 1879.

Stem stout, 3–4 dm. high, usually branched. Flowers 2–5, erect, open-campanulate; sepals oblong-lanceolate, greenish tinged with lilac on the inner surface, and with a yellowish hairy spot above the base; petals lilac, barred with yellow below, fan-shaped, 39–40 cm. long, strongly pitted and arched, thinly villous on the upper part of the inner surface, densely so on the lower with yellow hairs; gland densely villous above the broad transverse laciniate scale; anthers obtuse, 10 mm. long; capsule 25 mm. long, 15 mm. broad, attenuate at apex into a stout beak, 8–12 mm. wide.

A little known species of northern California. Collected in Siskiyou and Modoc Counties. Type locality: Siskiyou County, California.

22. Calochortus howéllii S. Wats.
Howell's Mariposa. Fig. 1078.

Calochortus howellii S. Wats. Proc. Am. Acad. **23**: 266. 1888.

Stem erect, 3–4 dm. high, often branched. Basal leaf about equalling the stem, 8–12 mm. wide; cauline leaf solitary, narrowly lanceolate, attenuate, about 5 cm. long; bracts 2–4 cm. long; pedicels erect, 5–10 cm. long; flowers open-campanulate, erect; sepals ovate, acuminate, 20 mm. long; petals white, rounded at apex, denticulate, 25 mm. long, the margin slightly ciliate toward the base, inner surface covered with short crisped hairs, those immediately above the gland denser and dark greenish; gland transversely oblong, densely clothed with short yellow hairs; anthers oblong, acute and spiculate, 6–7 mm. long; capsule elliptic, acute, 18–20 mm. long.

Transition Zone; southern Oregon, from the vicinity of Roseburg to the Siskiyou Mountains. Type locality: near Waldo, Oregon.

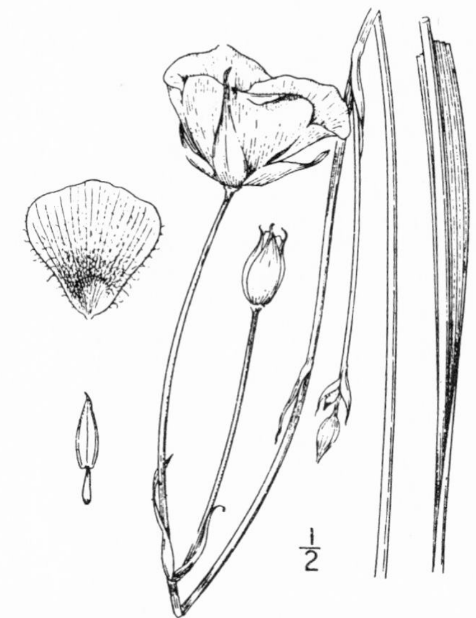

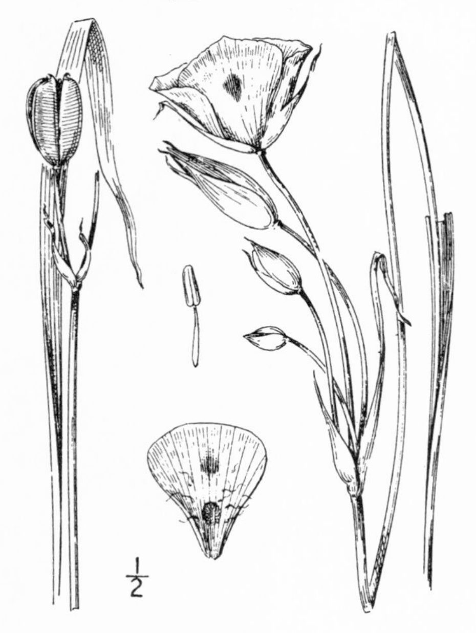

23. Calochortus eurycàrpus S. Wats.
Big-pod Mariposa. Fig. 1079.

Calochortus eurycarpus S. Wats. Bot. King Expl. 348. 1871.

Calochortus nitidus eurycarpus Henderson, Bull. Torrey Club **27**: 356. 1900.

Stem erect, stiff 25–45 cm. high. Umbels 1–4-flowered; sepals ovate-lanceolate, long-acuminate, usually exceeding the petals; petals white to lavender, with a conspicuous orbicular indigo spot in the middle, 25–45 mm. long, sparingly long-villous around the small round gland, this covered with short yellow hairs completely concealing the narrow scale at the base; filaments dilated below, 10 mm. long; anthers oblong 6–8 mm. long; capsule elliptic, 20–25 mm. long, broadly winged, abruptly rounded at apex to the short beak.

Open meadows, of the Canadian Zone; Blue Mountains, eastern Oregon, to Montana and northern Nevada. Type locality: "East Humboldt Mountains, Nevada, 7,000–9,500 feet altitude."

24. Calochortus catalìnae S. Wats.
Catalina Mariposa. Fig. 1080.

Calochortus catalinae S. Wats. Proc. Am. Acad. **14**: 268. 1879.
Calochortus lyoni S. Wats. Proc. Am. Acad. **21**: 455. 1886.

Stems erect, 25–70 cm. high, branching, leafy, often bulbiferous. Leaves flat; flowers 1–several, terminating the branches; sepals green, spotted with purple near the base within; petals 3–5 cm. long, cuneate, longer than wide, white tinged with lilac, to lilac-purple, with an obovate maroon-colored area surrounding the gland and extending to the base, naked except for a few scattered hairs above the gland; gland covered with rather long reflexed hairs; filaments deep purple, slender, dilated below, 10–12 mm. long; anthers narrowly oblong, obtuse, 6–7 mm. long; capsule linear-oblong, obtuse at both ends, 25–35 mm. long, 8–10 mm. wide.

Grassy hills and mesas usually in heavy soil, Upper Sonoran Zone; southern California from Santa Barbara to San Diego Counties, also on the Channel Islands. Type locality: Santa Catalina Island.

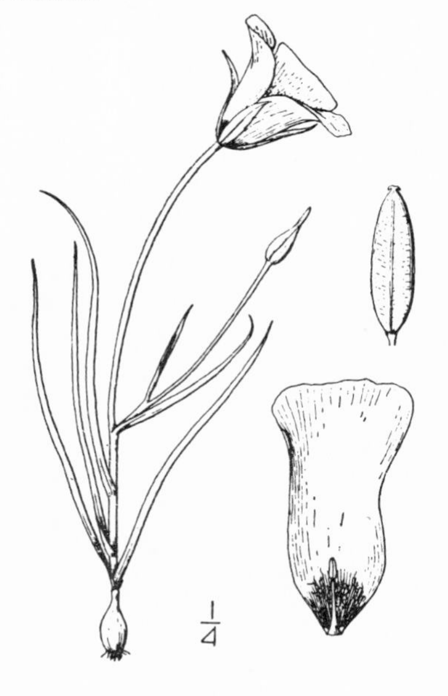

25. Calochortus clavàtus S. Wats. Club-haired Mariposa. Fig. 1081.

Calochortus clavatus S. Wats. Proc. Am. Acad. **14**: 265. 1879.

Stem stout, stiff, simple or usually branched, 4–18 dm. high, not bulbiferous. Leaves deeply channelled; flowers in a 2–5-flowered terminal umbel scattered along the stem on the shorter branches; sepals ovate, yellowish within; petals 34–45 mm. long, broadly fan-shaped and strongly arched, with a broad claw, nearly truncate at apex, deep yellow with a usually reddish brown claw, and a reddish brown band above the hairy zone, the sides of the gland and the lower half of the petals hairy with yellow club-shaped hairs; gland deep, covered with short matted hairs; anthers oblong, obtuse, 10–12 mm. long, red-brown; capsule lanceolate, long-attenuate, 6–7 cm. long.

Wooded cañon slopes, Upper Sonoran Zone; San Benito County to the Santa Monica Mountains, Los Angeles County, California; also in Mariposa County, according to Purdy. Type locality: San Luis Obispo, California.

26. Calochortus cóncolor (Baker) Purdy. Golden-bowl Mariposa. Fig. 1082.

Calochortus luteus concolor Baker, Garden **48**: 440. 1895.
Calochortus concolor Purdy, Proc. Calif. Acad. III. **2**: 135. 1901.

Stems stout, 3–5 dm. high, mostly simple. Leaf linear-lanceolate, glaucous, deeply channelled; cauline leaves 1–5, linear-attenuated, involute, 6–10 cm. long; sepals 25–30 mm. long, ovate-lanceolate, acute, broadly scarious-margined, yellow within and usually with a purple spot near the base; petals a deep yellow, sometimes tinged with purple, with or without a purple area above and below the gland, fan-shaped, 25–35 mm. long and as broad, the lower third covered with erect yellow hairs; gland small, oblong; anthers yellow, 8–10 mm. long, nearly twice the length of the broad filaments; capsule strongly 3-angled with thickened margins, attenuate at apex.

Dry hillsides, Upper and Lower Sonoran Zones; southern slopes of the San Bernardino Mountains to northern Lower California. Type locality: Laguna, San Diego County, California.

27. Calochortus kennédyi Porter. Red or Kennedy's Mariposa. Fig. 1083.

Calochortus kennedyi Porter, Bot. Gaz. **2**: 79. 1877.
Calochortus speciosus Jones, Contr. West. Bot. **14**: 27. 1912.

Stems simple, stout, glaucous, 10–25 cm. high. Basal leaves narrowly linear, deeply channelled, glaucous, about equalling the stem; stem leaves, 1 or 2 or sometimes wanting, 3–5 cm. long, thickish and recurved, broadly scarious-margined; umbels 2–4-flowered; sepals ovate, acute, 20–30 mm. long, orange-red within, usually with a conspicuous brown spot at base; petals cuneate-obovate, apiculate at apex, 25–35 mm. long, bright orange-red, the claws usually brownish-purple, naked except for a few scattered hairs about the gland, this small, round, covered with very short matted hairs; anthers oblong-ovate, obtuse, purple, 6–8 mm. long; filaments dilated, half the length of the anthers; capsule stout, linear-lanceolate, 4–5 cm. long.

Dry gravelly hills and mesas of the desert region, Sonoran Zones; Mount Pinos, Ventura County, and the desert slopes of the Tehachapi Mountains to southern Nevada and Arizona. Type locality: Kern County, California.

28. Calochortus macrocàrpus Dougl.
Green-banded Star Tulip. Fig. 1084.

Calochortus macrocarpus Dougl. Trans. Hort. Soc. 7: 276. pl. 8. 1830.
Calochortus cyaneus A. Nels. Bot. Gaz. 53: 219. 1912.
Calochortus maculosus A. Nels. & Macbr. Bot. Gaz. 56: 471. 1913.

Stems rather stout, strict, 3–5 dm. high, glaucous. Leaves several, linear, revolute becoming recurved, 5–10 cm. long; bracts similar, equalling the very stout pedicels; flowers usually 2, campanulate; sepals lanceolate long-acuminate, 35–45 mm. long, with broad scarious margins, tinged with purple within; petals obovate, 30–45 mm. long, acute or acuminate, purple with yellow base, with a prominent green band down the middle, sparsely clothed with yellow hairs on both sides of the gland and for a short distance above; anthers lanceolate, but obtuse, 10–12 mm. long; capsule lanceolate, 45 mm. long, 8 mm. thick, very narrowly winged.

Dry plains and hillsides, Upper Sonoran and Transition Zones; British Columbia to Montana and northern California. In the Pacific States it is east of the Cascade Mountains extending from Washington to Modoc County, California. Type locality: "dry barren grounds around the Great Falls of the Columbia and on the summit of the low hills between them and the Grand Rapids."

29. Calochortus nuttàllii Torr. Sego Lily. Fig. 1085.

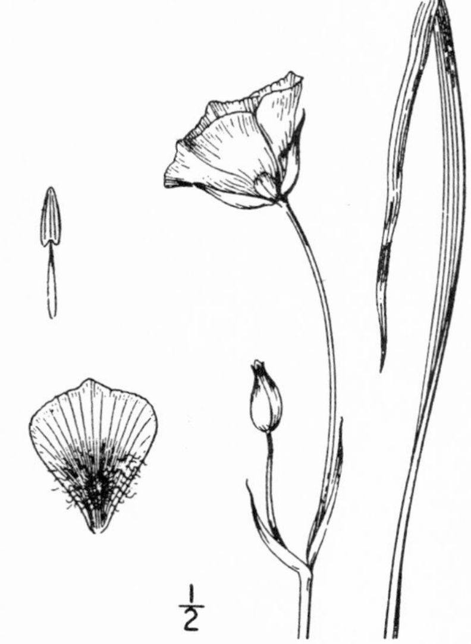

Calochortus nuttallii Torr. Stansbury Expl. 397. 1855.
Calochortus excavatus Greene, Pittonia 2: 71. 1890.
Calochortus discolor Davids. Bull. S. Calif. Acad. 14: 11. 1902.
Calochortus campestris Davids. Bull. S. Calif. Acad. 14: 12. 1915.

Stem simple, 2–4 dm. high. Leaves very glaucous, somewhat fleshy, narrowly linear, 10–15 cm. long, the cauline 1 or 2, deeply channelled and more or less recurved; umbel 2–4-flowered; sepals ovate-lanceolate, about two-thirds the length of the petals, erect or the tips slightly recurved, broadly scarious-margined, usually with a conspicuous dark purple spot (sometimes two) near the base; petals cuneate-obovate, 25–35 mm. long, obtuse or abruptly acute at apex, white tinged with purple above and often down the middle, dark purple around the gland and on the claw; sparsely hairy around the glands or entirely naked, the hairs yellowish with dark purple bases; gland rounded, yellow, deeply impressed; anthers yellow, oblong, obtuse, 10 mm. long, exceeding the filaments; capsule lanceolate, attenuate above the middle, 4–5 cm. long.

Dry gravelly slopes and ridges, Upper Sonoran and Transition Zones; eastern Washington to the eastern slopes of the Sierra Nevada and the mountains of southern California, east to South Dakota and New Mexico. Type locality: Salt Lake Valley, Utah. A variable and perhaps composite species as here treated. The southern California plants are more slender with a less depressed gland and yellow anthers. *Calochortus excavatus* Greene has a deeply depressed purple gland and purple anthers, and is probably worthy of at least varietal distinction.

30. Calochortus davidsoniànus Abrams.
Davidson's Mariposa. Fig. 1086.

Stem simple or usually branched, 3–6 dm. high, not bulbiferous at base. Basal leaves usually 2 or 3, about 10 cm. long, 4–6 mm. wide; flowers terminating the branches, or appearing in 2-flowered umbels; calyx lanceolate, long-acuminate, recurved, usually nearly equalling the petals, lavender on the inner surface with purple spots near the base; petals broadly fan-shaped, 25–45 mm. long, rounded at apex, lilac-purple to the very short deep purple claw, sparsely hairy over the lower third with yellowish somewhat tangled hairs; gland (sometimes wanting) obovate, covered with short glutinous hairs; anthers oblong, obtuse, purple, 4–6 mm. long, shorter than the dilated deep purple filaments; capsule linear 5–6 mm. long, 5–6 mm. wide, attenuate at apex.

On grassy or chaparral-covered hillsides, Upper Sonoran Zone; Santa Barbara to San Diego Counties. Type: grassy slopes between the Onofre Mountains and the sea, San Diego County, *Abrams 3275.* 1903.

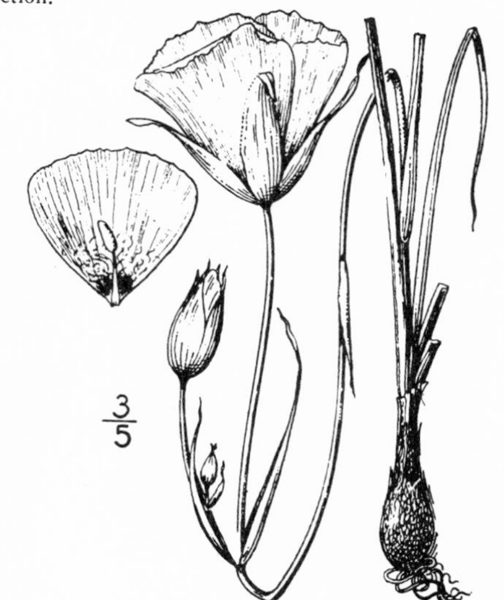

31. Calochortus spléndens Dougl.
Splendid Mariposa. Fig. 1087.

Calochortus splendens Dougl.; Benth. Trans. Hort. Soc. II. 1: 411. *pl. 15.* 1835.

Stems branching, 4–6 cm. high, not bulbiferous at base. Flowers terminating the branches or often only 2 and then umbellate; sepals lanceolate, long-acuminate, usually recurved, with purple spot near the base, about two-thirds the length of the petàls; petals 3–5 cm. long, broadly fan-shaped, rounded at apex, lilac purple to the very short dark purple claw, sparsely hairy with long tangled hairs from near the middle to about 5 mm. from the gland; gland covered with very short glutinous hairs, or these sometimes wanting; anthers lilac to blue, narrowly linear-lanceolate, acute, 8–10 mm. long, equalling the dilated filaments; capsule linear, attenuate at apex, 2–6 cm. long.

Dry gravelly hills, usually in chaparral, Upper Sonoran Zone; California Coast Ranges, Santa Clara County to San Luis Obispo. Type locality: California, probably near the San Antonio Mission, Monterey County.

32. Calochortus pàlmeri S. Wats.
Palmer's Mariposa. Fig. 1088.

Calochortus palmeri S. Wats. Proc. Am. Acad. 14: 266. 1879.
Calochortus striatus Parish, Bull. S. Calif. Acad. 1: 122. 1902.

Stem very slender, lax and flexuous, 3–5 dm. high, glaucous, not bulbiferous near the base, arising from an oblong membranously coated corm. Basal leaves 1 or 2, narrowly linear, about 10 cm. long; bracts linear, scarious-margined; sepals oblong, short-acuminate, the tip at length recurved; petals rather narrowly obovate-cuneate, 20–25 cm. long, white to very light purple, the claw brown, naked except for a few scattering hairs surrounding the gland, this large, undefined, covered with short linear hairs; anthers oblong, obtuse, 6 mm. long; capsule very narrow, about 25 mm. long or more.

Alkaline meadows, Lower Sonoran Zone; Mojave Desert, southern California. Type locality: "near the Mojave River," probably Las Flores Rancho, San Bernardino County.

33. Calochortus dúnnii Purdy.
Dunn's Mariposa. Fig. 1089.

Calochortus dunnii Purdy, Proc. Calif. Acad. III. 2: 147. 1901.

Calochortus palmeri dunnii Jepson & Ames, Fl. Calif. 1: 294. 1921.

Stem slender, erect, branching, 3–5 dm. high, not bulbiferous. Basal leaves usually 2, linear, deeply channelled, 10–20 cm. long; the stem leaves bract-like, 3–5 cm. long; flowers terminating the branches, rather open-campanulate; sepals ovate, acute, 15 mm. long, greenish white within and faintly spotted, the tips not recurving; petals broadly cuneate, 20 mm. long, as broad as long, rounded at apex, white with a reddish brown band above the gland, naked except for short scattering hairs on either side of the gland; gland small, round, densely covered with short matted hairs; anthers oblong, obtuse, mucronulate, 3–4 mm. long, white; filaments dilated below, 5 mm. long; capsule linear, attenuate at apex, 3 cm. long.

Dry gravelly hillsides in open chaparral, Upper Sonoran Zone; Cuyamaca Mountains, San Diego County, California. Type locality: "near Julian, San Diego County, California."

34. Calochortus paludicòlus Davidson.
Meadow Mariposa. Fig. 1090.

Calochortus splendens montanus Purdy, Proc. Acad. III. **2**: 143. 1901. In part.
Calochortus paludicolus Davidson, Bull. S. Calif. Acad. **9**: 53. 1910.
Calochortus palmeri paludicolus Jepson & Ames, Fl. Calif. **1**: 294. 1921.

Stem 4–6 dm. high, bulbiferous at base. Basal leaves 2 or 3, 15–20 cm. long, 4–5 mm. wide; stem leaves 2 or 3, 25–50 mm. long; flowers usually 2–4, terminating the branches, never umbellate, sepals oblong-lanceolate, acuminate, 15 mm. long, yellowish within, with a conspicuous oblong brown spot toward the base, recurved in anthesis; petals fan-shaped, rounded at apex, or somewhat apiculate, 25 mm. long, rose-colored, the claw brownish purple, hairy on the lower third with yellowish hairs; gland triangular, covered with short hairs; filaments brownish purple, slender, 7–8 mm. long; anthers oblong, obtuse, 4 mm. long, white or pinkish; capsule 5 cm. long, attenuate at apex.

Mountain meadows and stream banks, Transition Zone: San Bernardino and San Jacinto Mountains, southern California. Type locality: meadows, Bear Valley, San Bernardino Mountains, California.

Calochortus flexuòsus S. Wats. Am. Nat. **7**: 393. 1873. Stems decumbent or ascending more or less flexuous and branching; petals broadly cuneate-obovate, purple or white, with yellowish base; gland rounded, clothed with simple linear hairs. Southern Nevada, and possibly adjacent California, east to southern Utah and Arizona. Type locality: Las Vegas, Nevada.

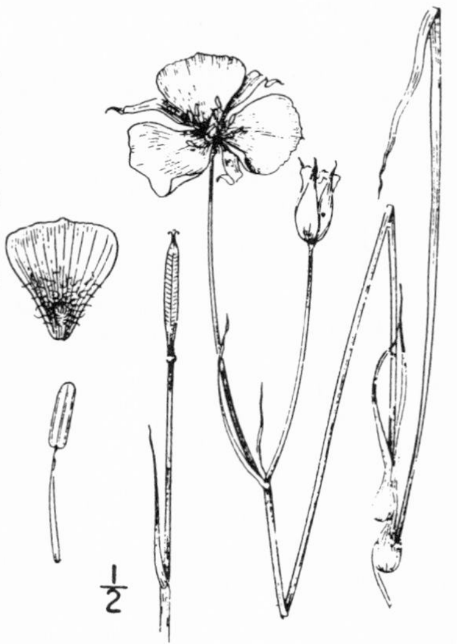

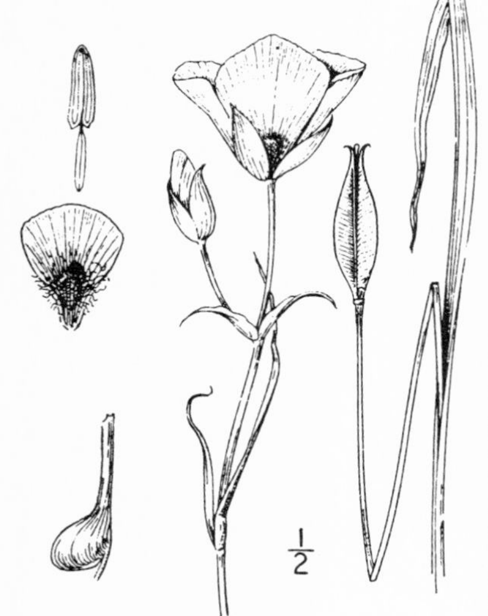

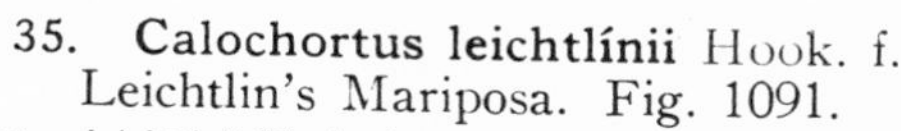
35. Calochortus leichtlínii Hook. f.
Leichtlin's Mariposa. Fig. 1091.

Calochortus leichtlinii Hook. f. Bot. Mag. **96**: *pl. 5862.* 1870.
Calochortus nuttallii subalpinus M. E. Jones, Contr. West. Bot. **12**: 78. 1908.

Stem erect, simple, 2–4 dm. high, bulbiferous at base. Leaves narrowly linear, strongly revolute, the basal 10–15 cm. long, the 1–3 cauline similar but shorter; umbels usually 2-flowered; sepals ovate-lanceolate, about two-thirds the length of the petals, yellowish within and usually with a deep purple sometimes hairy spot below the middle; petals cuneate-obovate, 25–35 mm. long, obtuse or abruptly acute, white above tinged with greenish yellow or lilac, and with a dark greenish purple spot above the yellow gland, claw yellow, sparsely hairy around the gland, with yellow hairs, otherwise naked; gland somewhat concealed by long reflexed yellow hairs, covered by a mat of tangled whitish hairs; filaments about equalling or shorter than the anthers, pale yellow; anthers linear-oblong, more or less conspicuously sagittate at base, 6–8 mm. long; capsule lanceolate-attenuate, above the middle, 3–5 cm. long.

Open gravelly situations, mainly Arid Transition Zone; eastern Oregon to the southern Sierra Nevada, eastward to South Dakota and New Mexico. Type locality: Sierra Nevada, California.

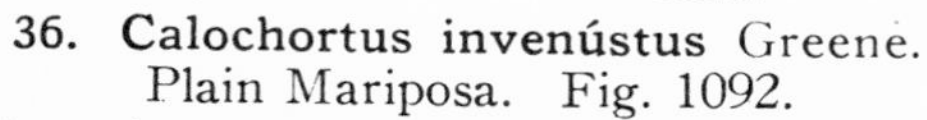
36. Calochortus invenústus Greene.
Plain Mariposa. Fig. 1092.

Calochortus invenustus Greene, Pittonia **2**: 71. 1890.
Calochortus invenustus montanus Parish, Bull. S. Calif. Acad. **1**: 124. 1902.
Calochortus montanus Davidson, Bull. S. Calif. Acad. **9**: 54. 1910.

Stems simple, 1–6 dm. high, bulbiferous at base, glaucous. Basal leaves 1 or rarely 2, narrowly linear, 15–25 cm. long, deeply channelled, glaucous; stem leaves 1 or rarely 2, 5–10 cm. long, sometimes wanting; umbel 2–4-flowered, sepals ovate-lanceolate, acute, about two-thirds the length of the petals, yellowish green within, the tips rarely recurving; petals narrowly obovate-cuneate, 25–35 mm. long, prominently apiculate at apex, white, more or less tinged with green, the short claw purplish, sparsely white hairy just above the gland, otherwise naked; gland small, lanceolate, purple, covered with short matted hairs; anthers oblong-linear, obtuse, 7–8 mm. long; filaments 5–6 mm. long; capsule linear, attenuate at apex, 4 cm. long.

Dry gravelly ridges, Transition Zone; Santa Lucia Mountains, Monterey County, and the Tehachapi Mountains, south to the Cuyamaca Mountains, San Diego County, California. Type locality: "higher mountains to the westward of the Mojave Desert," probably the Tehachapi Mountains.

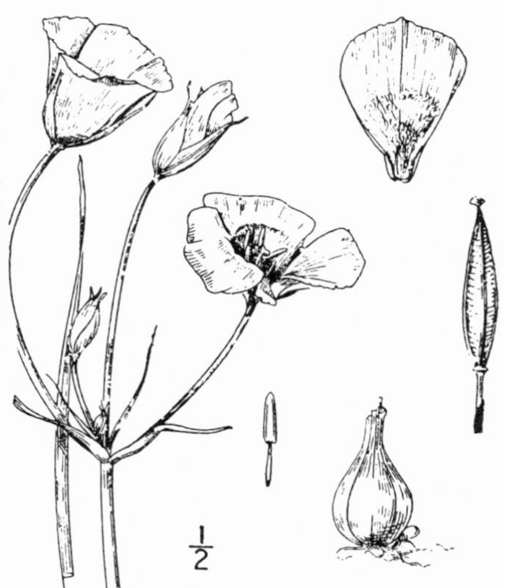

37. **Calochortus lùteus** Dougl. Yellow Mariposa. Fig. 1093.

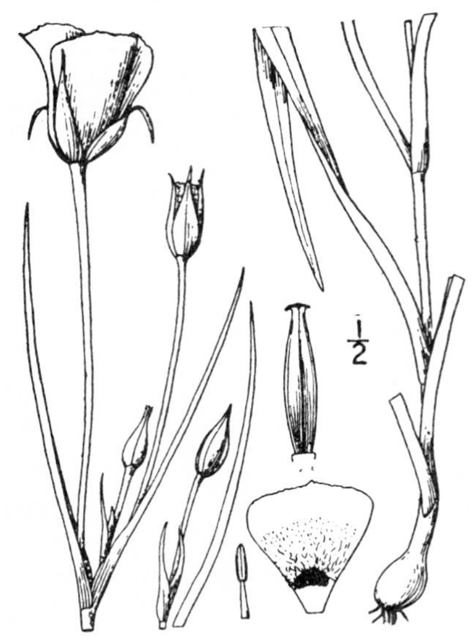

Calochortus luteus Dougl.; Lindl. Bot. Reg. **19**: *pl. 1567*. 1833.
Calochortus venustus citrinus Baker, Journ. Linn. Soc. **14**: 310. 1875.
Calochortus luteus citrinus S. Wats. Proc. Am. Acad. **14**: 265. 1879.
Calochortus luteus oculatus S. Wats. Proc. Am. Acad. **14**: 265. 1879.
Calochortus luteus robusta Purdy, Proc. Calif. Acad. III. **2**: 139. 1901.

Stem stiffly erect, slender, frequently branching 1.5–5 dm. high, bulbiferous at base. Basal leaves 1 or 2, narrowly linear, 10–20 cm. long, 2–5 mm. wide; cauline leaves 1 or 2, attenuate, 4–8 cm. long; flowers in 2–3-flowered umbels or in robust plants a few scattered down the stem on lateral branches; bracts 3–10 cm. long; pedicels 6–15 cm. long; sepals about equalling the petals, oblong-lanceolate acuminate, the tips at length more or less recurved, yellow within; petals fan-shaped, 25–35 mm. long and as broad, deep yellow and more or less penciled with brown, hairy over the lower third of the inner surface with yellow hairs; gland broad, lunate, densely covered with matted ascending yellow hairs; anthers oblong, obtuse, pale yellow, 8–10 mm. long; filaments slender scarcely dilated below, equalling or little exceeding the anthers; capsule linear-lanceolate, attenuate at apex, 35–50 mm. long.

Open hillsides in dry or gravelly ground, Upper Sonoran Zone; California Coast Ranges from Mendocino to San Luis Obispo Counties, also in the foothills of the Sierra Nevada, but here it is usually replaced by the varieties. Type locality: California, probably Monterey.

Calochortus luteus citrìnus S. Wats. Petals deep yellow with a prominent dark brown spot near the base, not oculated. This variety is best developed in the north Coast Ranges of California, in Mendocino and Sonoma Counties, but it also occurs less clearly defined in the foothills of the Sierra Nevada.

Calochortus luteus oculàtus S. Wats. Petals pale creamy white to lilac, with a prominent central dark spot surrounded by a band of yellow; sepals also with an eye-spot. Best developed in the Sierra Nevada foothills from Shasta to Butte Counties, but also found in the north Coast Ranges, California.

38. **Calochortus venústus** Dougl. Butterfly Mariposa. Fig. 1094.

Calochortus venustus Dougl.; Benth. Trans. Hort. Soc. II. **1**: 412. *pl. 15*. 1835.
Calochortus venustus lilacinus Baker, Gard. Chron. II. **8**: 70. 1877.
Calochortus venustus purpurea Baker, Gard. Chron. II. **8**: 70. 1877.
Calochortus venustus purpurascens S. Wats. Proc. Am. Acad. **14**: 266. 1879.
Calochortus vesta Purdy, Proc. Calif. Acad. III. **2**: 139. 1901.
Calochortus venustus eldorado Purdy, Proc. Calif. Acad. III. **2**: 141. 1901.
Calochortus venustus roseus Purdy, Proc. Calif. Acad. III. **2**: 141. 1901.
Calochortus venustus sulphureus Purdy, Proc. Calif. III. **2**: 141. 1901.
Calochortus venustus caroli Cockerell, Nature **67**: 235. 1903.

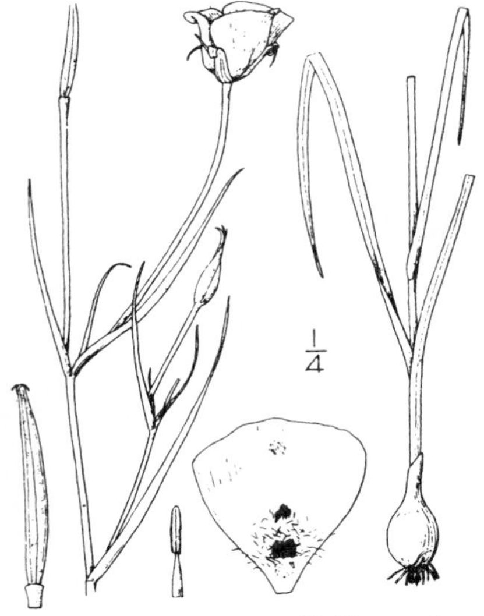

Stem stiffly erect, usually branching, 1–8 dm. high, bulbiferous at base with 1–4 bulblets. Basal leaves 1 or 2, linear, 10–20 cm. long, 2–6 mm. wide, glaucous; cauline leaves 1 or 2, or sometimes none, 3–8 cm. long; flowers in a terminal 1–4-flowered umbel, and in large specimens a few scattered down the stem on lateral branches; bracts 3–7 cm. long, scarious-margined below; pedicels rather stout, 5–20 cm. long; sepals oblong-lanceolate, long-acuminate, the tips at length revolute, 25–35 mm. long with a brownish spot surrounded by yellow below the middle within; petals broadly obovate-cuneate, 30–40 mm. long, slightly rounded and erose at apex, white to lilac, with a prominent eye-spot in the middle and commonly with a reddish blotch near the apex, frequently penciled toward the base, with scattered hairs over the lower one-third of the inner surface; gland lunate, roundish or oblong, densely covered with ascending matted yellow hairs; filaments slightly dilated below; anthers oblong, obtuse, 7–8 mm. long, usually lilac; capsule linear, 5–7 cm. long.

Usually in light sandy soil, but also found in adobe or even alkaline situations, Upper Sonoran and Transition Zones; California Coast Ranges from Humboldt to Los Angeles Counties, also in the foothills of the Sierra Nevada, and the Sacramento-San Joaquin Valley.

The species is extremely variable in coloration and markings. Some color forms will appear fairly constant in a colony or over considerable range, only to break up into a veritable riot of color forms elsewhere. A number of these color strains have been given varietal names:

Calochortus venustus purpuráscens S. Wats. The Sierra Nevada plants differ from the Coast Range forms in having relatively narrow and more deeply colored petals, the terminal reddish blotch is more often absent, the variation in color especially in the southern Sierra Nevada is often to deep purple. *Calochortus venustus eldorado* Purdy was proposed as a name applying to the entire Sierran variation, while *purpurascens* S. Wats. was applied merely to the purple forms of the strain.

Calochortus vésta Purdy is the larger flowered variation of the north Coast Ranges, marked with a broad reddish or dark brown band across the middle, instead of with an eye-spot.

Calochortus venustus càroli Cockerell (*Calochortus venustus purpurascens* Purdy, not S. Wats.). Flowers lilac to purple, with a definite eye-spot, but without the terminal reddish blotch. This color variant is common in the Coast Ranges from San Francisco Bay to San Luis Obispo County.

Calochortus venustus sulphùreus Purdy. Flowers a light yellow, with a central eye-spot and a terminal rose-colored blotch. Has been found at Alcalde, Kern County, and at Newhall, Los Angeles County, the southern limit of the species.

39. **Calochortus plúmmerae** Greene. Plummer's Mariposa. Fig. 1095.

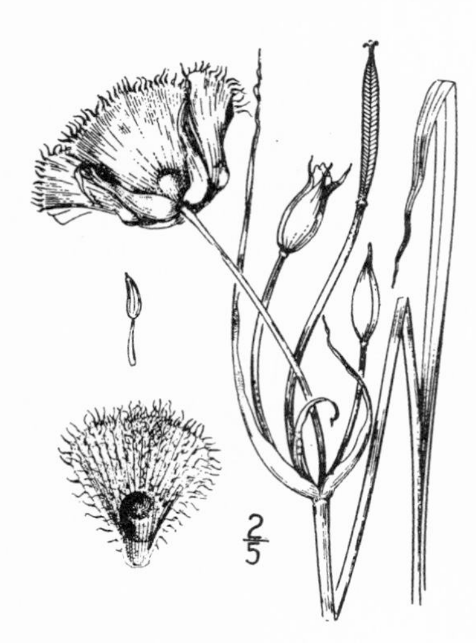

Calochortus weedii purpurascens S. Wats. Proc. Am. Acad. **14**: 265. 1879, in part.

Calochortus plummerae Greene, Pittonia **2**: 70. 1890.

Stem simple or branching, 4–8 dm. high; basal leaves 1 or more, linear, convolute, attenuate above; cauline leaves usually several, often 10 cm. long or more, narrowly attenuate; flowers several and solitary on the racemose branches, or when reduced to 2 appearing umbellate; sepals lanceolate, acuminate, about equalling the petals, tinged with purple; petals fan-shaped, somewhat rounded at apex, 30–40 mm. long, purple, the margin entire or sometimes ciliate, clothed to near the apex with orange-colored hairs; gland orbicular, densely clothed with short hairs; anthers oblong-lanceolate, acutish, 10–14 mm. long, equalling the dilated filaments; capsule linear, acuminate, 5 cm. long, 6 mm. thick.

Dry gravelly hillsides, Upper Sonoran Zone; Los Angeles to Riverside and Orange Counties. Type locality: Mill Creek Cañon, San Bernardino Mountains, California.

40. **Calochortus weèdii** Wood. Weed's Mariposa. Fig. 1096.

Calochortus weedii Wood, Proc. Acad. Philad. **1868**: 169. 1868.

Calochortus luteus weedii Baker, Journ. Linn. Soc. **14**: 306. 1874.

Calochortus citrinus Baker, Bot. Mag. **101**: *pl. 6200.* 1875.

Stem simple or commonly branching, 3–5 dm. high. Leaves linear, long-attenuate, the basal nearly equalling the stem; flowers 1–4, terminating the stiff branches; sepals ovate-lanceolate, or lanceolate-acuminate, shorter than or exceeding the petals, yellow within and penciled with brown; petals deep yellow, penciled and often bordered with orange-brown, fan-shaped, 25-35 mm. long, truncate-rounded at apex, and often apiculate, strongly ciliate on the summit, the entire inner surface above the gland villous with yellow hairs; gland orbicular, 4 mm. wide covered with reflexed light yellow hairs; filaments 12–15 mm. long, dilated below; anthers oblong, obtuse, 8–10 mm. long; capsule linear, acuminate, 5 cm. long.

Dry gravelly hills and mesas, Upper Sonoran Zone; western San Diego County, California. Type locality: San Diego, California.

41. **Calochortus obispoénsis** Lemmon. Lemmon's Mariposa. Fig. 1097.

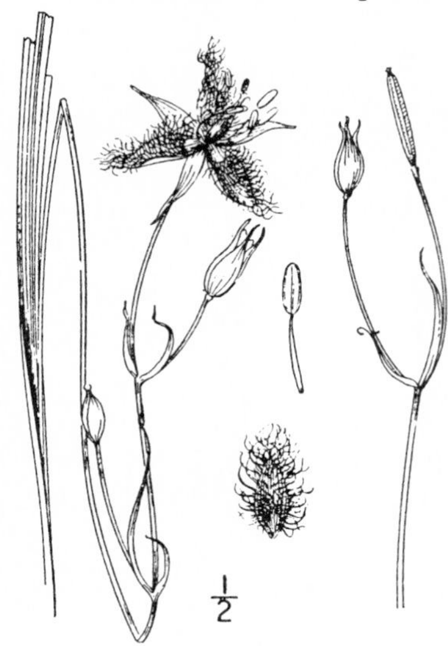

Calochortus obispoensis Lemmon, Bot. Gaz. **11**: 180. 1886.
Calochortus weedii obispoensis Purdy, Proc. Calif. Acad. III. **2**: 133. 1901.

Stems simple or branched, from a deep-seated coarsely fibrous-coated bulb. Leaves linear, long-attenuate; flowers usually several terminating the racemose branches; sepals lanceolate, long-acuminate, 20 mm. long, orange within; petals oblong-ovate, 10–15 mm. long, deep orange, tinged with brown, ciliate on the margin, the region of the gland covered with long spreading orange-colored hairs, forming a conspicuous tuft, upper half of petal but sparsely hairy; stamens equalling the petals; filaments slender, 8 mm. long; capsule linear-lanceolate, attenuate at apex.

Rocky hillsides, Upper Sonoran Zone; San Luis Obispo County. Type locality: on dry stony hills, Steele's Ranch, near San Luis Obispo, California.

22. **YÚCCA** L. Sp. Pl. 319. 1753.

Acaulescent or arborescent plants with a woody simple or branched caudex, and soft or usually rigid leaves, entire, minutely denticulate or filiferous on the margins. Flowers in large panicles, their perianth–segments open-campanulate to incurved, thin or thickish. Stamens 6, filaments enlarged above; anthers versatile. Stigmas sessile or on a short style; 6-lobed or notched, with the stigmatic surface on the wall of the central cavity. Fruit a dehiscent loculicidal capsule, or indehiscent and dry and spongy, or fleshy and berry-like. Seeds numerous, thin and flat, usually lustrous black. [The Haytien name.]

An American genus of about 30 species. Most abundant in southwestern United States and the Mexican Plateau region. Pollination is affected by the Yucca moth, which deposits its eggs in the cavities of the ovary and then places pollen within the stigmatic tube of the stigma. Type species, *Yucca aloifolia* L.

Leaves minutely denticulate; perianth-segments more or less incurved; stigma sessile; fruit dry and spongy. 1. *Y. brevifolia.*
Leaves filiferous on the margin; perianth-segments open-campanulate; style evident; fruit fleshy.
  Perianth-segments spreading from the base; style very short. 2. *Y. mohavensis.*
  Perianth-segments appressed to above the ovary, then spreading; style elongated, 2–3 cm. long. 3. *Y. baccata.*

1. **Yucca brevifòlia** Engelm. Joshua Tree. Fig. 1098.

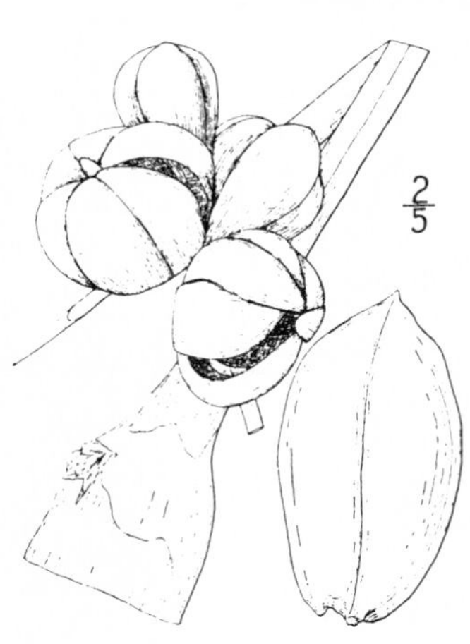

Yucca draconis arborescens Torr. Pacif. R. Rep. **4**: 147. 1857.
Yucca brevifolia Engelm. Bot. King. Expl. 496. 1871.
Yucca arborescens Trelease, Rep. Mo. Bot. Gard. **3**: 163. 1892.
Cleistoyucca arborescens Trelease, Rep. Mo. Bot. Gard. **13**: 43. 1902.

Tree 8-12 m. high, becoming much branched and forming a rounded open head, with a roughened trunk, 5–8 dm. in diameter, young trees unbranched until 2–3 m. high and clothed nearly to the ground with spreading or reflexed leaves. Leaves very rigid and sharp-pointed, forming large tufts at the ends of the branches, spreading in all directions, 15–25 cm. long, 10–15 mm. wide, more or less 3-sided, minutely denticulate; panicle sessile, 25–35 cm. long, dense, often short hispid; flowers greenish white, oblong or globose; perianth-segments 25–30 mm. long; fruit ovoid, 5–10 cm. long, dry and spongy, indehiscent.

One of the most striking features of the Mojave Desert, Lower Sonoran Zone; Antelope Valley, Los Angeles County, California to Detrital Valley, northern Arizona, and the Beaver Dam Mountains, southern Utah. Type locality: not indicated.

2. Yucca mohavénsis Sarg. Mojave Yucca. Fig. 1099.

Yucca mohavensis Sarg. Gard. & Forest. **9**: 104. 1896.

A small tree or shrub, 5 m. high or less, with a simple or branched trunk, 15–20 cm. in diameter, the surface dark brown and scaly or often clothed to the ground with living leaves. Leaves 45–50 cm. long, about 3–4 cm. wide, thin and concave, smooth, with a stout sharp-pointed apex, the margins separating into long pale filaments; flowers in dense sessile or short-stalked panicles 30–50 cm. long; perianth-segments creamy white or the outer surface often purplish, 25–45 mm. long; filaments more or less hairy; style short; fruit 7–10 cm. long, about half as broad, baccate.

Arid desert regions, Lower Sonoran Zone; southern Nevada and northwestern Arizona, to western edge of the Mojave Desert and northern Lower California, extending into the coastal region in the San Gorgonio Pass and the mesas and foothills of Riverside and San Diego Counties, California. Type locality: only the general distribution given.

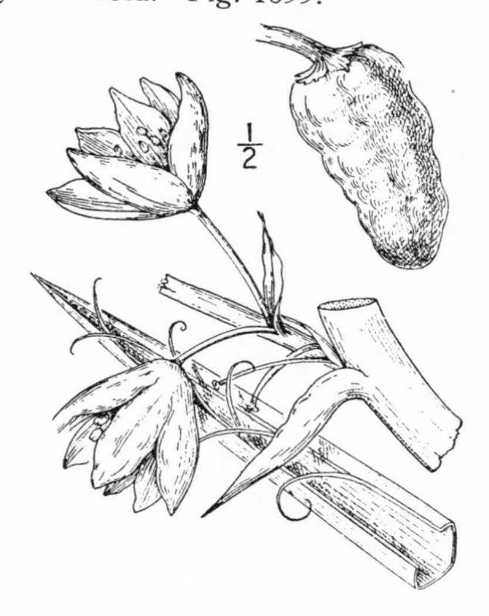

3. Yucca baccàta Torr. Fleshy-fruited Yucca. Fig. 1100.

Yucca baccata Torr. Bot. Mex. Bound. 221. 1859.

Caudex very short, usually appearing acaulescent, bearing a tuft of leaves on the ground or but little above. Leaves pale bluish green, 4–6 dm. long, rigid, with a few coarse fibres on the margin; panicle 5–9 dm. long; perianth-segments 6–8 cm. long, appressed to above the ovary then spreading; filaments papillate; style 2–3 cm. long; fruit conical, very large, often 15–20 cm. long.

Rocky hillsides in the piñon belt, Upper Sonoran Zone; Providence Mountains, southeastern California to southern Colorado and New Mexico. Type locality: "high tablelands between the Rio Grande and the Gila."

23. HESPEROYÚCCA (Engelm.) Baker, Kew Bull. **1892**: 8. 1892.

Low plants with a short woody caudex, densely clothed with straight needle-pointed rough-margined flat leaves. Flowers borne in large panicles on an elongated erect flowering stem; perianth-segments thin concave. Stamens 6; filaments enlarged above; anthers versatile. Style short, slender; stigma capitate, long-papillate. Fruit an incompletely 6-celled capsule, loculicidal; seeds thin flat, lustrous black. [Greek, meaning western yucca.]

A monotypic genus peculiar to southern California

1. Hesperoyucca whípplei (Torr.) Baker. Spanish Bayonet. Fig. 1101.

Yucca whipplei Torr. Bot. Mex. Bound. 222. 1859.
Yucca graminifolia Wood, Proc. Acad. Philad. **1868**: 167. 1868.
Hesperoyucca whipplei Baker, Kew Bull. **1892**: 8. 1892.

Subacaulescent with a short woody caudex, densely clothed by the needle-pointed leaves, forming a rounded clump about 50 cm. broad, or sometimes broader with surrounding daughter clumps formed from suckers. Leaves spreading in all directions, 20–45 cm. long, 15–20 mm. wide, rigid, glaucous, finely but sharply denticulate, tipped with a slender very sharp spine; flowering stalk 2–4 m. high, often 7–10 cm. in diameter, glabrous; panicle 1–1.5 m. long, oblong; flowers pendent, fragrant, with creamy white perianth-segments, 35–50 mm. long; capsule 3–4 cm. long, 25 mm. broad.

On chaparral-covered hills and mesas, Upper Sonoran Zone; Monterey and San Benito Counties in the Coast Ranges, and Tule River in the Sierra Nevada to northern Lower California. Type locality: near San Pasqual, California.

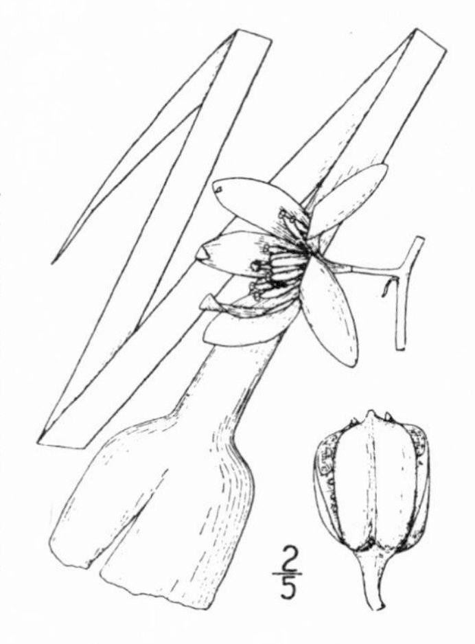

24. NOLÌNA Michx. Fl. Bor. Am. 1: 207. 1803.

Perennial with a thick woody trunk, often much dilated at the base. Leaves numerous, narrowly linear, rigid, finely serrate. Flowers borne on a short nearly naked flowering stem in a compound racemose panicle, the main branches subtended by foliaceous long-pointed bracts. Pedicels jointed usually near the base, subtended by minute scarious bracts. Perianth small, persistent, its segments 6, 1-nerved. Stamens 6, usually abortive in the fertile flowers; filaments very short, slender. Ovary sessile, deeply 3-lobed; style very short, recurved; ovules 2 in each cell. Fruit a capsule, with a thin inflated wall, usually bursting irregularly, exposing the rounded light colored seeds. [Named in honor of C. P. Nolin, French scientist of the 18th century.]

About 24 species native of southern United States and Mexico. Type species, *Nolina georgiana* Michx.

Leaves with merely roughish margins; fruit less than 12 mm. broad; seeds ovate-oblong. 1. *N. bigelovii.*
Leaves strongly serrate; fruit 12 mm. broad; seeds subglobose. 2. *N. parryi.*

1. Nolina bigelòvii (Torr.) S. Wats.
Bigelow's Nolina. Fig. 1102.

Dasylirion bigelovii Torr. Pacif. R. Rep. 4: 151. 1857.
Nolina bigelovii S. Wats. Proc. Am. Acad. 14: 247. 1879.

Caudex stout 6–10 dm. high. Leaves flat, 8–12 dm. long, about 2 cm. wide, above the dilated base, roughish on the margins; scape 5–6 dm. long; branches of the inflorescence slender, 5–10 cm. long, jointed near the middle; perianth-segments 2 mm. long, oblong-linear; capsule 6 mm. long and about as broad, its walls very thin; seeds ovate oblong, 3 mm. long, whitish gray, distinctly wrinkled.

Arid gravelly slopes,. Lower Sonoran Zone; western Arizona and the desert slopes of the Cuyamaca Mountains, southern California and northern Lower California. Type locality: mountain sides, Williams River, Arizona.

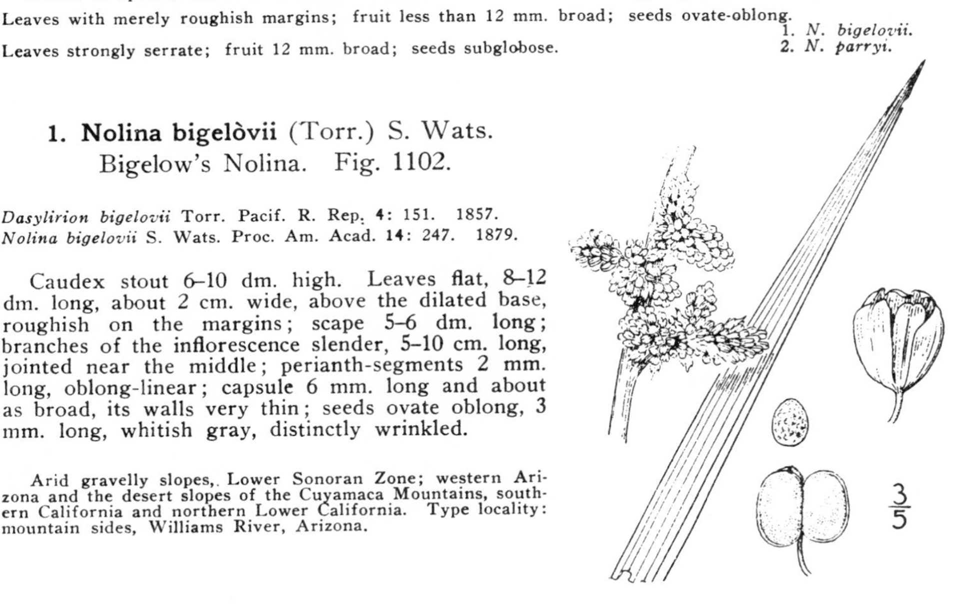

2. Nolina párryi S. Wats.
Parry's Nolina. Fig. 1103.

Nolina parryi S. Wats. Proc. Am. Acad. 14: 247. 1879.

Caudex unbranched, usually 1–2 m. high. Leaves forming a dense crown at the apex of the trunk, spreading in all directions, rigid, thick and slightly concave, often 1 m. long, about 2 cm. wide above the dilated base, strongly serrate; flowering stalk erect, 5–6 dm. long; pedicels and branches of the inflorescence stout; perianth-segments 4 mm. long; capsule 15 mm. long, and about as broad, its walls thin; seeds subglobose, light brown, with very thin transparent finely and irregularly wrinkled seed-coat.

Arid gravelly slopes, Lower Sonoran Zone; Morongo Cañon, desert base of the San Bernardino Mountains to the desert slopes of the San Jacinto Mountains, and the interior mesas and hills of the coastal region in Riverside and San Diego Counties, California. Type locality: near Whitewater, desert slopes of the San Bernardino Mountains.

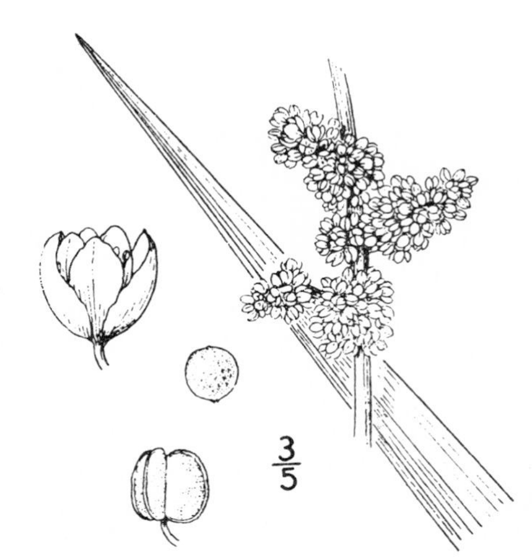

Family 19. CONVALLARIÀCEAE.

LILY-OF-THE-VALLEY FAMILY.

Scapose or leafy-stemmed herbs, with simple or branched rootstocks, never with bulbs or corms. Flowers solitary, racemose, paniculate or umbellate, regular and perfect. Leaves broad, parallel-veined and sometimes with cross-veinlets, the cauline alternate or verticillate, or in *Asparagus* and its allies, reduced to scales bearing filiform or flattened branchlets in their axils. Perianth 4–6-parted with separate segments, or oblong, cylindric or urn-shaped and 6-lobed or 6-toothed. Stamens 3 or 6, rarely 4, hypogynous or inserted on the perianth; anthers introrse or extrorse, or sometimes laterally dehiscent. Ovary superior, 1–3-celled; ovules anatropous or amphitropous; style slender or short; stigma mostly 3-lobed. Fruit a fleshy berry; rarely a capsule. Seeds few or numerous. Embryo small.

About 23 genera and 215 species, widely distributed.

Stamens 3; fruit a 1-celled capsule. — 1. *Scoliopus.*
Stamens 6; fruit a 3-celled berry.
  Leaves 3, whorled at the summit of the stem. — 2. *Trillium.*
  Leaves not whorled at the summit of the stem.
    Leaves reduced to scales; branchlets filiform, green. — 3. *Asparagus.*
    Leaves broad; stems simple or somewhat branched.
      Leaves basal; flowers solitary or umbellate. — 4. *Clintonia.*
      Leaves alternate, or solitary and rising from the rootstock on a long slender petiole.
        Flowers in terminal racemes or panicles.
          Perianth-segments 6. — 5. *Vagnera.*
          Perianth-segments 4. — 6. *Unifolium.*
        Flowers solitary or in few-flowered umbels, drooping.
          Flowers terminal. — 7. *Disporum.*
          Flowers axillary, usually solitary.
            Perianth narrowly campanulate. — 8. *Streptopus.*
            Perianth rotate. — 9. *Kruhsea.*

1. SCOLIÒPUS Torr. Pacif. R. Rep. 4: 145. *pl. 22.* 1857.

Nearly acaulescent glabrous perennial herbs, with a short fibrous-rooted rootstock. Stem very short, subterranean. Leaves 2, oval to lanceolate, sessile or subsessile, many-nerved with transverse veinlets, more or less punctuate with purple dots. Flowers in a sessile umbel, on elongated, more or less recurved and tortuous pedicels, ill-scented. Perianth-segments 6, distinct, deciduous; the outer 3 spreading, lanceolate, several-nerved, punctate; the inner erect and converging over the pistil, linear, 3-nerved. Stamens 3, attached at the base of the outer segments; filaments filiform-subulate; anthers oblong, 2-celled, extrorse, attached above the middle. Ovary sessile, narrow, attenuate at apex, strongly 3-angled, 1-celled, with the placentae in the thickened angles; style very short; stigmas 3, linear, recurved, deeply channelled on the inner surface, persistent; ovules about 10, in 2 rows on each placenta. Capsule oblong-lanceolate, strongly 3-angled, thin-membranous, dehiscing irregularly. Seeds oblong, slightly curved, sulcate-striate longitudinally. [Greek meaning, crooked foot or stalk, in allusion to the tortuous pedicels.]

Two species of the Pacific States. Type species, *Scoliopus bigelovii* Torr.

Perianth-segments about 15 mm. long; stigmas 5 mm. long. — 1. *S. bigelovii.*
Perianth-segments about 8 mm. long; stigmas 2 mm. long. — 2. *S. hallii.*

1. Scoliopus bigelòvii Torr. California Fetid Adder's-tongue. Fig. 1104.

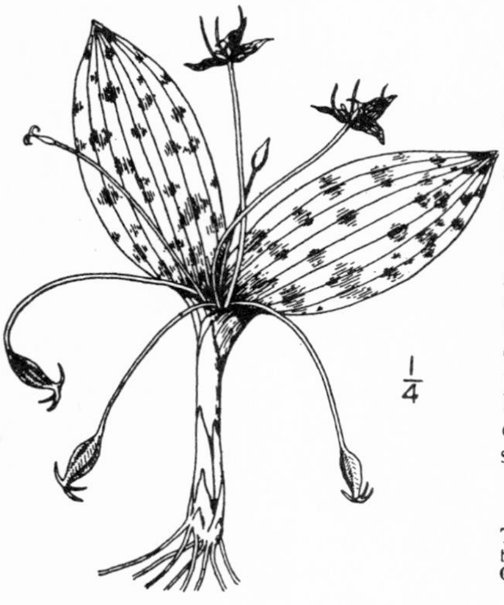

Scoliopus bigelovii Torr. Pacif. R. Rep. 4: 145. *pl. 22.* 1857.

Stem subterranean, 2–5 cm. long. Leaves elliptic to oblong, 10–20 cm. long, 5–10 cm. wide, obtuse at apex, sheathing at base, often mottled with dark splotches, peduncle very short; pedicels 3–12, elongated and scape-like, shorter than or exceeding the leaves, more or less tortuous and becoming strongly recurved in fruit, flowers mottled with green and purple, fetid; outer perianth-segments ovate-lanceolate, about 15 mm. long, spreading or recurving; inner segments linear-subulate, about equalling the outer segments, scarcely over 1 mm. wide, converging over the pistil; stamens 5–6 mm. long; stigmas 5–6 mm. long, becoming strongly recurved; capsule 15–18 mm. long, beaked by the persistent style and stigmas.

Moist wooded slopes, especially among redwoods, Humid Transition Zone; Humboldt County to the Santa Cruz Mountains, California. Type locality: "Tamul Pass, Marin County, California."

1/4

2. Scoliopus hállii S. Wats.
Oregon Fetid Adder's-tongue. Fig. 1105.

Scoliopus hallii S. Wats. Proc. Am. Acad. **14**: 272. 1879.

Stem very short, subterranean, sheathed by the old leaf-bases. Leaves 2, oblong-elliptic, 8–15 cm. long, 3–5 cm. wide, obtuse at apex, sheathing at base, and somewhat petioled; pedicels 1–8, very slender, 3–6 cm. long, ascending in flower, recurved in fruit; perianth-segments about 8 mm. long; the outer segments mottled with yellowish green and purple, lanceolate or oblancolate, spreading; the inner segments linear-spatulate, arched over the pistil; stamens about 4 mm. long; stigmas recurved, linear, 2 mm. long; capsule including the beak about 15 mm. long, brownish purple.

On moist mossy banks along mountain streams, Humid Transition Zone; western Oregon. Type locality: Cascade Mountains, Oregon.

2. TRÍLLIUM L. Sp. Pl. 339. 1753.

Glabrous erect unbranched herbs, with short scarred rootstocks and 3 leaves whorled at the summit of the stem, subtending the sessile or peduncled solitary bractless flower. Solitary long-petioled leaves are sometimes borne on the rootstock. Perianth of 2 distinct series of segments, the outer 3 (sepals) green, persistent, the inner 3 (petals) white, pink, purple or sometimes greenish, deciduous or withering. Stamens 6, hypogynous; filaments short; anthers linear, mostly introrse. Ovary sessile, 3–6-angled or lobed, 3-celled; ovules several or numerous in each cavity; styles 3, stigmatic along the inner side. Berry many-seeded. Seeds horizontal. [Latin in allusion to the 3-parted flowers and the 3 leaves.]

About 25 species of North America and Asia, besides the following there are 12 in eastern North America. Type species, *Trillium cernuum* L.

Flower distinctly pedicelled.
  Capsule winged; stems 3–6 dm. high. — 1. *T. ovatum.*
  Capsule globose, scarcely angled; stems 1–2 dm. high. — 2. *T. rivale.*
Flower sessile.
  Leaves sessile. — 3. *T. chloropetalum.*
  Leaves long-petioled. — 4. *T. petiolatum.*

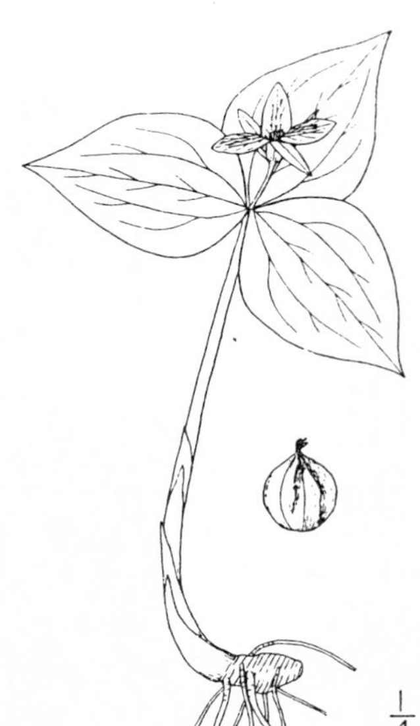

1. Trillium ovàtum Pursh.
Western Wake-robin or Trillium. Fig. 1106.

Trillium ovatum Pursh, Fl. Am. Sept. 245. 1814.
Trillium californicum Kellogg, Proc. Calif. Acad. **2**: 50. *f. 2.* 1860.
Trillium crassifolium Piper, Erythea **7**: 104. 1899.
Trillium scouleri Rydb. Bull. Torrey Club **33**: 394. 1906.
Trillium venosum Gates, Ann. Mo. Bot. Gard. **4**: 66. *pl. 6, f. 1.* 1917.

Stem 3–5 dm. high, from a horizontal or suberect rootstock. Leaves rhombic-ovate, 5–15 cm. long, acute or short-acuminate, narrowed or somewhat rounded to the sessile or shortly petiolate base; pedicel erect, 25–75 mm. long; sepals broadly oblong-lanceolate to narrowly lanceolate 25–50 mm. long; petals equalling or little exceeding the sepals, 6–25 mm. wide, white soon turning to deep rose color; anthers 8–15 mm. long, longer than the filaments; berry broadly ovate, somewhat winged, 12–18 mm. long.

Wooded slopes, Transition and Canadian Zones; British Columbia and Montana, southward through Oregon and the Coast Ranges of California to the Santa Cruz Mountains. Type locality: Cascades of the Columbia River. Considerable variation occurs throughout the range of the species and several of the more striking forms have been designated by name, but all are inconstant even in any given region so that separation seems impracticable. But to the plant-breeder and floral gardener the variations in size, color, and shape of the flowers in this species and *chloropetalum* offer opportunities for the development of many garden varieties.

2. **Trillium riväle** S. Wats.

Brook Trillium. Fig. 1107.

Trillium rivale S. Wats. Proc. Am. Acad. **20**: 378. 1885.

Stem slender, 10–20 cm. high. Leaves ovate-lanceolate, rounded or subcordate at base, acute or acuminate, 25–45 mm. long; petioles 5–25 mm. long; pedicels slender, erect or at length declinate, a little shorter or longer than the leaves; petals 15–20 mm. long, white or marked with purple, rhombic-ovate, acute; filaments about equalling the anthers; berry globose, 10–15 mm. broad, scarcely angled.

Stream banks, Humid Transition Zone; Coast Ranges and the Siskiyou Mountains of southern Oregon and adjacent California. Type locality: Big Flat, 30 miles east of Crescent City, California.

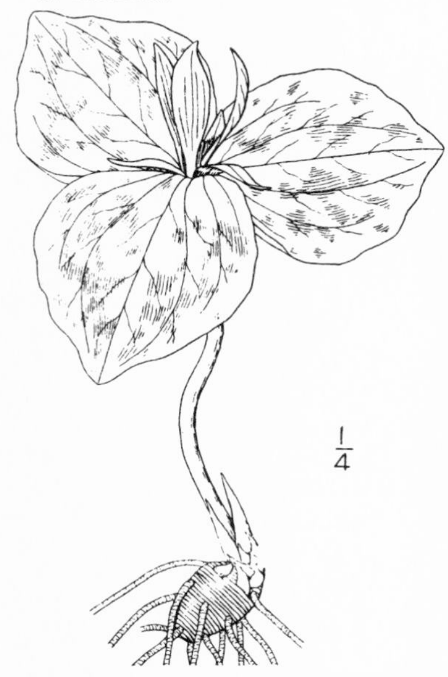

3. **Trillium chloropétalum** (Torr.) Howell.

Giant Wake-robin or Trillium. Fig. 1108.

Trillium sessile giganteum Hook. & Arn. Bot. Beechey, 402. 1841.
Trillium sessile chloropetalum Torr. Pacif. R. Rep. **4**: 151. 1857.
Trillium sessile angustipetalum Torr. Pacif. R. Rep. **4**: 151. 1857.
Trillium sessile californicum S. Wats. Proc. Am. Acad. **14**: 273. 1879.
Trillium chloropetalum Howell, Fl. N. W. Am. 1: 661. 1902.
Trillium giganteum Heller, Bull. S. Calif. Acad. Sci. **2**: 67. 1903.
Trillium giganteum chloropetalum Gates, Ann. Mo. Bot. Gard. **4**: 50. 1917.
Trillium giganteum angustipetalum Gates, Ann. Mo. Bot. Gard. **4**: 51. 1917.

Stems stout, one or more arising from the very stout rootstock, 3–5 dm. high. Leaves round-ovate, 8–15 cm. long, and nearly or quite as wide, usually conspicuously mottled with dark splotches, obtuse to somewhat accuminate at apex, rounded to the sessile base; flowers sessile; sepals narrowly to rather broadly oblong-lanceolate, 3–5 cm. long, petals obovate to linear-lanceolate, 4–9 cm. long, 8–35 mm. wide, deep maroon varying to greenish yellow or white; anthers 15–25 mm. long; berry strongly winged, green or red.

Wooded slopes, Transition Zone; western Washington, where it has been collected at Roy, southward to San Luis Obispo and Mariposa Counties, California. Type locality: "Redwoods," probably in the vicinity of Point Reyes, California, where the greenish yellow form is prevalent. From Mendocino and Napa Counties, California, northward, the prevailing form is white or yellowish white, often with a purple base. In the San Francisco region south to Monterey the maroon form is prevalent; this was originally described as *sessile giganteum*. The narrow-petaled form was described as *sessile angustipetalum* Torr., from specimens collected in the Mariposa Big Tree Grove. It is the prevailing type in the Sierra Nevada, but also occurs in the San Francisco Bay region.

4. **Trillium petiolàtum** Pursh.

Petioled Wake-robin or Trillium. Fig. 1109.

Trillium petiolatum Pursh, Fl. Am. Sept. 1: 244. 1914.

Stems from a short stout rootstock, 5–15 cm. long, but mainly subterranean, usually only 2–3 cm. above the ground. Leaves round-ovate 6–15 cm. long, 5–10 cm. wide, rounded or obtuse at apex, rounded at base to a stout petiole, 5–12 cm. long; flowers sessile in the axils of the petioles; sepals oblong-linear, 15–35 mm. long, 5–7 mm. wide, erect; petals narrowly oblanceolate, 25–45 mm. long, 5–10 mm. wide, maroon color; anthers 10–15 mm. long, exceeding the stigmas, on very short filaments; stigmas slender, 10 mm. long, but little spreading.

Moist rich soil usually in copses, Arid Transition Zone; eastern Washington and Idaho south to eastern Oregon. Type locality: Lolo River, Idaho.

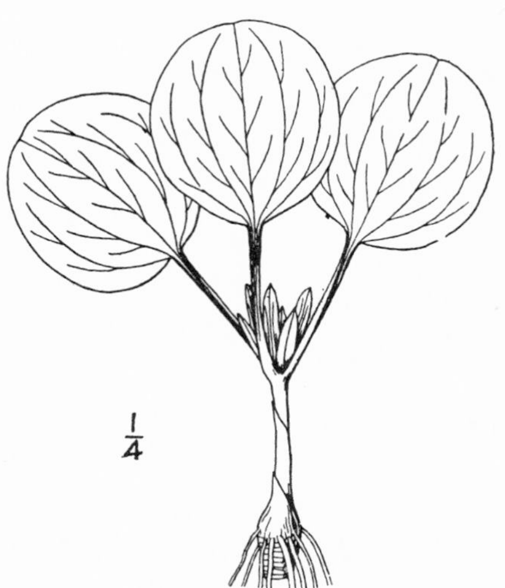

3. ASPÀRAGUS L. Sp. Pl. 313. 1753.

Stem at first simple, fleshy, scaly, at length much branched; the branchlets filiform and mostly clustered in the axils of the scales, or in some species flattened and linear to ovate. Flowers small solitary, racemose or umbellate. Perianth-segments equal, separate or slightly united at the base. Stamens inserted at the base of the perianth-segments; filaments usually filiform; anthers ovate or oblong, introrse. Ovary sessile, 3-celled; ovules 2 in each cavity; style slender, short; stigma 3, short, recurved. Fruit a globose berry. Seeds few, rounded. [Ancient Greek name.]

About 100 species, natives of the Old World. Type species, *Asparagus officinalis* L.

1. Asparagus officinális L.
Asparagus. Fig. 1110.

Asparagus officinalis L. Sp. Pl. 313. 1753.

Roostock much branched. Young stems succulent, edible, stout, later branching, and becoming 1–2 m. high, the filiform branchlets 1–2 cm. long, mostly clustered in the axils of minute scales; flowers mostly solitary at the nodes, green, drooping on jointed filiform pedicels; perianth campanulate, about 6 mm. long, the segments linear, obtuse; stamens shorter than the perianth; berry red, about 8 mm. in diameter.

Escaped from cultivation, and naturalized especially in subsaline or alkaline soils. Native of Europe.

4. CLINTÒNIA Raf. Journ. Pys. 89: 102. 1819.

Pubescent, or nearly glabrous, scapose herbs, with slender rootstocks, erect simple scapes, and few broad petioled sheathing basal leaves. Flowers bractless, terminal, solitary, or umbellate. Perianth-segments separate, equal or nearly so, erect-spreading. Stamens 6, inserted at the bases of the perianth-segments, filaments filiform; anthers oblong, laterally dehiscent. Ovary 2–3-celled; ovule 2–several in each cavity; style stout or slender; stigma obscurely 2–3 lobed. Fruit an oval or globose, blue or black berry. [Name in honor of DeWitt Clinton, 1769–1828, American naturalist, Governor of New York.]

Six species, 2 native of western North America, 2 of eastern North America, and 2 of eastern Asia.

Flowers 1 or 2 white. 1. *C. uniflora.*
Flowers umbellate, deep red. 2. *C. andrewsiana.*

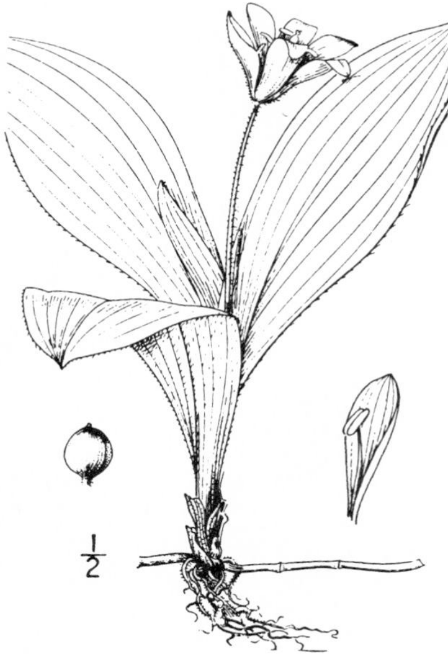

1. Clintonia unifl̀ora (Schult.) Kunth.
Single-flowered Clintonia. Fig. 1111.

Smilacina borealis uniflora Schult. in Roem & Schult. Syst. 7: 307. 1829.
Smilacina uniflora Menzies, Hook. Fl. Bor. Am. 2: 175. *pl. 190.* 1839.
Clintonia uniflora Kunth, Enum. 5: 159. 1850.

Stem only 2–5 cm. long, mostly subterranean, arising from a slender creeping rootstock. Leaves 2 or 3, rarely 4 or 5, oblanceolate to obovate, 8–15 cm. long, 2–7 cm. wide, acute, tapering at base, more or less villous-pubescent; peduncle slender, half to two-thirds the length of the leaves, naked or with 1 or 2 small bracts; flowers 1 or rarely 2, erect, white; perianth-segments 18–25 mm. long, 5–7 mm. wide, pubescent; filaments pubescent; anthers linear, 4–5 mm. long, sagittate; berry blue, globose or pyriform, 8–12 mm. in diameter, the cavities 6–10 seeded.

Wooded mountain slopes, Transition and (mainly) Canadian Zones; southern Alaska to Montana, southward through the Cascade-Sierra Nevada Ranges to Mariposa County, California. Type locality: "in ora occidentali Americae borealis", collected by Menzies.

2. Clintonia andrewsiàna Torr.
Red Clintonia. Fig. 1112.

Clintonia andrewsiana Torr. Pacif. R. Rep. **4**: 150. 1857.

Stem rather stout, 5–15 cm. long, mainly subterranean. Leaves mostly 5 or 6, oval to oblanceolate, 15–20 cm. long, acute or abruptly short-acuminate, sparsely villous-pubescent on the margins and midrib toward the base, otherwise nearly or quite glabrous; peduncle stout, 25–50 cm. long, usually with 1 or rarely 2 or 3 foliaceous bracts; inflorescence villous-pubescent, of a terminal umbel and one or more lateral umbellate fascicles; pedicels slender, unequal, 10–25 mm. long; perianth deep rose-purple, the segments gibbous at base 8–15 mm. long; filaments pubescent; anthers oblong, 2 mm. long; berry metallic blue, ovoid, 8–10 mm. long, the cavities 8–10-seeded.

Shady woods near the coast, Humid Transition Zone; California Coast Ranges, Humboldt County to Monterey. Type locality: "probably not far from San Francisco."

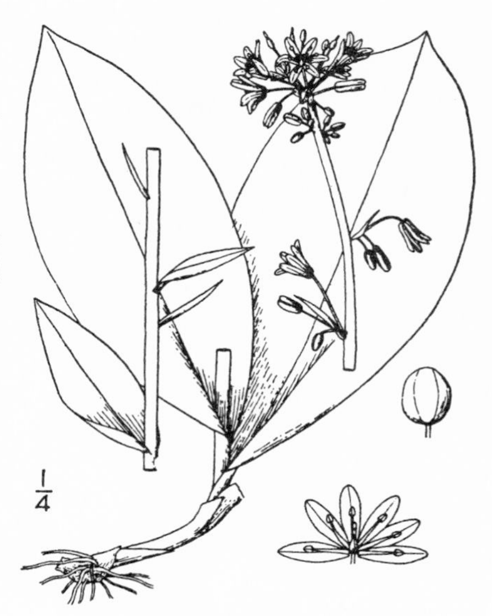

5. SMILACÌNA Desf. Am. Mus. Paris 9: 51. 1807.

[VAGNERA Adans. Fam. Pl. **2**: 496. 1763.]

More or less pubescent herbs, with slender, or short and thick rootstocks, simple stems, scaly below, leafy above. Leaves alternate, short-petioled or sessile, ovate, lanceolate or oblong. Flowers in a terminal raceme or panicle, small, white or greenish-white. Perianth of 6 separate spreading equal segments. Stamens 6, inserted at the base of the perianth-segments; filaments filiform or subulate; anthers ovate, introrse. Ovary 3-celled, sessile, subglobose; ovules 2 in each cavity; style short or slender; stigma 3-grooved or 3-lobed. Berry globular. Seeds usually 1 or 2, subglobose. [Name diminutive of *Smilax*.]

About 25 species, natives of North America, Central America, and Asia. Besides the following, two others occur in eastern North America. Type species, *Smilacina stellata* (L.) Desf.

Flowers paniculate, subsessile, numerous. — 1. *S. amplexicaulis.*
Flowers racemose, few; pedicels equalling or longer than the flowers.
  Leaves flat and spreading; raceme zigzag; pedicels widely spreading. — 2. *S. sessilifolia.*
  Leaves folded, ascending; raceme seldom zigzag; pedicels ascending. — 3. *S. liliacea.*

1. Smilacina amplexicaùlis Nutt. Western Solomon's Seal. Fig. 1113.

Smilacina amplexicaulis Nutt. Journ. Acad. Philad. **7**: 58. 1834.
Vagnera amplexicaulis Greene, Man. Bay Region 316. 1894.
Smilacina racemosa brachystyla Henderson, Bull. Torrey Club **27**: 357. 1900.
Vagnera brachypetala Rydb. Bull. Torrey Club **28**: 268. 1901.

Rootstock stout, fleshy, elongate. Stem 3–10 dm. high, erect or ascending, slender or stout, somewhat angled, more or less puberulent with spreading hairs. Leaves ovate to broadly lanceolate, 5–15 cm. long, sessile and amplexicaul or with a short dilated clasping petiole, acute or abruptly short-acuminate at apex, rounded at base, more or less pubescent with short stiff hairs, panicle sessile or short-peduncled, 3–15 cm. long, oblong-ovate; pedicels 1–3 mm. long; perianth-segments linear-lanceolate to linear-oblong, 1.5–2.5 mm. long; filaments subulate usually a third to a half longer than the segments; style 1–2 mm. long; berry bright red, sprinkled with reddish purple dots, 4–5 mm. in diameter; seed usually solitary, whitish, 3 mm. broad.

On rich wooded slopes, Transition Zone; British Columbia and Montana south to New Mexico and the San Bernardino Mountains, California. Type locality: "in the valleys of the Rocky Mountains about the sources of the Columbia River." Considerable variation in relative length of petals, stamens and style have led to several segregations, but these characters appear inconstant often on the same individual.

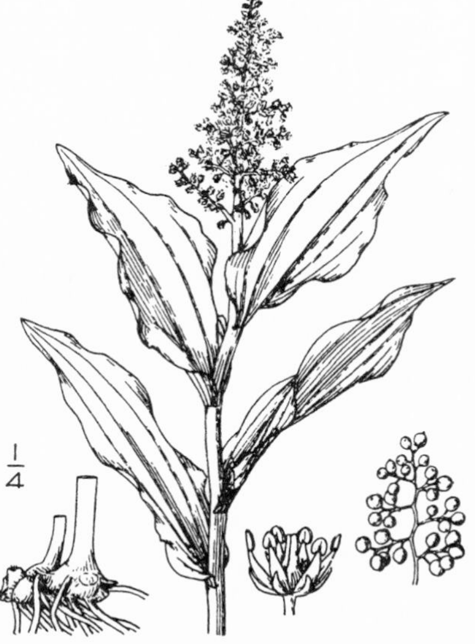

Smilacina amplexicaulis glàbra Macbr. Contr. Gray Herb. II. **41**: 18. 1918. A glabrate plant which seems fairly constant over a well defined range. It occurs in the southern Cascade Mountains and extends southward, mainly in the Canadian Zone, through the Sierra Nevada to the San Bernardino Mountains.

2. Smilacina sessilifòlia (Baker) Nutt. Nuttall's Solomon's Seal. Fig. 1114.

Tovaria sessilifolia Baker, Journ. Linn. Soc. **14**: 566. 1875.
Smilacina sessilifolia Nutt.; S. Wats. Proc. Am. Acad. **14**: 245. 1879.
Vagnera sessilifolia Greene, Man. Bot. Bay Region 316. 1894.
Smilacina stellata sessilifolia Henderson, Bull. Torrey Club **27**: 358. 1900.

Stem arising from a slender rootstock, 3–6 dm. high, erect, usually flexuous above. Leaves ovate-lanceolate, 5–15 mm. long, sessile and clasping, acute or acuminate, flat and spreading, more or less pubescent; raceme sessile or short-peduncled, 2–5 cm. long; mostly 5–8-flowered, rachis often zigzag; pedicels slender, the lower 10–15 mm. long, spreading; perianth-segments 4–7 mm. long, lanceolate, rarely 3-nerved; stamens half to two-thirds as long as segments; style about equalling the ovary; berry globose, 6–10 mm. in diameter, dark-purple; seeds 1–3, 3 mm. broad, brown.

On moist slopes and stream banks, Transition Zone; British Columbia to Montana and Utah, south through western Oregon and the Coast Ranges of California to Monterey County.

3. Smilacina liliàcea (Greene) Wynd. Lily-leaved Solomon's Seal. Fig. 1115.

Unifolium liliaceum Greene, Pittonia **1**: 280. 1889.
Vagnera liliacea Rydb. Mem. N. Y. Bot. Gard. **1**: 101. 1900.
Vagnera pallescens Greene, Proc. Acad. Sci. Philad. **1895**: 551. 1896.
Smilacina liliacea Wynd, Amer. Midland Nat. **17**: 902. 1936.

Stem arising from a rather stout fleshy rootstock, glabrous, 2–6 dm. high, straight and strict throughout, leafy. Leaves oblong-lanceolate or lanceolate, 5–15 cm. long, 15–30 mm. wide, acute or acuminate at apex, sessile or somewhat clasping at base, minutely pubescent beneath, or glabrate; raceme sessile or short-peduncled, 3–10 cm. long, 3–10-flowered, rarely slightly zigzag; pedicels, at least the lower, often 2 or 3 times as long as the flowers; perianth-segments oblong, obtuse, 5–7 mm. long; stamens half as long; style 1 mm. long; berry reddish purple, 7–10 mm. in diameter.

Wooded slopes, Arid Transition and Canadian Zones; British Columbia to New Mexico and southern California. Closely related to the eastern *S. stellata* (L.) Desf. to which it has been referred. Type locality: "the higher Sierra in California."

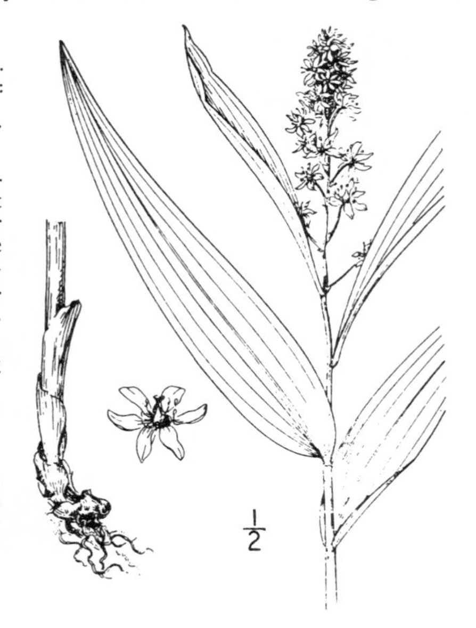

6. MAIÁNTHEMUM Weber in Wigg. Prim. Fl. Hols. 14. 1780.

[UNIFOLIUM Haller; Zinn, Cat. Plant Hort. Goett. 104. 1757.]

Glabrous or pubescent herbs, with slender rootstocks and simple erect few-leaved stems. Leaves alternate, solitary, petioled or sessile, cordate. Flowers small, white, in a terminal raceme, the pedicels often 2 or 3 together. Perianth of 4 separate spreading segments. Stamens 4, inserted at the bases of the perianth-segments, filaments filiform; anthers introrse. Ovary 2-celled, sessile, globose; ovules 2 in each cavity; style slender, 2-lobed or 2-cleft. Berry globose, 1–2-seeded. [Name Greek, meaning May and flower, in reference to the blooming period.]

A genus of two or three species inhabiting Asia and western North America. Type species *Maianthemum convallaria* Weber.

1. Maianthemum dilatàtum (Wood) Abrams
Two-leaved Solomon's Seal. Fig. 1116.

Maianthemum bifolium dilatatum Wood, Proc. Acad. Sci. Philad. **1868**: 174. 1868.
Unifolium dilatatum Greene, Man. Bay Reg. 316. 1894.
Unifolium bifolium kamtschaticum Piper, Contr. Nat. Herb. **11**: 200. 1906, in part.

Stem arising from a slender rootstock, erect, glabrous, 15–35 cm. high. Leaves of the sterile shoots round-cordate, 5–10 mm. wide, on a slender petiole 5–15 cm. long, the flowering stems with a few sheathing scarious bracts at base, or the uppermost bract produced into a small ovate leaf; leaves of the fertile stems usually 2, cordate to sagittate with large rounded auricles, 5–10 cm. long, 5–8 cm. wide, glabrous; petioles slender, 15–45 mm. long; raceme rather short-peduncled 25–75 mm. long, many-flowered; pedicels spreading 2–4 mm. long, usually in fascicles of 2–4; perianth-segments 2–3 mm. long, oblong, obtuse, becoming reflexed; stamens shorter than the segments; style stout, 1 mm. long, berry red, globose, 6 mm. in diameter, 1–4-seeded; seeds ovate, brown, 3 mm. long.

On moist shaded banks, Humid Transition Zone; Alaska to Idaho, and in the Coast Ranges to Marin County, California. Type locality: "from Astoria to The Dalles," Oregon.

7. DÍSPORUM Salisb. Trans. Hort. Soc. 1: 331. 1812.

[PROSARTES D. Don, Trans. Linn. Soc. 1: 48. 1839.]

More or less pubescent herbs, with slender rootstocks, and branching stems, scaly below, leafy above. Leaves alternate, somewhat inequilateral, sessile or clasping. Flowers terminal, drooping, solitary, or few in a cluster, whitish or greenish yellow. Perianth of 6 narrow equal separate deciduous segments. Stamens 6, hypogynous; filaments filiform or flattened; anthers oblong or linear, extrorse. Ovary 3-celled; ovules 2 or sometimes several in each cavity; style slender; stigma 3-cleft or entire. Berry ovoid or oval, obtuse. [Greek referring to the 2 ovules in each cavity of the ovary.]

About 15 species, natives of North America and Asia. Besides the following, 2 others occur in eastern North America. Type species, *Disporum pullum* Salisb.

Stigma 3-lobed.
  Style woolly-pubescent to the summit. — 1. *D. smithii.*
  Style glabrous or only slightly pubescent toward the base. — 2. *D. trachycarpum.*
Stigma entire.
  Anthers glabrous.
    Style glabrous. — 3. *D. hookeri.*
    Style pubescent below the middle. — 4. *D. oreganum.*
  Anthers pubescent. — 5. *D. trachyandrum.*

1. Disporum smíthii (Hook.) Piper.
Large-flowered Fairy Bell. Fig. 1117.

Uvularia smithii Hook. Fl. Bor. Am. **2**: 174. *pl. 189.* 1838.
Prosartes menziesii D. Don, Trans. Linn. Soc. **1**: 48. 1839.
Disporum menziesii Britton, Bull. Torrey Club **15**: 188. 1888.

Short-wooly pubescent or almost glabrous, 3–9 dm. high. Leaves ovate to ovate-lanceolate, 5–15 cm. long, narrowly acuminate, or the lowest acute, rounded or slightly cordate at base, 3–5-nerved; flowers 1–5; perianth gibbously truncate and broad at base, the segments 6, 12–20 mm. long, acute, nearly erect, creamy-white or somewhat greenish; stamens two-thirds the length of the segments; anthers glabrous; style about equalling the stamens, wholly-pubescent, 3-cleft at the apex; ovary glabrous; berry bright salmon red, 12–16 mm. long, oblong-obovoid, usually narrowed at apex into a short wooly beak, triangular, 3–6 seeded.

Moist wooded banks, Humid Transition Zone; Coast Ranges from British Columbia to Santa Cruz County, California. Type locality: "Nutka Sound."

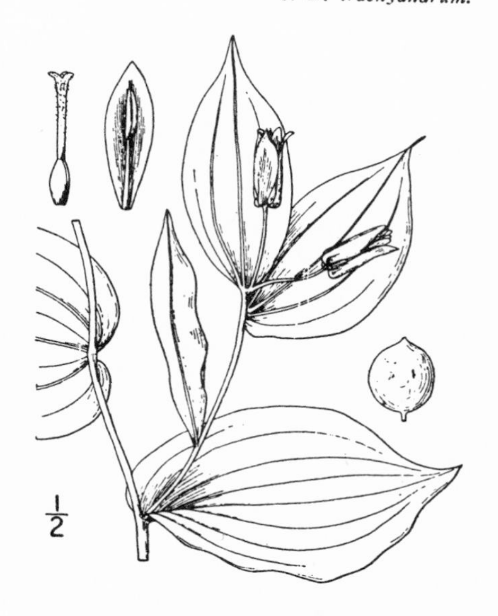

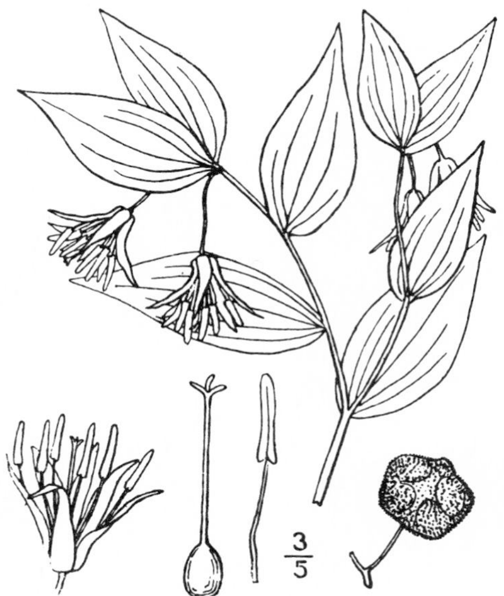

2. Disporum trachycàrpum (S. Wats. Benth. & Hook.

Rough-fruited Fairy Bell. Fig. 1118.

Prosartes trachycarpa S. Wats. Bot. King. Expl. 344. 1871.
Disporum trachycarpum Benth. & Hook. Gen. Pl. **3**: 832. 1883.
Disporum majus Britton, Bull. Torrey Club **15**: 188. 1888.

Short-woolly pubescent at least when young, 3–6 dm. high. Leaves ovate to oblong-lanceolate, 8 cm. long, acute or short-acuminate at the apex, rounded or subcordate at base, 5–11-nerved; flowers solitary or 2 or 3 together, creamy-white, 8–14 mm. long; perianth narrowly campanulate, the segments narrowly oblong or oblanceolate, acute, little spreading; stamens slightly exceeding the segments; style slender nearly equalling the stamens, or only slightly pubescent toward base, 3-lobed at apex; ovary depressed-globose; berry depressed-globose or somewhat obovoid, 8–10 mm. in diameter, roughened, 4–18-seeded.

Moist wooded slopes, Arid Transition Zone; British Columbia and Alberta south to Colorado, northern Arizona, eastern Washington and the Blue Mountains, Oregon. Type locality: "Wasatch and Uintahs," Utah.

3. Disporum hoòkeri (Torr.) Britt.

Hooker's Fairy Bell. Fig. 1119.

Prosartes hookeri Torr. Pacif. R. Rep. **4**: 144. 1857.
Disporum hookeri Britt. Torrey Club **15**: 188. 1888.

More or less roughish pubescent, with short usually spreading hairs, 3–6 dm. high. Leaves ovate to oblong-lanceolate, 3–8 cm. long, 2–4 cm. wide, acute or acuminate at apex, mostly deeply cordate at base, the uppermost very oblique; flowers 1–6 in a cluster, greenish white, narrowly campanulate, 10–12 mm. long; perianth-segments mostly oblanceolate, spreading at the tip; stamens about equalling or a little exceeding the segments; anthers obtuse, 4–5 mm. long, style equalling the stamens, entire, glabrous; capsule broadly obovoid, obtuse, about 8 mm. in diameter, 6-seeded.

On shady wooded slopes, Humid Transition and Upper Sonoran Zones; Coast Ranges of California from Mendocino and Sonoma Counties south to Santa Cruz and Santa Clara Counties. Type locality: "mountains near Oakland, California."

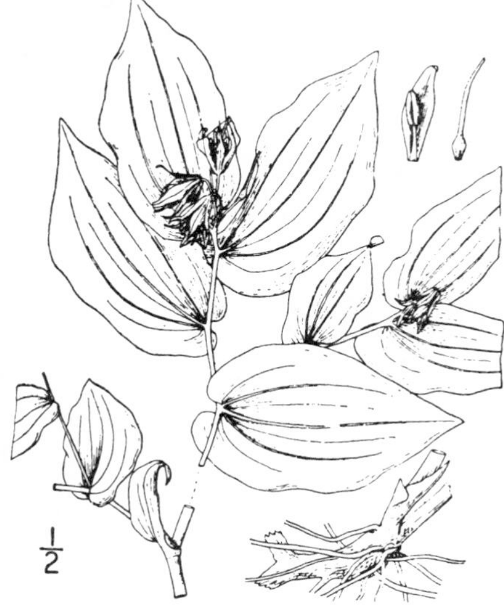

4. Disporum oregànum (S. Wats.) Benth. & Hook.

Oregon Fairy Bells. Fig. 1120.

Prosartes oregana S. Wats. Proc. Am. Acad. **14**: 271. 1879.
Disporum oreganum Benth. & Hook.; Howell, Fl. N. W. Am. 1: 659. 1902.

More or less soft-woolly pubescent, 3–8 dm. high. Leaves ovate to oblong-lanceolate, 3–9 cm. long, 1.5–6 cm. wide, mostly long-acuminate, deeply cordate at base; flowers 1–4 in a cluster, creamy-white often tinged with green, narrowly campanulate, 10–12 mm. long; perianth-segments oblanceolate acute and spreading at apex; stamens mostly well-exserted; anthers glabrous, oblong, rounded at apex, 2–3 mm. long; styles very slender, equalling the stamens, entire or obscurely 3-lobed, glabrous to near the base; ovary oblong, narrowed to the style, pubescent; capsule ovoid, acute, somewhat pubescent, about 8–16 mm. in diameter, 6-seeded.

In deep shaded coniferous woods, Transition Zone; British Columbia to Idaho, Blue Mountains, eastern Oregon and the Siskiyou Mountains, southern Oregon. Type locality: not designated.

5. **Disporum trachyándrum** (Torr.) Britt. Sierra Fairy Bells. Fig. 1121.

Prosartes trachyandra Torr. Pacif. R. Rep. 4: 144. 1857.
Prosartes lanuginosa trachyandra Baker, Journ. Linn. Soc. 14: 587. 1875.
Disporum trachyandrum Britton, Bull. Torrey Club 15: 188. 1888.

Minutely and somewhat rough-pubescent, 3–6 dm. high. Leaves ovate to oblong-lanceolate, 3–8 cm. long, 15–50 mm. wide, acute or short-acuminate, shallowly cordate at base or the upper not at all cordate; flowers 1–4 in a cluster, narrowly campanulate, creamy-white, often tinged with green, about 10 mm. long; perianth-segments oblanceolate, acute and spreading at apex; stamens about two-thirds the length of the segments; anthers oblong, obtuse, minutely hispid; style slender, equalling the stamens, entire, glabrous; ovary glabrous, narrowly oval; berry broadly ovoid, with a short stout beak, about 8 mm. in diameter, smooth and glabrous, 6-seeded.

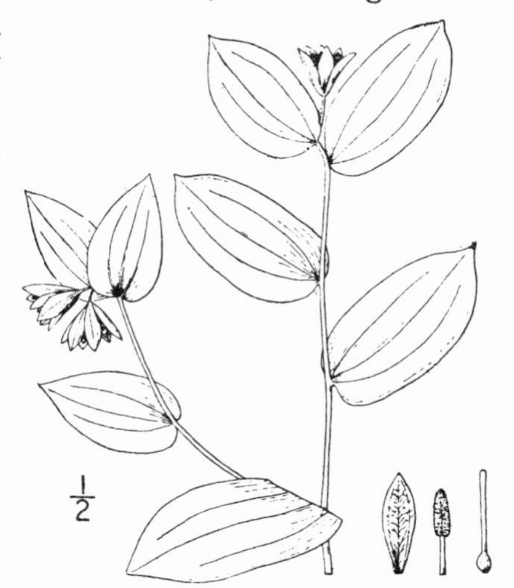

Coniferous forests, Arid Transition Zone; southern Oregon, south to Tulare County, in the Sierra Nevada, and to Trinity County in the Coast Ranges. Type locality: "hillsides, Duffield's Ranch, Sierra Nevada," California.

Disporum parvifòlium (S. Wats.) Britton, Bull. Torrey Club 15: 188. 1888. (*Prosartes parvifolia* S. Wats. Bot. Calif. 2: 179. 1880.) Distinguished by its nearly sessile acute anthers, but probably only a sterile form of *D. oreganum*. Similar sterile flowers with short stamens are frequent on normal plants of *D. hookeri*. It was originally collected by Rattan in the "Siskiyou Mountains between Happy Camp and Waldo," Oregon.

8. **STRÉPTOPUS** Michx. Fl. Bor. Am. 1: 200. 1803.

Glabrous or sparingly pubescent branching herbs, with stout or slender rootstocks. Leaves alternate, sessile or clasping, thin, many-nerved. Flowers small, somewhat campanulate, solitary or 2 together, extra-axillary, slender-peduncled. Peduncles bent or twisted at about the middle. Perianth-segments 6, separate, recurved or spreading, deciduous, the outer flat, the inner keeled. Stamens 6, hypogynous; filaments short; flattened; anthers sagittate, extrorse. Ovary 3-celled, sessile; slender, 3-cleft, 3-lobed or entire; ovules numerous, in 2 rows in each cavity. Berry globose or oval, red, many-seeded. [Greek, meaning twisted-stalk, in reference to the bent or twisted peduncle.]

About 5 species, natives of the north temperate zone. The following and the type are the only North American species. Type species, *Streptopus roseus* Michx.

Stems branched; pedicels geniculate; berry white. 1. *S. amplexifolius.*
Stems simple; pedicels not geniculate; berry red. 2. *S. curvipes.*

1. **Streptopus amplexifòlius** (L.) DC. Clasping-leaved Twisted-stalk. Fig. 1122.

Uvularia amplexifolia L. Sp. Pl. 304. 1753.
Streptopus amplexifolius DC. Fl. France 3: 174. 1805.

Rootstock short, stout, horizontal, covered with thick fibrous roots. Stem usually branching below the middle, 4–10 dm. high, glabrous. Leaves ovate to lanceolate, 5–15 cm. long, 2.5–5 cm. wide, acuminate at apex, cordate clasping at base, glabrous, glaucous beneath; peduncles 25–50 mm. long, 1–2-flowered; flowers 8–12 mm. long, greenish white; perianth-segments narrowly lanceolate, acuminate, widely spreading or recurved above; anthers subulate-pointed; stigma entire, obtuse or truncate; berry oval, 10–18 mm. long.

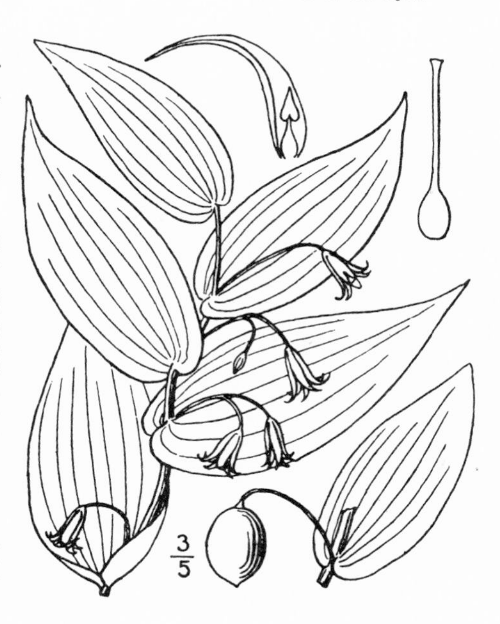

In moist woods, Transition Zone; Alaska and Labrador, south to Pennsylvania, Arizona, and Mendocino County, California, also in Europe and Asia. Type locality: Europe.

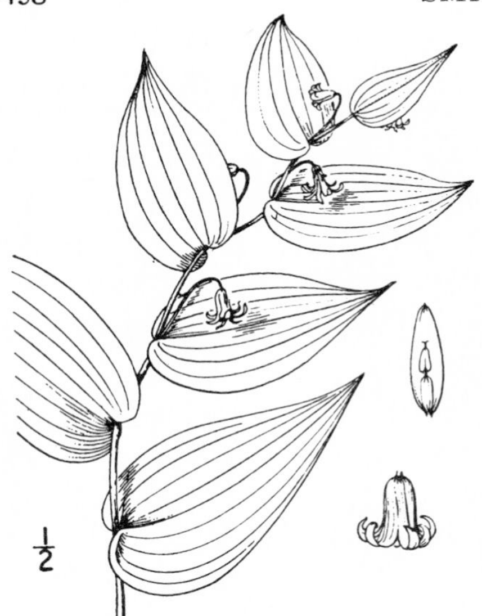

2. Streptopus cúrvipes Vail.
Simple-stemmed Twisted-stalk. Fig. 1123.

Streptopus curvipes Vail, Bull. Torrey Club **28**: 267. 1902.

Rootstock slender, covered with few fibrous rootlets. Stem simple, 1–3 dm. high, glabrous. Leaves ovate or oblong-lanceolate, 3–8 cm. long, acuminate at apex, rounded and slightly clasping at base, the margins finely glandular-ciliate; flowers few usually 3–5 solitary; peduncles not geniculate, 5–15 mm. long, glandular-pubescent; flowers rose-colored, 5–7 mm. long, campanulate; perianth-segments lanceolate, scarcely spreading at the apex, minutely glandular-pubescent on the inner surface; anthers 2-beaked; style 3-cleft; berry globose, 7–9 mm. in diameter.

On moist wooded slopes, Humid Transition Zone; Alaska to southern Oregon. Type locality: Asulkan Pass, altitude 4400 ft., British Columbia. Closely related to the eastern *Streptopus roseus* Michx., but distinguished by the small simple stem, more elongate rootstock and shorter peduncles.

9. KRÙHSEA Regel, Nouv. Mem. Soc. Nat. Moscow. 11: 122. 1859.

Glabrous simple stemmed herbs, with slender rootstocks. Leaves alternate, broad, sessile or slightly clasping, several-nerved. Flowers usually solitary, extra axillary, nodding on slender pedicels. Perianth-segments 6, equal, rotate spreading with reflexed tips. Stamens 6, hypogynous; filaments slender, very short; anthers acute, opening by lateral slits. Ovary 3-celled; ovules numerous in 2 rows in each cavity; style none; stigma sessile. Berry globose, red. [Name in honor of Dr. Krushe, of Siberia.]

A monotypic genus of western North America and Siberia.

1. Krushea streptopoìdes (Ledeb.) Kearney.
Kruhsea. Fig. 1124.

Smilacina streptopoides Ledeb. Fl. Ross. **4**: 128. 1853.
Kruhsea tilingiana Regel, Nouv. Mem. Soc. Nat. Mosc. **11**: 122. 1859.
Streptopus brevipes Baker, Journ. Linn. Soc. **14**: 592. 1875.
Kruhsea streptopoides Kearney in Herron, Expl. Alaska. Adj. Gen. Off. **31**: 74. 1901.
Streptopus streptopoides Nelson & Macb. Bot. Gaz. **61**: 30. 1916.

Rootstock very slender, creeping. Stem simple, 10–20 cm. high, slender, glabrous or obscurely scabrous; leaves ovate-lanceolate, 3–5 cm. long, acuminate at apex, rounded and slightly clasping at base; flowers few, 1–5, solitary in the axils; pedicels recurved, about 10 mm. long in fruit; perianth-segments ovate-lanceolate, 2 mm. long; stamens less than half the length of the segments; anthers scabrous; berry red, about 4 mm. in diameter.

In Alpine forests, Canadian Zone; Alaska to northern Washington; also in Siberia. Type locality: Siberia.

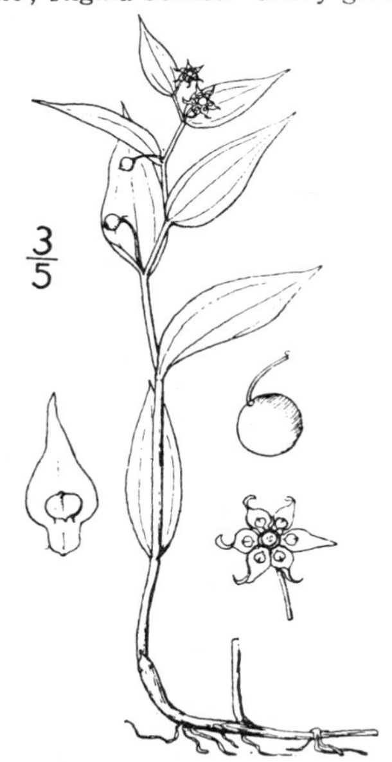

Family 20. SMILACÀCEAE.

SMILAX FAMILY

Mostly vines, with woody or herbaceous, often prickly stems arising from rootstocks. Leaves alternate, petioled, net-veined, several-nerved, usually punctate, deciduous from the persistent sheathing petioles, these bearing at the base a pair of tendril-like appendages. Flowers small, in axillary umbels, usually green, dioecious. Perianth-segments 6, the two sets similar. Stamens mostly 6, distinct; filaments ligulate; anthers basifixed, 2-celled, introrse. Ovary 3-celled, ovules 1 or 2 in each cavity, orthotropus, suspended; style very short or none; stigmas 1–3. Fruit a globose berry containing 1–6 seeds. Seeds brownish; endosperm horny, copious; embryo small, oblong, remote from the hilum.

Three genera and about 235 species, widely distributed in temperate and tropical zones; only the following genus in North America. The roots of many species are employed in medicine, under the name of Sarsaparilla.

1. SMÌLAX L. Sp. Pl. 1028. 1753.

Rootstock usually large and tuberous. Stems usually twining and climbing by means of the spirally coiling tendril-like appendages of the petioles. Lower leaves reduced to scales; the upper entire or lobed. Pedicels borne on the enlarged globose or conic summit of the peduncle, inserted in small pits, generally among minute bractlets. Pistillate flowers usually smaller than the staminate, with an ovary and usually 1–6 abortive stamens. Staminate flowers without an ovary. Perianth-segments deciduous. Stamens inserted on the bases of the perianth-segments. Berry black, red or purple, rarely white, with 3 strengthening bands of tissue running through the outer part of the pulp, connected at the base and apex. Embryo lying under a tubercle at the upper end of the seed. [Ancient Greek name, perhaps not originally applied to these plants.]

About 225 species widely distributed, most abundant in tropical America and Asia. Besides the following there are about 22 species in the eastern and southern United States. Type species, *Smilax aspera* L.

1. Smilax califórnica (A. DC.) A. Gray. California Smilax. Fig. 1125.

Smilax rotundifolia californica A. DC. Monog. Phaner. 1: 75. 1878.

Smilax californica A. Gray; S. Wats. Bot. Calif. 2: 186. 1880.

Stem woody, terete or somewhat angled, more or less covered with slender straight spreading brownish prickles, sometimes nearly or quite naked. Leaves broadly ovate, 5–10 cm. long, 25–70 mm. wide, abruptly acute and apiculate at apex, sub-cordate at base, rather thin, glabrous, deciduous; petioles 10–15 mm. long; peduncles slender, flat, 2–5 cm. long; pedicels 8–20, 6–8 mm. long; staminate flowers with ligulate perianth-segments about 5 mm. long, 1 mm. wide; widely spreading and recurved from below the middle; stamens 3 mm. long; berry black, 6 mm. in diameter.

Stream banks, climbing bushes and small trees often to a height of several feet. Upper Sonoran and Transition Zones; southern Oregon to Butte County, California. Type locality: near Chico, California.

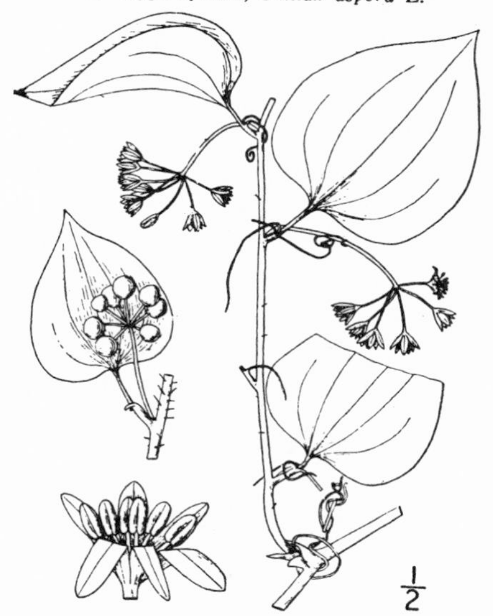

Family 21. AMARYLLIDÀCEAE.

AMARYLLIS FAMILY.

Perennial herbs or some tropical species woody and arboreous, with bulbs or rootstocks, scapose or sometimes leafy stems and narrow and entire leaves, or, in ours, thick and often spiny-toothed. Flowers perfect, regular or nearly so. Perianth 6-parted or 6-lobed, the segments distinct or united below, into a tube adnate to the ovary. Stamens 6, inserted on the bases of the perianth-segments, or in the throat opposite the lobes. Anthers versatile or basifixed, 2-celled, the sacs usually longitudinally dehiscent. Ovary wholly or partly inferior, usually 3-celled. Style filiform, entire, 3-lobed, or divided into 3 stigmas at the summit. Ovules numerous, rarely only 1 or 2 in each cavity, anatropous. Fruit capsular, rarely fleshy. Seeds generally black, the embryo small, enclosed in fleshy endosperm.

About 70 genera and 800 species, principally natives of tropical and subtropical regions.

1. AGÀVE L. Sp. Pl. 323. 1753.

Short-stemmed or usually acaulescent perennial from a thick fibrous-rooted crown. Leaves clustered, thick, rigid and fleshy, spiny pointed and entire or often spiny tufted. Flowers numerous, spicate or paniculate upon a tall bracteate woody scape. Pedicels short, jointed, bracteate. Perianth 6-cleft, the tube tubular or campanulate, thick, fleshy and somewhat persistent, the lobes valvate similar and nearly equal. Stamens adnate to the perianth tube, geniculately inflexed in the bud, at length exserted; anthers linear, versatile, style stout and elongated; stigma thickened; capsule coriaceous, with numerous flattened horizontal black seeds.

An American genus of about 100 species, chiefly Mexican. Type species, *Agave americana* L.

Flower clusters about 4-flowered, racemosely distributed on the main flower stalk. 1. *A. utahensis.*
Flower clusters many-flowered terminating the branches of the panicle.
- Apparently acaulescent, the stout trunk subterranean; stamens inserted at the mouth of the tube.
  - Ovary flask-shaped, scarcely longer than the perianth-segments. 2. *A. deserti.*
  - Ovary fusiform, about twice as long as the segments. 3. *A. consociata.*
- Caulescent, forming a leafy ellipsoid head; stamens inserted at the middle of the tube. 4. *A. shawii.*

1. **Agave utahénsis** Engelm. Utah Agave. Fig. 1126.

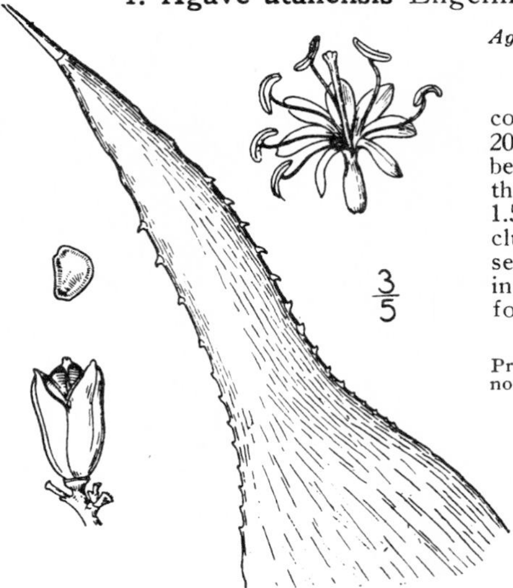

Agave utahensis Engelm.; S. Wats. Bot. King. Expl. 497. 1871.

Acaulescent, the trunk subterranean. Leaves concave, thick, hard, glaucous, 12–20 cm. long, 20–25 mm. wide, the margins more or less repand, beset with white deltoid teeth about 2 mm. long, the terminal spine 20–35 mm. long; flower stalk 1.5–2.5 m. high; panicle very narrow; flowers in clusters of 2, 4, or rarely 6, yellow; perianth-segments about as long as the ovary; stamens inserted near the middle of the broadly funnel-form tube; capsule ovoid, 2–3 cm. long.

Desert slopes, Lower Sonoran Zone; Death Valley and Providence Mountains, California, to southern Utah and northern Arizona. Type locality: St. George, Utah.

2. **Agave déserti** Engelm. Desert Agave. Fig. 1127.

Agave deserti Engelm. Trans. Acad. St. Louis **3**: 310. 1875.

Densely cespitose, acaulescent. Leaves triangular-lanceolate, gradually acute, 15–30 cm. long, 5 cm. wide, openly concave becoming channelled near the apex, falcately ascending, gray-green, sometimes transversely banded; spine dull brown becoming gray, 3–30 mm. long, grooved or involute below the middle; prickles dull brown, triangular, 3 or 4 mm. long, mostly reflexed, 5–10 mm. apart, the intervening leaf-margin nearly straight, herbaceous; scape slender, 2–3 m. high, bracts separated, narrowly lanceolate, ascending; panicle 3–6 dm. long; branches few, ascending, compactly short-branched; pedicels slender, about 5 mm. long, fascicled; flowers chrome-yellow, rather fetid, about 35 mm. long; ovary flask-shaped, with a distinct neck, 15–20 mm. long, about equalling the perianth; perianth-tube funnel-form, 3 or 4 mm. long; segments about 15 mm. long, 4 mm. wide; stamens inserted nearly in the throat, 25–30 mm. long; anthers 10–15 mm. long; capsule 45 mm. long, 15 mm. thick, nearly sessile, beaked; seeds 4–5 mm. in diameter.

Gravelly or rocky desert slopes, Lower Sonoran Zone; western borders of the Colorado Desert near the bases of the San Jacinto and Cuyamaca Mountains, California. Type locality: east of San Felipe, California.

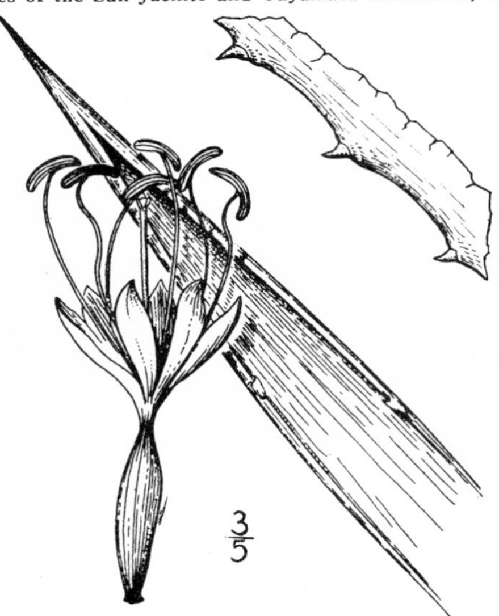

3. **Agave consocìata** Trel. Coahuilla Agave. Fig. 1128.

Agave consociata Trel. Rep. Mo. Bot. Gard. **22**: 53. 1911.

Densely cespitose. Leaves oblong-lanceolate, nearly smooth, 20–30 cm. long, about 6 cm. wide, the marginal prickles 10–25 or 30 mm. apart, 4–8 mm. long, triangular, the intervening margin straight or slightly undulated; flower stalk 5–7 m. high; flower clusters as in the preceding; ovary fusiform, 25–30 mm. long, nearly twice the length of the perianth-segments; stamens inserted at the mouth of the tube; capsule 15–35 mm. long, beaked.

Gravelly desert slopes associated with *A. deserti*, Lower Sonoran Zone; eastern base of the San Jacinto Mountains, southern California, to northern Lower California. Type locality: San Felipe, California.

4. **Agave shàwii** Engelm. Shaw's Agave. Fig. 1129.

Agave shawii Engelm. Trans. Acad. St. Louis 3: 314. *pl.* 2–4. 1875.

Cespitose, shortly caulescent, scarcely 1 m. high. Leaves ovate to lanceolate-ovate, acuminate, 20–50 cm. long, 6–12 cm. wide, smooth, glossy green, openly concave, stiffly erect compact, and rather closely imbricated; spine thick, 2–4 cm. long, 3–6 cm. wide, reddish, very openly grooved almost to the end; prickles stout 10–15 mm. long, abruptly widened to about 10 mm. at base, the intervening space repand, usually continuously horny; scape about 3 m. high, stout, the upper third or less ovoid-paniculate; bracts deltoid, crowded and subimbricate, appressed; branches few, stout, nearly horizontal, simple or three-branched with almost capitate bracted flower-clusters; pedicels very stout, much thickened in fruit; flowers 7–9 cm. long, greenish yellow, slightly fetid; ovary subfusiform, 3–4 cm. long; segments attenuate from a wide base, 20 mm. long, 8 mm. wide; stamens inserted at about the middle of the tube, 55–60 mm. long; capsule oblong, 5–7 cm. long, 20–25 mm. thick, with thick exocarp; seed 7–9 mm. in diameter.

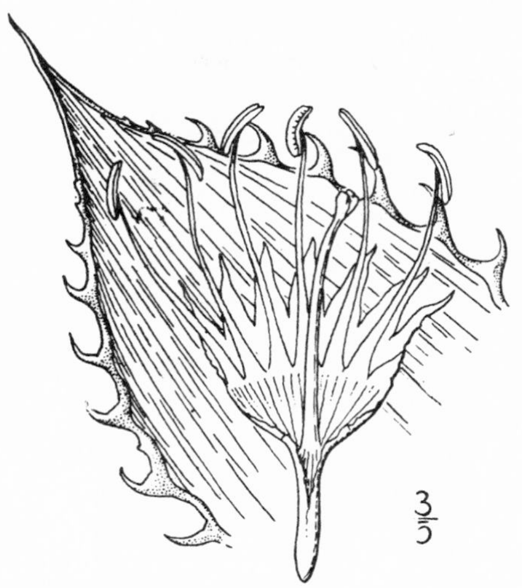

Dry hillsides near the sea, south of San Diego, California. Type locality: on the international boundary, near the Initial Boundary Monument. The only locality known for this rare plant.

Family 22. IRIDÀCEAE.

Iris Family.

Perennial herbs arising from bulbs or commonly from rootstocks, with narrow equitant 2-ranked leaves and perfect regular or irregular bracted flowers. Perianth of 6 segments or 6-lobed, its tube adnate to the ovary, the segments or lobes in two series of 3 each, convolute in the bud, withering-persistent. Stamens 3, inserted opposite the outer segments or lobes; filaments filiform separate or united; anthers 2-celled extrorse. Ovary inferior, mostly 3-celled; ovules usually numerous in each cavity, anatropous; style 3-cleft, its lobes sometimes divided. Fruit a 3-celled, 3-angled or 3-lobed (sometimes 6-lobed), loculicidally dehiscent capsule. Seeds numerous in 1 or 2 rows in each cavity of the capsule. Endosperm fleshy or horny; embryo straight, small.

About 57 genera and 1000 species, widely distributed in temperate and tropical regions of both hemispheres.

Styles petal-like; flowers very large, the two series of perianth-segments unlike. — 1. *Iris.*
Styles filiform; the two series of perianth-segments similar.
  Filaments united to the top; flowers usually blue. — 2. *Sisyrinchium.*
  Filaments not united to the top; flowers not blue.
    Filaments united only at the base; flowers reddish purple. — 3. *Olsynium.*
    Filaments united for more than half their length; flowers yellow. — 4. *Hydastylus.*

1. ÌRIS [Tourn.] L. Sp. Pl. 38. 1753.

Perennial herbs with creeping, often woody and sometimes tuber-bearing rootstocks, erect stems, erect or ascending equitant leaves, and large regular flowers in terminal racemes or sometimes panicles. Perianth of 6 clawed segments, united below into a tube, the 3 outer segments broad, spreading or reflexed, the 3 inner usually narrower and erect. Stamens inserted at the base of the outer perianth-segments; anthers linear or oblong. Ovary 3-celled; divisions of the style petal-like, arching over the stamens, bearing the stigmatic surface on the inner surface on a flat scale immediately below the usually 2-lobed tip; style-base adnate to the perianth-tube. Capsule oblong or oval, 3–6-angled or lobed, mostly coriaceous. Seeds numerous, vertically compressed, in 1 or 2 rows in each cavity. [Greek, rainbow, referring to the variegated flowers.]

About 100 species, mostly in the north temperate zone. Besides the following about 15 species occur in the eastern United States. Type species, *Iris germanica* L.

Perianth-tube very short, less than 1 cm. long.
Bracts contiguous or rarely separate; stems naked or nearly so; rootstock stout; palludose species.
Bracts scarious; interior species. 1. *I. missouriensis.*
Bracts foliaceous; coastal species. 2. *I. longipetala.*
Bracts separate and often distant; rootstock slender.
Stem-leaves foliaceous.
Stems 1-flowered; flowers deep violet-purple. 3. *I. tenax.*
Stems 2-flowered.
Bracts scarious; flowers white veined with lilac; capsule depressed-globose; pedicel scarcely equalling the ovary. 4. *I. tenuis.*
Bracts not scarious; flowers yellow veined with lilac, or lilac-purple; capsule oblong; pedicel 2–3 times the length of the ovary. 5. *I. hartwegi.*
Stem-leaves bract-like, the basal solitary, elongated; flowers yellow. 6. *I. bracteata.*
Perianth-tube elongated 2 cm. long or more.
Perianth-tube rather stout, 15–20 mm. long; leaves in age arched, dark glossy green on upper surface and pale beneath, reddish at base. 7. *I. douglasiana.*
Perianth-tube very slender, 25–80 mm. long; leaves pale glaucous green.
Stems clothed with more or less overlapping inflated sheathing bract-like leaves with very short attenuate blades. 8. *I. purdyi.*
Stems with 1–3 leaves, their sheathing bases not inflated, their blades 5 cm. long or more.
Flowers cream-color, with lilac veins; stem 10–30 cm. tall; perianth-tube 25–40 mm. long. 9. *I. californica.*
Flowers lilac-purple; stems very short; perianth-tube filiform, 4–8 cm. long. 10. *I. macrosiphon.*

1. **Iris missouriénsis** Nutt. Western Blue-flag or Iris. Fig. 1130.

Iris missouriensis Nutt. Journ. Acad. Philad. **7**: 58. 1834.
Iris tolmieana Herbert, Hook. & Arn. Bot. Beechey 396. 1839.
Iris caurina Herbert, Hook. Fl. Bor. Am. **2**: 206. 1839.

Rootstock stout. Stem rather slender, usually simple, terete, 2–5 dm. high, 1–2-flowered; leaves mostly basal, green sometimes purplish below, shorter than or about equalling the stem, 4–10 mm. wide; bracts dilated, scarious, 4–7 cm. long, acute or acuminate; flowers pale blue, variegated; pedicels 15–50 mm. long; perianth-tube 6–8 mm. long, narrowed below; sepals 5–6 cm. long, glabrous and crestless; petals a little shorter; capsule oblong, 3–5 cm. long, 10–15 mm. broad, obtusely 6-angled; seeds obovate, 4 mm. long.

Wet meadows, Arid Transition Zone; British Columbia to Dakota, south to Arizona and southern California. In the Pacific Coast it is mainly east of the Cascade-Sierra Nevada Divide, extending as far south as San Diego County, California. Type locality: "toward the sources of the Missouri."

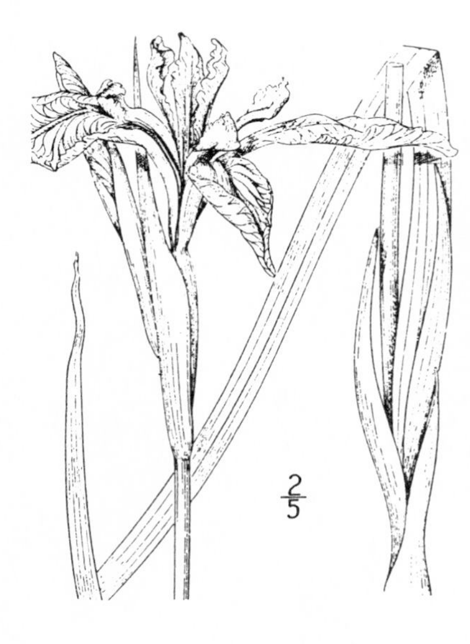

2. **Iris longipétala** Herbert. Long-petaled Iris. Fig. 1131.

Iris longipetala Herbert, Hook. & Arn. Bot. Beechey 395. 1841.

Rootstock stout. Stem stout, 3–5 dm. high, 3–5-flowered; leaves mainly basal, about equalling or longer than the stem, 5–10 mm. broad; bracts approximate, foliaceous, 7–10 cm. long, acuminate; pedicels stout, 25–50 mm. long; flowers bright lilac; sepals 6–8 cm. long, 25–30 mm. broad, narrowed to a short claw, white below and veined with violet, the midvein yellow; petals 5 cm. long, oblanceolate; perianth-tube about 6 mm. long; anthers 18 mm. long; styles broadly crested; capsule oblong, 5 cm. long, narrowed at each end; seeds flattened, 6 mm. long.

In meadows or on moist grassy slopes, Humid Transition and Upper Sonoran Zones; coastal region of central California, Contra Costa County and San Francisco, to Monterey. Type locality: not given, but probably Monterey.

3. Iris tènax Dougl.

Tough-leaved Iris. Fig. 1132.

Iris tenax Dougl.; Lindl. Bot. Reg. **15**: *pl. 1218.* 1829.

Rootstock slender, usually short and forming dense tufts. Stems slender 15–30 cm. high, bearing several short bract-like leaves; basal leaves numerous, much longer than the stem, 3–4 mm. wide, acuminate; bracts foliaceous, lanceolate, acute 3–5 cm. long; flowers solitary bright lilac-purple, or sometimes white; pedicels 15–25 mm. long; sepals 4–6 cm. long, 20–35 mm. wide, with a broad claw; petals nearly as long as the sepals, spatulate; perianth-tube about 5 mm. long; anthers 15–18 mm. long; capsule 20-25 mm. long, oblong, obtuse at both ends; seeds scarcely flattened, obtusely angled.

Open grassy places, Humid Transition Zone; Washington to southern Oregon west of the Cascade Mountains. Type locality: "a common plant in north California and along the coast of New Georgia, in dry soils or open parts of woods." The tough fibrous leaves of this species, and perhaps of some of the *macrosiphon* group are used by the Indians in making nets and snares.

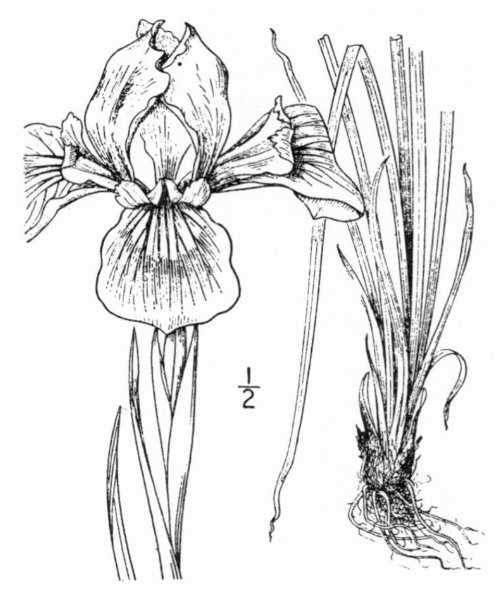

4. Iris ténuis S. Wats.

Clackamas Iris. Fig. 1133.

Iris tenuis S. Wats. Proc. Am. Acad. **17**: 380. 1882.

Rootstock very slender, only 2–4 mm. thick. Stems 20–25 cm. high, with 2 or 3 thin and scarious bract-like leaves 5–7 cm. long; basal leaves very thin, equalling or surpassing the stems, 8–12 mm. wide; bracts contiguous, 5–10 cm. long; flowers 2, on slender pedicels equalling or exceeding the bracts, white lightly striped and blotched with pale yellow and purple; sepals oblong-spatulate, about 3 cm. long; the petals a little shorter, emarginate; perianth-tube 4–6 mm. long; capsule depressed-globose, 12 mm. in diameter.

Open wooded slopes, Humid Transition Zone; Cascade Mountains of northern Oregon, along the Clackamas River. Type locality: "Eagle Creek, a branch of Clackamas River," Oregon.

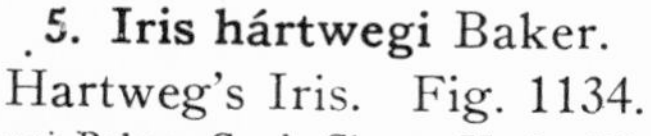

5. Iris hártwegi Baker.

Hartweg's Iris. Fig. 1134.

Iris hartwegi Baker, Gard. Chron. II. **5**: 323. 1876.

Rootstock slender, forming small tufts. Stems slender and flattened, 10–30 cm. high, with 1–3 leaves; basal leaves longer than or often much surpassing the stem, 4–6 mm. wide, acuminate; bracts rarely nearly contiguous, linear-lanceolate 5–7 cm. long; pedicels 2–7 cm. long; flowers 2, yellow with lavender veins, or pale lilac with deeper colored veins and yellow medial portion; sepals 3–5 cm. long, about 12 mm. wide; petals a little shorter and narrower; perianth-tube stout, 6–7 mm. long; anthers 12 mm. long, equalling the filaments; capsule oblong, 2–3 cm. long, acute at both ends, 3-angled; seeds flattened and angled.

Open coniferous forests, Arid Transition Zone; Siskiyou Mountains south through the Sierra Nevada to Kern County. Type locality: Sierra Nevada, probably on the American River.

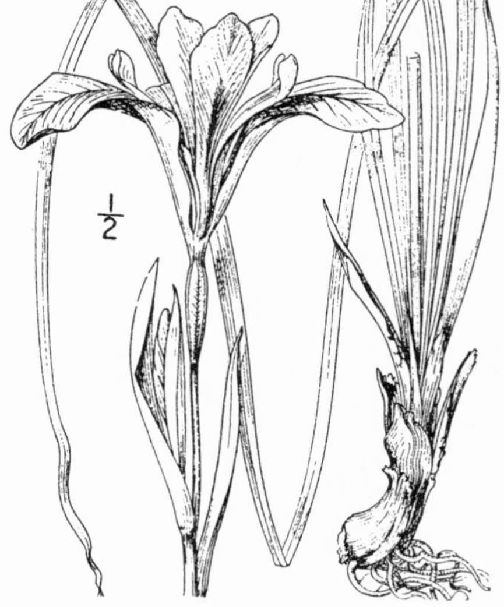

Iris hartwegi austrális Parish, Erythea **6**: 86. 1898. Usually larger than the typical species; flowers lilac-purple; sepals mostly 5–6 mm. long, 20 mm. wide; anthers 20 mm. long, a little exceeding the filaments.

Open coniferous forests in dry soils, Arid Transition Zone; mountains of southern California, from the San Gabriel to the San Jacinto. Type locality: dry ridges of the San Bernardino Mountains at 4000 ft. altitude.

6. Iris bracteàta S. Wats.

Siskiyou Iris. Fig. 1135.

Iris bracteata S. Wats. Proc. Am. Acad. **20**: 375. 1885.

Rootstock very slender, forming tufts. Stems slender, simple, flattened, 2–3 dm. high; basal leaf solitary, usually much longer than the stem, 6–10 mm. wide, glossy green above, pale beneath, firm and persistent; stem-leaves 2–3, bract-like, the sheathing bases somewhat inflated and often colored, the blades about 25 mm. long; bracts somewhat inflated and tinged with red, 5–8 cm. long, 5–7 mm. wide, attenuate at apex; flowers 2, yellow; pedicels slender, 30–35 mm. long; perianth-tube funnel-form, 6 mm. long; sepals oblong, 5–7 cm. long; petals oblanceolate, somewhat shorter than the sepals; capsule ovate-oblong, 3 cm. long; anthers 12 mm. long.

Open coniferous forests, Transition Zone; Siskiyou Mountains, Oregon. Type locality: near Waldo, Josephine County.

7. Iris douglasiàna Herbert.

Douglas' Iris. Fig. 1136.

Iris douglasiana Herbert; Hook. & Arn. Bot. Beechey 395. 1841.
Iris douglasiana bracteata Herbert; Hook. & Arn. Bot. Beechey 395. 1841.
Iris douglasiana nuda Herbert; Hook. & Arn. Bot. Beechey 395. 1841.
Iris beecheyana Herbert; Hook. & Arn. Bot. Beechey 395. 1841.
Iris watsoniana Purdy; Erythea **5**: 128. 1897.

Rootstock rather stout, elongated, clothed by the persistent leaf-base, these not becoming fibrous. Stem flattened, 25–50 cm. high, with 2 or 3 leaves, simple and 2–3-flowered, or rarely branched above and each branch 2–5 flowered. Basal leaves equalling or usually much exceeding the stem, 7–15 cm. long, acuminate, bright glossy green, reddish at base; bracts separate, foliaceous, lanceolate, 6–8 cm. long; pedicels 2–3 cm. long; flowers bright or pale lilac-purple varying to white or cream with lilac veins; sepals 5–7 cm. long, 15–25 cm. wide; petals oblanceolate, equaling the sepals; perianth-tube 15–20 cm. long; style-branches with long toothed crests; ovary narrowly oblong, 3–4 cm. long; capsule oblong, 4–7 cm. long, short-acuminate, or acute at apex; seeds nearly globose.

Open woods, especially in the redwood belt, forming tufts or small colonies, but on open hillsides near the coast often forming extensive colonies, Humid Transition (mainly) and Upper Sonoran Zones; Coast Ranges from Curry County, Oregon, to Monterey County, California. Type locality: California, probably Monterey.

8. Iris púrdyi Eastw.

Purdy's Iris. Fig. 1137.

Iris purdyi Eastw. Proc. Calif. Acad. II. Bot. **1**: 78. *pl. 7, fig. 2.* 1897.

Rootstock slender, about 3 mm. in diameter. Stems 15–30 cm. high, somewhat flattened. Basal leaves exceeding the stems, 6–7 mm. wide, erect or laxly spreading, long-acuminate, pale green; cauline leaves several, more or less overlapping, bract-like, the sheathing base inflated, tinged with red, the leaf-blade attenuate, scarcely 2 cm. long; bracts inflated, broadly lanceolate, 5–7 cm. long, about 7 mm. wide, abruptly acuminate, tinged with rose; flowers 2, short-pediceled; sepals 5–7 cm. long, 20 mm. wide, cream-color, with yellow lines on the claw and purple veins on the spreading blades; petals linear-oblong, somewhat shorter than the sepals; perianth-tube slender, 4–5 cm. long; anthers 15–18 mm. long; capsule oblong-ovoid, 25–30 cm. long, acute at each end.

Hillsides in open forests, Upper Sonoran and Humid Transition Zones; Siskiyou Mountains, southern Oregon to Mendocino County, California. Type locality: "redwood region of Mendocino County, around Ukiah."

9. **Iris califórnica** Leicht.
California Iris. Fig. 1138.

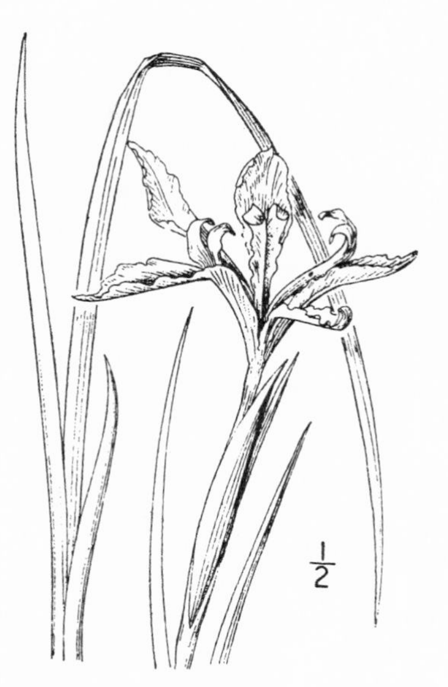

Iris californica Leicht. Garden **52**: 126. 1897.

Rootstock slender, about 5 cm. in diameter. Stem slender, somewhat flattened, 15–40 cm. high. Leaves of the sterile stems surpassing the flowers, 4–6 mm. wide, long-acuminate, pale green; those of the flower-bearing stem 2 or 3 about 10–20 cm. long, the upper sheathing for half their length; bracts lanceolate acuminate, 7–10 cm. long, 6–9 mm. wide; flowers usually 2, cream color, veined with lilac, very short-pedicelled; perianth-tube slender, 4–6 cm. long; sepals 5–6 cm. long, 15–18 mm. wide; petals spatulate, about equalling the sepals; anthers about 15 mm. long; style-branches with oblong crests, 8–10 mm. long; capsule broadly oblong, abruptly acute at each end, 25–40 mm. long, on pedicels 10–15 mm. long; seeds compressed.

Grassy hillsides and open woods in dry soil, Transition and (mainly) Upper Sonoran Zones; California Coast Ranges from Lake and Glenn Counties to Santa Clara County, also in the Sierra foothills of Butte County. Type locality: California, but locality not given.

10. **Iris macrosìphon** Torr.
Slender-tubed Iris. Fig. 1139.

Iris macrosiphon Torr. Pacif. R. Rep. **4**: 144. 1857.

Rootstocks slender, forming small tufts. Stems slender 3–20 cm. high, somewhat flattened. Basal leaves much exceeding the stem, 3–4 mm. rarely wider; cauline leaves 1 or 2, 5–10 cm. long; bracts linear-lanceolate, long-acuminate, 6–10 cm. long, 4–6 mm. wide; flowers 1 or 2, short-pedicelled, bright lilac-purple; sepals 35–50 mm. long, 12–15 mm. wide; petals oblanceolate, about equalling the sepals; perianth-tube very slender, 4–8 cm. usually about 6 cm. long; anthers about 12 mm. long; capsule oblong-ovoid, 25–30 mm. long, shortly acute at each end; seeds compressed, angled.

On grassy hillsides, or in open woods, Upper Sonoran and Humid Transition Zones; California Coast Ranges from Humboldt and Siskiyou Counties south to Marin County, and the Mount Hamilton Range, Santa Clara County. Type locality: "hillsides, Corte Madera," Marin County, California.

Iris chrysophýlla Howell, Fl. N. W. Am. **1**: 633. 1902. Probably only a form of *I. californica*, the chief difference being the elongated (15–20 mm.) lanceolate stigma-crests. Open pine forest, Arid Transition Zone, southern Oregon.

Iris amàbilis Eastw. Bull. Torrey Club **30**: 484. 1903. Clearly a member of the *macrosiphon* group and doubtfully distinct from *I. californica*, the chief distinction being a somewhat shorter (25–35 mm.) perianth-tube. Open pine forests, Nevada City, California.

Iris tenuíssima Dykes, Gard. Chron. III. **51**: 18. 1912. An imperfectly known relative of *I. macrosiphon*, differing, probably inconstantly, in the narrower, tapering perianth-segments; stigma-crests about as long as the stigmas. Described from specimens collected at the Pitt River Ferry, Shasta County, California.

2. **SISYRÍNCHIUM** L. Sp. Pl. 954. 1753.

Perennial tufted herbs, with short rootstocks, slender compressed more or less winged stems, linear grass-like leaves. Flowers blue, borne in a few-flowered terminal umbel from a pair of erect green bracts. Perianth-tube short or none, the 6 spreading segments oblong or obovate, similar, aristulate. Filament united to near the apex. Ovary 3-celled, each cavity several-ovuled. Style branches filiform, undivided, alternate with the stamens. Capsule globose or obovoid, loculicidally 3-valved. Seeds subglobose or ovoid, smooth or pitted. [Name of Theophrastus for a bulbous plant allied to Iris.]

About 60 species, natives of North America and the West Indies.

Stems simple and leafless with a sessile terminal spathe, rarely with a terminal leaf-bearing node and a solitary peduncle.
  Bracts of the spathe linear, equally narrow, the inner exceeding the pedicels, the outer much longer. — 1. *S. sarmentosum.*
  Bracts of the spathe dissimilar, the inner broader than the outer and shorter than the flowers.
    Pedicels glandular-pubescent; capsules pale not brownish. — 2. *S. halophyllum.*
    Pedicels glabrous; capsules brownish. — 3. *S. idahoense.*
Stems bearing 1, or more leaf-bearing nodes, each node with 2 or more peduncles. — 4. *S. bellum.*

1. Sisyrinchium sarmentòsum Suksdorf.
Suksdorf's Blue-eyed Grass. Fig. 1140.

Sisyrinchium sarmentosum Suksdorf, Erythea **3**: 121. 1895.
Sisyrinchium septentrionale Bick. Bull. Torrey Club **26**: 452. 1899.

Stems growing in small tufts, 10–20 cm. high, simple and leafless, bearing a sessile spathe, or sometimes with a leaf-bearing node and a solitary peduncle, about 1.5 mm. wide, narrowly winged, the margins smooth. Leaves about half to nearly equalling the stems, 2 mm. wide, attenuate at apex; spathe very narrow; outer bract linear, 40–50 mm. long, scarcely 2 mm. wide; inner bract similar, 25–30 mm. long, exceeding the slender glabrous pedicels; flowers usually 1–3; perianth-segments about 8 mm. long, pale violet, not emarginate, slenderly aristulate; capsule about 3 mm. high, glabrous.

Margins of wet meadows and streams, Arid Transition and Canadian Zones; British Columbia to Manitoba, eastern Washington and North Dakota. Type locality: Skamania County, Washington.

2. Sisyrinchium halophýllum Greene.
Nevada Blue-eyed Grass. Fig. 1141.

Sisyrinchium halophyllum Greene, Pittonia **4**: 34. 1899.
Sisyrinchium leptocaulon Bick. Bull. Torrey Club **26**: 451. 1899.

Stems arising from rather coarse fibrous roots, simple, bearing a solitary sessile spathe, 10–30 cm. high, smooth, the margins narrowly winged. Leaves about half the length of the stems, 1.5–3 mm. wide, pale green, their margins smooth; spathe narrow, the outer bract 12–30 mm. long, lanceolate, acute or attenuate at apex, broadly scarious-margined; pedicels glandular-pubescent; perianth violet-purple, 9–10 mm. long, somewhat cyathiform, prominently aristulate at apex; capsule about 4 mm. high, pale grayish green, pubescent.

Meadows, especially in alkaline soil, Upper Sonoran and Arid Transition Zones; Idaho to Utah west to the eastern slopes of the Sierra Nevada, California. Type locality: Humboldt Wells, near Wells, Nevada.

3. Sisyrinchium idahoénse Bick.
Idaho Blue-eyed Grass. Fig. 1142.

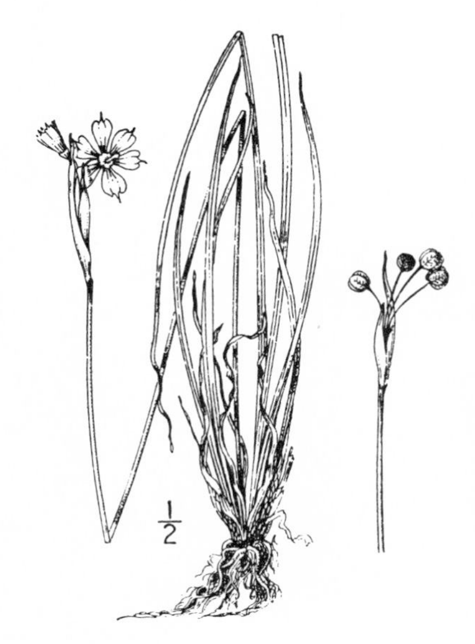

Sisyrinchium idahoense Bick. Bull. Torrey Club **26**: 445. 1899.
Sisyrinchium segetum Bick. Bull. Torrey Club **26**: 449. 1899.
Sisyrinchium oreophilum Bick. Bull. Torrey Club **31**: 381. 1904.

Stems arising from fibrillose roots, simple and leafless or occasionally a single terminal peduncle with subtending leaf, 10–45 cm. high, 1–2 mm. wide, distinctly winged, the margins smooth or more or less ciliolate. Leaves about half the length of the stems, 1–3.5 mm. wide, smooth on the margins; spathe narrow, outer bract 2–6 cm. long, narrowly attenuate; pedicels glabrous; perianth violet-purple 10–15 mm. long, the segments slightly if at all emarginate, mucronulate-aristulate; ovary glandular-pubescent; capsule globose, 3–6 mm. long, brown, sparsely puberulent.

Moist grassy places, Arid Transition and Canadian Zones; Idaho and Washington to the Sierra Nevada, California. Type locality: Nez Perces County, Idaho.

4. **Sisyrinchium béllum** S. Wats. California Blue-eyed Grass. Fig. 1143.

Sisyrinchium bellum S. Wats. Proc. Am. Acad. **12**: 227. 1876.

Bermudiana bella Greene, Man. Bay Region 308. 1894.

Sisyrinchium greenei Bick. Bull. Torrey Club **31**: 383. 1904.

Sisyrinchium eastwoodiae Bick. Bull. Torrey Club **31**: 385. 1904.

Sisyrinchium hesperium Bick. Bull. Torrey Club **31**: 390. 1904.

Sisyrinchium maritimum Heller, Muhlenbergia **1**: 48. 1904.

Stems 10–50 cm. high or more, smooth or ciliolate-scabrous on the narrowly winged margins, with 1–3 nodes, each node bearing 1–4 peduncles. Leaves about half the length of the stem, 2–4 mm. wide; peduncles equalling or surpassing the nodal leaf; perianth violet-purple, 10–15 mm. long; the segments truncate-emarginate and aristulate; stamineal column 3–5 mm. long; capsule 3–4 mm. long, brownish green.

Common on grassy hillsides either in heavy or sandy soil, Upper Sonoran and Transition Zones; throughout California from Shasta and Humboldt Counties to Lower California. Type locality: not indicated.

Sisyrinchium macoùnii Bick. Bull. Torrey Club **27**: 245. 1900. Stems simple, leafless, with a sessile spathe, or rarely with a leaf-bearing node and a solitary peduncle, 30–50 cm. high, very slender and scarcely winged, 1–1.5 mm. wide. Leaves mostly 20–30 mm. long, attenuate at apex; pedicles slender and glabrous; perianth segments 20 mm. long, emarginate and slenderly aristulate; anthers 2 mm. long.

Known only from the type locality, Comax, Vancouver Island, British Columbia.

Sisyrinchium funèreum Bick. Bull. Torrey Club **31**: 387. 1904. Stem pale glaucous green, stiffly erect, 50–60 cm. high, 2–4.5 mm. wide, narrowly margined, smooth, bearing a solitary node. Leaves about half the length of the stem, 2–3.5 mm. wide, the one at the node stiff, 4–10 cm. long; peduncles 2 or 3; spathes 18–25 mm. long, the outer bract scarcely equalling the inner; flowers often 12–18; perianth-segments 12–14 mm. long, mucronate at apex; capsules light colored, 3–6 mm. broad.

A little known species, based on a single collection, Furnace Creek Cañon, Funeral Mountains, California.

3. **OLSỲNIUM** Raf. New Fl. Am. **1**: 72. 1836.

Perennial tufted herbs with the general habit of *Sisyrinchium;* rootstocks very short, bearing coarse fibrous roots. Stems simple erect with a few leaves toward the base, compressed but not winged. Basal leaves with much reduced blades and bract-like. Spathe solitary and terminal with two very unequal bracts, the outer much elongated and foliaceous. Flowers usually 1 or 2, reddish purple. Perianth-segments similar, obtuse or apiculate, neither aristulate nor emarginate. Stamens with their filaments united only at the base. Styles cleft at the apex, much exceeding the stamens. Capsule broadly oblong or subglobose. Seeds many rounded, dark colored, finely pitted.

A monotypic genus.

1. **Olsynium douglásii** (A. Dietr.) Bick. Purple-eyed Grass. Fig. 1144.

Sisyrinchium grandiflorum Dougl.; Lindl. Bot. Reg. **16**: *pl. 1364.* 1830. not Cav. 1790.

Sisyrinchium douglasii A. Dietr. Sp. Pl. **2**: 504. 1833.

Olsynium grandiflorum Raf. New. Fl. Am. **1**: 72. 1836.

Olsynium douglasii Bick. Bull. Torrey Club **27**: 237. 1900.

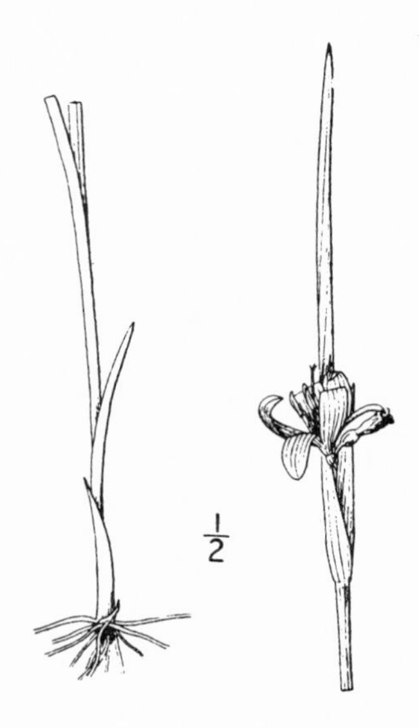

Stems 15–30 cm. high, 2–3 mm. wide, naked above the middle. Basal leaves bract-like with much reduced blades, 1 or 2 with blades 2–15 mm. long, stiffly erect, the long sheathing base scarious on the margins; spathe solitary, terminal; outer bract long-attenuate, 5–12 cm. long, far surpassing the flowers or rarely reduced and shorter than the pedicels; inner bracts shorter than the pedicels or but slightly exceeding them, obtuse; perianth-segments 15–20 mm. long, oblong-obovate, 5-nerved; stamens about two-thirds the length of the segments; anthers orange-colored, 3–4 mm. long; capsule 6–8 mm. long, often tinged with purple; seeds 2 mm. long.

Open grassy places, mainly Arid Transition Zone; British Columbia to northern Nevada and northern California. Type locality: Celilo Falls, Columbia River, Washington.

4. HYDÁSTYLUS Salisb. Trans. Hort. Soc. Lond. 1: 310. 1812.

Annual or perennial herbs with the habit and appearance of *Sisyrinchium,* staining purple and usually turning black on drying; rootstocks obscure or poorly developed, the roots fibrous pale and slender. Leaves narrowly linear, conduplicate at base. Stems ancipital, simple, scapose, terminated by a solitary spathe. Bracts of the spathe two, conduplicate, enclosing membranous scales. Pedicels slender often well exserted. Perianth yellow with dark or orange veins, not aristulate or emarginate. Filaments free to below the middle, somewhat spreading above. Anthers narrowly linear, versatile. Style branches slender, divergent. Capsule oblong to globose, more or less 3-lobed. Seeds few to many, rounded, distinctly pitted.

About 12 species, confined to western United States and Mexico. Type species, *Hydastylus californicus* (Ker.) Salisb.

Herbage turning black in drying and leaving a purple stain; coastal species.
  Anthers 3–5 mm. long; perianth-segments 12–18 mm. long.
    Pedicels 2–4 cm. long, mostly longer than the bracts; seeds 1.25–1.5 mm. in diameter. 1. *H. californicus.*
    Pedicels 1–2 cm. long, mostly shorter than the bracts; seeds .75–1 mm. in diameter. 2. *H. brachypus.*
  Anthers 2–2.5 mm. long; perianth-segments 8–10 mm. long. 3. *H. borealis.*
Herbage not ordinarily turning dark in drying, and not staining purple; montane species. 4. *H. elmeri.*

1. Hydastylus califórnicus (Ker.) Salisb.
California Golden-eyed Grass. Fig. 1145.

Marica californica Ker. Bot. Mag. *pl. 983:* 1806.
Sisyrinchium californicum Dryand. Hort. Kew. ed. 2. 4: 135. 1812.
Hydastylus californicus Salisb. Trans. Hort. Soc. Lond. 1: 310. 1812.
Sisyrinchium flavidum Kell. Proc. Calif. Acad. 2: 50. 1863.
Bermudiana californica Greene, Man. Bay Region 308. 1894.

Stems 15–60 cm. long, dull green and glaucescent, broadly winged, 2–6 mm. wide. Leaves mostly about half the length of the stem, 2–6 mm. wide; outer bract 2–5 cm. long, usually slightly surpassing the inner; flowers bright yellow, 3–7; pedicels 2–4 cm. long; perianth-segments 12–18 mm. long, obtuse or acutish, with 5–7 black nerves; anthers 3–5 mm. long; capsule ellipsoid or obovoid, 7.5–12 mm. long; seeds 1.25–1.5 mm. in diameter.

Moist meadows near the coast, Humid Transition Zone; Curry County, Oregon, to Monterey, California. Type locality: "northwest coast," Menzies.

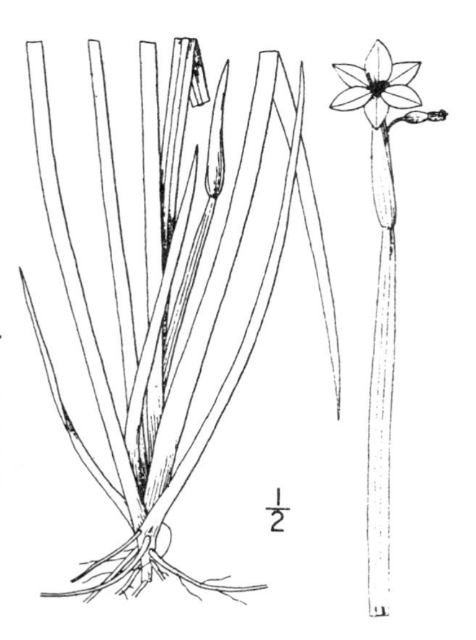

2. Hydastylus bráchypus Bick.
Oregon Golden-eyed Grass. Fig. 1146.

Hydastylus brachypus Bick. Bull. Torrey Club 27: 379. 1900.

Stems mostly low, 8–15 cm. high, or sometimes twice as high, stout, broadly winged, 2–5 mm. wide. Leaves about half the length of the stems or longer, short-pointed; outer bracts 15–25 mm. long, subequal or slightly surpassing the inner, 3–4 mm. wide, obtuse, united clasping for 2–5 mm.; pedicels usually shorter than the inner bract, 10–15 mm. long, obtuse; anthers 3–4 mm. long; capsule oblong or ellipsoid, 7–10 mm. long; seeds .75–1 mm. in diameter.

Open grassy places near the coast, Humid Transition Zone; Washington to central Oregon. Type locality: Newport, Oregon.

3. Hydastylus boreàlis Bick.

Northern Golden-eyed Grass. Fig. 1147.

Hydastylus borealis Bick. Bull. Torrey Club **27**: 378. 1900.

Stems 6–30 cm. high, wing-margined. Leaves about half the length of the stem, 1–3 mm. wide, often with an abrupt linear tip; outer bract 15–20 mm. long, slightly surpassing the inner; pedicels erect, slightly exserted, 15–20 mm. long; perianth-segments yellow, 8–10 mm. long, mostly 5-nerved; anthers 2–2.5 mm. long; capsule ellipsoid, 6–8 mm. long; seeds .75–1 mm. in diameter, finely pitted.

Moist situations, Humid Transition Zone; Vancouver Island and Whatcom County, Washington. Type locality: Whatcom County, Washington. More extended field work and garden culture may prove this species and *Hydastylus brachypus* to be mere forms of *Hydastylus californica*.

4. Hydastylus élmeri (Greene) Bick.

Drew's Golden-eyed Grass. Fig. 1148.

Sisyrinchium elmeri Greene, Pittonia **2**: 106. 1890.
Hydastylus elmeri Bick. Bull. Torrey Club **27**: 380. 1900.
Hydastylus rivularis Bick. Bull. Torrey Club **27**: 380. 1900.

Stems 10–30 cm. high, very slender, 1–2 mm. wide, narrowly winged. Leaves half the length of the stem, 1–3.5 mm. wide; outer bract 12–25 mm. long, united clasping for 5–6 mm., narrowed to an obtuse apex, scarcely exceding the inner; pedicels usually well exserted and spreading; perianth-segments 8–12 mm. long, orange-yellow with 5 dark purple veins, obtusely pointed; anthers 3–5 mm. long; capsule oblong, 7 mm. long.

Borders of mountain streams and meadows, Arid Transition and Canadian Zones; Sierra Nevada from Plumas County to the San Bernardino Mountains, California. Type locality: Lake Eleanor, Sierra Nevada, California.

Family 23. ORCHIDÀCEAE.

Orchid Family.

Perennial herbs with corms, bulbs or tuberous roots, sheathing entire leaves, sometimes reduced to scales, the flowers perfect, irregular, bracted, solitary, spiked or racemed. Perianth superior, of 6 segments, the 3 outer (sepals) similar or nearly so, 2 or the inner ones (petals) lateral, alike; the third inner one (lip) dissimilar, often markedly so, usually larger, often spurred, sometimes inferior by torsion of the ovary or pedicel. Stamens variously united with the style into pear-shaped, usually stalked masses (pollinia), united by elastic threads, the masses waxy or powdery, attached at the base to a viscid disk (gland). Style often terminating in a beak (rostellum) at the base of the anther or between its sacs. Stigma a viscid surface, facing the lip beneath the rostellum, or in a cavity between the anther-sacs (clinandrium). Ovary inferior, usually long and twisted, 3-angled, 1-celled; ovules numerous, anatropous, on 3 parietal placentae. Capsule 3-valved. Seeds very numerous, minute, mostly spindle-shaped, the loose coat hyaline, reticulated; endosperm none; embryo fleshy.

About 430 genera and over 5000 species, of wide distribution, most abundant in the tropics, many of those of warm regions epiphytes.

Anthers 2; lip a large inflated sac. (CYPRIPEDIEAE.) 1. *Cypripedium.*
Anthers solitary.
Pollinia with a caudicle, which is attached at the base to a viscid disk or gland; our genera with a spur. (ORCHIDEAE.)
Stems leafy. 2. *Limnorchis.*
Stems scapose with two basal leaves.
Anther sacs parallel or nearly so; lateral sepals adnate at base to the claw of the lip. 3. *Piperia.*
Anther sacs widely divergent; lateral sepals free at base. 4. *Lysias.*
Pollinia not produced into a caudicle..
Pollinia granulose or powdery. (NEOTTIEAE.)
Plants with ordinary green foliage.
Stems leafy.
Leaves broad; anthers operculate. 5. *Epipactis.*
Leaves narrow; anthers not operculate; spike often twisted. 7. *Spiranthes.*
Stems scapose; leaves basal or nearly so.
Leaves 2, opposite, not white-veined. 8. *Ophrys.*
Leaves several, white-veined. 9. *Peramium.*
Plants without green foliage, white, leaves reduced to alternate scales. 6. *Eburophyton.*
Pollinia smooth and waxy. (EPIDENDREAE.)
Leaves 1 or 2; plants bulbous.
Leaves 2, opposite; lip flat. 10. *Liparis.*
Leaf 1.
Flower solitary; lip saccate. 11. *Calypso.*
Flowers racemose; lip not saccate. 12. *Aplectrum.*
Leaves reduced to scales; plants without green herbage, from coralloid roots. 13. *Corallorrhiza.*

1. CYPRIPÈDIUM L. Sp. Pl. 975. 1753.

Glandular-pubescent herbs, with leafy stems and coarsely fibrous roots. Leaves large, broad, many-nerved. Flowers solitary or several, drooping, large and showy. Sepals spreading, separate, or usually two of them united into one under the inflated sac-like lip. Column declined, bearing a sessile or stalked fertile anther on each side and a dilated petaloid but thickish sterile stamen above, which covers the summit of the style. Pollinia granular, without a caudicle or glands. Stigma terminal, broad, obscurely 3-lobed. [Name incorrectly Latinized from Κύπρις, *Venus,* and πέδιλον, a shoe, therefore by some authors spelled *Cypripedilum.*]

About 20 species, natives of the north temperate zone. Besides the following 4 other species occur in eastern North America, 2 in the Rocky Mountains and 1 in Alaska. Type species, *Cypripedium calceolus* L.

Leaves several, alternate.
Sepals obtuse, shorter than the lip; flowers 3–7, racemose. 1. *C. californicum.*
Sepals acuminate, longer than the lip; flowers solitary.
Flowers yellow; lip 15–30 mm. long. 2. *C. parviflorum.*
Flowers brown, with a white lip 30–50 mm. long. 3. *C. montanum.*
Leaves 2, opposite, broadly oval; flowers greenish, fasciculate. 4. *C. fasciculatum.*

1. Cypripedium califórnicum A. Gray. California Lady's Slipper. Fig. 1149.

Cypripedium californicum A. Gray, Proc. Am. Acad. 7: 389. 1867.

Stems leafy, 3–7 dm. high, glandular-pubescent. Leaves 7–10 cm. long, ovate-lanceolate, acute, the upper becoming smaller, lanceolate and acuminate; flowers 3–7, solitary in the axils of the foliaceous bracts, and much shorter than them; sepals broadly oval, the lower wholly united, 15 mm. long; petals oblong-linear, equalling the sepals; ovary twisted, acutish, greenish-yellow; lip obovate-globose, white or often rose-color, spotted with brown, 18–20 mm. long, pubescent within at the base; sterile anther rounded and arching, nearly sessile, 4 mm. long; stigma roughened 4 mm. long; capsule reflexed, oblong 20 mm. long.

In damp soils of open woods, Transition Zone; Siskiyou Mountains, southern Oregon, south to Marin and Placer Counties, California. Type locality: swamps on Red Mountain, Mendocino County, California.

2. Cypripedium parviflòrum Salisb. Yellow Lady's Slipper. Fig. 1150.

Cypripedium parviflorum Salisb. Trans. Linn. Soc. 1: 77. 1791.

Cypripedium pubescens Willd. Sp. Pl. 4: 143. 1805.

Cypripedium bulbosum flavescens Farwell, Rep. Mich. Acad. Sci. 15: 170. 1913.

Cypripedium hirsutum parviflorum Rolfe, Orchid Rev. 15: 184. 1907.

Stems leafy, 3–8 dm. high. Leaves oval or elliptic, 5–15 cm. long, acute or acuminate; flowers 1–3, racemose in the axils of the bracts; sepals ovate-lanceolate, 3–5 cm. long, longer than the lip, yellowish or greenish, striped with purple; petals linear-lanceolate, equalling the sepals or longer; lip much inflated, 2–5 cm. long, pale yellow with purple veins, with a tuft of white jointed hairs near the top; sterile stamen triangular; stigma thick, somewhat triangular, incurved.

In moist woods, mainly Transition Zone; British Columbia to Nova Scotia, south to Nebraska and Georgia. In the Pacific States it has been found only in the vicinity of Spokane, Washington. Type locality: Virginia.

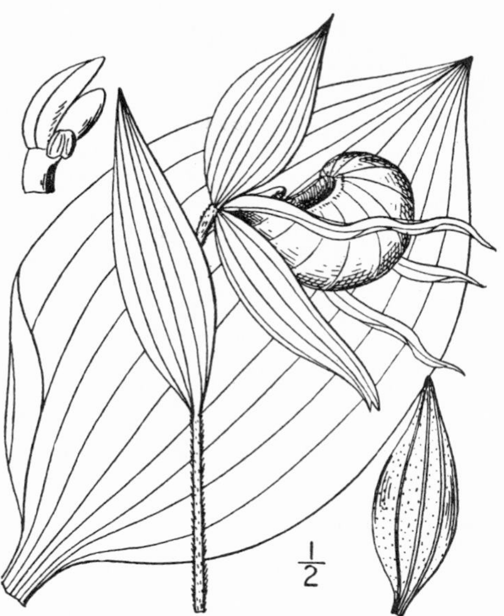

3. Cypripedium montànum Dougl. Mountain Lady's Slipper. Fig. 1151.

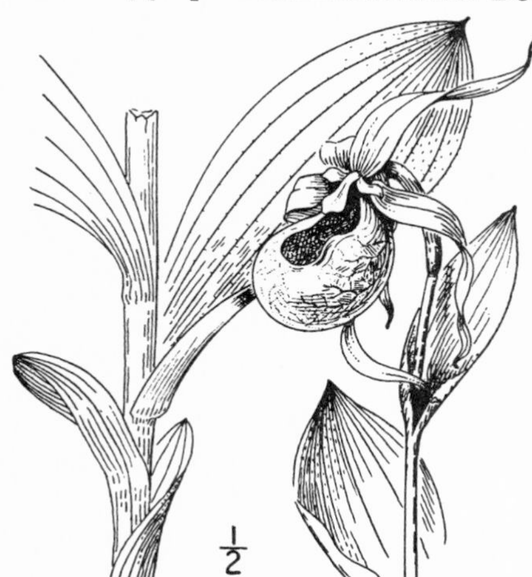

Cypripedium montanum Dougl.; Lindl. Gen. & Sp. Orchid. 528. 1840.

Cypripedium occidentale S. Wats. Proc. Am. Acad. 11: 147. 1876.

Stems stout, leafy, 3–7 dm. high. Leaves ovate to broadly lanceolate, 8–16 cm. long, 4–8 cm. wide, acute or abruptly short-acuminate. Flowers 1–3, racemose in the axils of the bracts; sepals lanceolate, acuminate, 4–6 cm. long, brownish purple, the lower united to near the tip; petals similar to the sepals in color, but narrower; lip oblong, 2–3 cm. long, white tinged and veined with purple; sterile stamen oblong-ovate, 8–10 mm. long, yellow with purple spots, deeply channelled above; stigma somewhat shorter than the sterile stamen; capsule erect or nearly so, oblong, 2 cm. long.

In moist open woods, Transition Zone; Vancouver Island to Saskatchewan and Wyoming, south to Mariposa and Santa Cruz Counties, California. Type locality: western North America.

4. Cypripedium fasciculàtum Kell. Clustered Lady's Slipper. Fig. 1152.

Cypripedium fasciculatum Kell. S. Wats. Proc. Am. Acad. 17: 380. 1882.

Cypripedium fasciculatum pussillum Hook. f. Bot. Mag. 119: *pl. 7275.* 1893.

Stems woolly-pubescent, 10–40 cm. high, with 1 or 2 scarious bracts at base. Leaves 2, usually in about the middle of the stem and appearing opposite, broadly oval, 5–10 cm. long, 4–7 cm. wide, obtuse or acutish; the peduncle-like stem above the leaves bearing 1 or 2 lanceolate bracts near the middle; flowers solitary or usually several in a cluster; bracteate, greenish; sepals lanceolate, acuminate, 15–25 mm. long, brown-veined, the lower pair wholly united or the very tips separate; petals similar but narrower; lip depressed-ovate, greenish yellow with brown-purple veins, 8–10 mm. long; sterile stamen oblong-ovate, obtuse, 3 mm. long, stigma equalling the sterile anther; capsule oblong, 2 cm. long.

Moist woods, Transition Zone; eastern Washington in the Wenatche Mountains to Plumas County in the Sierra Nevada and the Santa Cruz Mountains in the California Coast Ranges. Type locality: White Salmon River, above the Falls, Washington.

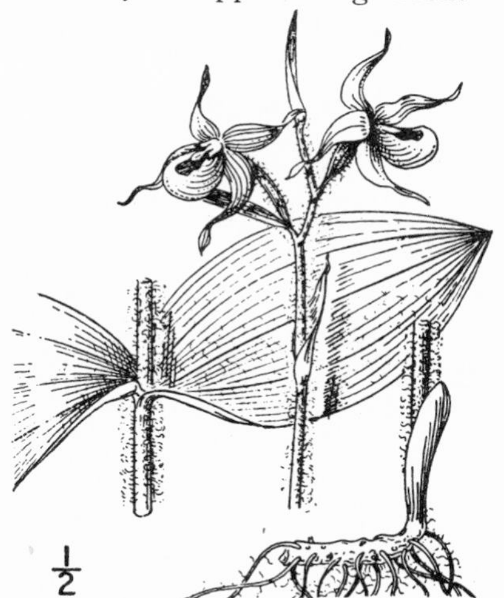

2. LIMNÓRCHIS Rydb. Mem. N. Y. Bot. Gard. 1: 104. 1900.

Leafy plants with thick fleshy roots and small greenish or whitish flowers in a long spike. Upper sepal ovate to suborbicular, erect, 3–7-nerved; lateral sepals from linear to ovate-lanceolate, free from the lip, usually 3-nerved, spreading or somewhat reflexed. Lateral petals erect, lanceolate, 3-nerved. Lip entire, flat or slightly concave, reflexed, free, linear to rhombic-lanceolate. Column short and thick; anther-sacs parallel, opening in front. Glands naked. Pollinia granular. [Greek, meaning marsh-orchis.]

A North American genus of about 15 species. Type species, *Limnorchis hyperborea* (L.) Rydb.

Flowers greenish or purplish; lip linear or lanceolate, not rhombic-dilated at base.
  Spur decidedly clavate, shorter than the lip.
    Petals purplish; spur one-half to two-thirds as long as the lip. — 1. *L. stricta.*
    Petals green; spur about equalling the lip. — 2. *L. viridiflora.*
  Spur filiform, exceeding the lip. — 3. *L. sparsiflora.*
Flowers white; lip rhombic-dilated at base.
  Spur usually shorter than or about equalling the lip; often somewhat clavate. — 4. *L. dilatata.*
  Spur slender, one-third to two-thirds longer than the lip.
    Dilated portion of lip oval; spur two-thirds longer than lip. — 5. *L. thurberi.*
    Dilated portion of lip rhombic; spur scarcely one-half longer than lip. — 6. *L. leucostachys.*

1. Limnorchis strícta (Lindl.) Rydb. Slender Bog Orchid. Fig. 1153.

Platanthera stricta Lindl. Gen. & Sp. Orchid. 288. 1836.
Habenaria gracilis S. Wats. Proc. Am. Acad. **12**: 277. 1877.
Limnorchis stricta (Lindl.) Rybd. Mem. N. Y. Bot. Gard. **1**: 105. 1900.

Stems 3–10 dm. high, strict. Lower leaves oblanceolate, obtuse, 5–12 cm. long, 15–25 mm. wide; the upper lanceolate acute; spike lax, 1–3 dm. long; bracts linear-lanceolate, the lower much longer than the flowers, often 25–40 mm. long; flowers 2–14 mm. long; sepals green, the upper ovate, erect, 4–5 mm. long, the lateral lanceolate, obtuse, 5-6 mm. long; petals purplish, lanceolate, acute; lip linear, 5–7 mm. long, obtuse, purple; spur one-half to two-thirds as long as the lip, purplish, very saccate.

In swampy places, Transition and Canadian Zones; Alaska to the Cascade Mountains of northern Oregon, east to Wyoming and Montana. Type locality: western North America.

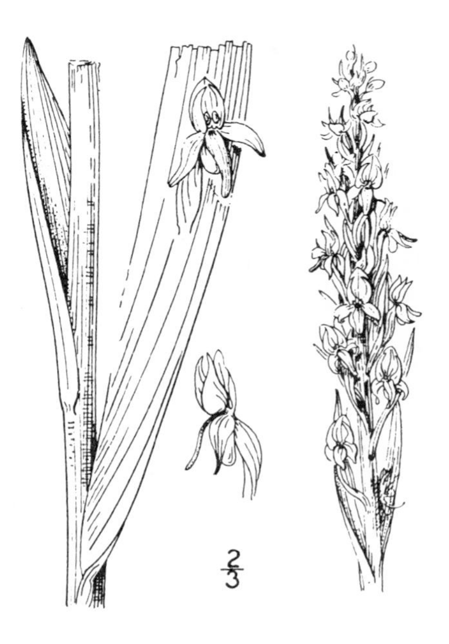

2. Limnorchis viridiflòra (Cham.) Rydb. Green-flowered Bog Orchid. Fig. 1154.

Habenaria borealis viridiflora Cham. Linnaea **3**: 28. 1828.
Limnorchis viridiflora Rybd. Bull. Torrey Club **28**: 616. 1901.

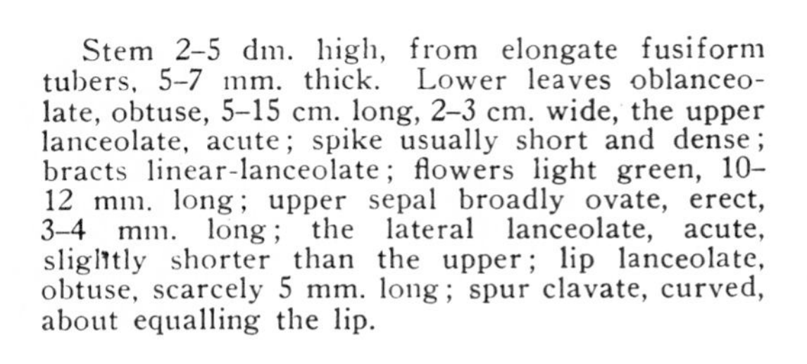

Stem 2–5 dm. high, from elongate fusiform tubers, 5–7 mm. thick. Lower leaves oblanceolate, obtuse, 5–15 cm. long, 2–3 cm. wide, the upper lanceolate, acute; spike usually short and dense; bracts linear-lanceolate; flowers light green, 10–12 mm. long; upper sepal broadly ovate, erect, 3–4 mm. long; the lateral lanceolate, acute, slightly shorter than the upper; lip lanceolate, obtuse, scarcely 5 mm. long; spur clavate, curved, about equalling the lip.

In bogs, Canadian Zone; Alaska, through the Rocky Mountains to North Dakota and Colorado, and in the Pacific States to the Olympic Mountains and Mount Rainier, Washington. Type locality: "Unalaschca."

3. **Limnorchis sparsiflòra** (S. Wats.) Rydb.
Sparsely Flowered Bog Orchid. Fig. 1155.

Habenaria sparsiflora S. Wats. Proc. Am. Acad. **12**: 276. 1877.
Limnorchis laxiflora Rydb. Bull. Torrey Club **28**: 630. 1901.
Limnorchis sparsiflora Rydb. Bull. Torrey Club **28**: 631. 1902.

Stem slender, 4–6 dm. high. Lower leaves oblanceolate, obtuse, 8–10 cm. long, 10–15 mm. wide, the upper linear-lanceolate, acute; spike slender and lax, 1–2 dm. long, few-flowered; bracts linear-lanceolate; flowers greenish, 10–12 mm. long; upper sepals broadly obovate, obtuse, about 4 mm. long, the lateral broadly lanceolate, acutish; lip linear, obtuse, about 6 mm. long; spur slightly clavate, a little longer than the lip.

Wet springy places, Transition and Canadian Zones; Washington, Oregon to northern Lower California and New Mexico. Type locality: near Mariposa Grove, Yosemite National Park, California.

4. **Limnorchis dilatàta**(Pursh)Rydb.
Boreal Bog Orchid. Fig. 1156.

Orchis dilatata Pursh, Fl. Am. Sept. 588. 1814.
Habenaria dilatata Hook. Exot. Fl. *pl. 95.* 1825.
Limnorchis dilatata Rydb. Bull. Torrey Club **28**: 622. 1901.
Habenaria borealis Cham. Linnaea **3**: 28. 1828.
Limnorchis borealis Rydb. Bull. Torrey Club **28**: 621. 1901.

Stems leafy, 4–8 dm. high, from thick, elongated, fusiform tubers. Leaves lanceolate, the lower obtuse, the upper acute, 7–20 cm. long; spike 1–2 dm. long, often rather densely flowered; flowers white, 10–14 mm. long; upper sepal ovate, obtuse, 4–5 mm. long, the lateral oblong-lanceolate; lip rhombic-lanceolate, obtuse, about 5 mm. long; spur usually shorter than the lip, more or less clavate.

Bogs and springy places, Canadian Zone; Alaska to Mount Hood, Oregon, east to New Foundland, New York and Nebraska. Type locality: Labrador.

5. **Limnorchis thúrberi** (A. Gray) Rydb.
Thurber's Bog Orchid. Fig. 1157.

Habenaria thurberi A. Gray, Proc. Am. Acad. **7**: 389. 1868.
Limnorchis thurberi Rydb. Bull. Torrey Club **28**: 624. 1901.

Stems stout and leafy, 4–6 dm. high, from elongated fusiform tubers. Leaves lanceolate to linear-lanceolate, acute, 1–2 dm. long; spike dense, 10–20 cm. long; flowers white 16–20 mm. long; upper sepal ovate, obtuse, the lateral lanceolate, acute; lip 7–8 mm. long, lanceolate, obtuse, the dilated portion ovate; spur filiform, curved, about two-thirds longer than the lip.

Bogs and borders of streams, Transition and Canadian Zones; vicinity of Mount Shasta to northern Lower California and New Mexico. Type locality: "Arizona."

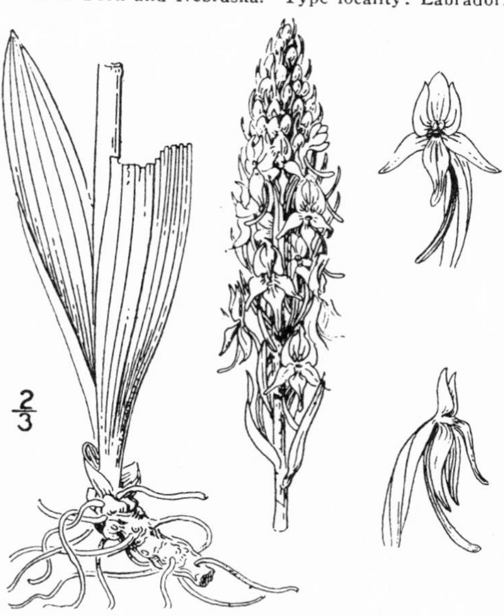

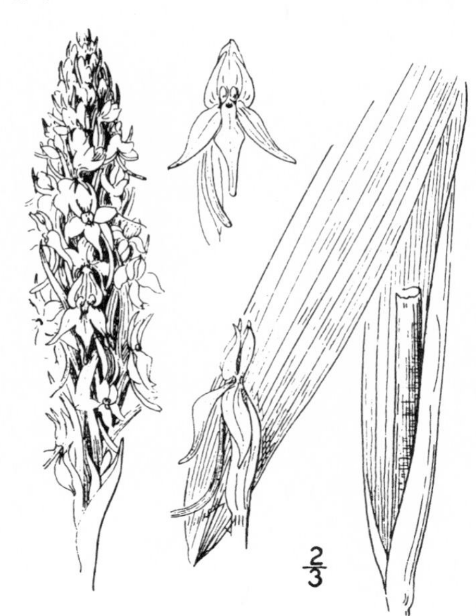

6. Limnorchis leucostáchys (Lindl.) Rydb.

White-flowered Bog Orchid. Fig. 1158.

Platanthera leucostachys Lindl. Gen. & Sp. Orchid. 288. 1835.
Habenaria leucostachys S. Wats. Bot. Calif. **2**: 134. 1880.
Limnorchis leucostachys Rydb. Mem. N. Y. Bot. Gard. **1**: 106. 1900.

Stem stout 6–10 dm. high, from thick elongated fusiform tubers. Lower leaves oblanceolate, 1–2 dm. long, the upper lanceolate acute; spike more or less densely flowered, 1–3 dm. long; flowers white, 15–20 mm. long; upper sepal ovate, obtuse, about 5 mm. long, the lateral lanceolate, acute, 7–8 mm. long; lip lanceolate, decidedly rhombic at base, about 8 mm. long; spur filiform, clavate, acutish, about one half longer than the lip.

Bogs and springs, Transition and Canadian Zones, Alaska to southern California, Nevada and Utah. Type locality: mountains of western North America.

Limnorchis leucostachys robústa Rydb. Bull. Torrey Club **28**: 626. 1901. Stouter plant with denser spike; spur clavate and obtuse. British Columbia to Oregon and Idaho.

3. PIPERÌA Rydb. Bull. Torrey Club 28: 269. 1901.

Simple-stemmed perennial herbs from spherical or broadly ellipsoid tubers. Leaves usually near the base and withering at or before anthesis. Flowers small, greenish or white, spicate; sepals and petals 1-nerved or obscurely 3-nerved; the upper erect, the lateral spreading. Lateral petals free, lanceolate or linear-lanceolate, oblique. Lip linear-lanceolate to ovate, concave, united with the bases of the lower sepals. Capsule ellipsoid. [Name in honor of Charles Vancouver Piper.]

About nine species natives of western North America. Type species, *Piperia élegans* (Lindl.) Rydb.

Spur shorter than the ovary or nearly equalling it in *lancifolia*.
  Lip oblong; stem leafy only at base. — 1. *P. unalaschensis.*
  Lip ovate; stem more or less leafy.
    Spur very saccate, slightly exceeding the lip. — 2. *P. cooperi.*
    Spur only slightly clavate, nearly twice as long as the lip. — 3. *P. lancifolia.*
Spur decidedly longer than the ovary, filiform.
  Lip linear to lanceolate.
    Spike very lax; lip 4–5 mm. long; spur about 8–10 mm. long. — 4. *P. leptopetala.*
    Spike dense; lip about 6 mm. long; spur 15–18 mm. long. — 5. *P. multiflora.*
  Lip ovate or ovate-lanceolate.
    Spike elongated, lax; leaves withering at anthesis.
      Bracts linear lanceolate; stem leafy only at base. — 6. *P. elegans.*
      Bracts ovate lanceolate; stem usually leafy. — 7. *P. longispica.*
    Spike short and very dense; leaves withering before anthesis.
      Petals and sepals about 4 mm. long; petals purplish or green. — 8. *P. michaeli.*
      Petals and sepals about 5 mm. long; petals white. — 9. *P. maritima.*

1. Piperia unalaschénsis (Spreng.) Rydb.

Alaska Piperia. Fig. 1159.

Sperianthes unalaschensis Spreng. Syst. **3**: 307. 1826.
Habenaria schischmareffiana Cham. Linnaea **3**: 29. 1828.
Habenaria foetida S. Wats. Bot. King. Expl. 341. 1871.
Piperia unalaschensis Rydb. Bull. Torrey Club **28**: 270. 1901.

Stem slender, 3–5 dm. high, leafy only near the base. Basal leaves oblanceolate, obtuse or acutish 10–15 cm. long, 10–35 mm. wide, withering at anthesis or soon after; stem leaves bract-like, attenuate; spike lax, 1–3 dm. long; bracts lanceolate to ovate-lanceolate, about equalling the ovary; flowers greenish, 8–10 mm. long; sepals and petals 2–4 mm. long; upper sepal ovate acutish, the lateral oblong, lanceolate, obtuse; lip oblong, obtuse, slightly hastate at base; spur filiform or slightly clavate, a little longer than the lip, but shorter than the ovary.

Dry ridges in open forests, Upper Sonoran and Transition Zones; Alaska and Quebec to Colorado and southern California. **Type locality**: Aleutian Islands.

2. Piperia coòperi (S. Wats.) Rydb.
Cooper's Piperia. Fig. 1160.

Habenaria cooperi S. Wats. Proc. Am. Acad. **12**: 276. 1876.

Stem strict, 3–10 dm. high, leafy below. Basal leaves oblong lanceolate, acute, about 10 cm. long, the lower stem leaves lanceolate, attenuate, 10–15 cm. long, the upper reduced and bract-like; spike lax, 10–30 cm. long; flowers yellowish green; sepals and petals about 4 mm. long, obtuse; upper sepal ovate, the lateral oblong-lanceolate; lip ovate, rounded at the apex, somewhat hastate at base; spur thick, conspicuously clavate, about equalling the lip.

Dry mesas and hills, in chaparral, Lower Sonoran Zone; vicinity of San Diego. Type locality: clay hills near San Diego, California.

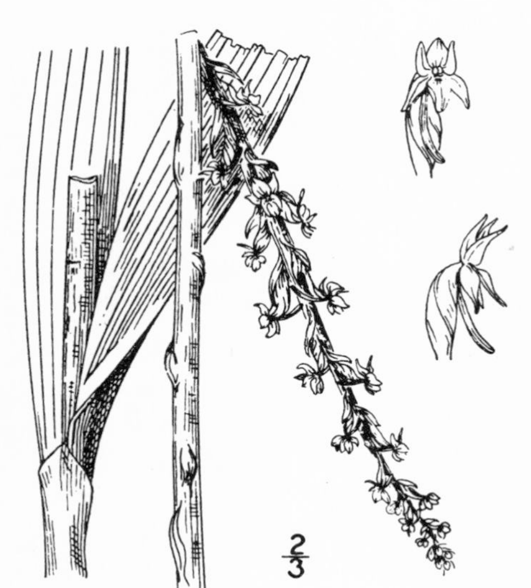

3. Piperia lancifòlia Rydb.
Lance-leaved Piperia. Fig. 1161.

Piperia lancifolia Rydb. Bull. Torrey Club **28**: 637. 1901.

Stem stout, 3–6 dm. high, the lower portion leafy. Basal leaves and lower stem leaves lanceolate, attenuate, 10–15 mm. long, 1–2 cm. wide, withering after anthesis; spike lax, many-flowered, 2–3 dm. long; flowers greenish; upper sepal ovate, obtuse, about 4 mm. long, the lateral slightly longer, oblong-lanceolate; lip about 4 mm. long, broadly ovate, rounded at apex, scarcely hastate at base; spur slightly clavate, almost twice as long as the lip.

Dry wooded slopes, Upper Sonoran Zone; Santa Monica and San Gabriel Mountains, southern California. Type locality: Santa Monica Mountains, California.

4. Piperia leptopétala Rydb.
Narrow-petaled Piperia. Fig. 1162.

Piperia leptopetala Rydb. Bull. Torrey Club **28**: 637. 1901.

Stem slender, 3–4 dm. high, leafy only at base. Basal leaves 2, oblong-lanceolate, obtuse or acutish, about 1 dm. long, 15–25 mm. wide, withering at anthesis; stem leaves few and bract-like; spike slender, lax, 1–2 dm. long; flowers greenish; upper sepal lanceolate, obtuse, about 4 mm. long, the lateral narrowly lanceolate, acute, about 5 mm. long; lip lanceolate, obtuse, hastate at base; spur filiform, about twice as long as the lip and longer than the ovary.

Dry wooded slopes. Transition Zone; Washington to southern California. Type locality: mountains east of San Diego.

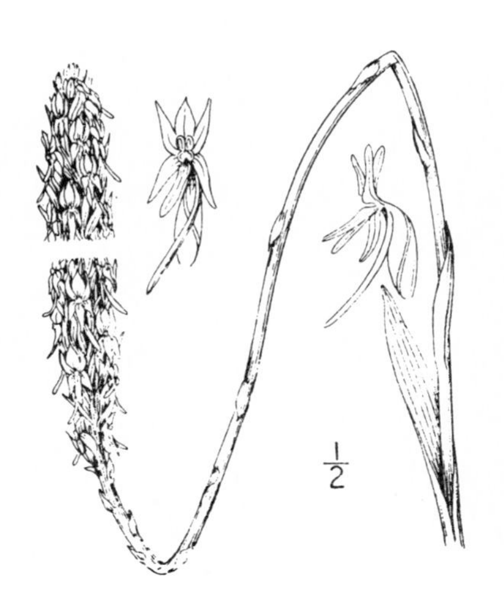

5. **Piperia multiflòra** Rydb.

Many-flowered Piperia. Fig. 1163.

Piperia multiflora Rydb. Bull. Torrey Club **28**: 638. 1901.

Stem stout 4–6 dm. high, leafy only near the base. Basal leaves 3–4, oblong or oblanceolate, obtuse or acutish, 10–15 cm. long, 2–3 cm. wide, withering at anthesis; lower stem leaves lanceolate, acute, the upper bract-like; spike very dense 1–2 dm. long; flowers greenish white, upper sepals lanceolate, acute, 4–5 mm. long, the lateral linear-lanceolate; lip nearly linear, obtuse, only slightly hastate at base, 6 mm. long; spur filiform about 15 mm. long.

Wooded slopes, Transition Zone. Washington and Montana to central California. This and *leptopetala* may be only forms of the common *elegans*. Type locality: Gray's Harbor, Washington.

6. **Piperia élegans** (Lindl.) Rydb.

Elegant Piperia. Fig. 1164.

Platanthera elegans Lindl. Gen. & Sp. Orchid. 285. 1835.

Habenaria elegans Boland. Cat. Pl. San Franc. 29. 1870.

Piperia elongata Rydb. Bull. Torrey Club **28**: 270. 1901.

Stem slender, 4–7 dm. high, leafy only at base. Basal leaves 2 or 3, lanceolate to ovate, acute or obtuse, 8–15 cm. long, 1–5 cm. wide; stem leaves reduced to bracts, 5–10 mm. long; spike usually lax 15–30 cm. long; flowers greenish white; upper sepal lanceolate, acute, 5 mm. long, the lateral linear-oblong, acute; lip broadly lanceolate, slightly hastate at base; spur filiform 10–12 mm. long.

Open woods, Transition Zone; British Columbia and Idaho to southern California. Type locality: western North America.

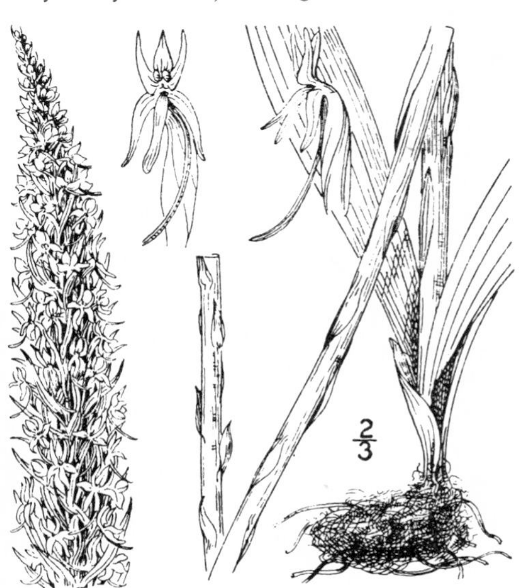

7. **Piperia longispìca** Durand.

Long-spiked Piperia. Fig. 1165.

Gymnandenia longispica Durand, Journ. Acad. Sci. Philad. II. **3**: 101. 1855.

Piperia longispica Rydb. Bull. Torrey Club **28**: 639. 1901.

Stem stout, 3–7 dm. high, more or less leafy towards the base. Basal leaves and lower stem leaves 2–4, lanceolate, acute, 10–15 cm. long, 2–4 mm. wide, withering at or soon after anthesis; upper stem leaves bract-like; spike many-flowered, but not dense. 1–3 dem. long; flowers greenish; upper sepal ovate, obtuse, 5 mm. long, the lateral oblong-lanceolate, obtuse; lip ovate, hastate at base, 5 mm. long; spur filiform, about 12 mm. long.

Open woods, Upper Sonoran and Transition Zones; central Sierra Nevada to the Santa Monica and San Bernardino Mountains. Type locality: near Nevada City, Nevada County, California.

8. Piperia michaèli (Greene) Rydb.
Purple-flowered Piperia. Fig. 1166.

Habenaria michaeli Greene, Man. Bay Reg. 306. 1894.
Piperia michaeli Rydb. Bull. Torrey Club **28**: 640. 1901.

Stem stout, 2–3 dm. high, leafless at flowering time. Basal leaves, elliptic or oblanceolate, about 15 cm. long, 4 cm. wide, withering before anthesis; stem leaves numerous, all bract-like; spike very dense, 5–15 cm. long; flowers greenish; upper sepal ovate, 4 mm. long, the lateral oblong lanceolate obtuse; lip ovate, obtuse, scarcely hastate at base; spur filiform, two and a half times the lip.

Wooded hillsides; Upper Sonoran and Transition Zones; Washington south through the Coast Ranges to San Luis Obispo County, California. Type locality: San Luis Obispo County, California.

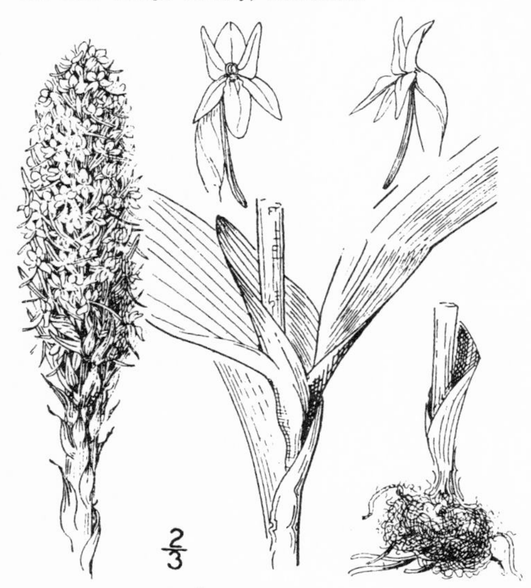

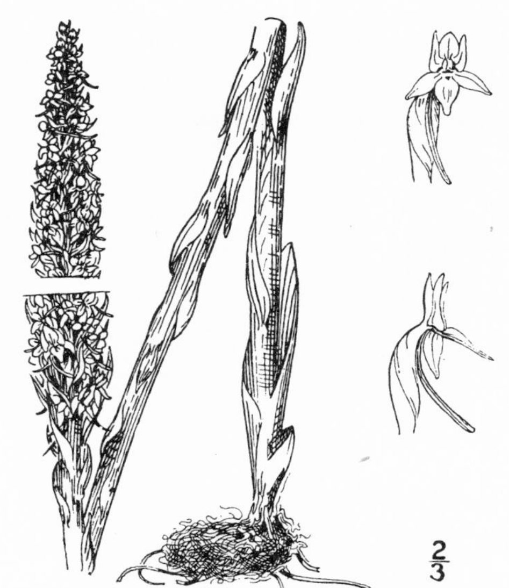

9. Piperia marítima (Greene) Rydb.
Coast Piperia. Fig. 1167.

Habenaria maritima Greene, Pittonia **2**: 298. 1892.
Piperia maritima Rydb. Bull. Torrey Club **28**: 641. 1901.

Stem stout, 2–3 dm. high, leafless at flowering time. Basal leaves 2 or 3, oblong, obtuse or acute, about 1 dm. long, 3–4 cm. wide, withering before anthesis; stem leaves all bract-like, numerous; spike very dense, 4–10 cm. long; flowers white; upper sepal ovate, obtuse, 4 mm. long, the lateral oblong lanceolate; lip elliptic, obtuse, slightly hastate, white and thin; spur filiform, about two and a half times longer than the lip.

Dry hills along the coast, Upper Sonoran and Transition Zones; San Francisco to Monterey County. Type locality: Point Lobos, San Francisco, California.

4. LỲSIAS Salisb. Trans. Hort. Soc. London 1: 288. 1812.

Stem scapose, from tubers or fleshy roots. Leaves two, basal. Flowers spicate, greenish or white. Sepals free, large and spreading. Petals small and narrow; lip entire, linear or nearly so; spur slender, usually longer than the ovary. Beak of the stigma without appendages. Anther sacs widely diverging, their narrow beak-like bases projecting forward. Capsule cylindric-clavate, distinctly stipitate. [Named for Lysias, an Attic orator.]

A circumboreal genus of about 6 species. Type species, *Lysias bifolia* (L.) Salisb.

1. Lysias orbiculàta (Pursh) Rydb.
Large Round-leaved Orchid. Fig. 1168.

Orchis orbiculata Pursh, Fl. Am. Sept. 588. 1826.
Habenaria macrophylla Goldie, Edinb. Phil. Journ. **6**: 331. 1822.
Habenaria orbiculata Torr. Comp. 318. 1826.
Lysias orbiculata Rydb. in Britton, Man. 294. 1901.

Scape short, bracted, 3–6 dm. high, occasionally bearing a small leaf. Basal leaves 2, orbicular, spreading flat on the ground, silver beneath, shining above, 5–8 cm. broad, raceme loosely many-flowered; pedicles about 1 cm. long, erect in fruit; flowers greenish white; upper sepal short, rounded, the lateral falcate-ovate, 8–10 mm. long, spreading; lip oblong-linear, obtuse, entire, white, 12 mm. long; spur longer than the ovary, often 3 cm. long.

In rich woods, Canadian Zone; British Columbia and Cascade Mountains of northern Washington east to Newfoundland and Virginia. Heal-all. July–Aug. Type locality: "on the mountains of Pennsylvania and Virginia."

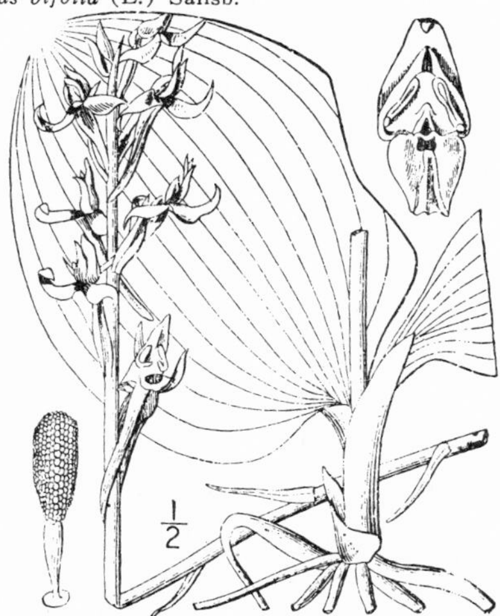

5. EPIPÁCTIS L. C. Rich. Mém. Mus. Par. **4**: 51, 60. 1818.

[Serapias L. Sp. Pl. 949. 1753.]

Tall short simple-stemmed leafy herbs from creeping rootstocks. Leaves ovate to lanceolate, plicate, clasping. Flowers few, in a terminal leafy-bracted raceme. Sepals and petals similar, all separate. Lip free, sessile, broad, concave below, constricted near the middle, the upper portion dilated and petal-like. Spur none. Column short and erect. Anther one, sessile, back of the broad truncate stigma. Pollinia 2-parted granulose, becoming attached to the glandular beak of the stigma. Capsule oblong, beakless. [Name from the Greek word *epipeqnuo,* because it curdles milk.]

About 10 species, widely distributed. Besides the following the type species, *Serapias helleborine* L., occurs in the eastern United States.

1. Epipactis gigántea Dougl.

Giant Helleborine. Fig. 1169.

Epipactis gigantea Dougl.; Hook. Fl. Bor. Am. **2**: 202. *pl. 202.* 1839.

Serapias gigantea A. A. Eaton, Proc. Biol. Soc. Wash. **21**: 67. 1908.

Stem erect, simple 3–10 dm. high, sparsely pubescent. Lower leaves ovate, the upper lanceolate, acute, 5–15 cm. long; flowers 3–15, greenish or purplish, on pedicels 4–8 mm. long; sepals 15 mm. long, the upper concave; the lateral spreading and acute; petals orange-purple, purple-veined; lip ovate, lanceolate, with short erect lobes and many callous tubercles near the base; capsule reflexed, 2–3 cm. long.

Stream banks and springs, Transition and Upper Sonoran Zones; British Columbia to Lower California, Montana and Texas. Type locality: "N. W. America, on the subalpine regions of the Blue and Rocky Mountains." Also called False Lady's Slipper.

6. EBURÓPHYTON Heller, Muhlenbergia **1**: 48. 1904.

Saprophytic perennial herb, with branched, creeping rootstock, whole plant white. Leaves reduced to scarious sheathing bracts. Flowers in a terminal raceme, nearly sessile, subtended by scarious bracts. Lateral sepals spreading, strongly keeled and somewhat concave. Upper sepal and petals erect and somewhat connivent. Lip free, the saccate base with broad wing-like margin, articulate at the middle with a callosity on each side; stigma beakless; anther stipitate. Pollinia distinct, not connected to a gland, granulose.

A monotypic genus of western North America.

1. Eburophyton aùstinae (A. Gray) Heller.

Phantom Orchid. Fig. 1170.

Chloroea austinae A. Gray, Proc. Am. Acad. **12**: 83. 1876.

Cephalanthera oregona Reichenb. Linnaea **41**: 53. 1877.

Cephalanthera austinae Heller, Cat. N. Am. Pl. ed. 2. **4**. 1900.

Eburophyton austinae Heller, Muhlenbergia 1: 49. 1904.

Scape stout, 2–5 dm. high, clothed with 3–4 scarious sheaths below and usually with 1 or 2 long free linear-lanceolate bracts above. Raceme 5–15 cm. long, about 5–20-flowered; sepals and petals oblong-lanceolate, nearly equal, 12–15 mm. long; lip a little shorter, the saccate base with a broad wing-like margin, the nerves in the center wavy-crested; column 4 mm. long, about twice as long as the anther.

In coniferous forests; Transition and Canadian Zones; Washington and Idaho south to Mariposa and Monterey Counties, California. Type locality: near Quincy, Plumas County, California.

7. SPIRÁNTHES L. C. Rich. Mém. Mus. Paris 4: 50. 1818.

Stems erect, leafy, from a cluster of tuberous roots. Flowers in a twisted spike, white, spurless. Sepals and petals narrow, erect or more or less connivent; lip oblong sessile or nearly so, the base embracing the column, with a callous protuberance on each side, the dilated summit spreading and usually entire. Column very short oblique, terminating in a short terete spike. Stigma ovate, with an acuminate bifid beak, covering the anther and stigmatic only beneath. Anther without a lid, borne on the back of the column, erect. Pollinia two, one in each sac, powdery. Capsule ovoid or oblong erect. [Name Greek, meaning spiral and flower, in reference to its twisted inflorescence.]

About 55 species, widely distributed over temperate and tropical regions. Besides the following about a dozen other species inhabit eastern North America. Type species, *Spiranthes aestivalis* L. C. Rich.

Callosities at the base of the lip obsolete. 1. *S. romanzoffiana.*
Callosities at the base of the lip nipple-shaped, directed downward. 2. *S. porrifolia.*

1. Spiranthes romanzoffiàna Cham. & Schl.

Hooded Ladies' Tresses. Fig. 1171.

Spiranthes romanzoffiana Cham. & Schl. Linnaea 3: 32. 1828.
Gyrostachys romanzoffiana McM. Met. Minn. 171. 1892.
Orchiastrum romanzoffianum Greene, Man. Bay Region 306. 1894.
Gyrostachys stricta Rydb. Mem. N. Y. Bot. Gard. 1: 299. 1901.
Ibidium romanzoffianum House, Muhlenbergia 1: 129. 1906.
Ibidium strictum House, Bull. Torrey Club 32: 381. 1905.

Stem glabrous, 15–45 cm. high. Lower leaves linear to linear-lanceolate, 5–35 cm. long; spike 5–15 cm. long, dense; flowers white or greenish, 6–8 mm. long; lip oblong, contracted below the cusped dilated apex, the callosities at base obsolete.

Wet meadows, Upper Sonoran to Canadian Zones; Alaska to New Foundland, south to southern California, Colorado, and Pennsylvania, May–Sept. Type locality: Unalaska.

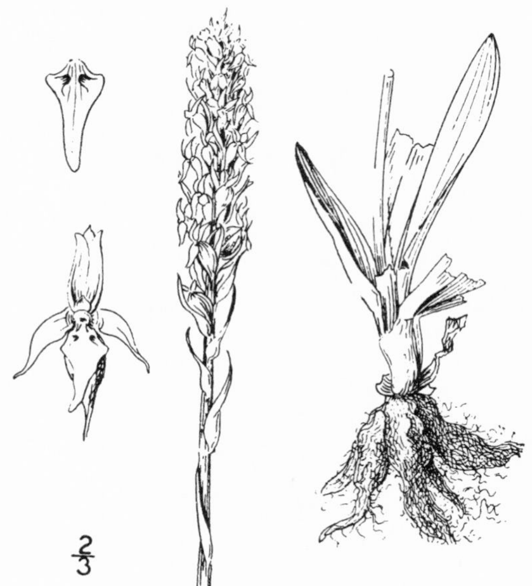

2. Spiranthes porrifòlia Lindl.

Western Ladies' Tresses. Fig. 1172.

Spiranthes porrifolia Lindl. Gen. & Sp. Orchid 467. 1840.
Gyrostachys porrifolia Kuntze, Rev. Gen. Pl. 2: 664. 1891.
Orchiastrum porrifolium Greene, Man. Bay Region 306. 1894.
Ibidium porrifolium Rydb. Bull. Torrey Club 32: 610. 1905.

Stem slender, 2–4 dm. high, glabrous. Leaves narrowly oblanceolate or linear; 1–2 dm. long; spike 5–10 cm. long, rather densely flowered; flowers greenish white; lip lanceolate, scarcely dilated at the apex; with 2 nipple-shaped callosities at the base.

Bogs and marshes, Upper Sonoran and Transition Zones; Washington and Idaho to southern California and Colorado. June–Aug. Type locality: northwest America.

8. ÓPHRYS [Tourn.] L. sp. Pl. 945. 1753.

Small perennial herbs with roostocks and fleshy-fibrous roots. Stem slender, bearing 1 or 2 sheathing scales towards the base, and 2 opposite leaves near the middle. Flowers in a terminal raceme, spurless. Sepals and petals similar, spreading or reflexed, free. Anther without a lid, jointed to the column. Pollinia 2, powdery, united to a minute gland. Capsule ovoid or obovoid. [Greek, the eyebrow.]

About a dozen species native of the north temperate and arctic regions. Besides the following two others occur in eastern North America. Type species, *Orphys ovata* L.

Lip broadly wedge-shaped, retuse or 2-lobed at apex.
Lip 9 mm. long; ovary glandular. 1. *O. convallarioides.*
Lip 5 mm. long; ovary glabrous. 2. *O. caurina.*
Lip oblong or linear, 2-cleft to near the middle. 3. *O. cordata.*

1. Ophrys convallarioìdes (Sw.) W. F. Wight.

Broad-lipped Twayblade. Fig. 1173.

Epipactis convallarioides Sw. Kongl. Vet. Acad. Handl. II. **21**: 232. 1800.

Listera convallarioides Torr. Comp. 320. 1826.

Ophrys convallarioides W. F. Wight, Bull. Torrey Club **32**: 380. 1905.

Stem slender, 10–25 cm. high, glandular-pubescent above the leaves. Leaves round oval or ovate, 3–5 cm. long, obtuse or cuspidate at apex, slightly cordate at base; glabrous, 3–9-nerved; raceme 3–7 cm. long, densely 3–15-flowered; flowers greenish yellow, on slender bracted pedicels 6–8 mm. long; sepals and petals linear lanceolate, much shorter than the lip; lip broadly wedge-shaped about 9 mm. long, with 2 obtuse lobes at the dilated apex, generally with a tooth on each side at base; column elongated, 5 mm. long, with 2 short rings above the middle; ovary glandular, pubescent; capsule about 6 mm. long.

Moist shady places, Canadian Zone; Alaska to Newfoundland, south to southern California, Michigan and Vermont. June–Aug. Type locality: "E. Terra Nova Amer. Sept."

2. Ophrys caurìna (Piper) Rydb.

Northwestern Twayblade. Fig. 1174.

Listera caurina Piper, Erythea **6**: 32. 1898.

Listera retusa Suksdorf, Deutsch. Bot. Monatss. **18**: 155. 1900.

Ophrys caurina Rydb. Bull. Torrey Club **32**: 610. 1905.

Stem slender, 10–30 cm. high, glandular-pubescent above the leaves. Leaves sessile by a clasping base, ovate, obtuse or acutish, 3–5 cm. long, glabrous; racemes 5–40-flowered; pedicels 5–7 mm. long; sepals and petals lanceolate spreading, not reflexed in anthesis, greenish; lip 4 mm. long, cuneate, obovate, with an inconspicuous tooth between the 2 shallow notches at apex, and a small tooth on each side near the base; column stout, 2 mm. long; ovary glabrous; capsule glabrous, ovoid, 5–6 mm. long.

Moist shady places, Canadian Zone; British Columbia to the Siskiyou Mountains, southern Oregon, east to Idaho and Montana. June–Aug. Type locality: "Cascade Mountains at about 3000 ft. altitude," Washington.

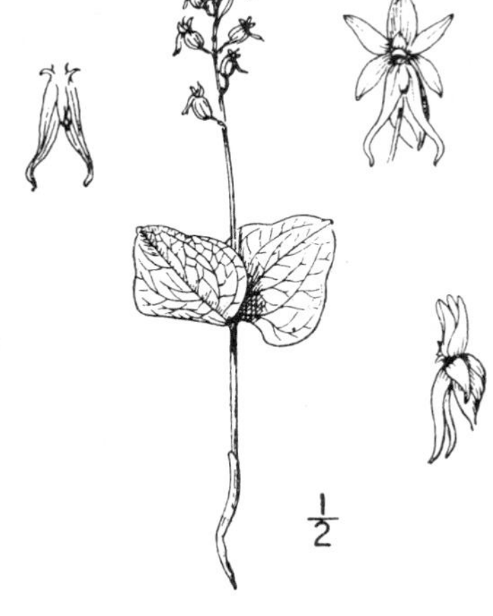

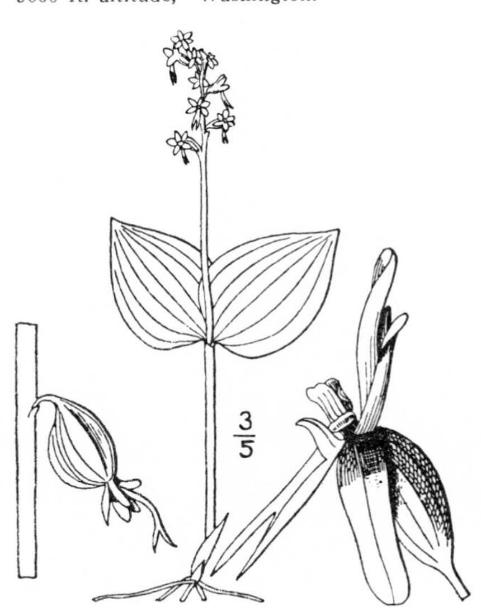

3. Ophrys cordàta L.

Heart-leaved Twayblade. Fig. 1175.

Ophrys cordata L. Sp. Pl. **2**: 946. 1753.

Listera cordata R. Br. in Ait. Hort. Kew. ed. 2. **5**: 201. 1813.

Listera nephrophylla Rydb. Mem. N. Y. Bot. Gard. **1**: 108. 1900.

Stem slender, glabrous or sparsely glandular-pubescent above the leaves. Leaves sessile, ovate, cordate, mucronate, 15–30 mm. long; racemes loosely 4–20-flowered; pedicels slender, 2 mm. long, twice the length of the bracts; sepals and petals greenish purple, oblong-linear, scarcely 2 mm. long; lip narrow, often with subulate teeth on each side at base, the segments setaceous and ciliolate; column scarcely over 1 mm. long; ovary glabrous; capsule ovoid, 4 mm. long.

Moist woods, Humid Transition and Canadian Zones; Alaska to Labrador, south to Oregon, Colorado, Michigan and New Jersey, also in Europe and Asia. June–Aug. Type locality: Europe.

9. PERÀMIUM Salisb. Trans. Hort. Soc. Lond. 1: 301. 1812.

Perennial herbs with bracted erect scapes, and thick fleshy fibrous roots. Leaves basal, tufted, often blotched with white. Flowers medium-sized, in a terminal bracted spike. Lateral sepals free, the upper united with the petals into a galea. Lip sessile, entire, round-ovate, concave or saccate, reflexed at apex, without callosities. Anther without a lid; attached to the column by a short stalk. Pollinia 2, one in each sac, attached to a small disk, granular. Column straight, rather short. [Greek, referring to the pouch-like lip.]

About 25 species, widely distributed in temperate and tropical regions. Type species, *Peramium repens* (L.) Salisb.

1. Peramium decípiens (Hook) Piper. Menzies' Rattlesnake Plantain. Fig. 1176.

Spiranthes decipiens Hook. Fl. Bor. Am. 2: 202. 1839.
Goodyera menziesii Lindl. Gen. & Sp. Orchid 492. 1840.
Peramium menziesii Morong, Mem. Torrey Club 5: 124. 1894.
Peramium decipiens Piper, Contr. Nat. Herb. 11: 208. 1906.
Epipactis decipiens Ames, Orchid. 2: 261. 1908.

Scape stout, 2–4 dm. high, glandular pubescent. Leaves ovate-lanceolate; 3–6 cm. long, with or without whitish blotches, acute at apex, narrowed at base to winged petiole; spike 5–15 cm. long, many-flowered, somewhat 1-sided; flowers greenish white; petals and sepals about 8 mm. long; galea ovate-lanceolate, concave, with an elongated usually recurved tip; lip with a long narrow recurved or spreading apex, and a swollen base; anther attached to the base of the column, ovate, acute; column prolonged above the stigma with an awl-shaped beak.

In forests, Transition Zone; British Columbia to Quebec, south to Marin County and the Sierra Nevada, California, Arizona, Michigan and New Hampshire. Type locality: Lake Huron.

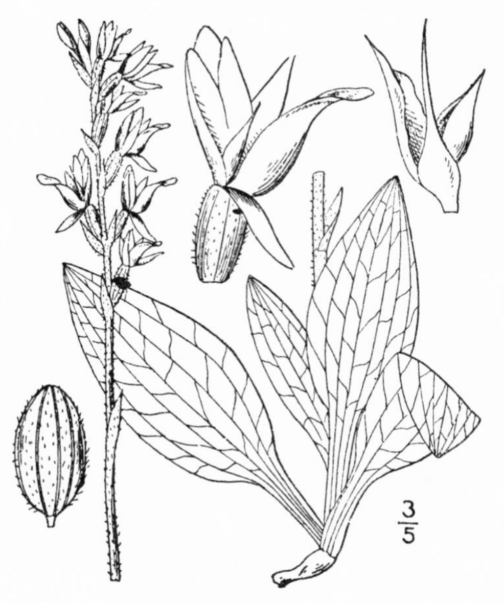

10. LIPÀRIS L. C. Richard, Mem. Mus. Paris 4: 43, 60. 1817.

Low scapose herbs, with solid bulbs, the base of the stem sheathed by several scales and 2 broad shining leaves. Flowers in terminal racemes. Sepals and petals nearly equal, linear, the petals usually very narrow. Column elongated, incurved, thickened and margined above. Pollinia 2 in each anther sac, waxy, slightly united, without stalks, threads or glands. Lip nearly flat, often bearing 2 tubercles above the base. [Greek, fat, referring to the texture of the leaves.]

About 100 species, widely distributed in temperate and tropical regions; only the following and *Liparis lilifolia* (L.) L. C. Rich. of the eastern states are known to occur in North America. Type species, *Liparis loeselii* (L.) L. C. Rich.

1. Liparis loesèlii (L.) L. C. Rich. Loesel's Twayblade. Fig. 1177.

Ophrys loeselii L. Sp. Pl. 947. 1753.
Liparis loeselii L. C. Rich. Mem. Mus. Paris 4: 60. 1817.
Leptorchis loeselii MacM. Met. Minn. 173. 1892.

Scape 5–20 cm. high, 5–7-ribbed, glabrous. Leaves elliptic-lanceolate or oblong, 5–15 cm. long, 15–45 mm. wide, keeled, obtuse; raceme few-flowered; flowers greenish, 4–6 mm. long, on stout pedicels nearly as long as the ovary; sepals narrowly lanceolate, spreading; petals linear, somewhat reflexed; lip obovate or oblong, 5 mm. long, yellowish green, the tip incurved; column half as long as its lip or less; capsule about 5 mm. long, narrowly winged.

In wet places in woods and thickets, Manitoba and Washington to Saskatchewan, south to Alabama and Missouri. Also in Europe. Known in the Pacific States only from Falcon Valley, Washington, *Suksdorf*. Type locality: Switzerland.

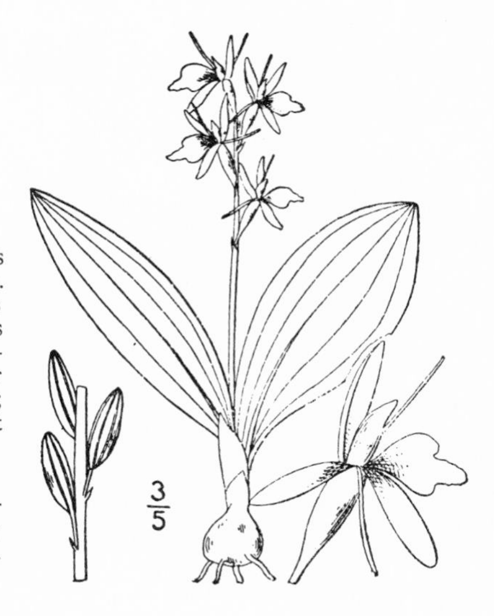

11. CALYPSO Salisb. Parid. Lond. *pl. 89.* 1807.

Low herb with a solid bulb and coralloid roots, the 1-flowered scape sheathed by 2 or 3 loose scales and a solidly petioled basal leaf, arising separately from the corm. Flowers large, showy, terminal. Sepals and petals similar. Lip large, saccate or swollen, 2-parted below. Column dilated, petal-like, bearing the lid-like anther just below the apex. Pollinia 2, waxy, each 2-parted, sessile on a thick gland. Stigma at the base. [The Greek goddess Calypso, whose name signifies concealment.]

A monotypic species in the cooler parts of the north temperate zone.

1. Calypso bulbòsa (L.) Salisb. Calypso. Fig. 1178.

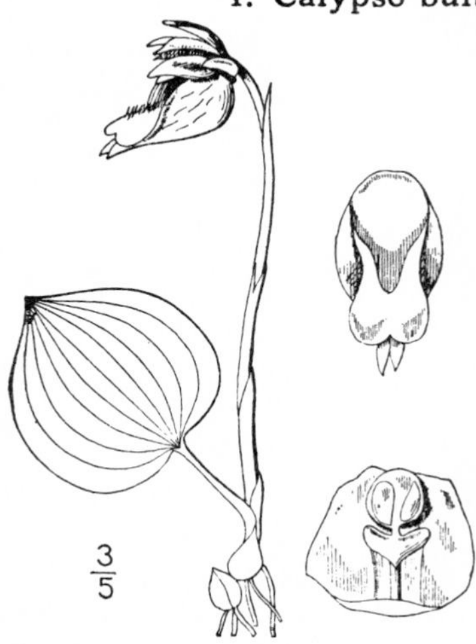

Cypripedium bulbosum L. Sp. Pl. 951. 1753.
Calypso borealis Salisb. Par. Lond. *pl. 89.* 1807.
Calypso bulbosa Oakes, Cat. Vermont Pl. 28. 1842.
Calypso bulbosa f. *occidentalis* Holz. Contr. Nat. Herb. **3**: 251. 1895.
Calypso occidentalis Heller, Bull. Torrey Club **25**: 193. 1898.
Cytherea bulbosa House, Bull. Torrey Club **32**: 382. 1905.

Scape 1–3 dm. high, smooth, with membranous sheathing bracts. Leaf appearing usually some time after anthesis, broadly oval or ovate, 25–35 mm. long, rounded or subcordate at base; petiole 25–50 mm. long; peduncle subtended by a petaloid bract; sepals and petals linear, erect or spreading, 10–14 mm. long, with 3 longitudinal lines of purple; lip large, saccate, spotted with purple, spreading or drooping with a patch of white woolly hairs; column shorter than the petals, erect; capsule about 1 cm. long, many-nerved.

In rich moist woods, especially around decayed logs, Transition and Canadian Zones; Alaska to Labrador, south to northern California, Arizona, Michigan and Maine. Also in Europe. June–July. Type locality: Europe.

12. APLÉCTRUM Nutt. Gen. 2: 197. 1818.

Scapose herbs, from a corm, produced on slender naked rootstocks or offshoots sometimes with coralloid fibers, the scape clothed with several sheathing scales. Leaf solitary, basal, petioled, developed in autumn or late summer and persisting through the winter. Flowers in terminal racemes, pedicellate and subtended by small bracts, sepals and petals similar, narrow. Lip clawed, somewhat 3-ridged, spur none. Column free, the anther borne a little below its summit. Pollinia 4, lense-shaped, oblique. [Greek, meaning without a spur.]

A monotypic North American genus.

1. Aplectrum spicàtum (Walt.) B. S. P. Adam-and-Eve Putty-root. Fig. 1179.

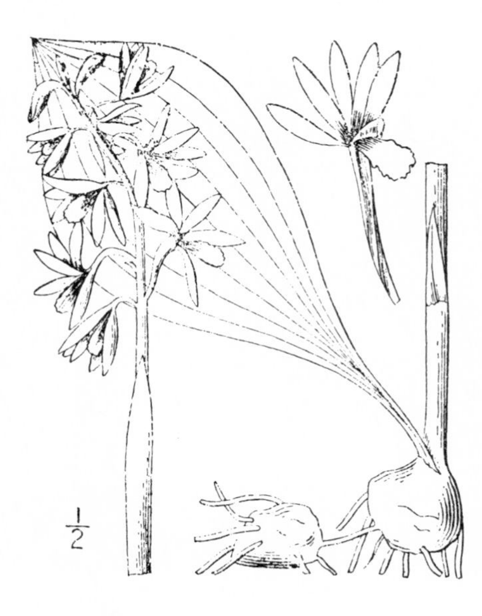

Arethusa spicata Walt. Fl. Car. 222. 1788.
Cymbidium hyemale Muhl.; Willd. Sp. Pl. **4**: 107. 1805.
Aplectrum hyemale Torr. Compend. 322. 1826.
Aplectrum spicatum B. S. P. Prel. Cat. N. Y. 51. 1888.
Aplectrum shortii Rydb. in Britton Man. 305. 1901.

Scape 3–6 dm. high, glabrous, bearing about 3 sheathing membranous bracts. Leaf arising from the corm, elliptic or ovate, 10–15 cm. long, 2–7 cm. wide; raceme 5–10 cm. long, loosely several-flowered; flowers dull yellowish brown mixed with purple, short-pedicelled; sepals and petals linear lanceolate, 12–15 mm. long; lip shorter than the petals, obtuse, somewhat 3-lobed and undulate; column slightly curved, shorter than the lip; capsule oblong-ovoid, angled, 2 cm. long.

Moist woods and swamps, Transition Zone; Saskatchewan to Ontario, south to Missouri and Georgia. Said to have been collected by Nuttall in Oregon, but not since reported from west of the Rocky Mountains. May–June. Type locality: Carolina.

13. CORALLORRHÌZA (Haller) Chatelain, Spec. Inaug. 8. 1760.

Scapose, yellowish or purplish herbs, saprophytes or root-parasites, with large masses of coralloid branching rootstocks. Leaves all reduced to sheathing scales. Flowers in terminal racemes, subtended by minute bracts. Sepals nearly equal, the lateral united at base with the foot of the column, forming a short spur or gibbous protuberance, the other one free, the spur adnate to the summit of the ovary. Petals about equalling the sepals, 1–3-nerved. Lip 1–3-ridged, column slightly incurved, somewhat 2-winged. Anther terminal, operculate. Pollinia 4, in 2 pairs, oblique, free, waxy. [Greek, from the coral-like roots.]

About 15 species, widely distributed in the north temperate regions; about 9 are North American. Type species, *Corallorrhiza trifida* Chatelain.

Spur present, adnate to the ovary.
  Sepals and petals 1-nerved, often very small. — 1. *C. corallorrhiza.*
  Sepals and petals 3-nerved; spur prominent.
    Lip 3-lobed, spotted; spur adnate its whole length. — 2. *C. maculata.*
    Lip entire, not spotted; spur free below the middle. — 3. *C. mertensiana.*
Spur none; lateral sepals oblique and gibbous at base.
  Lip deeply and narrowly concave, 6–8 mm. long. — 4. *C. bigelovii.*
  Lip shallowly and broadly concave, 10–12 mm. long. — 5. *C. striata.*

1. Corallorrhiza corallorrhìza (L.) Karst. Early Coral-root. Fig. 1180.

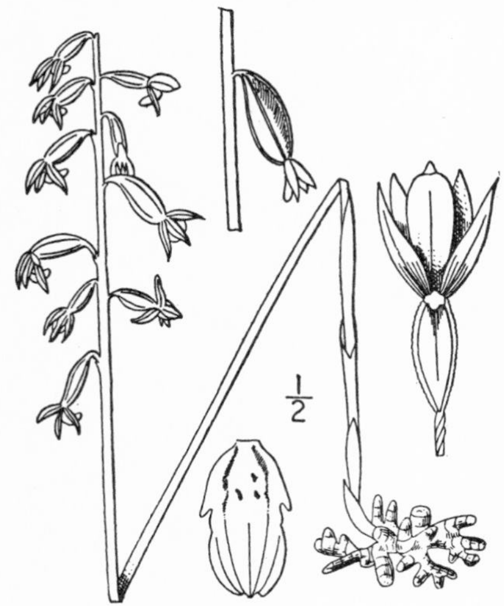

Ophrys corallorrhiza L. Sp. Pl. 945. 1753.
Corallorrhiza trifida Chatelain, Spec. Inaug. 8. 1760.
Corallorrhiza neottia Scop. Fl. Corn. ed. 2. 2: 207. 1772.
Corallorrhiza innata R. Br. in Ait. Hort. Kew. ed. 2. 5: 209. 1813.
Corallorrhiza corallorrhiza Karst. Deutsch Fl. 448. 1880–83.

Scape rather slender, 1–3 dm. high, clothed with 2–5 closely sheathing scale bracts. Raceme 3–7 cm. long, 3–12-flowered; bracts minute; flowers dull purple, very short pedicelled; sepals and petals about 5 mm. long, narrow, 1-nerved; lip shorter than the petals, oblong, whitish, 2-lobed or 2-toothed toward the base; spur a small protuberance adnate to the summit of the ovary; capsule 8–12 mm. long, oblong or somewhat obovoid.

On forest floors, Transition and Canadian Zones; Alaska to Newfoundland south to Washington, Colorado, Ohio and Georgia. May–June. Type locality: Europe.

2. Corallorrhiza maculàta Raf. Spotted Coral-root. Fig. 1181.

Corallorrhiza maculata Raf. Am. Month. Mag. 2: 119. 1817.
Corallorrhiza multiflora Nutt. Journ. Acad. Phila. 3: 138. *pl. 7.* 1823.
Corallorrhiza multiflora occidentalis Lindl. Gen. & Sp. Orchid. 534. 1840.

Scape stout, 2–5 dm. high, purplish, clothed with usually 3 sheathing appressed scales. Raceme 5–20 cm. long, 10–35-flowered; bracts minute; pedicels about 2 mm. long; flowers mainly brownish purple; sepals and petals somewhat connivent at base, linear-lanceolate, 7–10 mm. long, 3-nerved; lip white, spotted and lined with crimson, oval to ovate in outline, deeply 3-lobed, the middle lobe broader than the lateral, crenulate, bearing 2 narrow lamellae; spur prominent, yellowish; capsule ovoid or oblong, 10–15 mm. long, drooping.

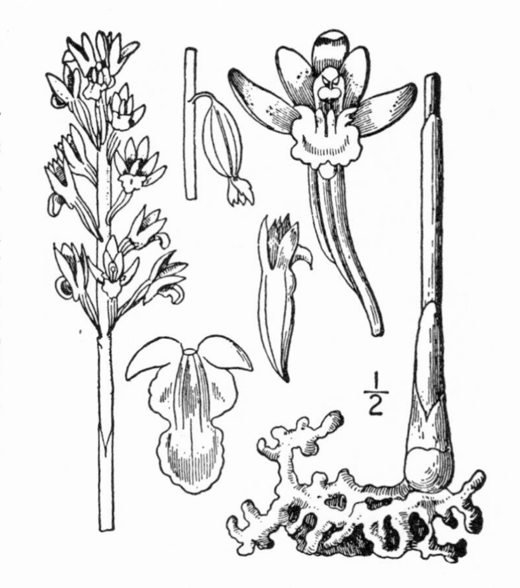

In woods, Transition and Canadian Zones; British Columbia to Nova Scotia, south to southern California, New Mexico and Florida. June–Sept. Type locality: Philadelphia.

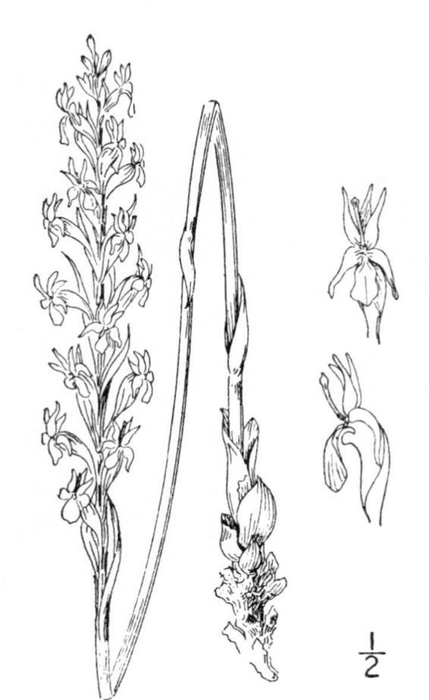

3. **Corallorrhiza mertensiàna** Bong.
Merten's Coral-root. Fig. 1182.

Corallorrhiza mertensiana Bong. Mem. Acad. St. Petersb. VI. **2**: 165. 1832.

Scape stout, 2–5 dm. high, brownish purple, clothed with usually 3 scarious sheathing bracts. Raceme 10–40-flowered; pedicels slender, 2–3 mm. long; sepals and petals narrowly linear-lanceolate, 6–8 mm. long, dull purple, 3-nerved; sepals reflexed; lip broadly oblong, narrowed at base, entire, thin and concave, reddish purple, not spotted; spur prominent, the lower half free from the ovary; column slender, nearly equalling the petals; capsule 15–18 mm. long, narrowed at base to the slender pedicel; lip reddish purple, not spotted.

Coniferous forests, Canadian Zone; Alaska south to western Montana and Humboldt County, California. Type locality: Sitka, Alaska.

4. **Corallorrhiza bigelòvii** S. Wats.
Biglow's Coral-root. Fig. 1183.

Corallorrhiza bigelovii S. Wats. Proc. Am. Acad. **12**: 275 1877.

Scape stout, 2–4 dm. high, brownish purple, clothed with 3 or 4 scarious sheathing scales. Raceme 5–25 cm. long, 10–45-flowered; pedicels very short; sepals and petals yellowish purple, 3-nerved, striped with dark purple lines, or the petals with 2 additional fainter stripes, 6–10 mm. long, oblong to elliptic, the petals a little broader than the sepals; lip reflexed, elliptic, 5–6 mm. long, deeply concave, auriculate at base; spur none, but sepals and petals gibbous at base; column about half the length of the petals, rather slender, broadly winged below; capsule oblong-ovate, 12–18 mm. long.

Usually in dark woods, Transition Zone; Coast Ranges and Sierra Nevada, northern and central California, extending south to the Santa Cruz Mountains, and the southern Sierra Nevada. Doubtfully distinct from the next. April–July. Type locality: Marin County, California.

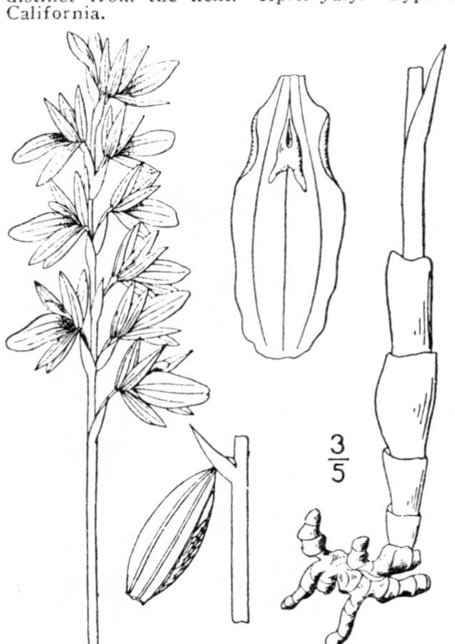

5. **Corallorrhiza striàta** Lindl.
Striped Coral-root. Fig. 1184.

Corallorrhiza striata Lindl. Gen. & Sp. Orchid. 534. 1840.
Corallorrhiza macraei A. Gray, Man. ed. 2. 453. 1856.

Scape stout, purplish, 2–5 dm. high. Raceme 5–15 cm. long, 10–25-flowered; flowers dark purple; sepals and petals narrowly elliptic, 3-striped with deeper purple lines; lip oval or obovate, striate-veined, entire or a little undulate, slightly auriculate at base, about equalling the petals; spur none, but sepals and petals gibbous at base; column slender, about half the length of the petals.

On the forest floor, Transition Zone; British Columbia to Ontario, south to northern California, Wyoming, Michigan and northern New York. Type locality: western North America.

Sub-class 2. *DICOTYLÉDONES.*

Stem exogenous, of pith, wood and bark, the wood in one or more layers surrounding the pith, traversed by medullary rays, and covered by the bark (endogenous in structure in *Nymphaeaceae*). Leaves usually pinnately or palmately veined, the veinlets forming a network. Parts of the flower usually in 4's or 5's, rarely in 3's or 6's. Embryo of the seed with two cotyledones (one only in *Cyclamen, Nymphaeaceae*, and some species of *Ranunculaceae*; in *Quercus* and a few other genera 3 sometimes occur, and in some species of *Amsinckia* 4), the first leaves of the germinating plantlet opposite.

Dicotyledonous plants are first definitely known in Cretaceous time. They constitute between two-thirds and three-fourths of the living angiosperms.

Series 1. *Choripétalae.*

Petals separate and distinct from each other or wanting (somewhat united in certain species of *Portulacaceae, Caryophyllaceae, Fumariaceae, Crassulaceae, Malvaceae,* and *Fabaceae*).

Family 24. SAURURÀCEAE.

LIZARD'S-TAIL FAMILY.

Perennial herbs with broad entire alternate petioled leaves and small perfect bracteolate flowers in peduncled spikes. Perianth none. Stamens 6–8, or sometimes fewer, hypogynous; anthers 2-celled, the sacs longitudinally dehiscent. Ovary 3–4-carpelled, the carpels distinct or united, 1–2-ovuled; ovules orthotropus. Fruit capsular or berry-like, composed of 3–4, mostly indehiscent carpels. Seeds globose or ovoid; endosperm copious, mealy; embryo minute, cordate, borne in a small sac near the end of the endosperm.

Three genera and four species, natives of North America and Asia. Represented in North America by the following and the Lizard's-tail (*Saururus cernuus* L.) of the eastern states.

1. ANEMÓPSIS Hook. Ann. Nat. Hist. 1: 136. 1838.

Stems nodose, scape-like, stoloniferous from aromatic creeping rootstocks. Leaves mostly radical, minutely punctate. Flowers in a compact spike surrounded at the base by a persistent colored involucre of 5–8 bracts; each flower except the lowest also surrounded by a small colored bract. Stamens 6–8. Ovary sunk in the rachis of the spike, 1-celled; stigmas 3–4. Capsule dehiscent at the apex.

A monotypic genus native of southwestern United States and northern Mexico.

1. Anemopsis califórnica Hook.
Yerba Mansa. Fig. 1185.

Anemopsis californica Hook. Ann. Nat. Hist. 1: 136. 1838.
Anemopsis bolanderi A. DC. Linnaea 37: 333. 1871.
Houttuynia californica Benth. & Hook.; S. Wats. Bot. Calif. 2: 483. 1880.

Stem 15–50 cm. long, with a broadly ovate clasping leaf above the middle and a fascicle of 1–3 small petioled leaves in the axil. Basal leaves elliptic-oblong, rounded above, more or less narrowed toward the cordate base, 5–15 cm. long; petioles 10–20 cm. long; spikes 1.5–4 cm. long; involucral bracts white, often reddish beneath, oblong, 1–3 cm. long; floral bracts white, obovate, unguiculate, 5–6 mm. long; ovules 6–10 on each placenta.

Low, usually alkaline ground, Upper and Lower Sonoran Zones; Sacramento Valley and Santa Clara County, south to northern Lower California, and Arizona. Mar.–Aug. Type locality: Santa Barbara and San Diego, California.

Family 25. **SALICÀCEAE.**

WILLOW FAMILY.

Trees or shrubs with simple deciduous alternate stipulate leaves. Flowers dioecious, in terminal aments, appearing with or before the leaves, maturing in a few weeks and the entire ament falling off as a single flower. Each flower subtended by a scale-like bract, without a perianth and with or without a cup-shaped glandular disk. Stamens 2 to many, rarely 1. Ovary 1-celled, ovoid or globose; stigmas 2–4. Fruit a 2–4-valved capsule, with numerous minute comose seeds.

Two genera, widely distributed, but most abundant in the north temperate and subarctic regions.

Bracts fimbriate, caducous; flowers on a cup-shaped disk; stamens numerous; buds with several imbricated scales. 1. *Populus.*
Bracts entire or merely toothed, persistent or tardily dehiscent; disk reduced to a gland; stamens 2–3, rarely 1; buds with a single scale. 2. *Salix.*

1. **PÓPULUS** L. Sp. Pl. 1034. 1753.

Trees with soft usually whitish wood, resinous buds, terete or angled branchlets, broad or narrow usually petioled leaves and minute fugacious stipules. Bracts of the flowers fimbriate or incised, narrowed at the base. Disk cup-shaped, oblique, lobed or entire. Staminate aments pendulous; stamens 12–60, or rarely reduced to 4, their filaments distinct. Pistillate aments pendulous, erect or spreading; ovary sessile; style short; stigmas 2–4, entire or 4-lobed. Capsule 2–4-valved. Seeds with long copious coma. [Ancient Latin name.]

About 30 species widely distributed over the northern hemisphere. Type species, *Populus alba* L.

Capsule oblong-conical, thin-walled; stigmas 2, 2-lobed, their lobes slender; buds slightly resinous. 1. *P. tremuloides.*
Capsule globose or oblong-globose, thick-walled; stigmas 2–4-lobed, their lobes dilated; buds conspicuously resinous.
  Petioles terete; leaves dark green on the upper surface, glaucous beneath.
    Ovary glabrous; petiole glabrous. 2. *P. balsamifera.*
    Ovary and capsule tomentose; petioles usually pubescent. 3. *P. trichocarpa.*
  Petioles strongly flattened laterally; leaves broadly deltoid, bright yellow-green on both surfaces. 4. *P. fremontii.*

1. **Populus tremuloìdes** Michx.

American Aspen. Fig. 1186.

Populus tremuloides Michx. Fl. Bor. Am. **2**: 243. 1803.
Populus vancouveriana Trel.; Tidestr. in Piper & Beattie, Fl. N. W. Coast 118. 1915.
Populus tremuloides vancouveriana Sarg. Bot. Gaz. **57**: 208. 1919.

Tree with a maximum height of 30–35 m., and a trunk 4–6 dm. in diameter, branches slender, often drooping. Bark smooth and greenish white, becoming fissured at the base of old trunks; leaves 3–5 cm. long, broadly ovate to nearly orbicular, crenate-serrate with small incurved glandular teeth except at the shallowly cordate to cuneate base, dark green above, pale yellow green beneath; petioles slender, laterally flattened, about equalling the leaves; aments 3–6 cm. long; bracts 3–5-lobed, fringed with long hairs; stamens 6–12; stigma lobes linear; capsule conical, 3 mm. long, glabrous.

Borders of streams, lakes and meadows, throughout the Canadian Zone of North America; Labrador to the Yukon, south to Pennsylvania, Missouri, Arizona and northern Lower California. Type locality: Canada and New England. The western plants have been segregated out as a species or variety by some, but the characters used to distinguish the two seem trivial and inconstant.

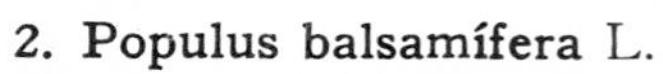

2. **Populus balsamífera** L.

Balsam Poplar. Fig. 1187.

Populus balsamifera L. Sp. Pl. 1034. 1753.

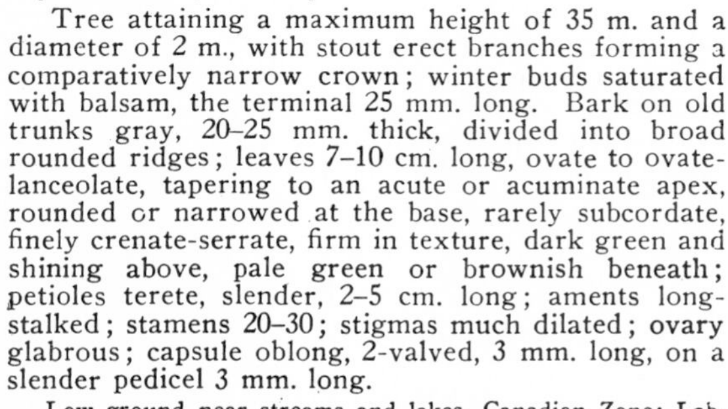

Tree attaining a maximum height of 35 m. and a diameter of 2 m., with stout erect branches forming a comparatively narrow crown; winter buds saturated with balsam, the terminal 25 mm. long. Bark on old trunks gray, 20–25 mm. thick, divided into broad rounded ridges; leaves 7–10 cm. long, ovate to ovate-lanceolate, tapering to an acute or acuminate apex, rounded or narrowed at the base, rarely subcordate, finely crenate-serrate, firm in texture, dark green and shining above, pale green or brownish beneath; petioles terete, slender, 2–5 cm. long; aments long-stalked; stamens 20–30; stigmas much dilated; ovary glabrous; capsule oblong, 2-valved, 3 mm. long, on a slender pedicel 3 mm. long.

Low ground near streams and lakes, Canadian Zone; Labrador to the Alaskan coast, south to New York, Michigan, South Dakota, Wyoming, Idaho and eastern Oregon. Type locality: eastern North America.

3. Populus trichocàrpa Torr. & Gray.
Black Cottonwood. Fig. 1188.

Populus trichocarpa Torr. & Gray; Hook. Icon. Pl. **9**: *pl. 878.* 1852.

The largest tree in the genus, attaining a maximum height of 60–70 m. and a diameter of nearly 3 m., branches long, forming a broad open crown, winter buds very resinous, long-pointed, the terminal 2 cm. long. Bark ashy gray, deeply fissured on old trunks, 3–6 cm. thick; leaves 3–5 cm. long, ovate, or on young trees or offshoots often oblong-lanceolate, narrowed to the acute apex, rounded, cordate or sometimes cuneate at the base, finely crenate-serrate, more or less pubescent when young, becoming glabrous, dark green and shining above, pale and usually glaucous beneath; petioles 2–6 cm. long, terete; stamens 40–60; ovary ovoid, pubescent; capsule nearly sessile, subglobose, about 4 mm. thick, pubescent.

Stream banks, Upper Sonoran, Transition and Canadian Zones; southern Alaska south on the Pacific Slope to the San Bernardino Mountains, California, east to northern Idaho. Type locality: Santa Clara River, Ventura County, California.

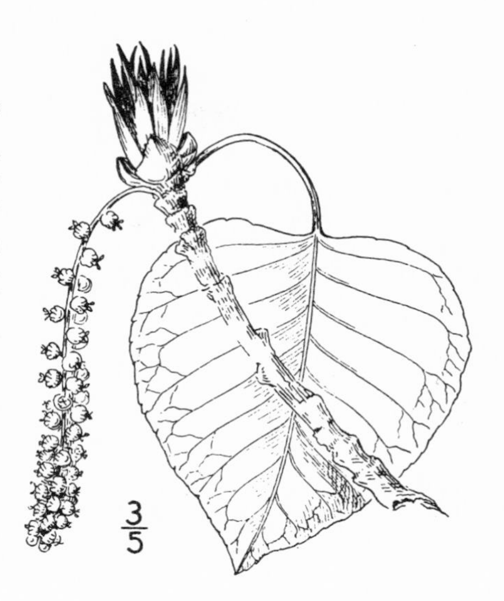

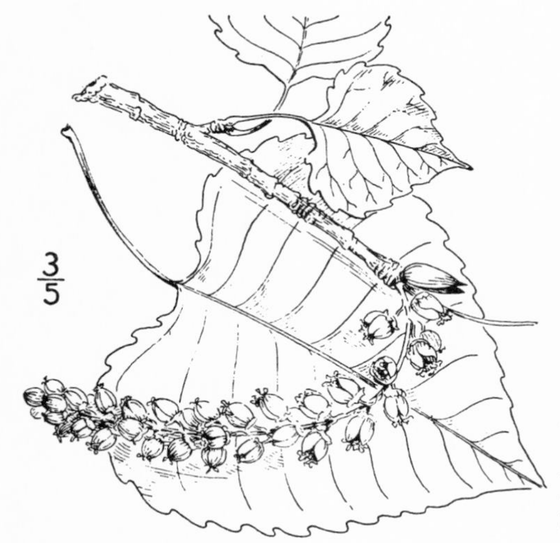

4. Populus fremóntii S. Wats.
Fremont's Cottonwood. Fig. 1189.

Populus fremontii S. Wats. Proc. Am. Acad. **10**: 350. 1875.

Tree attaining 35 m. in height with a trunk 1–2 m. in diameter, branches large and spreading, forming a broad open crown. Winter buds 4 mm. long, ovate, sharp-pointed, with light green scales; bark 3–5 cm. thick, light gray, becoming furrowed toward the base of old trunks; leaves 4–7 cm. long, deltoid, abruptly sharp-pointed at the apex, truncate or cordate at the base, coarsely serrate-dentate, glabrous or somewhat pubescent, bright green on both surfaces; petioles flattened laterally, yellowish, 4–8 cm. long; stamens 60 or more; anthers dark red; ovary glabrous; capsule 8–12 mm. long, 3-valved; pedicels about 3 mm. long.

Stream banks and bottom lands, mainly Lower Sonoran Zone; Upper Sacramento Valley to northern Lower California, southern Nevada and Arizona. Type locality: Deer Creek, Tehama County, California. A pubescent form in southern California has been named var. *pubescens* Sarg.

Populus macdougálii Rose, Smithson. Misc. Coll. **61**[12]: 1. *pl. 1.* 1913. This is the tree on the Colorado River delta. It differs from *fremontii* in its pubescent twigs and leaves, and smaller (3 mm. wide) fruiting disk, but these are not constant characters, and it is doubtful if the species is distinct.

2. SÀLIX* L. Sp. Pl. 1051. 1753.

Creeping shrubs to tall trees, with single bud scales lined with an adherent membrane, usually narrow and short-petioled leaves, and persistent to deciduous large to minute stipules (or these wanting). Aments sessile to pedunculate, erect or spreading, appearing before (precocious), with (coetaneous), or after (serotinous) the leaves; bracts of the aments entire or rarely shallowly dentate at apex; each flower accompanied by 1 or sometimes 2 small glands. Stamens 1–10, usually 2 or 5, the filaments distinct or sometimes more or less united. Ovary and capsule 2-valved, sessile or short-pedicelled, glabrous or hairy; style wanting to elongate, entire or bifid; stigmas short to long, entire to divided.

About 300(?) species, mostly of the north temperate and arctic zones, rare in the tropics and southward. Useful for wood, posts, poles, furniture, whistles, holding stream banks, browse for stock, tanning, medicine, etc. Type species, *Salix pentandra* L.

---

*Text contributed by Dr. Carleton R. Ball, U. S. Department of Agriculture.

Scales yellow, deciduous; leaves elongated, usually 4–5 or more times longer than wide; styles very short.
Leaves linear-lanceolate to lanceolate, closely serrulate; stamens 3–8 or 10.
Petioles glandular above.
Leaves glaucous beneath. 1. *S. lasiandra.*
Leaves green beneath. 2. *S. caudata.*
Petioles not glandular.
Leaves green beneath. 3. *S. gooddingii.*
Leaves glaucous beneath.
Young twigs yellowish, drooping; petioles slender. 4. *S. amygdaloides.*
Young twigs brown, spreading; petioles stout. 5. *S. laevigata.*
Leaves linear, or linear-lanceolate (except No. 7), entire or remotely denticulate; stamens 2.
Stigma lobes long, linear or oblong.
Leaves and capsules densely pilose.
Blades linear-lanceolate. 6. *S. hindsiana.*
Blades lanceolate to broadly elliptical. 7. *S. sessilifolia.*
Leaves and capsules thinly pilose to glabrate.
Blades linear-lanceolate or oblanceolate. 8. *S. fluviatilis.*
Blades linear. 9. *S. parishiana.*
Stigma lobes short.
Leaves and capsules white pilose. 10. *S. argophylla.*
Leaves pubescent to glabrate; capsules glabrous.
Blades linear, mostly entire, pubescent. 11. *S. exigua.*
Blades lanceolate, denticulate, glabrate. 12. *S. melanopsis.*
Scales brownish to nearly black (yellow in No. 42); more or less persistent; leaf-length rarely more than 3–4 times the width; styles 0.5–2 mm. long.
Capsule glabrous (pubescent in Nos. 19a, 23b, and 27a, and sometimes in 22).
Leaf blades glabrous (usually hairy beneath in No. 19, and thinly so above in No. 22).
Styles very short (0.2–0.3 mm. long).
Blades 2–4 cm. long, entire, finely reticulate. 13. *S. pedicellaris.*
Styles 0.4–0.7 mm. long.
Blades glaucous beneath.
Blades lanceolate to ovate-lanceolate, mostly serrulate.
Pedicels 1–2 mm. long (2.5 in var. *platyphylla*). 14. *S. lutea.*
Pedicels 2.5–4 mm. long. 15. *S. mackenziana.*
Blades oblanceolate, mostly entire.
Blades 3–5 cm. long; scales glabrate. 16. *S. farrae.*
Blades 6–10 cm. long; scales densely woolly. 19. *S. lasiolepis.*
Blades green beneath, serrulate.
Blades ovate-lanceolate to obovate, thin. 17. *S. monochroma.*
Blades lanceolate, thick, firm. 18. *S. pseudomyrsinites.*
Styles 1–1.5 mm. long; blades broad, glaucous beneath.
Aments subsessile.
Blades elliptic-ovate to ovate. 21. *S. pseudomonticola.*
Blades elliptic to oblanceolate. 26. *S. piperi.*
Aments on leafy peduncles 1–3 cm. long. 22. *S. barclayi.*
Leaf blades hairy on at least one side.
Aments sessile or subsessile.
Blades oblanceolate to narrowly obovate; styles 0.5 mm. long. 19. *S. lasiolepis.*
Blades broadly oval; styles 1 mm. long. 27. *S. hookeriana.*
Aments on leafy peduncles; styles 1–1.5 mm. long.
Blades thinly tomentose above. 22. *S. barclayi.*
Blades tomentose on both sides. 23. *S. commutata.*
Capsules hairy (see also 19a, 22, 23b, and 27a).
Styles distinct, 0.5–1 or 1.5 mm. long.
Leaves more or less glabrate at maturity.
Prostrate creeping shrubs, blades dull.
Blades glabrous, green beneath. 28. *S. cascadensis.*
Blades hairy on margins, glaucescent beneath. 29. *S. petrophila.*
Erect or ascending shrubs; blades bright green.
Blades broadly elliptic-ovate, 1–3.5 cm. long. 30. *S. monica.*
Blades oblanceolate to obovate, 3–6 cm. long. 31. *S. pennata.*
Blades lanceolate to narrowly elliptical, 4–9 cm. long. 32. *S. lemmoni.*
Leaves densely hairy (at least beneath) at maturity.
Branchlets always pruinose; blades silvery pubescent beneath.
Pubescence short; scales thinly pilose. 33. *S. subcoerulea.*
Tomentum long; scales densely lanate. 34. *S. bella.*
Branchlets never pruinose.
Styles 0.7–1.5 mm. long; stamens 2.
Leaves silky villous on both sides. 20. *S. wolfii idahoensis.*
Leaves tomentose on both sides.
Aments leafy-pedunculate. 24. *S. eastwoodiae.*
Aments subsessile. 25. *S. orestera.*
Leaves densely hairy beneath.
Floral gland short (less than ½ mm.).
Blades oval or elliptic. 35. *S. drummondiana.*
Blades narrowly oblanceolate. 36. *S. jepsoni.*
Floral gland long.
Blades linear lanceolate. 37. *S. breweri.*
Blades oblong-obovate. 38. *S. delnortensis.*
Styles 0.5–0.7 mm. long; stamen 1.
Blades cuneate-obovate, short-pubescent. 39. *S. sitchensis.*
Blades elliptic to oblong-ovate, densely opaque-tomentose. 40. *S. coulteri.*
Styles obsolete or very short (0–0.3 mm. long).
Erect shrubs 1 m. high or more, subalpine or lower; leaves lanceolate to obovate; capsules 5–8 mm. long.
Aments dense; scales black; stigmas long.
Blades oblanceolate to obovate. 41. *S. scouleriana.*
Aments lax; scales yellowish; stigmas short.
Blades broad; pedicels 3–5 mm. long. 42. *S. bebbiana.*
Blades narrow; pedicels 2 mm. long. 43. *S. geyeriana.*
Low or prostrate alpine shrubs, 0.5–5 dm. high; leaves oblong-oval to orbicular; capsules 3–5 mm. long.
Blades glabrous.
Blades 0.5–1 cm. long. 44. *S. nivalis.*
Blades 1.5–3 cm. long. 45. *S. saximontana.*
Blades long-villous, especially beneath.
Blades 3–5 cm. long. 46. *S. vestita.*

1. Salix lasiándra Bentham. Red Willow. Fig. 1190.

S. lasiandra Benth. Pl. Hartweg. 335. 1857.

S. speciosa Nutt. N. Am. Sylva 1: 58. *pl. 17.* 1843, not Host, 1828, or Hook. and Arn. 1832.

S. arguta lasiandra Anderss. Svensk. Vetensk. Akad. Handl. (Monog. Sal.) 6: 33. 1867.

S. lasiandra lyallii Sarg. Gard. & For. 8: 463. 1895.

S. lyallii (Sarg.) Heller, Bull. Torrey Club 25: 580. 1898.

A tree, 5–15 m. high; bark rough; twigs deep red, lustrous; leaves lanceolate to broadly lanceolate or sometimes oblanceolate, acuminate at the apex, acute to rounded at the base, closely gladular crenate-serrulate, dark green and shining above, glaucous beneath, 6–10 cm. long, 1.5–3.5 cm. wide, up to 5x20 or 30 cm. on sprouts; stipules small, acute, glandular; petioles stout, glandular above near base of blade; aments pedunculate; staminate aments 2–6 cm. long, 1–1.3 cm. wide; pistillate aments 3–10 cm. long, 1.2–2 cm. wide; scales lanceolate to ovate, usually dentate, sometimes glandular at the apex; capsule pale straw-color or light brown, lanceolate, 5–7 mm. long; pedicel 1.5–2 mm. long; style short.

Stream banks, Upper Sonoran and Transition Zones; British Columbia and Alberta to southern California and central New Mexico. Type locality: "ad flumen Sacramento," California.

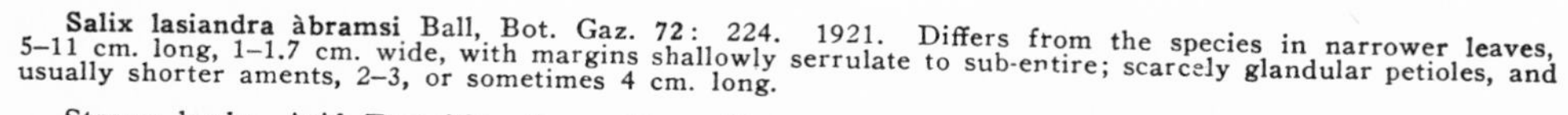

Salix lasiandra àbramsi Ball, Bot. Gaz. 72: 224. 1921. Differs from the species in narrower leaves, 5–11 cm. long, 1–1.7 cm. wide, with margins shallowly serrulate to sub-entire; scarcely glandular petioles, and usually shorter aments, 2–3, or sometimes 4 cm. long.

Stream banks, Arid Transition Zone; Sierra Nevada Mountains from Plumas County south to Fresno County, California. Type locality: "near Sentinel Hotel, Yosemite Valley, Yosemite National Park, altitude 4000–4500 feet."

Salix lasiandra lancifòlia (Anderss.) Bebb. Willows Calif. (repr. S. Wats. Bot. Calif. 2: 84.) 1879. (*Salix lancifolia* Anderss. Svensk. Vetensk. Akad. Handl. (Monog. Sal.) 6: 34. *pl. 2, f. 23.* 1867.) Young branchlets more or less densely pubescent-pilose. With the species, especially in the northern part of its range. Type locality: "America septentrionali-occidentali in insula Vancouver."

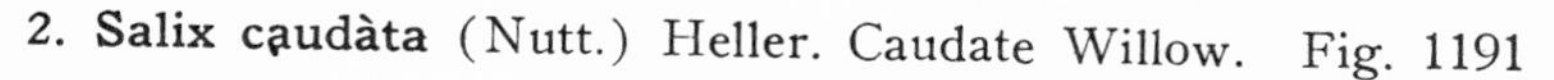

2. Salix caudàta (Nutt.) Heller. Caudate Willow. Fig. 1191

Salix pentandra caudata Nutt. Sylva. 1: 61. *pl. 18.* 1842.

Salix lasiandra caudata (Nutt.) Sudw. Bull. Torrey Club 20: 43. 1893.

Salix caudata (Nutt.) Heller Muhlenb. 2: 186. 1906.

Salix fendleriana of various authors, not Anderss.

Salix lasiandra fendleriana (Anderss.) Bebb. Willows Calif. (repr. S. Wats. Bot. Calif. 2: 84.) 1879.

A cespitose shrub, 2–5 m. high; twigs long, reddish to chestnut, shining; leaves narrowly lanceolate to lanceolate, long-acuminate, dark green on both sides or somewhat paler but never glaucous beneath, 6–13 cm. long, 1.2–3 cm. wide, closely glandular-serrulate; stipules usually none, if present, small, semicordate to reniform; staminate aments 2–4 cm. long, 1–1.2 cm. wide; pistillate 2–5.5 cm. long, 1.5 cm. wide; scales lanceolate to oblanceolate, glabrate outside; capsule pale straw-color or brownish, 5–7 mm. long; pedicels 1–1.5 mm. long; styles 0.5–0.7 mm. long.

Mountain streams and meadows, to 9,000 feet elevation, Arid Transition and Canadian Zones; from British Columbia and Washington, east of the Cascade Mountains, east to Alberta and the Black Hills of South Dakota, and south to east central California, southern Utah, and northern New Mexico. Type locality: "by streams in the valleys of the Rocky Mountains, towards their western slope, in Oregon, and also the Blue Mountains of the same territory."

Salix caudata parvifòlia Ball, Bot. Gaz. 72: 225. *f. 1.* 1921. Leaves smaller, 5–8 cm. long, 7–12 mm. wide. With the species from Wasco County, Oregon, to the Yellowstone Park, and Alberta in the Rocky Mountains at Banff. Type locality: "along Swiftcurrent Creek, below Lake McDermott," Glacier National Park, Montana.

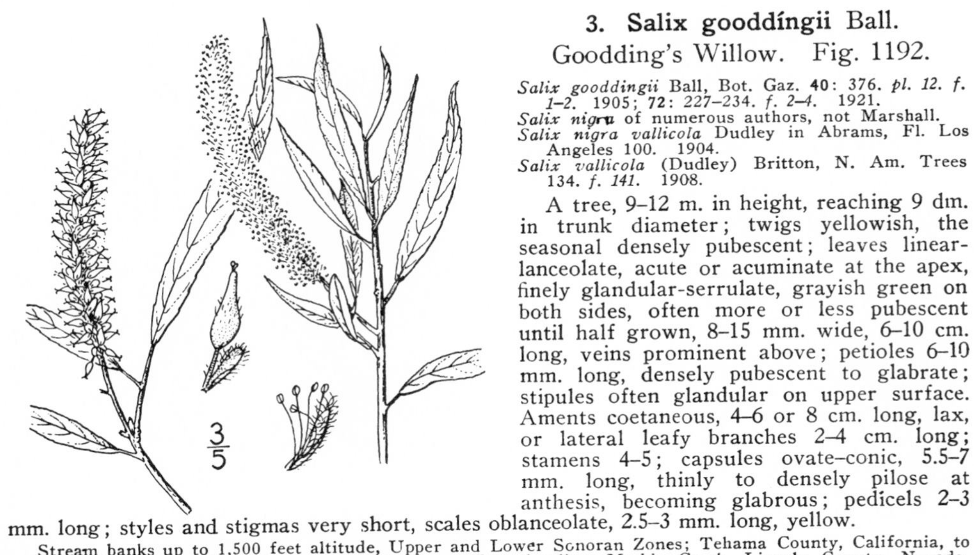

3. Salix gooddíngii Ball.
Goodding's Willow. Fig. 1192.

Salix gooddingii Ball, Bot. Gaz. **40**: 376. *pl. 12. f. 1–2.* 1905; **72**: *227–234. f. 2–4.* 1921.
Salix nigra of numerous authors, not Marshall.
Salix nigra vallicola Dudley in Abrams, Fl. Los Angeles 100. 1904.
Salix vallicola (Dudley) Britton, N. Am. Trees 134. *f. 141.* 1908.

A tree, 9–12 m. in height, reaching 9 dm. in trunk diameter; twigs yellowish, the seasonal densely pubescent; leaves linear-lanceolate, acute or acuminate at the apex, finely glandular-serrulate, grayish green on both sides, often more or less pubescent until half grown, 8–15 mm. wide, 6–10 cm. long, veins prominent above; petioles 6–10 mm. long, densely pubescent to glabrate; stipules often glandular on upper surface. Aments coetaneous, 4–6 or 8 cm. long, lax, or lateral leafy branches 2–4 cm. long; stamens 4–5; capsules ovate-conic, 5.5–7 mm. long, thinly to densely pilose at anthesis, becoming glabrous; pedicels 2–3 mm. long; styles and stigmas very short, scales oblanceolate, 2.5–3 mm. long, yellow.

Stream banks up to 1,500 feet altitude, Upper and Lower Sonoran Zones; Tehama County, California, to Lower California, southern Nevada and Arizona. Type locality: Muddy Creek, Lincoln County, Nevada.

4. Salix amygdaloìdes Anderss.
Peach-Leaved Willow. Fig. 1193.

S. amygadaloides Anderss. in Ofv. Svensk. Vetensk. Akad. Forh. **15**: 114. 1858.

A tree, 3–12 m. high, yellowish-green in mass-color, youngest twigs yellow, slender, somewhat drooping; leaves lanceolate to ovate-lanceolate, acuminate, closely serrulate, light green above, paler and usually glaucous beneath, 5–12 cm. long, 1.5–3 cm. wide; petioles slender, 5–15 mm. long, glabrous; aments appearing with the leaves on leafy branches, the staminate slender, yellow, 3–5 cm. long, the pistillate lax, 4–8 cm. long in fruit; scales yellow, lanceolate or broader, glabrous outside, crisp villous inside; stamens 5–7; capsules lanceolate, glabrous, 4–5 mm. long; pedicels filiform, 2 mm. long; style short, stigmas distinct.

Transition and Canadian Zones; southeastern British Columbia and Washington and Oregon east of the Cascade Mountains, south to west central Nevada and southern Utah and east to Quebec. Type locality: Missouri River, near Fort Pierre, South Dakota.

5. Salix laevigàta Bebb.
Red or Polished Willow. Fig. 1194.

Salix laevigata Bebb. Am. Nat. **8**: 202. 1874.

A tree, 5–13 m. in height; branchlets yellowish to reddish-brown. Leaves lanceolate (often oblanceolate to obovate while expanding, with long yellowish hairs beneath, especially at the abruptly-cuspidate apex), 7–15 cm. long, 1.5–3 cm. wide, acutish to rounded or subcordate at base, acute at the apex, subentire to shallowly serrulate, coriaceous, glabrous, shining above, glaucous beneath; petioles 5–10 mm. long, stout, stipules small, glandular-serrate to dentate; aments 3–10 cm. long, on short (1–3 cm.) leafy peduncles; staminate broad; stamens 4–6; pistillate lax in fruit, 1–1.5 cm. wide; capsules ovate, acute, 3.5–5 mm. long, glabrous; pedicels slender, 1.5–3 mm. long; style and stigmas very small; scales oblanceolate, obtuse or truncate, with 2–4 teeth, yellow.

Along streams, Upper Sonoran and Transition Zones; from Mendocino and Siskiyou Counties, California, south to the southern border, southern Nevada and western Arizona. Type locality: Santa Cruz, Santa Cruz County, California.

Salix laevigata araquipa (Jepson) Ball, Bot. Gaz. **72**: 234. 1921. (*Salix laevigata*, forma *araquipa* Jepson Fl. Calif. 339. 1909.) Small tree; seasonal twigs, petioles, and basal portion of midrib beneath, densely pilose.

Sparingly in the northern and central part of the range of the species, abundant in southern California and western Arizona. Type locality: "dry gulches, Araquipa Hills, Solano County," California.

6. Salix hindsiàna Bentham. Valley Willow. Fig. 1195.

Salix hindsiana Benth. Pl. Hartw. 335. 1857.
Salix sessilifolia hindsiana (Benth.) Anderss. in Oefv. K. Vet.-Akad. Forh. **15**: 117. 1858; in Proc. Am. Acad. **4**: 56. (Sal. Bor.-Am.) 1858.

Shrub or tree 3–9 m. high, and 2.5 or 3 dm. in diameter, with gray furrowed bark; young twigs densely silvery tomentose; stipules minute or wanting; leaves linear to linear-lanceolate, tapering to both ends, 4–8 cm. long, 3–6 or 10 mm. wide, usually entire, silky-villous or subtomentose, sometimes becoming glabrate; aments serotinous (on leafy peduncles 2–5 cm. long), numerous, solitary or in pairs, 2–4 or 5 cm. long; capsule subsessile, 5–6 mm. long, lanceolate, villous-tomentose; style about 0.5 mm. long, stigmas divided, 1 mm. long, often revolute; stamens 2, filaments pubescent; scales oblanceolate, acute, villous.

River banks up to 3,000 feet elevation, Upper and Lower Sonoran Zones; southwestern Oregon (Douglas, Jackson, Josephine and Klamath Counties) and California to Lower California. Type locality: "ad ripas fluvii Sacramento."

A dubious variety, *leucodendroides*, described as having larger, denticulate leaves, is recognized by Rowlee (*S. macrostachya leucodendroides*, Bull. Torrey Club **27**: 250. 1900) and Schneider (*S. hindsiana leucodendroides* (Rowl.) Schn. Bot. Gaz. **65**: 26. 1918) as replacing the species in southern California, from Monterey and Kern Counties southward.

7. Salix sessilifòlia Nutt. Soft-leaved Willow. Fig. 1196.

Salix sessilifolia Nutt. Sylva **1**: 68. 1843.

Shrub 2–7 m. high; branchlets villous-tomentose; stipules ovate-lanceolate; leaves narrowly to broadly lanceolate or elliptic-lanceolate, 2.5–5 cm. long, 8–15 mm. wide (the sprout leaves larger) acute to rounded at the base, acute or short-acuminate and subaristate at the apex, spinulose-denticulate, especially on the outer half, green and densely to thinly lanate-tomentose on both sides; aments serotinous, solitary or in twos or threes, terminal on leafy peduncles 1–5 cm. long; aments 4–6 cm. long, the deciduous yellow scales broadly elliptic-lanceolate, densely villous; pedicel 0.5 mm. long, villous; capsul lanceolate, 5–6.5 mm. long, villous; style about 0.5 mm. long, divided; stigmas nearly 1 mm. long, divided; stamens 2; filaments pubescent.

River banks, Upper Sonoran and Transition Zones; Vancouver Island, British Columbia, and the Puget Sound district, to the upper valley of the Willamette River, and on the Umpqua River, in Oregon. Type locality: on the "rocky borders of the Oregon (Columbia) at the confluence of the Wahlamet" (Willamette).

8. Salix fluviátilis Nutt.
River Willow. Fig. 1197.

Salix fluviatilis Nuttall, Sylva 1: 73. 1843.

Shrub 2–5 m. high; young twigs tomentose, leaves narrowly elliptical, linear-lanceolate, or linear-oblanceolate, acute, 5–7 cm. long, 8–15 mm. wide, the larger 8–10 cm. by 15–20 mm. remotely spinulose-denticulate, green above, paler or occasionally subglaucous beneath, thinly silky-villous, becoming glabrate; aments serotinous, 3–7 cm. long, usually 2–5 together, on long leafy branches; capsule lanceolate, 5–7 mm. long, subsessile, villous to nearly glabrous in age; style short, stigmas divided, 1 mm. long; scales elliptic-lanceolate, 3 mm. long, yellow, glabrate; stamens 2, filaments pubescent.

Transition Zone; on the lower Columbia River from The Dalles westward, and southward up the Willamette River in Multnomah County, Oregon. Type locality: the "border of the Oregon (Columbia) a little below its confluence with the Wahlamet" (Willamette).

9. Salix parishiàna Rowlee.
Parish Willow. Fig. 1198.

Salix parishiana Rowlee, Bull. Torrey Club **27**: 249. *pl. 9, f. 3.* 1900.

Shrub 1–3 m. high; young twigs silky tomentose; leaves exstipulate, narrowly linear to linear-lanceolate, 4–7 cm. long, 2–5 mm. wide, gray silky-pubescent, becoming glabrate, nearly entire; aments serotinous, 2–3 cm. long, on long leafy peduncles; capsules 5 mm. long, densely sericeus, becoming glabrate in age; pedicel very short; style very short; stigmas linear, about 1 mm. long; scales elliptical, yellow, villous; stamens 2, filaments pubescent.

Stream banks, Upper and Lower Sonoran Zones; California, southern portion from Tulare and Los Angeles Counties southward. Type locality: Matilija Cañon, Ventura County.

10. Salix argophýlla Nuttall.
Silver-Leaved Willow. Fig. 1199.

Salix argophylla Nutt. Sylva 1: 71. *pl. 20.* 1843.
Salix malacophylla Nutt. ined. (Specimens in Kew and Gray Herbaria.)

Shrub 2–4 m. high; young twigs densely white villous-tomentose; leaves stipulate (occasionally), linear to linear-oblanceolate, 4–6 or 8 cm. long, 5–8 or 12 mm. wide, acute at both ends, aristate at apex, nearly entire, densely villous with shining white hairs; aments serotinous, terminating leafy shoots, 1–3 together, 2–4 cm. long; scales lanceolate, acute, villous; capsules lanceolate, 5–6 mm. long; densely villous, pedicel about 0.5 mm.; style obsolete, stigmas divided, about 0.5 mm. long.

Stream banks, mainly Upper Sonoran Zone; Washington and Oregon, east of the Cascade Mountains to northern Nevada, east throughout Idaho to northeastern Utah and northwestern Montana. Type locality: "the river Boiseé, toward its junction with the Shoshone" (Snake), in west central Idaho.

11. **Salix exígua** Nuttall. Slender Willow. Fig. 1200.

Salix exigua Nuttall, Sylva 1: 75. 1843.
Salix longifolia exigua (Nutt.) Bebb, Willows, Calif. (repr. S. Wats. Bot. Calif. 2: 85.) 1879.
Salix fluviatilis exigua (Nutt.) Sarg. Silva N. Am. 9: 124. 1896, in part.

Shrub 2–4 m. high; color effect grayish; leaves linear, acute at both ends, 5–12 cm. long, 2–10 mm. wide, entire or remotely denticulate, opaque, more or less canescent on both surfaces or silky with a fine silvery tomentum beneath, or on both surfaces on young shoots (then frequently mistaken for *S. argophylla*); peduncles sometimes 7 cm. long; staminate aments 2–4 cm. long, the pistillate 3–6 cm. long; scales lanceolate, mostly acute, glabrate or more or less white-pilose; capsules narrowly lanceolate, 4–6 (mostly 5) mm. long, sessile or short-pedicelled, especially at the base of the ament, glabrous or thinly villous; gland about 0.5 mm. long; stigmas sessile, divided, about 0.5 mm. long.

Very common along mountain streams and wet places from sea level to 6,000 feet elevation, Sonoran and Transition Zones; Fraser River in British Columbia southward, east of the Cascade and Sierra Nevada ranges, throughout southern California, and east to Alberta and New Mexico or west Texas. Type locality: probably "on the banks of Lewis River of the Shoshone" (the Snake River, in Idaho).

Variety *virens* Rowlee (Bull. Torrey Club 27: 255. 1900) apparently represents merely the large greenish, glabrate leaves found commonly with normal foliage.

Salix exigua nevadénsis (S. Wats.) Schneider, Bot. Gaz. 67: 331. 1919. (*Salix nevadensis* S. Wats. Am. Nat. 7: 302. 1873.) Capsule (and ovary) glabrous, short-pedicelled, with both dorsal and ventral glands; leaves may be somewhat shorter and broader.

With the species, east central California (Nevada County) southward to the border and east to southeastern Idaho and southern Utah.

12. **Salix melanópsis** Nuttall. Dusky Willow. Fig. 1201.

Salix melanopsis Nutt. Sylva 1: 78. *pl. 21.* 1843.

Dark green shrub or small tree, 3–5 m. high, more divaricately branched than the related species; twigs brown to blackish, often lustrous; leaves oblanceolate to elliptical, acute at both ends, 4–6 or 8 cm. long, 6–15 mm. wide, rather closely denticulate, often spinulose-denticulate, especially near the apex, or subentire, dark green and glabrous above, paler to glaucescent beneath; stipules lanceolate to semicordate, dentate; aments slender, 3–4 cm. long, the pistillate 7–8 mm. wide; scales oblong to obovate, sometime erose at the apex, glabrous to thinly pilose outside, the pistillate often plainly striate with 3–5 nerves; capsules ovate-lanceolate, glabrous, sessile or subsessile, 4–5 mm. long.

Stream banks, Upper Sonoran and Transition Zones; Rocky Mountains of Alberta and British Columbia south to southern California, east to western Montana and Wyoming and northeastern Utah. Type locality: "Fort Hall, on the alluvial lands of Lewis River of the Shoshonee" (Snake), in southeastern Idaho.

Salix melanopsis bolanderiàna (Rowlee) Schneider, Bot. Gaz. 67: 338. 1919. (*Salix bolanderiana* Rowlee, Bull. Torrey Club 27: 257. *pl. 9, f. 9*(?). 1900.) Leaves more or less densely pubescent to densely gray-tomentose beneath. The characters given by Rowlee do not separate his species in any way from true *melanopsis*. The characters assigned by Schneider (longer leaves, wider aments, and longer capsules) are not shown by the type (*Bolander 4958*), which is pubescent beneath, or by the topotype (*Bolander 5031*).

With the species throughout; abundant in Glacier National Park. Type locality: "Yosemite Valley in California (Bolander 4958, 5031)."

13. **Salix pedicellàris** Pursh. Bog Willow. Fig. 1202.

Salix pedicellaris Pursh, Fl. Am. Sept. **2**: 611. 1814.

Salix pedicellaris hypoglauca Fernald, Rhodora **11**: 161. 1909.

Salix myrtilloides of American authors, not L.

Shrub, 1–2 m. high, glabrous throughout; leaves elliptic to oblanceolate, ovate-oblong or obovate, 2–4 cm. long, margins entire and revolute, mostly obtuse, narrowed at the base, short-petioled, firm in texture, green above and glaucous beneath, finely and closely reticulate on both sides; aments 1–2 cm. long, appearing with the leaves and borne on leafy-bracted peduncles; stamens 2; capsules narrowly conic, 5–7 mm long, glabrous; pedicels slender, 2–3 cm long, longer than the persistent acutish thinly villous scale; style very short.

In sphagnum bogs and wet meadows, Transition and Canadian Zones; southern British Columbia, Washington west of the Cascades, and northern Idaho; east to Quebec and New Jersey. Type locality: "on the Catskill Mountains, New York."

14. **Salix lùtea** Nutt. Yellow Willow. Fig. 1203.

Salix lutea Nutt. N. A. Sylva **1**: 63. *pl. 19.* 1843.

Salix cordata lutea (Nutt.) Bebb, Gard. & For. **8**: 473. 1895.

Salix cordata watsoni Bebb. Willows Calif. (repr. S. Wats. Bot. Calif. **2**: 86.) 1879.

Salix flava Rydb. Bull. Torrey Club **28**: 273. 1901, not Schoepf, 1796.

Salix watsoni (Bebb) Rydb. Bull. Torrey Club **33**: 137. 1906.

A clustered shrub 2–5 m. high; twigs yellow, or reddish brown on the sunny side, glabrous or those of the season puberulent; leaves narrowly to broadly lanceolate, acute to short-acuminate at the apex, rounded to cordate at the base, 4–8 or 10 cm. long, 1 5–4 cm. wide, serrulate to entire, mostly yellowish-green, glaucous beneath; stipules ovate to lunate, serrulate to entire; aments nearly sessile, on very short bracted peduncles; staminate 2–3 cm. long; pistillate 2–4 cm. long, 1 cm. wide; scales oblanceolate, acute to obtuse, tawny, thinly crisp pilose; capsules ovate-conic, 4–5 mm. long; glabrous; pedicels 0.7–2 mm. long; styles less than 0.5 mm. long; filaments free.

Along streams and in wet places up to 7,000 feet. Transition and Canadian Zones; Washington and Oregon east of the Cascades, east to northern Alberta and Manitoba, south to Colorado, northern Arizona, and (sparingly) in the Sierra Nevada to the San Bernardino Mountains, California. Type locality: "Rocky Mountains westward to the Oregon" (Columbia).

Salix lutea ligulifòlia Ball, Bot. Gaz. **71**: 428. 1921. Twigs mostly dark, sometimes yellowish: leaves narrowly lanceolate or oblong, strap-like, 5–10 by 1–2 cm., subentire, dark green; capsule 4–5.5 mm. long, pedicel about 1 mm.

With the species, Yosemite Valley, California, and western Nevada to the Black Hills, South Dakota, and southern New Mexico.

Salix lutea platyphýlla Ball, Bot. Gaz. **71**: 430. 1921. (*Salix cordata* var. Bebb, in King, U. S. Explor. 40th Par. **5**: 325. 1871.) Differs from the species in its longer and more slender petioles, 8–15 mm. long, and broader and shorter leaves, elliptic-obovate to ovate, 1.5–3 cm. wide, 4–8 cm. long, commonly 2x4, 2–2.5x5–7, or on shoots, 4.5x9 cm.; pedicels longer, 1–2.5 or 3 mm. long.

Mountains, northern Oregon, east of the Cascades, to southwestern Wyoming, south to southern Utah, with the species.

15. **Salix mackenziàna** (Hooker) Barratt. Mackenzie Willow. Fig. 1204.

Salix cordata mackenziana Hook. Fl. Bor. Am. **2**: 149. 1838.

Salix mackenziana (Hook.) Barratt, in Anderss. Svensk. Vetensk. Akad. Handl. (Monog. Sal.) **6**: 160. 1867.

A shrub or small tree, 2–6 m. tall, with elongated dark brown or yellowish branchlets; leaves lanceolate or ovate-lanceolate, to conic-lanceolate (often obovate while expanding), short-acuminate, rounded to cordate at the base, glandular-serrulate, 6–10 cm. long, 2–3.5 cm. wide, glabrous, glaucous beneath; aments appearing with the leaves, on short (4–8 mm.) peduncles bearing 2–3 small leaves; pistillate rather lax, 2.5–6 cm. long; scales oblanceolate, obtusish, drying brown, thinly tomentose on the outside, densely so within; capsule 4.5–5.5 mm. long, glabrous; pedicel 2.5–4 mm. long; style about 0.5 mm. long; staminate aments shorter and more dense; stamens 2; filaments 2, glabrous.

Sea level to 5,000 feet elevation, Transition and Canadian Zones: southern British Columbia to Lake County, California, and Washoe County, Nevada, east to northern Utah, western Montana, and Mackenzie (Great Slave Lake). Type locality: "Great Slave Lake and Mackenzie River."

Salix mackenziana macrogémma Ball, in Piper and Beattie, Flora N. W. Coast 116. 1915. Branchlets densely pubescent-tomentose with gray hairs; buds elongated, 10–16 mm. long, lanceolate, acuminate, densely pilose-tomentose with long gray hairs; petioles and basal portion of midrib on upper surface pubescent, otherwise as in the species.

Low ground, sea level, to 500 feet. Transition Zone; Puget Sound from Seattle, Washington, south up the Willamette Valley to Eugene, Oregon. Type locality: "Seattle."

16. **Salix fàrrae** Ball. Farr's Willow. Fig. 1205.

Salix farrae Ball, in Standley, Contrib. U. S. Nat. Herb. **22**: 321. 1921.

Small shrub, probably 3–6 dm. tall; branchlets glabrous, bright red, lustrous (often drying dull brown), seasonal twigs yellowish; stipules obsolete or small; petioles slender, twisted; blades elliptical-oblanceolate to elliptic-oval or broadly oval-lanceolate, acute at apex, acute to usually rounded at base, entire or nearly so, 1–1.8 x 3.5–5 cm.; glabrous, glaucous or subglaucous beneath, loosely reticulate with slender raised veins; aments coetaneous, short-pedunculate; pistillate 2–4 cm. long in fruit; capsule lanceolate, often acute at base, 4–5 cm. long, glabrous; pedicel 1–1.5 mm. long; style 0.3–0.5 mm. long; stigmas short; scales oblong or oblanceolate, obtuse, fuscous at apex, paler below, glabrous outside.

Rocky Mountains of southern British Columbia and Alberta, and northwestern Montana to Yellowstone Park, and on the Imnaha, Wallowa County, Oregon. Type locality: "on the flats along the Kicking Horse River, Field, B. C., Can."

17. Salix monóchroma Ball. One-colored Willow. Fig. 1206.

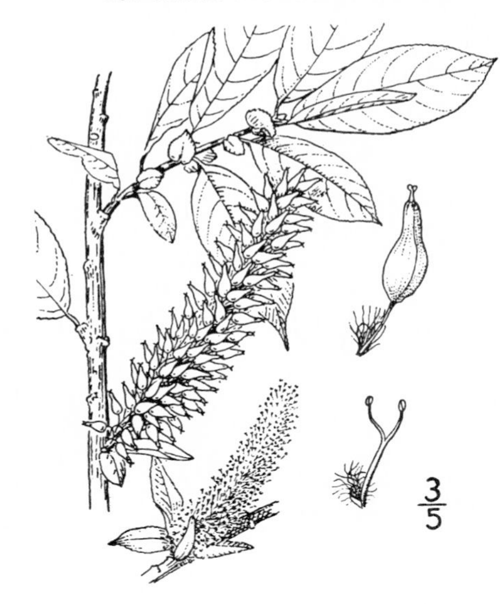

Salix monochroma Ball, Bot. Gaz. **71**: 431. *f. 1.* 1921.
Salix pyrifolia as interpreted by Ball in Coulter and Nelson, Man. Rocky Mtn. Botany 131. 1915, not Anderss.

Shrub 1–3 m. high; twigs glabrous, bright chestnut to brown, lustrous; leaves ovate-lanceolate, ovate, or obovate-oval, abruptly cuspidate to short-acuminate at the apex, rounded or cordate at the base, 3.5–7 cm. long, 1.5–3.5 cm. wide, shallowly glandular-serrulate, thin, translucent, pure dark green and reticulate with slender veins on both sides; stipules lunate or broadly ovate; peduncles short, leafy; aments coetaneous, long, lax; the staminate slender, flexuous, 4–6 cm. long, 8–10 mm. wide; filaments glabrous, united for one-third to three-fourths their length; pistillate aments 3–6 cm. long, 1.2–1.8 mm. wide; capsules 4–7 mm. long, glabrous; pedicels 2.5–4 mm. long, about three times as long as the scales; styles 0.3–0.7 mm long; stigmas usually deeply divided.

Mountains, Transition and Canadian Zones; southern Alberta to Oregon, east of the Cascade Mountains (sparingly in the Willamette Valley of western Oregon), east to western Montana and Colorado, up to elevations of about 6,000 feet. Type locality: "valley of Hatwai Creek, Nez Perce County, Idaho."

18. Salix pseudomyrsinìtes Anderss. Firm-leaf Willow. Fig 1207.

Salix pseudomyrsinites Anderss. Oefv. Vet. Akad. Foerhandl **15**: 129. 1858. (Nord. Am. Pil.); Sal. Bor. Am. 25. 1858.
Salix pseudocordata, in part, of authors.

Shrub 1–3 m. high; twigs mostly divaricate, leafy, bright chestnut to dark brown, lustrous; leaves elliptic-oblanceolate to lanceolate-oblong, acute to acuminate at the apex, mostly rounded at the base (or oval-oblong and subcordate at base); 3–6 cm. long, 1–2 cm. wide, shallowly glandular-serrulate or subentire, dark green above, pure green, scarcely paler and coarsely reticulate with broad veins beneath; stipules lanceolate to ovate; peduncles 0.5–1 cm. long, leafy; aments coetaneous, small, 2–3 cm. long in both sexes; scales oblong or oblanceolate, acutish to truncate, thinly white-pilose; capsules greenish, 4–5 mm. long; pedicels 1–1.5 mm. long, usually pubescent; styles 0.5–0.7 mm. long; stigmas thick, entire; stamens 2, filaments glabrous, free.

Southern British Columbia through Washington and Oregon, east of the Cascade Mountains, and in the high Sierras of California, east to Saskatchewan and south in the Rockies to New Mexico. Up to 10,000 or 12,000 feet elevation in the south, and perhaps a distinct variety, smaller in every way, with shorter, broader leaves. Type locality: "on the Grand Rapid of the Saskatchewan."

19. Salix lasiolèpis Benth.
Arroyo Willow. Fig. 1208.

Salix lasiolepis Benth. Pl. Hartwg. 335. 1857.

Shrub or small tree 2–12 m. high, stems clustered, branchlets brownish-black, sometimes yellowish, usually pubescent; leaves narrowly to broadly oblanceolate, 6–10 cm. long, 1–2 cm. wide, pubescent to glabrate and glaucous beneath, margins nearly entire and somewhat revolute; aments precocious, subsessile, 3–7 cm. long; scales obovate, dark brown, densely pilose-tomentose; capsule 4–5.5 mm. long, glabrous; pedicel 0.5–1 mm. long, style 0.5 mm. long; stamens 2, filaments united at base.

Rocky stream banks, up to about 5,000 feet, Upper Sonoran and Transition Zones; eastern Washington and adjacent Idaho to California, western Nevada, Arizona, western New Mexico and Mexico. Type locality: "ad ripas fluviorum Salinas et Carmel prope Monterey," California.

Salix lasiolepis bàkeri (v. Seem.) Ball, Bot. Gaz. **71**: 436. 1921. (*Salix bakeri* v. Seemen, Bull. Torrey Club **30**: 635. 1903.) As the species except capsule thinly pubescent, at least near the apex, and apparently slightly larger, 5–6 mm. long.

Sparingly about San Francisco Bay. Type locality: "foothills near Stanford University, Santa Clara County, California."

Salix lasiolepis bigelòvii (Torr.) Bebb, Willows Calif. (repr. S. Wats. Bot. Calif. **2**: 86.) 1879. (*Salix bigelovii* Torr. Pac. R. Rep. **4**: 139. 1857. *Salix franciscana* von Seemen, Bull. Torrey Club **30**: 634. 1903.) Leaves broadly cuneate-oblanceolate or spatulate to obovate, pubescent beneath, often densely so; capsule to 6 mm. long; pedicel 1–2 mm. long; otherwise as the species.

With the species in the northern and central parts of its range. Type locality: "near San Francisco," California.

20. **Salix wolfii idahoénsis** Ball. Idaho Willow. Fig. 1209.

Salix wolfii idahoensis Ball, Bot. Gaz. **40**: 378. 1905.

Low shrub 6–10 dm. high; twigs chestnut, leafy, the younger yellowish, lustrous; leaves oblanceolate or rhomboid-oblanceolate, or elliptical, acute at the apex, rounded at the base, 3–6 cm. long, 1–1.8 cm. wide, entire, dull green and silky-villous with shining hairs on both sides; stipules small or wanting; peduncles short, bearing several leaf-like bracts; pistillate aments 1–3.5 cm. long, subglobose to oblong, densely flowered; staminate aments 0.5–2 cm. long; scales dark, obovate, sparsely long-villous; capsules 4.5–5 mm. long, thinly to densely tomentose; pedicels about 0.5 mm. long; styles 0.8–1.2 mm. long. (The species with smaller leaves and aments, and glabrous capsules).

In the higher mountains, above 6,000 feet, Boreal Zone; northeastern Oregon (Wallowa County) to Yellowstone Park, Wyoming, adjacent Montana and sparingly in Colorado. (The species from Yellowstone Park to southern Colorado; up to 12,000 feet.) Type locality: "forks of Wood River, altitude 6,000 feet" (Blaine County, Idaho).

21. **Salix pseudomontícola** Ball. False Mountain Willow. Fig. 1210.

Salix pseudomonticola Ball, in Standley, Contrib. U. S. Nat. Herb. **22**: 321. 1921.

Shrubs 1–3 m. high; branchlets yellowish to red or brown, shining, usually glabrous or the youngest occasionally pubescent; stipules large; blades elliptic-ovate or narrowly to broadly ovate (young obovate) 4–6 cm. long, 1.5–3 cm. wide, common sizes 1.8x4.5, 2x4, 2.5x5, and 3x6 cm., rounded to mostly cordate at base, acute to abruptly acuminate at apex, coarsely glandular crenate-serrate, green above, glaucous beneath, glabrous and strongly veined on both surfaces; aments coetaneous, subsessile, pistillate 3–7 cm. long; scales broadly oblanceolate, obtuse, brown, clothed densely within and thinly outside with long hairs; capsules 6–8 mm. long, glabrous; pedicel 1–1.5 mm. long, glabrous; style 0.6–1 mm. long; stamens 2, filaments glabrous, free.

Rocky Mountains, 4,000–6,000 feet, Boreal Zone; in Alberta (Jasper Park) and British Columbia, southward to western Montana and northwest Wyoming; doubtfully to Cascade Mountains in Washington. Type locality: Rocky Mountain Park, Banff, Alberta, Canada.

22. **Salix bàrclayi** Anderss. Barclay's Willow. Fig. 1211.

Salix barclayi Anderss. Oefvers. Svensk. Vet. Akad. Foerhandl. **15**: 125. 1858 (Nord.-Am. Pil.).
Salix conjuncta Bebb, Bot. Gaz. **13**: 111. 1888.

Shrub 1–4 m. high; twigs stoutish, blackish, tomentose to glabrate; leaves ovate-lanceolate to oval or obovate, acute or cuspidate at the apex, cuneate to rounded at the base, 3–8 cm. long, 1.5–2.5 or more cm. wide, crenate-serrulate, the glandular teeth inflexed, or subentire, thinly tomentose above, especially on the midrib and veins, becoming glabrate, glabrous, reticulate and mostly glaucous beneath; stipules large; aments stout, on leafy peduncles, 1–3 cm. long; staminate aments 1–3, and the pistillate 2–6 cm. long, 1.5 cm. wide; scales fuscous, lanceolate, mostly acute; thinly to densely villous with long hairs; capsules glabrous, or sometimes thinly sericeous, greenish, 5–8 mm. long; pedicels 1 mm. long; styles 1–1.5 mm. long; stigmas long.

Boreal Zones; from Unalaska and southern Yukon southward to Washington, northeastern Oregon, and western Montana, only at higher elevations in the southern part of its range. Type locality: near Cape Greville, Kadiak Island, Alaska.

23. **Salix commutàta** Bebb. Fig. 1212.

Salix commutata Bebb, Bot. Gaz. **13**: 110. 1888.

Shrub 1–3 m. high, branchlets as in *barclayi;* leaves elliptical to broadly oblanceolate or obovate, 6–8 cm. long, entire or nearly so, cuspidate, densely tomentose on both sides when young, becoming glabrate in age, not paler beneath; aments appearing with the leaves, stout, 2–5 cm. long, on leafy peduncles, as in *S. barclayi;* scales brownish, oblanceolate, obtuse, woolly; capsule 5–7 mm. long, glabrous; pedicels 1 mm. long, usually pubescent; style 1–1.5 mm. long; stigmas short; stamens 2; filaments glabrous, free.

Mountains, at high elevations, 4,000–7,000 feet, Boreal Zones; southern Alaska and adjacent Yukon to southern Oregon (doubtfully to northern California and Nevada) and east to Alberta, western Montana and northern Wyoming. Type locality: Eagle Creek (or Wallowa) Mountains, Wallowa County, Oregon.

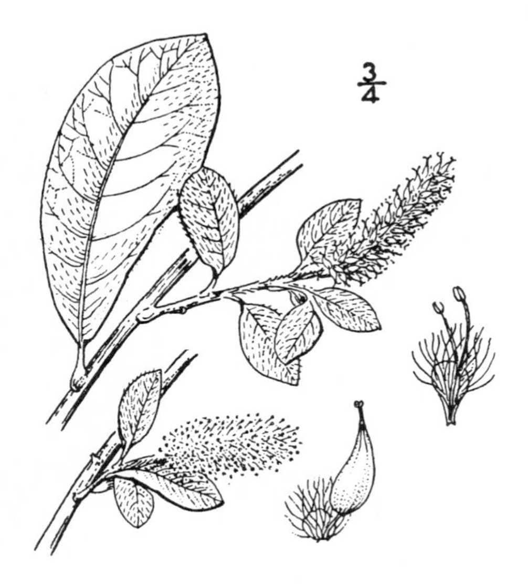

Salix commutata denudàta Bebb, Bot. Gaz. **13**: 111. 1888. Leaves becoming glabrate to glabrous, and the margins subserrate.

With the species, in Oregon, Washington and British Columbia. Type locality: Eagle Creek Meadows, Wallowa County, Oregon.

Salix commutata pubèrula Bebb, Bot. Gaz. **13**: 111. 1888. "Capsules thinly puberulous. Transition to *S. californica.*"

With the species, sparingly, in Oregon and Washington.

24. Salix eastwoòdiae Cockerell. Eastwood's Willow. Fig. 1213.

Salix californica Bebb, Willows Calif. (repr. S. Wats. Bot. Calif. 2: 89.) 1879. not Lesquereux, 1878.

Salix eastwoodiae Cockerell, in Heller, Cat. N. Am. Pl. ed. 2, 89. 1910.

Low shrub, 0.5–2 m. high; branchlets usually tomentose, dark brown; leaves elliptic-lanceolate or elliptical, to elliptic-obovate, 4–6 cm. long, 1–2 cm. wide, acute at both ends or rounded at base, green and grayish tomentose on both sides, somewhat glandular-denticulate, especially the lower; aments coetaneous, leafy-pedunculate, pistillate 2–5, the staminate 1–3 cm. long; scales broadly lanceolate, brownish, tomentose; capsule 5–6.5 mm. long, gray tomentose; pedicels 0.7–1 mm. long, tomentose; style 1–1.5 mm. long; stamens 2; filaments pilose at base, free.

Mountains, 6,000 to 10,000 feet, Boreal Zones; in the Cascade and Sierra Nevada ranges from central Washington to Tulare County, California, east to western Idaho and western Nevada. Type locality: Sierra Nevada Mountains, 8,000–9,000 feet.

25. Salix oréstera Schneider. Sierra Willow. Fig. 1214.

Salix orestera Schneider, Jour. Arnold Arb. 1: 164. 1920.

Salix glauca villosa of California authors, not Anderss.

Shrub 1–3 m. high; branchlets yellowish-brown, more or less tomentose; leaves oblanceolate, lanceolate or narrowly elliptic-lanceolate, 4–6.5 cm. long, 0.8–2 cm. wide, mostly acute at both ends, entire, or rarely sparsely glandular-serrate, densely to thinly silky-villous on both sides, paler to glaucescent and glabrescent beneath; aments coetaneous, dense, subsessile; 1–2.5 cm. long (2–5 cm. in fruit); scales narrowly oblong to obovate-oblong and obtuse, dark brown, silky-pilose; capsule 7–8.5 mm. long, densely to thinly silky-villous; pedicel 1–1.5 mm. long; style 1 mm. long, entire; stigmas short, bifid; stamens 2; filaments free or partly united, pilose below.

Stream banks, Boreal Zones; Sierra Nevada and the San Bernardino Mountains, California, and in Elko County, northeastern Nevada. Type locality: "Mount Goddard, Fresno County, California.

26. Salix pìperi Bebb. Piper's Willow. Fig. 1215.

Salix piperi Bebb, Gard. & For. 8: 482. 1895.

Large shrub, 5–6 m. tall; branchlets stout, glabrous, shining; leaves broadly elliptical or oblanceolate, or obovate, 6–12 cm. long, 2–5 cm. wide, obliquely acute or acuminate, serrulate, glabrous, shining green above, densely glaucous beneath; stipules rare, if present, semicordate or reniform; aments precocious or appearing with the leaves; staminate cylindrical, 3–5 cm. long, short-peduncled; scales brown, long-pilose; stamens 2, the glabrous filaments united for one-third their length; pistillate aments 4–10 cm. long, dense, on short leafy peduncles; capsules 6–7 mm. long, glabrous or thinly pubescent above; pedicel and style each 1 mm. long.

Along streams and in swamps, Humid Transition Zone; Puget Sound near Seattle, Washington, southward in Oregon in the Willamette Valley and along the coast. Type locality: Lake Washington, Seattle, Washington.

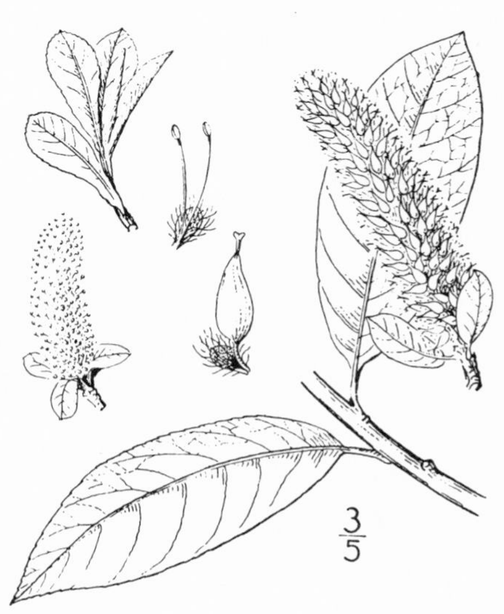

27. Salix hookeriàna Barratt.

Coast Willow. Fig. 1216.

Salix hookeriana Barratt., in Hooker, Fl. Bor.-Am. 2: 145. *pl. 18.* 1839.

Shrub or tree, 4–8 or 10 m. high, with rough dark-gray bark; twigs densely pubescent and very brittle; leaves broadly lanceolate or elliptic-oval, acute, crenate-serrate to subentire, dark green and glabrous above, densely white- (becoming rusty-) tomentose beneath, 3–12 cm. long, 1–4 cm. wide; aments appearing before the leaves; staminate aments stout, short-cylindric, densely flowered, 2–5 cm. long, the obovate scales black, densely covered with long white hairs; pistillate aments 4–7 cm. long, stout; capsules glabrous, 6–8 mm. long; pedicels 1 mm. long; style 1 mm. long; stigmas short.

A species occurring mainly near the ocean beach, Humid Transition and Canadian Zones; from Vancouver Island to Humboldt County, California (Eureka, *Piper 5027*); rare along the shores of Puget Sound. Type locality: "N. W. Coast of America," probably Vancouver Island.

Salix hookeriana tomentòsa Henry, in Schneider, Jour. Arnold Arb. 1: 220. 1920. As the species except capsules wholly or mostly tomentose.

With the species. Type locality: Vancouver Island, British Columbia.

28. Salix cascadénsis Cockerell.

Cascade Willow. Fig. 1217.

Salix cascadensis Cockerell, Muhlenb. 3: 9. 1907.

Salix tenera Anderss. in DC. Prod. 16²: 288. 1868, not Braun, 1850.

Prostrate creeping shrub about 5 cm. high; leaves narrowly elliptical to obovate, broadest at or above the middle, acute at each end or rarely obtuse at the apex, entire, green and shining above, pale beneath, glabrous, 8–15 mm. long, strongly veined, the old leaves persistent; aments small, subglobose, 5–20 mm. long, few flowered; scales brownish, villous; capsules sessile, 4–5 mm. long, tomentose; styles elongate, 1–1.5 mm. long; stamens 2.

Rare; on alpine summits, Boreal Zones; from southern British Columbia through Washington to Wyoming. Type locality: eastern summits of "Cascade Mountains, lat. 49°, 7,000 feet."

29. Salix petróphila Rydberg.

Alpine Willow. Fig. 1218.

Salix arctica petrea Andersson, in DC. Prod. 16²: 287. 1868.

Salix pertophila Rydberg, Bull. N. Y. Bot. Gard. 1: 268. 1899.

Salix tenera as interpreted by Jepson, Fl. Calif. 344. 1909, not Anderss. 1868.

Stems horizontal, creeping; branches erect, glabrous, brown or yellowish, 5–10 cm. high; leaves elliptical or broadly elliptical to obovate, abruptly acute to obtusish, entire, deep green above, distinctly paler to subglaucous beneath, rather strongly veined, 1.5–4 cm. long, 1–1.5 cm. wide, glabrate or glabrous; petioles slender, yellow; aments coetaneous, terminating leafy branches; the staminate 1–2 cm. long; stamens

2, filaments glabrous, free; the pistillate 2–4 cm. long, slender, lax at base; capsules lanceolate, sessile, gray-tomentose, 4–6 mm. long; styles 1–1.5 mm. and stigmas 0.5 mm. long; scales broadly oblanceolate to obovate, often somewhat erose, brown to blackish, long-pilose within, thinly so without.

Alpine summits, 8,000 to 14,000 feet, Boreal Zones; Selkirk Mountains, British Columbia, and Rocky Mountains of Alberta, south in the Rocky and Wasatch systems to northern New Mexico and Utah. Rare in northeastern Oregon. Type locality: "in summits Rocky Mountains," probably Alberta.

Salix petrophila caespitòsa (Kennedy) Schneider, Bot. Gaz. **66**: 136. 1918. (*Salix caespitosa* Kennedy, Muhlenb. 7: 135. *pl. 9,* 1912.) Leaves acute at both ends, or apex subacuminate, more or less densely pilose-tomentose above, glabrate beneath, scales narrower.

Alpine summits, 9,000–11,000 feet, Sierra Nevada Mountains, Sierra to Fresno Counties, California, and adjacent Washoe County, Nevada. Type locality: "Mount Rose, Washoe County, Nevada."

30. Salix monìca Bebb.
Mono Willow. Fig. 1219.

Salix monica Bebb, Willows Calif. (repr. S. Wats. Bot. Calif. 2: 90.) 1879.

Salix chlorophylla of various western authors, for the most part; not Anderss. 1867.

Salix planifolia monica (Bebb) Schneider, Jour. Arnold Arb. 1: 78. 1919.

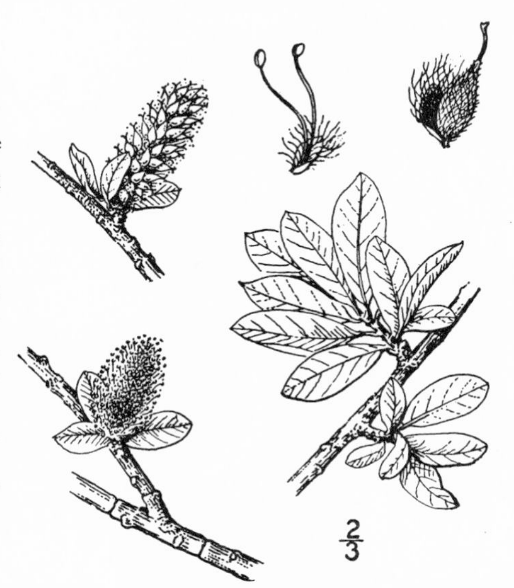

Twigs glabrous, bright chestnut to deep rich brown; leaves broadly elliptic-ovate or obovate, 2–3.5 cm. long, 0.8–2 cm. wide, obtuse at both ends or acutish at the base and abruptly acute at the apex, entire or nearly so; buds large, chestnut or darker; aments appearing with the leaves, 1–3 cm. long, 1–1.3 cm. wide, sessile; scales ovate, acute, clothed on both sides with long yellowish-white hairs; capsules ovate-conic, 4–6 mm. long; styles 1–1.5 mm. long, somewhat pubescent at the base; stigmas thick, mostly entire, 0.3–0.7 mm. long; stamens 2, filaments free.

Apparently differs from *S. planifolia* Pursh in smaller leaves; sessile, denser, smaller aments; and shorter styles and stigmas. Until the identity of *S. chlorophylla* Anderss. is settled I prefer to retain Bebb's name.

Alpine situations, 8,000–13,000 feet, Boreal Zones; central Sierra Nevada, western Montana to northern New Mexico, in the Rocky and Wasatch systems. Type locality: "Mono Pass Summit," Mono County, California.

31. Salix pennàta Ball. Feather-vein Willow. Fig. 1220.

Salix pennata Ball, Bot. Gaz. **60**: 45. *f. 1.* 1915.

Low shrub with dark, divaricate, stoutish, glabrous branchlets and large chestnut-colored buds; leaves obovate or elliptic-obovate, 3–6 cm. long, 1.5–3.5 cm. wide (4x2, 5x2.5), acute, narrowed at base, entire, very dark green above, glaucous beneath, the raised midrib and parallel primary veins conspicuous beneath, glabrous; aments sessile, stout, pistillate 2.5–7 cm. long; capsules subsessile, 6–8 mm. long, densely silvery pubescent; style about 1.4 mm. long; stigmas 0.5 mm. long; scales obovate, acute, black, densely pilose; stamens 2; filaments glabrous, free.

At 4,000 to 6,000 feet on slopes of the high peaks of the central Cascade Mountains, Mount Hood, Mount Adams, Mount Rainier, Boreal Zones. Type locality: "Mount Paddo [Adams]," Pierce County, Washington.

32. **Salix lémmoni** Bebb. Lemmon's Willow. Fig. 1221.

Salix lemmoni Bebb, Willows Calif. (repr. S. Wats. Bot. Calif. **2**: 88.) 1879.
Salix austinae Bebb (in part), *ibid.* **2**: 88.
Salix lemmoni austinae (Bebb) Schneider, Jour. Arnold Arb. **2**: 79. 1920.

Shrub 1–5 m high, young branches thinly silky to glabrous, brownish-black, shining, sometimes thinly pruinose; leaves stipulate, lanceolate, oblanceolate or narrowly elliptic–lanceolate, acute, 4–9 cm. long, 1–1.5 cm. wide, firm, entire or sparsely denticulate, more or less silky when young, becoming glabrous, deep green and shining above, glabrescent and glaucescent beneath at maturity; aments coetaneous, short-leafy-pedunculate; staminate cylindric, 1–3 cm. long; stamens 2, filaments usually free, pilose; pistillate 2–4 cm. long, 1.5–2 cm. wide; capsule 6–8 mm. long, silky tomentose; pedicel 1.5–2 mm. long; style 0.5–0.8 mm. long; stigmas short, bifid; scales narrowly obovate, obtuse, thinly silky-pilose, dark brown to black.

Mountains, 6.000–10,000 feet, Boreal Zones; southwest Oregon (Jackson and Klamath Counties), south to San Bernardino County, California, east to Ormsby and Washoe Counties, Nevada, and northeastern Oregon. Type locality: "Sierra County," California.

33. **Salix subcoerùlea** Piper. Bluish Willow. Fig. 1222.

Salix subcoeruiea Piper, in Bull. Torrey Club **27**: 400. 1900.
Salix covillei Eastwood, Zoe **5**: 80. 1900.
Salix pachnophora Rydberg, Bull. Torrey Club **31**: 403. 1904.
Salix pellita of western authors, not Anderss.

Shrub 1–3 m. high; twigs glabrous, brown to blackish, sublustrous, pruinose; leaves narrowly to broadly oblong-lanceolate or oblanceolate, acute or acuminate at both ends, 3–5 or 7 cm. long, 0.8–2.5 cm. wide, entire or slightly crenulate, green and sparsely pubescent above, densely covered with a short silvery pubescence beneath; midveins beneath conspicuous, yellowish, glabrate; stipules none, or occasional on vigorous shoots; aments coetaneous, sessile or very short-pedunculate, 1–4 cm. long, densely flowered; capsules ovate or ovate-lanceolate, subsessile or short (1 mm.)-pediceled, 3.5–5 mm. long, silvery sericeous; styles 1–1.5 mm. long, yellowish or brown; stigmas 0.2–0.5 mm., entire; stamens 2, filaments glabrous, free; scales oblong, ovate or obovate, acute or obtuse, brown to black, thinly pilose.

Mountains 3,000–10,000 feet, Boreal Zones; eastern Washington and eastern Oregon, east to southern Alberta and south to southeastern Utah, and northern New Mexico, and locally in Fresno and Tulare Counties, California. Type Locality: "Powder River Mountains, in wet meadows near the head of Eagle Creek," Wallowa County, Oregon.

34. Salix béllá Piper. Beautiful Willow. Fig. 1223.

Salix bella Piper, Bull. Torrey Club **27**: 399. 1900.

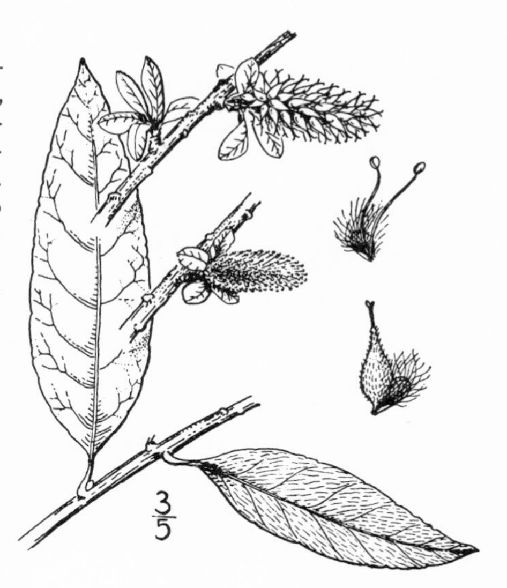

Shrub, 2–4 m. high; branchlets brittle, yellowish to purplish black, somewhat pruinose, glabrate; buds large, pruinose; leaves lanceolate to oblong-lanceolate or oblong-oblanceolate, acute at apex, rounded at base, 4–7 cm. long, 1.5–2.5 cm. wide (sprout leaves up to 12x3.4 cm.), entire or the larger repand-crenate, green and glabrate (at maturity) above, densely clad with opaque silvery tomentum beneath, the yellowish midrib glabrous; stipules lanceolate to semi-cordate, acute; aments precocious, stout, densely flowered, sessile or the pistillate short-pedunculate, 3–6 cm. long; capsule 4.5–5.5 mm. long, gray silky; pedicel 1 mm. long; style 1–1.5 mm. long; stigmas entire or bifid; staminate aments 2–3 cm. long; stamens 2; filaments glabrous, free; anthers yellow; scales oblong-ovate, black, densely lanate.

Mountains, at elevations of 3,000–5,000 feet, Transition and Canadian Zones; southeastern Washington (Whitman and Spokane Counties); northeastern Oregon (Wallowa County); northern Idaho (Bonner, Kootenai, Shoshone, and Latah Counties); northwestern Montana (Flathead County, Glacier National Park); and Alberta near Banff. Type locality: "6 miles east of Pullman, near Garrison, Whitman County, Washington."

35. Salix drummondiàna Barratt. Drummond Willow. Fig. 1224.

Salix drummondiana Barratt, in Hooker Fl. Bor.-Am. **2**: 144. 1838.

Salix sitchensis (in part) or *S. subcoerulea* (in part) of various Rocky Mountain authors, not Sanson or Piper.

Shrub 1–3 (?) m. high, older branchlets rich dark brown, glabrous, lustrous, never pruinose, younger light brown to yellowish, pubescent to puberulent; stipules small; leaves broadly oblong, elliptical, obovate or oval, 3–6 cm. long, 1–3 cm. wide, commonly 3x1.3, 4x1.7, 5x2 or 2.5, and 6x2.2 or 2.5 cm., on sprouts 7–10x 3.5–4 cm., acute to obtusish at base, mostly obtuse and plicate-apiculate at apex, entire or sparingly serrulate near apex, revolute, green and puberulent to glabrous and somewhat shining, with impressed veins above, densely clad beneath with a somewhat lustrous white tomentum partly obscuring the raised veins, the midvein yellowish and pubescent; aments precocious (?) or coetaneous, sessile or subsessile, dense; 1–3 (?) cm. long; stamens 2, filaments glabrous, free; capsules about 5 mm. long, silvery sericeous; pedicels 0.5–1 mm. long; styles 0.5–1 mm. long.

Mountains, elevations of 3,000–7,000 feet, Transition and Canadian Zones; Jasper Park, Alberta, south to northern Idaho (Bonner County) and northwestern Montana (Glacier National Park). Type locality: "marshes and prairies of Rocky Mountains," probably Alberta, Canada.

36. **Salix jépsoni** Schneider. Jepson's Willow. Fig. 1225.

Salix jepsoni Schneider, Journ. Arnold Arb. 1. 89. 1919.

Shrub, probably 1–2 m. high; branchlets chestnut to dark purple, shining, the youngest pubescent, not pruinose; leaves narrowly oblanceolate, from a cuneate base, obtuse or acute at apex, 3–6 or 7 cm. long, 0.6–1.2 cm. wide (sprout leaves to 9x2 cm), firm, entire, margins somewhat revolute, sparsely short-silky above, densely silky tomentose beneath, including raised yellow midvein; aments subprecocious or coetaneous, masculine subsessile, 1–2 cm. long, feminine 2–4 cm. long, on bracted peduncles 0.5–2 cm. long; scales oblong-obovate, pale brown, densely pilose; stamens 2; filaments glabrous, free or sometimes united at base, or for one-half their length; capsule 4.5–5.5 mm. long, densely silky-pubescent; pedicel 1–1.3 mm. long; style 0.7–1 mm. long; stigmas short.

High elevations, 7,000–10,000 feet, Boreal Zone; in the central Sierra Nevada (Placer to Madera Counties), California. Type locality: "high mountains near Donner Pass, Placer County, California."

37. **Salix brèweri** Bebb. Brewer's Willow. Fig. 1226.

Salix breweri Bebb, Willows Calif. (repr. S. Wats. Bot. Calif. 2: 89.) 1879.

Shrub (probably low); branchlets slender, dark brown, glabrous or the younger pubescent; stipules small, ovate-lanceolate, or wanting; leaves oblong or linear-lanceolate to narrowly lanceolate, acute at both ends, 3–6 cm. long; 5–10 mm. wide, common size 5 cm. by 8 mm., entire, dull green and sparingly tomentose (on midrib especially) above, gray tomentose and distinctly reticulate with raised veins beneath; aments precocious or coetaneous, subsessile, linear-cylindric; staminate 1.5–2 cm. long, 0.5 cm. wide; stamens 2, filaments free, glabrous or glabrate; pistillate 2–3.5 cm. long, 1 cm. wide; capsules sessile, about 5 mm. long, short-tomentose; style 1 mm. long, often bifid; stigmas short, bifid or entire; scales oblanceolate, yellowish, or reddish at apex, pilose; gland linear, about 1 mm. long.

Upper Sonoran Zone; inner Coast Ranges, San Benito, Santa Clara, Napa, Lake, and Colusa Counties, California. Type locality: "on San Carlos Mountain, in a dry ravine, at 3,500 feet altitude," San Benito County, California.

38. **Salix delnorténsis** Schneider. Del Norte Willow. Fig. 1227.

Salix delnortensis Schneider, Jour. Arnold Arb. 1: 96. 1919.

Shrub, probably as *S. breweri,* young branches usually gray-tomentose; leaves obovate-oblong to obovate or broadly obovate, cuneate at base, obtuse or rounded or abruptly apiculate at apex, 1.5–2.5 cm. long, 1–1.5 cm. wide, entire, tomentose to glabrescent and dull green above, densely tomentose and rugose-reticulate beneath; aments coetaneous, subpedunculate; masculine 1.5–2.5 cm. long, 0.5–0.7 cm. wide; stamens 2, filaments free, glabrous; anthers violet (?); pistillate aments 3–5 cm. long 0.8–1 cm. wide, slender; capsule 5.5 mm. long, pilose-tomentose; style 1 mm. long, entire or subbifiid; stigmas short, bifid; scales oblong, obtuse, light brown (reddish in maturity?), densely silky-pilose; gland narrowly conic, 0.6 mm. long. (Description after Schneider; specimens not seen).

Coast Range, Humid Transition Zone; Del Norte County, California. Type locality: "Gasquets, Del Norte County, California."

39. Salix sitchénsis Sanson.
Sitka Willow. Fig. 1228.

Salix sitchensis Sanson, in Bongard, Mem. Acad. Imp. Sci. St. Petersb. VI. 2: 162. 1833.

Shrub, 2–7 m. high, branchlets slender, dark brown to black, dull to sublustrous, glabrous or rarely pubescent; leaves spatulate-obovate or the upper oblong-obovate, cuneate at base, mostly acute at apex, entire or obscurely crenate, 4–7cm. long, 1.5–2.5 cm. wide; dull green (the young pubescent) above, with impressed veins, covered beneath with a short appressed satiny pubescence; aments coetaneous, long, slender, densely flowered, subsessile to short-pedunculate, 2–8 cm. long; capsules ovate-conical, acute, subsessile, 4–6 mm. long, silky-pubescent; style 0.5–0.7 mm. long; stigmas short, entire, erect; stamen 1, filament glabrous, anthers violet (?); scales oblanceolate, brown, thinly villous.

Along streams, low elevations, Transition and Canadian Zones; southern Alaska to southwestern Oregon (and northern California?), east to northeastern Oregon and western Montana. Type locality: "Indian River, near Sitka, Alaska."

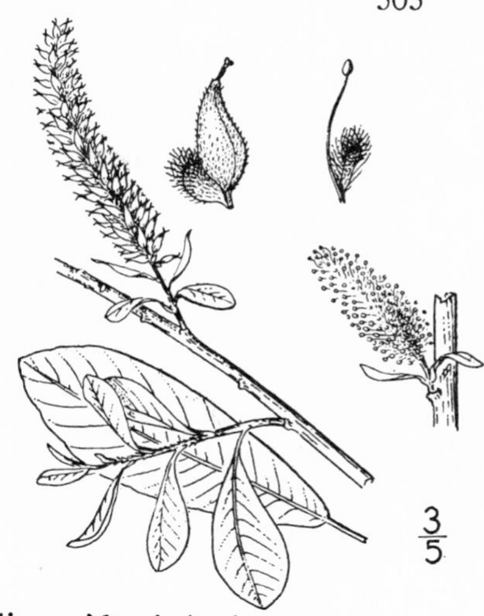

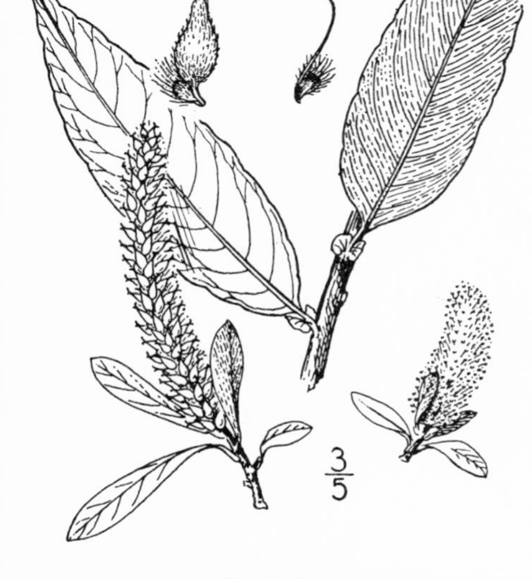

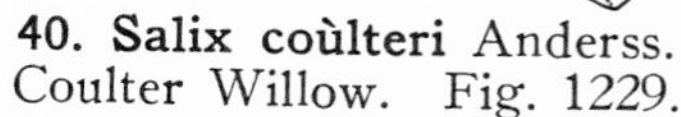

40. Salix coùlteri Anderss.
Coulter Willow. Fig. 1229.

Salix coulteri Anderss. in Oefv. Svensk. Vet. Akad. Förh. 15: 119. 1858.
Salix sitchensis as interpreted by various California authors, not Sanson.
Salix sitchensis forma *coulteri* Jepson, Fl. Calif. 342. 1909.

Shrub, 2–4 m. high; branchlets stoutish, brown, pubescent to densely tomentose; leaves oblanceolate to oblong-obovate or obovate (on same plant), 4–8 or 10 cm. long, 1.5–3 cm. wide, acute to obtuse at apex, acute to cuneate at base, entire or nearly so, coriaceous, dark green and glabrate (except the midrib) with impressed veins above, densely white opaque tomentose beneath; stipules reniform; aments coetaneous, slender, subsessile or short-pedunculate; staminate 3–7 cm. long; stamen 1, filament glabrous, anthers yellow; pistillate 5–10 cm. long; capsules ovoid, 4-4.5 mm. long, subsessile, silvery tomentose; style about 0.5 mm. long; stigmas short; scales oblanceolate, tawny, or fuscous at tip, densely white lanose.

Stream banks, mainly Humid Transition Zone; sparingly in Washington (Cowlitz County, *Coville;* Klickitat County, on the Columbia River, *Suksdorf*), southward more commonly through Oregon, west of Cascade Mountains, and western California to Santa Barbara County. Type locality: "in California," probably Monterey.

41. Salix scouleriàna Barratt. Scouler Willow. Fig. 1230.

Salix scouleriana Barratt, in Hooker, Fl. Bor.-Am. 2: 145. 1839.
Salix flavescens Nutt. N. A. Sylva 1: 65. 1843, not Host. 1828.
Salix stagnalis Nutt. N. A. Sylva 1: 66. 1843.
Salix brachystachys Benth. Pl. Hartweg. 336. 1857.
Salix capreoides Andersson, in Oefv. Svensk. Vet. Akad. Förhandl. 15: 120. 1858.
Salix nuttallii Sarg. Gard. & For. 8: 463. 1895.

Shrub or small tree, 4–10 m. tall, with dull gray bark; leaves very variable, mostly oblanceolate to obovate, sometimes oblong or elliptical, mostly obtuse or abruptly acute at the apex and cuneate at base, 3–10 or 12 cm. long, entire to shallowly crenate-serrulate, thick, green and glabrate above, silvery to (in age) rusty-pubescent to glabrous and glaucous and reticulate beneath; stipules minute to large, ear-shaped, denticulate; aments precocious, densely flowered, sessile or the pistillate very short peduncled, 2–5 cm. long; capsules long-beaked, 7–9 mm. long, tomentose; pedicels pubescent, 1.5 mm. long; stigmas sessile, long; scales obovate, black, long-hairy; stamens 2; filaments glabrous, free.

The commonest willow in our limits, occurring both in dry uplands and swamps, Transition and Canadian Zones. The young leaves and bark have a peculiar fetid odor. An immensely variable species especially as to foliage. To attempt to distinguish varieties here would add to confusion.

Southern Alaska and Yukon, to southern California and New Mexico, up to 10,000 feet. Type locality: "on the Columbia" River.

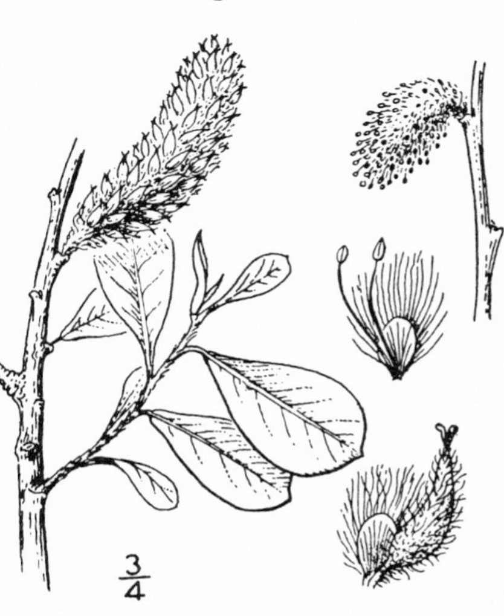

42. **Salix bebbiàna** Sarg. Bebb Willow. Fig. 1231.

Salix bebbiana Sarg. Gard. & For. **8**: 463. 1895.
Salix rostrata Richardson, Bot. App. in Franklin, Narr. Jour. Polar Sea 753. 1823.

Shrub with few stems, or a small tree, 2–5 m. high; twigs short, divaricate, brown or darker, pubescent to glabrate; leaves narrowly elliptical and acute at both ends to broadly oblanceolate or obovate-oval and abruptly short-acuminate, 2–4 or 5 cm. long, 1–2 or 2.5 cm. wide, entire or subentire, dull green above, paler or subglaucous and reticulate with raised veins beneath, more or less pubescent on both surfaces or glabrate in age; stipules none, or small, dentate; staminate aments subsessile or on short peduncles, slender, 1–2 cm. long, yellowish; filaments capillary; pistillate aments (on peduncles 0.5–2 cm. long) 2–4 or 6 cm. long, 1–2 cm. wide, very lax in fruit; scales in both sexes lanceolate-oblong, 1–2 mm. long, acute, yellowish; capsules 6–9 or 10 mm. long, thinly pubescent; pedicels slender, pubescent, 2–5 mm. long; stigmas nearly sessile, deeply divided.

Some of the specimens may represent var. *perrostràta* (Rydb.) Schneider, with smaller, narrower, more glabrate and less deeply reticulate leaves, but it is difficult to distinguish from the species.

Boreal Zones; Alaska and Yukon south to central California, Arizona and New Mexico, east to Labrador, Newfoundland and New Jersey. Rare west of the Cascades. Type locality: "wooded country from lat. 54° to 64° north," Northwest Territory, Canada.

43. **Salix geyeriàna** Anderss. Geyer Willow. Fig. 1232.

Salix macrocarpa Nutt. N. Am. Sylva **1**: 67. 1843 (in part); not Trautvetter. 1832.
Salix geyeriana Anderss. in Oefv. Vet. Akad. Förhandl. **15**: 122. 1858.

Shrub 1 m. high; twigs glabrous, very leafy, black with a bluish bloom; leaves small, linear-oblanceolate or elliptical, acute or short-acuminate at apex, acute at base, 2–4 or 6 cm. long, 5–10 mm. wide, dark green above, more or less glaucous beneath, thinly silky-pilose on both surfaces, margins entire, revolute; stipules none; aments short-leafy-pedunculate, the staminate 1 cm. or less, oblong; the pistillate subglobose, numerous, 1 or occasionally 2 cm. long, 1–1.3 cm. wide; scales linear-oblong, acute, thinly pilose, the base tawny, the tip red; capsules 5–7 mm. long, rostrate; pedicels stout, pubescent, 2–2.5 mm. long; styles short or none; stigmas short, divided.

At middle altitudes, Transition and Canadian Zones; southwest Oregon (Jackson and Klamath Counties), south through the Sierra Nevada, east to Montana, South Dakota and Nebraska, and south to Arizona and Colorado. Type locality: "clumps in wet places where the water is stagnant," probably Oregon (*Nuttall*); "Missouri and Oregon, Rocky Mountains" (*Andersson.*)

Salix geyeriana argéntea (Bebb) Schneider, Jour. Arnold Arb. **2**: 74. 1920. (*Salix macrocarpa argentea* Bebb, Bot. Gaz. **10**: 223. 1885.) Differs from the species in pubescent twigs and rather densely and permanently silvery pilose leaves. Sparingly, with the species in the central half of its range, California, to Wyoming and Colorado. Type locality: "Sierra County, Plumas County," California.

Salix geyeriana meleína Henry, Fl. So. Brit. Columbia, 98. 1915. Taller; twigs not pruinose; leaves oblong, becoming glabrous above, very glaucous beneath and somewhat brownish-pubescent.

Southern British Columbia to northwestern Oregon, west of the Cascade Mountains; also (according to Schneider) in northern Idaho and western Montana. Type locality: "Shawnigan, Victoria, New Westminster," British Columbia.

44. Salix nivàlis Hooker.
Snow Willow. Fig. 1233.

Salix nivalis Hooker, Fl. Bor.-Am. 2: 152. 1839.

Creeping shrub, 2–3 cm. high, from buried stems; leaves elliptic, oblong–obovate, or suborbicular, acute, obtuse or retuse, 7–12 mm. long, 4–8 mm. wide, green and shining above, glaucous and strongly reticulate beneath, glabrous, margins entire and revolute; petioles 2–7 mm. long, yellowish; aments serotinous, globose or oblong, 3–6-flowered, 1 cm. or less in length; scales oblong–obovate, obtuse, yellowish, nearly glabrous; capsules 2.5–3 mm. long, tomentose; style very short; stigmas short, divided.

Boreal Zones; high peaks of the Cascade, Rocky and Wasatch Mountain systems from British Columbia and Alberta south to Washington, Utah and Colorado. Type locality: "near the summits of the peaks in the Rocky Mountains," lat. 52°–56°, Alberta or British Columbia.

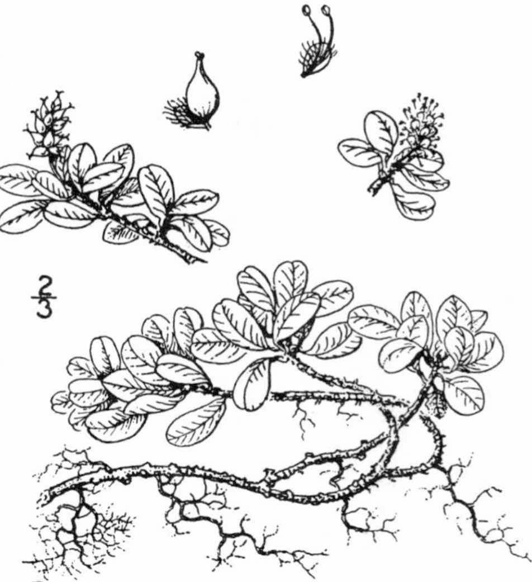

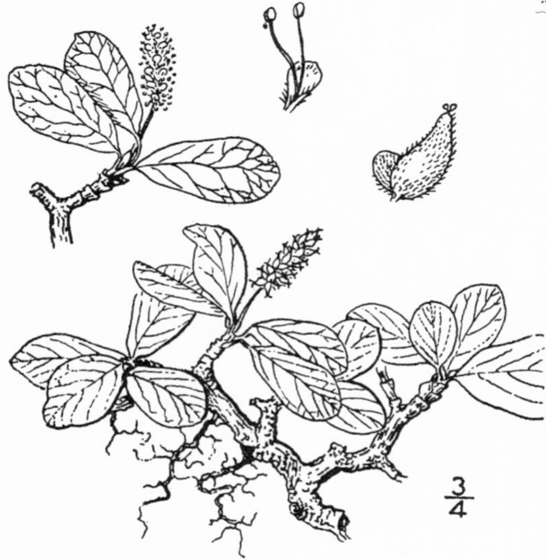

45. Salix saximontàna Rydberg.
Rocky Mountain Willow.
Fig. 1234.

Salix saximontana Rydb. Bull. N. Y. Bot. Gard. 1: 261. 1899.
Salix aemulans von Seemen, Bot. Jahrb. 29: Beibl. 65: 28. 1900.
Salix reticulata of Am. authors (in part), not L.
Salix nivalis saximontana Schneider, Bot. Gaz. 67: 47. 1919.

Densely cespitose and much branched, 3–6 cm. high; leaves elliptic-oblong to suborbicular, obtuse or abruptly acute at the apex, 1.5–3 cm. long, 1–2 cm. wide, with entire and revolute margin, light green above, glaucous and strongly net-veined beneath, glabrous throughout; aments serotinous, 1–2 cm. long, many-flowered; scales broadly obovate, yellowish, nearly glabrous; capsules 3–4 mm. long, sessile, densely white-tomentose; style very short; stigmas short, divided.

Boreal Zones; high elevations, southern British Columbia and Alberta, south to northeast Nevada, and northern New Mexico. Type locality: "Gray's Peak (Clear Creek(?) County), Colorado."

46. Salix vestìta Pursh.
Rock Willow. Fig. 1235.

Salix vestita Pursh, Fl. Am. Sept. 2: 610. 1814.
Salix fernaldii Blankinship, Mont. Agr. Coll. Sci. Studies Bot. 1: 46. 1905.

A shrub 1–5 dm. high, with glabrous, angular twigs; leaves elliptic to oblong-obovate or suborbicular, rounded to retuse at the apex, acute to obtuse at the base, 3–4.5 cm. long, 2–3.5 cm. wide, obscurely crenulate, deep green and glabrous above, coarsely reticulate and long-pilose beneath, especially on the midrib and veins; petioles 2–10 mm. long; aments serotinous on villous peduncles; the staminate linear, 2–3.5 cm. long, 0.5 cm. wide; pistillate aments 2–4.5 cm. long, 1 cm. wide; scales oblong-ovate, tomentose; capsules ovate-conic, 4–5 mm. long; style very short.

Boreal Zones; rocky places, mountains of southern British Columbia and Alberta south to northeastern Oregon, and northwestern Montana, also Hudson Bay south to Quebec and Newfoundland. Type locality: "in Labrador."

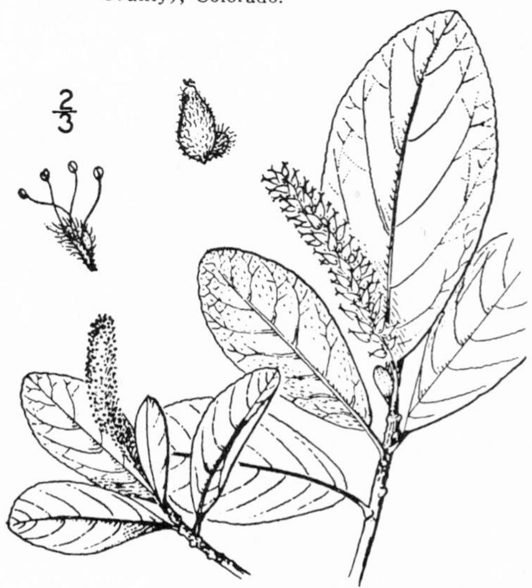

Family 26. MYRICÀCEAE.

Bayberry Family.

Trees or shrubs with simple alternate deciduous or evergreen leaves, without stipules. Flowers monoecious or dioecious, in short aments borne in the axils of the leaves of the same year. Perianth none. Staminate flowers with 2–16, usually 4–8 stamens, with short distinct or united filaments and ovate 2-celled anthers, dehiscent by a longitudinal slit. Pistillate flowers with a solitary, 1-celled ovary, subtended by 2–8 bractlets; ovule 1, orthotropus; styles 2, linear. Fruit a small oblong drupe or nut, the exocarp often waxy. Seed erect; endosperm none.

A family of two genera with about 40 species, widely distributed through the temperate and tropical regions.

1. MÝRICA L. Sp. Pl. 1024. 1753.

Small trees or shrubs with entire dentate or lobed, usually resinous dotted deciduous or evergreen leaves. Staminate aments oblong or narrowly cylindrical; stamens 4–16. Pistillate aments avoid or subglobose; ovary subtended by 2–4 short bractlets. [Ancient name of the Tamarisk.]

About 30 species, widely distributed through the temperate and tropical regions. Type species, *Myrica gale* L.

Flowers dioecious; leaves appearing after the flowers, deciduous.
  Leaves sometimes puberulent below, otherwise glabrous; cuneate at base to the very short petiole; nutlet waxy coated. 1. *M. gale.*
  Leaves more or less pubescent, at least on the margins, attenuate at base to a slender petiole 6–12 mm. long; nutlets not wavy coated. 2. *M. hartwegi.*
Flowers monoecious; leaves evergreen. 3. *M. californica.*

1. Myrica gàle L. Sweet Gale. Fig. 1236.

Myrica gale L. Sp. Pl. 1024. 1753.

A low shrub, 5–15 dm. high, with dark brown glabrous branches. Leaves deciduous, appearing after the flowers, serrate toward the obtuse apex, narrowed to a cuneate subsessile base, firm in texture, dark green and glabrous above, pale and puberulent or glabrous below, 25–60 mm. long, 12–16 mm. wide; staminate aments linear-oblong, 12–18 mm. long, crowded; pistillate aments ovoid-oblong, 8 mm. long in fruit; nutlet waxy coated, scarcely equalling the persistent bractlets which are adnate to its base and clasping.

In low moist situations, Humid Transition and Canadian Zones; **Alaska to Labrador, south on Pacific Coast to southern Oregon, Michigan and Virginia; also in Eurasia.** Type locality: Europe.

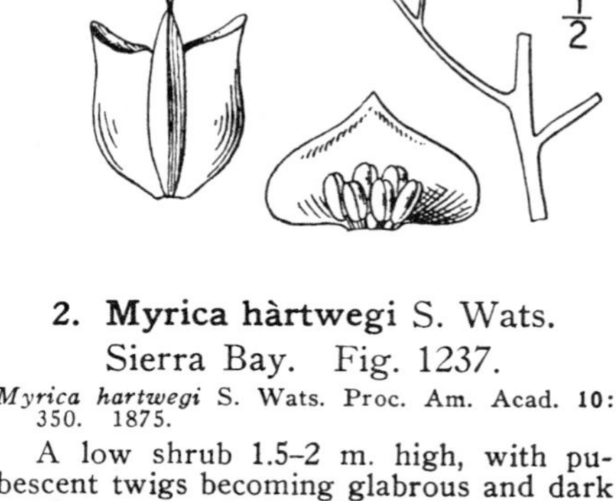

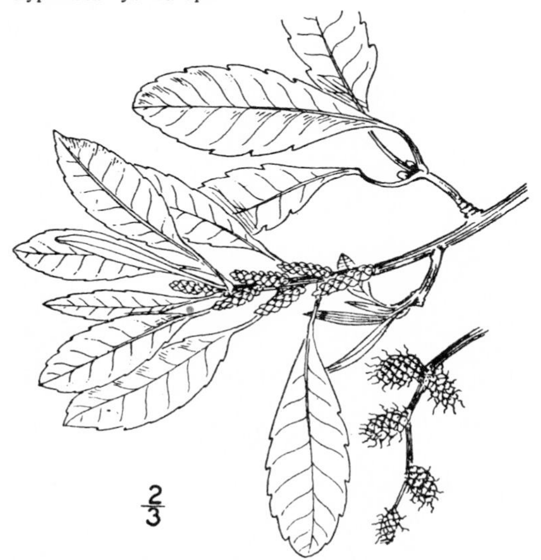

2. Myrica hàrtwegi S. Wats. Sierra Bay. Fig. 1237.

Myrica hartwegi S. Wats. Proc. Am. Acad. **10**: 350. 1875.

A low shrub 1.5–2 m. high, with pubescent twigs becoming glabrous and dark red-brown, marked with small oval lenticels in the second or third year. Leaves deciduous, appearing after the flowers, broadly oblanceolate, 4–8 cm. long, coarsely serrate toward the acute or obtuse apex, attenuate at base to a slender petiole 6–12 mm. long, thin, light green and minutely pubescent or glabrate above, paler below and permanently pubescent especially on the margins and midvein; staminate aments cylindric, 12–18 mm. long; bracts glabrous, light brown; pistillate aments subglobose, about 8 mm. long in fruit; nutlets smooth or with scattered resinous globules, 2 mm. long, compressed and winged by the thickened, acute bractlets.

A restricted species inhabiting stream banks on the western slope of the Sierra Nevada in the Lower Transition Zone; Yuba to Fresno Counties. Type locality: probably on the San Joaquin River.

3. Myrica califórnica Cham. & Sch.
California Wax-myrtle. Fig. 1238.

Myrica californica Cham. & Sch. Linnaea 6: 535. 1831.

An arborescent shrub or tree occasionally 18 m. high with a trunk 3–4 dm. in diameter; bark smooth, thin, dark grey or light brown; wood close-grained, heavy, hard and strong with a slightly reddish tinge. Leaves evergreen, 5–10 cm. long, glabrous, thick, oblong to oblanceolate, acute at apex, narrowed at base to a short petiole, remotely serrate or nearly entire; flowers monoecious; staminate aments borne below the pistillate, 12–16 mm. long; stamens 7–16, united by their filaments; pistillate aments in the axils of the upper leaves, 8–12 mm. long; ovary ovate; bractlets minute; fruit brownish purple, covered with a whitish wax, 6–8 mm. in diameter.

Cañons and moist hillsides of the Coastal region, Humid Transition Zone; Puget Sound to the Santa Monica Mountains, California. Type locality: San Francisco.

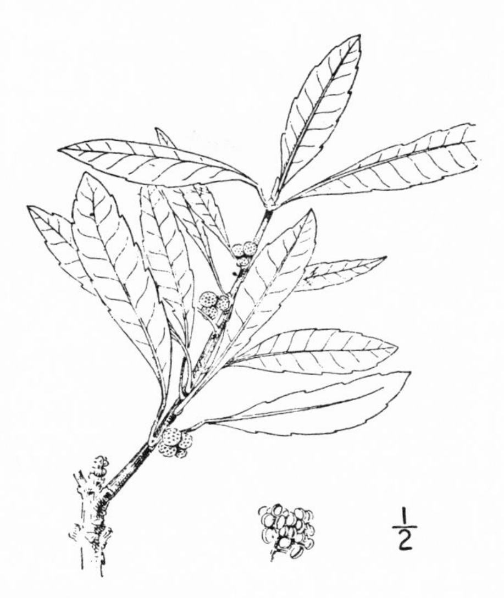

Family 27. JUGLANDÀCEAE.
WALNUT FAMILY.

Trees and shrubs with alternate pinnate leaves, without stipules, and monoecious, bracteolate flowers, the staminate in long drooping aments on twigs of the previous year, the pistillate solitary or clustered on peduncles at the end of the shoots of the season. Staminate flowers consisting of 3 to many stamens, with or without an irregularly lobed calyx adnate to the bractlet. Anthers splitting longitudinally, erect on short filaments. Pistillate flowers bracted and usually with 2 bracteoles and a 3–5-lobed calyx. Petals sometimes present. Ovary inferior, 1-celled or incompletely 2–4-celled. Ovules solitary, erect; styles 2. Fruit a drupe, with fibrous somewhat fleshy exocarp and a bony, rugose or sculptured endocarp. Seed large, 2–4-lobed, without endosperm; cotyledons 2-lobed, corrugated, fleshy and oily.

A family of six genera, two of which, *Juglans* and *Hicoria* (Hickory), occur in North America.

1. JÚGLANS L. Sp. Pl. 997. 1753.

Trees or shrubs with resinous aromatic bark and foliage, superposed buds and deciduous odd-pinnate leaves. Staminate flowers with 4–40 stamens, in 2 or more series. Pistillate flowers with a 4-lobed calyx and 4 petals; styles short, fimbriate. Drupe with an indehiscent somewhat fleshy exocarp. [Name Latin, meaning the nut of Jupiter.]

About 10 species, distributed over the temperate regions of the northern hemisphere and extending into the mountains of tropical America and the Andes. Type species, *Juglans regia* L.

1. Juglans califórnica S. Wats.
California Walnut. Fig. 1239.

Juglans californica S. Wats. Proc. Am. Acad. 10: 349. 1875.

An arborescent shrub, 5–10 m. high or sometimes a tree attaining a height of 15–18 m. with a trunk 6 dm. in diameter; young branchlets with brownish tomentum; bark, on old trunks, dark, 8–12 mm. thick, divided into broad irregular ridges. Leaves 15–20 cm. long, on slender puberulous petioles; leaflets 9–17, ovate-lanceolate, often long-pointed, 35–75 mm. long, coarsely serrate except at base, glabrous above, with tufts of hairs in the axils beneath; staminate aments slender, 5–7 cm. long, brownish pubescent; calyx-lobes acute; stamens 30–40; pistillate flowers in small spicate clusters, ovoid, with recurved yellow stigmas,

10–12 mm. long; fruit globose, 20–30 mm. in diameter, with a thin dark brown soft pubescent husk; nut globose, slightly compressed, with a few remote shallow grooves, thin-walled.

In cañons, especially on north slopes, Upper Sonoran Zone; coast mountains of Ventura County to the Santa Ana and San Bernardino Mountains, California. Type locality: not indicated.

Juglans californica hindsii Jepson, Fl. Calif. 1: 365. 1909. Tree 15–25 m. high with a trunk 3–15 dm. in diameter; leaflets mostly lanceolate and acuminate occasionally oblong-lanceolate, 5–7 cm. long; fruit 3–5 cm. in diameter. As pointed out by Jepson these northern trees are found about ancient Indian village sites and were probably introduced from the southern part of the state. Walnut and Lafayette Creeks, Contra Costa County; near Walnut Grove, and on the east slope of the Napa Range near Wooden Valley. Often planted as a shade tree, and used as a stock for the English walnut.

Family 28. BETULÀCEAE.

Birch Family.

Trees and shrubs without terminal buds, the branches prolonged by one of the upper naked axillary buds. Leaves deciduous, alternate, simple and usually deeply serrate. Flowers monoecious, both kinds in aments, usually appearing before the leaves. Staminate aments pendulous, with 1–3 flowers in the axils of each bract. Calyx membranous, 2–4-parted or none, stamens 1–10. Pistillate aments erect or drooping, spike-like or capitate. Calyx when present adnate to the 1–2-celled ovary. Style 2-cleft. Ovules 1 or 2 in each cell, pendulous. Fruit a small 1-seeded winged nutlet in spike-like clusters, or the nut larger, enclosed by an involucre formed by the enlarged bracts, and solitary or few in a cluster. Endosperm none; cotyledons fleshy.

Six genera confined to the northern hemisphere.

Fruit a nut enclosed in an involucre; staminate flowers solitary in the axils of the ament scales. 1. *Corylus.*
Fruit a winged nutlet in the axils of the ament scales; staminate flowers usually three in the axils of the ament scales.
  Pistillate aments solitary, their scales thin, 3-lobed, deciduous. 2. *Betula.*
  Pistillate aments racemose, their scales erose, persistent, becoming thick and woody. 3. *Alnus.*

1. CÓRYLUS [Tourn.] L. Sp. Pl. 998. 1753.

Shrubs with thin double-toothed leaves folded lengthwise in the bud, flowering in early spring before the leaves. Staminate flowers in drooping aments at the ends of the branches of the preceding year; the flowers solitary in the axils of the bracts, of about 4 stamens and 2 bractlets. Calyx none. Pistillate flowers several in a rounded scaly bud terminating the branchlet, 2 to each bract, with 2 bractlets. Calyx adherent to the ovary; style short; stigmas 2, elongated, bright red. Fruit a smooth nut enclosed by an involucre formed by the enlarged and united bractlets. [Name Greek, from the helmet-like involucre.]

About 12 species distributed over the cooler parts of the north temperate zone. Besides the following 2 others occur in eastern North America. Type species, *Corylus arellana* L.

1. Corylus califórnica (A. DC.) Rose. California Hazelnut. Fig. 1240.

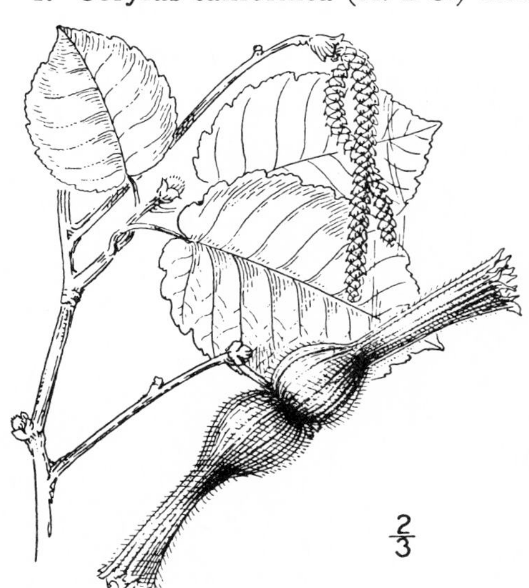

Corylus rostrata californica A. DC. Prod. 16²: 133. 1864.

Corylus californica Rose, Gard. & Forest, 8: 263. 1895.

A loose spreading shrub, 2–3.5 m. high, with smooth bark, and glandular-pubescent branchlets becoming glabrous the second year. Leaves rounded to obovate, 3.5–8 cm. long, obtuse or abruptly acute, doubly serrate, sometimes obscurely 3-lobed, glandular-hairy, pale beneath; anthers with a sparse tuft of hairs at the apex; involucre hispid, prolonged above the nut into a lacinate tube, 15–25 mm. long; nut ovoid, 12–15 mm. long.

Moist cañon slopes or along streams, Transition Zone; British Columbia to the southern Sierra Nevada and the Santa Cruz Mountains, California. Type locality: Santa Cruz, California.

2. BÉTULA [Tourn.] L. Sp. Pl. 982. 1753.

Trees or sometimes shrubs with close-grained hard wood, smooth aromatic resinous bark, marked by long longitudinal lenticels, often separating into thin papery sheets. Leaves deciduous, simple, toothed, petioled; stipules scarious, caducous. Aments elongated and solitary or the staminate sometimes clustered, blooming in early spring before the leaves. Flowers 3 in the axil of the bract, the two lateral subtended by bractlets adnate to the base of the bract. Calyx in the staminate flowers 2–4-lobed, wanting in the pistillate. Stamens 2, 2-branched above, each branch bearing a pollen sac. Ovary sessile, compressed. Bracts accrescent, 3-lobed, deciduous with the nutlet in fruit. Nutlet minute, oval or obovate, compressed and scariously winged. [The ancient name.]

About 30 species inhabiting the subarctic and the cooler parts of the north temperate regions. Type species, *Betula alba* L.

Low shrubs; leaves usually less than 25 mm. wide, simply serrate or crenate, distinctly reticulate; wings much narrower than the nutlets.
  Twigs not pubescent, thickly resinous glandular; leaves mostly rounded. 1. *B. glandulosa.*
  Twigs pubescent or hairy; leaves obovate-cuneate. 2. *B. pumila glandulifera.*
Trees or large shrubs; leaves usually over 25 mm. long, indistinctly reticulate and usually more or less doubly serrate; wings usually broader than the nutlet.
  Bark separable into thin layers. 3. *B. papyrifera occidentalis.*
  Bark not separable into thin layers, reddish or yellowish brown. 4. *B. fontinalis.*

1. Betula glandulòsa Michx. Glandular or Scrub Birch. Fig. 1241.

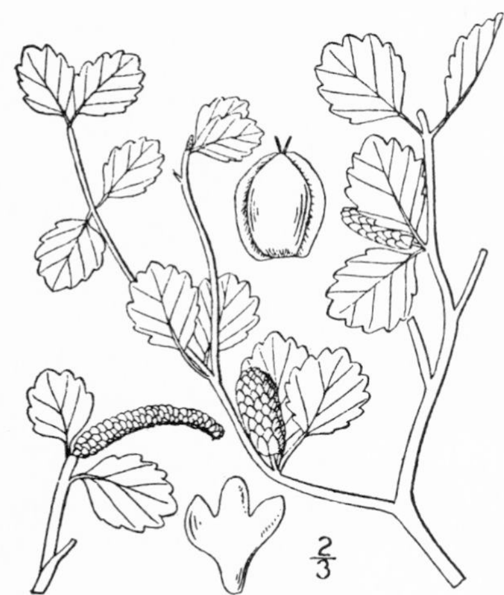

Betula glandulosa Michx. Fl. Bor. Am. 2: 180. 1803.

Shrub 3–12 dm. high, the twigs densely dotted with wart-like resinous glands, otherwise glabrous. Leaves broadly obovate to orbicular, 12–25 mm. long, rounded at apex, usually cuneate at base, crenate-dentate, reticulated, pale green and glandular-dotted beneath, firm and subcoriaceous; petioles 4–6 mm. long; fruiting aments 12–25 mm. long, 3–4 mm. thick, mostly erect, sessile or short-stalked; bractlets with divergent lobes rounded at the apex; nutlets with a narrow or obscure wing.

Usually in sphagnum bogs, Canadian Zone; Arctic regions of North America and Asia, south to Oregon, Colorado and New York. Type locality: about lakes, from Hudson Bay to Lake Mistassini.

2. Betula pùmila glandulífera Regel. Western Low Birch. Fig. 1242.

Betula pumila glandulifera Regel, Bull. Soc. Nat. Moscou 38[2]: 410. 1866.
Betula hallii Howell, Fl. N. W. Am. 1: 614. 1902.
Betula glandulifera Butler, Bull. Torrey Club 36: 425. 1909.

Shrub, 1–2 m. high, the slender reddish brown branchlets slightly resiniferous or glandular-dotted, clothed with long scattered hairs. Leaves 2–4 cm. long, 15–30 mm. wide, obovate to nearly orbicular, acute or obtuse at apex, cuneate at base, serrate to crenate-serrate, dark green and shining above, resinous beneath, hairy on the veins when young; fruiting aments 15–20 mm. long, 3–5 mm. thick, erect; bractlets puberulent, middle lobe truncate, longer than the ascending rounded lateral ones; wings much narrower than the oval body.

In bogs, especially sphagnum bogs, Humid Transition and Canadian Zones; Alaska to Marion County, Oregon, east to Minnesota. Type locality: not indicated.

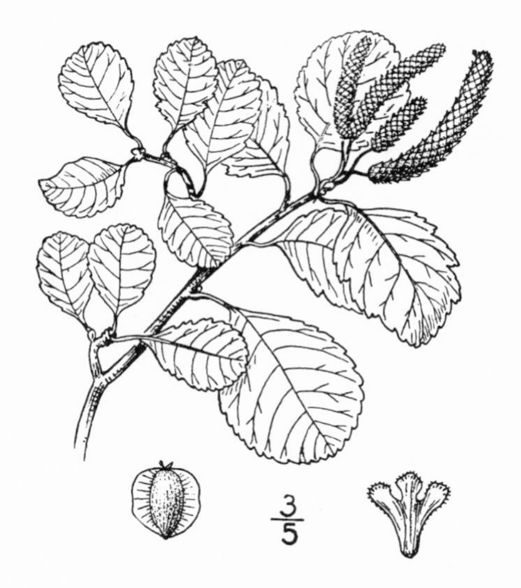

3. **Betula papyrífera occidentàlis** (Hook) Sarg. Western Paper Birch. Fig. 1243

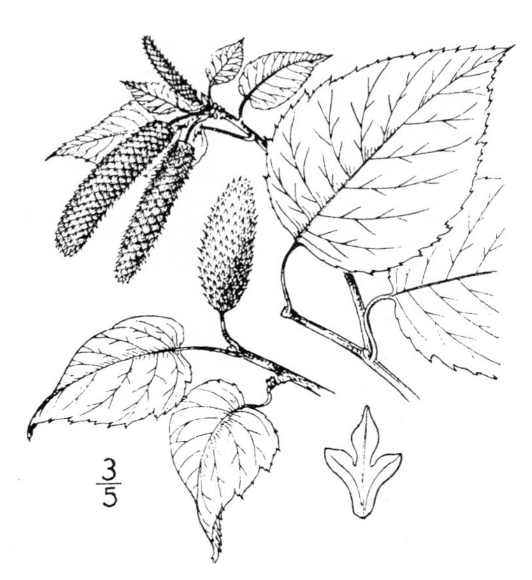

Betula occidentalis Hook. Fl. Bor. Am. **2**: 155. 1839.

Betula papyracea occidentalis Dippel, Hanb. Laubholzk. **2**: 177. 1892.

Betula microphylla occidentalis M. E. Jones, Contr. West. Bot. **12**: 77. 1908.

Betula papyrifera occidentalis Sarg. Journ. Arnold Art. **1**: 63. 1919.

A large tree attaining a maximum height of 40 m. with a trunk diameter of 10–13 dm. Bark orange-brown or white, shining, peeling off readily; the inner bark bright orange-yellow; young twigs loosely covered with long hairs, glandular. Leaves 6–10 cm. long, ovate, acute or acuminate at apex, subcordate to subcuneate at base, doubly serrate with irregular long-pointed teeth, glabrous with age except for a tuft of hairs in the axils of the veins; petioles about 15 mm. long; fruiting aments cylindric, spreading, 3–4 cm. long, 10–12 mm. thick; bractlets ciliate, 5–6 mm. long, middle lobe elongated, awl-shaped, the lateral lobes ovate or rhombic, widely spreading; wings but little wider than the nutlet.

Low ground bordering streams and lakes, Transition Zone; southwestern British Columbia and northwestern Washington to eastern Washington, northern Idaho, and northern Montana, west of the continental divide. Best developed along the Fraser River, British Columbia, and on the islands of Puget Sound. This and its eastern relative, the typical *Betula papyrifera,* have been considered by some botanists as varieties of the Old World *Betula alba.* Type locality: "Straits of Juan de Fuca."

Betula papyrifera subcordàta (Rydb.) Sarg. Journ. Arnold Arb. **1**: 63. 1919. (*Betula subcordata* Rydb. Bull. Torrey Club **36**: 436. 1909.) A small tree, 8–15 m. high with a trunk 3–5 dm. in diameter. Bark white or sometimes orange-brown, separating freely into thin layers. Leaves broadly ovate, 5–7 cm. long, acute or short-acuminate at apex, rounded or subcordate at base, finely and often doubly serrate, glabrous; fruiting aments drooping on slender peduncles, cylindric, 25–35 mm. long, 8 mm. thick; bractlets puberulous, ciliate on the margins, the middle lobe acute, longer than the broad truncate lateral lobes; wings broader than the body of the seed.

Stream banks, Transition and Canadian Zones; British Columbia and western Washington to Alberta, northern Montana, Idaho, and northeastern Oregon (Minum River Valley). Type locality: Hatwai Creek, Nez Perce County, Idaho.

4. **Betula fontinàlis** Sarg. Spring Birch. Fig. 1244.

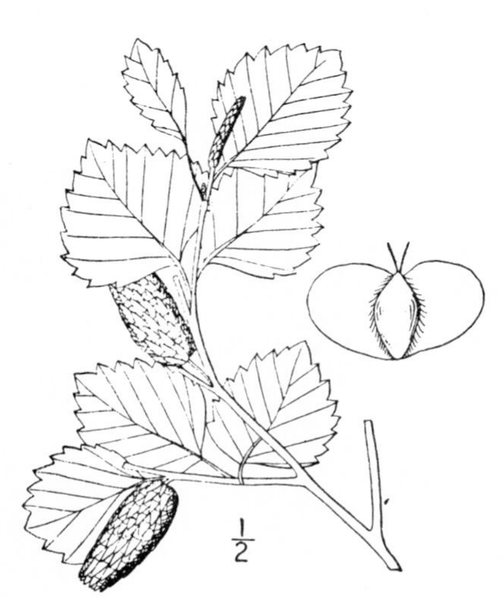

Betula fontinalis Sarg. Bot. Gaz. **31**: 239. 1901.

Betula piperi Britton, Bull. Torrey Club **31**: 165. 1904.

Betula microphylla fontinalis M. E. Jones, Contr. West. Bot. **12**: 77. 1908.

Betula occidentalis, f. *inopina* Jepson, Fl. Calif. **1**: 349. 1909.

A shrub or sometimes a small tree 10 m. high, with a trunk 1–4 dm. in diameter. Bark dark bronze and shining, not separable into layers; young twigs roughened by large resinous glands. Leaves broadly ovate, 3–5 cm. long, acute at apex, cuneate to subcordate at base, sharply and often doubly serrate, resinous glandular on the upper surface when young, becoming dull dark green above, paler and shining beneath and covered with minute glandular dots; fruiting aments 25–30 mm. long, spreading or pendulous on slender grandular peduncles; bractlets glabrous or puberulent, ciliate on the margins, lateral lobes erect, much shorter than the middle one; wing wider than the nutlet.

Along mountain streams, Canadian Zone; eastern British Columbia to South Dakota, south to New Mexico and Inyo County, California. The California specimens have the wings equalling or narrower than the nutlet. Type locality: Sweetwater River, Wyoming.

3. ÁLNUS [Tourn.] Hill, Brit. Herb. 510. 1756.

Trees and shrubs with astringent scaly bark, soft close-grained wood, naked leaf buds and deciduous dentate or serrate, petioled leaves. Flowers monoecious, both kinds in aments; the staminate aments clustered at the ends of the branchlets, long and pendulous, the pistillate clustered usually near the base of the staminate cluster, erect. Staminate flowers 3–6 in each axil, consisting of a 4-parted perianth and 1–4 stamens, subtended by 2–4 bractlets. Pistillate flowers 2 in the axils, without perianth, subtended by 2–4 bractlets, adnate to the scales and becoming woody in fruit, forming an ovoid or subglobose cone. Nutlet flat, ovate or oblong, with lateral wings or membranous border. [The ancient Latin name.]

About 15 species inhabiting the northern hemisphere and the Andes of South America. Type species, *Alnus vulgaris* Hill.

Flowers developed with the leaves on the twigs of the season; fruiting peduncles slender, longer than the cones; leaves shining; nutlets with a broad membranous wing. 1. *A. sinuata.*
Flowers developed on last year's twigs, opening before the leaves in early spring or winter; fruiting peduncles shorter than the cones; leaves dull.
  Leaves rusty pubescent beneath at least on the veins, revolute on the margins; stamens 4; nutlets with a narrow somewhat membranous wing. 2. *A. oregona.*
  Leaves green beneath not revolute on the margins; nutlets merely margined.
    Leaves mostly cuneate at base, doubly serrate, not lobed except on vigorous shoots; stamens 2 or 3. 3. *A. rhombifolia.*
    Leaves rounded or cordate at base, doubly serrate and lobed; stamens 4. 4. *A. tenuifolia.*

1. **Alnus sinuàta** (Regel) Rydb. Wavy-leaved or Sitka Alder. Fig. 1245.

Alnus viridis sinuata Regel, in DC. Prod. **26**: 183. 1868.
Alnus sinuata Rydb. Bull. Torrey Club **24**: 190. 1897.
Alnus sitchensis Sarg. Silva N. Am. **14**: 61. 1902.

A small tree attaining a maximum height of 15 m. and a trunk diameter of 20 cm., or more often a shrub forming thickets. Leaves ovate, 7–15 cm. long, acute at the apex, rounded or cuneate at base, shallowly lobed and doubly glandular-serrate, yellow-green above, pale and shining beneath, glabrous or long-hairy on the mid-rib and axils of the veins; petioles 12–18 mm. long; staminate aments with ovate apiculate scales; cones 12–15 mm. long, their scales truncate and thickened at apex; nutlet oval, about as wide as the wings.

Alpine streams, Canadian and Hudsonian Zones; Alaska to Oregon and Colorado, also Siberia. Type locality: Kamchatka.

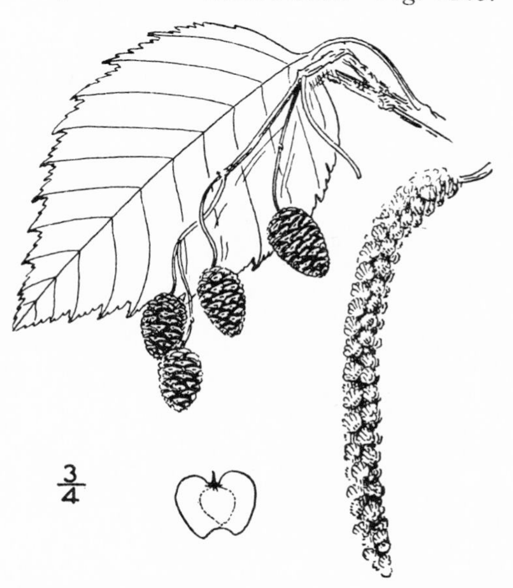

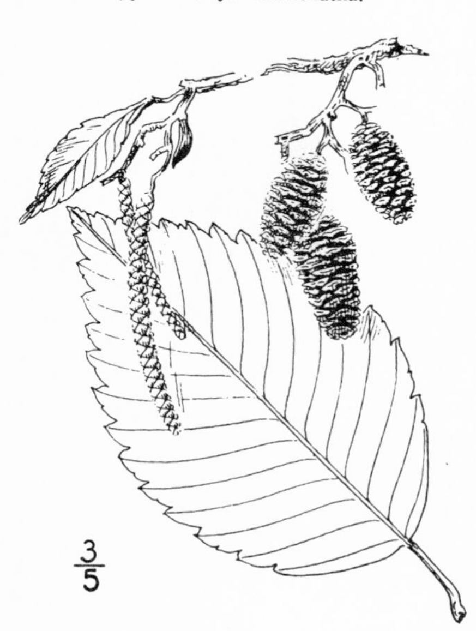

2. **Alnus oregòna** Nutt. Oregon or Red Alder. Fig. 1246.

Alnus rubra Bong. Mem. Acad. St. Petersb. VI. **2**: 162. 1837, not Marsh. 1785.
Alnus oregona Nutt. Sylva **1**: 28. 1842.

Tree 15–25 m. high, with a trunk 6–12 dm. in diameter; bark thin, pale gray or nearly white, the inner red-brown. Leaves ovate 7–12 cm. long, acute at apex, cuneate at base, crenately lobed and minutely glandular-dentate, tomentose when young, becoming dark green and glabrous or sparsely pubescent on the upper surface, rusty pubescent on the lower; petioles 15–20 mm. long; staminate aments 10–15 cm. long, their scales ovate, acute, glabrous; cone ovate to oblong, 20–25 mm. long; scales truncate, much thickened toward the apex; nutlet ovate, much broader than the membranous wing.

Stream banks, Humid Transition and Canadian Zones: southern Alaska south in the coastal region to Santa Barbara County, California. Type locality: probably western Oregon.

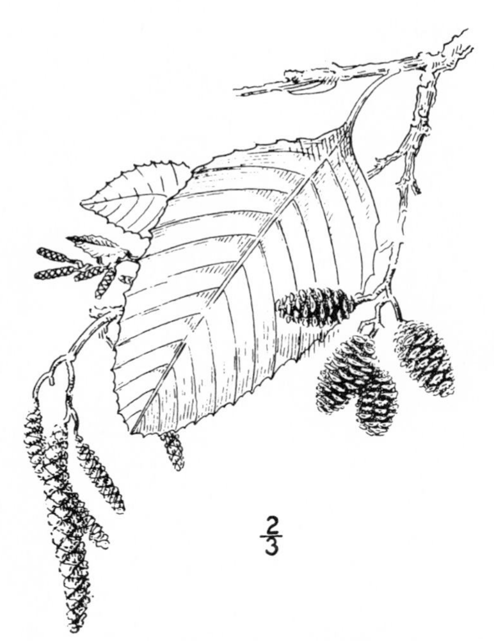

3. Alnus rhombifòlia Nutt. White Alder. Fig. 1247.

Alnus rhombifolia Nutt. Sylva 1: 33. 1842.

A tree, often 25 m. high, with a straight trunk 6–10 dm. in diameter. Leaves ovate, 5–7.5 cm. long, rounded or acute at apex, cuneate at base, finely or coarsely and doubly serrate, becoming dark green, glabrous or minutely pubescent above, light yellow-green and puberulent beneath; petioles pubescent, flattened; cones oblong, 8–12 mm. long, their scales slightly thickened and lobed at apex; nutlets broadly ovate, with a narrow thin margin.

Along streams, Transition and Upper Sonoran Zones; southern British Columbia and northern Idaho south through the Pacific States to northern Lower California. Type locality: Monterey, California.

4. Alnus tenuifòlia Nutt. Thin-leaved Alder. Fig. 1248.

Alnus tenuifolia Nutt. Sylva 1: 32. 1842.
Alnus incana virescens S. Wats. Bot. Calif. 2: 81. 1880.
Alnus occidentalis Dippel, Handb. Laubh. 2: 158. 1892.

Shrub 1–2 m. high or sometimes a small tree 10 m. high with a trunk 20 cm. in diameter. Leaves 5–10 cm. long, ovate to oblong, acute at apex, rounded to cordate or sometimes cuneate at base, usually acutely and deeply lobed and doubly serrate, pubescent when young, becoming dark green and glabrous above, yellow-green and glabrous or finely pubescent beneath; cones ovoid to oblong, 8–12 mm. long, their scales truncate and 3-lobed at the much thickened apex; nutlets nearly orbicular, with a very narrow membranous border.

Borders of streams, lakes and mountain meadows, Arid Transition and Canadian Zones; Yukon Territory and British Columbia to New Mexico and Lower California. The common alder of the Sierra Nevada and the eastern slopes of the Cascades. Type locality: Rocky Mountains, and the Blue Mountains.

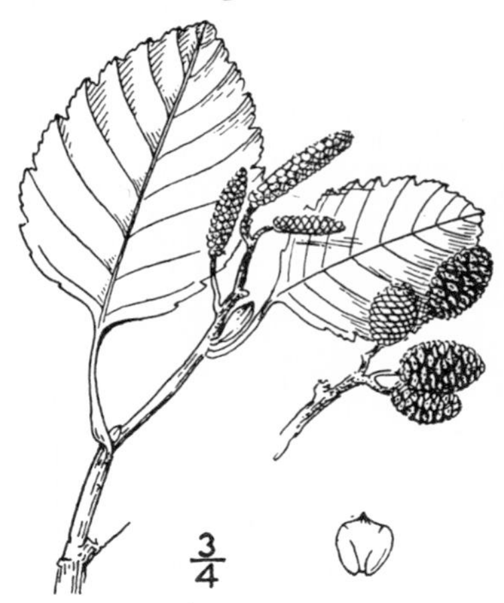

Family 29. **FAGÀCEAE.**

Beech Family

Trees and shrubs with deciduous or evergreen alternate petioled leaves and small usually deciduous stipules. Flowers monoecious, without petals, the staminate in pendulous or erect aments or sometimes in peduncled heads, the pistillate solitary or in small clusters, subtended by an involucre which becomes a bur or cup. Staminate flowers with 4–7-lobed calyx and 4–20 stamens. Pistillate flowers with a 4–8-lobed calyx, adnate to the 3–7-celled ovary. Ovules 1–2 in each cell, pendulous, only 1 in each ovary maturing. Styles as many as the cells of the ovary, linear. Fruit a 1-seeded nut with bony exocarp. Endosperm none; cotyledons large and fleshy.

A family of six genera, mainly confined to the northern hemisphere, and all represented in North America except *Nothofagus,* which is restricted to the southern hemisphere.

Fruit composed of 1–3 nuts, enclosed in a spiny bur-like involucre. 1. *Castanopsis.*
Fruit a single nut surrounded at base by a scaly cup-like involucre.
  Staminate aments densely flowered, erect, their flowers clustered in the axils of persistent bracts; pistillate flowers clustered at the base of the staminate aments. 2. *Lithocarpus.*
  Staminate aments loosely flowered, spreading or drooping, their flowers solitary in the axils of the thin caducous bracts; pistillate flowers in axillary clusters. 3. *Quercus.*

1. CASTANÓPSIS Spach, Hist. Veg. Phan. 11: 185. 1842.

Trees and shrubs with scaly bark, astringent wood, evergreen leaves, and buds with imbricated scales. Staminate aments erect, densely flowered, the flowers in clusters of 3 in the axils of bracts; stamens 10–12; calyx 5–6-lobed. Pistillate flowers usually on the base of the staminate ament, 1–3 in the involucre; styles 3. Fruit maturing the autumn of the second year; nut ovoid, more or less angled, 1–3 in the spiny bur-like involucre. [Name Greek, in allusion to the resemblance of these plants to *Castanea.*]

Approximately 30 species, of which all but the following are Asiatic. Type species, *Castanopsis armata* (Roxb.) Spach.

Tree or sometimes a shrub, forming a conical crown; bark thick; leaves pointed. 1. *C. chrysophylla.*
Shrub low and spreading; bark thin; leaves usually obtuse. 2. *C. sempervirens.*

1. Castanopsis chrysophýlla (Dougl.) A. DC. Giant Chinquapin. Fig. 1249.

Castanea chrysophylla Dougl.; Hook. Fl. Bor. Am. 2: 159. 1839.

Castanopsis chrysophylla A. DC. Seem. Journ. Bot. 1: 182. 1863.

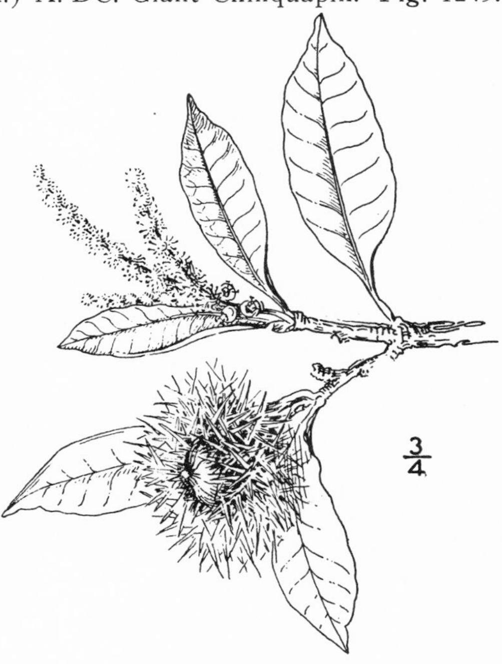

Tree 15–45 m. high, with a trunk 1–2 m. in diameter, and stout spreading branches forming a conical crown; bark thick, broken into longitudinal furrows; wood light reddish brown, light and soft. Leaves 5–15 cm. long, lanceolate, gradually narrowed at the ends, entire, thick and coriaceous, very dark green above, golden tomentose beneath, falling the second or third year; petioles 6–8 mm. long; aments clustered at the ends of the branches; burs chestnut-like, 4-valved; nuts 1 or rarely 2, 8–12 mm. long.

Wooded slopes, mainly Humid Transition Zone; Skamania County, Washington, south on the western slope of the Cascade Mountains and the Coast Ranges to Mendocino County California. Type locality: "Grand Rapids of the Columbia."

Castanopsis chrysophylla minor (Benth.) A. DC. A small tree or shrub with a conical crown; leaves revolute, densely golden brown beneath, very dark green above. Usually on gravelly or rocky ridges, Humid Transition and Upper Sonoran Zones; Mendocino County to the Santa Lucia Mountains, California. Type locality: Monterey.

2. Castanopsis sempérvirens (Kell.) Dudley. Sierra Chinquapin. Fig. 1250.

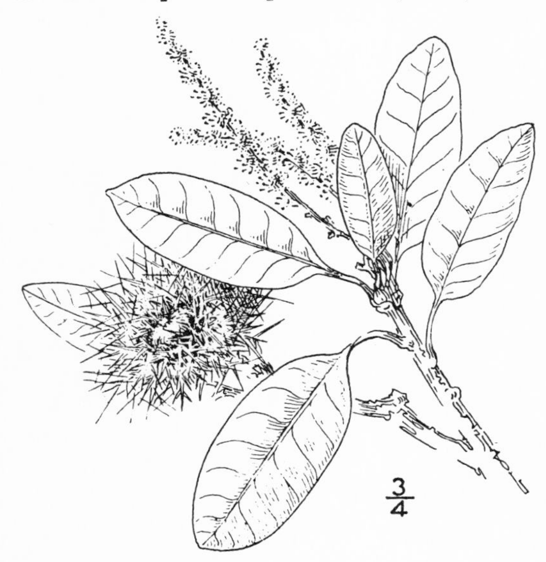

Castanea sempervirens Kell. Proc. Calif. Acad. 1: 71. 1855.

Castanopsis sempervirens Dudley; Merriam, N. Am. Fauna. 16: 142. 1899.

Low spreading round-topped shrub, 1–2.5 m. high, with smooth brown bark. Leaves oblong, obtuse or sometimes acute at apex, 3.5–7.5 cm. long, on petioles about 15 mm. long, light yellowish gray-green above, golden or pale rusty tomentose beneath; flowers and burs similar to the preceding species.

Dry ridges in open coniferous forests, Arid Transition and Canadian Zones; Cascade Mountains of southern Oregon through the Sierra Nevada to the San Bernardino Mountains, southern California. Type locality: near Mariposa, California.

2. LITHOCÀRPUS Blume, Bijdr. 526. 1825.

Trees with astringent deeply furrowed bark, stellate pubescence, hard close-grained wood, petioled evergreen leaves and persistent stipules. Staminate aments erect, dense; the flowers in clusters of 3, in the axils of ovate rounded bracts with 2 lateral bractlets; calyx tomentose, 5-lobed; stamens 10. Pistillate flowers usually at the base of the staminate ament, solitary in the axils of acute bracts and minute lateral bractlets; calyx 6-lobed; styles 3. Fruit an acorn, maturing in the autumn of the second year. Nut 1-seeded, surrounded at base by the cup-like involucre. [Greek, meaning stone fruit.]

About 100 species confined to southeastern Asia, with the exception of one California species. Type species, *Lithocarpus javensis* Blume.

1. Lithocarpus densiflòra (Hook. & Arn.) Rehd. Tan-bark Oak. Fig. 1251.

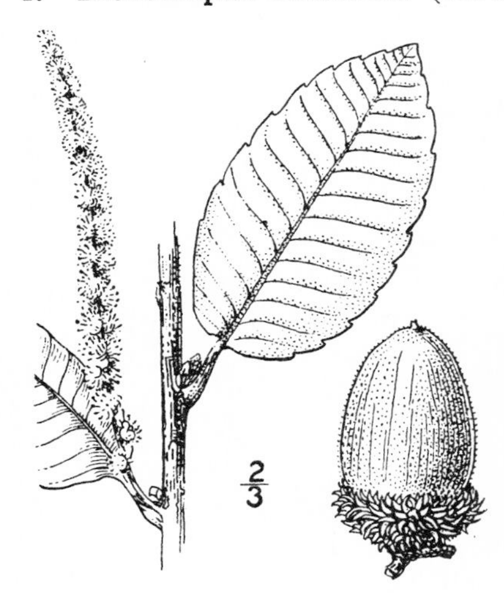

Quercus densiflora Hook. & Arn. Bot. Beech. 391. 1841.
Pasania densiflora Oerst. in Kloeb. Vidensk. Middel. 84. 1866.
Lithocarpus densiflora Rehd. in Bailey, Cycl. Hort. **6**: 3569. 1917.

Forest tree 25–35 m. high, with a trunk 1–2 m. in diameter, and a narrow conical crown; young twigs tomentose; bark 2–4 cm. thick, fissured. Leaves persisting 3–4 years, oblong, 5–12 cm. long, acute or sometimes rounded at both ends, lateral veins parallel, prominent, ending in sharp teeth; petioles 15–20 mm. long; staminate aments densely flowered, erect, 7–10 cm. long, ill-scented; nut rounded at base, gradually tapering to the apex, the shell thick; cup shallow with linear stiff spreading scales, tomentose on the inner surface.

In forests of the Humid and Arid Transition Zones; ranging from the Umpqua River, Oregon, to Mariposa County, in the Sierra Nevada and the Santa Ynez Mountains in the Coast Ranges of California. Best developed in the redwood belt. The bark is extensively used for tanning. The wood is hard and strong, suitable for implements. Type locality: California, locality not indicated.

Lithocarpus densiflora echinoìdes (R. Br.) Abrams. (*Quercus echinoides* R. Br. Campst. Ann. & Mag. Nat. Hist. IV. 7: 251. 1871.) Low shrub 0.5–3 m. high. Leaves entire, 25–50 mm. long, their veins inconspicuous; cups with slender awl-shaped recurved and uncinate scales. This shrubby variety, or possibly distinct species inhabits the Canadian Zone in the Mt. Shasta and Siskiyou region, extending to elevations of 8,000 ft. The Sierran form of the species is more or less intermediate between the two, but may prove a distinct geographical variety.

3. QUÉRCUS [Tourn.] L. Sp. Pl. 994. 1753.

Trees and shrubs with persistent or deciduous leaves, hard wood with astringent properties, and slender angled twigs marked by lenticels. Staminate aments slender drooping or spreading, the flowers solitary in the axils of caducous bracts; calyx usually 6-lobed; stamens 5–12; pistillate solitary in many-bracted involucres; style 3, short. Fruit an acorn consisting of the imbricated and more or less united bracts of the cup-like involucre subtending and partly enclosing the 1-seeded coriaceous nut. [The classical Latin name.]

A large genus of approximately 200 species, inhabiting the temperate regions of the northern hemisphere and extending into the tropics in mountainous districts. Type species, *Quercus robur* L.

Bark not scaly, smooth or on old trunks irregularly ruptured; stigmas on slender styles; scales of the involucre thin and closely imbricated; nut tomentose on the inner surface. BLACK OAKS
- Leaves deciduous, large and divided into bristle-tipped lobes. 1. *Q. kelloggii.*
- Leaves evergreen, small and coriaceous.
  - Acorns maturing the second autumn; leaves plane, bright yellow-green and glabrous below. 2. *Q. wislizenii.*
  - Acorns maturing the first autumn; leaves convex, pale beneath usually with tufts of hairs in the axils of the principal veins. 3. *Q. agrifolia.*

Bark scaly, and on large trees usually furrowed; stigmas broad and nearly or quite sessile; cups usually with tuberculate scales. WHITE OAKS
- Acorns maturing the first autumn; nut glabrous on the inner surface.
  - Leaves deciduous.
    - Leaves not blue-green, distinctly lobed.
      - Acorn cup deep hemispheric; nut tapering at apex.
        - Branchlets drooping; bark deeply fissured; leaves not prominently reticulate-veined beneath. 4. *Q. lobata.*
        - Branchlets not drooping; bark scaly; leaves prominently reticulate-veined beneath. 9. *Q. macdonaldi.*
      - Acorn cup shallow; nut broadly oblong, rounded at the apex.
        - Tree or rarely shrubby; winter buds grayish and densely villous-tomentose, 8–12 mm. long. 5. *Q. garryana.*
        - Shrub; winter buds brown, puberulent becoming glabrate, 3–5 mm. long. 6. *Q. breweri.*
    - Leaves blue-green above, merely wavy-margined or sometimes toothed; bark finely checked. 7. *Q. douglasii.*

Leaves evergreen.
Leaves not chestnut-like.
Tree; leaves oblong, entire or toothed, blue-green above; bark finely checked. 8. *Q. engelmanni.*
Shrubs; bark covered with loose scales.
Twigs rusty-tomentose.
Leaves brittle, plane or undulate, bright glossy green above. 10. *Q. dumosa.*
Leaves leathery, convex, dull green and stellate pubescent above. 11. *Q. durata.*
Twigs grayish tomentose; branches and leaves very rigid. 12. *Q. alvordiana.*
Leaves chestnut-like, 2–4 in. long, toothed, lateral veins parallel and prominent. 13. *Q. sadleriana.*
Acorns developing the second autumn; nut tomentose on the inner surface.
Cup large, thick-walled, densely tomentose.
Leaves with prominent parallel lateral veins, 2–4 in. long. 14. *Q. tomemtella.*
Leaves with lateral veins not prominent nor parallel, dull green above, pale or golden beneath. 15. *Q. chrysolepis.*
Cup smaller, thin-walled, bowl-shaped; shrubs or small trees.
Small tree, with short stiff branches; leaves rigid, prickly dentate, pale beneath. 16. *Q. palmeri.*
Low spreading montane shrub, with slender pliable branches. 17. *Q. vaccinifolia.*

1. **Quercus kellóggii** Newb. Kellogg's or California Black Oak. Fig. 1252.

Quercus tinctoria californica Torr. Pacif. R. Rep. **4**: 138. 1857.

Quercus kelloggii Newb. Pacif. R. Rep. **6**: 28, *f. 6.* 1857.

Quercus californica Cooper, Smiths. Rep. 261. 1858.

Quercus sonomensis Benth. in DC. Prod. **16²**: 62. 1864.

A tree, 10–25 or rarely 35 m. high, with a trunk 6–13 dm. in diameter; bark smooth becoming 25–35 mm. thick, divided in broad ridges near the base of old trunks. Leaves deciduous, 8–20 cm. long, deeply sinuate-lobed into about 3 lobes on each side, each lobe with 1–4 bristle-tipped teeth, bright green and glabrous or sometimes pubescent above; petioles 25–50 cm. long; acorns solitary or clustered; cup 15–25 mm. deep, puberulent within, chestnut brown without, the scales thin, with a membranous margin; nut oblong, 25–30 mm. long.

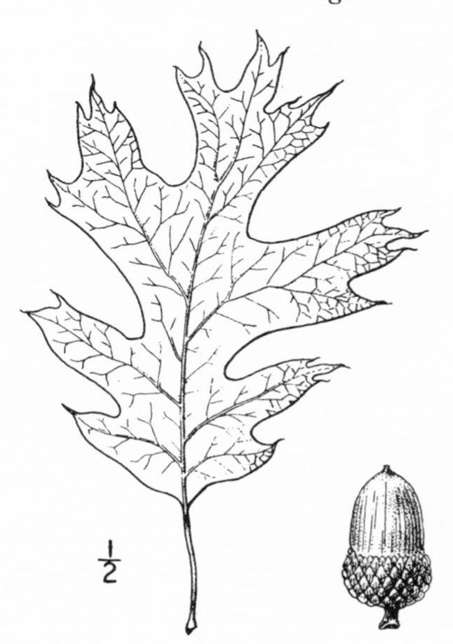

Interior foothills and mountains, Arid Transition Zone; Mackenzie River, Oregon, through the Sierra Nevada and the Coast Ranges, to the Cuiamaca Mountains, San Diego County, California. Type locality: "south and north of San Francisco in the Coast Mountains," and "between Fort Redding and Lassen's Butte," California.

2. **Quercus wislizénii** A. DC. Sierra Live Oak. Fig. 1253.

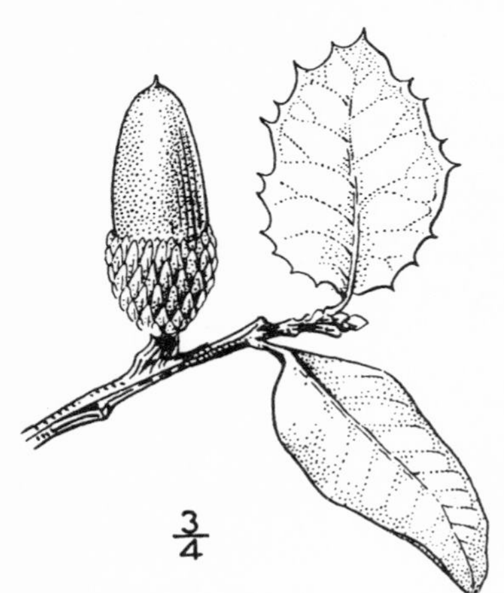

Quercus wislizenii A. DC. Prod. **16²**: 67. 1864.

A round-topped tree, 15–25 m. high, with a trunk 1–2 m. in diameter; bark smooth, 5–7 cm. thick and divided into broad ridges toward the base of old trunks. Leaves evergreen, 25–35 mm. long, lanceolate to elliptic, entire or spiny-toothed, thick and coriaceous, plane, glabrous and shining on both surfaces, dark green above, yellow-green beneath; acorns maturing the second autumn; cup turbinate, 12–20 mm. deep, or sometimes cup-shaped, its scales thin, light brown, often ciliate; nut slender, oblong-ovate, abruptly narrowed at base, pointed at apex, 20–30 mm. long.

Mountain slopes, foothills and valleys, Upper Sonoran and Transition Zones; Mt. Shasta southward through the Coast Ranges, the Sacramento Valley, and the Sierra foothills to northern Lower California. The variety *frutescens* Engelm. is a scrubby form found in the chaparral formations of the Coast Ranges and southern California. Type locality: American River, California.

Quercus morehus Kell. Proc. Calif. Acad. **2**: 36. 1863. This widely distributed but nowhere common oak is a supposed hybrid between the two preceding species. It has evergreen leaves a little smaller and less deeply lobed than those of *Quercus kelloggii.* Individual specimens have been found here and there from Lake and Tehama Counties to the San Bernardino Mountains, California.

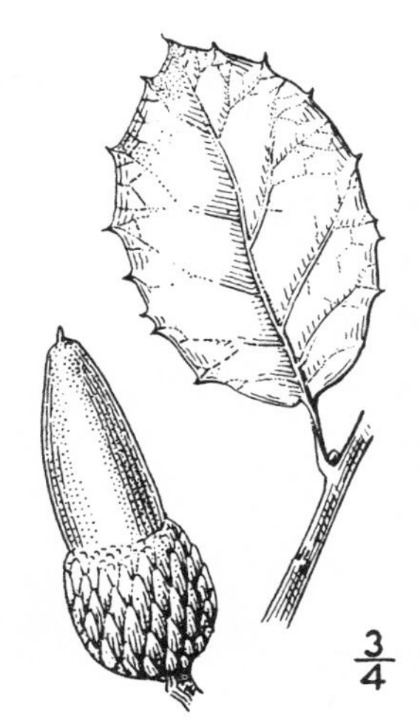

3. Quercus agrifòlia Nee. Encina or California Live Oak. Fig. 1254.

Quercus agrifolia Nee, Anal. Cienc. Nat. 3: 271. 1801.
Quercus oxyadenia Torr. Sitg. Rep. 172. *pl. 17.* 1853.
Quercus acroglandis Kell. Proc. Calif. Acad. 1: 25. 1855.

Tree 10–25 m. high, with a short trunk 1–1.5 m. or rarely 2.5 m. in diameter, and a low dense rounded head sometimes 50 m. across; bark smooth, becoming divided into broad ridges, checking into plates on old trunks. Leaves evergreen but falling as the new leaves appear in the spring, oval to oblong, usually 25–50 m. long, rounded or acute at the apex, rounded at the base, convex, finely spiny-toothed, becoming firm-coriaceous, dull green above, paler beneath, usually with tufts of hairs in the axils of the veins, acorns maturing the first autumn, sessile or nearly so; nut slender, pointed, 25–35 mm. long, the shell densely tomentose within; cup shallow or sometimes deeper, silky-pubescent within, its scales thin, rounded at the apex, puberulent especially toward the base of the cup.

Mountain slopes, foothill cañons and well-watered valleys of the Upper Sonoran and Transition Zones. Mendocino County, California, southward through the Coast Ranges to San Pedro Martir Mountain, Lower California. Type locality: Monterey, California.

4. Quercus lobàta Nee. Roble or California White Oak. Fig. 1255.

Quercus lobata Nee, Anal. Cienc. Nat. 3: 277. 1801.
Quercus hindsii Benth. Bot. Sulph. 55. 1844.

A stately tree and the largest of all the American oaks, often 35 m. high, with a trunk up to 4 m. in diameter, divided 5–10 m. above the ground into great spreading branches; young twigs silky pubescent, becoming gray and glabrous the second year; bark 6–8 cm. thick or less, divided into broad ridges by longitudinal fissures and covered by loosely appressed light gray scales. Leaves deciduous, oblong to obovate, 5–10 cm. long, deeply divided into 3–5 pairs of obtuse lobes, dark yellow-green and stellate pubescent above, pale and pubescent below, yellow-veined; petioles stout, hirsute, 6–12 mm. long; acorns maturing the first autumn; nut conical, pointed, 3–5 cm. long; cup usually enclosing about ⅓ the nut; upper scales elongated into ciliate free tips, the lower much thickened into prominent wart-like processes.

Rich loam soils of the valleys and foothills, Upper Sonoran Zone; extending from the head of Sacramento Valley southward through the Great Valley, its surrounding foothills and the valleys and foothills between the inner and outer Coast Ranges to Alhambra, Los Angeles County. An especially large handsome tree on the Bidwell Ranch at Chico has been named in honor of the noted English botanist, Sir Joseph Dalton Hooker. Another in Priests Valley, San Benito County, is the largest of all known oaks, 36 feet 7 inches in circumference.

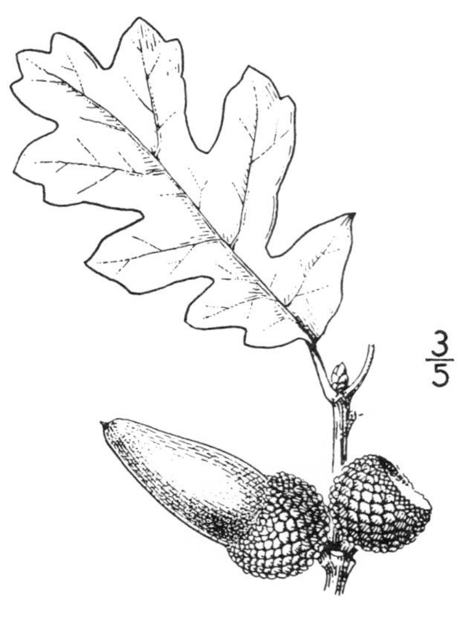

5. Quercus garryàna Dougl. Garry's or Oregon Oak. Fig. 1256.

Quercus garryana Dougl.; Hook. Fl. Bor. Am. 2: 159. 1839.
Quercus oerstediana R. Br. Campst. Ann. & Mag. Nat. Hist. IV. 7: 250. 1871.

A tree, commonly 15–20 m. occasionally 35 m. high, with a trunk 0.5–1 m. in diameter, and stout spreading branches forming a broad rounded compact head; young twigs rufous pubescent, becoming glabrous and reddish brown the second year; bark 25 mm. thick or less, shallowly divided into broad ridges and covered by grayish scales. Leaves obovate to oblong, 10–15 cm. long, coarsely pinnatifid, at maturity subcoriaceous, dark green shining and glabrous above, pale green or yellowish pubescent below; petioles pubescent, 15–25 mm. long; nut oval, obtuse, 25–30 mm. long; cup shallow or slightly turbinate, puberulent within, its scales ovate, acute, usually thickened toward the base and more or less united, pubescent or tomentose.

Upper Sonoran and mainly Humid Transition Zones; Lower Frazer River and Vancouver Island, British Columbia, southward through western Washington, Oregon and the Coast Ranges of California to Marin County. It is the only oak north of the Columbia River. Type locality: "on the plains near Fort Vancouver," along the Columbia River.

6. Quercus bréweri Engelm.
Brewer's Oak. Fig. 1257.

Quercus lobata fruticosa Engelm. Trans. St. Louis Acad. 3: 389. 1876.
Quercus breweri Engelm. Bot. Calif. 2: 96. 1880.
Quercus garryana breweri Jepson, Fl. Calif. 1: 354. 1909.

A shrub, 1–5 m. high, with smoothish gray bark, and slightly pubescent twigs; winter buds brown, ovate, 2–4 mm. long, sparsely puberulent. Leaves 5–12 cm. long, pinnatifid to the middle, the lobes obtuse or acute, entire or toothed, finely pubescent, at length nearly glabrous and shining above, pale beneath and often velvety pubescent; cup shallow, somewhat tuberculate, 12–20 mm. broad; nut oval, obtuse, about 20 mm. long.

Mountain slopes, Upper Sonoran to Canadian Zones; the typical form inhabits the high altitudes of the Siskiyou and Mt. Shasta regions. On the western slopes of the Sierra Nevada at middle elevations a form with less tuberculate scales occurs from Tuolumne County southward to the Kaweah Basin. This has been described as (*Quercus garryana senota* Jepson, *loc. cit.*) and may prove distinct. Type locality: "mountains west of Shasta."

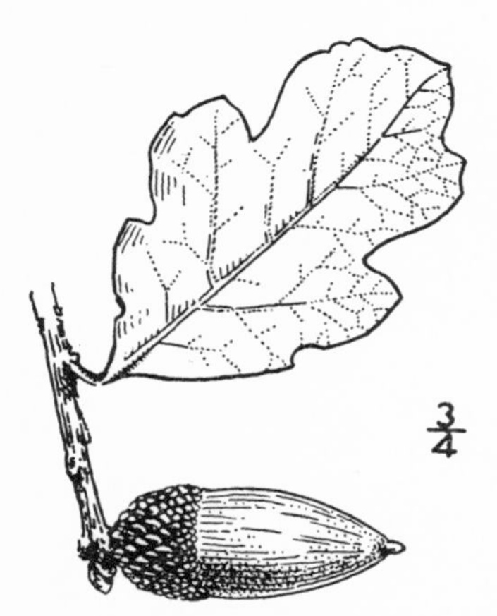

7. Quercus douglásii Hook. & Arn.
Blue Oak. Fig. 1258.

Quercus douglasii Hook. & Arn. Bot. Beech. 391. 1841.
Quercus ransomi Kell. Proc. Calif. Acad. 1: 25. 1855.

A tree usually 10–20 m. high, with a trunk 5–12 dm. in diameter, and a dense round-topped crown; young twigs hoary tomentose; bark shallowly checked into light gray scales. Leaves deciduous, 5–12 cm. long, oblong, with broad shallow lobes appearing like undulations on the margin, or sometimes rather deeply lobed, blue-green above, pale beneath, prominently veined and pubescent; acorn cups shallow, 10–12 mm. broad, the scales thickened; nut oval, 20–25 mm. long, narrow at base, acute at apex, tomentose.

Dry rocky hillsides, Upper Sonoran Zone; head of Sacramento Valley south through the Sierra Nevada foothills and the Coast Ranges to the Sierra Liebre Mountains, Los Angeles County. Type locality: California, locality not indicated.

8. Quercus engelmànni Greene. Englemann's Oak. Fig. 1259.

Quercus engelmanni Greene, W. Am. Oaks 33. 1889.
Quercus macdonaldi elegantula Greene, W. Am. Oaks 26. 1889.

A tree 15–20 m. high, with a trunk 0.5–1 m. in diameter, and spreading branches forming a broad rounded or irregular head; young twigs hoary tomentose; bark divided by narrow fissures and covered by thin appressed scales. Leaves evergreen, falling after the appearance of the new leaves in spring, oblong to obovate, 3–7 mm. long usually obtuse at apex, entire, often undulate or sinuate toothed, firm in texture, dark blue-green and glabrous or sparsely pubescent above, pale beneath and pubescent or glabrous; cup usually enclosing nearly half the nut, light brown and puberulent within, its scales light brown, tomentose, tuberculate toward the base, the uppermost acute and ciliate; nut oblong to oval, dark chestnut brown, 20–25 mm. long.

Foothills and valleys, Upper Sonoran Zone; extending from Pasadena southward through the foothills of San Diego County, California. Type locality: "mountains of southern California, from the mesas east of San Diego northward to Kern County."

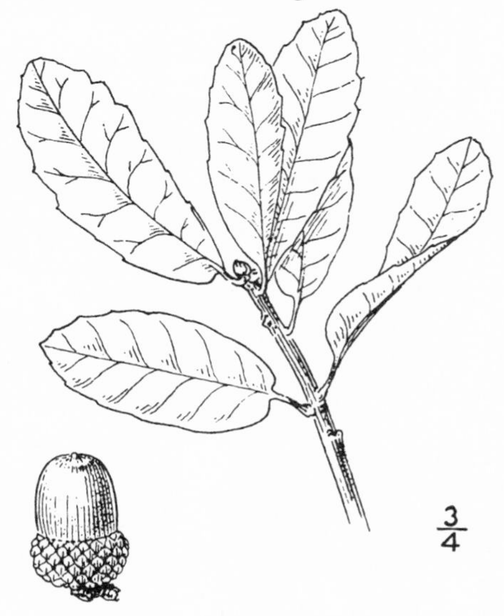

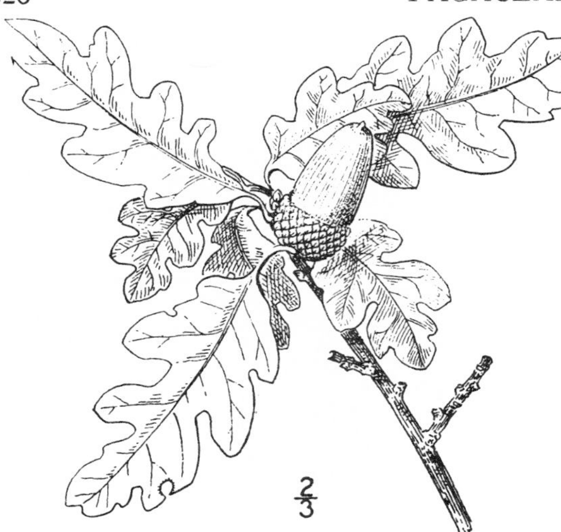

9. **Quercus macdónaldi** Greene. MacDonald's Oak. Fig. 1260.

Quercus macdonaldi Greene, W. Am. Oaks 25. 1889.

A small tree, 8–12 m. high with a trunk 5 dm. in diameter, and a compact rounded crown; young twigs densely pubescent; bark scaly. Leaves deciduous, oblong to obovate, 5–7 cm. long, with 2–4 lobes on a side, these blunt or sharp-pointed and bristle-tipped, narrowed and entire at base, thick and firm, glabrous above, densely pubescent below becoming less so and prominently reticulate-veined; acorns sessile; cup deeply hemispherical, its scales strongly tuberculate; nut oblong-conical acute 20–35 mm. long.

Cañon slopes, Upper Sonoran Zone; Santa Cruz and Santa Catalina Islands, off the coast of southern California. Type locality: Santa Cruz Island.

10. **Quercus dumòsa** Nutt. California Scrub Oak. Fig. 1261.

Quercus dumosa Nutt. Sylva 1: 7. 1842.
Quercus acutidens Torr. Bot. Mex. Bound. 207. *pl. 51.* 1859.

A shrub 1–3 m. high, with stout short stiff branches; young twigs rusty-tomentose; bark scaly, light grayish brown. Leaves evergreen but falling as the new leaves appear in the spring, oblong to roundish, 15–25 mm. long, entire or usually irregularly spinose-serrate, or sinuate-lobed, coriaceous, plane or undulate, dark green and shining above, pale beneath and pubescent; cup deep cup-shaped, enclosing about one-half the nut, light brown and pubescent within, its scales ovate, united and tuberculate toward the base, more or less rusty-tomentose; nut oval, broad at base, rounded or acute at apex, 1–3 cm. long.

Chaparral covered foothills and mesas, Upper Sonoran Zone; northern Coast Ranges of California and the western slope of the central Sierra Nevada, southward to northern Lower California. Type locality: Santa Barbara. A very common species in the southern part of California and remarkably variable in foliage and fruit. Apparent hybrids between it and *Quercus engelmanni* sometimes occur.

Quercus dumosa turbinélla (Greene) Jepson, Fl. Calif. 1: 356. 1909. (*Quercus turbinella,* Greene, W. Am. Oaks 37. 1889.) Leaves oblong or elliptic, bright green above, pale beneath; cup turbinate, its scales thin not tuberculate; nut slender acute, about 25 mm. long. Mountain slopes, Upper Sonoran Zone; San Jacinto Mountains, southern California, to northern Lower California. Type locality: mountains of Lower California, 20 to 30 miles below the international boundary.

11. **Quercus duràta** Jepson. Leather Oak. Fig. 1262.

Quercus dumosa bullata Engelm. Bot. Calif. 2: 96. 1880.
Quercus durata Jepson, Fl. Calif. 1: 356. 1909.

A shrub 1–3 m. high, with rigid branches and branchlets; young twigs rusty-tomentose; bark scaly, light grayish brown. Leaves evergreen, persisting well after the new leaves are formed, 15–25 mm. long, oval, thick and coriaceous, spinose dentate, revolute, dark dull green and stellate-pubescent above, paler and tomentose below; cup bowl-shaped, enclosing about one-half the nut, its scales thick and tuberculate, grayish tomentose.

Chaparral covered hills and mountains, Upper Sonoran Zone; Coast Ranges from Lake to San Luis Obispo Counties, California. Type locality: San Carlos Range, California.

12. Quercus alvordiàna Eastw.
Alvord's Oak. Fig. 1263.

Quercus alvordiana Eastw. Handb. Trees Calif. 48. *pl. 27, fig. 4.* 1905.

Shrub 2–3 m. high, with stiff brittle almost divaricate branches, young twigs grayish tomentose. Leaves evergreen, broadly oval to oblong, 15–40 mm. long, entire or commonly spinulose dentate, grayish green above and sparsely stellate-pubescent, paler and stellate-tomentose beneath, prominently reticulate, thick, coriaceous and often undulate; cups shallow, 10–15 mm. broad, the scales grayish tomentose somewhat tuberculate; nut slender acute, 20–30 mm. long.

Dry interior hills or desert slopes, Upper Sonoran Zone; south Coast Ranges (San Carlos Range) to northern Lower California. Referred in the Bot. Calif. to *Querus undulata pungens* Engelm. Type locality: San Emigdo Cañon, Kern County, California.

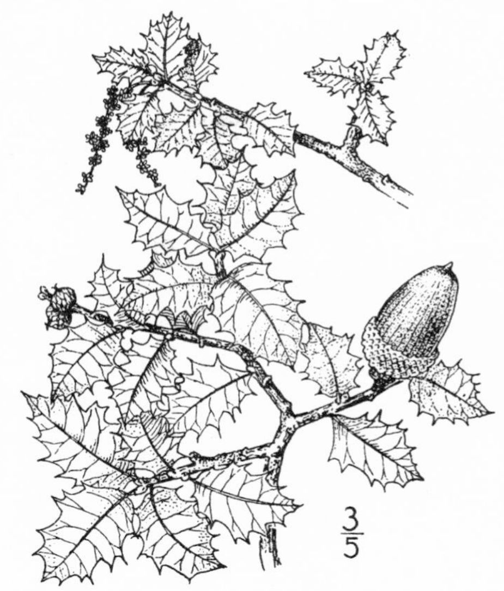

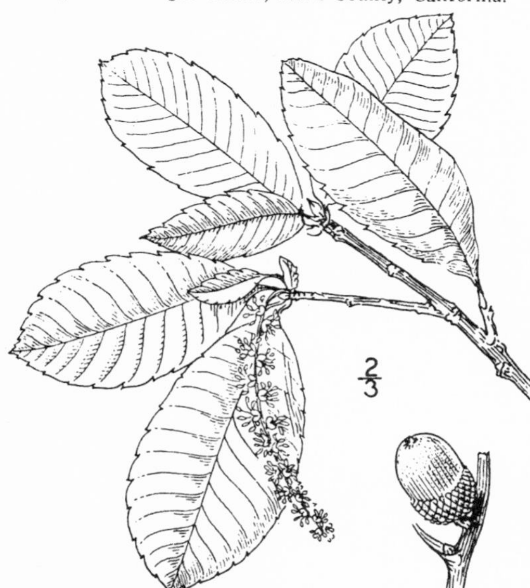

13. Quercus sadlerìana R. Br.
Deer or Sadler's Oak. Fig. 1264.

Quercus sadleriana R. Br. Campst. Ann. & Mag. Nat. Hist. IV. 7: 249. 1871.

Low shrub, 0.5–2.5 m. high branching from the base, young twigs glabrous or nearly so. Leaves evergreen, falling in the summer after the appearance of the new leaves, oblong-ovate to broadly ovate, 5–14 cm. long, coarsely serrate, prominently pinnate-veined glabrous or nearly so; stipules 10–15 mm. long, oblanceolate, densely silky; acorns solitary or several crowded in the axils of the upper leaves, maturing the first autumn; cup cup-shaped, thin, the scales thin, scarcely tuberculate; nut ovate, 15–20 mm. long.

Dry slopes and ridges, at high altitudes, Canadian Zone; Josephine and Curry Counties, Oregon, south through the Siskiyou and Klamath Mountains to Trinity County, California. Type locality: Crescent City trail between Sailors' Diggings, Oregon, and Smith Creek, California.

14. Quercus tomentélla
Engelm. Island Oak.
Fig. 1265.

Quercus tomentella Engelm. Trans. St. Louis Acad. **3**: 393. 1877.

A small round-topped tree, 8–12 m. high, with a trunk 3–6 dm. in diameter; young twigs hoary tomentose becoming brown; bark thin, reddish brown, covered with loosely appressed scales. Leaves persisting until the third year, 5–10 cm. long, oblong lanceolate, usually acute at the apex, rounded or wedge-shaped at base, crenate-dentate with callous-tipped teeth or sometimes entire, stellate-pubescent beneath; acorns maturing the second autumn; cup shallow, pubescent within, its scales ovate, acute, more or less hidden in the thick hoary tomentum; nut ovate, 15–20 mm. long.

Peculiar to the islands off the coast of southern California and Lower California, where it has been found on Santa Cruz, Santa Rosa, Santa Catalina, San Clemente and Guadaloupe. Type locality: Guadaloupe Island.

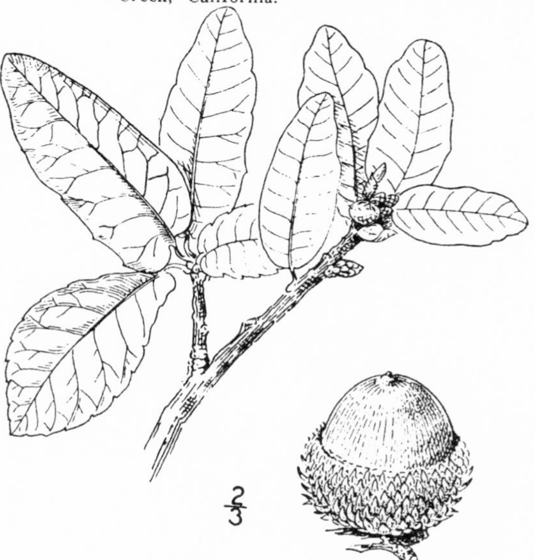

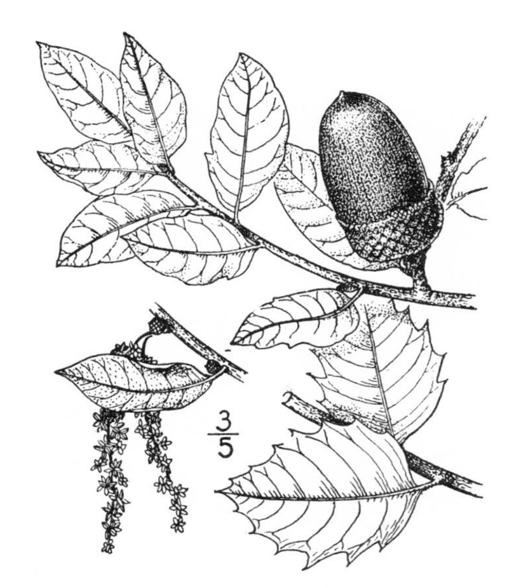

15. **Quercus chrysolèpis** Liebm. Cañon or Maul Oak. Fig. 1266.

Quercus chrysolepis Liebm. Dansk. Vidensk. Forhandl. **1854**: 173. 1854.

Quercus fulvescens Kell. Proc. Calif. Acad. **1**: 67. 1855.

Quercus crassipocula Torr. Pacif. R. Rep. **4**: 137. 1857.

A tree 10–20 m. high, with a trunk 0.5–2 m. in diameter or sometimes reduced to a mere shrub; young twigs hoary-tomentose; bark ashy gray and smooth, covered with closely appressed gray-brown scales. Leaves evergreen 1.5–6 cm. long, usually oblong-ovate, entire or on vigorous shoots or young trees often spiny-toothed, thick coriaceous, becoming dark green above, grayish or yellowish tomentose or ultimately becoming glabrous and glaucous beneath; acorns maturing the second autumn; cup shallow, with usually very thick walls, silky within, its scales mostly hidden by a dense coat of rusty tomentum; nut oval or oblong, 25–30 mm. long, nearly as broad, the shell with scanty tomentum within.

Cañons and moist mountain slopes, Upper Sonoran and Transition Zones; southern Oregon southward along the western slopes of the Sierra Nevada and the Coast Ranges to San Pedro Martir Mountain, Lower California, and also in the mountains of Arizona, New Mexico and Sonora. Type locality: California, locality not designated.

16. **Quercus pàlmeri** Engelm. Palmer's Oak. Fig. 1267.

Quercus palmeri Engelm. Trans. St. Louis, Acad. **3**: 393. 1877.

Quercus dunnii Kell. Pacif. Rural Press. **17**: 371. 1879.

A shrub or small tree, 3–5 m. high, with short rigid branches and finely checked dark grayish brown bark. Leaves evergreen 2–3 cm. long, stiff coriaceous, elliptic to rounded, undulate, spinose, dull green above, when young densely white or yellowish tomentose becoming nearly smooth but pale or whitish beneath; acorns maturing the second autumn; cup shallow, subturbinate, 10–12 mm. broad, its walls rather thin; scales densely clothed with tomentum; nut ovoid, tomentose on the inner surface of the shell.

Chaparral covered mountains, Upper Sonoran Zone; extending from the San Jacinto Mountains, southern California, southward into northern Lower California. Type locality: northern Lower California.

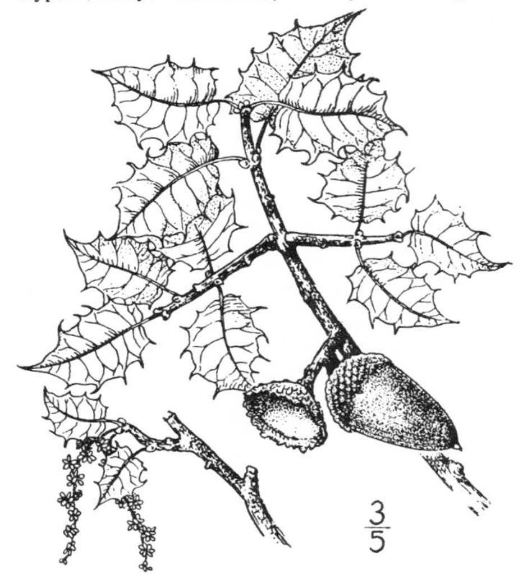

17. **Quercus vaccinifòlia** Kell. Huckleberry Oak. Fig. 1268.

Quercus vaccinifolia Kell. Proc. Calif. Acad. **1**: 96. 1855.

Quercus chrysolepis vaccinifolia Engelm. Trans. St. Louis Acad. **3**: 393. 1877.

Low spreading often prostrate shrub, with slender flexible branchlets, simulating the habit of the huckleberry. Leaves evergreen, about 15–25 mm. long, oblong-ovate, usually entire, pale shiny green above, tomentose beneath; acorns maturing the second year; cup shallow, bowl-shaped, usually pubescent, its walls thin; nut rounded-ovate, 10–15 mm. long.

Dry subalpine ridges often forming thickets, Canadian Zone; extending from the Siskiyou Mountains, southern Oregon, to the Trinity Mountains and the southern Sierra Nevada. Type locality; Trinity, Scott and Siskiyou Mountains.

Family 30. **ULMÀCEAE.**

Elm Family.

Trees with watery juice, terete branchlets arising from an upper lateral scaly bud. Leaves alternate, simple, serrate, deciduous, petioled and 2-ranked, oblique at the base; stipules caducous. Flowers perfect, polygamous or monoecious, clustered or the pistillate solitary. Calyx 4–9-parted or lobed. Stamens 4–6; styles 2; ovary usually 1-celled, with a single suspended ovule. Fruit a samara, nut or drupe; seed with little or no endosperm.

A family of thirteen genera widely distributed over the temperate and tropical regions. Several species of *Ulmus* (elm) are frequently planted in the Pacific States, and in some localities have become more or less established by seeding or suckering.

1. **CÉLTIS** [Tourn.] L. Sp. Pl. 1043. 1753.

Trees with thin smooth bark, serrate or entire leaves and usually scarious stipules. Flowers monoecious or polygamo-monoecious, appearing with the leaves on branches of the season, minute, pedicellate; the staminate cymose or fascicled, the pistillate solitary or in few-flowered clusters in the axils of the upper leaves. Calyx deeply 4–5-lobed, greenish yellow, deciduous. Ovary ovoid, with a short sessile style. Fruit a drupe with a thick-walled smooth or rugose nutlet. [Ancient Latin name.]

About 60 species widely distributed over the temperate and tropical regions of both hemispheres. Type species, *Celtis australis* L.

1. **Celtis douglásii** Planch. Douglas's Hackberry. Fig. 1269.

Celtis douglasii Planch. Ann. Sci. Nat. III. **10**: 293. 1848.

Shrub or small tree 2–3.5 m. high, with rough grayish brown bark and glabrous or puberulent branchlets. Leaves ovate or ovate-lanceolate, acute or acuminate, 10 cm. long or less, serrate to nearly or quite entire, rough above, strongly netted-veined beneath; petioles mostly not over 6–8 mm. long, pubescent; pedicels well exceeding the petioles, pubescent; fruit globose, about 8 mm. in diameter, dark brown when mature.

Cañon floors and bottom lands along streams, Upper Sonoran Zone; eastern Washington and Oregon to Utah, also Kern and San Diego Counties, California. Type locality: "the arid interior region, along the Columbia River."

Family 31. **URTICÀCEAE.**

Nettle Family.

Annual or perennial herbs (some tropical species trees or shrubs), with alternate or opposite, mostly stipulate leaves. Flowers small, greenish, dioecious, monoecious or polygamous, in cymose racemes or fascicles. Calyx 2–5-cleft or of separate sepals. Petals none. Stamens as many as calyx-lobes or sepals and opposite them, the filaments inflexed. Ovary superior, 1-celled; style simple with a capitate or filiform stigma; ovule solitary, erect or ascending, orthotropous. Fruit an achene. Endosperm oily, usually not copious; embryo straight.

About 40 genera and 550 species of wide geographic range.

Herbs with stinging hairs; leaves opposite, stipulate.
- Calyx deeply 4-parted, the inner much the longer in the pistillate flower and inclosing the achene. — 1. *Urtica.*
- Calyx 4-parted in the staminate flower, tubular and 2–4-toothed in the pistillate. — 2. *Hesperocnide.*

Herbs without stinging hairs; leaves alternate, without stipules. — 3. *Parietaria.*

1. ÚRTICA [Tourn.] L. Sp. Pl. 983. 1753.

Annual or perennial, simple or branching herbs, with stinging hairs, and opposite 3–7-nerved petioled serrate or dentate stipulate leaves. Flowers clustered in axillary geminate racemes or heads. Staminate flowers with a deeply parted calyx and 4 stamens. Pistillate calyx with unequal sepals, the inner larger and at length enclosing the flattened achene. Stigma sessile, tufted. [The ancient Latin name.]

About 30 species of wide geographic distribution. Type species, *Urtica dioica* L.

Perennials; pistillate and staminate flowers in separate clusters.
  Stem and leaves densely pubescent.
    Stem and leaves coarsely velvety pubescent; seeds tan-colored, smooth. 1. *U. holosericea.*
    Stem finely strigose; lower surface of leaves short pubescent; seed olive-brown, finely tuberculate. 2. *U. breweri.*
  Stem nearly glabrous; leaves dark green above and glabrate in age; seeds tan-colored, smooth.
    Leaves grayish pubescent below, deeply cordate, not with a prolonged entire apex. 3. *U. californica.*
    Leaves green on both surfaces, with an entire caudate apex. 4. *U. lyallii.*
Annual; pistillate and staminate flowers mixed in same cluster. 5. *U. urens.*

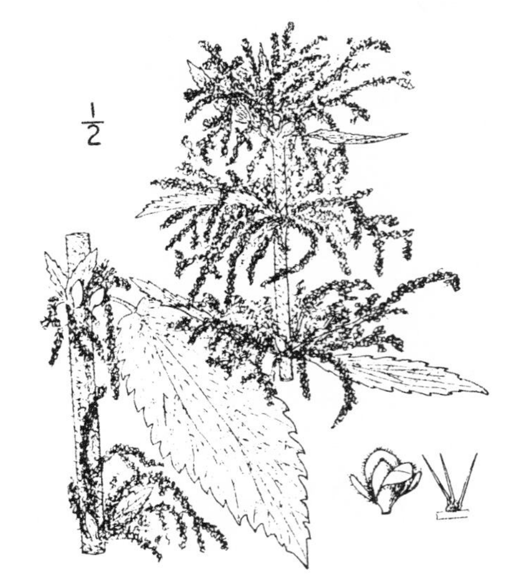

1. Urtica holosericea Nutt. Hoary Nettle. Fig. 1270.

Urtica holosericea Nutt. Journ. Acad. Phila. II. 1: 183. 1847.
Urtica gracilis holosericea Jepson, Fl. Calif. 1: 367. 1909.

Stems simple, stout, 1–3 m. high or more, more or less bristly and finely pubescent. Leaves cinereous densely pubescent beneath and soft velvety, less so above or with only a few scattering bristles, ovate to lanceolate, 5–10 cm. long, the upper much shorter, on petioles one-fourth as long, coarsely serrate; stipules narrowly oblong, acute or obtuse, 6–10 mm. long; staminate flower clusters rather loose, nearly equalling the leaves; pistillate denser and shorter; inner sepals ovate, densely hispid, 1 mm. long, about equalling the broadly ovate achene.

Low ground and stream banks, Transition and Upper Sonoran Zones; Washington and Idaho, south to northern Lower California. The plants of the Sierra Nevada and southern California mountains are greener, with the stems sometimes nearly glabrous. July–Sept. Type locality: Monterey, California.

2. Urtica bréweri S. Wats. Brewer's Nettle. Fig. 1271.

Urtica breweri S. Wats. Proc. Am. Acad. 10: 348. 1875.

Stem 1–2 m. high, stout, nearly glabrous or grayish with a short strigose pubescence and scattered bristles. Leaves 4–10 cm. long, ovate to ovate-lanceolate, rounded or subcordate at base, coarsely toothed, sparsely strigose above, becoming glabrate, velvety pubescent beneath, with a short fine grayish pubescence; petioles usually strongly bristly, about one-third the length of the blades; flower clusters dense, much shorter than the leaves; achenes usually much shorter than the inner sepals, dull olive brown, sparsely papillate-roughened.

Low ground and stream banks, Upper Sonoran Zone; southern California and Utah to western Texas and adjacent Mexico. June–Oct. Type locality: near Los Angeles.

3. Urtica califórnica Greene. California or Coast Nettle. Fig. 1272.

Urtica californica Greene, Pittonia 1: 281. 1889.
Urtica lyallii californica Jepson, Fl. W. Mid. Calif. 147. 1901.

Stem simple or branched from the base, 6–10 dm. high, hispid and finely (often sparsely) strigose. Leaves broadly ovate, deeply cordate, coarsely serrate, 7–10 cm. long or the lower often 15 cm. long, short-pubescent beneath and often gray, dark green above and nearly glabrous; petioles 3–7 cm. long; stipules oblong, obtuse, 8–12 mm. long; flower clusters simple or paniculately branched, mostly exceeding the petioles; achene ovate, puncticulate, about equalling the inner sepals.

Low ground near the coast, Transition Zone; Marin to San Mateo Counties, California. June–Sept. Type locality: Point Pietras, San Mateo County.

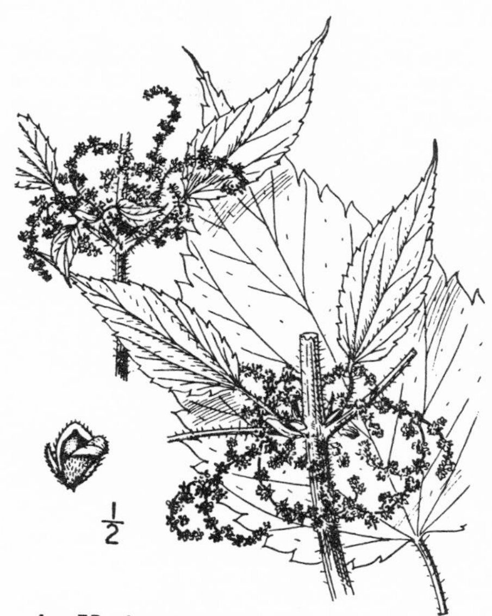

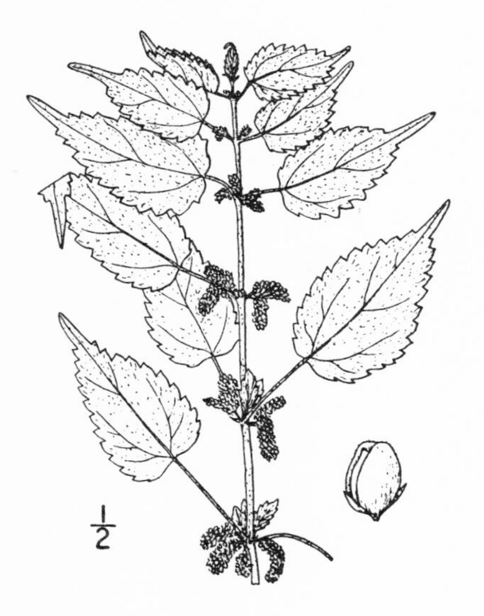

4. Urtica lyállii S. Wats. Lyall's Nettle. Fig. 1273.

Urtica lyallii S. Wats Proc. Am. Acad. 10: 281. 1875.

Stem branching from the base, erect, 1–2 m. high, sparsely bristly or entirely glabrous. Leaves 3–15 cm. long, ovate or shallowly cordate, thin, coarsely toothed, with an entire elongate acumination at apex, green and nearly glabrous or minutely pubescent; petioles usually about one-third the length of the blade; stipules oblong, obtuse, about 1 cm. long; staminate flower clusters paniculate, well exceeding the petioles; the pistillate unbranched, scarcely exceeding the petioles; achene broadly ovate; smooth, equalling the inner sepals.

Low ground, Canadian and Transition Zones; Alaska south to the Coastal region of Oregon. May–Sept. Type locality: Vancouver Island.

5. Urtica ùrens L. Small or Dwarf Nettle. Fig. 1274.

Urtica urens L. Sp. Pl. 984. 1753.

Erect, annual, stem branching from the base or sometimes simple, 15–50 cm. high, stinging bristly. Leaves ovate or oblong-ovate, thin, glabrous or nearly so, 3–5-nerved, deeply and sometimes doubly serrate, 1–4 cm. long; on slender petioles of about the same length; stipules 4 mm. long; flower clusters rather dense, mostly shorter than the petioles; flowers androgynous, mainly pistillate, both kinds mixed in the same clusters; fruiting calyx with hispid-ciliate margins; achene 2 mm. long.

In waste places, and a garden and orchard weed. Naturalized from Europe and widely distributed over North America.

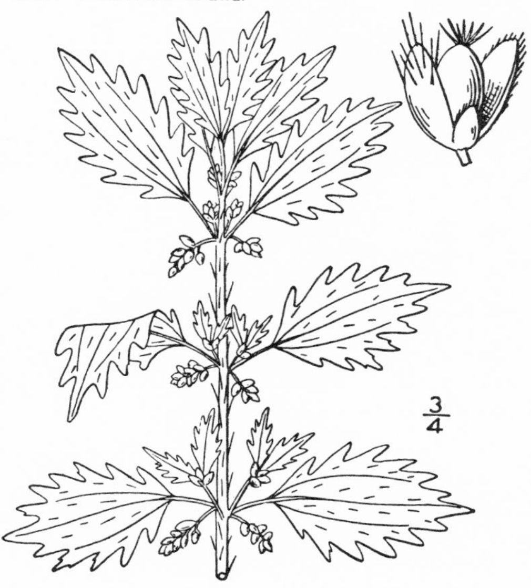

2. HESPEROCNÌDE Torr. Pacif. R. Rep. 4: 139. 1857.

Annual herbs with stinging hairs, and opposite petioled serrate or dentate leaves. Flowers in clustered loose axillary heads, monoecious, the staminate and pistillate in the same clusters. Calyx of the staminate flowers 4-parted, the lobes equal, that of the pistillate flowers tubular, 2–4-toothed, beset with uncinate hairs. Stamens 4, opposite the calyx-lobes, inflexed. Stigma terminal, nearly sessile, tufted. Achene enclosed by the thin membranous calyx. [Greek, meaning western nettle.]

Two species, the other occurring on the Hawaiian Islands. Type species, *Hesperocnide tenella* Torr.

1. Hesperocnide tenélla Torr. Western Nettle. Fig. 1275.

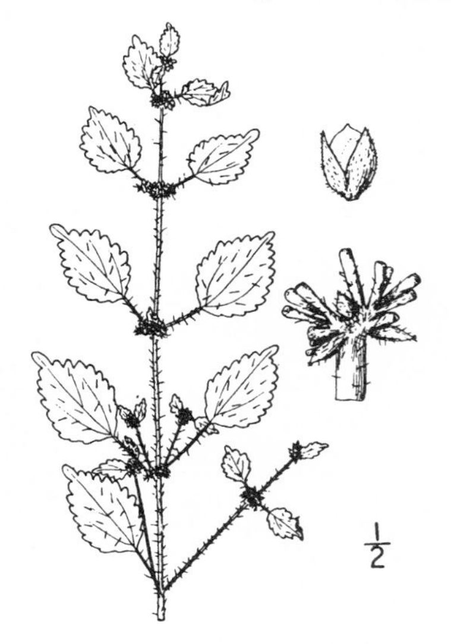

Hesperocnide tenella Torr. Pacif. R. Rep. 4: 139. 1857.

Stems slender and weak, 25–50 cm. high, simple or branched, somewhat hispid with branching hairs and bristly. Leaves 1–3 cm. long, thin, ovate, obtusely serrate; petioles slender, half as long as the leaf blades; flower clusters rather dense, nearly glomerate, shorter than the petioles; calyx thin, hispid, with hooked hairs, in fruit 1–1.5 mm. long; achene membranous, striately tuberculate with minutely rough points.

On shaded cañon slopes, and hillsides, Upper Sonoran Zone; Coast Ranges from Napa County and the central Sierra Nevada, south to San Diego County. March–July. Type locality; Napa Valley, California.

3. PARIETÀRIA L. Sp. Pl. 1052. 1753.

Annual or perennial herbs (some tropical species shrubs or trees) without stinging hairs. Leaves alternate, entire, 3-nerved, petioled, without stipules. Flowers in axillary glomerate clusters, polygamous, subtended by leafy bracts. Calyx of the perfect flowers 4-parted, of the pistillate tubular-ventricose and 4-cleft with connivent lobes. Style slender or none; stigma spatulate, recurved, densely tufted. Achene ovoid, smooth and shining, enclosed in the dry brownish nerved calyx. [Ancient Latin name, referring to the growth of some species on walls.]

About 7 species, widely distributed. Besides the following two others inhabit the eastern states. Type species, *Parietaria officinalis* L.

Involucre 2 or 3 times as long as the sepals; stems simple or sparingly branched. 1. *P. pennsylvanica.*
Involucre very slightly surpassing the sepals; stems much branched. 2. *P. floridana.*

1. Parietaria pennsylvànica Muhl. Pennsylvania Pellitory. Fig. 1276.

Parietaria pennsylvanica Muhl.; Willd. Sp. Pl. 4[3]: 955. 1805.

Parietaria occidentalis Rydb. Bull. Torrey Club 39: 306. 1912.

Stems slender erect; simple or branched at base, gibbous, pubescent, 1–4 dm. long. Leaves lanceolate or oblong-lanceolate, acuminate at the apex, narrowed at the base, 25–75 mm. long; petiole slender, usually much shorter than the leaves; flowers glomerate in all but the lowest axils; bracts of the involucre linear, 2–3 times as long as the sepals; sepals linear-lanceolate; achene 1 mm. long.

Moist shaded places, Transition Zone; Washington, British Columbia and Maine, south to Tennessee, Mexico and northern California. May–Aug. Type locality: Pennsylvania.

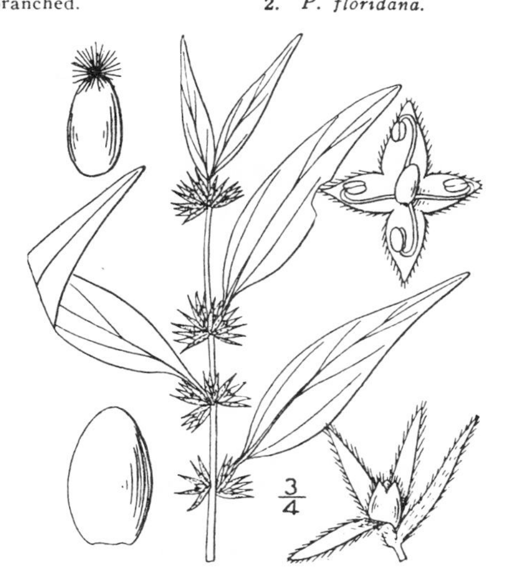

2. Parietaria floridàna Nutt. Florida Pellitory. Fig. 1277.

Parietaria floridana Nutt. Gen. **2**: 208. 1818.

Stems very slender usually diffusely branching from the base, 10–25 cm. long, somewhat hispid. Leaves 5–10 mm. long or more, broadly ovate, obtuse, rounded at base or abruptly cuneate; bracts linear or linear-lanceolate acute about 3 mm. long; sepals lanceolate acute, nearly equalling the bracts; achene 1 mm. long.

Moist shady places, Upper Sonoran Zone; central California to Lower California, also in the southeastern states. March–May. Type locality: "near St. Mary's, west Florida." *Parietaria debilis* Forst, to which the species has been referred is a closely related old world species.

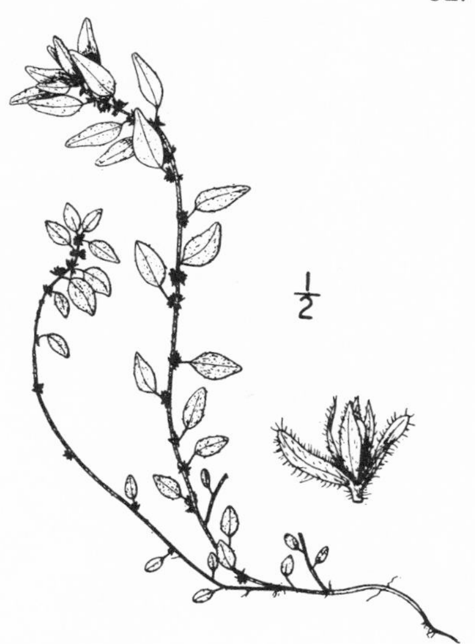

Family 32. SANTALÀCEAE.

Sandalwood Family.

Herbs, shrubs or trees, with entire exstipulate leaves. Flowers, perfect or unisexual, axillary or terminal. Calyx adnate to the base of the ovary or disk, 4–5-cleft, valvate in the bud. Petals none. Stamens as many as the calyx-lobes and opposite them, inserted on the fleshy disk. Style 1, or sometimes none; stigma capitate. Ovary 1-celled; ovules 2–4, suspended from the summit of the central placenta. Fruit a drupe or nut. Seed 1, ovoid or globose; testa none; endosperm copious; embryo small, apical.

About 26 genera and 250 species mostly in the tropical regions.

1. COMÁNDRA Nutt. Gen. 1: 157. 1818.

Smooth perennial herbs (sometimes parasitic on roots) with erect stems from a woody base. Leaves alternate, entire, nearly sessile. Flowers perfect. Calyx campanulate, 5-lobed, the tube lined with a 5-lobed disk. Stamens 5, inserted between the lobes of the disk; anthers attached to the calyx-lobes by a tuft of hairs. Fruit drupe-like, crowned by the persistent calyx. [Greek referring to the hairy attachment of the anthers.]

Four species, 3 in North America and 1 in Europe. Type species, *Comandra umbellata* (L.) Nutt.

Flowers in terminal corymbose or paniculate cymes.
- Leaves oblong, green; calyx-tube produced above the globose fruit. 1. *C. umbellata.*
- Leaves lanceolate, very pale; calyx-tube not produced above the ovoid fruit. 2. *C. pallida.*

Flowers on axillary 2–3-flowered peduncles. 3. *C. livida.*

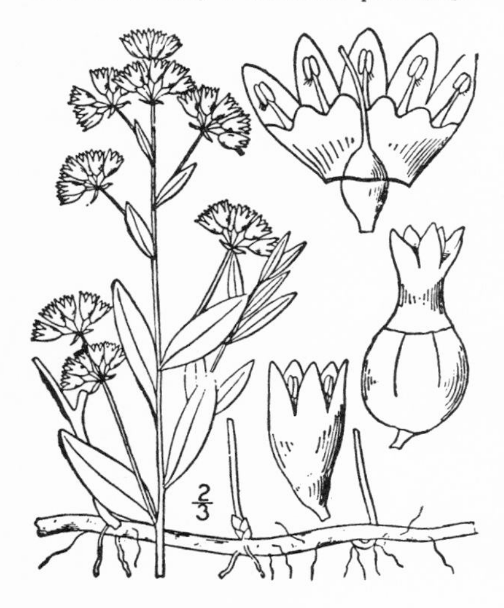

1. Comandra umbellàta (L.) Nutt. Bastard Toad-flax. Fig. 1278.

Thesium umbellatum L. Sp. Pl. 208. 1753.
Comandra umbellata Nutt. Gen. 1: 157. 1818.

Stems slender, very leafy, usually branched, 15–45 cm. high. Leaves sessile, oblong or oblong-lanceolate, 15–30 mm. long, acute or subacute at both ends; cymes corymbose at the summit or more commonly ellipsoid paniculate, several-flowered, calyx greenish white or purplish, about 4 mm. long, the lobes oblong; drupe globose, 5–6 mm. in diameter, the calyx-tube produced above the drupe 1 mm. or more.

Dry soil, Transition Zone; British Columbia to Labrador south to western Oregon and Georgia. Type locality: Virginia.

2. **Comandra pállida** A. DC. Pale Comandra. Fig. 1279.

Comandra pallida A. DC. Prodr. **14**: 636. 1857.

Comandra umbellata pallida Jones, Proc. Calif. Acad. II. **5**: 722. 1895.

Comandra californica Eastw.; Heller, Muhlenbergia **2**: 20. 1905.

Similar to the preceding species, but paler and glaucous, usually much branched. Leaves sessile, linear or linear-lanceolate, or the lower oblong-elliptic, 15–35 mm. long, acute; cymes few–several-flowered, corymbosely clustered at the summit; pedicels 2–4 mm. long; calyx purplish, about 4 mm. long, the lobes oblong; drupe ovoid-oblong, 6–8 mm. long, 4–5 mm. in diameter, the calyx-tube scarcely produced above the body of the fruit.

Dry soils, arid Transition Zone; British Columbia to Manitoba south to the Sierra Nevada, California, Arizona, Texas, and Kansas. Type locality: near Clearwater, Idaho. The fruits are sweet and edible.

3. **Comandra lívida** Richards. Northern Comandra. Fig. 1280.

Comandra livida Richards. App. Frank. Journ. 734. 1823.

Stems slender, usually simple, 10–30 cm. high. Leaves 10–25 mm. long, oval, obtuse at apex, narrow at base, thin, green; short-petioled; cymes axillary, few, or often only 1, 1–5-flowered; peduncle shorter than the leaves, filiform; flowers sessile; calyx-lobes ovate; style very short; drupe oblong-globose, about 6 mm. in diameter, red and edible, the calyx-tube not produced above the body of the fruit.

In moist soils, Canadian and Hudsonian Zones; Alaska to Newfoundland, south to Washington, Idaho, Michigan and New Hampshire. Type locality: "not seen to the northward of Great Slave Lake."

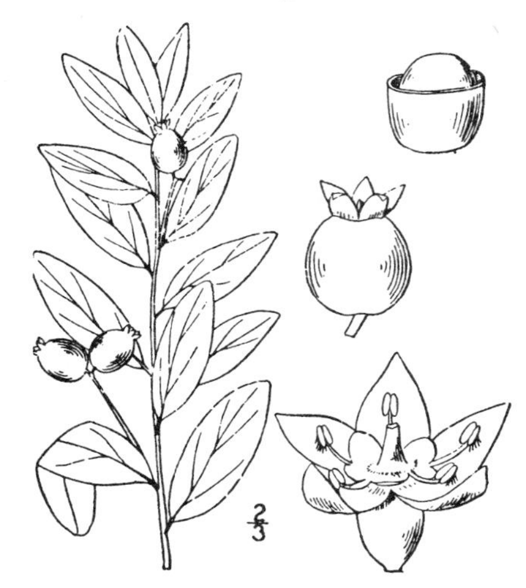

Family 33. LORANTHÀCEAE.

Mistletoe Family.

Evergreen shrubs or herbs, parasitic on shrubs or trees and absorbing food from their sap through specialized roots (haustoria). Stems dichotomously branched, swollen at the joints and bearing opposite thick coriaceous entire exstipulate leaves, foliaceous or reduced to connate scales. Flowers dioecious, regular, clustered or solitary, small and greenish. Petals none. Calyx-tube adnate to the ovary, 2–5-lobed. Stamens equalling the calyx-lobes and inserted upon them; anthers 2-celled or confluently 1-celled. Ovary inferior, 1-celled, 1-ovuled; style simple or none; stigma 1. Fruit a berry. Seed solitary with glutinous testa and copious endosperm; embryo straight, terete or angled.

About 21 genera and 500 species, widely distributed, most abundant in tropical regions, where some are terrestrial.

Anthers 1-celled; pistillate calyx 2-lobed; berry compressed. 1. *Arceuthobium.*
Anthers 2-celled; pistillate calyx 3-lobed; berry globose. 2. *Phoradendron.*

1. ARCEUTHÒBIUM Marsch.-Bieb. Fl. Tour. Cauc. Suppl. 629. 1819.

Plants yellow or greenish brown with fragile jointed angled stems. Leaves reduced to opposite connate scales. Flowers solitary or several from the same axil. Staminate flowers mostly 3-parted, compressed. Anthers sessile on the calyx-lobes, circular, 1-celled, dehiscent at the base by a circular slit; pollen-grains spinulose. Pistillate flowers ovate, compressed, 2-toothed, subsessile, at length exserted on reflexed pedicels. Berry fleshy, compressed, dehiscing elastically at the circumscissile base. Cotyledons very short. [Name Greek, meaning juniper and life, in reference to the host of original species.]

About 10 species; all North American except the type. Type species, *Arceuthobium Oyxcedri* Marsch.-Bieb.

Staminate flowers paniculate, nearly all terminal on distinct peduncle-like joints. 1. *A. americanum.*
Staminate flowers in simple or compound spikes, 2–4 mm. thick.
  Stems stout, 6–12 cm. long.
    Staminate spikes stout, 3–4 mm. thick, usually over 1 cm. long; plants parasitic on *Pinus.* 2. *A. campylopodum.*
    Staminate spikes slender, 2 mm. thick, less than 1 cm. long.
      Plants parasitic on piñon pines. 3. *A. divaricatum.*
      Plants parasitic on *Abies,* 4. *A. abietinum.*
  Stems slender, 1–2 mm. thick, 1–4 cm. long.
    Stems well branched, 2–4 cm. long.
      Plants parasitic on *Larix.* 5. *A. larace.*
      Plants parasitic on *Tsuga.* 6. *A. tsugense.*
    Stems nearly simple, 1–2 cm. long; parasitic on *Pseudotsuga.* 7. *A. douglasii.*

1. Arceuthobium americànum Nutt. American Dwarf Mistletoe. Fig. 1281.

Arceuthobium americanum Nutt.; Engelm. Bost. Journ. Nat. Hist. **6**: 214. 1850.

Razoumofskya americana Coville, Contr. Nat. Herb. **4**: 192. 1893.

Stems slender, dichotomously or verticillately much branched, greenish yellow, the staminate plants often 6–10 cm. long, with stems 1–2 mm. thick at base, the pistillate plants much shorter, 15–30 mm. long; staminate flowers terminal on distinct peduncle-like joints, paniculate, 2 mm. broad; calyx-lobes 3, round-ovate, acutish; pistillate flowers smaller; fruit 2 mm. long, bluish.

Parasitic on *Pinus contorta* and *P. divaricata;* British Columbia to Colorado and California. Type locality: on *Pinus contorta,* Oregon.

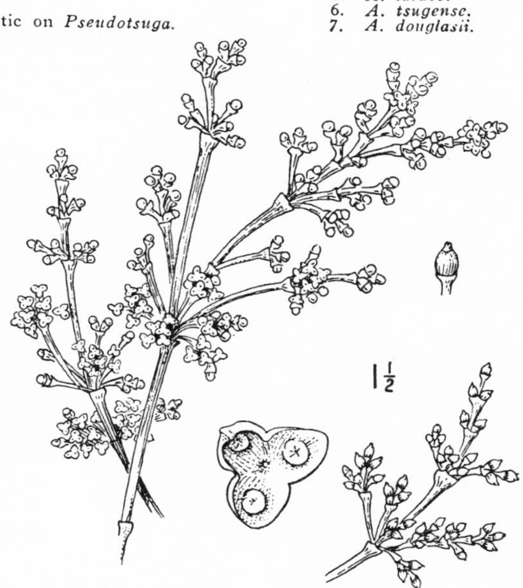

2. Arceuthobium campylopòdum Engelm. Western Dwarf Mistletoe. Fig. 1282.

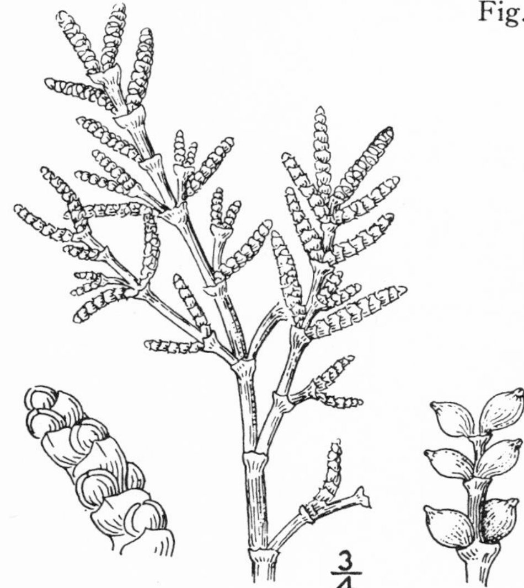

Arceuthobium campylopodum Engelm. Bost. Journ. Nat. Hist. **6**: 214. 1850.

Arceuthobium occidentale Engelm. U. S. Geol. Surv. West 100th Merid. **6**: 254. 1878.

Razoumofskya occidentalis Kuntze, Rev. Gen. Pl. **2**: 587. 1891.

Razoumofskya campylopoda Kuntze, Rev. Gen. Pl. **2**: 587. 1891.

Staminate plants stout, dichotomously branched, 6–12 cm. high, greenish brown, the stems 4 mm. thick at base, the pistillate plants often higher, but usually more slender and darker. Spikes of the staminate flowers 2 to a joint, the pairs usually well separated or sometimes crowded, simple, mostly over 1 cm. long, 2.5–3 mm. thick, with mostly 7–8 joints; calyx-lobes of the staminate flowers 4, rarely 3 or 5, oblong-ovate, acutish, about 3 mm. long; fruit 4–5 mm. long, bluish, drooping on the curved pedicels.

Parasitic on *Pinus ponderosa, P. jeffreyi, P. radiata, P. coulteri,* and *P. sabiniana;* Idaho, south to southern California. The plants on *P. radiata* and *P. sabiniana* have the staminate spikes more crowded and often compound. Type locality: "Oregon, on *Pinus ponderosa.*"

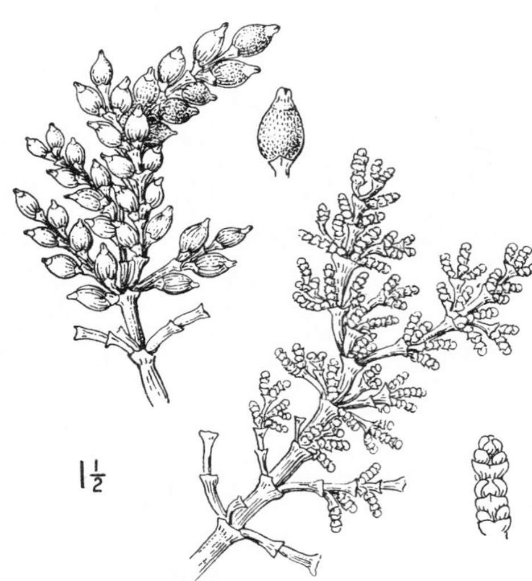

3. Arceuthobium divaricàtum

Engelm. Piñon Dwarf Mistletoe.
Fig. 1283.

Arceuthobium divaricatum Engelm. U. S. Geol. Surv. 100th Merid. **6**: 253. 1878.

Razoumofskya divaricata Coville, Contr. Nat. Herb. **4**: 192. 1893.

Staminate plants rather stout, much branched, olive green, or brown, 5–10 cm. high, the stem, 2–3 mm. thick at base, the branches spreading, often flexuous or recurved; the pistillate plants similar. Staminate spike very slender, scarcely 2 mm. thick, mostly 6–8 mm. long, with 5–6 joints; calyx-lobes of staminate flowers 3, ovate acute; fruit 3–4 mm. long.

Parasitic on Piñon pines; eastern slopes of the Sierra Nevada, California to Colorado, northern Mexico, and Lower California. Found in California on *Pinus monophylla* and to be expected on *Pinus quadrifolia*. Type locality: on *Pinus edulis* and *P. monophylla*, "from southern Colorado through New Mexico to Arizona."

4. Arceuthobium abietìnum

Engelm. Fir Dwarf Mistletoe.
Fig. 1284.

Arceuthobium abietinum Engelm.; Proc. Am. Acad. **8**: 401. 1872, *nomen nudum.*

Arceuthobium douglasii abietinum Engelm. in S. Wats. Bot. Calif. **2**: 106. 1880.

Arceuthobium occidentale abietinum Engelm. in S. Wats. Bot. Calif. **2**: 107. 1880.

Razoumofskya douglasii abietina Greene, Fl. Fran. 341. 1892.

Razoumofskya abietina Abrams, Ill. Fl. Pacif. Sts. **1**: 530. 1923.

Staminate plant rather stout, spreading, greenish brown or tinged with yellow, 5–10 cm. high; the stems 2–3 mm. thick at base; pistillate plants usually a little higher and darker; staminate spikes not crowded on the branches, 6–8 mm. long, 2–2.5 mm. thick, 6–7-jointed; calyx-lobes 7; staminate flowers usually 4, ovate, 2 mm. long; fruit 3–4 mm. long.

Parasitic on species of *Abies;* Washington to California and Utah. Type locality: Sierra Valley, California, on *Abies concolor*. There are no essential differences between the plants on *Abies concolor,* which were described by Engelmann as a variety of *A. douglasii* and those on *Abies grandis,* which he first considered a species but later published as a variety of *A. occidentalis.*

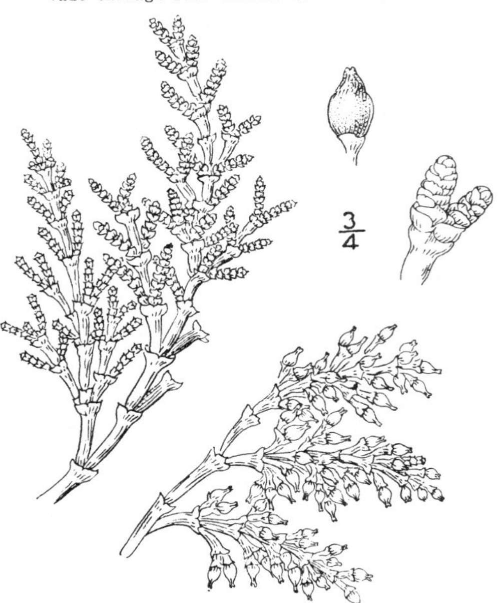

5. Arceuthobium lárice (Piper)

St. John. Larch Dwarf Mistletoe.
Fig. 1285.

Razoumofskya douglasii laricis Piper, Contr. Nat. Herb. **11**: 223. 1906, *nomen nudum.*

Razoumofskya laricis Piper, Piper & Beattie, Fl. Southeast. Washington. 80. 1914.

Arceuthobium douglasii laricis M. E. Jones, Bull. Univ. Montana Biol. ser. **15**: 25. 1910.

Arceuthobium larice St. John, Fl. Southeast Wash. 115. 1937

Staminate plants divaricately branched, greenish yellow, 3–5 cm. high, the stems about 2 mm. thick at base; scales very acute; calyx-lobes of the staminate flowers 3, ovate, acutish, 2 mm. long; fruit about 4 mm. long.

Parasitic on *Larix occidentalis.* Washington to Montana, south to Oregon. Type locality: Mt. Adams, Washington.

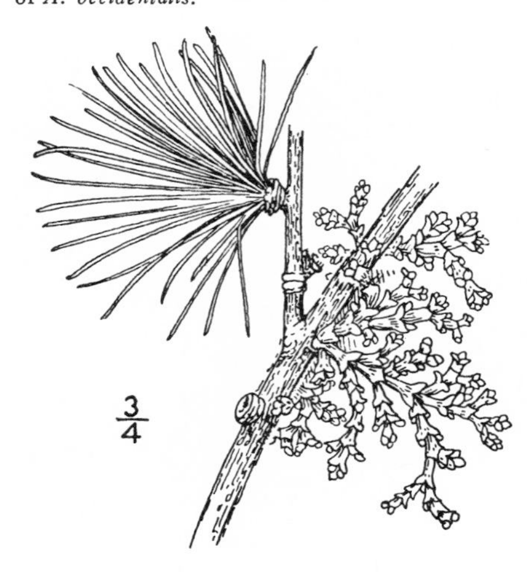

6. Arceuthobium tsugénse (Rosend.) G. N. Jones. Hemlock Dwarf Mistletoe. Fig. 1286.

Razoumofskya tsugensis Rosend. Minn. Bot. Studies III. **2**: 272. 1903.

Razoumofskya douglasii tsugensis Piper, Contr. Nat. Herb. **11**: 222. 1906.

Arceuthobium douglasii tsugensis M. E. Jones, Bull. Univ. Montana Biol. ser. **15**: 25. 1910.

Arceuthobium tsugense G. N. Jones, Univ. Wash. Publ. Biol. **5**: 139. 1936.

Staminate plants paniculately branched, or dichotomously branched above, 4–10 cm. high, brownish yellow, the stems about 1.5 mm. thick at base; pistillate plants shorter, usually less branched and darker; calyx-lobes of the staminate flowers 4, suborbicular, abruptly acute at apex; fruit 4–5 mm. long.

Parasitic on *Tsuga;* western British Columbia and Washington. Type locality: on *Tsuga heterophylla* (Raf.) Sarg. Vancouver Island, British Columbia.

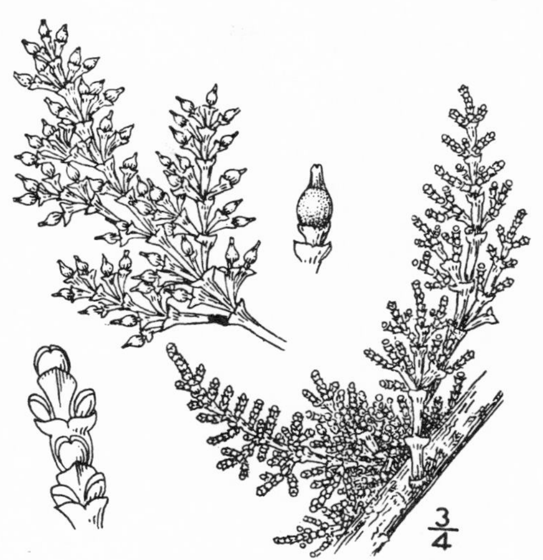

7. Arceuthobium douglásii Engelm. Douglas's Dwarf Mistletoe. Fig. 1287.

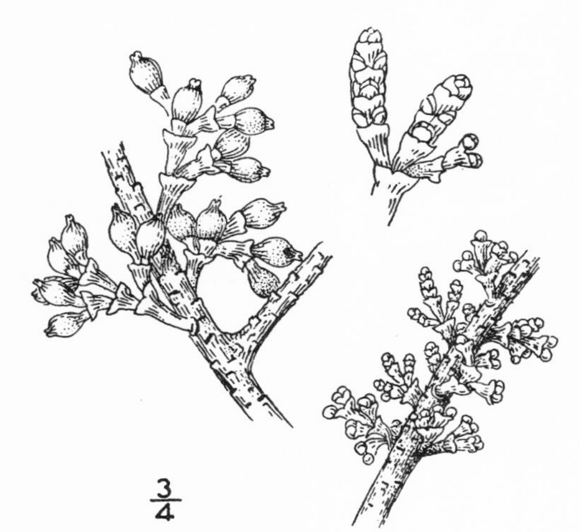

Arceuthobium douglasii Engelm. U. S. Geol. Surv. 100th Merid. **6**: 253. 1878.

Razoumofskya douglasii Kuntze, Rev. Gen. Pl. **2**: 587. 1891.

Staminate plant slender, 1–3 cm. high, greenish yellow; stems about 2 mm. thick at base, dichotomously branched, the branches simple or with accessary branches behind the first; calyx-lobes of the staminate flowers round-ovate, 2 mm. long; fruit 5 mm. long.

Parasitic on *Pseudotsuga taxifolia.* Washington to Montana south to California, Arizona and New Mexico. It has not been collected on *Pseudotsuga macrocarpa.* Type locality: Santa Fe River, New Mexico.

Arceuthobium cyanocarpum (A. Nels.) Abrams. *Razoumofskya cyanocarpa* A. Nels.; Rydb. Fl. Colo. 100, 101. 1906. Stems greenish yellow, 2–3 cm. high, much branched; calyx-lobes usually 3, ovate. Parasitic on *Pinus flexilis.* Wyoming and Colorado. To be expected on this host in California, but not yet reported.

2. PHORADÉNDRON Nutt. Journ. Acad. Phila. II. 1: 185. 1848.

Woody plants with terete usually jointed and brittle stems. Leaves foliaceous, entire, faintly nerved, or reduced to connate scales. Flowers sunk in the jointed rachis, usually several in the axil of each bract. Staminate flowers with a mostly 3-lobed globose calyx, bearing a sessile transversely 2-celled anther at the base of each lobe. Pistillate flowers with a similar calyx-adnate to the inferior ovary. Berry sessile, ovoid, or globose, fleshy. [Greek meaning tree-thief, from the parasitic habit.]

About 100 species, all American and chiefly tropical. Type species, *Phoradendron californicum* Nutt.

Pistillate flowers 2 on each joint.
  Leaves represented by short thin scales.
    Herbage canescent; spikes several-jointed. — 1. *P. californicum.*
    Herbage glabrous; spikes 1-jointed.
      Scales obscurely constricted but not grooved at base. — 2. *P. libocedri.*
      Scales conspicuously constricted, grooved at base as if by a string. — 3. *P. ligatum.*
  Leaves oblanceolate-spatulate.
    Leaves usually 12–20 mm. long; on *Sabina* and *Cupressus.* — 4. *P. densum.*
    Leaves usually 25–35 mm. long; on *Abies.* — 5. *P. pauciflorum.*
Pistillate flowers 6 or more on each joint.
  Joints of staminate spikes 6–15 mm. long, 20–30-flowered.
    Herbage glabrous or very sparsely short-villous. — 6. *P. coloradense.*
    Herbage canescently short-villous. — 7. *P. longispicum.*
  Joints of staminate spikes 5 mm. long, about 12-flowered; on *Quercus.* — 8. *P. villosum.*

1. **Phoradendron califórnicum** Nutt. California Mistletoe. Fig. 1288.

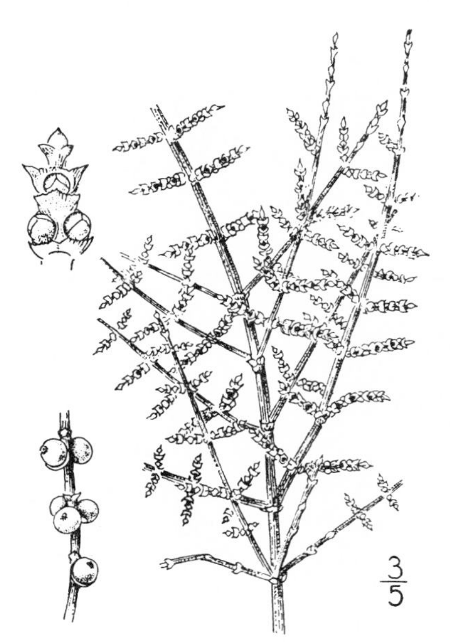

Phoradendron californicum Nutt. Journ. Acad. Phila. II. 1: 185. 1848.

Plants with elongated branching terete stems, 4–8 mm. long, rather slender, puberulent, becoming glabrous in age, internodes 15–30 cm. long. Leaves reduced to short thin acute scales, not disarticulating from the stem; spikes axillary, mostly solitary, 5–10 mm. long, with about 4 joints; fruit reddish pink, fragrant.

Parasitic on various desert shrubs, but most common on *Prosopis* and never on conifers. Lower Sonoran Zone; Mohave Desert, southern California and Arizona south to Sonora and Lower California. Type locality: "mountains of Upper California. Parasitic on the trunks and branches of a *Strombocarpus*."

Pharadendron californicum dístans Trel. Monogr. Phoradendron, 21, 1916. Differs from the type in its elongated fruiting spikes with distinctly separated whorls of fruit. Perhaps only a form having a similar range and occurring on the same hosts.

2. **Phoradendon libocèdri** (Engelm.) Howell. Libocedrus Mistletoe. Fig. 1289.

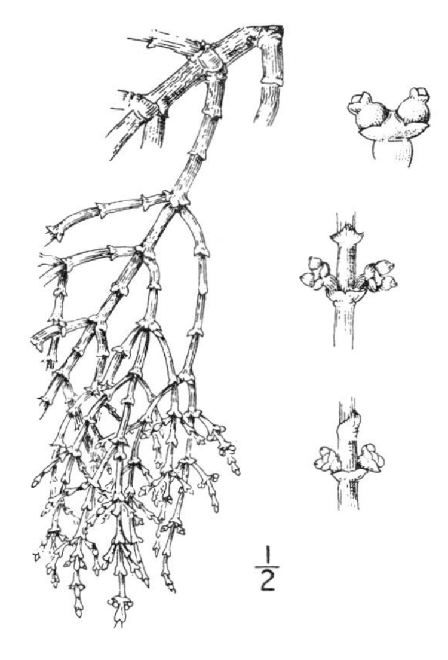

Phoradendron juniperinum libocedri Engelm. in S. Wats. Bot. Calif. 2: 105. 1880.
Phoradendron libocedri Howell Fl. N. W. Am. 1: 608. 1902.

Plants much branched, pendent, yellowish green, 3–5 cm. long, glabrous, the branches somewhat quadrangular, the internodes 10–15 mm. long, cellular-granulated. Leaves reduced to scales, thick, half-ovate, obtuse, about 1 mm. long, obscurely restricted at base; spikes solitary, 2–3 mm. long, 1-jointed, scales and calyx-lobes sparsely short ciliate, otherwise glabrous, fruit straw-colored, 3 mm. in diameter.

Parasitic on *Libocedrus;* southern Oregon to northern Lower California. Type locality: Duffield's Ranch, Sierra Nevada, California.

3. **Phoradendron ligàtum** Trel. Constricted Mistletoe. Fig. 1290.

Phoradendron ligatum Trel. Monogr. Phoradendron. 24. 1916.

Plants branched and compact, 15–25 cm. high; yellowish green, glabrous, the branches somewhat quadrangular, the internodes 10–15 mm. long cellular-granular. Leaves reduced to scales nearly semiorbicular, sharply constricted-grooved at base, about 1 mm. long; spikes solitary in the axils, about 2 mm. long, 1-jointed, glabrous except for the short cilia on the scales and calyx-lobes; fruit not known.

Parasitic on *Sabina;* eastern Oregon to southern California, also in the plateau region of northern Mexico. Closely related to the Rocky Mountain *P. juniperinum* Engelm., from which it differs primarily by the constricted bases of the scales. Type locality: Crook County, Oregon.

4. Phoradendron dénsum Torr. Dense Mistletoe. Fig. 1291.

Phoradendron densum Torr.; Trel. Monogr. Phoradendron 27. 1916.

Plants much branched and dense, 10–25 cm. high, the stem often 5–7 mm. thick at base, glabrous, the internodes 10–15 mm. long, terete or nearly so. Leaves obanceolate usually very obtuse, 9–15 mm. long, 4–6 mm. wide; spikes sometimes clustered, about 3 mm. long, 1–2-jointed, the staminate about 12-flowered; fruit straw-colored, subglobose, about 4 mm. in diameter.

Parasitic on *Sabina* and *Cupressus;* Lake County, Oregon, south to San Diego County, California, and Arizona. This species has been confused with the Mexican species *P. bolleanum* Eichler. Type locality: Mt. Shasta, California. The plants on *Cupressus* usually have larger leaves than those on *Sabina* and are more or less intermediate between this species and *P. pauciflorum* Torr. to which Trelease refers them.

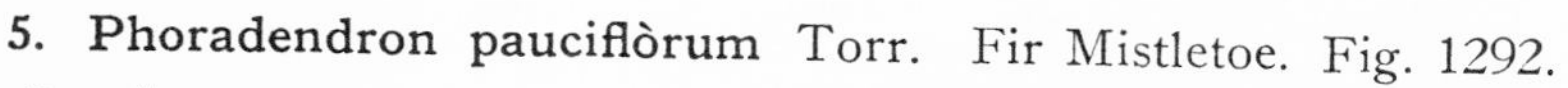

5. Phoradendron pauciflòrum Torr. Fir Mistletoe. Fig. 1292.

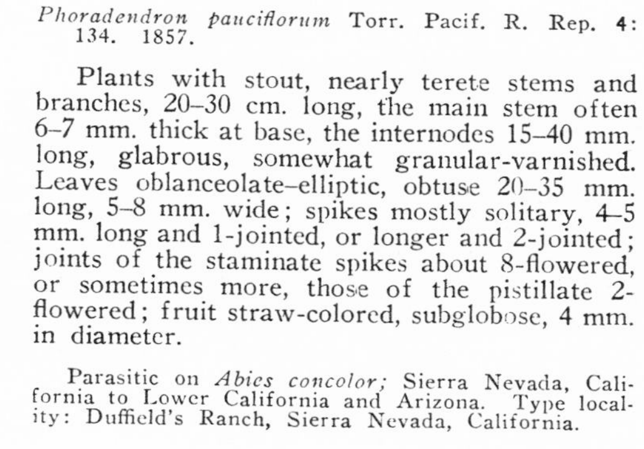

Phoradendron pauciflorum Torr. Pacif. R. Rep. **4**: 134. 1857.

Plants with stout, nearly terete stems and branches, 20–30 cm. long, the main stem often 6–7 mm. thick at base, the internodes 15–40 mm. long, glabrous, somewhat granular-varnished. Leaves oblanceolate–elliptic, obtuse 20–35 mm. long, 5–8 mm. wide; spikes mostly solitary, 4–5 mm. long and 1-jointed, or longer and 2-jointed; joints of the staminate spikes about 8-flowered, or sometimes more, those of the pistillate 2-flowered; fruit straw-colored, subglobose, 4 mm. in diameter.

Parasitic on *Abies concolor;* Sierra Nevada, California to Lower California and Arizona. Type locality: Duffield's Ranch, Sierra Nevada, California.

6. Phoradendron coloradénse Trel. Colorado Desert Mistletoe. Fig. 1293.

Phoradendron coloradense Trel. Monogr. Phoradendron 39. 1916.

Plants stout, 3–6 dm. high, with the internodes 3–4 cm. long, 5–6 mm. thick, minutely canescent, soon becoming glabrous. Leaves oblanceolate-obovate, very obtuse, 3–6 cm. long, 15–25 mm. wide, glabrous, cuneately petioled; spikes 3–5 cm. long, glabrous, with about 3 joints, the pistillate joints 6-flowered, the staminate 20–30-flowered; fruit white, subglobose, about 5 mm. in diameter.

Parasitic on *Prosopis;* desert regions of southern California and Arizona, from The Needles to Yuma. Type locality: Fort Yuma, Arizona.

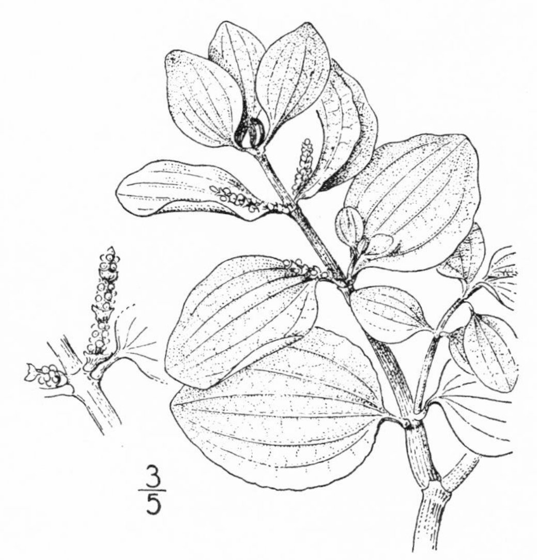

7. **Phoradendron longispìcum** Trel. Long-spiked Mistletoe. Fig. 1294.

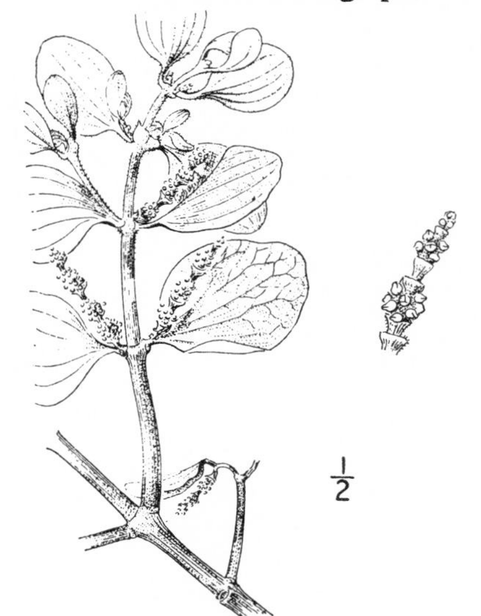

Phoradendron longispicum Trel. Monogr. Phoradendron 39. 1916.

Phoradendron longispicum cyclophyllum Trel. *loc. cit.* 40. 1916.

Plants stout, 4–6 cm. long, the internodes canescently velvety-tomentose, 3–5 cm. long. Leaves elliptic-obovate to oblanceolate 4–6 cm. long, 2–3 cm. wide, velvety-tomentose to nearly glabrous and glossy, cuneately petioled; spikes 15–30 mm. long, short-villous, the pistillate joints 12-flowered, the staminate about 20-flowered; fruit white, tinged with pink, about 4 mm. in diameter, in rather distinct whorls.

Parasitic on various deciduous trees, *Salix, Populus, Juglans, Platanus, Fraxinus;* Sacramento Valley south to San Diego County, California and Arizona. Type locality: Sacramento Valley.

8. **Phoradendron villòsum** Nutt. Hairy Mistletoe. Fig. 1295.

Phoradendron villosum Nutt. Journ. Acad. Philad. II. 1: 185. 1848.

Phoradendron flavescens villosum Engelm. U. S. Geol. Surv. 100th Merid. 6: 252. 1878.

Plants stout, 3–6 cm. high, its internodes 2–3 cm. rarely 4 cm. long, densely short-villous. Leaves oblanceolate-obovate, 3–4 cm. or rarely 5 cm. long, 15–20 mm. or rarely 30 mm. wide, short-villous; petioles cuneate, 3–5 mm. long; spikes often clustered, 10–15 mm. long, short-villous 3-jointed, the pistillate joints 6-flowered, the staminate 12-flowered; fruit pinkish white, somewhat short-villous at apex, 4 mm. in diameter, in close whorls.

Parasitic on various species of *Quercus*, but also sometimes occurring on *Umbellularia, Arctostaphylos,* and *Aesculus;* Willamette Valley, Oregon, south to Lower California. Type locality: Willamette Valley, Oregon.

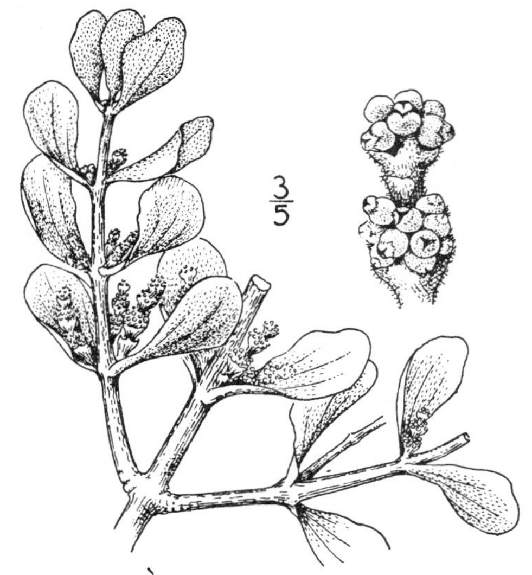

Family 34. **ARISTOLOCHIÀCEAE.**

Birthwort Family.

Low caulescent or often acaulescent herbs, or twining shrubs with bitter tonic or stimulating properties and sometimes aromatic. Leaves alternate or basal, petioled, mostly cordate or reniform and entire. Flowers axillary or terminal, solitary or clustered, perfect, regular or irregular. Calyx-tube mostly adnate to the ovary; calyx-lobes 3 or 6. Petals none. Stamens 5–12, more or less united to the style, anthers 2-celled, adnate, extrorse. Ovary wholly or partly inferior, 6-celled, with parietal placentae; ovules numerous in each cavity, anatropous. Fruit a many-seeded mostly 6-celled capsule. Seeds angled or compressed, smooth or wrinkled, usually with a large fleshy raphe; embryo minute; endosperm fleshy.

Six genera and about 200 species, of wide distribution. The affinity of the family is uncertain, but probably it is more akin to the *Cucurbitaceae,* than the families with which it is associated in prevailing schemes of classification.

Acaulescent herbs; calyx regular, 3-lobed; stamens 12. — 1. *Asarum.*
Caulescent herbs or twining shrubs; perianth irregular; stamens 6. — 2. *Aristolochia.*

1. ÁSARUM [Tourn.] L. Sp. Pl. 442. 1753.

Acaulescent perennial herbs, with aromatic creeping rootstocks bearing 2 or 3 scales and thick fleshy roots. Leaves basal, long-petioled, cordate, mostly ovate or orbicular. Flowers solitary large, borne in the lower axils very near the ground on a short peduncle. Calyx regular, 3-cleft or 3-parted, campanulate. Stamens 12, with short stout filaments, the connectives of the anthers prolonged into a point. Capsule rather fleshy, crowned by the persistent calyx, globose, bursting irregularly or loculicidally dehiscent; seeds large, compressed. [The ancient name, of obscure derivation.]

About 20 species, natives of the North Temperate Zone. Besides the following, 4 others occur in extreme eastern North America. Type species, *Asarum europaeum* L.

Calyx-lobes long-attenuate.
  Terminal appendages longer than the pollen-sacs. — 1. *A. hartwegi.*
  Terminal appendages much shorter than the pollen-sacs. — 2. *A. caudatum.*
Calyx-lobes obtuse or merely acute, not attenuate. — 3. *A. lemmoni.*

1. **Asarum hàrtwegi** S. Wats. Hartweg's Wild Ginger. Fig. 1296.

Asarum hookeri major Ducharte, in DC. Prodr. 15[1]: 424. 1864.

Asarum hartwegi S. Wats. Proc. Am. Acad. 10: 463. 1875.

Asarum major Coville, Contr. Nat. Herb. 4: 192. 1893.

Rootstock often much branched and tufted, rather stout. Leaves evergreen, cordate with large rounded auricles, 5–15 cm. long, and nearly as broad, usually acute, glabrous and mottled above or rarely sparsely pubescent on the veins, the margins ciliate; petioles floccose; peduncles stout, 15–25 mm. long; calyx brownish purple, floccose without, pubescent within, the tube wholly adnate to the ovary, about 12 mm. broad, the lobes ovate, narrowed to an elongated linear apex, 25–65 mm. long; the prolonged free apex of the connective as long or twice as long as the anther; styles shorter than the stamens, nearly distinct; seeds ovate, 4 mm. long.

Coniferous forests, Arid Transition Zone; Siskiyou Mountains, southern Oregon to the southern Sierra Nevada, not found in the California Coast Ranges. Type locality: central Sierra Nevada, probably on the American River.

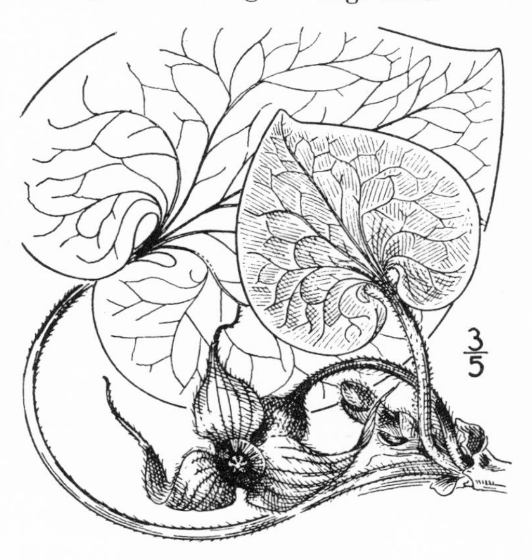

2. **Asarum caudàtum** Lindl. Long-tailed Wild Ginger. Fig. 1297.

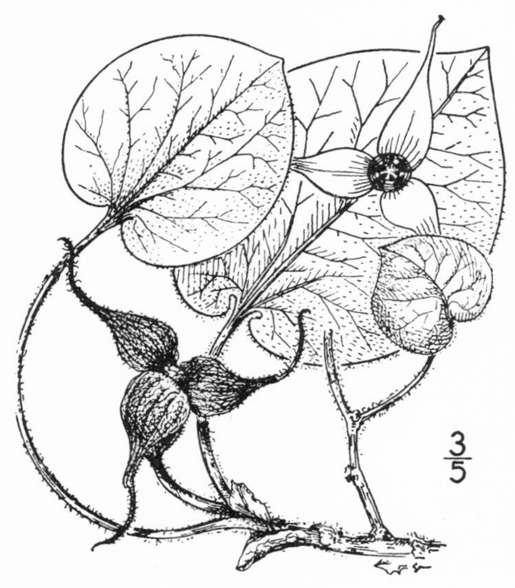

Asarum caudatum Lindl. Bot. Reg. 17: under *pl. 1399.* 1831.

Asarum hookeri Field. & Gardn. Sert. Plant. 1: *pl. 32.* 1844.

Rootstocks slender, elongated, branching. Leaves 2 at a node, cordate-reniform, evergreen, 4–12 cm. long, acute or acutish or sometimes obtuse, dark green above and sparingly short-pubescent at least on the veins; petioles floccose-woolly to glabrous; peduncles slender, 15–30 mm. long; calyx brownish purple, floccose woolly to nearly glabrous without, purplish pubescent within; calyx-tube entirely adnate to the ovary, about 1 cm. broad; calyx-lobes oblong, more or less long-attenuate, 25–85 mm. long; the prolonged apex of the connective much shorter than the anther; styles united, equalling the stamen; seeds ovate, 3 mm. long.

Moist shaded woods, Transition Zone; British Columbia and Idaho south to the Blue Mountains, and in the Coast Ranges to the Santa Cruz Mountains, California. Type locality: Fort Vancouver, Washington.

3. **Asarum lémmoni** S. Wats. Lemmon's Wild Ginger. Fig. 1298.

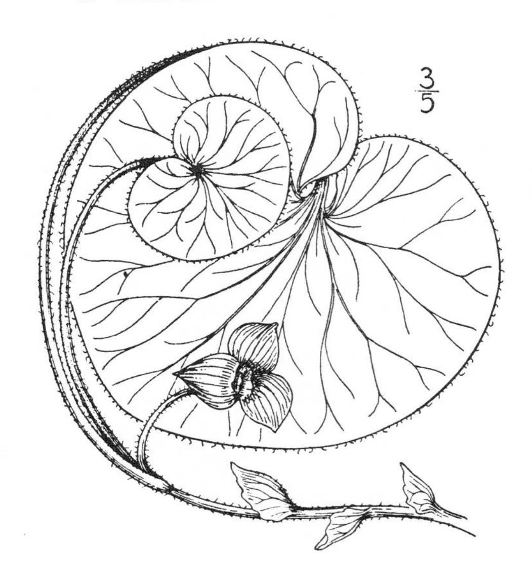

Asarum lemmoni S. Wats. Proc. Am. Acad. **14**: 294. 1879.

Rootstocks slender, elongated, bearing 2 leaves at a joint. Leaves cordate-reniform, 4–8 cm. long, and usually a little broader, rounded at the apex, dark green and glabrous above, ciliate on the margins, the lower surface and the petioles more or less floccose; peduncle slender, about 2 cm. long; calyx about 8 mm. broad, brownish purple; the lobes ovate, merely acute, 8–10 mm. long; the prolonged free apex of the connective much shorter than the anther; styles united.

Coniferous forests, Transition Zone; Sierra Nevada, from Plumas to Tulare Counties, California. Much less common than *A. hartwegi*. Type locality: Sierra County, California.

2. **ARISTOLÒCHIA** [Tourn.] L. Sp. Pl. 960. 1753.

Perennial erect climbing or twining herbs or shrubs. Leaves alternate, cordate, entire or lobed, palmately nerved. Flowers solitary or clustered, irregular, greenish or lurid purple. Calyx more or less adnate to the ovary, the tube usually inflated around the style and contracted at the throat, deciduous, the limb spreading or reflexed entire, 3–6-lobed or appendaged. Stamens usually 6; anthers sessile adnate to the short style or stigma. Ovary 6-celled with 6-parietal placentae. Styles 3–6-lobed. Capsule naked, septicidally 6-valved. Seeds compressed, their sides flat or concave. [Named for its reputed medicinal properties.]

About 200 species, widely distributed in tropical and temperate regions. Besides the following about 9 other species inhabit eastern and southern United States. Type species, *Aristolochia rotunda* L.

1. **Aristolochia califórnica** Torr. California Pipe-vine. Fig. 1299.

Aristolochia californica Torr. Pacif. R. Rep. **4**: 128. 1857.

Stems woody, climbing to a height of 3–4 m., the slender branches more or less pubescent or short-tomentose. Leaves ovate-cordate, 4–15 cm. long acutish or obtuse, tomentose on both surfaces; petioles slender, 15–45 mm. long; peduncles slender, solitary, with a small foliaceous bract near the middle; calyx-tube closely doubled on its self near the middle, strongly saccate, greenish purple, 20–40 mm. to the top of the curviture, 10–15 mm. broad, the limb 2-lipped; ovary linear-clavate; capsule obovate, narrow to a slender base, 30–40 mm. long, 6-winged; seeds deeply concave.

Stream banks and borders of lakes, Upper Sonoran Zone; foothills at the head of Sacramento Valley south to Eldorado and Monterey Counties, California. Type locality: near Corte Madera, California.

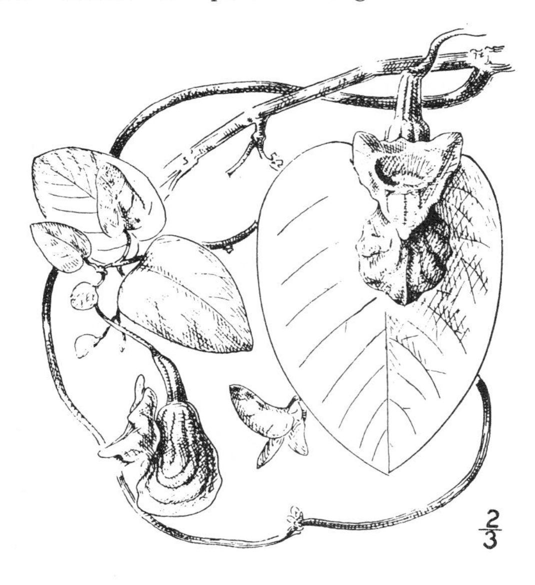

INDEX OF GENERA AND FAMILIES

[Phyla, classes, and families in SMALL CAPITALS; genera in Roman; synonyms and casual references in *Italic*.]